Biochemistry, Cell and Molecular Biology for Pharmacists and Health Professionals

Audrey Minden

Biochemistry, Cell and Molecular Biology for Pharmacists and Health Professionals

From Biomolecules to Drug Development

Springer

Audrey Minden
Department of Chemical Biology
Ernest Mario School of Pharmacy, Rutgers University
Piscataway, NJ, USA

ISBN 978-3-032-20528-5 ISBN 978-3-032-20529-2 (eBook)
https://doi.org/10.1007/978-3-032-20529-2

This Springer imprint is published by the registered company Springer Nature Switzerland AG
The registered company address is: Gewerbestrasse 11, 6330 Cham, Switzerland

Preface

This book discusses fundamental concepts in biochemistry, molecular biology, and cell biology in a context relevant to pharmacy students and other health professional students. It is designed to help readers understand how these topics have formed the foundation of modern medicine and drug development. The book explores the basics of cell biology and the fundamental biomolecules in biochemistry, including proteins, nucleic acids, lipids, and carbohydrates. It delves into biochemical pathways such as cellular respiration pathways and amino acid metabolism. Students will learn how improper regulation of these pathways can lead to disease, and how this understanding is important for the development of pharmaceutical interventions. The book will discuss various types of drugs that have been developed based on our understanding of biochemistry and cell biology. These include drugs that are enzyme inhibitors, a traditional mechanism for drug development that has been used for many years. The book will also explore more modern technologies, such as biologics, which have gained prominence in recent years. It will also discuss newer mechanisms for treating diseases, such as gene editing and antisense technologies, which are revolutionizing the development of drugs and patient treatment. The book also covers advanced topics, including signal transduction, oncogenes, and tumor suppressor genes, and introduces technologies such as CRISPR and Omics technologies, as well as new drug delivery mechanisms, including the use of liposomes and nanotechnology. The reader will gain a better understanding of the importance of these topics in modern medicine and pharmaceutical development and be encouraged to explore these subjects in greater depth.

Piscataway, NJ, USA — Audrey Minden

Acknowledgments

This book is dedicated to the loving memory of my father Percy Minden, whose long career and love of science inspired my own curiosity. His influence continues to shape how I approach science, writing, mentoring, and teaching.

I am deeply grateful to Yiling Wang for her valuable assistance in creating many of the figures. I am grateful to Oyinkansolami Babalola for her assistance with research, organizing charts and tables, and helping with figures. I also thank Yasmeen Sakr, Dmytro Sosnovskyy, Kyle McNally, Monica Ragheb, Noorien Ali, John Salib, Naml Ahmad, George Ebeid, and Ghadir Esayed for helping with research, proofreading, editing, organizing, creating practice questions, and for bringing the perspective of PharmD students to the development of this book. I am grateful to Polly Minden for helping create some of the figures, organize some of the references, and for making delicious meals to sustain me while I wrote. I would like to thank Eden Kim for discussions and assistance with research on CRISPR technologies.

I would like to acknowledge my colleagues in the Department of Chemical Biology in the Ernest Mario School of Pharmacy for advice, guidance, encouragement, and thoughtful discussions, especially Dr. Suzie Chen, Dr. Wei-Xing Zong, Dr. Philip Furmanski, Dr. Nanjoo Suh, Dr. Christina Glytsou, Dr. Mathew McBride, Dr. Luca Cartegni, Dr. Fang Liu, Renping Zhou, and Dr. Jessie Yanxiang. I would also like to thank our staff, Deborah Stalling, Aleesha Mahajan, Roin-Lyn Willamson, and Angelica Arroyo, for their invaluable support.

Finally, I thank my family for their constant support, patience, and encouragement: my mother, Doris Minden, who always encouraged me, my children, Polly Minden and Daniel Minden, who have been a constant source of encouragement, joy, and perspective throughout this project, my sisters, Judy Sylvan, Claudia Minden, Shelley Minden, nieces, nephews, and cousins, who always inspire me.

Contents

About the Author

Audrey Minden is an Associate Professor in the Department of Chemical Biology in the Ernest Mario School of Pharmacy at Rutgers, the State University of New Jersey. She received her BA from Colorado College and her PhD from the University of Illinois, followed by postdoctoral training at the University of California at San Diego (UCSD) under the guidance of Dr. Michael Karin. Her research explores how signaling pathways regulate cell growth and how such pathways can be improperly activated in cancer cells. Her lab focuses on small GTP-binding proteins (such as Cdc42, Rac, and Rho) and their downstream effectors, particularly the p21-activated kinases (Paks). A major focus of her work is studying the roles of these signaling proteins in breast cancer cells, and recent work from her lab has revealed that targeting these protein kinases may be promising for reducing the growth of triple-negative breast cancer cells. In addition to research, she is committed to teaching and mentoring students and postdoctoral fellows in the lab and in the classroom. She developed and teaches a course for the PharmD program at Rutgers entitled "Biochemical Foundations of Pharmaceutical Biotechnology." This course is designed to strengthen students' understanding of biochemistry and to help them understand how biochemical concepts are critical for drug discovery and for the practice of pharmacy.

Part I

Introduction and Cell Biology

This textbook covers the topics of biochemistry and cell and molecular biology, tailored toward pharmacy students and other health professional students. These fields are essential for understanding how the body functions under normal conditions and what problems arise when cells and biochemical pathways are not functioning properly. The book is divided into seven parts. This part includes three chapters: Chapter 1 provides an overall introduction to the textbook, explaining its organization, scope, and the background knowledge students are expected to have. It also includes an introduction to the evolving mechanisms by which drugs are developed, for example, it discusses how drugs are often designed to target enzymes, but newer approaches are also being developed, such as targeting metabolic pathways. Therapeutic treatment options are also expanding to nucleotide-based treatments such as those involving CRISPR and personalized medicine techniques. Chapter 2 gives an overview of cell biology. While students reading this book should have some fundamental knowledge of cell biology, this chapter reviews the basics, discussing cell function, cell organelles, the cell cycle, the evolution of cells, and some history of cell biology. Chapter 3 discusses the cell membrane and transport proteins. While most of this book focuses on molecular biology and biochemistry, this is the only chapter that is entirely dedicated to a single aspect of cell biology. This is because the cell membrane and transport proteins are often critical drug targets, and they are also often critical for regulating the transport of drugs into cells.

1 Introduction to Molecular, Cell Biology, Biochemistry, and Their Pharmaceutical Applications

Abstract

This chapter serves as an introduction to the textbook. First, it introduces the basics of cell biology, molecular biology, and biochemistry. Next, since this book is geared toward pharmacy students, it then provides an overview of drug development, considerations in drug design and use, and the ways in which pharmacy and therapeutics are grounded in cell and molecular biology and biochemistry. Core biochemical concepts that students should already be familiar with are also introduced. Finally, this chapter concludes with an overview of the book's organization and structure.

Keywords

Cell biology · Molecular biology · Biochemistry · Drug action · Drug development

What You Will Learn in This Chapter

- An introduction to this textbook and its basic organization
- The basic definitions of cell biology, molecular biology, and biochemistry, and how they interconnect
- Basic mechanisms of drug action
- Different categories of drugs, such as small molecules and biologics
- How new technologies such as PROTACs and CRISPR have expanded the way patients are treated
- The importance of pharmacokinetics and pharmacodynamics in determining how drugs will behave in the body

This textbook covers the topics of biochemistry and cell and molecular biology, tailored toward pharmacy students and other health professional students. Such students have a strong interest in disease, pharmacological treatments, and the mechanisms through which drugs exert their effects. Because pharmaceuticals interact with cells, molecules, and biochemical pathways, a strong background in these fields is essential for understanding how drugs work.

The goal of this book is not only to provide an understanding of biochemistry and molecular and cell biology, but also to **highlight the direct relevance of these topics to drug action and development**. For example, some pharmaceuticals target cell surface receptors, others may be transported into the cell and target DNA molecules, and many drugs target enzymes. This book aims to establish a connection between fundamental biological concepts and their practical applications in pharmacy. Students are expected to have taken at least a basic course in biology and cell biology, which will be helpful

A. Minden, *Biochemistry, Cell and Molecular Biology for Pharmacists and Health Professionals*,
https://doi.org/10.1007/978-3-032-20529-2_1

before reading this book. However, certain key topics in cell biology and biology are reviewed here, and references at the end of the chapter can be helpful for students who need a refresher on these topics.

This chapter serves as an introduction to the remainder of the textbook. It is divided into four main parts. **Part 1** introduces the basics of the fields of cell biology, molecular biology, and biochemistry. Next, since this book is geared toward pharmacy students, **part 2** provides a brief overview of the different mechanisms that are used for drug development, various considerations for the design and use of drugs, and how the study of pharmacy and drug development is linked to our understanding of cell and molecular biology and biochemistry. Next, **part 3** covers basic concepts in biochemistry, such as specific bonds and functional groups, that students should review before delving into the subject. Finally, **Part 4** discusses the general organization of the book's chapters.

1 What Are the Main Areas of Cell Biology, Molecular Biology, and Biochemistry?

Living organisms grow, move, metabolize nutrients, use energy, respond to stimuli, and reproduce. Many types of living organisms exist on Earth, but all share a basic feature: they are made up of cells. Some organisms consist of a single cell. Examples include yeast, amoeba, and bacteria. Other, more complex organisms, such as animals and plants, are made up of many cells. In fact, animals and plants can consist of trillions of cells, each with specialized functions, which must work together and communicate to form the whole organism [1].

Cells are the basic units of life. In multicellular organisms, cells interact in a coordinated way to carry out complex biological functions. Eukaryotic cells consist of organelles and a nucleus, all within a plasma membrane. The field of **cell biology** focuses on the structure and function of cells, as well as the process of understanding how cells communicate with one another and with their environment [2].

At the molecular level, cells are made up of complex molecules. Examples of complex molecules within cells are nucleic acids and proteins. The field of **molecular biology** involves studying these large molecules and their functions. Examples are the study of how DNA is replicated or how gene expression is regulated.

These complex molecules within cells are, in turn, made up of simpler biomolecules. These can include nucleotides, amino acids, carbohydrates, and lipids. The field of **biochemistry** encompasses these types of molecules and their interactions with one another and the environment. An important topic in biochemistry is the role these biomolecules play in cellular metabolism, which involves the use and generation of energy. These biomolecules are composed of atoms, which in turn contain protons, neutrons, and electrons, topics that are covered in **chemistry and organic chemistry** courses (see Fig. 1.1). Finally, while it is helpful to define these basic fields of biology, it is also important to note that these fields are often interconnected and frequently overlap. Some concepts do not fit neatly into a single category [3–6].

2 Mechanisms of Drug Action

Many drugs are designed to interact with cellular and biochemical targets, such as enzymes, transport proteins, or metabolic intermediates. The following is an introduction to some types of medications used today, along with an overview of their functions and interact with biomolecules and cellular components. Many of these and other drugs are discussed in more detail throughout the text.

2.1 Some History of Drug Development

Drug development has come a long way over the course of human history. In ancient times, some medicines were extracted from plants. Examples are tree bark extract used to treat pain, which later became the source of aspirin (salicylic acid), and poppies, which were a source of opium, also used

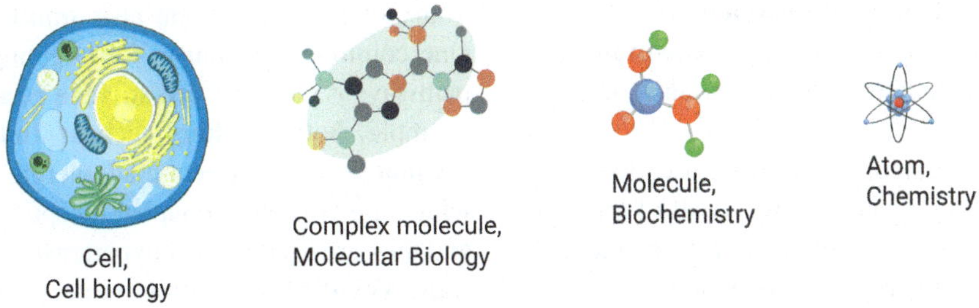

Fig. 1.1 Cell biology, molecular biology, and biochemistry. These topics deal with the study of cells, complex molecules, and simpler molecules. (Source: Illustrations from Vecteezy.com. (Pro license))

for treating pain. With the increasing understanding of cell biology and biochemistry, more rational methods were developed for drug creation, allowing scientists to target specific enzymes or biological processes. Currently, drug development employs a range of different strategies. Historically, many drugs have been developed to target and inhibit enzymes, and most drugs are still enzyme inhibitors. Increasingly, however, drugs are also being developed using a variety of other strategies. These include targeting genes, targeting signaling or metabolic pathways, and using monoclonal antibodies or gene therapy mechanisms. Understanding how drugs work and developing new treatments relies more than ever on a strong understanding of biochemical processes and cell biology [7].

2.2 Many Pharmaceuticals Are Either Small Molecules or Biologics

Today's drugs usually fall into the categories of small molecules or biologics, although some drugs don't fit neatly into either of these categories. **Small molecules** are typically synthesized in the lab and are usually under 900 Daltons in size, composed of up to about 100 atoms. Some can be taken orally, and others are administered by injection. **Biologics** were developed more recently, and they are larger and more complex than small molecules. They may contain hundreds to thousands of atoms and are over 10,000 Daltons in size. They generally consist of components that are produced normally in living organisms, such as proteins or nucleic acids. They are often delivered by injection [8–10].

Aside from their origins and sizes, small molecules and biologics have several significant differences. Small molecules are stable at room temperature, whereas biologics are often unstable and require storage in cold temperatures. Small molecules can often permeate the cell membrane and enter cells, whereas biologics usually cannot cross the plasma membrane. Biologics often target cell surface proteins, such as receptors, instead. Small molecules can be distributed throughout the body via circulation, but the circulation of biologics can be more limited. Biologics often cost more than small molecules, but they also offer several advantages. They can be highly specific to their targets; in many cases, they can enhance or suppress the immune system and may have a longer half-life in the body compared to small molecules. **Many biologics are antibodies** or modifications of antibodies, but other types include cytokines, growth factors, and cellular therapy [9].

2.3 Other Categories of Pharmaceuticals

Most drugs can be classified as small molecules or biologics, but some categories of drugs fall outside of these two groups. For example, **peptidomimetics** are drugs that contain small synthetic peptides, which are intermediate in size between small molecules and biologics. Some drugs are considered **natural products**, while others are **synthetic or semisynthetic**

derivatives of natural products. Another category of drug constitutes **oligonucleotide therapeutics**, which consist of short strands of nucleic acids [11].

Different strategies have been used to develop drugs, and these can encompass all of the above subtypes of drugs. Some examples of these different strategies are described below.

2.4 Different Mechanisms of Action of Pharmaceuticals

2.4.1 Enzyme Inhibitors

Enzyme inhibitors were among the earliest drugs developed and have been available for many years. Enzyme inhibitors are still commonly used, and inhibiting enzymes has historically been the most widely used strategy for drug development. Enzymes catalyze biological reactions that could not occur efficiently without them, and many enzyme inhibitors are **small molecules**. **Aspirin** is an example of a widely used enzyme inhibitor. It inhibits the **enzyme cyclooxygenase (COX)**. Consequently, it inhibits the synthesis of prostaglandins, and this in turn reduces inflammation and pain. **Statins** (such as **Lipitor (atorvastatin)** [12]) inhibit **HMG-CoA reductase**, the rate-limiting enzyme in cholesterol biosynthesis, and statins are widely used to treat high cholesterol. **ACE inhibitors** block angiotensin-converting enzyme (ACE) and are used to lower blood pressure [3, 7, 13–16].

Some enzyme inhibitors specifically target enzymes that are mutated in diseases such as cancer, rather than the wild-type form. An example is **Gleevec (imatinib)** [17]. This drug targets **Bcr–Abl**, a cancer-specific fusion protein with tyrosine kinase activity, commonly linked to chronic myelogenous leukemia (CML) [18].

2.4.2 Drugs that Target Signaling Pathways

Some drugs **target signaling pathways** that are important for cell communication with the environment or other cells. These types of drugs may **inhibit, mimic, or activate** signaling pathways. Many of these drugs are also **small molecules**. One common mechanism for targeting signaling pathways is to target **cell surface receptors**. For example, **beta-blockers** such as **Inderal (propranolol)** [19] target β-adrenergic receptors, which are G protein-coupled receptors (GPCRs), and they are used to treat hypertension. **Nolvadex (tamoxifen)** [20] is a small molecule that targets the estrogen receptor. It is a class of drugs called **selective estrogen receptor modulators (SERMs)** and is used to treat estrogen-positive breast cancer. Other drugs have been developed that target other pathways, such as the MAP Kinase pathway and PI3-Kinase pathway. Some components of signaling pathways are enzymes, particularly protein kinases. Therefore, some drugs in this category are also enzyme inhibitors, highlighting the significant overlap between these various strategies [6, 7, 14, 21–23].

2.4.3 Drugs That Target Metabolic Pathways

Metabolic pathways include those that involve the breakdown and synthesis of carbohydrates, fats, and amino acids, as well as the use and expenditure of energy. Metabolic pathways are a major target for drugs that treat metabolic diseases such as diabetes and obesity. **Ozempic (semaglutide)** is an important example of a drug that mimics a metabolic pathway. It is a peptide and therefore considered a biologic. It **mimics the GLP-1 hormone and binds to the GLP-1 receptor,** leading to increased insulin secretion in a glucose-dependent manner and increased satiety [24, 25].

2.4.4 Drugs That Target Channels and Transporters

Some drugs **target channels and transporters,** thereby regulating what can be transported in or out of the cell. For example, SSRIs (selective serotonin reuptake inhibitors) such as **Lexapro (escitalopram)** [26] and **Prozac (fluoxetine)** [27] inhibit the serotonin transporter, and they are used to treat depression. **Xylocaine (lidocaine)** [28] targets voltage-gated sodium channels and is used as a local anesthetic [6, 7].

2.4.5 Drugs That Involve the Immune System, Antibodies, or Antibody Fusions

Manipulating the immune system to target abnormal cells has seen significant progress in recent years. An important example is the use of **monoclonal antibodies**. Monoclonal antibodies are antibodies developed in the lab that recognize a specific antigen, and they are considered **biologics**. Monoclonal antibodies are derived from a single clone of B cells and recognize a specific epitope of a single antigen. Monoclonal antibodies bind to a target, often on the surface of a cell, and either inactivate the target or, in some cases, lead to an immune response against it [7].

Monoclonal antibodies are large and cannot enter the cell, so they are usually generated against extracellular targets such as receptors. **Herceptin (trastuzumab)** [29] and **Humira (adalimumab)** [30] are two examples of **monoclonal antibodies** that recognize cell surface receptors or ligands. Herceptin (trastuzumab) is a monoclonal antibody that targets Her2, a member of the EGF receptor family. It is used to treat HER2-positive breast cancer. Humira (adalimumab) targets TNFα, a cytokine involved in inflammation, and is used to treat inflammatory diseases such as arthritis [6, 31, 32].

Engineered fusion proteins are a newer category of biologics. They are formed by fusing two or more protein domains together. Often, one component is part of an antibody, usually the Fc region, which is modified to extend its half-life by binding to the neonatal Fc receptor (FcRn). The other part of the fusion is usually a receptor for the target. These fusion proteins can be used to neutralize their targets. An example is **Enbrel (etanercept)** [33]. It contains the extracellular domain of the human TNF receptor fused to the Fc portion of an antibody (IgG1). Thus, it can bind, sequester, and neutralize TNFα by serving as a type of decoy TNF receptor. It reduces inflammation in autoimmune diseases [34, 35].

Antibody drug conjugates (ADCs) are multicomponent drugs that consist of a monoclonal antibody linked to a drug, typically a cytotoxic chemotherapy agent. They are designed to deliver the cytotoxic drug specifically to a tumor cell. An example is **Kadcyla (ado-trastuzumab emtansine)** [36], which contains a monoclonal antibody that binds to Her2 and is linked to a highly toxic chemotherapeutic substance. The goal is targeted delivery of the chemotherapy drug while reducing side effects due to the specific nature of the delivery [7, 37].

Immunotherapy, including **checkpoint inhibitors,** does not involve administering antibodies to the patient but involves stimulating the patient's own immune system. This type of treatment has gained importance in the treatment of certain cancers. Examples include **Keytruda (pembrolizumab)** and **Opdivo (nivolumab)** [38, 39]. They are designed to help the normal immune cells target and kill cancer cells in a tumor. They do so by blocking the PD-1 receptor on T cells, which releases an inhibitory effect on the immune cells, allowing them to kill cancer cells more efficiently [40].

Cancer vaccines are another approach to treating cancer. These include peptide-based vaccines, in which a peptide corresponding to a small portion of the tumor is injected to stimulate a specific immune response. Peptide vaccines are promising and relatively straightforward to develop, but other types of cancer vaccines include whole-protein vaccines, dendritic cell vaccines (such as **Provenge (sipuleucel-T)**) [41], tumor cell-based vaccines, and nucleic acid-based vaccines. All of these involve stimulating the immune system to target the cancer cells [42].

2.4.6 Therapies That Involve Nucleic Acids

mRNA therapies are **biologics** that have received a great deal of attention in recent years. They are designed to encode a protein that is produced in the body and are usually administered by injection. The **COVID-19 mRNA vaccine** brought this type of technology into the spotlight in 2020 [43]. It encodes a protein produced by the COVID-19 virus, called the spike protein. The mRNA is delivered within nanoparticles made up of lipids. After the injection, the mRNA is translated into protein, and the immune system makes antibodies against the resulting protein. This will trigger an immune reaction if the virus infects the individual in the future. This technology has advantages because it does not involve the injection of live or inactivated virus, and because new vaccines can be made relatively quickly if the virus sequence changes over time [44].

The COVID-19 vaccine is a preventive rather than a therapeutic measure. Several mRNA drugs are currently in clinical trials for the treatment of actual diseases. However, for diseases such as cancer and autoimmune disease, none are yet FDA-approved [45].

RNA interference **(siRNA (small interfering RNA) or shRNA (short hairpin RNA)**) is also an RNA-based treatment. This type of treatment is designed to target specific mRNAs in the body, resulting in their degradation. The first FDA-approved siRNA drug is **Onpattro (patisiran)** [46]. It uses siRNA to promote the degradation of the transthyretin (TTR) transcript, thereby treating the disease hereditary ATTR (hereditary transthyretin-mediated amyloidosis) [47].

Antisense oligonucleotides (ASOs) use short DNA or RNA sequences that are complementary to a specific mRNA, which block its translation or alter its splicing pattern. An example is **Spinraza (nusinersen)** [48]. It treats spinal muscular atrophy by interfering with the splicing of the gene encoding SMN2. Both siRNA and ASOs are biologics [49].

Gene therapy involves the correction of a gene or the insertion of a new gene. An example is **Luxturna (voretigene neparvovec)**. It is used to treat an inherited retinal disease caused by mutations in the RPE65 gene. This drug introduces a functional copy of the RPE65 gene into the retinal cells using an adenovirus delivery system. As a result, the cells in the retina can produce a functional RPE65 enzyme. This treatment slows down the progression of the disease and improves vision [50].

The **CRISPR-Cas9** system is one of the most recent technologies for altering genes and a promising mechanism for gene therapy. The first FDA-approved CRISPR-Cas9 treatment is **Casgevy (exagamglogene autotemcel)** (FDA-approved in 2023) [51]. This drug edits regulatory elements of the gene encoding BCL11A, which has an important role in hemoglobin synthesis. Casgevy was developed to treat patients with sickle cell disease, a condition caused by abnormalities in hemoglobin. CRISPR and other gene therapy technologies involve the modification of genes and are considered **biologics** [52].

2.4.7 Proximity Inducing Small Molecules

This is a new class of drugs that brings two proteins into proximity that are not necessarily normally found together.

PROTACs (proteolytic targeting chimeras) are important members of this group. They target a particular protein for destruction, rather than merely inactivating it like an enzyme inhibitor. They achieve this by directing their target protein to the ubiquitin-proteasome system, which results in its degradation. PROTACs are **small molecules,** but they have more complex structures than other small molecules. They consist of multiple parts. Part of the PROTAC targets the protein to be destroyed. Another part contains a ligand for the E3 ligase. E3 ligase plays a crucial role in the ubiquitination and degradation of its target. An example of a PROTAC is **ARV-110**. It targets the androgen receptor, and it is in clinical trials for the treatment of prostate cancer. Currently, no PROTACs have been FDA-approved; however, this area of research is actively being pursued [53, 54].

Molecular glues and **protein stabilizers** also bring proteins into proximity, but with a different outcome. They promote or stabilize interactions between two proteins. These can lead to degradation, similar to PROTACs, but they can also have other effects, such as altering the function of a protein. **Vyndaqel (tafamidis)** [55] is considered a protein stabilizer. It is a drug that treats **transthyretin amyloidosis**. It binds to transthyretin (TTR) (a type of transport protein) and stabilizes it. This prevents it from breaking apart and forming aggregates, a characteristic of the disorder. It was approved in 2019 [56].

2.4.8 Cellular Therapy

Cellular therapy is another important concept in disease treatment. Rather than treating patients with a drug, it involves transplanting cells into patients to replace or repair damaged cells or to combat disease. This type of therapy is considered a biologic, as it uses cells as therapy. Cellular therapy is a broad category encompassing several subtypes. **Hematopoietic stem cell therapy (HSCT)**, for example, has been used for many years to treat certain types of blood cancer, and more recently, it has been explored for the treatment of other disorders, such as multiple sclerosis (MS). Various treatments involving different types of **stem cells** are being used or are under investigation for regenerative medicine [57].

CAR T-cell therapy (chimeric antigen receptor T-cell therapy) is also a type of cellular therapy. It involves modifying T cells from a patient with cancer and reintroducing the cells into the patient with the goal of targeting the cancer cells [58].

2.4.9 Antivirals

Some drugs are **antivirals,** and they can block or slow the progression of a virus that has infected cells. Examples include **Biktarvy** [59], an HIV capsid inhibitor; **Veklury (remdesivir)** [60], an RNA polymerase inhibitor targeting the SARS-CoV-2 (COVID-19) virus; and **Tamiflu (oseltamivir)** [61], which inhibits the influenza virus. All three of the above are small molecules [6, 7].

2.5 Vaccines

Some treatments are not really drugs designed to treat disease, but rather to prevent it. These usually include **vaccines,** which can be considered **preventable biologics**. In addition to the COVID-19 vaccine, described above, more traditional vaccines have been used for many years. Examples of vaccines are the **polio vaccine, MMR vaccine, hepatitis vaccine,** and **influenza vaccines**. Vaccines work by injecting an antigen, such as an inactivated virus, that stimulates the body's immune response, enabling the patient's immune system to respond to future infections [62].

2.6 Combination Therapies

Combination therapy refers not to a new type of drug, but to combining more than one drug together. In many cases, single-agent treatments (using just one drug) are less effective than combination approaches. This is especially true in cancer treatment, where patients are often given a cocktail or combination of different drugs that target different signaling pathways or target the cancer in different ways. This is generally more effective than single-agent treatments. Combination treatments are also effective in treating other disorders, such as diabetes and other metabolic diseases. The use of combination treatments is promising, but it also increases the complexity of determining the most effective treatment strategy for patients and managing potential side effects. Many of the upcoming chapters in this book discuss individual drugs, but it is important to consider that combinations of multiple treatments are often used and can sometimes be the most effective approach [63, 64].

2.7 Considerations About Pharmacokinetics and Pharmacodynamics

When prescribing or administering a drug, another important consideration is how the drug functions in the body. Two important aspects of this are **pharmacokinetics (PK)** and **pharmacodynamics (PD).**

PK describes a drug's absorption, distribution, metabolism, and excretion (ADME). In other words, it is associated with how **the body interacts with the drug**. **PK** is often measured by taking blood or plasma samples from patients at various time points after drug administration to determine the concentration of the drug in the body. These types of studies involve the measurement of specific PK parameters such as Cmax (measured in ng/mL), which is the maximum plasma concentration, Tmax, the time to reach Cmax, and the half-life of the drug (the time it takes for the drug level to be reduced by half in the body, usually measured in hours) [65].

PD has a different focus; its focus is on **what the drug does to the body**. PD involves studying the mechanism of action (MOA) of the drug and assessing its potency and effectiveness. These factors are often measured through biomarkers and clinical endpoints. One measurement examined is the EC50, which represents the concentration of a drug required to achieve 50% of its maximum effect. This can be measured in nM or μM. Emax is another measurement, representing the maximum possible effect the drug can have. Both are measured using dose-response curves. For example, if studying a blood pressure drug, one might measure the maximum change in blood pressure obtainable [66].

Together, PK and PD are used to determine factors such as what dose to use, and how and when to administer the drug. While PK and PD will not be a focus of this textbook, it is important to note, when reading about drugs, that these factors are essential for optimizing their safety and effectiveness [6, 7, 67, 68].

2.8 A Broad Picture of Molecular Cell Biology, Biochemistry, and Drug Action

The field of drug development has expanded rapidly in recent years, and numerous strategies are now employed to design pharmaceuticals. Above, we discuss some of the important strategies for designing drugs. Some drugs can fit into multiple categories (such as enzyme inhibitors that are also signal transduction inhibitors), while others do not fit perfectly into any category.

Advances in pharmaceutical development have significantly improved the treatment of many disorders, and continued research holds promise for the development of even more effective therapies. Such advances would not be possible without a thorough understanding of molecular and cell biology and biochemistry. For example, a solid background in biochemistry is essential for understanding how enzymes function, and understanding enzymes is crucial for developing effective enzyme inhibitors. Similarly, insight into metabolic pathways that regulate energy consumption and production has proven critical for developing drugs like **Ozempic (semaglutide)** [69], which treat diabetes, obesity, and other metabolic disorders. A clear understanding of how proteins are degraded in cells has led to research into the field of PROTACs, and familiarity with nucleic acid structure and function is critical for gene therapy treatments. Likewise, understanding membrane transport mechanisms is important for identifying new drug targets, such as drugs that inhibit transporters and channels. It is also crucial for understanding how drugs get transported into and throughout cells. A strong understanding of intracellular signaling pathways is essential for recognizing how their disruption can lead to disease and the importance of treatments that can either disrupt or activate these pathways.

A solid foundation in cell biology and biochemistry enables pharmacists and healthcare professionals to understand how drugs function at a biochemical and molecular level. It also helps them make informed decisions in clinical practice and counsel patients about the medicines they are prescribed. For those pursuing research, this knowledge is fundamental to developing new pharmaceuticals and technologies for treating and preventing disease.

Understanding cell and molecular biology and biochemistry has transformed medicine in numerous ways, paving the way for the development of drugs that have saved many lives and improved the quality of life for many others. At the same time, it is important to remember that the same drugs that target biochemical pathways to treat disease may sometimes also have unwanted side effects. This can happen because biochemical pathways and systems in the body are interconnected. For example, consider digoxin, which can be lifesaving to a patient with heart failure as it enhances cardiac contractility by inhibiting a sodium-potassium ATPase pump.

At the same time, however, disrupting the normal distribution of ions can also lead to other health issues, including other heart problems. Along the same lines, drugs that inhibit enzymes or receptors involved in inflammation may benefit patients with autoimmune disease but could negatively affect wound healing or the ability to fight infections. For these reasons, the side effects of any drugs must be monitored, and the dosage must be adjusted as necessary [6, 70].

It is also important to recognize that drug responses vary among different individuals. Many factors, such as age, genetics, metabolism, and other drugs the patient may be taking, can affect whether and how a patient responds to a drug. As a result, a drug that is highly effective in one patient may be less effective or ineffective in another. Due to this problem, the field of personalized medicine is receiving significant attention. The fields of "omics", such as genomics and proteomics, are beneficial for tackling these issues and for the development of more personalized disease treatments [71, 72]. These topics are also briefly introduced later in this book.

3 The Organization of the Textbook

3.1 The Seven Sections in the Book

This book is divided into **seven main sections**, each with multiple chapters. **Section one** introduces foundational concepts about cell biology. It is assumed that readers already have a foundational knowledge of cell biology and various basic biological concepts. As a refresher, this section therefore includes an overview of the structure and function of eukaryotic cells, as well as a chapter that focuses on the composition of the cell membrane and the transport of molecules across the membrane. **Section two** introduces proteins, **one of the four main macromolecules in cells**. After introducing basic protein structure, this section focuses primarily on an important category of proteins: enzymes. Enzymes are introduced early in the book because they are necessary for nearly all the biochemical processes and pathways discussed in the rest of the chapters. **Section three** discusses the major biomolecules and processes that are important for the flow of genetic information. Genetic information generally flows from DNA to RNA (both nucleic acids) to protein. This section covers the basic structure and function of DNA, as well as the process of DNA replication. It then covers transcription, where mRNA is synthesized from a DNA template, and translation, the synthesis of proteins from an RNA template. This section also discusses proteins in more depth, including protein modification, folding, and transport. **Section four** examines two additional key biomolecules, carbohydrates and lipids. Their basic structure and roles in key metabolic pathways are covered, with an emphasis on how they serve as energy sources and how they are stored during periods of low energy requirements. Topics include the fed and fasting states, the maintenance of energy homeostasis, and metabolic diseases that occur when these pathways are not functioning properly. **Section five** continues the theme of metabolism, focusing on the metabolism of amino acids and nucleotides. One-carbon metabolism, which involves the transfer of single-carbon units between different molecules, is also covered, and one chapter covers vitamins, many of which also have important roles as metabolic cofactors. **Section six** covers some advanced topics. These include cell signaling, cell cycle regulation, and topics that are important in cancer, such as tumor suppressor genes and oncogenes. Cell death and autophagy are also covered. A brief overview of immunology and virology is also included because, while this is not a major topic of the book, many drugs involve the immune system or viral infections. Finally, **section seven** introduces some advanced technologies. These include the use of stem cells and CRISPR-Cas9 technology. The use of omics studies to investigate DNA and gene expression, as well as their contributions to personalized medicine, is also discussed.

Importantly, throughout the text, the connections between these concepts and health, disease, and therapeutic strategies are addressed. Examples of topics include drugs that target cell surface receptors and transporters, as well as those that target specific enzymes. Drugs that have roles in altering DNA replication, gene expression, or the modification and translation of proteins are also covered. Many current therapies target metabolic pathways, such as those involving carbohydrate metabolism, and these will also be examined. Treatment of disorders that are associated with improper cell proliferation, such as cancer or autoimmune disease, will also be covered. Drugs that have been developed or are in the research phase, utilizing new technologies such as CRISPR, will also be addressed.

3.2 General Notes About the Composition of the Chapters

Many different drugs are mentioned in the text. Most drugs have a brand name and a generic name. Throughout the text, the first reference to all drugs is as follows: Commonly used brand name (generic name). For example: **Ozempic (semaglutide).** After this first reference, drugs may be referenced by their most commonly used names.

Enzymes play critical roles in biochemistry and are often important targets for drugs. Throughout the text, some important enzymes are highlighted in bold, as are drugs and some key concepts.

A series of three appendices is included at the end of the book. Appendix A provides a review of selected key biochemical concepts, including some basic structures of functional groups and molecule classes. It is assumed that students already have a fundamental background in chemistry and basic biology; however, this section may be useful for students as a review. It may be helpful to consult this section either before reading the book or when encountering unfamiliar concepts and structures throughout the text. Appendix B provides a review of shuttle systems and cycle systems covered in the metabolism sections, and Appendix C describes certain transamination reactions relevant to the study of protein metabolism.

Finally, at the end of each chapter, you will find:

- A table summarizing FDA-approved drugs that were discussed in the text, including their FDA approval dates. Some drugs have more than one FDA approval date as they are approved for different treatments. Generally, only the first approval date is included here.
- A list of definitions of key terms used in the chapter.
- A list of abbreviations used in the chapter.
- Practice questions to test your knowledge of the material.

FDA-Approved Drugs Covered in This Chapter

Commonly used brand name	Generic name	First FDA approval year	Main condition(s)	Main target
Aspirin	*Acetylsalicylic acid*	Pre-FDA	Pain and inflammation	Cox-1/2
Biktarvy	*Bictegravir/emtricitabine/tenofovir alafenamide*	2018	HIV	HIV integrase and reverse transcriptase
Casgevy	*Exagamglogene autotemcel*	2023	Sickle cell disease	β-Globin gene
Enbrel	*Etanercept*	1998	Autoimmune diseases	TNFα
Gleevec	*Imatinib*	2001	Chronic myelogenous leukemia (CML)	BCR–Abl
Herceptin	*Trastuzumab*	1998	HER2-positive breast cancer	HER2 receptor
Humira	*Adalimumab*	2002	Rheumatoid arthritis and other inflammatory diseases	TNFα
Inderal	*Propranolol*	1967	Hypertension and anxiety	β-Adrenergic receptors

(continued)

Commonly used brand name	Generic name	First FDA approval year	Main condition(s)	Main target
Kadcyla	*Ado-trastuzumab emtansine*	2013	HER2-positive breast cancer	HER2 receptor
Keytruda	*Pembrolizumab*	2014	Various cancers	PD-1 receptor
Lexapro	*Escitalopram*	2002	Depression and anxiety	Serotonin transporter (SERT)
Lipitor	*Atorvastatin*	1996	High cholesterol	HMG CoA reductase
Luxturna	*Voretigene neparvovec*	2017	Inherited retinal disease	RPE65 gene
Nolvadex	*Tamoxifen*	1977	Estrogen-positive breast cancer	Estrogen receptor
Onpattro	*Patisiran*	2018	Hereditary ATTR amyloidosis	Transthyretin mRNA
Opdivo	*Nivolumab*	2014	Various cancers	PD-1 receptor
Ozempic	*Semaglutide*	2017	Type 2 diabetes, obesity	GLP-1 receptor
Provenge	*Sipuleucel-T*	2010	Prostate cancer	PAP antigen
Prozac	*Fluoxetine*	1987	Depression	SERT
Spinraza	*Nusinersen*	2016	Spinal muscular atrophy	SMN2 pre-mRNA splicing
Tamiflu	*Oseltamivir*	1999	Influenza	Influenza neuraminidase
Veklury	*Remdesivir*	2020	COVID-19	Sars-Cov-2 RNA polymerase
Vyndaqel	*Tafamidis*	2019	Transthyretin amyloidosis	TTR tetramer
Xylocaine	*Lidocaine*	1948	Local anesthesia	Voltage-gated sodium channels

References

1. Alberts B, Bray D, Hopkin K, Johnson A, Lewis J, Raff M, Roberts K, Walter P. Essential cell biology. Norton and Company; 2023.
2. Cooper G, Adams K. The cell: a molecular approach. Sinauer Associates; 2022.
3. Berg JM, Tymoczko JL, Gatto GJ Jr, Stryer L. Biochemistry. 9th ed. New York: MacMillan Learning; 2019.
4. Alberts B, Heald R, Johnson A, Morgan D, Raff M, Roberts K, Walter P, Wilson J, Hunt T. Molecular biology of the cell. W.W. Norton and Company; 2022.
5. Atkins P, de Paula J, Keeler J. Atkins' physical chemistry. 11th ed. Oxford University Press; 2018.
6. Whalen K, Lerchenfeldt S, Giordano C. Lippincott illustrated reviews: pharmacology. Wolters Kluwer; 2021.
7. Ritter JM, Flower RJ, Henderson G, Loke YK, MacEwan DJ, Rang HP. Rang and dale's pharmacology. 10th ed; 2023.
8. FDA. What are "biologics" questions and answers. 2018. Available from: https://www.fda.gov/about-fda/center-biologics-evaluation-and-research-cber/what-are-biologics-questions-and-answers.
9. Singh N, Vayer P, Tanwar S, Poyet J-L, Tsaioun K, Villoutreix BO. Drug discovery and development: introduction to the general public and patient groups. Front Drug Discov. 2023:31201419. https://doi.org/10.3389/fddsv.2023.1201419.
10. Southey MW, Brunavs M. Introduction to small molecule drug discovery and preclinical development. Front Drug Discov. 2023;29 https://doi.org/10.3389/fddsv.2023.1314077.
11. Moumné L, Marie AC, Crouvezier N. Oligonucleotide therapeutics: from discovery and development to patentability. Pharmaceutics. 2022;14(2) https://doi.org/10.3390/pharmaceutics14020260.
12. FDA. LIPITOR (atorvastatin) approval overview. 1996. Available from: https://www.accessdata.fda.gov/scripts/cder/daf/index.cfm?event=overview.process&ApplNo=20702.
13. Miner J, Hoffhines A. The discovery of aspirin's antithrombotic effects. Tex Heart Inst J. 2007;34(2):179–86.
14. Katzung BG, Vanderah TW, Trevor AAJ. Katzung's basic and clinical pharmacology. 16th ed. McGraw-Hill Education; 2024.

15. Chen W-H, Chen C-H, Hsu M-C, Chang R-W, Wang C-H, Lee T-S. Advances in the molecular mechanisms of statins in regulating endothelial nitric oxide bioavailability: interlocking biology between eNOS activity and L-arginine metabolism. Biomed Pharmacother. 2024;171:116192. https://doi.org/10.1016/j.biopha.2024.116192.
16. Silva-Velasco DL, Cervantes-Pérez LG, Sánchez-Mendoza A. ACE inhibitors and their interaction with systems and molecules involved in metabolism. Heliyon. 2024;10(2):e24655. https://doi.org/10.1016/j.heliyon.2024.e24655.
17. FDA. Gleevec(imatinib): drugs@FDA approval overview Silver Spring, MD: U.S. Food and Drug Administration. 2001. Available from: https://www.accessdata.fda.gov/scripts/cder/daf/index.cfm?event=overview.process&ApplNo=021588.
18. Lin YL, Meng Y, Jiang W, Roux B. Explaining why Gleevec is a specific and potent inhibitor of Abl kinase. Proc Natl Acad Sci USA. 2013;110(5):1664–9. https://doi.org/10.1073/pnas.1214330110.
19. FDA. INDERAL (propranolol) approval overview. 1967. Available from: https://www.accessdata.fda.gov/scripts/cder/daf/index.cfm?event=overview.process&ApplNo=016418.
20. FDA. NOLVADEX (tamoxifen) approval overview. 1977. Available from: https://www.accessdata.fda.gov/scripts/cder/daf/index.cfm?event=overview.process&ApplNo=17970.
21. do Vale GT, Ceron CS, Gonzaga NA, Simplicio JA, Padovan JC. Three generations of β-blockers: history, class differences and clinical applicability. Curr Hypertens Rev. 2019;15(1):22–31. https://doi.org/10.2174/1573402114666180918102735.
22. Goodsell DS. The molecular perspective: tamoxifen and the estrogen receptor. Stem Cells. 2002;20(3):267–8. https://doi.org/10.1634/stemcells.20-3-267.
23. Jordan VC. Selective estrogen receptor modulation: concept and consequences in cancer. Cancer Cell. 2004;5(3):207–13. https://doi.org/10.1016/s1535-6108(04)00059-5.
24. Nauck MA, Quast DR, Wefers J, Meier JJ. GLP-1 receptor agonists in the treatment of type 2 diabetes - state-of-the-art. Mol Metab. 2021;46:101102. https://doi.org/10.1016/j.molmet.2020.101102.
25. Tamayo-Trujillo R, Ruiz-Pozo VA, Cadena-Ullauri S, Guevara-Ramírez P, Paz-Cruz E, Zambrano-Villacres-R, et al. Molecular mechanisms of semaglutide and liraglutide as a therapeutic option for obesity. Front Nutr. 2024;11 https://doi.org/10.3389/fnut.2024.1398059.
26. FDA. LEXAPRO (escitalopram) approval overview. 2002. Available from: https://www.accessdata.fda.gov/scripts/cder/daf/index.cfm?event=overview.process&ApplNo=021323.
27. FDA. Prozac (fluoxetine) approval overview. 1987. Available from: https://www.accessdata.fda.gov/scripts/cder/daf/index.cfm?event=overview.process&ApplNo=18936.
28. FDA. Xylocaine (lidocaine) approval overview. 1948. https://www.accessdata.fda.gov/scripts/cder/daf/index.cfm?event=overview.process&ApplNo=006488
29. FDA. HERCEPTIN (trastuzumab) approval overview. 1998. Available from: https://www.accessdata.fda.gov/scripts/cder/daf/index.cfm?event=overview.process&ApplNo=103792.
30. FDA. USFaDA. HUMIRA (adalimumab) approval overview 1001. Available from: https://www.accessdata.fda.gov/scripts/cder/daf/index.cfm?event=overview.process&ApplNo=125057.
31. Maadi H, Soheilifar MH, Choi WS, Moshtaghian A, Wang Z. Trastuzumab mechanism of action; 20 years of research to unravel a dilemma. Cancers (Basel). 2021;13(14) https://doi.org/10.3390/cancers13143540.
32. Evangelatos G, Bamias G, Kitas GD, Kollias G, Sfikakis PP. The second decade of anti-TNF-a therapy in clinical practice: new lessons and future directions in the COVID-19 era. Rheumatol Int. 2022;42(9):1493–511. https://doi.org/10.1007/s00296-022-05136-x.
33. FDA. Enbrel (etanercept): drugs@FDA approval overview Silver Spring, MD: U.S. Food and Drug Administration. 1998. Available from: https://www.accessdata.fda.gov/scripts/cder/daf/index.cfm?event=overview.process&ApplNo=103795.
34. Amgen. Enbrel®(etanercept) Mechanism of action: ENBREL is a soluble form of the tumor necrosis factor (TNF) receptor. 2024. Available from: https://www.enbrelpro.com/mechanism-of-action#:~:text=ENBREL%20is%20a%20dimeric%20fusion,1.
35. Rath T, Baker K, Dumont JA, Peters RT, Jiang H, Qiao SW, et al. Fc-fusion proteins and FcRn: structural insights for longer-lasting and more effective therapeutics. Crit Rev Biotechnol. 2015;35(2):235–54. https://doi.org/10.3109/07388551.2013.834293.
36. FDA. KADCYLA (ado trastuzumab emtansine) approval overview. 2013. Available from: https://www.accessdata.fda.gov/scripts/cder/daf/index.cfm?event=overview.process&ApplNo=125427.
37. Kadcyla. Proposed mechanism of action (MOA) for KADCYLA. Available from: https://www.kadcyla-hcp.com/early-breast-cancer/about/how-kadcyla-works.html?c=kad-17fb7730ed5&gad_source=1&gclid=CjwKCAjw%2D%2DK_BhB5EiwAuwYoyvgldsrmzgerEPD0MzVXPL8YGFldraHb0-okTnK26FpzwM5S3SgBixoCnOQQAvD_BwE&gclsrc=aw.ds.
38. FDA. KEYTRUDA (pembrolizumab) approval overview. 2014. Available from: https://www.accessdata.fda.gov/scripts/cder/daf/index.cfm?event=overview.process&ApplNo=125514.
39. FDA. OPDIVO (nivolumab) approval overview. 2014. Available from: https://www.accessdata.fda.gov/scripts/cder/daf/index.cfm?event=overview.process&ApplNo=125554.

40. Momtaz P, Postow MA. Immunologic checkpoints in cancer therapy: focus on the programmed death-1 (PD-1) receptor pathway. Pharmgenomics Pers Med. 2014;7:357–65. https://doi.org/10.2147/pgpm.S53163.
41. FDA. Provenge (sipuleucel-T) approval overview. 2010. Available from: https://www.fda.gov/vaccines-blood-biologics/cellular-gene-therapy-products/provenge-sipuleucel-t
42. Saxena M, van der Burg SH, Melief CJM, Bhardwaj N. Therapeutic cancer vaccines. Nat Rev Cancer. 2021;21(6):360–78. https://doi.org/10.1038/s41568-021-00346-0.
43. FDA. FDA approves first COVID-19 vaccine: U.S. Food and Drug Administration. 2021. Available from: https://www.fda.gov/news-events/press-announcements/fda-approves-first-covid-19-vaccine.
44. Leong KY, Tham SK, Poh CL. Revolutionizing immunization: a comprehensive review of mRNA vaccine technology and applications. Virol J. 2025;22(1):71. https://doi.org/10.1186/s12985-025-02645-6.
45. Shi Y, Shi M, Wang Y, You J. Progress and prospects of mRNA-based drugs in pre-clinical and clinical applications. Signal Transduct Target Ther. 2024;9(1):322. https://doi.org/10.1038/s41392-024-02002-z.
46. FDA. ONPATTRO (patisiran) approval overview. 2018. Available from: https://www.accessdata.fda.gov/scripts/cder/daf/index.cfm?event=overview.process&ApplNo=210922.
47. Mansouri M, Mansouri K, Taheri Z, Hossaini Alhashemi S, Dehshahri A. The Fomivirsen, Patisiran, and Givosiran odyssey: how the success stories may pave the way for future clinical translation of nucleic acid drugs. BioDrugs. 2025; https://doi.org/10.1007/s40259-025-00711-7.
48. FDA. Spinraza (nusinersen) approval overview. 2016. Available from: https://www.accessdata.fda.gov/scripts/cder/daf/index.cfm?event=overview.process&ApplNo=209531.
49. Ottesen E. ISS-N1 makes the first FDA-approved drug for spinal muscular atrophy. Transl Neurosci. 2017;8:1–6. https://doi.org/10.1515/tnsci-2017-0001.
50. Sparks Therapeutics. About Luxturna. 2024. Available from: https://luxturna.com/about-luxturna/#about-luxturna.
51. FDA. Casgevy. 2024. Available from: https://www.fda.gov/vaccines-blood-biologics/casgevy.
52. Hoy SM. Exagamglogene Autotemcel: first approval. Mol Diagn Ther. 2024;28(2):133–9. https://doi.org/10.1007/s40291-024-00696-z.
53. Li K, Crews CM. PROTACs: past, present and future. Chem Soc Rev. 2022;51(12):5214–36. https://doi.org/10.1039/d2cs00193d.
54. Chen QH, Munoz E, Ashong D. Insight into recent advances in degrading androgen receptor for castration-resistant prostate cancer. Cancers (Basel). 2024;16(3) https://doi.org/10.3390/cancers16030663.
55. FDA. Vyndaqel (Tafamidis) approval overview. 2019. Available from: https://www.accessdata.fda.gov/scripts/cder/daf/index.cfm?event=overview.process&ApplNo=211996.
56. Powers ET, Amass L, Baylor L, Fernández-Arias I, Riley S, Kelly JW. Transthyretin kinetic stabilizers for ATTR amyloidosis: a narrative review of mechanisms and therapeutic benefits. Cardiol Ther. 2025;14(3):333–50. https://doi.org/10.1007/s40119-025-00423-7.
57. Hoang DM, Pham PT, Bach TQ, Ngo ATL, Nguyen QT, Phan TTK, et al. Stem cell-based therapy for human diseases. Signal Transduct Target Ther. 2022;7(1):272. https://doi.org/10.1038/s41392-022-01134-4.
58. Sun D, Shi X, Li S, Wang X, Yang X, Wan M. CAR-T cell therapy: a breakthrough in traditional cancer treatment strategies (Review). Mol Med Rep. 2024;29(3) https://doi.org/10.3892/mmr.2024.13171.
59. FDA. Biktarvy(bictegravir/emtricitabine/tenofovir): Drugs@FDA approval overview. 2018. Available from: https://www.accessdata.fda.gov/scripts/cder/daf/index.cfm?event=overview.process&ApplNo=210251.
60. FDA. Velkury (remdesivir) approval overview. 2020. Available from: https://www.accessdata.fda.gov/scripts/cder/daf/index.cfm?event=overview.process&ApplNo=214787.
61. FDA. Tamiflu (oseltamivir) approval overview. 1999. Available from: https://www.accessdata.fda.gov/scripts/cder/daf/index.cfm?event=overview.process&ApplNo=021246.
62. Orenstein WA, Offit PA, Edwards KM, Plotkin SA. Plotkin's vaccines. 8th ed. Elsevier; 2023.
63. Wang Y, Minden A. Current molecular combination therapies used for the treatment of breast cancer. Int J Mol Sci. 2022;23(19) https://doi.org/10.3390/ijms231911046.
64. Damodaran C, Cho J-Y, Güngör C. Therapeutic resistance and combination therapy for cancer: recent developments and future directions. Sci Rep. 2025;15(1):26881. https://doi.org/10.1038/s41598-025-12725-w.
65. Grogan S, Preuss C. Pharmacokinetics. In: StatPearls. Treasure Island (FL): StatPearls Publishing; 2023.
66. Holford N. Pharmacodynamic principles and the time course of immediate drug effects. Transl Clin Pharmacol. 2017;25(4):157–61. https://doi.org/10.12793/tcp.2017.25.4.157.
67. Bio K. Pharmacodynamics: key differences in PK vs PD. 2025. Available from: https://kcasbio.com/blogs/pharmacokinetics-vs-pharmacodynamics/#:~:text=Pharmacokinetics%20and%20pharmacodynamics%20are%20two,efficacy%20and%20minimal%20side%20effects.
68. Glassman PM, Muzykantov VR. Pharmacokinetic and pharmacodynamic properties of drug delivery systems. J Pharmacol Exp Ther. 2019;370(3):570–80. https://doi.org/10.1124/jpet.119.257113.

69. FDA. Ozempic (semaglutide) approval overview. 2017. Available from: https://www.accessdata.fda.gov/scripts/cder/daf/index.cfm?event=overview.process&ApplNo=209637.
70. Cleveland Clinic. Immunosuppressants. 2023. Available from: https://my.clevelandclinic.org/health/treatments/10418-immunosuppressants.
71. Sadee W, Wang D, Hartmann K, Toland AE. Pharmacogenomics: driving personalized medicine. Pharmacol Rev. 2023;75(4):789–814. https://doi.org/10.1124/pharmrev.122.000810.
72. National Human Genome Research Institute. Personalized Medicine. 2025. Available from: https://www.genome.gov/genetics-glossary/Personalized-Medicine.

Introduction to Cells in Health and Disease

2

Abstract

Understanding basic cell biology concepts is crucial for comprehending biochemistry, molecular biology, and drug development. This chapter provides an overview of cell biology, including some historical context for the field, and highlights the relevance of cellular function to human health and disease. The evolution of prokaryotic and eukaryotic cells is introduced, followed by a discussion of the basic components of eukaryotic cells, their organelles, and how malfunctioning cellular components can lead to disease. The cell cycle and the concept of cell differentiation are introduced, along with a discussion of the differences between single-cell and multicellular organisms. This chapter also introduces cellular communication via signal transduction.

Keywords

Cell biology · Organelles · Cell structure · Cell cycle · Cell signaling

Abbreviations

ATP	Adenosine triphosphate
CFTR	Cystic fibrosis transmembrane conductance regulator
DNA	Deoxyribonucleic acid
ECM	Extracellular matrix
EGF	Epidermal growth factor
ER	Endoplasmic reticulum
G0	Quiescent phase
G1/G2	Gap 1/Gap 2 phases
HMG-CoA	3-Hydroxy-3-methylglutaryl-coenzyme A
LDL	Low-density lipoprotein
M phase	Mitotic phase
mRNA	Messenger RNA
PCM	Pericentriolar material
RNA	Ribonucleic acid
rRNA	Ribosomal RNA
S phase	Synthesis phase
tRNA	Transfer RNA

What You Will Learn

- Why is it important to study cells
- How most disorders are linked to cell behavior
- The basic tenets of cell theory
- How cells evolved
- The difference between prokaryotic and eukaryotic cells
- The basic components of eukaryotic cells, including organelles, ribosomes, the cytoskeleton, and the plasma membrane
- How different cellular components are associated with disease when they are not functioning properly
- An introduction to the cell cycle
- Single-cellular versus multicellular organisms
- An introduction to cell signaling

A. Minden, *Biochemistry, Cell and Molecular Biology for Pharmacists and Health Professionals*,
https://doi.org/10.1007/978-3-032-20529-2_2

1 Introduction: Why Is It Important to Study Cells?

The cell is the basic unit of life. All of us begin from a single cell, which, during the course of development, divides and gives rise to all the specialized cells in the body. Cells are fundamental to life, and without them, life as we know it could not exist. At the same time, most diseases and disorders arise from problems within our cells. Therefore, understanding how cells work is critical not only for understanding normal development but also for learning how to treat disorders arising from abnormal cell function.

Understanding some basic cell biology concepts is crucial for comprehending biochemistry, molecular biology, and drug development. Though students are expected to have some familiarity with basic cell biology, this chapter provides an overview of the subject. It discusses why cell biology is relevant to human health and disease and provides some historical context for the study of cell biology. This chapter also covers the evolution of cells, followed by a discussion of the basic components of eukaryotic cells, their organelles, and their relevance to human health. This chapter also provides a brief overview of the cell cycle, the process of cell differentiation, the differences between single-cell and multicellular organisms, and signal transduction. The following chapter will focus on membranes and transport across membranes, as these topics are especially important in the fields of pharmacology and for understanding how drugs function.

2 Disorders Linked to Abnormal Cell Behavior

Abnormalities in cell function can lead to health problems. Abnormal cell behavior can take on various forms. These can include too much cell division, too little cell division, or abnormalities in cell function that may not necessarily be directly related to cell division, as discussed below.

2.1 Some Disorders Are Caused by Too Much Cell Division

To renew themselves, cells must periodically divide, a process in which one cell divides into two daughter cells. This is the process of cell division. Cell division is a normal part of the cell life cycle, and normal cells respond to cues that dictate when they should divide. Cell division can go awry when these normal regulatory mechanisms are not functioning, which can sometimes lead to health problems. One problem that can lead to serious consequences is excessive or unregulated cell division.

An important example is cancer. In **cancer**, the regulatory mechanisms that normally control cell division can be disrupted, resulting in uncontrolled cell division. As a result, an excessive number of cells are produced, which can lead to the formation of a tumor. As the cells continuously divide, they may also develop gene mutations, leading to more abnormalities. Unchecked cell division is also sometimes linked to inappropriate cell migration. Cancer cells that move inappropriately from their origins to invade distant tissues and organs are the basis of metastasis, which is difficult to treat.

Tumor cells that migrate from their origin and invade tissues and organs are considered malignant. Some tumor cells do not become invasive, and these are often benign tumors. Benign tumors are also associated with dysregulated cell division. While benign tumors are usually not life-threatening, they can, in some cases, lead to health problems such as pain, pressure, or physical disfiguration, and if they disrupt vital tissues and organs, they can lead to severe problems [1, 2].

Dysregulated cell division can also lead to other health problems beyond cancer. For example, in some inflammatory and **autoimmune diseases**, abnormal cell division and activation of immune cells contribute to inflammation and tissue damage. In psoriasis [3], for example, skin cells (keratinocytes) divide excessively. Additionally, some T cells involved in the inflammatory response undergo abnormal cell division, resulting in immune cell expansion that contributes to the disease's inflammatory response. **Psoriasis** is characterized by thickened, scaly patches of skin, which can cause discomfort, itching, and bleeding.

Uterine **fibroids** are another example of a problem that stems from unregulated cell division. Uterine fibroids are noncancerous growths in the uterus. They result from the uncontrolled growth of cells in smooth muscle and connective tissue. They can lead to pain, heavy menstrual bleeding, pressure on nearby organs, and sometimes to fertility issues. These are just a few examples of the potential dangers of excessive cell division [4].

2.2 Some Disorders and Health Problems Are Caused by Too Little Cell Division

Conversely, in contrast to too much cell division, other health problems are caused by insufficient cell division [5].

Heart cells, particularly cardiomyocytes, are an example of a cell type with very limited capacity to divide after birth. In some **heart conditions**, such as heart failure or after a heart attack, there is a weakening of these heart muscle cells. Because the adult cardiomyocytes do not regenerate by cell division, the heart's ability to repair itself is inefficient. This makes these conditions challenging to treat.

Similarly, nerve cells do not divide efficiently, and most do not divide at all after birth. For this reason, spinal cord injury can be difficult or impossible to treat. After a severe accident in which nerve cells in the spinal cord are damaged, a lack of division of these cells can lead to paralysis.

Aplastic anemia is another disorder characterized by inadequate cell division. In this rare condition, reduced cell division of certain blood cells produced in the bone marrow leads to severe anemia [6].

Muscular dystrophies are characterized by progressive muscle degeneration and weakness. In some cases of muscular dystrophy, insufficient cell division and regeneration contribute to the loss of muscle tissue over time [7].

2.3 Some Disorders Are Caused by Abnormal Cell Function

Some disorders are caused by abnormalities in cell function, sometimes independently of the cell division process.

In **hemophilia**, for example, the production of specific clotting factors is deficient. The clotting factors that are deficient in hemophilia are normally synthesized by hepatocytes (liver cells) or endothelial cells within the liver. Patients with hemophilia, therefore, suffer severe consequences when they experience bleeding, because their blood cannot clot normally [8].

In **cystic fibrosis**, a defective protein (cystic fibrosis transmembrane conductance regulator (CFTR)) leads to disruption of ion transport across the membranes of epithelial cells that rely on this transporter. This is especially problematic in the epithelial cells lining the lungs, where disrupted chloride and water movement across membranes leads to the buildup of a thick mucus. This impairs normal lung function and leads to infections and inflammation [9].

2.4 Some Health Problems Are Caused by Infection with Foreign Cells

A different type of health problem involves not only our own cells but **infection by foreign cells**. The infectious cells are often single-celled prokaryotes such as *Escherichia coli* (*E. Coli*). Although large amounts of bacteria live normally in our bodies and have necessary and beneficial functions, such as aiding in digestion, some bacteria cause infections and lead to tissue damage. An important milestone in treating human disease has been the development of antibiotics to combat such infections. The development of antibiotics required a deep understanding of the biology of prokaryotic cells [10, 11].

Since cells play such important roles in human health and disease, studying them and understanding their functions is critical for identifying disease causes and developing treatments. A deeper understanding of the mechanisms that regulate cell division and function may help scientists develop ways to reestablish normal processes and combat diseases.

3 Cell Theory

Before understanding that cells were the fundamental units of life, scientists proposed various theories about the origins and basis of life, such as spontaneous generation and the concept of vitalism (the idea that organisms require a vital force) [12]. In the nineteenth century, cell theory became recognized as a basic principle in biology, describing cells as the basic units of life. This theory remains a foundation of modern biology.

3.1 The Basics of Cell Theory

The basic tenets of cell theory are:

1. Every living thing on Earth is made of one or more cells.
2. The cell is the basic unit of structure and organization in living organisms.
3. Cells arise from preexisting cells.

These basic tenets of cell theory have been revised and expanded over time as scientific methods and tools improved. Understanding the concept of cells was impossible until mechanisms existed to visualize small, microscopic organisms. Therefore, a key to discovering cells and understanding their importance was the invention and improvement of the microscope. This led to many studies and discoveries that were fundamental for the development of cell theory. Some of the key discoveries are discussed below. To read more about these discoveries, see [12–16].

3.2 Early Discoveries

1590: The first compound microscopes were made, but these early microscopes were inefficient.

1655: **Robert Hooke** of England improved the microscope. Using a microscope that he made, he observed cork plants. He saw that they contained what appeared to be individual **compartments**, which he named **"cells,"** which is Latin for "small rooms" [17, 18]. This finding can mark the beginning of the field of cell biology as we know it.

Late 1600s: **Antonie van Leeuwenhoek** also built a microscope with superior lenses compared to the previous one. He used his microscope to observe bacteria and protozoa [18].

1838: **Matthias Jakob Schleiden**, a botanist from Germany, observed plant tissue under the microscope and concluded that plants were made of cells.

Early 1830s: German physiologist **Theodore Schwann** observed animal tissue under the microscope. He reached a conclusion similar to Schleiden's, observing that animals are also made of cells. An important step in formulating cell theory was when **Schleiden** and **Schwann** compared their observations. Upon doing so, they realized that cells could be found in plants as well as animals, thus suggesting that cells were the basic units of all life forms [14].

3.3 Expansion of the Theory

In the mid-1800s, **Robert Remak,** a German biologist, proposed that cells arose from pre-existing cells, which is another important concept in cell theory. However, his idea was not initially accepted. Shortly after this, German pathologist **Rudolf Virchow** published an article stating that "Omnis cellula e cellula (Latin for "every cell comes from a cell"), indicating that every cell is derived from cells. This formed the basis for the third tenet of cell theory [12, 14, 19].

The three tenets of cell theory have since become foundational for understanding the molecular and cellular basis of life, paving the way for the development of treatments for disorders linked to cell function.

4 Evolution of Cells

4.1 How Did Cells First Evolve?

While it took many years of studying, discussion, and research to determine how the first cells evolved, it is now widely accepted that cells evolved via abiogenesis. This is the idea that life originated from nonlife spontaneously.

4.1.1 The Miller–Urey Experiment

The now-famous Miller–Urey experiment was pivotal in strengthening the theory that the first cells arose by abiogenesis. In this experiment, conducted in 1953, Stanley Miller and Harold Urey simulated early Earth conditions in the laboratory. They combined water, ammonia, methane, hydrogen, heat, and electricity. These were all believed to be components of the primitive Earth, and the electricity was meant to mimic lightning. Under these conditions, amino acids were found to have formed spontaneously after about a week. This was a groundbreaking result because it demonstrated that amino acids, major components of cells and life, could form spontaneously under the conditions found on primitive Earth. Subsequent experiments showed that nucleotides could also form under similar conditions. This is an even more significant finding because nucleotides comprise DNA and RNA, the molecules that enable life to replicate.

According to the theory of abiogenesis, monomeric biomolecules, such as amino acids and nucleotides, formed spontaneously in primitive Earth conditions. Over time, these biomolecules polymerized with each other to form macromolecules, specifically proteins and RNA. These macromolecules eventually became encapsulated within a phospholipid bilayer, marking the first step toward the evolution of cells [20].

4.1.2 Criticism of the Miller–Urey Experiment

While the Miller–Urey experiment was groundbreaking, it faced some scientific criticism over time. One criticism is that the experiment may not have precisely mimicked Earth's early conditions, and it remained unclear whether amino acids and nucleic acids could form spontaneously under the actual conditions of primitive Earth. Thus, the exact mechanism by which molecules formed spontaneously under Earth's primitive conditions is still not completely understood, but abiogenesis is still widely considered to be the most plausible explanation for the origin of life [21].

4.2 The First Protocell

Among the organic molecules thought to have formed in the early atmosphere are simple amphiphilic molecules, such as fatty acids, which may have been precursors to more modern phospholipids [1]. Like phospholipids, fatty acids have an amphipathic structure, meaning that they are made up of a hydrophilic head group (water-attracting) and a hydrophobic tail (water-repelling). This structure was foundational to the eventual development of cells as we know them today. Due to their structure, fatty acids spontaneously form a bilayer in aqueous conditions at the right pH and concentrations. The hydrophobic ends interact with each other and cluster away from the aqueous environment, while the hydrophilic heads face outward, interacting with the aqueous environment. This arrangement typically leads to the spontaneous formation of a sphere, which results in all the hydrophobic groups being completely shielded from the aqueous environment [22] (Fig. 2.1). Although modern cells have phospholipid bilayers, the earliest protocells are thought to have been bound by simple fatty acid membranes. Complex cells of today's multicellular organisms are encapsulated in a phospholipid bilayer with features resembling this ancient structure [1].

The first protocells that appeared on Earth were likely simpler than today's cells. They were likely simple fatty acid bilayers that encapsulated

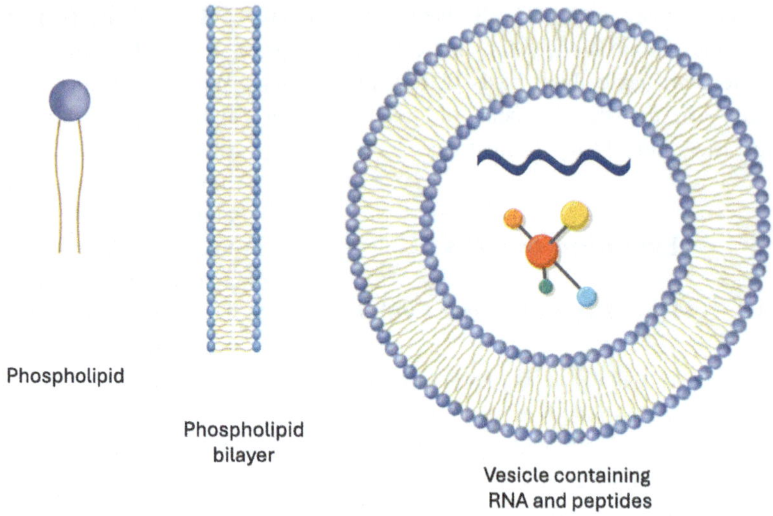

Fig. 2.1 The basic structure of the first protocell. The first Protocell likely consisted of a simple fatty acid bilayer that enclosed RNA molecules, and possibly some small peptides. (*Source*: Phospholipids and molecule illustrations from Freepik (premium license); RNA illustration adapted from Servier Medical Art (https://smart.servier.com/), licensed under the Creative Commons Attribution 4.0 (CC BY 4.0) license)

RNA molecules and lacked any organelles. According to the RNA world hypothesis [23, 24], RNA, rather than DNA, is thought to have carried the first genetic information, and it could also catalyze chemical reactions by acting as ribozymes (RNA molecules that function as enzymes). The bilayers on these protocells may have grown in size upon incorporating additional fatty acids from the environment, and when they became too large, they could split into two smaller vesicles. At the same time, some RNA molecules possessed the remarkable ability to catalyze their own replication, allowing genetic information to be transmitted to daughter cells/vesicles. This mechanism probably formed the basis of the first cells and laid the foundation for replication and early life.

RNA is thought to be the first genetic material. Over time, however, there was a transition from RNA to DNA as the genetic material of most organisms. The transition to DNA was a huge evolutionary advantage and an important step. DNA is more stable than RNA and is therefore ideal for long-term storage. DNA is also double-stranded, which has multiple advantages, including providing a backup template that can aid in DNA repair. DNA replication is more accurate than the RNA-based process that likely existed in the RNA world. Importantly, however, DNA replication is not perfectly accurate and is sometimes associated with occasional mutations. This ability to occasionally mutate was essential for the development of more complex cells. As DNA mutates, cells undergo gradual changes. The changes that are most beneficial to survival are those with a lasting impact. DNA's ability to replicate, combined with a low level of mutation, is a critical factor in the diversity of life on Earth and remains important today [1, 12, 25–27].

5 Prokaryotic Cells Compared with Eukaryotic Cells

5.1 Prokaryotic Cells

The first cells on Earth formed about 3.5 billion years ago. These first cells were primitive prokaryotic cells, which, over time, gave rise to the current prokaryotes: bacteria and archaea. These prokaryotic organisms were unicellular, consisting of a single cell. Prokaryotes are still unicellular today, although archaea can form multicellular aggregates.

Today's prokaryotic cells share certain common features. A major factor that distinguishes prokaryotes from eukaryotes is the absence of organelles and a nucleus. Prokaryotes have a plasma membrane, and most but not all bacteria have a cell wall outside the membrane. Some prokaryotic cells also have a capsule outside their cell walls. The capsule, usually made up of polysaccharides (sugar), mainly provides protection and adhesion, while the cell wall provides protection and structural support. The composition of the cell wall varies, but usually includes peptidoglycans (composed of polysaccharides and proteins) and lipopolysaccharides (composed of lipids and polysaccharides) [28]. In prokaryotes, the DNA is generally circular, and since prokaryotes lack a nucleus, they do not have a separate compartment for this genetic material.

5.2 Eukaryotic Cells

Eukaryotic cells include plant cells and animal cells (Fig. 2.2). The first eukaryotic cells appeared on Earth about 1.5 billion years ago. The main difference between eukaryotes and prokaryotes is that eukaryotes have a nucleus and membrane-bound organelles. The evolution of eukaryotic cells and their organelles is thought to have involved endosymbiosis.

Mitochondria, often referred to as the powerhouse organelles of eukaryotic cells, play a central role in the endosymbiosis theory. According to this theory, mitochondria evolved through a symbiotic relationship between two types of prokaryotic organisms: anaerobic bacteria and aerobic bacteria. Aerobic bacteria can carry out aerobic respiration, utilizing oxygen to produce ATP, the primary energy source in cells. In contrast, anaerobic bacteria are at a disadvantage because they cannot utilize oxygen for energy and instead rely on anaerobic respiration, which yields a significantly lower amount of ATP.

As the Earth became more oxygen-rich due to the presence of photosynthetic organisms, aerobic cells (those that utilize oxygen for energy production) gained a significant advantage over anaerobic cells, which cannot use oxygen. According to the endosymbiotic theory, some ancestral anaerobic cells are hypothesized to have engulfed aerobic prokaryotic cells. Instead of digesting them, a symbiotic relationship formed. The engulfed aerobic cells helped the host cell generate energy more efficiently by utilizing oxygen. The new combined cell now had an evolutionary advantage. Over many years of evolution, the engulfed aerobic bacteria developed into permanent cell structures, which are now known as the mitochondria. This theory is supported by evidence that, even now, mitochondria possess some of their own DNA, which is circular, like prokaryotic DNA, and is separate from the nucleus.

Other organelles probably did not evolve from the direct engulfment of other cells, but they likely developed from internal folding and protrusions in the host cell membranes, some of which resulted from the engulfment of anaerobic bacteria [1, 29, 30].

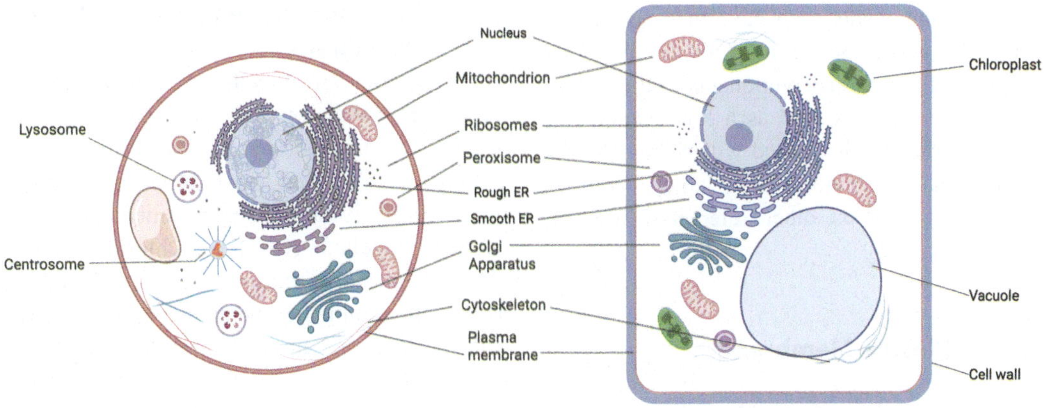

Fig. 2.2 Eukaryotic cells. Eukaryotic cells include animal cells and plant cells. They are membrane bound, and they contain a nucleus and membrane-bound organelles. Plant cells also contain a cell wall. (*Source*: Created with Biorender.com)

6 Eukaryotic Cell Structures, Functions, and Disorders

The cell membrane, also known as the plasma membrane, surrounds eukaryotic cells and acts as a barrier to control what enters and exits the cell. Inside the plasma membrane is the cytoplasm. The cytoplasm contains the cytosol, a gel-like fluid. The cytoplasm also contains the cytoskeleton and the cell's organelles. Many of the cell's organelles are part of the endomembrane system. This system consists of organelles that are connected by shared membranes. This network includes the outer nuclear membrane (ONM), rough and smooth endoplasmic reticulum (ER), Golgi apparatus, lysosomes, endosomes, and the plasma membrane. These structures are connected, either directly or indirectly, by membrane flow, which involves budding and fusion of vesicles that shuttle material between the different organelles or the cell surface. Other important cellular components, such as ribosomes, the cytoskeleton, mitochondria, and the centrosome, are not part of the endomembrane system. The cytoskeleton is not membrane-bound and is not an organelle. The ribosomes are also not membrane-bound but are usually classified as organelles. The centrosome is also classified as an organelle, despite not being membrane-bound. The mitochondria are membrane-bound organelles, but they are not part of the endomembrane system. The various cellular components each have their specialized functions, which are essential for the life of the cell [31]. The various components of the cell are described below, as well as a few examples of disorders that are associated with them. The discussion in this section draws largely from the [1, 31] as well as other references listed in the text.

6.1 Plasma Membrane

6.1.1 Function of the Plasma Membrane

The organelles and all components of the cell are found within the plasma membrane. The primary function of the plasma membrane is to act as a gatekeeper. The membrane helps determine what can enter or exit the cell. It is semipermeable, meaning only certain things can pass through it without the use of transport proteins or channels [1].

6.1.2 Basic Structure of the Plasma Membrane

The plasma membrane comprises a phospholipid bilayer with proteins and other lipids embedded within it. The phospholipid bilayer structure is also the basic structure of all membrane-bound organelles within the cell. The membrane's fundamental component is the phospholipid. Phospholipids, as discussed above, are amphipathic molecules (see Fig. 2.3). They have polar (hydrophilic) head groups and nonpolar (hydrophobic) tails [1]. Within the membrane, components can move around within their individual layer. This is called the fluid-mosaic model and was first proposed in 1972 [32]. Within the plasma membrane are embedded proteins that have many functions. These proteins include channels, pores, transport proteins, pumps, and signaling molecules such as receptors that bind to hormones and growth factors. Other lipids, such as cholesterol, are also embedded in the membrane, and they have roles such as regulating membrane fluidity. The plasma membrane is discussed in more detail in Chap. 3.

6.1.3 Diseases That Involve the Plasma Membrane

Many diseases are directly associated with the plasma membrane or proteins embedded within it, such as transporters. **Cystic Fibrosis**, for example, as discussed above, involves an abnormality in the cystic fibrosis transmembrane conductance regulator (CFTR), a chloride channel that regulates the transport of water and ions across the epithelial membranes of epithelial cells. A deficiency in this transporter leads to the buildup of thick mucus in the tissues surrounding the lungs [9].

Duchenne Muscular Dystrophy is caused by mutations in the gene for the protein dystrophin. Dystrophin is a cytosolic protein rather than a

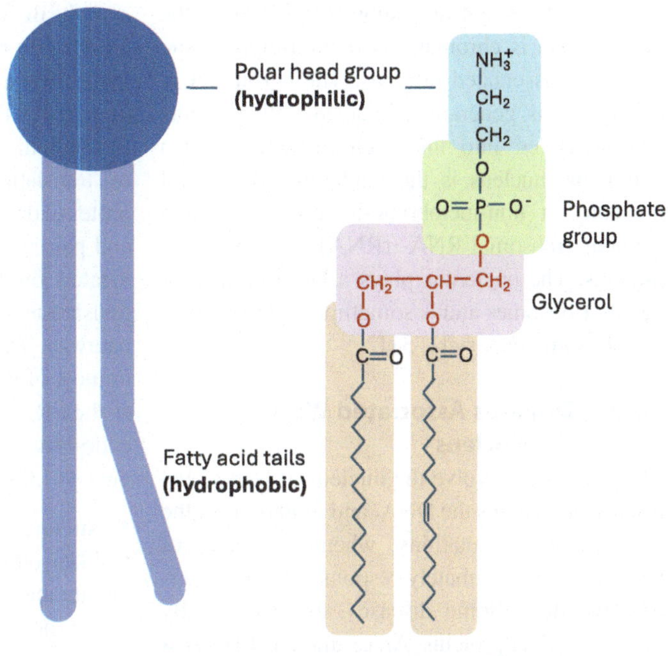

Fig. 2.3 The basic structure of phospholipids. Phospholipids contain fatty acid side chains, glycerol, a phosphate group, and a polar head group

membrane protein, but it is closely associated with the membrane. It binds to the cytosolic side of muscle cell membranes by interacting with other membrane-bound proteins, like β-dystroglycan, a transmembrane protein that binds to the extracellular matrix (ECM) on its extracellular domain. As a result of this interaction, dystrophin has a role in linking the cytoskeleton of muscle cells to the extracellular matrix. It thereby provides mechanical stability in muscle cells. Consequently, abnormalities in dystrophin are associated with progressive muscle weakness [7].

Hypercholesterolemia (high cholesterol) often involves a defect in the LDL receptor. The LDL receptor is a transmembrane protein that binds to LDL (low-density lipoprotein) on its extracellular domain and transports cholesterol [33]. LDL receptors on liver cells play a crucial role in clearing cholesterol from the bloodstream. However, abnormalities in these receptors can lead to hypercholesterolemia, which in turn increases the risk of cardiovascular disease.

6.2 Nucleus

6.2.1 Function of the Nucleus

The nucleus is a defining characteristic of eukaryotic cells. In eukaryotes, the genetic material, the DNA, is protected within the nucleus. In contrast, prokaryotes lack a nucleus. The nucleus in eukaryotic cells is vital for storing genetic information. It is also the site of DNA replication and transcription, as well as the location of the major steps in ribosome formation [1, 31].

6.2.2 The Structure of the Nucleus

The nucleus is enclosed in a nuclear envelope consisting of a double membrane, composed of phospholipid bilayers. This double membrane consists of an outer nuclear membrane (ONM), which faces the cytosol, and an inner nuclear membrane (INM), which faces the inside of the nucleus. Between these two membranes is the perinuclear space. The nuclear envelope contains pores that allow certain substances to flow in and out of the nucleus. The material inside the nucleus, surrounded by the INM, is the nucleoplasm, a gel-like substance. Within the nucleoplasm is chromatin, which comprises the DNA and associated proteins. The chromatin in eukaryotes is generally packaged with proteins and organized into linear chromosomes. Also within the nucleus is the nucleolus (or several nucleoli), a non-membrane-bound region that contains ribosomal RNA (rRNA) and ribosomal proteins. The nucleolus plays a key role in producing ribosomes and is sometimes referred to as the ribosome factory [1, 31].

6.2.3 Diseases Associated With the Nucleus

Many diseases involve the nucleus at some level because it contains the DNA, and mutations in the DNA (genetic mutations, whether somatic or hereditary) are ultimately responsible for numerous disorders. Some diseases also specifically involve nuclear proteins. An example is **Progeria** (Hutchinson–Gilford Progeria Syndrome), a rare disease that occurs in children and leads to premature aging. It is caused by a mutation in the gene that codes for Lamin A, a protein that provides structural support to the nuclear envelope. The result is a misshapen and unstable nucleus. The nucleus has vital roles in the cell, and abnormalities in its structure lead to impaired DNA repair, abnormal gene expression, increased cell death and stress, and reduced cell division and regeneration. These defects are also associated with normal aging, and patients with Progeria experience premature aging at an unusually early age [34].

6.3 Ribosomes

6.3.1 Eukaryotic Ribosomes

- The Function of Ribosomes

The primary function of ribosomes is the synthesis of proteins from mRNA during the process of translation, which is the formation of proteins (peptide chains) using mRNA as a template. While a nucleosome can be referred to as a ribosome factory, ribosomes can be considered protein factories. Some ribosomes are located in the cytoplasm, unbound to any membranes. These are the free ribosomes. Others are attached to the cytosolic surface of the rough endoplasmic reticulum (ER). Both categories have essential roles in translation. Those in the cytoplasm are involved in translating cytosolic proteins, as well as proteins destined for the nucleus, mitochondria, and peroxisomes. In contrast, ribosomes that are located on the rough ER are involved in the translation of proteins that enter the secretory pathway. These are proteins that may be destined to most of the membrane-bound organelles (such as the ER, the Golgi apparatus, the lysosome, and endosomes), the cell surface, or secretion from the cell [35].

- Strcture of Ribosomes

Unlike other organelles, ribosomes are not membrane-bound; instead, they are considered to be non-membrane-bound organelles. Ribosomes are made of a combination of ribosomal RNA (rRNA) and protein. Eukaryotic ribosomes are 80S in size (S stands for sedimentation coefficient and is a unit of size that reflects the size and shape of a particle during sedimentation in a centrifugal field). Ribosomes consist of two main subunits: a large and a small subunit. The large subunit is 60S (containing 28S rRNA, 5.85S rRNA, 5S rRNA, and 40 proteins), and the small subunit is 40S (containing 18S rRNA and 30 proteins). The ribosomal subunits remain separate until they are needed for translation, at which point they interact with each other and with the mRNA being translated [28, 35].

6.3.2 Bacterial Ribosomes and Antibiotics

Bacteria also have ribosomes, which are essential for the translation of bacterial mRNA and thus for the production of bacterial proteins. Bacterial ribosomes differ in size compared with eukaryotic ribosomes. They are 70S in size, containing a 30S and 50S subunit. Bacterial ribosomes are significant for pharmacists because they are frequent targets of antibiotics that are used to combat bacterial infections. By targeting ribosomes, bacterial protein synthesis (translation) can be blocked, thus blocking the viability of the bacteria.

Inhibitors of bacterial ribosomes have achieved significant success in combating infections. Some of the **many antibiotics that target bacterial ribosomes** include **streptomycin, tetracyclines,** and **puromycin** (puromycin is generally used in biological research, rather than for treating infections) [36]. While such antibiotics have been used successfully to combat infection for years, some bacteria have also developed resistance mechanisms. This can occur by either resisting the uptake of the antibiotic, inactivating the antibiotic, or altering the target (in this case, components of the ribosome), so that the antibiotic is no longer effective. Combatting the problem of antibiotic resistance is a critical, ongoing, and vital area of research in the field of infectious disease [37].

6.4 Endoplasmic Reticulum (ER)

6.4.1 Structure of the ER

The endoplasmic reticulum (ER) is a large, interconnected membrane-bound network within the cell. It covers a significant amount of area, making it the largest organelle by surface area in most eukaryotic cells. The two types of ER are the rough ER and the smooth ER. The rough ER is studded with ribosomes, giving it a "rough" appearance, unlike the smooth ER, which is not bound to ribosomes. Muscle cells have a specialized smooth ER called the sarcoplasmic reticulum (SR). The ER consists of a network of membranes. Inside the membranes of the ER is the internal space, called the ER lumen. Within the lumen are cisternae, a type of flattened vesicle [38].

6.4.2 Function of the ER

- Function of the Rough ER

The rough ER is studded with ribosomes. The primary functions of the rough ER include processing and transporting certain polypeptides (proteins), specifically those translated via the ribosomes located on its surface. These are primarily proteins that are destined for the secretory pathway (described below). The polypeptides that are synthesized by ribosomes on the rough ER are either fully translocated into the ER lumen or partially inserted into the ER membrane. This depends on the protein's ultimate destination. Transmembrane proteins, for example, are embedded in the ER membrane, while proteins destined for the lumens of organelles or for secretion are fully translocated into the ER lumen.

Within the ER lumen, the proteins undergo processing. In the case of transmembrane proteins, only the regions exposed to the lumen are processed within the ER. Processing of proteins within the ER involves folding and some posttranslational modifications, such as glycosylation, which is the process of attaching carbohydrates to the proteins, forming glycoproteins. This processing is catalyzed by enzymes within the ER lumen.

The rough ER is important for protein synthesis and processing, but it also has another function, which involves protein sorting. Specifically, it is the starting point of the secretory pathway. Proteins that enter the secretory pathway are en route to specific destinations within the cell or are destined for secretion. Some proteins that enter the secretory pathway are transported to organelles such as the Golgi apparatus, lysosomes, endosomes, or transport vesicles, while others may remain in the ER. Other proteins may be transported to the cell surface, where they can become membrane proteins or be secreted from the cell. All these destinations along the secretory pathway are connected via the endomembrane system.

Transport along the secretory pathway involves vesicular trafficking. In this system, proteins are packaged into membrane-bound transport vesicles that bud off from the ER membrane. These vesicles can then fuse with other membranes that are part of the endomembrane system. For example, they may fuse with the plasma membrane and release their contents from the cell. This is the process of exocytosis. Alternatively, they may deliver membrane proteins to the cell surface, which become part of the plasma membrane. Alternatively, vesicles can fuse with cellular organelles within the endomembrane system. In this way, they deliver their cargo into the organelle lumen, or onto the organelle membrane.

Regardless of the destination, the transport of the proteins along the secretory pathway generally involves the Golgi apparatus as an intermediary, where proteins are further processed before reaching their final destination [1].

Not all proteins are processed within the ER. The proteins processed by the ER are those destined for locations along the secretory pathway. Cytoplasmic proteins are synthesized on ribosomes that are not attached to the ER, and they are not processed within the ER. Likewise, proteins destined for the nucleus, the mitochondria, or the peroxisomes are synthesized on cytoplasmic ribosomes (although the mitochondria can also synthesize some of their own proteins on mitochondrial ribosomes).

- Function of the Smooth ER

The smooth ER is connected to the rough ER but is not covered with ribosomes. The smooth ER has different functions from the rough ER. It plays an essential role in the synthesis and metabolism of lipids, including phospholipids for membrane construction, and steroids such as cholesterol. The smooth ER is also involved in the detoxification of harmful substances. This involves adding hydroxyl groups to toxins to enhance their water solubility, allowing them to be flushed from the body more readily. The smooth ER also stores calcium ions. This is important for cell signaling, and in muscle cells, it is also important for contraction [38].

6.4.3 Some Drugs Target ER Proteins

Statins are medications used to **lower cholesterol levels**. They target **HMG-CoA-Reductase**, an integral membrane protein and enzyme found in both smooth and rough ER. HMG-CoA reductase has an important role in cholesterol synthesis, and targeting this enzyme can effectively lower cholesterol levels in patients with high cholesterol levels [39]. It will be discussed in more detail in Chap. 5.

6.5 The Golgi Apparatus

6.5.1 Function of the Golgi Apparatus

A major function of the Golgi apparatus is to modify and sort proteins after they have left the rough ER and before they reach their final destination. This includes the addition of various sugars or molecular tags to target proteins to certain parts of the cell [40].

6.5.2 Structure of the Golgi Apparatus

The Golgi comprises a series of sacs called cisternae. The cisternae contain enzymes that are involved in modifying proteins. The Golgi has a cis face and a trans face. Vesicles arrive at the cis face of the Golgi from the ER and depart from the trans face on their way to other parts of the cell [1, 40].

6.5.3 Abnormalities Associated with the Golgi Apparatus

Under some conditions, **Golgi fragmentation** can occur in cells. Fragmentation of the Golgi is often associated with cells undergoing cell death by apoptosis. Apoptosis can be a normal process, but Golgi fragmentation can also be associated with disease. For example, it is seen in some types of **cancer**. Golgi fragmentation is also common in some neurodegenerative diseases, such as **Parkinson's**. The exact mechanism and clinical relevance of Golgi fragmentation during disease are still not entirely clear [40, 41].

6.6 Lysosome

6.6.1 Structure and Function of the Lysosome

The lysosome is a membrane-bound sac containing an acidic environment. One of its major functions is digesting cellular waste. The lysosome contains enzymes that break down large molecules through hydrolysis. They also digest molecules that are no longer needed, as well as damaged organelles, as part of a normal cellular process called autophagy [1, 42].

6.6.2 Lysosomal Storage Diseases

Lysosomal storage diseases are a group of diseases that involve the misfunction of lysosomes. An example of a lysosomal storage disease is Gaucher disease. This disease is associated with abnormalities in the gene for the lysosomal enzyme **glucocerebrosidase** (also called **acid-β-glucosidase**), which breaks down lipids, mainly glucocerebroside. As a result, glucocerebroside accumulates in the lysosomes of cells of various organs. Macrophages that accumulate this lipid are called Gaucher cells. The normal functions of these cells are disrupted, leading them to carry out abnormal functions, including chronic cytokine release, which can cause disruption of normal organ function, inflammation, and tissue damage. Some types of Gaucher disease are treatable, but others, which occur in infants, are usually fatal. The differences lie in the different mutations in the gene for the enzyme [43].

6.7 Mitochondria

6.7.1 Function of the Mitochondria

The primary function of mitochondria (singular: mitochondrion) is cellular respiration to produce energy. Almost all eukaryotic cells have mitochondria (with some exceptions, such as mature red blood cells, which do not have organelles). Some cells can have hundreds of mitochondria and, therefore, generate large amounts of energy. Mitochondria are the sites where the Krebs cycle and the electron transport chain occur. These major metabolic pathways result in the generation of ATP through oxidative phosphorylation, the primary energy-producing process in eukaryotic cells.

6.7.2 Structure of the Mitochondria

The mitochondria contain two membranes. The outer membrane is smooth, and the inner membrane is made of folds called cristae. The intermembrane space is the area between these two membranes. The mitochondrial matrix is the innermost part of the mitochondria, within the inner membrane. The mitochondria have some of their own DNA, which is circular, and it is the only organelle in animal cells besides the nucleus to contain DNA. Mitochondria also have their own ribosomes, which are similar to those found in bacteria. Mitochondrial DNA, ribosomes, and enzymes are in the matrix, although the inner membranes also have some membrane-bound enzymes [1, 44] See Fig. 2.4.

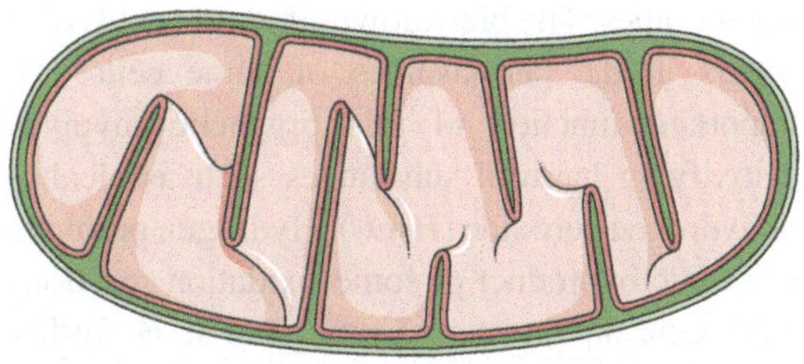

Fig. 2.4 The basic structure of the mitochondria. A mitochondrion contains a double membrane, with a smooth outer membrane and a highly folded inner membrane. It is a critical organelle for energy production. (*Source*: Image adapted from Servier Medical Art (https://smart.servier.com/), licensed under the Creative Commons Attribution 4.0 (CC BY 4.0) license)

6.7.3 Mitochondrial Disease

Leigh Syndrome is an example of a mitochondrial disease. It involves mutations in a component of the electron transport chain. These mutations can occur in either mitochondrial DNA or nuclear DNA, which encodes mitochondrial proteins. Symptoms of Leigh syndrome include overall weakness, tremors, and neurological symptoms resulting from energy deficits in cells, especially in the brain and nervous system. The disease is progressive and usually fatal [44].

6.8 Peroxisome

6.8.1 Structure and Function of the Peroxisome

The peroxisome is another membrane-bound organelle. The primary functions of the peroxisome are metabolism and detoxification. Fatty acid oxidation (breakdown) occurs mostly in the mitochondria, but some long-chain fatty acids and some other lipids are broken down in the peroxisomes. The breakdown of amino acids also occurs in the peroxisomes of some cells. An important function of this organelle involves detoxifying harmful substances such as alcohol or hydrogen peroxide (H_2O_2). Hydrogen peroxide is a toxic byproduct of some oxidation reactions that occur in the peroxisome, but it is further broken down into water and oxygen by the enzyme **catalase** within the peroxisome to avoid toxicity (**Catalase** catalyzes the reactions $2H_2O_2 \rightarrow 2H_2O + O_2$) [1].

6.8.2 Peroxisome Disorders

Zellweger Syndrome is a disorder in which peroxisome function is severely impaired. It is caused by mutations in genes that encode proteins necessary for the proper assembly of peroxisomes. Symptoms include impaired development and neuronal abnormalities. These can result from abnormal buildup of toxic compounds, including long-chain fatty acids. Due to the absence of functional peroxisomes, these long-chain fatty acids cannot be broken down, and they accumulate in tissues, including the brain. This causes neurodegeneration and impaired neuronal function. Other molecules also accumulate. Patients usually survive only up to one year of age [45].

6.9 Endosomes and Vesicles

Endosomes and vesicles are transient membrane-bound structures. They have roles in transport and sorting of solutes (such as proteins, lipids, or other cellular molecules). Endosomes are involved in endocytosis, importing material from the outside of the cell. Vesicles are involved in shuttling solutes between organelles and to the cell surface [46].

6.10 Cytoskeleton

6.10.1 Structure of the Cytoskeleton

The cytoskeleton is not an organelle, but it has crucial functions. It consists of a network of fibers throughout the cell (Fig. 2.5a). The three major components are microtubules, microfilaments (actin), and intermediate filaments [1]. Microtubules are discussed in more detail in Chap. 20.

6.10.2 Functions of the Cytoskeleton

The cytoskeleton's major functions include maintaining cell shape, integrity, and motility. Some cytoskeleton components, specifically microtubules, can also provide tracks for motor proteins, which facilitate the accurate movement of vesicles throughout the cell [47]. The microtubules also have crucial roles in segregating the chromosomes during mitosis, when the cell divides into two daughter cells. In fact, **microtubules are targets for certain anti-cancer drugs** because inhibiting their function prevents chromosome segregation and, consequently, cell division. Examples are **Taxol (paclitaxel)**, **Velban (vinblastine)**, and **Oncovin (vincristine)** [48–51] (While the original brand names of these drugs are discontinued, the main ingredients of paclitaxel and vincristine are still used for cancer treatment [52]). These are discussed in more detail in Chaps. 21 and 23.

6.11 Centrosome

6.11.1 Structure and Function of the Centrosome

The centrosome (Fig. 2.5b) is considered an organelle even though it is not membrane-bound. The centrosome contains a pair of centrioles (which are made up of microtubules) surrounded by Pericentriolar Material (PCM; a protein-rich matrix). Resting cells have one centrosome, and dividing cells have two. The primary functions of the centrosomes are their critical roles in cell division. The centrosome functions as a microtubule-organizing center, and it organizes microtubule assembly during cell division. During cell division, microtubules polymerize and extend from the two centrosomes, and they play essential roles in proper chromosome alignment and segregation during mitosis. These microtubules form the mitotic spindle, and they contain the polar, astral, and kinetochore

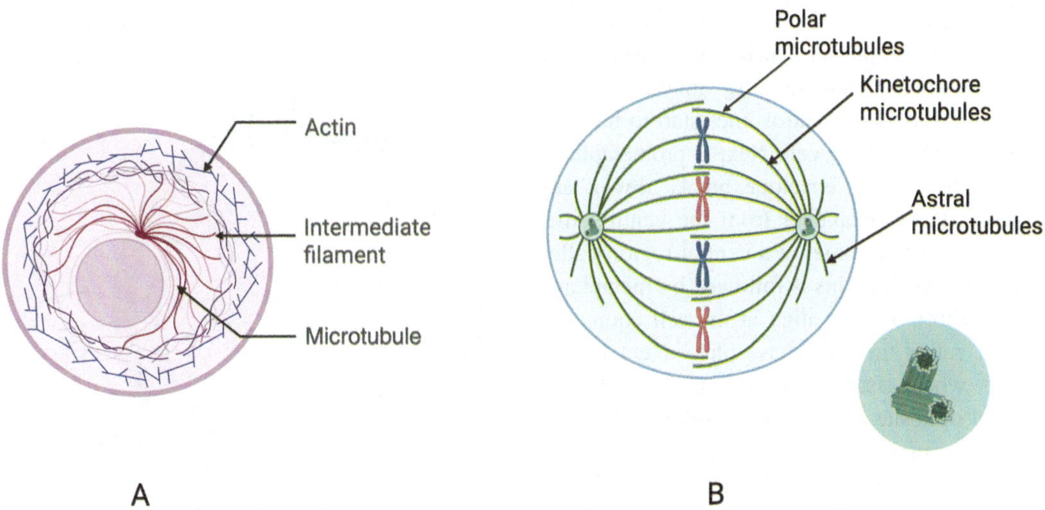

Fig. 2.5 (**a**) The cytoskeleton. The cytoskeleton consists of actin, intermediate filaments, and microtubules. (**b**) The centrosome and the mitotic spindle. The centrosome (lower right in panel **b**) consists of a pair of centrioles surrounded by pericentriolar material (PCM). It serves as a microtubule-organizing center (MTOC). The centrosome doubles during S phase and remains double during M phase. During M phase, microtubules radiate from the centrosome, forming the mitotic spindle. This spindle contains microtubules, including the kinetochore microtubules that attach to metaphase chromosomes. (*Source*: Created with Biorender.com)

microtubules [1, 53]. This is discussed in more detail in Chap. 20.

6.11.2 Diseases Involving the Centrosome

Due to their essential roles in chromosome segregation during cellular division, centrosome abnormalities can have devastating effects. Abnormal chromosome number, associated with chromosome instability, is a feature of many **cancer** cells. This can be caused by abnormalities in the centrosomes or an improper number of centrosomes, resulting in an unbalanced distribution of chromosomes to daughter cells when the cell divides [54].

6.12 Extracellular Matrix

The extracellular matrix (ECM) is not a part of the cell or an organelle; rather, it is the area outside the cell. It is composed largely of fibrous proteins, such as collagen and elastin, as well as polysaccharides and adhesive molecules like fibronectin and laminin. Although it is outside the cell, it is critical for cell function. It provides structural support and helps anchor cells in place. It contributes to the structure of organs composed of cells. Some body parts, such as bone and connective tissue, are especially rich in ECM content. It is also an important mediator of communication between cells. It serves as a reservoir for signaling molecules that bind to cell surface receptors. The proteins in the ECM are synthesized in the cell and secreted into the extracellular space via the secretory pathway [1, 55].

7 The Cell Cycle

Every time a cell divides, it splits into two daughter cells. In the process, it goes through one round of the cell cycle. The cell cycle is discussed in more detail in Chap. 20, but is introduced briefly here. The cell cycle can be divided into major phases [56]: G1 phase is a gap phase where cells are active but not dividing. This is followed by, S phase, where DNA replication occurs, then G2 phase, another gap phase, and then M phase, where the cell splits into two daughter cells.

M phase is divided into several subphases. It starts with prophase. During this phase, the chromosomes condense, the mitotic spindle begins to form, and centrosomes start to move to opposite poles of the cell. Next is prometaphase, where the nuclear envelope breaks down, and spindle fibers emanating from the centrosomes attach to chromosomes at a central region called the kinetochores. This is followed by metaphase, where chromosomes align at the cell equator, a region called the metaphase plate. Next is anaphase, where sister chromatids are pulled apart toward opposite poles. This is followed by telophase, during which the chromatids arrive at their opposite poles and the nuclear envelope reforms. The chromosomes begin to decondense. This is the end of mitosis, followed by cytokinesis, where the cytoplasm divides, resulting in two new daughter cells, which marks the end of the M phase. The phases of the cell cycle can now start again, at G1 phase [1].

During G1 phase, there is also a quiescent phase called G0, which takes place in many cells. Many cells that are not actively dividing are in G0 phase [57]. Some cells remain in the G0 phase for their entire life in the adult, such as many neurons, which do not divide. Other cells divide when more cells are needed. Once these cells receive signals to divide, they reenter G1 phase, then transition from G1 phase to S phase, followed by the rest of the phases of the cell cycle. G1, S, and G2 phases are all part of interphase (I) (see Fig. 2.6).

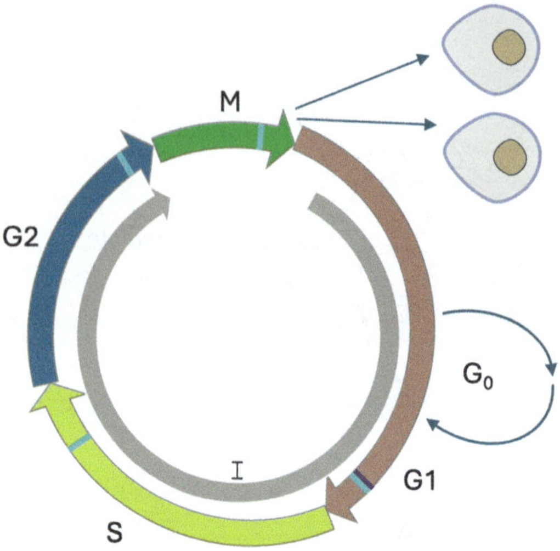

Fig. 2.6 The phases of the cell cycle. The cell cycle consists of G1 phase, S phase, and G2 phase (which are all part of interphase (I), followed by M phase. DNA replicates in S phase and the cell splits in M phase

8 Single Cellular Versus Multicellular Eukaryotes

While the first cells to evolve were single-celled organisms, today's eukaryotic cells can be either single-celled or multicellular [1]. Examples of single-celled eukaryotes include yeast and algae. Following the evolution of single-celled eukaryotes approximately 1.5 billion years ago, multicellular eukaryotes emerged about 600 million years ago. Multicellular eukaryotes include animals and plants. In multicellular animals, groups of cells form tissues, which then form organs, and ultimately, these organs comprise the entire organism. Animals can have trillions of cells, and different cells have various specialized functions [58] (Fig. 2.7). Cells that have taken on their specialized functions are called differentiated cells. The process of developing into a specialized cell is called cell differentiation.

There are many types of cells in an animal's body. After birth, nearly all cells are differentiated (although adult stem cells remain partly undifferentiated). Differentiated cells after birth are organized into four categories of tissues. The four major tissue types are epithelial tissue (sheets of cells that line organs, glands, and exterior surfaces), connective tissue (tissues such as tendons, ligaments, cartilage, and bone, which are found between other tissues), muscle tissue, and nervous tissue [1]. Within each of these four major categories are various specific tissues, each composed of cells that are similar in structure and function, working together to perform specialized tasks. Cells within these tissues must maintain their differentiated state, and at times they may need to replicate to replace dead or damaged cells. This involves the regulation of gene expression, cell cycle regulation, signal transduction, and, in some cases, replenishment by stem cells [59], all of which will be discussed in upcoming chapters.

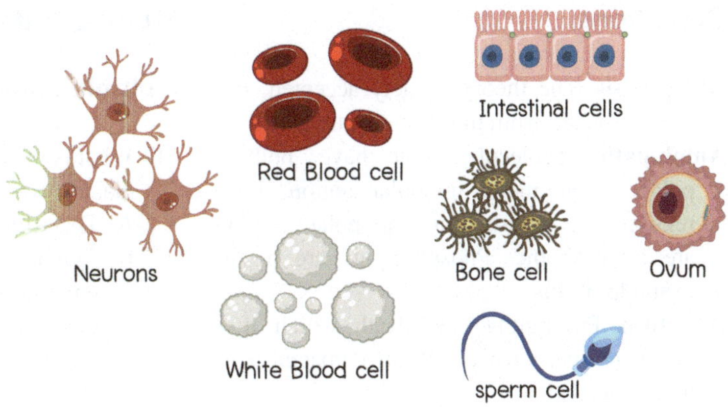

Fig. 2.7 Differentiated cells. Multicellular organisms consist of multiple types of specialized cells with specific functions, also called differentiated cells. (Illustrations are schematic. They are meant to emphasize that cells differ from, each other and not meant to depict the actual shape, size, or morphology of these cells types). (*Source*: Illustrations from Vecteezy.com (Pro license))

9 Signal Transduction Pathways

All cells need a way to respond to extracellular cues. These can come from other cells and from the environment. Chapter 18 is dedicated to the discussion of signal transduction, which is also briefly introduced here. Cells respond to signals in several different ways [60]. Commonly, they respond to signals that are in the form of small peptides, such as growth factors or cytokines, which are broadly classified as ligands. These are called extracellular signals because they are outside the cells in the surrounding space and tissues. Many ligands cannot cross the phospholipid bilayer, and instead, they bind to cell surface receptors. Receptor tyrosine kinases are an example of a type of cell surface receptor to which growth factors bind. Ligand binding activates these receptors, triggering a cascade of events within the cell. This can be referred to as a signaling pathway or a signaling cascade. These signaling pathways may include a series of enzymes that each activate the next one in the pathway. Often, this type of pathway leads to the activation of transcription factors in the nucleus, which then activate the expression/transcription of specific genes. For example, a growth factor such as epidermal growth factor (EGF) may activate a signaling pathway that leads to increased transcription of genes that have roles in cell growth and proliferation.

In contrast to extracellular signals, some ligands do enter the cell. These are small ligands, such as some hormones, that bind and activate intracellular receptors, potentially leading to cellular changes like the regulation of gene expression [1, 60].

FDA-Approved Drugs Discussed in This Chapter

Commonly used brand name	Generic name	First FDA approval year	Common conditions	Main targets
(No common brand name)	*Streptomycin*	1944	Antibiotics for the treatment of bacterial infections, primarily tuberculosis, and plague	30S ribosomal subunit, bacterial
(No common brand name)	*Tetracycline*	1950s	Antibiotics for treatment of bacterial infections, including acne, respiratory infections, and sexually transmitted infections	30S ribosomal subunit, bacterial
Oncovin	*Vincristine*	1963	Cancer, including leukemia, lymphoma, and neuroblastoma	Tubulin (microtubules)
Taxol	*Paclitaxel*	1992	Cancer, including ovarian, breast, lung, and pancreatic cancers	Tubulin (microtubules)
Velban	*Vinblastine*	1965	Cancer, including Hodgkin's lymphoma, testicular cancer, and breast cancer	Tubulin (microtubules)

Definitions

Abiogenesis The theory, widely accepted, that life originated from non-life spontaneously.

Amphipathic Molecules that have both a polar region and a nonpolar region. Polar regions are hydrophilic, nonpolar regions are hydrophobic. Phospholipids are a key example of this.

Apoptosis Programmed cell death, which often occurs normally when cells are damaged or no longer needed.

Cisternae A type of flattened membrane-bound vesicle.

Differentiated cell A cell that has specific, specialized functions. Most adult cells are differentiated, although stem cells remain largely undifferentiated.

Endocytosis A process where substances are taken into the cell via vesicles.

Endosymbiosis The generally accepted theory that during evolution, eukaryotic cells obtained organelles, specifically the mitochondria, by engulfing bacteria.

Exocytosis The process by which molecules are exported from the cells by vesicle transport, which involves vesicle fusion with the plasma membrane.

Glycosylation The enzymatic process that attaches sugars to proteins or lipids.

Lumen The internal space or cavity within a membrane-bound organelle.

Mutation A change in the DNA sequence that can affect gene function.

Nucleoplasm The gel-like substance in the nucleus, contained within the nuclear membrane.

Permeability In cell biology usually refers to the ability of substances to flow through a membrane.

Quiescent state A state in which a cell is not dividing or progressing through the cell cycle, but may be healthy and metabolically active. This is often considered a G_0 stage.

Vesicle A small, membrane-bound sac used by cells to transport molecules within the cell or to the outside of the cell for secretion.

Study Questions

Multiple-Choice Questions

1. What is the main function of the rough endoplasmic reticulum (rough ER)?
   A. Detoxification of harmful substances
   B. Synthesis and processing of proteins entering the secretory pathway
   C. Generation of ATP in the electron transport chain
   D. Both B and C
2. Which of the following best describes the endosymbiotic theory?
   A. Prokaryotes evolved from eukaryotic cells.
   B. Cells arise spontaneously from nonliving matter.
   C. Eukaryotic organelles like mitochondria evolved from engulfed prokaryotes.
   D. Prokaryotes used RNA as the genetic material, and eukaryotes used DNA.
3. Which of the following is NOT part of the endomembrane system?
   A. Endoplasmic reticulum
   B. Golgi apparatus
   C. Ribosomes
   D. The mitochondria
   E. C and D
4. Which disease results from a mutation affecting the LDL receptor on cell membranes?
   A. Hypocholesterolemia
   B. Hypercholesterolemia
   C. Progeria
   D. Gaucher disease
5. Which organelle is primarily responsible for generating cellular energy via the electron transport chain?
   A. Lysosome
   B. Nucleus
   C. Mitochondrion
   D. Golgi apparatus
   E. Endoplasmic reticulum
6. Which of the following is NOT a principle of the cell theory?
   A. All living organisms are composed of cells.

B. Cells are the basic units of structure and function in living organisms.
C. All cells arise from preexisting cells.
D. All cells have a nucleus.
E. None of the above.

7. Which evolutionary process is thought to be important for the origin of organelles in eukaryotic cells?
A. Endosymbiosis
B. Binary fission
C. Conjugation
D. Transformation
E. Glycolysis

8. Which of the following is a characteristic of prokaryotic cells?
A. They have a nucleus.
B. They lack a plasma membrane.
C. They are typically larger than eukaryotic cells.
D. They lack membrane-bound organelles.

9. Which of the following is most directly responsible for protein synthesis?
A. Nucleus
B. Golgi apparatus
C. Mitochondrion
D. Ribosome
E. Lysosome

10. Which organelle is responsible for generating ATP through oxidative phosphorylation?
A. Lysosome
B. Mitochondrion
C. Chloroplast
D. Nucleus

11. In a multicellular organism, cells of similar structure and function form a:
A. Tissue
B. Organ
C. Organism
D. Organ system

12. Which of the following proteins is a component of the cytoskeleton?
A. DNA polymerase
B. Actin
C. Transferrin
D. Collagen

Fill-in-the-Blank Questions

13. The scientist who coined the term "cell" after observing cork tissue under a microscope was __________.
14. The plasma membrane is composed primarily of a __________ bilayer.
15. The cytoskeleton consists of intermediate filaments, actin, and __________.

Case Studies

16. Case Study [A]
A patient presents with chronic respiratory infections and thick mucus buildup in the lungs. Genetic analysis reveals a mutation in the gene coding for the CFTR protein.
Question:
Which cellular structure or function is most directly impaired in this patient?
A. DNA replication in the nucleus
B. Ion transport across the plasma membrane
C. Detoxification in the peroxisome
D. Energy production by mitochondria
E. Protein folding in the endoplasmic reticulum

17. Case Study [B]:
A newborn is diagnosed with Zellweger syndrome, a genetic condition associated with impaired development and neuronal abnormalities. Biochemical analysis of cells from the newborn revealed accumulation of very long chain fatty acids and toxic compounds in cells.
Question:
Which organelle is most likely affected in this condition?
A. Golgi apparatus
B. Mitochondria
C. Peroxisome
D. Centrosome

Answer Key

Multiple-Choice Questions

1. **Answer**: B
2. **Answer**: C
3. **Answer**: E
4. **Answer**: B
5. **Answer**: C
6. **Answer**: D
7. **Answer**: A
8. **Answer**: D
9. **Answer**: D
10. **Answer**: B
11. **Answer**: A
12. **Answer**: B

Fill-in-the-Blank Questions

13. **Answer**: Robert Hooke
14. **Answer**: phospholipid
15. **Answer**: microtubules

Case Studies

16. **Case Study A**
    Answer: B
    Explanation:
    Cystic fibrosis involves an abnormality in the cystic fibrosis transmembrane conductor (CFTR), a chloride channel that regulates the transport of water and ions across epithelial membranes of epithelial cells. Deficiency in this transporter leads to the buildup of a thick mucus in the tissues surrounding the lungs
17. **Case Study B**
    Answer: C
    Explanation:
    Zellweger syndrome involves abnormalities in or the absence of peroxisomes. Because of the lack of functional peroxisomes, these long-chain fatty acids cannot be broken down, and they accumulate in tissues, including the brain. This causes neurodegeneration and impaired neuronal function.

References

1. Alberts B, Heald R, Johnson A, Morgan D, Raff M, Roberts K, Walter P, Wilson J, Hunt T. Molecular biology of the cell. W.W. Norton and Company; 2022.
2. Boutry J, Tissot S, Ujvari B, Capp J-P, Giraudeau M, Nedelcu AM, et al. The evolution and ecology of benign tumors. Biochim Biophys Acta (BBA) – Rev Cancer. 2022;1877(1):188643. https://doi.org/10.1016/j.bbcan.2021.188643.
3. Boehncke WH, Schön MP. Psoriasis. Lancet. 2015;386(9997):983–94. https://doi.org/10.1016/s0140-6736(14)61909-7.
4. National Library of Medicine. Overview: uterine fibroids. 2025. Available from: https://www.ncbi.nlm.nih.gov/books/NBK279535/#:~:text=It%20is%20not%20known%20exactly,weight%20itself%20is%20the%20cause.
5. Wiman KG, Zhivotovsky B. Understanding cell cycle and cell death regulation provides novel weapons against human diseases. J Intern Med. 2017;281(5): 483–95. https://doi.org/10.1111/joim.12609.
6. Benson-Quarm N, Gandhi S, Kulasekararaj A, Marsh J. Aplastic anaemia. Hematology. 2014;19(1):60–1. https://doi.org/10.1179/1024533213Z.000000000233.
7. Duan D, Goemans N, Takeda S, Mercuri E, Aartsma-Rus A. Duchenne muscular dystrophy. Nat Rev Dis Primers. 2021;7(1):13. https://doi.org/10.1038/s41572-021-00248-3.
8. Berntorp E, Fischer K, Hart DP, Mancuso ME, Stephensen D, Shapiro AD, et al. Haemophilia. Nat Rev Dis Primers. 2021;7(1):45. https://doi.org/10.1038/s41572-021-00278-x.
9. Ong T, Ramsey BW. Cystic fibrosis: a review. JAMA. 2023;329(21):1859–71. https://doi.org/10.1001/jama.2023.8120.
10. Fleming A. On the antibacterial action of cultures of a penicillium, with special reference to their use in the isolation of B. influenzæ. Br J Exp Pathol. 1929;10(3): 226–36.
11. Hutchings MI, Truman AW, Wilkinson B. Antibiotics: past, present and future. Curr Opin Microbiol. 2019;51:72–80. https://doi.org/10.1016/j.mib.2019.10.008.
12. Mukherjee S. The song of the cell: an exploration of medicine and the new human. Scribner; 2022.
13. Alberts BM. Cell theory. Enchclopaedia Britannic, Inc.; 2025. Available from: https://www.britannica.com/science/cell-theory.
14. Mazzarello P. A unifying concept: the history of cell theory. Nat Cell Biol. 1999;1(1):E13–E5. https://doi.org/10.1038/8964.
15. Müller-Wille S. Cell theory, specificity, and reproduction, 1837–1870. Stud Hist Philos Biol Biomed Sci. 2010;41(3):225–31. https://doi.org/10.1016/j.shpsc.2010.07.008
16. Cell Theory. National Geographic Education, National Geographic Society; 2023. https://education.nationalgeographic.org/resource/cell-theory/

17. Hooke R, Allestry J, Martyn J. Micrographia, or, some physiological descriptions of minute bodies made by magnifying glasses:with observations and inquiries thereupon. London: Printed by Jo. Martyn and Ja. Allestry, printers to the Royal Society; 1665.
18. Gest H. The discovery of microorganisms by Robert Hooke and Antoni Van Leeuwenhoek, fellows of the Royal Society. Notes Rec R Soc Lond. 2004;58(2): 187–201. https://doi.org/10.1098/rsnr.2004.0055.
19. Ribatti D. An historical note on the cell theory. Exp Cell Res. 2018;364(1):1–4. https://doi.org/10.1016/j.yexcr.2018.01.038.
20. Miller SL. THe formation of organic compounds on the primitive earth. Ann N Y Acad Sci. 1957;69:260–75.
21. Ruiz-Mirazo K, Briones C, de la Escosura A. Prebiotic systems chemistry: new perspectives for the origins of life. Chem Rev. 2014;114(1):285–366. https://doi.org/10.1021/cr2004844.
22. Schrum JP, Zhu TF, Szostak JW. The origins of cellular life. Cold Spring Harb Perspect Biol. 2010;2(9): a002212. https://doi.org/10.1101/cshperspect.a002212.
23. Gilbert W. Origin of life: the RNA world. Nature. 1986;319(6055):618. https://doi.org/10.1038/319618a0.
24. Joyce GF. The antiquity of RNA-based evolution. Nature. 2002;418(6894):214–21. https://doi.org/10.1038/418214a.
25. Mansy SS, Schrum JP, Krishnamurthy M, Tobé S, Treco DA, Szostak JW. Template-directed synthesis of a genetic polymer in a model protocell. Nature. 2008;454(7200):122–5. https://doi.org/10.1038/nature07018.
26. Black RA, Blosser MC. A self-assembled aggregate composed of a fatty acid membrane and the building blocks of biological polymers provides a first step in the emergence of protocells. Life. 2016;6(3):33. https://doi.org/10.3390/life6030033.
27. Joyce GF, Szostak JW. Protocells and RNA self-replication. Cold Spring Harb Perspect Biol. 2018;10 (9) https://doi.org/10.1101/cshperspect.a034801.
28. Cooper G, Adams K. The cell: a molecular approach. Sinauer Associates; 2022.
29. Archibald JM. Endosymbiosis and eukaryotic cell evolution. Curr Biol. 2015;25(19):R911–R21. https://doi.org/10.1016/j.cub.2015.07.055.
30. Vosseberg J, van Hooff JJE, Köstlbacher S, Panagiotou K, Tamarit D, Ettema TJG. The emerging view on the origin and early evolution of eukaryotic cells. Nature. 2024;633(8029):295–305. https://doi.org/10.1038/s41586-024-07677-6.
31. Alberts B, Bray D, Hopkin K, Johnson A, Lewis J, Raff M, Roberts K, Walter P. Essential cell biology. Norton and Company; 2023.
32. Singer SJ, Nicolson GL. The fluid mosaic model of the structure of cell membranes. Science. 1972;175(4023): 720–31. https://doi.org/10.1126/science.175.4023.720.
33. Abifadel M, Boileau C. Genetic and molecular architecture of familial hypercholesterolemia. J Intern Med. 2023;293(2):144–65. https://doi.org/10.1111/joim.13577.
34. Vidak S, Foisner R. Molecular insights into the premature aging disease progeria. Histochem Cell Biol. 2016;145(4):401–17. https://doi.org/10.1007/s00418-016-1411-1.
35. Wilson DN, Doudna Cate JH. The structure and function of the eukaryotic ribosome. Cold Spring Harb Perspect Biol. 2012;4(5) https://doi.org/10.1101/cshperspect.a011536.
36. Lin J, Zhou D, Steitz TA, Polikanov YS, Gagnon MG. Ribosome-targeting antibiotics: modes of action, mechanisms of resistance, and implications for drug design. Annu Rev Biochem. 2018;87:451–78. https://doi.org/10.1146/annurev-biochem-062917-011942.
37. Wilson DN. Ribosome-targeting antibiotics and mechanisms of bacterial resistance. Nat Rev Microbiol. 2014;12(1):35–48. https://doi.org/10.1038/nrmicro3155.
38. Voeltz GK, Rolls MM, Rapoport TA. Structural organization of the endoplasmic reticulum. EMBO Rep. 2002;3(10):944–50. https://doi.org/10.1093/embo-reports/kvf202.
39. Buhaescu I, Izzedine H. Mevalonate pathway: a review of clinical and therapeutical implications. Clin Biochem. 2007;40(9–10):575–84. https://doi.org/10.1016/j.clinbiochem.2007.03.016.
40. Li J, Ahat E, Wang Y. Golgi structure and function in health, stress, and diseases. Results Probl Cell Differ. 2019;67:441–85. https://doi.org/10.1007/978-3-030-23173-6_19.
41. Makhoul C, Gosavi P, Gleeson PA. Golgi dynamics: the morphology of the mammalian Golgi apparatus in health and disease. Front Cell Dev Biol. 2019;7:112. https://doi.org/10.3389/fcell.2019.00112.
42. Zhang Z, Yue P, Lu T, Wang Y, Wei Y, Wei X. Role of lysosomes in physiological activities, diseases, and therapy. J Hematol Oncol. 2021;14(1):79. https://doi.org/10.1186/s13045-021-01087-1.
43. Platt FM, d'Azzo A, Davidson BL, Neufeld EF, Tifft CJ. Lysosomal storage diseases. Nat Rev Dis Primers. 2018;4(1):27. https://doi.org/10.1038/s41572-018-0025-4.
44. Nunnari J, Suomalainen A. Mitochondria: in sickness and in health. Cell. 2012;148(6):1145–59. https://doi.org/10.1016/j.cell.2012.02.035.
45. Wanders RJA, Baes M, Ribeiro D, Ferdinandusse S, Waterham HR. The physiological functions of human peroxisomes. Physiol Rev. 2023;103(1):957–1024. https://doi.org/10.1152/physrev.00051.2021.
46. Huotari J, Helenius A. Endosome maturation. EMBO J. 2011;30(17):3481–500. https://doi.org/10.1038/emboj.2011.286.
47. Hohmann T, Dehghani F. The cytoskeleton-A complex interacting meshwork. Cells. 2019;8(4) https://doi.org/10.3390/cells8040362.

48. FDA,. Taxol (paclitaxel) Approval overview. 1992. Available from: https://www.accessdata.fda.gov/scripts/cder/daf/index.cfm?event=overview.process&ApplNo=20262.
49. FDA. Oncovin (vincristine) approval overview. 1963. Available from: https://www.accessdata.fda.gov/scripts/cder/daf/index.cfm?event=overview.process&ApplNo=75493.
50. Ritter JM, Flower RJ, Henderson G, Loke YK, MacEwan DJ, Rang HP. Rang and Dale's pharmacology. 10th ed; 2023.
51. Jordan MA, Wilson L. Microtubules as a target for anticancer drugs. Nat Rev Cancer. 2004;4(4):253–65. https://doi.org/10.1038/nrc1317.
52. Hertz DL, Joerger M, Bang YJ, Mathijssen RH, Zhou C, Zhang L, et al. Paclitaxel therapeutic drug monitoring – international association of therapeutic drug monitoring and clinical toxicology recommendations. Eur J Cancer. 2024;202:114024. https://doi.org/10.1016/j.ejca.2024.114024.
53. Farina F, Gaillard J, Guérin C, Couté Y, Sillibourne J, Blanchoin L, et al. The centrosome is an actin-organizing centre. Nat Cell Biol. 2016;18(1):65–75. https://doi.org/10.1038/ncb3285.
54. Piemonte KM, Anstine LJ, Keri RA. Centrosome aberrations as drivers of chromosomal instability in breast cancer. Endocrinology. 2021;162(12) https://doi.org/10.1210/endocr/bqab208.
55. Hynes RO. The extracellular matrix: not just pretty fibrils. Science. 2009;326(5957):1216–9. https://doi.org/10.1126/science.1176009.
56. Morgan DO. The cell cycle: principles of control. New Science Press; 2007.
57. Coller HA. Cell biology. The essence of quiescence. Science. 2011;334(6059):1074–5. https://doi.org/10.1126/science.1216242.
58. Brunet T, King N. The origin of animal multicellularity and cell differentiation. Dev Cell. 2017;43(2):124–40. https://doi.org/10.1016/j.devcel.2017.09.016.
59. Blanpain C, Fuchs E. Stem cell plasticity. Plasticity of epithelial stem cells in tissue regeneration. Science. 2014;344(6189):1242281. https://doi.org/10.1126/science.1242281.
60. Marks F, Klingmüller U, Müller-Decker K. Cellular signal processing: an introduction to the Molecular mechanisms of signal transduction. 2nd ed. Garland Science; 2017.

3 Membranes and Membrane Transport in Cell Biology and Drug Action

Abstract

The cell membrane (the plasma membrane) is the main barrier that regulates the movement of substances in and out of the cell. The organization of the membrane as a phospholipid bilayer containing embedded proteins, glycoproteins, glycolipids, and cholesterol is described, with emphasis on the major phospholipid components and their structures. Mechanisms of solute transport across the membrane are covered, highlighting the important roles of transport proteins and channels. Cell surface receptors that mediate responses to extracellular stimuli, such as growth factors and cytokines, are also introduced. The functional and clinical importance of membrane proteins is emphasized by discussing their roles in disease and their frequent targeting by therapeutic drugs, many of which either act directly on membrane proteins or enter cells through transporters or passive diffusion. While eukaryotic cell membranes are the main focus, bacterial membranes and cell walls are also discussed, as they are the targets of many drugs, such as antibiotics.

Keywords

Phospholipid bilayer · Transporters · Channels · Vesicular transport · Channel blockers

Abbreviations

ABC	ATP-binding cassette (transporter)
ACh	Acetylcholine
ADP	Adenosine diphosphate
ASM	Acid sphingomyelinase
ASMD	Acid sphingomyelinase deficiency
ATP	Adenosine triphosphate
Ca^{2+}	Calcium ion
CFTR	Cystic fibrosis transmembrane conductance regulator
Cl^{-}	Chloride ion
CNS	Central nervous system
ER	Endoplasmic reticulum
GLUT	Glucose transporter
GPCR	G protein-coupled receptor
K^{+}	Potassium ion
Km	Michaelis constant (affinity of a transporter or enzyme)
mAChR	Muscarinic acetylcholine receptor
MDR	Multidrug resistance protein
Na^{+}	Sodium ion
nAChR	Nicotinic acetylcholine receptor
Pgp/P-gp	P-Glycoprotein
Pi	Inorganic phosphate
PNS	Peripheral nervous system
RME	Receptor-mediated endocytosis
SERCA	Sarcoplasmic/endoplasmic reticulum Ca^{2+} ATPase

A. Minden, *Biochemistry, Cell and Molecular Biology for Pharmacists and Health Professionals*,
https://doi.org/10.1007/978-3-032-20529-2_3

SERT	Serotonin reuptake transporter
SGLT	Sodium–glucose linked transporter
SNARE	Soluble NSF attachment protein receptor
SR	Sarcoplasmic reticulum
SSRI	Selective serotonin reuptake inhibitor
VMAT	Vesicular monoamine transporter
Vmax	Maximum rate of transport

What You Will Learn

- The major components of the eukaryotic plasma membrane include phospholipids, proteins, and carbohydrates
- How solutes are transported across membranes
- The difference between passive transport and active transport
- Different types of transporters and channels in the membrane
- Examples of drugs that target membrane transport proteins, such as calcium channel blockers (that inhibit voltage-gated Ca^{2+} channels), and are used to treat hypertension
- Some transporters are associated with disease, such as the MDR transporters that may be overproduced in cancer, and CFTR, which is mutated in cystic fibrosis
- Vesicular transport
- Transport in neuronal cells and the role of ion transport in the action potential and neuronal cell signaling
- Drugs that modulate neuronal communication by targeting transporters, such as SSRIs that inhibit the serotonin reuptake transporter (SERT), and that may be used to treat depression and anxiety
- The importance of understanding the bacterial plasma membrane
- Bacteria often also have a cell wall
- Antibiotics that inhibit the bacterial cell wall are used to combat bacterial infections

This chapter is dedicated to the cell membrane (the plasma membrane) because it is the primary barrier that regulates the movement of substances in and out of the cell. Transport proteins and channels in the membrane regulate the flow of material in and out of the cell, while cell surface receptors mediate responses to extracellular stimuli, such as growth factors, that can regulate cell function. Many drugs either target membrane proteins or enter the cell through membrane transporters. This chapter is organized into two sections. Section 1 focuses on the basic structural features of the membrane, and Sect. 2 focuses on the transport of material across the membrane.

1 Membrane Structure

1.1 Introduction to Membranes

The plasma membrane separates the inside of the cell from the outside of the cell. It is semipermeable, meaning that only certain substances can enter or exit the cell without the assistance of transport proteins or channels. The defining component of the plasma membrane is the phospholipid bilayer, which consists of two layers, or leaflets, of phospholipids. One of the two leaflets faces inward, toward the inside of the cell, and the other faces outward. A hydrophobic part of the membrane, which is resistant to aqueous solutions, is buried in the section between the two layers, while the outer parts of the membrane can interact with the aqueous environment (see Chap. 2, Fig. 2.1). The major components of the plasma membrane that give it its unique characteristics are the phospholipids, which are amphipathic, having both hydrophobic and hydrophilic ends. The membrane also contains glycolipids (lipids attached to a sugar), sterols (such as cholesterol), other lipids, and proteins. The membrane is not a rigid structure. Instead, the different components of the plasma membrane can all move around within their layer. This is referred to as the fluid-mosaic model. While the plasma membrane surrounds the entire cell, the organelles within the cell are also surrounded by a membrane. The membranes of organelles are composed of a phospholipid bilayer. In some cases, such as the nucleus and mitochondria, they contain two phospholipid bilayers, forming a double membrane. Much of the discussion of basic membrane structure and transport in this chapter is drawn from [1–3] as well as other references cited in the text.

1.2 Lipids Are Fundamental Components of the Membrane

1.2.1 The Major Types of Lipids in Membranes

Phospholipids are the central components of membranes, giving them their characteristic bilayer structure. Phospholipids are amphipathic molecules (Fig. 3.1). They have two fatty acid (hydrocarbon) side chains, known as fatty acid tails, which are nonpolar and hydrophobic. They are usually attached to a glycerol, a phosphate group, and a molecule that contains an amino group (such as serine, choline, or ethanolamine). One of the two fatty acid tails is saturated, and the other is unsaturated. These lipids are technically glycerophospholipids due to their glycerol backbones [1].

There are four major phospholipids in membranes. These are phosphatidylethanolamine, phosphatidylserine, phosphatidylcholine, and sphingomyelin. For the first three, the major differences are the head group: ethanolamine, serine, or choline. Sphingomyelin differs from the other three. Its head group is choline, but it has a sphingosine backbone instead of a glycerol backbone and has one fatty acid side chain instead of two, although the sphingosine contributes a second hydrophobic tail (Fig. 3.1). The amphipathic nature of phospholipids is key to the special structure of the membrane. Amphipathic phospholipids spontaneously form bilayer and spherical structures (Fig. 3.2) when they are in an aqueous environment [1]. This is because the nonpolar (hydrophobic) tails tend to cluster together and face away from the aqueous environment. In contrast, the polar (hydrophilic) head groups face outward, where they can interact with the aqueous environment. As a result, the phospholipids interact to form a bilayer, and this bilayer folds into a spherical shape in an aqueous environment, so the hydrophobic tails are entirely sheltered from any aqueous solution. This is the basis for the overall characteristic of the plasma membrane [4].

In addition to the four main phospholipids, other less abundant lipids are present in membranes. These include glycolipids, such as glycosphingolipids, sterols, ceramides, and gangliosides.

1.2.2 Glycolipids

Glycolipids consist of a carbohydrate or sugar attached to a sphingoid base with no phosphate group. They have various roles, including roles in cell signaling, cell–cell recognition, and membrane stability. Ceramides consist of a sphingoid base linked to a fatty acid. They have several different roles. For example, they provide structural support and can have roles in cell signaling and apoptosis. Gangliosides are a type of glycosphingolipid with a complex sugar attached. They are most common in the nervous system. They have roles in signal transduction and synaptic transmission [1].

1.2.3 Sterols

Sterols are rigid ring structures containing four fused hydrocarbon rings. They have a polar hydroxyl group, while the rest of the structure is mostly hydrophobic. An example of a sterol is cholesterol. Cholesterol has an important role in maintaining membrane fluidity at both high and low temperatures. It prevents phospholipids from packing too tightly together at low temperatures, and at high temperatures, it limits phospholipid movement, thus helping to maintain a stable membrane structure [2].

1.2.4 Lipid Rafts

The membrane is composed of lipids, cholesterol, and proteins as described above. In some areas of the membrane, these components are clustered into domains that are more ordered than other parts of the membrane. These areas are called lipid rafts. The functions of these structures are not fully understood, but they can modulate membrane fluidity in specific areas and play crucial

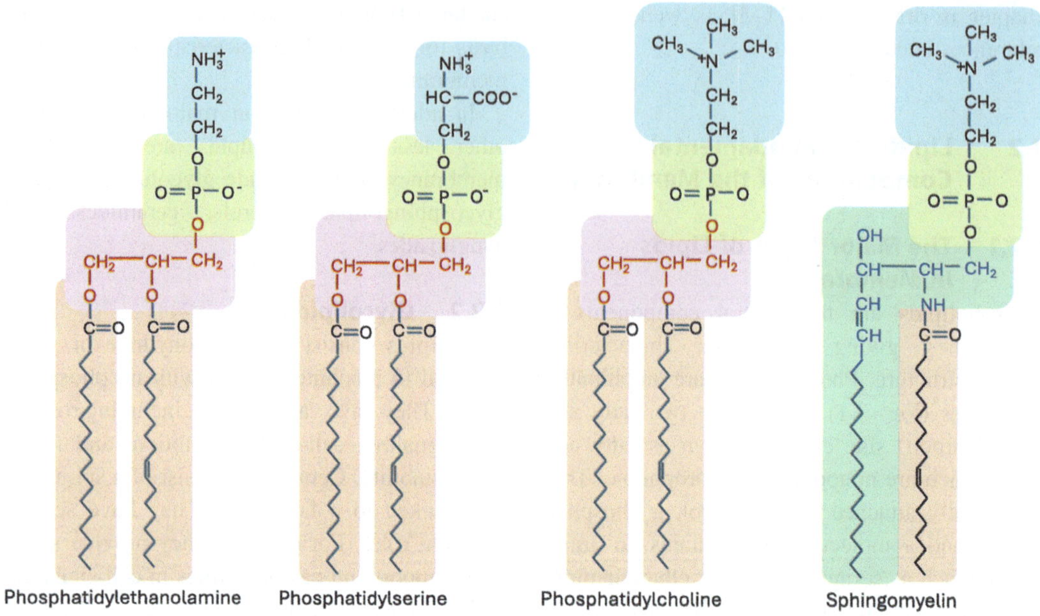

Fig. 3.1 The four main phospholipids in cell membranes. The four main phospholipids are phosphatidylethanolamine, phosphatidylserine, phosphatidylcholine, and sphingomyelin. Sphingomyelin differs from the other three because it has a sphingomyelin backbone and one fatty acid tail, rather than two fatty acid tails

Fig. 3.2 Phospholipids form a bilayer due to their amphipathic nature. The hydrophobic domain faces the interior of the layer, and the hydrophilic domains face the lumen of the structure or the outside of the structure, where they are in contact with the aqueous environment. (*Source*: Image adapted from Servier Medical Art (https://smart.servier.com/), licensed under the Creative Commons Attribution 4.0 (CC BY 4.0) license)

roles in various cellular processes, including cell signaling, membrane trafficking, and membrane organization [1, 5].

1.2.5 Disruption of Membrane Lipids or Their Metabolism Can Contribute to Disease

Membrane components, like all parts of the cell, are not static. Instead, they undergo a regular cycle of turnover. When cells are old or damaged, they can, for example, be phagocytosed by macrophages. This is a normal biological process that removes unnecessary cells from the body. Within these macrophages, the large molecules that make up the phagocytosed cells are transported to lysosomes, where enzymes break them down. This process includes the breakdown of membrane components.

Acid sphingomyelinase deficiency (ASMD), also referred to as Niemann–Pick disease types A and B, is a disorder characterized by a deficiency in the breakdown of sphingomyelin from cellular membranes. This disorder is caused by a mutation in the gene encoding the enzyme **acid**

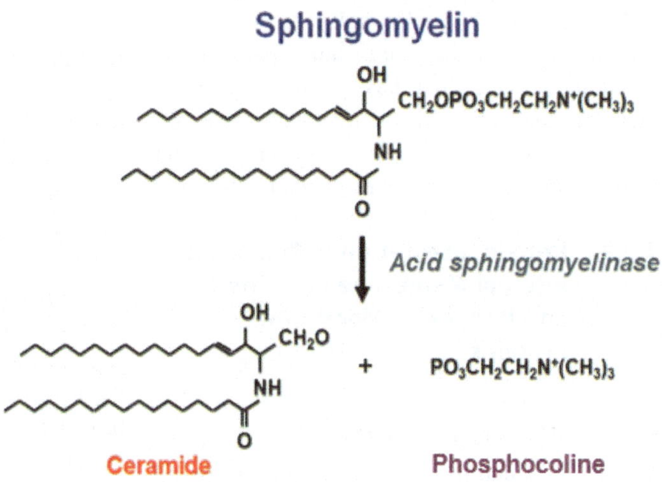

Fig. 3.3 Acid sphingomyelinase breaks down sphingomyelin into ceramide and phosphocholine. In most cells, there is a balance between sphingomyelin and its breakdown products. In ASMD, sphingomyelin cannot be broken down in the lysosomes. (*Source*: Reproduced from Mauhin et al. [142]. Licensed under the Creative Commons Attribution 4.0 (CC BY 4.0))

sphingomyelinase (**ASM**), which normally breaks down sphingomyelin in the acidic conditions specific to the lysosome (Fig. 3.3). ASMD is a rare and severe lysosomal storage disease.

Because of the deficiency in ASM, abnormally high levels of sphingomyelin accumulate within the lysosomes of the macrophages. The macrophages that accumulate sphingomyelin are referred to as foam cells due to their characteristic foamy appearance under the microscope. These cells travel throughout the body, causing damage to many tissues for several reasons. For example, they can lead to abnormal inflammatory effects. This is partly due to immune responses to these abnormal cells. The resulting chronic inflammation leads to widespread tissue damage, which can ultimately result in organ dysfunction. Symptoms include enlarged liver and spleen, difficulty eating, loss of muscle tone, lung disease, eye problems, low platelet counts, and delayed growth. The disease can show up in infants and has a poor prognosis; symptoms can be milder when it begins in adulthood [6].

Xenpozyme (olipudasa-alfa-rpcp) is a drug approved (FDA approved 2022) [7] for the treatment of ASMD. It is a replacement therapy for ASM. This drug is a recombinant, lab-made form of human acid sphingomyelinase. It is administered by repeated IV infusions. It received breakthrough therapy designation, which expedites the development of drugs to treat severe or life-threatening conditions. Small early clinical trials showed improvement in several symptoms. It was FDA-approved on August 31, 2022. It is the first drug approved for the treatment of the disease [8, 9].

1.3 Membrane Proteins and Carbohydrates

1.3.1 Overview of Proteins and Carbohydrates in Membranes

In addition to lipids, the membrane contains proteins embedded into the bilayer. Proteins can interact with the phospholipid bilayer in different ways. Some proteins span the entire membrane. These are called membrane-spanning proteins. Some of these proteins span the membrane multiple times, while others span the membrane only once. Some proteins may embed into only one layer of the phospholipid bilayer. They may penetrate this layer, or they may be attached to various membrane components by covalent attachments. Some membrane proteins can be modified or attached to other molecules. For example, some membrane proteins are glycoproteins, which have a sugar (carbohydrate) covalently attached to them. Some of the lipids in membranes are also attached to a sugar component. These lipids are called glycolipids. Due to the presence of large amounts of glycolipids and glycoproteins, many cells contain a cell coat called a glycocalyx. This cell coat is rich in carbohydrates on the extracellular side of the membranes. This coat has various functions, including facilitating interactions with the extracellular matrix, protecting cells from damage, and maintaining a distance between cells [1, 10].

1.3.2 Protein Integration in Membranes: Integral Membrane Proteins and Peripheral Membrane Proteins

- Integral Membrane Proteins

Membrane-bound proteins are categorized based on how they interact with the membrane. **Integral membrane proteins** are embedded in the membrane. There are two main categories of integral membrane proteins: **transmembrane proteins and monotopic proteins**. Transmembrane proteins span both leaflets of the membrane, thus spanning the entire bilayer, whereas **monotopic proteins** are embedded in only one leaflet (see Fig. 3.4).

Transmembrane proteins can cross the membrane once (single-pass) or multiple times (multi-pass). Examples include receptor tyrosine kinases (RTKs), which span the membrane once, and G protein-coupled receptors (GPCRs), which span the membrane seven times (also known as seven-pass transmembrane proteins). Ion channels and transporters are also usually multi-pass proteins.

- Peripheral Membrane Proteins

Peripheral membrane proteins do not penetrate either of the bilayers. Instead, they are associated with the membrane surface through noncovalent interactions with membrane lipids or with integral membrane proteins. Peripheral membrane proteins can be more readily dissociated from the membrane.

Peripheral membrane proteins attach to the membrane in different ways. Some, for example, can interact via electrostatic interactions with negatively charged lipid head groups. Some proteins are **lipid-anchored proteins**. These proteins are covalently attached to lipid molecules that insert into one leaflet of the membrane, and they are considered a type of peripheral membrane protein.

2 Transport Across the Membrane

2.1 The Plasma Membrane is Semipermeable

Transport in and out of the cell membrane is key to the cell's ability to interact with its surroundings while maintaining a stable internal environment. The molecules that are transported across the cell membrane, such as proteins, lipids, gases, or ions, are referred to as solutes. The membrane is semipermeable, and although most solutes can-

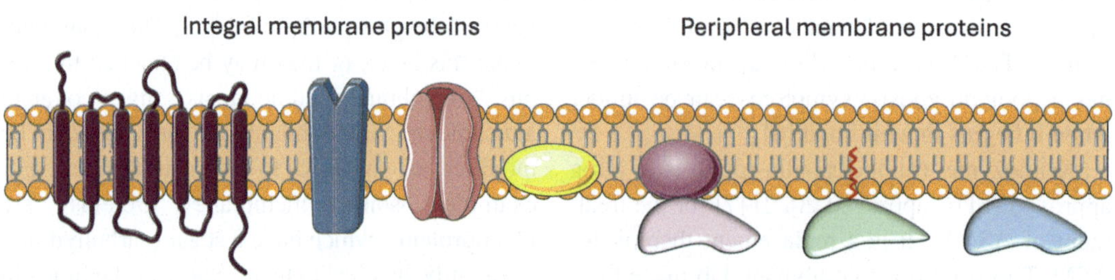

Fig. 3.4 Proteins are embedded in the plasma membrane. Proteins include integral membrane proteins and peripheral membrane proteins. (*Source*: Image adapted from Servier Medical Art (https://smart.servier.com/), licensed under the Creative Commons Attribution 4.0 (CC BY 4.0) license)

not cross the membrane without assistance, some small nonpolar molecules (such as O_2 and CO_2) and even some uncharged polar molecules can diffuse across the membrane without the need for any transporters. They can simply pass through the phospholipid bilayer. Most molecules, including larger ones, as well as many ions and polar molecules, cannot pass through the membrane without the help of transport proteins, discussed below [1, 2]. Understanding how molecules cross the membrane is critical for the study of pharmacy because many drugs either cross the plasma membrane to enter cells or interact with transporters and channels, thereby influencing the transport of other molecules.

2.2 Introduction to Membrane Transport Proteins

Most molecules require transport proteins to cross the membrane. The two main types of transport proteins are **transporters** (which include carriers and pumps) and channels.

Transporters are usually specific as to what they transport. They bind their solute, then undergo conformational changes to expose the solute to the other side of the membrane. Transporters can function in different ways: They can be uniports, symports, or antiports. Uniports allow one type of solute to pass in one direction across the membrane (either into or out of the cell). Symports allow two types of solutes to pass together in the same direction, and antiports transport one type of solute in one direction while the other type of solute is transported in the other direction (i.e., one type of solute is transported into the cell while the other is transported out of the cell). Symports and antiports can both be referred to as coupled transporters.

Channels are different from transporters. They can open or close due to conformational changes. When they are open, they form pores across the phospholipid bilayer. Transport through channels is usually faster than transport via transporters. Channels can sometimes also be specific as to which solute can pass through [11].

2.3 The Two Main Types of Transport Are Passive Transport and Active Transport

Passive transport does not require energy: It usually occurs when solutes flow down their concentration gradients (downhill). This generally means that the solutes flow from a region of high concentration to a region of low concentration due to the natural tendency to reach equilibrium. Additionally, the charge of the molecule can also impact transport. For example, a positively charged ion may be driven into a cell that has a negatively charged internal environment. A molecule is said to be moving down its electrochemical gradient if it is moving from a region of high concentration to a region of low concentration, as well as from an area of like charge to opposite charge (such as positive to negative).

In contrast, active transport requires energy: It is usually used when solutes flow against their concentration or electrochemical gradients (uphill). Active transport can use energy directly or indirectly, as discussed further below. Overall, passive transport is the more common mechanism for the transport of drugs [1] (see Fig. 3.5).

2.3.1 Passive Transport

There are different mechanisms by which passive transport can occur, as described below.

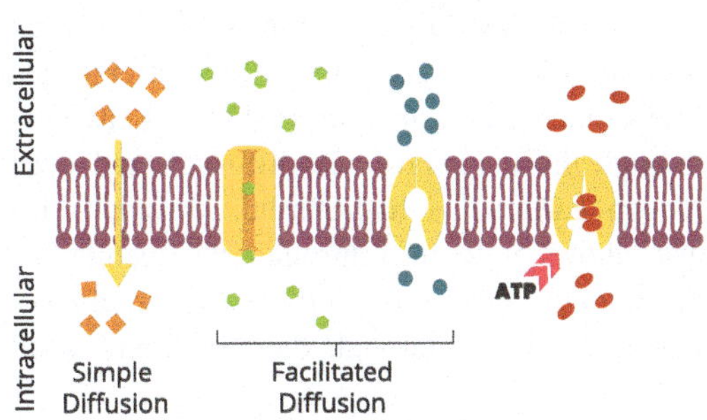

Fig. 3.5 Different types of transport mechanisms across the plasma membrane. Transport across the membrane can occur via simple diffusion (left) in some cases, but most solutes cross the membrane via facilitated diffusion. This involves the use of transport proteins such as channels (second to left) carriers (third from left). Active transport (right) requires energy). (*Source*: Illustration from Freepik.com (premium license))

- Passive Diffusion
  - Some solutes pass through the membrane by passive diffusion

A few types of solutes can diffuse freely through the phospholipid bilayer, down their concentration or electrochemical gradients, without requiring energy, channels, or transporters. This process is known as diffusion (or passive diffusion). Some small gases, small lipids, and some small uncharged molecules can pass through the membrane in this way.

- Some drugs can pass through the membrane by passive diffusion

Nonsteroidal anti-inflammatory drugs (NSAIDs), such as **Advil (Ibuprofen)** [12], **Aleve (naproxen)** [13], **and Aspirin (acetylsalicylic acid),** are examples of drugs that can pass through the membrane by passive diffusion, although some NSAIDs also use transporters [14]. These drugs are used to **relieve pain and inflammation**. Other examples of drugs that can diffuse through the membrane include some antiviral drugs, such as **Retrovir (Azidothymidine (AZT), or Zidovudine)** [15] **and Epivir (lamivudine)** [16, 17]. These drugs are both reverse transcriptase inhibitors that inhibit the activity of some viruses. Both can be used for the treatment of HIV AIDs. **Ethanol and Caffeine** are also examples of substances that impact cell function and can pass through membranes via diffusion. Ethanol is a central nervous system depressant, while caffeine can lead to increased wakefulness. Ethanol can pass through the membrane via diffusion, and even though **caffeine** is hydrophilic, its small size also allows it to pass through the membrane. Some of its effects on cells, however, do not require it to pass through the membrane at all [2, 18].

- Facilitated Diffusion

Most molecules cannot pass freely through the membrane; instead, they need a protein to assist in their transport. The types of proteins that assist in transport are usually channels or transporters (Fig. 3.5). Even molecules that travel down their concentration or electrochemical gradient usually need a transporter to cross the membrane. This is because most solutes are either too big or too polar to diffuse across the lipid bilayer. This type of transport is called facilitated diffusion. As long as the solute is flowing down its concentration or electrochemical gradient, it is still a type of passive transport because it does not require ATP. Examples of substances that cross the membrane via facilitated diffusion include some nucleotides, amino acids, glucose, and ions. Nearly all channels and some transporters operate by passive transport [1, 19–21]. Both channels and transporters are discussed here.

- Channels

Channels usually do not use energy and are thus generally used for passive transport, specifically in facilitated diffusion [1, 2]. Solutes can be transported down their concentration or electrochemical gradients through channels (Fig. 3.5).

Channels are **transmembrane proteins** that allow certain solutes to pass through the membrane. Often, these are ions that cannot pass through the lipid bilayer due to their charges. Channels are generally made of **polypeptides that fold into structures** that create a hydrophilic pore passing all the way through the membrane. Ions or some small molecules can pass through this core. Often, channels are selective as to what can pass through; this can be due to the shape and size of the pore and sometimes charge distribution. Many channels are **gated**, which means that only certain solutes can flow through them. Many gated channels are **ligand-gated**. They open when a certain ligand binds to a binding site on the channel. An example of a ligand that may bind a channel and cause it to open is a neurotransmitter. Some channels are **voltage-gated**. These channels may open in response to a change in membrane potential, which is the voltage difference across the membrane caused by the distribution of ions inside and outside the membrane. Some important channels are discussed below:

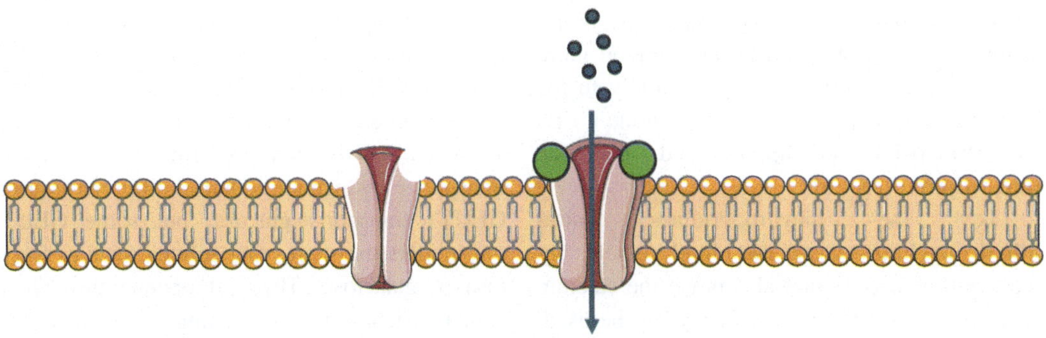

Fig. 3.6 Ligand-gated channels. Ligand-gated ion channels open for the passage of ions when a ligand binds. (*Source*: Image adapted from Servier Medical Art (https://smart.servier.com/), licensed under the Creative Commons Attribution 4.0 (CC BY 4.0) license)

Aquaporins
Aquaporins are a type of channel through which water molecules can pass. Although some water can flow through membranes without the aid of channels, some cells also have aquaporins. Cells, such as kidney cells, have an especially high level of aquaporins, which help them transport large amounts of water. This type of channel is typically not gated [1, 21, 22].

Ion Channels
Ion channels are another important example of channels. Ion channels transport ions such as Na^+, K^+, Ca^{2+}, and Cl^- from one side of the membrane to the other. Ion channels are usually selective as to which ions they transport. Many ion channels are gated (Fig. 3.6). **Gated ion channels** can open briefly in response to a stimulus, allowing ions to flow through. After the stimulus, the channels close again [1, 19].

Calcium Channels: Calcium channels are an important type of ion channel. There are multiple types of calcium channels. They are found in various cell types, including muscle cells, neurons, endocrine cells, and immune cells. Calcium influx into cells via calcium channels elicits different responses, depending on the cell type. In smooth and skeletal muscle cells, calcium influx leads to muscle contraction. In cardiac muscle cells, calcium plays a role in regulating heart contraction and relaxation. In neurons, the flow of calcium is essential for neurotransmitter release. In endocrine cells, such as cells of the pancreas or adrenal glands, calcium influx triggers the release of hormones into the bloodstream. In some immune cells, such as T cells and mast cells, calcium influx regulates immune responses. Most calcium channels on the plasma membrane are voltage-gated. Other types of channels, such as IP3 receptors (discussed in Chap. 18), are found on intracellular membranes such as the ER or the SR [1, 21].

Calcium channel blockers (CCBs) constitute a group of drugs that reduce the flow of calcium ions by blocking calcium channels. They generally function by binding to the channels and stabilizing them in a **closed** conformation. The result is inhibition of calcium flow through the channels. CCBs often block calcium channels in cardiac muscle or smooth muscle lining the blood vessels. This reduces intracellular calcium levels because inward calcium flow is reduced. This can lead to relaxation of vascular smooth muscle, vasodilation, and a decrease in cardiac contractility. These drugs are often used to manage **hypertension and some heart arrhythmias**. Two major classes of CCBs are **dihydropyridine CCBs and non-dihydropyridine CCBs** [19].

Dihydropyridines target primarily **calcium channels** in **smooth muscle** rather than cardiac muscle. They are good drugs for treating hypertension, which is associated with a high level of vascular resistance (the resistance of blood vessels to blood flow, which may be due to constriction [23]). Dihydropyridines, such as

Norvasc (amlodipine) [24], **relax the blood vessels**, allowing the heart to pump blood more easily. It also increases the blood supply to the heart. Other examples of Dihydropyridine CCBs are **Procardia (nifedipine)** and **Cardene (nicardipine)** [25, 26].

Like the dihydropyridines, **non-dihydropyridines** target the blood vessels, but the non-dihydropyridines also target the calcium channels in the heart muscle. They can be used, therefore, not only for treating high blood pressure but also for controlling heart arrhythmias, although they may have undesirable side effects in some patients with heart conditions. Examples of non-dihydropyridine CCBs are **Cardizem (diltiazem)** [27] and **Calan (verapamil)** [21, 28–32].

– Transporters (Carriers) in Passive Transport

How Transporters Work

Transporters are different from channels. Rather than forming a passage for solutes to go through, transporters physically transport the solute. They bind the solute on one side of the membrane via solute binding sites. After binding their solute, the transporters undergo a conformational change, allowing them to transport the solute across the membrane. The conformation of the transporter changes throughout the transportation process. When transporting a solute into the cell, the transporter changes from outward open (facing the outside of the cell) to occluded (where the solute is now buried in the lipid bilayer and shielded from either side of the membrane) to inward open, where the solute is released into the cell (Fig. 3.7). Many transporters participate in active transport, but some are also involved in passive transport. One example of a transporter that operates via passive transport is the Glucose transporter (Glut), which transports glucose down its concentration gradient. The Glut transporter is a uniport. There are 14 different isoforms of the GLUT transporters, which are found in different tissues. GLUT1 transports glucose, but some GLUT transporters transport other solutes, such as fructose or galactose. They all operate through a similar mechanism, functioning as uniports that facilitate the transport of their solutes down their concentration gradients [1, 33]. GLUT transporters become especially important when discussing metabolism, and therefore, a list of different GLUT transporters is shown in Chap. 12.

When studying the kinetics of transport via transporters, Vmax and Km are important concepts. They represent mechanisms for expressing the kinetics of the transport process and the efficiency of the solute-binding properties of the transporters. **Vmax** refers to the maximum rate of transport across the membrane when all the transport binding sites are saturated with solute. Thus, it reflects how efficiently the transporter operates when the conditions are optimal. **Km** is a measure of the transporter's affinity for its solute. Each carrier has a specific affinity for its solute. Km is the solute concentration when the transport rate is half its maximum value. A lower Km indicates higher affinity. In other words, the transporter can achieve half its maximal transport rate at lower solute concentrations. In contrast, a higher Km means that higher solute concentrations are required for the transporter to achieve its maximum transport rate [1, 34]. Vmax

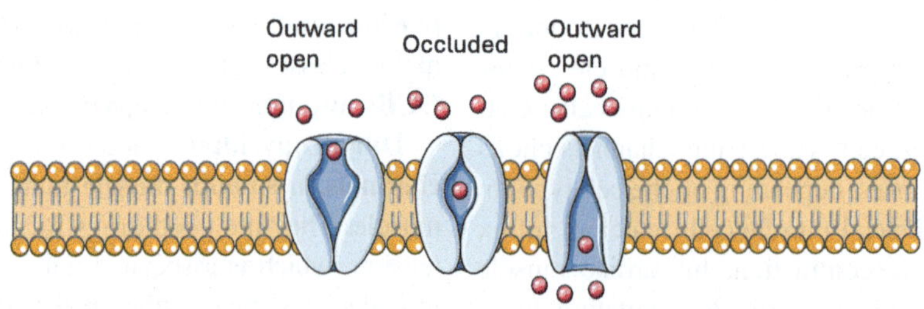

Fig. 3.7 Transporters. Transporters in the membrane change their conformation to transport cargo. (*Source*: Image adapted from Servier Medical Art (https://smart.servier.com/), licensed under the Creative Commons Attribution 4.0 (CC BY 4.0) license)

and Km are especially important concepts in enzyme kinetics and will be discussed in more detail in Chap. 5.

Inhibitors of Transporters

The binding of solute to the transporter can potentially be blocked by inhibitors. Such inhibitors are usually competitive inhibitors or noncompetitive inhibitors. Competitive inhibitors compete for the same binding site as the solute, although the inhibitor itself may or may not be transported. Noncompetitive inhibitors bind elsewhere in the transporter rather than at the actual solute binding site, but they may alter the transporter's structure, affecting its ability to bind or transport the solute [1].

2.3.2 Active Transport and Pumps

Here, we switch from the discussion of passive transport to active transport. Unlike passive transport, active transport requires energy to move solutes against their concentration or electrochemical gradients. There are two main types of active transport: primary and secondary active transport. Primary active transport directly uses energy, often in the form of ATP hydrolysis, to move molecules against their gradients. Active transporters that use primary active transport are commonly referred to as pumps. A well-known example is the Na^+/K^+ pump, which plays an important role in establishing gradients across cell membranes.

Secondary active transport, in contrast, harnesses energy indirectly, often via coupled transport. This can occur when a concentration gradient is separately set up, via a pump that utilizes active transport. In this type of coupled transport mechanism, one solute (from the pre-established gradient) moves down its concentration gradient, releasing energy. This energy is used to drive the transport of another solute against its concentration gradient. For example, in sodium–glucose cotransporters, Na^+ ions moving down their gradient provide energy for transporting glucose into the cell, relying on a pre-established gradient set up by the Na^+/K^+ pump. It is worth noting that some sources may label secondary transport as passive transport because it relies on a previously established gradient, but technically, this process qualifies as active transport.

In addition to these mechanisms, other transport mechanisms exist in specialized systems such as bacteria, archaea, mitochondria, and chloroplasts. These systems can use light or redox reactions to drive transport, but those are not discussed here [34].

Some types of active transporters are described below:

- Primary Active Transport
  - P-Type Pumps

Transport proteins, which are considered pumps, hydrolyze ATP to yield ADP and utilize the energy released from ATP hydrolysis to drive transport. An important type of pump is the P-type pump, also known as a P-type ATPase. The P-type pump [35] gets its name from the fact that it becomes phosphorylated. P-type pumps autophosphorylate themselves during the pumping cycle. The phosphorylation is linked to the hydrolysis of ATP via their intrinsic ATPase activity. Phosphorylation occurs on a conserved aspartate found on all P-type pumps. This phosphorylation results in a conformational change. This type of pump includes many ion pumps that are responsible for setting up ion gradients. Some examples are the sodium–potassium (Na^+/K^+) pump and some calcium (Ca^{2+}) pumps, as discussed below:

The Sodium (Na^+) Potassium (K^+) Pump

The Na^+/K^+ pump is a P-type pump that plays a crucial role in maintaining the electrochemical gradient across cells (see Fig. 3.8). This pump transports three Na+ ions out of the cell and two K^+ ions into the cell. The energy is derived from the hydrolysis of ATP and the subsequent phosphorylation of the pump, as described below. This pump has the very important function of establishing the electrochemical gradient across cells, in which the inside of the cell has low Na^+ and high K^+ levels, while the outside has high Na^+ and low K^+ levels [1, 36].

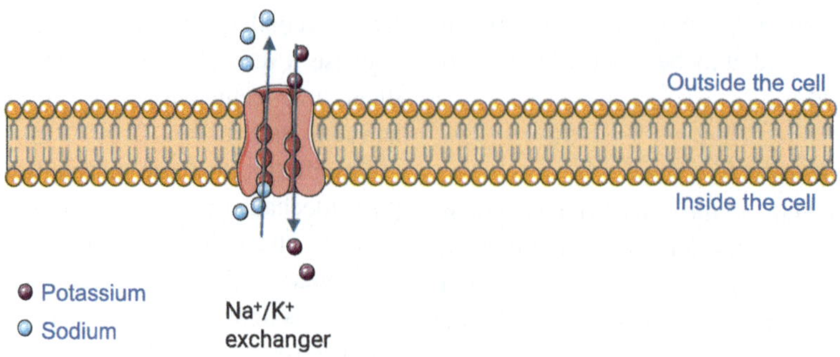

Fig. 3.8 Sodium–potassium pump. The sodium–potassium pump pumps sodium out of the cell and potassium into the cell, creating a gradient. (*Source*: Image adapted from Servier Medical Art (https://smart.servier.com/), licensed under the Creative Commons Attribution 4.0 (CC BY 4.0) license)

The pump works by first binding 3 Na^+ ions present in the cytoplasm to their binding sites on the cytoplasmic side of the pump. As the pump functions, it undergoes conformational changes starting with the E1 conformation, followed by E1–P, then E2–P, then E2, and back to E1, as follows. First, the binding of the three Na^+ ions bind to the cytoplasmic side of the pump. This binding trigger hydrolysis of ATP to yield ADP and inorganic phosphate (Pi), leading to phosphorylation of a site on the pump, where the released Pi binds to a conserved aspartate on the pump. This phosphorylation leads to a conformational change from an E1 state, where the Na^+ binding sites face the cytoplasm, to an E1–P state, followed by an E2–P state, where the Na^+ binding site faces the extracellular space. After the conformational change, the Na^+ ions are released into the extracellular space (outside the cell). Two binding sites for K^+ are now present on the extracellular side of the pump. The binding of K^+ to these sites triggers the release of the phosphate group that had been added to the pump. This leads to another conformational change, to an E2 state and back to the E1 state, which now exposes the K^+ ions to the inside of the cell, where they are released. This cycle repeats over and over again, establishing a gradient in the cell where there are high K^+ levels inside the cell and high Na^+ levels outside the cell. This type of mechanism, including the different states (E1→E1–P→E2–P→E2→E1), is common to all P-type pumps.

This electrochemical gradient involving Na^+ and K^+ is established across nearly all cells. It has several crucial functions. Importantly, the energy harnessed from the electrochemical gradient can drive the transport of certain molecules across the membrane. This is a prime example of indirect active transport, where the energy for transporting certain molecules is derived from the electrochemical gradient. One example of a transporter that relies on the Na^+/K^+ gradient for energy is a type of glucose transporter known as the SGLT (the sodium–glucose cotransporter, also called the sodium-glucose linked transporter), also referred to as the sodium–glucose cotransporter. This pump moves glucose into the cell against its concentration gradient, using the Na^+ gradient to facilitate the process. It couples the inward flow of Na^+ down its electrochemical gradient with the inward flow of glucose. (It is a symporter, transporting both Na^+ and glucose in the same direction). It is found in intestinal cells and renal tubule cells, where it plays a role in absorbing glucose from the lumens of these tissues into the cells lining them. It is a target of some diabetes medications, to be discussed later in this chapter.

The Na^+/K^+ gradient also has a role in neuronal signaling. It is the basis for transmitting electrical signals in neurons and other electrically excitable cells. Another function of the gradient is maintaining cell volume; it helps regulate the movement of water into and out of cells, thereby maintaining the correct cell volume. It also plays a key role in maintaining membrane potential. The membrane potential, which reflects the difference in electrical charge across the membrane, is important for many cellular processes, such as the uptake of nutrients and the release of neurotransmitters [1, 37].

Digoxin, a drug used to treat heart conditions, functions by inhibiting the Na^+/K^+ pump.

Digoxin is a **small-molecule drug** derived from the plant Digitalis. It is a steroidal glycoside, meaning that it has a steroid-like core with sugar residues [38]. It is used to treat heart conditions such as heart failure. Digoxin **binds directly to Na^+/K^+ pumps on cardiac cells and inhibits their activity**. By inhibiting the Na^+/K^+ pumps on cardiac cells, digoxin disrupts the Na^+/K^+ gradient. This, in turn, disrupts another transporter, the Na^+/Ca^{2+} exchanger, which depends on the Na^+/K^+ gradient to function. (The Na^+/Ca^{2+} exchanger is a classic example of secondary active transport because it depends on the gradient set up by the Na+/K+ pump for its function.) This Na^+/Ca^{2+} exchanger normally has the function of pumping Ca^{2+} out of the cell (Fig. 3.9). Disrupting the exchanger thus results in an increased concentration of intracellular calcium. Increased intracellular calcium has several beneficial effects. For example, it strengthens heart muscle contractions, which can be beneficial for patients with heart failure.

Digoxin also has another function in the nervous system, affecting the vagus nerve, a parasympathetic nerve that sends signals to the heart. Activation of this nerve by digoxin slows down the heart rate, but the exact mechanism is not completely understood [38].

Digoxin is helpful for some patients with **arrhythmias** because it improves contraction and controls certain types of irregular heartbeats [39, 40]. Digoxin can be lifesaving to heart patients, but it can also cause serious side effects, some of which may be related to disrupting the Na^+/K^+ balance. Side effects include gastrointestinal tract effects, central nervous system abnormalities, eye problems, and muscle weakness. It can also occasionally lead to an abnormal rise of K^+ in the bloodstream because K^+ leaks out of the cells due to their abnormally high level inside the cell. High blood potassium (hyperkalemia) can lead to further heart problems. Therefore, while this drug has many benefits, patients must be closely monitored and the dosage optimized [40].

The P-Type Calcium Pump

There are two major types of calcium transporters involved in the removal of calcium from the cytosol. There is the P-type Ca^{2+} ATPase [1, 35, 41], which is found on the plasma membrane (PMCA) and on the membranes of the endoplasmic or sarcoplasmic reticulum (SRCA). The other is the

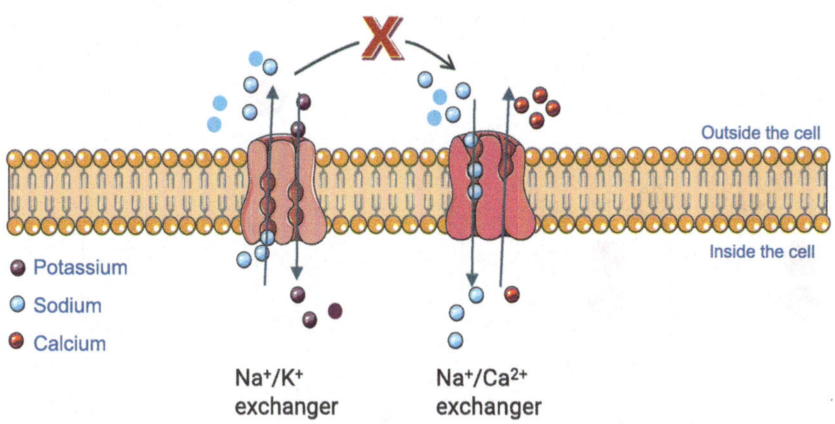

Fig. 3.9 Digoxin. By inhibiting sodium–potassium pumps, the drug digoxin disrupts a sodium–calcium exchanger. (*Source*: Image adapted from Servier Medical Art (https://smart.servier.com/), licensed under the Creative Commons Attribution 4.0 (CC BY 4.0) license)

Na^+/Ca^{2+} exchanger (an antiporter), driven by the Na^+ gradient (described above as the target for digitalis) [42]. The P-type Ca^{2+} ATPase pump is important in establishing the Ca^{2+} gradient across membranes. The SERCA transporter is discussed below.

The SERCA Calcium Pump is a P-Type Calcium Pump

Sarcoplasmic/endoplasmic reticulum Ca^{2+} ATPase (SERCA) is a P-type Ca^{2+} pump (It is sometimes referred to as simply the Ca^{2+} ATPase). SERCA is found in high levels in the membrane of the sarcoplasmic reticulum (SR) in skeletal muscle cells, but it is also found on the ER membrane in non-muscle cells. The SERCA pumps Ca^{2+} into the SR or ER. This results in high levels of Ca^{2+} stored in the SR or ER and correspondingly low levels in the cytoplasm.

Due to the SERCA, the calcium level is higher inside the SR or ER than in the cytoplasm. This calcium gradient is important because small influxes of calcium into the cytoplasm (from the ER or the SR) in response to various signals can lead to important cellular changes. In **muscle cells**, upon stimulation by various intracellular signals such as action potentials, the calcium stored in the SR can be rapidly released into the cytosol to trigger **contraction**. Afterward, SERCA resequesters calcium back into the SR, allowing the muscle to **relax.** In **non-muscle cells**, the stored calcium can be released into the cytosol upon activation of certain signaling pathways, often involving second messengers. The released calcium is involved in various **calcium-dependent functions**, such as regulation of various signaling pathways, gene regulation, secretion, or metabolism [1, 43].

The SERCA operates via a mechanism that involves phosphorylation, driven by ATP, and multiple conformational states, following a pattern similar to the sodium potassium pump (E1→ E1–P → E2–P, → E2→ E1). In the E1 state, the Ca^{2+} binding sites on the pump are exposed to the cytosol, allowing two Ca^{2+} ions to bind to the SERCA. ATP then binds and is hydrolyzed to ADP, after transferring its γ-phosphate group to a conserved aspartate residue on the pump's P domain, forming the E1–P state. The ADP is then released. This phosphorylation triggers a conformational change leading to the formation of the E2–P state. In this state, the two Ca^{2+} binding sites now face the lumen of the SR or the ER, and two Ca^{2+} ions are then released from the pump into the SR or ER lumen. After this, two to three H^+ ions from the lumen bind to the pump, which is then dephosphorylated, forming the E2 state. This is followed by a return to the E1 state. This process allows Ca^{2+} to be transported into the SR or ER lumen as protons are transferred from the SR lumen to the cytosol [44] (Fig. 3.10).

SERCA in Disease

Disorders associated with aging have been shown to be linked to the suppression of SERCA, thus impairing calcium homeostasis. This may be

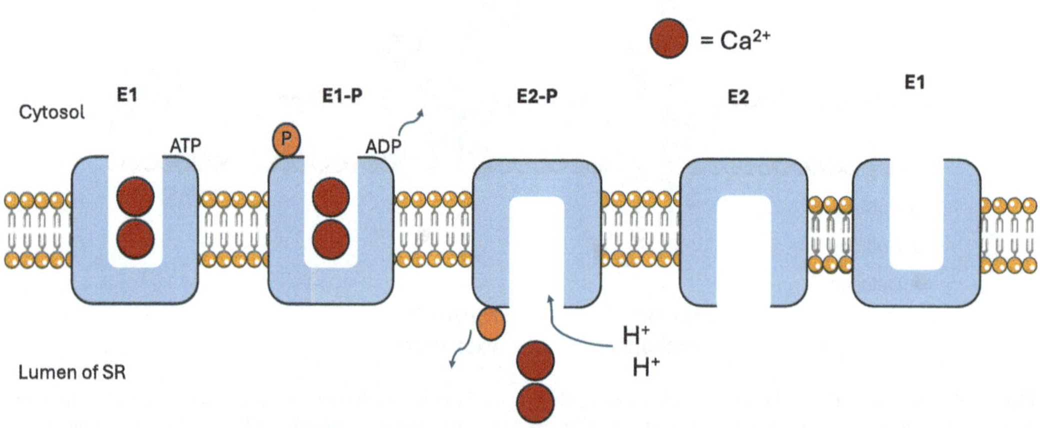

Fig. 3.10 The SERCA. Calcium pump (sarcoplasmic/endoplasmic reticulum calcium ATPase). The SERCA pump transports calcium into the sarcoplasmic reticulum. See text for details

linked to oxidative stress in response to aging. For example, some cases of muscular dystrophy and neurodegenerative disorders such as Alzheimer's and ALS have been linked to downregulation of SERCA, although these disorders can also have other causes and often have unknown causes. In muscular dystrophies, abnormal calcium levels can affect muscle function, and experimental studies show that SERCA dysfunction may be involved [45]. Disruption of calcium homeostasis in the nervous system may also be linked to diseases such as Alzheimer's disease, and this may also be associated with abnormalities in SERCA [43]. Impaired SERCA is also linked to diabetes because by impacting calcium homeostasis, leading to ER stress (a concept that will be discussed in Chap. 8), which is a major contributor to diabetes [46]. For these reasons, targeting SERCA, particularly its activation, is a promising therapeutic approach for treating these types of disorders.

Strategies to target SERCA may be an approach to treating these types of disorders. Approaches for activating SERCA involve using compounds that activate the pump or using genetic overexpression. For example, **CDN1163** is an organic compound that is an allosteric SERCA activator. This means that it binds to a site that is neither its ATP binding site nor its Ca^{2+} binding site. Instead, it leads to a conformational change that can increase SERCA's ability to pump Ca^{2+}. This compound has been shown to restore impaired SERCA activity in animal studies. This compound has led to beneficial changes in muscle function in mice, as well as reduced ER stress, improved insulin tolerance, and reduced fasting blood glucose [43, 46]. SERCA activators are primarily in the experimental stage and are being studied for potential clinical use.

– ABC Transporters

ABC transporters (**A**TP-**b**inding **c**assette transporters) have structures that differ from the P-type pumps. They primarily pump small organic molecules. While they are technically pumps, they have unique characteristics that differ from other pumps and are often referred to as transporters rather than pumps [47] (Fig. 3.11).

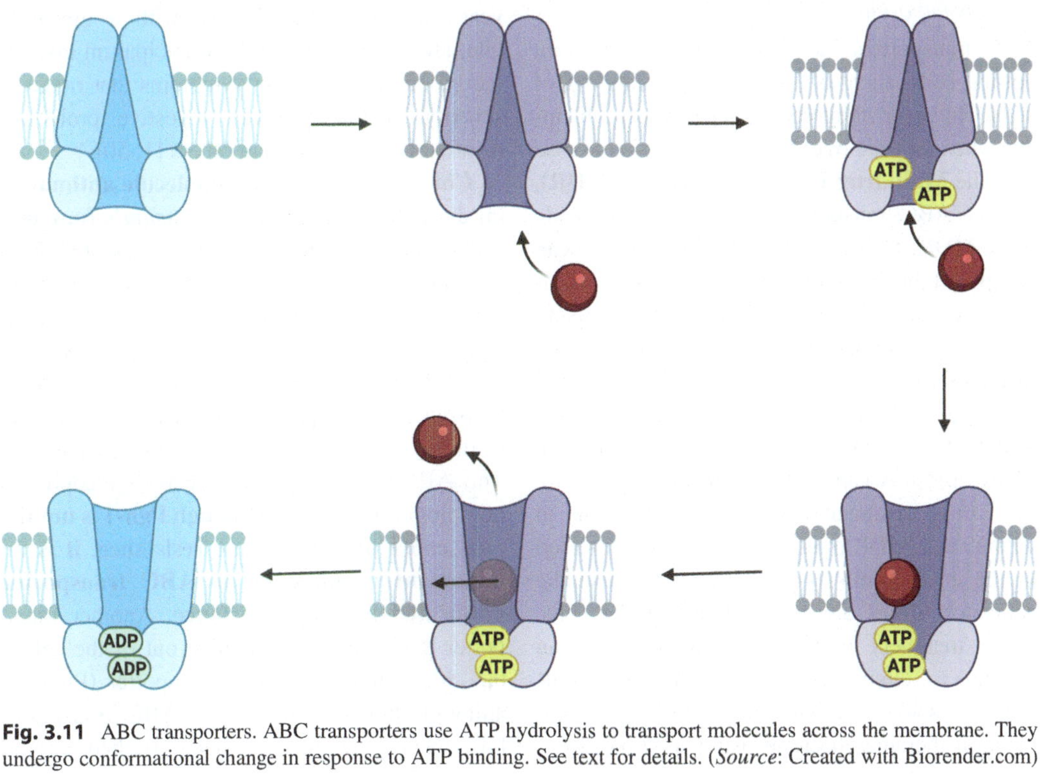

Fig. 3.11 ABC transporters. ABC transporters use ATP hydrolysis to transport molecules across the membrane. They undergo conformational change in response to ATP binding. See text for details. (*Source*: Created with Biorender.com)

ABC transporters are the largest family of membrane transport proteins. They consist of two transmembrane domains that form a channel and two ATP-binding domains (called ATP-binding cassettes) on the cytoplasmic side. The solute binds to its binding site on the transporter. ATP then binds to the binding cassettes, causing the two ATP-binding cassettes to come together. ATP is then hydrolyzed into ADP and inorganic phosphate (Pi), providing energy for transport. The transporter undergoes a conformational change, allowing the solute to be transported across the membrane to the other side. In eukaryotic cells, this type of transporter usually pumps its solute out of the cell. There are 49 different ABC genes in humans, which are categorized into seven distinct groups. They are involved in pumping various substrates out of the cell, including ions, peptides, metabolites, and certain lipids. They can be involved in detoxification by excreting toxins from cells. They are also known, however, to transport drugs out of cells, which can be a problem, especially during cancer treatment [48].

MDR is a member of the ABC transporter family and has important clinical significance.

ABC transporters have important functions in normal cells, such as transporting toxins out of cells. However, they can also become problematic when they are overexpressed. This is the case with the **Multidrug Resistance Protein (MDR)**, the first ABC transporter identified in eukaryotic cells. MDR, also known as P-glycoprotein, can pump certain drugs out of cells. MDR is commonly elevated in human cancer cells, causing drugs to be pumped out of the cells and thereby making the cells resistant to many drugs designed to treat the cancer. Treatment of cancer cells can lead to the selection of cells that overproduce MDR, and this can make MDR levels increase upon drug treatment. Treatment for patients in whom these MDR transporters are overexpressed in the cancer cells is challenging because the anticancer drugs are pumped out of the cells. Some treatment options are employed in an attempt to circumvent the problem of MDR overexpression. One possibility is the use of combination therapies because using multiple drugs can sometimes reduce the likelihood of the cells developing resistance. Drug sequencing is another option. Delivering drugs in a specific order has been shown to help avoid the development of resistance. Another option is to use certain drugs that have an inhibitory effect on MDR transporters. This can include targeted therapies that are directed at known MDR proteins. More recently, the use of nanotechnology has been explored. Nanoparticles delivering anticancer drugs can sometimes be designed in a way that they will be resistant to the MDR transporters [49].

P-glycoprotein homologue 1 (Pgh-1) is an ABC transporter found in the parasite Plasmodium falciparum, which has important implications in Malaria treatment

The ABC transporter P-glycoprotein homologue 1 (Pgh-1), found in *Plasmodium falciparum* (*P. falciparum*), is one mechanism that contributes to resistance to the antimalarial drug chloroquine. The protist Plasmodium falciparum is a parasite that infects humans and is transmitted by mosquitoes, causing malaria, an infectious disease that causes thousands of deaths each year. Infection often begins with a mosquito bite, after which Plasmodium falciparum invades red blood cells, causing symptoms that range in severity, including fever, digestive problems, respiratory problems, and others [1, 50].

Chloroquine is a **small-molecule antimalarial drug** that **disrupts *P. falciparum*'s function**. It does so by accumulating in the parasite's food vacuole, leading to toxicity [19]. However, in some instances, the protists can develop resistance to the drug. The resistant protists were found to have an amplified gene for an ABC transporter that pumps out the chloroquine, resulting in drug resistance in infected patients. This ABC transporter is known as P-glycoprotein homologue 1 (Pgh-1). **Although Pgh-1 is not the only cause of chloroquine resistance, it is an important example of an ABC transporter implicated in drug resistance.** Another transporter that pumps chloroquine out of the cell is the chloroquine resistance transporter (PfCRT), although this one is not an ABC transporter [50]. New strategies are being studied to find ways to block the actions of these types of transporters or to develop new drugs that are not pumped out of the *P. falciparum* [51].

CFTR is a Cl⁻ transporter that is mutated in cystic fibrosis (CF)

CFTR is the cystic fibrosis transmembrane conductance regulator protein. It is often considered to be an atypical ABC transporter. It functions slightly differently from other ABC transporters, functioning more like a channel than a typical transporter. CFTR is a Cl^- transport protein present on the surface of epithelial cells. It regulates the concentration of chloride ions in the extracellular fluid, particularly in the lungs, which indirectly affects the regulation of other ions as well. Mutations of both copies of the CFTR gene result in cystic fibrosis (CF).

Some drugs for CF are designed to potentiate the function of CFTR. One example of such a drug is **Kalydeco (ivacaftor)**. It is a **small-molecule drug that binds to the mutant CFTR** and increases its ability to open. It is effective in some patients and facilitates Cl^- flow [52, 53].

V-Type Pumps (V-ATPases)

Vacuolar-type proton pumps (V-type pumps) are another type of ATP-dependent proton pump. They are found primarily on the membranes of organelles and vesicles, such as lysosomes, endosomes, and the Golgi apparatus. They are important for regulating pH and promoting acidification within organelles. They use energy from ATP to transport protons (H^+) across membranes, thus contributing to an electrochemical gradient. They contain a domain for ATPase activity and a domain that forms a channel [54].

This type of transporter has an ATPase activity domain (V1 subunit), which faces the cytoplasm. It is responsible for hydrolyzing ATP for energy. The other domain is the proton channel domain (V0 subunit) through which the protons are transported [55].

Dysregulated V-type pumps are linked to certain diseases, such as lysosomal storage diseases. In some cases, they are involved in tumor progression because abnormal acidification can affect cell function and survival. Some types of experimental compounds are being tested to determine how drugs designed to regulate this type of pump may be effective in treating human disease [55, 56].

- Coupled Transport

Coupled transporters are a group of transporters that often carry out secondary transport. The energy source is usually the electrochemical gradient rather than the direct use of ATP. For example, various symporters utilize the electrochemical gradient established by the Na^+/K^+ pump [1]. Here are some examples:

– The Sodium–Glucose Cotransporter (SGLT)

The sodium–glucose co-transporter (SGLT) is a coupled transporter (see Fig. 3.12). It is a

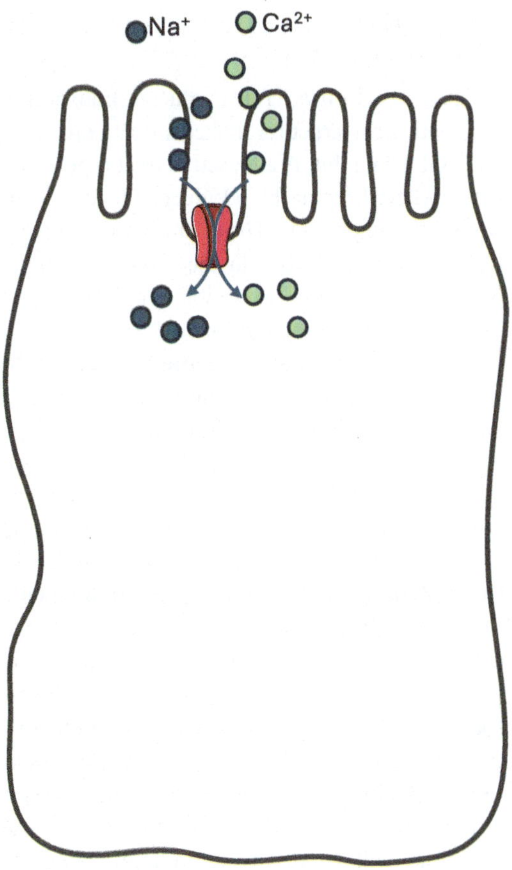

Fig. 3.12 SGLT: The sodium glucose cotransporter. Sodium enters the cell, going down its concentration gradient, driving the flow of glucose up its concentration gradient. (*Source*: Image adapted from Servier Medical Art (https://smart.servier.com/),

symporter that has a role in glucose transport across the membrane, especially in intestinal and kidney epithelial cells. (Note—this is different from the Glut transporter, the uniport described above in the passive transport section). There are multiple SGLTs. SGLT1 is located in the kidney and intestine, and SGLT2 is located in the kidney. SGLT2 is known to have a role in the reabsorption of glucose from the kidney filtrate into the blood. SGLTs use the energy stored in the sodium gradient (established by the Na^+/K^+ pump) to move glucose against its concentration gradient into cells. This is an example of secondary active transport. Sodium moves down its electrochemical gradient into the cell, and glucose is co-transported with it [1, 57].

- SGLT Inhibitors Are Used to Treat Metabolic Disease

Because SGLT plays a role in glucose transport, it is sometimes a target for diabetes treatment. Several SGLT2 inhibitors are used to treat type 2 diabetes, where keeping blood glucose levels under control is important. These inhibitors prevent reabsorption of glucose in the kidneys, thereby promoting glucose excretion in the urine [58]. **Jardiance (empagliflozin)** [59], for example, is a synthetic **small molecule SGLT2 inhibitor** used to treat **type 2 diabetes** [60]. It is a competitive inhibitor of SGLT2. It binds to the glucose-binding site on SGL2, thereby preventing glucose from binding. Consequently, it prevents glucose reabsorption, helping to lower blood sugar levels. **Farxiga (dapagliflozin) (AstraZeneca)** [61] is another **small molecule SGLT2 inhibitor**, also used to treat type 2 diabetes [62]. Other examples of small molecule inhibitors of SGLT are **Brenzavvy (bexagliflozin)** [63] and **Inpefa (sotogliflozin)** [64] (both FDA approved in 2023). While **Inpefa (Sotogliflozin)** is mainly an **SGLT2 inhibitor**, it also inhibits SGLT1 [65, 66]. Some of these inhibitors will be discussed in more detail in Chap. 13.

2.4 Summary of the Different Types of Glucose Transporters

Glucose is an important energy source in the body, and in this chapter, the two main categories of glucose transporters have been introduced: the GLUT transporter and the SGLT transporter. Glucose transport is critical for cellular metabolism, and it is therefore beneficial to review the different types of glucose transporters. Which type of glucose transporter used to transport glucose depends on whether glucose is flowing down or up its concentration gradient (passive transport or active transport).

The GLUT transporters transport glucose down its concentration gradient. These transporters use facilitated diffusion and transport glucose from areas of high concentration to areas of low concentration, down their concentration gradients. Different GLUT isoforms function in different tissues to either import or export glucose depending on the concentration gradient. The four most common and best characterized are GLUT 1—4. GLUT1 is found in the cells of many different tissues, GLUT2 is found in the liver, kidney, and pancreatic beta cells, GLUT3 is found in neurons, and GLUT4 is found in muscle and adipose tissue. GLUT4 is an important example of the GLUT transporters. It functions when glucose levels are in the bloodstream, such as after a meal, and the glucose needs to be transported down its concentration gradient into cells, such as in muscle or adipose tissue. This will be discussed in more detail in Sect. 2.4, where metabolism is covered.

The SGLT transporters, in contrast, use secondary active transport to move glucose against its concentration gradient. This type of transporter relies on the sodium gradient maintained by the Na^+-K^+ ATPase pump, which utilizes ATP as its energy source. SGLT transporters are primarily found in epithelial tissues, specifically absorptive tissues such as the proximal tubule of the kidney and the enterocytes of the small intestine, where glucose needs to be absorbed even when intracellular concentrations are already high [1, 33].

2.5 Vesicular Transport

We have discussed solutes crossing the plasma membrane via passive diffusion, channels, or carrier proteins. Relatively small molecules can be transported in this way. Larger molecules, such as proteins, can also be transported into cells, but this requires different mechanisms. One mechanism, called the Sec system [67], is not discussed in this chapter. The other is vesicular transport [1, 2], discussed here. In vesicular transport, vesicles pinch off a membrane (such as the plasma membrane or an organelle membrane), and in the process, they trap their cargo. Later, they fuse with a target membrane and release their cargo. Vesicular transport plays roles in exocytosis (the secretion of substances from the cell), endocytosis (the uptake of substances into the cell), and intracellular trafficking (the movement of substances between organelles). Although vesicular transport requires energy, it is often categorized separately from other types of active transport systems because it uses a distinct mechanism that involves membrane-bound vesicles. Below are some major examples of vesicle transport:

2.5.1 Receptor-Mediated *Endocytosis* (RME)

In RME, certain molecules, often cell surface receptors, bind to receptors on the cell surface. This triggers the formation of a vesicle. Often this is a clathrin-coated vesicle (a vesicle coated with the protein clathrin). This vesicle buds off from the membrane (invagination), and the whole vesicle and its contents enter the cell. The vesicle then uncoats. Commonly, the vesicle then usually fuses with other vesicles called early endosomes, which can sort their contents for delivery to the lysosome for degradation. They may also recycle contents back to the membrane. This type of transport does not depend on the concentration of the transported molecule, but it does require ATP [1].

2.5.2 Pinocytosis

Unlike RME, pinocytosis is the nonspecific uptake of fluid and small solutes. These may be taken up from outside the cell when vesicles pinch off from the plasma membrane (membrane invagination), bringing the contents into the cell.

2.5.3 Phagocytosis

Phagocytosis is the process of engulfing larger molecules or even whole cells, such as bacteria. This type of transport is only carried out in some types of cells. For example, in some cells of the immune system, it is common for cells to extend pseudopodia around a target and internalize it into large vesicles called phagosomes, which usually fuse with lysosomes for degradation. This system can be receptor-mediated.

2.5.4 Vesicle-Mediated Exocytosis

Exocytosis involves molecules leaving the cell, such as secretion. This also occurs after the solutes are packaged into vesicles, often originating from the Golgi apparatus. These vesicles are then transported to the plasma membrane. Exocytosis via vesicle transport is always nonspecific (no receptor binds to the solute to be transported). This type of transport can be either constitutive or regulated. Constitutive exocytosis occurs due to some level of continuous vesicle transport in which vesicles merge with the cell membrane. Regulated exocytosis is regulated by a stimulus. We see this example later in this chapter, where we discuss neurotransmitter release (such as acetylcholine) from a neuron in response to a stimulus (in this case, calcium) into the space between the neuron and another neuron or muscle [1, 68].

2.5.5 Intracellular Vesicle Transport

Vesicular transport not only involves moving solutes in and out of the cell, but it also happens within the cell. It is especially important as part of the secretory pathway, where newly synthesized proteins made in the ER are transported to the Golgi apparatus and then sorted into different organelles. This pathway is discussed in more detail in Chap. 8.

2.6 Transport and Signaling Systems in Neurons

Transport across neurons is a complex process that requires a combination of different types of transporters and channels. Neuron transport mediates communication between neurons as well as between neurons and muscle cells. Neuron signaling provides a good example of how various types of transporters discussed in this chapter can operate together to mediate cell function and respond to a signal.

2.6.1 Action Potentials in Neurons

Action potentials are examples in which ion transport plays an important role. Additionally, the action potential serves as an example of the coordination of multiple types of transport, particularly the combination of ion transport and vesicle transport.

- Overview of Action Potentials, Synaptic Potentials, and Acetylcholine Receptors

Several types of ion channels are important in neurons during an action potential (Fig. 3.13), including Na^+ and K^+ channels, Ca^{2+} channels, and acetylcholine receptors (AChRs), which are transmitter-gated ion channels. These channels are central to the generation of action potentials and synaptic potentials, which are fundamental in neuron communication. Action potentials are an essential mechanism by which signals travel along the length of a neuron. A synaptic potential is a response in a postsynaptic neuron or muscle that results from neurotransmitters, such as acetylcholine, binding to ligand-gated ion channels.

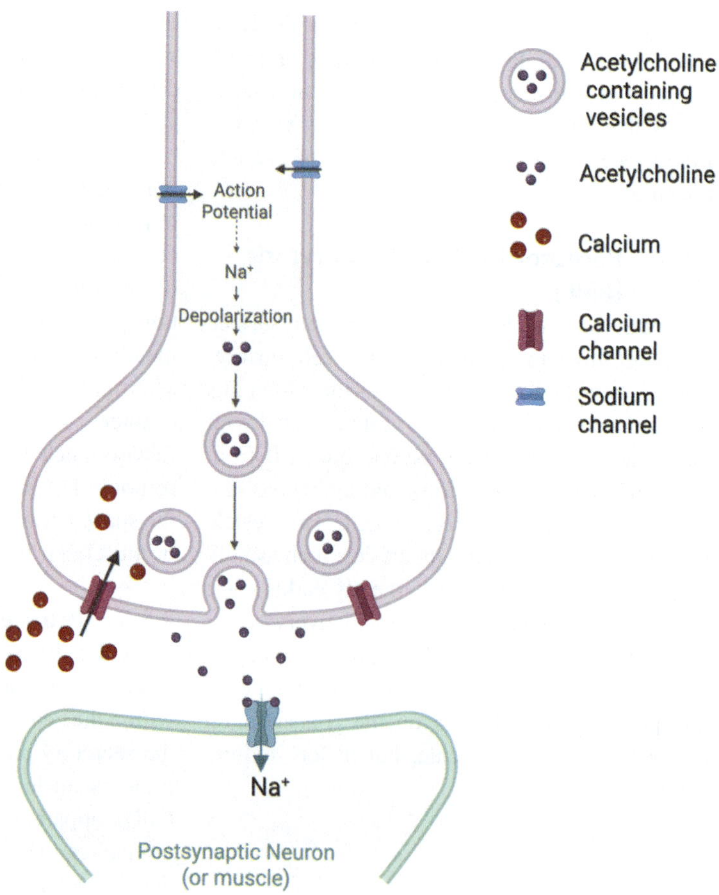

Fig. 3.13 The main components of an action potential. Membrane depolarization is driven by sodium influx. Vesicles containing neurotransmitters such as acetylcholine fuse with the membrane. This can lead to the opening of channels on the postsynaptic neuron. See text for details. (*Source*: Created with Biorender.com)

- The steps involved in an action potential

1. The action potential is triggered by a change in electric charge (changes in voltage) caused by the movement of ions (such as Na^+ and K^+) across the membrane.
2. When a neuron's membrane reaches a specific voltage called a threshold potential, this causes voltage-gated Na^+ channels in the cell membrane to open. Consequently, Na^+ ions flow into the neurons via passive transport (down their concentration gradient) through these Na^+ channels. This inward flow of Na^+ ions leads to membrane depolarization, which is a change in electrical potential across the membrane from negative to more positive.
3. When the action potential arrives at the presynaptic terminal, the membrane depolarization at the presynaptic terminal leads to the opening of voltage-gated Ca^{2+} channels in the presynaptic neuron. This causes Ca^{2+} to enter the cell.
4. This influx of Ca^{2+} causes vesicles that are filled with neurotransmitters (especially acetylcholine) to fuse with the membrane. These vesicles are called synaptic vesicles. The fusion of these synaptic vesicles with the membrane requires calcium and involves several steps.
   - First, Ca^{2+} binds to a calcium-sensing protein called synaptotagmin on the membrane of the synaptic vesicle.
   - This causes the synaptotagmin to undergo a conformational change so that it interacts with a complex called the SNARE complex.
   - The SNARE complex is made up of several proteins, including synaptobrevin on the vesicle's membrane, and syntaxin and SNAP25 on the plasma membrane. This complex plays an important role in bringing the two membranes close together (the vesicle membrane and the plasma membrane), ultimately leading to the fusion of the vesicle with the membrane.
   - At this point, the proteins that make up the SNARE complex (synaptobrevin, syntaxin, and SNAP25) dock together in a coiled structure that primes the vesicle for fusion. When calcium binds to synaptotagmin, it triggers the final fusion step.
   - Once the vesicle and the membrane fuse, acetylcholine is released from the cell into the synaptic cleft [1, 21, 69].
5. Once acetylcholine (ACh) is released, it diffuses across the synapse and binds to acetylcholine receptors (AChRs) on the postsynaptic membrane, which may be on the membrane of a neuron or a skeletal muscle cell. Signaling mechanisms such as this that use ACh as a neurotransmitter are referred to as cholinergic signaling.
6. AChRs are ligand-gated ion channels, and they open transiently in response to acetylcholine binding to them. The opening of these channels allows ions, especially Na^+ and K^+, to pass through, with Na^+ flowing inward and K^+ flowing outward. This generates the synaptic potential in the target (postsynaptic) cell.
7. This leads to a response in the postsynaptic cells. In neurons, it may trigger further signaling, and in muscle cells, it may lead to **contraction** [1, 21, 69–72]. (Fig. 3.14).

There Are Two Main Types of Acetylcholine Receptors

Above, we discussed acetylcholine receptors (AChRs) located on the surface of postsynaptic neurons or muscle cells. It is important to note that there are two types of AChRs. The type of Acetylcholine receptor (AChR) described above is the Nicotinic acetylcholine receptor (nAChR). nACHRs are ligand-gated ion channels that are activated by acetylcholine, but they can also be activated by nicotine. They are found in the central nervous system (CNS) and the peripheral nervous system (PNS). In the CNS, they are mainly at synapses between neurons, where they mediate excitatory neurotransmission. In the PNS, they are at neuromuscular junctions and mediate the transmission of signals from motor neurons to skeletal muscle, as described above. This type of AChR is not to be confused with a different kind of AChR, the Muscarinic acetylcholine receptors (mAChRs). This is a different category of AChRs, which differ from the

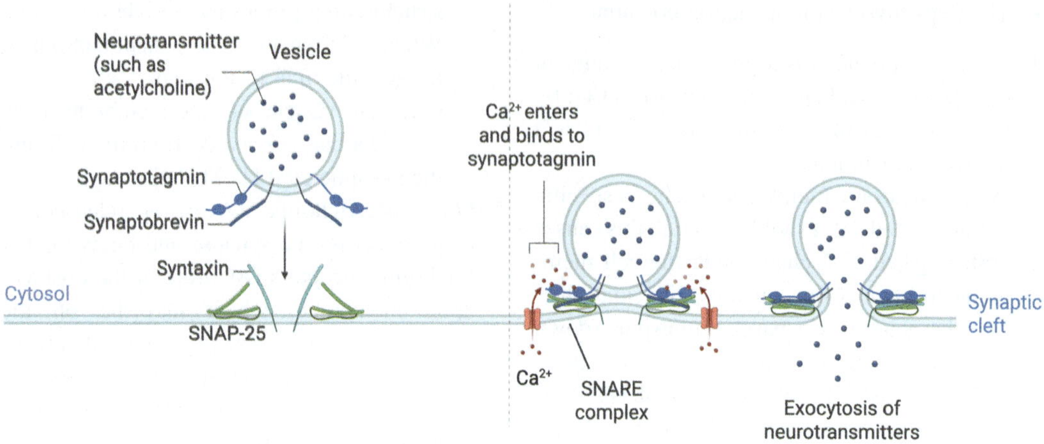

Fig. 3.14 The SNARE complex facilitates vesicle fusion. The complex includes SNAP-25 and syntaxin. Vesicle fusion allows exocytosis of molecules such as neurotransmitters from cells. See text for details. (*Source*: Created with Biorender.com)

receptors described above. Rather than being channels, they are G protein-coupled receptors (GPCRs) that are activated by acetylcholine and certain other agonists. They are located in the CNS and some peripheral tissues, where they play diverse functions [1, 71, 73], and are not covered in this chapter.

2.6.2 Clinical Examples of and Drug Targets Involving Neurotransmission, Ion Channels, and Cholinergic Signaling

- Botulinum toxin blocks acetylcholine release

Daxxify (daxibotulinumtoxinA-lanm) is an example of a drug that blocks acetylcholine signaling) is a **cosmetic drug** used to **reduce wrinkles**. Daxxify is related to the well-known drug **Botox (onabotulinumtoxinA)** [74]. Both of these drugs contain the same basic core enzyme: **botulinum toxin A**, a neurotoxin produced by the bacterium Clostridium botulinum. Daxxify and Botox are considered biologics as they comprise large proteins made naturally by bacteria [75]. Botulinum toxin is a protease (an enzyme that cleaves other proteins) that blocks the release of acetylcholine from the presynaptic neuron. Specifically, it can **cleave SNAP25** (part of the SNARE complex described above), a protein necessary for docking of the synaptic vesicle containing acetylcholine, to the membrane (see Fig. 3.14). Consequently, when SNAP25 is blocked, vesicle fusion is inhibited, and acetylcholine release is blocked. This leads to decreased muscle function, and this localized relaxation of the muscles in response to the drug reduces wrinkles. This effect is not permanent because, over time, there is a gradual recovery of SNAP25. Daxxify and Botox are made of the same neurotoxin, but Daxxify includes a synthetic protein that increases its stability and may help it enter neurons more efficiently. Therefore, its effects last a bit longer. **Daxxify** was FDA-approved in 2022, while Botox was approved for cosmetic use in 2002 and for treating other disorders, such as **neuromuscular disorders**, in 1989 [75]. **LETYBO (letibotulinumtoxinA-wlbg)**, approved in 2024, operates by a similar mechanism, and it is also made of botulinum toxin A but differs in its formulation. It is also used for reducing wrinkles. Likewise, several other drugs have also been formulated based on this same neurotoxin [75, 76].

While these drugs are popular for wrinkle treatment, they also serve other functions. Some drugs in this family are used to treat multiple different health problems (such as migraines, spasticity after a **stroke or in MS or cerebral palsy, overactive bladder**, and others). For migraines, the drugs are effective because they reduce the release of neurotransmitters that are associated with pain. Relaxation of muscle activity reduces contractions associated with plasticity, and blocking the release of acetylcholine at the bladder wall can improve urinary incontinence. These drugs are quite versatile, and other possible uses are also under investigation [77].

- Drugs that Activate or Inhibit nAChRs

Several other drugs affect AChRs, and some of these are summarized Table at the end of the chapter. One example is **Nicotine**. Nicotine is an alkaloid found naturally in tobacco. Although it is found naturally in tobacco, it is classified as a small molecule due to its size and structure. Alkaloids are small molecules that typically form ring structures and contain nitrogen and are found mostly in plants [78]. Nicotine can bind to **nACHRs on dopamine neurons** and activate them. Activation of nACHRs on these neurons by nicotine or acetylcholine enables calcium entry, which promotes vesicle fusion and release of neurotransmitters such as dopamine. Thus, unlike the other drugs discussed above, this action stimulates the nACHRs, rather than inhibiting them. The release of neurotransmitters like dopamine produces pleasurable effects and is addictive.

Another example of an nAChR activator is **Chantix (varenicline)** [79], a medication used to aid **smoking cessation**. This is also a **small molecule,** synthesized in the lab, and it is a **partial agonist of nAChRs**. By activating nAChRs, it provides a milder stimulation than nicotine, reducing nicotine cravings. Anesthetics such as **Anectine (succinylcholine)** [80] also activate nAChRs. Anectine is also a lab-made small molecule. It binds to nAChRs at the neuromuscular junction. It initially activates the receptor, but its prolonged binding leads to depolarization and muscle relaxation during surgical procedures.

Curare (**also called D-tubocurarine**) is an **antagonist of nAChRs** at neuromuscular junctions. This plant-derived drug is a small molecule that binds to and blocks acetylcholine receptors on skeletal muscle cells, resulting in **muscle relaxation**. It is considered a neuromuscular blocker. It was previously used as a muscle relaxant and was the first drug identified to work as a neuromuscular blocker by antagonizing the nAChRs. It is now mostly replaced by newer nAChR antagonists, but this drug paved the way for the generation of this class of drugs [19, 30, 81].

- Myasthenia Gravis (MG) involves destruction of nAChRs, potential treatments.

Myasthenia gravis (MG) (Fig. 3.15) is an A clinically relevant example of a disorder involving nAChRs is Myasthenia gravis (MG), which occurs at the neuromuscular junction. MG is an autoimmune disease where the nACHRs are attacked by the immune system. When these receptors (located on muscle cells at the junctions between motor neurons and muscles) are destroyed, the muscles do not receive signals properly, and the influx of Na^+ is inhibited. Consequently, certain muscles become weak. Difficulty speaking, swallowing, and chewing can result. There is **no cure, but regulating the destruction of nAChR is a promising mechanism for treatment**. In the disease, antibodies bind to the nAChR, leading to its destruction. This also activates the complement system, which is known to play an important role in enhancing the immune response, leading to further damage at the neuromuscular junction (The immune system and complement system are discussed in Chap. 24). The complement system plays a key role in the destruction of nACHRs in the disease, making it an important target for drug development to treat it [82, 83].

- Nav1.8 sodium channels as drug targets for pain treatment

Nav1.8 is a type of voltage-gated sodium channel found in some neurons. Voltage-gated sodium channels are a large family of channels, but this one is particularly interesting as a drug target. Nav1.8 is primarily found on peripheral sensory neurons, particularly those involved in pain sensation. It mediates action potentials in these neurons, especially in response to inflammation

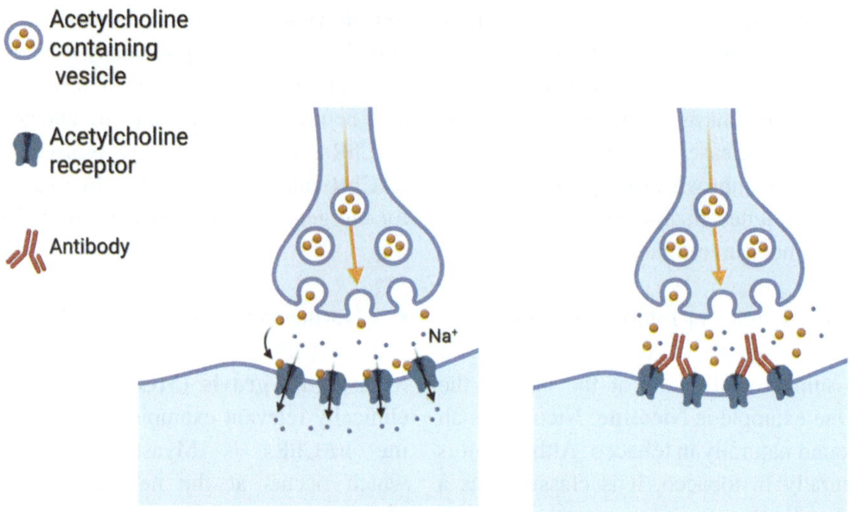

Fig. 3.15 Myasthenia Gravis. In myasthenia gravis, antibodies interfere with sodium channels. (*Source*: Created with Biorender.com)

or injury. Blocking this channel is often an effective way to achieve pain relief. Importantly, it can be blocked by drugs that do not contain opioids. This is important due to the addictive nature of opioids.

Journavx (suzetrigine) is a new drug for the treatment of (FDA approved in 2025) [84] that targets this type of channel. It is a small-molecule drug used for the **treatment of pain**. It specifically binds to the **NAV1.8 channel** and blocks sodium influx. This, in turn, prevents the action potential so that pain signals do not propagate. It is specific as to which sodium channel it inhibits, which is important because other sodium channels play important roles in both the nervous system and the heart. [85, 86].

2.6.3 GABA Receptors

• Overview of GABA Receptor Signaling

GABA (gamma-aminobutyric acid) is an inhibitory neurotransmitter. It helps to balance excitatory and inhibitory activity in the brain. GABA can bind to two main receptors, the $GABA_A$ receptor ($GABA_AR$), and the $GABA_B$ receptor type ($GABA_BR$). The $GABA_AR$ is an ionotropic receptor; it is a ligand-gated ion channel. The $GABA_BR$ is metabotropic; it is a G protein-coupled receptor. Both receptors regulate sleep, anxiety, and muscle tone, but they function differently. Here, the focus is on $GABA_AR$. This receptor is a ligand-gated chloride (Cl^-) channel. GABA binding to $GABA_AR$ on a postsynaptic neuron causes an influx of Cl^- ions into the cell. This influx of negatively charged ions reduces the likelihood of an action potential occurring, thereby exerting an inhibitory effect. The disruption of $GABA_AR$ is associated with various neurological and psychiatric disorders, including epilepsy, anxiety disorders, and schizophrenia [19, 21, 87–89] (Fig. 3.16).

GABA is synthesized from glutamate in certain neurons called GABAergic neurons, only in a reaction catalyzed by glutamate decarboxylase (GAD). The resulting GABA is packaged into vesicles and released at the synaptic cleft.

After GABA is released, its action is terminated primarily by GAT1 (GABA transporter 1). GAT1 is a transporter and takes GABA back up into neurons. This reuptake process is important and helps limit the inhibitory effect.

GABA is made from glutamate, and interestingly, glutamate is also a neurotransmitter. In fact, it is the major excitatory neurotransmitter. It binds to receptors such as AMPA and NMDA receptors, allowing sodium (Na^+) and calcium

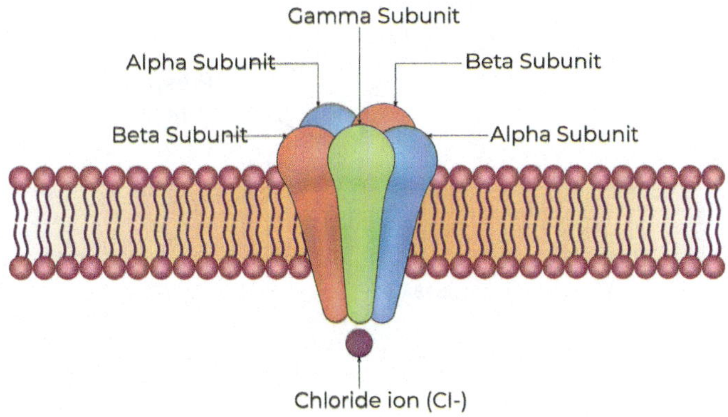

Fig. 3.16 The GABA receptor. The GABA receptor is an ion channel with a pore through which chloride ions flow. (*Source*: Illustration from Vecteezy.com (Pro license))

(Ca^{2+}) ions to enter postsynaptic neurons. These neurons, in turn, are more likely to fire an action potential. Thus, glutamate causes excitation of neurons, but when it is converted into GABA, it can also lead to inhibitory effects [90].

- Many Drugs Act on the GABA Receptor

– The Benzodiazepine and Barbiturate Family of Drugs

Drugs in the **benzodiazepine** and **barbiturate** families are examples of **small-molecule drugs** that bind to the **GABA receptor**. **Benzodiazepines** include drugs such as **Valium (diazepam)** and **Ativan (lorazepam)** [91]. These drugs **enhance the inhibitory effects of GABA** by increasing the receptor's response to GABA. They are often used as **anxiolytics, sedatives, and anticonvulsants**. **Barbiturates** include drugs such as **Luminal (phenobarbital)** (no longer in use), and **Sezaby (phenobarbital sodium)** (FDA approved **2022**) [92]. These drugs also enhance GABA's inhibitory effects. Sezaby has sedative, hypnotic, and anticonvulsant properties and is used for the treatment of seizures [88, 93–95] (see Fig. 3.17a).

Zurzuvae (zuranolone) was FDA-approved in 2023 [96] as a treatment for women with **postpartum depression**. It is a synthetic s**mall-molecule** drug that acts as a positive **allosteric regulator of the GABA receptor**. This means that the drug enhances the effects of GABA by binding to the receptor. As an allosteric regulator, it binds to a region on the receptor other than its active site (the site where GABA binds) and increases the receptor's affinity for GABA. This amplifies the GABA receptor's inhibitory effects on neuronal activity. The drug is thought to have a calming effect on the nervous system, thus alleviating some symptoms of depression and anxiety [97, 98].

In addition to drugs that target the GABA receptor, some drugs function by regulating GABA reuptake instead. The GABA transporter GAT1, on presynaptic neurons. GAT1 is a Na+/Cl- dependent symporter that transports GABA back into presynaptic neurons. GABA flows inward through this transporter, while Na^+ ions and Cl^- ions flow out. It clears GABA from the synaptic cleft, thereby terminating its inhibitory signaling. Drugs such as **Gabitril (tiagabine) inhibit GAT1** and, consequently, increase extracellular GABA concentrations, prolonging GABA signaling. **Tiagabine** is used to **treat seizures** in patients with epilepsy and has anticonvulsant effects [99].

- Ztalmy (ganaxolone) is an anticonvulsant drug that acts on the GABA receptor

Ztalmy (ganaxolone) [100] is **a small-molecule drug used as a treatment for seizures**, especially in children with CDD (CDKL5 deficiency disorder). It was FDA-approved in 2022. CCD is a disorder caused by a mutation in the gene encoding CDKL5 (cyclin-dependent kinase-like 5). CDKL5 is a cyclin-dependent kinase (CDK) important for normal brain development (CDKs will be discussed in more detail in Chap. 20). The CDKL5 mutation can lead to hyperexcitability of

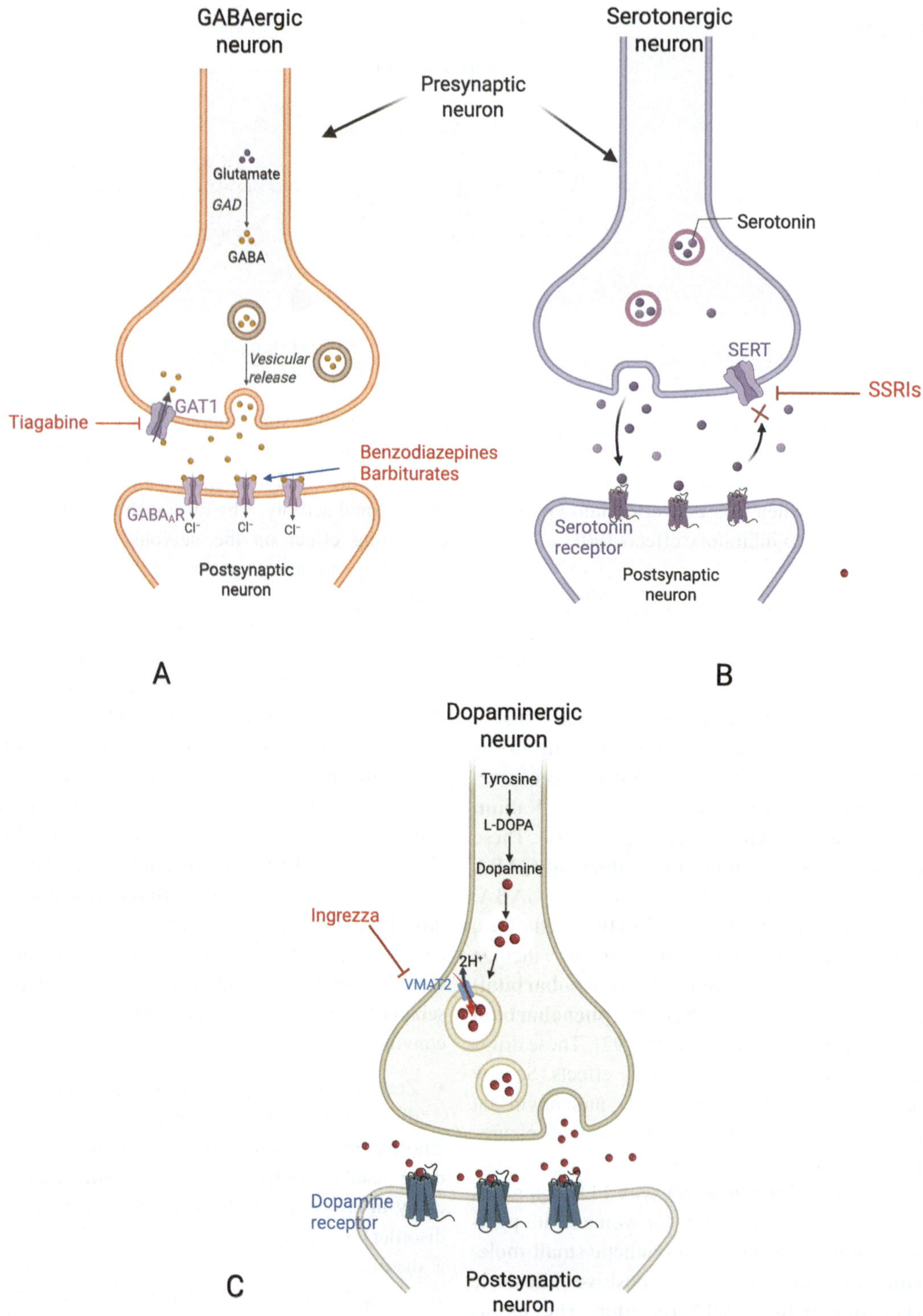

Fig. 3.17 (**A**) Neurotransmission and some drugs that target neurotransmission. . Panel **A** shows GABA neurotransmission and shows how drugs such as Tiagabine, Benzodiazepines, and Barbiturates act on GABAergic neurons and postsynaptic neurons (although they can also act on other types of neurons). (Panel **B** shows serotonin neurotransmission and illustrates how SSRIs are drugs that target SERT (serotonin transporter). Panel **C** shows doamine neurotransmission. In this figure VMATs are also shown, which are antiporters on synaptic vesicle membranes that transport monoamines. Ingrezza is an example of a drug that targets VMATs. . (*Source*: Created with Biorender.com)

synaptic signaling. The disorder is associated with early onset seizures.

Although the GABA receptor is not directly affected by CDD, targeting it is a promising treatment approach because its activation has a calming effect on the nervous system, thereby compensating for the mutation in CDKL5. **Ztalmy binds to and thereby activates the GABA receptor** (it is a positive allosteric modulator). Ztalmy can function differently from some other GABA receptor activators, however, because it can act on extrasynaptic GABA receptors. These include receptors that are not located precisely at the contact area between the pre and postsynaptic cells. This additional activation provides a more sustained and widespread effect than some of the previous GABA receptor activators [101–103].

Xdemvy (lotilaner ophthalmic solution) (FDA approved 2023) [104] is a **small-molecule** drug used to treat **demodex blepharitis**, a condition characterized by inflammation of the eyelids. This condition can be caused by Demodex mites, which can overgrow in the eyelids. These mites are a normal part of the skin's microbiome, but their overpopulation in the eye can lead to problems, including itchy, red, and inflamed eyes, sometimes accompanied by crusty discharge. **Xdemvy targets the mites by inhibiting their GABA receptors**, leading to their paralysis and death. **Xdemvy** does not target mammalian GABA receptors and thus only targets the mites, thereby treating the condition without significant side effects [105, 106].

2.6.4 Serotonin Transporter (SERT)

- Mechanism of Action of SERTs

The serotonin transporter (SERT) is a Na^+-dependent neurotransmitter transporter, also known as a reuptake transporter. It is found in neuronal membranes and uses the Na^+ concentration gradient to transport serotonin into the cell. SERT is primarily located on presynaptic neurons in the central nervous system (CNS), specifically on serotonin neurons. It is responsible for the reuptake of serotonin (5-HT) from the synaptic cleft between two neurons, returning it to the presynaptic neuron. This regulates the concentration of serotonin in the synaptic cleft and ultimately terminates the signal between the two neurons [21, 107, 108] (Fig. 3.17b).

- Selective Serotonin Reuptake Inhibitors (SSRIs) Block the SERTs

SSRIs are synthetic small-molecule drugs used to treat depression and anxiety. They function by **binding to SERT and preventing serotonin from binding to it**. They thereby block SERT function, preventing serotonin reuptake and thereby increasing synaptic serotonin levels. As a result, these drugs impact mood and relieve anxiety. They are highly selective for SERTs rather than other neurotransmitters. SSRIs are often considered first-line treatments for depression as they have fewer side effects than older antidepressants [109]. Some examples of SSRIs are **Prozac (fluoxetine)** [110], **Zoloft (sertraline)** [111], **Lexapro (escitalopram)** [112], **Celexa (citalopram)** [113], **Paxil (paroxetine)** [114], and **Luvox (fluvoxamine)** [21, 115, 116] (Fig. 3.17b).

2.6.5 The Vesicle Monoamine Transporters (VMATs)

Vesicle monoamine transporters (VMAT1 and VMAT2) are antiporters found on synaptic vesicle membranes, and they transport monoamines from the cytoplasm into the synaptic vesicles for storage and later release. Monoamines consist of neurotransmitters such as dopamine and serotonin. Dopamine is made from the amino acid tyrosine, via the intermediate L-DOPA, and serotonin is made from the amino acid tryptophan. As antiporters, the VMATs couple the transport of monoamines to the transport of protons (H^+) in the opposite direction. They differ in their expression patterns: VMAT1 is primarily found in peripheral tissues, particularly neuroendocrine cells, and VMAT2 is the main form of VMAT in the CNS [21, 117]. Neurotransmitters bind the VMAT transporter on the cytoplasmic side of a vesicle. On the other side of the VMATs (the luminal side, or inside, of the vesicle), protons bind the VMAT, and energy from a proton gradient causes the VMAT to change conformation so that the neurotransmitter now faces the inside of the vesicle. There, it will be released from the VMAT and stored in the vesicle until it is needed for release at a synapse. The proton is released into the cell's cytoplasm, outside the vesicle lumen. The VMAT then returns to its original conformation and can pick up more neurotransmitters [118].

VMAT2 has become an important drug target. VMAT2 has a key role in packaging monoamines (such as dopamine, serotonin, norepinephrine, and histamine) into vesicles in presynaptic neurons. As with all neurotransmitters, packing into vesicles helps control their release into the synaptic cleft, thereby influencing their levels in the synapse and regulating neurotransmission. Some diseases associated with abnormal body movements are linked to disruptions in monoamine signaling, particularly dopamine. Examples are **Tardive Dyskinesia**, which may be caused by drugs that are dopamine inhibitors, and **Huntington's disease**, which is associated with altered dopamine signaling [21, 119, 120]. VMAT2 inhibitors have shown promise in relieving symptoms of these disorders. Examples of **drugs that inhibit VMAT2** are **Xenazine (tetrabenazine)**, **Austedo (deutetrabenazine)**, and **Ingrezza (valbenazine)** [117, 121–123]. These **small-molecule** drugs are reversible competitive inhibitors. **They bind directly to the VMAT2 transporter** at or near the site where the monoamines normally bind, thereby preventing the monoamines from entering the vesicles. This results in reduced dopamine release into the synapse, which helps to decrease excessive dopamine signaling associated with involuntary movements [117, 124–126] (see Fig. 3.17c). They are used to treat movement disorders such as **Tardive dyskinesia** and **Huntington's Disease.**

2.7 Asymmetric Distribution of the Various Transporters and Channels in Cells

In this chapter, we discussed multiple types of transporters. Different types of transporters may be found on different types of cells and tissues. It is important to note that even within a single cell type, different types of transporters can be found on different parts of the cell. In gut epithelial cells, for example, transporters at the apical side, considered the absorptive side, transport nutrients into the cell. These types of transporters are typically Na^+-linked symporters. In contrast, at the basolateral part of the membrane, uniporters allow nutrients to leave the cell and travel down their concentration gradients through passive transport. These nutrients eventually enter the bloodstream for distribution throughout the body [1, 127].

3 The Bacterial Cell Wall and Cell Membrane Are Targets for Some Antibiotics

3.1 Bacterial Envelopes

Until now, we have discussed the membranes of eukaryotic cells, but the membranes and cell walls of bacteria are also important for the study of pharmacy. A major reason is that cell walls and cell membranes of bacteria are targets for many antibiotics (although antibiotics can also have other targets, such as ribosomes). Antibiotics are one of the greatest achievements in medicine, having saved millions of lives by combating infections. Antibiotics remain lifesaving drugs, despite the emergence of new strains of antibiotic-resistant bacteria, which pose new challenges.

Understanding the construction of the bacterial membrane and cell wall is important for understanding how many antibiotics work. Bacteria are surrounded by a cell envelope, which consists of multiple layers and typically includes a cell membrane and a cell wall. All bacteria have a cell membrane, and most bacteria also have a cell wall. Some bacteria, such as mycoplasma, not addressed here, do not have a cell wall. Bacteria are often classified as Gram-positive or Gram-negative, based on the construction of their cell envelopes [1].

The cell walls of Gram-positive bacteria are thick and are rich in peptidoglycan.

Peptidoglycan is a molecule containing polysaccharide chains cross-linked with short peptides, providing mechanical support for the bacteria and preventing them from lysing in response to osmotic pressure. The formation of crosslinks is catalyzed by enzymes, including **DD-transpeptidases**, which belong to a family of proteins known as **penicillin-binding proteins (PBPs)**. The cell wall of these Gram-positive bacteria surrounds an inner membrane made of a phospholipid bilayer.

The cell envelopes of Gram-negative bacteria are more complex. They contain an inner membrane, a thin cell wall, and an outer membrane. The cell wall of the Gram-negative bacteria also contains peptidoglycan, but it is much thinner than the cell wall of the Gram-positive bacteria, and it is located in the periplasmic space between the two membranes [21, 128, 129].

3.2 Antibiotics: Some Antibiotics Inhibit Cell Wall Synthesis

Antibiotics are usually derived from microorganisms, such as bacteria and molds, although they are occasionally synthesized in the laboratory. Many antibiotics come from soil microbes. It is not clear why these microbes produce antibiotics, but they may serve as a self-defense mechanism against other life forms or competitors in the soil [130, 131]. There are many types of antibiotics.

Although the use of microbes to treat wounds dates back thousands of years, the history of antibiotics really starts with the discovery of penicillin by Alexander Fleming in the 1920s [131]. This discovery was made unexpectedly when he noticed that a mold had contaminated a Petri dish of *Staphylococcus* bacteria in his lab. He noticed that this mold was surrounded by a clear zone where the bacteria had been killed. He identified the mold and found that it was producing a substance that inhibited bacterial growth. He named the mold penicillin. This eventually led to the development of penicillin as the first widely used antibiotic [132, 133].

3.3 Major Categories of Antibiotics

Some of the main categories of antibiotics that interfere with the bacterial envelope are described here:

3.3.1 β-Lactams

A common mechanism for antibiotics is the inhibition of bacterial cell wall synthesis. The cell wall of Gram-positive bacteria contains a thick layer of peptidoglycan. Staphylococcus aureus (commonly referred to as staph) is a Gram-positive bacterium and a leading cause of staph infections. **Penicillin** is an antibiotic that is highly effective in combating staph infections. It does so by inhibiting the **transpeptidases (the PBPs)**, the enzyme that is important for the production of the peptidoglycan layer. Penicillin is a member of the **β-lactam** family, which is made up of β-lactam rings (four-membered rings which contain three carbons and a Nitrogen) [130]. In the presence of penicillin, the staph bacteria are unable to form their cell walls effectively, and as a result, they cannot survive. **Penicillin** is highly effective in combating various bacterial infections [1]. Penicillin and other antibiotics are discussed in more detail in Chap. 5. Penicillin was originally identified in fungi. It is still isolated from fungi but is also modified in the lab [134].

3.3.2 Glycopeptides

Other types of antibiotics include **glycopeptides,** such as **vancomycin**. A glycopeptide is a short stretch of amino acids with a sugar molecule attached. Vancomycin also inhibits peptidoglycan synthesis, but it does so by a different mechanism. This category of antibiotics **binds to an amino acid sequence of a peptidoglycan precursor**. This prevents crosslinking and the formation of mature peptidoglycan in the cell wall. This ultimately leads to the destruction of the cell wall and, consequently, the destruction of the bacteria [21, 130, 135, 136]. Due to their size, glycopeptides are not considered small molecules, but do not fit into the category of biologics. They are often considered natural products or semisynthetic derivatives of natural products because they are modified in the lab.

3.3.3 Cyclic Lipopeptide Antibiotics

Cyclic lipopeptides are a class of antibiotics that are made of a cyclic peptide core linked to a fatty acid tail. They come from bacteria and are usually considered natural products. They are also considered small-molecule drugs, although they are a bit larger than most small molecules.

This category of antibiotics includes **Polymyxins**. They are used to treat **infections caused by Gram-negative bacteria**. An example is colistin (polymyxin E). This type of antibiotic **binds to lipopolysaccharides (LPS),** a part of the outer membrane of Gram-negative bacteria. Consequently, the membrane is disrupted, and the bacteria eventually die [103, 104].

Cubicin (Daptomycin) is also a cyclic lipopeptide antibiotic [137], but it is used to treat **Gram-positive infections**, such as those caused by MRSA (methicillin-resistant Staphylococcus aureus) or VRE (vancomycin-resistant enterococcus). Daptomycin also contains a cyclic peptide, but with a different type of modification. It binds to the **bacterial cell membrane** and becomes inserted into the membrane, forming channels. This causes ions to leak through the channels, leading to cell death [130, 138, 139].

Orlynvah (sulopenem etzadroxil and probenecid) is a drug that combines an antibiotic with another treatment, approved by the FDA in 2024. This drug is **a combination of sulopenem, etzadroxil, and probenecid**, all of which are small molecules. It is used to treat urinary tract infections (UTIs) in adult women. **Sulopenem etzadroxil** is a prodrug, meaning it does not become activated until it undergoes cleavage in the body. Once cleaved, it functions as an antibiotic, inhibiting bacterial cell wall synthesis by binding to penicillin-binding proteins (PBPs). The other component of the drug is **Probenecid**. This is a renal tubular inhibitor. It inhibits a **transporter** in the **kidney** called **OAT3**. Inhibiting this transporter slows down the excretion of drugs in the renal system. Thus, this drug prolongs the effectiveness of the antibiotic [140, 141].

3.4 Antibiotic Resistance

An increasingly common problem is antibiotic resistance in bacteria. Antibiotic resistance can occur in different ways. For example, the PBP proteins can be modified so that antibiotics of the β−lactam family cannot target them. In the case of glycopeptides, resistant bacteria may synthesize an altered peptidoglycan precursor so that it cannot be targeted by the antibiotic. Modification of LPS on bacterial membranes can lead to resistance to antibiotics in the polymyxin family. Care must be taken to avoid overusing antibiotics in both human patients and agriculture, as this can increase the likelihood of antibiotic resistance. Research into new types of antibiotics is also important for treating antibiotic-resistant infections [130].

FDA-Approved Drugs Mentioned in This Chapter

Commonly used brand name	Generic name	First FDA approval year	Common conditions	Main targets
Advil	*Ibuprofen*	1984	Pain and inflammation	Cox-1/2
Aleve	*Naproxen*	1994	Pain and inflammation	Cox-1/2
Anectine and Quelicin	*Succinylcholine chloride*	1953	Anesthetic: Muscle relaxation during surgery	Nicotinic ACh receptor
Aspirin	*Acetylsalicylic acid*		Pain and inflammation	Cox-1/2
Ativan	*Lorazepam*	1977	Anxiety, seizures	GABA receptor
Austedo	*Deutetrabenazine*	2017	Movement disorders	VMAT2
Botox	*OnabotulinumtoxinA*	1989	Cosmetic use and neuromuscular disorders	SNAP25

(continued)

Commonly used brand name	Generic name	First FDA approval year	Common conditions	Main targets
Brenzavvy	*Bexagliflozin*	**2023**	**Type 2 diabetes**	**SGLT-2**
Calan	*Verapamil hydrocholoride*	1984	Hypertension and arrhythmias	Ca2+ channels
Cardene	*Nicardipine hydrochloride*	1988	Hypertension	Ca2+ channels
Cardizem	*Diltiazem hydrochloride*	1982	Hypertension and arrhythmias	Ca2+ channels
Celexa	*Citalopram*	1998	Depression and anxiety	SESRT
Chantix	*Varenicline*	2006	Smoking cessation	Nicotinic ACh receptor
Daxxify	*DaxibotulinumtoxinA-lanm*	**2022**	Wrinkles and neuromuscular disorders	SNAP-25
Digoxin	*Digoxin*	1954	Heart conditions	Na+/K+-ATPase
Epivir	*Lamivudine*	1995	HIV/AIDS	HIV reverse transcriptase
Farxiga	*Dapagliflozin*	2014	Type 2 diabetes	SGLT-2
Ingrezza	*Valbenazine*	2017	Movement disorders	VMAT2
Gabitril	*Tiagabine*	1997	Seizures	GAT1
Inpefa	*Sotogliflozin*	**2023**	Type 2 diabetes	SGLT-2
Jardiance	*Empagliflozin*	2014	Type 2 diabetes	SGLT-2
Journavx	*Suzetrigine*	**2025**	**Pain**	**Immune system**
LETYBO	*LetibotulinumtoxinA-wlbg*	**2024**	Wrinkle reduction	SNAP-25
Lexapro	*Escitalopram*	2002	Depression, anxiety	SERT
Luminal (brand name discontinued)	*Phenobarbital*		Seizures	GABA receptor
Luvox	*Fluvoxamine*	1994	Depression and anxiety	SERT
Norvasc	*Amlodipine*	1996	Hypertension	Ca2+ channel
Paxil	*Paroxetine*	1992	Depression and anxiety	SERT
Procardia	*Nifedipine*	1982	Hypertension	Ca2+ channel
Prozac	*Fluoxetine*	1987	Depression and anxiety	SERT
Retrovir	*Azidothymidine (AZT), now called zidovudine*		HIV/AIDS	HIV reverse transcriptase
Sezaby	*Phenobarbital sodium*	**2022**	**Seizures**	**GABA receptor**
Tubarine	*Curare (d-tubocurarine)*	1945	Muscle relaxation. Largely replaced by newer drugs	Nicotinic ACh receptor (
Valium	*Diazepam*	1963	Anxiety and seizures	GABA receptor
Xdemvy	*Lotilaner ophthalmic solution*	**2023**	**Demodex blepharitis**	**Demodex mite**
Xenazine	*Tetrabenazine*	2008	Movement disorders	VMAT2
Xenpozyme	*Olipudasa alfa-rpcp*	2022	ASMD	Acid sphingomyelinase
Zoloft	*Sertraline*	1991	Depression and anxiety	SERT
Ztalmy	*Ganaxolone*	2022	CDKL5 deficiency disorder, which is characterized by seizures	GABA receptor
Zurzuvae	*Zuranolone*	2023	Postpartum depression	GABA receptor

Antibiotics Mentioned in This Chapter

Commonly used brand Name	Generic name	First FDA approval year	Major conditions	Main targets
(no common brand name)	*Penicillin*		Bacterial infections	Penicillin-binding proteins (PBPs)
(no common brand name)	*Vancomycin*		Bacterial infections	a peptidoglycan precursor
Coly-Mycin E	*Colistin (polymyxin E)*	1970	Gram-negative infections	LPS on Gram-negative bacteria
Cubicin	*Daptomycin*	2003	MRSA, VRE	Bacteria cell membrane
Orlynvah	*Sulopenem etzadroxil and probenecid*	2024	Urinary tract infections	PBP and the kidney transporter OAT3

Definitions

Active transport A type of transport across membranes that requires energy.

Antiport A transporter that transports two different molecules in opposite directions across the membrane

Apical side of an epithelial cell The side of the cell that faces the lumen that the cell is lining, often the interior of the organ, such as the small intestine. The other surface of the cell is the basolateral side

Endocytosis The process by which a cell takes materials in by engulfing them in vesicles formed at the plasma membrane.

Exocytosis The process where materials are exported from the cell via vesicles.

Facilitated diffusion Passive transport of molecules across a membrane via transport proteins or channels.

Fluid mosaic model A widely accepted model in which the membrane is not a rigid structure. Instead, the different components of the plasma membrane can all move around within their layer.

Glycolipid A lipid molecule with a carbohydrate group attached.

Integral membrane protein A membrane protein that is embedded in the membrane and spans the membrane.

Ligand-gated ion channel These channels open when a certain ligand binds to a binding site on the channel. An example of a ligand that may bind a channel and cause it to open is a neurotransmitter.

Lipid raft Areas of the membrane where cholesterol, sphingolipids and proteins are clustered into domains that are more ordered than other parts of the membrane.

Na^+/K^+ ATPase (sodium-potassium pump) An active transporter that moves 3 Na^+ ions out and 2 K^+ ions into the cell using ATP as an energy source. This pump is important for maintaining electrochemical gradients across membranes.

Passive transport The transport of molecules down their concentration gradient or electrochemical gradient, in a way that does not require energy

Peripheral membrane protein A protein attached to the membrane surface, often via interactions with integral membrane proteins or lipids.

Phospholipid bilayer A double layer of phospholipids that makes up the plasma membrane and organelle membranes.

Secondary active transport This type of transport harnesses energy indirectly, often via coupled transport. It usually couples the movement of one solute down its concentration gradient to drive the transport of another molecule.

Snare complex A group of proteins (including SNAP25, synaptobrevin, and syntaxin) that mediate the fusion of vesicles with membranes during neurotransmitter release.

Sterol A type of lipid that is a rigid ring structure with four linked rings.

Symport A type of cotransport where two molecules move in the same direction across a membrane.

Vesicular transport Transport of molecules via membrane-bound vesicles, this can include travel within the cell, or endocytosis or exocytosis.

Study Questions

Multiple-Choice Questions

1. What is the function of cholesterol in the plasma membrane?
   A. Serves as a receptor for ligands
   B. Stabilizes channels and transporters
   C. Helps maintain membrane fluidity
   D. Inhibits membrane protein movement
2. Which of the following drugs is an inhibitor of SGLT2, and used to treat type 2 diabetes?
   A. Digoxin
   B. Jardiance (empagliflozin)
   C. Verapamil
   D. Ztalmy
3. Which transporter carries out active transport of sodium and potassium ions across the membrane?
   A. GLUT1
   B. Na+/K+ ATPase
   C. CFTR
   D. SGLT1
4. Which of the following statements about ion channels is TRUE?
   A. They undergo a conformational change when they bind to their ion.
   B. They always require ATP for transport.
   C. They allow passive movement of ions down their electrochemical gradients.
   D. They are only found in organelles, rather than the plasma membrane.
   E. Both C and D.
5. A solute that crosses the membrane, it goes down its concentration gradient, and it needs a channel to cross the membrane. The mechanism it uses is:
   A. MDR transporter
   B. Active transport
   C. Facilitated diffusion
   D. A membrane-bound pump
   E. B and C
6. Nicotinic acetylcholine receptors (nAChRs) are:
   A. Ligand-gated ion channels
   B. MDR pumps
   C. Sodium–potassium pumps
   D. Neurotransmitters
   E. Calcium channels
7. Daxxify is a cosmetic drug that reduces wrinkles. It functions by inhibiting SNAP25. This leads to what response?
   A. It blocks the release of acetylcholine from the presynaptic neuron.
   B. It blocks the release of calcium from the presynaptic neuron.
   C. It blocks the uptake of acetylcholine to the postsynaptic neuron.
   D. It blocks the uptake of sodium from the postsynaptic neuron.
   E. It stimulates the activity of the postsynaptic muscle.
8. Activation of the GABA receptor leads to:
   A. Excitation of nerve cells
   B. Inhibition of the activity of nerve cells
   C. Increased action potential
   D. Increased sodium transport
   E. B and C
9. What is the major function of acid sphingomyelinase?
   A. Facilitates the synthesis of sphingomyelin
   B. Facilitates in the breakdown of sphingomyelin
   C. Involved in the breakdown of ceramides
   D. Causes Niemann–Pick disease when it is in excess
   E. B and D
10. The sodium potassium pump is an example of a?
    A. Uniport
    B. Symport
    C. Antiport
    D. Channel
    E. B and D

True/False Question

11. CFTR is a transporter that functions as a chloride channel. ___________.

Fill-in-the-Blank Questions

12. The _________ transporter is a family of transporters that carries glucose down its concentration gradient.
13. Acid sphingomyelinase deficiency (ASMD) leads to the accumulation of _________ in lysosomes.
14. The _________ pump transports calcium into the sarcoplasmic reticulum promotes muscle relaxation.
15. A complex that has an important role in vesicle fusion and release of acetylcholine from vesicles into the synaptic cleft is the:
16. The category of transporter that transports one molecule down its gradient and another molecule up its gradient but both in the same direction is a ____________ . (the broad category, not the name of a specific transporter.

Short-Answer Questions

17. Briefly explain how P-glycoprotein (MDR) can lead to cancer drug resistance.

Matching Question

18. **Match the Drug to its Target**
    Drugs:
    (a) Digoxin
    (b) Xenpozyme
    (c) Chantix
    (d) Botox
    Targets:
    (i) SNAP25
    (ii) Na+/K+ ATPase
    (iii) Acid sphingomyelinase
    (iv) Nicotinic acetylcholine receptor

Answer Key

Multiple-Choice Questions

1. **Answer**: C
2. **Answer**: B
3. **Answer**: B
4. **Answer**: C
5. **Answer**: C
6. **Answer**: A
7. **Answer**: A
8. **Answer**: B
9. **Answer**: B
10. **Answer**: C

True/False Question

11. **Answer**: True

Fill-in-the-Blank Questions

12. **Answer**: GLUT
13. **Answer**: sphingomyelin
14. **Answer**: SERCA (Sarcoplasmic Reticulum Ca^{2+} ATPase)
15. **Answer**: Snare complex
16. **Answer**: Symporter

Short-Answer Questions

17. **Answer**: This type of transporter can transport cancer drugs out of the cell, thereby leading to resistance

Matching Question

18. **Answer**:
    (a) ii, (b) iii, (c) iv, (d) i

References

1. Alberts B, Heald R, Johnson A, Morgan D, Raff M, Roberts K, Walter P, Wilson J, Hunt T. Molecular biology of the cell. W.W. Norton and Company; 2022.
2. Cooper G, Adams K. The cell: a molecular approach. Sinauer Associates; 2022.
3. Edidin M. Lipids on the frontier: a century of cell-membrane bilayers. Nat Rev Mol Cell Biol. 2003;4 (5):414–8. https://doi.org/10.1038/nrm1102.
4. van Meer G, Voelker DR, Feigenson GW. Membrane lipids: where they are and how they behave. Nat Rev Mol Cell Biol. 2008;9(2):112–24. https://doi.org/10.1038/nrm2330.
5. Simons K, Gerl MJ. Revitalizing membrane rafts: new tools and insights. Nat Rev Mol Cell Biol. 2010;11(10):688–99. https://doi.org/10.1038/nrm2977.
6. Schuchman EH, Desnick RJ. Types A and B Niemann-Pick disease. Mol Genet Metab. 2017;120 (1–2):27–33. https://doi.org/10.1016/j.ymgme.2016.12.008.
7. FDA. Xenpozyme(olipudasa alfa-rpcp) approval overview. 2022. Available from: https://www.accessdata.fda.gov/scripts/cder/daf/index.cfm?event=overview.process&ApplNo=761261.
8. Keam SJ. Olipudase Alfa: first approval. Drugs. 2022;82(8):941–7. https://doi.org/10.1007/s40265-022-01727-x.
9. Wasserstein MP, Diaz GA, Lachmann RH, Jouvin MH, Nandy I, Ji AJ, et al. Olipudase alfa for treatment of acid sphingomyelinase deficiency (ASMD): safety and efficacy in adults treated for 30 months. J Inherit Metab Dis. 2018;41(5):829–38. https://doi.org/10.1007/s10545-017-0123-6.
10. Vinothkumar KR, Henderson R. Structures of membrane proteins. Q Rev Biophys. 2010;43(1):65–158. https://doi.org/10.1017/S0033583510000041.
11. Locher KP, Bass RB, Rees DC. Breaching the barrier. Science. 2003;301(5633):603–4. https://doi.org/10.1126/science.1088621.
12. FDA. Advil (ibuprofen) Approval overview. 1974. Available from: https://www.accessdata.fda.gov/scripts/cder/daf/index.cfm?event=overview.process&ApplNo=201803.
13. FDA. Aleve (naptroxin) Approval overview. 1994. Available from: https://www.accessdata.fda.gov/scripts/cder/daf/index.cfm?event=overview.process&ApplNo=21076.
14. Loetsch J, Oertel B. NSAIDs, pharmacokinetics. In: Gebhart GF, Schmidt RF, editors. Encyclopedia of pain. Berlin/Heidelberg: Springer Berlin Heidelberg; 2013. p. 2350–61.
15. FDA. Retrovir/zidovudine (azidothymidine (AZT)) approval overview. 1987. Available from: https://www.accessdata.fda.gov/scripts/cder/daf/index.cfm?event=overview.process&ApplNo=020518
16. FDA. Epivir(lamivudine) approval overview. 1995. Available from: https://www.accessdata.fda.gov/scripts/cder/daf/index.cfm?event=overview.process&ApplNo=020564.
17. PharmGKB. Zidovudine pathway, pharmacokinetics/pharmacodynamics. 2024. Available from: https://www.pharmgkb.org/pathway/PA165859361.
18. National Library of Medicine. Pharmacology of Caffeine. In: Caffeine for the sustainment of mental task performance: formulations for military operations [Internet]. Washington, DC: National Academies Press (US); 2001. Available from: https://www.ncbi.nlm.nih.gov/books/NBK223808/.
19. Katzung BG, Vanderah TW, Trevor AAJ. Katzung's basic and clinical pharmacology. 16th ed. McGraw-Hill Education; 2024.
20. Alberts B, Bray D, Hopkin K, Johnson A, Lewis J, Raff M, Roberts K, Walter P. Essential cell biology. Norton and Company; 2023.
21. Ritter JM, Flower RJ, Henderson G, Loke YK, MacEwan DJ, Rang HP. Rang and Dale's pharmacology. 10th ed; 2023.
22. Magouliotis DE, Tasiopoulou VS, Svokos AA, Svokos KA. Chapter 5: Aquaporins in health and disease. In: Makowski GS, editor. Advances in clinical chemistry, vol. 98. Elsevier; 2020. p. 149–71.
23. Hall JE, Ha ME. Guyton and Hall textbook of medical physiology. 14th ed. Elsevier; 2020.
24. FDA. Norvasc(amlodipine): approval overview. 1996. Available from: https://www.accessdata.fda.gov/scripts/cder/daf/index.cfm?event=overview.process&ApplNo=019787.
25. FDA. Cardene(nicardipine hydrochloride) approval overview. 1988.
26. Plus M. Amlodipine. 2024. Available from: https://medlineplus.gov/druginfo/meds/a692044.html.
27. FDA. Cardizem(diltiazem hydrochloride) approval overview. 1982. Available from: https://www.accessdata.fda.gov/scripts/cder/daf/index.cfm?event=overview.process&ApplNo=018622.
28. FDA. Calan(verapamil hydrocholoride) approval overview. 1984. Available from: https://www.accessdata.fda.gov/drugsatfda_docs/label/2017/018817s033lbl.pdf
29. Marks FK, Muller-Decker, K. Cellular signal processing: an introduction to the Molecular mechanisms of signal transduction. 1st: Garland Science; 2008.
30. Coventry House Publishing. Top 300 drugs: pocket reference guide. 3rd ed. Coventry House Publishing; 2024.
31. Elliott WJ, Ram CV. Calcium channel blockers. J Clin Hypertens (Greenwich). 2011;13(9):687–9. https://doi.org/10.1111/j.1751-7176.2011.00513.x.
32. Jones KE, Hayden SL, Meyer HR, Sandoz JL, Arata WH, Dufrene K, et al. The evolving role of Calcium Channel blockers in hypertension management: pharmacological and clinical considerations. Curr Issues

Mol Biol. 2024;46(7):6315–27. https://doi.org/10.3390/cimb46070377.
33. Navale AM, Paranjape AN. Glucose transporters: physiological and pathological roles. Biophys Rev. 2016;8(1):5–9. https://doi.org/10.1007/s12551-015-0186-2.
34. Stein WD, Litman T. Channels. Carriers, and pumps: an introduction to membrane transport. 2nd ed. Academic; 2015.
35. Dyla M, Kjærgaard M, Poulsen H, Nissen P. Structure and mechanism of P-type ATPase ion pumps. Annu Rev Biochem. 2020;89:583–603. https://doi.org/10.1146/annurev-biochem-010611-112801.
36. Scheiner-Bobis G. The sodium pump. Its molecular properties and mechanics of ion transport. Eur J Biochem. 2002;269(10):2424–33. https://doi.org/10.1046/j.1432-1033.2002.02909.x.
37. Pivovarov AS, Calahorro F, Walker RJ. Na(+)/K(+)-pump and neurotransmitter membrane receptors. Invertebr Neurosci. 2018;19(1):1. https://doi.org/10.1007/s10158-018-0221-7.
38. Fender J, Klöcker J, Boivin-Jahns V, Ravens U, Jahns R, Lorenz K. "Cardiac glycosides"-quo vaditis?-past, present, and future? Naunyn Schmiedeberg's Arch Pharmacol. 2024;397(12):9521–31. https://doi.org/10.1007/s00210-024-03285-3.
39. Khandelwal R, Vagha JD, Meshram RJ, Patel A. A comprehensive review on unveiling the journey of digoxin: past, present, and future perspectives. Cureus. 2024;16(3):e56755. https://doi.org/10.7759/cureus.56755.
40. Gheorghiade M, Adams KF, Colucci WS. Digoxin in the Management of Cardiovascular Disorders. Circulation. 2004;109(24):2959–64. https://doi.org/10.1161/01.CIR.0000132482.95686.87.
41. Bublitz M, Morth JP, Nissen P. P-type ATPases at a glance. J Cell Sci. 2011;124(15):2515–9. https://doi.org/10.1242/jcs.088716.
42. Ottolia M, John S, Hazan A, Goldhaber JI. The cardiac Na(+) -Ca(2+) exchanger: from structure to function. Compr Physiol. 2021;12(1):2681–717. https://doi.org/10.1002/cphy.c200031.
43. Xu H, Van Remmen H. The SarcoEndoplasmic Reticulum Calcium ATPase (SERCA) pump: a potential target for intervention in aging and skeletal muscle pathologies. Skelet Muscle. 2021;11(1):25. https://doi.org/10.1186/s13395-021-00280-7.
44. Das A, Rui H, Nakamoto R, Roux B. Conformational transitions and alternating-access mechanism in the sarcoplasmic reticulum calcium pump. J Mol Biol. 2017;429(5):647–66. https://doi.org/10.1016/j.jmb.2017.01.007.
45. Mareedu S, Million ED, Duan D, Babu GJ. Abnormal calcium handling in Duchenne muscular dystrophy: mechanisms and potential therapies. Front Physiol. 2021;12 https://doi.org/10.3389/fphys.2021.647010.
46. Kang S, Dahl R, Hsieh W, Shin A, Zsebo KM, Buettner C, et al. Small Molecular allosteric activator of the Sarco/endoplasmic reticulum Ca2+-ATPase (SERCA) attenuates diabetes and metabolic disorders. J Biol Chem. 2016;291(10):5185–98. https://doi.org/10.1074/jbc.M115.705012.
47. Rees DC, Johnson E, Lewinson O. ABC transporters: the power to change. Nat Rev Mol Cell Biol. 2009;10(3):218–27. https://doi.org/10.1038/nrm2646.
48. Vasiliou V, Vasiliou K, Nebert DW. Human ATP-binding cassette (ABC) transporter family. Hum Genomics. 2009;3(3):281–90. https://doi.org/10.1186/1479-7364-3-3-281.
49. Ahmad J, Akhter S, Greig NH, Kamal MA, Midoux P, Pichon C. Engineered nanoparticles against MDR in cancer: the state of the art and its prospective. Curr Pharm Des. 2016;22(28):4360–73. https://doi.org/10.2174/1381612822666160617112111.
50. van Schalkwyk DA, Egan TJ. Quinoline-resistance reversing agents for the malaria parasite plasmodium falciparum. Drug Resist Updat. 2006;9(4):211–26. https://doi.org/10.1016/j.drup.2006.09.002.
51. Saliba KJ, Lehane AM, Kirk K. A polymorphic drug pump in the malaria parasite. Mol Microbiol. 2008;70(4):775–9. https://doi.org/10.1111/j.1365-2958.2008.06451.x.
52. Lee AG. A calcium pump made visible. Curr Opin Struct Biol. 2002;12(4):547–54. https://doi.org/10.1016/s0959-440x(02)00360-3.
53. Vertex Pharmaceuticals Incorporated. Kalydeco® (ivacaftor)—patient information. 2024. Available from: https://www.kalydeco.com/.
54. Beyenbach KW, Wieczorek H. The V-type H+ ATPase: molecular structure and function,physiological roles and regulation. J Exp Biol. 2006;209(4):577–89. https://doi.org/10.1242/jeb.02014.
55. Chen F, Kang R, Liu J, Tang D. The V-ATPases in cancer and cell death. Cancer Gene Ther. 2022;29(11):1529–41. https://doi.org/10.1038/s41417-022-00477-y.
56. Holliday LS. Vacuolar H(+)-ATPases (V-ATPases) as therapeutic targets: a brief review and recent developments. Biotarget. 2017;1 https://doi.org/10.21037/biotarget.2017.12.01.
57. Sano R, Shinozaki Y, Ohta T. Sodium–glucose cotransporters: functional properties and pharmaceutical potential. J Diabetes Invest. 2020;11(4):770–82. https://doi.org/10.1111/jdi.13255.
58. Vallon V, Thomson SC. Targeting renal glucose reabsorption to treat hyperglycaemia: the pleiotropic effects of SGLT2 inhibition. Diabetologia. 2017;60(2):215–25. https://doi.org/10.1007/s00125-016-4157-3.
59. FDA. Jardiance(empagliflozin) approval overview. 2014. Available from: https://www.accessdata.fda.gov/scripts/cder/daf/index.cfm?event=overview.process&ApplNo=204629.

60. Frampton JE. Empagliflozin: a review in type 2 diabetes. Drugs. 2018;78(10):1037–48. https://doi.org/10.1007/s40265-018-0937-z.
61. FDA. Farxiga(dapagliflozin) approval overview. 2014. https://www.accessdata.fda.gov/drugsatfda_docs/appletter/2014/202293Orig1s000ltr.pdf
62. Dhillon S. Dapagliflozin: a review in type 2 diabetes. Drugs. 2019;79(10):1135–46. https://doi.org/10.1007/s40265-019-01148-3.
63. FDA. Brenzavvy(bexagliflozin) approval overview. 2023. Available from: https://www.accessdata.fda.gov/scripts/cder/daf/index.cfm?event=overview.process&ApplNo=214373.
64. FDA. Inpefa(sotogliflozin) approval overview. 2023. https://www.accessdata.fda.gov/scripts/cder/daf/index.cfm?event=overview.process&ApplNo=216203
65. Hoy SM. Bexagliflozin: first approval. Drugs. 2023;83(5):447–53. https://doi.org/10.1007/s40265-023-01848-x.
66. Lexicon Pharmaceuticals, Inc. INPEFA® (sotagliflozin) HCP. 2024. Available from: https://www.inpefahcp.com/.
67. Itskanov S, Park E. Mechanism of protein translocation by the Sec61 Translocon complex. Cold Spring Harb Perspect Biol. 2023;15(1) https://doi.org/10.1101/cshperspect.a041250.
68. Südhof TC, Rizo J. Synaptic vesicle exocytosis. Cold Spring Harb Perspect Biol. 2011;3(12) https://doi.org/10.1101/cshperspect.a005637.
69. Rizo J. Molecular mechanisms underlying neurotransmitter release. Annu Rev Biophys. 2022;51:377–408. https://doi.org/10.1146/annurev-biophys-111821-104732.
70. Mamrack M. Exercise and sport pharmacology. Tayler and Francis; 2020.
71. Marks F, Klingmüller U, Müller-Decker K. Cellular signal processing: an introduction to the Molecular mechanisms of signal transduction. 2nd ed. Garland Science; 2017.
72. Luo L. Principles of neurobiology. 2nd ed. Garland Science; 2020.
73. Hille B. Ion channels of excitable membranes. 3rd ed; 2001.
74. FDA. Botox(onabotulinumtoxinA) approval overview. 1989. Available from: https://www.accessdata.fda.gov/scripts/cder/daf/index.cfm?event=overview.process&ApplNo=103000.
75. Salame N, Eber AE, Dover J. DaxibotulinumtoxinA-lanm (Daxxify™): a comprehensive overview. Skin Therapy Lett. 2023;28(4):1–3.
76. Bennek M, Rudowitz D, Kerscher M. Diffusion characteristics of LetibotulinumtoxinA, OnabotulinumtoxinA, and AbobotulinumtoxinA and its impact on muscle relaxation: a randomized Split-face clinical trial. Dermatol Ther (Heidelb). 2025;15(8):2147–58. https://doi.org/10.1007/s13555-025-01458-3.
77. Ayoub N. Botulinum toxin therapy: a comprehensive review on clinical and pharmacological insights. J Clin Med. 2025;14(6) https://doi.org/10.3390/jcm14062021.
78. Letchuman S, Madhuranga HDT, Madhurangi BLNK, Premarathna AD, Saravanan M. Alkaloids unveiled: a comprehensive analysis of novel therapeutic properties, mechanisms, and plant-based innovations. Intell Pharm. 2024; https://doi.org/10.1016/j.ipha.2024.09.007.
79. FDA. Chantix(varenicline) approval overview. 2006. Available from: https://www.accessdata.fda.gov/scripts/cder/daf/index.cfm?event=overview.process&ApplNo=021928
80. FDA. Anectine (succinylcholine chloride) approval overview. 1953. Available from: https://www.accessdata.fda.gov/scripts/cder/daf/index.cfm?event=overview.process&ApplNo=008453.
81. Changeux JP. Discovery of the first neurotransmitter receptor: the acetylcholine nicotinic receptor. Biomolecules. 2020;10(4) https://doi.org/10.3390/biom10040547.
82. Bubuioc AM, Kudebayeva A, Turuspekova S, Lisnic V, Leone MA. The epidemiology of myasthenia gravis. J Med Life. 2021;14(1):7–16. https://doi.org/10.25122/jml-2020-0145.
83. San PP, Jacob S. Role of complement in myasthenia gravis. Front Neurol. 2023;14:1277596. https://doi.org/10.3389/fneur.2023.1277596.
84. FDA. Journavx(suzetrigine): approval overview. 2025. Available from: https://www.accessdata.fda.gov/scripts/cder/daf/index.cfm?event=overview.process&ApplNo=219209.
85. FDA. FDA approves novel non-opioid treatment for moderate to severe acute pain. 2025. Available from: https://www.fda.gov/news-events/press-announcements/fda-approves-novel-non-opioid-treatment-moderate-severe-acute-pain#:~:text=In%20this%20section-,Press%20Announcements,of%20non%2Dopioid%20pain%20treatment.
86. Vertex Pharmaceuticals. JOURNAVX is an oral nonopioid indicated for the treatment of moderate-to-severe acute pain in adults. 2024. Available from: https://www.journavxhcp.com/.
87. Jembrek MJ, Vlainic J. GABA receptors: pharmacological potential and pitfalls. Curr Pharm Des. 2015;21(34):4943–59. https://doi.org/10.2174/1381612821666150914121624.
88. Whalen K, Lerchenfeldt S, Giordano C. Lippincott illustrated reviews: pharmacology. Wolters Kluwer; 2021.
89. Han W, Shepard RD, Lu W. Regulation of GABAaRs by transmembrane accessory proteins. Trends Neurosci. 2021;44(2):152–65. https://doi.org/10.1016/j.tins.2020.10.011.
90. Kandel ER, Koester JD, Mack SH, Koester J. Principles of neural science. 6th ed. McGraw Hill/Medical; 2021.

91. FDA. Ativan(lorazepam) approval overview. 1977. Available from: https://www.accessdata.fda.gov/scripts/cder/daf/index.cfm?event=overview.process&ApplNo=017794.
92. FDA. Sezaby(phenobarbital sodium) approval overview. 2022. Available from: https://www.accessdata.fda.gov/scripts/cder/daf/index.cfm?event=overview.process&ApplNo=215910.
93. McGovern G. The role of pharmacists in optimizing care for seizures in neonatal patients. 2025. Available from: https://www.pharmacytimes.com/view/the-role-of-pharmacists-and-sezaby-in-optimizing-care-for-seizures-in-neonatal-patients.
94. Park B. Sezaby approved to treat seizures in term and preterm infants. 2022. Available from: https://www.empr.com/home/news/sezaby-approved-to-treat-seizures-in-term-and-preterm-infants/.
95. Ghit A, Assal D, Al-Shami AS, Hussein DEE. GABAA receptors: structure, function, pharmacology, and related disorders. J Genet Eng Biotechnol. 2021;19(1):123. https://doi.org/10.1186/s43141-021-00224-0.
96. FDA. Zurzuvae(zuranolone) approval overview. 2023. Available from: https://www.accessdata.fda.gov/scripts/cder/daf/index.cfm?event=overview.process&ApplNo=217369.
97. Deligiannidis KM, Meltzer-Brody S, Maximos B, Peeper EQ, Freeman M, Lasser R, et al. Zuranolone for the treatment of postpartum depression. Am J Psychiatry. 2023;180(9):668–75. https://doi.org/10.1176/appi.ajp.20220785.
98. Biogen and Sage Therapeutics I. ZURZUVAE™ (zuranolone) C-IV | official HCP website. 2024. Available from: https://www.zurzuvaehcp.com/.
99. Meldrum BS, Chapman AG. Basic mechanisms of gabitril (tiagabine) and future potential developments. Epilepsia. 1999;40(Suppl 9):S2–6. https://doi.org/10.1111/j.1528-1157.1999.tb02087.x.
100. FDA. Ztalmy(ganaxolone) approval overview. 2022. Available from: https://www.accessdata.fda.gov/scripts/cder/daf/index.cfm?event=overview.process&ApplNo=215904.
101. Reddy DS, Estes WA. Clinical potential of Neurosteroids for CNS disorders. Trends Pharmacol Sci. 2016;37(7):543–61. https://doi.org/10.1016/j.tips.2016.04.003.
102. Lamb YN. Ganaxolone: first approval. Drugs. 2022;82(8):933–40. https://doi.org/10.1007/s40265-022-01724-0.
103. Immedica, Pharma U.S. Inc. ZTALMY® (ganaxolone) official site. Available from: https://www.ztalmy.com/.
104. FDA. Xdemvy(lotilaner ophthalmic solution) approval overview. 2023. Available from: https://www.accessdata.fda.gov/scripts/cder/daf/index.cfm?event=overview.process&ApplNo=217603.
105. Tarsus Pharmaceuticals, Inc. XDEMVY® (lotilaner opthalmic solution). Available from: https://xdemvy.com
106. Davey PG, Farid M, Karpecki P, Gaddie IB, Chan A, Mun J, et al. Lotilaner ophthalmic solution, 0.25%, for the treatment of Demodex blepharitis. Healthcare. 2024;12(15):1487.
107. Baudry A, Pietri M, Launay JM, Kellermann O, Schneider B. Multifaceted regulations of the serotonin transporter: impact on antidepressant response. Front Neurosci. 2019;13:91. https://doi.org/10.3389/fnins.2019.00091.
108. Yang D, Gouaux E. Illumination of serotonin transporter mechanism and role of the allosteric site. Sci Adv. 2021;7(49):eabl3857. https://doi.org/10.1126/sciadv.abl3857.
109. Gabriel FC, de Melo DO, Fráguas R, Leite-Santos NC, Mantovani da Silva RA, Ribeiro E. Pharmacological treatment of depression: a systematic review comparing clinical practice guideline recommendations. PLoS One. 2020;15(4):e0231700. https://doi.org/10.1371/journal.pone.0231700.
110. FDA. Prozac (fluoxetine) approval overview. 1987. Available from: https://www.accessdata.fda.gov/scripts/cder/daf/index.cfm?event=overview.process&ApplNo=18936.
111. FDA. Zoloft(sertraline) approval overview. 1991. Available from: https://www.accessdata.fda.gov/scripts/cder/daf/index.cfm?event=overview.process&ApplNo=019839.
112. FDA. LEXAPRO (escitalopram) approval overview. 2002. Available from: https://www.accessdata.fda.gov/scripts/cder/daf/index.cfm?event=overview.process&ApplNo=021323.
113. FDA. Celexa(citalopram) approval overview. 1998. Available from: https://www.accessdata.fda.gov/scripts/cder/daf/index.cfm?event=overview.process&ApplNo=020822
114. FDA. Paxil(paroxetine): approval overview. 1992. Available from: https://www.accessdata.fda.gov/scripts/cder/daf/index.cfm?event=overview.process&ApplNo=020031.
115. FDA. Luvox(fluvoxamine): approval overview. 1994. Available from: https://www.accessdata.fda.gov/scripts/cder/daf/index.cfm?event=overview.process&ApplNo=021519.
116. Edinoff AN, Akuly HA, Hanna TA, Ochoa CO, Patti SJ, Ghaffar YA, et al. Selective serotonin reuptake inhibitors and adverse effects: a narrative review. Neurol Int. 2021;13(3):387–401. https://doi.org/10.3390/neurolint13030038.
117. Harriott ND, Williams JP, Smith EB, Bozigian HP, Grigoriadis DE. VMAT2 inhibitors and the path to Ingrezza (Valbenazine). Prog Med Chem. 2018;57(1):87–111. https://doi.org/10.1016/bs.pmch.2017.12.002.
118. Yaffe D, Forrest LR, Schuldiner S. The ins and outs of vesicular monoamine transporters. J Gen Physiol. 2018;150(5):671–82. https://doi.org/10.1085/jgp.201711980.

119. Savitt D, Jankovic J. Tardive syndromes. J Neurol Sci. 2018;389:35–42. https://doi.org/10.1016/j.jns.2018.02.005.
120. Schwab LC, Garas SN, Drouin-Ouellet J, Mason SL, Stott SR, Barker RA. Dopamine and Huntington's disease. Expert Rev Neurother. 2015;15(4):445–58. https://doi.org/10.1586/14737175.2015.1025383.
121. FDA. Ingrezza(valbenazine) approval overview. 2017. Available from: https://www.accessdata.fda.gov/scripts/cder/daf/index.cfm?event=overview.process&ApplNo=209241.
122. Floridia Rietmann LM, Romano C, Beltrán Covarrubias SA, Gomez Miranda JA, Briceño Cardeña OE, Shenod S, et al. Efficacy and safety of VMAT-2 inhibitors and dopamine stabilizers for Huntington's chorea: a systematic review, meta-analysis, and trial sequential analysis. Med Sci. 2025;13(3):201.
123. FDA. Austedo (deutetrabenazine) approval overview. 2017. Available from: https://www.accessdata.fda.gov/scripts/cder/daf/index.cfm?event=overview.process&ApplNo=209885.
124. Bashir HH, Jankovic J. Treatment of tardive dyskinesia. Neurol Clin. 2020;38(2):379–96. https://doi.org/10.1016/j.ncl.2020.01.004.
125. Anderson KE, Stamler D, Davis MD, Factor SA, Hauser RA, Isojärvi J, et al. Deutetrabenazine for treatment of involuntary movements in patients with tardive dyskinesia (AIM-TD): a double-blind, randomised, placebo-controlled, phase 3 trial. Lancet Psychiatry. 2017;4(8):595–604. https://doi.org/10.1016/s2215-0366(17)30236-5.
126. Heo YA, Scott LJ. Deutetrabenazine: a review in chorea associated with Huntington's disease. Drugs. 2017;77(17):1857–64. https://doi.org/10.1007/s40265-017-0831-0.
127. Caplan MJ. Membrane polarity in epithelial cells: protein sorting and establishment of polarized domains. Am J Physiol-Renal Physiol. 1997;272(4):F425–F9. https://doi.org/10.1152/ajprenal.1997.272.4.F425.
128. Salton MRJ, Kim KS. Structure. In: Baron S, editor. Medical microbiology. 4th ed. Galveston: University of Texas Medical Branch at Galveston; 1996.
129. Silhavy TJ, Kahne D, Walker S. The bacterial cell envelope. Cold Spring Harb Perspect Biol. 2010;2(5):a000414. https://doi.org/10.1101/cshperspect.a000414.
130. Baran A, Kwiatkowska A, Potocki L. Antibiotics and bacterial resistance—a short story of an endless arms race. Int J Mol Sci. 2023;24(6):5777. https://doi.org/10.3390/ijms24065777.
131. Hutchings MI, Truman AW, Wilkinson B. Antibiotics: past, present and future. Curr Opin Microbiol. 2019;51:72–80. https://doi.org/10.1016/j.mib.2019.10.008.
132. Chain E. The early years of the penicillin discovery. Trends Pharmacol Sci. 1979;1(1):6–11. https://doi.org/10.1016/0165-6147(79)90004-X.
133. Fleming A. On the antibacterial action of cultures of a Penicillium, with special reference to their use in the isolation of B. Influenzæ. Br J Exp Pathol. 1929;10(3):226–36.
134. Miller EL. (2010) The penicillins: a review and update. J Midwifery Women's Health. 2002;47(6):426–34. https://doi.org/10.1016/S1526-9523(02)00330-6.
135. Reynolds BDo. History of vancomycin. 2024. Available from: https://doseme-rx.com/vancomycin/articles/vancomycin-mechanisms-of-action#:~:text=Vancomycin%20is%20an%20antibacterial%20medication,10%5D.
136. National Cancer Institute. Glycopeptide antibiotic. Available from: https://www.cancer.gov/publications/dictionaries/cancer-drug/def/glycopeptide-antibiotic.
137. FDA. Cubicin (daptomycin) approval overview. 2003. Available from: https://www.accessdata.fda.gov/scripts/cder/daf/index.cfm?event=overview.process&ApplNo=021572.
138. El-Sayed Ahmed MAE, Zhong LL, Shen C, Yang Y, Doi Y, Tian GB. Colistin and its role in the Era of antibiotic resistance: an extended review (2000–2019). Emerg Microbes Infect. 2020;9(1):868–85. https://doi.org/10.1080/22221751.2020.1754133.
139. Bialvaei AZ, Samadi KH. Colistin, mechanisms and prevalence of resistance. Curr Med Res Opin. 2015;31(4):707–21. https://doi.org/10.1185/03007995.2015.1018989.
140. FDA. FDA approves new treatment for uncomplicated urinary tract infections in adult women who have limited or no alternative oral antibiotic treatment options. 2024. Available from: https://www.fda.gov/drugs/news-events-human-drugs/fda-approves-new-treatment-uncomplicated-urinary-tract-infections-adult-women-who-have-limited-or-no.
141. Rusu A, Oancea O-L, Donici E, Uncu L. Recent developments in Penem antibiotics: structural and therapeutic perspectives. Molecules. 2025;30(10):2126. https://doi.org/10.3390/molecules30102126.
142. Mauhin W, et al. Acid sphingomyelinase deficiency: sharing experience of disease monitoring and severity in France. J Clin Med. 2022;11:920. https://doi.org/10.3390/jcm11040920.

Part II

Introduction to Proteins and Enzymes

This part introduces proteins. This part covers the overall structure of proteins and the properties of amino acids, providing an overview of the diverse functions proteins play in the cell. Chapter 4 discusses the basic structure of proteins. It covers amino acids, the bonds that link amino acids, and the hierarchical structure of proteins. Chapter 5 covers enzymes. It covers the concepts of a catalyst, the different types of enzymes, and how enzyme kinetics and thermodynamics are defined. The chapter is dedicated to enzymes because enzymes have key roles in nearly every biological process discussed throughout the rest of the text. Enzymes are important drug targets, and multiple examples of pharmaceuticals that target enzymes are included in this part.

Amino Acids and Protein Structure 4

Abstract

Proteins are one of the four major macromolecules in cells, and they carry out a wide range of essential functions. This chapter introduces protein structure and function, beginning with a discussion of amino acids, which have an amino group, a carboxyl group, an R group, and an alpha carbon. Amino acids are linked together via peptide bonds to form polypeptide chains. Protein structure is prevented as hierarchical, including primary, secondary, tertiary, and sometimes quaternary structures. Each level of structure builds upon the level before. Proteins have many important functions in our cells. Some examples of proteins, their names, and their functions are addressed. Examples are transport proteins, signaling proteins, enzymes, and others. Protein interactions are discussed, and the concepts of equilibrium constants and dissociation constants are introduced.

Keywords

Amino acids · Polypeptide chains · Proteins · Alpha helix · Beta pleated sheet · Protein interactions

Abbreviations

EGFR	Epidermal growth factor receptor
GPCR	G protein-coupled receptor
Kd	Dissociation constant
Km	Michaelis constant
mAb	Monoclonal antibody
RTK	Receptor tyrosine kinase
Vmax	Maximum velocity

What You Will Learn

- Proteins are made of amino acids.
- The general structure of an amino acid, including the amino group, carboxyl group, R group, and alpha carbon.
- Polypeptide chains and peptide bonds.
- Protein structure hierarchy, including primary, secondary, tertiary, and quaternary structure.
- Proteins can have different domains with different functions.
- Examples of protein functions such as transport proteins, signaling proteins, chaperones, immune proteins, and enzymes.
- Proteins can interact with other proteins or molecules, and this interaction can be quantified using equilibrium constants such as the dissociation constant.

A. Minden, *Biochemistry, Cell and Molecular Biology for Pharmacists and Health Professionals*,
https://doi.org/10.1007/978-3-032-20529-2_4

- The importance of understanding proteins in pharmacy. Some proteins, such as cytochrome CYP450 enzymes, are important for metabolizing drugs, and most drugs target proteins.

This chapter introduces the structure and function of proteins. Proteins are one of the four major macromolecules in cells, as illustrated in Fig. 4.1. This chapter begins with an overview of protein structure. This includes the amino acids that constitute proteins, the bonds that link these amino acids together, and the secondary, tertiary, and quaternary structures that proteins form. This chapter then explores the diverse functions of proteins. Some key examples of the various types of proteins present in cells, along with their functions, are discussed. One especially important group of proteins, enzymes, will be discussed in more detail in the next chapter.

1 Structure of Proteins

Like other major macromolecules, proteins are chains of monomers that are linked together to form a polymer. The monomers that make up proteins are the **amino acids**. Chains of amino acids form polypeptide chains that fold into various three-dimensional structures [1, 2] (Fig. 4.2).

1.1 Amino Acids

Amino acids are the basic building blocks of proteins. The basic structure of an amino acid (a free amino acid that is not in a chain) includes **a central carbon (the α-carbon),** and around this carbon are **an amino group** ($-NH_2-$ which is usually protonated as $-NH_3^+$ at physiological pH)**,** a hydrogen atom (H), **a carboxyl group** ($-COOH$; which is usually deprotonated as $-COO^-$ at physiological pH), and a **side chain (R)** (Fig. 4.3). The side chain (R) is the part that varies from one amino acid to another. Twenty different amino acids make up proteins in all plants and animals (Fig. 4.4). Amino acids can have a charge, depending on the pH in the environment. At low pH (acidic), they are fully protonated and carry a net positive charge (cationic form). At neutral pH (around pH 7), they exist mainly as zwitterions with both positive and negative charges. At high pH (basic), they are deprotonated and carry a net negative charge (anionic form) [1, 3, 4] (Fig. 4.5).

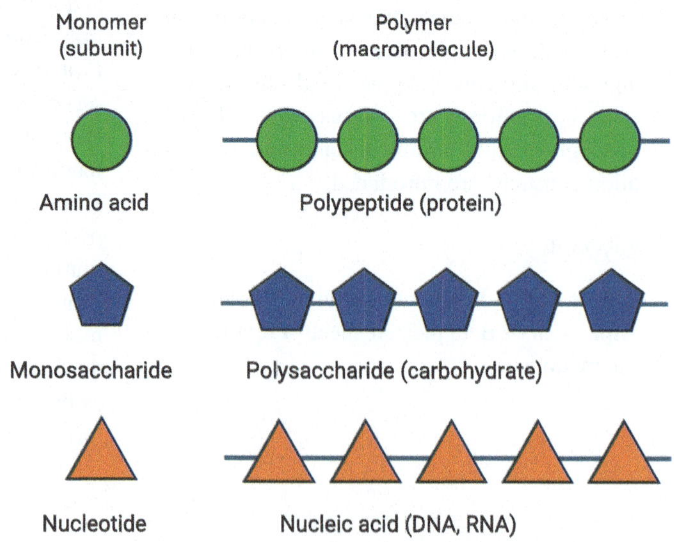

Fig. 4.1 Biomolecules. The four biomolecules in cells are polypeptides (proteins), polysaccharides (carbohydrates), nucleic acids (DNA and RNA), and lipids. Polypeptides, polysaccharides, and nucleic acids are polymers made up of monomers

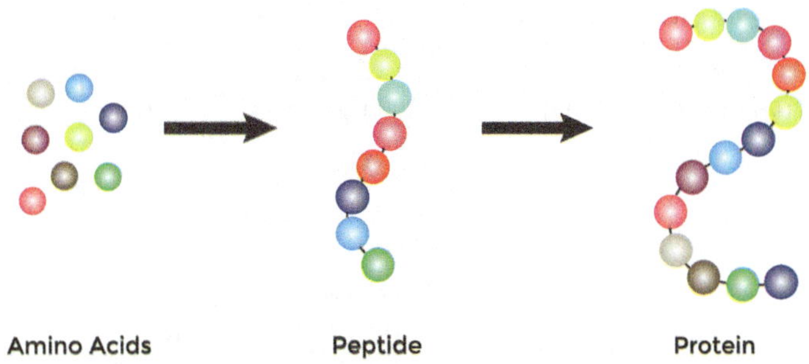

Fig. 4.2 Proteins: Monomers (amino acids) make up peptides, which are folded into various shapes to make proteins. (*Source*: Illustration from Freepik.com (premium license))

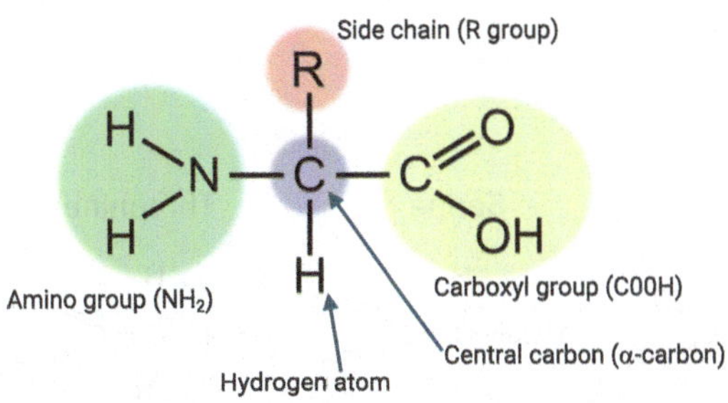

Fig. 4.3 Basic structure of amino acids. Amino acids have a central carbon, an amino group, a carboxyl group, and a side chain (R group). Amino groups can exist as NH_2 as shown here, or as NH_3^+, while carboxyl groups can exist as COOH as shown here or as COO^- (*Source*: Image from OpenClipart.org (openclipart.org/61279), Public Domain (CC0 1.0))

1.1.1 Categorization of Amino Acids by Their Side Chains

There are different ways that amino acids can be categorized. One way to categorize them is by their **side chains**, the **R groups**, as follows:

- Amino Acids with Nonpolar Side Chains

The side chains on this category of amino acids generally have no electrically charged regions. These side chains are hydrophobic and tend not to form hydrogen bonds with water or other molecules. These amino acids, including alanine, glycine, valine, leucine, isoleucine, proline, phenylalanine, methionine, and tryptophan, are often found in the interior of proteins and are away from aqueous environments. Cysteine is sometimes included in this group as well, though its side chain can form polar interactions and disulfide bonds (Fig. 4.4).

- Amino Acids with Polar side Chains

This group consists of amino acids with polar side chains. The side chains of these amino acids are hydrophilic. This category of amino acids can be subdivided further:

– Amino Acids with Polar Side Chains That Are Uncharged

These amino acids have uncharged side groups, but their side groups can still form hydrogen bonds with water. This is because they contain groups such as hydroxyl (–OH) or amide (–$CONH_2$) groups that have electronegative atoms. The amino acids in this group include asparagine, glutamine, serine, threonine, cysteine, and tyrosine.

– Amino Acids with Polar Side Chains That Are Charged

These can include amino acids with negatively charged side chains, which are polar and acidic, or positively charged side chains, which are polar and basic. These amino acids can interact with water, engage in electrostatic interactions with other molecules, and form hydrogen bonds. These include aspartate, glutamate, arginine, lysine, and histidine.

1.1.2 Categorization of Amino Acids by Whether They Are Essential or Nonessential Amino Acids

Another way amino acids can be categorized is based on whether they are essential, meaning we need to obtain them from our diet, or nonessential, meaning we produce them and therefore do

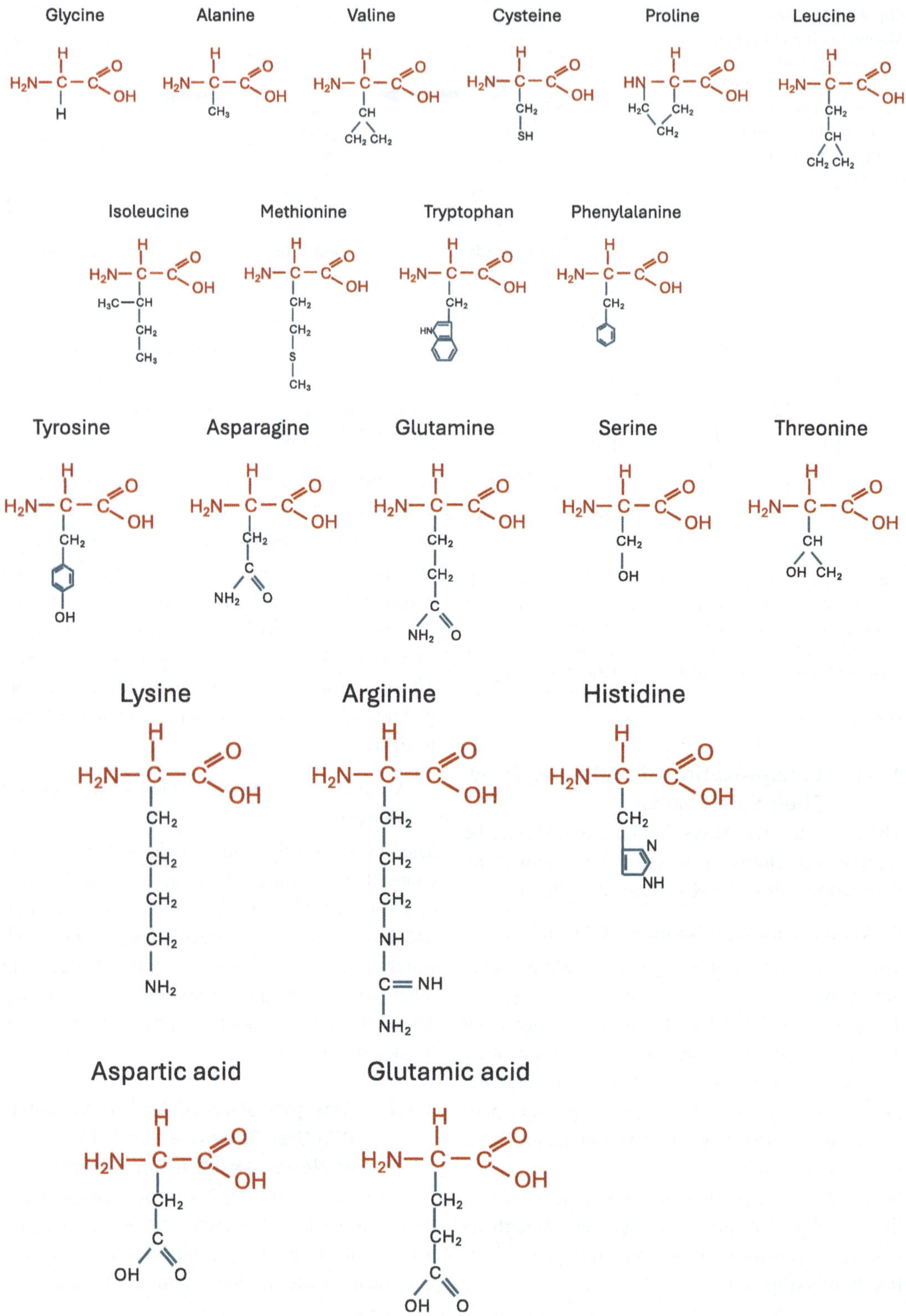

Fig. 4.4 (**a**) Structures of amino acids. Nonpolar amino acids. (**b**) Structures of amino acids. Polar uncharged (neutral) amino acids. (**c**) Structures of amino acids. Polar charged amino acids. For simplicity, all amino groups are shown as NH_2 and all carboxyl groups are shown as COOH

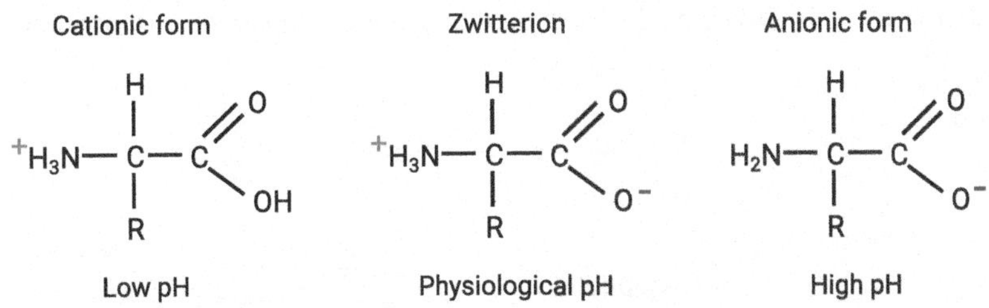

Fig. 4.5 The charge of an amino acid depends on the pH. At low pH (high H^+ concentration), the amino acid tends to gain protons and the amino group group is, present at NH_3^+ while the carboxyl group is generally present as COOH. At high pH (low H^+ concentration), amino acids tend to lose protons. Under these conditions the amino group will be present as NH_2 and the carboxyl group as COO^-. At neutral pH amino acids are commonly present as zwitterions (NH_3^+/COO^-) . Amino and carboxyl groups have different PKAs, and thus tend to gain or lose protons at different pH values. (pKa quantifies the strength of an acid and its tendency to donate a proton in water). For simplicity, the amino groups and carboxyl groups are shown in

not need to obtain them from our diet. The essential amino acids are histidine, isoleucine, leucine, lysine, methionine, phenylalanine, threonine, tryptophan, and valine. Clinically, understanding essential amino acids is important when evaluating protein malnutrition, where essential amino acid deficiency impairs protein synthesis. We can obtain essential amino acids from various food sources, including meat, dairy products, fish, eggs, legumes, nuts, and seeds. The nonessential amino acids are alanine, arginine, asparagine, aspartic acid, cysteine, glutamic acid, glutamine, glycine, proline, serine, and tyrosine [1, 3]. These amino acids can be synthesized in the body.

1.2 Polypeptide Chains

Amino acids linked together make up polypeptide chains. Chains of amino acids can vary in length. Dipeptides are made of two amino acids, oligopeptides are made of a small number of amino acids, usually 10 or fewer, and amino acid chains that are longer than oligopeptides are polypeptides (Fig. 4.6). All of these, however, are also sometimes lumped into the broad category of polypeptides [1, 5].

1.3 Peptide Bonds Link Amino Acids

Amino acids in polypeptide chains are linked together by bonds. The type of bond that links the amino acids in the chain is the peptide bond (Fig. 4.7). This bond links the carbon of the carboxyl group (–COOH) of one amino acid with the nitrogen on the amino group (–NH_2) of its neighboring amino acid. This results in an amide bond (C–N), also known as the peptide bond. In the process, an amide group forms, which consists of a carbonyl group (C=O) linked to a nitrogen atom that is linked to one hydrogen (N–H). This reaction is a dehydration reaction, releasing one water molecule. These repeated linkages form the backbone of the protein [1, 3, 4].

Proteins are typically composed of long polypeptides. Due to the way amino acids are linked together, polypeptide chains have an amino (N) terminus and a carboxyl (C) terminus. At the N terminus, the amino group of an amino acid is exposed to the end of the chain, and at the C terminus, the carboxyl group of an amino acid is exposed to the end. The chains have directionality, and when writing a protein sequence, we write it from the N terminus to the C terminus (See Fig. 4.8).

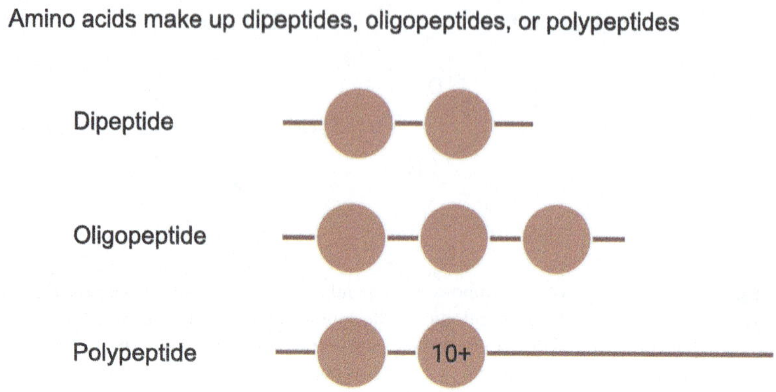

Fig. 4.6 Peptide chains. Peptide chains can be classified into dipeptides, oligopeptides, and polypeptides

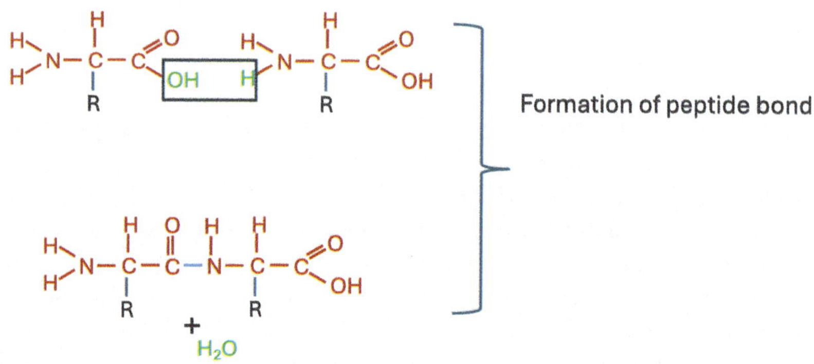

Fig. 4.7 Peptide bonds. Peptide bonds link the amino acids in polypeptide chains

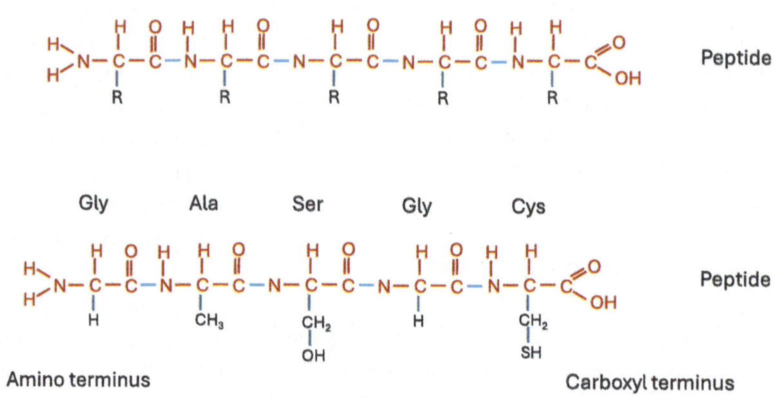

Fig. 4.8 Peptides have directionality. The end with the amino group is the N-terminal end, and the other end is the C-terminal end. Amino acids are linked with peptide bonds

1.4 Sizes of Proteins

A protein is made up of one or more polypeptides. Most proteins range in size from 50 to 2000 amino acids. Smaller proteins often function in signaling (e.g., peptide ligands or hormones), while larger ones more often have structural or mechanical roles (such as the cytoskeleton or collagen). A protein's molecular weight (grams per mole) can be represented in Daltons or Kilodaltons (one Dalton is equivalent in molecular weight to 1/12 the molecular weight of a carbon atom, or 1 gram per mole; this is also approximately the molecular weight of a hydrogen atom, and a mole is 6.02×10^{-23} copies). An average amino acid has a molecular weight of 110 Daltons (Da), or 110 grams/mole. Thus, if most proteins average in size from 50 to 2000 amino acids in length, they are 5500 to 220,000 Daltons in molecular weight ($50 \times 110 = 5500$

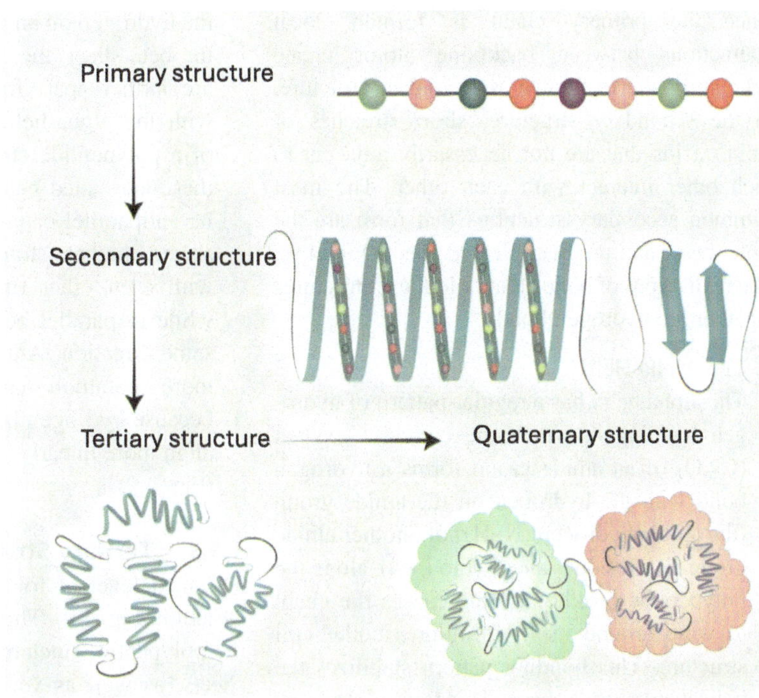

Fig. 4.9 Protein hierarchy. Primary, secondary, tertiary, and quaternary structures. (*Source*: Illustration from Freepik.com (premium license))

Daltons, 2000 × 110 = 220,000 Daltons). This would usually be expressed as 5.5 to 220 Kilodaltons (5.5 to 220 kD) [1, 5, 6].

1.5 Protein Structure Hierarchy

Proteins consist of polypeptides that are usually folded into various conformations. This folding is hierarchical, from simple linear sequences to complex multi-subunit arrangements. Proteins have primary, secondary, tertiary, and sometimes quaternary structures (Fig. 4.9). Each level of structure builds upon the level before. The structural hierarchy discussed below is also described in detail in [1, 3, 5].

1.5.1 Primary Structure

The primary structure is the basic linear sequence of the amino acids in the polypeptide chain (Fig. 4.9). Alterations in the primary structure can lead to abnormal protein function and disease. Examples include sickle cell anemia and cystic fibrosis. Both disorders exemplify how a single point mutation or deletion can disrupt protein structure and function. In sickle cell anemia, one glutamate (at position 6) is changed to the amino acid valine on the β-chain of hemoglobin. Hemoglobin is the protein that carries oxygen in red blood cells. This single amino acid change is sufficient to render hemoglobin defective and distort the shape of red blood cells, thereby disrupting oxygen flow and causing pain [3, 7].

Cystic fibrosis (CF) is caused by mutations in the CFTR gene (cystic fibrosis transmembrane conductance regulator), which regulates the flow of ions, specifically chloride (Cl–), across membranes. There are many mutations that can cause the disease, but a common one is a small mutation in which a phenylalanine (F) at position 508 in the CFTR protein is missing. This one change is enough to disrupt the protein's function. As a result, Cl– transport is disrupted, resulting in CF, which is associated with thick mucus accumulation, especially in certain organs such as the lungs [8].

1.5.2 Secondary Structure

For most proteins, the chains of amino acids in the primary structure fold to make specific shapes.

Once the primary chain is formed, local interactions between backbone atoms create structural patterns known as secondary structure. In the secondary structure, short stretches of amino acids that are not necessarily adjacent to each other interact with each other. The most common secondary structures that form are the alpha helix and the beta pleated sheet (Fig. 4.9). The main types of bonds that hold these structures together are hydrogen bonds.

- The Alpha Helix
  The alpha helix has a regular pattern of hydrogen bonding in which the carbonyl oxygen (C=O) of an amide group forms a hydrogen bond with the hydrogen on the amide group (the amide hydrogen (N–H)) of another amide group four residues ahead (I to i + 4), along the polypeptide backbone. This causes the chain to wrap around itself, forming a coiled-coil structure. This bonding pattern stabilizes the helical structure's shape. This type of structure is a right-handed spiral (Fig. 4.9).
- The Beta-Sheet
  The beta-sheet is the other major type of secondary structure. Beta sheets form when a polypeptide chain folds back on itself. In the beta-sheet, hydrogen bonds form between amino acids that may be far from each other in the primary sequence but are brought close to each other in the folded protein. In this structure, hydrogen bonds also form between the oxygen on the carbonyl groups (C=O) and the hydrogen on an amide group (N–H), but in the beta sheet, the amide groups that interact are farther apart from each other compared with the alpha-helix. Usually, only portions of a polypeptide chain form beta sheets, and these are called beta strands. Beta sheets can be antiparallel or parallel. In antiparallel beta sheets, the beta strands that come into contact with each other run in opposite directions, while in parallel beta sheets, they run in the same direction. Antiparallel arrangements are more common and usually more stable because hydrogen bonds in antiparallel sheets align more linearly, strengthening the network (Fig. 4.9).

1.5.3 Tertiary Structure

The next level of folding results in the tertiary structure (Fig. 4.9). This is the full 3D structure of the polypeptide, including alpha helices and beta sheets, as well as various folds between them. Tertiary structures are quite varied and can contain globular regions and many kinds of bonds. For example, myoglobin is a compact globular protein whose tertiary structure allows it to store oxygen efficiently in muscle tissue. They can contain bonds that link amino acids that are quite a bit further apart than those in the secondary structure. In the tertiary structure of a protein, residues with hydrophobic side chains tend to be found on the inside of a structure, while hydrophilic residues tend to be on the outside. This hydrophobic core helps minimize exposure of

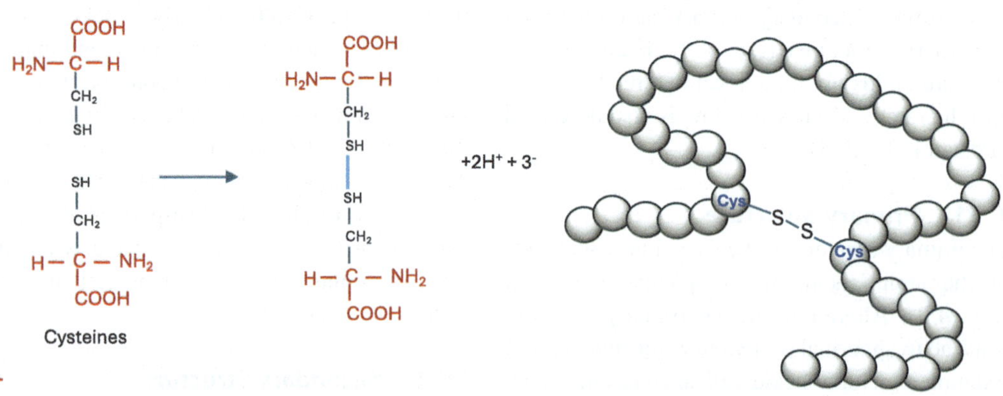

Fig. 4.10 Disulfide bonds. Disulfide bonds can occur between cysteine residues in proteins

nonpolar groups to water, stabilizing the protein. There are multiple types of bonds in tertiary structures. One important bond is the disulfide bond. Disulfide bonds are a type of covalent bond that can link amino acids, specifically cysteines, that are not necessarily adjacent to each other. Cysteine residues have a thiol (–SH) group on their side chain. When disulfide bonds form between two cysteines, the cysteines lose the hydrogen atom (H) from the thiol group (an oxidation reaction), and the two remaining sulfur (S) atoms form covalent bonds with each other. This disulfide bond helps stabilize the protein (Fig. 4.10). Other important bonds in the tertiary structure are hydrophobic interactions, ionic bonds, hydrogen bonds, and Van der Waals Interactions.

1.5.4 Quaternary Structure

Finally, some proteins consist of multiple polypeptides that combine to form one protein, often referred to as a protein complex. This is the quaternary structure (Fig. 4.9). In these protein complexes, each polypeptide chain is called a subunit. The simplest type of quaternary structure is the dimer, which consists of two subunits. A dimer may be a homodimer, where both subunits are identical, or a heterodimer, where they are both different. Quaternary structures can also be larger. A good example is Hemoglobin, which is made of four polypeptides. Other examples include DNA polymerase and ATP synthase. The bonds found in quaternary structures include hydrogen bonds, ionic bonds, disulfide bonds, and hydrophobic interactions.

Finally, it is important to remember that protein structure is not rigid, but dynamic. Proteins can fold and unfold and can sometimes convert to different conformations.

1.6 Domains of Proteins

Large proteins, even if they do not have separate subunits, are often made up of different domains. A domain is a modular unit within the rest of the protein, and it folds independently. Domains of proteins may have different functions. For

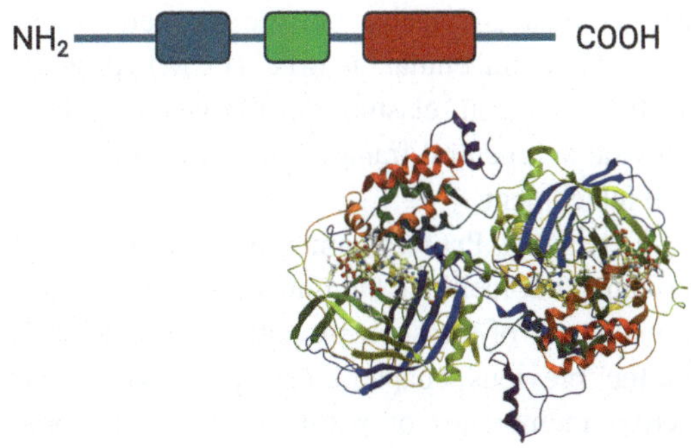

Fig. 4.11 Protein domains. Proteins can have multiple domains that can have specific functions, illustrated in the top panel in different colors. A folded protein is modeled below. (catalase). (*Source*: Protein structure from Freepik.com (premium license))

example, one domain may have catalytic activity if it is an enzyme, another domain may have membrane-spanning sequences, and another may have a protein interaction domain that facilitates an interaction with another protein. A protein may have one domain, or it may have multiple domains, all connected via mostly unstructured polypeptide sequences [9]. Protein domains can all be from a single polypeptide, or they can be from different polypeptides in the quaternary structure of the protein. Proteins containing their various structural elements are often depicted as 'ribbon diagrams' where the different structures can be identified visually (Fig. 4.11).

2 Proteins Have a Variety of Functions

Proteins have many different functions, and together with other biomolecules, they carry out nearly all the cell's activities. Below are some examples of protein categories and their different roles, but keep in mind that there is often considerable overlap between these categories. These protein functions are also described in [1, 5, 9, 10].

Structural Proteins provide support and help maintain the shapes of cells and tissues. Some structural proteins are in the cell, such as the cytoskeletal proteins actin and tubulin. Others are secreted and reside outside the cell. These may be extracellular matrix (ECM) proteins such as collagen, elastin, and fibronectin, which provide a structural framework and a surface for cell attachment.

Transport Proteins are a broad category of proteins that facilitate the transport of molecules. Some types of transport proteins were described in the previous chapter. Transport may occur across membranes or within cells, or in some cases throughout the body. Transport proteins, including ion channels, carrier proteins, and pumps, facilitate the transport of molecules across membranes. They also include proteins involved in vesicle transport as well as motor proteins, which use energy to move molecules or organelles along cytoskeletal tracks (mainly microtubules) within the cell. Proteins that transport molecules throughout the body can also be considered transport proteins. Examples include transferrin, which transports iron in the bloodstream, and hemoglobin, which transports oxygen. Defects in these proteins can lead to disorders such as iron deficiency, anemia, or sickle cell disease.

Transmembrane Proteins span both leaflets of cell membranes and play roles in cell signaling, adhesion, and transport. Examples include cell surface receptors such as receptor tyrosine kinases (RTKs), ion channels, and integrins, which link cytoplasmic proteins with the ECM. Many transmembrane proteins also serve as transport proteins, such as channels and pumps.

Signal Transduction Proteins mediate responses to signals and regulate cell communication. Many types of proteins fall into this category. Examples include cell surface receptors and extracellular ligands that bind to receptors, as well as proteins inside the cell that mediate a response to a stimulus. Ligands may include small peptides or proteins such as insulin, cytokines, and growth factors. Cell surface receptors include the EGF receptor and the TNF receptor. Cytoplasmic and nuclear proteins have roles in signaling include protein kinases, phosphatases, G-proteins, and others. Many drugs target signaling proteins. For example, therapeutic monoclonal antibodies can target TNF and control pathways involved in inflammation.

Motor Proteins generate mechanical force and movement within cells. They can do so by interacting with cytoskeletal elements and have especially important roles in muscle. Examples include myosin, kinesin, and dynein. Many proteins in this category also overlap with transport proteins and have roles in vesicle transport.

Immune Proteins are involved in defense. They defend the body against foreign invaders. Examples include antibodies, which bind to foreign antigens, cytokines, which can mount an inflammatory response, major histocompatibility complex (MHC) proteins, which are involved in the recognition of self-versus non-self, and components of the complement system, which is a part of the immune system that helps eliminate pathogens.

Chaperones are proteins that assist in the folding and assembly of other proteins. A classic example is the heat shock protein family. Chaperones ensure proper folding and prevent protein aggregation. Misfolding and aggregation of proteins are associated with disorders such as Parkinson's disease and Alzheimer's disease.

Storage Proteins store certain molecules, such as amino acids, ions, or metal ions. Examples include ferritin and albumin.

Enzymes are catalysts that accelerate chemical reactions in cells. The enzyme category can overlap with nearly any of the above categories. They are involved in almost every aspect of cell function, including metabolism, intracellular signaling, DNA synthesis, transcription, and protein synthesis. Some examples are DNA polymerase, which catalyzes DNA synthesis, and various kinases and proteases, which regulate the phosphorylation of proteins or lipids. Enzymes are especially critical for biochemistry and pharmacy, and many drugs target enzymes. The next chapter, therefore, is dedicated to the discussion of enzymes.

3 Proteins Can Interact with Other Proteins or Molecules

3.1 Introduction to Protein Interactions

Rather than functioning alone, proteins usually act together with other proteins or other molecules within the cell, often via direct interactions [9, 10], which will be discussed in this section. These interactions can be critical to protein function. Some interactions are transient, lasting only a short time, while others are more stable. There are many examples of proteins that bind to other proteins or molecules. For example, antibodies bind to invaders such as viruses and bacteria, marking them for destruction. Tubulin units bind to other tubulin units to form microtubules, and enzymes bind to their substrates to catalyze various reactions [9].

When two or more proteins bind, they are referred to as binding partners or interactors. The region of a protein that binds to another protein is called the protein binding site. The interactions between proteins can depend on conformational shapes that allow the proteins to fit together. Various chemical bonds mediate the binding of a protein to its binding partner. These can be covalent bonds or noncovalent interactions, such as hydrogen bonds, van der Waals interactions, and electrostatic attractions.

In addition to interacting with other proteins, proteins can interact with small molecules (such as ions, metabolites, and hormones), nucleic acids, lipids, carbohydrates, metal ions (such as iron, zinc, magnesium, and copper), vitamins and cofactors, and gases. When bound to proteins, these molecules are commonly referred to as **ligands** (although in some cases, proteins themselves, such as growth factors, can also be considered ligands), and the region of the protein that binds the ligand is referred to as the ligand binding site. The ligand binding site often comprises a cavity, into which the ligand can fit. Understanding these mechanisms is crucial for rational drug design, since most drugs act as ligands that modulate protein function.

3.2 Some Examples of Proteins That Interact with Other Proteins or Other Molecules

3.2.1 Growth Factors and Growth Factor Receptors

The interaction between a cell surface receptor and its ligand is a good example of a protein–protein interaction. For example, epidermal growth factor receptor (EGFR) is a growth factor receptor. It spans the cell membrane, and on the extracellular side, it binds to growth factors such as epidermal growth factor (EGF). EGF is a small peptide. When EGF binds to the EGFR, it causes two EGFRs to bind to each other, forming a dimer. Thus, interactions occur between the EGF and the EGFR, as well as between the two EGFRs. This interaction triggers a cascade of events within the cell, culminating in a response to the growth factor signal. The ligand–receptor interaction is transient and involves various types of noncovalent bonds, including hydrogen bonds, ionic bonds, and others. The dimerization between two EGFRs is also a transient interaction and involves noncovalent bonds [11, 12].

3.2.2 Hormone Receptors and Hormone

Hormone receptors, such as the estrogen receptor and the androgen receptor, also bind to their respective ligands. However, in this case, their ligands are not proteins but steroid hormones, which are lipids. These receptors are not transmembrane proteins; instead, they are inside the cell. Their ligands, such as estrogen or testosterone, diffuse through the phospholipid bilayer and bind to the receptors inside the cell. The estrogen receptor is found in tissues such as the mammary glands and the endometrium. When estrogen binds, the receptors form dimers, resulting in transcriptional changes within the cell. Signaling by estrogens has important roles in the development of the female reproductive system and secondary sex characteristics [13]. Testosterone functions similarly and binds to the androgen receptor. It is associated with the male reproductive system and secondary sex characteristics. Testosterone is present in both males and females, but females have fewer androgen receptors than males. Testosterone can also bind to the estrogen receptor, but this interaction is weak [14].

3.2.3 Antibodies and Antigen

Antibodies bind to an antigen (which is generally a foreign substance). This interaction is an essential part of the immune system. Antibodies, also called immunoglobulins, are Y-shaped glycoproteins (proteins with a carbohydrate attached), and they bind to an antigen via a specific region of the antibody called the variable region. This interaction is highly specific and involves multiple noncovalent interactions [15]. Even a single amino acid change in an antigen can strongly reduce antibody binding affinity, illustrating the precision of immune recognition [16, 17] (see Chap. 24).

3.2.4 Hemoglobin and Myoglobin

Hemoglobin binds oxygen and transports it throughout the bloodstream. Adult hemoglobin is made up of four proteins, two alpha chains and two beta chains. These are the four subunits of hemoglobin. Each subunit contains a heme group, which is a nonprotein component. The heme contains iron, and the iron binds to O_2. The protein subunits of hemoglobin form interactions with the heme group. Hemoglobin also consists of protein-protein interactions, as the four polypeptide chains bind to each other. Myoglobin, found in muscle cells, also has a heme group that binds to oxygen and iron. Myoglobin has a different type of structure, though, as it consists of only one subunit [18–20].

4 Understanding Proteins Is Important for the Study of Pharmacy

Understanding how proteins function is essential for pharmacy because proteins have various functions that are directly relevant to human health and drug development. For example, enzymes, in addition to being frequent targets of drugs, can also be involved in drug metabolism. This dual role means that enzymes are not only drug targets but also determine how drugs are activated, inactivated, or cleared from the body. **Cytochrome P450 (Cyp450)** is an example of a family of enzymes found in the liver and some other organs. They have essential roles in **metabolizing drugs** into a form that is more easily secreted (detoxification). Their activity can also be affected by drugs, however. Some drugs can activate the Cyp450 enzymes, while others can inhibit them. It is, therefore, vital to understand how drugs and drug combinations impact Cyp450 activity. This is especially important when prescribing combinations of drugs to patients.

Many proteins, such as some signaling proteins, are effective drug targets [21–23]. G protein-coupled receptors (GPCRs), for example, are cell surface receptors that mediate responses to extracellular stimuli. **Beta-blockers** are a family of drugs that bind to beta-adrenergic receptors, which are GPCRs. These receptors mediate responses to hormones, mainly epinephrine and norepinephrine. By binding to the receptors, beta blockers reduce their signaling. They are used to treat **hypertension, angina, and other conditions**. They are discussed in more detail in Chap. 18.

Proteins with roles in defense and protection are also important for pharmacy and medicine [11]. **Antibodies** are proteins that play a crucial role in immune responses. Antibodies can also be produced in the lab using biotechnology techniques and used to treat human diseases [10, 15]. This is usually done by injecting mice or rabbits with an antigen and triggering them to make specific antibodies, which can, in turn, sometimes be used as drugs. These are usually monoclonal antibodies (mAbs; discussed in more detail in Chap. 24). mAbs come from a single clone of B cells and recognize a single epitope of an antigen. Monoclonal antibodies used therapeutically fall into the category of biologics, and many promising drugs are composed of them. For example, **Humira (adalimumab)** [24] is **a monoclonal antibody** (mAb) that binds to the cytokine TNF, preventing it from binding to its receptor. It is used to treat **inflammatory disorders such as arthritis** [25]. **Xolair (omalizumab)** [26] binds to IgE and is used to treat **asthma and allergies**. IgE is also an antibody, and it is involved in allergic reactions by triggering the release of histamine and other chemicals from immune cells. Inhibiting IgE can reduce the symptoms of allergies. **Herceptin (trastuzumab)** [27] binds to the cell surface receptor HER2. HER2 is a cell surface receptor that is overexpressed in some types of breast cancer, specifically HER2-positive breast cancer. Herceptin is used to treat **HER2-positive breast cancer** [21].

Endogenous antibodies, those we make ourselves, can also be manipulated by some pharmaceutical treatments. Endogenous antibody production can be stimulated as a way to prevent or treat disease by using vaccines. **Vaccines** are designed to prime the body to mount an immune response against infectious diseases. They can lead to activation of immune cells and the production of antibodies against a specific invader, such as a virus [21]. Vaccines are also discussed in Chap. 24.

5 Equilibrium

Proteins often bind to other proteins or molecules, but the strength of their binding can vary. Sometimes, proteins bind tightly to their interacting partners or ligands, and the interaction may last a long time or be relatively permanent. Other times, the interactions are weak and transient. These various binding properties can be defined using terms such as equilibrium, the equilibrium constant, and the dissociation constant.

Equilibrium refers to the state in which binding (association) and unbinding (dissociation) are balanced. For two molecules that interact with each other (molecule *A* and molecule *B*), we can express them as:

$$A + B \rightleftharpoons AB$$

where *A* and *B* are the individual molecules (the reactants), and *AB* represents the interaction or complex between the two (the product).

In solution or in a cell, *A* and *B* can interact with each other, and over time, the *AB* interaction (association) increases. Also, over time, however, some *AB* complexes come apart (dissociation), producing *A* + *B*. Over time, the system reaches equilibrium, where the rate of association (the formation of *AB*) equals the rate of dissociation (the breakdown of *AB*). At this point, there is no net change in the concentrations of *A*, *B*, and *AB* over time, even though the reactions continue. These concepts, also discussed in [1, 5] are important for understanding the equilibrium constant and dissociation constant.

5.1 The Equilibrium Constant (K)

At equilibrium, the forward and reverse reactions happen at the same rate, so the concentrations of *A*, *B*, and *AB* (described above) remain constant. In this type of binding reaction, the equilibrium constant is commonly referred to as the association constant (Ka), and it describes how tightly two molecules bind. It is defined by the ratio of product concentrations to reactant concentrations when the reaction is at equilibrium. It is useful for understanding how tightly molecules interact [1, 28, 29]. *K* can be expressed as follows:

$$\mathrm{Ka} = \frac{[AB]}{[A][B]}$$

Where:

[*AB*] is the concentration of the complex at equilibrium.

[*A*] and [*B*] are the concentrations of the unbound proteins *A* and *B* at equilibrium (*K* can also be referred to as *K*eq).

This ratio indicates how likely the complex [*AB*] is to form compared to the unbound proteins [*A*] and [*B*]. A large Ka means [*AB*] is favored. In other words, most of *A* and *B* are bound together. A small Ka means the unbound forms [*A*] and [*B*] are favored. This indicates that the two molecules do not bind each other strongly. (In this example, the units of *K* are 1/molar or (M^{-1}), and the Kd, discussed below, has units of molar (M)).

5.2 The Dissociation Constant

In addition to the association constant (Ka) discussed above, another term that is especially helpful is the dissociation constant (Kd). The Kd is the reciprocal of Ka. Kd can be expressed as follows:

$$\mathrm{Kd} = \frac{1}{K} = \frac{[A][B]}{[AB]}$$

Kd can give us useful information. The Kd can be used to determine the concentration of free *B* at which half of the *A* molecules are bound to *B*. A lower Kd means tighter binding, where it takes less *B* to occupy half of *A*.

The reaction described above is a relatively simple 1:1 interaction, where one protein binds to one other protein. Some interactions, however, can be more complex. For example, a single protein might have multiple sites for a binding partner, or a protein may even have more than one type of binding partner. In more complex situations, a reaction may even involve cooperative binding where one binding site affects another binding site.

In these cases, the form of the equilibrium constant changes, and the units also change to match the reaction stoichiometry (the ratio and number of molecules involved in the interaction). For example, in a situation where *A* binds to two molecules of *B*, the following equation and reaction would apply:

$$A + 2B \rightleftharpoons AB_2$$

$$K = \frac{[AB_2]}{[A][B]^2}.$$

FDA-Approved Drugs Mentioned in This Chapter

Commonly used brand name	Generic name	First FDA approval year	Common conditions	Main targets
Herceptin	*Trastuzumab*	1998	Her2 positive breast cancer	Her2
Humira	*Adalimumab*	2002	Arthritis and other inflammatory disorders	TNFα
Xolair	*Omalizumab*	2003	Asthma and allergies	IgE

Definitions

Amino acids The basic building blocks of proteins. The basic structure of an amino acid (a free amino acid that is not in a chain) includes a central carbon (C), and around this carbon are an amino group, a hydrogen atom, a carboxyl group (COOH), and a side chain.

Chaperone proteins Proteins that assist in the folding and assembly of other proteins. A classic example is the heat shock family of proteins.

Disulfide bond A covalent bond formed between two cysteine residues, helping stabilize protein structure.

Enzyme A biological catalyst, usually a protein that lowers the activation energy of the reaction, thereby accelerating the reaction.

Hydrophobic interactions The tendency of nonpolar (hydrophobic) groups to cluster together away from an aqueous environment.

Peptide bond The type of covalent bond that links amino acids together. The bond links the carboxyl group of one amino acid to the amino group of another,

Primary structure The linear sequence of amino acids in a polypeptide chain.

Quaternary structure The arrangement of multiple polypeptide chains (subunits) into a protein complex.

Secondary structure Short stretches of amino acids that are not necessarily adjacent to each other interact with each other. Includes structures such as α-helices and β-sheets, they contain hydrogen bonds.

Tertiary structure The full 3D shape of a polypolypeptide. These structures contain globular regions and many kinds of bonds

Study Questions

Multiple-Choice Questions

1. Which level of protein structure is primarily responsible for the overall 3D shape of a single polypeptide chain?
   A. Primary
   B. Secondary
   C. Tertiary
   D. Quaternary
2. Which of the following amino acids can form disulfide bonds?
   A. Methionine
   B. Tyrosine
   C. Cysteine
   D. Serine
   E. C and D
3. What is the correct direction in which protein sequences are written?
   A. From the C-terminus to the N-terminus.
   B. Random order.
   C. From the N-terminus to the C-terminus.
   D. It varies by protein type.
4. Which of the following protein types is a transmembrane receptor?
   A. Hemoglobin
   B. Estrogen receptor
   C. EGFR
   D. Actin
   E. Both B and C
5. Beta-blockers generally targeting which type of protein?
   A. Glucocorticoid receptors
   B. Receptor tyrosine kinases
   C. SNAP25
   D. G protein-coupled receptors (GPCRs)
   E. C and D
6. Which level of protein structure involves interactions between multiple polypeptide chains?
   A. Primary structure
   B. Secondary structure
   C. Quaternary structure
   D. Alpha helix
   E. C and D
7. Which of the following is not a component of an amino acid?
   A. Amino group
   B. Carboxyl group
   C. Phosphate group
   D. R group

Fill-in-the-Blank Questions

8. The primary structure of a protein is determined by the specific sequence of__________.
9. The family of proteins that assist in the folding and assembly of other proteins are the __________ proteins.
10. A bond between the carboxyl group of one amino acid and the amino group of another is called a __________ bond.
11. In a beta sheet, the type of bonds that can form between distant amino acids are __________ bonds.
12. Amino acids are made of a central carbon, an amino group, a side chain, a hydrogen, and a _________.

True/False Question

13. Transmembrane proteins are always transport proteins.

Short-Answer Questions

14. **Case Study**
    A patient with cystic fibrosis has a mutation causing the deletion of phenylalanine at position 508 in CFTR. What is the clinical significance of this mutation?

15. Describe how alpha helices and beta sheets differ and how they are similar.
16. Why are monoclonal antibodies (mAbs) important in pharmacy?

Answer Key

Multiple-Choice Questions

1. **Answer**: C
2. **Answer**: C
3. **Answer**: C
4. **Answer**: C
5. **Answer**: D
6. **Answer**: C
7. **Answer**: C

Fill-in-the-Blank Questions

8. **Answer**: amino acids
9. **Answer**: chaperone
10. **Answer**: peptide
11. **Answer**: hydrogen
12. **Answer**: carboxyl group

True/False Question

13. **True or False**:
    Answer: False.

Short-Answer Questions

14. **Case Study**
    Answer:
    It affects protein folding and prevents proper transport of Cl^- across membranes, leading to thick mucus accumulation and CF symptoms.
15. **Answer**: They are similar in that they are both secondary structures that form in proteins. They are also similar in that they both are stabilized by hydrogen bonds. However, alpha helices form bonds every four residues ($i \rightarrow i + 4$) within a single chain, creating a coiled shape, while beta sheets form bonds between amino acids that can be farther apart.
16. **Answer**: They can be designed to target specific antigens. Drugs that consist of mAbs can inactivate a target or in some cases activate an immune response to lead to destruction of the target. In other cases, mAbs can act as agonists.

References

1. Berg JM, Tymoczko JL, Gatto GJ Jr, Stryer L. Biochemistry. 9th ed. New York: MacMillan Learning; 2019.
2. Fersht AR. Structure and mechanisms in protein science: a guide to enzyme catalysis and protein folding. Singapore: World Scientific Publishing Co.; 2017.
3. McKee T, McKee JR. Biochemistry: the molecular basis of life. 7th ed. New York: Oxford University Press; 2023.
4. Stollar EJ, Smith DP. Uncovering protein structure. Essays Biochem. 2020;64(4):649–80. https://doi.org/10.1042/ebc20190042.
5. Alberts B, Heald R, Johnson A, Morgan D, Raff M, Roberts K, Walter P, Wilson J, Hunt T. Molecular biology of the cell. New York: W.W. Norton and Company; 2022.
6. Pauling L. General chemistry. 3rd ed. Newburyport: Dover Publications; 2014.
7. Inusa BPD, Hsu LL, Kohli N, Patel A, Ominu-Evbota-K, Anie KA, et al. Sickle cell disease-genetics, pathophysiology, clinical presentation and treatment. Int J Neonatal Screen. 2019;5(2):20. https://doi.org/10.3390/ijns5020020.
8. Ong T, Ramsey BW. Cystic fibrosis: a review. JAMA. 2023;329(21):1859–71. https://doi.org/10.1001/jama.2023.8120.
9. Morris R, Black KA, Stollar EJ. Uncovering protein function: from classification to complexes. Essays Biochem. 2022;66(3):255–85. https://doi.org/10.1042/ebc20200108.
10. Alberts B, Bray D, Hopkin K, Johnson A, Lewis J, Raff M, Roberts K, Walter P. Essential cell biology. New York: Norton and Company; 2023.
11. Marks F, Klingmüller U, Müller-Decker K. Cellular signal processing: an introduction to the molecular mechanisms of signal transduction. 2nd ed. New York: Garland Science; 2017.
12. Westermarck J, Ivaska J, Corthals GL. Identification of protein interactions involved in cellular signaling. Mol Cell Proteomics. 2013;12(7):1752–63. https://doi.org/10.1074/mcp.R113.027771.

13. Fuentes N, Silveyra P. Estrogen receptor signaling mechanisms. Adv Protein Chem Struct Biol. 2019;116:135–70. https://doi.org/10.1016/bs.apcsb.2019.01.001.
14. Rochefort H, Garcia M. Androgen on the estrogen receptor. I – Binding and in vivo nuclear translocation. Steroids. 1976;28(4):549–60. https://doi.org/10.1016/0039-128x(76)90023-4.
15. Chiu ML, Goulet DR, Teplyakov A, Gilliland GL. Antibody structure and function: the basis for engineering therapeutics. Antibodies (Basel). 2019;8(4). https://doi.org/10.3390/antib8040055.
16. Oraby AK, Stojic A, Elawar F, Bilawchuk LM, McClelland RD, Erwin K, et al. A single amino acid mutation alters multiple neutralization epitopes in the respiratory syncytial virus fusion glycoprotein. NPJ Viruses. 2025;3(1):33. https://doi.org/10.1038/s44298-025-00119-8.
17. Alegre ML, Collins AM, Pulito VL, Brosius RA, Olson WC, Zivin RA, et al. Effect of a single amino acid mutation on the activating and immunosuppressive properties of a "humanized" OKT3 monoclonal antibody. J Immunol. 1992;148(11):3461–8.
18. Gell DA. Structure and function of haemoglobins. Blood Cell Mol Dis. 2018;70:13–42. https://doi.org/10.1016/j.bcmd.2017.10.006.
19. CDC. Sickle Cell Disease (SCD). Centers for Disease Control and Prevention. 2023. Available from: https://www.cdc.gov/sickle-cell/index.html.
20. Olivé M, Engvall M, Ravenscroft G, Cabrera-Serrano-M, Jiao H, Bortolotti CA, et al. Myoglobinopathy is an adult-onset autosomal dominant myopathy with characteristic sarcoplasmic inclusions. Nat Commun. 2019;10(1):1396. https://doi.org/10.1038/s41467-019-09111-2.
21. Katzung BG, Vanderah TW, Trevor AAJ. Katzung's basic and clinical pharmacology. 16th ed. New York: McGraw-Hill Education; 2024.
22. Waring RH. Cytochrome P450: genotype to phenotype. Xenobiotica. 2020;50(1):9–18. https://doi.org/10.1080/00498254.2019.1648911.
23. do Vale GT, Ceron CS, Gonzaga NA, Simplicio JA, Padovan JC. Three generations of β-blockers: history, class differences and clinical applicability. Curr Hypertens Rev. 2019;15(1):22–31. https://doi.org/10.2174/1573402114666180918102735.
24. FDA. HUMIRA (adalimumab) Approval Overview 1001. Available from: https://www.accessdata.fda.gov/scripts/cder/daf/index.cfm?event=overview.process&ApplNo=125057.
25. Horiuchi T, Mitoma H, Harashima S, Tsukamoto H, Shimoda T. Transmembrane TNF-alpha: structure, function and interaction with anti-TNF agents. Rheumatology (Oxford). 2010;49(7):1215–28. https://doi.org/10.1093/rheumatology/keq031.
26. FDA. Xolair (omalizumab) approval overview. 2003. Available from: https://www.accessdata.fda.gov/scripts/cder/daf/index.cfm?event=overview.process&ApplNo=103976.
27. FDA. HERCEPTIN (trastuzumab) Approval Overview. 1998. Available from: https://www.accessdata.fda.gov/scripts/cder/daf/index.cfm?event=overview.process&ApplNo=103792.
28. LibreTexts. 6:3 ligand binding: biology LibreTexts. Available from: https://bio.libretexts.org/Bookshelves/Biochemistry/Supplemental_Modules_%28Biochemistry%29/6._Lab_Notes_Part_2/6.3%3A_Ligand_binding?utm.
29. Pääkkönen J, Jänis J, Rouvinen J. Calculation and visualization of binding equilibria in protein studies. ACS Omega. 2022;7(12):10789–95. https://doi.org/10.1021/acsomega.2c00560.

Enzymes, their Catalytic Properties, and Enzymes as Drug Targets

5

Abstract

Enzymes are biological catalysts essential for nearly all cellular processes, including DNA replication, transcription, nutrient metabolism, and cell division. Enzymes are catalysts. They help speed up reactions that may otherwise not occur or occur inefficiently. Fundamental principles of enzyme kinetics are presented, including the Michaelis–Menten constant (K_m) and maximum reaction rate (V_{max}). Enzyme thermodynamics is also addressed, and the different strategies used for enzymes in catalysis are discussed. This chapter introduces the seven major categories of enzymes, along with several important examples. Enzyme regulation is explored, including the roles of cofactors and allosteric regulators. Many drugs function by inhibiting enzymes, and several are introduced here. Some drugs inhibit bacterial enzymes rather than mammalian enzymes, and these are frequently used to treat bacterial infections.

Keywords

Enzymes · Catalysts · Activation energy · Transition state · Enzyme inhibitors · Antibiotics

Abbreviations

ADP	Adenosine diphosphate
ATP	Adenosine triphosphate
FAD	Flavin adenine dinucleotide
$FADH_2$	Reduced flavin adenine dinucleotide
K_{cat}	Turnover number (catalytic rate constant)
K_m	Michaelis constant
LDH	Lactate dehydrogenase
NAD^+	Nicotinamide adenine dinucleotide (oxidized form)
NADH	Nicotinamide adenine dinucleotide (reduced form)
pH	Potential of hydrogen (measure of acidity/basicity)
v	Reaction velocity
V_{max}	Maximum reaction velocity

What You Will Learn

- Enzymes are catalysts, they help speed up a chemical reaction
- The basic components of an enzymatic reaction, including the enzyme, substrate, product, and active site
- Enzymes may contain regulatory sites
- The concept of activation energy

A. Minden, *Biochemistry, Cell and Molecular Biology for Pharmacists and Health Professionals*,
https://doi.org/10.1007/978-3-032-20529-2_5

- Basic concepts of enzyme kinetics, including V_{max} and K_m
- Thermodynamic principles that determine whether a reaction can proceed
- The seven major categories of enzymes
- The enzyme commission (EC) number for categorizing enzymes
- Different strategies for catalysis utilized by enzymes, such as covalent catalysis or acid–base catalysis
- Some important enzymes and enzyme families, such as proteases and protein kinases
- Enzymes often use cofactors and allosteric regulators
- Mechanisms by which enzymes may be regulated
- Different types of enzyme inhibitors: competitive, noncompetitive, and uncompetitive inhibitors
- Many drugs are enzyme inhibitors. Examples include statins (HMG-CoA reductase inhibitors), ACE inhibitors, MAO inhibitors, aspirin (a COX inhibitor), and certain anticoagulants
- Antibiotics often inhibit bacterial enzymes

1 Introduction to Enzymes

There are thousands of enzymes in eukaryotic cells. Enzymes are essential for all the reactions that occur in cells, including DNA replication, transcription, nutrient metabolism, and cell division, among others. Enzymes are also targets for many drugs, and some enzymes play a crucial role in the metabolism of the drugs we take. For these reasons, it is essential to understand how enzymes function before delving into any of the other basic concepts in biochemistry and cell biology.

1.1 Enzymes Are Catalysts

In biochemistry, an enzyme is a biomolecule that serves as a catalyst. A catalyst is a molecule that can speed up the rate of a reaction without being consumed in the reaction. In this way, the catalyst makes the reaction more efficient. In cells, catalysts can perform various functions, such as regulating the making or breaking of covalent bonds in chemical reactions. After a reaction is complete, the catalyst remains intact and unchanged and available to catalyze another reaction. A catalyst, by definition, lowers the activation energy of a reaction, thereby speeding it up. Enzymes are abundant and necessary in cells because they catalyze biochemical reactions essential for the cell's survival. An enzyme can catalyze a reaction that might be nearly impossible under physiological conditions, making it much more likely to occur without the enzyme. There are many enzymes in cells that catalyze a vast number of reactions. In general, one can identify an enzyme by its name, which usually (but not always) ends in the suffix "ase." The rest of the name usually refers to what the enzyme acts on. For example, lactase is an enzyme that breaks down lactose. This chapter covers basic mechanisms of enzyme action and how enzymes are regulated. This chapter draws on the following references, as well as references in the text, which also include further information about enzymes [1–10].

1.2 Basic Components and Functions of Enzymes and Substrates

1.2.1 Introduction to the Enzyme Reaction

- The Enzyme, Substrate, and Active Site

Every enzyme recognizes a specific substrate. The substrates bind to a specific region of the enzyme known as the active site. This is the site on the enzyme where the chemical reaction will occur, catalyzed by the enzyme. The enzyme and substrate temporarily form a complex, called the enzyme–substrate complex. The substrate and active site interact precisely due to favorable interactions, such as hydrogen bonds, hydrophobic interactions, ionic interactions, and van der Waals forces. Some enzymes may bind to more than one substrate. In this case, the active site can accommodate all the substrates, either together or sequentially [7, 9, 10].

- The Product

Once the substrate interacts with the active site of the enzyme, the reaction occurs. The end result is the formation of a product, which is then released from the enzyme. The enzyme remains unchanged and free to catalyze another reaction.

- The Reaction

The reaction that an enzyme catalyzes can be depicted by this equation or a variation of it:

$$E + S \leftrightharpoons ES \leftrightharpoons EP \rightarrow E + P$$

Where E is the enzyme, S is the substrate, ES is the enzyme/substrate complex, EP is the enzyme/product complex, and P is the product. The above equation shows that the enzyme (E) and substrate (S) interact to form the enzyme–substrate complex (ES). Within the enzyme's active site, the product (P) is made as a result of the enzymatic reaction. At this point, the enzyme product complex (EP) is formed (EP is often omitted from equations, but is included here for clarity). After the reaction, the enzyme (E) and product (P) separate. After the reaction is complete, the enzyme is left unchanged and can catalyze another reaction [5, 7, 11, 12].

In the reaction represented here, the substrate consists of one or more reactants in a chemical reaction, and the terms "reactant" and "substrate" are sometimes used interchangeably when discussing enzymatic reactions (the equation above typically represents one substrate and product, but the equation can be expanded to accommodate multiple substrates and/or products) (see Fig. 5.1).

- Lactase is a Classic Model of Enzyme Terminology

The enzyme terminology described above can be clearly illustrated with the enzyme **lactase**. **Lactase** is an enzyme that breaks down the sugar lactose. Lactose (found in milk) is composed of two sugars, glucose and galactose. Lactase works by catalyzing the cleavage of the bond between these two sugars. In this reaction, lactase is the enzyme (E), lactose is the substrate (S), which binds to the active site on **lactase**, and glucose and galactose are the products (P) [1, 7].

1.2.2 Models for Enzyme–Substrate Binding

Two key models have been used to describe enzyme–substrate binding: (1) The **lock and key model**. It suggests that the substrate must have a shape that matches the enzyme's active site. (2) **The induced fit model**. This is now the more widely accepted model for how most enzymes work. According to this model, the substrate does not always fit readily into the active site, but it can induce a conformational change in the enzyme, resulting in a strong interaction. Thus, the enzyme has some flexibility in terms of what substrates it can interact with [9, 10, 13].

1.2.3 Regulatory Sites

Many enzymes also have additional important sites besides the active site. For example, many enzymes have one or more regulatory sites. Various molecules or ions can bind to these regulatory sites, modulating the enzyme's activity. Some regulatory sites are allosteric sites. The binding of molecules to allosteric sites changes the

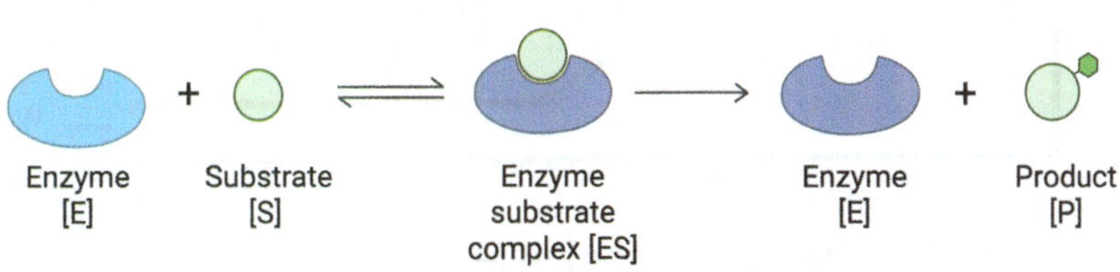

Fig. 5.1 Enzyme–substrate interactions. In a basic enzymatic reaction, the enzyme and substrate interact. The substrate is converted to product, and the enzyme remains unchanged

conformation of the enzyme. These changes can enhance or inhibit the enzyme's catalytic efficiency [4, 5, 7, 14].

2 Activation Energy

2.1 What Is Activation Energy

Activation energy (E_a) is an essential concept for understanding enzymes because enzymes function by lowering the activation energy of a reaction. Activation energy refers to the minimum amount of energy required for a reaction to occur. Activation energy can be thought of as the "effort" needed to initiate an event. Imagine you are riding an E-bike (electric-assist bike), but you are not using the electric option because you want to get a good workout. You are relying on manual pedaling. You approach a steep hill. You find it challenging to get up the hill. Your muscles begin to ache. The effort you are putting into climbing this hill represents the activation energy required to reach the top. The next time you go up the same hill, you engage the electric option on your e-bike, which provides a pedal assist function. The motor kicks in. You still need to pedal, but the effort required is reduced. This time, therefore, you get up the same hill much more easily and more quickly. Here, your e-bike's electric pedal assist option serves as a catalyst. It makes going up the hill easier because you use less energy. In other words, it lowers the activation energy [8, 11], but the height of the hill, as well as the starting and ending points of the ride, remain the same.

In cells, enzymes lower the activation energy in various ways. For example, when an enzyme causes two molecules to interact, it can function by bringing these two reactant molecules close to each other, allowing them to interact more readily. It can also act as a template to correctly orient the molecules and encourage interaction. An enzyme can also provide an environment that can stabilize charged or unstable intermediates, making an interaction more likely. Enzymes can also participate directly in a reaction by donating or accepting chemical groups to facilitate the formation or breakage of chemical bonds. However, by the end of the reaction, the enzyme must return to its original state. All of these strategies can enhance the efficiency and likelihood of the reaction [5, 7, 11, 12] (Fig. 5.2).

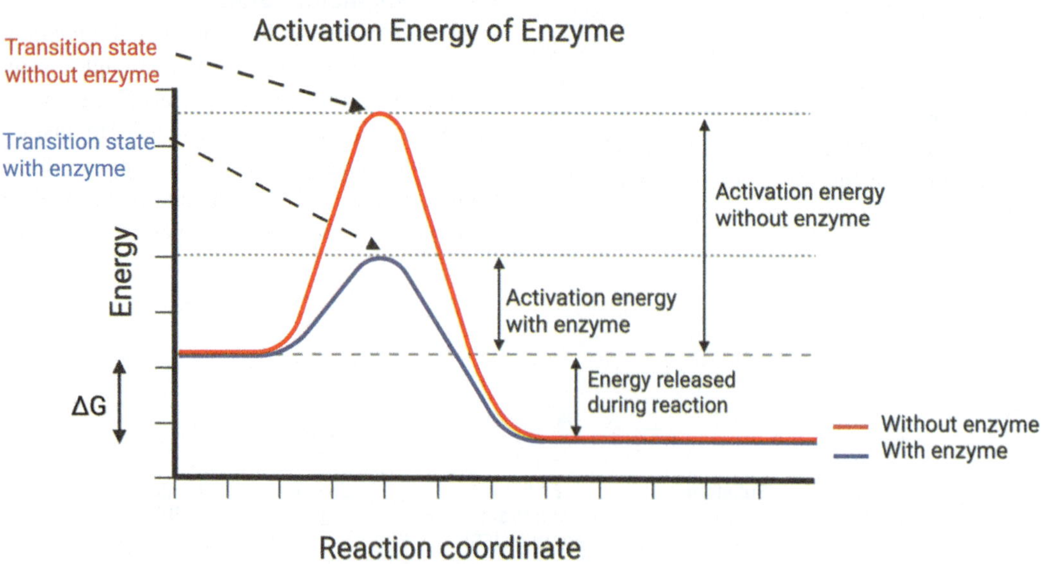

Fig. 5.2 Activation energy. The enzyme lowers the activation energy of a reaction. (*Source*: Created with Biorender.com)

2.2 The Transition State and Its Relationship to the Activation Energy

In any enzymatic reaction, a transition state (‡) takes place as reactants are converted to products, before the reaction is complete. The transition state is a short-lived, high-energy state in which bonds are partially formed or broken. It represents a point at which the reaction is most likely to proceed. The transition state is highly unstable, and a certain amount of free energy is required to reach this state [9]. The energy required is the activation energy for a reaction. This is where enzymes play such an important role. Enzymes help because they have a higher affinity for the transition state than for the stable, unreacted substrate (also called the ground form). Consequently, the enzyme binds tightly to the transition state, reducing the activation energy barrier and facilitating a faster reaction. It thus lowers the energy required for reaching the transition state [5, 7, 11, 12] (see Fig. 5.2).

3 Enzyme Kinetics and Thermodynamics Determine Whether a Reaction Can Occur

Enzyme kinetics is the study of the rates and efficiency of the reactions catalyzed by an enzyme and the factors that influence them. Enzyme thermodynamics deals with the direction and favorability of a reaction. Together, enzyme kinetics and thermodynamics provide an understanding of the speed, efficiency, and energetic favorability of a reaction. Information on enzyme kinetics is drawn from the following references, which also contain additional valuable information about the topic [1, 5, 7, 11, 12, 15–17].

3.1 Enzyme Kinetics

In this section, several important concepts will be introduced that are important for understanding Michaelis–Menten kinetics, a model that describes how enzyme-catalyzed reaction rates vary with substrate concentration. In the sections below, several important terms will be described, including V_{max}, the Michaelis–Menten plot, K_m (the Michaelis–Menten constant), v (reaction velocity), the Michaelis–Menten equation, the Lineweaver–Burk plot, K_{cat} (the turnover number), catalytic efficiency, and rate constants.

3.1.1 V_{max} and the Michaelis–Menten Plot

V_{max} (maximum velocity) represents the maximum rate at which an enzyme can catalyze a reaction when all of its active sites are saturated with substrate. V_{max} depends on both enzyme concentration and the enzyme's catalytic efficiency. The V_{max} can be represented on a Michaelis–Menten plot (Fig. 5.3). In this type of plot, the enzyme's reaction rate (reaction velocity) (V) is plotted on the Y-axis against the substrate concentration ($[S]$) on the X-axis. The plot represents how initially, as the substrate concentration increases, the reaction rate also increases. However, as the enzyme approaches saturation (when all of the enzyme's active sites are occupied by substrate), the reaction rate plateaus at V_{max}. This plateau can be seen on the Michaelis–Menten plot, indicating that the

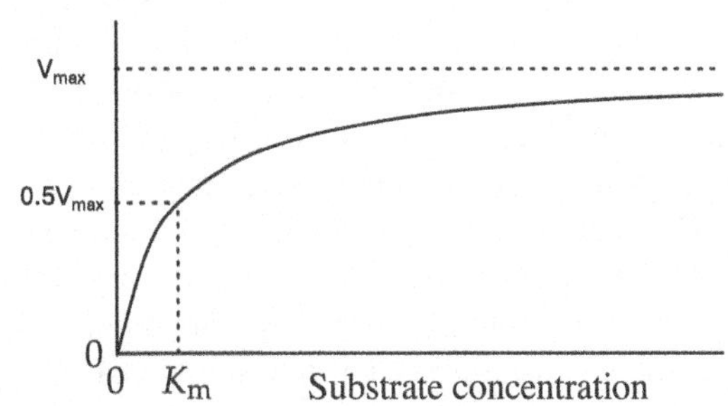

Fig. 5.3 Michaelis–Menten plot. The Michaelis–Menten plot shows the relationship between reaction velocity and substrate concentration. In the Michaelis–Menten plot, reaction velocity (v) is often expressed as moles of product per liter per second, and $[S]$ in moles of substrate per liter (molarity). From this type of plot, one can see how the reaction rate plateaus at a substrate concentration increases. At this point, the enzyme cannot work any faster and is working at its maximum rate. This is the V_{max}. (*Source*: Adapted from Vickers [116])

enzyme cannot catalyze the reaction any faster, even if more substrate is added.

In summary, the V_{max} can be understood by thinking of an enzyme and substrate together, where an increasing amount of substrate is added to the system. As more substrate is added, the increase in substrate concentration will raise the rate at which the enzyme produces the product. However, the enzyme will reach a limit where it cannot catalyze any faster. Therefore, the enzyme's rate eventually reaches its maximum even if more substrate is available. At that point, the enzyme is considered to be saturated with substrate; it cannot work any faster even if there is more substrate, and a plateau is reached. The rate at which the enzyme functions at this plateau is known as the V_{max}.

3.1.2 The K_m: The Michaelis–Menten Constant

The Michaelis–Menten plot can be used to determine another important number, the Michaelis–Menten Constant (K_m). The K_m represents the concentration of substrate when the reaction rate is half maximal (half the V_{max}, or $V_{max}/2$). The K_m can be determined from the Michaelis–Menten plot by identifying the substrate concentration on the *x*-axis that corresponds to half of V_{max} on the *y*-axis. The K_m also provides an indication of the enzyme's affinity for its substrate. A low K_m means the enzyme binds strongly to its substrate; thus, the enzyme has a higher substrate affinity (because less substrate is needed to reach half-maximum velocity due to this strong binding). A high K_m, in contrast, corresponds to weak substrate binding. K_m values vary widely for different enzymes [4, 11, 17]. K_m is useful for comparing enzymes or understanding how environmental conditions can affect an enzyme's properties.

3.1.3 V and the Michaelis–Menten Equation

The relationship between the reaction rate catalyzed by the enzyme and the substrate concentration can be described by the Michaelis–Menten equation (5.1):

$$v = \frac{V_{max}[S]}{K_m + [S]} \tag{5.1}$$

In the above Eq. (5.1), v is the initial rate (also called initial velocity) of the reaction at a given substrate concentration $[S]$. V_{max}, as defined above, is the maximum reaction rate achieved when the enzyme is saturated with substrate. K_m, also defined above, is the Michaelis–Menten constant, representing the substrate concentration at which the reaction rate is half the V_{max}.

The Michaelis–Menten equation explains the saturation behavior of enzymes. At low substrate concentration $[S]$, the reaction rate $[v]$ increases nearly linearly with $[S]$ since many enzymes' active sites are available. At high substrate $[S]$ concentrations, the reaction rate approaches V_{max} as substrate concentration increases and the enzyme becomes saturated, meaning that most active sites are occupied.

3.1.4 The Lineweaver-Burk Plot

The Lineweaver–Burk plot is another way to express V_{max}. The plot linearizes the Michaelis-Menten equation, making it easier to estimate kinetics graphically. The Lineweaver–Burk plot is linear (whereas the Michaelis–Menten plot is hyperbolic).

The formula for the Lineweaver–Burk plot is below (see Eq. 5.2):

$$\frac{1}{v} = \frac{K_m}{V_{max}[S]} + \frac{1}{V_{max}} \tag{5.2}$$

This formula is derived from the reciprocal of the Michaelis–Menten equation. The Lineweaver–Burk plot is a double reciprocal plot. An example of a Lineweaver-Burk plot is shown in Fig. 5.4 below. In this graph, the *Y* intercept is $1/V_{max}$ (the reciprocal of V_{max}), and the *X* intercept is $-1/K_m$. The slope is K_m/V_{max}. From this plot, we can determine the V_{max} by looking at the *Y* intercept. The Lineweaver–Burk plot is often used to experimentally determine the value of V_{max} and K_m by measuring the reaction velocities of an enzyme at various substrate concentrations and plotting the results.

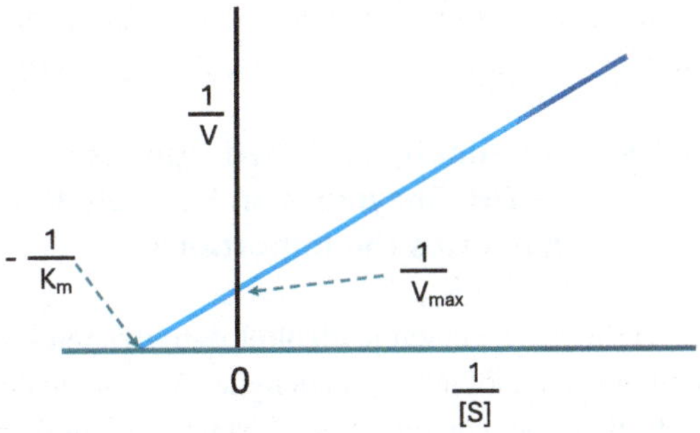

Fig. 5.4 Lineweaver-Burk plot. This plot shows the double-reciprocal relationship between *1/v* and $1/[S]$. It allows the determination of V_{max} and K_m. (*Source*: Created with Biorender.com)

3.1.5 K_{cat}: Turnover Number

Other measurements can also be calculated to tell us more about an enzyme. For example, the turnover number (K_{cat}) describes the catalytic rate of an enzyme. It represents the rate of substrate turnover. Specifically, it is expressed as the number of substrate molecules converted to product by one enzyme molecule per unit of time (in seconds) when the enzyme is saturated with substrate [12]. (The units for K_{cat} are 1/s). K_{cat} is represented as follows in the Eq. (5.3):

$$K_{cat} = \frac{V_{max}}{E_{total}} \tag{5.3}$$

Here, E_{total} is the total enzyme concentration, and as described above, V_{max} is the rate of the reaction when the enzyme is saturated with substrate. The higher the K_{cat}, the faster the enzyme can convert substrate molecules into products, and K_{cat} is usually expressed as s^{-1} (per second). For example, if the K_{cat} is 100, then the enzyme converts 100 molecules of substrate to product per second. K_{cat} can therefore be measured experimentally by knowing the V_{max} and the concentration of active enzyme. The higher the K_{cat}, the more substrates are converted into product per second.

More information can be obtained from the turnover number (K_{cat}), such as catalytic efficiency, which measures how effectively an enzyme converts a substrate into a product. Catalytic efficiency is measured as shown in Eq. (5.4):

$$\text{Catalytic efficiency} = \frac{K_{cat}}{K_m} \tag{5.4}$$

A high catalytic efficiency (K_{cat}/K_m ratio) indicates an enzyme that is both efficient at binding its substrate and at converting it to product.

Knowing the K_{cat} also provides another way to determine the reaction velocity (v) of a reaction because under saturating conditions, **v** is equal to the product of K_{cat} and the total enzyme concentration (E_{total}) (see Eq. 5.5).

Speed of a reaction at saturating substrate concentration:

$$v = K_{cat}[E_{total}] \tag{5.5}$$

3.1.6 Rate Constants

The reaction between an enzyme and its substrates can be broken down into steps, and **rate constants** describe the rate of each specific step. For example, see Eq. (5.6) [5, 18].

$$\text{Rate constants:} \quad E + S \underset{K_{-1}}{\overset{K_1}{\rightleftharpoons}} ES \underset{K_{-2}}{\overset{K_2}{\rightleftharpoons}} EP \overset{K_3}{\rightarrow} E + P \tag{5.6}$$

Where:

K_1 is the rate constant for the **binding** of the substrate (S) to the enzyme (E) to form ES.

K_{-1} is the rate constant for the **dissociation** of ES back to free E and S.

K_2 is the rate constant for the **conversion** of ES into EP (the term $[EP]$ is not always included in a typical equation, but is shown here for completeness).

K_{-2} is the rate constant for the conversion of EP to ES.

K_3 is the rate constant for the **release** of the final product (P) and regeneration of free enzyme (E).

The units of rate constants depend on the molecularity, which refers to how many molecules are involved in each step. This determines whether the step is a first-order or second-order reaction.

For example, for K_1, E, and S must come together, meaning that two molecular species (E and S) participate in the reaction step. This is thus a bimolecular reaction and is therefore classified as a **second-order** reaction. The rate depends on the concentration of both reactants and the rate constant and has units of $M^{-1}s^{-1}$. In this case, a rate constant of, for example, $10^6\ M^{-1}s^{-1}$ means that at a concentration of 1 M of both reactants (E and S), the reaction would proceed at a rate of 10^6 M/s.

For K_2, only one molecule must change (ES becomes EP). This is therefore a unimolecular reaction, which is a first-order reaction. In this case, the rate depends only on the concentration of ES, and the rate constant has units of s^{-1}. For example, a rate constant of 100 s^{-1} means that each molecule of ES will, on average, convert to EP at a rate of 100 events per second per molecule.

The rate constants are typically measured experimentally or, in some cases, through mathematical modeling studies. Rate constants can provide a way of understanding enzyme kinetics. For example. The K_m can be derived from the rate constants in the following reaction, see Eq. (5.7):

$$K_m = \frac{K_{-1} + K_2}{K_1} \tag{5.7}$$

3.1.7 Effects of Different Conditions on Enzyme Efficiency

Finally, it is important to note that the results obtained from the equations and calculations described above may differ under different conditions. In particular, the rates of enzyme-catalyzed reactions may increase with increasing temperature. This trend increases until the temperature is so high that the proteins are denatured, and catalytic activity consequently ceases.

3.1.8 The Aldehyde Dehydrogenase (ALDH) Enzyme Is an Example that Illustrates the Importance of the K_m

In the liver, the enzyme **alcohol dehydrogenase** converts ethanol into acetaldehyde. Acetaldehyde will then be converted to acetate by another enzyme, **aldehyde dehydrogenase (ALDH)** (Fig. 5.5).

Most people have two forms of **ALDH**: a **mitochondrial ALDH**, which has a low K_m, and a **cytoplasmic ALDH**, which has a high K_m (remember—a low K_m represents a strong ES interaction while a high K_m represents a weak interaction). Some people have a polymorphism in the gene encoding **mitochondrial ALDH**. A polymorphism is the occurrence of different forms of an allele at a gene locus. This particular polymorphism results in an amino acid substitution that makes the mitochondrial ALDH significantly less active, and in individuals who are homozygous for this polymorphism, it is nearly completely inactive. For this reason, in these individuals, acetaldehyde must be processed primarily by the **cytoplasmic ALDH**. However, since the **cytoplasmic ALDH** enzyme has a high K_m, it does not catalyze the reaction efficiently unless acetaldehyde concentrations are high. Consequently, some acetaldehyde does not get converted to acetate at all. This incomplete conversion of acetaldehyde to acetate leads to the accumulation of acetaldehyde in the blood. This leads to physiological effects such as flushing and discomfort. People with this polymorphism, therefore, have side effects when drinking alcohol and sometimes may avoid it [5, 16].

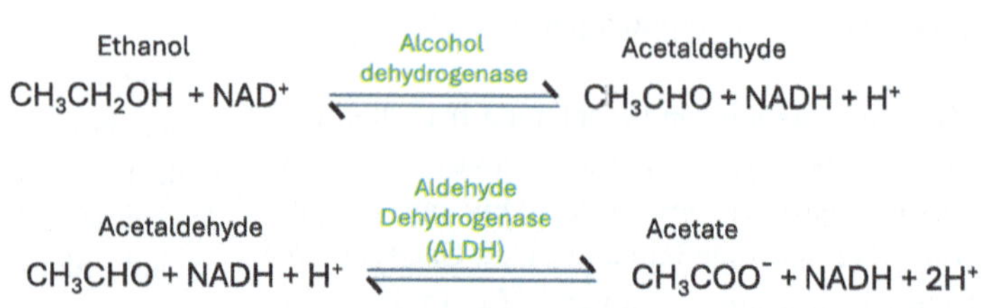

Fig. 5.5 Alcohol dehydrogenase. Chemical reactions catalyzed by alcohol dehydrogenase and aldehyde dehydrogenase

3.2 Enzyme Thermodynamics

Enzyme thermodynamics deals with the direction and favorability of a reaction as determined by Gibbs free energy (ΔG).

3.2.1 Gibbs Free Energy and ΔG

Gibbs free energy is a concept in thermodynamics that measures the maximum reversible work a system can perform under constant temperature and pressure. Gibbs free energy is denoted as G, and the change in Gibbs free energy as a system progresses from one state to another is denoted as ΔG. Every molecule has a certain free energy, which is determined by its chemical makeup, bonds, and structure. This free energy state can change. For example, Adenosine Triphosphate (ATP) has a high free energy state, but when it is hydrolyzed, and one phosphate group is removed, the resulting Adenosine Diphosphate (ADP) has a lower free energy state.

In an enzyme-catalyzed reaction, the free energy state of the reactants changes as they change to products. This change is the ΔG, which is equivalent to $G_{\text{products}} - G_{\text{reactants}}$ as in:

$$\Delta G = G_{\text{products}} - G_{\text{reactants}}$$

If $\Delta G < 0$, the reaction is spontaneous and thermodynamically favorable because the products have lower free energy than the reactants. This reaction is exergonic.

If $\Delta G = 0$, the reactants and products have the same free energy. This system is at equilibrium, and the forward and reverse reactions can continue at equal rates (regardless of the presence of the enzyme). This is called dynamic equilibrium because reactions may still occur in both directions, but there is no net change in the concentration of either reactants or product over time.

If $\Delta G > 0$, then the product has a higher free energy than the reactants. This reaction cannot occur spontaneously. Instead, energy is required for the reaction to proceed. This reaction is endergonic.

It should be noted that ΔG only refers to the free energy states of the products minus the free energy of the reactants. It has does not depend on the enzyme or its catalytic mechanism. Enzymes lower the activation energy required to reach the transition state but do not alter the Gibbs free energy of the reaction (see Fig. 5.2). An uncatalyzed reaction has the same ΔG as a catalyzed reaction. Finally, ΔG does not provide information about the rate of a reaction, only whether or not the reaction is thermodynamically favorable ($\Delta G < 0$) or unfavorable ($\Delta G > 0$) [5, 11].

3.2.2 Activation Energy and the Transition State in Thermodynamic Terms

Now, let us revisit the activation and the transition states to discuss how free energy applies. The transition state represents the highest free energy state during the reaction, and it is therefore the least stable structure in the reaction. The energy difference between the reactants (or substrate) and the transition state is called the Gibbs free energy of activation ($\Delta G‡$). This value represents the activation energy in thermodynamic terms. This energy barrier must be overcome for the reaction to proceed [5].

Enzymes function by lowering the Gibbs free energy of activation ($\Delta G‡$), thereby stabilizing the transition state and making it easier for the substrate to be converted into product. Thus, enzymes increase the reaction rate. It is important to note that although enzymes change the Gibbs free energy of activation ($\Delta G‡$), they do not change the overall Gibbs free energy change (ΔG) of the reaction or affect whether the reaction is thermodynamically favorable.

3.3 Bidirectionality of Enzymatic Reactions

Many enzymatic reactions are bidirectional. This means that the same enzyme can catalyze the reaction in both the forward and reverse directions (bidirectional reactions, sometimes called reversible reactions, are represented by bidirectional arrows). The direction in which a reaction proceeds at any given time depends on various factors, including the concentrations of substrates and products, as well as the cell's needs. Both the forward and reverse directions have an activation energy. In the forward direction, the activation energy is the energy needed to reach the transition state from the substrate(s), while in the reverse reaction, it is the energy needed to reach the transition state from the product(s). This bidirectionality or reversibility is especially important in metabolic pathways. When we discuss metabolism, we will discuss how the direction of an enzymatic reaction can change depending on the cell's metabolic needs and the availability of various metabolites.

4 There Are Seven Major Categories of Enzymes

Now that we have introduced how enzymes work, we will discuss the different types of enzymes in cells. There are thousands of enzymes in most cells, which can be classified into various categories. There were originally six categories, but a seventh category was added in recent years. The information from this section is drawn from the following references [4, 5, 11, 19–21] and additional references included below. The seven categories are below and also outlined in Table 5.1.

1. **Transferase**: Enzymes in this group transfer a functional group (which in this case can range from a small chemical group to a larger one like a peptide) from one molecule (the donor) to another (the acceptor) [4, 5, 11, 17, 22]. This can be represented by:

$$A + BX \rightarrow AX + B$$

   where A and B are molecules and X is a functional group.
   - Some Examples of Transferases
   Peptidyl transferase catalyzes the transfer of the amino acid from the tRNA to the growing amino acid chain (see Chap. 8) during translation.
   DNA polymerase catalyzes the transfer of nucleotide monophosphates to the 3′ end of a growing DNA chain during DNA replication, repair, and recombination.
   Kinases are a large family of enzymes. Kinases transfer a phosphate group onto their substrates; they phosphorylate them. Most kinases are highly specific and phosphorylate only certain substrates. Many kinases are protein kinases that phosphorylate proteins, but some can also phosphorylate other types of substrates, such as lipids.
2. **Ligases**: Ligases join two molecules together [19–21]. This requires energy and can be represented by:

$$A + B \rightarrow AB$$

   DNA ligase is an example of a ligase. DNA ligase links DNA fragments together. For example, during DNA replication, Okazaki fragments form on the lagging strand. **DNA ligase** connects these strands by catalyzing the formation of bonds between adjacent fragments.
3. **Oxidoreductases**: Enzymes in this group transfer electrons from one molecule to another.

Table 5.1 Seven main categories of enzymes

	Category	Mechanism	Example
1	Transferase	$A + BX \rightarrow AX + B$	Peptidyl transferase
2	Ligase	$A + B \rightarrow AB$	DNA ligase
3	Oxidoreductase	$A + B{:} \rightleftharpoons A{:} + B$	Lactate dehydrogenase
4	Isomerase	$A \rightarrow B$	Phosphoglucose isomerase
5	Hydrolase	$A + H_2O \rightarrow B + C$	Chymotrypsin
6	Lyase	$A \rightarrow B + C$	Arginosuccinate lyase
7	Translocase	$A \rightleftharpoons A$	Na+/K+ ATPase (also a hydrolase)

This can be represented as:

$$A + B :\rightleftharpoons A : +B$$

These enzymes are involved in redox reactions. They catalyze reactions in which one molecule is oxidized while the other is reduced. They can be more specifically named, such as oxidases, reductases, or dehydrogenases. An oxidase is involved in oxidizing, or removing electrons from, a molecule. A reductase is involved in reducing a molecule by giving electrons to it. Oxidoreductases can usually both oxidize and reduce their substrate [11].

- Examples of Enzymes That Are Oxidoreductases
  Lactate Dehydrogenase is an oxidoreductase. It catalyzes the following reaction that occurs during lactic acid fermentation:

pyruvate + NADH $\rightleftharpoons$ lactic acid + NAD+

  In this reaction, NADH is oxidized (it loses electrons), and pyruvate is reduced (it gains electrons). The enzyme transfers hydride (H^-; a hydrogen with two electrons) from NADH to pyruvate. In this process, the NADH is oxidized to NAD^+ and the pyruvate is reduced to lactate. This transfer of electrons is characteristic of redox reactions and a hallmark of dehydrogenase enzymes (which transfer hydrogen, in the form of a hydride (H^-) ion).
  Catalase is also an oxidoreductase. Catalase is a peroxisomal enzyme that breaks down hydrogen peroxide (H_2O_2) in the reaction $2H_2O_2 \rightarrow 2H_2O + O_2$. H_2O_2 is common in cells but is toxic, so it must be broken down quickly. This involves a reaction where H_2O_2 acts as both an electron donor and an electron acceptor. Catalase breaks down H_2O_2 at a very high rate, but without catalase, H_2O_2 would take years to break down.

4. **Isomerases**: Enzymes in this category catalyze changes in the spatial rearrangement of substrates. In other words, a molecule is converted into one of its isomers, and the bonds within a molecule may be rearranged, but the chemical formula remains the same. This reaction can be represented as:

$$A \rightarrow B$$

Phosphoglucose isomerase is an example of an isomerase. It catalyzes the reaction:

Glucose-6-P $\rightarrow$Fructose-6-P,

which is a step in glycolysis.

Glucose-6-P and Fructose-6-P are isomers (they have the same chemical formulas, but the atoms are arranged differently).

5. **Hydrolases**: This category of enzymes catalyzes the hydrolysis of chemical bonds, which causes them to break apart [5, 11, 23]. This means that they use H_2O to cleave a molecule into smaller parts, which is also called hydrolytic cleavage. The chemical reaction they carry out can be summarized like this:

$$A + H_2O \rightarrow B + C$$

- Examples of Enzymes That Are Hydrolases
  Proteases are a family of enzymes that break down proteins by breaking bonds between amino acids. Examples of proteases include pepsin, trypsin, and chymotrypsin. These enzymes are found in the digestive system and are important for digesting proteins that come from food.
  Nucleases are *hydrolases* that break down nucleic acids (DNA and RNA) by hydrolyzing the bonds between the nucleotides. Nucleases include endonucleases and exonucleases. Endonucleases cleave nucleic acids from within polynucleotide chains, and exonucleases cleave them from the ends of the chains.
  Lipases are also *hydrolases*. Lipases break down lipids. They are important in the small intestine, where they can break the ester bonds found in triacylglycerols

(TAGs) into their building blocks: fatty acids and glycerol [24].

Amylases are *hydrolases* that break down carbohydrates, specifically starches. An important amylase is **salivary amylase**, which breaks down starch in the mouth at the beginning of the digestion process [25].

Phosphatases catalyze the hydrolytic removal of phosphate groups from proteins or other molecules.

ATPases hydrolyze ATP, resulting in ADP and inorganic phosphate. This cleavage is a source of energy for many reactions. Myosin is an example of an ATPase, but many other ATPases also exist in cells.

GTPases hydrolyze GTP to form GDP; GTPases, such as Ras, are important in many signaling pathways and in the regulation of cell division [26].

6. **Lyase**: Enzymes in this category catalyze the cleavage of covalent bonds without the involvement of water or redox reactions. The reaction usually results in the formation of a double bond or ring structure. Many lyases catalyze reactions that can be summarized like this:

$$A \rightarrow B + C$$

Lyases often involve the removal of functional groups such as CO_2 or NH_3, leaving behind a double bond [27]. In the reverse direction, lyases can also catalyze addition reactions, where a molecule is added across a double bond.

- Examples of Lyases

  Argininosuccinate Lyase is an example of a lyase [5, 28]. It catalyzes the reaction:

  $$\text{Arginosuccinate} \rightarrow \text{arginine} + \text{succinate}$$

  This reaction takes place during the urea cycle.

  Carbonic Anhydrase is a lyase, though it does not fit perfectly with this or other categories. Carbonic anhydrase catalyzes the conversion of carbon dioxide (CO_2, which is an end result of aerobic metabolism) and water to bicarbonate (HCO_3^-) and protons in RBCs. This is a bidirectional reaction, and the enzyme can also dehydrate bicarbonate in the blood to form CO_2, which eventually gets exhaled. Carbonic anhydrase has an essential role in regulating the blood pH.

7. **Translocases**: This is the newest family. Some members of this family were previously assigned to different categories. Translocases consist of enzymes that couple a reaction with the movement of proteins or ions, usually across membranes [29].

5 Enzyme Commission (EC) Number for Categorizing Enzymes

Since so many enzymes exist, not only in humans but also in all life forms, scientists need a way to name them accurately. The enzyme commission (EC) number was developed by the International Union of Biochemistry and Molecular Biology (IUBMB) to establish a universal system, allowing all scientists to refer to enzymes in a consistent manner. It is a numerical classification system for enzymes based on the chemical reactions they catalyze. This system is based on the seven main categories of enzymes described above. The EC number consists of four parts, separated by periods. The first number refers to the main class, one of the seven major categories: (1. Oxidoreductases, 2. Transferases, 3. Hydrolases, 4. Lyases, 5. Isomerases, 6. Ligases, and 7. Translocases). The second and third digits represent the subclass and the sub–subclass. These can refer to different things, depending on the main class. For example, the subclass might refer to the type of compound or bond the enzyme acts on. The fourth digit is a serial number identifying the specific enzyme within that sub-subclass. The EC number format is: EC X.X.X.X [1, 20, 21, 30–32].

As an example, the EC number of alcohol dehydrogenase is EC 1.1.1.1. Here, EC1 stands for Oxidoreductase (the class), EC 1.1 refers to the fact that this enzyme acts on the CH–OH group of donors (this is the subclass), EC 1.1.1 refers to NAD+ or NADP+ acting as the acceptor (sub–subclass), and finally, EC 1.1.1.1 refers to the specific name, Alcohol dehydrogenase [32]. Every enzyme has its own unique EC number. This EC numbering system allows scientists to report and share results in a standardized and universally understood way. This system has many advantages. For example, it allows scientists to clearly identify the enzyme used in a published study, making it easier to interpret and replicate experiments in various systems.

6 Catalytic Strategies

There are different mechanisms by which the active site of an enzyme can interact with a substrate and catalyze the enzymatic reaction. These mechanisms are often referred to as **catalytic strategies**. Seven major strategies are discussed here. These mechanisms are not mutually exclusive; rather, enzymes can use combinations of the different mechanisms, see also the following resources for more information [5, 7, 14, 33, 34] as well as the additional references below.

6.1 Covalent Catalysis

In covalent catalysis, a temporary covalent bond forms between the enzyme and the substrate. This can occur when a **nucleophilic group** in the active site reacts with an electrophilic center in the substrate (An **electrophilic center** is an electron deficient atom, often a carbon atom, that is susceptible to attack by a nucleophile). The nucleophilic group is usually a functional group on an amino acid, such as the hydroxyl group on a serine. Other amino acids that may have a nucleophilic group include cysteine, lysine, and histidine. The covalent linkage forms a transient intermediate that modifies the substrate. This can be an important step in a cleavage reaction, leading to the substrate's cleavage.

Serine proteases such as **chymotrypsin** are examples of enzymes that use this type of catalytic mechanism. The serine residue in the active site of chymotrypsin (in the catalytic triad, discussed more below) forms a covalent bond with the substrate, facilitating its cleavage. This temporary covalent intermediate lowers the activation energy of the reaction [5].

6.2 Electrostatic Catalysis

In this type of catalysis, the enzyme stabilizes charged intermediates or transition states by providing a complementary electrostatic environment. This often involves charged amino acid side chains at the active site, such as lysine, arginine, or aspartate. These help stabilize positive or negative charges that develop during the reaction. For example, a negatively charged transition state may be stabilized by nearby positively charged residues in the active site [14, 35].

6.3 Catalysis by Approximation

In this model, two or more substrates may bind the active site of an enzyme in a reaction. This is also known as the proximity and orientation effects. Here, the enzyme brings substrates into close proximity with each other and into the proper orientation [36]. The enzyme may act as a sort of mold, with a configuration that orients and stabilizes the substrates, encouraging a reaction between them. An example is DNA polymerase, which positions the incoming nucleotide and the growing DNA strand in the proper orientation to facilitate formation of the phosphodiester bond. **carbonic anhydrase** also partially uses this mechanism. It brings together CO_2 and water to form bicarbonate (HCO_3^-) and a proton (H^+), or it aligns bicarbonate and a proton for the reverse reaction, although it also depends on metal ion catalysis [4, 37].

6.4 Acid–Base Catalysis

Acids donate protons (H^+) and bases accept protons. In acid–base catalysis, the enzyme functions as an acid or base by donating or accepting protons. This is due to the side chains of amino acids in their active sites. Histidine is an example of an amino acid that may accept a proton, thus acting as a base. On the other hand, aspartate or glutamate can function as acids and donate a proton. The result is the transfer of protons between the enzyme and the substrate, which can stabilize intermediates and promote the breakage of bonds in the substrate. This type of catalysis can be used in enzymes that cleave polypeptide bonds. Some examples of enzymes that use acid-base catalysis are **chymotrypsin, carbonic anhydrase**, and **myosin**. These enzymes also use other mechanisms. For example, chymotrypsin also uses covalent catalysis and catalysis by approximation.

6.5 Metal Ion Catalysis

Enzymes that use this type of catalysis have metal ions (such as Zn^{2+}, Mg^{2+}, Fe^{2+}/Fe^{3+}, and Mn^{2+}) in their active sites, and these metal ions have a role in the enzyme's function. For example, they can stabilize negatively charged substrates or intermediates, play roles in oxidation-reduction reactions, or position the substrate in the correct orientation. For example, magnesium ion (Mg^{2+}) is important for the catalytic activity of some restriction enzymes, such as **EcoRV**, and ATP-dependent motors like myosin require magnesium ions [14, 38].

6.6 Transition State Stabilization

This is also sometimes called preferential binding of the transition state. This refers to the fact that the enzyme generally works by binding the transition state more tightly than the substrate or product. This helps stabilize the transition state and lower the activation energy of the reaction, helping it to proceed.

6.7 Desolvation

Desolvation refers to removing the water molecules from the active site of an enzyme. In this strategy, the enzyme shields its substrate from water or polar molecules. This promotes interactions between the substrate and catalytic groups in the enzyme's active site (Desolvation in some textbooks is lumped together with other categories such as electrostatic catalysis or transition state stabilization) [39].

7 Lysozyme as a Classic Model for How an Enzyme Uses Various Mechanisms

Lysozyme is a member of the ***hydrolase*** family. It is an enzyme that catalyzes the cleavage of glycosidic bonds in the polysaccharides of bacterial cell walls, thereby leading to the destruction of the bacteria. It is found in secretions such as saliva and tears and provides protection from bacterial infection. Lysozyme specifically cleaves the $\beta(1\rightarrow4)$ linkage between N-acetylmuramic acid (NAM) and N-acetylglucosamine (NAG). These two sugars alternate in the bacterial peptidoglycan. The peptidoglycan is a structure based on polysaccharides (sugars, specifically NAG and NAM) and peptides, which form the bacterial cell wall. Gram-positive bacteria are especially susceptible to lysozyme because their cell walls consist of a thick, exposed peptidoglycan layer (see Fig. 5.6).

Polysaccharide molecules within the peptidoglycan have glycosidic bonds between individual sugars (NAG and NAM), which are kinetically stable and difficult to rapidly cleave without the aid of an enzyme [4]. **Lysozyme** solves this problem by stabilizing the transition state, thus lowering the activation energy for bond cleavage. The active site of lysozyme forms a cleft that positions six linked sugars (consisting of repeating units of NAM and NAG, often labeled sugars A–F), on a region of a polysaccharide. This conformation causes two adjacent sugars to be distorted. It is this distortion that stabilizes the transition state, resulting in the bond connecting them being bent into an unstable configuration that is more readily cleaved.

For the cleavage reaction, the enzyme employs a combination of two main catalytic strategies:

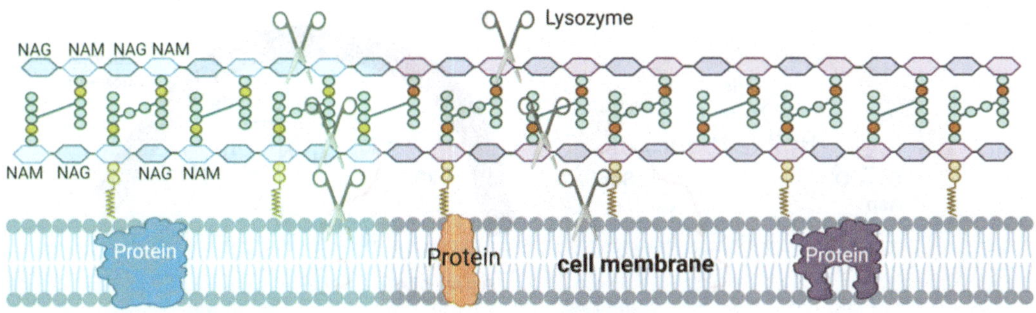

Fig. 5.6 Lysozyme. Lysozyme (indicated as scissors) cleaves β(1→4) glycosidic bonds between N-acetylmuramic acid (NAM) and N-acetylglucosamine (NAG) in the cell wall of bacteria, shown here external to the bacterial membrane. (*Source*: Reproduced from Miao et al. [117]. Licensed under the Creative Commons Attribution 4.0 (CC BY 4.0) license)

acid–base catalysis and covalent catalysis, to facilitate a hydrolysis reaction that cleaves glycosidic bonds between two sugars.

First, in acid–base catalysis, a glutamate (Glu35) in the active site of **lysozyme** is a proton donor. Glu35 remains protonated in the largely hydrophobic environment of the active site, allowing it to donate a proton to the oxygen in the glycosidic bond of the substrate (the polysaccharide). This is the first step in facilitating cleavage of the glycosidic bond.

At the same time, covalent catalysis takes place. Here, an aspartic acid (Asp52) in the active site has a key role. This ASP is usually in its deprotonated form and has a negatively charged carboxylate group (COO^-) on its side chain. It acts as a nucleophile (a species that is electron-rich and can donate electrons), interacting with one of the carbons on the sugar to form a covalent intermediate, which consists of the enzyme and substrate.

Next, hydrolysis takes place. Here, a water molecule, activated by Glu35, attacks the covalent bond between ASP52 and the polysaccharide, hydrolyzing the bond. This results in cleavage and release of the two cleaved sugar molecules. Because it combines multiple strategies, lysozyme is considered a classic model for understanding enzymatic catalysis [4, 11, 14].

8 Some Important Enzymes and Enzyme Families

There are far too many enzymes in cells to discuss all of them here. Here are a few examples of important enzymes and enzyme families. Enzymes employ multiple catalytic strategies, but the primary mechanisms of action for these enzymes are described below. The purpose of this chapter is to provide a foundational understanding of how these enzymes function at a molecular level. However, many of the enzymes discussed here have important roles in biochemistry and cell biology. For example, many of these enzymes have critical roles in signal transduction, DNA replication, and metabolism. Therefore, while this chapter focuses on catalytic mechanisms, the functional significance of many of these enzymes, as well as additional enzymes, will be addressed in more detail in later chapters.

8.1 Proteases

Proteases are a family of enzymes in the **hydrolase family**. Some proteases, especially serine proteases, rely on a catalytic triad. A catalytic triad is a group of three amino acids in the active site of an enzyme that act together to facilitate the reaction. Catalytic triads commonly consist of

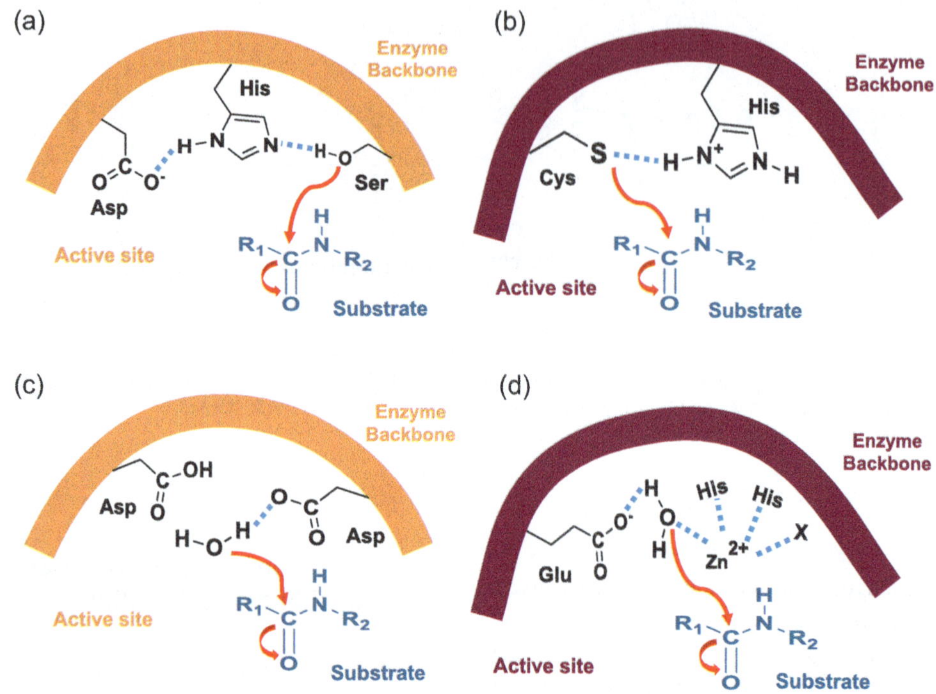

Fig. 5.7 Proteases. Proteases fall into four main categories. The active sites are shown here. (**a**) Serine protease, (**b**) cysteine protease, (**c**) aspartyl proteas, (**d**) metalloprotease. See text for details. (*Source*: Yiling Wang, used with permission)

Ser-His-Asp, although some other combinations are also possible. Some examples of serine proteases will be described below, and also see references [5, 40, 41].

8.1.1 Serine Proteases

Chymotrypsin is an example of a **serine protease**. It catalyzes the breakdown of proteins in the digestive system. It cleaves peptide bonds on the carboxyl-terminal sides of large hydrophobic residues (such as those containing phenylalanine, tryptophan, and tyrosine). Serine proteases get their name not because they cleave their substrates at serine residues, but because they use a serine residue in the catalytic triad in their active site to catalyze the cleavage of their substrates. In the case of **chymotrypsin**, the serine is located at residue 195.

- The Catalytic Triad in Serine Proteases
  The catalytic triad in **serine proteases** typically consists of serine (Ser), histidine (His), and aspartate (Asp) (see Fig. 5.7).
  - The serine acts as a nucleophile. It directly attacks the substrate to form a covalent intermediate (A nucleophile tends to donate electrons in the process of forming a bond).
  - The histidine acts as a general base (A general base accepts a proton from a nucleophile). It has the role of helping to activate the nucleophile (serine) by accepting a proton from it. This increases the nucleophile's (serine) reactivity.
  - The aspartate is an acid/base stabilizer. It will stabilize the charge on histidine, increasing its ability to act as a general base.
- How the Serine Protease Enzyme Works
  Cleavage works as follows:
  - The substrate to be cleaved binds to the enzyme's active site, at a region by the catalytic triad.
  - The histidine at the catalytic triad will act as a general base and accept a proton from the serine's hydroxyl group, increasing its reactivity.
  - The aspartate will stabilize this histidine.

- Next, the nucleophilic attack occurs. The serine in the catalytic triad, which functions as a nucleophile, attacks the carbonyl carbon of the peptide bond in the substrate peptide at the location that is to be cleaved. This results in the formation of a tetrahedral intermediate, a temporary covalent structure in which the carbon is bonded to four atoms in a specific 3D shape.
- Next, this transient tetrahedral intermediate will collapse, followed by cleavage of the targeted bond.
- One fragment of the target protein is released, and the rest remains bound to the enzyme.
- Finally, water enters the active site. Histidine deprotonates the water, and the resulting hydroxide ion (OH^-) attacks the intermediate that is still bound to the enzyme. (This is acid–base catalysis, where histidine attracts a proton to generate the nucleophilic hydroxide ion). This leads to the release of the second part of the product [5, 40, 41].

Note that this type of enzyme uses multiple catalytic strategies. The serine, which is a nucleophile, forms a covalent bond with the substrate peptide bond to form an intermediate. This is covalent catalysis. The histidine residue acts as a general base, representing acid-base catalysis. The aspartate residue stabilizes the histidine through electrostatic interactions, representing electrostatic catalysis. The serine protease stabilizes an unstable intermediate, representing transition-state stabilization. The enzyme binds the substrate in a way that the bond to be cleaved is aligned with the catalytic triad, which contributes to catalysis by approximation [5, 42].

8.1.2 Different Types of Proteases

Not all proteases are serine proteases. Some, for example, are **cysteine proteases**. Cysteine proteases rely on a Cys residue within their active site to serve as the nucleophile instead of serine.

Aspartyl proteases do not use a catalytic triad at all. Instead of a catalytic triad, they utilize a pair of Aspartate (Asp) residues in their active site that work together to enable a water molecule to attack the peptide bond in the substrate (see Fig. 5.7). In this system, one of the Asp residues is deprotonated and acts as a general base. It activates the bound water molecule by removing a proton. This generates a hydroxide ion (OH^-) from the bound water molecule. The other Asp residue is protonated and acts as a general acid. It helps stabilize the reaction intermediates. The activated OH^- attacks a carbonyl carbon of the peptide bond on the substrate, and as with the other proteases, an intermediate forms, and one part of the cleaved substrate is released. The other fragment is protonated by the other aspartate, and it is also released [43].

Metalloproteases also do not use a catalytic triad. They use a mechanism involving a bound metal ion (typically zinc) that activates a water molecule to act as a nucleophile, attacking the peptide carbonyl group of the peptide substrate [5, 44] (see Fig. 5.7).

8.1.3 Proteases Have Diverse Functions

Proteases have many different functions and have crucial roles in cells. Some proteases include **caspases** (which are cysteine proteases), which play key roles in programmed cell death (apoptosis), **thrombin** (a serine protease), which has an important role in blood coagulation, and **angiotensin converting enzyme (a metalloprotease)**, which has an important role in blood pressure regulation [45–47]. Proteases are found in the body, but they are also used in various industrial products. For example, they are included in detergents, laundry products, pest control products, and contact lens cleaners [48].

8.2 Restriction Enzymes

Restriction Enzymes (also called **restriction endonucleases**) are a group of enzymes that belong to the ***hydrolase family***. Restriction enzymes are good examples of endonucleases, which cleave nucleic acids by hydrolyzing the bonds between the nucleotides. Restriction enzymes are found in bacteria, where they destroy the DNA of bacteriophage (viruses that infect

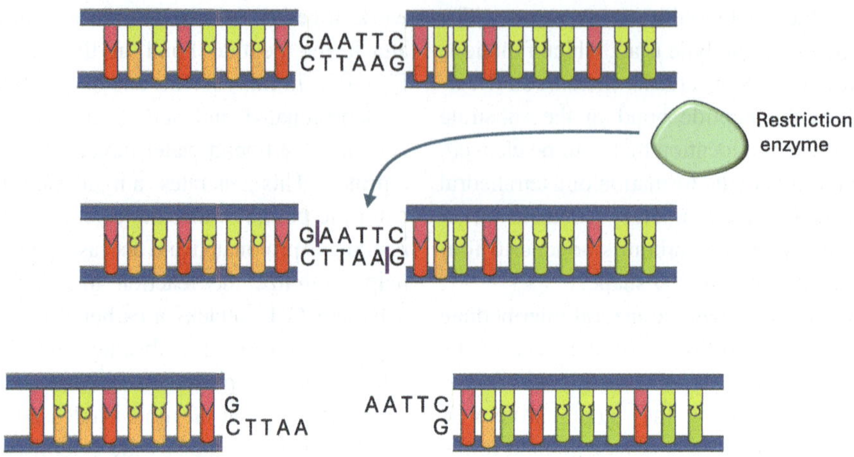

Fig. 5.8 Restriction enzymes. Restriction enzymes cleave DNA at a consensus sequence. Depending on their cleavage sites, they can leave sticky ends or blunt ends. Sticky ends are shown here. (*Source*: Adapted from Servier Medical Art (https://smart.servier.com/), licensed under the Creative Commons Attribution 4.0 (CC BY 4.0) license)

bacteria). They recognize certain DNA sequences (recognition sequences) within the bacteriophage DNA, and they cleave the DNA at these sites. In this way, restriction enzymes serve a protective role in bacteria [5, 11].

Restriction enzymes catalyze the hydrolysis of the bond between an oxygen and a phosphorus in DNA. They require Mg^{2+} to catalyze the reaction. Mg^{2+} forms a bridge between the enzyme and the DNA substrate. Mg^{2+} is essential for this reaction for multiple reasons. For example, it stabilizes the negatively charged transition state that results from cleavage of the phosphodiester bond. In bacteria, only the viral DNA is cut. The host cell DNA is protected because methyl groups at the recognition site prevent it from being cleaved. (The addition of these methyl groups to the DNA also involves an enzyme, **DNA methylase**) [5, 11, 49, 50].

While they have important functions in bacteria, **restriction enzymes** are valuable tools for molecular biology labs, where they are used to cleave DNA at specific sites (see Fig. 5.8). Genetic engineering techniques almost always involve the use of restriction enzymes, which can be used to make alterations in various genes and study their functions. Genetic engineering techniques involving restriction enzymes are also used to develop drugs or study human disease. This is due to their ability to precisely cut DNA at specific sequences, enabling scientists to manipulate genes and insert them into target vectors for protein production or preparation for research studies and sequencing analysis. For example, the human insulin gene has been inserted into a bacterial vector, allowing for the production of large amounts of insulin for patients. This was a significant improvement over using purified insulin extracted from animals, which was much less efficient. Similar methods have been used to produce growth hormone, albumin, monoclonal antibodies, and antigens used in vaccines [51, 52].

Restriction enzymes are also used in DNA profiling, where they are used to cleave DNA to observe the resulting banding patterns on a gel (corresponding to the sizes of DNA fragments). This method can be used to study genetic differences between different individuals, with the goal of developing personalized medicine treatments. This type of DNA profiling is also valuable for forensic science [5, 11, 49].

8.3 Myosin

Myosin is a molecular motor protein (It is an enzyme even though its name does not follow

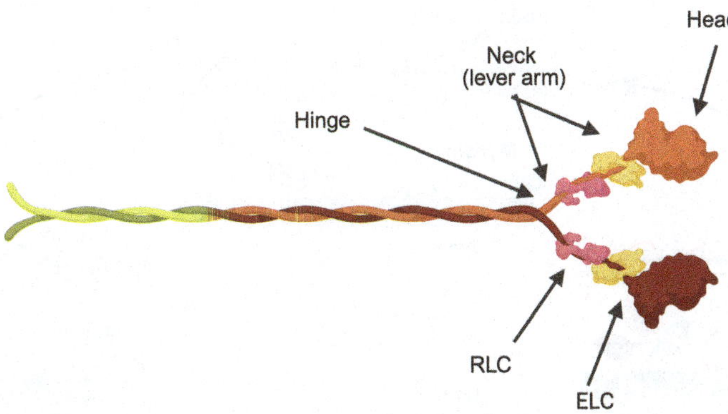

Fig. 5.9 Myosin. Myosin consists of myosin light chains (MLCs) and myosin heavy chains (MHCs). The MLC includes the regulatory light chain (RLC) and essential light chain (ELC). The head region of MHC forms the motor domain where ATP hydrolysis and actin binding occur. The neck region, which acts as a lever arm, binds the light chains (RLC and ELC). The hinge region connects the neck with the tail region, the long dimerized coiled region of the MHC. (*Source*: Created with Biorender.com)

the typical naming pattern of most enzymes). Specifically, myosin is an ATPase that powers muscle contraction and movement. It is a multi-domain protein (see Fig. 5.9), and the ATPase activity is in one of its domains, called the head domain. **Myosin**, like other ATPases, is a **hydrolase**. It catalyzes the cleavage of ATP into ADP and inorganic phosphate (Pi) in a reaction that uses H_2O. This cleavage provides the energy for its motor activity.

Myosin is actually a family of enzymes, and there are three different **myosins** in mammalian cells. Conventional **myosins**, described here, have a role in the movement and contraction of muscle cells. Myosin consists of a head and a tail domain. The head domains of myosin enzymes bind both actin and ATP. The tail regions can dimerize with the tails of other myosin enzymes. Multiple dimers of tails can coil around each other to form a filament called the thick filament. Between the head and tail domains, there are other domains, including a flexible domain called the neck, which contains the lever arm (Fig. 5.9). **Myosin** activity requires Mg^{2+} or Mn^{2+} ions, which bind to **ATP**. The ATP-Mg^{2+} complex is the physiological substrate of this enzyme. The transition state leads to a large conformational change in myosin that will drive the power stroke [5, 10], described next.

The overall mechanism by which Myosin works is as follows: Myosin begins in the **rigor state**, tightly bound to actin. ATP (in complex with Mg^{2+}) binds to the myosin head. The ATP binding causes a conformational change in myosin, reducing its affinity for actin. This leads to the release of myosin from the actin. ATP hydrolysis then occurs, and myosin is now bound to ADP and inorganic phosphate (Pi). This hydrolysis provides the energy to change the myosin head into a high-energy state and leads to a change in the orientation of the lever arm. Now, myosin binds actin further along the actin filament, and the Pi is released. This release causes the “power stroke” and the lever arm of myosin to change position again, pulling on the actin filament. The ADP is then released, and myosin returns to the rigor state. The cycle can start again with the binding of another ATP. In the process, myosin moves (“walks”) along the actin filament. Each myosin makes only a small movement during this process, but our muscles have trillions of myosins, and there are about 300 myosins bound together in each thick filament in the muscle. The coordination of many myosin molecules in these thick filaments can generate enough force to lead to muscle contraction and thereby power muscle cells. This cycle is known as the cross-bridge cycle [5, 10] (see Fig. 5.10).

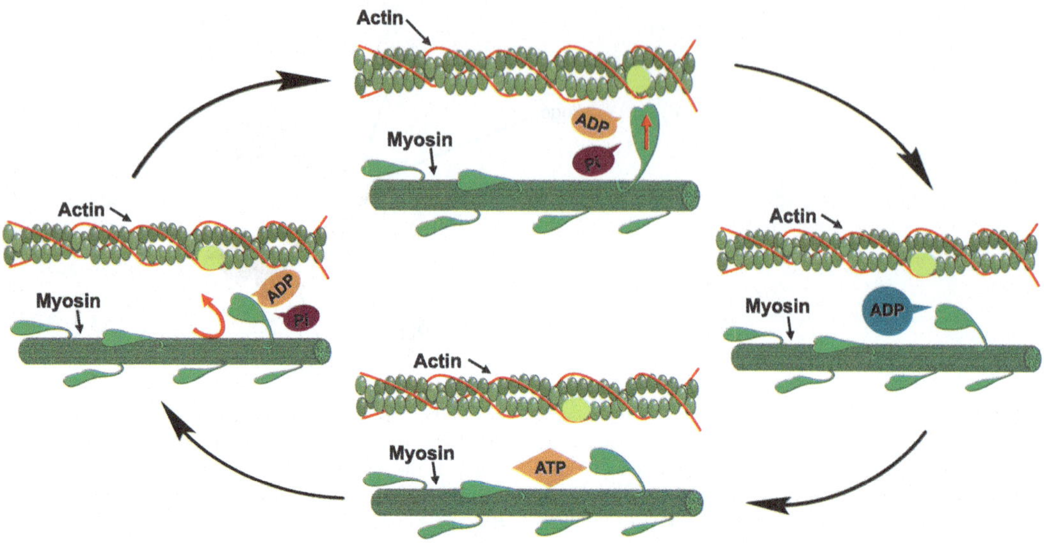

Fig. 5.10 The interaction between actin and myosin. The myosin head binds to actin. Making multiple small movements, it leads to contraction. (*Source*: Made by Yiling Wang, with permission, adapted from Servier Medical Art (https://smart.servier.com/), licensed under the Creative Commons Attribution 4.0 (CC BY 4.0) license)

8.4 Protein Kinases

Protein kinases have roles in almost every aspect of cell function, including signaling, metabolism, cell cycle regulation, transport, and more. There are hundreds of **protein kinases** in cells. Protein kinases (and other kinases, such as lipid kinases) are members of the ***transferase family***. They phosphorylate their substrates, meaning that they catalyze the covalent addition of a phosphate group onto them. The three major categories of protein kinases are tyrosine kinases (which phosphorylate amino acids on tyrosine residues), serine/threonine kinases (which phosphorylate serine and threonine residues), and dual-specificity kinases (that phosphorylate serine, threonine, and tyrosine). Kinases have a variety of functions and substrates, and although they are all quite diverse in sequence, they also have certain general features in common. They have a **catalytic domain**, which is the part of the protein that is most conserved; a **regulatory domain**; and a **substrate recognition domain**.

The **catalytic domain** is responsible for the enzyme's reaction. It is also called the kinase domain. It plays an important role in transferring a phosphate group from ATP to the substrate. Within this domain are an N-terminal lobe and a C-terminal lobe. The N-terminal lobe is smaller, and it plays a role in correctly positioning the ATP for catalysis. The **N-terminal lobe** is made largely of β-sheets along with one α-helix. The **C-terminal lobe** is made mostly of α-helices.

Between these lobes is a **hinge region**, a flexible segment that connects the lobes and helps stabilize ATP in the active site. The active site spans the cleft between the two lobes, and it includes the ATP binding pocket, where ATP binds, and the substrate recognition site, where the substrate's target residue (the residue that gets phosphorylated) is positioned. This is a critical part of the kinase. When a protein kinase operates, it breaks a phosphate bond in this ATP molecule and transfers the terminal phosphate group of this ATP to its substrate. A large amount of free energy is released when the terminal phosphate of the ATP is released and the bond between the second and third phosphate (the β and γ phosphate) of ATP is broken [11].

Within the C-terminal lobe, which is the larger lobe, is the **catalytic loop**. The catalytic loop has residues that are important for facilitating the transfer of a phosphate group onto its substrate. Also within the catalytic domain is the activation

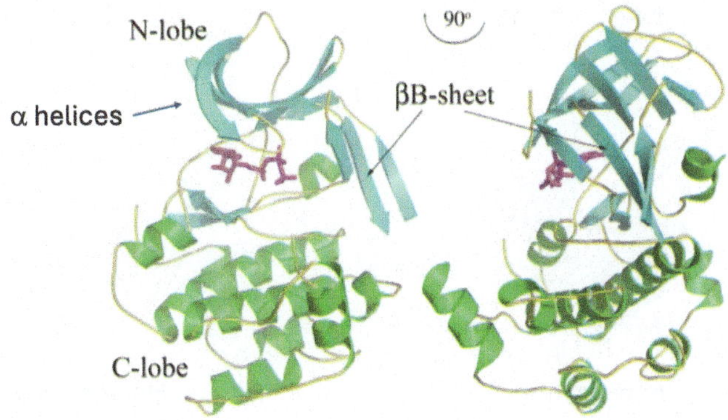

Fig. 5.11 Structure of a protein kinase. Ribbon diagram. The protein kinase RSK, shown from two angles, is a model for protein kinase structure. The C lobe, N lobe, and beta pleated sheets (blue) and alpha helices (cyan) are also shown. (*Source*: Adapted from Malakhova et al. [118]. Licensed under the Creative Commons Attribution 4.0 (CC BY 4.0) license)

loop. This activation loop has a role in regulating the activity of the kinase by regulating the conformation of the active site. In some protein kinases, this region must itself be phosphorylated in order for the kinase to be active.

Outside of the catalytic domain is the regulatory domain. This domain does not have catalytic activity, but it regulates the activity of the catalytic domain. It can either inhibit or activate its activity. Some kinases also contain an additional substrate recognition domain (separate from the above-mentioned substrate recognition site). This domain ensures that the kinase recognizes the correct substrate.

Although kinases tend to have these different domains, their overall amino acid sequences largely differ from each other. Overall, kinases tend to share a common sequence of about 250–300 amino acids within their catalytic domains, but differ in amino acid sequence outside this domain. The above information about protein kinases can also be found in the following resources [4, 53–56]. (For an example of protein kinase structure, see Fig. 5.11).

9 Cofactors and Allosteric Regulators

Many enzymes require other factors to function. These can include different subtypes of cofactors and allosteric regulators. When enzymes require a cofactor, the enzyme is considered an apoenzyme when the cofactor is not bound, but a holoenzyme once the cofactor binds (Fig. 5.12). Cofactors and allostesric regulators are also found in [5, 7, 40].

9.1 What Are Cofactors

Often, an enzyme requires a cofactor to function. When an enzyme that requires a cofactor is in an inactive state and not bound to the cofactor, it is called an apoenzyme. When such an enzyme binds its cofactor, the enzyme is referred to as the holoenzyme. Cofactors can be metal ions or chemicals, but not proteins. There are various types of cofactors, which can be categorized as follows.

9.1.1 Inorganic Cofactors

These are often metal ions such as Mg^{2+}, Fe^{2+}, Zn^{2+}, or other nonprotein molecules. They are involved in metal ion catalysis. They are considered essential minerals that we must obtain through our diet.

9.1.2 Organic Cofactors

These can be further subcategorized as follows:

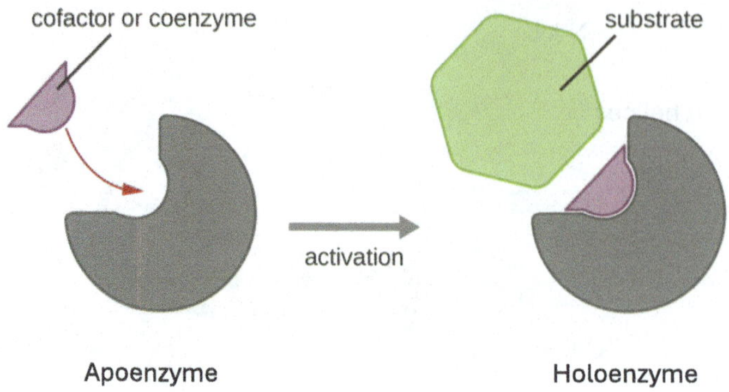

Fig. 5.12 Apoenzyme and holoenzyme. The apoenzyme is not active until a cofactor or coenzyme binds. Once the cofactor binds, the enzyme is considered a holoenzyme. It can interact with a substrate and catalyze a reaction. (*Source*: Adapted from *OpenStax Microbiology*, licensed under the Creative Commons Attribution 4.0 (CC BY 4.0) license)

- Coenzymes

Coenzymes are usually derived from vitamins. They often serve as carriers for chemical groups or electrons. Examples of coenzymes are NAD^+, derived from niacin, and FAD, derived from riboflavin.

- Prosthetic Groups

Unlike the other groups, prosthetic groups are usually permanently bound to the enzyme. Some examples of prosthetic groups are heme or biotin. Heme is a ring structure with an iron ion in the center, and biotin is a vitamin.

Note on classification. Above, coenzymes and prosthetic groups are grouped into a subcategory of organic cofactors. This is the widely used classification system, but some forms of classification categorize cofactors, coenzymes, and prosthetic groups as completely distinct categories.

9.2 Allosteric Regulators

Allosteric regulators can enhance or inhibit the activity of some enzymes, but unlike cofactors, they are not required by the enzyme. Also, unlike cofactors, they usually bind to the regulatory sites on enzymes, specifically the allosteric sites. The binding of an allosteric regulator to an enzyme changes the enzyme's conformation, which can impact its function and its ability to interact with its substrate [5, 7]. This type of regulation is called allosteric regulation.

9.3 Some Examples of Enzymes and Enzyme Families That Use Cofactors or Prosthetic Groups

Carboxypeptidases are enzymes that cleave the carboxyl ends of peptides. Typically, they cleave one peptide, but some carboxypeptidases cleave more. Many carboxypeptidases, such as pancreatic carboxypeptidases, are bound to a zinc (Zn^{2+}) ion at their active sites and are part of the metalloprotease family (see Fig. 5.7). In these enzymes, zinc forms a bond with the substrate and helps stabilize the negatively charged transition states that occur during hydrolysis. It is also necessary for structural integrity and proper orientation within the active site [37, 57, 58].

DNA Polymerase catalyzes the synthesis of DNA by adding nucleotides to the 3′ end of the growing DNA strand. All eukaryotic DNA polymerases are bound to two magnesium (Mg^{2+}) ions at their active sites. The magnesium ions stabilize the negative charges on the phosphate groups of the incoming nucleotide and facilitate the nucleophilic attack of the 3′ OH group on the nucleotide. They also promote the release of the pyrophosphate (PPi) as the phosphodiester bonds are formed [5, 11, 59].

Acetyl-CoA Carboxylase (ACC) catalyzes the carboxylation of acetyl-CoA during fatty acid metabolism. Biotin is a cofactor for this enzyme, forming a transient bond with the carboxylate group that gets transferred [37, 60].

DNA Methyltransferases (DNMTs) catalyze the methylation of substrates, DNA. They use SAM (S-adenosylmethionine) (produced in the folate pathway) as a coenzyme [61].

The **Pyruvate Dehydrogenase Complex (PDH)** is an enzyme complex that plays a crucial role in metabolism. It catalyzes the oxidative decarboxylation of pyruvate to acetyl-CoA. This complex uses lipoamide as a cofactor. Lipoamide is a small organosulfur compound (an organosulfur compound is an organic molecule that contains sulfur covalently bound to carbon). It serves as a flexible structure that transfers groups between different parts of the enzyme complex [5, 61].

10 Various Mechanisms by Which Enzyme Activity May Be Regulated

There are many enzymes in the cell, but they do not all operate simultaneously. In fact, cells would not survive if all the enzymes were active at the same time. Instead, each enzyme must be regulated so that it is only active when it is needed. There are mechanisms that control when, where, and to what extent different enzymes are activated. Enzyme activity can be both positively and negatively regulated [5, 11]. Below are some of the many mechanisms that can regulate whether specific enzymes are active. Enzymes can be regulated by one or a combination of these or other mechanisms. These regulatory mechanisms are distinct from cofactors, which affect enzyme activity by enabling catalysis rather than controlling when or where enzymes are active.

10.1 Enzymes Can Be Regulated by Their Amounts in the Cell

The amount of an enzyme present in a cell or tissue is tightly regulated, often at the level of transcription of the gene that codes for it. The tissue-specific expression pattern of an enzyme determines where it is produced, because different cells and tissues express their own unique set of genes. For example, muscle Myosin II (one subtype of myosin) is expressed primarily in muscle cells, ensuring that the enzyme is specifically active in those cells. Additionally, gene expression can vary across different developmental stages, with some enzymes being expressed in the embryo and others in the adult. Beyond transcription regulation, enzymes can also be controlled through RNA degradation, translation efficiency, and posttranslational modifications. These layers of regulation help control when and where enzymes are present [11]. Transcription and gene expression are discussed in more detail in Chap. 7.

10.2 Enzymes Can Be Regulated by Their Cellular Locations

Some enzymes might be limited to certain parts of the cell. For example, they may be membrane-bound and only act on substrates that are near the membrane. Alternatively, they may be located only in a certain organelle, such as the lysosome, and only act on substrates in that organelle [62].

10.3 Some Enzymes May Be Regulated by Environmental Conditions

Enzymes may be regulated by environmental conditions [7]. Digestive enzymes, for example, function only at low pH. If the conditions are not optimal, they may be inactive or even destroyed [63].

10.4 Enzymes Can Be Regulated by Covalent Modifications

Covalent modification is a common method by which many enzymes are regulated. Covalent modifications can either positively or negatively regulate the activity of certain enzymes. A common modification is phosphorylation, but other modifications can include methylation, acetylation, ubiquitination, sumoylation, glycosylation, or lipidation. Some of these modifications are long-term or permanent, but others may be transient and reversible. Transient and reversible changes are important for enzyme regulation. These types of changes can occur quickly, allowing enzymes to be inactivated or activated efficiently and rapidly in response to the cell's needs. Some enzymes are regulated by proteolytic cleavage. This often leads to enzyme activation, and unlike phosphorylation, it is a permanent change. Enzymes that are regulated by covalent modifications are sometimes modified by other enzymes that can also be modified by similar modifications. Below, we discuss phosphorylation and cleavage as two common enzyme modifications [11].

10.4.1 Regulation of Enzyme Activity by Phosphorylation

The addition of a phosphate group (phosphorylation) to an enzyme is a common and important mechanism by which many enzymes are transiently activated (and sometimes inactivated). Many thousands of proteins, including enzymes, are regulated by phosphorylation. These enzymes are regulated by other enzymes, **protein kinases**, which add a phosphate group to them, or by **phosphatases**, which remove a phosphate group.

There are different ways in which phosphorylation can regulate an enzyme's activity. The phosphate group generally contains two negative charges. This can sometimes cause a conformational change. For example, if the phosphate groups attract positively charged amino acids on another part of the enzyme, this may lead to a conformational change that may affect the enzyme's activity [10, 44].

Another way a phosphate group may affect the activity of an enzyme is by causing the enzyme to interact with other proteins. Some proteins bind specifically to phosphorylated sites and, therefore, the presence of a phosphate group may lead to protein-protein interactions that impact enzyme function. In contrast, the presence of a phosphate group might inhibit protein–protein interactions. In this case, the phosphate group may mask the site where two proteins would otherwise interact. All of these types of changes may positively or negatively regulate the activity of the targeted enzyme [10].

10.4.2 Enzyme Regulation by Irreversible Proteolytic Cleavage

Some enzymes, called **zymogens**, are inactivated in their full-length forms. In their inactive forms, they are considered proenzymes, and they remain inactive until they are cleaved at specific sites, which activates them [64]. Unlike phosphorylation, cleavage is irreversible, and it does not require an energy source. Some examples of zymogens are the digestive enzymes **pepsinogen, chymotrypsinogen, trypsinogen**, and **procarboxypeptidase**. These enzymes get activated by cleavage. For example, **chymotrypsinogen** gets cleaved by and thereby activated by **trypsin** to become **chymotrypsin**. In their active forms, digestive enzymes catalyze the cleavage of proteins from digested food. For more information about zymogens see also [5, 11, 64, 65].

Another family of **zymogens** is the **caspases**. **Caspases** are also **proteases**. They are activated by cleavage, and in turn, they can cleave and activate other caspases. This leads to the activation of pathways necessary for apoptosis, although these pathways can also have other functions. The blood clotting cascade also involves a series of enzymes that are regulated by proteolytic cleavage [11, 65] and is discussed further later in this chapter.

10.5 Enzymes Can Be Regulated by Cell Signaling Pathways

Signal transduction pathways are well known to influence the activity of enzymes. Activation of growth factor receptors, for example, can lead to a

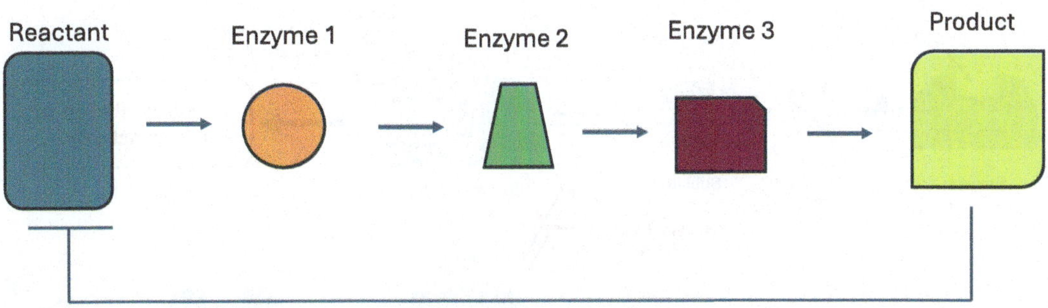

Fig. 5.13 Feedback inhibition. Feedback inhibition can occur when one enzyme in a pathway negatively regulates an earlier enzyme in the pathway. Here the product (yellow box) is an enzyme that inhibits the reactant (blue rectangle), which is also an enzyme (*Source*: Yiling Wang, used with permission)

cascade of kinase activity where one kinase regulates the next in the cascade. This is also an example of regulation by posttranslational modification, because many enzymes in signaling pathways are regulated by modifications such as phosphorylation [10].

10.6 Enzymes Can Be Regulated by Intramolecular Interactions or Protein–Protein Interactions

Some enzymes may adopt an open or closed configuration due to intramolecular interactions, which regulate their activity. A closed configuration may mask the catalytic site, inhibiting enzymatic activity, while an open configuration may stimulate enzymatic activity. The activity of some enzymes can also be impacted by protein–protein interactions. These interactions can include dimerization of two identical enzymes or interactions between enzymes and other proteins [10].

10.7 Enzymes Can Be Regulated by Feedback Inhibition

Feedback inhibition can be a mechanism for turning off (negative regulation) an enzyme's activity when it is no longer needed. A good example of feedback inhibition happens during signal transduction cascades. In signal transduction cascades (see Chap. 18), one enzyme may activate another, which in turn activates another, and so on, in a cascade of enzymatic activity. Often, these enzymes are protein kinases that activate one another through phosphorylation. They may lead to the phosphorylation of a final protein in the cascade, which is often involved in regulating transcription. Such cascades are advantageous because they can amplify a reaction, but eventually, enough substrate will be phosphorylated. At this point, one of the substrates in the cascade may act as an inhibitor, to inhibit the first enzyme in the cascade. For example, one of the last kinases to be activated may phosphorylate the first enzyme on an inhibitory site and inhibit its activity. In this way, when large quantities of a final product accumulate, the feedback inhibition mechanism prevents the generation of more product [10] (Fig. 5.13).

10.8 Enzymes Can Be Regulated by Inhibitors

Different types of inhibitors can inhibit enzyme activity. These include allosteric regulators, competitive inhibitors, and noncompetitive inhibitors. Inhibitors can inhibit enzymes by reversible and irreversible mechanisms. They can be naturally occurring, but many drugs are also designed to be enzyme inhibitors. Inhibitors will be discussed in more depth in the next section and are also discussed in [1, 5, 7, 11].

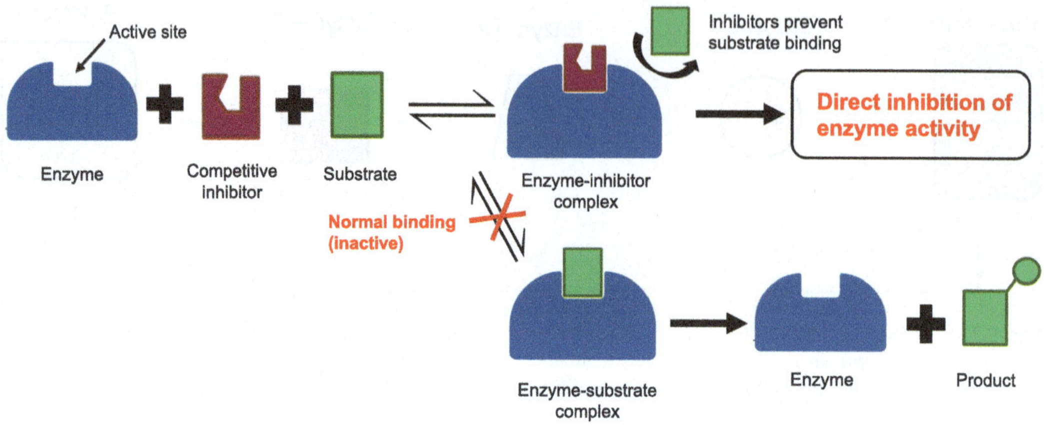

Fig. 5.14 Enzyme inhibition: competitive inhibition. In competitive inhibition, the inhibitor binds to the enzyme's active site, directly blocking the substrate's interaction with the enzyme. Normal substrate binding, which is inactive in the presence of the inhibitor, is shown at the bottom. (*Source*: Yiling Wang, used with permission)

11 Enzyme Inhibitors and Drugs That Are Enzyme Inhibitors

Enzyme inhibitors can inhibit the catalytic activity of enzymes [1]. They can be reversible or irreversible, though reversible inhibitors are more common. Irreversible inhibitors involve covalent interactions with the enzymes, while reversible inhibitors involve noncovalent interactions and can dissociate from the enzyme. Inhibitors are naturally present in the cell, but many drugs are also designed to act as enzyme inhibitors. In fact, the majority of drugs are enzyme inhibitors, although other mechanisms for drug development are continually being explored. The widespread use of enzyme inhibitors to treat disease makes sense, since enzymes control so many biological activities in the body. For this reason, it is especially important for pharmacists to have an understanding of how enzyme inhibitors work. Here, we discuss how enzyme inhibitors work and focus on those that are used as pharmaceuticals.

11.1 Reversible Inhibitors

Reversible inhibitors typically fall into two categories: competitive inhibitors and noncompetitive inhibitors. Slightly less common are uncompetitive inhibitors [5, 7, 11].

11.1.1 Competitive Inhibitors

Competitive inhibitors bind to the active site of an enzyme (see Fig. 5.14). They have a structure similar to that of the normal substrate and can often be considered substrate analogs. They compete directly with the substrate for the same binding site. In the case of competitive inhibitors, adding more natural substrate can overcome the inhibitor's effects. Competitive inhibitors increase the K_m of the enzyme, but the V_{max} remains unchanged. This type of enzyme inhibition is known as competitive inhibition. Some examples of competitive inhibitors are discussed below.

Statins, such as **Lipitor (atorvastatin)**, are **small-molecule** drugs that are competitive inhibitors of the enzyme **HMG-CoA reductase**, which is the rate-limiting enzyme in cholesterol biosynthesis [66]. HMG-CoA reductase catalyzes the conversion of **HMG-CoA** to **mevalonate**. The reaction is HMG-CoA + 2 NADPH + 2 H^+ → mevalonate + 2 $NADP^+$ + CoA-SH, in a three-step reaction. **Mevalonate** is an important precursor in the pathway leading to the synthesis of cholesterol. **Statins** mimic HMG-CoA and compete with it for binding to the active site of HMG-CoA reductase. This inhibits the synthesis of cholesterol, making statins effective **cholesterol-lowering drugs**. **Lipitor** has a structure similar to HMG-CoA, and thus can compete with it for binding to **HMG-CoA reductase** [67, 68].

ACE inhibitors are **small-molecule drugs** that are **competitive inhibitors of angiotensin-converting enzyme (ACE)**. **ACE** is a **zinc metalloprotease** that hydrolyzes peptides. **ACE** (specifically ACE 1) converts the peptide angiotensin I (ang I; inactive) to angiotensin II (ang II; active) by removing a dipeptide (2 amino acids) from **ang I**. **Ace** is a **carboxypeptidase**, but unlike classic carboxypeptidases, it cleaves two amino acids from the carboxyl-terminal end rather than a single amino acid. Angiotensin II is a hormone that circulates in the blood. Its main role is to constrict blood vessels. High levels of angiotensin II correlate with cardiovascular disease and can cause high blood pressure (hypertension). **ACE inhibitors** are used to **manage hypertension** by blocking **ACE**. They bind the active site of **ACE** because they have structures that mimic angiotensin I (the natural substrate of **ACE**). **ACE inhibitors** typically have groups that can bind the zinc in the active site of **ACE**, helping to prevent the enzyme from functioning. As a result, **ACE inhibitors** help relax the veins and arteries, thereby **lowering blood pressure**. They prevent heart failure, reduce heart attacks in patients with high blood pressure, and protect kidney function in diabetic patients or those with mild kidney disease [69]. There are many **ACE inhibitors**. Some examples are **Capoten (captopril)** and **Vasotec (enalapril)** [47, 70].

11.1.2 Noncompetitive Inhibitors

Noncompetitive inhibitors, also described in bind to a site that is not the active site [5, 7, 11, 37, 65]. They cause a conformational change in the enzyme, which inhibits its activity. Some noncompetitive inhibitors bind to an allosteric site, and these are referred to as allosteric inhibitors. They can bind to the enzyme alone or to the enzyme-substrate complex. This lowers the concentration of functional enzyme in the cell. In the presence of noncompetitive inhibitors, the V_{max} increases, but the K_m remains unchanged, and enzymatic activity cannot be overcome by increasing the substrate concentration. This type of inhibition is called noncompetitive inhibition (Fig. 5.15).

Noncompetitive inhibitors can be pure noncompetitive inhibitors or mixed noncompetitive inhibitors. Pure noncompetitive inhibitors bind equally well to the enzyme alone and to the enzyme with substrate bound. In this case, only the turnover number is decreased in response to the inhibitor. Mixed noncompetitive inhibitors have a preference. They either bind preferentially to the free enzyme or to the enzyme-substrate complex (although they maintain the ability to bind to either) [5].

- Examples of Drugs That are Noncompetitive Inhibitors

Doxycycline [71] is an antibiotic that functions by inhibiting protein synthesis in bacteria (see Chap. 8). However, it also has another function: acting as a mixed **noncompetitive inhibitor** of matrix **metalloproteinases (MMPs)**, particularly **collagenase (MMP-1)**, an enzyme that breaks down collagen, the main structural protein in connective tissue. **Collagenase** cleaves collagen into peptide fragments [72]. **Doxycycline** can bind to the **collagenase** enzyme noncompetitively, by binding to a region other than its active site. This inhibits the activity of the collagenase enzyme, thus blocking its ability to break down collagen [73].

By inhibiting **collagenase**, **Doxycycline** helps preserve collagen in tissues. This makes it helpful for treating conditions that involve **tissue breakdown**. For this reason, it is used to treat periodontal disease, which can be caused by collagen degradation in the gums. It is also sometimes effective at treating other disorders where tissue breakdown occurs. These include diseases such as **osteoarthritis** and **rheumatoid arthritis**, where it helps preserve cartilage and treat chronic wounds by stabilizing the collagen in the extracellular matrix, thereby aiding in healing. Additionally, Doxycycline is also studied in the field of **cancer treatment**, due to its role in controlling tissue invasion via MMP inhibition [33, 74–78].

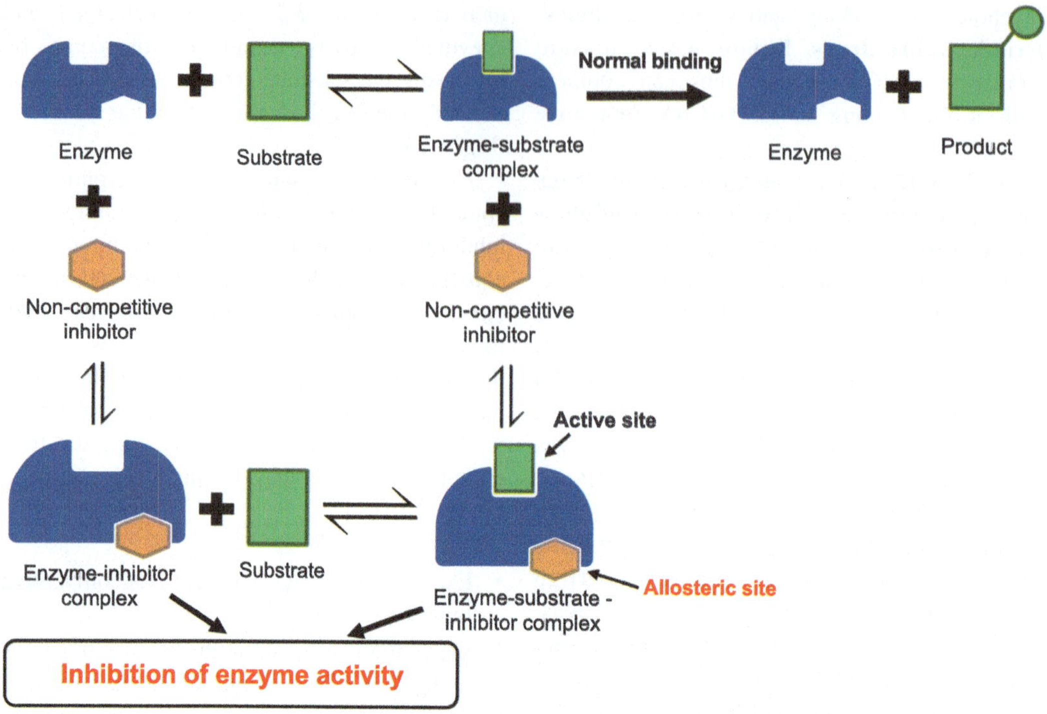

Fig. 5.15 Enzyme inhibitors: noncompetitive inhibitor. A noncompetitive inhibitor does not bind the active site. It binds to another site on the enzyme but inhibits the enzyme's activity. (*Source*: Yiling Wang, used with permission)

Monoamine oxidase inhibitors (MAOIs) are primarily noncompetitive inhibitors of **monoamine oxidases (MAOs)**, although some may act as **competitive inhibitors**. **MAOs** are enzymes needed to break down certain neurotransmitters, specifically those in the monoamine family, such as norepinephrine, epinephrine, and dopamine. **MAOs** oxidize the neurotransmitters, which is the first step in their degradation [79]. Their roles in breaking down neurotransmitters help the body maintain the right balance of neurotransmitter signaling [79]. Thus, **MAOIs** increase neurotransmitter levels and can enhance neurotransmission. **MAOIs** are used **to treat depression, anxiety, and some neurodegenerative diseases**. Commonly used **MAOIs** include the small molecule drugs **Nardil (phenelzine)** [80], **Emsam (selegiline)** [81], **Parnate (tranylcypromine)** [47, 82, 83]. Although they are effective, these drugs have been largely replaced by newer antidepressant drugs [84, 85].

11.1.3 Uncompetitive Inhibitors

This type of inhibitor is less common. It binds the enzyme–substrate complex, rather than to either one alone (Fig. 5.16). It can prevent catalysis rather than preventing substrate binding to the enzyme. For this type of inhibitor, adding more substrate will not increase enzyme activity. In response to this type of inhibitor, both the K_m and the V_{max} decrease [5, 7, 40].

Camptothecin (CPT) is an example of an uncompetitive inhibitor. It is a small molecule and natural compound that is a topoisomerase I inhibitor. Derivatives of CPT, such as **Topotecan** [86] (also sold under the brand name Hycamtin) and **irinotecan** [87, 88], are FDA-approved for use in treating several **cancers**. Although approved in 1996, they are still used today. As an uncompetitive inhibitor, this type of inhibitor binds to the topoisomerase-I-DNA complex. **Topoisomerase** is an enzyme that has an important role in proper DNA replication (see Chap. 6), and therefore, disrupting its function inhibits

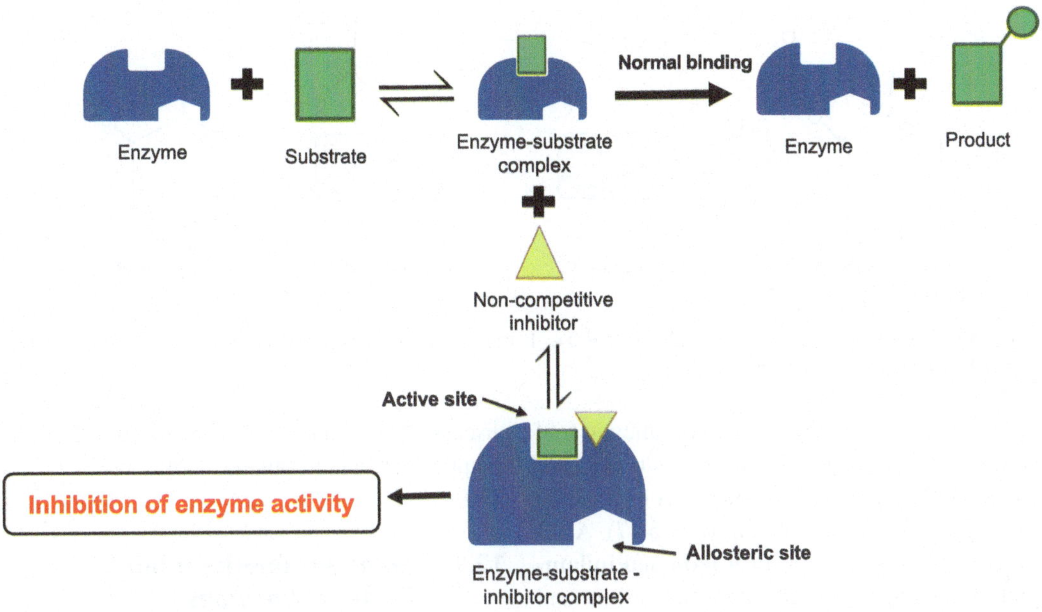

Fig. 5.16 Enzyme inhibition: uncompetitive inhibitor. The uncompetitive inhibitor binds the enzyme/substrate complex, preventing catalysis. (*Source*: Yiling Wang, used with permission)

DNA synthesis. Inhibition of DNA synthesis in turn inhibits cell proliferation and can lead to cell death. This compound is used to treat several types of **cancer** [89–93].

11.2 Irreversible Inhibitors

11.2.1 How Irreversible Inhibitors Work

An irreversible inhibitor acts as an analog of the substrate and binds to the active site of the enzyme. Unlike the competitive inhibitor, however, it forms a covalent bond with the enzyme. This type of inhibitor is also called a **suicide inhibitor** because it permanently inactivates the enzyme [5, 17].

11.2.2 Examples of Irreversible Inhibitors

Aspirin (acetylsalicylic acid) is a **small molecule** and an irreversible inhibitor of the enzyme **cyclooxygenase (COX)**. Aspirin is an interesting compound because of its long history, dating back well before its functions were understood. As far back as 400 BC, willow bark was used to relieve **pain**. The active compound in willow tree bark was identified as **salicin** in the 1800s, and salicylic acid was later generated from salicin. Salicylic acid was effective as an analgesic (a substance that reduces pain), but it also caused gastric irritation. To reduce this, it was made into **acetylsalicylic acid**, which was less irritating. This is today's **aspirin**.

There are two **COX** enzymes: COX1 and COX2. COX is an enzyme that converts arachidonic acid into prostaglandin H (PGH2) in two steps. First, it adds two oxygen molecules to arachidonic acid, forming prostaglandin G_2 (PGG_2). PGG_2 is then converted to PGH_2, also catalyzed by **COX**, in a reaction where an –OOH group is reduced to an –OH (hydroxyl) group. PGH_2 is then further metabolized by other enzymes into prostacyclin, thromboxanes, and other prostaglandins.

Prostaglandins are important products of these reactions. Prostaglandins promote the inflammatory response, which includes swelling, pain, and fever. Inhibiting COX thus reduces this inflammatory and painful response. **Aspirin** (acetylsalicylic acid) inhibits COX by transferring its own acetyl group to the side chain of a serine residue on **COX's active site**. This results in a permanent covalent bond with **COX**, irreversibly blocking its activity. In the process, aspirin

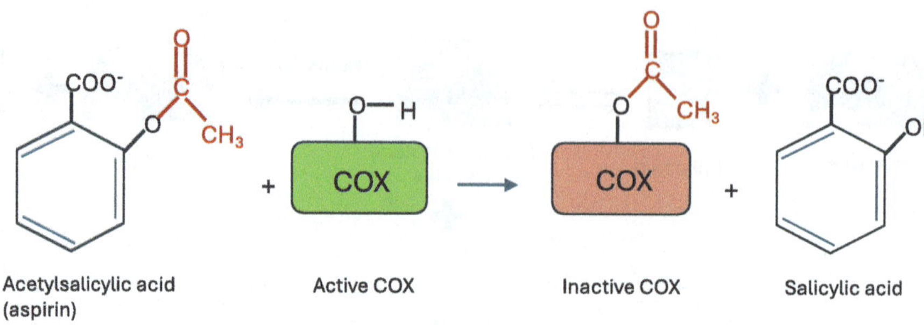

Fig. 5.17 Aspirin is an irreversible inhibitor of COX. It transfers its own acetyl group to the active site of COX, irreversibly inhibiting its activity

(acetylsalicylic acid) is converted to salicylic acid (which still has some biological activity and can contribute to aspirin's side effects, which include some residual gastric irritation) (Fig. 5.17). Aspirin is still widely used to treat **pain** and **inflammation**, among other uses. Aspirin's history and mechanism of action are also described in the following sources [5, 94].

Penicillin is another example of a **suicide inhibitor**. **Penicillin** is an **antibiotic** in the β-lactam family. Like other antibiotics, it is produced by some bacteria and fungi to kill or inhibit other microbes. Penicillium mold (a type of fungus) naturally produces the antibiotic penicillin. When used as a drug, penicillin is effective against many bacterial infections caused by staphylococci and streptococci species. It is used to treat a wide range of **infections**, including strep throat, some skin infections, syphilis, and others. Over the years, it has saved millions of lives. Penicillin irreversibly inactivates the bacterial enzyme **DD-transpeptidase** (which is one of several proteins known as **penicillin-binding proteins (PBPs)**). It does so by reacting with a serine at the enzyme's active site and forming a covalent bond with it. This enzyme catalyzes the last step in the synthesis of the bacterial cell wall and therefore causes the bacteria to burst [5, 95–97].

12 Other Drugs That Inhibit Enzymes

Enzymes are a common drug target. A few more of the many drugs that target enzymes are discussed here, and others are discussed in the chapters throughout this textbook.

12.1 Some Antibiotics Inhibit Bacterial Enzymes

Antibiotics are used to treat **bacterial infections**, and they kill bacteria in various ways. Some of them function by disrupting enzymes [98]. **Penicillin**, a member of the **β-lactam** family of antibiotics, was discussed above. **Fluoroquinolones**, another class of antibiotics, kill bacteria by a different mechanism. They are inhibitors of **DNA gyrase** and **topoisomerase IV**, which are **bacterial type II topoisomerases** that play a role in DNA replication [99]. **Rifamycins** (such as rifampin) are another type of antibiotic, and they inhibit bacterial **RNA polymerase**, preventing transcription [100]. Rifamycin is also discussed in Chap. 7.

Sulfonamides (such as **Bactrim (sulfamethoxazole)**) [101] are **antibiotics** that inhibit **dihydropteroate synthase**, an enzyme involved in bacterial folate synthesis. In bacteria, as in mammalian cells, folate is important for the synthesis of nucleic acids and the metabolism of amino acids, and is therefore required for bacterial cell growth and viability. Bacteria synthesize their folate, while animals must obtain it from the diet. **Dihydropteroate synthase (DHPS)** is a bacterial enzyme that is necessary for folate production by bacteria. It catalyzes the reaction that leads to dihydropteroate, an intermediate in the synthesis of folate. **Sulfonamides** are competitive inhibitors of DHPS. They inhibit the enzyme by binding to its active site [102]. Human cells do not have DHPS, and therefore these antibiotics do not impact folate synthesis in humans. **Trimethoprim** is also an antibiotic that inhibits folate synthesis in bacteria. It inhibits another enzyme, **dihydrofolate reductase (DHFR)**, which also has a role in folate metabolism and synthesis. While mammalian cells also have DHFR, the bacterial DHFR is inhibited more strongly by the antibiotic. Trimethoprim is also a competitive inhibitor [102].

Rezzayo (rezafungin acetate) is not an antibiotic but an inhibitor of **fungal infections**, specifically the fungus Candida. It inhibits the fungal enzyme **1,3-β-D glucan synthase**. This enzyme belongs to the transferase family and is essential for forming 1,3-β-D-glucan, a crucial component of the cell walls of many fungi, including Candida. **Rezzayo** is a noncompetitive inhibitor of this enzyme. It does not bind to the enzyme's active site, but to a different part of the enzyme, which impacts its activity. The enzyme is present in fungi but not in mammalian cells, and thus, the drug specifically targets fungi [103].

12.2 Anticoagulant Drugs

The rest of the inhibitors discussed below inhibit enzymes in mammalian cells. **Anticoagulant drugs**, also called **blood thinners**, are good examples of enzyme inhibitors in medicine. Anticoagulant drugs reduce **blood clot formation**. They are used to prevent **stroke, myocardial infarction** (heart attack; the death of heart muscle tissue), and pulmonary embolism (a blockage of a pulmonary artery in the lung caused by a blood clot) in high-risk patients such as those with atrial fibrillation. They can also reduce the risk of clots during certain types of surgery and can be used to treat blood clots in the legs [68, 104].

Anticoagulant drugs function by inhibiting the action of **clotting factors**. Clotting factors are part of a cascade of **serine proteases** that consist of **zymogens**. This cascade is called the coagulation cascade or the blood clotting cascade. Each protease in the cascade cleaves and thereby activates the next one. At each step, the signal is amplified, forming the coagulation cascade. This cascade is divided into three pathways: the **intrinsic pathway**, the **extrinsic pathway**, and the **common pathway**. Different stimuli activate these pathways, which all converge on the cleavage of **factor X** (an inactive serine protease) to form **factor Xa** (an active serine protease). Each activated enzyme in the cascade cleaves and activates the next downstream zymogen, leading to a rapid amplification of the signal. Consequently, even a small trigger (such as a vessel injury) can generate a fast and large response.

Once factor Xa is formed, it cleaves prothrombin (factor II) to generate **thrombin** (also called factor IIa) (together with factor Va, calcium ions, and phospholipids). **Thrombin** is also an active serine protease, and once activated, it has several roles. Importantly, it converts **fibrinogen** into **fibrin**, a fibrous protein that is important for clot formation. It also activates factor XIII, which cross-links fibrin strands to stabilize the clot. It also further amplifies coagulation by activating upstream factors in the cascade, creating positive feedback loops (Fig. 5.18) [68, 105]. Some important anticoagulant drugs are discussed below.

12.2.1 Direct Thrombin Inhibitors (DTIs) and Factor Xa Inhibitors

Direct Thrombin Inhibitors (DTIs) are **anticoagulant drugs** that bind to the active site on **thrombin** (Fig. 5.18). An example is **Acova (Argatroban)** [106], a **DTI** that is a monovalent inhibitor (meaning it binds to a single site) at the **thrombin active site**. By binding, it inhibits thrombin's activity and thus inhibits the production of fibrin. **Angiomax (bivalirudin)** [107] is also a **DTI**, but it is a bivalent inhibitor [108]. This means that it binds to two sites: an active site and a regulatory site [68, 109].

While DTIs directly inhibit thrombin, factor Xa inhibitors act one step up in the coagulation cascade. **Eliquis (apixaban)** [110] is a competitive inhibitor of the enzyme **factor Xa**, the enzyme responsible for the cleavage of prothrombin (factor II) to form thrombin (factor IIa). This reduces the formation of thrombin indirectly, interfering with clot formation at an earlier stage [68, 111].

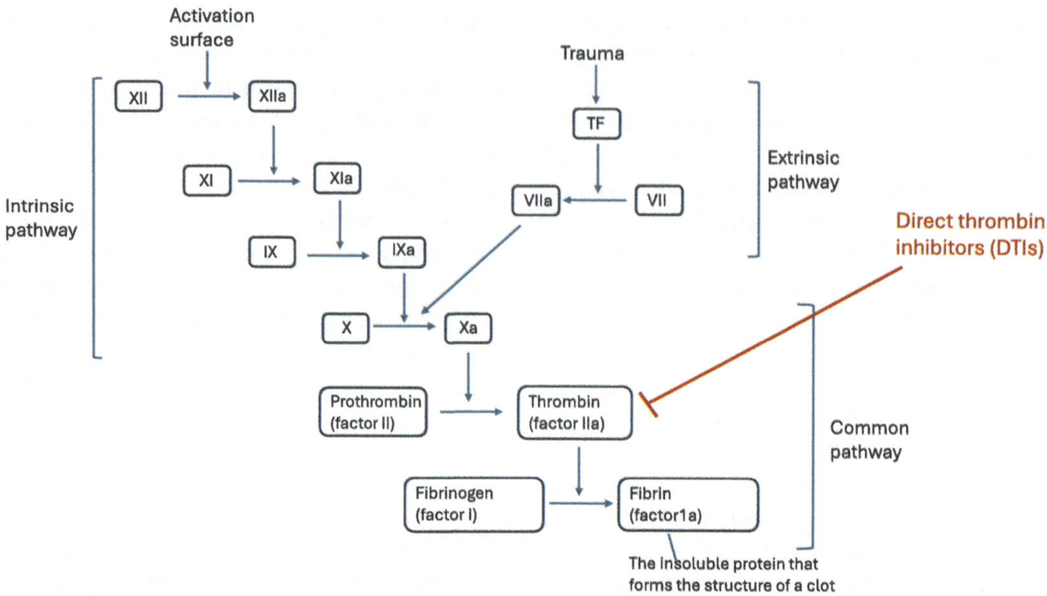

Fig. 5.18 Coagulation cascade. The coagulation cascade is described in the text. Direct Thrombin Inhibitors (DTIs) can block the pathway and reduce blood clotting

12.2.2 Coumadin (Warfarin)

Coumadin (warfarin) is also an **anticoagulant drug**, but it operates differently. It inhibits the enzyme **vitamin K Epoxide Reductase (VKOR)**. This has an important impact on clotting because of the role of vitamin K in the clotting cascade. Several factors in the clotting cascade require a posttranslational modification (γ carboxylation) in order to be active. This reaction is catalyzed by the enzyme **gamma-glutamyl carboxylase (GGCX). GGCX** requires a reduced form of vitamin K as a co-enzyme (vitamin K hydroxyquinone), but vitamin K gets oxidized in the process. **VKOR** regenerates the active, reduced form of vitamin K, allowing it to remain available for clotting factor activation. These types of inhibitors are described in more detail in the following sources [68, 112, 113].

12.3 Drugs That Are Kinase Inhibitors

Kinase inhibitors are an especially important family of enzyme inhibitors. Many drugs are kinase inhibitors, which are used to treat various disorders where kinases are improperly regulated or mutated. Dysregulated kinase activity is often associated with diseases, particularly those caused by uncontrolled cell growth, which can lead to cancer. However, kinases are also linked to other diseases, such as diabetes, Parkinson's disease, and autoimmune diseases.

A common mechanism by which kinase inhibitors work is through competitive inhibition, in which the inhibitor binds to the active site on the protein kinase, specifically to the ATP-binding pocket. This prevents ATP from binding, and consequently, the kinase is unable to phosphorylate its substrate. These ATP-competitive inhibitors often mimic ATP and bind in its place to the ATP-binding pocket [114].

Not all kinase inhibitors are competitive inhibitors. Some bind to other regions of a kinase. For example, some may be allosteric inhibitors, which bind to regulatory regions rather than the kinase domain, thereby inhibiting the enzyme's activity.

Some kinase inhibitors have been approved for treating various conditions, including cancer. An important challenge remains: developing inhibitors that are highly specific to the targeted kinase and do not inhibit other kinases. This is important for avoiding unwanted side effects

[115]. ATP competitive inhibitors, as described above, are strongly inhibitory toward protein kinases, but their use can also be challenging because ATP-binding regions are highly conserved among different kinases. This means that one kinase inhibitor may inadvertently inhibit other kinases as well, leading to unwanted side effects. Using allosteric inhibitors may be advantageous because allosteric regions are less highly conserved. Nevertheless, allosteric inhibitors are not very common and can sometimes be more challenging to develop. Other methods, such as the use of PROTACs, are being developed to destroy kinases rather than inhibit their activity. These and other methods are discussed in Chap. 12. More detailed examples of various kinase inhibitors are described in Chaps. 18 and 19.

A notable example of a kinase inhibitor used successfully to treat cancer is **Gleevec (imatinib)**, which is actually the first kinase inhibitor approved for cancer treatment. **Gleevec** was FDA-approved in 2001 for the treatment of **chronic myelogenous leukemia (CML)**. It is an inhibitor of the tyrosine kinase **BCR–ABL**, an abnormal fusion protein that has constitutive **tyrosine kinase activity**. Other kinase inhibitors have been developed against **receptor tyrosine kinases (RTKs)**, such as **HER2**, which is overexpressed in certain types of breast cancer. Efforts continue to develop inhibitors of other kinases. These and other kinase inhibitors are discussed in more detail in Chaps. 18 and 19.

FDA-Approved Drugs Mentioned in This Chapter

Brand name	Generic name	First FDA approval year	Main conditions	Main target
Acova	*Argatroban*	2000	Anticoagulation	Thrombin
Angiomax	*Bivalirudin*	2000	Anticoagulation	Thrombin
Aspirin	*Acetylsalicylic acid*	1899	Pain and inflammation.	Cyclooxygenase (COX)
Bactrim	*Sulfamethoxazole—trimethoprim*	1973	Bacterial infections	Dihydropteroate synthase (DHPS) and dihydrofolate reductase (DHFR).
Capoten	*Captopril*	1981	Hypertension	Angiotensin-converting enzyme (ACE)
Coumadin	*Warfarin*	1954	Anticoagulation	Vitamin K epoxide reductase (VKOR)
Doxycycline	*Doxycycline*	1967	Periodontal disease, acne, and arthritis. Also, an antibiotic.	Matrix metalloproteinases (MMPs), also 30S subunit of bacterial ribosomes.
Eliquis	*Apixaban*	2012	Anticoagulation	Factor Xa
Emsam	*Selegiline*	2006	Depression and Parkinson's	Monoamine oxidase (MAO)
Gleevec	*Imatinib*	2001	Chronic myelogenous leukemia	BCR–ABL tyrosine kinase
Irinotecan	*Irinotecan*	1996	Cancer	Topoisomerase I
Lipitor	*Atorvastatin*	1996	High cholesterol	HMG-CoA reductase
Nardil	*Phenelzine*	1959	Depression	Monoamine oxidase (MAO-A and MAO-B)
Parnate	*Tranylcypromine*	1961	Depression	Monoamine oxidase (MAO)
Penicillin	*Penicillin*	1943	Bacterial infections	DD-transpeptidase (PBP)
Rezzayo	*Rezafungin acetate*	2023	Fungal infections	1,3-b-D glucan synthase
Rifadin	*Rifampin*	1971	Tuberculosis, bacterial infections	RNA polymerase
Topotecan	*Topotecan*	1996	Cancer	Topoisomerase I
Trimethoprim	*Trimethoprim*	1979	Bacterial infections	Dihydrofolate reductase (DHFR)
Vasotec	Enalapril	1985	Hypertension	Angiotensin-converting enzyme (ACE)

Definitions

Activation energy The minimum energy required for a reaction to occur.

Active site The site on an enzyme where the reaction occurs, which is the site where substrate binds.

Allosteric regulation Regulation by allosteric regulators, which bind to regulatory sites of an enzyme rather than the active site; allosteric sites. Allosteric regulators can lead to conformational changes in the enzyme which impact its function.

Apoenzyme An enzyme without its cofactor or coenzyme, this form is generally inactive.

Catalytic efficiency Measures how effectively an enzyme converts a substrate into a product. It is represented by the K_{cat} divided by the K_m.

Coenzyme Coenzymes are organic cofactors, usually derived from vitamins. They often serve as carriers for chemical groups or electrons.

Cofactor A nonprotein chemical compound required for enzyme activity.

Competitive inhibition In enzyme inhibition, this type of inhibition occurs when the inhibitor binds to the active site and competes for substrate binding.

Electrophilic center The electron deficient atom within the electrophilic group that accepts electrons from a nucleophile.

Electrophilic group A electron-deficient atom or group that attracts electrons from nucleophiles.

Enzyme A biological catalyst that accelerates chemical reactions without being consumed.

Feedback inhibition Inhibition of an enzyme by target proteins, often the endpoint of a metabolic reaction, that inhibits an early enzyme in the pathway to limit its activity.

Free energy refers to the **Gibbs free energy (ΔG)**, which is a measure of the energy in a system available to do useful work in a chemical reaction. If ΔG is positive the reaction is non-spontaneous and requires energy input. If it is negative, the reaction is spontaneous.

Holoenzyme An enzyme that requires a cofactor, bound to its cofactor.

Lineweaver–Burk plot This plot is derived by the equation representing the double reciprocal of the Michaelis–Menton equation.

Michaelis constant (K_m) The substrate concentration at which the reaction rate is half of V_{max}. It reflects half the V_{max} and it also reflects enzyme affinity.

Michaelis–Menten kinetics A model describing how the reaction rate depends on substrate concentration.

Noncompetitive inhibition Inhibition of an enzyme where the inhibitor does not bind the active site and does not compete with the substrate.

Nucleophile (Nucleophilic group) An atom, ion, or molecule that donates a pair of electrons to form a chemical bond in a reaction. Nucleophiles are attracted to electron-deficient atoms or groups called electrophiles.

Oxidation The process where a substance loses electrons. The substance that loses electrons is said to be oxidized (In terms of hydrogen atoms, oxidation is defined as the loss of hydrogen atoms. When a substance loses hydrogen atoms, it is typically losing electrons because hydrogen atoms (H) are often accompanied by their electrons when they are transferred).

pH scale Measures the concentration of hydrogen ions in a solution. As positive H+ ions decrease, the pH goes up.

Product The molecule (or molecules) that results from an enzyme reaction.

Rate constants Define the speed at which individual steps in a reaction occur.

Reaction velocity (V) The reaction rate at a given substrate concentration. This rate increases until it reaches V_{max}.

Reduction The process where a substance gains electrons. The substance that gains electrons is said to be reduced. (In terms of hydrogen atoms, it can be defined as the gain of hydrogen atoms, because hydrogen atoms bring electrons with them).

Substrate The molecule (or molecules) upon which an enzyme acts.

Transition state A high-energy intermediate in a reaction. Enzymes stabilize the transition state to lower the activation energy of a reaction.

Turnover number (K_{cat}) The turnover number describes the number of substrate molecules converted to product by the enzyme per unit of time in seconds, when enzyme is saturated with substrate.

Vitamin An organic compound that is essential for physiological function, but is either not synthesized in sufficient quantities or not synthesized at all in the body. Most vitamins must be obtained through the diet, though some, such as vitamin D, can be produced in other ways, such as through exposure to sunlight. Many vitamins are precursors to coenzymes.

V_{max} The maximum rate of an enzyme-catalyzed reaction when the enzyme is saturated with substrate.

Zymogen (proenzyme) An inactive precursor of an enzyme. It requires a change in order to become activated. Often, this can be cleavage of the proenzyme.

Study Questions

Multiple-Choice Questions

1. Which of the following best describes a transition state in enzyme catalysis?
   A. The point at which the enzyme binds irreversibly to the product
   B. The lowest energy state of the enzyme–substrate complex
   C. A high-energy intermediate between substrate and product
   D. The enzyme-product complex just before release
   E. The turnover number.
2. Which of the following enzyme types catalyzes the transfer of a phosphate group to its substrate?
   A. Hydrolases
   B. Ligases
   C. Oxidoreductases
   D. Kinases
   E. Nucleases
3. In competitive inhibition, the inhibitor:
   A. Binds to a site other than the active site
   B. Forms a covalent bond with the enzyme
   C. Competes with the substrate for the active site
   D. Binds to the cofactor of the enzyme
   E. Both C and D.
4. Protein kinases are:
   A. Enzymes that catalyze the removal of phosphate groups from proteins
   B. Enzymes that catalyze the addition of phosphate groups onto proteins
   C. Enzymes that catalyze hydrolysis of GTP
   D. Enzymes that bind to SH2 domains
   E. B and D

Fill-in-the-Blank Question

5. The Michaelis constant (K_m) is defined as the substrate concentration at which the reaction rate is at ______ ofV_{max}.

Short-Answer Question

6. What does the enzyme lactase do in the body in response to the ingestion of milk?

Matching Question

7. Match each enzyme class with its primary function:
   Enzyme Class
   A. Oxidoreductase
   B. Hydrolase
   C. Ligase
   Function
   1. Ligate two molecules
   2. Transfer electrons
   3. Break bonds using water

Answer Key

Multiple-Choice Questions

1. **Answer**: C
2. **Answer**: D
3. **Answer**: C
4. **Answer**: B

Fill-in-the-Blank Question

5. Half

Short-Answer Question

6. **Answer**: It catalyzes the cleavage of the bond between glucose and galactose, in the disaccharide lactose.

Matching Question

7. **Answer**:
   A → 2, B → 3, C → 1

References

1. Palmer T, Bonner PL. Enzymes: biochemistry, biotechnology, clinical chemistry. 2nd ed. Woodland Publishing; 2007.
2. Nelson DL, Cox MM, Hoskins AA. Lehninger principles of biochemistry. 8th ed. Macmillan Learning; 2021.
3. GeorgiaTech Bioprinciples. Energy and enzymes. Available from: https://bioprinciples.biosci.gatech.edu/04-energy-and-enzymes/.
4. Alberts B, Bray D, Hopkin K, Johnson A, Lewis J, Raff M, Roberts K, Walter P. Essential cell biology. Norton and Company; 2023.
5. Berg JM, Tymoczko JL, Gatto GJ Jr, Stryer L. Biochemistry. 9th ed. New York: MacMillan Learning; 2019.
6. Thompson RHS. Classification and nomenclature of enzymes. Science. 1962;137(3528):405–8. https://doi.org/10.1126/science.137.3528.405.
7. Robinson PK. Enzymes: principles and biotechnological applications. Essays Biochem. 2015;59:1–41. https://doi.org/10.1042/bse0590001.
8. Engel PC. Enzymes: a very short introduction. Oxford University Press; 2020.
9. Fersht AR. Structure and mechanisms in protein science: a guide to enzyme catalysis and protein folding. World Scientific Publishing Co.; 2017.
10. Cooper G, Adams K. The cell: a molecular approach. Sinauer Associates; 2022.
11. Alberts B, Heald R, Johnson A, Morgan D, Raff M, Roberts K, Walter P, Wilson J, Hunt T. Molecular biology of the cell. W.W. Norton and Company; 2022.
12. Marangoni AG. Enzyme kinetics: a modern approach. Hoboken: Wiley-Interscience; 2003.
13. Koshland DE. Application of a theory of enzyme specificity to protein synthesis. Proc Natl Acad Sci USA. 1958;44(2):98–104. https://doi.org/10.1073/pnas.44.2.98.
14. Kessel A, Ben-Tal N. Introduction to proteins: structure, function, and motion. 2nd ed. Boca Raton: CFC Press; 2018.
15. Srinivasan B. A guide to the Michaelis-Menten equation: steady state and beyond. FEBS J. 2022;289(20):6086–98. https://doi.org/10.1111/febs.16124.
16. Fry M. Essential biochemistry for medicine. 1st ed. Wiley; 2010.
17. Smith JJ, Simons C. Enzymes and their inhibitors: drug development. Boca Raton: Taylor and Francis; 2004.
18. Kaleeg A. Kinetics and thermodynamics. 2006. Available from: https://web.stanford.edu/~kaleeg/chem32/kinT/#4.
19. CreativeEnzymes. Enzyme definition and classification. Creative Enzymes. Available from: https://www.creative-enzymes.com/resource/enzyme-definition-and-classification_18.html.
20. McDonald AG, Tipton KF. Enzyme nomenclature and classification: the state of the art. FEBS J. 2023;290(9):2214–31. https://doi.org/10.1111/febs.16274.
21. Ann Benore M. What is in a name? (or a number?): the updated enzyme classifications. Biochem Mol Biol Educ. 2019;47(4):481–3. https://doi.org/10.1002/bmb.21251.
22. Beringer M, Rodnina MV. The ribosomal peptidyl transferase. Mol Cell. 2007;26(3):311–21. https://doi.org/10.1016/j.molcel.2007.03.015.
23. CreativeEnzymes. Hydrolases. Available from: https://www.creative-enzymes.com/cate/hydrolases_75.html#:~:text=Hydrolases%20are%20a%20large%20and,food%20production%2C%20and%20environmental%20management.
24. Lim SY, Steiner JM, Cridge H. Lipases: it's not just pancreatic lipase! Am J Vet Res. 2022;83(8):ajvr.22.03.0048. https://doi.org/10.2460/ajvr.22.03.0048.
25. Peyrot des Gachons C, Breslin PA. Salivary amylase: digestion and metabolic syndrome. Curr Diab Rep. 2016;16(10):102. https://doi.org/10.1007/s11892-016-0794-7.

26. Reiner DJ, Lundquist EA. Small GTPases. WormBook. 2018;2018:1–65. https://doi.org/10.1895/wormbook.1.67.2.
27. CreativeEnzymes. Lyases. Available from: https://www.creative-enzymes.com/cate/lyases_76.html.
28. Brosnan ME, Brosnan JT. Renal arginine metabolism. J Nutr. 2004;134(10):2791S–5S. https://doi.org/10.1093/jn/134.10.2791S.
29. CreativeEnzymes. Translocase introduction. Available from: https://www.creative-enzymes.com/resource/translocase-introduction_42.html#:~:text=Nowadays%2C%20these%20enzymes%20have%20been%20classified%20under,across%20membranes%20or%20their%20separation%20within%20membranes.
30. Nomenclature Committee of the International Union of Biochemistry and Molecular Biology. Enzyme nomenclature. Queen Mary University of London; 2024. Available from: https://iubmb.qmul.ac.uk/enzyme/.
31. TheBumblingBiochemist. Enzyme classification, EC (enzyme commission) numbers, and exploring the ExplorEnz database. Youtube 2022. Available from: https://www.youtube.com/watch?v=8yO1XEzoVIE
32. International Union of Biochemistry and Molecular Biology. ExplorEnz: the Enzyme Database. 2024. Available from: https://www.enzyme-database.org.
33. Purich DL. Chapter 2: Active sites and their chemical properties. In: Purich DL, editor. Enzyme kinetics: catalysis & control. Boston: Elsevier; 2010. p. 53–169.
34. CreativeEnzymes. Catalytic modes of enzymes. Available from: https://www.creative-enzymes.com/resource/catalytic-modes-of-enzymes_36.html.
35. Fried SD, Boxer SG. Electric fields and enzyme catalysis. Annu Rev Biochem. 2017;86:387–415. https://doi.org/10.1146/annurev-biochem-061516-044432.
36. Jencks WP. Catalysis in chemistry and enzymology. New York: McGraw Hill; 1969.
37. Campbell MK, Farrell SO. Biochemistry. Pacific Grove: Brooks/Cole; 2007.
38. Baldwin GS, Sessions RB, Erskine SG, Halford SE. DNA cleavage by the EcoRV restriction endonuclease: roles of divalent metal ions in specificity and catalysis. J Mol Biol. 1999;288(1):87–103. https://doi.org/10.1006/jmbi.1999.2672.
39. Reuben J. Substrate anchoring and the catalytic power of enzymes. Proc Natl Acad Sci USA. 1971;68(3):563–5. https://doi.org/10.1073/pnas.68.3.563.
40. McKee JR, McKee T. Biochemistry: the molecular basis of life. 7th ed. New York: Oxford University Press; 2023.
41. Radisky ES, Lee JM, Lu C-JK, Koshland DE. Insights into the serine protease mechanism from atomic resolution structures of trypsin reaction intermediates. Proc Natl Acad Sci USA. 2006;103(18):6835–40. https://doi.org/10.1073/pnas.0601910103.
42. Neitzel JJ. Enzyme catalysis: the serine proteases. Nat Educ. 2010;3(9):21.
43. Hofer F, Kraml J, Kahler U, Kamenik AS, Liedl KR. Catalytic site pKa values of aspartic, cysteine, and serine proteases: constant pH MD simulations. J Chem Inf Model. 2020;60(6):3030–42. https://doi.org/10.1021/acs.jcim.0c00190.
44. Chakraborti S, Dhalla NS, Mandal M, editors. Proteases in physiology and pathology. 1st ed. Singapore: Springer; 2017.
45. López-Otín C, Bond JS. Proteases: multifunctional enzymes in life and disease. J Biol Chem. 2008;283(45):30433–7. https://doi.org/10.1074/jbc.R800035200.
46. Marks D. Marks' essentials of medical biochemistry: a clinical approach. Lippincott Williams & Wilkins (LWW); 2017.
47. Ritter JM, Flower RJ, Henderson G, Loke YK, MacEwan DJ, Rang HP. Rang and Dale's pharmacology. 10th ed. London: Elsevier; 2023.
48. Singh S, Bajaj BK. Potential application spectrum of microbial proteases for clean and green industrial production. Energy Ecol Environ. 2017;2(6):370–86. https://doi.org/10.1007/s40974-017-0076-5.
49. Adams RLP, Burdon RH. The function of DNA methylation in bacteria and phage. In: Molecular biology of DNA methylation. New York: Springer; 1985. p. 73–87.
50. Zahran M, Berezniak T, Imhof P, Smith JC. Role of magnesium ions in DNA recognition by the EcoRV restriction endonuclease. FEBS Lett. 2011;585(17):2739–43. https://doi.org/10.1016/j.febslet.2011.07.036.
51. Loenen WA, Dryden DT, Raleigh EA, Wilson GG, Murray NE. Highlights of the DNA cutters: a short history of the restriction enzymes. Nucleic Acids Res. 2014;42(1):3–19. https://doi.org/10.1093/nar/gkt990.
52. Absolute Antibody. What is recombinant DNA technology. Available from: https://absoluteantibody.com/general/recombinant-dna-technology/#:~:text=Recombinant%20Proteins,-Much%20like%20recombinant&text=The%20ability%20to%20customize%20and,been%20developed%20to%20treat%20diabetes.
53. Taylor SS, Kornev AP. Protein kinases: evolution of dynamic regulatory proteins. Trends Biochem Sci. 2011;36(2):65–77. https://doi.org/10.1016/j.tibs.2010.09.006.
54. Arter C, Trask L, Ward S, Yeoh S, Bayliss R. Structural features of the protein kinase domain and targeted binding by small-molecule inhibitors. J Biol Chem. 2022;298(8):102247. https://doi.org/10.1016/j.jbc.2022.102247.
55. Huse M, Kuriyan J. The conformational plasticity of protein kinases. Cell. 2002;109(3):275–82. https://doi.org/10.1016/S0092-8674(02)00741-9.

56. Modi V, Dunbrack RL. Defining a new nomenclature for the structures of active and inactive kinases. Proc Natl Acad Sci USA. 2019;116(14):6818–27. https://doi.org/10.1073/pnas.1814279116.
57. Wouters MA, Husain A. Changes in zinc ligation promote remodeling of the active site in the zinc hydrolase superfamily. J Mol Biol. 2001;314(5): 1191–207. https://doi.org/10.1006/jmbi.2000.5161.
58. Morrison H. Chapter 8: Carboxypeptidase A. In: Morrison H, editor. Enzyme active sites and their reaction mechanisms. Academic Press; 2021. p. 37–40.
59. Beard WA, Wilson SH. Structure and mechanism of DNA polymerase β. Biochemistry. 2014;53(17): 2768–80. https://doi.org/10.1021/bi500139h.
60. Lee C-K, Cheong H-K, Ryu K-S, Lee JI, Lee W, Jeon YH, et al. Biotinoyl domain of human acetyl-CoA carboxylase: structural insights into the carboxyl transfer mechanism. Proteins. 2008;72(2):613–24. https://doi.org/10.1002/prot.21952.
61. Chandel NS. Navigating metabolism. 2nd ed. Cold Spring Harbor: Cold Spring Harbor Laboratory Press; 2020.
62. Blanco A, Blanco G. Chapter 8: Enzymes. In: Blanco A, Blanco G, editors. Medical biochemistry. Academic Press; 2017. p. 153–75.
63. CreativeEnzymes. Effect of pH on enzymatic reaction. Available from: https://www.creative-enzymes.com/resource/effect-of-ph-on-enzymatic-reaction_51.html#:~:text=For%20example%2C%20the%20optimum%20pH,slows%20down%20and%20then%20stops.
64. Neurath H, Walsh KA. Role of proteolytic enzymes in biological regulation (a review). Proc Natl Acad Sci USA. 1976;73(11):3825–32. https://doi.org/10.1073/pnas.73.11.3825.
65. Hames BD, Hooper NM. BIOS instant notes in biochemistry. Taylor and Francis; 2011.
66. Istvan ES, Deisenhofer J. Structural mechanism for statin inhibition of HMG-CoA reductase. Science. 2001;292(5519):1160–4. https://doi.org/10.1126/science.1059344.
67. Friesen JA, Rodwell VW. The 3-hydroxy-3-methylglutaryl coenzyme-A (HMG-CoA) reductases. Genome Biol. 2004;5(11):248. https://doi.org/10.1186/gb-2004-5-11-248.
68. Katzung BG, Vanderah TW, Trevor AAJ. Katzung's basic and clinical pharmacology. 16th ed. McGraw-Hill Education; 2024.
69. Cushman DW, Ondetti MA. Design of angiotensin converting enzyme inhibitors. Nat Med. 1999;5(10): 1110–2. https://doi.org/10.1038/13423.
70. U.S. Food and Drug Administration. Vasotec(Enalapril). Approval overview. 1985. https://www.accessdata.fda.gov/scripts/cder/daf/index.cfm?event=overview.process&ApplNo=018998
71. U.S. Food and Drug Administration. Doxycycline capsules, USP. 1967. https://www.accessdata.fda.gov/scripts/cder/daf/index.cfm?event=overview.process&ApplNo=050641
72. Wu S, Zhou X, Jin Z, Cheng H. Collagenases and their inhibitors: a review. Collagen Leather. 2023;5(1):19. https://doi.org/10.1186/s42825-023-00126-6.
73. Nordström D, Lindy O, Lauhio A, Sorsa T, Santavirta S, Konttinen YT. Anti-collagenolytic mechanism of action of doxycycline treatment in rheumatoid arthritis. Rheumatol Int. 1998;17(5): 175–80. https://doi.org/10.1007/s002960050030.
74. Ghasemi K, Ghasemi K. A Brief look at antitumor effects of doxycycline in the treatment of colorectal cancer and combination therapies. Eur J Pharmacol. 2022;916:174593. https://doi.org/10.1016/j.ejphar.2021.174593.
75. Golub LM, Lee H-M. Periodontal therapeutics: current host-modulation agents and future directions. Periodontol 2000. 2020;82(1):186–204. https://doi.org/10.1111/prd.12315.
76. Smith GN Jr, Mickler EA, Hasty KA, Brandt KD. Specificity of inhibition of matrix metalloproteinase activity by doxycycline: relationship to structure of the enzyme. Arthritis Rheum. 1999;42(6):1140–6. https://doi.org/10.1002/1529-0131(199906)42:6<1140::AID-ANR10>3.0.CO;2-7.
77. Shanmugasundaram S, Solanki K, Saseendar S, Chavada VK, D'Ambrosi R. Role of doxycycline as an osteoarthritis disease-modifying drug. J Clin Med. 2023;12(8):2927. https://doi.org/10.3390/jcm12082927.
78. Xu DH, Zhu Z, Fang Y. The effect of a common antibiotics doxycycline on non-healing chronic wound. Curr Pharm Biotechnol. 2017;18(5):360–4. https://doi.org/10.2174/1389201018666170519095339.
79. Bortolato M, Chen K, Shih JC. Monoamine oxidase inactivation: from pathophysiology to therapeutics. Adv Drug Deliv Rev. 2008;60(13–14):1527–33. https://doi.org/10.1016/j.addr.2008.06.002.
80. U.S. Food and Drug Administration. Nardil. 1959. https://www.accessdata.fda.gov/drugsatfda_docs/label/2007/011909s038lbl.pdf
81. U.S. Food and Drug Administration. Emsam (Selegiline). Approval overview. 2006. Available from: https://www.accessdata.fda.gov/scripts/cder/daf/index.cfm?event=overview.process&ApplNo=021336.
82. U.S. Food and Drug Administration. Parnate tablets. 1961. https://www.accessdata.fda.gov/drugsatfda_docs/label/2007/011909s038lbl.pdf
83. Cleveland Clinic. MAOIs (Monoamine oxidase inhibitors). Available from: https://my.clevelandclinic.org/health/treatments/25220-maois-monoamine-oxidase-inhibitors.
84. Pinder RM. New antidepressants or more of the same? Neuropsychiatr Dis Treat. 2007;3(5): 519–20.
85. Gaweska H, Fitzpatrick PF. Structures and mechanism of the monoamine oxidase family. Biomol

Concepts. 2011;2(5):365–77. https://doi.org/10.1515/bmc.2011.030.
86. U.S. Food and Drug Administration. Topotecan (Topotecan). Approval overview. 1996. Available from: https://www.accessdata.fda.gov/scripts/cder/daf/index.cfm?event=overview.process&ApplNo=022453.
87. U.S. Food and Drug Administration. Irinotecan (Irinotecan). Approval overview. 1996. https://www.accessdata.fda.gov/scripts/cder/daf/index.cfm?event=overview.process&ApplNo=020571
88. Khaiwa N, Maarouf NR, Darwish MH, Alhamad DWM, Sebastian A, Hamad M, et al. Camptothecin's journey from discovery to WHO essential medicine: fifty years of promise. Eur J Med Chem. 2021;223:113639. https://doi.org/10.1016/j.ejmech.2021.113639.
89. Yakkala PA, Penumallu NR, Shafi S, Kamal A. Prospects of topoisomerase inhibitors as promising anti-cancer agents. Pharmaceuticals (Basel). 2023;16(10):1456. https://doi.org/10.3390/ph16101456.
90. Bailly C. Irinotecan: 25 years of cancer treatment. Pharmacol Res. 2019;148:104398. https://doi.org/10.1016/j.phrs.2019.104398.
91. Li F, Jiang T, Li Q, Ling X. Camptothecin (CPT) and its derivatives are known to target topoisomerase I (Top1) as their mechanism of action: did we miss something in CPT analogue molecular targets for treating human disease such as cancer? Am J Cancer Res. 2017;7(12):2350–94.
92. Hertzberg RP, Caranfa MJ, Hecht SM. On the mechanism of topoisomerase I inhibition by camptothecin: evidence for binding to an enzyme-DNA complex. Biochemistry. 1989;28(11):4629–38. https://doi.org/10.1021/bi00437a018.
93. Pommier Y. Topoisomerase I inhibitors: camptothecins and beyond. Nat Rev Cancer. 2006;6(10):789–802. https://doi.org/10.1038/nrc1977.
94. Miner J, Hoffhines A. The discovery of aspirin's antithrombotic effects. Tex Heart Inst J. 2007;34(2):179–86.
95. Kenakin TP. Chapter 6: Enzymes as drug targets. In: Kenakin TP, editor. Pharmacology in drug discovery. Boston: Academic Press; 2012. p. 105–24.
96. Lobanovska M, Pilla G. Penicillin's discovery and antibiotic resistance: lessons for the future? Yale J Biol Med. 2017;90(1):135–45.
97. Bush K, Bradford PA. β-Lactams and β-lactamase inhibitors: an overview. Cold Spring Harb Perspect Med. 2016;6(8):a025247. https://doi.org/10.1101/cshperspect.a025247.
98. Hutchings MI, Truman AW, Wilkinson B. Antibiotics: past, present and future. Curr Opin Microbiol. 2019;51:72–80. https://doi.org/10.1016/j.mib.2019.10.008.
99. Spencer AC, Panda SS. DNA gyrase as a target for quinolones. Biomedicine. 2023;11(2):371. https://doi.org/10.3390/biomedicines11020371.
100. Mosaei H, Zenkin N. Inhibition of RNA polymerase by rifampicin and rifamycin-like molecules. EcoSal Plus. 2020;9(1). https://doi.org/10.1128/ecosalplus.ESP-0017-2019.
101. U.S. Food and Drug Administration. Bactrim(Sulfamethoxazole). Approval overview. 1973. Available from: https://www.accessdata.fda.gov/scripts/cder/daf/index.cfm?event=overview.process&ApplNo=021105.
102. Capasso C, Supuran CT. Sulfa and trimethoprim-like drugs—antimetabolites acting as carbonic anhydrase, dihydropteroate synthase and dihydrofolate reductase inhibitors. J Enzyme Inhib Med Chem. 2014;29(3):379–87. https://doi.org/10.3109/14756366.2013.787422.
103. Garcia-Effron G. Rezafungin-mechanisms of action, susceptibility and resistance: similarities and differences with the other echinocandins. J Fungi (Basel). 2020;6(4):262. https://doi.org/10.3390/jof6040262.
104. Cleveland Clinic. Anticagulants. Available from: https://my.clevelandclinic.org/health/treatments/22288-anticoagulants.
105. Milling TJ Jr, Ziebell CM. A review of oral anticoagulants, old and new, in major bleeding and the need for urgent surgery. Trends Cardiovasc Med. 2020;30(2):86–90. https://doi.org/10.1016/j.tcm.2019.03.004.
106. U.S. Food and Drug Administration. Acova (Argatroban). Approval overview. 2020. Available from: https://www.accessdata.fda.gov/scripts/cder/daf/index.cfm?event=overview.process&ApplNo=020883.
107. U.S. Food and Drug Administration. Angiomax (Bivalirudin). Approval overview. 2000. Available from: https://www.accessdata.fda.gov/scripts/cder/daf/index.cfm?event=overview.process&ApplNo=020873.
108. Warkentin TE, Koster A. Bivalirudin: a review. Expert Opin Pharmacother. 2005;6(8):1349–71. https://doi.org/10.1517/14656566.6.8.1349.
109. Warkentin TE, Greinacher A, Koster A. Bivalirudin. Thromb Haemost. 2008;99(5):830–9. https://doi.org/10.1160/th07-10-0644.
110. U.S. Food and Drug Administration. Eliquis (Apixaban). Approval overview. 2012. Available from: https://www.accessdata.fda.gov/scripts/cder/daf/index.cfm?event=overview.process&ApplNo=202155.
111. Shantsila E, Lip GY. Apixaban, an oral, direct inhibitor of activated factor Xa. Curr Opin Investig Drugs. 2008;9(9):1020–33. https://doi.org/10.1007/s12325-012-0003-2.
112. Hirsh J, Fuster V, Ansell J, Halperin JL. American Heart Association/American College of Cardiology Foundation guide to warfarin therapy. Circulation. 2003;107(12):1692–711. https://doi.org/10.1161/01.CIR.0000063575.17904.4E.

113. Berkner KL. Vitamin K-dependent carboxylation. In: Vitamins & hormones, vol. 78. Academic Press; 2008. p. 131–56.
114. Olivieri C, Li GC, Wang Y, Manu VS, Walker C, Kim J, et al. ATP-competitive inhibitors modulate the substrate binding cooperativity of a kinase by altering its conformational entropy. Sci Adv. 2022;8(30):eabo0696. https://doi.org/10.1126/sciadv.abo0696.
115. Attwood MM, Fabbro D, Sokolov ́AV, Knapp S, Schiöth HB. Trends in kinase drug discovery: targets, indications and inhibitor design. Nat Rev Drug Discov. 2021;20(11):839–61. https://doi.org/10.1038/s41573-021-00252-y.
116. Vickers T. MM curve. Wikimedia Commons, public domain. 2006. https://commons.wikimedia.org/wiki/File:MM_curve.png.
117. Miao X, Zhao Y, Zhu L, Zeng Y, Yang C, Zhang R, Lund AK, Zhang M. The equilibrium of bacterial microecosystem: probiotics, pathogenic bacteria, and natural antimicrobial substances in semen. Microorganisms. 2024;12(11):2253. https://doi.org/10.3390/microorganisms12112253.
118. Malakhova M, et al. Structural diversity of the active N-terminal kinase domain of p90 ribosomal S6 kinase 2. PLoS One. 2009;4(11):e8044. https://doi.org/10.1371/journal.pone.0008044.

Part III

The Flow of Genetic Information

This part covers the biochemical processes involved in the flow of genetic information. According to the central dogma of molecular biology, genetic information flows primarily from DNA to RNA to protein. This part covers the structures and mechanisms that are essential for the flow of genetic information. Chapter 6 starts with DNA. This chapter discusses the structure of DNA and the mechanism of DNA replication. Chapter 7 discusses RNA. It covers RNA structure and its differences from DNA. This is followed by a discussion of transcription, which is the synthesis of RNA using DNA as a template. Next is a discussion of the regulation of gene transcription, also known as the regulation of gene expression, which involves different transcription factors that can regulate various genes in different cells or under different conditions. Chapter 8 explores translation, the process by which proteins are synthesized using DNA as a template. This part also discusses how proteins are processed, folded, transported, and, in some cases, destroyed when they are no longer needed by the cell. Multiple therapeutic strategies are also discussed in this part, including targeting DNA synthesis in cancer cells, utilizing RNA- and DNA-based therapies to treat diseases, and targeting specific proteins for destruction.

DNA and DNA Replication in Normal Cells and in Disease

6

Abstract

DNA contains the genetic information that underlies the central dogma of molecular biology, in which information flows from DNA → RNA → protein. This chapter begins with an examination of DNA structure, including the nucleotides that constitute DNA and the bonds (phosphodiester and hydrogen) that link them. In the cell, DNA takes on the form of chromatin, and in this form, it is bound to proteins. An important protein that binds chromatin is histone. Histones and other DNA-binding proteins are discussed. The second section of this chapter covers DNA replication. Important proteins involved in DNA replication include DNA polymerase and topoisomerase. Then, the replication fork, the leading and lagging strands, and Okazaki DNA polymerase. This chapter also discusses why studying DNA is important for pharmacists and health science professionals. It addresses the importance of understanding genetic disease and how mutations in a single gene can lead to disease in some cases. This chapter covers several examples of therapeutic strategies that target DNA or its replication to treat various conditions.

Keywords

DNA · Central dogma · Nucleic acids · Chromatin · Histones

Abbreviations

AML	Acute myeloid leukemia
ATP	Adenosine triphosphate
Bp	Base pairs
cAMP	Cyclic adenosine monophosphate
CDK	Cyclin-dependent kinase
CHD	Chromodomain helicase DNA-binding
CRCs	Chromatin remodeling complexes
CTCF	CCCTC-binding factor
DC	Dyskeratosis congenita
dCTP	Deoxycytidine triphosphate
DDK	Dbf4-dependent kinase
DNA	Deoxyribonucleic acid
DNMT	DNA methyltransferase
dNTP	Deoxyribonucleoside triphosphate
DOX	Doxorubicin (also referred to by its brand name Adriamycin)
ESCO2	Establishment of sister chromatid cohesion N-acetyltransferase 2
Gb	Gigabase pair (billion base pairs)
HAT	Histone acetyltransferase
HDAC	Histone deacetylase
INO80	INO80 chromatin remodeling complex
ISWI	Imitation Switch complex
Kb	Kilobase pair (thousand base pairs)
Mb	Megabase pair (million base pairs)
MCM	Minichromosome maintenance
mRNA	Messenger RNA

A. Minden, *Biochemistry, Cell and Molecular Biology for Pharmacists and Health Professionals*,
https://doi.org/10.1007/978-3-032-20529-2_6

NIPBL	Nipped-B-like (cohesin loading factor)
ORI	Origin of replication
PDS5A/ PDS5B	Protein partners in cohesin regulation
Pi	Inorganic phosphate
PPi	Pyrophosphate
Pre-IC	Pre-initiation complex
RAD21	Cohesin subunit (gene/ protein name)
RBS	Roberts syndrome
RNA	Ribonucleic acid
SA	Stromal antigen
SAM	S-Adenosyl methionine
SMC	Structural maintenance of chromosomes
SWI/SNF	Switch/sucrose nonfermentable complex
TAD	Topologically associating domain
TERC	Telomerase RNA component
TERT	Telomerase reverse transcriptase
TET	Ten-eleven translocation enzyme
tRNA	Transfer RNA
VCP	Valosin-containing protein

What You Will Learn

- The central dogma of molecular biology: DNA → RNA → protein.
- The basic structure of DNA, including nucleic acids, hydrogen bonds, and phosphodiester bonds.
- Why DNA matters in pharmacy and healthcare: understanding diseases that have a genetic basis, drugs that interfere with DNA, and gene editing technologies.
- Genes and the genome, and the concepts of genes, loci, alleles, and zygosity.
- Chromatic structure and DNA-binding proteins such as histones.
- Cohesins and condensins and how they organize, condense, and compact DNA.
- DNA and histones can be modified by epigenetic modifications such as methylation or acetylation.
- The mechanisms of DNA replication, including the replication bubble, licensing, pre-initiation, initiation, elongation, and termination.
- Telomeres and telomerase, how telomeres protect the DNA from shortening, how abnormalities in telomeres can be linked to cancer, and how telomeres are associated with health.
- Topoisomerases and their roles in relieving supercoiling during DNA replication.
- Topoisomerase inhibitors are sometimes used to treat diseases involving too much cell replication, especially cancer. Some antibiotics are also topoisomerase inhibitors.

This chapter covers DNA and is divided into two sections. Section 1 covers the basic structure of DNA. This includes a discussion of the nucleotides in DNA and the bonds that link them in the DNA strands. It discusses proteins that bind to DNA and the structure of chromatin. Section 2 focuses on DNA replication. It covers the proteins involved in DNA replication, including DNA polymerase. This chapter also discusses why studying DNA is important for pharmacists and health science professionals, how abnormalities in DNA structure and replication can lead to disease, and provides examples of therapeutic strategies that target DNA or its replication for treating various conditions.

1 DNA Structure

1.1 Introduction to Nucleic Acids

Nucleic acids include deoxyribonucleic acid (DNA) and ribonucleic acid (RNA). The basic building blocks of nucleic acids are nucleotides. DNA has the vital roles of coding, storing, and transmitting genetic information. RNA is necessary for the transcription of genes and protein translation. While these roles in genetic information storage and expression are primary functions of DNA and RNA, nucleic acids and nucleotides also have other functions. For example, the nucleotide ATP is the major energy currency of the cell. Cyclic AMP (cAMP), derived from ATP, is a second messenger with important roles in cell signaling. Ribozymes are examples of RNA molecules that function as enzymes, catalyzing various chemical reactions [1–3]. This chapter will focus on DNA, its basic structure, and the process of DNA replication. Much of the discussion in this chapter is drawn from [2], which also provides additional information about the topic, as well as other references cited throughout the text.

1.2 The Central Dogma and DNA Basics

In eukaryotic cells, DNA is found primarily in the nucleus and contains the genetic code. DNA is fundamental to the central dogma of molecular biology. The central dogma describes the flow of genetic information from DNA to RNA to proteins, which occurs in all living organisms (see Fig. 6.1). It involves the processes of transcription (the synthesis of mRNA using DNA as a template), mRNA processing, translation (the synthesis of proteins using mRNA as a template), and protein processing.

The **flow of genetic information** begins with the DNA. Specific regions of DNA contain genes, including coding regions that encode proteins. The genetic code is a mostly universal code that is the basis of the coding regions. It consists of codons, which are three-nucleotide sequences that each code for an amino acid. Coding regions of DNA get transcribed into messenger RNA (mRNA) in the nucleus, with the DNA serving as the template. This is the process of *transcription*. The resulting mRNA gets processed and shuttled out of the nucleus through pores and into the cytoplasm. There, it interacts with ribosomes, and transfer RNA (tRNA) delivers the correct amino acids to each codon in the mRNA molecule during translation (see Fig. 6.2). This results in the synthesis of proteins. These proteins then get folded, processed, modified, and delivered to their correct locations [2, 4, 5]. In this chapter, we will focus on DNA, and in subsequent chapters, we will discuss RNA, transcription, and protein synthesis (translation).

DNA's unique structure makes it optimal for encoding genetic information and for replicating to pass that information to subsequent generations. DNA is a polymer made of long sequences of monomers. The genetic information

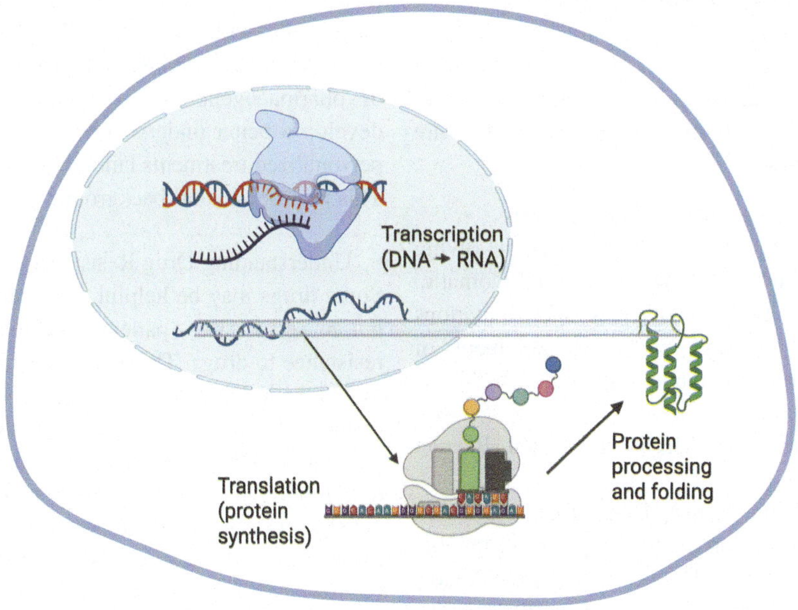

Fig. 6.1 The central dogma of molecular biology. Information flows from DNA to RNA to protein. (*Source*: Created with BioRender.com)

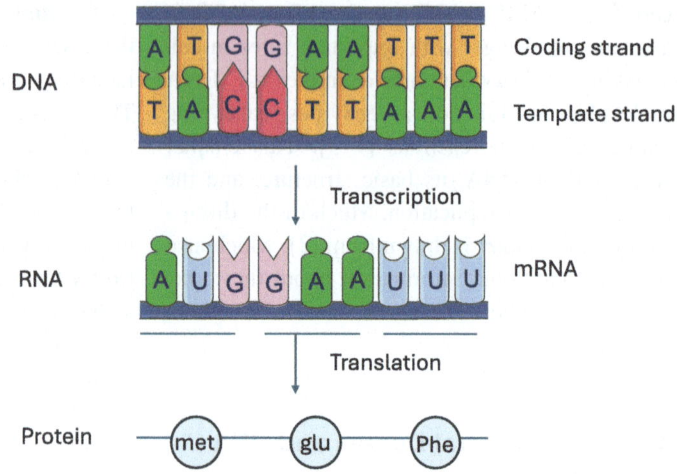

Fig. 6.2 Double-stranded DNA, mRNA, and protein. mRNA is transcribed from the DNA template strand. (*Source*: Adapted from Servier Medical Art (https://smart.servier.com/), licensed under the Creative Commons Attribution 4.0 (CC BY 4.0) license)

is encoded in the specific sequence of four nucleotides. The vast number of possible combinations of these nucleotides allows for tremendous variability, enabling the encoding of a wide variety of genetic traits. Furthermore, DNA's double-stranded complementary structure aids in accurate replication and allows it to be preserved over the generations as it replicates [2].

1.3 Understanding the Structure and Function of DNA Is Important for Pharmacists

There are several reasons why having a strong foundation in the basics of DNA and its structure is important for pharmacists; a few of these are outlined below.

- Many Diseases Have a Genetic Basis

Most diseases have a genetic basis, involving either inherited (germline) or acquired (somatic) mutations in DNA. Identifying such mutations has paved the way to the development of therapies for a wide range of disorders [4, 6].

- Development of Mechanisms That Interact with DNA

The elucidation of the DNA structure has led scientists to develop multiple types of drugs that target DNA directly. Drugs that target DNA have been especially prevalent for cancer treatments. For example, some cancer drugs intercalate into DNA. This interferes with the normal structure of DNA and consequently inhibits its replication. The result is cell death, especially in rapidly dividing cells such as cancer cells [7].

- Pharmacogenomics and Personalized Medicine

Every individual has a unique genetic makeup, and this can impact how they respond to drugs. This interplay between genetics and the response to drugs is covered in the field of pharmacogenomics. While traditionally, two individuals with the same disorder would be given the same treatment, the growth of the field of pharmacogenomics is leading scientists to develop a better understanding of how to create personalized treatments tailored to the individual. This requires a solid background in genetics [8].

- Understanding Drug Resistance

Some drugs may be helpful for a period of time, but in some cases, patients eventually develop resistance to drugs. This resistance can be due to mutations in DNA or changes in gene expression patterns. The ability to identify specific mutations that may lead to drug resistance is also crucial for determining the most effective treatment for a patient and conducting follow-up studies when patients use medications for extended periods [9, 10].

- Gene Editing Technologies

Modern disease treatments do not only involve the use of pharmaceuticals, but increasingly they can also involve the manipulation of genes. This may involve replacing abnormal genes or inserting new genes. Technologies such as the CRISPR-Cas9 system have made this even more feasible, requiring a strong knowledge of DNA structure and function [11].

For all of these reasons and more, a solid background in DNA structure, gene function, and DNA replication is essential for students in pharmaceutical sciences and health sciences.

1.4 The Basic Structure of DNA

This section discusses the basic structure of DNA. DNA is made of nucleotides. RNA is also made of nucleotides, although there are some differences. Therefore, the basic components of DNA and RNA are described in this section, and subsequent sections in this chapter will focus on DNA.

1.4.1 DNA and RNA Are Made Up of Nucleotides

Like the other major macromolecules, DNA is a polymer made up of monomers. The monomers in DNA are called nucleotides. Likewise, the monomers in RNA are nucleotides. **Nucleotides** consist of three components: **a nitrogenous base, a sugar, and a phosphate group**. A typical DNA molecule is double-stranded; it is composed of two strands of nucleotides, and the bases on opposing strands form hydrogen bonds with each other. In contrast, RNA is generally single-stranded. DNA is made from four different nucleotides, distinguished by their nitrogenous bases: adenine, thymidine, cytosine, and guanine (A, T, C, and G), while RNA is made from adenine, uracil, cytosine, and guanine (A, U, C, and G) [2]. The basic structures of these nucleotides are described below. Although this chapter focuses on DNA, the nucleotides that make up RNA are also briefly described, as they share a structural similarity.

- Basic Structure of Nucleotides

Nucleotides in DNA (Fig. 6.3) consist of **five-carbon sugars** that bind to a phosphate group and a nitrogenous base (also called a nucleobase or referred to simply as a base). The nitrogenous base binds to carbon 1 (1′) of the sugar, while the phosphate groups are bound to carbon 5 (5′) of the sugar. Free nucleotides (not bound to DNA or RNA) usually have three phosphate groups, but they can also have only one or two phosphate groups.

- The Sugar in Nucleotides

The sugar in the nucleotides in DNA is deoxyribose. "Deoxy" refers to the absence of an oxygen attached to the 2′ carbon of the sugar. In other words, this site has a hydrogen (H) instead of a

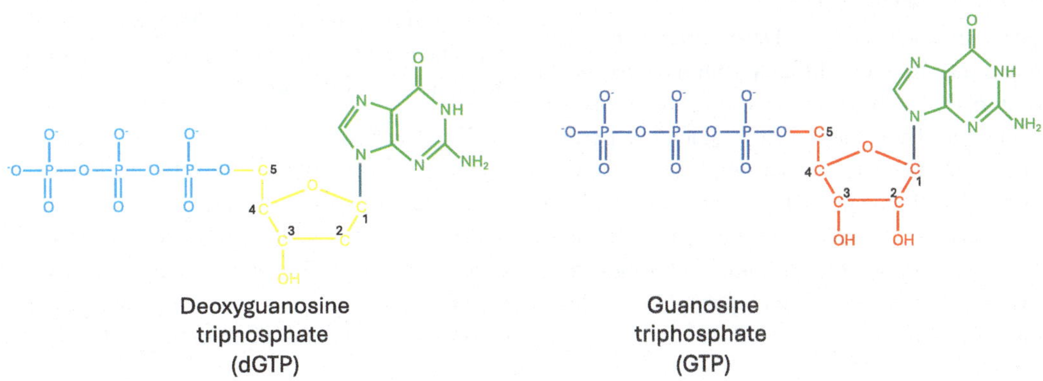

Fig. 6.3 dGTP and GTP. dGTP is a deoxyribonucleotide in DNA, while GTP is a ribonucleotide in RNA. Deoxynucleotides, in DNA, lack a hydroxyl group at carbon 2 of the sugar residue

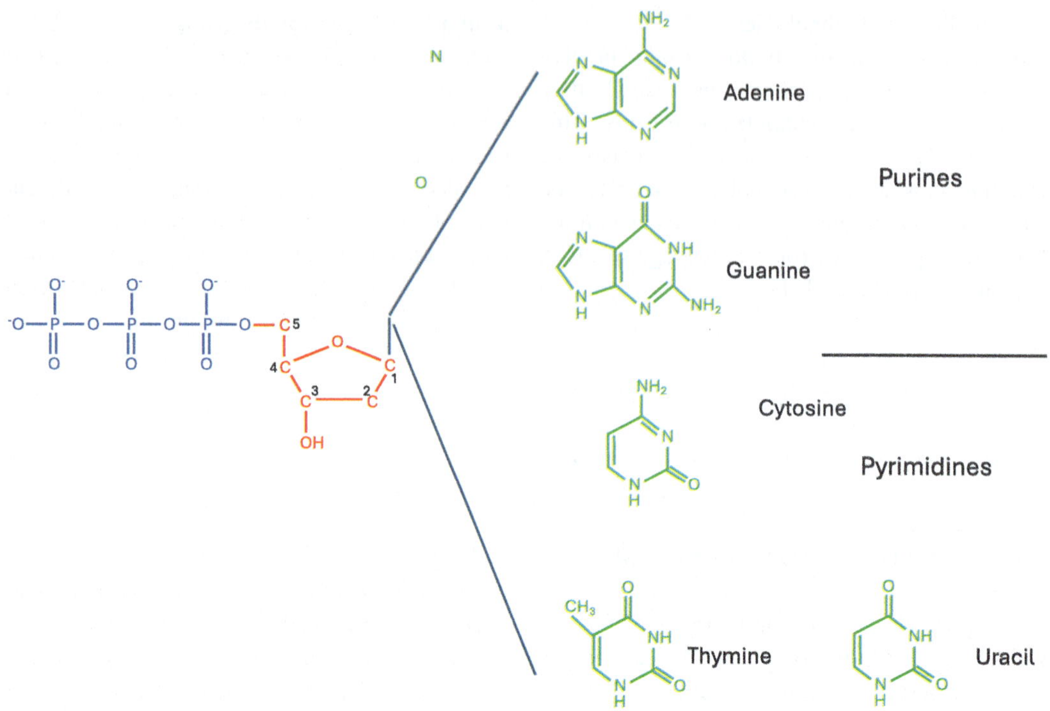

Fig. 6.4 Nucleotides. In nucleotides that comprise DNA the deoxyribose sugar may be bound to adenine, guanine, cytosine, or thymine in DNA. In RNA, the ribose sugar may be bound to adenine, guanine, cytosine, or uracil

hydroxyl group (OH). Only the 3′ carbon of this sugar has an OH group. This is different from ribose, the sugar in RNA, which has an OH group at both the 2′ and 3′ carbons. For this reason, the nucleotides in DNA are technically referred to as deoxyribonucleotides, and nucleotides in RNA are ribonucleotides (Fig. 6.3).

- The Nitrogenous Bases

The four nucleotides in DNA differ from each other due to their different nitrogenous bases (Fig. 6.4). The bases all have ring structures and a nitrogen. These bases are cytosine (C: a pyrimidine), guanine (G: a purine), adenine (A: a purine), and thymine (T: a pyrimidine). RNA has uracil (U: a pyrimidine) instead of thymine. Bases are purines or pyrimidines. Purines have two-ring structures, and pyrimidines have one-ring structures.

- Nomenclature

The nomenclature for the different nucleotides can vary depending upon whether one is referring to the base alone, the base together with the sugar, or the complete nucleotide consisting of the base, sugar, and phosphate group [1, 2, 6, 12, 13], as discussed below.

– The Base Alone

The base alone is referred to as the nitrogenous base or simply the base. In DNA, the bases are cytosine, guanine, adenine, or thymine. RNA has uracil instead of thymine.

– The Nucleoside

The base plus the sugar is called the nucleoside. A nucleoside consists of the nitrogenous base covalently linked to a sugar (deoxyribose for DNA or ribose for RNA). The nucleosides in DNA are deoxycytidine, deoxyguanosine, deoxyadenosine, and thymidine. In RNA, the nucleosides are cytidine, guanosine, adenosine, and uridine. The deoxy prefix is typically used for DNA nucleotides, although it is usually omitted before thymidine, since thymidine is only found in DNA and not RNA. (Uridine is only found in RNA and not DNA.)

– The Nucleotide

The base + the sugar + the phosphate group(s): The base plus the sugar plus one or more phosphate groups is called the nucleotide. Most free nucleotides have three phosphate groups. These are called nucleoside triphosphates. They are named deoxycytidine triphosphate (dCTP), and so on. If they have two phosphates, they are diphosphates, and if they have one phosphate, they are monophosphates.

– Alternative Nomenclature for Monophosphate Nucleotides

When a nucleotide has only one phosphate group, it can be called by its monophosphate name, such as cytidine monophosphate (CMP). However, free monophosphates (not part of a DNA or RNA chain) can also be referred to by an alternative name that ends with the suffix -ylate: cytidylate, guanylate, adenylate, and thymidylate (or uridylate for RNA). These -ylate names are more commonly used in biochemical reactions such as those involved in metabolic pathways [14, 15].

This nomenclature for nucleotides is summarized *in* Table 6.1, along with the commonly used abbreviations.

1.4.2 DNA Is Double-Stranded

DNA is a double-stranded molecule made up of two chains of nucleotides, and each chain has a sugar-phosphate backbone. These two backbones coil around each other to form a double helix. The two strands of the helix are antiparallel, running in opposite directions from each other. The two strands of DNA wind around each other in a right-handed helix. Each turn of the helix has 10.4 nucleotide pairs. The coiling of the two strands around each other forms two grooves, the major groove and the minor groove (see Fig. 6.5a).

The bases face the inside of the double helix, and they bind together via complementary base pairing. A is complementary to T, while G is complementary to C. Thus, A binds to T while G binds to C via hydrogen bonds (see Fig. 6.6b). In both cases, the bulkier base (the purine, which has two rings) pairs with the less bulky base (the pyrimidine, which has one ring) [1, 12, 13].

The length of a DNA strand is usually measured in base pairs (bps) because of the double-stranded nature of DNA. For longer sequences, kilobase pairs (kbs) are used, which measure thousands of base pairs (a kilobase pair

Table 6.1 Nomenclature for nucleotides

Nitrogenous base	Nucleoside: the base + the sugar	Nucleotide: base + sugar + 3 phosphate groups	Nucleotide abbreviation	Alternative name for the monophosphates (1 phosphate group)
Cytosine (C)	Deoxycytidine (DNA) Cytidine (RNA)	Deoxycytidine triphosphate Cytidine triphosphate	dCTP CTP	Cytidylate (same as cytidine monophosphate: CMP)
Guanine (G)	Deoxyguanosine (DNA) Guanosine (RNA)	Deoxyguanosine triphosphate Guanosine triphosphate	dGTP GTP	Guanylate (guanosine monophosphate: GMP)
Adenine (A)	Deoxyadenosine (DNA) Adenosine (A)	Deoxyadenosine triphosphate Adenosine triphosphate	dATP ATP	Adenylate (adenosine monophosphate: AMP)
Thymine (T)	Thymidine (DNA)	Thymidine triphosphate	TTP	Thymidylate (thymidine monophosphate: TMP)
Uracil	Uridine (RNA)	Uridine triphosphate	UTP	Uridylate (uridine monophosphate: UMP)
General (any nucleotide		Nucleotide triphosphate	dNTP NTP	

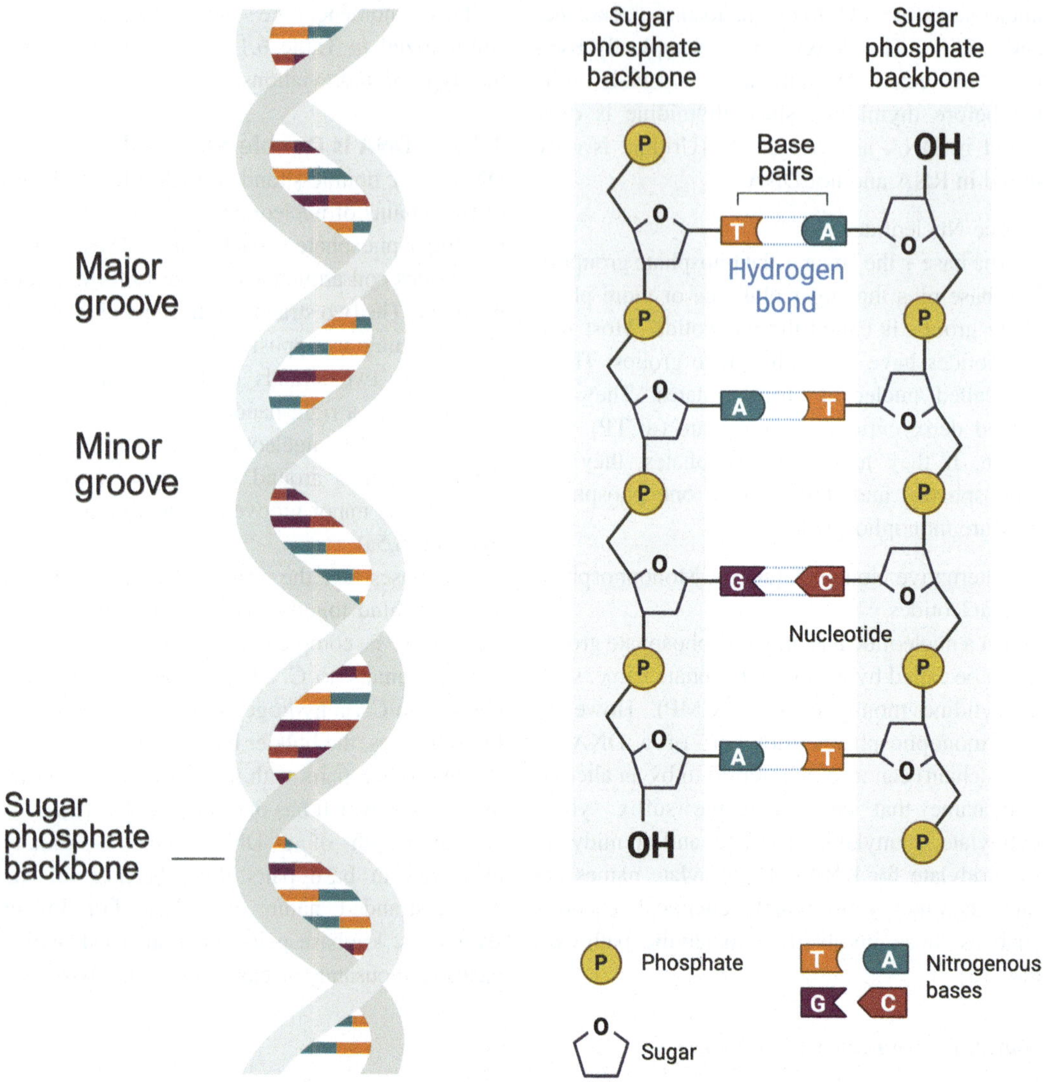

Fig. 6.5 DNA structure. Left: DNA forms a double helix, with a major groove and a minor groove. Right: The two strands are antiparallel, and the 3′ end of each strand has a hydroxyl group. (*Source*: Created with BioRender.com)

is 1000 base pairs). Megabase pairs (Mbs) are used for millions of bases. The entire human genome contains approximately 3.2 billion base pairs, equivalent to 3.2 gigabase pairs (Gb).

1.4.3 Chemical Bonds in DNA

Different kinds of bonds link the nucleotides in DNA [2]. The primary bonds in DNA are the hydrogen bonds and phosphodiester bonds.

- Hydrogen Bonds

Hydrogen bonds are the bonds between complementary nucleotides that base pair with each other (see Fig. 6.6b). Hydrogen bonds are noncovalent interactions. They occur when a hydrogen atom covalently bonded to an electronegative atom, such as a nitrogen or oxygen in a base, is attracted to another electronegative atom on its complementary base. Two hydrogen

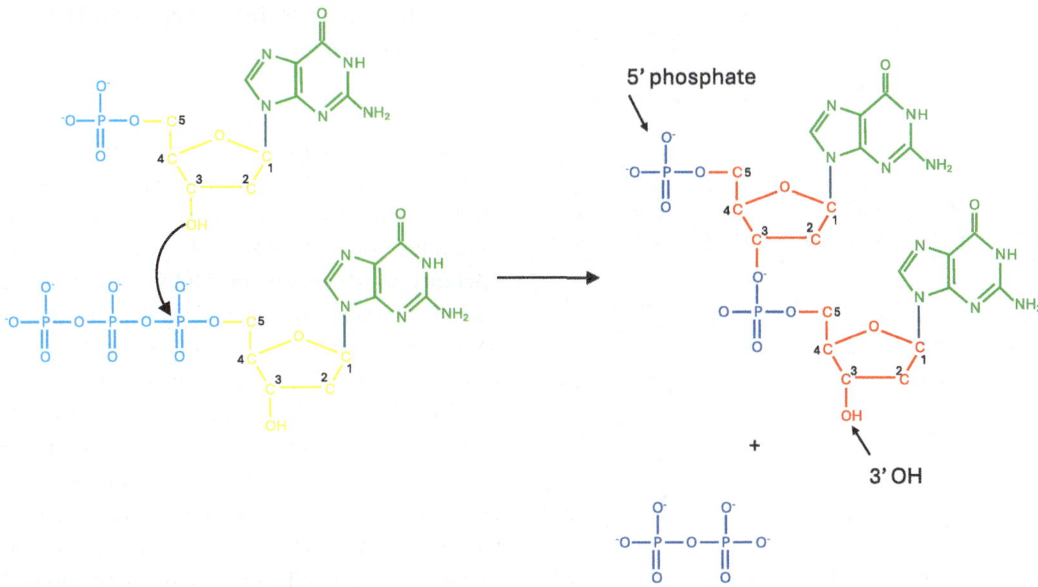

Fig. 6.6 Phosphodiester bonds. The phosphodiester bond connects the sugar and phosphate groups in the DNA backbone

bonds form between A and T, and three hydrogen bonds form between C and G. Individually, hydrogen bonds are weak bonds, but as a group, they form a strong interaction. These bonds therefore stabilize base pairing between complementary base pairs. Nevertheless, these bonds can be broken each time the DNA replicates or during transcription.

- Phosphodiester Bonds

In contrast to the hydrogen bonds, phosphodiester bonds are the bonds that hold the nucleotides together along the backbone (the sugar phosphate backbone). Unlike hydrogen bonds, these are covalent bonds. They link the sugars from one nucleotide with the phosphate in another nucleotide. Free nucleotides generally have three phosphate groups, but when incorporated into DNA, the nucleotides in the DNA backbone have one phosphate group. When each nucleotide is incorporated into the growing chain of DNA, the two terminal phosphate groups are released as pyrophosphate (PPi: two linked phosphate groups). This released pyrophosphate is hydrolyzed to two inorganic phosphate ions (Pi) by the enzyme **pyrophosphatase**. This hydrolysis releases energy, which helps drive the formation of the phosphodiester bond in the DNA backbone (Figs. 6.5b and 6.6). In this bond, the phosphate attached to the 5′ carbon of the sugar on one nucleotide forms a bond with the 3′ OH group on the sugar molecule on its neighboring nucleotide.

1.4.4 Polarity of DNA

The DNA double helix has polarity. The 5′ end of DNA contains a phosphate group (which is bound to the 5′ carbon on the sugar at that end), while at the 3′ end is a hydroxyl group on carbon 3 of the sugar at the other end. When DNA is replicated, new nucleotides can only be added to the 3′ end, where the 3′ OH group is free (Figs. 6.5b and 6.6). The strands in the double-stranded DNA structure are antiparallel. In this configuration, the 5′ end of one strand faces the 3′ end of the other strand [12, 13].

1.4.5 Chromosomes

The DNA in eukaryotic cells is not one continuous molecule. Rather, it is divided and packaged among multiple chromosomes. Chromosomes contain DNA bound to multiple DNA-binding

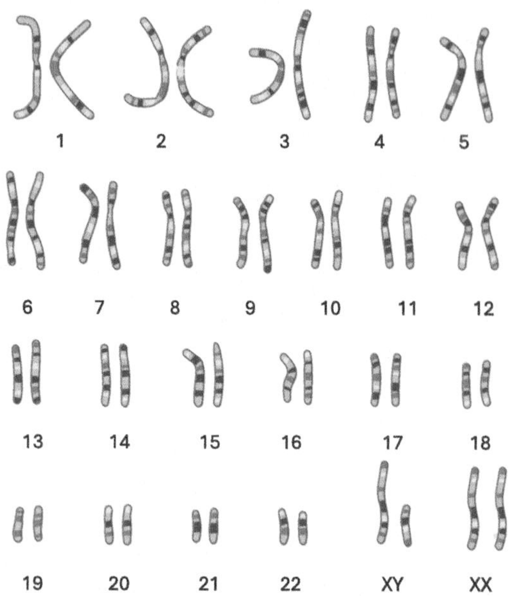

Fig. 6.7 The karyotype. The karyotype is a visualization of all the chromosomes at metaphase. Humans have 23 pairs of chromosomes. (*Source*: Adapted from Servier Medical Art (https://smart.servier.com/), licensed under the Creative Commons Attribution 4.0 (CC BY 4.0) license)

proteins. Humans have 22 pairs of autosomes and 1 pair of sex chromosomes (X or Y), totaling 23 pairs, or 46 chromosomes in all. A karyotype is the display of all chromosomes, which can be visualized by microscopy, at M phase (Fig. 6.7). Chromosomes can be stained with dyes that reveal a pattern of bands specific to each chromosome; this process is known as chromosome banding. Chromosomes can also be analyzed by chromosome painting, a technique in which different-colored fluorescent probes recognize different chromosomes. Chromosomes are the most condensed during M phase. During the rest of the cell cycle, they form longer threads, which are less condensed and not readily visualized and distinguished from each other [12, 16]. During metaphase, chromosomes can be observed, with each sister chromatid having a long arm, referred to as the q arm, a short arm, called the p arm, and a centromere that holds the sister chromatids together.

1.5 Drugs That Interfere with DNA Structure

DNA is crucial for cell survival, and it must maintain its normal structure to function correctly. Several drugs have been developed to interfere with DNA's structure, with the goal of inhibiting the growth of **cancer** cells. **Adriamycin (doxorubicin; DOX)** [17, 18] is an example of a small-molecule drug that can impact DNA structure. DOX is an anticancer drug that operates via multiple mechanisms. One of its major mechanisms is intercalating into DNA. It inserts itself in between the base pairs of the double helix, distorting the DNA's normal structure. DNA is essential for cell survival, and by interfering with DNA, DOX can inhibit the growth of cancer cells, ultimately leading to cell death.

Doxorubicin was FDA-approved in 1974, but it is still one of the most widely used drugs for treating multiple types of cancer (although the brand name Adriamycin is no longer available in the United States, the generic counterpart doxorubicin remains). It is administered intravenously, either alone or in combination with other treatments. Dox is lipophilic (fat-soluble), so it can diffuse across the plasma membrane. It can then enter the nucleus via nuclear pores. Although DOX is effective at targeting cancer cells, it also has considerable side effects due to its role in interfering with DNA and inhibiting the growth of normal cells. Another problem is that cancer cells can develop resistance over time [19–21].

1.6 Genes and the Genome

1.6.1 Some Key Terms

The DNA in a cell contains all the genes that make up the genome. Some key terms that are important for understanding genes and the genome are described below, and also see [4, 22] for further discussion.

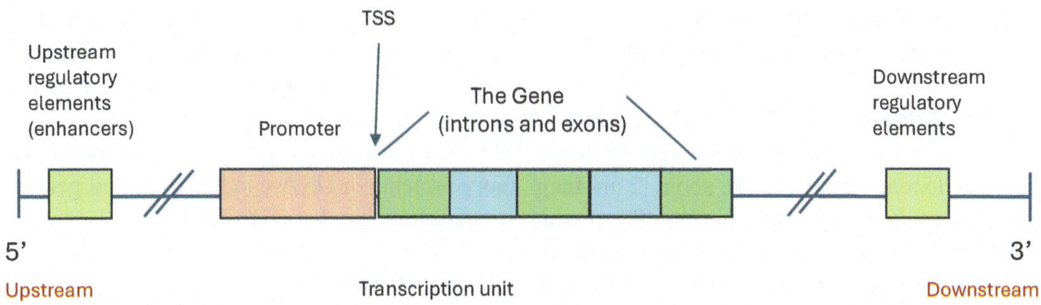

Fig. 6.8 Components of a typical gene locus. TSS is the transcription start site. Upstream and downstream regulatory sequences contain noncoding sequences that play a role in gene regulation. The gene itself contains introns and exons

- Genes

Each chromosome can contain up to thousands of genes. Genes are segments of DNA that contain instructions for making a specific protein. We each have two copies of every gene, one maternal and one paternal.

- Loci

Genes are on a specific region of DNA that is conserved among individuals, called the gene locus (plural: loci) [4, 22].

- Alleles

Different individuals have essentially the same genes, but they have slight variations in each gene. The different forms of a gene are called alleles. An example is the gene OCA2, which codes for melanin and determines the color of certain tissues, such as the eye [23]. One allele of this gene might lead to brown eyes in one person, while another allele may lead to blue eyes in another person. These two individuals carry different alleles of the same gene [4, 22].

- Zygosity

We all have two copies of each gene. If both alleles are the same, then we are homozygous for that gene. If both alleles are different, then we are heterozygous for that gene [4, 22].

1.6.2 Basic Structure of a Gene Locus

A gene typically consists of several sequences in addition to the coding sequence. These include upstream regulatory elements (enhancers and a promoter) and downstream regulatory elements. Between these regions lies the actual gene, which contains the exons and introns. The exons contain the coding sequences, and the introns, which are between the exons, contain noncoding sequences.

The DNA sequence exhibits directionality; the 5′ end of the gene is considered the upstream end, while the 3′ end is the downstream end [4, 22, 24] (Fig. 6.8).

1.6.3 The Genome

The genome refers to the complete DNA sequence of an organism. Every cell has a complete copy of the genome, with a few exceptions, such as mature red blood cells, which lack a nucleus. The human genome has approximately 3 billion base pairs, which comprise about 20,000 protein-coding genes. These genes can give rise to even more proteins due to mechanisms such as alternative splicing, which allows one gene to code for more than one variant of a protein. In addition to genes, the genome also contains regulatory elements, such as promoters and enhancers, as well as noncoding sequences, including introns. A large amount of DNA that does not code for proteins at all or even known gene regulatory elements. In fact, the largest proportion of DNA is noncoding, and some of it still lacks a clearly defined function [25, 26].

1.7 Chromatin Structure and Condensation of DNA

1.7.1 Overview of DNA Condensation

If naked DNA in eukaryotic cells were stretched out end to end, it would be about 2 m long [27, 28], way too long to fit into cells. The reason DNA can fit into our cells is that it rarely exists in isolation. Instead, it binds to and wraps around proteins, causing it to become highly condensed (or compacted). DNA packaged with proteins is called chromatin.

Chromatin structure is dynamic. Although there is always some level of compaction, the level of condensation varies throughout the cell cycle. During interphase, chromatin is more relaxed, allowing for increased gene expression. During the M phase, it becomes significantly more condensed, allowing for more efficient chromosome segregation.

Various proteins are involved in the condensation of DNA. The major proteins that play a role in condensing and organizing DNA include histones, as well as the cohesin and condensin complexes. Histones bind directly to DNA and form the basic chromatin structure. During interphase, cohesin complexes form ring structures that loop around histone-bound chromatin, further condensing the DNA and forming domains called topologically associating domains (TADs). Cohesins also have another function, which is to link the sister chromatids together after S phase and until M phase. During M phase, when the DNA is most condensed, the condensins play a role in further condensing the DNA [2, 4, 22]. The roles of these proteins and various accessory proteins in DNA condensation and genome organization are discussed below.

1.7.2 Histones

The main proteins that bind DNA are histones, but numerous nonhistone proteins also bind DNA. Histones help organize and condense DNA into chromatin. Histones are a family of proteins that tend to be positively charged because they are rich in arginine and lysine. This allows them to interact with the negatively charged backbone of DNA. There are five major types of histones. These are the core histones: histone 2A (H2A), histone 2B (H2B), histone 3 (H3), histone 4 (H4), and the linker histone, histone 1 (H1). There are also some histone variants. Histones are found in all cell types that have a nucleus, and they are among the most evolutionarily conserved of all proteins. Histones are the basis of the nucleosome core, which will be discussed next.

- The Nucleosome Core

The basic structure formed by DNA wrapped around histones is the nucleosome core [2, 12, 29, 30]. In this complex, DNA wraps around a

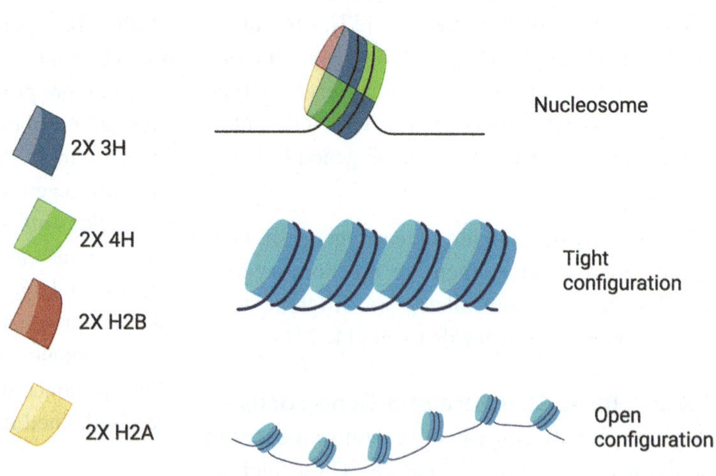

Fig. 6.9 Histones. Eight histones, made up of histone H3, H4, H2A, and H2B make up the nucleosome. DNA can be packaged with nucleosomes in tight or open configurations. (*Source*: Created with BioRender.com)

complex of histones, consisting of eight proteins, forming an octamer. This octamer consists of two dimers consisting of H2A and H2B, and one tetramer consisting of two H3 proteins and two H4 proteins (see Fig. 6.9). This octamer, with DNA wrapped around it, is the nucleosome core.

- The Complete Nucleosome

In addition to the core histone proteins in the octamer, a fifth histone, histone 1 (H1), also has an important role. H1 is a linker histone. It is less well-conserved in sequence than the other histones. Histone H1 is generally located between nucleosomes on the DNA (Fig. 6.9). Together, the nucleosome core and H1 make up the nucleosome.

- Variant Histones

In addition to the core histones and H1, eukaryotes also have some variant histones, which can also assemble onto the nucleosomes. There are fewer of them, and they are less well-conserved. Some of these include H2A.Z, H2A.X, MacroH2A, H3.3, H2B.3, and others. Variant histones create additional variability in chromatin structure [12, 31].

- The Interactions Between the DNA and the Nucleosomes

Nucleosomes are found at about every 200 nucleotide pairs along the DNA. Approximately 147 nucleotides of DNA wrap around each core in about 1.65 turns of the DNA strand. Over 20% of the amino acids in histones are lysine or arginine. Their positive charges can help them interact with the DNA backbone, which is negatively charged due to its phosphate groups. The DNA is associated with the nucleosomes via hydrogen bonds and electrostatic interactions, usually between the histones and the sugar-phosphate backbone of the DNA. Hydrophobic and salt linkages are also present [13, 32].

- The Overall Beads-on-a-String Structure

The basic level of DNA condensation is often referred to as the beads-on-a-string model (see Fig. 6.9). DNA in this configuration is about 10 nM thick, and nucleosomes are connected by segments of linker DNA. The linker DNA is the only portion of the DNA not bound to nucleosomes. This is the basic level of DNA condensation. This structure predominates during interphase (I). The most actively transcribed regions are called euchromatin and are typically in a condensed yet open configuration, allowing for the binding of transcription factors and gene expression [2, 27, 32]. Modifications to histones (discussed later) help regulate how tightly histones are packed. More tightly packed chromatin is called heterochromatin and is generally not actively transcribed. It involves further levels of condensation, which will be discussed in the following section.

1.7.3 Further Condensation by Loop Extrusion: Cohesins and Condensins

The beads on a string structure are present throughout interphase. During this phase, the DNA also tends to form a higher-level structure, organizing into topologically associating domains (TADs). These domains are formed by loops, which bring distant regions of the genome into proximity (Fig. 6.10a, b). This organization plays an important role not only in DNA condensation but also in gene regulation, as it can bring DNA regulatory elements, such as enhancers and promoters, into proximity with the genes they regulate. Once cells reach M phase, transcription largely stops, TADs are not necessary, and the DNA becomes even more compact.

These further levels of chromatin condensation involve protein complexes called cohesins and condensins. **Cohesins** organize DNA in interphase into loop structures that form TADs, and condensins lead to further compaction of the DNA, mostly in M phase. Cohesins also have another role, besides compacting DNA and forming TADs, which is to bind sister chromatids together, which will be discussed in more detail below [33].

- The Basic Functions of Cohesins and Condensins

Cohesins and condensins interact with chromatin, which is DNA bound to histones, contributing to

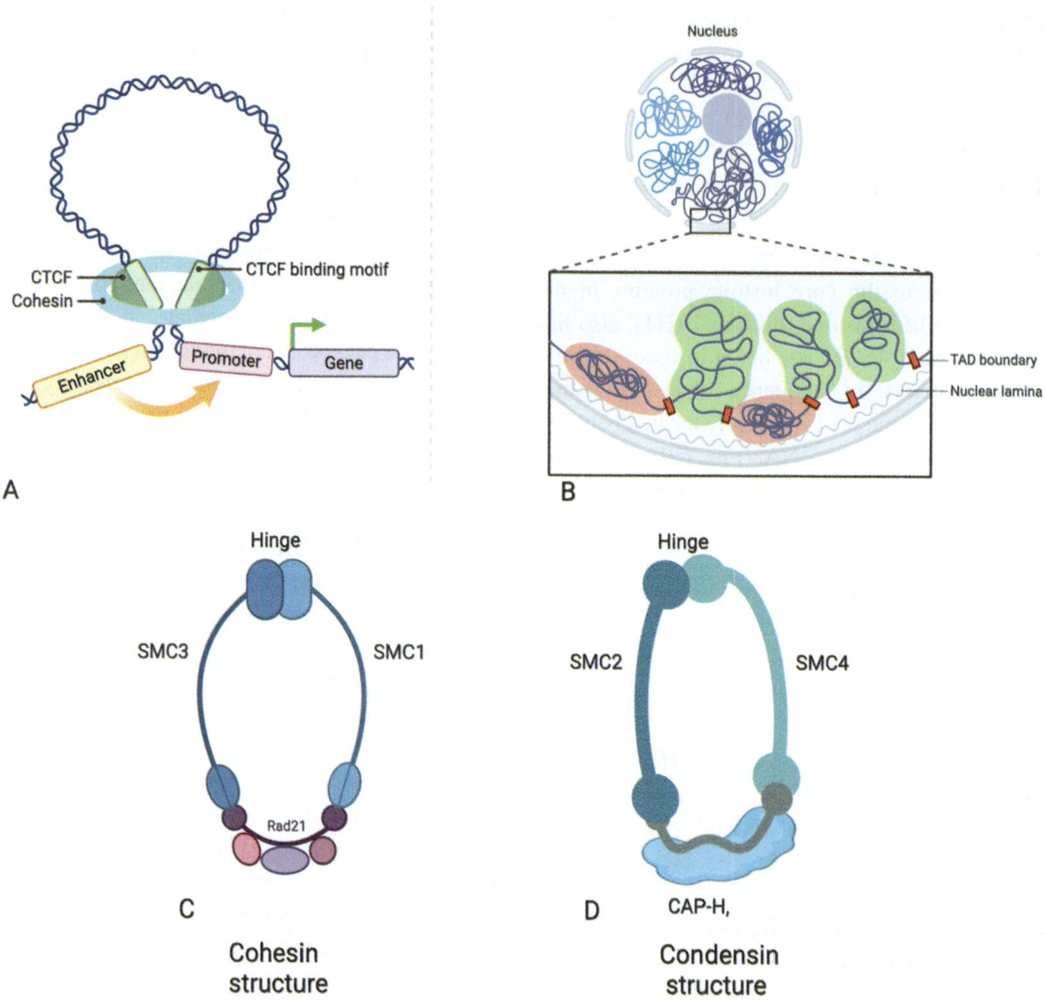

Fig. 6.10 Cohesins and condensins in higher order chromatin structure. **A**. Cohesin encircling DNA regions. **B**. Cohesin is important for the formation of Topologically Associating Domains (TADs). **C**. Basic cohesin structure, **D**. Basic condensin structure. (*Source*: Created with BioRender.com)

even higher-order chromatin structure. They are both members of the structural maintenance of chromosomes (SMC) family of complexes. The SMCs are ATPases that use ATP hydrolysis to facilitate interactions with DNA and regulate chromatin architecture. They form ring-like structures and help organize the DNA into loops (TADs) [2, 12, 25, 34, 35].

Cohesins play roles in both intra-molecular crosslinking (looping within the same chromatid) throughout interphase and inter-molecular crosslinking (between sister chromatids) from the S phase to the M phase. Condensins have roles only in intramolecular crosslinking, and their main functions are during M phase [33]. Both are described below. First, we discuss their roles in organizing and condensing DNA, and then we delve into the various proteins that comprise the cohesin and condensin complexes in more detail.

- Cohesins

– Cohesins and Intramolecular Crosslinking, the Formation of TADs

Cohesins are important for the formation of topologically associated domains (TADs) throughout interphase. Cohesins are loaded onto DNA after M phase, in early G1 phase. They contribute to TAD formation through a process called loop extrusion, in which cohesin complexes extrude loops of chromatin through their ring-like structures [36].

This cohesion-mediated loop extrusion is a regulated process. It is regulated by a DNA-binding protein called CTCF (CCCTC-binding factor), which binds to a specific DNA sequence (that usually contains the sequence CCCTC or a related sequence). This binding serves as an anchor or barrier that prevents further cohesion-mediated loop extrusion, thereby allowing defined loops to form [37]. Thus, although cohesins do not bind DNA directly, the formation of loop structures is organized by CTCF and several other associated proteins, which define the beginning and end points of the loops (see Fig. 6.10a).

– Cohesins and Intermolecular Crosslinking Between Sister Chromatids

In addition to their roles in DNA condensation, cohesins also have another important function. This is the linking of sister chromatids. This involves intermolecular interactions that hold replicated chromatids together until the onset of anaphase. After DNA replication, a subset of the cohesins will trap sister chromatids due to the cohesin's ring-like structure. In this way, they hold the sister chromatids together. This is important in early mitosis as it helps ensure proper chromosome segregation. During prophase, the first subphase of mitosis, most of these cohesins (those involved in sister chromatid adhesion) are released from chromosome arms, while some cohesins remain at the centromeres. A protein called Shugoshin (Sgo1) protects cohesins at centromeres. Later, at the onset of anaphase, a protease called separase cleaves a subunit of the cohesion complex called kleisin. This releases all remaining cohesins that bind sister chromatids together. Now the sister chromatids can separate [2, 38] (see Fig. 6.11).

- Condensins

Condensins have similar ring-shaped structures to condensins, but they are important for even further compaction of the DNA. During M phase, DNA becomes even more compact. This highly condensed state is optimal for accurate segregation during cell division. This is where condensins have a key role. Transcription generally does not take place during M phase, and thus, the TAD domains are not necessary at this point. Condensins largely replace the cohesins at this stage, TADs are mostly eliminated, and through further loop extrusion, the DNA becomes organized into even more condensed arrays of loops. Most condensins are removed from DNA after M phase [2, 39, 40].

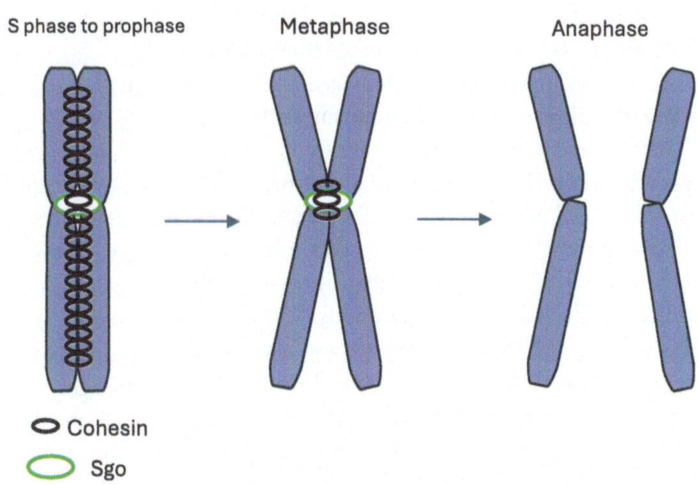

Fig. 6.11 Cohesins and sister chromatids. One function of cohesins, along with Shugoshin (Sho1) is to link sister chromatids after S phase and throughout M phase until anaphase

Condensins are especially important in M phase, but even during interphase, some condensins (condensing II, discussed below) may remain and contribute to chromatin organization. They are involved in chromosome condensation in specific genomic regions or cell types. However, overall, chromatin remains more relaxed and less condensed in interphase compared to mitosis, and the condensins now play a lesser role.

- Cohesins and Condensins Are Made of SMC Complexes

– Cohesin SMC Complexes

Cohesin is made up of SMC3 and SMC1 that bind together to form a V-shaped dimer. The kleisin subunit Rad21 forms a bridge between the two ATPase head domains of the SMCs. Thus, the kleisin closes the ring structure, forming a ring shape that can encircle the DNA. Additional regulatory subunits, including SA1/SA2 (also known as STAG1/STAG2), PDS5A/B, and WAPL, regulate the stability and dynamics of the ring. SA1/SA2 helps stabilize cohesin on the DNA, while PDS5 and WAPL regulate its release. The basic structure of cohesin is seen in Fig. 6.10c [41, 42].

– The Condensin SMC Complexes

Condensin is made of SMC2 and SMC4, which also come together to form a V shape. There are two condensins, condensin I and condensin II. They both use different kleisin subunits. One is called CAP-H, and it functions on condensin I (see Fig. 6.10c), and the other is CAP-H2, which functions on condensin II. In both cases, they also close the ring structure. Condensin I also has CAP-D2 and CAP-G, which are proteins that play a role in chromatin binding and regulation of condensin activity, while condensin II has CAP-D3 and CAP-G2, which serve similar roles in condensin II. The two condensin complexes differ in their localization and timing. Condensin I is cytoplasmic during interphase and associates with chromosomes after the breakdown of the nuclear envelope. Condensin II remains nuclear throughout the cell cycle and initiates chromatin condensing earlier, in prophase [2, 41, 43, 44].

The ATPase activity of the SMC proteins is essential because ATP binding and hydrolysis control the opening and closing of the cohesion and condensin rings. This is necessary for driving the loop extrusion process, compacting chromatin, and regulating the overall dynamics of chromatin organization [12, 25, 35].

- Some Disorders Involving Cohesins and Condensins

Cohesins and condensins play crucial roles in regulating DNA structure, and therefore, mutations in the genes encoding them can lead to developmental disorders and diseases. Mutations in cohesin subunits can lead to a group of developmental disorders known as cohesinopathies [45]. More broadly, such disease falls into the category of chromatinopathies, which are diseases involving the regulation of chromatin structure. An example is **Cornelia de Lange syndrome (CdLS)**. This is a rare developmental disorder characterized by growth delays, intellectual disability, and distinctive facial features. CdLS is considered to be a spectrum of disorders, and it can be caused by mutations in one of several different genes. Mutations in genes encoding cohesin subunits or proteins that regulate cohesins are thought to be involved in the disease. The symptoms can be variable and range in severity, depending on the specific mutations. Some genes that have been identified as mutated in the disease include SMC1A, SMC3, NIPBL, RAD21, and HDAC8. SMC1A and SMC3 are the basic proteins that make up cohesins, as described above. The NIBPL gene is a cohesion-loading factor. It codes for delangin, which has a role in facilitating the interaction between the cohesion complex and chromatin [46]. Rad21 is a cohesion subunit [47], and HDAC8 deacetylates a subunit of cohesion (SMC3) and thus has a role in regulating cohesin's activity [48]. Cohesins are involved not only in chromatin structure and linking chromatids together, but also in regulating gene expression. Gene expression abnormalities are seen in the disease, particularly in genes involved in the development of the face and limbs

[49, 50]. So far, treatment of the disease has focused on addressing its symptoms, although understanding the molecular basis of this type of disorder may hold promise for more effective treatments in the future.

Roberts syndrome (RBS) is also a disease associated with cohesins [51]. It is also a rare disease caused by a mutation in a gene called ESCO2 (establishment of sister chromatid cohesion N-acetyltransferase 2). This gene encodes a protein that is important for acetylation of cohesion, which is necessary for its interaction with chromatids. RBS leads to severe prenatal growth retardation, limb malformations, and facial abnormalities. Some, but not all, symptoms of RBS overlap with those of CdLS [52].

Mutations in cohesin are also implicated in certain cancers. For example, mutations in cohesin components are found in some cases of **acute myeloid leukemia (AML)** [53]. Likewise, cohesin mutations are common in bladder cancer [54]. Overexpression of a type of condensin is sometimes seen in breast cancer and may be associated with genomic instability and tumor progression [55].

There are various types of mutations in genes encoding cohesion or cohesin regulators that are associated with cancer. Many of these are inactivating mutations, but sometimes overexpression is seen instead. There are currently no approved treatments for cancer that target cohesins; however, studies are underway to develop methods that specifically target cohesins or those cancer cells with cohesion mutations [56, 57].

1.7.4 The Chromatin Structure at the Centromere

We have discussed some of the important features of the chromatin, but the centromere of chromosomes has some unique features. The centromere is the region of DNA near the center of each chromosome that links the sister chromatids during prophase. The proteins that bind this region of the DNA have some differences compared with the proteins that bind the rest of the DNA. On either side of the centromere are the arms of the chromosomes. These are the short arms (the p arms) and the long arms (the q arms). The chromatin structure at the centromeres is very compact. It contains a variant of histone H3 called CENP-A (centromere protein A).

The centromere provides a site of attachment for the complex of proteins that constitute the kinetochore. The kinetochore, in turn, links to the spindle microtubules during mitosis. The kinetochore consists of three major parts. The inner kinetochore is the part that binds directly to the centromere. The fibrous layer is the outermost part and binds the microtubules. In between the inner kinetochore and the fibrous layer is the outer kinetochore [2, 12, 58] (see Fig. 6.12).

In addition to these features, condensins are also present on the centromere regions. They are responsible for holding sister chromatids together until after anaphase, as discussed above. Condensin complexes are also found at the centromere, contributing to the high level of compaction in this region [59, 60].

1.8 Epigenetic Modifications on DNA and Histones

1.8.1 What Are Epigenetic Modifications?

Chromatin, composed of DNA and associated proteins, can undergo various chemical modifications. When these modifications are stable and passed on through cell division, they are considered epigenetic modifications. These modifications do not affect nucleotide sequence of the DNA. Instead, the nucleotide sequence stays the same, but the nucleotides have a structural modification. Additionally, the associated proteins, especially histones, may be modified. The modifications may include acetylation, phosphorylation, methylation, or others [2] (see Fig. 6.13a).

1.8.2 Modifications of Histones

Histones are subject to various modifications. This plays an important role in regulating the level of DNA condensation and the accessibility of other proteins, such as transcription factors that bind DNA. Each core histone has an amino

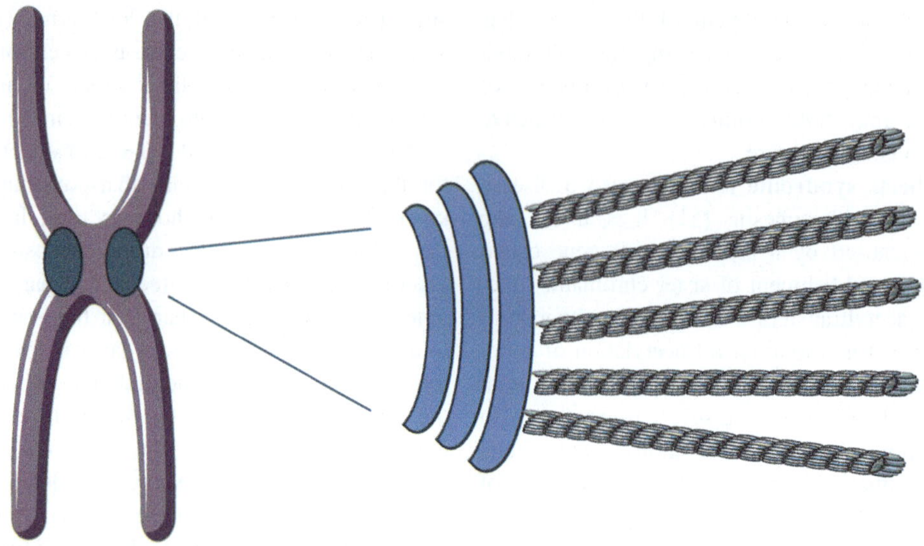

Fig. 6.12 The kinetochore. The kinetochore contains highly compact DNA, and the DNA in this region contains a variant of histone H3. (*Source*: Adapted from Servier Medical Art (https://smart.servier.com/), licensed under the Creative Commons Attribution 4.0 (CC BY 4.0) license)

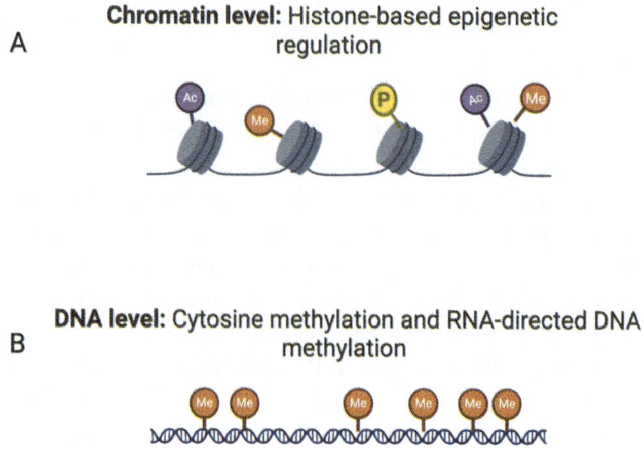

Fig. 6.13 Epigenetic modifications. Histones and DNA can be modified. For example, histones can be modified by acetylation, methylation, or phosphorylation. CpG regions in DNA are sometimes modified by methylation. (*Source*: Created with BioRender.com)

(N) terminal "tail" that extends out from the DNA-histone core. These tails are often the regions that are modified [2, 61]. Histones and their various modifications are discussed below.

Histone Modifications and *Histone Modifiers*

Histones are often modified by acetylation, which is the addition of acetyl groups (-C-CH_3), usually on the ε amino group (the terminal amino group on the side chain) of the amino acid lysine. The enzyme that catalyzes the addition of acetyl groups to histones is **histone acetyltransferase (HAT)**. The acetyl groups can also be removed. This is catalyzed by a different enzyme, **histone deacetylase (HDAC)** [2] (Fig. 6.14).

Histones can also be modified by methylation, usually on lysine residues. The enzymes that catalyze the addition and removal of methyl groups are histone methyltransferases and histone demethylases, respectively. Histones may

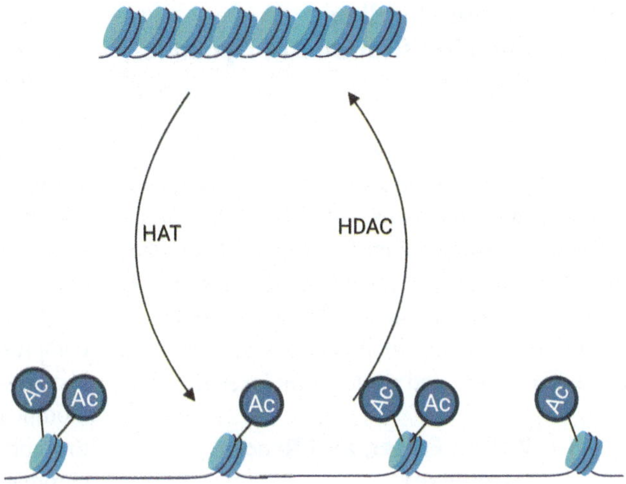

Fig. 6.14 Regulation of histone acetylation by HAT and HDACs. Histone acetylation is regulated by HAT, which catalyzes the addition of acetyl groups onto histones, and HDACs, which remove them. (*Source*: Created with BioRender.com)

sometimes be modified by phosphorylation. Phosphate groups are added to serines by kinases and removed by phosphatases [1, 61]. The types of enzymes that modify histones, such as HDACs, are members of the family of histone modifiers.

Histone Codes Can Describe the Histone Modifications

Histone modifications can be referred to by a specific type of "histone code." These codes follow a pattern as follows: (1) the abbreviation for the histone (such as H2A, H2B, H3, H4); (2) the marked (modified) amino acid side chain on that histone; (3) the distance of the modified amino acid from that histone's amino terminus; and (4) the type of modification. For example, H3K4me3 refers to Histone H3, lysine (K), position 4, with the modification being trimethylation. Alternatively, H3K27AC means that the lysine on position 27 of histone H3 is modified by acetylation (Ac) [2, 12, 62, 63].

Histone Modifications Have Various Functions

The number of possible combinations of histone modifications is extensive, and each can have a specific function. Modifications can alter chromatin configuration by inducing the chromatin to adopt a more open or closed conformation. They can also play a role in recruiting other DNA-binding proteins to the chromatin. These can include transcription activators or repressors, and thus, they have roles in regulating gene expression. They can also function in signaling that a stretch of chromatin is newly replicated or that the DNA in a particular chromatin region is damaged and needs repair [64, 65].

1.8.3 Modification of DNA: Methylation

Epigenetic changes can also be found directly on the DNA (see Fig. 6.13b). Importantly, DNA can be methylated. Most commonly, this happens on cytosine (C) residues. This is particularly relevant for C residues followed by a guanine (G). **DNA methyltransferases (DNMTs)** are enzymes that have a key role in methylating DNA. They transfer a methyl group from s-adenosyl methionine (SAM) to cytosine. (This methylation process is also discussed in Chap. 16.). In contrast to DNMTs, **ten-eleven translocation (TET)** enzymes are involved in demethylation, or the removal of methyl groups from DNA [6, 66, 67].

Modifications to both histones and DNA play important roles. Among their important functions are their roles in regulating gene expression. These functions have led to the development of drugs that can modulate these epigenetic changes, such as HDAC inhibitors. These topics will be discussed in more detail in Chap. 7, which covers RNA and transcription.

1.9 Histone Binding and Modification Are Dynamic Processes

Modifications on histones are dynamic. They can be made anew and can also be removed. They can also spread to other parts of the chromatin. This is especially important after DNA replication, when modifications must spread across the newly synthesized DNA strand. These processes of adding or removing modifications are regulated by writer, eraser, and reader complexes [2].

1.9.1 Writer, Eraser, and Reader Complexes

- Basic Functions of the Complexes

Writer complexes, "write" or add modifications onto the histones. They contain enzymes that can add modifications such as methyl groups and acetyl groups onto the histones. An example is the histone modifier **histone acetyltransferase (HAT)**.

In contrast, eraser complexes can remove specific modifications. Histone deacetylases (discussed above) and histone demethylases, which remove acetyl groups and methyl groups, respectively, are examples of eraser complexes.

Reader complexes are different than writer and eraser complexes. Rather than adding or removing modifications, they "read" the modifications that are already present on histones. In return, they may carry out specific functions encoded by the modifications.

Reader complexes contain protein domains that can bind to specific modifications and recruit other proteins to the DNA in response to these modifications. For example, they may contain a domain that binds to an acetylated lysine residue on histone tails, and another domain that interacts with a transcriptional activator. In this way, they may recruit transcriptional activators to specific places on the genome, thus regulating gene expression [2, 68].

- Spreading of Histone Modifications

When new histone modifications are made, they usually need to be spread throughout the nearby regions of DNA. For example, during DNA replication, when new DNA is copied, new modifications need to be added to histones on the duplicated DNA and then spread to neighboring chromatin regions. This spreading also occurs during interphase, when chromatin modifications take place in response to various signals and the need for the regulated transcription of specific genes.

During the spreading of modifications, writer complexes add specific modifications to the histones. Reader complexes then recognize these modifications and recruit more writer complexes to replicate the modifications in neighboring chromatin regions. This becomes a cycle of reading and writing, which can occur over long stretches of DNA, although there is a limit as to how far the spreading can occur [2, 12].

1.9.2 Chromatin Remodeling Complexes

Histones bind to DNA and form nucleosomes. Histones can also be removed, moved, or added. Enzymes that catalyze these changes are ATP-dependent complexes called chromatin remodeling complexes (CRCs). (In some cases, they are also called nucleosome remodeling complexes.) Rather than modifying histones, CRCs reposition or remove histones. One type of change that can occur is the eviction of histones. During this process, certain histones are completely removed from the DNA, and consequently, the entire nucleosome is disassembled. This can cause the DNA to adopt a more open configuration, enabling the binding of proteins such as transcription factors, which, in turn, facilitates gene transcription. Another type of change that can occur is histone replacement, where one type of histone may be replaced by a different type. For example, in some situations, histone H3 gets replaced with the variant H3.3. Histone H3.3 is more suitable for active transcription compared with H3, because it can undergo modifications that lead to a more open chromatin configuration.

In some cases, nucleosome sliding occurs. In this situation, the nucleosomes "slide" along the DNA but are not removed. As a result, certain regions of DNA become exposed, allowing transcription factors to bind more readily to the DNA. There are two major ways that nucleosome sliding can occur. The first is spontaneous sliding, which may sometimes occur without any particular stimulus. The second mechanism is ATP-dependent sliding. Here, ATP-dependent chromatin remodeling complexes can actively move nucleosomes from one location to another along the DNA [2, 12, 61, 69].

Examples of CRCs include proteins in the SWI/SNF family (also called BAF/BRG1/BRM-associated factor complexes) and those in the ISWI family. These complexes are composed of several proteins and peptides, and they include an ATPase motor protein. The SWI/SNF complex has a role in disrupting histone–DNA interactions. It can induce nucleosome sliding, ejection, and unwrapping, making the DNA more accessible for transcription factors to bind. The complex is often associated with activated transcription, and in some cancers, it has been linked to tumor suppressor activity. Imitation Switch (ISWI) is usually involved in nucleosome sliding but not in nucleosome eviction. They may also be involved in regulating transcription by influencing chromatin structure. Other examples of CRCs include the CHD family and the INO80 family, which are also involved in nucleosome sliding and exchange [2, 70–74].

1.10 Organization of DNA Within the Nucleus

DNA within the nucleus can be visualized by microscopy after staining with certain dyes. These staining techniques have revealed that within the nucleus, most cells have a combination of heterochromatin and euchromatin. Heterochromatin is highly condensed, while euchromatin has a more open, less condensed form. Some of the euchromatin is active chromatin. This is the DNA that is being actively transcribed. This is the most open form of DNA because DNA needs to be in an open configuration during transcription. The rest of the euchromatin is inactive; it is not actively transcribed, but it remains more open than the heterochromatin. Heterochromatin can exist in two major states: constitutive heterochromatin, which is permanently condensed, and facultative heterochromatin, where the condensation level can be regulated to control gene expression. The overall level of DNA packaging also depends on the stage of the cell cycle. During the M phase, all the DNA takes on the most highly condensed form, which is why chromosomes can be visualized by microscopy during this phase.

Whether the DNA is in a heterochromatin or euchromatin state is related to multiple factors, including histones and their modifications in the bound nucleosomes. In the heterochromatin, the nucleosomes are less acetylated, while the histones on euchromatin are more likely to be acetylated. Phosphorylation and methylation of certain sites on the histones are also more likely to lead to euchromatin (a more open configuration), but methylation of certain other sites on histones may lead to condensation (heterochromatin). There are different histone variants in the different classes of chromatin as well, which can impact how compact the chromatin is. DNA methylation also contributes to methylation. More methylation of the DNA is more likely to be associated with euchromatin rather than heterochromatin.

Cohesin and condensin complexes also have a role. Cohesins, which help organize the chromatin into TADs, are more often associated with euchromatin, which is found in transcriptionally active regions. Cohesins contribute to higher levels of compaction and are associated with heterochromatin during mitosis.

The proportion of DNA in the different states changes during different phases of the cell cycle. During M phase, most of the DNA is in the form of heterochromatin. During interphase, where transcription occurs, a mixture of heterochromatin and euchromatin is present. Both heterochromatin and euchromatin can replicate during S phase [2, 12, 75].

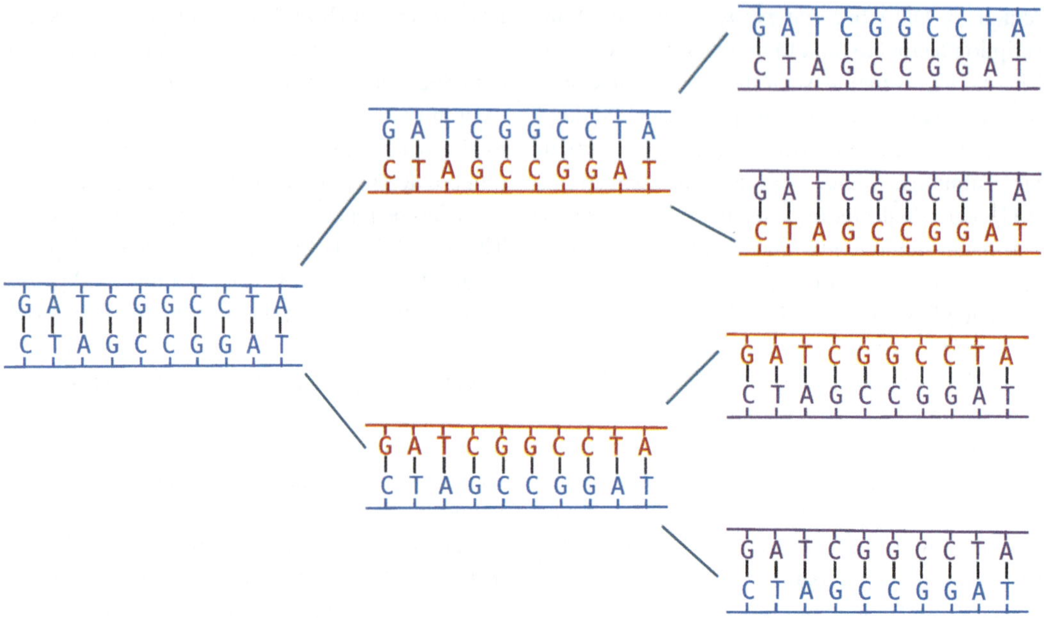

Fig. 6.15 Semiconservative nature of DNA replication. When DNA replicates, each new molecule consists of one original (parental) strand and one newly synthesized strand

2 DNA Replication

2.1 DNA Replication: Overview

DNA replication is essential for passing down genetic material from one generation to the next, and consequently, it is essential for life. DNA replication is also called DNA synthesis because it involves the synthesis of an entire new copy of the DNA. It occurs during S phase of the cell cycle so that a complete new copy of DNA is made each time the cell goes through a round of the cell cycle. When DNA replicates, each strand of the original double-stranded DNA serves as a template for the synthesis of a new complementary strand. These newly synthesized complementary strands are called daughter strands, and paired with their template strands, they form two identical double-stranded DNA molecules. The daughter strands are made by complementary base pairing, where new nucleotides are added to the daughter strands as they form hydrogen bonds with the complementary nucleotide in the template strands (A pairs with T and C pairs with G). DNA replication is semi-conservative. This means that each new double-stranded DNA molecule is composed of one original (parental) strand and one newly synthesized (daughter) strand (Fig. 6.15).

DNA replication always occurs in the 5′ to 3′ direction. This means that new nucleotides get added to the 3′ hydroxyl group of the previous nucleotide on a growing chain. Thus, it is the 3′ end of the chain that elongates. When nucleotides are added to a growing DNA chain, they start off as deoxyribonucleoside triphosphates (dNTPs), which contain three phosphate groups. Base pairing occurs between an incoming dNTP and its complementary nucleotide on the template strand. As the base pairing proceeds, two of the phosphate groups from the dNTP are released in the form of pyrophosphate (PPi), and one phosphate group remains as part of the growing DNA chain. **DNA polymerase** is the enzyme that catalyzes the addition of nucleotides onto the growing chain. It is actually a family of proteins that can function under different situations [2, 76, 77], as discussed later in this chapter.

2.2 DNA Replication Basics

2.2.1 DNA Replication Occurs During S Phase

The amount of DNA differs in different phases of the cell cycle. The amount of DNA is represented by "N." In this situation, "amount" refers to the number of copies of each chromosome. When a typical diploid cell is in G1 phase, the amount of DNA is 2N during G1 phase—two of each chromosome (one of maternal and one paternal). After the DNA replicates in S phase, the amount of DNA is 4N because an exact replica of each chromosome has been produced. The steps of DNA replication that we discuss below occur in S phase, during which the DNA content increases from 2N to 4N, although some of the initial steps preparing for DNA replication begin to occur in G1 phase [2, 4].

2.2.2 Parental and Daughter Strands

When DNA replicates, the two strands separate, allowing them to unwind and break the hydrogen bonds between the complementary base pairs. These two original strands are referred to as the template (or parental) strands. They will each be used as a template for replication, and two new complementary strands will be synthesized. The result is a parental strand and a daughter strand, each times two (due to the double-stranded nature of DNA).

2.2.3 Formation of Bonds and Elongation of the Chain

The process of DNA replication involves complementary base pairing between nucleotides, which is critical for accurately copying the genetic material. This process involves the formation of hydrogen bonds between the nitrogenous bases and the establishment of phosphodiester bonds within the sugar-phosphate backbone of the DNA molecule. Specifically, the nitrogenous base of the incoming nucleotide forms a hydrogen bond with its complementary base on the template strand, while the sugar of the incoming nucleotide forms a phosphodiester bond with the sugar of the neighboring nucleotide in the growing chain, linking the 3′ OH group of one sugar to the 5′ phosphate group of the next.

The nucleotides are always added in a 5′ to 3′ direction. This means that the 5′ phosphate group (the phosphate that remains attached to the 5′ carbon of the sugar) binds the 3′ hydroxyl (OH) group (the OH group on the 3′ C of the sugar) of the last nucleotide in the existing DNA strand. The addition of nucleotides to the growing strand is catalyzed by the enzyme **DNA polymerase** [2] (Fig. 6.16). More details about the different steps involved in DNA synthesis, including DNA polymerases, primers, and replication bubbles, will be discussed in the following sections.

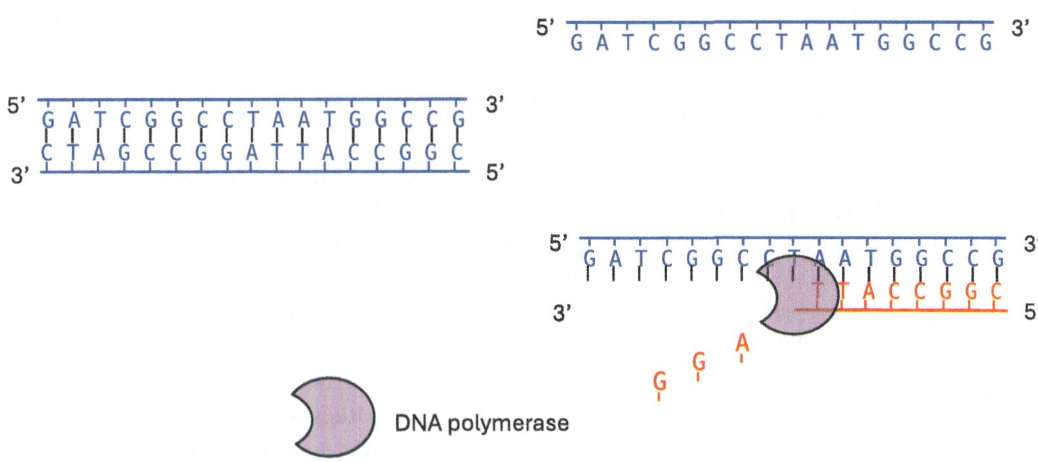

Fig. 6.16 DNA synthesis direction. DNA synthesis, catalyzed by DNA polymerase, occurs in the 5′ to 3′ direction

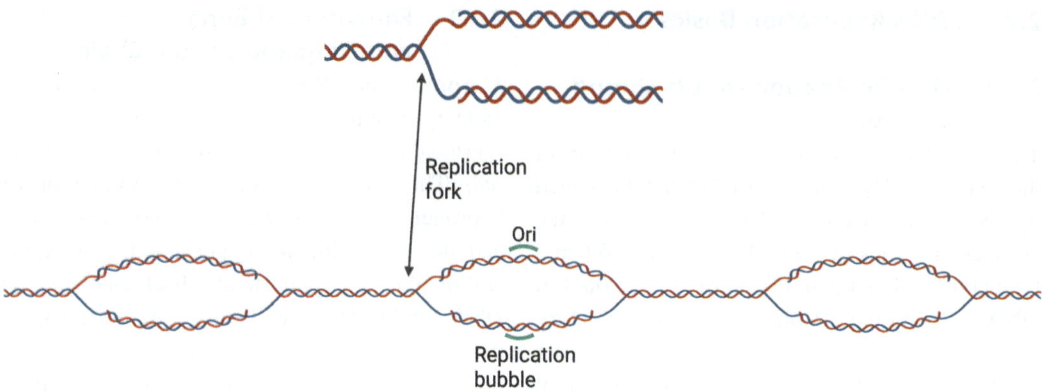

Fig. 6.17 Replication bubbles and replication forks. DNA replication occurs within replication bubbles, beginning at the origin of replication (ORI). (*Source*: Created with BioRender.com)

2.3 Key Structures That Form in DNA Replication

2.3.1 Replication Forks and Replication Bubbles

The task of replicating all the DNA in a eukaryotic cell at once is a big one. Rather than being replicated from the beginning to the end of each chromosome, DNA replication in eukaryotes occurs at multiple sites along the chromosomes. These sites are replication bubbles. Replication bubbles are formed by two replication forks. Replication forks are the structures that form when the double DNA helix unwinds (Fig. 6.17). The replication forks are on either side of the replication bubble. As DNA replicates, the bubbles expand until a replication fork meets a fork from another bubble. Eventually, the entire DNA molecule on a chromosome is replicated [2, 4].

2.3.2 Origin of Replication (ORI)

At the center of each replication bubble is an origin of replication (ORI) (see Fig. 6.17). There are many ORIs throughout the eukaryotic genome (prokaryotes, which have circular DNA, have only one). The ORI is the region where the two strands of DNA first separate. The ORI is distinguished by specific DNA sequences. These sequences are usually rich in As and Ts. These regions separate more easily because A and T are held together by two hydrogen bonds, unlike G and C, which are held together by 3 hydrogen bonds. DNA replication begins at each ORI, and the whole stretch of DNA that is replicated beginning from a particular ORI is called a replicon [2, 78].

2.4 DNA Replication: Initiation, Elongation, Termination

The separation of DNA at the ORI, the formation of the replication bubble, and the synthesis of the replicon and the entire DNA sequence occur in a specific series of steps. DNA replication is a complex process, but it can be broken down into the licensing stage followed by three overall stages. These phases are initiation, elongation, and termination. While the first steps may begin in G1 phase or even at the end of M phase, DNA replication occurs during S phase. The different stages are described below and also described in detail in [2, 77–80].

2.4.1 Licensing, Pre-initiation, and Initiation

- Licensing and Binding of Helicase

This stage begins in late M phase and continues into early G1 phase before the actual initiation of DNA replication takes place in S phase. During late M phase, a complex of proteins called the **origin recognition complex (ORC)** binds to the ORI. This marks the start site of DNA replication (see Fig. 6.18).

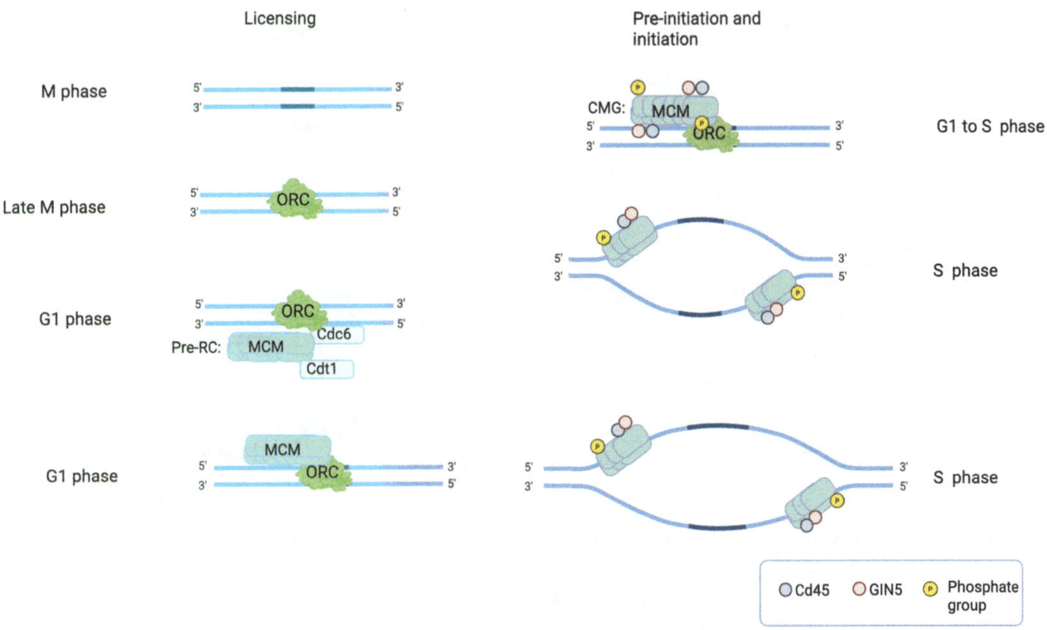

Fig. 6.18 DNA replication: licensing, pre-initiation, and initiation. See text for details. (*Source*: Created with BioRender.com)

Next, during G1 phase, other proteins, **CDC6** and **Cdt1**, bind the ORC. These proteins are licensing proteins, which play a role in marking the DNA for replication, ensuring that DNA replication occurs only once per cell cycle.

These licensing proteins recruit a double **MCM** (minichromosome maintenance) complex (a double hexamer) to the rest of the complex at the **ORI**. **MCM** is a replicative **DNA helicase**, consisting of a hexamer composed of six proteins: MCM2–7. **Helicases** are unwinding proteins. DNA helicases break the hydrogen bonds between complementary base pairs, allowing the two strands to separate and initiate the unwinding process. This **helicase** uses energy from ATP hydrolysis to unwind DNA at the ORI and separate the two strands, but at this point, it remains inactive. The assembled **ORC-Cdc6-Cdt1-MCM complex** is referred to as the pre-replication complex (pre-RC). The two MCMs bind to the Ori as a head-to-head double-hexamer that encircles double-stranded DNA (only one is shown in the figure, for simplicity). This process is called licensing because it marks the origins that are permitted to initiate replication at the upcoming S phase.

After MCM loading, **Cdt1** becomes inhibited by the protein geminin and/or gets targeted for degradation. Cdt1 inactivation and degradation are important because they prevent the licensing of new origins. Geminin is a protein that is regulated in the cell cycle. Its levels are low in the G1 phase, allowing licensing to occur, but they rise during the S and subsequent phases. Once its levels rise, it directly binds to Cdt1, inactivating it and preventing its recruitment to the helicase. In addition to being inactivated, Cdt1 is also degraded by an E3 ubiquitin ligase. Cdc6 is also removed from the complex [2, 77, 78, 81] (see Fig. 6.19).

MCM is only activated once it gets phosphorylated, which occurs at the beginning of S phase, as discussed next. Activation of MCM and its regulation are discussed in the the following reference [2] and in next steps below.

- Pre-initiation and Initiation; Activation and Regulation of the Helicases

At the G1 to S phase transition, the **DNA helicases** (the MCM complexes, called the MCM helicases) become phosphorylated, which leads to their activation. They get phosphorylated

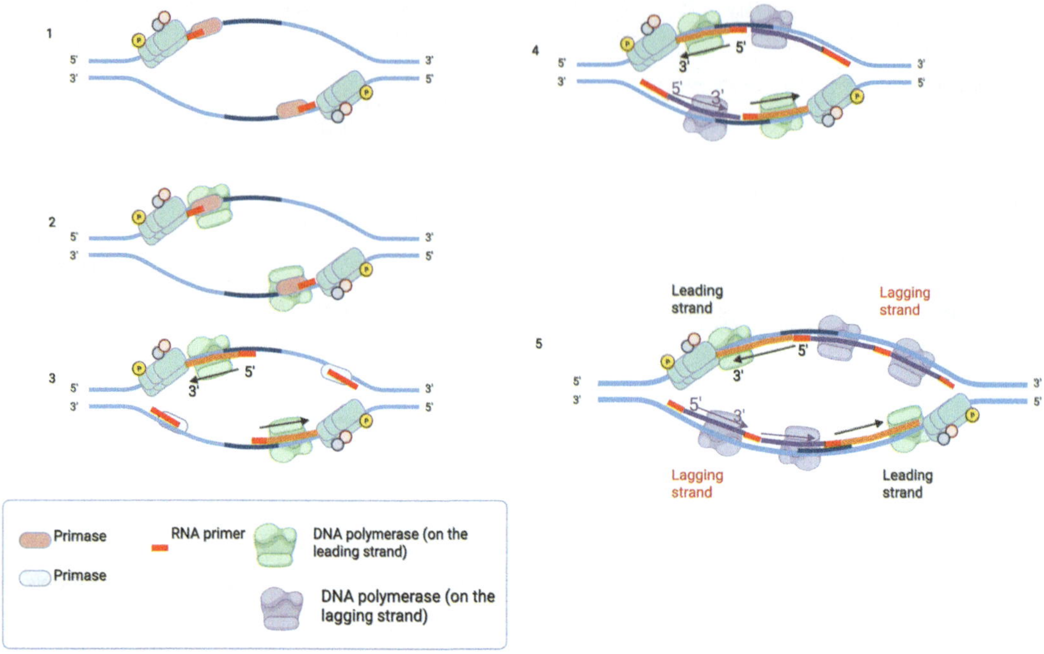

Fig. 6.19 DNA replication; elongation. Leading strand and lagging strand are shown. (*Source*: Created with BioRender.com)

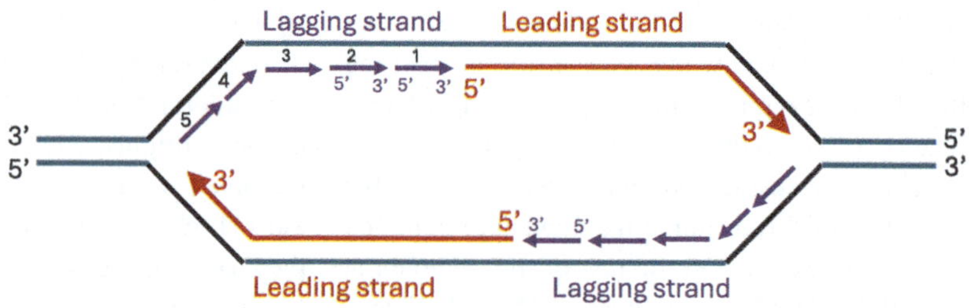

Fig. 6.20 Leading and lagging strand. In the diagram, the fragements in the top lagging strand are numbered for clarity. Strand 1 was made first, strand 2 was made next, and strand 5 was made last in this sequence

by kinases that include **CDKs** and **DDK** (Dbf4-dependent kinase). This phosphorylation is necessary for MCM's activation. It also leads to the recruitment of other proteins, including CDC45 and the GINS complex. Together, MCM2–7, CDC45, and GINS form the CMG complex, which is the fully active **helicase**. This active helicase can unwind the DNA, marking the start of the S phase.

The assembly of this complex is known as the formation of the pre-initiation complex (pre-IC), and it marks the commitment to initiate DNA replication. Once activated, this CMG helicase uses energy from ATP hydrolysis to disrupt the hydrogen bonds between the complementary base pairs at the ORI, thereby initiating the unwinding process. During this process, the **MCM hexamer**, which initially surrounds the double helix, separates into two individual helicase complexes, each encircling one strand of DNA, and these complexes will move in opposite directions [2, 77–80] (Fig. 6.20).

- Formation and Expansion of the Replication Bubble, Marking the Transition to Elongation

As the helicases move apart in opposite directions, the DNA continues to unwind, and the area of unwound DNA expands. The result is an expanding replication bubble. Within this bubble, both DNA strands are separated. At either side of this bubble is a replication fork. The helicases remain associated with the replication forks, facilitating the continued unwinding as the replication bubble expands (see Fig. 6.20).

2.4.2 Elongation

- The Replisome, DNA Primase, and the RNA Primer

As the replication bubble expands, the DNA separates during S phase, and a protein complex called the replisome assembles at each fork. The replisome includes the **MCM helicase** structure described above, along with **DNA primase, DNA polymerase**, and several other proteins necessary for the DNA replication process, some of which will be discussed below. DNA synthesis has an important requirement. It can only occur if a primer, containing nucleotides, is already in place, to which new nucleotides are then added. To circumvent this problem, the cells make use of the enzyme **DNA primase**. DNA primase synthesizes short RNA sequences called RNA primers after unwinding begins (a process aided by the various accessory proteins). DNA primase starts this process by adding ribonucleotides, which bind to their complementary nucleotides on the DNA strand via hydrogen bonds. This results in the synthesis of RNA primers, bound to their complementary DNA strand. These RNA primers are approximately ten nucleotides long. They are essential for replication because, unlike DNA, RNA can be synthesized de novo, without the need for a pre-existing primer. This RNA primer provides a 3′ hydroxyl (−OH) group to which the deoxyribonucleotides that make up DNA can be added later. One RNA primer is made on the leading strand, and multiple RNA primers are made on the lagging strand. The leading and lagging strands are discussed in more detail later [2, 77–80, 82] (see Fig. 6.19).

- DNA Polymerase

The next step involves **DNA polymerase**, the enzyme that synthesizes DNA by adding deoxyribonucleotides to the growing strand. There are multiple different DNA polymerases in eukaryotic cells. Among these, Pol α, Pol δ, and Pol ε are important for DNA strand elongation. Pol α is the least efficient but is the first to function in the elongation process. Pol α forms a complex with the **primase**, which by this point has laid down the RNA primer (see Fig. 6.19). Now, Pol α will add deoxyribonucleotides (dNTPs) to the RNA primer's free 3′ OH group to begin synthesizing DNA. It synthesizes a short strand of DNA, approximately 20 nucleotides long, before DNA synthesis is taken over by Pol δ δ or Pol ε, which are more efficient polymerases. Pol ε replicates the leading-strand DNA. Thus, it synthesizes long stretches at a time. Pol δ is mostly involved in replicating the lagging strand and produces the Okazaki fragments [2, 77–80, 82] (see below).

- Leading and Lagging Strands and Okazaki Fragments

DNA is always synthesized in the 5′ to 3′ direction. This creates a potential problem in the replication bubbles, leading to what is known as semi-discontinuous replication. The leading strand is synthesized continuously, while the lagging strand is synthesized in short discontinuous segments, as follows. DNA polymerase on the leading strand can synthesize DNA continuously as the replication bubble continues to open. As it is synthesized, the new DNA follows in the path of the helicase as it expands the replication bubble. The DNA on the other strand also starts with a primase, near the helicase. But this DNA cannot continue its synthesis in the 5′ to 3′ direction as the replication bubble opens; instead, it "runs into" the DNA from the other strand. Therefore, DNA on this strand (the lagging strand) is replicated in short sequences of about 200 nucleotides called Okazaki fragments [83]. Each of these fragments is initiated with its own RNA primer in the region of the replication fork, followed by the addition of the DNA sequence. For each fragment, synthesis stops when the polymerase reaches another fragment. As a result, multiple small fragments are made, one after the other, as the replication bubble expands. Thus, the leading strand consists of one long fragment of DNA that is being

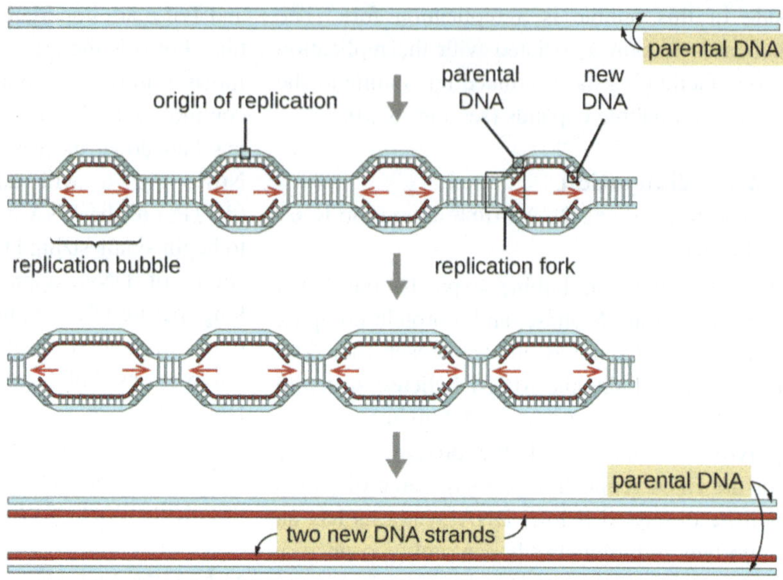

Fig. 6.21 Termination of DNA replication. DNA replication terminates when adjacent replication bubbles meet. (*Source*: Reproduced from *OpenStax Microbiology* (by OpenStax), licensed under the Creative Commons Attribution 4.0 (CC BY 4.0) license)

replicated, while the lagging strand consists of the Okazaki fragments. These fragments will eventually be joined together (*see* Fig. 6.20). Later, nucleases will remove the RNA primers, and a DNA polymerase will replace these primers with deoxyribonucleotides (DNA). Finally, the enzyme DNA ligase will join the multiple fragments together. For each Okazaki fragment, the process of removing the RNA primer, replacing the RNA with DNA, and ligating the fragments together is called maturation of the Okazaki fragments [2, 77–80, 82].

2.4.3 Termination

Thousands of terminations happen each time the DNA replicates. Termination is not associated with a specific DNA sequence; instead, it occurs when a replication bubble encounters another bubble moving in the opposite direction (see Fig. 6.21). At this stage, the last Okazaki fragment in the replication bubble undergoes the maturation process [2].

When two forks meet, the **CMG complex** (the helicase, composed of Cdc45, MCM, and GINS) at the fork is modified by ubiquitin on its **MCM** subunits. Ubiquitin can have various functions, but in this case, ubiquitination serves as a signal for the regulated disassembly of the helicase complex from the DNA. This disassembly process is facilitated by additional factors, including the segregase enzyme p97/VCP, which helps extract **CMG** from the DNA and marks it for disassembly. As a result of this process, the **CMG helicase** is disassembled, and a new complex must be assembled for future rounds of replication. At the end of the process, there is now a new complementary strand of each DNA strand [2, 77–79, 82, 84, 85].

2.5 Telomeres

2.5.1 Telomeres and Termination of Replication at the End of the Chromosome

Most termination events happen within replication bubbles. However, termination also occurs at the ends of the chromosomes. When the termination process occurs at the end of a chromosome rather than in a replication bubble, the process is more complex. This is an issue for eukaryotic cells, unlike prokaryotes, which have circular DNA. In eukaryotic cells, the **CMG helicase** is released from the end of the chromosome. However, the lagging strand encounters a problem because no new DNA can be made to replace the primers on it. DNA will therefore shorten each time it replicates (see Fig. 6.22). For this

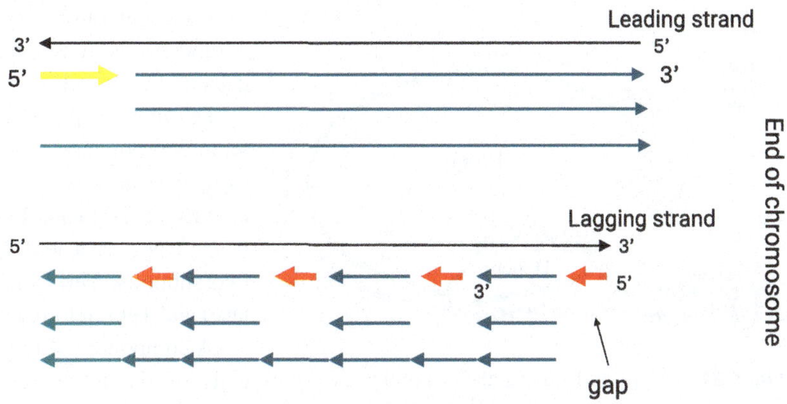

Fig. 6.22 The gap at the laggind strand of DNA. At the end of the chromosome the lagging strand cannot be fully copied, leaving a gap at the chromosome end. For this reason, DNA becomes slightly shorter every time DNA replicates

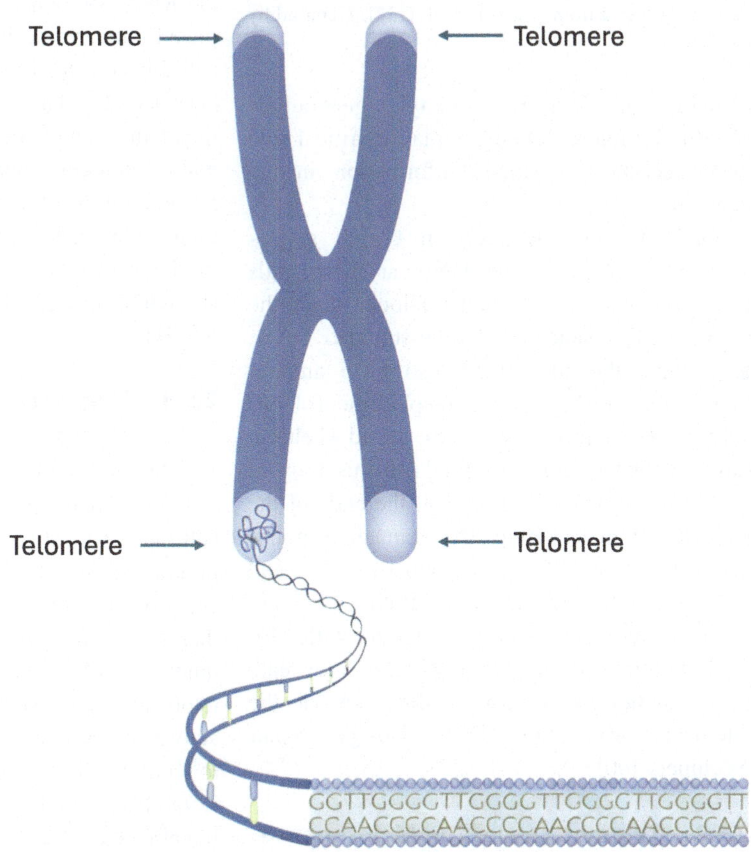

Fig. 6.23 Telomeres. Telomeres are located at the ends of chromosomes. They contain repeated sequences. They shorten every time the DNA replicates and help protect the coding sequences. (*Source*: Illustration from Freepik.com (premium license))

reason, telomeres play a vital role in preventing the loss of important genetic material. Telomeres are at the ends of chromosomes, and they contain thousands of copies of a highly conserved sequence (GGGTTA in Humans), thus forming the sequence GGGTTAGGGTTAGGGTTAGG GTTA and so on. (Sometimes the repeats are written as TTAGGG, depending on where in the sequence of repeats one starts; see Fig. 6.23.) These repeats shorten every time the chromosome replicates, but the genetic material that codes for proteins remains untouched. Once telomeres become critically short, cells lose the ability to divide, but the telomere sequences are long

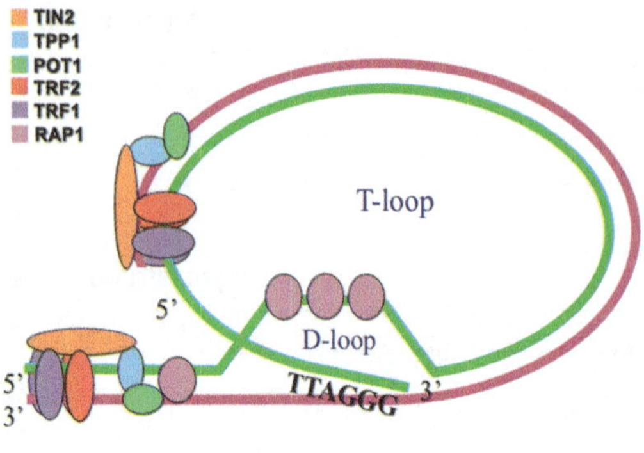

Fig. 6.24 T loop and D loop at the end of the telomere. This region is protected by the shelterin complex, which includes TIN1, TPP1, POT1, TRF2, TRF1, and RAP1. (*Source*: Adapted from Ivancich et al. [132]. Licensed by CC BY)

enough to survive many generations [2, 86]. Therefore, telomeres play a critical role in preserving the genetic information in the genome.

The telomere ends contain G-rich single-stranded overhangs. The DNA at these ends forms a loop structure called a T-loop, where the G-rich ends invade the double-stranded DNA, and within this loop, they also form another loop called a D-loop (the displacement loop) (Fig. 6.24). A protein complex called shelterin (also called the telosome) binds to this loop to protect the structure and prevent the ends of the telomeres from activating DNA damage response pathways [2, 87–89]. These loop structures and shelterin binding are not static. Rather, they are dynamic and therefore do not interfere with telomere shortening during DNA replication. Instead, their function is to protect the ends of the telomeres from the DNA damage repair machinery [90].

2.5.2 Telomerase and the Maintenance of Telomeres

In most normal somatic cells, the telomeres shorten each time the cell divides, and eventually telomeres will become critically short. In some cells, such as stem cells and germ cells, however, telomere length is maintained. This is due primarily to the enzyme **telomerase**. In most normal adult cells, telomerase is inactive, absent, or expressed at only very low levels. In contrast, telomerase is functional in cells such as stem cells and germ cells.

Telomerase plays a crucial role in replacing telomere sequences that are lost during cell division. It is an enzyme with reverse transcriptase activity. This means that it can synthesize DNA using RNA as a template. Telomerase contains two subunits: Tert (telomerase reverse transcriptase) and Terc (telomerase RNA component). The RNA component of telomerase (TERC) binds the ends of the telomeres. In addition to helping maintain the shape of the telomerase complex, the RNA component of TERC serves as a template for the synthesis of new telomere DNA after cell division, facilitated by the reverse transcriptase activity. Thus, it enables the telomere to maintain its length rather than shortening. In summary, telomerase attaches its RNA sequence to the end of the telomere, then uses its reverse transcriptase activity to extend the DNA sequence of the telomere. This prevents telomeres from shortening in cells that have telomerase [77, 86, 87, 91].

2.5.3 Telomere Length and Human Health

- Telomere Length and Cell Fate

Telomeres play an important role in determining cell fate. In most cells that lack or have minimal telomerase activity, telomeres shorten with every cell division. In cells that are growing rapidly, therefore, the telomeres shorten especially quickly. After many rounds of shortening, telomeres may become critically short. This activates DNA damage responses that can lead to cellular senescence, apoptosis, or genomic instability. In senescence, cells stop dividing but remain metabolically active. This occurs naturally in many cells, but it can also be a contributing factor to tissue aging and eventual loss of function. Apoptosis is a type of programmed cell death that eliminates cells that are damaged or no longer functional. Genomic instability can be associated with chromosomal fusions and rearrangements that may result from the damaged ends of the chromosomes. This can limit cell proliferation, but under some circumstances, it can lead to uncontrolled growth and cancer.

- Telomere Length, Aging, and Disease

– Telomeres and Aging

A strong link exists between telomere length and aging. Older individuals typically have shorter telomeres in most cell types compared to younger individuals, although telomere length varies across tissues. Shortened telomeres are associated with an increased risk of age-related diseases, including heart disease and infectious illnesses. Certain lifestyle factors, such as smoking, obesity, dietary factors, and exposure to environmental toxins, may increase the rate of telomere shortening, and this may be linked to diseases such as heart disease and other age-related problems.

Due to the strong links between telomeres and disease, studying telomere length is an important topic in the field of aging and the treatment of age-related health problems. It should be noted, however, that many other factors aside from telomeres also contribute to aging and related disease [86, 92, 93].

– Telomeropathies: Genetic Disorders of Telomeres

Telomeropathies are a group of genetic disorders caused by abnormalities in telomeres. Dyskeratosis congenita (DC) is an example of a telomeropathy. It is often characterized by accelerated telomere shortening. Patients with DC often have bone marrow failure, aplastic anemia, and an elevated risk of leukemia. Mutations in genes like *TERT* and *TERC*, which encode components of telomerase, are common in DC. Treatment options include steroids, stem cell transplantation, and supportive care [94, 95].

While there are no targeted therapies for telomeropathies, researchers recently found that adding thymidine to cells from patients with DC lengthened their telomeres. It is unclear why thymidine, in particular, rather than other nucleotides, has this effect on telomeres; however, the results are promising for the treatment of telomere disorders. Scientists are working on developing clinical trials to further explore the utility of using thymidine to benefit patients with DC [95, 96].

– Cancer

The effects of telomere length on human health are complex. Telomere dynamics play a dual role in cancer. Telomere shortening can inhibit cell growth by activating pathways that respond to DNA damage, specifically the ATM/ATR pathways (discussed in Chap. 23). However, critically short telomeres can also lead to cancer because the loss of telomere protection can result in chromosomal fusions and other abnormalities associated with genomic instability, leading to increased uncontrolled cell growth and cancer. **Telomerase** maintains telomere length in some types of cells, but it is downregulated in most normal adult diploid cells. This is not the case in cancer cells, however. In about 90% of cancer cells, the telomerase gene is abnormally activated. This reactivation allows telomere lengthening and can encourage unchecked cell growth [86, 89, 92, 93].

– Myelodysplastic Syndromes (MDS)

Myelodysplastic syndromes (MDS) are a group of blood cell disorders affecting the bone marrow. This disease may have many causes, but in some patients, **telomerase** is upregulated in their hematopoietic stem cells. Telomerase overexpression in MDS is associated with poor survival and may contribute to tumor growth.

The link between telomerase activity and MDS is complex, and telomeres are not always longer in this disease. In fact, they are often shorter. There are several reasons for this. For example, high levels of telomerase can lead to increased cell proliferation, and this rapid cell division can actually result in shorter telomeres and a large number of cells. Nevertheless, the finding that telomerase is overexpressed in some MDS patients led researchers to develop a telomerase inhibitor for the treatment of the disease. **Rytelo (imetelstat)** [97, 98] is a **telomerase inhibitor**, and it was FDA-approved in 2024 for treating some patients with **MDS**. It is an oligonucleotide that can bind to the RNA portion of telomerase and act as a competitive inhibitor of its activity. It is attached to a lipid moiety, which

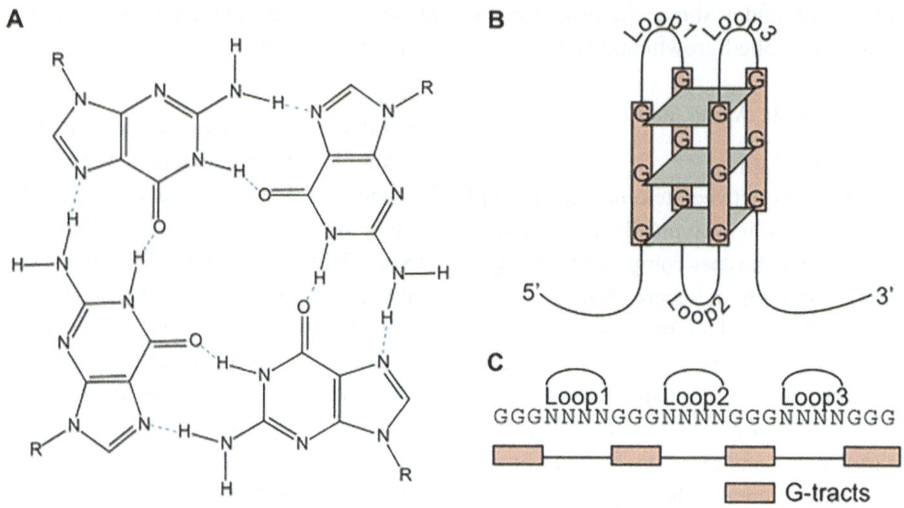

Fig. 6.25 G quadruplexes. Diagram of G-quadruplexes, a type of structure present in telomeres. (*Source*: Adapted from Capra et al. [103]. Licensed under Creative Commons Attribution (CC BY))

enhances cellular uptake and stability. Rytelo (imetelstat) falls into the category of oligonucleotide therapeutics. It is administered by IV infusion every 4 weeks [99–101].

– Telomestatin is a Telomerase Inhibitor

Another compound that affects telomere function is an experimental compound called **telomestatin**. Telomestatin functions by stabilizing **G-quadruplexes** (G4), which are secondary structures on the 3′ overhangs on telomeres [102, 103]. Telomeres contain G-rich sequences, and under some conditions, they tend to form G-quadruplex (G4) complexes. These complexes contain stacked complexes of four guanines stabilized by Hoogsteen hydrogen bonding (see definition at the end of the chapter) (Fig. 6.25). They naturally inhibit telomerase by blocking its access to the 3′ telomere overhangs that they normally bind to. Because they inhibit telomerase, they may be promising drug targets for cancer cells where telomerase levels are often upregulated [87]. Telomestatin is a natural product and a small molecule. It is isolated from the bacterium *Streptomyces annulatus*, and it binds to and stabilizes G4 complexes. By binding to these complexes, telomestatin stabilizes them. This, in turn, inhibits the activity of telomerase. This is because G4 complexes prevent telomerase from binding to the telomere DNA. Thus, stabilizing the G4 complex interferes with telomere elongation. Over time, this drug can therefore lead to telomere shortening. Telomestatin is an experimental compound, but it has been shown to have antiproliferative effects in several types of cancer. The potential for using this type of compound for cancer treatment is an ongoing area of research [104, 105].

– POT1-Associated Cancers

Shelterin is made of six different protein subunits. Abnormalities in any of these proteins can affect shelterin activity, thereby interfering with telomere function. One example is POT1 (Protection of telomeres 1), an essential subunit of shelterin. Mutations in POT1 have been identified in various cancers, including **melanoma and other malignancies**. POT1 mutations lead to telomere dysfunction and consequently to chromosomal instability. When POT1 is mutated, the result can be uncontrolled access of telomerase to the telomeres. This can, in some cases, lead to cell immortalization and uncontrolled cell growth. There is currently no targeted treatment for cancers with POT1 mutations, but standard cancer therapies are often used. Development of specific inhibitors is an important area of research [106, 107].

2.6 Additional Proteins with Roles in DNA Replication

This chapter has covered some important proteins involved in DNA replication, such as **DNA polymerase, DNA primase,** and **helicase**. However, DNA replication is a complex process, and multiple other proteins are involved [2, 77–80, 82]. Some of these are discussed below and illustrated in Fig. 6.26.

2.6.1 Single-Stranded DNA-Binding (SSB) Protein

When the replication bubble opens, some single-stranded DNA is exposed before it gets replicated. This single-stranded DNA needs to be protected, or it will fold back on itself, forming hairpin loops. This problem is solved by SSB proteins, which coat the single-stranded DNA and are removed once the DNA is replicated. In eukaryotes, the SSB protein is made of three different subunits [2, 77–80, 82].

2.6.2 Sliding Clamp and Clamp Loader

The sliding clamp (called **PCNA**, proliferating cell nuclear antigen) is a protein complex that helps position **DNA polymerase** on the DNA during DNA synthesis. The sliding clamp (PCNA) is ring-shaped, and it can slide along the DNA behind the **DNA polymerase**, helping it stay attached to the DNA template during elongation. The clamp loader has a role in loading the clamp onto the DNA. The sliding clamp can be seen in Fig. 6.26.

The sliding clamp (PCNA) requires the **clamp loader (RFC: replication factor C),** an enzyme complex that uses energy from ATP hydrolysis to open and close the clamp ring structure as needed. The clamp loader binds to a free clamp molecule and, using ATP, forces it to adopt an open ring conformation. This enables the clamp to fit around DNA. The loader then positions the clamp onto the DNA, usually at the 3′ end of a primer, where DNA synthesis will begin. The loader then dissociates, and DNA polymerase binds to the clamp, allowing DNA synthesis to occur.

As DNA synthesis proceeds, the clamp moves along with the DNA polymerase, keeping it stably attached to the DNA. When the **DNA polymerase** reaches a double-stranded DNA sequence (such as at the beginning of another Okazaki fragment), it detaches from the clamp and from the DNA, and DNA synthesis stops. The clamp later dissociates as well, but will be available to recruit another clamp loader [2, 108, 109].

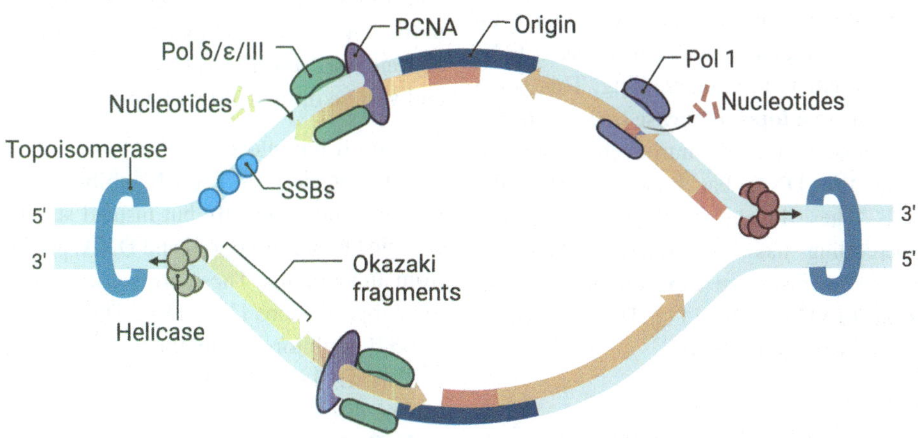

Fig. 6.26 Replication bubble. Summary of proteins that bind to the replication bubble. See text for details. (*Source*: Created with BioRender.com)

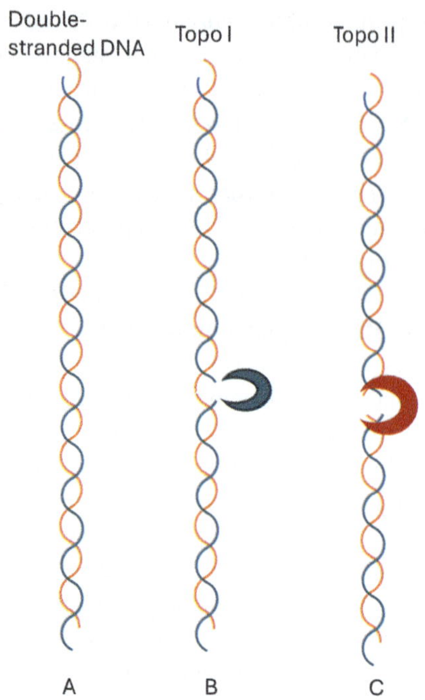

Fig. 6.27 Topoisomerase. Topoisomerase 1 makes single-stranded cuts in DNA, and topoisomerase II makes double-stranded cuts

2.6.3 DNA Topoisomerase

Since DNA is a double helix, it can become supercoiled when the replication fork opens. You can think of this as someone pulling on a portion of two strings that are wound around each other. When one portion is pulled apart, the remaining strings wind around each other even more tightly. Therefore, a mechanism needs to be in place to regulate this supercoiling. This is the job of the enzyme **topoisomerase** (see Fig. 6.27). Topoisomerase works by introducing transient breaks in the DNA backbone to relax the supercoils. These breaks are then resealed after the supercoiling has been relieved. The two main topoisomerases are **topoisomerase I** and **topoisomerase II** [2, 110, 111]. There are multiple subtypes of both of these.

- Topoisomerase I (Topo I)
  Topo I makes single-stranded breaks in DNA. It attaches to DNA, breaking a phosphodiester bond. This produces transient single-stranded breaks in the phosphodiester backbone. The DNA on either side of the break can now rotate through the break, and after this, the bond re-forms [2, 110].
- Topoisomerase II (Topo II)
  Topo II makes double-stranded breaks in DNA. It can be activated at sites where two strands of a double helix cross over or where two entire helices cross over each other. This can occur at a replication fork. Topo II binds to these sites and then breaks one complete double helix to create an opening. This allows a nearby helix to pass through the opening. The break then re-seals [2, 110].

- Drugs That Are Topoisomerase Inhibitors

Topoisomerases are essential for maintaining the integrity of DNA. Without topoisomerase, the DNA would eventually be destroyed during cell division. The topoisomerases are especially important for rapidly dividing cells. Consequently, they are compelling **targets for drugs that inhibit the growth of cancer cells**. The idea is that targeting topoisomerases will interfere with normal DNA replication, thereby leading to cell death. Topoisomerase inhibitors are among the earliest drugs used for cancer treatment. They are nonspecific inhibitors of rapidly dividing cells and therefore have many side effects, but are still used, often in combination with other treatments [112]. Both topoisomerase I (Topo I) and topoisomerase II (Topo II) inhibitors are used, but Topo II inhibitors were used earlier, and are used for a broad range of cancers [113] and will be discussed first below.

- Topo II Inhibitors

One major class of Topo II inhibitors does not inhibit Topo II directly but instead stabilizes the complex between Topo II and DNA, preventing it from functioning. These are sometimes referred to as topo poisons [114–116]. The other class of Topo II inhibitors can inhibit the catalytic activity of Topo II [114, 117]. Examples of both of these will be discussed here.

VePesid (Etoposide) [118, 119] is a commonly used **small-molecule topoisomerase II inhibitor** for the treatment of several types of **cancer**. It does not act by inhibiting the catalytic activity of **Topo II**. Instead, it stabilizes the complex made between Topo II and DNA, thus "trapping" Topo II in this complex. Normally, Topo II makes a double-stranded break in DNA and temporarily binds to the DNA at the site of the break. This interaction between DNA and Topo II is the Topo II covalent cleavage complex. Etoposide acts by binding to this complex. This stabilizes the Topo II-DNA cleavage complex, preventing the repair of the break from occurring. Consequently, the DNA break becomes permanent. The result is an accumulation of breaks, which leads to genetic instability and subsequently cell death [114, 120–122].

Doxorubicin, also used to treat several types of **cancer**, interferes with DNA by intercalating between the bases. But it also interferes by inhibiting **Topo II**. When it intercalates into DNA, it can stabilize the Topo II-DNA covalent cleavage complex. This is similar to etoposide, but the mechanism is somewhat different because it relies on the drug's ability to intercalate into DNA. Doxorubicin also has other functions besides inhibiting Topo II. For example, it produces reactive oxygen species, which also has a cytotoxic effect. Thus, doxorubicin is a powerful drug that can kill cells by multiple mechanisms. **Novantrone (Mitoxantrone)** [123] is also a **Topo II inhibitor**. Like doxorubicin, it intercalates into DNA and binds the Topo II-DNA covalent cleavage complex, inhibiting Topo II activity. It has significant side effects, and it is recommended only if the benefits outweigh the risks [124], and newer treatments may be preferable.

Bisdioxopiperazines are a family of small-molecule drugs that inhibit **Topo II** by the second mechanism. They can inhibit Topo II by binding to it and inhibiting its catalytic activity. Drugs in this category are used to treat certain **cancers such as leukemias, lymphomas, sarcomas, and breast cancer** [117].

– Topo I Inhibitors

Some drugs are **Topo I inhibitors**. These include **camptosar (irinotecan)** and **hycamtin (topotecan)**, both **small-molecule** drugs. They are derivatives of **camptothecin (CPT)**, which is a natural alkaloid derived from the bark of the *Camptotheca acuminata* tree. Thus, they are both derivatives of natural products. They are used for the treatment of **colorectal and ovarian cancer, and others**. They are often used in combination therapies. Camptothecin and its derivatives have multiple functions, but an important function is inhibiting Topo I. They do so by stabilizing the transient complex formed between Topo I and DNA, also called a cleavage complex. The inhibitors stabilize the complex and interfere with the function of the Topo I enzyme [112, 125–127].

– Antibiotics That Inhibit Bacterial Topoisomerases

Bacteria also have topoisomerases, which serve a similar purpose as eukaryotic topoisomerases. Two important topoisomerases in bacteria are **DNA gyrase** and **topoisomerase IV**. Both introduce double-stranded breaks in the DNA and prevent DNA tangling and supercoiling, similar to mammalian topoisomerases. Both of these processes result in double-stranded breaks in DNA. **Fluoroquinolones** are a widely used class of **broad-spectrum antibiotics for treating bacterial infections**. An example is **Cipro (ciprofloxacin)** [128]. **Fluoroquinolones inhibit both DNA gyrase and topoisomerase IV** by stabilizing the enzyme-DNA cleavage complex. This prevents re-ligation after cleavage, leading to bacterial death. Cipro is used to treat a wide range of **bacterial infections, including pneumonia and skin infections** [129, 130].

2.7 Replication of Histones as DNA Replicates

2.7.1 Replication and Reassembly of Nucleosome Core Proteins

DNA replication requires not only the replication of the DNA but also the replication and reassembly of all the protein complexes associated with the DNA. This includes histones, which bind along the length of DNA. New histones must be synthesized and added to the newly formed DNA strand. Most eukaryotes have multiple copies of the histone gene. Histones are synthesized mainly during the S phase, unlike most other proteins, which can be made throughout the cell cycle. Histone proteins are very stable, unlike the histone RNA, which is relatively transient.

As the replication fork proceeds, the histones on the DNA are transiently displaced, leaving a stretch of free unbound nucleotides. The histones that make up the nucleosome cores are replaced as follows. The nuclear core contains a histone octamer, which is made of one H3–H4 tetramer and two H2A–H2B dimers (see Fig. 6.9). Parent H3–H4 tetramers (those that were on the original DNA molecule pre-replication) remain near the replication fork that forms as the DNA unwinds. Once the DNA begins to replicate, these complexes are redistributed at random to the two daughter DNA strands. H2A–H2B proteins are moved completely from the fork but are added back to both the DNA strands as the DNA replicates, along with newly made H2A-H2B complexes [2, 131].

2.7.2 The Roles of Chaperone Proteins in Histone Reassembly

Chaperone proteins facilitate the release and reassembly of histones onto the newly synthesized DNA. Chaperone proteins have roles in the folding, unfolding, assembly, disassembly, or transport of various proteins without becoming a part of the final structure of the proteins. In the case of histones, chaperones play important roles in protein folding, ensuring that newly synthesized histones fold correctly. Chaperones are also involved in the assembly of histone octamers onto DNA to form nucleosomes, and they are involved in histone disassembly, especially during DNA replication, DNA repair, and transcription.

Some chaperone proteins important in regulating the duplication and reassembly of histones include FACT, NAP1, and CAF1. FACT is the chaperone protein that facilitates the release of H2A and H2B proteins from the nucleosomes at the replication fork. It has several domains that interact with nucleosomes, facilitating their release and disassembly. NAP1 and CAF1 are chaperones involved in the reassembly of the histones. They help restore all the histones to the new and template DNA [2, 131].

FDA-Approved Drugs Covered in This Chapter

Commonly used brand name	Generic name	First FDA approval year	Common conditions	Main target
Adriamycin	*Doxorubicin*	1974	Cancer	DNA (intercalation), topoisomerase II
Camptosar	*Irinotecan*	1996	Colorectal cancer	Topoisomerase I-DNA cleavage complex
Cipro	*Ciprofloxacin*	1987	Bacterial infections	Bacterial DNA gyrase, topoisomerase IV
Hycamtin	*Topotecan*	1996	Ovarian and other cancers	Topoisomerase I-DNA cleavage complex
Novantrone	*Mitoxantrone*	1987	Cancer, multiple sclerosis	Topoisomerase II-DNA cleavage complex
Rytelo	*Imetelstat*	2024	Myelodysplastic syndromes (MDS)	Telomerase (RNA component)
VePesid	*Etoposide*	1983	Cancer	Topoisomerase II-DNA cleavage complex

Definitions

Acetylation Addition of an acetyl group to histones, often associated with a more open chromatin configuration and active transcription.

Allele A variant form of a gene located at a specific position (locus) on a chromosome. Different alleles may correspond to different traits, such as blue or brown eyes color.

Antiparallel The two strands in a DNA helix are antiparallel, one runs 5′ to 3′. The other, complementary, strand is also 5′ to 3′ but runs in the opposite direction.

Central dogma of molecular biology The dogma that the flow of genetic information runs from DNA to RNA to protein.

Chromatin The DNA in complex with proteins, many of which are histones. The chromatin can be densely or loosely packed.

Cohesin Cohesins are protein complexes that have important roles in chromatin organization. They interact with the DNA and have roles in both inter-molecular crosslinking (crosslinking between sister chromatids) and intra-molecular crosslinking (crosslinking within the same chromatid).

Electronegative atom An atom with a strong tendency to attract electrons toward itself.

Electronegativity The measure of an atom's ability to attract electrons.

Epigenetics Changes in gene expression that do not involve alterations to the DNA sequence, but rather modification of DNA or histones. Modifications may include the addition of methyl groups to the DNA or histones, or an acetyl group to the histones.

Histones A family of basic proteins that DNA wraps around to form nucleosomes, which are the main protein/DNA structures in chromatin.

Hoogsteen hydrogen bonding A type of non-canonical base pairing between nucleotides that differs from Watson and Crick base pairing. It can form in DNA structures that differ from the standard double helix.

Hydrogen bonds Hydrogen bonds form when a hydrogen atom covalently bonded to an electronegative atom (like oxygen or nitrogen) is attracted to another electronegative atom.

Locus/loci The location or position of a gene or genetic marker. (Loci is plural for locus.)

Methylation The addition of a methyl group (CH_3). In the case of DNA, the methyl group is commonly added to cytosine.

Nitrogenous base A chemical structure that contains nitrogen and has the chemical properties of a base.

Nucleosome In chromatin, DNA wraps around histones to form the nucleosome, which is the basic unit of chromatin. Eight histones in the nucleosome form the histone octamer.

Nucleotide The basic building block of DNA and RNA, consisting of a nitrogenous base, a sugar (deoxyribose or ribose), and one or more phosphate groups.

Okazaki fragments Short stretches of DNA that form along the lagging strand as DNA replicates. They will ultimately be ligated together by the enzyme ligase.

Origin of replication (ORI) A specific DNA sequence where replication begins. Eukaryotic chromosomes have multiple ORIs. When the replication bubble opens, the ORI will be at its center.

Pharmacogenomics The study of how individual genetic makeups influence how individuals respond to drugs.

Phosphate group A phosphate ion (PO_4^{3-}) covalently attached to a carbon-containing molecule, such as the sugar in nucleotides. Nucleotides may have 1, 2, or 3 phosphate groups. Sometimes the phosphate group is referred to simply as "phosphate," but it is technically a phosphate group when attached to a molecule.

Phosphate ion A phosphate ion has a central phosphorus atom and four oxygens bound to it. It usually has a charge of −3 when it is in its fully deprotonated form (PO_4^{3-}).

Phosphodiester bonds The bonds that hold the nucleotides together along the backbone (the sugar phosphate backbone) of DNA and RNA.

Phosphorus (P) The atom at the center of a phosphate ion. Its atomic number is 15; it has 15 protons and 15 electrons distributed equally across three electron shells.

Purine A type of nitrogenous base with a double-ring structure; includes adenine (A) and guanine (G).

Pyrimidine A type of nitrogenous base with a single-ring structure; includes cytosine (C), thymine (T), and uracil (U).

Replication bubble An open region of DNA where DNA replication takes place. On either side is a replication fork.

Replication fork A Y-shaped structure formed when DNA unwinds to allow the formation of the replication bubble, where DNA replication can take place.

Sugar A sugar is a compound with the formula CnH_2On. Nucleotides contain sugars that a can form a ring structure. The important sugars in nucleotides are deoxyribose sugars in DNA and ribose sugars in RNA.

Telomerase An enzyme with reverse transcriptase activity that adds sequences to the ends of the telomeres, helping maintain their length in some cells.

Topologically associating domains (TADS) Genomic regions that bring DNA regions into proximity with each other that are not in proximity linearly. For example, they can bring regulatory elements into proximity to coding regions.

Zygosity The amount of similarity or differences in alleles of genes. Genes may be homozygous or heterozygous.

Study Questions

Multiple-Choice Questions

1. Which of the following nitrogenous bases is found only in RNA and not DNA?
   A. Adenine
   B. Cytosine
   C. Uracil
   D. Thymine
2. DNA replication is described as
   A. Conservative
   B. Transcription
   C. Semi-conservative
   D. Epigenetic
   E. C or D
3. Which enzyme catalyzes the synthesis of the RNA primer during DNA replication?
   A. RNA polymerase
   B. DNA ligase
   C. DNA primase
   D. tRNA primase
4. What type of bond links the nucleotides along the backbone of a DNA strand?
   A. Hydrogen bond
   B. Phosphodiester bond
   C. Peptide bond
   D. Ionic bond
   E. Glycosidic bond
5. Which histone is considered a linker histone?
   A. H2A
   B. H3
   C. H1
   D. H4
6. Which of the following drugs intercalates into DNA to inhibit replication?
   A. Methotrexate
   B. Doxorubicin
   C. Etoposide
   D. Vincristine
7. What is the function of telomerase?
   A. Degrades damaged DNA
   B. Binds DNA polymerase to enhance DNA replication
   C. Repairs damaged DNA
   D. Maintains the length of telomeres
8. Which of the following unwinds DNA?
   A. DNA primase
   B. Helicase
   C. DNA ligase
   D. DNA polymerase
   E. RNA primer
9. Which nucleotide pair forms three hydrogen bonds in the double-stranded DNA molecule?
   A. A-T
   B. A-U
   C. G-C
   D. T-G

10. Which of the following is not a component of a nucleotide?
    A. Nitrogen base
    B. Phosphate group
    C. Ribose group
    D. Peptide bond
    E. All of the above
11. The 3′ end of the DNA contains the
    A. Hydroxyl group
    B. Phosphate group
    C. Ligase group
    D. Histone group
12. An example of a regulatory region of a gene is
    A. The coding sequence of the gene
    B. The promoter sequence upstream to a gene
    C. An enhancer sequence upstream to the gene
    D. B and C
    E. All of the above
13. The region of the chromosome that the kinetochore binds to is called the
    A. Histone
    B. Centromere
    C. Microtubule
    D. Centrosome
    E. H1
14. In eukaryotic cells, where does DNA replication primarily occur?
    A. Nucleus
    B. Cytoplasm
    C. Mitochondria
    D. Ribosome
15. DNA polymerase synthesizes new strands of DNA in which direction?
    A. 5′ to 3′
    B. 3′ to 5′
    C. Bidirectional
    D. Random
    E. Depends on whether it is the lagging or the leading strand
16. What is the purpose of DNA primase in DNA replication?
    A. To add nucleotides to the growing DNA strand
    B. To proofread and correct errors in the DNA sequence
    C. To create RNA primers for DNA synthesis
    D. To create DNA primers for DNA synthesis
    E. To repair damaged DNA strands
17. Which of the following is responsible for joining the Okazaki fragments on the lagging strand during DNA replication?
    A. RNA polymerase
    B. Helicase
    C. Ligase
    D. Primase
    E. DNA polymerase
18. What type of bond connects the nitrogen bases in a double helix in DNA?
    A. Hydrogen bond
    B. Covalent bond
    C. Ionic bond
    D. Peptide bond
    E. All of the above

Fill-in-the-Blank Questions

19. Histones are rich in the amino acids________and arginine.
20. In eukaryotic cells, DNA replication occurs during_______of the cell cycle.
21. The nitrogenous base that pairs with guanine in DNA is_________.
22. The___________________is the region of DNA where replication begins.
23. The enzyme_____________joins Okazaki fragments on the lagging strand.
24. The most condensed form of DNA is found during the__________phase of mitosis.
25. Methylation of nucleotides is a type of ______________ modification of DNA. (one word)
26. An octamer of four types of histones make up the ______________.

Short-Answer Question

27. The nucleotides that base pair in the double helix of DNA are held together by what kind of bonds?

Matching Question

28. Match the DNA replication enzymes with their functions:
    Enzymes:
    A. Helicase
    B. Primase
    C. DNA polymerase
    D. Ligase
    Functions:
    1. Joins Okazaki fragments
    2. Synthesizes RNA primer
    3. Unwinds DNA
    4. Adds nucleotides

Ordering Question

29. Place the following events of DNA replication in the correct order:
    1. Binding of helicase
    2. Formation of RNA primer
    3. DNA strand elongation
    4. Activation of helicase

Answer Key

Multiple-Choice Questions

1. **Answer**: C
2. **Answer**: C
3. **Answer**: C
4. **Answer**: B
5. **Answer**: C
6. **Answer**: B
7. **Answer**: D
8. **Answer**: B
9. **Answer**: C
10. **Answer**: D
11. **Answer**: A
12. **Answer**: D
13. **Answer**: B
14. **Answer**: A
15. **Answer**: A
16. **Answer**: C
17. **Answer**: C
18. **Answer**: A

Fill-in-the-Blank Questions

19. **Answer**: Lysine
20. **Answer**: S phase
21. **Answer**: Cytosine
22. **Answer**: Origin of replication
23. **Answer**: DNA ligase
24. **Answer**: Metaphase
25. **Answer**: Epigenetic
26. **Answer**. Nucleosome

Short-Answer Question

27. **Answer**: Hydrogen

Matching Question

28. Match the DNA replication enzymes with their functions:
    Answer: A-3, B-2, C-4, D-1

Ordering Question

29. Place the following events of DNA replication in the correct order:
    Answer: 1 → 4 → 2 → 3

References

1. McKee T, McKee J. Biochemistry: the molecular basis of life. 7th ed. New York: Oxford University Press; 2023.
2. Alberts B, Heald R, Johnson A, Morgan D, Raff M, Roberts K, Walter P, Wilson J, Hunt T. Molecular biology of the cell. W.W. Norton and Company; 2022.
3. Watson JD, Crick FHC. Molecular structure of nucleic acids: a structure for deoxyribose nucleic acid. Nature. 1953;171(4356):737–8. https://doi.org/10.1038/171737a0.
4. Hartwell LH, Goldberg M, Fischer JA, Hood L, Reynolds AE, Aquadro CF, Silver LM. Genetics: from genes to genomes. 6th ed. New York: McGraw Hill; 2017.
5. Dai X, Shen L. Advances and trends in omics technology development. Front Med (Lausanne). 2022;9:911861. https://doi.org/10.3389/fmed.2022.911861.
6. Latchman DS. Gene regulation: a eukaryotic perspective. New York: Taylor and Francis; 2010.
7. Venugopal S, Sharma V, Mehra A, Singh I, Singh G. DNA intercalators as anticancer agents. Chem Biol Drug Des. 2022;100(4):580–98. https://doi.org/10.1111/cbdd.14116.
8. Aneesh TP, Sonal Sekhar M, Jose A, Chandran L, Zachariah SM. Pharmacogenomics: the right drug to the right person. J Clin Med Res. 2009;1(4):191–4. https://doi.org/10.4021/jocmr2009.08.1255.
9. Munita JM, Arias CA. Mechanisms of antibiotic resistance. Microbiol Spectr. 2016;4(2). https://doi.org/10.1128/microbiolspec.VMBF-0016-2015.
10. Vasan N, Baselga J, Hyman DM. A view on drug resistance in cancer. Nature. 2019;575(7782):299–309. https://doi.org/10.1038/s41586-019-1730-1.
11. Li T, Yang Y, Qi H, Cui W, Zhang L, Fu X, et al. CRISPR/Cas9 therapeutics: progress and prospects. Signal Transduct Target Ther. 2023;8(1):36. https://doi.org/10.1038/s41392-023-01309-7.
12. Alberts B, Bray D, Hopkin K, Johnson A, Lewis J, Raff M, Roberts K, Walter P. Essential cell biology. Norton and Company; 2023.
13. Brown TA. Genomes. 5th ed. CRC Press; 2023.
14. Nelson DL, Cox MM, Hoskins AA. Lehninger principles of biochemistry. 8th ed. Macmillan Learning; 2021.
15. Berg JM, Tymoczko JL, Gatto GJ Jr, Stryer L. Biochemistry. 9th ed. New York: MacMillan Learning; 2019.
16. Carter NP. Cytogenetic analysis by chromosome painting. Cytometry. 1994;18(1):2–10. https://doi.org/10.1002/cyto.990180103.
17. U.S. Food and Drug Administration. Doxycycline capsules, USP. 1967. https://www.accessdata.fda.gov/scripts/cder/daf/index.cfm
18. National Cancer Institute. Doxorubicin Hydrochloride. Bethesda (MD) National Cancer Institute. Available from: https://www.cancer.gov/about-cancer/treatment/drugs/doxorubicin-hydrochloride-liposome
19. Agudelo D, Bourassa P, Bérubé G, Tajmir-Riahi H-A. Intercalation of antitumor drug doxorubicin and its analogue by DNA duplex: structural features and biological implications. Int J Biol Macromol. 2014;66:144–50. https://doi.org/10.1016/j.ijbiomac.2014.02.028.
20. Pfizer. Doxorubicin dosage and administration. Available from: https://www.pfizermedical.com/doxorubicin/dosage-admin
21. Sritharan S, Sivalingam N. A comprehensive review on time-tested anticancer drug doxorubicin. Life Sci. 2021;278:119527. https://doi.org/10.1016/j.lfs.2021.119527.
22. Krebs JE, Goldstein ES, Kilpatrick ST. Lewin's genes XII. 12th ed. Jones and Bartlett Learning; 2017.
23. MedlinePlus. OCA2 gene: OCA2 melanosomal transmembrane protein. Available from: https://medlineplus.gov/genetics/gene/oca2/#:~:text=The%20OCA2%20gene%20(formerly%20called,necessary%20for%20most%20biological%20processes.
24. Genomics and Your Health. Genetics basics. Available from: https://www.cdc.gov/genomics-and-health/about/index.html.
25. Strachan T, Read A. Human molecular genetics. 5th ed. New York: Garland Science; 2019.
26. Pagni S, Mills JD, Frankish A, Mudge JM, Sisodiya SM. Non-coding regulatory elements: potential roles in disease and the case of epilepsy. Neuropathol Appl Neurobiol. 2022;48(3):e12775. https://doi.org/10.1111/nan.12775.
27. Vagnarelli P. Chromatin reorganization through mitosis. Adv Protein Chem Struct Biol. 2013;90:179–224. https://doi.org/10.1016/b978-0-12-410523-2.00006-7.
28. Federal Judicial Center. DNA basics: nucleotides, genes, and the genome. Available from: https://www.fjc.gov/content/361230/DNA-basics-nucleotides-genes-genome.
29. Luger K, Mäder AW, Richmond RK, Sargent DF, Richmond TJ. Crystal structure of the nucleosome core particle at 2.8 Å resolution. Nature. 1997;389(6648):251–60. https://doi.org/10.1038/38444.
30. Cutter AR, Hayes JJ. A brief review of nucleosome structure. FEBS Lett. 2015;589(20 Pt A):2914–22. https://doi.org/10.1016/j.febslet.2015.05.016.
31. Talbert PB, Henikoff S. Histone variants at a glance. J Cell Sci. 2021;134(6):jcs244749. https://doi.org/10.1242/jcs.244749.
32. Mariño-Ramírez L, Kann MG, Shoemaker BA, Landsman D. Histone structure and nucleosome stability. Expert Rev Proteomics. 2005;2(5):719–29. https://doi.org/10.1586/14789450.2.5.719.
33. Brooker AS, Berkowitz KM. The roles of cohesins in mitosis, meiosis, and human health and disease. Methods Mol Biol. 2014;1170:229–66. https://doi.org/10.1007/978-1-4939-0888-2_11.
34. Uhlmann F. A unified model for cohesin function in sister chromatid cohesion and chromatin loop formation. Mol Cell. 2025;85(6):1058–71. https://doi.org/10.1016/j.molcel.2025.02.005.

35. Skibbens RV. Condensins and cohesins – one of these things is not like the other! J Cell Sci. 2019;132(3):jcs220491. https://doi.org/10.1242/jcs.220491.
36. Hoencamp C, Rowland BD. Genome control by SMC complexes. Nat Rev Mol Cell Biol. 2023;24(9):633–50. https://doi.org/10.1038/s41580-023-00609-8.
37. Agarwal A, Korsak S, Choudhury A, Plewczynski D. The dynamic role of cohesin in maintaining human genome architecture. Bioessays. 2023;45(10):e2200240. https://doi.org/10.1002/bies.202200240.
38. Morales C, Losada A. Establishing and dissolving cohesion during the vertebrate cell cycle. Curr Opin Cell Biol. 2018;52:51–7. https://doi.org/10.1016/j.ceb.2018.01.010.
39. Gibcus JH, Samejima K, Goloborodko A, Samejima I, Naumova N, Nuebler J, et al. A pathway for mitotic chromosome formation. Science. 2018;359(6376):eaao6135. https://doi.org/10.1126/science.aao6135.
40. Houlard M, Godwin J, Metson J, Lee J, Hirano T, Nasmyth K. Condensin confers the longitudinal rigidity of chromosomes. Nat Cell Biol. 2015;17(6):771–81. https://doi.org/10.1038/ncb3167.
41. Yuen KC, Gerton JL. Taking cohesin and condensin in context. PLoS Genet. 2018;14(1):e1007118. https://doi.org/10.1371/journal.pgen.1007118.
42. Makrantoni V, Marston AL. Cohesin and chromosome segregation. Curr Biol. 2018;28(12):R688–R93. https://doi.org/10.1016/j.cub.2018.05.019.
43. Onn I, Aono N, Hirano M, Hirano T. Reconstitution and subunit geometry of human condensin complexes. EMBO J. 2007;26(4):1024–34. https://doi.org/10.1038/sj.emboj.7601562.
44. Ono T, Fang Y, Spector DL, Hirano T. Spatial and temporal regulation of Condensins I and II in mitotic chromosome assembly in human cells. Mol Biol Cell. 2004;15(7):3296–308. https://doi.org/10.1091/mbc.e04-03-0242.
45. Banerji R, Skibbens RV, Iovine MK. How many roads lead to cohesinopathies? Dev Dyn. 2017;246(11):881–8. https://doi.org/10.1002/dvdy.24510.
46. MedlinePlus. NIPBL gene: NIPBL cohesin loading factor. Available from: https://medlineplus.gov/genetics/gene/nipbl/.
47. Cheng H, Zhang N, Pati D. Cohesin subunit RAD21: from biology to disease. Gene. 2020;758:144966. https://doi.org/10.1016/j.gene.2020.144966.
48. Milazzo G, Mercatelli D, Di Muzio G, Triboli L, De Rosa P, Perini G, et al. Histone deacetylases (HDACs): evolution, specificity, role in transcriptional complexes, and pharmacological actionability. Genes (Basel). 2020;11(5):556. https://doi.org/10.3390/genes11050556.
49. Boyle MI, Jespersgaard C, Brøndum-Nielsen K, Bisgaard AM, Tümer Z. Cornelia de Lange syndrome. Clin Genet. 2015;88(1):1–12. https://doi.org/10.1111/cge.12499.
50. Avagliano L, Parenti I, Grazioli P, Di Fede E, Parodi C, Mariani M, et al. Chromatinopathies: a focus on Cornelia de Lange syndrome. Clin Genet. 2020;97(1):3–11. https://doi.org/10.1111/cge.13674.
51. Xu B, Lu S, Gerton JL. Roberts syndrome: a deficit in acetylated cohesin leads to nucleolar dysfunction. Rare Dis. 2014;2:e27743. https://doi.org/10.4161/rdis.27743.
52. Piché J, Van Vliet PP, Pucéat M, Andelfinger G. The expanding phenotypes of cohesinopathies: one ring to rule them all! Cell Cycle. 2019;18(21):2828–48. https://doi.org/10.1080/15384101.2019.1658476.
53. Cuartero S, Innes AJ, Merkenschlager M. Towards a better understanding of cohesin mutations in AML. Front Oncol. 2019;9:867. https://doi.org/10.3389/fonc.2019.00867.
54. Waldman T. Emerging themes in cohesin cancer biology. Nat Rev Cancer. 2020;20(9):504–15. https://doi.org/10.1038/s41568-020-0270-1.
55. Cai X, Gao J, Shi C, Guo W, Guo D, Zhang S. The role of NCAPG in various of tumors. Biomed Pharmacother. 2022;155:113635. https://doi.org/10.1016/j.biopha.2022.113635.
56. Antony J, Chin CV, Horsfield JA. Cohesin mutations in cancer: emerging therapeutic targets. Int J Mol Sci. 2021;22(13):6788. https://doi.org/10.3390/ijms22136788.
57. National Cancer Institute. Synthetic lethality. Available from: https://www.cancer.gov/publications/dictionaries/cancer-terms/def/synthetic-lethality.
58. Stirpe A, Heun P. The ins and outs of CENP-A: chromatin dynamics of the centromere-specific histone. Semin Cell Dev Biol. 2023;135:24–34. https://doi.org/10.1016/j.semcdb.2022.04.003.
59. Samoshkin A, Arnaoutov A, Jansen LE, Ouspenski I, Dye L, Karpova T, et al. Human condensin function is essential for centromeric chromatin assembly and proper sister kinetochore orientation. PLoS One. 2009;4(8):e6831. https://doi.org/10.1371/journal.pone.0006831.
60. Andrade Ruiz L, Kops G, Sacristan C. Vertebrate centromere architecture: from chromatin threads to functional structures. Chromosoma. 2024;133(3):169–81. https://doi.org/10.1007/s00412-024-00823-z.
61. Baldi S, Korber P, Becker PB. Beads on a string-nucleosome array arrangements and folding of the chromatin fiber. Nat Struct Mol Biol. 2020;27(2):109–18. https://doi.org/10.1038/s41594-019-0368-x.
62. Ewan's Blog. Bioinformation at large: a cheat's guide to histone modifications. 2014. Available from: https://ewanbirney.com/2014/09/a-cheats-guide-to-histone-modifications.html?utm_source=chatgpt.com.
63. Bannister AJ, Kouzarides T. Regulation of chromatin by histone modifications. Cell Res. 2011;21(3):381–95. https://doi.org/10.1038/cr.2011.22.

64. Suganuma T, Workman JL. Signals and combinatorial functions of histone modifications. Annu Rev Biochem. 2011;80:473–99. https://doi.org/10.1146/annurev-biochem-061809-175347.
65. Kouzarides T. Chromatin modifications and their function. Cell. 2007;128(4):693–705. https://doi.org/10.1016/j.cell.2007.02.005.
66. Lyko F. The DNA methyltransferase family: a versatile toolkit for epigenetic regulation. Nat Rev Genet. 2018;19(2):81–92. https://doi.org/10.1038/nrg.2017.80.
67. Moore LD, Le T, Fan G. DNA methylation and its basic function. Neuropsychopharmacology. 2013;38(1):23–38. https://doi.org/10.1038/npp.2012.112.
68. Hyun K, Jeon J, Park K, Kim J. Writing, erasing and reading histone lysine methylations. Exp Mol Med. 2017;49(4):e324. https://doi.org/10.1038/emm.2017.11.
69. Becker PB. Nucleosome sliding: facts and fiction. EMBO J. 2002;21(18):4749–53. https://doi.org/10.1093/emboj/cdf486.
70. Pazin MJ, Kadonaga JT. SWI2/SNF2 and related proteins: ATP-driven motors that disrupt-protein-DNA interactions? Cell. 1997;88(6):737–40. https://doi.org/10.1016/S0092-8674(00)81918-2.
71. Bieluszewski T, Prakash S, Roulé T, Wagner D. The role and activity of SWI/SNF chromatin remodelers. Annu Rev Plant Biol. 2023;74:139–63. https://doi.org/10.1146/annurev-arplant-102820-093218.
72. Narlikar GJ, Fan H-Y, Kingston RE. Cooperation between complexes that regulate chromatin structure and transcription. Cell. 2002;108(4):475–87. https://doi.org/10.1016/S0092-8674(02)00654-2.
73. Reyes AA, Marcum RD, He Y. Structure and function of chromatin remodelers. J Mol Biol. 2021;433(14):166929. https://doi.org/10.1016/j.jmb.2021.166929.
74. Tyagi M, Imam N, Verma K, Patel AK. Chromatin remodelers: we are the drivers!! Nucleus. 2016;7(4):388–404. https://doi.org/10.1080/19491034.2016.1211217.
75. Morrison O, Thakur J. Molecular complexes at euchromatin, heterochromatin and centromeric chromatin. Int J Mol Sci. 2021;22(13):6922. https://doi.org/10.3390/ijms22136922.
76. Albrecht LV, Tejeda-Muñoz N, De Robertis EM. Cell biology of canonical Wnt signaling. Annu Rev Cell Dev Biol. 2021;37:369–89. https://doi.org/10.1146/annurev-cellbio-120319-023657.
77. Cooper G, Adams K. The cell: a molecular approach. Sinauer Associates; 2022.
78. Ekundayo B, Bleichert F. Origins of DNA replication. PLoS Genet. 2019;15(9):e1008320. https://doi.org/10.1371/journal.pgen.1008320.
79. DePamphilis ML, Bell SD. Genome duplication. Garland Science; 2010.
80. O'Donnell M, Langston L, Stillman B. Principles and concepts of DNA replication in bacteria, archaea, and eukarya. Cold Spring Harb Perspect Biol. 2013;5(7):a010108. https://doi.org/10.1101/cshperspect.a010108.
81. Zhang H. Regulation of DNA replication licensing and re-replication by Cdt1. Int J Mol Sci. 2021;22(10):5195. https://doi.org/10.3390/ijms22105195.
82. Burgers PMJ, Kunkel TA. Eukaryotic DNA replication fork. Annu Rev Biochem. 2017;86:417–38. https://doi.org/10.1146/annurev-biochem-061516-044709.
83. Okazaki R, Okazaki T, Sakabe K, Sugimoto K, Sugino A. Mechanism of DNA chain growth. I. Possible discontinuity and unusual secondary structure of newly synthesized chains. Proc Natl Acad Sci USA. 1968;59(2):598–605. https://doi.org/10.1073/pnas.59.2.598.
84. Tarcan Z, Poovathumkadavil D, Skagia A, Gambus A. The p97 segregase cofactor Ubxn7 facilitates replisome disassembly during S-phase. J Biol Chem. 2022;298(8):102234. https://doi.org/10.1016/j.jbc.2022.102234.
85. Xia Y. The fate of two unstoppable trains after arriving destination: replisome disassembly during DNA replication termination. Front Cell Dev Biol. 2021;9:658003. https://doi.org/10.3389/fcell.2021.658003.
86. Turner KJ, Vasu V, Griffin DK. Telomere biology and human phenotype. Cells. 2019;8(1):73. https://doi.org/10.3390/cells8010073.
87. Lu W, Zhang Y, Liu D, Songyang Z, Wan M. Telomeres-structure, function, and regulation. Exp Cell Res. 2013;319(2):133–41. https://doi.org/10.1016/j.yexcr.2012.09.005.
88. Hu H, Yan HL, Nguyen THD. Structural biology of shelterin and telomeric chromatin: the pieces and an unfinished puzzle. Biochem Soc Trans. 2024;52(4):1551–64. https://doi.org/10.1042/bst20230300.
89. Maciejowski J, de Lange T. Telomeres in cancer: tumour suppression and genome instability. Nat Rev Mol Cell Biol. 2017;18(3):175–86. https://doi.org/10.1038/nrm.2016.171.
90. de Lange T. Shelterin: the protein complex that shapes and safeguards human telomeres. Genes Dev. 2005;19(18):2100–10. https://doi.org/10.1101/gad.1346005.
91. Fletcher H, Hickey I. BIOS instant notes in genetics. Abingdon: Garland Science; 2012.
92. Shammas MA. Telomeres, lifestyle, cancer, and aging. Curr Opin Clin Nutr Metab Care. 2011;14(1):28–34. https://doi.org/10.1097/MCO.0b013e32834121b1.
93. Savage SA. Telomere length and cancer risk: finding Goldilocks. Biogerontology. 2024;25(2):265–78. https://doi.org/10.1007/s10522-023-10080-9.
94. MedlinePlus. Dyskeratosis congenita. Available from: https://medlineplus.gov/genetics/condition/dyskeratosis-congenita/#causes.
95. Boston Children's Hospital. Dyskeratosis congenita. Available from: https://www.childrenshospital.org/conditions/dyskeratosis-congenita.
96. Mannherz W, Agarwal S. Thymidine nucleotide metabolism controls human telomere length. Nat Genet. 2023;55(4):568–80. https://doi.org/10.1038/s41588-023-01339-5.

97. U.S. Food and Drug Administration. Rytelo (Imetelstat). Approval overview. 2024. Available from: https://www.accessdata.fda.gov/scripts/cder/daf/index.cfm?event=overview.process&ApplNo=217779.
98. Keam SJ. Imetelstat: first approval. Drugs. 2024;84(9):1149–55. https://doi.org/10.1007/s40265-024-02080-x.
99. Geron Corporation. RYTELO® (imetelstat) Injection. Available from: https://www.rytelohcp.com/
100. Platzbecker U, Santini V, Fenaux P, Sekeres MA, Savona MR, Madanat YF, et al. Imetelstat in patients with lower-risk myelodysplastic syndromes who have relapsed or are refractory to erythropoiesis-stimulating agents (IMerge): a multinational, randomised, double-blind, placebo-controlled, phase 3 trial. Lancet. 2024;403(10423):249–60. https://doi.org/10.1016/s0140-6736(23)01724-5.
101. Bazinet A, Bravo GM. New approaches to myelodysplastic syndrome treatment. Curr Treat Options in Oncol. 2022;23(5):668–87. https://doi.org/10.1007/s11864-022-00965-1.
102. Bryan TM. G-quadruplexes at telomeres: friend or foe? Molecules. 2020;25(16):3686. https://doi.org/10.3390/molecules25163686.
103. Capra JA, Paeschke K, Singh M, Zakian VA. G-quadruplex DNA sequences are evolutionarily conserved and associated with distinct genomic features in Saccharomyces cerevisiae. PLoS Comput Biol. 2010;6(7):e1000861. https://doi.org/10.1371/journal.pcbi.1000861.
104. Teng FY, Jiang ZZ, Guo M, Tan XZ, Chen F, Xi XG, et al. G-quadruplex DNA: a novel target for drug design. Cell Mol Life Sci. 2021;78(19–20):6557–83. https://doi.org/10.1007/s00018-021-03921-8.
105. Kim MY, Vankayalapati H, Shin-Ya K, Wierzba K, Hurley LH. Telomestatin, a potent telomerase inhibitor that interacts quite specifically with the human telomeric intramolecular g-quadruplex. J Am Chem Soc. 2002;124(10):2098–9. https://doi.org/10.1021/ja017308q.
106. Wu Y, Poulos RC, Reddel RR. Role of POT1 in human cancer. Cancers (Basel). 2020;12(10):2739. https://doi.org/10.3390/cancers12102739.
107. Baptista Freitas M, Desmyter L, Badoer C, Smits G, Vandernoot I, t'Kint de Roodenbeke D. POT1 tumour predisposition: a broader spectrum of associated malignancies and proposal for additional screening program. Eur J Hum Genet. 2024;32(8):980–6. https://doi.org/10.1038/s41431-024-01611-0.
108. De March M, Merino N, Barrera-Vilarmau S, Crehuet R, Onesti S, Blanco FJ, et al. Structural basis of human PCNA sliding on DNA. Nat Commun. 2017;8(1):13935. https://doi.org/10.1038/ncomms13935.
109. Kelch BA, Makino DL, O'Donnell M, Kuriyan J. Clamp loader ATPases and the evolution of DNA replication machinery. BMC Biol. 2012;10(1):34. https://doi.org/10.1186/1741-7007-10-34.
110. Bush NG, Evans-Roberts K, Maxwell A. DNA topoisomerases. EcoSal Plus. 2015;6(2). https://doi.org/10.1128/ecosalplus.ESP-0010-2014.
111. Wang JC. Cellular roles of DNA topoisomerases: a molecular perspective. Nat Rev Mol Cell Biol. 2002;3(6):430–40. https://doi.org/10.1038/nrm831.
112. Bailly C. Irinotecan: 25 years of cancer treatment. Pharmacol Res. 2019;148:104398. https://doi.org/10.1016/j.phrs.2019.104398.
113. Bondarev AD, Jonsson J, Chubarev VN, Tarasov VV, Lagunas-Rangel FA, Schiöth HB. Recent developments of topoisomerase inhibitors: clinical trials, emerging indications, novel molecules and global sales. Pharmacol Res. 2024;209:107431. https://doi.org/10.1016/j.phrs.2024.107431.
114. Nitiss JL. Targeting DNA topoisomerase II in cancer chemotherapy. Nat Rev Cancer. 2009;9(5):338–50. https://doi.org/10.1038/nrc2607.
115. Le TT, Wang MD. Topoisomerase II and etoposide – a tangled tale. Nat Chem Biol. 2023;19(5):546–7. https://doi.org/10.1038/s41589-022-01236-8.
116. Pendleton M, Lindsey RH Jr, Felix CA, Grimwade D, Osheroff N. Topoisomerase II and leukemia. Ann N Y Acad Sci. 2014;1310(1):98–110. https://doi.org/10.1111/nyas.12358.
117. Hevener K, Verstak TA, Lutat KE, Riggsbee DL, Mooney JW. Recent developments in topoisomerase-targeted cancer chemotherapy. Acta Pharm Sin B. 2018;8(6):844–61. https://doi.org/10.1016/j.apsb.2018.07.008.
118. U.S. Food and Drug Administration. VePesid (Etoposide) capsules. Approval overview. 1983. Available from: https://www.accessdata.fda.gov/scripts/cder/daf/index.cfm?event=overview.process&ApplNo=075635.
119. U.S. Food and Drug Administration. VePesid (Etoposide) injection. Approval overview. 1983. Available from: https://www.accessdata.fda.gov/scripts/cder/daf/index.cfm?event=overview.process&ApplNo=019557
120. Baldwin EL, Osheroff N. Etoposide, topoisomerase II and cancer. Curr Med Chem Anticancer Agents. 2005;5(4):363–72. https://doi.org/10.2174/1568011054222364.
121. Montecucco A, Zanetta F, Biamonti G. Molecular mechanisms of etoposide. EXCLI J. 2015;14:95–108. https://www.excli.de/vol14/Montecucco_19012015_proof.pdf
122. Agudelo D, Bourassa P, Bérubé G, Tajmir-Riahi HA. Review on the binding of anticancer drug doxorubicin with DNA and tRNA: structural models and antitumor activity. J Photochem Photobiol B Biol. 2016;158:274–9. https://doi.org/10.1016/j.jphotobiol.2016.02.032.
123. U.S. Food and Drug Administration. Novantrone (Mitoxantrone). Approval overview. 1987. https://

www.accessdata.fda.gov/drugsatfda_docs/label/2012/019297s035lbl.pdf

124. Today MN. Uses and risks of mitoxantrone. 2025. Available from: https://www.medicalnewstoday.com/articles/322211?utm.
125. U.S. Food and Drug Administration. Camptosar (Irinotecan). Approval overview. 1996. Available from: https://www.accessdata.fda.gov/scripts/cder/daf/index.cfm?event=overview.process&ApplNo=020571.
126. Yakkala PA, Penumallu NR, Shafi S, Kamal A. Prospects of topoisomerase inhibitors as promising anti-cancer agents. Pharmaceuticals (Basel). 2023;16(10):1456. https://doi.org/10.3390/ph16101456.
127. Li F, Jiang T, Li Q, Ling X. Camptothecin (CPT) and its derivatives are known to target topoisomerase I (Top1) as their mechanism of action: did we miss something in CPT analogue molecular targets for treating human disease such as cancer? Am J Cancer Res. 2017;7(12):2350–94.
128. U.S. Food and Drug Administration. Cipro(Ciprofloxacin). Approval overview. 1987. Available from: https://www.accessdata.fda.gov/scripts/cder/daf/index.cfm?event=overview.process&ApplNo=019537.
129. Katzung BG, Vanderah TW, Trevor AAJ. Katzung's basic and clinical pharmacology. 16th ed. McGraw-Hill Education; 2024.
130. MedlinePlus. Ciprofloxacin. 2021. Available from: https://medlineplus.gov/druginfo/meds/a688016.html.
131. Masai H, Foiani, M. (eds). DNA replication: from old principles to new discoveries: Springer; 2018.
132. Ivancich M, et al. Treating cancer by targeting telomeres and telomerase. Antioxidants. 2017;6(1):15. https://doi.org/10.3390/antiox6010015.

RNA and Transcription, Abnormalities in Gene Expression in Disease

7

Abstract

RNA plays multiple essential roles in the flow of genetic information. There are multiple types of RNA in cells, and the major focus of this chapter is messenger RNA (mRNA). mRNA is the crucial intermediate that carries information from DNA to the encoded proteins. This is the process of transcription, and it is a key step in the transmission of information according to the central dogma. Other types of RNA that play key roles in the central dogma include transfer RNA (tRNA), which is important in the process of translation, and rRNA, a component of ribosomes. Other types of RNA include siRNA, microRNA, lncRNA, and snRNA. This chapter discusses the structure of the ribonucleotides that constitute RNA. It also discusses the major steps in transcription, including the role of RNA polymerase and other important enzymes. Regulation of transcription, also referred to as gene expression, is also covered, including the cis-regulatory elements and transcription factors that are critical for this process. The importance of differential expression is covered; the process by which different cells express different sets of genes, giving specialized cells their own unique characteristics and functions. Various therapies targeting RNA, as well as RNA-based therapies and drugs that modulate chromatin remodeling, are also discussed. Bacterial gene regulation is briefly covered, as well as the use of pharmaceuticals that may target bacterial transcription.

Keywords

Ribonucleotides · Template strand · Coding strand · Gene locus · Transcription · mRNA processing

Abbreviations

AML	Acute myeloid leukemia
APL	Acute promyelocytic leukemia
ATP	Adenosine triphosphate
CBC	Cap-binding complex
CBP	CREB-binding protein
CRCs	Chromatin remodeling complexes
CTCL	Cutaneous T-cell lymphoma
CTD	C-terminal domain
DNMT	DNA methyltransferase
EJC	Exon junction complex
ELK1	ETS-like gene 1 protein
ERE	Estrogen response element
GMP	Guanosine monophosphate
GTP	Guanosine triphosphate
HAT	Histone acetyltransferase
HDAC	Histone deacetylase
lncRNA	Long noncoding RNA
MBD	Methyl-CpG-binding domain
MD	Muscular dystrophy

A. Minden, *Biochemistry, Cell and Molecular Biology for Pharmacists and Health Professionals*,
https://doi.org/10.1007/978-3-032-20529-2_7

MDM2	Mouse double minute 2 homolog
MDS	Myelodysplastic syndrome
MEF2	Myocyte enhancer factor 2
miRNA	MicroRNA
NCoR	Nuclear receptor co-repressor
NPC	Nuclear pore complex
NXF1	Nuclear RNA export factor 1
NXT1	Nuclear transport factor 2-like export factor 1
p53	Tumor protein p53
PABP	Poly(A)-binding protein
PAP	Poly(A) polymerase
PIC	Pre-initiation complex
Pol II	RNA polymerase II
PRC2	Polycomb repressive complex 2
RAR	Retinoic acid receptor
rRNA	Ribosomal RNA
RXR	Retinoid X receptor
siRNA	Small interfering RNA
snRNA	Small nuclear RNA
SRF	Serum response factor
SWI/SNF	Switch/sucrose nonfermentable (chromatin remodeling complex)
T3	Triiodothyronine
TAF	TBP-associated factor
TBP	TATA-binding protein
TET	Ten-eleven translocation (enzymes involved in demethylation)
TFIID	Transcription factor II D
THR	Thyroid hormone receptor
tRNA	Transfer RNA
TSS	Transcription start site

What You Will Learn

- The basic structure of RNA, ribonucleotides, and phosphodiester bonds, and how RNA differs from DNA.
- The different types of RNA, with a focus on messenger RNA (mRNA).
- mRNA is synthesized in the process of transcription. mRNA is complementary to the template strand of DNA. The DNA template strand is complementary to the DNA coding strand.
- The organization of a gene locus.
- The proteins involved in transcription, including RNA polymerase and transcription regulators, and how some drugs can target these proteins to treat disease.
- The basic steps in transcription include initiation, elongation, and termination.
- After transcription, mRNA is processed, which involves 5′ capping, addition of a poly-A tail, and splicing, the final product is exported from the nucleus.
- The concept of differential expression: different genes are expressed in different cells, giving different cells their specialized characteristics.
- Differentiation therapy in cancer treatment, to restore the pattern of gene expression associated with differentiated cells. An example is all trans retinoic acid (ATRA).
- Transcription regulators can regulate the expression of different genes in different cells or under different conditions.
- Chromatin structure and modification can regulate gene expression.
- Some important examples of transcription factors and their functions, such as p53 and MyoD.
- Chromatin remodelers can impact gene expression. Some drugs impact chromatin remodeling, such as SAHA, an HDAC inhibitor used in cancer treatment, and Duvyzat, which is used to treat muscular dystrophy.
- The genes for transcription factors can also be regulated, leading to networks of transcription factors.
- Genes can be regulated by alternative splicing.
- RNA-based vaccines; the SARS-CoV-2 vaccine.
- Antisense technology to Spinraza, to treat spinal muscular atrophy (SMA).
- Drugs that use small interfering RNA (siRNA) technology to silence disease genes.
- Bacterial genes are organized by operons.
- Important bacterial operons include the lac operon and the trp operon.
- Bacterial RNA polymerase as a target to combat bacterial infections.

1 RNA Structure and Transcription

This chapter focuses on RNA, specifically messenger RNA (mRNA) and the process of transcription. Transcription is the synthesis of mRNA from a DNA template. Transcription is often referred to as gene expression, even though gene expression technically encompasses the entire process, from the transcription of DNA into RNA to the subsequent translation of RNA into protein. In this chapter, we will often use the term gene expression when referring to transcription. This chapter is divided into three sections. Section 1 discusses the structure of RNA, with a focus on mRNA. This chapter covers the basic steps involved in transcription. The coverage of transcription involves further discussion of both DNA as well as mRNA, since mRNA is synthesized from a DNA template. Section 2 also discusses DNA, mRNA, and transcription, but it focuses on regulated transcription. Regulation of transcription refers to the process by which different genes are transcribed at varying levels in different cells or under different conditions. This is sometimes referred to as differential gene expression. Section 3 discusses other topics related to RNA. For instance, Sect. 3 introduces other types of RNA, such as rRNA and tRNA. Other topics in this section include RNA-based therapies and bacterial RNA.

This chapter also emphasizes the importance of regulated transcription in maintaining controlled cell growth and cell differentiation, which are critical for health. This chapter also discusses how abnormalities in RNA and transcription can lead to disease. Examples are provided of drugs designed to treat abnormalities associated with RNA or transcription. RNA-based vaccines and drug treatments are also discussed, along with antisense technologies for treating disease.

1.1 Introduction to RNA

There are multiple types of RNA in cells. Messenger RNA (mRNA) is the crucial intermediate that carries information from the DNA to the encoded proteins. It has a central role in the DNA → RNA → protein pathway of the central dogma, and it will be the main focus of this chapter. There are also other types of RNA in the cell, however. Transfer RNA (tRNA) plays a key role in translation, where proteins are synthesized using mRNA as a template, and rRNA is a component of ribosomes. Thus, these two types of RNA also have direct roles in the central dogma. Other types of RNA have more indirect roles. They include siRNA, microRNA, lncRNA, and snRNA. These are noncoding RNAs but also have important roles in regulating genetic information, such as regulating gene expression and RNA processing. The discussion in this chapter is drawn from the following source [1] as well as references indicated in the text below.

1.2 Ribonucleotides

While there are different categories of RNA molecules, all of them are made of ribonucleotides (see Chap. 6, Fig. 6.3). Ribonucleotides have the same basic structure as deoxyribonucleotides that make up DNA, with certain important differences. First, in contrast to the deoxyribonucleotides, ribonucleotides are made with ribose as the sugar rather than deoxyribose. This means that, unlike the sugars in DNA, the sugars in RNA have an –OH (hydroxyl) group at the $2'$ carbon (as well as the $3'$ carbon). Second, while ribonucleotides have nitrogen bases similar to those of deoxyribonucleotides, there is one key difference. While the bases on deoxyribonucleotides are cytosine, guanine, adenine, or thymine, ribonucleotides have uracil instead of thymine. Thus, the nucleotides in RNA are cytidine (C), guanosine (G), adenosine (A), and uridine (U) [1, 2] (see Chap. 6, Fig. 6.4).

1.3 Messenger RNA (mRNA)

Messenger RNA (mRNA) is the main intermediate between the template DNA and protein

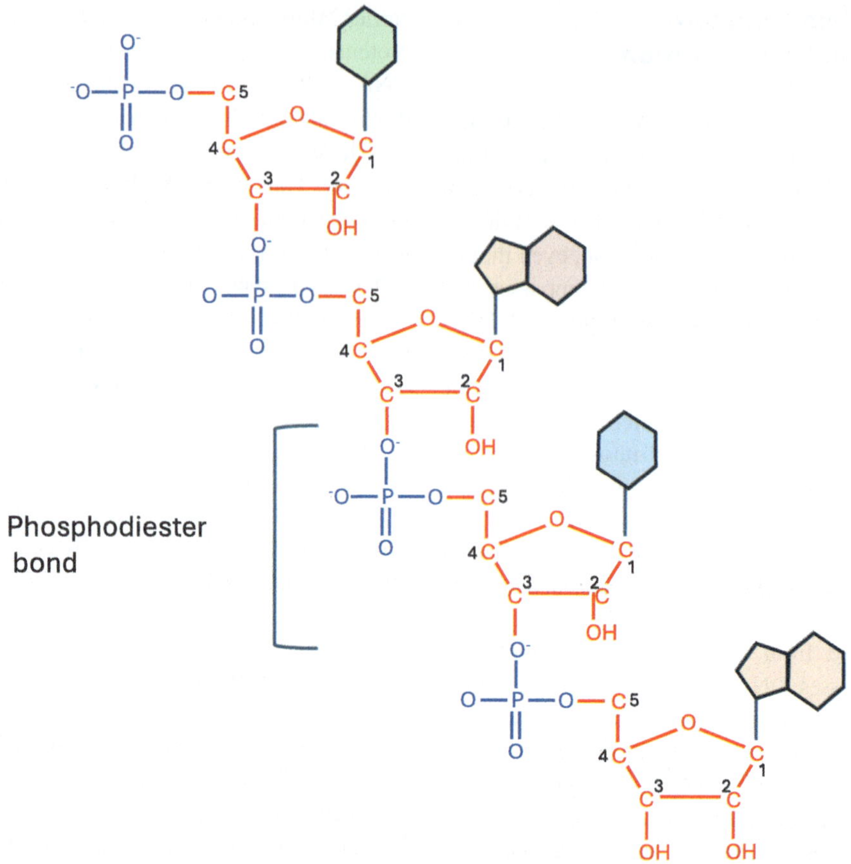

Fig. 7.1 Messenger RNA (mRNA) mRNA is usually single stranded. The ribonucleotides are held together by phosphodiester bonds. (*Source*: Created with BioRender.com)

synthesis. It is the category of RNA made during transcription, and it is the main focus here. mRNA is a polymer of ribonucleotides. The types of bonds in mRNA exhibit similarities and some differences compared to those in DNA. Like DNA, the nucleotides in RNA are linked by phosphodiester bonds, forming the backbone (Fig. 7.1). The hydrogen bonds that stabilize complementary base pairs in DNA, however, are absent in mRNA because it is single-stranded. During transcription, however, hydrogen bonds do temporarily form between the complementary nucleotides of the DNA template and the growing RNA strand (see Fig. 7.2), allowing the mRNA sequence to be accurately synthesized. Once the RNA strand separates from the DNA, these bonds are broken. Additionally, single-stranded RNA molecules sometimes fold back on themselves, forming secondary structures, such as hairpin loops. These structures, which are sometimes temporary, are stabilized by hydrogen bonds between nucleotides [1].

1.4 An Overview of Transcription

Transcription of coding genes is the process of transcribing the template strand of DNA into a single-stranded mRNA strand. After transcription, the mRNA will be released from the DNA, and it will become the template for translation during protein synthesis (Chap. 6, Figs. 6.1 and 6.2).

Similar to DNA synthesis, the process of transcription involves complementary base pairing and the formation of hydrogen bonds between

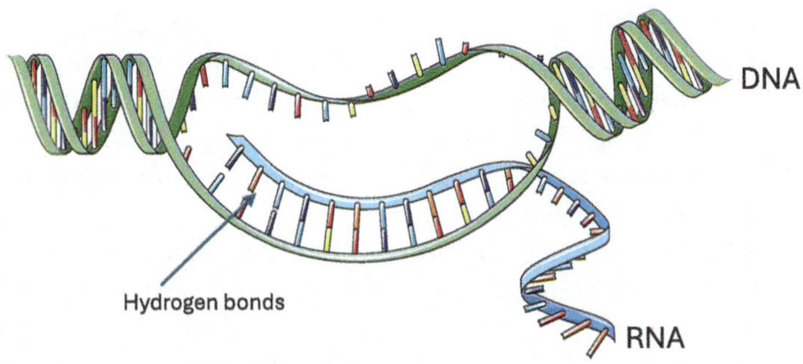

Fig. 7.2 Transcription. When transcription takes place, transient hydrogen bonds are formed between deoxyribonucleotides on DNA and ribonucleotides on the newly synthesized mRNA. (*Source*: Adapted from Servier Medical Art (https://smart.servier.com/), licensed under the Creative Commons Attribution 4.0 (CC BY 4.0) license)

DNA Strand 1		DNA Strand 2	DNA Strand		RNA
A	::	T	A	::	U
T	::	A	T	::	A
G	:::	C	G	:::	C
C	:::	G	C	:::	G

Fig. 7.3 Base pairing. Base pairing between two DNA strands, and base pairing between the DNA template strand and RNA. RNA contains U instead of T. Double and triple hydrogen bonds are indicated by dots

bases on RNA and their complementary bases on DNA. **RNA polymerase** is the enzyme that catalyzes the synthesis of the mRNA strand. It reads the DNA template in the 3′ to 5′ direction and thus builds the mRNA strand in the 5′ to 3′ direction. It adds the nucleotides one at a time to the growing mRNA strand, based on complementary base pairing. Base pairing is as follows: C on DNA base pairs with G on RNA, G on DNA base pairs with C on RNA, T on DNA base pairs with A on RNA, and A on DNA base pairs with U on RNA. A::U and T::A pairs have double hydrogen bonds, while G:::C and C:::G pairs have triple hydrogen bonds (Fig. 7.3). As with DNA, each nucleotide triphosphate is added to the 3′ OH group of the last nucleotide in the growing chain. This leads to hydrogen bonding between the nucleotides on the RNA and those on the DNA, and it also leads to the formation of a phosphodiester bond between the 3′ OH of the ribose sugar and the 5′ phosphate group of the incoming nucleotide, with the release of pyrophosphate. After transcription of the mRNA strand, the hydrogen bonds are broken, and single-stranded mRNA is released (Chap. 6, Fig. 6.1).

Transcription shares some features in common with DNA replication. For example, as with DNA replication, during transcription, DNA unwinding occurs, followed by base pairing (in this case, deoxyribonucleotides pair with ribonucleotides). Furthermore, a bubble forms in the DNA during transcription in the region of DNA that is being actively transcribed (see Fig. 7.2). However, in the case of transcription, only a portion of the DNA is transcribed, and the RNA is released from the template strand after the transcription process is complete. Unlike DNA replication, which occurs in S phase of the cell cycle, transcription can occur during any cell cycle phase, although almost all transcription occurs in interphase (I) [1–3].

1.5 The Nucleotide Strands Involved in Transcription

The region of DNA that codes for a gene contains the template from which the mRNA is transcribed. DNA is a double-stranded molecule, consisting of a coding strand and a template strand. These two strands, relative to the mRNA strand that will result from transcription, are described below and in the following sources [1, 2, 4, 5] (Chap. 6, Fig. 6.2).

- The Coding Strand of the DNA

The coding strand of the double-stranded DNA molecule is the strand that actually has the genetic code for the protein that will ultimately be produced. For example, ATG codes for methionine, GCC codes for alanine, and so forth. (The genetic code will be discussed in more detail in Chap. 8.) This strand can also be called the nontemplate strand or the sense strand.

- The Template Strand

The template strand of the double-stranded DNA has the sequence that is complementary to the coding strand. This will be the template from which mRNA will be transcribed. It can also be called the anti-sense strand. **RNA polymerase** will read this strand in the 3′ to 5′ direction, synthesizing the mRNA in the 5′ to 3′ direction.

- The mRNA

The mRNA is transcribed using the template strand as its template. Thus, the sequence of the mRNA will be identical to the coding strand sequence, except that it will have Us instead of Ts. It will be complementary to the template strand [1, 5].

1.6 The Gene Locus

Only a portion of DNA is transcribed at any one time and in any particular cell. Specifically, it is the protein-coding regions of the DNA that are transcribed, and only according to the needs of the cell. The location of the coding regions, along with the surrounding cis-regulatory elements, is the gene locus (see Chap. 6). This region of DNA generally contains certain key features in its coding strand, which will be discussed below. In the following, remember, upstream refers to regions of DNA toward the 5′ end, and downstream refers to 3′. The transcription start site is at the 5′ end of any gene, and the termination site is at the 3′ end [2].

1.6.1 The Transcription Unit

The part of the DNA that contains the gene to be transcribed is the transcription unit [1, 2]. It contains the following (*see* Chap. 6, Fig. 6.8).

- The Gene Coding Sequence

The gene coding sequence is the sequence that encodes the exons (which contain the coding sequence) and the introns (which contain noncoding sequence between the exons).

- The Transcription Start Site (TSS)

Upstream to the first exon is a DNA sequence that gets transcribed but does not get translated. Within this sequence is the transcription start site (TSS).

- The Promoter

The promoter is a sequence of DNA that lies upstream of the coding sequence. It is called a cis-regulatory element because it is located outside the coding region but has a role in regulating transcription. Usually, it is located approximately 30 nucleotides upstream from the transcription start site (TSS). The promoter is almost always necessary for gene transcription to take place. It is the site where RNA polymerase and key transcription factors bind to initiate transcription.

- Downstream Untranslated Sequence

This consists of some sequence downstream of the last exon. This part of the sequence will be transcribed but will not be translated into protein later. Toward the end of this sequence is a polyadenylation signal, and shortly downstream to this is a termination signal, which signals for transcription to stop.

1.6.2 Other cis-Regulatory Elements in the DNA

Some cis-regulatory elements can be found in distant regions, both upstream and sometimes downstream, of the gene coding sequence. Since they can be quite distant from the coding sequence, they can influence transcription without being part of the transcription unit. Cis-regulatory elements outside the promoter are often called enhancers. Proteins that bind these sequences may act as transcription activators, leading to an increased level of transcription. When bound by inhibitory transcription factors (repressors), these regions may also be referred to as silencers. Cis-regulatory elements can be quite far from the coding sequence, but they may also be closer or even within an intron [1, 2].

1.7 Proteins That Are Involved in Transcription

Various proteins have roles in transcription, including **RNA polymerase** and transcription regulators, discussed below and also described in the following resources [1, 2].

1.7.1 RNA Polymerase

RNA polymerase is the enzyme that catalyzes the synthesis of RNA from the DNA template in the transcription process. There are three different RNA polymerases in eukaryotes. RNA **polymerase II (RNA Pol II)** transcribes protein-coding genes and synthesizes mRNA, which will be the main focus here. **RNA Pol II** moves along the DNA, unwinds part of the double helix, and facilitates the incorporation of ribonucleotides and the formation of phosphodiester bonds on the RNA backbone. As transcription proceeds, the ribonucleotides are hydrogen-bonded to their complementary nucleotides on the template DNA strand. Unlike DNA polymerase, RNA polymerase can start transcription without a primer. As the new RNA chain grows, it extends in the 5′ to 3′ direction (moving along the template in the 3′ to 5′ direction). Transcription can occur quickly, and multiple RNA polymerases can transcribe a single gene, producing many RNA copies in a short time [1, 2, 6].

1.7.2 Transcription Regulators

Transcription regulators (Fig. 7.4) regulate gene expression by binding directly or indirectly to cis-regulatory sequences on DNA. Transcription regulator is a broad term that includes transcription factors (which bind to DNA directly) and other factors such as co-regulators, mediator complexes, and others. Transcription factors can include both transcription activators, which enhance transcription, and transcription repressors, which inhibit transcription. They are sometimes also referred to simply as activators and repressors. Transcription factors have a DNA-binding domain, which allows them to bind directly to DNA, and they may also have an activation domain, enabling interaction with other proteins to help regulate transcription. They do not bind DNA at random, but each type of transcription factor has a specific sequence or set of sequences that it will bind to. This is called its consensus sequence.

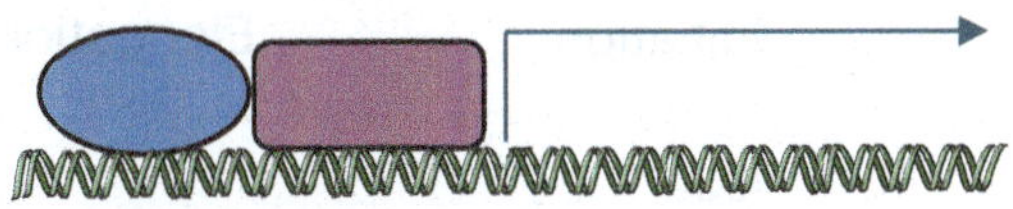

Fig. 7.4 Transcription regulators. Transcription regulators bind to regulatory elements (cis regulatory sequences) such as promoters or enhancers and regulate transcription. (*Source*: Adapted from Servier Medical Art (https://smart.servier.com/), licensed under the Creative Commons Attribution 4.0 (CC BY 4.0) license)

While transcription factors bind directly to DNA at promoter or cis-regulatory regions, co-regulators are proteins that interact with transcription factors. These can include co-activators and co-repressors, and they do not necessarily bind DNA directly. Some chromatin modifiers are examples of co-regulators. They are proteins that are recruited to transcription factors and that also have a domain that modulates the structure of the chromatin, thereby making the DNA either more or less accessible to other transcription factors [1, 2, 6, 7].

1.8 The Basic Steps in Transcription

Broadly, the major steps of transcription are initiation, elongation, termination (see Fig. 7.5), and processing, all of which take place in the nucleus. During initiation, **RNA polymerase** binds to the promoter region and forms the transcription initiation complex. During elongation, **RNA polymerase** unwinds the DNA and creates a ribonucleotide chain that is complementary to the DNA template strand. Termination happens when the RNA polymerase reaches a termination signal and stops transcribing the DNA. At this

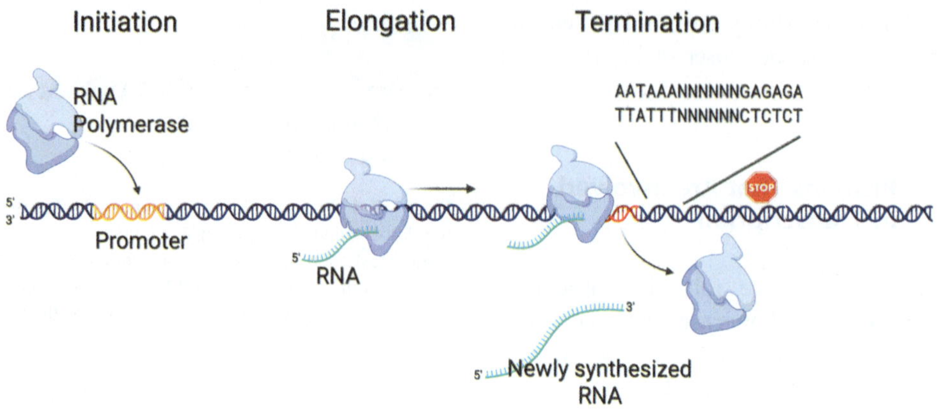

Fig. 7.5 Transcription. The basic steps of transcription are initiation, elongation, and termination. See text for details. (*Source*: Created with BioRender.com)

point, RNA polymerase and the RNA chain will dissociate from the DNA. Once the RNA is made, it still needs to undergo processing. This involves the addition of a poly-A tail and a 5′ cap, as well as splicing. Once processed, the mRNA is ready to be transported out of the nucleus and into the cytoplasm, where translation will occur. The basic steps of initiation, elongation, and processing are described in more detail below, and also described in [1, 2, 6, 8, 9].

1.8.1 Transcription Initiation

- Binding of the *General Transcription Factors* to the Promoter

The steps that occur during the initiation of transcription are summarized in Fig. 7.6. The first step of transcription initiation in most protein-coding genes is the binding of the protein complex **TFIID (transcription factor II D)** to the gene's promoter (Fig. 7.6a). TFIID is a large multiprotein complex that has a subunit called **TBP** (TATA-binding protein) and multiple proteins called TAFs (TBP-associated factors). The TBP subunit recognizes and binds to promoters with TATA boxes. **TATA boxes** are sequences rich in As and Ts, present within the promoters of many genes. This binding leads to a change in DNA conformation that is important for subsequent steps of initiation, and it helps mediate interactions with other transcription factors and regulatory [1].

After TFIID binds (via TBP), other proteins bind to this region of the promoter, either directly or indirectly. These include **TFIIA, TFIIB, TFIIF, TFIIE, and TFIIH**. Together, these comprise the general transcription factors (and sometimes called the initiation factors). **RNA Pol II** typically binds the promoter as part of a complex with TFIIF, and this stabilizes its interaction with the DNA (Fig. 7.6b). Together with RNA Pol II, **the general transcription factors** form the **pre-initiation complex (PIC)**, which is essential for the initiation of nearly all genes transcribed by **RNA Pol II**. All these components together are sometimes referred to as the general transcription machinery [1, 10].

While many genes contain a TATA box in their promoters, some genes do not. Nevertheless, TFIID plays a key role in the initiation of transcription from most genes. It can often bind even to "TATA-less" promoters via TAFs (TBP-associated factors), followed by the recruitment of the other general transcription factors and **RNA Pol II** [11].

- Upstream cis-Regulatory Elements Play a Role in Positioning the Initiation Complex

In addition to the promoter, many genes also contain cis-regulatory regions, such as enhancers, upstream (5′) to the promoter. The binding of transcription activators to these elements can help stabilize the PIC at the promoter. These cis-regulatory elements are not required for the

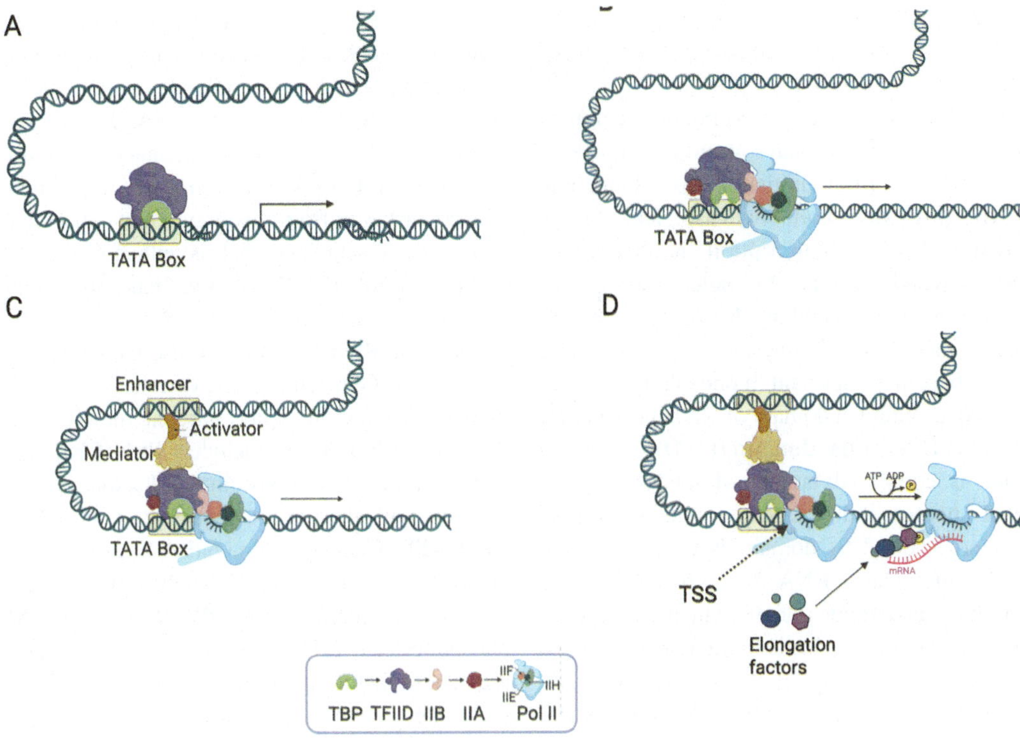

Fig. 7.6 Transcription. Steps involved in the initiation of transcription. Important elements are the TATA box, often present within the promoter, the TSS, and enhancers. The general transcription factors along with RNA polymerase make up the general transcription machinery. (*Source*: Created with BioRender.com)

transcription initiation of all genes, but they enhance the transcription efficiency of many genes.

Transcription activators bound to enhancers can usually interact with a **mediator complex** (Fig. 7.6c). This mediator complex is a multiprotein complex that mediates an interaction between transcription activators and the components of the general transcription machinery. For example, the mediator plays a role in positioning TFIIH near the tail of **RNA polymerase II (RNA Pol II)**, facilitating its phosphorylation, an important step in the elongation process, as discussed below. The mediator complex can also recruit chromatin-modifying enzymes, including those that modulate histones, to make the DNA more accessible to DNA-binding proteins.

Although these interactions between enhancer binding proteins and the general transcription machinery can support the transcription initiation process, they are not always necessary for the basal level of transcription of all genes. However, these types of interactions become especially important when discussing gene regulation and differential gene expression, as presented in Sect. 2 of this chapter.

1.8.2 Transcription Elongation

The elongation phase begins when RNA polymerase starts to add the ribonucleotides, one by one, to the growing mRNA chain. The bases on the ribonucleotides form hydrogen bonds with the complementary bases on the template DNA strand. The addition of ribonucleotides starts with the one that is complementary to the nucleotide at the **transcription start site (TSS)**. As the ribonucleotides are added, a phosphodiester bond forms between each new ribonucleotide on the growing mRNA strand. The hydrogen bonds between RNA and DNA are transient, and as the RNA chain grows, the growing RNA strand separates from the DNA.

TFIIH is part of the pre-initiation complex (described above) and plays an important role in the transition from initiation to elongation. **TFIIH** possesses both helicase and kinase activities. During transcription initiation, the DNA unwinds at the promoter through its helicase activity, enabling RNA synthesis to occur. Once RNA synthesis starts, its kinase activity becomes important. It phosphorylates serine residues within a domain of RNA polymerase called the **C-terminal domain (CTD)**. This CTD region is composed largely of tandem repeats called heptapeptide repeats, which contain 7 amino acids. This domain is also called the "tail" of the RNA polymerase. The phosphorylated serines are within these repeats. This phosphorylation leads to the release of most of the general transcription factors, which are not needed for the elongation phase. Now **RNA Pol II** can proceed down the DNA beyond the promoter region, leading to **elongation** of the new mRNA chain. During elongation, ribonucleotides (ATP, CTP, UTP, and GTP) from the cell's nucleotide pool are added to the growing mRNA chain (see Fig. 7.6d).

Phosphorylation of the CTD also provides docking sites for a set of proteins that are important for elongation. These proteins include elongation factors. They are important for elongation of transcription and also help ensure that **RNA Pol II** will not dissociate before reaching the end of the gene. Proteins that bind to **RNA Pol II** also have other functions, such as RNA processing, histone modification, and chromatin remodeling. As elongation proceeds, another site on CTD gets phosphorylated, further facilitating the elongation process by recruiting additional elongation factors and chromatin-modifying enzymes that help RNA pol II function [1, 5, 12–15].

1.8.3 Transcription Termination

Once the RNA-Pol II reaches the end of the gene's coding region and encounters the 3′ noncoding region, transcription will terminate. This requires the **RNA Pol II** to encounter certain signals. There are different sequences that can signal the RNA Pol II to stop, but an important termination signal is the polyadenylation (poly-A) signal at the 3′ end of the DNA. This region often encodes (is complementary to) **the sequence AAUAAA** or a similar sequence. Further downstream from this region is a region that codes for a sequence that is either G-rich or GU-rich. RNA Pol II will terminate after transcribing through these regions.

Once the **RNA Pol II** has transcribed the poly-A and G or GU-rich signals, certain factors will bind to this region of the nascent (newly transcribed) mRNA. These include **CPSF** (**Cleavage** and **P**olyadenylation **S**pecificity **F**actor), which interacts with the poly-A sequence of the RNA, and **CSTF** (Cleavage **S**timulation **F**actor), which interacts with the G or GU-rich region. These factors are recruited to the RNA via the phosphorylated CTD domain of RNA polymerase. These factors then recruit additional factors, including cleavage factors and poly **(A) polymerase**.

The cleavage factors **cleave** RNA at a specific site downstream to the poly-A signal but upstream of the G/GU-rich sequence. This results in a sequence upstream and one downstream to the cleavage site. The upstream sequence is the pre-mRNA. At this point, transcription of the pre-mRNA has completed. It will be further processed as discussed below. The downstream fragment, both RNA and DNA, remains attached to the **RNA-Pol II** briefly.

The **RNA-Pol II** still needs to be released from the DNA. This is thought to occur via the "torpedo model," where an **exonuclease** (such as **XRN2**) degrades the "uncapped" RNA that is still associated with the **RNA Pol II**. As this exonuclease "catches up" with the RNA-Pol II as it degrades, it breaks the RNA-DNA hybrid within the Pol II, and this releases the Pol II from the DNA. The hydrogen bonds that remain between this downstream RNA strand and the DNA template strand are broken. At this point, transcription is terminated [7, 16].

This sequence of events is also important for the addition of the poly-A tail, which will be discussed in more detail below. It is also

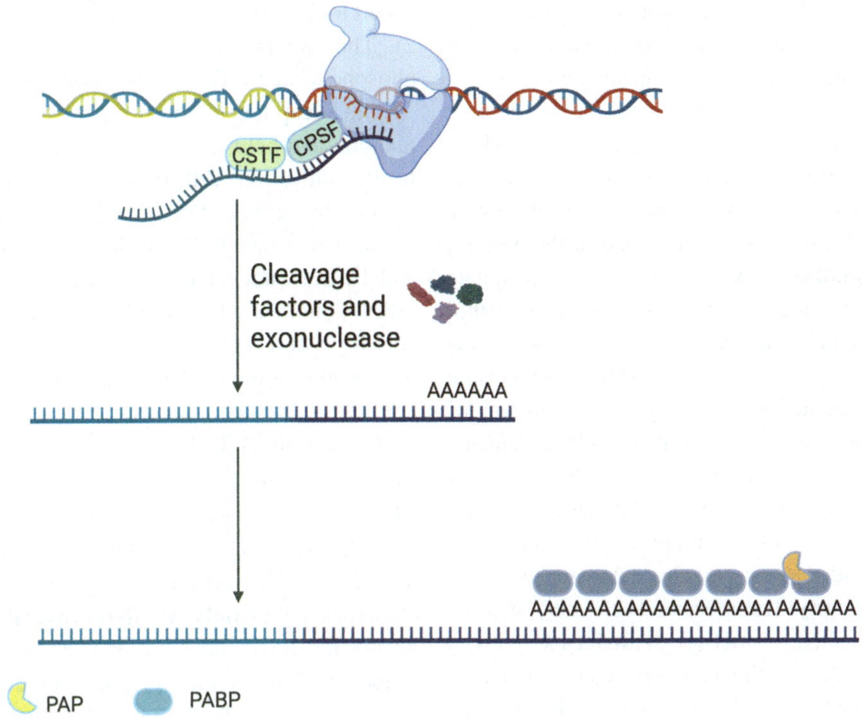

Fig. 7.7 Transcription termination and Polyadenylation. CPSF and CSTF interact with the 3′ end of the mRNA, followed by cleavage of the mRNA. PAP adds Poly(A) to make the Poly(A) tail. See text for details. (*Source*: Created with BioRender.com)

Fig. 7.8 mRNA processing. The complete mRNA processing pathway involves splicing, polyadenylation (Poly A tail), and 5′ capping

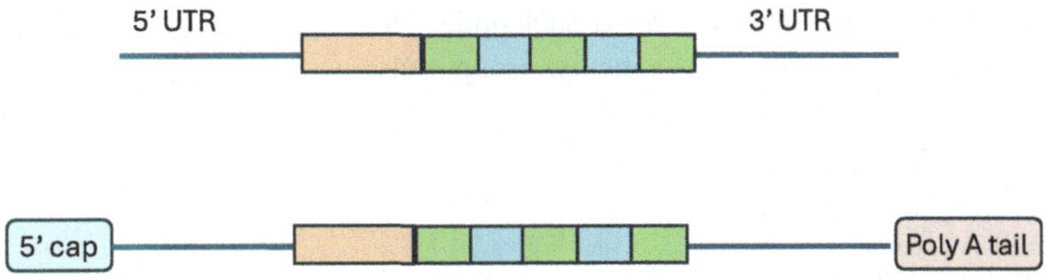

important to note that although the AAUAAA signal is crucial for termination and polyadenylation in many genes, some genes use other termination mechanisms, and CPSF can recognize alternative polyadenylation signals (Fig. 7.7).

1.8.4 Processing of the Pre-mRNA

After transcription is finished, the mRNA is considered to be pre-mRNA. This pre-mRNA still requires further processing before it is mature and can exit the nucleus. Three important modifications of the pre-mRNA take place, usually in this order: 5′ capping, splicing, then 3′ polyadenylation (poly-A tail addition), followed by splicing. After these modifications, the RNA is considered mature mRNA and ready to be transported out of the nucleus and into the cytoplasm (Fig. 7.8) [11].

• The 5′ Cap

The 5′ cap of mRNA is a guanosine nucleotide that is modified by methylation, as described below and also discussed in [1, 2, 4, 6, 17]. This modified guanosine is called 7-methylguanosine (m7G). The addition of the cap does not wait until after transcription is terminated; it occurs soon after transcription begins. It remains on the mRNA after transcription is terminated.

5′ RNA capping starts as soon as the first 20 nucleotides of RNA are synthesized. There are several steps to the capping process. First, as part of the RNA synthesis process, a triphosphate group will be present at the 5′ end of the mRNA. A **phosphatase** removes one of the phosphate groups. This leaves a diphosphate group rather than a triphosphate group at the 5′ end of the RNA. Next, a free GMP (guanosine monophosphate) is formed by the removal of two phosphate groups from a GTP (guanosine **tri**phosphate; from the intracellular pool of nucleotides). This GMP is then added to the diphosphate group remaining at the 5′ end of the RNA strand. The enzyme that catalyzes this reaction is a **guanyl transferase**. In the next step, an enzyme called a **methyl transferase** adds a methyl group to the new guanosine, which is now located at the 5′ end of the mRNA. This methyl group is added to position 7 of the guanosine, forming the basic cap structure (Cap 0) (see Fig. 7.9).

In mammalian cells, additional methyl groups are common. A methyl group is usually added to the ribose of the first transcribed nucleotide. This is called Cap 1. Sometimes a methyl group is also added to the ribose of the second nucleotide (Cap 2). The capping enzymes are recruited by the phosphorylated CTD of RNA polymerase II.

The 5′ cap helps protect the mRNA from degradation and helps promote translation initiation. It also allows the cell to distinguish the mRNA (made by RNA Pol II) from other RNA molecules, since RNA made from RNA Pol I and III does not have a cap. The 5′ cap on the mRNA binds to a Cap Binding Complex (CBC), which plays a role in protecting the mRNA and transporting it out of the nucleus [1, 18].

- The 3′ Poly-A Tail

The Poly-A tail is a strand of up to 250 or more adenine nucleotides added to the 3′ end of the mRNA after transcription has completed. Once the RNA is cleaved during the termination process, the enzyme **poly-A polymerase** (**PAP**) adds adenosine nucleotides to the 3′ end of the sequence. This is the beginning of the poly-A tail. Proteins called poly-A-binding proteins (PABPs) bind to the tail and help protect and stabilize it as it lengthens to its final size. The poly-A tail has multiple functions, such as helping protect the mRNA from being destroyed by exonucleases, facilitating mRNA export from the nucleus, and regulating gene expression (Fig. 7.7) [1–3, 6, 19].

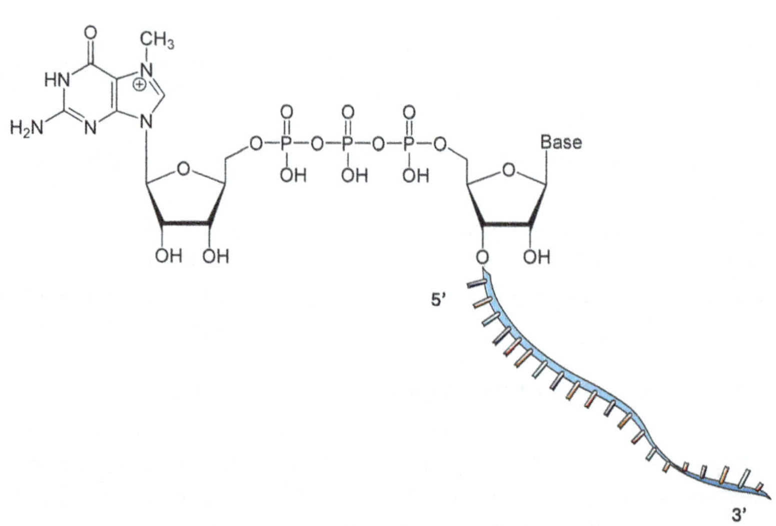

Fig. 7.9 5′ Capping. The 5′ CAP gets added as the mRNA is being synthesized. It consists of a modified (usually methylated) guanosine added at the 5′ end of the mRNA. (*Source*: Adapted from "Cap structure.svg" by Yikrazuul, Wikimedia Commons (public domain). mRNA strand is adapted from Servier Medical Art (https://smart.servier.com/), licensed under the Creative Commons Attribution 4.0 (CC BY 4.0) license)

• Splicing

Splicing is the process by which introns are removed from the mRNA, and exons are joined together at the sites where the introns are removed. Splicing is a complex process that depends on the presence of certain **consensus sequences**, described here and also covered in [1, 2, 4, 6, 17, 20]. Here, we will consider only two exons (which we will call the 5′ exon and the 3′ exon), with the intervening intron to be spliced out. The 5′ end of the intron is called the 5′ splice site, and the 3′ end of the **intron** is called the **3′ splice site**. In the sequence just downstream of the **5′ splice site**, there is the dinucleotide sequence **GU**. Just upstream of the **3′ splice site**, there is another dinucleotide sequence, **AG**. Upstream to this AG sequence, also within the intron, there is a sequence called the **branch point** sequence, which contains an adenine (**A**).

The process of intron removal during RNA splicing involves the formation of a loop structure called a lariat. A complex of proteins called the spliceosome is necessary for the formation of this structure. This complex consists of SNRPs (small nuclear riboprotein molecules), which are composed of small RNA molecules and proteins. The different SNRPs are the U1, U2, U4, U5, and U6 SNRPs. Various proteins, including BBP (branchpoint-binding protein), U2AF (U2 auxiliary factor), also play key roles [17].

At the beginning of the splicing process, the U1 SNRP binds to the 5′ splice junction at the GU sequence. It does so due to its RNA component, which can base pair with the RNA at this region. At the same time, BBP and U2AF bind at the branch point (at the A) and the 3′ splice site (at the AG sequence). U2 SNRP then displaces both BBP and U2AF and remains bound at the branch point (A). A protein complex called the triple complex (tr-SNRP), which contains U4, U5, and U6 SNRPs, now binds with U2 SNRP at the branch point. At this point, a loop structure begins to form in the RNA. U6 SNRP (part of the triple complex) now displaces U1 SNRP at the 5′ splice site. The NTC/NTR protein complex now binds to this complex (containing the U2, U4, U5, and U6 SNRPs) on the RNA and positions the complex and the RNA so that the A at the branch point comes close to the GU at the 5′ splice site. Contact between the A at the branch site and the GU at the 5′ splice site will eventually lead to a cut in the sugar-phosphate backbone of the RNA at the 5′ splice junction. This leaves a free OH group at the 3′ end of the 5′ exon. This step is known as the first transesterification reaction, where the 2′ hydroxyl group of the branch point adenine attacks the phosphate at the 5′ splice site.

Now, the end of the intron, which is released from the 5′ exon, covalently binds to the A at the branch point via its 5′ phosphate, forming a 2′–5′ phosphodiester bond. This leads to the beginning of the lariat structure, but it is still attached to the 3′ splice site.

The RNA structure now rearranges so that the free 3′ OH group of the 5′ exon can interact with the phosphate group at the 5′ end of the 3′ exon, forming a phosphodiester bond. This is the second transesterification reaction, in which the 3′ hydroxyl of the 5′ exon attacks the phosphate at the 3′ splice site, joining the two exons together and releasing the lariat.

The lariat is released and eventually degraded, and the exons are joined together. This entire process requires energy in the form of ATP hydrolysis. An outline of the basic splicing steps and the locations of the important consensus sequences is summarized in Fig. 7.10.

1.8.5 Export from the Nucleus

Finally, once the mRNA has been made and processed (the 5′ cap is added, the poly-A tail is added, and splicing has taken place), it needs to be shuttled out of the nucleus and into the cytoplasm, where translation will occur. The mature mRNA binds to various proteins in the nucleus, forming an mRNA-protein complex (mRNP). These proteins include the exon junction complex (EJC), which gets bound to exon–exon junctions during splicing. The mRNP then interacts with nuclear pore complexes (NPCs), which form channels on the nuclear membrane. The mRNP is transported through these channels. Nuclear transport receptors, such as NXF1/TAP and NXT1, mediate the interaction between the mRNP and the NPCs. In the cytoplasm, the EJC is released from the mRNA, and translation can begin. This entire process requires energy. Any

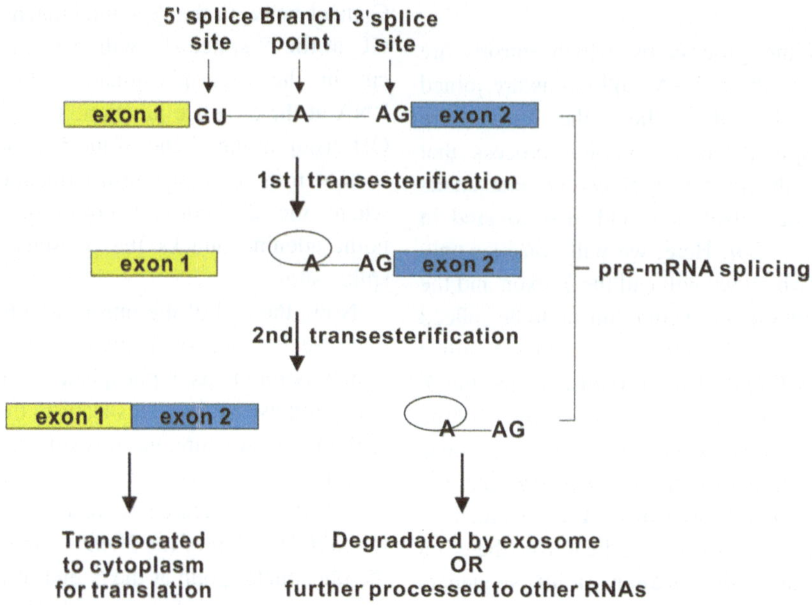

Fig. 7.10 Splicing. The main consensus sequences and transesterification steps are shown. See text for more details. (*Source*: From Pan et al. [149], licensed under the Creative Commons Attribution 4.0 (CC BY 4.0) license)

RNA that is no longer needed, such as spliced-out introns or incompletely processed RNA, remains in the nucleus and gets degraded [1, 7, 21, 22].

2 Regulation of Transcription

2.1 Mechanisms in Transcription Regulation

Up until now, we have focused on the basic mechanisms of gene transcription. While these basic mechanisms are essential, the transcription of genes within a cell, especially in a complex multicellular organism, is subject to additional levels of regulation. A remarkable and seemingly amazing feat of nature is that, despite every cell in the body containing the same DNA, different cell types vary vastly from one another and can carry out distinct functions. This is due to differences in gene expression, which determine the proteins present in each cell type. In other words, not all genes are transcribed in every cell or at all times. Instead, gene expression is tightly regulated so that only specific genes are expressed in a particular cell type, or under certain conditions or developmental stages. This is known as differential gene expression. It explains how the same genome can give rise to numerous different cell types, each with distinct functions. Even within a single cell type, external signals can regulate the expression of specific sets of genes. For example, the transcription of some genes may be activated in response to growth factors, while the transcription of others may be turned on or off in response to cellular stress [11, 23]. Here, we will explore how transcription is regulated beyond the basal level to understand how gene transcription can be differentially regulated in different cell types and different conditions. This discussion is drawn from [1, 2] and from sources within the text.

2.1.1 Regulation of Transcription by *Transcription Regulators* That Bind cis-Regulatory Elements

- Introduction to Transcription Regulators

Regulation of gene expression involves the activities of different combinations of transcription regulators that bind to cis-regulatory elements (see Fig. 7.11a). Transcription regulators that bind directly to DNA are transcription factors. Transcription regulators can be transcription activators or repressors, which positively or negatively regulate transcription, respectively. The cis-regulatory elements that transcription factors bind to are usually the promoter of a gene or enhancer. An enhancer is a region that is often located upstream of the promoter, but it can also be found downstream of the coding sequence, or sometimes within introns. Transcription regulators may include transcription activators or transcription repressors that bind directly to DNA. Co-regulators (co-activators or co-repressors) can bind DNA indirectly and also regulate transcription. Some transcription regulators were introduced in Sect. 1, which focused on those that are part of the general transcription machinery. Many more transcription regulators are involved in regulated transcription, however, and some of these are discussed here.

- Different Categories of Transcription Regulators

There are multiple subcategories of transcription factors. Transcription factors can be categorized into different subtypes based on their DNA-binding domains. They can include, for example, zinc finger, leucine zipper, helix-turn-helix (HTH), and helix-loop helix (HLH) transcription factors. Zinc finger transcription factors have finger-like structures that insert into the major groove of DNA, and they are stabilized by one or more zinc ions, which interact with the protein. HTH transcription factors have two alpha helices connected by a short turn sequence. One of the helices fits into the major groove of DNA. Leucine zipper transcription factors dimerize via a leucine-rich region, and they bind DNA through basic regions. HLH transcription factors also dimerize. They have a loop that separates two helices [24, 25].

- Transcription Activators and Transcription Repressors

Transcription activators can stimulate gene expression through various mechanisms. For example, they may facilitate the binding of **RNA polymerase II** or promote the assembly of the general transcription machinery to the promoter. They may also stabilize the transcription machinery involved in transcription elongation. Transcription activators may sometimes bind to regulatory elements that are far from the general transcription machinery. This can cause the intervening DNA to loop out of the way. In these cases, the mediators complex, discussed in Sect. 1, can play a role in linking activators at enhancers to the transcription machinery at the promoter. Transcription activators can also act cooperatively with other transcription activators, and sometimes they can recruit co-activators. Co-activators do not bind DNA directly but enhance transcription by interacting with other transcription factors or modifying chromatin structure (discussed in more detail below), helping the local DNA form a more open configuration.

Transcription repressors repress transcription rather than activate it. They may block the binding of **RNA polymerase** or important components of the general transcription machinery, such as the initiation factors, to the DNA. They may also block the binding of transcription activators, and a balance between activators and repressors can determine whether and to what extent a gene is expressed. They may interact with co-repressors and modulate the chromatin, causing the DNA to be in a closed configuration and inaccessible to the transcription machinery [1, 2, 6, 11]. Refer to Fig. 7.11b, c for co-activators and co-repressors.

- Tissue-Specific Promoters and Enhancers

Transcription factors bind to cis-regulatory regions and, in turn, regulate the transcription of genes. These cis-regulatory regions are usually promoters or enhancers. Some promoters and enhancers are active only in certain tissues or specific types of cells. These are known as tissue-specific promoters or tissue-specific enhancers. Genes regulated by these types of tissue-specific promoters or enhancers are expressed only if the tissue-specific factors are

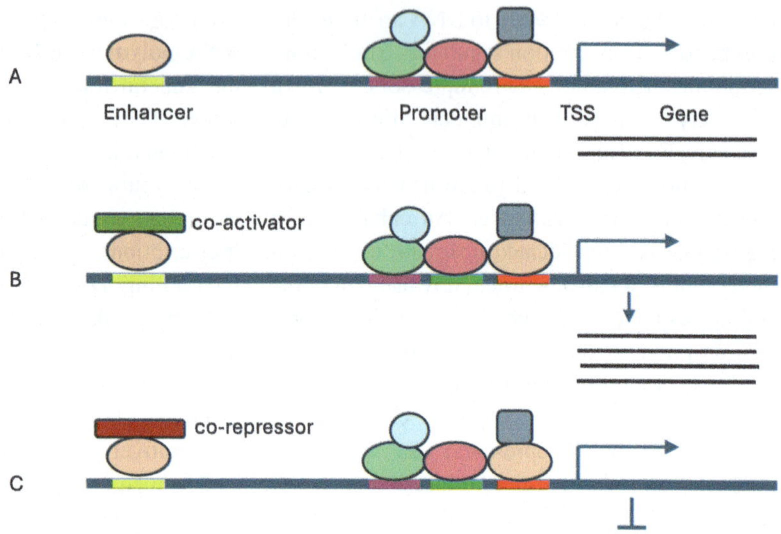

Fig. 7.11 Transcription regulation. A combination of transcription factors control gene expression. Shown here are multiple different transcription regulators, represented by various shapes and colors that bind to promoter and enhancer regions and lead to different outcomes. The relative amount of mRNA, representing the level of gene expression (high or low or none, is represented by solid lines under the gene

present. For example, a muscle-specific enhancer may be bound and activated only by transcription factors like MyoD, which is present in muscle precursors, but not in other cell types. The gene that is controlled by this promoter would therefore only be expressed (transcribed) in muscle cells. This is an important regulatory mechanism for controlling differential gene expression [1].

2.1.2 Histone Positioning and Modifications Impact Gene Expression

Along with transcription factors binding to cis-regulatory elements, other important considerations in gene regulation are chromatin structure and epigenetic changes such as DNA methylation [1].

- An Open Chromatin Configuration Encourages Transcription

DNA structure in eukaryotes is complex because DNA is always in the form of chromatin, where it is bound to proteins such as histones, which form the nucleosome structures. When the chromatin is in an open configuration, there is space containing free DNA between the nucleosomes. This open configuration favors transcription because much of the DNA is accessible to transcription factors. In contrast, if the nucleosomes are tightly packed, the chromatin is more condensed and considered to be in a closed configuration. In this situation, there is not much free DNA, transcription factors cannot access the DNA, and transcription is repressed. Whether the chromatin is in an open or closed configuration can depend on various factors, including histone modification.

- Modification of Histones and Gene Expression

Histone modifications were introduced in Chap. 6. Histones can undergo various types of post-translational modifications.

As discussed in Chap. 6, an important modification of histones is acetylation (see Chap. 6, Fig. 6.13). Acetyl groups add "bulk" and create space between the histones, thereby keeping them some distance from each other. The presence of **acetyl groups** is therefore important for keeping the chromatin in an **open, active conformation**, where histones are separated from each other, making the DNA **accessible to transcription factors**. Removal or addition of acetyl groups can, therefore, determine whether the chromatin is in a more open or more closed configuration.

- Some Transcription Regulators Can Interact with Histone Modifiers

Some transcription regulators play important roles in determining whether chromatin is in an open or closed configuration. One way they can do so is by regulating the acetylation of the histones through interaction with histone modifiers such as **histone acetyltransferase (HATs)** and **histone deacetylases (HDACs)**. Co-activators can be involved in this process. For example, a co-activator may have one domain that binds to transcription factors, and another that promotes acetylation of histones. This may be due to interacting with **HATs**, which acetylate histones. This will promote an open configuration in the local DNA, thus encouraging the binding of transcription factors and active transcription.

In contrast, co-repressors might recruit **HDACs**, which remove acetyl groups, thereby decreasing acetylation. This results in a closed chromatin configuration, which reduces transcription [1, 7, 26].

- Positioning of Histones

Chromatin remodeling complexes (CRCs) (which were discussed in Chap. 6) are also involved in modifying chromatin structure by repositioning or removing histones. CRCs will change the extent to which DNA is accessible to transcription factors, potentially increasing or decreasing transcription. Chromatin remodeling complexes often collaborate with histone modifiers, leading to intricate levels of transcription regulation [27, 28].

Some transcription factors may function by recruiting or interacting with CRCs. For example, the transcription factor p53 can interact with SWI/SNF complexes and guide them to the regulatory elements of target genes. SWI/SNF complexes (which were discussed in Chap. 6) use ATP to move nucleosomes or to eject them from the DNA, thereby making the genes more accessible to the transcription machinery [29, 30].

SWI/SNF plays important roles in transcription regulation. Mutations in some of their subunits are often observed in melanoma and several other cancers. Such mutations are associated with tumor progression, metastasis, and resistance to treatment. These mutations are typically loss-of-function mutations that reduce their chromatin remodeling activity. For example, reduced chromatin accessibility can result from the loss of function of SWI/SNF, which can suppress the expression of some tumor suppressor genes (genes that encode proteins that negatively regulate cell division), thereby leading to excessive cell proliferation. For these reasons, these complexes are of interest to researchers seeking new approaches to treating melanoma patients. Dealing with a target that is already inactivated is a complex problem in cancer treatment, but treating cancers with SWI/SNF mutations is an important area of research [31, 32].

2.1.3 Modification of DNA Also Impacts Gene Expression

Methylation of cytosines (C) within CpG-rich regions in DNA, particularly in gene promoters, can suppress gene transcription. Regions that are rich in CpG (cytosine followed by a guanine) are sometimes called CpG islands (or CG islands). They can be found anywhere in the DNA, but they are primarily relevant in promoters. The reason methylated CpG regions in promoters can inhibit gene expression is that they lead to the deacetylation of local histones bound to the DNA. This is due to proteins that bind to the methylated CpG regions. A family of proteins with domains called methyl-CpG-binding domains (MBDs) can bind to the methylated CpGs. Examples of proteins that have MBDs are MECP2, MBD1, MBD2, and MBD4. These MBD proteins recruit other proteins, including a protein called SIN3A, which in turn recruits **HDACs**. **HDACs**, in turn, remove acetyl groups from local histones, leading to a closed chromatin configuration and resulting in transcription inhibition (Fig. 7.12) [1, 17, 33].

Some transcription factors can modulate methylation. **DNMT** and **TET** are enzymes that have important roles in adding or removing methyl groups from DNA, respectively. Some transcription factors can directly or indirectly influence the recruitment of **DNMTs** or **TETs** to DNA, thereby modulating methylation and regulating transcription [34, 35]. For example, the transcription factor E2F can interact with a **DNMT** to facilitate its recruitment to specific promoters, leading to the silencing of certain genes [34]. (E2F and its role in regulating transcription of genes that are important for S phase and DNA synthesis are discussed in Chap. 20).

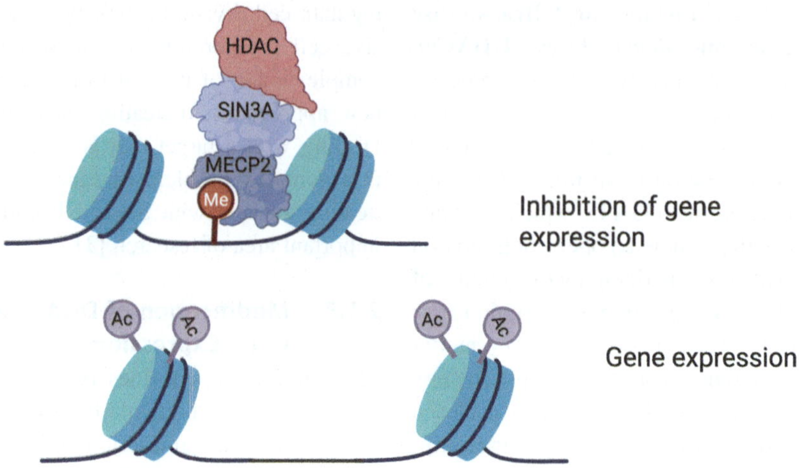

Fig. 7.12 DNA methylation. DNA methylation can lead to the regulation of histone acetylation, thereby impacting gene expression. The roles of the proteins such as MECP2, SIN3A, and HDAC are discussed in the text. (*Source*: Created with BioRender.com)

2.2 MyoD and p53 Are Two Examples of Transcription Factors That Regulate Gene Expression Via Multiple Mechanisms

2.2.1 MyoD

MyoD (myogenic differentiation 1) is a transcription factor that plays a role in the differentiation of muscle cells and some other types of cells. It is discussed here because it provides a good model system to illustrate how some of the different mechanisms described above can work together to regulate gene expression.

MyoD is a transcription factor that binds to DNA but is not part of the general transcription machinery. Instead, it is a tissue-specific transcription regulator found only in certain tissues, primarily in muscle precursors. MyoD binds to the regulatory elements of various genes, including those that control muscle development. The cis-regulatory elements it usually binds to are called E boxes (enhancer boxes), which have the consensus sequence CANNTG (where N can be any nucleotide). E-boxes are found only in enhancers and promoters of certain genes, including many genes involved in muscle development. They are also only active in cells where the transcription factors needed to activate them, such as MyoD, are present. Once bound to these E-box elements, MyoD can bind to other transcription factors. One example of a transcription factor that MyoD can interact with is MEF2. MEF2 binds to MyoD and also to its own consensus site on the DNA. The combination of MyoD and MEF2 leads to a synergistic effect, resulting in strong activation of transcription of genes whose promoters and enhancers they bind to.

How does this synergistic activation work? MyoD and MEF2 can help recruit each other to the DNA and stabilize each other's binding. The combination of MyoD and MEF2 also facilitates the recruitment of **RNA Pol II** to the promoter, thus encouraging transcription initiation. Furthermore, MyoD and MEF2 can recruit co-activators. For example, they can recruit p300/CBP and SWI/SNF. SWI/SNF, as described above, is a chromatin remodeling complex that can reposition the histones on the chromatin to create a more open configuration. p300/CBP is a histone modifier. It has histone acetyltransferase (HAT) activity. It increases histone acetylation, resulting in a more open chromatin configuration, which encourages transcription. Genes that are regulated by the combination of MyoD and MEF2 include those encoding myosin, actin, and various muscle-specific proteins [1, 11, 36, 37].

2.2.2 p53

p53 is another example of a transcription factor that regulates gene expression via multiple mechanisms. It can activate some target genes, but it can also repress others by binding to specific DNA response elements. It can have different functions depending on the conditions. It can promote cell cycle arrest under some conditions, and under more severe conditions, it can lead to cell death by apoptosis. In both cases, it functions by regulating the expression of target genes.

Under certain conditions, such as in response to DNA damage, p53 activates the expression of genes that lead to cell cycle arrest. Under these conditions, p53 binds its consensus p53 response elements (p53REs) in promoters or enhancers of target genes. For example, it binds to the promoter of the gene encoding p21 (the CDKN1A gene) and activates its transcription. P21 is a cyclin-dependent kinase inhibitor that puts a brake on the cell cycle. When p53 binds to DNA, it may also recruit co-activators, such as p300/CBP, and chromatin remodelers, like SWI/SNF. This contributes to its role in activating transcription of target genes [38]. P53 can also interact with the components of the pre-initiation complex [39]. Once DNA damage is repaired or other stresses are alleviated, then the cell cycle can continue.

Under other conditions, such as under more prolonged or severe DNA damage that may be irreparable, p53 may lead to apoptosis instead of cell cycle arrest. Under these conditions, p53 binds to different DNA response elements, called noncanonical or variant response elements. Under these conditions, p53 can **repress gene expression** instead of activating it. This can involve the recruitment of co-repressors or other transcription regulators. An example of a gene whose transcription is repressed under these conditions is the Survivin gene. Repressing its expression encourages apoptosis [40].

p53 itself is also regulated. An important regulator of p53 is MDM2 (mouse double minute 2 homologue). MDM2 is a ubiquitin ligase that binds p53 and targets it for degradation. By doing so, it inhibits p53 from regulating its target genes. Interestingly, p53 can activate expression of the gene encoding MDM2, thus activating a negative feedback loop that keeps p53 levels in check. This inhibition is relieved during cellular stress or when DNA damage occurs. Thus, under these conditions, p53 can remain active for a longer period and regulate target genes [41].

2.3 Chromatin Remodelers and Chromatin Modifiers and Human Health and Medications

2.3.1 HDAC Inhibitors

- HDAC Inhibitors and Cancer Treatment

Since histone acetylation and deacetylation can regulate gene expression, **HDAC inhibitors** are considered promising for regulating gene expression in some types of cancer. The rationale for using HDAC inhibitors to treat cancer is that the HDAC inhibitors may increase the expression of genes that may be inappropriately turned off in some types of cancer, particularly tumor suppressor genes. Furthermore, **HDAC** levels are increased in certain cancer types, which may be linked to decreased transcription of certain genes, most likely including some growth regulatory genes [42].

Studies have shown that **HDAC inhibitors** can reduce or **block the growth of certain cancer cells**, leading to their death through apoptosis. They exhibit significantly lower cytotoxicity in normal cells compared to cancer cells, thereby minimizing potential side effects. Some HDAC inhibitors have been approved for cancer treatment and have shown some success in the clinic. **The first FDA-approved HDAC inhibitor** was the small molecule drug **Zolinza (vorinostat)** [43]. The drug is frequently referred to by its chemical name, **Saha** (an abbreviation for suberoylanilide hydroxamic acid) [44]. Saha inhibits 11 different HDACs (there are 18 known HDACs in human cells) [45]. Saha has been shown to alter the transcription of multiple genes, including those involved in apoptosis and cell cycle regulation [45]. Saha is a competitive inhibitor of **HDACs**; it binds to their active sites and inhibits them. It also functions as a chelator for zinc ions. This is important because some HDACs are zinc-dependent, meaning that they require zinc binding for their deacetylase activity [46]. Saha is used to treat **cutaneous T-cell lymphoma (CTCL)** and is being tested for other types of **cancer** [47].

Other HDAC inhibitors include **Beleodaq (Belinostat)** [48] and **Istodax (Romidepsin)** [49]. Another drug, **Epidaza (Tucidinostat)**, is relatively new and not yet FDA-approved (as of 2025). These drugs are used to treat some types of **leukemia** and are under investigation for the treatment of **other types of cancer**, both alone and in combination with other therapies [42, 47].

- HDAC Inhibitor and Muscular Dystrophy

Muscular dystrophy (MD) is a group of diseases associated with muscle loss and weakness [50]. The disease is usually X-linked and most common in boys. The most common form of muscular dystrophy is Duchenne muscular dystrophy. It is caused by a mutation in the dystrophin gene. This is a large gene that encodes the dystrophin protein. Dystrophin is a protein that is necessary for normal muscle contraction. The lack of dystrophin affects muscle primarily through a mechanical mechanism, but it also leads to secondary effects, including alterations in gene expression. For example, lack of dystrophin has been shown to be associated with decreased expression of the gene encoding follistatin. Follistatin is a protein with numerous functions, including a crucial role in muscle mass development. Its altered expression contributes to the pathology of Duchenne muscular dystrophy. Recently, the lack of dystrophin has been shown to lead to an increased activation of HDAC. This is thought to be one of the causes of decreased expression of the follistatin gene [51]. This connection is the basis for a new drug, **DUVYZAT (givinostat)** (FDA approved in 2024) [52], which is also an **HDAC inhibitor**. **Duvyzat**, which causes increased acetylation of histones, leads to a more open chromatin configuration and enhances gene expression. This helps to restore the expression of genes such as follistatin. This, in turn, is designed to aid in the repair of muscle tissue, and treatment with **Duvyzat** is expected to slow the progression of **muscular dystrophy** by slowing muscle loss and inflammation associated with the disease [53, 54].

2.3.2 Demethylating Drugs

Vidaza (azacitidine) [55] is a **small-molecule** drug that **demethylates DNA**. It is used to treat **some cancers, such as MDS** (myelodysplastic syndromes: a group of cancers that start in the bone marrow [56], and acute myeloid leukemia (AML). It functions as an analogue of cytidine [57] and incorporates into DNA. Within the DNA, it functions as a false substrate for **DNMTs**. **DNMTs** are enzymes that add methyl groups to DNA. Once DNMT is bound to the drug, the DNMT becomes "trapped" and becomes degraded [58]. The result is undermethylation of the DNA. It functions as a type of enzyme inhibitor called a suicide inhibitor (discussed more in Chap. 5) because it irreversibly inhibits certain DNMTs (there are multiple DNMTs in human cells). Even if the DNA is already methylated, once it replicates, the new copies will not be methylated. By inhibiting DNA methylation, this drug can reactivate genes that were silenced due to hypermethylation. It has been shown to lead to the re-expression of growth-inhibitory genes, such as tumor suppressor genes [59]. **Onureg (azacitidine)** [60] and **Dacogen (decitabine)** [61], **two** formulations **of azacitidine**, function similarly and are also used to treat certain types of **cancer, as well as MDS** [62, 63].

2.4 Transcription Factors Are Often Regulated by Signal Transduction Pathways That Lead to the Expression of Specific Genes

Cells can respond to environmental cues by activating signal transduction pathways. These signal transduction pathways lead to a response to the stimulus, and one important way they do so is by regulating the activity of transcription factors. Depending on the stimulus and environmental conditions, these pathways can lead to both transient and long-term changes in transcription within the cell. Both of these types of responses are discussed below. The concepts of cell signaling and transcriptional regulation are closely linked because many signal transduction pathways lead to changes in gene expression. Therefore, while this chapter discusses some key signaling pathways with respect to their roles in regulating gene expression, most of these pathways are discussed in more detail in Chap. 18.

2.4.1 Gene Expression Changes That Are Transient

Some gene expression changes are transient, such as those that occur in response to specific environmental cues like growth factors or some hormones. For example, when cells are exposed to growth factors, they may respond by upregulating genes that lead to increased cell division. One way growth factors lead to responses in cells is by binding to cell surface receptors, which activate signaling pathways that, in turn, lead to the upregulation or activation of certain transcription factors. For example, some growth factors can bind to cell surface receptors, which are transmembrane proteins. These cell surface receptors become activated, and although they cannot enter the cell, they can activate a cascade of enzymes and other proteins within the cell. This is called a signaling pathway. This type of pathway can lead to the nucleus, where transcription factors become regulated. One example is the MAP Kinase pathway, which can lead to positive regulation of the transcription factors in response to stimulation by growth factors. For example, the MAP Kinase pathway can lead to activation of the transcription factor ELK1 (ETS-like gene-1). ELK1 can form a complex with another transcription factor, SRF (serum response factor), and activate the transcription of target genes that have roles in cell cycle progression, leading to cell division [64, 65]. This effect is transient because once the growth factors are no longer present, the MAP Kinase pathway becomes inactive, and cell cycle progression stops until more growth factors become available [11].

Another example of a transient change in gene expression is the response to glucocorticoids. Glucocorticoids are steroid hormones, such as cortisol, that play a crucial role in the body's response to stress. Glucocorticoids act by diffusing across the plasma membrane and entering the cell, where they bind to glucocorticoid receptors (GRs). GRs are nuclear hormone receptors (NHRs). NHRs are different than cell surface receptors because they function inside the cell. They are a type of intracellular receptor. The GR translocates from the cytoplasm to the nucleus once it is bound to glucocorticoids, and there it functions as a transcription factor. In the nucleus, the GR interacts with other transcription factors, and together, they bind to target genes, regulating their expression. GRs tend to regulate the expression of target genes that mediate responses to stress, such as those involved in glucose metabolism, blood pressure control, and the production of certain cytokines [1, 11].

2.4.2 Mechanisms of Gene Expression Regulation That Are Associated with Long-Term Changes and Cell Differentiation

While some transcriptional changes are transient, others lead to long-term cellular changes. These long-term cellular changes are important for cell differentiation and for maintaining a differentiated state. Cell differentiation into specialized cell types requires certain patterns of gene expression that must be maintained for the long term, in some cases for the cell's lifetime. Some transcription factors can regulate the expression and maintenance of genes crucial for cell specialization, whose expression needs to be regulated for extended periods. Some examples of signaling systems that lead to long-term gene expression changes associated with differentiation are discussed below, and some of the associated signaling pathways are discussed in more depth in Chap. 18.

- Notch Signaling

The Notch receptor is a transmembrane protein that can bind to extracellular signals. It is different than some other transmembrane receptors because it can be cleaved, and the intracellular domain can translocate to the nucleus. When the Notch receptor binds to its ligand, this triggers a series of events (described in more detail in Chap. 18) that lead to the release of the intracellular domain of Notch (the Notch intracellular domain or NICD), which then translocates to the nucleus. In the nucleus, NICD interacts with the transcription factor CSL (CBF1/RBO-Jk) [66], and together these two proteins form a complex. This complex binds to promoters of target genes and regulates their expression. Some of these target genes have roles in cell differentiation. Notch function is complex, however, because it regulates different target genes in different cell types, leading to different outcomes. While it promotes differentiation in some cells, it plays a role in maintaining a more undifferentiated state, such as a progenitor state, in others, depending on the cellular environment and the presence or absence of various co-regulators [67, 68].

- Estrogen Signaling

Estrogen signaling also leads to changes in gene expression associated with maintaining the characteristics of certain specialized cells. Estrogen, a steroid hormone, enters the cell and binds to the estrogen receptor (ER). The ER is a nuclear hormone receptor, and once bound to estrogen, it enters the nucleus (although in some cells, the ER is already in the nucleus, and estrogen binds to it there). The ER undergoes a conformational change, dimerizes, and functions as a transcription factor. It binds to the estrogen response element (ERE) in the promoters or enhancers of target genes. Some of these target genes have specialized functions in glands, uterine cells, and bone, or have metabolic functions [69]. Estrogen has a wide range of functions. In addition to these long-term effects, Estrogen can lead to more short-term changes in cells by regulating the expression of target genes involved in cell growth or apoptosis [70, 71].

- Networks of Transcription Factors and Master Transcription Regulators

Transcription factors play a crucial role in regulating gene expression, which is essential for cell differentiation and the maintenance of a differentiated state. Rather than a single transcription factor, however, networks of transcription factors are usually required to regulate gene expression and, consequently, cell differentiation. Such networks can also be associated with nondifferentiated cells, such as stem cells and progenitor cells, but here we focus on differentiated cells [1, 72, 73].

In these networks, transcription factors may activate the expression of genes that encode other transcription factors. These, in turn, regulate genes that control cell function and the expression of other transcription factor-encoding genes. These types of networks can become quite complex, leading to the regulation of various transcription factors. Ultimately, the set of transcription factors present in a given cell type, rather than a single transcription factor, will impact what genes are expressed and thus determine the cell's characteristics [1] (Fig. 7.13).

Some transcription factors in such networks are considered master transcription regulators. A master transcription regulator acts at the "top" of a hierarchy or network of transcription factors. A good example of a master regulator is MyoD. The transcription factor MyoD promotes muscle cell differentiation not only by regulating the expression of muscle cell-specific genes, but also by activating the expression of genes that encode other transcription factors, further driving the muscle differentiation program [1, 36, 74].

This role for MyoD in differentiation may have important clinical implications. When MyoD is introduced into nonmuscle cells in the laboratory, it activates events that drive the cells to express muscle-specific proteins and adopt muscle cell characteristics. This is called transdifferentiation. MyoD can also cause stem cells to differentiate into muscle cells (stem cells are discussed in more detail in Chap. 26). Studying MyoD is therefore valuable for understanding muscle development and regeneration, as well as

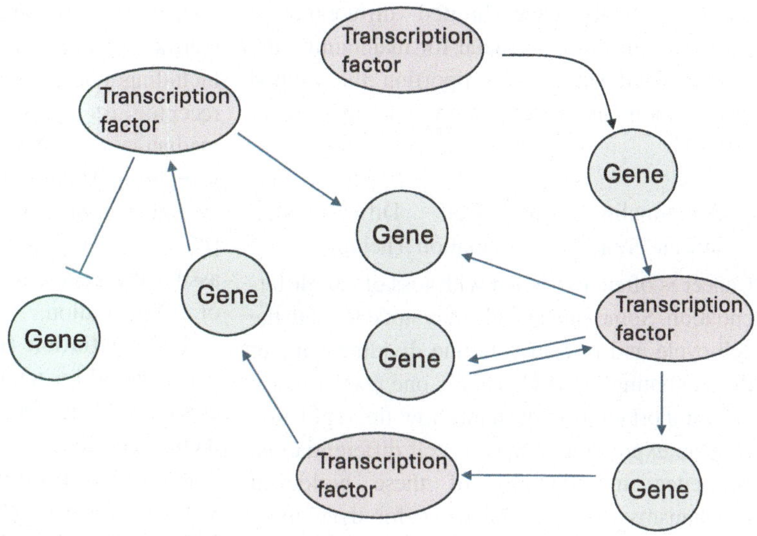

Fig. 7.13 Transcription networks. Genes encoding transcription factors can also be regulated by transcription factors, resulting in networks of transcription factors that control gene expression in different cells. An example of a hypothetical transcription network is shown here

muscular diseases, such as muscular dystrophies [36]. Understanding how MyoD works is considered important for the field of regenerative medicine, where scientists are investigating methods to repair or replace damaged muscle tissue. However, the use of MyoD in medical treatment is still in the research and preclinical stages [75–77].

- Maintaining a Differentiated State

Regulation of transcription plays a key role not only in establishing cell identity but also in maintaining cell memory of its identity. Once a cell has become a differentiated cell with its own specialized characteristics, it maintains that state. Even after a cell divides, the daughter cells must "remember" their differentiated characteristics. Furthermore, even cells that do not divide, such as neurons, must maintain their differentiated state throughout the cell's life. How do cells "remember" what kind of cell they are? One mechanism that can contribute is the positive feedback loop. In a positive feedback loop, a transcription regulator, especially a master regulator, not only regulates a gene or set of genes important for maintaining specific characteristics in a differentiated cell but also positively regulates the expression of its own gene. This process is called autoregulation. This is only one mechanism for maintaining a differentiated state. Other mechanisms include epigenetic modification and changes in chromatin organization [1, 11].

MyoD once again provides a good example. Once activated, MyoD not only regulates the expression of muscle-specific genes but also binds to regulatory elements within the MyoD gene itself. This leads to a positive feedback loop that ensures the long-term expression of MyoD and MyoD target genes [78]. This loop helps reinforce muscle identity in both developing and regenerating muscle. For example, during muscle regeneration, satellite cells (muscle stem cells) are activated, and they eventually differentiate into mature muscle cells. As they differentiate, MyoD drives the expression of muscle genes while also maintaining its own expression. In this way, it stabilizes the gene regulation program that is important for the differentiated muscle state [79].

Another example is **Pdx1** (pancreatic and duodenal homeobox 1). Pdx1 is a transcription factor that is important for pancreatic development and maintenance of pancreatic beta cell differentiation. In mature pancreatic beta cells, Pdx1 activates genes that are important for pancreatic function, and it also binds to the promoter of the Pdx1 gene, thus leading to a positive feedback loop. This enables it to maintain its own expression in addition to the expression of pancreatic beta cell-specific genes such as insulin. Mature pancreatic cells have limited differentiation potential, but Pdx1 is crucial for maintaining the differentiated state and supporting the limited proliferation and regeneration that does occur [80, 81].

- Abnormalities in Cell Differentiation in Cancer, and Differentiation Therapy

Cancer is often associated with loss of cell differentiation. Some cancer cells may not complete the cell cycle and instead continue dividing without differentiating [82–84]. This is one reason that it is so important to understand how the regulation of gene expression relates to cell differentiation. A better understanding of these biological mechanisms may provide clues for developing new cancer treatments, especially those that restore differentiation in undifferentiated cells.

Inhibition of differentiation is common in myeloid cancers. In **acute promyelocytic leukemia (APL)**, a chromosomal translocation between chromosomes 15 and 17 results in the production of a fusion protein called PML-RARα, where a protein called promyelocytic leukemia protein (PML) is fused to the retinoic acid receptor (RARα).

The normal RAR is a nuclear hormone receptor. Retinoic acid (RA) is a metabolite of vitamin A and functions similarly to a hormone by binding to a receptor and regulating gene expression. RA binds the RAR, and in response, the RAR forms a heterodimer with another nuclear hormone receptor called RXR (retinoid X receptor). This heterodimer (RAR/RXR) binds to retinoic acid response elements (RAREs) in the promoters of certain target genes. This leads to the release of any repressors that are associated with the gene, and to the recruitment of co-activators such as HATs. RAR/RXR thereby positively regulates the expression of target genes in response to RA. The products of some of these target genes promote differentiation in response to RA [85].

In contrast to the normal RAR, the fusion protein (PML-RAR) functions mostly as a transcription repressor instead of a transcription activator. It still binds the RAREs on target genes, but instead of activating transcription, it represses transcription of these genes. It does so by recruiting repressors to the promoter DNA, including nuclear co-receptor (NCoR) (nuclear receptor co-repressor) and SMRT (silencing mediator of retinoid and thyroid hormone receptors). This repression is stable and resistant to being replaced by the normal RAR/RXR [86, 87]. This affects certain differentiation-associated genes, thereby inhibiting the process of differentiation.

Differentiation therapy has become a standard method for treating **some myeloid cancers**, especially APL. The main type of differentiation therapy for this type of cancer involves treatment with **all-trans retinoic acid (ATRA)**. **ATRA** is a synthetic form of **RA**. **ATRA** can bind to the PML-RARα fusion proteins. This causes them to undergo a conformational change, thereby releasing their associated co-repressors, such as HDACs and NCoR. Instead, they can now recruit activator proteins such as **HATs**. **ATRA** can also lead to the degradation of the PML-RARα protein. The result is reactivation of genes that had been silenced by the fusion protein. This restoration of normal transcription can cause the undifferentiated cancer cells to differentiate. This type of treatment has increased the prognosis for patients with APL [88–90]. ATRA, used for differentiation therapy, is considered a small-molecule drug because of its low molecular weight and simple structure. The full name of this drug is **Vesanoid (tretinoin)** [91].

More recently, scientists have studied the possible use of differentiation therapy for the treatment of medulloblastoma, a malignant brain tumor in children. Researchers showed that treating medulloblastoma cells with the drug **Triiodothyronine (T3) (an active form of thyroid hormone) increased** their level of differentiation and slowed their growth. Researchers also gave the T3 to mice with medulloblastoma; as a result, tumor growth stopped, and survival increased. This raises the hope that differentiation therapy may be helpful for treating this type of cancer [84, 92].

The **T3** treatment may work by increasing the expression of the gene encoding NeuroD1, a ***transcription factor*** that promotes differentiation.

T3 binds to thyroid hormone receptors (THRs), which are nuclear hormone receptors. THRs function as transcription factors, which in turn regulate transcription of genes that have thyroid hormone response elements (TREs) in their promoters. TRα1 (thyroid hormone receptor alpha 1) is a thyroid hormone receptor isoform that predominates in the brain. It regulates the transcription of genes, including the one encoding NeuroD1, which promotes differentiation.

In the cancer cells, TRα1 functions differently than in normal cells. In cancer cells, it interacts with the gene encoding EZH2, but this interaction does not occur in normal cells. EZH2 is the catalytic subunit of an enzyme complex called **polycomb repressive complex 2 (PRC2)**. This enzyme methylates histones, leading to gene silencing. One of the genes it silences is the gene for NeuroD1. When medulloblastoma cells are treated with T3, the T3 enters the cell and disrupts the interaction between TRα1 and EZH2. Consequently, NeuroD1 can now be expressed. As a result, treating medulloblastoma cells with T3 increases transcription of the NeuroD1 gene, thereby promoting differentiation of the cancer cells. This study is in its experimental stages, but it may hold promise for the future development of differentiation therapy for this type of cancer [92].

2.5 Genes Can Be Regulated by Multiple Additional Mechanisms in Addition to the Regulation of Transcription

Regulated gene transcription is critical for determining a cell's characteristics. As described above, the regulation of gene expression involves various transcription regulators, interacting with DNA, other transcription regulators, and the protein components of the chromatin. However, there are also other mechanisms by which mRNA can be regulated. These can include alternative splicing, RNA editing, regulation of RNA export out of the nucleus, and variability in RNA stability, as discussed below.

2.5.1 Alternative Splicing

Once mRNA is synthesized, it undergoes processing, including splicing, where introns are removed. In some cases, some exons may also be removed along with the introns. This allows for alternative splicing, where different intron–exon combinations may be spliced out, changing the final mRNA structure (Fig. 7.14). The presence of multiple exons thus increases the possibility of producing a diversity of products from a single gene [17].

A classic example of alternative splicing is the CDKN2A gene locus. This locus encodes both p16INK4a and p14ARF. The final spliced p16INK4a and p14ARF share exons 2 and 3, but they have different first exons due to alternative splicing. This causes them to be translated via completely different reading frames, thus producing completely different proteins. P16INKa blocks cell cycle progression by inhibiting CDK4/6 (a cyclin-dependent kinase discussed more in Chap. 20). P14ARF has a different function. It binds and inhibits MDM2, a negative regulator of p53. P53, as discussed above, is a transcription factor that can also negatively regulate the cell cycle. Thus, p16INK4a and p14ARF can inhibit the cell cycle by different mechanisms. Which product is obtained is associated with many factors, including the cell context, availability of transcription factors, and epigenetic factors [93, 94].

Another example of a gene that undergoes alternative splicing is CD44. CD44 encodes a cell surface glycoprotein important for cell adhesion, migration, and signaling. It is expressed in most types of cells. The CD44 pre-mRNA contains 20 exons. Exons 1–5 and 16–20 are called the constant exons. The other exons are the variant exons. They are numbered V2-V10 in humans. The standard CD44 mRNA product contains all of the constant exons and is widely expressed in tissues. This is called CDC44s. Some alternatively spliced forms of CD44 are called CD44 variant isoforms (CD44v). They have different combinations of the variant exons. This can impact the translocated protein's functions and tissue-specific expression patterns.

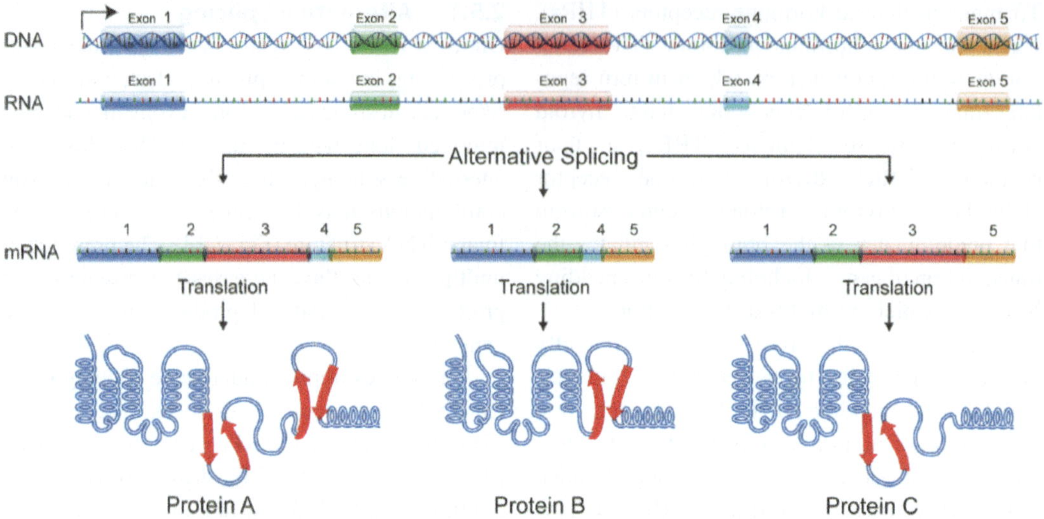

Fig. 7.14 Alternative splicing. The presence of multiple exons leads to the possibility of alternative splicing, where some exons can be spliced out. Alternatively spliced genes can have different functions. (*Source*: Public domain image from the National Human Genome Research Institute, via Wikimedia Commons: https://commons.wikimedia.org/wiki/File:DNA_alternative_splicing.gif)

Fig. 7.15 RNA modification. One example of RNA modification is the deamination of adenosine, which converts it to inosine

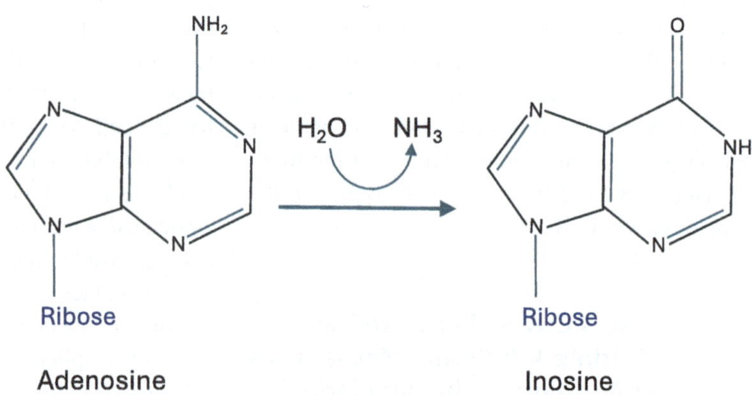

Some CD44 splice variants have been identified as being linked to some types of cancer, and they may have a role in tumor invasiveness and metastasis, and responsiveness to chemotherapy [95, 96].

2.5.2 RNA Editing

Sometimes, RNA may be modulated by RNA editing. RNA editing is a covalent modification of the bases on ribonucleotides in RNA that changes them to different bases. RNA editing occurs primarily in the nucleus. The most common modifications are deamination of the base cytosine to uracil (C to U editing), catalyzed by enzymes in the **APOBEC family**, or deamination of the base adenine to inosine (A to I editing), catalyzed by **ADAR (adenosine deaminase acting on RNA)** enzymes (*see* Fig. 7.15). (Inosine is a base found only in RNA, created by deamination of adenine.) Both reactions are forms of deamination. For example, in A to I editing, adenosine undergoes a deamination reaction with H_2O, releasing the NH_2 group as NH_3 and replacing it with a double-bonded oxygen, converting adenosine to inosine. RNA editing is catalyzed by enzymes that carry out deamination [97].

RNA editing within the coding region can lead to changes in the resulting amino acid sequence of the translated protein or the insertion of a stop codon, resulting in premature termination of translation and a truncated protein. Edits outside the coding region may also have an effect. For example, they may impact splicing, transcription factor binding, gene regulation, or transport out of the nucleus [7].

An example of a gene that is modified by C to U editing is the apolipoprotein B gene. In the intestine, the RNA transcript for this gene undergoes C to U editing. This creates a premature stop codon, resulting in a shorter form of the protein, known as ApoB-48. In contrast, C to U editing does not occur in liver cells. As a result, the full-length isoform, called Apo-100, is produced in the liver. Thus, two different protein isoforms are produced depending on the tissue type, and each has distinct properties. The isoform in the liver is essential for the secretion of VLDL (very low-density lipoprotein), while the intestinal isoform is involved in the transport of dietary lipids [1, 11, 98].

The GluA2 gene is an example of a gene that is regulated by A-to-I editing. The GluA2 protein is a subunit of the AMPA receptor. The AMPA receptor mediates excitatory synaptic transmission in the brain and functions as an ion channel. It is permeable to Na^+ and K^+. A to I editing at a single site in the GluA2 mRNA causes a codon change from glutamine (Q) to arginine (R). This editing occurs normally in neurons and is important for the normal function of AMPA receptors. The editing, leading to the Q/R substitution, renders the AMPA receptors impermeable to calcium (Ca^{2+}) ions, which is crucial for neuronal health. In contrast, the unedited mRNA allows calcium permeability. An excessive influx of calcium is toxic to the cells and can lead to neuronal damage. A reduction in Glu2 editing has been seen in the brains of patients with temporal lobe epilepsy, and it may contribute to seizure activity [99].

2.5.3 Changes in mRNA Stability

mRNA stability can be another mechanism that controls which genes get translated into proteins and how much protein is made. Eukaryotic mRNA varies in stability. Beta-globin mRNA is an example of a stable mRNA with a half-life of more than 10 h. Most mRNAs are less stable, with half-lives of less than 30 min. Genes for growth factors and transcription regulators tend to have especially short lives. Several mechanisms are involved in the eventual destruction of an mRNA molecule. One mechanism is poly-A shortening. The poly-A tail at the 3′ end of mRNA gradually shortens over time. As this shortening continues, the mRNA eventually becomes completely degraded from the 3′ end. Another mechanism is de-capping, where the 5′ cap is removed. Once the 5′ cap is removed, the mRNA is rapidly degraded from its 5′ end. Some mRNAs can also be degraded by other mechanisms, such as by endonucleases that cleave the mRNA internally. Certain nucleotide sequences can impact the efficiency of mRNA degradation and, therefore, influence its half-life. For example, AU-rich regions in the 3′ end can speed up mRNA degradation by recruiting certain proteins that are involved in its degradation (although regulation of degradation is complex, because in some cases other proteins can bind to these same regions and enhance RNA stability instead [100]) [101]. The half-life of the mRNA is also influenced by its translation efficiency. The factors involved in Poly-A shortening and de-capping can compete with the machinery that translates the mRNA, and any factors that increase translation efficiency will tend to reduce degradation [1, 17].

3 Other RNA Topics

3.1 Other Categories of RNA

In addition to mRNA, there are other categories of RNA in cells, also discussed in [1, 102, 103]. Some of these are discussed briefly here.

rRNA is ribosomal RNA. rRNA is an important component of ribosomes, which are made of both rRNA and protein. Ribosomes are the sites of protein synthesis, where proteins are translated. Among the different types of RNA, rRNA is actually the most abundant in cells, as large amounts of it are required to form all the ribosomes necessary for protein synthesis. There are different forms of rRNA. The main rRNAs are 18S, 5.8S, 28S, and 5S rRNA. The first three of these are transcribed from a single precursor rRNA molecule, while a separate gene encodes the 5S rRNA. In the process of ribosome biogenesis, ribosomal proteins are transported to the nucleolus within the nucleus, where they are assembled with rRNA to make the ribosomal subunits. The assembled ribosomal subunits are transported out of the nucleus via nuclear pores.

tRNA is transfer RNA. tRNA is essential for protein synthesis. It plays the critical role of linking the amino acids to the mRNA during translation. It will be discussed in more detail in Chap. 8. **Telomerase RNA** serves as the template for the **telomerase** enzyme, which maintains the length of telomeres in certain cells. **snRNAs** are small nuclear RNAs. They have various functions, including roles in the splicing of pre-mRNA. **snoRNAs** are small nucleolar RNAs. They have roles in modifying **rRNAs**. **IncRNAs** are long noncoding RNAs. Some lncRNAs have unknown functions, while others serve as scaffolds to regulate various processes, such as X-chromosome inactivation. Additionally, some have been found to function as antisense RNAs that can silence the expression of certain genes [102, 104, 105].

piRNAs are the Piwi-interacting RNAs. These RNAs have diverse functions. They bind to nuclear RNA-binding proteins called piwi (**P**-Element **I**nduced **Wi**mpi Testis) proteins. They play a role in the epigenetic modification of DNA. They are important for regulating the expression of germline cells. They also play a role in protecting germ cells from transposable elements, which are short pieces of movable and independently replicating DNA [106].

miRNA (microRNA) is a small noncoding RNA molecule. It is an especially interesting type of RNA molecule because it can regulate gene expression by inhibiting translation or causing the degradation of specific mRNA sequences. miRNAs are single-stranded RNA molecules that are naturally found in cells (i.e., they are endogenous). They have sequences that are complementary to various mRNAs. They originate from longer precursor molecules, known as primary miRNAs (pri-miRNAs).

The miRNAs result from pri-mRNAs and regulate specific genes as follows: The pri-miRNAs can fold into hairpin loops, resulting in a double-stranded region. These pri-mRNAs are processed by the enzyme **Drosha**, along with another protein, **DGCR8** (DiGeorge Syndrome Critical Region 8), in the nucleus. **Dosha** is a nuclease, and it trims the **pri-miRNA** into a smaller molecule called **pre-miRNAs**. This pre-miRNA is a small (about 70 nt) hairpin-shaped structure [107]. The pre-miRNA is then transported from the nucleus to the cytoplasm. There, it is processed further by another nuclease, **Dicer**, which cuts off the hairpin loop. This creates a short double RNA duplex, about 22 nucleotides in length. One strand of this duplex is the final mature **miRNA** [107]. One strand of this miRNA, called the guide strand, becomes incorporated into a complex called the **RNA-induced silencing complex (RISC)**. Within the **RISC** complex, the mature miRNA guides the complex to complementary mRNA sequences. Although the match is not necessarily perfect, the result is usually repression of translation due to interference with the translation machinery. Other times, the result can be mRNA degradation or cleavage due to the degree of complementarity with the mRNA [108] (Fig. 7.16).

siRNAs are small interfering RNAs. siRNAs differ from some of the other RNAs discussed above because they are often synthesized experimentally in the laboratory. miRNAs and siRNAs share the feature of being able to silence genes by interfering with mRNA. miRNAs are more common in mammalian cells compared with siRNAs. Some siRNAs exist naturally, but many siRNAs are made experimentally in the lab and designed to have sequence identity to target genes. They can be used to disrupt genes and, in some cases, to

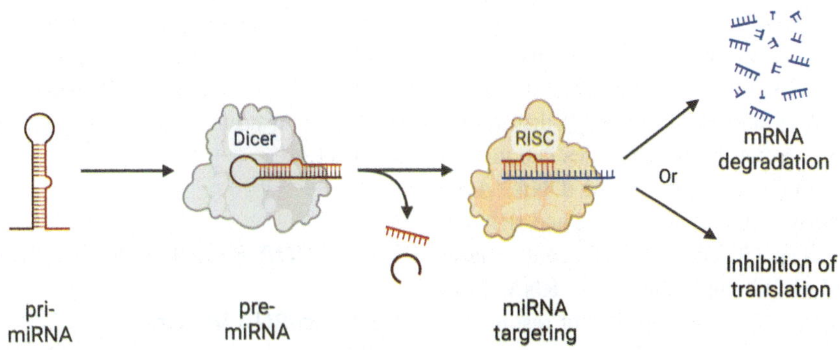

Fig. 7.16 microRNA (miRNA) processing. miRNA is a small noncoding single stranded RNA molecule that can regulate gene expression. miRNAs have sequences complementary to various mRNAs. See text for details. (*Source*: Created with BioRender.com)

treat genetic disorders. While miRNAs may bind imperfectly to their complementary mRNAs, siRNAs can usually be made to be a perfect match. siRNAs are usually double-stranded. One strand is called the guide RNA, which is used for targeting mRNA, and the other strand is usually degraded [109].

3.2 There Are Multiple RNA Polymerases

3.2.1 Overview of RNA Polymerases in Eukaryotic Cells

While we have focused on **RNA Pol II**, there are two other RNA polymerases in eukaryotes. These are **RNA polymerase I (RNA Pol I)** and **RNA polymerase III (RNA-Pol III)**. The main function of **RNA Pol I** is the transcription of 5.8S rRNA, 18S rRNA, and 28S rRNA from a single precursor. **RNA Pol III** transcribes 5S rRNA. It also transcribes tRNA and some snRNA. **RNA pol-II**, in addition to the transcription of mRNA, also has a role in transcribing other RNAs such as miRNA and siRNA [1, 6, 17, 110, 111].

3.2.2 RNA Polymerases in Human Health and as Targets for Drug Treatment

- Inhibition of RNA Polymerase II in Eukaryotic Cells Is Toxic

Alpha-amanitin is a **toxic** compound found in some very poisonous mushrooms, such as the "death cap mushroom." Ingestion of this toxin can be fatal. It specifically inhibits **eukaryotic RNA polymerase II** by binding to it, and it also inhibits **RNA polymerase III**. Without functional **RNA Pol II**, transcription cannot occur, and cells cannot survive. There are not many reliable treatments available once one has ingested this toxin. Some treatments include stomach pumping, blocking the toxin with activated charcoal, or liver transplants in cases where the liver has been destroyed in response to the toxin. Treatment usually needs to occur quickly to be effective. Some researchers have proposed directing modified forms of this and other mushroom toxins directly to cancer cells, although this type of treatment would be risky due to its high toxicity [11].

Alpha-amanitin is sometimes also used as a research tool. It can be used to turn off gene expression in cells to study which genes are transcribed under various conditions, the importance of various transcription factors, and the functions of specific genes. However, its high level of toxicity makes it dangerous to use, which is a significant drawback to using it as a research tool [112, 113].

- Inhibition of RNA Pol I in Cancer Treatment

Dactinomycin (**actinomycin D; ActD**) (also called **Cosmegen**) is a **small-molecule chemotherapy** drug [114]. It intercalates into DNA, especially at GC-rich regions. This interferes with **RNA polymerase**, causing it to pause, thereby blocking transcription. Due to its intercalation into DNA, the drug can indirectly inhibit both RNA Pol I and RNA Pol II. It distorts the double helix so that the polymerases can no longer pass through this part of the DNA. **ActD** binds primarily to GC-rich regions, and rRNA is especially rich in GC content. Cells contain large amounts of rRNA and, therefore, **RNA Pol-I** inhibition is an especially significant result from treatment with the drug. Furthermore, since rRNA is an important component of ribosomes, **ActD** interferes with ribosome production, which is essential for the synthesis of proteins and thereby for cell viability and duplication. This drug is therefore considered to be promising for inhibiting the growth of rapidly dividing cells. It is used to treat several types of cancer, including Wilms' tumor and Ewing sarcoma. While it can block the growth of cancer cells, it also has serious side effects because to affects the growth of healthy cells [115, 116].

- Bacterial RNA Polymerase as a Drug Target

Bacteria also have **RNA polymerase**. They have one RNA polymerase, and it is an important drug target, especially for combating bacterial infections. **Rifadin (rifampin)** [117] is an antibiotic that inhibits **RNA polymerase** in multiple types of bacteria. **Rifadin** is a semisynthetic derivative of **rifamycin B**, a small molecule found in certain types of bacteria as a defense mechanism against other bacterial species. **Aemcolo (rifamycin)** is an enzyme inhibitor. It binds to **bacterial RNA polymerase** and inhibits its activity, thereby blocking elongation of RNA [118, 119]. Due to the inhibition of bacterial RNA polymerase, the transcription of bacterial genes is suppressed, resulting in bacterial cell death. This drug is used to treat bacterial infections, such as tuberculosis (TB) [120, 121].

Xifaxan (rifaximin) [122] also works by inhibiting bacterial RNA polymerase. It is used to treat bacterial infections such as those that cause traveler's diarrhea, especially infections caused by *E. coli*. It is also used to treat hepatic encephalopathy by stopping a population of ammonia-producing gut bacteria that can worsen this liver disease. Another drug that inhibits bacterial RNA polymerase is **Priftin (rifapentine)** [123]. It is used to treat bacterial infections such as TB [124].

3.3 RNA-Based Drug Treatments

3.3.1 mRNA Vaccines

mRNA vaccines, which are considered biologics, are a relatively new method for immunization [125]. While this technology has been under investigation for some time, mRNA vaccines only became actively used during the COVID-19 pandemic. The **COVID-19 vaccine**, the first FDA-approved mRNA vaccine, became a successful and groundbreaking approach, saving many lives and helping to slow the spread of the virus worldwide. The idea behind the mRNA vaccine technology is to identify an antigen that can trigger an immune response. In the case of COVID-19, the spike protein of the SARS-CoV-2 virus was chosen as the antigen. A synthetic mRNA encoding this antigen was made in the lab and encapsulated in lipid nanoparticles for protection. This is important since mRNA is easily degraded. The complex is injected intramuscularly into patients, and it enters the cytoplasm of their cells. There, the mRNA is translated by the normal cellular mechanisms that utilize the host cell's ribosomes and tRNA. Cells that take up the mRNA include immune cells, such as antigen-presenting cells, which present the translated proteins on their surface. This triggers an immune response involving the production of antibodies and the activation of T cells and memory cells. If an individual is later infected with the actual virus, the immune system will now be prepared to mount a response. Due to its mechanism of action, it can reduce the risk of infection. More importantly, it prevents a serious reaction if an infection does occur by enabling the body to mount an immune response quickly. The commonly used COVID-19 vaccines are **Comirnaty** (BNT162b2) (Pfizer) [126] and **Spikevax** (mRNA-1273) (Moderna) [127, 128].

mRNA vaccines are a real breakthrough. Vaccines have been developed to protect against viral infections for many years; however, the use of mRNA vaccine technology is relatively new. This technology has several advantages over other approaches. Unlike more traditional methods, mRNA vaccines do not use live or attenuated viruses, and this makes them safer and easier to develop. This type of vaccine can be made more quickly than traditional methods. Speed is especially important when it comes to viruses because viruses mutate over time, necessitating the development of new viruses. New mRNA vaccines can be made relatively quickly to respond to these viral changes.

The use of mRNA rather than DNA for the vaccine is also important. Unlike DNA, mRNA does not need to enter the nucleus. This eliminates the risk that the nucleic acid will incorporate into the cell's genome. Additionally, the use of mRNA results in a rapid response. mRNA does not need to enter the nucleus, and transcription does not need to occur. Instead, the mRNA can be readily translated into protein.

This mRNA technology is quite promising, and the hope is that it can be applied to other types of vaccines beyond the COVID-19 vaccine. Nevertheless, mRNA vaccines also come with challenges. mRNA is not very stable and can potentially degrade in storage. This necessitates the use of freezers to store the vaccine until it is administered.

3.3.2 Antisense Drugs Using Antisense Oligonucleotides (ASOs)

Antisense technology is designed to target and degrade specific mRNA molecules, consequently inhibiting the expression of certain genes. One important antisense technology method involves designing short, synthetic strands of nucleotides (single-stranded) that are complementary to a specific mRNA sequence to be targeted. These strands are often chemically modified to increase their stability and affinity for binding to their complementary sequences. These oligonucleotides are referred to as antisense oligonucleotides (ASOs). They are usually made of DNA, but in some cases, they can be made of RNA. The idea is that the ASOs bind their complementary mRNA sequences in the cell. The resulting formation of a DNA/RNA hybrid leads to the recruitment of the enzyme **RNaseH**. **RNaseH** is an enzyme that has a high affinity for DNA/RNA hybrids and that degrades RNA. Once it binds, it degrades the RNA strand of the hybrid. Consequently, expression of the specific gene is inhibited or reduced. Thus, an important outcome of antisense oligonucleotides is the destruction of a target mRNA. In some cases, the ASOs can also lead to different outcomes, such as interfering with the normal splicing pattern or altering gene function [129].

Several drugs have been developed based on antisense technology, a few of them are described below.

- Spinraza (Nusinersen) for the Treatment of Spinal Muscle Atrophy (SMA)

Antisense drugs are therapies that primarily utilize oligonucleotides (ASOs). Antisense drugs are usually synthetic and fall into the category of oligonucleotide therapeutics. One example is an antisense drug to treat spinal muscular atrophy (SMA), a disease that leads to muscle atrophy and wasting. SMA is caused by a mutation or deletion in the SMN1 (survival of motor neuron 1) gene, so that the SMN1 protein is not produced in sufficient amounts. People also have a second gene, SMN2, which can also make SMN. The SMN2 gene cannot compensate for the lack of SMN1 protein, however, because its product is often truncated due to alternative splicing. This is because a single nucleotide change in exon 7 leads to alternative splicing, causing exon 7 to be spliced out from SMN2.

By knowing the sequence of the SMN2 gene, researchers were able to design an ASO that restores splicing, ensuring exon 7 is included in the final mature SMN2 mRNA (rather than being spliced out). This ASO is complementary to, and therefore binds to, a splicing silencer element called ISS-N1 (intron splicing silencer N1) on the SMN2 pre-mRNA. Splicing silencer elements are RNA sequences that recruit splicing repressor proteins, such as the ribonucleoprotein hnRNP

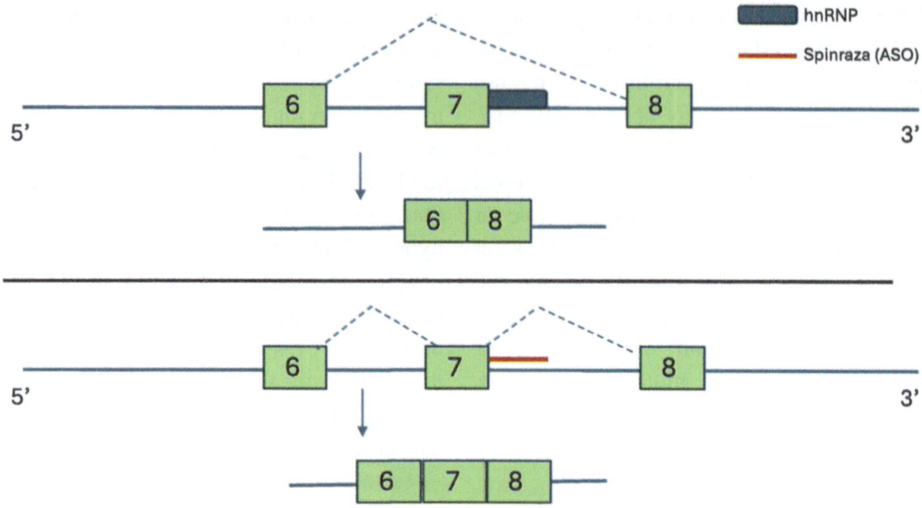

Fig. 7.17 Spinraza. Spinraza is an antisense oligonucleotide that binds to the SMN2 pre-mRNA prompting inclusion of exon 7. It does so by blocking the binding of hnRNP splicing repressors. See text for details

A1. Splicing repressor proteins are proteins that can normally bind to RNA and prevent the inclusion of certain exons (such as in this case, exon 7) in the final mature mRNA. By binding to the splicing silencer element, the ASO blocks the binding of hnRNP A1 and possibly other repressors to the RNA. This restores the production of full-length SMN protein from the SMN2 gene, and now exon 7 is not spliced out. This is an effective treatment because SMN2 can now help compensate for the lack of SMN1.

This oligonucleotide is now a drug: **Spinraza (nusinersen)**. This drug is injected into patients' cerebrospinal fluid (CSF) so it can reach the motor neurons [130]. **Spinraza** alleviates some symptoms of **SMA** and increases the survival rate [131, 132] (see Fig. 7.17). This is a good example of how antisense methods have the potential to treat genetic disorders. It also illustrates how understanding the concept of alternative splicing helped researchers to develop this drug.

- Wainua (Eplonterson), an Antisense Oligonucleotide Therapy for the Treatment of Transthyretin Amyloidosis (ATTR)

The drug **Wainua (eplonterson)** [133] also uses **antisense technology**. It is used to treat a disease called **hereditary transthyretin amyloidosis (hATTR)**. It was FDA-approved in 2023. ATTR is mentioned several times in this chapter because it is a disease that has responded particularly well to gene silencing technologies. This is due in part to the clear genetic target, the TTR gene.

hATTR involves the accumulation of an improperly folded protein called transthyretin (TTR; formerly called prealbumin). TTR is a transport protein that binds to two molecules, thyroxine (T4) and retinol-binding protein (RBP), and transports them to their target tissue. TTR is a mainly extracellular protein, and it normally circulates in the serum and cerebral spinal fluid (CSF) [134, 135]. The hereditary form of ATTR is associated with mutations in the TTR gene, which cause it to misfold and aggregate. As they build up, these aggregates can organize into amyloid fibrils. These fibrils deposit outside of cells, which can interfere with tissue structure and function. The accumulation of the unfolded form leads to neurological problems and tissue and organ damage. Symptoms include neuropathy, such as pain, numbness, and weakness, as well as cardiac symptoms [136]. Several treatments for the disease utilize antisense technology. Using antisense technology for this disorder has received a great deal of attention because it is a good example of a disease for which antisense technology has been quite successful at treating a severe disease.

Wainua (eplonterson) is an **antisense oligonucleotide**, made of DNA, that binds to the **TTR mRNA**. This leads to the degradation of some of the TTR mRNA, thus specifically lowering the levels of TTR protein. As a result, the lowered TTR levels alleviate some of the TTR aggregation. Wainua is designed to alleviate symptoms, although side effects must also be managed due to decreased normal TTR levels [137]. **Tegsedi (inotersen)** [138] is also a drug that targets TTR using an antisense oligonucleotide (ASO). Both drugs are self-administered by subcutaneous injection weekly. Both have shown improvement in symptoms in some, but not all, patients with **ATTR** who took the drug [139].

3.3.3 Drugs Made Using siRNA Technology

siRNA sequences are often synthetic and laboratory-made, designed to target specific sequences. siRNA sequences lead to the destruction of mRNA. siRNAs are usually double-stranded. They are usually about 20–25 nts in length, derived from larger double-stranded fragments processed by Dicer (Dicer is also used in miRNA processing). One strand of the siRNA (the guide strand) is incorporated into the RISC complex (as described earlier for miRNAs), and the other strand may be degraded. The siRNA RISC complex matches with its exact match in the cell's mRNA. This leads to cleavage of the mRNA. Cleavage is mediated by Ago2, a component of the RISC complex that possesses endonuclease activity. The result is degradation of the target mRNA. In general, the process of inhibiting mRNA using this type of small interfering RNA is often referred to as RNAi (RNA interference) [1, 140].

- siRNA Technology for the Treatment of Transthyretin Amyloidosis (ATTR)

Above, we discussed antisense technology for the treatment of hATTR. siRNA technology has been used to treat hATTR. In fact, the first FDA-approved siRNA technology was used to treat this disorder. **Onpattro (patisiran)** (FDA approved in 2018) [141] is the first FDA-approved siRNA drug, and it is designed to treat hereditary ATTR. It utilizes a synthetic double-stranded siRNA molecule in which one strand is complementary to the gene encoding transthyretin (TTR). This process involves Dicer and the RISC complex, reducing the production of the TTR protein, as described above. **Onpattro** is administered by intravenous infusion in lipid nanoparticles, which help deliver the siRNA to the liver, where TTR is produced [135, 142].

Amvuttra (vutrisiran) [143] is a newer **siRNA drug** (FDA approved 2022), also designed to treat **ATTR** by targeting the TTR mRNA. It uses a similar mechanism to the drugs described above, but the delivery mechanism differs. Amvuttra can be administered by subcutaneous injection instead of intravenous infusion, which is more convenient for patients [144].

- Givlaari (Givosiran) for the Treatment of Acute Hepatic Porphyria (AHP)

The second **siRNA** drug to be FDA approved is **Givlaari (Givosiran)**, approved in 2019 [145, 146]. It is designed for the **treatment of acute hepatic porphyria (AHP)**. This is a group of genetic disorders characterized by a deficiency in one of several enzymes involved in heme synthesis in the liver. This leads to a block in the heme production pathway and an accumulation of harmful intermediates, which can cause symptoms such as pain, vomiting, weakness, and others. One of the key enzymes in the pathway is ALAS1 (hepatic δ-aminolevulinic acid synthase), and it functions upstream to the blocked enzymes. The ALAS1 gene becomes overexpressed in response to low heme levels, leading to the accumulation of harmful intermediates. ALAS1 is the target of this drug. **Givlaari** is a siRNA drug that **targets the ALAS1** mRNA, leading to its degradation. It functions specifically in the liver. This is because the siRNA in this drug is conjugated with GalNAc (N-acetylgalactosamine), which directs the siRNA to the liver. There, it leads to degradation of ALAS1 mRNA, without affecting any genes outside the liver. Targeting ALAS1 reduces the accumulation of harmful intermediates, thereby decreasing symptoms.

3.4 Transcription Regulation in Bacteria

Much of what we learned about transcription regulation came from studies in bacteria. Bacteria also need to regulate gene expression according to their specific needs, but their genetic makeup is less complex, making it more straightforward to delineate the mechanisms involved. *E. coli* is a typical model for studying prokaryotes. It has a single circular DNA molecule that codes for approximately 4000–5000 proteins. Only some of its genes are expressed (transcribed) at any given time due to differential gene expression. One major difference between bacteria and eukaryotic cells is that in bacteria, operons regulate the transcription of clusters of genes controlled by a single promoter. Upstream to the genes in an operon is an operator, and upstream to the operator is a promoter sequence that controls all the genes in the operon. RNA polymerase can bind the promoter, and transcription repressors can bind the operator. The genes in these operons can be regulated as a group and can be subject to positive or negative regulation [1, 110]. Some examples of operons and their regulation are described below.

3.4.1 Tryptophan Operon

The tryptophan (trp) operon is an example of an operon where a single promoter controls a cluster of five genes, all of which can be negatively regulated by one transcription factor. These five genes code for enzymes that are necessary for the production of the amino acid tryptophan. The genes in this operon are transcribed (i.e., transcription is turned on) when tryptophan concentrations in the bacteria are low. However, their transcription is turned off when tryptophan levels are high, as more tryptophan is not needed under these conditions. This is controlled by a transcription repressor called the tryptophan repressor, which can bind to the tryptophan operator sequence located immediately downstream of the promoter.

When tryptophan concentrations in the bacteria are low, the repressor does not bind to the operator. Instead, RNA polymerase can bind, and the five genes are transcribed. Thus, the operon is on.

However, when tryptophan levels are high, tryptophan acts as a co-repressor by binding to the repressor. This causes the repressor to change its conformation, allowing it to bind to the operator. By binding to the operon, it prevents RNA polymerase from binding to the promoter, thereby blocking transcription initiation. The result is no transcription. Thus, the operon is off.

In this manner, the bacteria produce tryptophan only when needed, and the expression of the genes in this operon is turned **off** when there is already an abundance of tryptophan in the environment [1, 3, 17, 147] (Fig. 7.18).

3.4.2 The Lac Operon

Some operons also use transcription activators (Fig. 7.19). The Lac operon is an example of an operon controlled by both a repressor and an activator [148]. This operon contains three genes that encode proteins required for the import and digestion of lactose. This operon ensures that bacteria utilize the most efficient energy source available. The best energy source for bacteria is glucose, but when glucose is absent, the bacteria

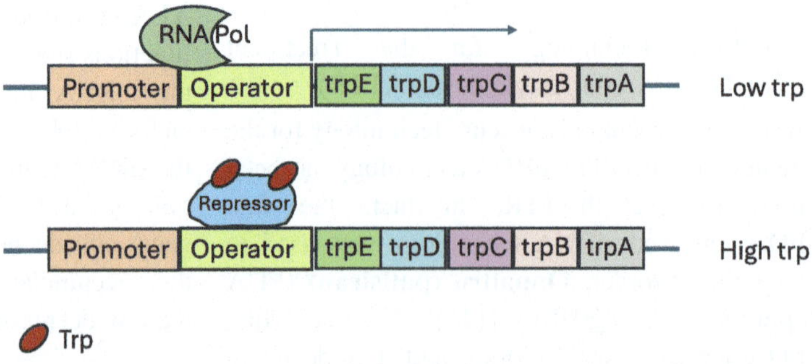

Fig. 7.18 The Trp operon in bacteria. The trp operon contains genes encoding enzymes required for Trp synthesis. It is regulated by the presence of Trp, as described in the text

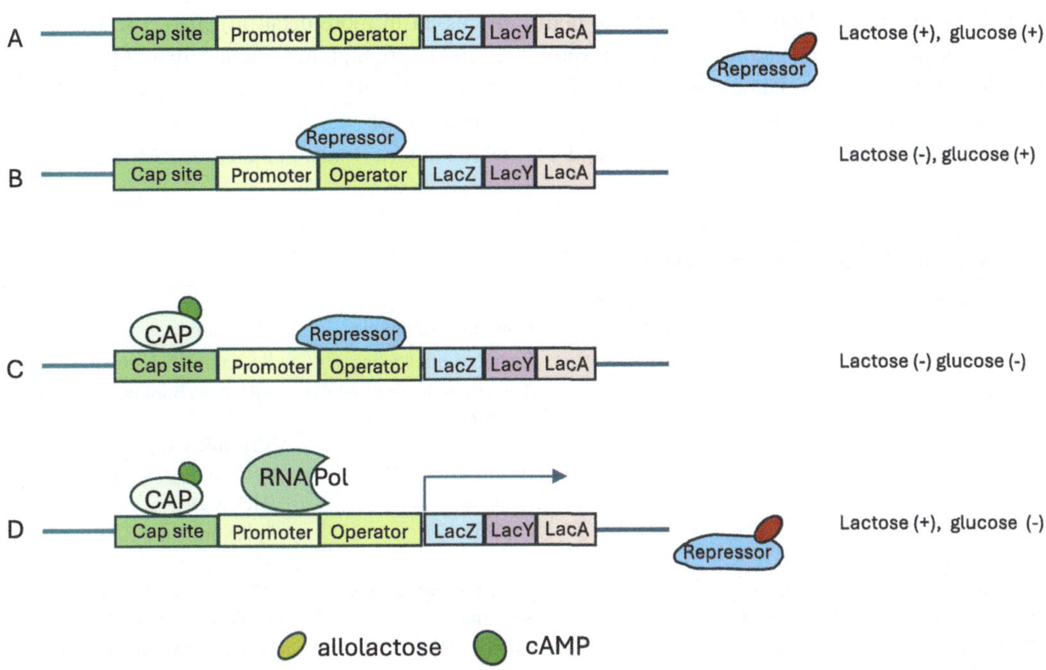

Fig. 7.19 The lac operon. The Lac operon in bacteria consists of genes that are important for importing and digesting lactose as an energy source. It is controlled by both a repressor and an activator, as described in the text

must rely on an alternative energy source. Under these conditions, the bacteria activate a pathway that turns on genes that allow the use of alternative energy sources, such as lactose. However, the genes needed for the utilization of lactose as an energy source will only be turned **on** (expressed) if lactose is present in the environment *and* if glucose is absent. Regulation of this operon involves both a repressor (the lac repressor) and an activator (the CAP protein) [4], as discussed below.

In the absence of lactose, a repressor (the lac repressor) binds to a sequence on the operator and prevents transcription of the lac genes.

In the presence of lactose, a lactose derivative called allolactose binds to the repressor, changing its conformation so that it can no longer bind the operator. However, this alone is insufficient to activate transcription of the lac genes because the absence of glucose is also required.

In the absence of glucose, a transcription activator called the CAP protein binds a sequence just upstream of the promoter. CAP only binds when it is activated by cyclic AMP (cAMP), whose levels rise when glucose levels are low. CAP can activate transcription, but only if the lac repressor is *not* also bound to the operator (in the presence of high lactose).

As a result, the lac operon is turned off under the following conditions: (a) when both glucose and lactose are present (CAP does not bind because cAMP levels are low, and the repressor is inactive). Transcription is low, (b) when glucose is present, but lactose is absent (the lac repressor binds the operator), and (c) when both glucose and lactose are absent (the lac repressor binds, and CAP is active but blocked by the repressor; the operon is turned off).

Only (d) when glucose is absent and lactose is present (CAP-cAMP binds upstream of the promoter, and the lac repressor is inactive) is the operon turned on, and the genes for import and utilization of lactose are transcribed [1, 3].

Some activators, such as CAP, activate transcription by regulating the binding of RNA polymerase to the DNA. In this example, RNA polymerase typically binds poorly to the promoter of this operon. However, when the activator is present, its binding is strengthened, allowing transcription to occur. The CAP protein, the activator, is itself subject to regulation. CAP stands for both cAMP receptor protein and catabolite activator protein. It receives its name because CAP is activated by cAMP, which is a second messenger in both prokaryotic and eukaryotic cells, and is discussed in more detail in Chap. 18. The levels of cAMP are high in bacteria when glucose levels are low because low glucose levels lead to the activation of the enzyme **adenylate cyclase**, which in turn results in increased cAMP levels [1].

FDA-Approved Drugs Mentioned in This Chapter

Commonly used brand name	Generic name	First FDA approval year	Common conditions	Main target
Aemcolo	*Rifamycin*	2018	Bacterial infections	Bacterial RNA polymerase
Amvuttra	*Vutrisiran*	2022	Transthyretin amyloidosis (ATTR)	siRNA technology TTR mRNA
Beleodaq	*Belinostat*	2014	Peripheral T-cell lymphoma	Histone deacetylases (HDACs)
Comirnaty	*BNT162b2 (Pfizer-BioNTech)*	**2021**	COVID-19	Spike protein mRNA (vaccine target)
Cosmegen	*Dactinomycin (Actinomycin D)*	1964	Wilms tumor, Ewing sarcoma	RNA polymerase I and II (via DNA intercalation)
Dacogen	*Decitabine*	2006	MDS, AML	DNA methyltransferases (DNMTs)
DUVYZAT	*Givinostat*	**2024**	Duchenne muscular dystrophy	Histone deacetylases (HDACs)
Givlaari	*Givosiran*	2019	Acute hepatic porphyria (AHP)	siRNA technology ALAS1 mRNA
Istodax	*Romidepsin*	2009	Cutaneous and peripheral T-cell lymphoma	Histone deacetylases (HDACs)
Onpattro	*Patisiran*	2018	Transthyretin amyloidosis (ATTR)	siRNA technology TTR mRNA
Onureg	*Azacitidine (oral)*	**2020**	MDS, AML	DNA methyltransferases (DNMTs)
Priftin	*Rifapentine*	1998	Tuberculosis	Bacterial RNA polymerase
Rifadin	*Rifampin*	1971	Tuberculosis	Bacterial RNA polymerase
Spikevax	*mRNA-1273 (Moderna)*	**2020**	COVID-19	Spike protein mRNA (vaccine target)
Spinraza	*Nusinersen*	2016	Spinal muscular atrophy (SMA)	ASO technology ISS-N1 on SMN2 pre-mRNA
Tegsedi	*Inotersen*	2018	Transthyretin amyloidosis (ATTR)	ASO technology TTR mRNA
Vesanoid	*Tretinoin (ATRA)*	1995	Acute promyelocytic leukemia (APL)	PML-RARa
Vidaza	*Azacitidine*	2004	MDS, AML	DNA methyltransferases (DNMTs)
Wainua	*Eplontersen*	2023	Transthyretin amyloidosis (ATTR)	TTR mRNA
Xifaxan	*Rifaximin*	2004	Traveler's diarrhea, hepatic encephalopathy	Bacterial RNA polymerase
Zolinza	*Vorinostat (SAHA)*	2006	Cutaneous T-cell lymphoma (CTCL)	Histone deacetylases (HDACs)

Definitions

5′ cap (m^{7}G cap) The 5′ cap of mRNA is a guanosine nucleotide on the 5′ end of mature mRNA that is modified by methylation. This modified guanosine is called 7-methylguanosine (m^{7}G).

Alternative splicing After transcription, introns are removed. In some cases, introns are removed along with exons, allowing different mature mRNA products to be produced from one gene.

Amyloid fibrils Insoluble protein aggregates that form fibrils with a specific structure that contains an ordered arrangement of beta strands and beta sheets.

Antisense (template) strand The DNA strand used as a template for mRNA synthesis; it is complementary to the mRNA sequence that is produced.

Chromatin remodeling complex (CRC) Protein complexes that reposition or restructure nucleosomes to regulate access to DNA for transcription.

Consensus sequence A DNA sequence, often one that is recognized by a protein, such as a transcription factor. The same or similar consensus sequence may be found in multiple genes.

Enhancer A cis-regulatory DNA element that is separate from the coding sequence or the promoter. Transcription factor binding to cis-regulatory elements can sometimes lead to the enhancement of gene expression (transcription).

Exon A segment of a gene that generally remains in the mature mRNA after RNA splicing takes place. The exon may encode amino acid sequences.

Intron A noncoding region of a gene whose corresponding mRNA gets spliced out during splicing. Introns are present between the exons.

Messenger RNA (mRNA) The RNA molecule that gets transcribed from the DNA sequence.

Poly-A tail An adenine-rich sequence added to the 3′ end of mRNA after transcription.

Promoter A DNA sequence upstream of the transcription start site. It provides a binding site for RNA polymerase and transcription factors.

RNA polymerase II (RNA Pol II) The enzyme that catalyzes the transcription of coding genes. It adds ribonucleotides to the growing mRNA strand.

Sense (coding) strand The DNA strand with a sequence identical to the mRNA (except T is replaced by U in RNA). It is not used as the template.

Splicing The process of removing introns and splicing together the remaining exons to form the mature mRNA after transcription.

TATA box A sequence in the promoters of many genes that is rich in As and Ts (adenine and thymine nucleotides). The transcription factor TFIID binds to this region via its TBP (TATA-binding protein) subunit.

TFIIH A general transcription factor that is part of the initiation complex. It has helicase and kinase activities and plays a role in unwinding DNA and phosphorylating the CTD (C-terminal domain) of RNA polymerase.

Transcription factor A protein that binds to specific DNA sequences and regulates transcription. It may either positively or negatively regulate transcription.

Transcription The process of synthesizing RNA from a DNA template, including messenger RNA (mRNA).

Study Questions

Multiple-Choice Questions

1. What is the main role of messenger RNA (mRNA)?
   A. To catalyze peptide bond formation
   B. To replicate DNA
   C. To carry genetic information from DNA to ribosomes for protein synthesis
   D. To maintain telomere length
   E. B and C

2. Which base is found in RNA but not in DNA?
   A. Thymine
   B. Adenine
   C. Uracil
   D. Guanine
3. Which enzyme is responsible for synthesizing RNA during transcription?
   A. DNA polymerase
   B. RNA polymerase
   C. Ligase
   D. Helicase
4. Where does transcription occur in eukaryotic cells?
   A. Cytoplasm
   B. Ribosome
   C. Endoplasmic reticulum
   D. Nucleus
5. When is the 5′ CAP added to the mRNA?
   A. During transcription, after transcription begins but before it is complete
   B. After transcription is complete
   C. After splicing has been completed
   D. After the poly-A tail has been added
   E. B and C
6. What signal initiates transcription termination?
   A. TATA box
   B. Splice site
   C. The 5′ CAP
   D. Polyadenylation signal (AAUAAA)
   E. A and C
7. What type of bond links ribonucleotides in RNA?
   A. Hydrogen bond
   B. Peptide bond
   C. Disulfide bond
   D. Phosphodiester bond
8. In eukaryotic cells, where does transcription occur?
   A. In the nucleus
   B. In the cytoplasm
   C. In the mitochondria
   D. In the endoplasmic reticulum
9. In eukaryotes, which of the following is a process that modifies pre-mRNA to form mature mRNA?
   A. Translation
   B. Replication
   C. Transcription
   D. RNA splicing
10. In bacteria, what is the primary function of an operon?
    A. To facilitate DNA replication of genes to work together
    B. To control protein translation of genes that work together
    C. To regulate the expression of a group of functionally related genes
    D. To regulate expression of one gene at a time
    E. B and C
11. In the regulation of the tryptophan operon in*E. coli*, which molecule acts as the co-repressor? That is, it binds to the repressor protein and facilitating its binding to the operator region to inhibit transcription.
    A. Tryptophan
    B. RNA polymerase
    C. Promoter
    D. Ribosome
    E. RNA polymerase
12. When combinations of transcription factors act synergistically, it means
    A. They inhibit each other's activity, leading to reduced gene expression.
    B. They have no effect on gene expression and remain independent.
    C. Their combined action enhances gene expression more than the sum of their individual effects.
    D. They compete for binding to the same DNA regulatory element.
    E. Their combined actions enhance gene expression to an amount equivalent to the sum of their individual effects.
13. Methylation of CpG regions on a promoter is associated with the repression of gene expression. How is this related to histone acetylation?
    A. Methylation and acetylation both promote gene expression.
    B. Methylation and acetylation are independent processes with no connection.

C. Methylation recruits histone acetyltransferases, leading to gene activation.
D. Methylation attracts histone deacetylases, resulting in gene silencing.
E. Methylation attracts histones, resulting in gene silencing.

Fill-in-the-Blank Questions

14. The coding strand of DNA is also called the ________ strand.
15. The sugar found in RNA is __________.
16. RNA uses the nitrogenous base _______ instead of thymine.
17. Coding regions on genes that remain in mRNA after splicing are the __________.

True/False Questions

18. Histone acetylation leads to a more open configuration helping transcription factors.
19. All genes require a TATA box for transcription to begin.

Short-Answer Question

20. What happens to introns during RNA splicing?

Matching Question

21. Match the RNA type with its function:
mRNA type
A. mRNA
B. tRNA
C. rRNA
Function:
1. Brings amino acids to the ribosome
2. Template for protein synthesis
3. Forms the core of ribosome's structure

Answer Key

Multiple-Choice Questions

1. **Answer**: C
2. **Answer**: C
3. **Answer**: B
4. **Answer**: D
5. **Answer**: A
6. **Answer**: D
7. **Answer**: D
8. **Answer**: A
9. **Answer**: D
10. **Answer**: C
11. **Answer**: A
12. **Answer**: C
13. **Answer**: D

Fill-in-the-Blank Questions

14. **Answer**: Sense
15. **Answer**: ribose
16. **Answer**: uracil
17. **Answer**: exons

True/False Questions

18. **Answer**: True
19. **Answer**: False

Short Answer Question

20. **Answer**: Introns are removed from the pre-mRNA and degraded, while exons are joined together.

Matching Question

21. **Answer**: A-2, B-1, C-3

References

1. Alberts B, Heald R, Johnson A, Morgan D, Raff M, Roberts K, Walter P, Wilson J, Hunt T. Molecular biology of the cell. W.W. Norton and Company; 2022.
2. Hartwell LH, Goldberg M, Fischer JA, Hood L, Reynolds AE, Aquadro CF, Silver LM. Genetics: from genes to genomes. 6th ed. New York: McGraw Hill; 2017.
3. Fletcher H, Hickey I. BIOS instant notes in genetics. Abingdon: Garland Science; 2012.
4. Hames BD, Hooper NM. BIOS instant notes in biochemistry. Taylor and Francis; 2011.
5. Lodish H, Berk A, Kaiser CA, Krieger M, Bretscher A, Ploegh H, Martin KC, Yaffe MB, Amon A. Molecular cell biology. 9th ed. Macmillan Learning; 2020.
6. Cramer P. Organization and regulation of gene transcription. Nature. 2019;573(7772):45–54. https://doi.org/10.1038/s41586-019-1517-4.
7. Latchman DS. Gene regulation: a eukaryotic perspective. New York: Taylor and Francis; 2010.
8. Kuehner JN, Pearson EL, Moore C. Unravelling the means to an end: RNA polymerase II transcription termination. Nat Rev Mol Cell Biol. 2011;12(5): 283–94. https://doi.org/10.1038/nrm3098.
9. Archuleta SR, Goodrich JA, Kugel JF. Mechanisms and functions of the RNA polymerase II general transcription machinery during the transcription cycle. Biomolecules. 2024;14(2):176. https://doi.org/10.3390/biom14020176.
10. Nikolov DB, Burley SK. RNA polymerase II transcription initiation: a structural view. Proc Natl Acad Sci USA. 1997;94(1):15–22. https://doi.org/10.1073/pnas.94.1.15.
11. Latchman DS. Gene control. New York: Garland; 2015.
12. Roeder RG. The role of general initiation factors in transcription by RNA polymerase II. Trends Biochem Sci. 1996;21(9):327–35.
13. Komarnitsky P, Cho EJ, Buratowski S. Different phosphorylated forms of RNA polymerase II and associated mRNA processing factors during transcription. Genes Dev. 2000;14(19):2452–60. https://doi.org/10.1101/gad.824700.
14. Egloff S. Role of Ser7 phosphorylation of the CTD during transcription of snRNA genes. RNA Biol. 2012;9(8):1033–8. https://doi.org/10.4161/rna.21166.
15. Phatnani HP, Greenleaf AL. Phosphorylation and functions of the RNA polymerase II CTD. Genes Dev. 2006;20(21):2922–36. https://doi.org/10.1101/gad.1477006.
16. Rodríguez-Molina JB, West S, Passmore LA. Knowing when to stop: transcription termination on protein-coding genes by eukaryotic RNAPII. Mol Cell. 2023;83(3):404–15. https://doi.org/10.1016/j.molcel.2022.12.021.
17. Brown TA. Introduction to genetics: a molecular approach. 1st ed. New York: Garland Science; 2011.
18. Ramanathan A, Robb GB, Chan SH. mRNA capping: biological functions and applications. Nucleic Acids Res. 2016;44(16):7511–26. https://doi.org/10.1093/nar/gkw551.
19. Rodríguez-Molina JB, Turtola M. Birth of a poly (A) tail: mechanisms and control of mRNA polyadenylation. FEBS Open Bio. 2023;13(7): 1140–53. https://doi.org/10.1002/2211-5463.13528.
20. Wilkinson ME, Charenton C, Nagai K. RNA splicing by the spliceosome. Annu Rev Biochem. 2020;89: 359–88. https://doi.org/10.1146/annurev-biochem-091719-064225.
21. Hocine S, Singer RH, Grünwald D. RNA processing and export. Cold Spring Harb Perspect Biol. 2010;2 (12):a000752. https://doi.org/10.1101/cshperspect.a000752.
22. De Magistris P. The great escape: mRNA export through the nuclear pore complex. Int J Mol Sci. 2021;22(21):11767. https://doi.org/10.3390/ijms222111767.
23. Lee TI, Young RA. Transcriptional regulation and its misregulation in disease. Cell. 2013;152(6):1237–51. https://doi.org/10.1016/j.cell.2013.02.014.
24. Struhl K. Helix-turn-helix, zinc-finger, and leucine-zipper motifs for eukaryotic transcriptional regulatory proteins. Trends Biochem Sci. 1989;14(4):137–40. https://doi.org/10.1016/0968-0004(89)90145-X.
25. Jones S. An overview of the basic helix-loop-helix proteins. Genome Biol. 2004;5(6):226. https://doi.org/10.1186/gb-2004-5-6-226.
26. Bannister AJ, Miska EA. Regulation of gene expression by transcription factor acetylation. Cell Mol Life Sci. 2000;57(8–9):1184–92. https://doi.org/10.1007/pl00000758.
27. Bieluszewski T, Prakash S, Roulé T, Wagner D. The role and activity of SWI/SNF chromatin remodelers. Annu Rev Plant Biol. 2023;74:139–63. https://doi.org/10.1146/annurev-arplant-102820-093218.
28. Narlikar GJ, Fan H-Y, Kingston RE. Cooperation between complexes that regulate chromatin structure and transcription. Cell. 2002;108(4):475–87. https://doi.org/10.1016/S0092-8674(02)00654-2.
29. Sudarsanam P, Winston F. The SWI/SNF family: nucleosome-remodeling complexes and transcriptional control. Trends Genet. 2000;16(8):345–51. https://doi.org/10.1016/S0168-9525(00)02060-6.
30. Mittal P, Roberts CWM. The SWI/SNF complex in cancer—biology, biomarkers and therapy. Nat Rev Clin Oncol. 2020;17(7):435–48. https://doi.org/10.1038/s41571-020-0357-3.
31. Mollapour Sisakht M, Amirkhani MA, Nilforoushzadeh MA. SWI/SNF complex, promising target in melanoma therapy: snapshot view. Front Med (Lausanne). 2023;10:1096615. https://doi.org/10.3389/fmed.2023.1096615.
32. Helming Katherine C, Wang X, Roberts Charles WM. Vulnerabilities of mutant SWI/SNF complexes

in cancer. Cancer Cell. 2014;26(3):309–17. https://doi.org/10.1016/j.ccr.2014.07.018.
33. Kundu TK, Rao MR. CpG islands in chromatin organization and gene expression. J Biochem. 1999;125(2):217–22. https://doi.org/10.1093/oxfordjournals.jbchem.a022276.
34. Hervouet E, Peixoto P, Delage-Mourroux R, Boyer-Guittaut M, Cartron P-F. Specific or not specific recruitment of DNMTs for DNA methylation, an epigenetic dilemma. Clin Epigenetics. 2018;10(1):17. https://doi.org/10.1186/s13148-018-0450-y.
35. Someya K, Nakatsukasa H, Ito M, Kondo T, Tateda KI, Akanuma T, et al. Improvement of Foxp3 stability through CNS2 demethylation by TET enzyme induction and activation. Int Immunol. 2017;29(8):365–75. https://doi.org/10.1093/intimm/dxx049.
36. Battistelli C, Garbo S, Maione R. MyoD-induced trans-differentiation: a paradigm for dissecting the molecular mechanisms of cell commitment, differentiation and reprogramming. Cells. 2022;11(21):3435. https://doi.org/10.3390/cells11213435.
37. Chan HM, La Thangue NB. p300/CBP proteins: HATs for transcriptional bridges and scaffolds. J Cell Sci. 2001;114(Pt 13):2363–73. https://doi.org/10.1242/jcs.114.13.2363.
38. Lee D, Kim JW, Seo T, Hwang SG, Choi E-J, Choe J. SWI/SNF complex interacts with tumor suppressor p53 and is necessary for the activation of p53-mediated transcription*. J Biol Chem. 2002;277(25):22330–7. https://doi.org/10.1074/jbc.M111987200.
39. Beckerman R, Prives C. Transcriptional regulation by p53. Cold Spring Harb Perspect Biol. 2010;2(8):a000935. https://doi.org/10.1101/cshperspect.a000935.
40. Hoffman WH, Biade S, Zilfou JT, Chen J, Murphy M. Transcriptional repression of the anti-apoptoticsurvivin gene by wild type p53*. J Biol Chem. 2002;277(5):3247–57. https://doi.org/10.1074/jbc.M106643200.
41. Sullivan KD, Galbraith MD, Andrysik Z, Espinosa JM. Mechanisms of transcriptional regulation by p53. Cell Death Differ. 2018;25(1):133–43. https://doi.org/10.1038/cdd.2017.174.
42. Ramaiah MJ, Tangutur AD, Manyam RR. Epigenetic modulation and understanding of HDAC inhibitors in cancer therapy. Life Sci. 2021;277:119504. https://doi.org/10.1016/j.lfs.2021.119504.
43. U.S. Food and Drug Administration. Zolinza [Vorinostat (SAHA)]. Approval overview. 2006. Available from: https://www.accessdata.fda.gov/scripts/cder/daf/index.cfm?event=overview.process&ApplNo=021991.
44. Marks PA. The clinical development of histone deacetylase inhibitors as targeted anticancer drugs. Expert Opin Investig Drugs. 2010;19(9):1049–66. https://doi.org/10.1517/13543784.2010.510514.
45. Marks PA. Discovery and development of SAHA as an anticancer agent. Oncogene. 2007;26(9):1351–6. https://doi.org/10.1038/sj.onc.1210204.
46. Porter NJ, Christianson DW. Structure, mechanism, and inhibition of the zinc-dependent histone deacetylases. Curr Opin Struct Biol. 2019;59:9–18. https://doi.org/10.1016/j.sbi.2019.01.004.
47. Parveen R, Harihar D, Chatterji BP. Recent histone deacetylase inhibitors in cancer therapy. Cancer. 2023;129(21):3372–80. https://doi.org/10.1002/cncr.34974.
48. U.S. Food and Drug Administration. Beleodaq (Belinostat). Approval overview. 2014. https://www.accessdata.fda.gov/scripts/opdlisting/oopd/detailedIndex.cfm?cfgridkey=289309
49. U.S. Food and Drug Administration. Istodax (Romidepsin). Approval overview. 2009. Available from: https://www.accessdata.fda.gov/scripts/cder/daf/index.cfm?event=overview.process&ApplNo=022393.
50. Mayo Clinic. Muscular dystrophy. Mayo Clinic. 2023. Available from: https://www.mayoclinic.org/diseases-conditions/muscular-dystrophy/symptoms-causes/syc-20375388.
51. Sandonà M, Cavioli G, Renzini A, Cedola A, Gigli G, Coletti D, et al. Histone deacetylases: molecular mechanisms and therapeutic implications for muscular dystrophies. Int J Mol Sci. 2023;24(5):4306. https://doi.org/10.3390/ijms24054306.
52. U.S. Food and Drug Administration. Duvyzat (Givinostat). Approval overview. 2024. https://www.fda.gov/news-events/press-announcements/fda-approves-nonsteroidal-treatment-duchenne-muscular-dystrophy
53. Lamb YN. Givinostat: first approval. Drugs. 2024;84(7):849–56. https://doi.org/10.1007/s40265-024-02052-1.
54. Muscular Dystrophy. Duvyzat (givinostat) for Duchenne muscular dystrophy. 2024. Available from: https://musculardystrophynews.com/givinostat-itf2357/.
55. U.S. Food and Drug Administration. Vidaza (Azacitidine). Approval overview. 2004. Available from: https://www.accessdata.fda.gov/scripts/cder/daf/index.cfm?event=overview.process&ApplNo=050794.
56. Memorial Sloan Kettering Cancer Center. Myelodysplastic syndrom (MDS). 2025. Available from: https://www.mskcc.org/cancer-care/types/myelodysplastic-syndrome.
57. Stomper J, Rotondo JC, Greve G, Lübbert M. Hypomethylating agents (HMA) for the treatment of acute myeloid leukemia and myelodysplastic syndromes: mechanisms of resistance and novel HMA-based therapies. Leukemia. 2021;35(7):1873–89. https://doi.org/10.1038/s41375-021-01218-0.
58. Thoms JAI, Yan F, Hampton HR, Davidson S, Joshi S, Saw J, et al. Clinical response to azacitidine in MDS is associated with distinct DNA methylation changes in HSPCs. Nat Commun. 2025;16(1):4451. https://doi.org/10.1038/s41467-025-59796-x.

59. Issa J-PJ, Kantarjian HM, Kirkpatrick P. Azacitidine. Nat Rev Drug Discov. 2005;4(4):275–6. https://doi.org/10.1038/nrd1698.
60. U.S. Food and Drug Administration. Onureg [Azacitidine (Oral)]. Approval overview. 2020. Available from: https://www.accessdata.fda.gov/scripts/cder/daf/index.cfm?event=overview.process&ApplNo=214120.
61. U.S. Food and Drug Administration. Dacogen (Decitabine). Approval overview. 2006. Available from: https://www.accessdata.fda.gov/scripts/cder/daf/index.cfm?event=overview.process&ApplNo=021790.
62. Yang Y, Li J, Geng Y, Liu L, Li D. Azacitidine regulates DNA methylation of GADD45γ in myelodysplastic syndromes. J Clin Lab Anal. 2021;35(2):e23597. https://doi.org/10.1002/jcla.23597.
63. Hackanson B, Daskalakis M. Decitabine. Recent Results Cancer Res. 2014;201:269–97. https://doi.org/10.1007/978-3-642-54490-3_18.
64. Krishna M, Narang H. The complexity of mitogen-activated protein kinases (MAPKs) made simple. Cell Mol Life Sci. 2008;65:3525–44. https://doi.org/10.1007/s00018-008-8170-7.
65. Marais R, Wynne J, Treisman R. The SRF accessory protein Elk-1 contains a growth factor-regulated transcriptional activation domain. Cell. 1993;73(2):381–93. https://doi.org/10.1016/0092-8674(93)90237-k.
66. Contreras-Cornejo H, Saucedo-Correa G, Oviedo-Boyso J, Valdez-Alarcón JJ, Baizabal-Aguirre VM, Cajero-Juárez M, et al. The CSL proteins, versatile transcription factors and context dependent corepressors of the notch signaling pathway. Cell Div. 2016;11(1):12. https://doi.org/10.1186/s13008-016-0025-2.
67. VanDussen KL, Carulli AJ, Keeley TM, Patel SR, Puthoff BJ, Magness ST, et al. Notch signaling modulates proliferation and differentiation of intestinal crypt base columnar stem cells. Development. 2012;139(3):488–97. https://doi.org/10.1242/dev.070763.
68. Marks F, Klingmüller U, Müller-Decker K. Cellular signal processing: an introduction to the molecular mechanisms of signal transduction. 2nd ed. Garland Science; 2017.
69. Findlay JK, Liew SH, Simpson ER, Korach KS. Estrogen signaling in the regulation of female reproductive functions. Handb Exp Pharmacol. 2010;(198):29–35. https://doi.org/10.1007/978-3-642-02062-9_2.
70. Clusan L, Ferrière F, Flouriot G, Pakdel F. A basic review on estrogen receptor signaling pathways in breast cancer. Int J Mol Sci. 2023;24(7):6834. https://doi.org/10.3390/ijms24076834.
71. Fuentes N, Silveyra P. Estrogen receptor signaling mechanisms. Adv Protein Chem Struct Biol. 2019;116:135–70. https://doi.org/10.1016/bs.apcsb.2019.01.001.
72. Weidemüller P, Kholmatov M, Petsalaki E, Zaugg JB. Transcription factors: bridge between cell signaling and gene regulation. Proteomics. 2021;21(23–24):2000034. https://doi.org/10.1002/pmic.202000034.
73. Benyoucef A, Haigh JJ, Brand M. Unveiling the complexity of transcription factor networks in hematopoietic stem cells: implications for cell therapy and hematological malignancies. Front Oncol. 2023;13:1151343. https://doi.org/10.3389/fonc.2023.1151343.
74. Fong AP, Tapscott SJ. Skeletal muscle programming and re-programming. Curr Opin Genet Dev. 2013;23(5):568–73. https://doi.org/10.1016/j.gde.2013.05.002.
75. Yagi M, Ji F, Charlton J, Cristea S, Messemer K, Horwitz N, et al. Dissecting dual roles of MyoD during lineage conversion to mature myocytes and myogenic stem cells. Genes Dev. 2021;35(17–18):1209–28. https://doi.org/10.1101/gad.348678.121.
76. Harvard Medical School. News and research: new hope for muscle-related diseases? 2021. Available from: https://hms.harvard.edu/news/new-hope-muscle-related-diseases.
77. Qabrati X, Kim I, Ghosh A, Bundschuh N, Noé F, Palmer AS, et al. Transgene-free direct conversion of murine fibroblasts into functional muscle stem cells. NPJ Regen Med. 2023;8(1):43. https://doi.org/10.1038/s41536-023-00317-z.
78. Tapscott SJ. The circuitry of a master switch: Myod and the regulation of skeletal muscle gene transcription. Development. 2005;132(12):2685–95. https://doi.org/10.1242/dev.01874.
79. Dumont NA, Wang YX, Rudnicki MA. Intrinsic and extrinsic mechanisms regulating satellite cell function. Development. 2015;142(9):1572–81. https://doi.org/10.1242/dev.114223.
80. Ebrahim N, Shakirova K, Dashinimaev E. PDX1 is the cornerstone of pancreatic β-cell functions and identity. Front Mol Biosci. 2022;9:1091757. https://doi.org/10.3389/fmolb.2022.1091757.
81. Fujimoto K, Polonsky KS. Pdx1 and other factors that regulate pancreatic beta-cell survival. Diabetes Obes Metab. 2009;11 Suppl 4(Suppl 4):30–7. https://doi.org/10.1111/j.1463-1326.2009.01121.x.
82. Cooper G, Adams K. The cell: a molecular approach. Sinauer Associates; 2022.
83. Weinberg RA. The biology of cancer. 3rd ed. Norton and Company; 2023.
84. Fox Chase Cancer Center Temple Health. Differentiation therapy a promising treatment for medulloblastoma, Fox Chase Cancer Center researchers show. 2024. Available from: https://www.foxchase.org/news/2024-08-12-differentiation-therapy-a-promising-treatment-for-medulloblastoma-fox-chase-cancer-center-researchers-show.
85. Rochette-Egly C, Germain P. Dynamic and combinatorial control of gene expression by nuclear retinoic

acid receptors (RARs). Nucl Recept Signal. 2009;7: e005. https://doi.org/10.1621/nrs.07005.
86. Minucci S, Maccarana M, Cioce M, De Luca P, Gelmetti V, Segalla S, et al. Oligomerization of RAR and AML1 transcription factors as a novel mechanism of oncogenic activation. Mol Cell. 2000;5(5):811–20. https://doi.org/10.1016/S1097-2765(00)80321-4.
87. Noguera NI, Catalano G, Banella C, Divona M, Faraoni I, Ottone T, et al. Acute promyelocytic leukemia: update on the mechanisms of leukemogenesis, resistance and on innovative treatment strategies. Cancers (Basel). 2019;11(10):1591. https://doi.org/10.3390/cancers11101591.
88. Stubbins RJ, Karsan A. Differentiation therapy for myeloid malignancies: beyond cytotoxicity. Blood Cancer J. 2021;11(12):193. https://doi.org/10.1038/s41408-021-00584-3.
89. Madan V, Koeffler HP. Differentiation therapy of myeloid leukemia: four decades of development. Haematologica. 2021;106(1):26–38. https://doi.org/10.3324/haematol.2020.262121.
90. Wang ZY, Chen Z. Acute promyelocytic leukemia: from highly fatal to highly curable. Blood. 2008;111(5):2505–15. https://doi.org/10.1182/blood-2007-07-102798.
91. U.S. Food and Drug Administration. Vesanoid [Tretinoin (ATRA)]. Approval overview. 1995. Available from: https://www.accessdata.fda.gov/scripts/cder/daf/index.cfm?event=overview.process&ApplNo=020438.
92. Yang Y, Valdés-Rives SA, Liu Q, Gao T, Burudpakdee C, Li Y, et al. Thyroid hormone suppresses medulloblastoma progression through promoting terminal differentiation of tumor cells. Cancer Cell. 2024;42(8):1434–49.e5. https://doi.org/10.1016/j.ccell.2024.07.008.
93. James MC, Peters G. Alternative product of the p16/CKDN2A locus connects the Rb and p53 tumor suppressors. Prog Cell Cycle Res. 2000;4:71–81. https://doi.org/10.1007/978-1-4615-4253-7_7.
94. Stott FJ, Bates S, James MC, McConnell BB, Starborg M, Brookes S, et al. The alternative product from the human CDKN2A locus, p14(ARF), participates in a regulatory feedback loop with p53 and MDM2. EMBO J. 1998;17(17):5001–14. https://doi.org/10.1093/emboj/17.17.5001.
95. Hassn Mesrati M, Syafruddin SE, Mohtar MA, Syahir A. CD44: a multifunctional mediator of cancer progression. Biomolecules. 2021;11(12):1850. https://doi.org/10.3390/biom11121850.
96. Jang BI, Li Y, Graham DY, Cen P. The role of CD44 in the pathogenesis, diagnosis, and therapy of gastric cancer. Gut Liver. 2011;5(4):397–405. https://doi.org/10.5009/gnl.2011.5.4.397.
97. Fisher AJ, Beal PA. Structural perspectives on adenosine to inosine RNA editing by ADARs. Mol Ther Nucleic Acids. 2024;35(3):102284. https://doi.org/10.1016/j.omtn.2024.102284.
98. Wang A-B, Liu D-P, Liang C-C. Regulation of human apolipoprotein B gene expression at multiple levels. Exp Cell Res. 2003;290(1):1–12. https://doi.org/10.1016/S0014-4827(03)00313-6.
99. Yang Y, Okada S, Sakurai M. Adenosine-to-inosine RNA editing in neurological development and disease. RNA Biol. 2021;18(7):999–1013. https://doi.org/10.1080/15476286.2020.1867797.
100. García-Mauriño SM, Rivero-Rodríguez F, Velázquez-Cruz A, Hernández-Vellisca M, Díaz-Quintana A, De la Rosa MA, et al. RNA binding protein regulation and cross-talk in the control of AU-rich mRNA fate. Front Mol Biosci. 2017;4:71. https://doi.org/10.3389/fmolb.2017.00071.
101. Dolicka D, Sobolewski C, Correia de Sousa M, Gjorgjieva M, Foti M. mRNA post-transcriptional regulation by AU-rich element-binding proteins in liver inflammation and cancer. Int J Mol Sci. 2020;21(18):6648.
102. Zhang P, Wu W, Chen Q, Chen M. Non-coding RNAs and their integrated networks. J Integr Bioinform. 2019;16(3):20190027. https://doi.org/10.1515/jib-2019-0027.
103. Chen L-L, Kim VN. Small and long non-coding RNAs: past, present, and future. Cell. 2024;187(23):6451–85. https://doi.org/10.1016/j.cell.2024.10.024.
104. George TP, Subramanian S, Supriya MH. A brief review of noncoding RNA. Egypt J Med Hum Genet. 2024;25(1):98. https://doi.org/10.1186/s43042-024-00553-y.
105. Jacobson EC, Pandya-Jones A, Plath K. A lifelong duty: how Xist maintains the inactive X chromosome. Curr Opin Genet Dev. 2022;75: 101927. https://doi.org/10.1016/j.gde.2022.101927.
106. Haase AD, Ketting RF, Lai EC, van Rij RP, Siomi M, Svoboda P, et al. PIWI-interacting RNAs: who, what, when, where, why, and how. EMBO J. 2024;43(22): 5335–9. https://doi.org/10.1038/s44318-024-00253-8.
107. He L, Hannon GJ. MicroRNAs: small RNAs with a big role in gene regulation. Nat Rev Genet. 2004;5(7):522–31. https://doi.org/10.1038/nrg1379.
108. Hammond SM. An overview of microRNAs. Adv Drug Deliv Rev. 2015;87:3–14. https://doi.org/10.1016/j.addr.2015.05.001.
109. Dana H, Chalbatani GM, Mahmoodzadeh H, Karimloo R, Rezaiean O, Moradzadeh A, et al. Molecular mechanisms and biological functions of siRNA. Int J Biomed Sci. 2017;13(2):48–57.
110. McLennan A, Bates A, Elcock L, Turner P. BIOS instant notes in molecular biology. 4th ed. New York: Garland Science; 2012.
111. Turner PC, McLennan AG, Bates AD, White MRH. BIOS instant notes in molecular Biology. Garland Science; 2007.
112. Xue J, Lou X, Ning D, Shao R, Chen G. Mechanism and treatment of α-amanitin poisoning. Arch Toxicol. 2023;97(1):121–31. https://doi.org/10.1007/s00204-022-03396-x.

113. Bensaude O. Inhibiting eukaryotic transcription: which compound to choose? How to evaluate its activity? Transcription. 2011;2(3):103–8. https://doi.org/10.4161/trns.2.3.16172.
114. U.S. Food and Drug Administration. Cosmegen [Dactinomycin (Actinomycin D)]. Approval overview. 1964. Available from: https://www.accessdata.fda.gov/scripts/cder/daf/index.cfm?event=overview.process&ApplNo=050682.
115. Tiburcio PDB, Chen K, Xu L, Chen KS. Actinomycin D and bortezomib disrupt protein homeostasis in Wilms tumor. bioRxiv. 2024. https://doi.org/10.1101/2024.06.11.598518.
116. Ferreira R, Schneekloth JS Jr, Panov KI, Hannan KM, Hannan RD. Targeting the RNA polymerase I transcription for cancer therapy comes of age. Cells. 2020;9(2):266. https://doi.org/10.3390/cells9020266.
117. U.S. Food and Drug Administration. Rifadin (Rifampin). Approval overview. 1989. Available from: https://www.accessdata.fda.gov/scripts/cder/daf/index.cfm?event=overview.process&ApplNo=050420.
118. Sudzinová P, Šanderová H, Koval T, Skálová T, Borah N, Hnilicová J, et al. What the Hel: recent advances in understanding rifampicin resistance in bacteria. FEMS Microbiol Rev. 2023;47(6):fuac051. https://doi.org/10.1093/femsre/fuac051.
119. U.S. Food and Drug Administration. Aemcolo (Rifamycin). Approval overview. 2018. Available from: https://www.accessdata.fda.gov/scripts/cder/daf/index.cfm?event=overview.process&ApplNo=210910.
120. Campbell EA, Korzheva N, Mustaev A, Murakami K, Nair S, Goldfarb A, et al. Structural mechanism for rifampicin inhibition of bacterial RNA polymerase. Cell. 2001;104(6):901–12. https://doi.org/10.1016/S0092-8674(01)00286-0.
121. Katzung BG, Vanderah TW, Trevor AJ. Katzung's basic and clinical pharmacology. 16th ed. McGraw-Hill Education; 2024.
122. U.S. Food and Drug Administration. Xifaxan (Rifaximin). Approval overview. 2004. Available from: https://www.accessdata.fda.gov/scripts/cder/daf/index.cfm?event=overview.process&ApplNo=021361.
123. U.S. Food and Drug Administration. Provenge (Sipuleucel-T). Approval overview. 2010. Available from: https://www.fda.gov/vaccines-blood-biologics/cellular-gene-therapy-products/provenge-sipuleucel-t
124. Rothstein DM. Rifamycins, alone and in combination. Cold Spring Harb Perspect Med. 2016;6(7):a027011. https://doi.org/10.1101/cshperspect.a027011.
125. Mir S, Mir M. The mRNA vaccine, a swift warhead against a moving infectious disease target. Expert Rev Vaccines. 2024;23(1):336–48. https://doi.org/10.1080/14760584.2024.2320327.
126. U.S. Food and Drug Administration. Comirnaty [BNT162b2 (Pfizer-BioNTech)]. Approval overview. 2021. https://www.fda.gov/news-events/press-announcements/fda-approves-first-covid-19-vaccine?utm
127. U.S. Food and Drug Administration. Spikevax [mRNA-1273 (Moderna)]. Approval overview. 2022. Available from: https://www.fda.gov/vaccines-blood-biologics/spikevax.
128. Gote V, Bolla PK, Kommineni N, Butreddy A, Nukala PK, Palakurthi SS, et al. A comprehensive review of mRNA vaccines. Int J Mol Sci. 2023;24(3):2700. https://doi.org/10.3390/ijms24032700.
129. Crooke ST, Baker BF, Crooke RM, Liang XH. Antisense technology: an overview and prospectus. Nat Rev Drug Discov. 2021;20(6):427–53. https://doi.org/10.1038/s41573-021-00162-z.
130. Biogen. Spinraza. SPINRAZA is delivered directly to the CSF at the site of motor neuron loss. 2025. Available from: https://www.spinraza.com/en_us/home/why-spinraza/how-spinraza-works.html.
131. Ottesen E. ISS-N1 makes the first FDA-approved drug for spinal muscular atrophy. Transl Neurosci. 2017;8:1–6. https://doi.org/10.1515/tnsci-2017-0001.
132. U.S. Food and Drug Administration. Spinraza (Nusinersen). Approval overview. 2016. Available from: https://www.accessdata.fda.gov/scripts/cder/daf/index.cfm?event=overview.process&ApplNo=209531.
133. U.S. Food and Drug Administration. Wainua (Eplontersen). Approval overview. 2023. Available from: https://www.accessdata.fda.gov/scripts/cder/daf/index.cfm?event=overview.process&ApplNo=217388.
134. Buxbaum JN, Reixach N. Transthyretin: the servant of many masters. Cell Mol Life Sci. 2009;66(19):3095–101. https://doi.org/10.1007/s00018-009-0109-0.
135. Luigetti M, Romano A, Di Paolantonio A, Bisogni G, Sabatelli M. Diagnosis and treatment of hereditary transthyretin amyloidosis (hATTR) polyneuropathy: current perspectives on improving patient care. Ther Clin Risk Manag. 2020;16:109–23. https://doi.org/10.2147/tcrm.S219979.
136. Ueda M. Transthyretin: its function and amyloid formation. Neurochem Int. 2022;155:105313. https://doi.org/10.1016/j.neuint.2022.105313.
137. Nie T. Eplontersen: first approval. Drugs. 2024;84(4):473–8. https://doi.org/10.1007/s40265-024-02008-5.
138. U.S. Food and Drug Administration. Tegsedi (Inotersen). Approval overview. 2018. Available from: https://www.accessdata.fda.gov/scripts/cder/daf/index.cfm?event=overview.process&ApplNo=211172.
139. Keam SJ. Inotersen: first global approval. Drugs. 2018;78(13):1371–6. https://doi.org/10.1007/s40265-018-0968-5.
140. Tijsterman M, Plasterk RHA. Dicers at RISC: the mechanism of RNAi. Cell. 2004;117(1):1–3. https://doi.org/10.1016/S0092-8674(04)00293-4.

141. U.S. Food and Drug Administration. Onpattro (Patisiran). Approval overview. 2018. Available from: https://www.accessdata.fda.gov/scripts/cder/daf/index.cfm?event=overview.process&ApplNo=210922.
142. Mansouri M, Mansouri K, Taheri Z, Hossaini Alhashemi S, Dehshahri A. The fomivirsen, patisiran, and givosiran odyssey: how the success stories may pave the way for future clinical translation of nucleic acid drugs. BioDrugs. 2025;39:359–71. https://doi.org/10.1007/s40259-025-00711-7.
143. U.S. Food and Drug Administration. Amvuttra (Vutrisiran). Approval overview. 2018. Available from: https://www.accessdata.fda.gov/scripts/cder/daf/index.cfm?event=overview.process&ApplNo=215515.
144. Alnylam Pharmaceuticals, inc. Amvuttra website. How Amvuttra works. Available from: https://www.amvuttrahcp.com/how-amvuttra-works.
145. Dickey AK, Leaf RK. Givosiran: a targeted treatment for acute intermittent porphyria. Hematology Am Soc Hematol Educ Program. 2024;2024(1):426–33. https://doi.org/10.1182/hematology.2024000663.
146. U.S. Food and Drug Administration. Givlaari (Givosiran). Approval overview. 2019. Available from: https://www.accessdata.fda.gov/scripts/cder/daf/index.cfm?event=overview.process&ApplNo=212194.
147. Merino E, Jensen RA, Yanofsky C. Evolution of bacterial trp operons and their regulation. Curr Opin Microbiol. 2008;11(2):78–86. https://doi.org/10.1016/j.mib.2008.02.005.
148. Istiqomah D, Joko T, Ogawa N. A brief overview of LacI-family transcriptional regulators in bacteria. Rev Agric Sci. 2023;11:310–25. https://doi.org/10.7831/ras.11.0_310.
149. Pan K, et al. Coupling and coordination in gene expression processes with pre-mRNA splicing. Int J Mol Sci. 2015;16(3):5682–96. https://doi.org/10.3390/ijms16035682.

Protein Synthesis, Folding, and Turnover 8

Abstract

Protein synthesis occurs through translation, a process that uses mRNA as a template. This chapter discusses the genetic code, degeneracy, and tRNA. The important role of ribosomes is discussed, including cytoplasmic ribosomes and those on the rough endoplasmic reticulum. The role of ribosomes as drug targets, especially bacterial ribosomes, is also discussed. Once proteins are synthesized, they become modified, and they must be properly folded. This chapter also delves into the unfolded protein response (UPR), a cellular response that occurs when an excessive number of proteins fold improperly. Some diseases are associated with improper protein folding or with the UPR. These disorders are also discussed, as well as current and potential therapeutic strategies. This chapter also explores the transport of proteins to their final destinations within the cell once they have been synthesized and folded properly. Over time, many proteins are degraded in the cell; therefore, this chapter also discusses protein turnover and degradation. Some drugs, such as PROTACs, also target proteins for degradation. Examples of such drugs are also discussed.

Keywords

Translation · Protein synthesis · Reading frame · Ribosomes · Protein turnover · Secretory pathway

Abbreviations

A site	Aminoacyl tRNA site
AAA+	ATPases associated with diverse cellular activities
AD	Alzheimer's disease
ALS	Amyotrophic lateral sclerosis
AMP	Adenosine monophosphate
APP	Amyloid precursor protein
ATF4	Activating transcription factor 4
ATF6	Activating transcription factor 6
BiP	Binding immunoglobulin protein
CH_3	Methyl group
CH_3CO	Acetyl group
CHOP	C/EBP homologous protein
CRBN	Cereblon (an E3 ubiquitin ligase substrate receptor)
E site	Exit site
E1	Ubiquitin-activating enzyme
E2	Ubiquitin-conjugating enzyme
E3	Ubiquitin ligase
eIF2α	Eukaryotic initiation factor 2 alpha
ER	Endoplasmic reticulum
ERAD	ER-associated degradation
GADD153	Growth arrest and DNA damage-inducible protein 153

A. Minden, *Biochemistry, Cell and Molecular Biology for Pharmacists and Health Professionals*,
https://doi.org/10.1007/978-3-032-20529-2_8

GRP78	Glucose-regulated protein 78
HIF-1α	Hypoxia-inducible factor 1-alpha
HR	Half-life reduction (sometimes used in degradation kinetics)
HSP	Heat shock protein
HSP60	Heat shock protein 60
HSP70	Heat shock protein 70
HSPA5	Heat shock protein family A member 5 (also known as BiP or GRP78)
IRE1	Inositol-requiring enzyme 1
MAPK	Mitogen-activated protein kinase
MEK	MAPK/ERK kinase
MM	Multiple myeloma (implied from Velcade/Kyprolis clinical context)
ORF	Open reading frame
P site	Peptidyl site
PD	Parkinson's disease
PERK	PKR-like ER kinase
PKR	Protein kinase R
PO_4^{3-}	Phosphate group
PROTAC	Proteolysis targeting chimera
rRNA	Ribosomal ribonucleic acid
S	Svedberg unit (used for ribosome subunit sizes, e.g., 40S, 60S, 80S)
tRNA	Transfer ribonucleic acid
Ub	Ubiquitin
UBE3A	Ubiquitin-protein ligase E3A (gene implicated in Angelman syndrome)
UPR	Unfolded protein response
UPS	Ubiquitin–proteasome system (commonly used term, may be implied)
XBP1	X-box binding protein 1

What You Will Learn

- How proteins are synthesized in the process of translation.
- The genetic code, codons, and degeneracy of the genetic code.
- Transfer RNA (tRNA).
- Ribosomes, their basic structure, and the A, P, and E active sites.
- Cytosolic ribosomes and ribosomes on the rough ER.
- The basic steps of translation: initiation, elongation, and termination.
- Ribosomes as drug targets: many antibiotics target bacterial ribosomes.
- Protein folding and post-translational modifications.
- Protein transport and the secretory pathway.
- The unfolded protein response.
- Misfolded proteins and their roles in disease, such as diabetes, Alzheimer's disease, and other neurological disorders, and potential strategies for treating these disorders.
- Protein turnover and the process of ubiquitination and the proteasome.
- Protein degradation in human health and disorders such as Angelman syndrome, autism spectrum disorder (ASD), and cancer.
- Drugs that either target or utilize the protein degradation process, such as proteasome inhibitors, PROTACs, and molecular glues.

This chapter brings together the topics of DNA, RNA, and protein to discuss translation, a key step in the central dogma of molecular biology, focusing primarily on proteins. This chapter is divided into three sections. Section 1 discusses translation, introduces protein folding, and explores the transport of proteins to their final destinations within the cell. Section 2 delves further into protein folding and discusses the unfolded protein response, a cellular response that occurs when an excessive number of proteins fold improperly. Finally, Section 3 discusses protein turnover and degradation.

1 Proteins—Translation, Assembly, Transport

1.1 Translation

Translation is the process by which proteins are synthesized, using mRNA as the template. Some of the basic components essential for translation, in addition to amino acids, include the mature mRNA, which is transcribed using DNA as a template, tRNA, and ribosomes. They are described below, followed by a more detailed description of the translation process [1, 2].

1.1.1 The DNA Template, Transcription, mRNA, and Codons

- The DNA Template

As discussed in previous chapters, the DNA double helix consists of two strands: the coding strand, which contains the genetic code, and the template strand, which is complementary to it. The mature mRNA will have the same sequence as the coding strand (except that it will have Us instead of Ts) [2] (see Chap. 6, Fig. 6.2).

- Codons and the Genetic Code

The DNA coding strand sequence is organized in triplets of nucleotides, and these triplets correspond to the codons in the mRNA. Each set of three nucleotides in the mRNA represents a codon. Each codon codes for an amino acid unless it is a stop codon, which signals the end of the protein. Given that there are four different bases, there are 64 possible combinations of codons (triplets of bases), but there are only 20 amino acids. Consequently, more than one codon will code for a particular amino acid. The list of codons, their abbreviations, and the amino acids they encode is shown in Fig. 8.1. Codons are typically identified by their mRNA sequence, which is essentially identical to the coding strand DNA sequence, except that U is substituted for T. The genetic code refers to the sequences of codons that make up the proteins in the cell, which the DNA encodes.

During translation, the reading frame is important (see Fig. 8.2). The reading frame depends on what is considered the first codon in the encoded protein, which is the start codon. The start codon in a protein-coding mRNA is AUG (and the corresponding DNA sequence on the coding strand is ATG), and it codes for methionine. The next triplet, after this, codes for the next amino acid, and so forth. In theory, each strand of RNA has three possible reading frames. However, the reading frame in which AUG is the first codon in the protein is the correct reading frame. At the end of the protein-coding region is an in-frame stop codon, which does not code for any amino acid. The stop codons are UAA, UAG, and UGA. The entire region on the DNA between the start codon (ATG) and the stop codon is called the open reading frame (ORF), and this is the part of the DNA that codes for a protein. Once this region is transcribed, the corresponding sequence on the

1 \ 2	U	C	A	G
U	UUU Phe	UCU Ser	UAU Tyr	UGU Cys
	UUC Phe	UCC Ser	UAC Tyr	UGC Cys
	UUA Leu	UCA Ser	UAA Stop	UGA Stop
	UUG Leu	UCG Ser	UAG Stop	UGG Trp
C	CUU Leu	CCU Pro	CAU His	CGU Arg
	CUC Leu	CCC Pro	CAC His	CGC Arg
	CUA Leu	CCA Pro	CAA Gln	CGA Arg
	CUG Leu	CCG Pro	CAG Gln	CGG Arg
A	AUU Ile	ACU Thr	AAU Asn	AGU Ser
	AUC Ile	ACC Thr	AAC Asn	AGC Ser
	AUA Ile	ACA Thr	AAA Lys	AGA Arg
	AUG **Met**	ACG Thr	AAG Lys	AGG Arg
G	GUU Val	GCU Ala	GAU Asp	GGU Gly
	GUC Val	GCC Ala	GAC Asp	GGC Gly
	GUA Val	GCA Ala	GAA Glu	GGA Gly
	GUG Val	GCG Ala	GAG Glu	GGG Gly

Alanine	Ala	A
Arginine	Arg	R
Asparagine	Asn	N
Aspartate	Asp	D
Cysteine	Cys	C
Glutamine	Gln	Q
Glutamate	Glu	E
Glycine	Gly	G
Histidine	His	H
Isoleucine	Ile	I
Leucine	Leu	L
Lysine	Lys	K
Methionine	Met	M
Phenylalanine	Phe	F
Proline	Pro	P
Serine	Ser	S
Threonine	Thr	T
Tryptophan	Trp	W
Tyrosine	Tyr	Y
Valine	Val	V

Fig. 8.1 Amino acids and corresponding codons and abbreviations

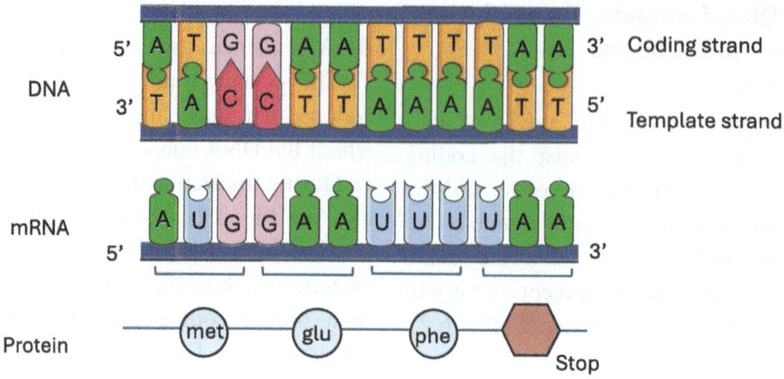

Fig. 8.2 The reading frame. The reading frame relates to the three nucleotide codons. The correct reading frame in a protein coding sequence starts with the start codon (AUG) and continues to the stop codon. (*Source*: Adapted from Servier Medical Art (https://smart.servier.com/), licensed under the Creative Commons Attribution 4.0 (CC BY 4.0) license)

mRNA is also called the ORF. The ORF on the mRNA is smaller than the full mRNA transcript because the mRNA includes noncoding sequences both upstream to the start codon and downstream to the stop codon [1, 3]. (At this point, introns have already been spliced out during mRNA processing.)

- Degeneracy of the Genetic Code

Degeneracy refers to the fact that codons have redundancy. This means that more than one codon can code for a single amino acid. For example, GCA, GCC, GCG, and GCU all code for alanine, and CGA, CGC, CGG, and CGU all encode arginine. In each case, it is the third nucleotide in the codon that is flexible, and this flexibility accounts for the degeneracy of the code. This is explained by the wobble hypothesis, described in more detail when translation and tRNA are discussed [4–6] (see Fig. 8.1).

1.1.2 Transfer RNA (tRNA)

After mRNA is synthesized (through transcription) and processed in the nucleus, it is transported to the cytoplasm, where it is directed to the ribosome. At the ribosome, another type of RNA plays a key role. This is the transfer RNA (tRNA). tRNA is a strand of RNA, 73–93 nucleotides long. It is clover-shaped and forms its unique shape because about half of its ribonucleotide bases pair with each other. One part of the tRNA is called the anticodon loop, which contains a 3-nucleotide antisense codon that is complementary to a specific codon. The terminus of the tRNA contains the sequence CCA. The hydroxyl (OH) group at the end of the CCA is attached to an activated amino acid, which corresponds to the anticodon (Fig. 8.3). There is a tRNA for every amino acid in the cell.

Each tRNA must be charged (activated) before it is ready to be used in translation. A charged tRNA is a tRNA that contains its attached amino acid. There are two steps that take place before the activated tRNA is formed.

First, before an amino acid can be added to a tRNA, it gets adenylated. This means that ATP reacts with the amino acid, releases two phosphate groups, and forms an aminoacyl-AMP (an amino acid attached to AMP, which contains one phosphate group). **Aminoacyl-tRNA synthase** is the enzyme that catalyzes the attachment of the AMP to the carboxyl group on the amino acid. This amino acid is now referred to as an aminoacyl-AMP.

$$\text{AA} + \text{ATP} \rightarrow \text{aminoacyl-AMP} + \text{PPi}$$

Second, this amino acid, which is now considered an activated amino acid, gets added to the tRNA at the 3′ hydroxyl group of the terminal adenosine on the tRNA's CCA tail, releasing the AMP. This is also catalyzed by the

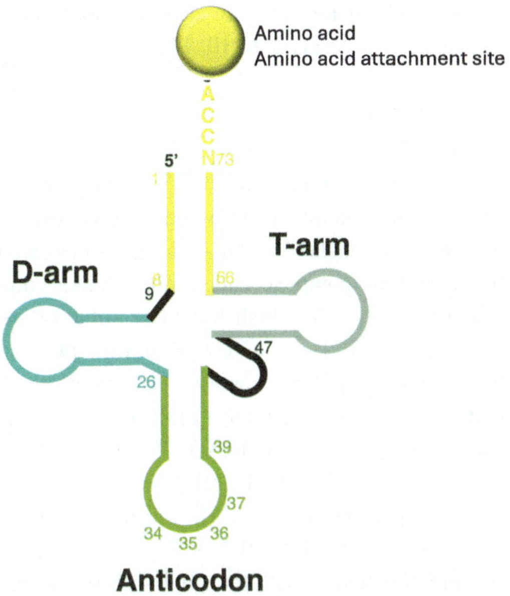

Fig. 8.3 tRNA. tRNA is the category of RNA that carries amino acids to their corresponding codons (complementary to the anticodon on the tRNA) on the mRNA during translation. (*Source*: Modified from Tamura [104]. Licensed under the Creative Commons Attribution 4.0 (CC BY 4.0) license)

aminoacyl-tRNA synthase enzyme. Most cells have a different aminoacyl-tRNA synthetase for each amino acid (Fig. 8.4). The tRNA that is bound to its activated amino acid can now be referred to as an aminoacyl-tRNA. It can also be named for the specific amino acid it is bound to, such as tRNA-Met or tRNA-Leu.

$$\text{aminoacyl-AMP} + \text{tRNA} \\ \rightarrow \text{aminoacyl-tRNA} + \text{AMP}$$

The amino acid attached to the tRNA is now also considered to be activated or charged because it is linked to the tRNA by a high-energy ester bond. The potential energy from this bond will later be used to form a peptide bond when translation occurs [1].

1.1.3 Ribosomes

Translation takes place at the ribosome. Ribosomes are made of protein and another type of RNA, **ribosomal RNA (rRNA)**. Ribosomes

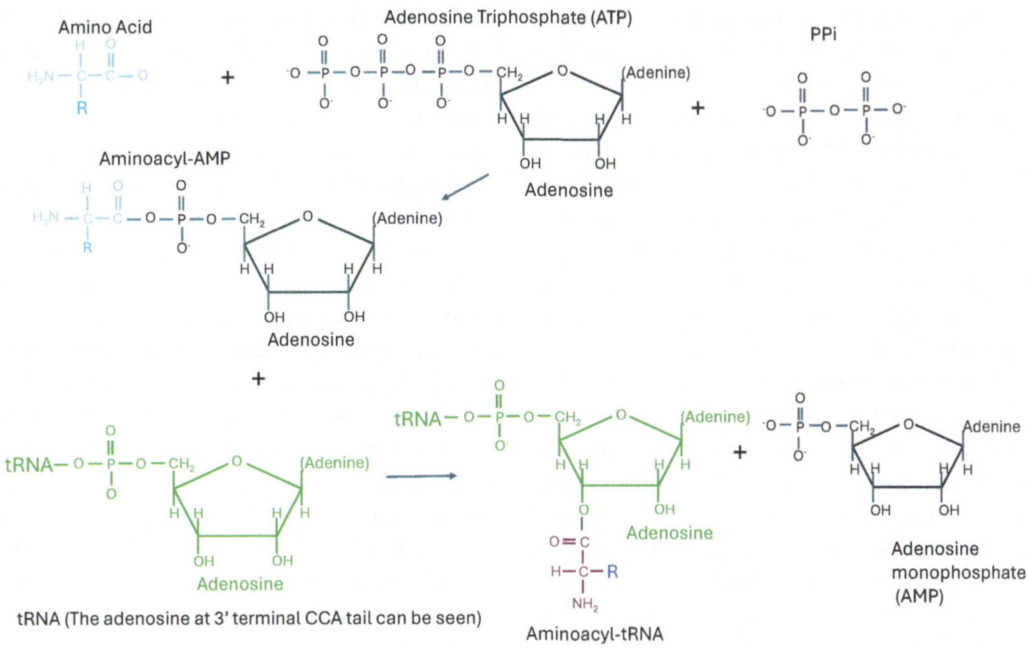

Fig. 8.4 Charging the tRNA. The tRNA gets charged by addition of its appropriate amino acid before it is used in translation. First the amino acid (purple) gets adenylated. In this process it gets added to Adenosine from ATP (black), with the release of pyrophosphate (two phosphate groups). Second, this activated amino acid (purple and black) gets added to the tRNA (green) specifically at the adenosine at the 3′ CCA tail of the tRNA. In the process, the original adenosine from the activated amino acid is released as AMP (black). Now the charged amino-acyl tRNA (green and black) is complete

have two subunits: a small (40S) and a large (60S) subunit, and the size of the full ribosome is 80S. S stands for Svedberg unit, also known as the sedimentation coefficient, and it measures the sedimentation rate during ultracentrifugation [7]. There are two major categories of ribosomes in the cell. They are structurally the same, but they differ in their locations. The first group is found free in the cytoplasm, and the second group is bound to the surface of the rough ER. They can be referred to as cytoplasmic ribosomes and ER-bound ribosomes, respectively. The difference lies in their involvement in the synthesis of proteins with different final destinations [1, 2, 8]. The different destinations will be discussed in more depth below.

1.1.4 The Steps of Translation

- Overview of Translation

During the translation process, each amino acid gets delivered to the mRNA via the appropriate aminoacyl-tRNA, which binds to its complementary codon on the mRNA. In this process, the anticodon of the tRNA base pairs with its complementary codon on the mRNA. Thus, the tRNAs deliver the correct amino acids to their positions along the mRNA. Each amino acid is delivered to the mRNA one by one, starting with the very first one. After an amino acid is in place, a peptide bond forms between this amino acid and the previous one. The tRNA then detaches from the mRNA and the amino acids, and this process continues until a stop codon is encountered, at which point the full-length polypeptide chain is formed [1, 6] (Fig. 8.5).

This process depends on the tRNA delivering the correct amino acid to its correct codon. The wobble hypothesis helps explain why the same tRNA molecule can recognize multiple codons (see above). According to this model, the first two bases of the codon form standard base pairs with the anticodon on the tRNA, but the third base can "wobble," forming nonstandard base pairing [2, 5, 9].

Translation takes place in three stages: **initiation**, **elongation**, and **termination**. These steps are described below, and are also discussed in the following sources[1, 2, 6, 10, 11].

- The Initiation Stage

In the initiation stage, the small ribosomal subunit binds to an initiator $tRNA_i^{Met}$ (a specialized tRNA that contains methionine and is always used to start translation). This binding is mediated by a protein complex called eIF2 (eukaryotic initiation factor 2), which has three subunits, α, β, and γ. eIF2 is a GTP-binding protein. (GTP-binding proteins are a class of proteins that can bind GTP or GDP.) eIF2 must be in its active form to bind to $tRNA_i^{Met}$. In its active form, it is bound to GTP; eIF2-GTP. The eIF2-GTP complex binds $tRNA_i^{Met}$, forming a ternary complex: e1F2-GTP+ $tRNA_i^{Met}$. This ternary complex associates with the small 40S ribosomal subunit, along with other initiation factors, to form a complex known as the 43S pre-initiation complex. This complex then attaches to the 5′ CAP structure at the 5′ end of an mRNA molecule. The 43S complex then scans this mRNA in the 5′ to 3′ direction until it reaches a start codon, **AUG**. Once this start codon is recognized, eIF2 hydrolyzes its bound GTP to GDP, stimulated by eIF5, and it is released. (eIF5, eukaryotic initiation factor 5, is a GAP; GTPase activating protein that catalyzes the hydrolysis of GTP to GDP.)

Now, the $tRNA_i^{Met}$ base pairs with the AUG codon via complementary base pairing between the anticodon on the tRNA and the codon on the mRNA. Several initiation factors are involved in this process. Importantly, the large subunit of the ribosome joins the complex at this point, while the mRNA remains in a region between the small and large ribosomal subunits. The now fully assembled ribosome has three active sites: **the E, P, and A sites** (see Fig. 8.5). **A** is the aminoacyl-tRNA site, **P** is the peptidyl-tRNA site, and **E** is the exit site. During initiation, the initiator tRNA-MET is uniquely positioned directly in the **P site** rather than the A site [1, 2, 4, 11, 12].

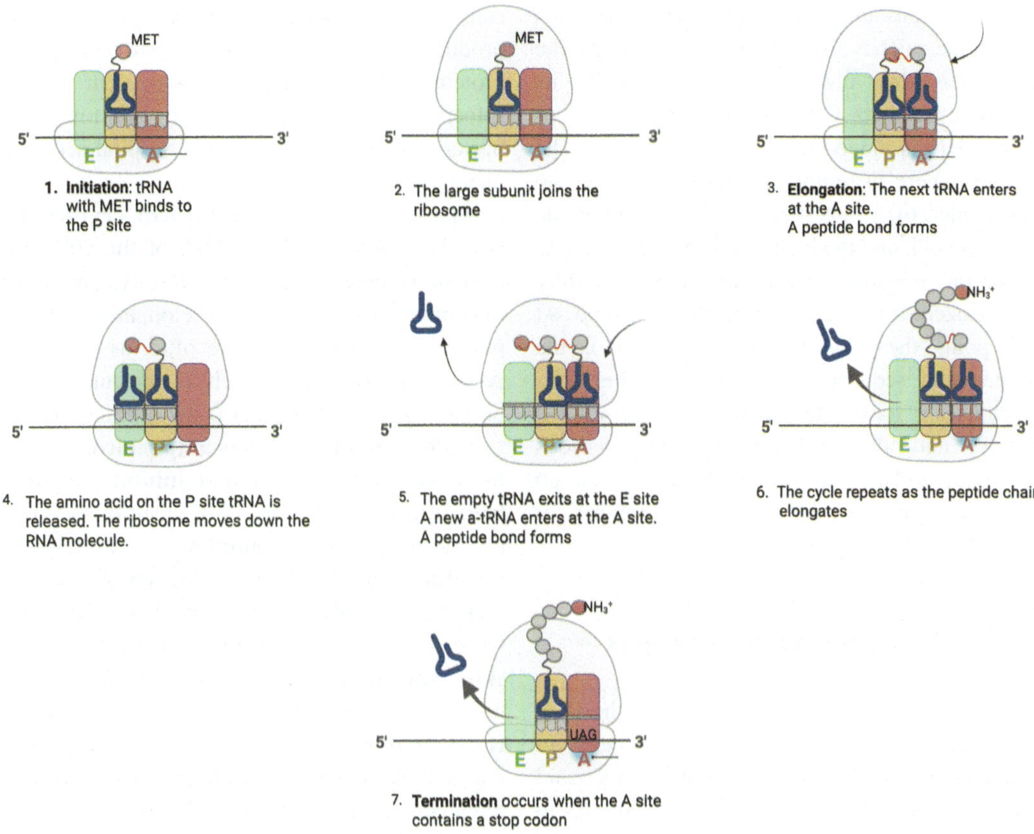

Fig. 8.5 The key steps in translation. The mRNA, ribosomes, and tRNA are shown. Initiation, elongation, and termination steps are indicated. See text for more details. (*Source*: Created with BioRender.com)

- The Elongation Stage

The addition of the second amino acid marks the beginning of the elongation stage. The second aminoacyl-tRNA enters the ribosome at the **A site**, and its anticodon base pairs with its complementary codon on the mRNA. Only the correct aminoacyl-tRNA can bind here because incorrect aminoacyl-tRNAs cannot base pair with the codon. A peptide bond now forms between the first amino acid (Met) and the second amino acid. This occurs at the peptidyl transferase center located in the large ribosomal subunit. The ribosome then slides along the mRNA in the 5′ to 3′ direction, causing the tRNA carrying the growing peptide chain to move from the **A site to the P site**, while the empty tRNA in the P site moves to the **E site**, where it exits the ribosome. The growing polypeptide chain emerges from the exit tunnel in the large ribosomal subunit. These steps—binding of aminoacyl-tRNA, peptide bond formation, ribosome translocation—are repeated in cycles until all the amino acids in the polypeptide chain have been [1, 2, 13].

- The Termination Stage

Finally, the termination stage occurs when the ribosome encounters a stop codon. When a stop codon enters the A site, a release factor also enters the A site. This leads to the separation of the large and small ribosomal subunits, the dissociation of mRNA from the ribosome, and the end of [1, 6].

- Summary of the Basic Translation Steps

The steps above can be briefly summarized in Fig. 8.5. During the entire translation process, each new amino acid is added to the C terminus of the previous amino acid as the ribosome scans across the mRNA from its 5′ to its 3′ end. The protein is thus always synthesized from its N-terminus to its C-terminus. Inside the ribosome, tRNAs move through three key sites. The A site, where incoming tRNAs carrying amino acids enter, the P site, where the growing peptide chain is held, and the E site, where empty tRNAs leave the ribosome. Once the protein is fully synthesized, it is released from the ribosome. At this point, the primary structure, which is the amino acid sequence, is complete. After this, proteins are folded and modified, as discussed below, and they proceed to their various destinations. They may stay in the cytoplasm, go to other organelles such as the mitochondria, or enter the secretory pathway.

1.2 Ribosomes as Drug Targets

Ribosomes are essential for protein synthesis in both eukaryotic cells and prokaryotic cells. In prokaryotes, ribosomes are smaller (70S) and are made up of 30S and 50S subunits [4]. As in eukaryotes, bacterial ribosomes are essential for protein synthesis. For this bacterial reason, ribosomes are important targets for some antibiotics. By inhibiting ribosomes, antibiotics can block protein synthesis, thereby killing bacteria. Since inhibition of bacterial ribosomes will not affect eukaryotic ribosomes, antibiotics only kill the bacteria, rather than the eukaryotic cells.

Many **antibiotics target either of the two subunits of bacterial ribosomes** [14]. **Doxycycline** [15] (which is a tetracycline), for example, binds to 16S RNA, which is part of the 30S subunit of bacterial ribosomes, and thereby prevents the attachment of aminoacyl-tRNA to the A site and blocks protein synthesis [16]. Doxycycline is commonly used to treat several types of infections, including respiratory tract infections, eye infections, and some sexually transmitted infections.

Several other **antibiotics** also **target ribosomes**. **Aminoglycosides** [such as **Gentak (gentamicin)** [17] **and Streptomycin**] also bind 16S rRNA of the 30S subunit. These antibiotics cause the misreading of mRNA and lead to the production of abnormal and dysfunctional proteins. They are often used to treat serious **infections** such as those caused by gram-negative bacteria, including sepsis and hospital-acquired pneumonia.

Macrolides [such as **Erythrocin (erythromycin)**] [18] **bind the 23S rRNA of the 50S subunit of ribosomes**. It blocks the exit tunnel of ribosomes, thus preventing elongation of the polypeptide chain. They are often used to treat **respiratory infections** such as pneumonia and bronchitis, as well as other types of infections.

Chloromycetin (chloramphenicol) also **binds the 50S subunit and inhibits peptidyl transferase activity**. It is infrequently used due to side effects. **Lincosamides** such as **Cleocin (clindamycin)** [19]. **Oxazolidinones**, such as **Zyvox (linezolid)** [20], also bind to the 50S ribosome subunit. Clindamycin is often used to treat **skin infections** caused by *Staphylococcus aureus*, while linezolid is used for certain **drug-resistant infections**, such as MRSA and VRE. All of these antibiotics are used to treat **various bacterial** infections by targeting bacterial ribosomes [16].

1.3 Protein Folding and Modification

1.3.1 Protein Folding

- Introduction to Protein Folding

After translation occurs and polypeptide chains are synthesized and released from the ribosomes, the polypeptides need to be folded into their correct structures. Protein folding begins as the protein is being synthesized, co-translationally, even as it is still emerging from the ribosome. The complete folding process, however, takes place after protein synthesis has been completed, post-translationally. Sometimes errors can occur, resulting in the presence of misfolded proteins in the cell, which can lead to cell damage. Protein folding can occur in the ER and in the cytoplasm, and in some cases, the mitochondria or nucleus, depending on the final destination of the protein [1].

- Protein Folding Happens Largely in the Cytosol or the ER

Protein folding occurs in different parts of the cell depending on the protein's destination. Proteins that are made on ER-bound ribosomes and destined for the secretory pathway are folded within the lumen of the endoplasmic reticulum (ER). In contrast, many proteins that are made on cytoplasmic ribosomes are folded in the cytosol, but some are only completely folded when they reach their final destination such as the mitochondria or the peroxisomes. Folding of proteins synthesized on ER-bound ribosomes and those synthesized on cytoplasmic ribosomes requires distinct but overlapping sets of chaperone proteins, which recognize misfolded or unfolded proteins and assist in proper folding. (The secretory pathway and the destination of proteins synthesized on cytoplasmic ribosomes are discussed later in this chapter.)

- Chaperones Have Key Roles in Protein Folding

Molecular chaperones are a highly abundant family of proteins that play a key role in helping proteins fold properly and correcting improperly folded proteins. Chaperones bind to unfolded or improperly folded proteins (the client proteins), either as they emerge from the ribosome or after they have completed the folding process. They usually recognize and bind to exposed hydrophobic regions that are to become buried in properly folded proteins. These client proteins can exist in various non-native forms. They may be completely unfolded, misfolded, or in intermediate states such as the molten globule form. The molten globule form is an intermediate stage with some structures, such as alpha helices or β-sheets, in place, but the overall tertiary form remains dynamic and not fully formed [21]. Chaperones help to guide these types of proteins to their correctly folded state. Chaperones hydrolyze ATP and use the resulting energy to fold proteins or to refold misfolded proteins properly. Chaperones undergo cycles of binding to proteins and releasing them, a process that also requires energy from ATP [22].

Protein folding in the cytosol and in the ER requires separate but related chaperone systems. In the cytosol, chaperones such as **HSP70** and **HSP60** are essential for co- and post-translational folding. HSP70, along with **co-chaperones like HSP40**, binds to polypeptides as they emerge from the ribosome. They prevent aggregation and help with folding. HSP60 functions differently. It is a chaperone in the chaperonin family. It acts on proteins that are already completely synthesized but that were improperly folded. It encloses the misfolded protein in its cavity, providing a protected environment for correct refolding. After refolding, the protein is released from the cavity. Within this cavity, the isolated environment helps the protein transition from the molten globule state to its correctly folded form. Some proteins synthesized at cytoplasmic ribosomes are folded in other organelles, specifically the mitochondria or peroxisomes. Folding in these organelles also involves HSP70, as well as other organelle-specific chaperones (such as mitochondrial HSP60 and mtHSP70) [1, 23–25].

In the ER, related chaperones perform similar functions for proteins entering the secretory pathway. The main chaperone is **HSPA5** (also known as BiP or GRP78). It binds to hydrophobic regions of newly translocated polypeptides, plays a role in their proper folding, and helps prevent protein aggregation. Like cytosolic HSP70, BiP requires ATP binding and hydrolysis. Additional ER chaperones, such as calnexin and calreticulin, also contribute to the folding of some proteins, especially glycoproteins [26, 27].

1.3.2 Formation of Quaternary Structures

Some proteins form quaternary structures after they are folded, where multiple proteins bind together. These can be permanent or transient structures. The result is a protein complex with multiple subunits. The subunits can be identical, or the complex can be made of different proteins. The subunits are held together by various noncovalent bonds, including hydrogen bonds, hydrophobic interactions, van der Waals forces, and ionic bonds. Proteins bind to each other through various regions that recognize each other and are complementary in terms of shape and electrostatic interactions. For example, hydrophobic regions tend to cluster together, charged amino acids interact with oppositely charged amino acids, and disulfide bonds connect proteins that contain cysteines. Chaperones may also play a role in the assembly of quaternary structures [1, 28].

1.3.3 Post-translational Modifications of Proteins

Even after a protein has been made and folded, modifications are often added. These can be permanent or transient and can change the functions of proteins. These modifications can occur in various places, such as the ER, the cytoplasm, or the nucleus. Some of the many types of modifications that can occur as follows, and also described in the following resources [1, 6, 29].

Phosphorylation involves adding a phosphate group (PO_4^{3-}) to serine, threonine, or tyrosine residues on a protein. Phosphate groups are added to proteins by **protein kinases**, but phosphorylation is reversible, and phosphate groups can be removed by another family of enzymes, the **phosphatases**. The activity of many proteins can be regulated by whether they are phosphorylated or not. For example, proteins within the MAP kinase pathway, such as MEK or MAPK, are activated by phosphorylation and inactivated by dephosphorylation (see Chap. 18).

Glycosylation involves the addition of carbohydrate chains to certain amino acids, usually serine, threonine, or asparagine. Glycosylation affects protein folding, stability, cellular location, and recognition by other molecules.

Acetylation is the addition of an acetyl group (CH_3CO) to the amino terminus of a protein or to lysine residues within the protein. Acetylation is reversible, and addition or removal of acetyl groups is catalyzed by **acetyltransferase** and **deacetylase** enzymes, respectively. Acetylation has various functions, such as the regulation of protein stability, DNA binding, and protein–protein interactions. As discussed in the previous chapter, the acetylation of histones, for example, has a significant impact on the density of nucleosomes, which in turn affects gene expression.

Methylation is the addition of methyl groups (CH_3) to amino acids like lysine or arginine on proteins. Histones, in addition to acetylation, can also be methylated, which also impacts DNA accessibility. Another example of a protein that can be methylated is p53, and methylation can affect its stability. Methylation of proteins can also affect protein–protein interactions, chromatin structure, gene expression, and signaling.

Ubiquitination involves the attachment of the small protein ubiquitin to lysine residues on proteins. Ubiquitination can lead to protein degradation via the ubiquitin–proteasome system, but it also has several other roles.

Proline hydroxylation involves the addition of a hydroxyl group (–OH) to certain proline residues. It has many different functions. For example, hydroxylation stabilizes collagen, a crucial connective tissue protein. Hydroxylation also has important roles in regulating some signaling pathways, such as the HIF-1α pathway, discussed in Chap. 21.

1.4 Protein Transport and Modification

Once they are synthesized, proteins are transported to their final destinations. Where they are transported to depends in part on whether they were translated on cytoplasmic ribosomes or ER-bound ribosomes [1].

1.4.1 Proteins Made on Cytoplasmic Ribosomes

Proteins translated on free cytosolic ribosomes can remain in the cytosol or be targeted to certain organelles such as the nucleus, mitochondria, or peroxisomes. Depending on where the proteins are destined, folding is coordinated with transport. For example, proteins destined for the mitochondria or peroxisomes are often kept partially or completely unfolded by chaperones, as this allows them to pass through the necessary

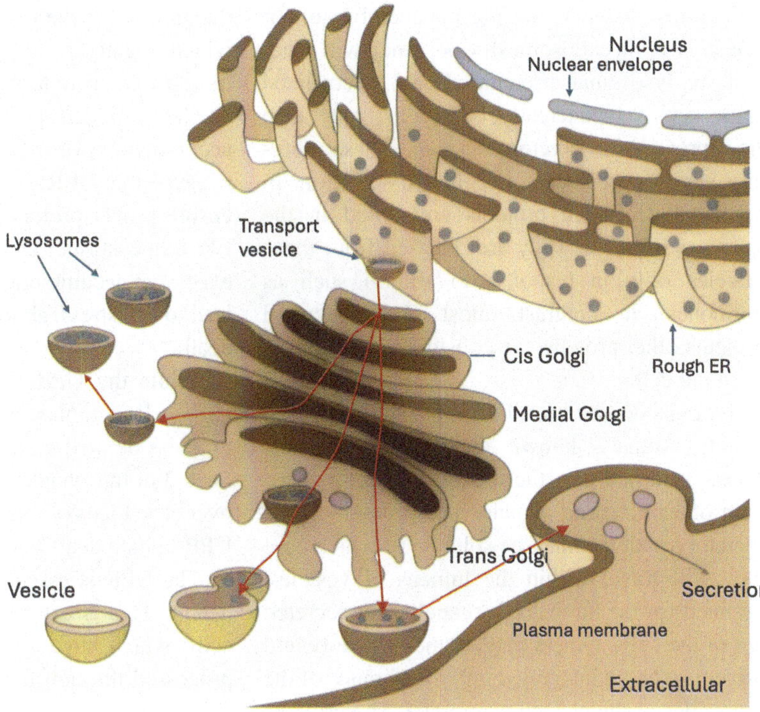

Fig. 8.6 The secretory pathway. Proteins generated at the rough ER (top right of the figure) enter the secretory pathway where they are destined to organelles within the pathway or to the plasma membrane or secreted. Proteins in the secretory pathway generally pass through the Golgi apparatus (middle of the figure) before reaching their final destinations. (*Source*: Illustration from Freepik.com (premium license))

channel. Proteins destined for the nucleus can enter the nucleus fully folded.

1.4.2 Proteins Synthesized on ER-Bound Ribosomes: The Secretory Pathway

Proteins that are synthesized on ribosomes bound to the rough ER enter **the secretory pathway** (Fig. 8.6). These proteins are folded in the ER and may also undergo some initial modifications. If they are not destined to stay in the ER, they are packaged into transport vesicles. Soluble proteins are enclosed within the vesicles, and membrane proteins are inserted into the vesicle membranes. These vesicles are transported to the Golgi apparatus. This process involves vesicle budding from the ER to the Golgi, as discussed below and in [1, 30].

The proteins enter the Golgi apparatus, which consists of a series of flattened, membrane-bound sacs called cisternae. It has a cis Golgi network (closest to the ER), a cis face, a medial region, and a trans face (closest to the plasma membrane). It also has multiple interconnecting vesicles. At the cis-Golgi network, closest to the ER, the Golgi receives newly synthesized proteins from the ER. (The Golgi also receives lipids from the smooth ER.) The protein then moves along the different stacks of the Golgi. This movement is thought to occur through a process called cisternal progression or cisternal maturation, in which the cisternae mature and move from the cis face to the trans face, transporting proteins within them. In the meantime, Golgi enzymes are transported backward through vesicles to maintain the identity of each cisterna.

The different sacks contain different enzymes that add modifications to proteins. These can include, for example, sugar chains (glycosylation), phosphate groups, and other modifications. Some of these modifications serve as a sort of "tag" or "address," indicating where the proteins are destined to go. From the medial part of the Golgi, the proteins are then sent to the trans face. In the trans face, the Golgi recognizes the molecular tags. The proteins are then sorted, packaged into vesicles, and sent to their destinations.

The main destinations for proteins in the secretory pathway are retention in the ER or the Golgi, the plasma membrane (either secreted from the cell or inserted into the plasma membrane), or lysosomes (via endosomes), where they will function as lysosomal proteins. Some specialized cells, such as pancreatic beta cells, may also store proteins in secretory vesicles. These secretory vesicles are used for regulated secretion. In regulated secretion, proteins are stored in the secretory vesicles. They are released from these vesicles only in response to stimuli such as hormones. In contrast, most other secreted proteins other proteins are constitutively secreted from the cell.

Proteins destined for the ER or the Golgi have specific signals, known as retention signals. Those destined for the lysosome are marked with other targeting signals. The proteins in the secretory pathway may be soluble, in which case they will travel within the lumens of vesicles, destined for the lumens of organelles, or secreted out of the cells. Others may be membrane-bound proteins that are integrated into membranes of the organelles or the plasma membrane. The proteins synthesized on ER-bound ribosomes are generally folded in the ER, even if they are destined for other locations in the cell [30].

2 The Unfolded Protein Response (UPR) and ERAD

2.1 Unfolded Protein Response (UPR) and ERAD

Protein folding is a critical part of the protein synthesis process. Most proteins can only function if they are folded correctly. Proper folding is vital, but sometimes proteins do not fold correctly. Therefore, cells sometimes need to respond to the accumulation of improperly folded proteins. Proteins synthesized on cytoplasmic ribosomes and those synthesized on ER-bound ribosomes are folded in different locations in the cell. Here, we focus on proteins synthesized on ER-bound ribosomes. Such proteins are folded in the ER and will then either remain there or be transported to organelles along the secretory pathway after they are folded. Accumulation of too many unfolded or misfolded proteins in the ER is dangerous for the cell. It can lead to ER stress, which negatively impacts cell function. The cell has a mechanism to address this problem, known as the unfolded protein response (UPR). The accumulation of unfolded proteins in the ER triggers the **UPR**. The goal of the UPR is to ensure proper protein folding and thus to restore ER homeostasis. This is important because an excessive accumulation of unfolded proteins can lead to cell stress, abnormal cell function, and cell death.

While the UPR addresses unfolded proteins in the ER, cytoplasmic protein folding is handled in a less defined system and will be discussed in Sect. 3 of this chapter, which covers protein turnover. The focus of this section of the chapter is the UPR, which deals with proteins folded in the ER.

The UPR is essential for maintaining cellular health. This is especially important in secretory cells, which have a high level of proteins being processed through the ER. During ER stress the UPR helps restore normal cell function by (a) increasing the production of molecular chaperone proteins to aid in proper protein folding, (b) slowing down overall protein translation to limit the influx of new proteins into the ER until proper folding of unfolded proteins can be restored, and (c) degradation of proteins whose misfolding could not be repaired, through ER-associated degradation (ERAD). If homeostasis cannot be restored, prolonged UPR activation can ultimately trigger apoptosis so that cells with abnormal proteins do not survive. The UPR is also described in [1, 4, 35–37] and other references indicated in the text below.

2.1.1 Overview of the Main Phases of the UPR

The UPR can be broadly divided into two parts. The different parts can be named in various ways, but we will refer to them here as the adaptive phase and the terminal phase.

- The Adaptive Phase
  This phase, which deals with maintaining cell viability while fixing the problem of unfolded proteins, can also be subdivided into two parts:

- Early adaptive phase—In the **first** part of this phase, the ER sensors are activated, global cell translation is decreased, and chaperones are upregulated. The goal is to give unfolded proteins a chance to refold properly, thereby restoring normal protein homeostasis without compromising the cell's health.
- ERAD activation—The **second** part of this phase (which can overlap with the first part) is activation of ERAD. This occurs if the initial part is insufficient, and it involves the destruction of unfolded proteins that cannot be folded or refolded properly.

Note on terminology: Some sources refer to ERAD as a distinct process from the UPR, while others include it as a component of the UPR. In this chapter, ERAD is treated as a downstream effector pathway of the UPR, activated when attempts to refold proteins are insufficient.

- The Terminal Phase

  The last part of the UPR is sometimes called the terminal or the pro-apoptosis phase. This may occur if the first parts of the UPR are overwhelmed, and the cell cannot restore homeostasis. Under these conditions, the cell may undergo apoptosis.

UPR will be discussed in more detail below.

2.1.2 ER Stress Sensors Are Critical Components of the UPR

A key component of the unfolded protein response (UPR) is a group of three stress sensors located on the ER membrane that detect the accumulation of unfolded or misfolded proteins in the ER. These proteins are **PERK** (protein kinase R–like endoplasmic reticulum kinase), **IRE1** (inositol-requiring enzyme 1), and **ATF6** (activating transcription factor 6). Under normal, nonstress conditions, these sensors remain inactive due to their association with the ER chaperone HSPA5 (also known as BiP **or** GRP78), which binds to their luminal domains. When unfolded proteins accumulate, HSPA5 dissociates to assist in protein folding, allowing the sensors to activate and trigger the UPR signaling pathways.

2.1.3 Activation of the Sensors in Response to ER Stress

When ER stress arises due to an excess of unfolded or improperly folded proteins in the ER lumen, cells initiate the UPR. In the early stages of the response, HSPA5 binds preferentially to the misfolded proteins in the ER lumen, at their exposed hydrophobic regions. This binding helps prevent protein aggregation, which facilitates the proper folding of the proteins. Importantly, as a consequence of binding to the misfolded proteins, HSPA5 dissociates from the luminal domains of the ER sensors, PERK, IRE1, and ATF6 (Fig. 8.7).

Now that these sensors (PERK, IRE1, and ATF6) are freed from inhibition by HSPA5, they undergo activation. They are activated by different mechanisms:

PERK and IRE1 oligomerize. This involves forming dimers or higher-order oligomers. This oligomerization promotes trans-autophosphorylation of their cytoplasmic kinase domains, leading to their activation.

ATF6 is activated differently. Membrane-bound ATF6 translocates from the ER to the Golgi apparatus, where it is cleaved by two proteases called the site-1 and site-2 proteases (S1P and S2P). This releases the cytosolic N-terminal domain of ATF6, which is a transcription factor [26, 31].

2.1.4 Downstream Signaling Pathways Activated by the Stress Sensors

Now that they are activated, these sensors activate different signaling pathways (in the cytoplasm and nucleus) that coordinate the UPR, as follows.

PERK phosphorylates eIF2α (the alpha subunit of eIF2), thereby effectively inactivating it. eIF2α is critical for translation, and its role is described in Sect. 1 of this chapter. e1F2 must be GTP-bound to be active, and phosphorylation on its α subunit prevents it from exchanging GDP for GTP once it is inactivated. Thus, phosphorylation

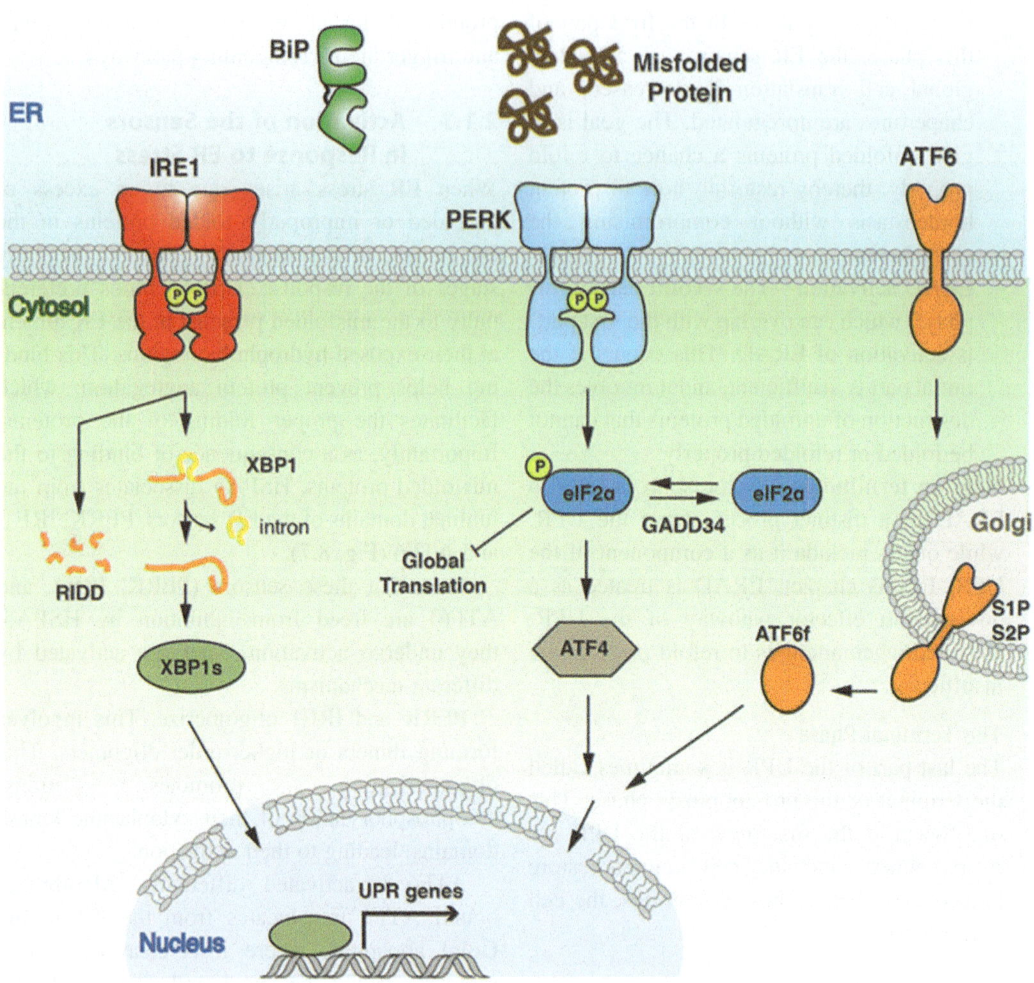

Fig. 8.7 ER stress is mediated by the ER stress sensors PERK, IRE1 and ATF6, which activate signaling pathways specifically in response to unfolded or misfolded proteins. See text for descriptions of these pathways. (*Source*: From Adams et al. [105]. Licensed under the Creative Commons Attribution 4.0 (CC BY 4.0) license)

of eIF2α inhibits its activity, leading to a global reduction in protein synthesis. This reduces the number of proteins entering the ER. This is important because it allows the ER to refold improperly folded proteins already present before obtaining new proteins to fold.

One important twist is that while overall protein synthesis is inhibited in response to eIF2α phosphorylation, some mRNAs escape this inhibition and, in fact, their translation increases rather than decreases. These are proteins encoded by genes that have a region called an upstream open reading frame (uORF), which is a short sequence within the 5′ untranslated region that can be translated into a small peptide [32]. Genes with a uORF can actually be translated more efficiently in times of cellular stress rather than less efficiently. An important example of a protein whose gene contains a uORF is ATF4. Translation of ATF4 is consequently **increased** rather than decreased in response to eIF2 inhibition [33]. ATF4 plays important roles in the adaptive phase of UPR. It upregulates several genes that are important for maintaining cell survival and viability during ER stress. Under prolonged ER stress, ATF4 can also lead to apoptosis, a topic that will be discussed later.

IRE1 has endoribonuclease activity, meaning that it can splice RNA. It specifically splices the mRNA for the gene XBP1 (X-box binding protein 1) and thus removes a short 26-bp intron. Unspliced XBP1 (which can also be called XBP1u) encodes a transcription factor with a short half-life and which is therefore essentially inactive. In contrast, the spliced form of XBP1, called XBP1s, encodes an active transcription factor. XBP1s translocates to the nucleus, upregulating the expression of several genes. These include genes that promote proper protein folding, prevent improperly folded proteins from leaving the ER, promote lipid biosynthesis, which is necessary for expanding the ER membrane, and genes that are involved in ERAD [34].

The cleaved **ATF6** translocates to the nucleus, where it serves as a transcription factor, inducing the expression of various genes. It leads, for example, to the expression of genes encoding ER chaperones, including HSPA5. It also leads to the expression of genes that encode other components of the protein folding and degradation machinery. Thus, it contributes to the cell's ability to handle the unfolded proteins.

In summary, together, these three sensors result in multiple responses that are crucial for the UPR. They reduce the amount of new proteins that form, enhance the degradation of misfolded proteins, and increase chaperone levels, thereby increasing proper folding. The goal of this UPR response is to restore homeostasis, thereby maintaining cell health and viability [35–37].

2.1.5 ERAD

As a result of the UPR process, misfolded proteins have a chance to be refolded properly, but in some cases, terminally misfolded proteins remain, and these are destroyed. Destruction of misfolded proteins is a result of the **ERAD (ER-associated degradation)** system. In this system, terminally misfolded proteins are recognized by ER chaperones and directed to the retrotranslocation complex, which facilitates the proteins' exit from the ER and entry into the cytosol. Two important proteins in this process are Derlin-1 and HRD1. Derlin-1 is a multi-pass ER membrane protein that acts as a channel that recognizes and initiates substrate export to the cytosol. HRD1 is an E3 ubiquitin ligase (ubiquitin ligases are discussed more in Sect. 3). It ubiquitinates its substrates (attaches ubiquitin protein), marking them for degradation. It also recruits cytosolic p97/VCP [38], which interacts with ubiquitinated proteins and facilitates their degradation in the proteasome. Once in the cytosol, these proteins, which are now ubiquitinated, will be degraded via the proteasome to prevent the accumulation of misfolded proteins [35–37]. (Ubiquitination and protein degradation are discussed in Sect. 3.)

2.1.6 Apoptosis Can Result When the UPR and ERAD Systems Become Overwhelmed

In some situations, the UPR, along with ERAD, can become overwhelmed by too much unfolded protein. This can trigger a pathway that leads to programmed cell death by apoptosis if the stress from the unfolded proteins is not relieved. In many types of cells, this involves the activation of a protein called CHOP (C/EBP homologous protein), also known as GADD153 (growth arrest and DNA damage-inducible 153). The gene for CHOP is upregulated in response to ATF4, which, as discussed above, is itself upregulated in response to PERK. To a lesser extent, CHOP is also upregulated by IRE1 signaling. CHOP is a transcription factor that increases the expression of several stress response genes that lead to apoptosis [37, 39] (see Fig. 8.8).

Before apoptosis is triggered, autophagy can be an alternative stress relief mechanism. Autophagy is a process that involves the degradation and recycling of damaged organelles and misfolded proteins. This response can protect cells against ER stress by reducing the accumulation of faulty proteins. Autophagy is also regulated by ATF4, which controls several genes involved in the autophagy process. If autophagy is not sufficient to restore or maintain

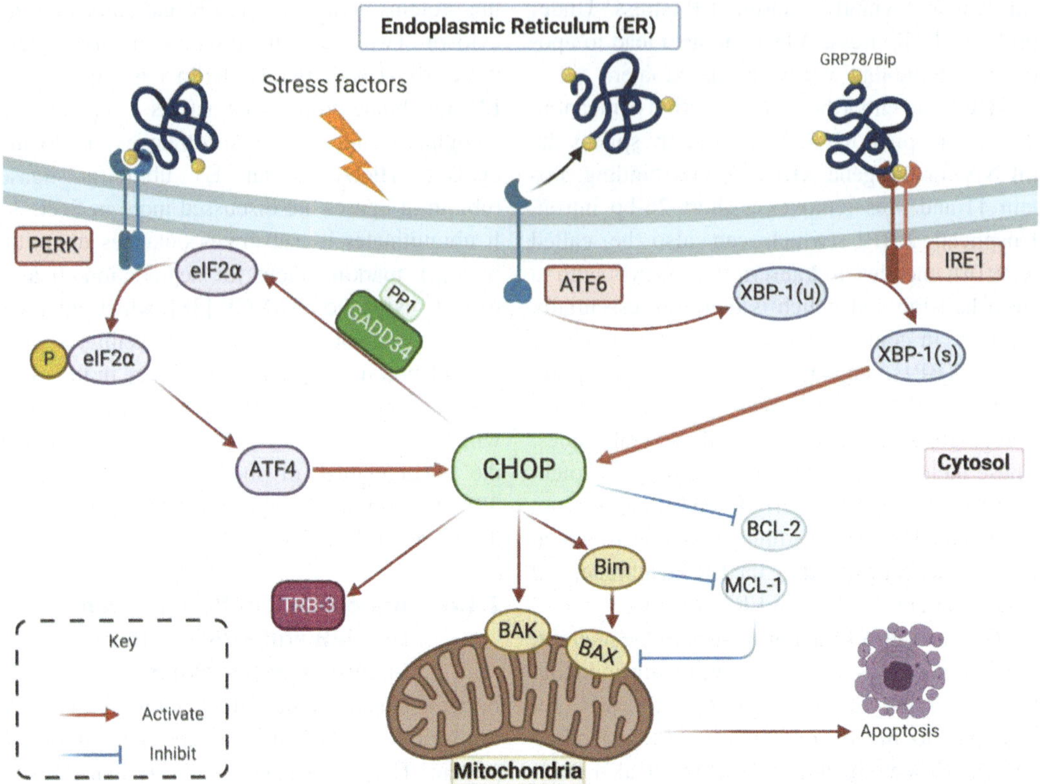

Fig. 8.8 CHOP-mediated apoptosis in response to ER stress. Apoptosis can result when the UPR and ERAD systems become overwhelmed. This is mediated by CHOP. See text for details. (*Source*: Reproduced from Turpin et al. [106]. Licensed under the Creative Commons Attribution 4.0 (CC BY 4.0) license)

normal cell function, then apoptosis will predominate. (Apoptosis and autophagy are described in more detail in Chap. 22.)

2.2 Misfolded Proteins, ER Stress, and Disease

Proper protein folding is critical for many biological processes. Consequently, protein misfolding can be detrimental. Below, several diseases that are specifically related to improper protein folding, and in some cases, the UPR, are discussed.

2.2.1 Diabetes

Insulin is produced in large amounts in the β-cells of the pancreas. The proper synthesis and folding of **insulin** are essential for **metabolism**, particularly in responding to changes in glucose levels, which will be discussed in Chap. 10. Insulin is initially synthesized as a long single amino acid chain called proinsulin [40]. Proinsulin is folded in the ER and then transported to the Golgi apparatus, where it is packaged into secretory granules. Secretory granules are a specialized type of vesicle that is found in secretory cells. They are used for regulated secretion, when secretion occurs in response to a signal. They can hold large amounts of cargo [4]. Within these secretory granules, proinsulin is cleaved by enzymes. Upon cleavage, insulin is now in its final form, known as mature insulin. Mature insulin is now stored in the granules and secreted into the bloodstream when it is needed, specifically in response to rising blood glucose levels.

Proper folding of insulin, which occurs within the ER, is important, although some amount of insulin misfolding is common. When there are too many misfolded proteins, UPR and ERAD can occur. The misfolded protein can either be folded correctly or destroyed, but the UPR and ERAD systems can become overwhelmed. This is even more likely to occur if large amounts of insulin are being produced, which can result from high blood glucose levels. Ultimately, this can result in increased production of CHOP, which plays a key role in leading to apoptosis of the β-cells. Once some of the β-cells undergo apoptosis, the remaining β-cells must produce even more insulin, leading to an even further increase in stress, which can result in further cell death. As a result, ER stress can kill many of the pancreatic β-cells. Too much pancreatic β-cell death contributes to overall β-cell failure, and this can increase the risk of diabetes. Too much insulin, which can occur, for example, in response to chronic high blood glucose levels, can increase the risk of ER stress and the destruction of β-cells. Keeping ER stress levels low is therefore important for preventing diabetes [41–43].

Some strategies under investigation for treating diabetes involve developing methods to encourage proper folding or to promote ERAD, thereby avoiding cell destruction. These types of strategies are mainly in experimental stages. They involve compounds that act as chaperones or other molecules that can enhance proper protein folding, or enhance the proteins involved in ERAD. Other mechanisms for enhancing β-cell survival are also important topics of investigation [31, 42].

2.2.2 Alzheimer's Disease (AD)

Protein misfolding and ER stress commonly occur in some neurodegenerative diseases [44]. ER stress, for example, **Alzheimer's disease (AD). AD is associated with the accumulation of unfolded proteins**. Specifically, it is characterized by the accumulation of β-amyloid (Ab) plaques on neurons, which are extracellular deposits made up mostly of misfolded β-amyloid peptides. These peptides are derived from the cleavage of the amyloid precursor protein (APP). APP is cleaved by **secretases**, including gamma-secretase. In Alzheimer's disease, APP is cleaved in the wrong place and misfolds. The misfolded peptides aggregate, and as a result, they form plaques that may contribute to neurodegeneration [45].

AD is also characterized by Tau tangles inside the neurons. Tau proteins normally stabilize microtubules, but in AD, they become hyperphosphorylated and misfolded. This causes them to detach from microtubules and aggregate into insoluble neurofibrillary tangles. These may impair neuronal transport systems and contribute to cell death.

While these misfolded proteins can lead to abnormalities in the nervous system, the high level of improperly folded proteins in the brains of AD patients may also lead to ER stress, causing more problems. UPR, and the death of cells in the nervous system. In fact, studies have shown that UPR markers, such as phosphor-IRE1 (p-IRE1), p-PERK, active ATF6a, and CHOP, are elevated in patients with AD and other neurodegenerative diseases [46]. These markers indicate that ER stress is occurring, and this may play a role in the death of neurons, especially when the UPR is unable to restore the normal structures of the unfolded proteins.

Some current ideas for treating AD involve developing compounds that may target UPR and relieve ER stress in the brains of AD patients. These include inhibitors of the IRE1, ATF6, and PERK signaling pathways. Multiple such compounds have been developed and are in either experimental stages or clinical trials [44].

Kisunia (donanemab-azbt) [47], a drug approved by the FDA in 2024, is the most recent FDA-approved treatment for AD (as of 2025). It is a **monoclonal antibody** (and thus a biologic), **directed against aggregated β-amyloid**. Kisunia binds specifically to the plaques. This causes immune cells to phagocytose the antibody-coated plaques. The goal is to slow the progression of the disease by breaking down the plaques. The drug is intended for **use by adults with early AD symptoms**. It has been shown to decrease progression of the disease in some patients, although it does not reverse the symptoms. It is delivered by intravenous (IV) infusion and is used monthly for a year [48–50].

In a final note about AD, AD is a complex disease. Despite the strong apparent link between ER stress, amyloid plaques, and neurofibrillary tangles and AD, it should be noted that the exact cause of AD is still incompletely understood, and the etiology of AD is an evolving field of study [46, 51].

2.2.3 Other Neurodegenerative Diseases

Some other neurodegenerative diseases have also been linked to protein misfolding [46]. Some examples are given below.

Parkinson's disease (PD) is characterized by the lack of dopamine neurons and the accumulation of Lewy bodies in the brain. Lewy bodies are abnormal clumps of proteins that consist of **misfolded aggregates of proteins called alpha-synuclein (α-SYN)**. These aggregates of misfolded proteins form fibrils within neurons. There is evidence that the accumulation of the α-SYN aggregates leads to ER stress and may contribute to the death of dopamine neurons [52].

NPT200–11 is a small-molecule, investigational compound known as **minzasolmin**. It targets α-SYN and helps prevent its misfolding. It was shown to **reduce the aggregation of α-SYN**. It **reduces some PD symptoms** in animal models and is being studied in clinical trials [52, 53].

Amyotrophic lateral sclerosis (ALS) is characterized by the degeneration of motor neurons and progressive weakness. The cause of the disease is unknown, and there is an interest in identifying genetic mutations that are associated with ALS. **Upregulation of UPR markers is often seen in neurons in the disease**. Certain proteins, such as superoxide dismutase 1 (SOD1) and TAR DNA-binding protein 43 (TDP-43), have been shown to misfold and aggregate within neurons in some ALS patients. TDP-43 has diverse cellular functions, including the regulation of metabolic pathways, and SOD1 is a protein that helps break down harmful superoxide radicals within cells. The improper folding of these proteins leads to the harmful accumulation of aggregates [54]. These misfolded proteins impair normal function and may also contribute to cell death due to causing prolonged ER stress responses [55, 56].

ALS is usually sporadic, but there is a rare genetic variant with a mutation in the gene for SOD1. **Mutations in SOD1** are thought to cause it to misfold. **Misfolded SOD** forms aggregates in the motor neurons, causing neuronal abnormalities [57]. **Qalsody (tofersan)** [58] is a drug based on **antisense technology** (thus it falls into the oligonucleotide therapeutics category of drugs) that **targets the mutated SOD1 gene**. It was FDA-approved in 2023. Qalsody consists of an oligonucleotide sequence that is complementary to part of the SOD1 mRNA. When it binds to the SOD1 mRNA, it leads to its degradation, thus reducing the amount of SOD1 protein produced. The drug is administered by injection into the cerebrospinal fluid (CSF). The goal is to reduce the production of the SOD1 protein in order to decrease the formation of aggregates. The drug lowers SOD1 protein and neurodegeneration markers, but its clinical benefits are still uncertain [57, 59, 60].

2.2.4 Cystic Fibrosis (CF)

Cystic fibrosis (CF) is caused by a **mutation in the CFTR (cystic fibrosis transmembrane conductance regulator) gene**, which encodes a chloride channel found in epithelial cells. CFTR's normal function is to regulate mucous consistency by regulating the transport of $Cl-$ ions across membranes. In CF patients, this process is disrupted due to mutations in the channel protein, leading to abnormal amounts of mucous accumulating in the extracellular space of epithelial tissues, such as in the airways. The most common CF mutation in the CFTR is the deletion of a phenylalanine (Phe) residue at position 508 in the CFTR. This leads to **improper folding of the protein**, and consequently, the CFTR does not reach the cell membrane once it is made. In addition, **improper folding of CFTR triggers the UPR and can eventually lead to degradation of the protein via ERAD**. The process can eventually lead to increased inflammation, impaired cell function, and cell death.

The connection between CF and ER stress forms the basis for developing new ways to treat the disease. One strategy is to try to restore proper protein folding via the use of **correctors**. Correctors are small **synthetic molecules that are synthesized to help misfolded proteins achieve a more functional conformation**. Correctors bind to misfolded CFTR and stabilize its structure, allowing it to fold correctly. Correctors can prevent aggregation, stabilize intermediates, and interact with chaperones [61]. **Orkambi (lumacaftor)** [62] is a **CFTR corrector**, and it is the first medicine to treat the underlying cause of **CF**. **Lumacaftor** is used in combination with another drug, **Kalydeco (ivacaftor)** (also discussed in Chap. 3), which supports the activity of the CFTR once it is properly folded. Lumacaftor and Ivacaftor both bind directly to CFTR. The two drugs together have the brand name **Orkambi**. **Symdeko (tezacaftor)** is also a **CFTR corrector** and is used in combination with **Kalydeco (ivacaftor)** in a single medicine called **Symdeko (tezacaftor/ivacaftor)** [63, 64].

2.2.5 Transthyretin Amyloidosis (ATTR)

ATTR was discussed in Chap. 7. It involves the accumulation of **improperly folded transthyretin (TTR)**. There is more than one category of ATTR, and it can be classified either by etiology or by clinical presentation. Regarding etiology, ATTR can be classified as either hereditary (variant, ATTRv) caused by mutations in the *TTR* gene, or wild-type (ATTRwt), which is associated with aging. Clinically, patients may present primarily with cardiomyopathy (ATTR-CM; a heart disease characterized by the myocardium becoming stiff or weakened), or they may have polyneuropathy (ATTR-PN; a disorder in which multiple peripheral nerves are damaged), or they may exhibit a mixed phenotype [65]. This distinction is important because current therapies are approved for specific subtypes.

We have discussed several treatments for this disorder. Here, we discuss a mechanism that is involved in stabilizing the TTR protein. In **ATTR**, the TTR protein becomes unstable, leading to its misfolding. This leads to the formation of amyloid fibrils that accumulate in certain tissues, such as the heart and nervous system, resulting in tissue damage. TTR normally forms a tetramer, consisting of four identical subunits. **Tafamidis** is a type of small molecule called a **stabilizer**. Tafamidis is designed to bind to the TTR tetramer and stabilize it. This prevents TTR from breaking apart into monomers, which are more prone to misfolding. **Tefamidis** is the active component of **Vyndaqel (tafamidis meglumine)** [66] and **Vyndamax (tafamidis-free acid)** [67], which are FDA-approved drugs for treating ATTR-CM [65]. These drugs both contain tafamidis in different formulations.

Another drug that treats **ATTR** and stabilizes TTR is **Attruby (acoramidis)** (FDA approved in 2024). **Attruby** binds more tightly to the TTR than the Tafamidis. It also differs in that it is designed to mimic a TTR variant (TTR (T119M)). In the TTR (T119M) variant, threonine is replaced with methionine at position 119. This is a rare natural variant that stabilizes the TTR, thus having a protective effect against the disease. My mimicking this variant, Attruby, stabilizes the TTR even more effectively than other stabilizers.

3 Protein Turnover

Proteins do not necessarily last for the lifetime of the cell. Instead, most proteins undergo periodic turnover. Short-lived or damaged proteins, or those no longer needed, are degraded, and new proteins are synthesized to replace them. Protein half-lives vary from a few minutes for some proteins to the organism's lifetime for others, and therefore some proteins turn over frequently, while others last longer. In addition to synthesis and degradation, proteins are also folded, modified, and transported to specific locations inside or outside the cell. The regulation of all of these factors, protein synthesis, folding, trafficking, and degradation to maintain a balanced proteome (the complete set of proteins in the cell), is known as proteostasis (protein homeostasis) [30, 68]. One important aspect of proteostasis is the regulation of protein degradation. For protein turnover to be regulated, some proteins are periodically and selectively degraded. This section of the chapter will focus on the systems of protein degradation.

3.1 The Ubiquitin–Proteasome System

3.1.1 Overview of the Ubiquitin–Proteasome System

An important part of protein turnover is proteolysis, which is the cleavage of peptide bonds by proteases, leading to protein degradation. There are different mechanisms by which proteins can be degraded, but an important mechanism is the ubiquitin–proteasome system (UPS). A high percentage of proteins that get degraded are destroyed by this mechanism. Ubiquitin is a small protein (76 kDa) that is abundant in most cells. Adding ubiquitin to a protein is called ubiquitination (sometimes called ubiquitylation). Ubiquitination tags a protein for destruction by targeting it to a proteasome. In the proteasome, proteins are broken down by proteolysis, a process that involves breaking peptide bonds. This is the ubiquitin–proteasome system, and it results in the permanent degradation of a protein. This differs from mechanisms that may temporarily inactivate a protein, such as certain post-translational [1, 30, 69, 70].

3.1.2 Ubiquitination

Proteins destined for destruction are ubiquitinated on lysine residues. The C terminus of ubiquitin contains a glycine tail, and it is this tail that becomes covalently attached to a lysine on the targeted protein. This linkage is between the carboxyl group of a glycine on the ubiquitin and the epsilon amino group of a lysine on the targeted protein. The bond that links them is an isopeptide bond. Lysine has two amino groups. The epsilon amino group is the amino group attached to the fourth carbon. Ubiquitin attachment is referred to as the conjugation of ubiquitin. Adding even one ubiquitin to a protein adds a significant amount of molecular weight to the protein. Sometimes, multiple ubiquitin monomers can be added, forming bonds with different lysine residues.

Ubiquitin itself can also be ubiquitinated. Ubiquitin has seven lysine residues that can be ubiquitinated, and this can result in polyubiquitin chains. There are eight types of polyubiquitin chains. Ubiquitin chains can be distinguished by which lysine the ubiquitins bind to. A polyubiquitin chain in which the ubiquitin units are linked by K48 (the 48th lysine on the ubiquitins) is well known for targeting proteins for destruction. In some cases, ubiquitination can have other functions besides proteolysis. For example, it can have roles in DNA repair, chromatin remodeling, immunity, membrane trafficking, and autophagy. Here, we focus on the role of ubiquitination in **proteolysis** [70–72].

3.1.3 The Enzymes Involved in Ubiquitination

Ubiquitin is added to proteins by a series of reactions that involve three families of enzymes: **E1**, **E2**, and **E3**, also discussed in [70–73] (Fig. 8.9). The reactions occur as follows:

1. **E1** is the **ubiquitin-activating enzyme**. In the first reaction, the carboxyl (C) terminus of a ubiquitin molecule is covalently attached to the active site of the **E1 enzyme** by a thioester bond at a cysteine residue in the enzyme. This is an ATP-dependent step. At this point, the ubiquitin is considered to be "activated."
2. **E2** is the **ubiquitin-conjugating enzyme**. In this step, the **E1 ubiquitin complex** (formed in step 1) binds to a cysteine on the **E2 enzyme**, transferring the ubiquitin to the cysteine on the E2's active site. The ubiquitin is now linked to the cysteine residue of **E2** via a thioester bond.
3. The **E3 enzyme** is a **ubiquitin ligase**. This enzyme transfers the ubiquitin from **E2** to its substrate. In this step, the **E2-ubiquitin** complex is transferred to the **E3 enzyme**. The **E3 enzyme** has binding sites for both the **E2-ubiquitin complex** and the **substrate protein** that will get ubiquitinated. Upon binding of both these groups to **E3**, the **E3 enzyme** catalyzes the transfer of ubiquitin from the cysteine of the **E2 enzyme** to the epsilon amino group of a lysine on the **substrate**

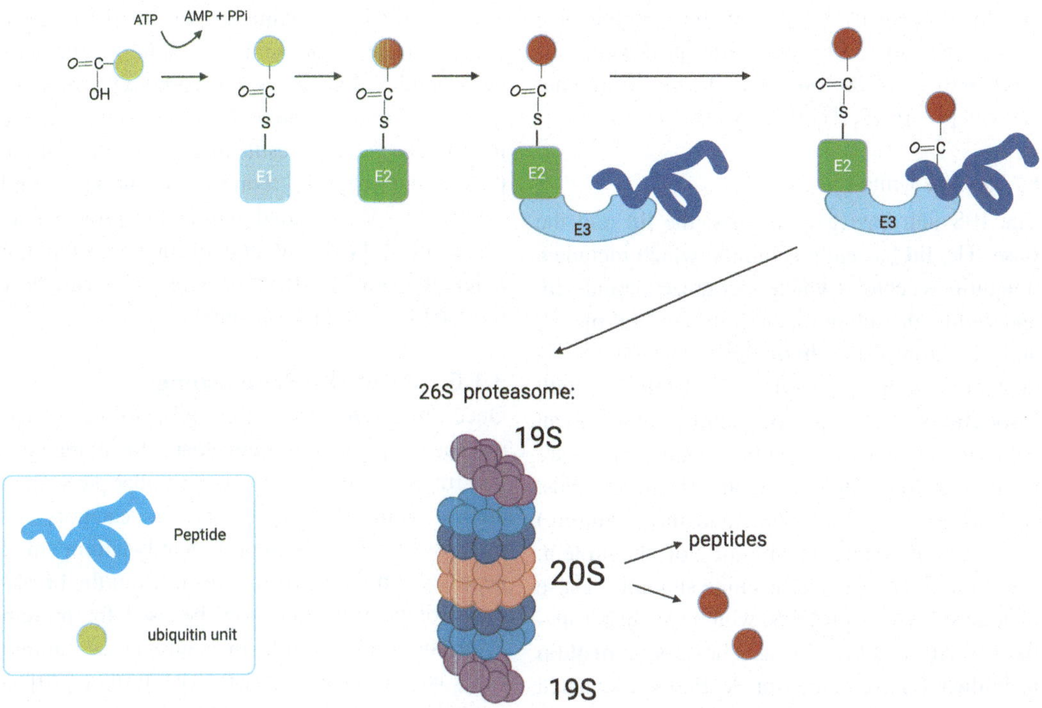

Fig. 8.9 The ubiquitin–proteosome system. Ubiquitination involve three families of enzymes: E1, E2, E3, and proteolysis via the proteosome. See text for details. (*Source*: Created with BioRender.com)

protein, forming a covalent bond between the protein and the ubiquitin.

Polyubiquitin chains are assembled the same way as described above, with each ubiquitin added to another via the above three enzymes.

E1, **E2**, and **E3** are families of enzymes. There are two types of **E1**, over 30 **E2 enzymes**, and **more than** 600 different **E3 enzymes** in human cells, all of which can target different proteins. This leads to diversity within the system and to the degradation of various types of proteins [74].

3.1.4 Degrons

The E3 enzyme usually adds ubiquitin to a region in the substrate protein that has a degron. A degron is an amino acid sequence that has a key role in targeting a protein for degradation. Sometimes, degrons are concealed and only become accessible once the protein undergoes certain changes. For example, they may become accessible once the protein has already undergone some initial degradation or as a result of cleavage, partial unfolding, or a conformational change. In unfolded or misfolded proteins hidden normally hidden degrons may become exposed, leading to their degradation; this is an important mechanism for the degradation of unfolded proteins in the cytosol [1, 70, 71, 75].

3.1.5 The Proteasome

Ubiquitin or polyubiquitin (on the protein to be destroyed) is recognized by the proteasome, and consequently, a ubiquitinated protein will bind to a proteasome. The proteasome (called the 26S proteasome) is a complex that digests ubiquitinated proteins, but it only digests those that are "admitted" into the proteasome. The proteasome consists of a barrel-shaped proteolytic chamber (the **20S unit**), which is the catalytic subunit, and two external cavities (the **19S units**), known as regulatory particles and sometimes referred to as caps. The 19S subunits can bind to either or both ends of the 20S units, and the 20S unit is only open if a 19S subunit is bound to it. The 19S unit can control whether the 20S subunit is open by regulating the opening of a "gate" into the 20S unit. The proteasome is discussed more below and in the following sources [4, 70, 76, 77] (Fig. 8.9).

- The 19S Units

The 19S unit has two subparts: the **lid** and the **base**. The **lid** has eight subunits, which include a ubiquitin receptor (which recognizes, binds to, and holds the ubiquitinated protein in place), and a **ubiquitin hydrolase**, which is a deubiquitinating enzyme (DUB) with **isopeptidase** activity. (An isopeptidase is an enzyme that cleaves isopeptide bonds, which are bonds that form on side chains of amino acids, such as between the protein and the ubiquitin.) This enzyme cleaves the ubiquitin off the protein. The base of the 19S unit has nine subunits. These include six AAA+ proteins, which are chaperone-like proteins that function as a hexameric **protein unfoldase** (AAA stands for **A**TPases **a**ssociated with diverse cellular **a**ctivities). AAA+ proteins have various functions, but in proteasomes, they can sense and bind ubiquitinated proteins. They subsequently help unfold the protein. After proteins bind the 19S unit and unfolding occurs, the 19S unit triggers a conformation change in the 20S subunit, causing it to open. Proteins can now travel into the 20S subunit.

- The 20S Unit

The 20S unit is the main catalytic unit, consisting of 28 subunits that form four rings. The outer two rings each have a row of 7 alpha-type subunits, and the two rings in the middle each have a row of **7 β-subunits**. The rings are thus heptameric and stacked on top of each other. The β-subunits are enzymes, with their active sites facing the inside of the proteosomes. The α-subunits have roles in controlling which substrates enter the catalytic cores of the β-subunits. There are three types of β-subunits that have catalytic activity. They have different types of active sites: caspase-like sites, which cleave proteins after acidic residues; trypsin-like, which cleave after basic residues; and chymotrypsin-like, which cleave after hydrophobic residues [78]. All of these β-subunits use an N-terminal threonine residue in their active site. These threonine residues become nucleophiles that attack the carbonyl groups of peptide bonds, thus forming acyl-enzyme intermediates. As a result of the functions of the β-subunits, peptide bonds are broken, and substrates are degraded within the proteasome. The result is the degradation of fragments ranging in length from 2 to 10 amino acids, which are now released from the proteasome.

3.1.6 After the Proteasome

Once released from the proteasome, other proteases, specifically **peptidases**, break the peptide fragments down into even smaller pieces and amino acids. Peptidases are a category of proteases that cleave peptide bonds within small peptides. Amino acids that result from the breakdown of peptides can now be used for protein synthesis (translation). In some cases, amino acids can be broken down even further [70], a process discussed in Chap. 15.

3.2 Why Are Proteins Degraded?

There are many situations in which proteins are targeted for proteolysis, resulting in their degradation. Proteins can get degraded, for example, when they are improperly folded. They also get degraded when they are damaged or no longer needed. Additionally, many normal cellular processes involve proteolysis and protein degradation. For example, some cell signaling pathways, such as the β-catenin pathway, involve proteolysis of various proteins. Proteolysis in a cell signaling pathway can regulate the activity of the pathway [71, 76]. Proteolysis also plays important roles in the cell cycle, apoptosis, and the regulation of transcription factors, such as p53 [70].

3.2.1 Protein Degradation and Human Health

- Angelman Syndrome

Abnormalities in **E3 ubiquitin ligase** have been associated with human health problems. Defective E3 leads to the accumulation of proteins that evade degradation, potentially resulting in disorders. **Angelman syndrome** is a severe neurological disorder characterized by cognitive problems, limited speech ability, and impaired coordination [70, 79]. It is caused by a deletion or abnormality in the gene for a member of the **E3 family**, **UBE3A** (also called **E6AP**). The lack of this E3 ligase leads to impaired protein regulation in neurons, which is believed to contribute to the symptoms of the disease.

The mechanism of UBE3A regulation is unusual and may make this disease potentially amenable to treatment in the future. In most tissues, both maternal and paternal copies of UBE3A are expressed. However, in the brain, only the maternal allele is expressed, while the paternal allele is silenced. This is a rare case of tissue-specific genomic imprinting. The silencing of the paternal allele is due to an lncRNA called UBE3A-ATS (ATS stands for antisense transcript). This lncRNA is controlled by a tissue-specific promoter, so it is only expressed in neurons. It is transcribed from the strand that is complementary to the UBE3A template strand. The two transcripts that result (UBE3A and UBE3A-ATS) are transcribed in opposite directions but overlap in sequence. Thus, in the region of overlap, they are antisense to each other. The UBE3A-ATS interferes with the transcription of the UBE3A allele by silencing it; this is known as transcriptional interference. This prevents the paternal allele from being translated and producing protein. This act of interference occurs only in cis—on the same allele. On the maternal allele, the gene for UBE3A-ATS is inactivated by methylation (epigenetic regulation). Therefore, the maternal UBE3A-ATS is not expressed, and UBE3A can still be transcribed and translated [80].

In Angelman syndrome, the maternal UBE3A allele is often nonfunctional, frequently deleted, mutated, or epigenetically silenced, and the paternal allele is normally silenced in neurons by a long noncoding antisense transcript, UBE3A. The result is that neurons lack functional UBE3A expression. To address this, researchers have explored "unsilencing" therapies aimed at suppressing UBE3A, thereby allowing the paternal allele to be expressed. In a recent study, a small molecule was identified through a high-content screen as a novel paternal UBE3A unsilencer. It downregulates Ube3a-ATS, increases paternal UBE3A mRNA and protein in neurons, and, when delivered by intraperitoneal injection in a mouse model, produces widespread neuronal unsilencing. The researchers propose that this molecule might have therapeutic potential for humans, though it remains an early-stage preclinical proof of concept [79, 81].

- Autism Spectrum Disorder (ASD)

Overexpression of E3 is also associated with health problems. For example, overexpression of **UBE3A3** is associated with a subset of autism spectrum disorder (ASD) cases. This overexpression is often due to copy number variations of the **UBE3A gene** that result in extra copies of the gene. A transgenic mouse model with extra copies of the gene also displays ASD behavioral characteristics [82].

- Cancer

Protein turnover is important for normal cell growth regulation and is linked to the regulation of some tumor suppressor genes and proto-oncogenes. Disruption of protein turnover, often due to abnormalities in the proteosome pathway, can lead to cancer. p53 is a tumor suppressor well known for its role in controlling cell growth, and it is inactivated or mutated in many types of cancer. Interestingly, **UBE3A** is involved in the degradation of p53. This may help explain why too much UBE3A can be associated with cancer [83].

3.2.2 Protein Degradation as Treatments for Disease

- Proteasome Inhibitors

Velcade (bortezomib) [84] is a **small-molecule** drug that is a **proteosome inhibitor**. It was the first proteasome inhibitor to be FDA-approved (2003) and destroys proteosomes. Proteosomes are necessary for cells because they prevent the buildup of unwanted proteins. The drug, therefore, is designed to **kill cancer cells** by causing the abnormal buildup of proteins within the cell.

It is used as a therapy for multiple myeloma [70, 85]. It is also being investigated for other purposes and for combination therapies [86].

Velcade works by targeting the β-subunits of the 20S catalytic core of proteasomes. It is a reversible inhibitor of the chymotrypsin-like activity of the β-subunits. It interacts with the N-terminal threonine residue in their active sites. Velcade works because it has a boron atom in its structure. This boron atom interacts with the N-terminal threonine residue and forms a bond, inhibiting the enzyme's activity [86].

Kyprolis (carfilzomib) [87] is also a **small-molecule proteasome inhibitor**, and it is used, in combination with other drugs, to treat refractory multiple myeloma [88]. Like other cancer drugs, these medications can cause side effects because they can target proteosomes in both normal cells and cancer cells. **Cancer** cells are generally fast-growing cells, but myeloma cells have an especially high turnover rate. Fast-growing cells rely on proteosomes more than slower-growing cells, and therefore, they are particularly susceptible to the drug [70, 76, 89].

- PROTACs

The PROTACs (proteolysis targeting chimeras) system is a new technology based on the ubiquitin–proteasome system [90] (see Fig. 8.10). This system is designed for the targeted degradation of proteins within cells. **PROTACs are molecules** that are designed to use the ubiquitin–proteasome system **to degrade specific proteins** within cells. PROTACs have two main parts. They have (**a**) a domain that binds the protein to be targeted, and (**b**) a domain

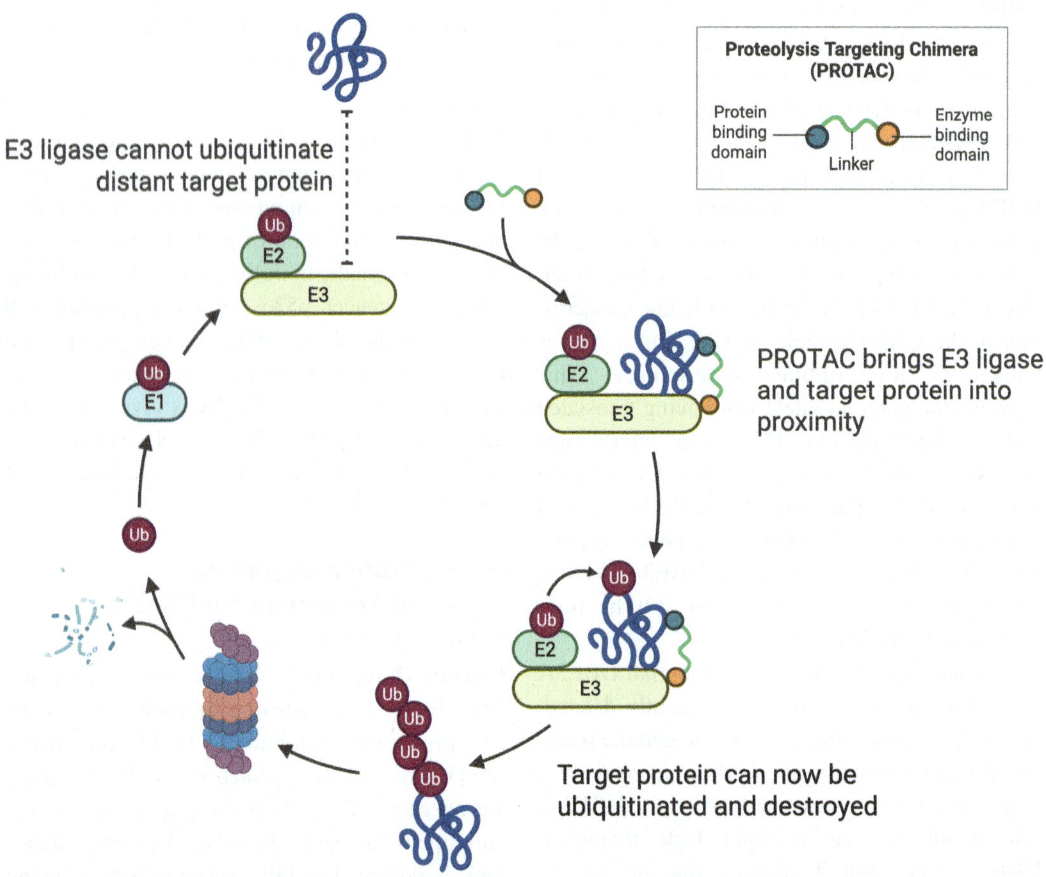

Fig. 8.10 Protacs. Protacs are designed to direct target proteins to the ubiquitin proteosome system, so that targets can be degraded. See text for details. (*Source*: Created with BioRender.com)

that binds to an E3 ubiquitin ligase. PROTACs are designed to enter cells, although this process can be complicated, especially for larger PROTACs [91]. Some are small molecules that can enter by passive diffusion, while others are designed to enter the cell via endocytosis. Because they bind to both an E3 and a target protein, the result is that the target protein is brought into proximity to the E3 ligase. The resulting complex is called a ternary complex. The E3 ligase then performs its function of ubiquitinating the target protein, even if it is not normally ubiquitinated. This targets the protein to the proteosome for its destruction.

PROTACs are especially promising for treating cancers for which specific proteins that drive uncontrolled growth have been identified and targeted. The PROTAC system offers benefits over other mechanisms, such as the use of inhibitors (like kinase inhibitors) because it leads to the total destruction of the target protein rather than merely inhibiting its activity, which can be transient. While there are no FDA-approved PROTAC-based treatments at this time (as of 2025), some are in clinical trials, and others have been given fast-track designation, a program to speed up the development of new and promising drugs.

ARV-110 is a **Protac degrader** being tested in **clinical trials** for the treatment of **metastatic prostate cancer**. It **targets the androgen receptor (AR) for degradation**. AR is a nuclear hormone receptor that plays a role in prostate cell growth. In many prostate cancers, AR is overexpressed or hyperactivated and contributes to disease progression. Targeting AR is a strategy for treating the disease. ARV-110 recruits AR to an E3 ubiquitin ligase, which leads to its degradation [92].

Vepdegestrant (ARV-471) from Pfizer and Arvinas was recently (2024) granted fast-track designation for treating patients with **ER$^+$/Her2$^-$ breast cancer**. This drug is a **PROTAC targeted degrader of the estrogen receptor (ER)**. The ER is overproduced in ER+ breast cancer and drives the expression of genes that promote cell proliferation and survival. Vepdegestrant is designed to overcome resistance to existing targeted ER therapies such as tamoxifen and aromatase inhibitors [93, 94].

KT-333 (developed by Kymera Therapeutics) is being fast-tracked as a treatment for **cutaneous T-cell lymphoma (CTCL) and peripheral T-cell lymphoma (PTCL)**. This drug is a **PROTAC** designed to **degrade STAT3**. STAT3 (part of the JAK/STAT pathway) is overproduced in several cancers, including both hematological tumors and solid tumors. STAT3 contributes to oncogenesis by promoting inflammation, survival, and immune evasion. By promoting its degradation, KT-333 is designed to disrupt cell growth-promoting pathways mediated by STAT3 [76, 95].

In addition to their promise in disease treatment, PROTACs are powerful research tools. Using PROTACs allows researchers to selectively degrade specific proteins. This provides scientists with a means to investigate the functions of these proteins in cells and animal models. This has certain advantages over other systems. For example, inhibiting proteins using inhibitors can be nonspecific, leading to the inhibition of a broad range of proteins, and inhibition can also be incomplete. Inhibition by gene knockout systems can be time-consuming and irreversible. By using PROTACs, genes can be destroyed relatively quickly, and the system can be reversible because once the PROTACs are removed, new protein can be synthesized. This gives the researcher control over the timing of protein degradation and allows scientists to study proteins that are difficult to knock out using other systems, especially proteins essential for cell viability [96].

- Molecular Glues

Molecular glues are another mechanism that can be used to target disease-causing proteins for degradation. Molecular glues are small molecules that also make use of the ubiquitin–proteasome system to promote the degradation of targeted proteins. Like PROTACs, they lead to the formation of a complex between a target protein and an **E3 ubiquitin ligase**, thus leading to ubiquitination and proteasomal degradation of the target protein.

Molecular glues also have some key differences compared with PROTACs, however. PROTACs are bifunctional molecules; they bind to two proteins: the target protein and the E3 ubiquitin ligase. In contrast, molecular glues are monovalent; they generally bind only to the E3 ligase. Instead of binding to another protein as well, they bind to the E3 ligase in a way that alters its substrate specificity (the substrate is the targeted protein). They create a new interface that enables the E3 ligase to interact with proteins that it does not normally ubiquitinate. For example, they may bind to the substrate receptor on the E3 ligase, which is the part of the E3 enzyme that recognizes and binds its specific target/substrate.

Iberdomide is an example of an **investigative drug** that falls into the **molecular glue** category. This drug is a modulator of an E3 ligase, specifically an E3 ligase complex called CRL4^{CRBN}. In this complex, **CRL4 (Cullin-Ring Ligase 4)** refers to the ligase complex, and **CRBN** is the **substrate receptor** component of this complex. The substrate receptor is the part of E3 that determines which substrates get targeted for ubiquitination.

The **CRL4 complex** has four main parts: (**a**) CUL4A, which functions as a scaffold protein (it holds together the E3 and its target); (**b**) DDB1, an adapter protein that links CRBN to CRL4; (**c**) CRBN (cereblon), the substrate receptor, which specifically recognizes and binds to its target; and (**d**) RBX1, a ring finger protein that recruits E2 ubiquitin for transfer of the ubiquitin from E2 to the target [97]. These individual components do not have intrinsic ubiquitin ligase activity, but the fully assembled complex enables targeted ubiquitination of its substrates.

Iberdomide binds to the cereblon (CRBN) component of this E3 ligase complex, and it is thus classified as a *c*ereblon *E*3 *l*igase *mo*dulator (CELMoD). Iberdomide, like other CELMoDs, contains a glutarimide ring (a 5-membered ring structure), which is the element that anchors the drug to the CRBN, and a substrate recognition domain. This binding to CRBN enables Iberdomide to change the complex's substrate specificity, allowing it to tag new proteins, known as neosubstrates, for ubiquitination and subsequent proteasomal degradation. Among these neosubstrates are the transcription factors Ikaros (IKZF1) and Aiolos (IKZF3), both of which are important for B cell development, survival, and plasma cell function (plasma cells are the terminally differentiated B cells). Degradation of these proteins inhibits the proliferation of immune cells. Because **multiple myeloma** is a cancer of plasma cells, this mechanism makes Iberdomide a promising therapeutic candidate currently under investigation for treating this malignancy [98, 99] (see Fig. 8.11).

Interestingly, **Iberdomide** is related to other drugs such as thalidomide, lenalidomide, and pomalidomide, all immunomodulatory drugs. However, the mechanism of action of these

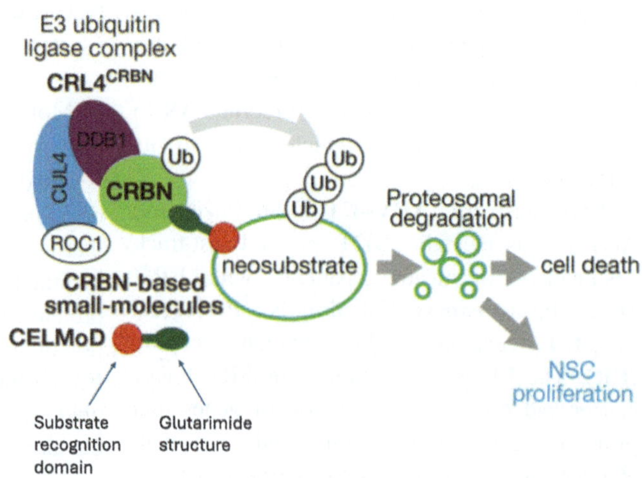

Fig. 8.11 CELMod. CELMod is an E3 ligase modulator. This system involves the reprogramming of E3 ligase to degrade a specific protein. See text for details. (*Source*: Adapted from Sato et al. [107]. Licensed under the Creative Commons Attribution 4.0 (CC BY 4.0) license)

other drugs were not understood when they were first developed, especially thalidomide which was developed in the 1950s.

Revlimid (lenalidomide) [100] is a similar drug and has been used for years [101]. It was first FDA-approved in 2005, and it is mainly used to treat multiple myeloma, but is sometimes used to treat other **blood cancers**. It was also used before its mechanism of action was understood. Later, it was found to function by inhibiting IKZF1 and IKZF3, and it is now also known to modulate the **E3 ligase CRBN** [102]. Although **Iberdomide** functions in the same way as **Revlimid**, it is expected to be more efficient due to several characteristics, including more efficient binding to **CRBN** [103].

FDA-Approved Drugs Mentioned in This Chapter

Commonly used brand name	Generic name	First FDA approval year	Common conditions	Main target
Attruby	*Acoramidis*	2024	Transthyretin amyloidosis (ATTR)	TTR tetramer (stabilizer; T119M mimic)
Chloromycetin	*Chloramphenicol*	1949	Serious bacterial infections	50S ribosomal subunit
Cleocin	*Clindamycin*	1970	Anaerobic infections, skin infections	50S ribosomal subunit
Erythrocin	*Erythromycin*	1952	Respiratory, skin, and other infections	50S ribosomal subunit (23S rRNA)
Gentak	*Gentamicin*	1970	Bacterial infections	30S ribosomal subunit (16S rRNA)
Kisunla	*Donanemab-azbt*	2024	Early Alzheimer's disease	Aggregated β-amyloid plaques
Kyprolis	*Carfilzomib*	2012	Relapsed refractory multiple melanoma	20S proteasome (chymotrypsin-like activity of β-subunits)
Orkambi	*Lumacaftor/ Ivacaftor*	2015	Cystic fibrosis (CF)	CFTR protein
Qalsody	*Tofersen*	2023	ALS with SOD1 mutation	SOD1 mRNA
Revlimid	*Lenalidomide*	2005	Multiple myeloma, other blood cancers	IKZF1, IKZF3, and E3 ligase CRBN
Streptomycin	*Streptomycin*	1944	Tuberculosis, bacterial infections	30S ribosomal subunit (16S rRNA)
Symdeko	*Tezacaftor/ Ivacaftor*	2018	Cystic fibrosis (CF)	CFTR protein
Velcade	*Bortezomib*	2003	Multiple myeloma	20S proteasome (chymotrypsin-like activity of β-subunits)
Vibramycin	*Doxycycline*	1967	Bacterial infections	30S ribosomal subunit (16S rRNA)
Vyndamax	*Tafamidis*	2019	Transthyretin amyloidosis (ATTR)	TTR tetramer (stabilizer)
Vyndaqel	*Tafamidis meglumine*	2019	Transthyretin amyloidosis (ATTR)	TTR tetramer (stabilizer)
Zyvox	*Linezolid*	2000	Resistant bacterial infections	50S ribosomal subunit

Definitions

19S regulatory particle A subunit of the proteasome. It contains a deubiquiting enzyme (removes ubiquitin) and a protein unfoldase.

20S core particle The main catalytic unit of the proteasome.

26S proteasome The proteasome complex that digests and degrades ubiquitinated proteins.

AAA+ ATPases Chaperone-like proteins with ATPase and unfoldase activity.

Activating transcription factor 6 (ATF6) A transcription factor that is an ER stress sensor, located in the ER membrane until it is activated.

Aminoacyl tRNA site (A site) A site on the ribosome. During translation, incoming aminoacyl-tRNA enters the ribosome at this site.

Aminoacyl-tRNA synthetase An enzyme that attaches the appropriate amino acid to its corresponding tRNA.

Aminoacyl-tRNA A tRNA molecule that has been activated, or "charged," with its specific amino acid.

Angelman syndrome A genetic disorder caused by a defect in the **UBE3A** gene, which encodes an E3 ubiquitin ligase.

Anticodon A three-nucleotide sequence on tRNA that base pairs with a corresponding mRNA codon.

Apoptosis Programmed cell death.

Autophagy A cellular process where cells degrade their own contents, including damaged organelles and some proteins.

Beta cells (β-cells) Insulin-producing cells in the pancreas

C/EBP homologous protein (CHOP) A transcription factor that promotes apoptosis if ER stress becomes overwhelmed. It is also called GADD153.

Cereblon (CRBN) An E3 ligase substrate receptor that is modulated by drugs like lenalidomide and iberdomide.

Codon A sequence of three nucleotides in the mRNA that corresponds to a specific amino acid or, in some cases, to a stop signal.

Endoplasmic reticulum (ER) stress Caused by the accumulation of too many unfolded or misfolded proteins in the ER.

Endoplasmic reticulum (ER)-associated degradation (ERAD) A cellular pathway where misfolded proteins are transported to the cytosol for proteasomal degradation.

Eukaryotic initiation factor 2 alpha (eIF2α) A protein with a role in translation. When it is phosphorylated, translation decreases globally.

Exit site (E site) A site on the ribosome. tRNAs exit the ribosome at this site after delivering their amino acid to the growing peptide chain.

GADD153 Another term for CHOP, see above.

Heat shock proteins (HSPs) A family of molecular chaperone proteins.

HSPA5 (BiP/GRP78) A chaperone protein that binds to unfolded proteins and regulates activation of ER stress sensors.

Inositol-requiring enzyme 1 (IRE1) An ER stress sensor. An ER transmembrane protein with kinase and endoribonuclease activity; splices XBP1 mRNA.

Isopeptidase An enzyme that cleaves isopeptide bonds

Isopeptide bond A type of covalent bond that forms between the side chains of amino acids, such as those that link ubiquitin to lysine residues on target proteins.

Molecular chaperones Proteins that assist in the proper folding of other proteins, preventing misfolding and aggregation.

Molecular glue Small molecules that enhance or create an interaction between two proteins. For example, a molecular glue can create a new interaction between an E3 ligase and its substrate, leading to protein degradation.

Neurodegenerative disease Diseases associated with the progressive loss of neurons or of neuron function. Some neurodegenerative diseases are linked to chronic ER stress and protein misfolding.

Open reading frame (ORF) The continuous sequence of a gene from the start codon to the stop codon, uninterrupted by other stop codons, that codes for a protein.

Peptidase An enzyme that breaks peptide bonds within peptides.

Peptidyl site (P site) One of the sites on the ribosome. During translation, the tRNA moves to the P site after the A site.

Peptidyl transferase center A region of the large ribosomal subunit where peptide bonds are formed between amino acids.

PKR-like ER kinase (PERK) An ER stress sensor and kinase that phosphorylates eIF2α when it is activated.

Polyubiquitin chain A chain of ubiquitin molecules on a protein that are linked to each other.

Post-translational modification Modifications of proteins. These can include the addition of phosphate groups, methyl groups, and others.

PROTACs (proteolysis targeting chimeras) Small molecules that target a particular protein for destruction, rather than merely inactivating it like an enzyme inhibitor. They can be designed to recruit E3 ligases to specific proteins, triggering their ubiquitination and degradation.

Proteasome A protein complex that degrades unwanted, damaged, or misfolded proteins.

Rough endoplasmic reticulum (rough ER) A large membrane-bound organelle. The rough ER is studded with ribosomes and is the site of translation of proteins that will enter the secretory pathway.

Start codon (AUG) The first codon of an mRNA transcript that signals the beginning of translation and codes for methionine.

Stop codon A codon (UAA, UAG, or UGA) that signals the termination of translation.

Svedberg unit (S) A measure of sedimentation rate during ultracentrifugation. Used to describe the size of ribosomal subunits (e.g., 40S, 60S).

Transfer RNA (tRNA) The RNA that delivers the amino acid to its appropriate codon on the mRNA.

Translation The process by which protein is synthesized from an mRNA template.

Ubiquitin ligase (E3 ligase) Recognizes specific substrate proteins and catalyzes the transfer of ubiquitin from E2 to the substrate.

Ubiquitin A small protein that is covalently attached to target proteins to mark them for degradation by the proteasome.

Ubiquitin-activating enzyme (E1 enzyme) An enzyme that catalyzes the first step of ubiquitin activation using ATP.

Ubiquitination The process of attaching ubiquitin to a substrate protein.

Ubiquitin-conjugating enzyme (E2 enzyme) An enzyme that transfers activated ubiquitin from E1 to E3.

Ubiquitin-protein ligase E3A (UBE3A) A gene encoding an E3 ubiquitin ligase; mutations in this gene are linked to Angelman syndrome.

Unfolded protein response (UPR) A cellular stress response triggered by the accumulation of unfolded or misfolded proteins in the endoplasmic reticulum.

Wobble hypothesis According to this model, the first two bases of the codon form standard base pairs with the anticodon on the tRNA, but the third base can "wobble," forming nonstandard base pairing.

X-box binding protein 1 (XBP1) A transcription factor produced by IRE1-mediated splicing that enhances the expression of UPR genes.

Study Questions

Multiple-Choice Questions

1. What is the function of tRNA in translation?
   A. It is the same as mRNA
   B. To synthesize RNA from DNA
   C. To deliver amino acids to the ribosome
   D. To synthesize the ribosomes
2. Which codon is the start codon for translation?
   A. UAA
   B. GUG

   C. UAG
   D. AUG
3. Where does translation occur in the cell?
   A. In the nucleus
   B. On the Golgi apparatus
   C. Inside lysosomes
   D. At the ribosome
   E. In the ER lumen
4. What happens when a stop codon is reached during translation?
   A. A tRNA adds a random amino acid.
   B. A ribosome initiates folding.
   C. mRNA is degraded.
   D. A release factor enters the A site and the subunits of the ribosome separate.
   E. A release factor binds the E site of the ribosome so that no new tRNAs can enter the ribosome.
5. Why is the ribosome a good target for antibiotics bacterial ribosomes?
   A. Because it kills the bacteria because this inhibits transcription
   B. Because it kills the bacteria because this inhibits DNA replication
   C. Because it kills the bacteria because this inhibits lysosomal function
   D. Because it kills the bacteria by inhibiting protein synthesis
6. What is the primary goal of the unfolded protein response (UPR)?
   A. To promote protein degradation only
   B. To increase translation of all mRNAs
   C. To stimulate mitochondrial energy production
   D. To restore normal protein folding and reduce ER stress
7. Which of the following proteins is an ER stress sensor?
   A. JAK
   B. BAX
   C. p53
   D. PERK
8. Which chaperone protein binds unfolded proteins in the ER?
   A. Calmodulin
   B. HSP90
   C. Tubulin
   D. HSPA5 (BiP/GRP78)
9. What happens when PERK is activated?
   A. It cleaves APP to form amyloid.
   B. It exports proteins to lysosomes.
   C. It activates mitochondria.
   D. It phosphorylates eIF2⍰ to reduce global protein synthesis.
10. How is ATF6 activated?
   A. It is activated by phosphorylation.
   B. It is activated by cleavage.
   C. It is activated by methylation.
   D. It is activated by a combination of A, B, C.
11. What is CHOP's main role in the context of ER stress?
   A. It assists in protein folding.
   B. It leads to autophagy and destruction of the ER.
   C. It promotes cleavage of ATF6.
   D. It promotes apoptosis if ER stress is unresolved.
12. What is the role of ubiquitin in protein turnover?
   A. To fold proteins into correct structures
   B. To direct proteins to the Golgi apparatus
   C. To mark proteins for degradation by the proteasome
   D. To mark proteins for exocytosis
   E. Either C or D
13. Which part of the proteasome is considered the catalytic core?
   A. The 19S subunit
   B. The 20S subunit
   C. The ATF6
   D. The AAA+
14. Velcade (bortezomib) is used to treat which condition?
   A. Parkinson's disease
   B. Multiple myeloma
   C. Diabetes
   D. Angelman syndrome

True/False Questions

15. All codons code for amino acids.
16. Post-translational modifications can affect a protein's function.
17. The 19S subunit of the proteasome is the site where peptide bonds are cleaved to break down the protein into smaller pieces.

Fill-in-the-Blank Questions

18. The start codon for translation is ________.
19. The ribosome is made of proteins and __________.
20. The site on the ribosome where the initiator tRNA-Met binds is the ________.
21. The three stages of translation are initiation, __________, and termination.

Short-Answer Questions

22. What are three examples of post-translational modification a protein might have?
23. What role does eIF2□ phosphorylation play during ER stress?
24. Describe one advantage of PROTACs over traditional small -olecule inhibitors.
25. Ibermide is an investigational drug for the treatment of multiple myeloma. It leads to the destruction of two transcription factors, IKZF1 and IKZF3. How does it do so?

Matching Questions

26. Match the enzyme to its function:
    Enzyme
    A. E1
    B. E2
    C. E3
    Function
    1. Transfers ubiquitin to substrate
    2. Activates ubiquitin
    3. Carries ubiquitin
27. Match each translation stage with its main feature:
    Stage
    A. Initiation
    B. Elongation
    C. Termination
    Feature
    1. Peptide bond formation
    2. Start codon recognition
    3. Release factor binding

Answer Key

Multiple-Choice Questions

1. **Answer**: C
2. **Answer**: D
3. **Answer**: D
4. **Answer**: D
5. **Answer**: D
6. **Answer**: D
7. **Answer**: D
8. **Answer**: D
9. **Answer**: D
10. **Answer**: B
11. **Answer**: D
12. **Answer**: C
13. **Answer**: B
14. **Answer**: B

True/False Questions

15. **Answer**: False
16. **Answer**: True
17. **Answer**: False

Fill-in-the-Blank Questions

18. **Answer**: AUG
19. **Answer**: Ribosomal RNA
20. **Answer**: P site
21. **Answer**: Elongation

Short-Answer Questions

22. **Answer**: Phosphorylation, glycosylation, acetylation, methylation, ubiquitination, proline hydroxylation (any three of these)
23. **Answer**: It decreases global translation while selectively allowing ATF4 translation.
24. **Answer**: They lead to destruction of the target protein rather than inactivation, which can be temporary or incomplete
25. **Answer**: It binds to cereblon (CRBN), a substrate receptor within an E3 ubiquitin ligase complex. This causes the complex to alter its substrate specificity and lead to the degradation of these two transcription factors.

Matching Questions

26. **Answer**: A-2, B-3, C-1
27. **Answer**: A-2, B-1, C-3

References

1. Alberts B, Bray D, Hopkin K, Johnson A, Lewis J, Raff M, Roberts K, Walter P. Essential cell biology. Norton and Company; 2023.
2. Hames BD, Hooper NM. BIOS instant notes in biochemistry. Taylor and Francis; 2011.
3. Saier MH Jr. Understanding the genetic code. J Bacteriol. 2019;201(15):e00091-19. https://doi.org/10.1128/jb.00091-19.
4. Alberts B, Heald R, Johnson A, Morgan D, Raff M, Roberts K, Walter P, Wilson J, Hunt T. Molecular biology of the cell. W.W. Norton and Company; 2022.
5. Krebs JE, Goldstein ES, Kilpatrick ST. Lewin's genes XII. 12th ed. Jones and Bartlett Learning; 2017.
6. McLennan ABA, Elcock L, Turner P. BIOS instant notes in molecular biology. 4th ed. New York: Garland Science; 2012.
7. Cleaves HJ. Svedberg unit. In: Gargaud M, Irvine WM, Amils R, Cleaves HJ, Pinti DL, Quintanilla JC, et al., editors. Encyclopedia of astrobiology. Berlin, Heidelberg: Springer Berlin Heidelberg; 2015. p. 2446.
8. Wilson DN, Doudna Cate JH. The structure and function of the eukaryotic ribosome. Cold Spring Harb Perspect Biol. 2012;4(5):a011536. https://doi.org/10.1101/cshperspect.a011536.
9. Agris PF, Eruysal ER, Narendran A, Väre VYP, Vangaveti S, Ranganathan SV. Celebrating wobble decoding: half a century and still much is new. RNA Biol. 2018;15(4–5):537–53. https://doi.org/10.1080/15476286.2017.1356562.
10. Ramakrishnan V. Ribosome structure and the mechanism of translation. Cell. 2002;108(4):557–72. https://doi.org/10.1016/s0092-8674(02)00619-0.
11. Fletcher H, Hickey I. BIOS instant notes in genetics. Abingdon: Garland Science; 2012.
12. Hinnebusch AG. The scanning mechanism of eukaryotic translation initiation. Annu Rev Biochem. 2014;83:779–812. https://doi.org/10.1146/annurev-biochem-060713-035802.
13. Kolář MH, McGrath H, Nepomuceno FC, Černeková M. Three stages of nascent protein translocation through the ribosome exit tunnel. Wiley Interdiscip Rev: RNA. 2024;15(6):e1873. https://doi.org/10.1002/wrna.1873.
14. Hauser AR. Antibiotic basics for clinicians. 3rd ed. LWW; 2018.
15. U.S. Food and Drug Administration. Doxycycline capsules, USP. 1967.
16. Lin J, Zhou D, Steitz TA, Polikanov YS, Gagnon MG. Ribosome-targeting antibiotics: modes of action, mechanisms of resistance, and implications for drug design. Annu Rev Biochem. 2018;87:451–78. https://doi.org/10.1146/annurev-biochem-062917-011942.
17. U.S. Food and Drug Administration. Gentak(Gentamicin). Approval overview. 1970. Available from: https://www.accessdata.fda.gov/scripts/cder/daf/index.cfm?event=overview.process&ApplNo=065024.
18. U.S. Food and Drug Administration. Erythrocin (Erythromycin). Approval overview. 1952. Available from: https://www.accessdata.fda.gov/scripts/cder/daf/index.cfm?event=overview.process&ApplNo=050609.
19. U.S. Food and Drug Administration. Cleocin (Clindamycin). Approval overview. 1970. Available from: https://www.accessdata.fda.gov/scripts/cder/daf/index.cfm?event=overview.process&ApplNo=050162.
20. U.S. Food and Drug Administration. Zyvox (Linezolid). Approval overview. 2000. Available from: https://www.accessdata.fda.gov/scripts/cder/daf/index.cfm?event=overview.process&ApplNo=021130.
21. Fersht AR. Structure and mechanisms in protein science: a guide to enzyme catalysis and protein folding. World Scientific Publishing Co.; 2017.
22. Liberek K, Lewandowska A, Zietkiewicz S. Chaperones in control of protein disaggregation. EMBO J. 2008;27(2):328–35. https://doi.org/10.1038/sj.emboj.7601970.

23. Dobson CM. Protein folding and misfolding. Nature. 2003;426(6968):884–90. https://doi.org/10.1038/nature02261.
24. Singh MK, Shin Y, Han S, Ha J, Tiwari PK, Kim SS, et al. Molecular chaperonin HSP60: current understanding and future prospects. Int J Mol Sci. 2024;25(10):5483. https://doi.org/10.3390/ijms25105483.
25. Garbuzynskiy SO, Marchenkov VV, Marchenko NY, Semisotnov GV, Finkelstein AV. How proteins manage to fold and how chaperones manage to assist the folding. Phys Life Rev. 2025;52:66–79. https://doi.org/10.1016/j.plrev.2024.12.006.
26. Wang J, Lee J, Liem D, Ping P. HSPA5 gene encoding Hsp70 chaperone BiP in the endoplasmic reticulum. Gene. 2017;618:14–23. https://doi.org/10.1016/j.gene.2017.03.005.
27. Adams BM, Oster ME, Hebert DN. Protein quality control in the endoplasmic reticulum. Protein J. 2019;38(3):317–29. https://doi.org/10.1007/s10930-019-09831-w.
28. Stollar EJ, Smith DP. Uncovering protein structure. Essays Biochem. 2020;64(4):649–80. https://doi.org/10.1042/ebc20190042.
29. Suskiewicz MJ. The logic of protein post-translational modifications (PTMs): chemistry, mechanisms and evolution of protein regulation through covalent attachments. Bioessays. 2024;46(3):e2300178. https://doi.org/10.1002/bies.202300178.
30. Cooper G, Adams K. The cell: a molecular approach. Sinauer Associates; 2022.
31. Chen X, Shi C, He M, Xiong S, Xia X. Endoplasmic reticulum stress: molecular mechanism and therapeutic targets. Signal Transduct Target Ther. 2023;8(1):352. https://doi.org/10.1038/s41392-023-01570-w.
32. Young SK, Wek RC. Upstream open reading frames differentially regulate gene-specific translation in the integrated stress response. J Biol Chem. 2016;291(33):16927–35. https://doi.org/10.1074/jbc.R116.733899.
33. Vattem KM, Wek RC. Reinitiation involving upstream ORFs regulates ATF4 mRNA translation in mammalian cells. Proc Natl Acad Sci. 2004;101(31):11269–74. https://doi.org/10.1073/pnas.0400541101.
34. Hetz C, Papa FR. The unfolded protein response and cell fate control. Mol Cell. 2018;69(2):169–81. https://doi.org/10.1016/j.molcel.2017.06.017.
35. Chakrabarti A, Chen AW, Varner JD. A review of the mammalian unfolded protein response. Biotechnol Bioeng. 2011;108(12):2777–93. https://doi.org/10.1002/bit.23282.
36. Read A, Schröder M. The unfolded protein response: an overview. Biology (Basel). 2021;10(5):384. https://doi.org/10.3390/biology10050384.
37. Hetz C, Zhang K, Kaufman RJ. Mechanisms, regulation and functions of the unfolded protein response. Nat Rev Mol Cell Biol. 2020;21(8):421–38. https://doi.org/10.1038/s41580-020-0250-z.
38. Meyer H, Weihl CC. The VCP/p97 system at a glance: connecting cellular function to disease pathogenesis. J Cell Sci. 2014;127(Pt 18):3877–83. https://doi.org/10.1242/jcs.093831.
39. Hu H, Tian M, Ding C, Yu S. The C/EBP homologous protein (CHOP) transcription factor functions in endoplasmic reticulum stress-induced apoptosis and microbial infection. Front Immunol. 2018;9:3083. https://doi.org/10.3389/fimmu.2018.03083.
40. Arunagiri A, Haataja L, Cunningham CN, Shrestha N, Tsai B, Qi L, et al. Misfolded proinsulin in the endoplasmic reticulum during development of beta cell failure in diabetes. Ann N Y Acad Sci. 2018;1418(1):5–19. https://doi.org/10.1111/nyas.13531.
41. Hotamisligil GS. Endoplasmic reticulum stress and the inflammatory basis of metabolic disease. Cell. 2010;140(6):900–17. https://doi.org/10.1016/j.cell.2010.02.034.
42. Scheuner D, Kaufman RJ. The unfolded protein response: a pathway that links insulin demand with beta-cell failure and diabetes. Endocr Rev. 2008;29(3):317–33. https://doi.org/10.1210/er.2007-0039.
43. Yong J, Johnson JD, Arvan P, Han J, Kaufman RJ. Therapeutic opportunities for pancreatic β-cell ER stress in diabetes mellitus. Nat Rev Endocrinol. 2021;17(8):455–67. https://doi.org/10.1038/s41574-021-00510-4.
44. Ajoolabady A, Lindholm D, Ren J, Pratico D. ER stress and UPR in Alzheimer's disease: mechanisms, pathogenesis, treatments. Cell Death Dis. 2022;13(8):706. https://doi.org/10.1038/s41419-022-05153-5.
45. Yuan S, She D, Jiang S, Deng N, Peng J, Ma L. Endoplasmic reticulum stress and therapeutic strategies in metabolic, neurodegenerative diseases and cancer. Mol Med. 2024;30(1):40. https://doi.org/10.1186/s10020-024-00808-9.
46. Ghemrawi R, Khair M. Endoplasmic reticulum stress and unfolded protein response in neurodegenerative diseases. Int J Mol Sci. 2020;21(17):6127. https://doi.org/10.3390/ijms21176127.
47. U.S. Food and Drug Administration. Kisunla (Donanemab-azbt). Approval overview. 2024. Available from: https://www.fda.gov/drugs/drug-approvals-and-databases/drug-trials-snapshots-kisunla?utm.
48. Kang C. Donanemab: first approval. Drugs. 2024;84(10):1313–8. https://doi.org/10.1007/s40265-024-02087-4.
49. Eli Lilly and Company. Kisunla™ (donanemab-azbt). 2024. Available from: https://kisunla.lilly.com/.
50. Khartabil N, Awaness A. Targeting amyloid pathology in early Alzheimer's: the promise of donanemab-azbt. Pharmacy (Basel). 2025;13(1):23. https://doi.org/10.3390/pharmacy13010023.
51. National Institute of Aging. Alzheimer's disease fact sheet. 2023. Available from: https://www.nia.nih.gov/health/alzheimers-and-dementia/alzheimers-

disease-fact-sheet#:~:text=body%20shuts%20down.-,What%20causes%20Alzheimer's%20disease?,differ%20from%20person%20to%20person.
52. Price DL, Koike MA, Khan A, Wrasidlo W, Rockenstein E, Masliah E, et al. The small molecule alpha-synuclein misfolding inhibitor, NPT200-11, produces multiple benefits in an animal model of Parkinson's disease. Sci Rep. 2018;8(1):16165. https://doi.org/10.1038/s41598-018-34490-9.
53. AlzForum Networking for a Cure. Therapeutics: minzasolmin. 2024. Available from: https://www.alzforum.org/therapeutics/minzasolmin.
54. Jeon YM, Kwon Y, Lee S, Kim HJ. Potential roles of the endoplasmic reticulum stress pathway in amyotrophic lateral sclerosis. Front Aging Neurosci. 2023;15:1047897. https://doi.org/10.3389/fnagi.2023.1047897.
55. Walker AK, Soo KY, Sundaramoorthy V, Parakh S, Ma Y, Farg MA, et al. ALS-associated TDP-43 induces endoplasmic reticulum stress, which drives cytoplasmic TDP-43 accumulation and stress granule formation. PLoS One. 2013;8(11):e81170. https://doi.org/10.1371/journal.pone.0081170.
56. de Mena L, Lopez-Scarim J, Rincon-Limas DE. TDP-43 and ER stress in neurodegeneration: friends or foes? Front Mol Neurosci. 2021;14:772226. https://doi.org/10.3389/fnmol.2021.772226.
57. ALS Association. Tofersen: current status. Available from: https://www.als.org/navigating-als/living-with-als/fda-approved-drugs/tofersen.
58. U.S. Food and Drug Administration. Qalsody (Tofersen). Approval overview. 2023. Available from: https://www.accessdata.fda.gov/scripts/cder/daf/index.cfm?event=overview.process&ApplNo=215887.
59. Blair HA. Tofersen: first approval. Drugs. 2023;83(11):1039–43. https://doi.org/10.1007/s40265-023-01904-6.
60. U.S. Food and Drug Administration. FDA approves treatment of amyotrophic lateral sclerosis associated with a mutation in the SOD1 gene. 2023. Available from: https://www.fda.gov/drugs/news-events-human-drugs/fda-approves-treatment-amyotrophic-lateral-sclerosis-associated-mutation-sod1-gene#:~:text=FDA%20approved%20Qalsody%20(tofersen)%20to,maintenance%20dose%20every%2028%20days.
61. McDonald EF, Meiler J, Plate L. CFTR folding: from structure and proteostasis to cystic fibrosis personalized medicine. ACS Chem Biol. 2023;18(10):2128–43. https://doi.org/10.1021/acschembio.3c00310.
62. U.S. Food and Drug Administration. Orkambi (Lumacaftor/Ivacaftor) (FDA). Approval overview. 2015. Available from: https://www.accessdata.fda.gov/scripts/cder/daf/index.cfm?event=overview.process&ApplNo=206038.
63. Fiedorczuk K, Chen J. Mechanism of CFTR correction by type I folding correctors. Cell. 2022;185(1):158–68.e11. https://doi.org/10.1016/j.cell.2021.12.009.
64. Symdeko. SYMDEKO® works on treating the underlying cause. 2023. Available from: https://www.symdeko.com/.
65. Poli L, Labella B, Cotti Piccinelli S, Caria F, Risi B, Damioli S, et al. Hereditary transthyretin amyloidosis: a comprehensive review with a focus on peripheral neuropathy. Front Neurol. 2023;14:1242815. https://doi.org/10.3389/fneur.2023.1242815.
66. U.S. Food and Drug Administration. Vyndaqel (Tafamidis). Approval overview. 2019. Available from: https://www.accessdata.fda.gov/scripts/cder/daf/index.cfm?event=overview.process&ApplNo=211996.
67. U.S. Food and Drug Administration. Clinical reviews. 2019. Available from: https://www.accessdata.fda.gov/drugsatfda_docs/nda/2019/211996Orig1s000%2C%202121610Orig1s000MedR.pdf.
68. Noormohammadi A, Calculli G, Gutierrez-Garcia R, Khodakarami A, Koyuncu S, Vilchez D. Mechanisms of protein homeostasis (proteostasis) maintain stem cell identity in mammalian pluripotent stem cells. Cell Mol Life Sci. 2018;75(2):275–90. https://doi.org/10.1007/s00018-017-2602-1.
69. Mayer RJ, Ciechanover AJ, Rechsteiner M, editors. Protein degradation: the ubiquitin-proteasome system and disease. 1st ed. Germany: Wiley; 2007.
70. Berg JM, Tymoczko JL, Gatto GJ Jr, Stryer L. Biochemistry. 9th ed. New York: MacMillan Learning; 2019.
71. Marks F, Klingmüller U, Müller-Decker K. Cellular signal processing: an introduction to the molecular mechanisms of signal transduction. 2nd ed. Garland Science; 2017.
72. Nandi D, Tahiliani P, Kumar A, Chandu D. The ubiquitin-proteasome system. J Biosci. 2006;31(1):137–55. https://doi.org/10.1007/bf02705243.
73. Gong X, Du D, Deng Y, Zhou Y, Sun L, Yuan S. The structure and regulation of the E3 ubiquitin ligase HUWE1 and its biological functions in cancer. Investig New Drugs. 2020;38(2):515–24. https://doi.org/10.1007/s10637-020-00894-6.
74. Toma-Fukai S, Shimizu T. Structural diversity of ubiquitin E3 ligase. Molecules. 2021;26(21):6682. https://doi.org/10.3390/molecules26216682.
75. Ravid T, Hochstrasser M. Diversity of degradation signals in the ubiquitin–proteasome system. Nat Rev Mol Cell Biol. 2008;9(9):679–89. https://doi.org/10.1038/nrm2468.
76. Schwab M. Encyclopedia of cancer. Berlin, Heidelberg: Springer; 2017.
77. Bard JAM, Goodall EA, Greene ER, Jonsson E, Dong KC, Martin A. Structure and function of the 26S proteasome. Annu Rev Biochem. 2018;87:697–724. https://doi.org/10.1146/annurev-biochem-062917-011931.

78. Zhang H, Zhou C, Mohammad Z, Zhao J. Structural basis of human 20S proteasome biogenesis. Nat Commun. 2024;15(1):8184. https://doi.org/10.1038/s41467-024-52513-0.
79. Camões Dos Santos J, Appleton C, Cazaux Mateus F, Covas R, Bekman EP, da Rocha ST. Stem cell models of Angelman syndrome. Front Cell Dev Biol. 2023;11:1274040. https://doi.org/10.3389/fcell.2023.1274040.
80. Wang T, Li J, Yang L, Wu M, Ma Q. The role of long non-coding RNAs in human imprinting disorders: prospective therapeutic targets. Front Cell Dev Biol. 2021;9:730014. https://doi.org/10.3389/fcell.2021.730014.
81. Vihma H, Li K, Welton-Arndt A, Smith AL, Bettadapur KR, Gilmore RB, et al. Ube3a unsilencer for the potential treatment of Angelman syndrome. Nat Commun. 2024;15(1):5558. https://doi.org/10.1038/s41467-024-49788-8.
82. Khatri N, Man HY. The autism and Angelman syndrome protein Ube3A/E6AP: the gene, E3 ligase ubiquitination targets and neurobiological functions. Front Mol Neurosci. 2019;12:109. https://doi.org/10.3389/fnmol.2019.00109.
83. Glickman MH, Ciechanover A. The ubiquitin-proteasome proteolytic pathway: destruction for the sake of construction. Physiol Rev. 2002;82(2):373–428. https://doi.org/10.1152/physrev.00027.2001.
84. U.S. Food and Drug Administration. Velcade (Bortezomib). Approval overview. 2003. Available from: https://www.accessdata.fda.gov/scripts/cder/daf/index.cfm?event=overview.process&ApplNo=021602.
85. Scott K, Hayden PJ, Will A, Wheatley K, Coyne I. Bortezomib for the treatment of multiple myeloma. Cochrane Database Syst Rev. 2016;4(4):CD010816. https://doi.org/10.1002/14651858.CD010816.pub2.
86. Sogbein O, Paul P, Umar M, Chaari A, Batuman V, Upadhyay R. Bortezomib in cancer therapy: mechanisms, side effects, and future proteasome inhibitors. Life Sci. 2024;358:123125. https://doi.org/10.1016/j.lfs.2024.123125.
87. U.S. Food and Drug Administration. Kyprolis (Carfilzomib). Approval overview. 2012. Available from: https://www.accessdata.fda.gov/scripts/cder/daf/index.cfm?event=overview.process&ApplNo=202714.
88. Kortuem KM, Stewart AK. Carfilzomib. Blood. 2013;121(6):893–7. https://doi.org/10.1182/blood-2012-10-459883.
89. Chauhan D, Hideshima T, Anderson KC. Proteasome inhibition in multiple myeloma: therapeutic implication. Annu Rev Pharmacol Toxicol. 2005;45:465–76. https://doi.org/10.1146/annurev.pharmtox.45.120403.100037.
90. Li K, Crews CM. PROTACs: past, present and future. Chem Soc Rev. 2022;51(12):5214–36. https://doi.org/10.1039/d2cs00193d.
91. Yokoo H, Naito M, Demizu Y. Investigating the cell permeability of proteolysis-targeting chimeras (PROTACs). Expert Opin Drug Discov. 2023;18(4):357–61. https://doi.org/10.1080/17460441.2023.2187047.
92. Chen QH, Munoz E, Ashong D. Insight into recent advances in degrading androgen receptor for castration-resistant prostate cancer. Cancers (Basel). 2024;16(3):663. https://doi.org/10.3390/cancers16030663.
93. Pfizer. Arvinas and Pfizer's vepdegestrant (ARV-471) receives FDA fast track designation for the treatment of patients with ER+/HER2− metastatic breast cancer. 2024. Available from: https://www.pfizer.com/news/announcements/arvinas-and-pfizers-vepdegestrant-arv-471-receives-fda-fast-track-designation.
94. Hamilton EP, Ma C, De Laurentiis M, Iwata H, Hurvitz SA, Wander SA, et al. VERITAC-2: a phase III study of vepdegestrant, a PROTAC ER degrader, versus fulvestrant in ER+/HER2− advanced breast cancer. Future Oncol. 2024;20(32):2447–55. https://doi.org/10.1080/14796694.2024.2377530.
95. Kymera Therapeutics. Release details. 2024. Available from: https://investors.kymeratx.com/news-releases/news-release-details/kymera-therapeutics-receives-us-fda-fast-track-designation-kt.
96. Hu Z, Crews CM. Recent developments in PROTAC-mediated protein degradation: from bench to clinic. Chembiochem. 2022;23(2):e202100270. https://doi.org/10.1002/cbic.202100270.
97. Neri P. CRISPRing the CRL4(CRBN) RING in multiple myeloma. Blood. 2018;132(12):1219–20. https://doi.org/10.1182/blood-2018-08-867069.
98. MultipleMyelomaHub. MultipleMyelomaHub. 2025. Available from: https://multiplemyelomahub.com/.
99. Lonial S, Dimopoulos MA, Berdeja JG, Richardson PG, Quach H, Rodríguez-Otero P, et al. EXCALIBER-RRMM: a phase III trial of iberdomide, daratumumab, and dexamethasone in relapsed/refractory multiple myeloma. Future Oncol. 2025;21(14):1761–9. https://doi.org/10.1080/14796694.2025.2501920.
100. U.S. Food and Drug Administration. Revlimid (Lenalidomide). Approval overview. 2005. Available from: https://www.accessdata.fda.gov/scripts/cder/daf/index.cfm?event=overview.process&ApplNo=021880.
101. Katzung BG, Vanderah TW, Trevor AJ. Katzung's basic and clinical pharmacology. 16th ed. McGraw-Hill Education; 2024.
102. Zhou N, Gutierrez-Uzquiza A, Zheng XY, Chang R, Vogl DT, Garfall AL, et al. RUNX proteins desensitize multiple myeloma to lenalidomide via protecting IKZFs from degradation. Leukemia. 2019;33(8):2006–21. https://doi.org/10.1038/s41375-019-0403-2.

103. Amatangelo M, Flynt E, Stong N, Ray P, Van Oekelen O, Wang M, et al. Pharmacodynamic changes in tumor and immune cells drive iberdomide's clinical mechanisms of activity in relapsed and refractory multiple myeloma. Cell Rep Med. 2024;5(6):101571. https://doi.org/10.1016/j.xcrm.2024.101571.
104. Tamura K. Origins and early evolution of the tRNA molecule. Life. 2015;5(4):1687–99. https://doi.org/10.3390/life5041687.
105. Adams CJ, et al. Structure and molecular mechanism of ER stress signaling by the unfolded protein response signal activator IRE1. Front Mol Biosci. 2019;6:11. https://doi.org/10.3389/fmolb.2019.00011.
106. Turpin J, El-Safadi D, Lebeau G, et al. CHOP pro-apoptotic transcriptional program in response to ER stress is hacked by zika virus. Int J Mol Sci. 2021;22(7):3750. https://doi.org/10.3390/ijms22073750.
107. Sato T, et al. Cereblon-based small-molecule compounds to control neural stem cell proliferation in regenerative medicine. Front Cell Dev Biol. 2021;9:629326. https://doi.org/10.3389/fcell.2021.629326.

Part IV

Carbohydrates and Lipids

The five chapters in this part cover carbohydrates and lipids, their basic structures, functions, and metabolism. Like proteins and nucleic acids, carbohydrates and lipids are macromolecules in biochemistry. They have especially important roles as energy sources and in metabolism. The chapters in this part are organized as follows: Chapter 9 covers the basics of carbohydrates and lipids. It covers the overall structure of these molecules and some of their essential functions and briefly introduces how they are metabolized. The next chapters in this part delve into the metabolism of carbohydrates and lipids. Chapter 10 begins with a general introduction to metabolism as a field within biochemistry, followed by a discussion of glucose metabolism, which lies at the heart of energy metabolism and is essential for understanding all other metabolic processes. Chapter 11 covers lipid metabolism, and Chap. 13 covers metabolic homeostasis, which is the body's ability to maintain a stable internal environment even as its energy needs fluctuate. Chapter 14 examines diseases associated with carbohydrate and lipid metabolism. Many human disorders are associated with disturbances in carbohydrate and lipid metabolism, including diabetes, obesity, and fatty liver disease. This chapter introduces various strategies for managing and treating these disorders.

9 Carbohydrates and Lipids

Abstract

Carbohydrates and lipids are essential biomolecules with diverse structural and metabolic roles. This chapter covers the basic structures and functions of carbohydrates and lipids. The basic structures of monosaccharides are described, and the differences between monosaccharides, disaccharides, oligosaccharides, and polysaccharides are discussed. Glycosidic bonds, which link monosaccharides together, are described. Some important polysaccharides, such as glycogen, are introduced as important energy sources. The basic structures of lipids are discussed, including fatty acids, the simplest forms of lipids. Other lipids, including triacylglycerols (TAGs), which are important storage forms, are also discussed. The differences between saturated and unsaturated fatty acids are described, as are the health impacts of dietary fatty acids. Cholesterol and the dangers of high low-density lipoprotein levels are discussed, as well as drugs designed to treat familial hypercholesterolemia. Metabolic disorders involving carbohydrates and lipids are introduced, along with some available treatments.

Keywords

Carbohydrates · Lipids · Fatty acids · Saturated fatty acids · Triacylglycerols (TAGs) · Steroids

Abbreviations

ApoC-III	Apolipoprotein C-III
ASO	Antisense oligonucleotide
ATP	Adenosine triphosphate
CH_3	Methyl group
CoA	Coenzyme A
COOH	Carboxylic acid group
CRISPR	Clustered regularly interspaced short palindromic repeats
ER	Endoplasmic reticulum
FA	Fatty acid
$FADH_2$	Flavin adenine dinucleotide (reduced form)
Fc	Fragment crystallizable region (of an antibody)
FFA	Free fatty acid
G6P	Glucose-6-phosphate
GalNAc	N-acetylgalactosamine
GD2	Disialoganglioside 2
GLUT	Glucose transporter
HDL	High-density lipoprotein
HMG-CoA	3-Hydroxy-3-methylglutaryl-coenzyme A
IDL	Intermediate-density lipoprotein
IgG	Immunoglobulin G
LDL	Low-density lipoprotein
LDLR	Low-density lipoprotein receptor
mRNA	Messenger RNA
MUFAs	Monounsaturated fatty acids
NADH	Nicotinamide adenine dinucleotide (reduced form)

A. Minden, *Biochemistry, Cell and Molecular Biology for Pharmacists and Health Professionals*,
https://doi.org/10.1007/978-3-032-20529-2_9

PCSK9	Proprotein convertase subtilisin/kexin type 9
PEG	Polyethylene glycol
PI	Phosphatidylinositol
PS	Phosphatidylserine
PUFAs	Polyunsaturated fatty acids
RME	Receptor-mediated endocytosis
RNAse H	Ribonuclease H
TAG	Triacylglycerol (also referred to as triglyceride)
UDP	Uridine diphosphate
VLDL	Very low-density lipoprotein

What You Will Learn

- The basic structure of carbohydrates, their basic chemical formula.
- Monosaccharides, their structure, and different stereoisomers, including D and L forms, aldoses and ketoses, linear and cyclic forms.
- The difference between monosaccharides, disaccharides, oligosaccharides, and polysaccharides.
- Glycosidic bonds that link monosaccharides together.
- Some important polysaccharides, such as glycogen, a major form of energy storage.
- The importance of carbohydrates in the field of pharmacy: their roles in metabolism and energy storage.
- Abnormalities in glycosylation in diseases like cancer.
- Lactose intolerance.
- The basic structure of lipids and some of their functions.
- The structures of fatty acids, the simplest lipid, and the difference between saturated and unsaturated fatty acids.
- Fatty acids in diet and health: the problem with trans-fatty acids.
- Triacylglycerols (TAGs) as storage forms of fatty acids.
- Diseases associated with TAGs and drugs designed to treat them.
- The structures and functions of other lipids, such as phospholipids, terpenes, and terpenoids.
- Steroids, including cholesterol, and their packaging into lipoprotein particles such as LDL and HDL. High cholesterol levels and human health, drugs such as statins, designed to treat elevated cholesterol.
- Metabolic disorders involving lipids, such as obesity.
- Lipids used in drug delivery, such as lipoproteins.

This chapter explores carbohydrates and lipids. It is divided into two sections. Section 1 covers carbohydrates, and Sect. 2 covers lipids.

1 Carbohydrates

1.1 Introduction to Carbohydrates

Carbohydrates are the primary and most readily available energy source for most animals. Carbohydrates provide energy in the form of calories (more precisely, kilocalories or Kcals), which are units of energy. Many of our carbohydrates are consumed in our diet. Most of these come from plants (which include fruits, vegetables, and grains such as flour) and are broken down and used for energy. While energy production is their primary role, carbohydrates also have other functions, such as providing structural support and participating in cell signaling. Carbohydrates get their name because they are hydrates of carbon; they are carbon molecules that have hydrogen and hydroxyl groups [1, 2].

Carbohydrates come in a range of sizes. They include monosaccharides (simple sugars), such as glucose and fructose; disaccharides (double sugars), such as sucrose and lactose; oligosaccharides, which are often found attached to proteins, and large complex carbohydrates, such as starch and glycogen. The discussion of carbohydrates and lipids in this chapter draws from the following sources [1, 3, 6], as well as references within the text.

1.2 Monosaccharides

1.2.1 Monosaccharide Structures

Like other macromolecules, carbohydrates are polymers that are made up of monomers. Monosaccharides are the monomers that make up carbohydrates, and carbohydrates that are made up of multiple monosaccharides are called disaccharides, polysaccharides, or oligosaccharides. Monosaccharides are also called sugars, or simple sugars. Monosaccharides are components of larger molecules, but individual monosaccharides also have many important functions. The names of monosaccharides (as well as disaccharides and some polysaccharides) usually end in the name "ose."

There are multiple different monosaccharides, and they can be categorized according to the number of carbons they have. For example, a monosaccharide can be a triose, tetrose, pentose, or hexose if it contains 3, 4, 5, or 6 carbons, respectively (Fig. 9.1). Most often in nature, we see pentoses and hexoses. The usual formula for a monosaccharide is $(CH_2O)n$, with n being the number of carbons they have. The carbons in a monosaccharide have names to identify them. The carbonyl carbon has a carbonyl group (C double-bonded to an O; C=O). The central carbons are called chiral centers. These carbons are bonded to each other and to an H and an OH. The terminal carbon is the one (in a linear molecule) that is farthest from the carbonyl carbon. It is bound to H_2OH. The penultimate carbon is the one right before the terminal carbon, and it is also a chiral center. The numbers can be referred to as C1, C2,..., starting with the carbonyl carbon [1, 3, 6].

1.2.2 Stereoisomers

- Stereoisomers Have Different Spatial Arrangements

Various characteristics can differentiate different monosaccharides from each other, aside from the number of carbons they have. For example, there can be stereoisomers, meaning that the molecules have the same chemical formulas but have different spatial arrangements. For monosaccharides, stereoisomers can vary based on the positions of the H and OH groups around the chiral carbons. Figure 9.2 shows an example of different pentoses (5 carbons), which are all stereoisomers that differ only by the positions of the H and OH groups [1, 3–5, 6].

- D and L Forms

In addition to the changes in the arrangements of the H and OH groups, stereoisomers can be either D forms or L forms, which are enantiomers, meaning that they are mirror images of each other (the chiral centers are mirror images of

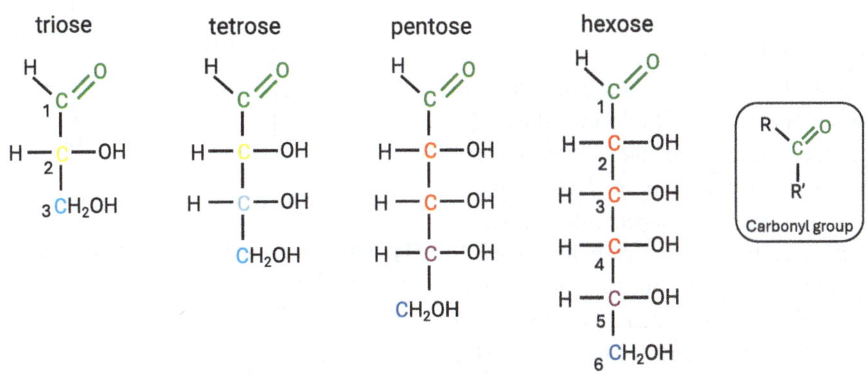

Fig. 9.1 The basic structure of monosaccharides, and some examples. The formula for monosaccharides is $(CH_2O)n$. These monosaccharides have a carbonyl group on one end, which contains a carbonyl carbon (C double-bonded to an O). Chiral centers, in red, are carbons that are attached to four different species. The penultimate carbon, in darker red, is the second-to-last carbon, and it is also a chiral center. Monosaccharides can be categorized based on the number of carbons they contain. In this example the triose is glyceraldehyde, the tetrose is erythrose, the pentose is ribose, hexose is aldohexose

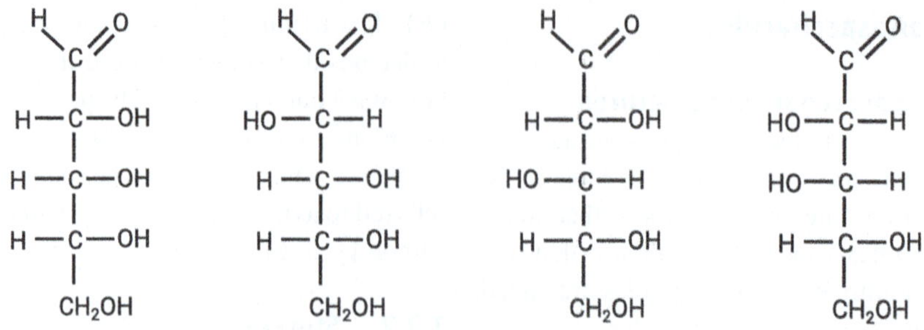

Fig. 9.2 Stereoisomers. In stereoisomers, the chemical formula stays the same, but the positions of the H and -OH groups vary. Some examples are shown here

each other). In nature, the D forms are more common. We use the penultimate carbon to identify whether a monosaccharide is in D or L form (Fig. 9.3) [1, 3, 6].

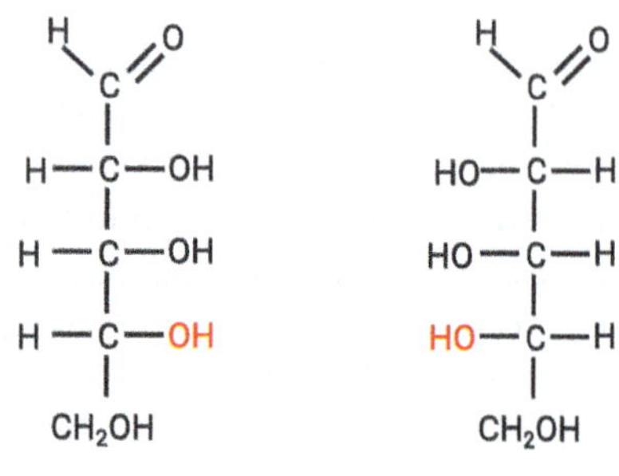

Fig. 9.3 Enantiomers. Enantiomers can be D or L forms. D and L forms are mirror images of each other

- Aldoses and Ketoses

Monosaccharides can also differ based on whether they are aldoses or ketoses (Fig. 9.4). In an **aldose**, the carbonyl group at the end of the carbon chain (C1) forms an aldehyde structure. In a ketose, the carbonyl group is located within the carbon chain (C2), forming a ketone structure (see *Appendix*). In an aldose, carbon 1 is called the **anomeric carbon**, and in a ketone, carbon 2 is the anomeric carbon. Glucose and fructose are examples of hexoses (6 carbons), where one is an aldose (glucose), and one is a ketose (fructose). Most carbohydrates are aldoses [1, 6].

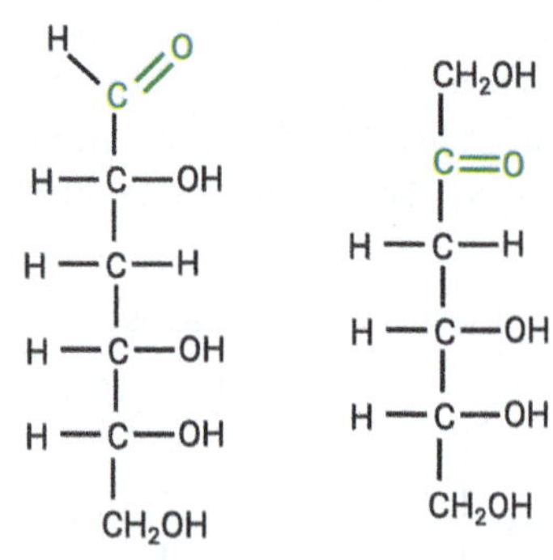

Fig. 9.4 Aldoses and ketoses. **Glucose** and **fructose** are examples of hexoses (6 carbons), where one is an aldose (glucose) and one is a ketose (fructose)

- Linear and Cyclic Forms

Depending on their sizes, monosaccharides can cycle between linear and cyclic forms. The cyclic form is produced due to an interaction between the anomeric carbon and one of the OH groups on a carbon in the same monosaccharide (generally the penultimate carbon or the one before that).

Here is how the cyclization reaction works, using the 6-carbon monosaccharide glucose as an example, where the bond forms between **C1** (the anomeric carbon) and the OH on **C5** (the penultimate carbon): When the monosaccharide is in an aqueous environment, the hydroxyl (OH) group on C5 performs a nucleophilic attack on the carbonyl carbon at C1 (the anomeric carbon). This converts the linear form of glucose into a ring structure. During this process, the OH on **C5** loses a proton (H^+) ion, often facilitated by solvent molecules like water. Meanwhile, the oxygen (O) on the carbonyl group on **C1** gains a proton (H^+), converting the carbonyl (C=O) into a hydroxyl group (C-OH). The resulting 6 cyclic molecule is a 6-membered ring. In this ring, C1, the anomeric carbon, is now also a new chiral center. This ring is considered a 6-membered ring because it has five Cs and one O. (C6, the terminal carbon, is not part of the actual ring structure). Both the linear and the cyclic forms of glucose are shown in Fig. 9.5. Monosaccharides can form both linear and cyclic forms, and they can interconvert, but the cyclic forms are favored in the body. It is usually the 5- or 6-carbon monosaccharides that take on cyclic forms.

Note that Fig. 9.5 also illustrates several common ways that monosaccharides can be depicted. The linear open-chain form is shown as a Fisher projection; the cyclic form is represented here as both a chair conformation and a Haworth projection. In all of these conformations, the carbons are explicitly labeled, but they are understood to be located at the vertices (bends) of the structure. Likewise, most hydrogens attached to carbons are omitted in the cyclic depictions. When all the carbons and hydrogens (Cs and Hs) are illustrated explicitly, as in Fig. 9.4, the depiction is referred to as the structural formula.

- Cyclic Monosaccharides Can Be α or β Anomers

Cyclic monosaccharides can be in the form of α or β anomer. When a cyclic monosaccharide forms, the hydroxyl group attacks the carbonyl carbon, meaning that it donates a pair of electrons to make a new covalent bond. This carbonyl carbon is now considered to be the anomeric carbon. As a result of the formation of this cyclic structure, two different **isomers** can form. In one isomer, the OH group attached to the anomeric carbon is positioned away from the CH_2OH group (the terminal carbon), and in the other, it faces the same direction. These are called the alpha and beta anomers, respectively (Fig. 9.6), and the anomeric carbons are referred to as alpha or beta anomeric carbons.

- Cyclic Monosaccharides Can Be Pyranose or Furanose Structures

Finally, a 6-membered monosaccharide (a hexose) that is an aldose can form either a pyranose or furanose structure. A pyranose contains a 6-membered ring, where the ring contains 5 carbons and an oxygen (C6, the terminal carbon, is not part of the ring structure), while a furanose is a 5-membered ring, containing 4 carbons and an oxygen. (The names are taken from the pyran and the furan, which are 6 and 5-membered rings, respectively). A monosaccharide with 6 carbons can form either of these, depending on which hydroxyl group (OH) interacts with the anomeric carbon (Fig. 9.7). A pyranose results when the OH on C5 attacks C1 to form the ring, while a furanose results when the OH group on C4 attacks C1 to

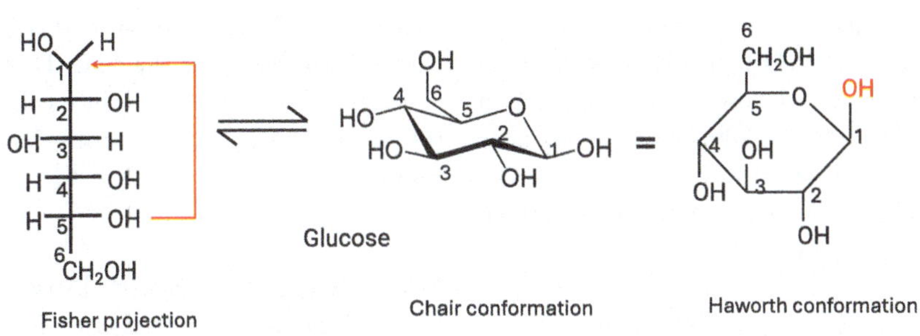

Fig. 9.5 Linear and cyclic forms of monosaccharides. Monosaccharides can cycle between a linear and cyclic form. Shown here are glucose in its linear and cyclic forms. Also shown are the different methods of presentation, the Fisher projection, chair conformation, and Haworth conformation. In aqueous environments, there is a balance between linear and cyclic forms of monosaccharides, though the cyclic forms predominate. Monosaccharides with fewer than five carbons generally do not form rings. (The cyclic form of glucose is also called glucopyranose). (*Source*: Chair conformation is modified from Jü (2012), reproduced under CC0 from Wikimedia Commons File:TetrahydroPYRANE_V.2.svg))

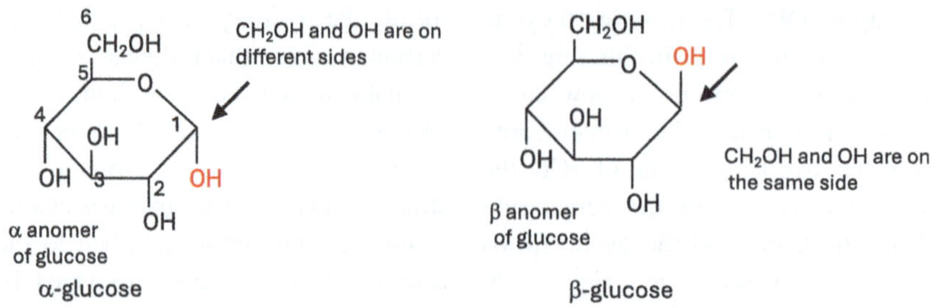

Fig. 9.6 Cyclic monosaccharides can exist in either alpha or beta anomer forms. Shown here are Haworth projections of alpha and beta forms of glucose. The alpha or beta forms depend on the arrangement of the OH around the carbon closest to the oxygen (O) in the ring (the anomeric carbon)

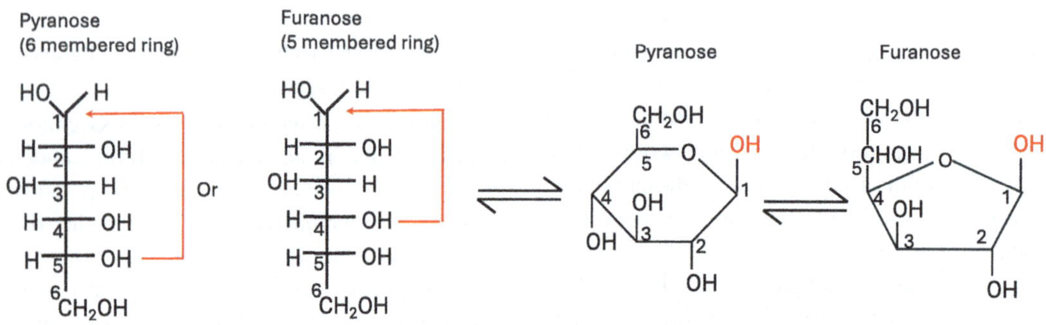

Fig. 9.7 Pyranose and furanose forms of cyclic monosaccharides. In the pyranose form of a hexose, the OH group on the fifth carbon (C-5) attacks the carbonyl carbon (C-1) to form a 6-membered ring. In the furanose, the OH group on the fourth carbon (C-4) attacks the carbonyl carbon (C-1), leading to a five-membered ring structure

form the ring. Glucose, for example, can potentially form a pyranose or a furanose structure, although it is generally found in the pyranose form, while fructose forms a furanose structure. Five carbon monosaccharides (pentoses) are usually aldoses, and they typically form only furanoses, where the OH on C4 interacts with C1. Six-membered ketoses can form both structures, but they typically make the furanose form. In the furanose form of a ketose, the OH group on C5 interacts with the anomeric carbon, which in this case is C2.

- Summary of Monosaccharide Structure

In summary, some characteristics that differentiate the various monosaccharides include **(1)** the **number of carbons** they have, **(2)** whether they are **aldoses** or **ketoses**, **(3)** whether they are in **linear** or **circular** forms, though they can cycle between the two forms and are usually in circular forms, **(4)** whether the circular form of a 6-carbon monosaccharide is a **pyranose or a furanose**, and **(5)** whether they differ in any of the following **stereoisomer** forms: **(a)** the **arrangement of hydrogens (H)** and **hydroxyls (OH)** around the carbons**,** **(b)** Whether they are **D or L forms** (usually in nature, they are in the D form), and **(c)** whether the circular forms have **α or β anomer** arrangements [1, 3, 5, 6].

1.3 Glycosidic Bonds Link Monosaccharides to Form Larger Molecules

Although monosaccharides can function independently, many carbohydrates consist of monosaccharides linked together.

Monosaccharides are linked together by covalent bonds called glycosidic bonds. They are formed by a dehydration reaction between hydroxyl (OH) groups on two adjacent monosaccharides (see Fig. 9.8). One of the OH groups is attached to the anomeric carbon. In the process of joining two OH groups, two Hs and an O are lost, resulting in the formation of H_2O, which is then lost. The result is the formation of the O-glycosidic bond [1, 6].

While all glycosidic bonds have the same chemical makeup, there are different variations. To identify them, they are numbered according to which OH groups are linked together. For example, the 1,4 glycosidic bond links carbon 1 of one monosaccharide with carbon 4 of the other monosaccharide (Fig. 9.8). Other common types of glycosidic bonds include 1,6 bonds and 2,6 bonds. They are also named alpha or beta, depending on whether the interacting carbons are alpha or beta anomers. Thus, an α-1,4 bond connects an α-anomeric carbon (C1) with C4 of its binding partner.

Carbohydrates are all made up of monosaccharides that are linked by these types of bonds. Monosaccharides bound together as pairs are disaccharides. 3-9 monosaccharides linked together are oligosaccharides, and 10 or more monosaccharides are considered polysaccharides. In a large molecule (such as a polysaccharide, discussed below), the type of bond will affect the molecule's structure.

1.4 Disaccharides

Disaccharides contain two monosaccharide units linked by a glycosidic bond. Some sugars are monosaccharides, and others are disaccharides. Some common disaccharides are **lactose** (made of β-galactose bonded to glucose (usually β-glucose), sucrose (α-glucose and β-fructose: table sugar; see Fig. 9.9), and maltose (made of two glucose units), which are all dietary carbohydrates. Disaccharides and all carbohydrates that have more than one monosaccharide have directionality. They have a reducing end and a nonreducing end. At the reducing end, there is a free anomeric carbon, which can potentially interact with another monosaccharide or participate in redox reactions. The other end is the nonreducing end [1, 6].

1.5 Oligosaccharides

Oligosaccharides usually contain 3–9 monosaccharides. Their functions are somewhat different from those of other carbohydrates in that they are not usually used for energy. They are more commonly used for other functions, such as cell signaling. Signaling molecules may be oligosaccharides alone, or they may be oligosaccharides attached to proteins. When oligosaccharides are attached to proteins, they are called glycoproteins. Glycoproteins are

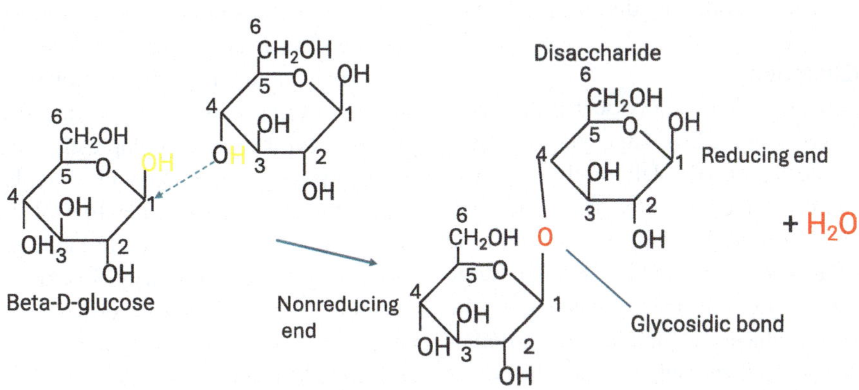

Fig. 9.8 Disaccharides and glycosidic bonds. Shown here is an example of a glycosidic bond in a disaccharide (two sugars). Formation of this bond involves a dehydration reaction. This example shows a 1,4 glycosidic bond

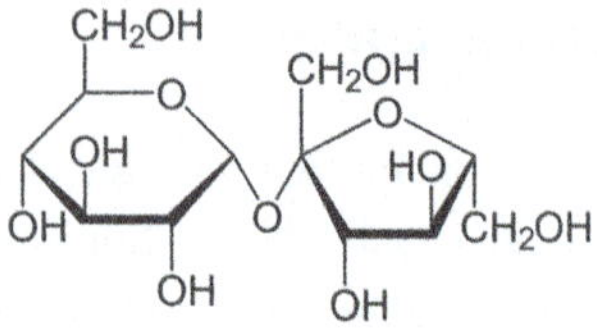

Fig. 9.9 Structure of sucrose. Sucrose is a disaccharide containing glucose and fructose. (*Source*: Public domain image by Calvero (2007), via Wikimedia Commons (File: Saccharose2.svg))

assembled on the rough ER and modified further in the Golgi. Oligosaccharides can also be attached to lipids, and then they are called glycolipids [1, 6].

1.6 Polysaccharides

Polysaccharides are the largest forms of carbohydrates. They are composed of 10 or more monosaccharides, but are often very large and can contain thousands of monosaccharides (see Fig. 9.10). Several important polysaccharides are made up of glucose, although polysaccharides can also be made of other monosaccharides. Even the different polysaccharides that are made up of glucose are different from each other. This is due to the types of glycosidic bonds and the branching in the structure. For example, an α-1,4 linkage favors a bend, resulting in a helical structure suitable for storage, while a β-1,4 bond favors straight chains, which are useful for structural purposes [1]. Some important polysaccharides are discussed below.

1.6.1 Glycogen

An important example of a polysaccharide is glycogen [1, 6] (see Fig. 9.10). Glycogen is a large polysaccharide composed solely of glucose. Glycogen is not obtained directly from the diet. Instead, glucose is obtained in the diet, and the body synthesizes glycogen by linking large amounts of glucose together. Glycogen is the storage form of glucose in animals. It is a homopolymer because all the subunits are the same: glucose. It is the most common homopolymer in animal cells. Most of the glucose units in

Fig. 9.10 Glycogen. Glycogen is an important example of a polysaccharide. Glycogen is shown here with the protein glycogenin, found within the glycogen molecule, which has a role in initiating the synthesis of glycogen. (Source: Mikael Häggström. Public domain (Wikimedia Commons). File: Glycogen_structure.jpg)

glycogen are linked by α-1,4 glycosidic bonds. Branches are formed by a-1,6 glycosidic bonds. There are many branches in glycogen, which increases the surface area to allow better access to enzymes that can rapidly break down glycogen when the body needs glucose. Glycogen is stored mainly in the liver and muscles. Glycogen is critical for energy production because it can be broken down to release glucose when the body needs energy.

Glycogen in the liver can have two outcomes. (a) It can be broken down via several steps into glucose. The resulting free glucose is exported from the liver and transported through the bloodstream to organs that need it. In these organs, glucose is broken down completely during cellular respiration, beginning with glycolysis. This leads ultimately to the generation of ATP as an energy source. (b) Glycogen in the liver can also serve as a source of energy for the liver. This glycogen is broken down via two enzyme-catalyzed steps to glucose-6-phosphate, and this glucose-6-phosphate can enter glycolysis, leading to energy production.

Likewise, in muscle, glycogen is broken down via two enzyme-catalyzed steps into glucose-6-phosphate, which enters glycolysis to provide energy for muscle. Muscle does not export glucose to the bloodstream but uses all of its glycogen for its own needs.

Glycogen is an essential but short-term form of energy storage compared to fat. The liver and muscle store enough glycogen to last up to a day, even during fasting, but fat stores last much longer, nearly a month [6, 7]. The breakdown and synthesis of glycogen, along with the enzymes involved, are discussed in detail in Chap. 10.

1.6.2 Starch

Another example of a polysaccharide is starch, a large, branched structure composed of glucose. Starch is not produced in the body, but it is obtained through the diet. It is found in foods like potatoes, wheat, rice, corn, and legumes. Starch is an important dietary source of glucose. Starch is broken down into glucose, which can enter glycolysis and participate in oxidative phosphorylation to produce ATP [1, 3, 8].

1.6.3 Cellulose

Cellulose is another large polysaccharide, composed entirely of glucose units, but it differs from other glucose-based polysaccharides in its structure. It consists of long unbranched chains of glucose linked by β-1,4 glycosidic bonds. Cellulose is found in the cell walls of plant cells, and we obtain it from our diet by eating plants, such as some vegetables, fruits, legumes, grains, and nuts and seeds. Cellulose has a different structure from glucose. It is more of a straight sequence of glucose molecules, rather than branched sequences. Interestingly, humans lack the enzymes to break the bonds in cellulose, despite having the enzymes to break down glycogen and starch. Other animals, such as cows, have the necessary enzymes and can, consequently, get most of their energy from eating grass. When humans consume foods such as lettuce that are rich in cellulose, we break down this indigestible cellulose into fiber. Fiber is an important part of the digestive process and can help bulk up the stool [1, 3, 9, 10].

1.7 Some Important Carbohydrates and Their Functions

Carbohydrates have many functions, with energy storage being a primary one, but they also serve other related purposes. Some examples of carbohydrates and their functions are discussed below.

1.7.1 Monosaccharides and Their Functions

Although monosaccharides can bind together to make complex molecules, many monosaccharides have important roles on their own. Some important monosaccharides and some of their functions are discussed and in [1, 3, 6].

Glucose (a hexose) is a monosaccharide, although it can also bind with other glucose or monosaccharide molecules to make large molecules. Glucose is the most important monosaccharide. It is a main source of energy; it can cross the blood–brain barrier and nourishes all tissues, including the brain. It forms the building block of more complex carbohydrates, such as starch, glycogen, and cellulose. Glucose is also a component of some disaccharides. For example, it can form maltose, a disaccharide, with another glucose molecule [1, 11].

Fructose (a hexose) is also a monosaccharide; it is sweet and is found in fruits and honey. Fructose is obtained mainly through food, such as fruits, honey, and root vegetables. It has a role in energy production. It is usually in the cyclic form [12].

Galactose (a hexose) is also found in various glycoproteins and has roles in cell signaling.

Ribose and deoxyribose (both pentoses) are the major components of nucleotides. Nucleotides are found in RNA and DNA. Ribose is also found in ATP. Ribose is also necessary for the biosynthesis of nucleotides.

Glyceraldehyde (a triose, usually in the D form) is an intermediate in carbohydrate metabolism, specifically in glycolysis (the breakdown of glucose to yield ATP) and gluconeogenesis (the synthesis of glucose from smaller molecules).

Dihydroxyacetone (a triose) is also an intermediate in glycolysis. In addition to its biological functions, it is also used in some types of cosmetics such as sunless tanning products [13].

Ribulose (a pentose) plays a central role in the pentose phosphate cycle, which is important for the synthesis of nucleotides and other important metabolic intermediates. It is also important in the Calvin cycle in plants and algae, and some bacteria [1].

1.7.2 Disaccharides and Their Functions

Sucrose is a disaccharide that contains fructose and glucose. Sucrose is also called table sugar. It is used for sweetening. It is also found naturally in sugar cane, sugar beets, various fruits, some vegetables, and other sources such as maple syrup [1, 14, 15].

Lactose is a disaccharide made up of galactose and glucose. It is the sugar found in mammalian milk [1].

1.7.3 Polysaccharides and Their Functions

Glycogen is the main energy storage form of glucose, made largely in the liver and muscle [7]. It is discussed in more detail above.

Starch: Starch is a polysaccharide made entirely of glucose, but the glucose can be in different forms. Part of the glucose is stored as amylose, which consists of linear unbranched chains of glucose units linked together by α-1,4-glycosidic bonds. The other, more significant portion is made up of Amylopectin. This is a highly branched structure consisting of glucose with α-1,6-glycosidic bonds forming branch points [8].

Cellulose is a polysaccharide of glucose found in plants such as leafy greens [1, 3, 9, 10].

Glycosaminoglycans are made of N-acetylglucosamine (a hexosamine; a hexosamine is a hexose with an amino group (NH_2). N-acetylglucosamines are derived from glucose. They are linear polysaccharides composed of various disaccharide units. They are abundant in connective tissues but are also found in other locations. They play roles in structural support, lubrication, and regulation of cellular activities [16].

1.7.4 Carbohydrates Attached to Other Molecules

The insulin receptor is a glycoprotein, which means it is a protein with carbohydrates attached. The insulin receptor is a transmembrane glycoprotein that is usually heavily glycosylated, meaning that it has numerous carbohydrate chains attached. The insulin receptor has important roles in various cell signaling pathways [17].

Mucins are also glycoproteins. They have a high proportion of carbohydrates and have important roles in lubrication [1].

1.8 The Importance of Carbohydrates in the Field of Pharmacy

1.8.1 Metabolism and Metabolic Diseases

One of the primary reasons carbohydrates are important to study is their critical role in metabolism. Metabolism refers to the process by which we break down the food we eat to produce energy and synthesize biomolecules when nutrient intake is insufficient. Glucose is the major source of energy in the body. It can be stored when energy levels are high and broken down when more energy is needed. Some organs, such as the brain, can only use glucose as their energy source. Metabolic diseases are a growing and significant global health concern. Diabetes mellitus, for example, is a metabolic disorder characterized by impaired regulation of blood glucose levels. Other metabolic disorders include obesity and glycogen storage disease. Several new and quite popular drugs have been made to combat some metabolic disorders, such as diabetes and obesity, but side effects are important considerations. The next four chapters are dedicated to a discussion of metabolism, metabolic disorders, and various treatment options. Below are some other reasons that studying carbohydrates is important for pharmacists.

1.8.2 Glycosylation and Cancer

Glycosylation is the addition of carbohydrates to either proteins, lipids, or other carbohydrates.

Glycosylation is a normal occurrence in cells, but alterations in glycosylation can be associated with diseases such as cancer. Tumor-associated glycoproteins (TAGs) are glycosylated proteins that are found specifically on some cancer cells. Since some of these modifications are specific to cancer cells, they are being tested as targets for some anticancer drugs [18, 19].

Disialoganglioside 2 (GD2) is a ganglioside. A ganglioside is a glycolipid that contains both a lipid component and a carbohydrate component. The carbohydrate component faces the exterior of the cell, and the lipid part is inserted into the plasma membrane. GD2 is found at low levels in some cells, especially in the nervous system, but it is upregulated in some types of cancer, including neuroblastoma [20]. It has therefore become a promising target for some types of cancer. **Unituxin (dinutuximab)** is an FDA-approved monoclonal antibody that binds to the carbohydrate portion of GD2 and is used to treat neuroblastoma in children. The binding of this monoclonal antibody to GD2 on neuroblastoma cells triggers an immune response and leads to the lysis of the cells. It is given to patients by infusion [19, 21]. The drug is still used but causes serious side effects, requiring careful patient selection and monitoring [22]. Another drug that functions similarly is **Danyelza (Naxitamab)** [23], a monoclonal antibody drug used to treat childhood neuroblastoma by binding to GD2 [24].

Carbohydrate antigen 19-9 (CA19-9) is a carbohydrate that is overproduced on pancreatic cancer cells as well as some other types of cancer cells. It is often used as a diagnostic tool and a marker to monitor disease progression. It also has the potential to be a drug target. **MVT-5873** is a **monoclonal antibody** that recognizes CA19-9, and it has been studied as an anticancer treatment for several types of cancer [19, 25, 26].

GP-100 (glycoprotein 100) is a glycoprotein found at high levels on the surface of melanocytes and melanoma cells [27]. It is a target of the drug **Kimmtrak (tabentafusp-tebn)**. **Kimmtrak**, FDA-approved in 2022, is used to treat uveal melanoma, which is a rare form of melanoma that affects the eye. It is delivered to the patient intravenously. **Kimmtrak** is a protein that attaches to GP-100 on melanoma cells and simultaneously binds to CD3, a receptor on the surface of T cells, which are a type of immune cell. By binding to both melanoma cells and T cells, the drug can bring the melanoma cells into close proximity to the T cells, thereby triggering an immune response to kill the cells. The medication is given by IV once per week [28].

1.8.3 Lactose Intolerance

When we digest disaccharides, an enzyme is needed to break the bonds between them because it is the monosaccharides that are digestible. Lactose (in milk) is a disaccharide consisting of β-galactose and glucose. Some people lack the enzyme required to break the bond that is formed between β-galactose and glucose. This enzyme is **lactase**. If the bonds are not broken apart, the **lactose** disaccharide cannot be digested. While children tend to have the enzyme, many adults experience a normal decline in **lactase** production, which is known as lactase nonpersistence. The degree of lactase loss varies among different individuals. Some people still maintain the enzyme at high levels in adults, while others experience a sharp decline in their **lactase** levels. This may have a genetic basis. If lactase is insufficient, lactose cannot be broken down, and therefore it cannot be absorbed from the intestine and taken to the bloodstream. This results in an osmotic effect. The disaccharide is larger than most molecules that are digested. Consequently, it goes through the intestine and pulls water toward it, causing water to accumulate in the gut. As a result, too much water in the intestines may lead to diarrhea. Additionally, bacteria in the colon can ferment undigested lactose, producing gas (including hydrogen and methane), which can cause symptoms such as bloating and flatulence [29]. Thus, the lack of the enzyme causes multiple symptoms leading to discomfort in people with lactose intolerance [1, 30]. To manage symptoms, some people with **lactose intolerance** (those with low levels of **lactase**) may limit their consumption of dairy products and opt for lactose-free milk, which has already had the lactose broken down. Some people also take lactase enzyme supplements to help digest dairy products. These supplements contain lactase, which can help break down lactose and aid digestion [31, 32].

1.8.4 Carbohydrates Can Be Used as Excipients

Carbohydrates are also useful as components of drug delivery systems. Drug developers can isolate and modify carbohydrates and use them as excipients. Excipients are inactive ingredients or vehicles for drug delivery. They can be used in tablets and pills, drugs dissolved in solution, and injectables. Carbohydrates can improve drug solubility and stability. Some carbohydrates can encapsulate drugs for better delivery. Carbohydrates can also be added to monoclonal antibodies (mAbs) to improve drug delivery and stability [33, 34].

2 Lipids

2.1 Introduction to Lipids

Lipids are hydrophobic molecules that usually consist of long hydrocarbon (carbon and hydrogen) chains or rings. Like carbohydrates, they are composed mostly of carbon and hydrogen. However, there are also big differences between lipids and carbohydrates. Many lipids have a **carboxyl group (COOH)** at one end. In contrast, carbohydrates have a carbonyl group (C=O) at one end of their linear forms, although this group becomes a hydroxyl (OH) group when the carbohydrate forms a ring structure. In lipids, the carbons are bonded to hydrogens while in carbohydrates, carbons are bonded to either hydrogens or hydroxyl (OH) groups. Carbohydrates form cyclic structures, while lipids are generally linear. Lipids have fewer oxygen atoms than carbohydrates. This makes them hydrophobic, allowing them to incorporate into membranes, and they can interact with each other to form globules or micelles in aqueous solution. Lipids have significant differences compared with other biomolecules. They are generally smaller, and they do not usually consist of repeating monomers. Some important types of lipids include fatty acids, phospholipids, triacylglycerols (TAGs), terpenes and terpenoids, and steroids [35, 36]. The fatty acid is the simplest form of a lipid [1, 35, 37, 38]. Its basic structure is seen in Fig. 9.11.

2.2 Some Functions of Lipids

One of the most important functions of lipids is energy storage. For example, fatty acids are stored in adipose tissue as triacylglycerols (also called triglycerides) (TAGs), and adipose tissue is a primary form of energy storage. While glycogen is also critical for energy storage, as discussed in the previous chapter, lipids are more suitable for long-term storage. Some lipids are also important for membrane structure. In particular, phospholipids are the major components of the plasma membrane and of organelle membranes. Some lipids provide insulation and protection by helping to regulate body temperature and by providing cushioning and padding. Lipids can also act as cell signaling molecules. For example, phosphatidylinositol (PI) is a membrane lipid. It can respond to extracellular signals and subsequently activate intracellular signaling pathways that can modulate cell function. Sphingolipids are also found in membranes; they are important for the production of signaling molecules that

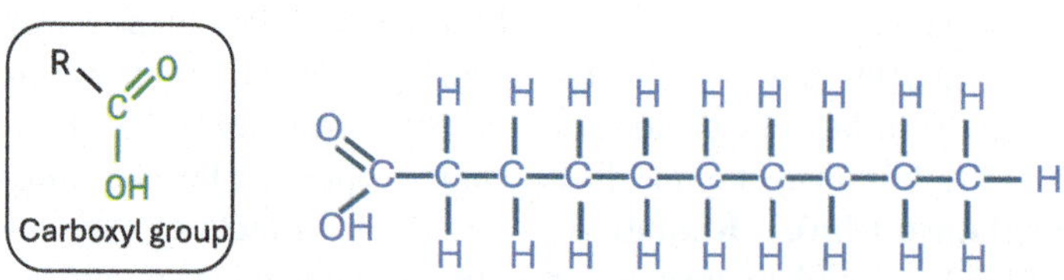

Fig. 9.11 The basic structure of a fatty acid. Fatty acids are the simplest forms of lipids. They generally contain a carboxyl group at one end of a hydrocarbon chain

regulate cell growth and apoptosis. Prostaglandins and leukotrienes are lipids that function as signaling molecules. They have roles in regulating inflammation, immune function, and other related processes. Lipids also have a role in the phagocytosis of dead cells. The lipid phosphatidylserine (PS), for example, accumulates on the outer membrane of apoptotic cells, which can target these cells for phagocytosis [1, 35, 37–39].

2.3 Fatty Acids

2.3.1 Overview of Fatty Acids

Fatty acids (FAs) are the simplest forms of lipids. Fatty acids generally contain a carboxyl group (COOH) at one end of a hydrocarbon chain. The end opposite the fatty acid contains a methyl group (a carbon with 3 hydrogens). All the rest of the Cs in the lipid are attached to two other carbons and 2 Hydrogens (Fig. 9.11). A major function of fatty acids is their role as "fuel molecules" due to their important roles in energy storage and production. Fatty acids are also important because they are building blocks for phospholipids and glycolipids. Fatty acids can attach covalently to some proteins, and this can cause these proteins to attach to the cell membrane because the fatty acid portion can insert into a leaflet of the phospholipid bilayer. Fatty acid derivatives can also act as hormones and second messengers [35, 40].

Each end of the fatty acid is referred to differently, and each carbon has a name. The carbon within the carboxyl group (COOH) is considered carbon 1 (C1). The next carbon (C2) is the **alpha (α) carbon.** The next carbon (C3) is the β carbon, and the next carbon (C4) is the γ carbon. This numbering continues in this way until the end of the fatty acid, which contains the methyl group (CH_3). This carbon, no matter what number it is, is called the **omega (ω) carbon**, and it marks the omega end of the fatty acid. The entire region between them is called the fatty acid backbone. Fatty acids are generally found as components of a larger molecule, the triacylglycerol (TAG) [35, 37, 40], which is discussed later.

2.3.2 Saturated and Unsaturated Fatty Acids, and Cis Versus Trans

There are hundreds of different kinds of fatty acids. Fatty acids can vary in length and saturation. Saturated fatty acids have no double bonds. Thus, they are saturated with hydrogen (H) atoms. Unsaturated fatty acids contain one or more double bonds. They can be monounsaturated fatty acids, which have one double bond, or polyunsaturated fatty acids, which have more than one double bond. Unsaturated fatty acids can be cis or trans (see Figs. 9.12 and 9.13). In cis fatty acids, the hydrogen atoms adjacent to the double bond are on the same side as each other. In trans fatty acids, the hydrogen atoms adjacent to the double bonds are on opposite sides of each other [1, 14].

Saturated fatty acids have a higher melting point and, thus, are more difficult to melt than unsaturated fatty acids. Unsaturated fatty acids are easier to melt because they do not pack together as well due to their shapes. Trans-unsaturated fatty acids have a more linear shape and a more rigid structure than cis-unsaturated fatty acids; therefore, they pack together more tightly and are more difficult to melt [35, 37, 38, 40] (Fig. 9.13).

2.3.3 Naming of Fatty Acids

Fatty acids can be named based on (a) the number of carbons they have (b) the number of double bonds they have, and (c) the positions of the double bonds. For example, a FA that has 10 carbons and no double bonds (it is saturated) can be depicted as C10:0. If there is a double bond, we look at the position of each double bond. Typically, we start with the COOH group as carbon 1 and use a Δ symbol to show the position. For example, an 18-carbon chain with 1 double bond at carbon #9 would be C18:1,Δ9.

Fatty acids can have varying numbers of carbons. Even-numbered fatty acids are more common, but some fatty acids are odd-numbered. Short-chain fatty acids have 2–6 carbons. An example is acetic acid (C2:0), which has two carbons. Medium-chain fatty acids range from 6 to 12 carbons. An example is lauric acid

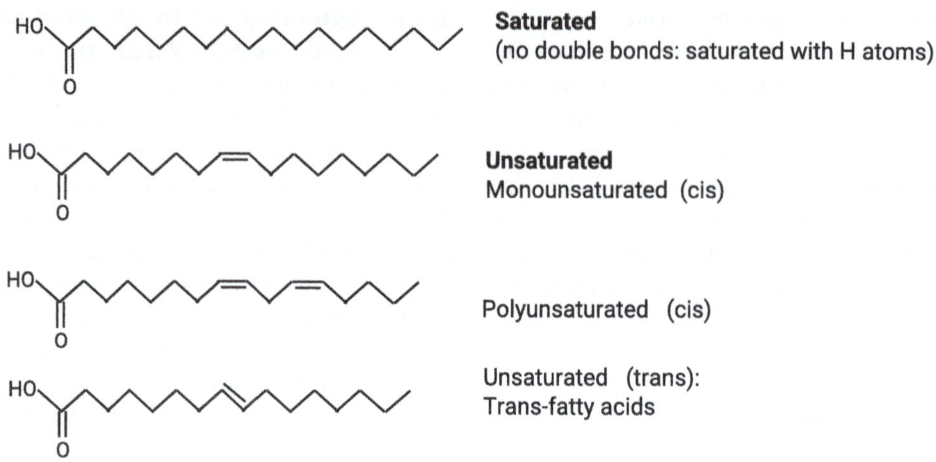

Fig. 9.12 Fatty acids can vary in saturation. Saturation is determined by how many hydrogens bind to each carbon. Saturated fatty acids have no double bonds, but unsaturated fatty acids have double bonds. Unsaturated fatty acids can be cis or trans

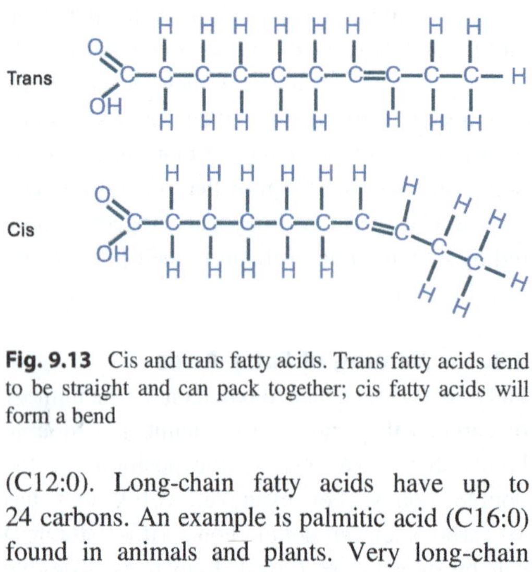

Fig. 9.13 Cis and trans fatty acids. Trans fatty acids tend to be straight and can pack together; cis fatty acids will form a bend

(C12:0). Long-chain fatty acids have up to 24 carbons. An example is palmitic acid (C16:0) found in animals and plants. Very long-chain fatty acids, over 24 carbons long, are less common but do exist in some tissues, such as the brain.

In addition to the nomenclature described above, fatty acids also have systematic names. These names adhere to the rules of the International Union of Pure and Applied Chemistry (IUPAC). As an example, the systematic name for palmitic acid is hexadecanoic acid. This name is based on the number of carbons and ends in anoic acid, which is the convention for fatty acids [1, 14]. Sometimes, the systemic names are written before the numerical nomenclature. An example is oleic acid (C18:1,Δ9). Other examples are palmitic acid (C16:0) and linoleic acid (C18: 2:Δ9,12) (linoleic acid has two double bonds, at positions 9 and 12). Finally, many fatty acids also have common names such as oleic acid, linoleic acid, and stearic acid.

2.3.4 Fatty Acids in Diet and Health

Saturated fatty acids are common in animal-based fats like butter, lard, fatty meat, and in oils such as coconut and palm oil. They are generally solid at room temperature. This is due to their linear structure, which allows them to pack together tightly. Excessive consumption of these fatty acids is associated with an increased risk of heart disease and elevated low-density Lipoprotein (LDL) cholesterol, considered the "bad cholesterol." Their relationship to heart disease is complex, however, and may depend on a combination of health factors and dietary habits.

Unsaturated fatty acids (cis) include monounsaturated and polyunsaturated fatty acids. Monounsaturated fatty acids (MUFAs) are found in olive oil, avocados, nuts, and some seeds. They may help lower LDL (low-density lipoprotein, considered the "bad" cholesterol) while maintaining high-density lipoprotein (HDL, the "good" cholesterol) [36, 41]. Polyunsaturated fatty acids (PUFAs) are found in fatty fish (such as salmon, mackerel, and sardines), some seeds, vegetable oils, and nuts. They play a role in reducing inflammation and improving brain function [42]. MUFAs and PUFAs are generally considered heart-healthy. This is largely due to their structures. The kinks in their structures make them less packed together and more likely to be in liquid form at room temperature [43].

Trans fatty acids, although unsaturated, are harmful to our health. They are generally artificially produced. They are made from partially hydrogenated oils [44] by the hydrogenation of naturally occurring cis-unsaturated fatty acids. Hydrogenation involves adding hydrogen gas (H_2) to the oil at high pressure and with heat (see Fig. 9.14). More than any other type of fatty acids, trans-fatty acids are associated with cardiovascular disease. This is because, due to their structures, they are more likely to form tightly packed complexes. The food industry has used trans fats for many years because they are generally less expensive than animal fats, have a longer shelf life, tend to be solid at room temperature, and are convenient for deep-frying due to their stability at high temperatures [45]. They also have a softer, spreadable texture than other oils. Nevertheless, because the dangers of these fats have been recognized, many countries, including the United States, have placed regulations and bans on them. This has significantly reduced the use of trans fats in foods [35, 37, 38, 40, 46].

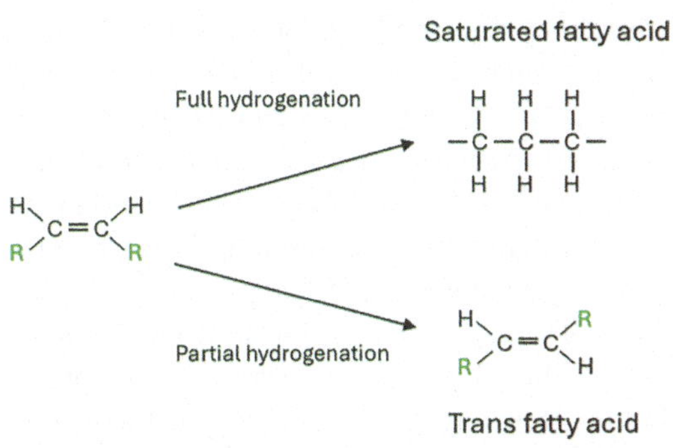

Fig. 9.14 Hydrogenation. Trans fatty acids are generally produced by hydrogenation. In this example full hydrogenation of a cis unsaturated fatty acid converts the double bond to a single bond and produces a saturated fatty acid. Partial hydrogenation can convert a cis double bond to a trans double bond. R represents the remainder of the hydrocarbon chain. (*Source*: Modified from Temkov and Muresan, Foods 2021,10,1376. https://doi.org/10.3390/foods10061376 licensed under the Creative Commons Attribution (CC BY) license)

2.3.5 Essential Fatty Acids

Linoleic acid, linolenic acid, and arachidonate (which is synthesized from linoleic acid) are essential fatty acids. Linolenic acid is often referred to as an omega-3 fatty acid, and linoleic acid as an omega-6 fatty acid. This is based on a nomenclature system that uses omega rather than delta. In the omega system, the position of the double bond is counted from the methyl group rather than from the carboxyl group. We lack the enzymes to synthesize linoleic acid and linolenic acid, and therefore, we must obtain them through our diet. This is because mammals lack the enzymes necessary to add the double bonds beyond the ninth carbon (C9) from the carboxyl group in fatty acids [1]. These essential fatty acids are necessary for various cellular processes, including maintaining cell membrane integrity, regulating inflammatory and immune responses, promoting cardiovascular health, and controlling various metabolic processes. Good sources of these fatty acids, especially omega-3 fatty acids, include some fish (such as salmon, mackerel, albacore, sardines, and halibut), certain vegetable oils, nuts, and seeds [1, 47].

2.4 Triacylglycerols (TAGs)

2.4.1 Basic Structure of TAGs

Triacylglycerols (TAGs) are also sometimes referred to as triglycerides. TAGs are the primary storage forms of fatty acids. They consist of three fatty acids (FAs) linked to a glycerol molecule (glycerol is a 3-carbon alcohol). The fatty acids bind to glycerol via an ester linkage (Fig. 9.15). There are many different TAGs. A simple TAG is one where the three fatty acids attached to the glycerol are identical. A mixed TAG is one where the fatty acids are different [1].

TAGs are largely consumed in our diet and are found in vegetable and animal fat, but we can also synthesize TAGs. Fatty acids and glycerol can be

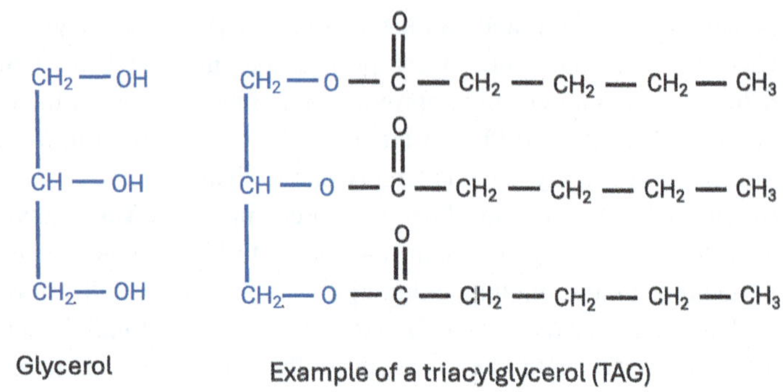

Fig. 9.15 Triacylglycerol (TAG) structure. TAGs consist of glycerol esterified to three fatty acids

synthesized, and fatty acids can be esterified to glycerol, as described in Chap. 11. TAGs are stored in adipocytes, the fat cells that make up adipose tissue, the main component of body fat. Adipose tissue is found throughout the body. It can be subcutaneous fat (under the skin) or visceral fat (which surrounds and protects the internal organs). Adipocytes are filled with TAGs. Within adipocytes, TAGs coalesce together to form large globules. A small amount of TAGs is also stored in skeletal muscle, which is available for energy during intense physical activity [1, 35, 37, 38, 40].

2.4.2 TAGs as an Energy Source

To be used as an energy source, the TAGs need to be broken down into fatty acids and glycerol. TAGs come from two main sources: dietary fat, which is packaged into chylomicrons and circulates in the blood, and stored fat, which is stored in adipose tissue. When the body needs energy, stored TAGs in adipocytes (fat cells) are mobilized through a process called lipolysis. Lipolysis is carried out by **lipases**, leading to the release of fatty acid chains from glycerol. There are three lipases involved in this; **adipose triglyceride lipase (ATGL)** removes the first fatty acid. **Hormone-sensitive lipase (HSL)** removes the second fatty acid (it is especially active when under the control of hormones such as adrenaline), and **monoglyceride lipase (MGL)** removes the final fatty acid, releasing glycerol. Next, the fatty acids and glycerol are also broken down, generating energy in the form of ATP [35, 37, 38, 40].

Glycerol can also be either broken down or reused. Fat is a highly efficient energy source due to the TAGs stored in adipose tissue. Oxidation of a gram of fat releases about twice as much energy as the same amount of carbohydrate [35]. These steps are described in some detail in Chap. 11, and an overview of the process is given below.

- Oxidation of Fatty Acids

FAs that are released from the TAGs enter the bloodstream, where they can bind serum albumin (which is abundant in the blood) and enter various organs. In muscle cells and other cells, FAs can be transported to the cytosol. There, they get activated by attachment to coenzyme A (CoA). They are then transported to the mitochondria, where they undergo β-oxidation, a stepwise process in which they are broken down into acetyl-CoA, 2C units, and NADH and $FADH_2$ are also produced. Acetyl CoA can then enter the Krebs cycle, and NADH and $FADH_2$ can be used in the electron transport chain, resulting in the generation of ATP for energy. Fatty acids are highly efficient energy sources for nearly all tissues except the brain, but in states of prolonged fasting, the acetyl CoA from fatty acid oxidation can also be made into ketone bodies, which can provide energy for multiple organs, even the brain [1, 35, 37, 38, 40].

- The Glycerol Released from TAGs

The glycerol released from the TAGs has a different outcome from the fatty acids. Glycerol can be converted to an intermediate of the glycolysis pathway in the liver. This can lead to the production of pyruvate, which can also feed into the Krebs cycle, ultimately leading to ATP production. Alternatively, depending on the cell's needs, pyruvate can lead to gluconeogenesis, resulting in the production of additional glucose [1, 48].

2.4.3 Disorders Associated with TAGs

TAGs that are not stored in fat are packaged in **chylomicrons** (discussed more below) after fat-containing foods are consumed. These lipoprotein particles transport dietary lipids, including TAGs and cholesterol, through the bloodstream. The enzyme **lipoprotein lipase (LPL),** which is attached to capillary walls in some tissues, including adipose tissue and muscle, hydrolyzes the TAGs within these chylomicrons. This process removes two fatty acids, leaving 2-monoacylglycerol and two free fatty acids for each TAG. These fatty acids then get taken up by cells in tissues. The process differs from the intracellular lipases discussed earlier, which act on stored fat [14, 49, 50], and will be discussed in more detail in Chap. 11.

Familial chylomicronemia syndrome (FCS) is a genetic disorder where TAG levels are especially high. This is a genetic abnormality involving **LPLs** or their cofactors. In this disorder, the TAGs in the chylomicrons are not efficiently broken down, and instead, chylomicrons with dangerously high levels of TAGs accumulate in the bloodstream. Patients need to eat a very low-fat diet in order to manage the disorder, but dietary adjustments are not always sufficient. Patients may suffer from abdominal pain and have other symptoms that can be dangerous, such as pancreatitis [51].

Tryngolza (olezarsen) [52] is a drug designed to treat **familial chylomicronemia syndrome (FCS)**. It is an antisense drug, FDA-approved in 2024, designed to inhibit or reduce the production of a protein called apolipoprotein-C-III (ApoC-III). ApoC-III inhibits TAG clearance from the bloodstream. **Elevated ApoC-III levels are strongly associated with reduced clearance of triglyceride-rich lipoproteins (TRLs).** However, with modern high-fat diets or in people with FCS, TAG clearance may be required consistently [53]. **Tryngolza** utilizes antisense oligonucleotide (ASO) technology, which is an RNA-based system. It is a modified single-stranded RNA antisense oligonucleotide (ASO), designed to bind to the ApoC-III mRNA. This triggers ApoC-III mRNA degradation through RNase H. (This type of technology is described in Chap. 6). The decrease in ApoC-III in patients lowers TAG levels because it improves the clearance of TAG from the blood [54]. ApoC-III is located in the liver, and Tryngolza is also targeted to the liver by being conjugated to N-acetylgalactosamine (GalNAc), which binds to hepatocyte-specific receptors and thus gets transported to the liver [51].

2.5 Phospholipids

Phospholipids are another important type of lipid. They contain two hydrocarbon chains (tails), glycerol (the backbone), a phosphate group, and a polar molecule (head group). Phospholipids are major and essential components of cell and organelle **membranes** and were discussed in more detail in Chap. 3 [35, 37, 39, 40] (see Chap. 3, Fig. 3.3).

2.6 Terpenes and Terpenoids

Terpenes and terpenoids are not commonly found in animal cells but can be obtained in the diet, especially from plants. Terpenes are made up of repeating units of isoprene. Isoprene is a 5-carbon structure with the formula C_5H_8. Some examples of terpenes include myrcene (found in mangoes), farnesene (found in apples), limonene (found in lemons), and beta-carotene (found in plants, carrots, and some fruits). Terpenoids are also made up of isoprenes but are often modified by oxidation. Terpenes and terpenoids are found in many fruits and other types of food. An example of a terpenoid is menthol (found in peppermint), which comprises two modified isoprenes (Fig. 9.16) [1, 55, 56].

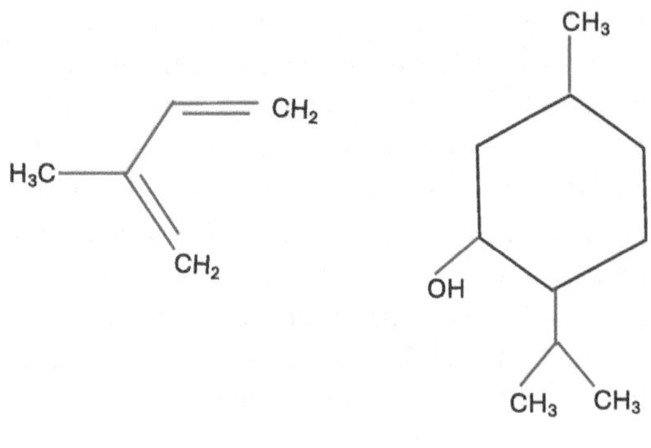

Isoprene Menthol

Fig. 9.16 Isoprenes and compounds that consist of isoprenes. Terpenes are made up of repeating units of isoprene, a 5-carbon structure. Terpenoids are also made up of isoprenes but are often modified by oxidation. Menthol is an example of a terpenoid; it is made of two isoprenes

2.7 Steroids

2.7.1 Overview of Steroids

Steroids are important lipids in mammalian cells. They include hormones, some vitamins, and cholesterol. The structure of a steroid is based on a 17-carbon ring structure with four rings and various branches and modifications (Fig. 9.17). Steroids have a hydrocarbon tail at one end and a hydroxyl (OH) group at the other end, where the rings are. An important example of a steroid is cholesterol. Cholesterol is essential for maintaining membrane integrity and is also a precursor to other steroids, including sex hormones and cortisol. Sex hormones (estrogens, androgens, progestins, and testosterone) have multiple functions. For example, they regulate various aspects of sexual development, sexual characteristics, and reproduction. Cortisol regulates activities such as metabolism and inflammation [1, 40].

2.7.2 Cholesterol

• Overview of Cholesterol

Cholesterol is a molecule that has 27 carbon atoms (Fig. 9.18). Cholesterol can be obtained from the diet, and it is also synthesized de novo. The liver and intestine are the main sites of cholesterol biosynthesis, where cholesterol is derived from acetyl-CoA. Cholesterol that is made or ingested is carried throughout the body by lipoproteins to various tissues. Cholesterol is essential. It is a key component of membranes and helps ensure membrane fluidity because of its bulky structure. The OH group of cholesterol interacts with the phospholipid head groups in the phospholipid bilayer, and the cholesterol is oriented parallel to the lipids in the bilayer. Cholesterol is also a precursor to bile acids. Bile acids are important because they act as detergents, breaking down large fat globules in the digestive system. They also help transport fatty acids and fat-soluble vitamins across the intestinal lining [57]. Cholesterol is also a precursor for the formation of steroid hormones, such as the sex hormones, as well as vitamin D. It also has important roles in cell signaling, including synaptic signaling [1, 40]

• Low-Density Lipoprotein Particles (LDLs)

LDL particles are lipoproteins and contain lipids and proteins. Their primary purpose is to carry lipids in the form of cholesterol, TAGs, and phospholipids throughout the body. LDL particles comprise a core of a large amount of cholesterol molecules esterified to fatty acids (commonly linoleate). They also have a shell of phospholipids, unesterified cholesterol, and a protein component called apolipoprotein B (apoB). This protein component is recognized by LDL receptors on target cells and is essential for directing LDL particles to their intended destinations. LDL uptake by various tissues is by receptor-mediated endocytosis [1, 58] (Fig. 9.18).

LDL delivers **cholesterol** from the **liver** to **peripheral tissues**. While the liver can metabolize and excrete cholesterol (discussed further below), most other tissues take up cholesterol from LDL and use it in membrane synthesis or hormone production, but they cannot degrade or excrete it. Thus, LDL leads to an increase in cholesterol levels in tissues. This is why it is often referred to as the "bad cholesterol."

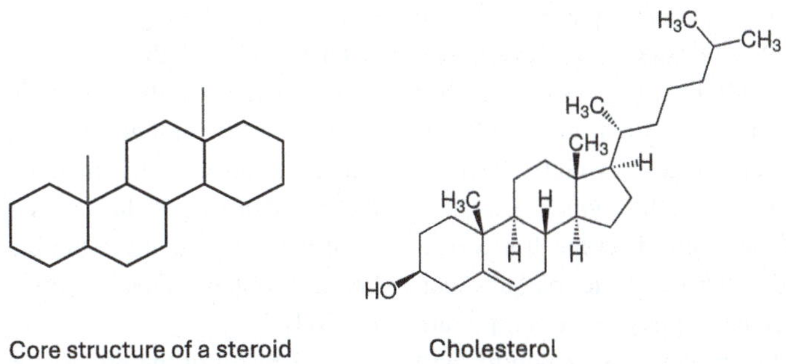

Fig. 9.17 Steroids. Steroids include hormones, some vitamins, and cholesterol. They are based on a ring structure, with branches and modifications. (*Source*: Cholesterol structure: Public domain image from Wikimedia Commons: File:Structural_formula_of_cholesterol.svg)

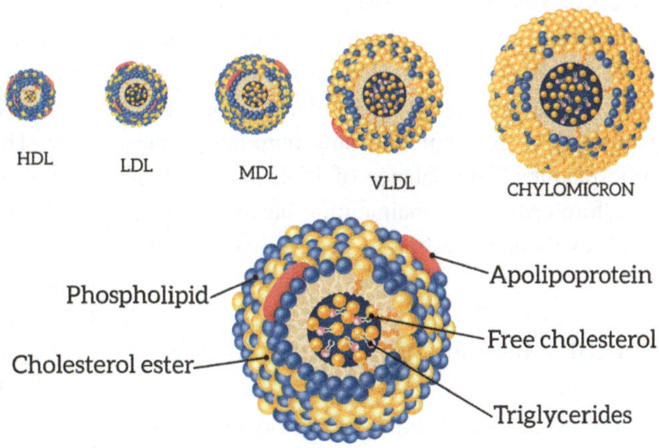

Fig. 9.18 Structures Compositions of different classes of lipoproteins. Lipoproteins package and transport cholesterol, TAGs, and other lipids. (*Source*: Illustration from Freepik.com (premium license))

- High-Density Lipoprotein Particles (HDLs)

HDL particles also carry cholesterol, but unlike LDL particles, they transport cholesterol in the opposite direction. HDL particles transport excess cholesterol from peripheral tissues to the liver, where the cholesterol is broken down and excreted. Thus, HDL removes cholesterol from tissues, and this is why HDL is often called the "good cholesterol" [1, 58].

Other particles that carry cholesterol in addition to LDL and HDL are chylomicrons, VLDL, and IDL, which have different proportions of proteins, cholesterol, phospholipids, and TAGs (see Fig. 9.18).

- Excretion of Cholesterol

Cells cannot degrade cholesterol, so cholesterol must either be used or excreted by the liver. For excretion, the liver converts some cholesterol into bile salts, thereby lowering the amount of cholesterol in the body. These bile salts are then secreted into the bile (digestive fluid produced by the liver) and released into the small intestine. There, they aid in the digestion of fat. Some cholesterol can also be secreted directly into the bile in its unmodified form. While a large portion of the bile salts is reabsorbed in the intestine and returned to the liver, a portion of it escapes reabsorption and is excreted in the feces. Excretion in the feces is the primary method by which the body eliminates excess cholesterol, as it cannot break it down enzymatically [1, 40, 59].

- Excess Cholesterol in Human Health

While a certain amount of cholesterol in the body is necessary, **excessive cholesterol** can lead to health problems. Excess cholesterol that is not used or excreted is transported in the bloodstream by low-density lipoprotein (LDL) particles. When LDL levels are high, cholesterol can accumulate in various tissues, including the walls of blood vessels. The cholesterol is then modified and taken up by macrophages, which become engorged and form foam cells, which are filled with cholesterol. Foam cells become trapped in blood vessels, and atherosclerotic plaques can form [1]. Atherosclerotic plaques contain foam cells, cholesterol, fat, blood cells, calcium, and fibrin (a type of insoluble protein) [60]. These plaques narrow the arteries and increase the risk of heart attacks and strokes. HDL plays a crucial role in regulating cholesterol levels by removing excess cholesterol from tissues and macrophages, transporting it back to the liver for excretion. If HDL levels are low or HDL does not function correctly, it fails to pick up enough cholesterol, leading to more foam cells and more atherosclerotic plaques. The balance of LDL and HDL is therefore critical for maintaining healthy cholesterol levels and preventing heart disease [1, 40, 61].

- PCSK9 Inhibitors for Lowering Cholesterol Levels

LDL receptors (LDLRs) are receptors on the surface of liver cells. LDLRs bind to LDL that is circulating in the bloodstream. Specifically, they bind to apolipoprotein B-100 (ApoB-100), a protein located on the outer surface of LDL particles. The LDLR-LDL complex is internalized into the liver cells via receptor-mediated endocytosis (RME). In this process, the complex is first internalized into a clathrin-coated vesicle. The vesicle then uncoats and fuses with an early endosome. This early endosome has an acidic environment, which causes the LDL to be released from the LDLR. The LDLR is recycled back to the cell surface, but the LDL remains in the endosome. The endosome fuses with a lysosome. Lysosomes contain enzymes that degrade LDL, and in this process, cholesterol can be released from the LDL particles. The released cholesterol may be used by the liver. For example, it plays a role in the synthesis of bile salts (see above), the maintenance of membranes, and the excretion of some bile salts. Having sufficient LDL receptors (LDLRs) thus helps to keep blood cholesterol levels low. Nevertheless, if cholesterol levels are too high, some of it gets packaged into lipoprotein particles and gets distributed to other tissues via the bloodstream. If this is excessive, then cholesterol may recirculate in the bloodstream, causing health problems such as **atherosclerotic plaques** [62–64].

As a result of the process described above, the LDLRs remove the LDL particles from the bloodstream. Having the correct balance of LDL receptors is important for maintaining a low level of LDL in the bloodstream. PCSK9 (proprotein convertase subtilisin/kexin type 9) is a protein that is produced in liver cells. It is a secreted protein, and most of it circulates in the bloodstream. PCSK9 binds to LDLRs on the surface of liver cells. This binding leads to internalization of the Pcsk9 LDLR complex into the cell. This prevents the LDLR from recycling and targets it for the lysosome, where it is destroyed. This decreases the number of LDLRs available to help clear LDL from the bloodstream (*see* Chap. 25).

The level of PCSK9 is thus also very important. If PCSK9 levels are too high, too many LDL receptors get internalized and destroyed in the liver cells, even when they are not bound to LDL. Consequently, LDL cannot be taken up into the liver in sufficient amounts, and cholesterol levels remain too high in the bloodstream (high circulating LDL levels). In contrast, when the levels of Pcsk9 are low, the LDL receptor is not destroyed. Instead, LDL receptors on the surface of liver cells bind and import LDL into the liver, where the LDL gets processed and removed from circulation. The LDL receptor is then recycled back to the cell surface, where it can bind to more LDL. This is preferable because it results in low circulating LDL levels. Pcsk9 is considered a promising drug target for treating high cholesterol. Lowering the amount of Pcsk9 would help reduce the amount of circulating LDL in the bloodstream [65].

Repatha (evolocumab) [66] is a drug designed to treat **high cholesterol**. It is a monoclonal antibody that binds to circulating PCSK9. This prevents PCSK9 from binding to LDLRs.

This increases the amount of LDLRs on liver cells that are available to clear LDL from the bloodstream. The drug is administered by injection every 2 weeks and can be used together with statins [67–70]. Repatha is a human monoclonal antibody in the IgG2 subclass. This mAb is less inflammatory than some other classes of antibodies [71], and the main goal of this drug is to prevent PCSK9 from binding to its receptor. (For a more detailed discussion about monoclonal antibodies, see Chap. 24).

- Statins

Cholesterol can be synthesized de novo, mostly in the liver. The rate of cholesterol synthesis is mediated by changes in the amount and activity of the enzyme **HMG-CoA reductase,** an integral membrane protein in the ER membrane that regulates the rate of cholesterol synthesis. This enzyme catalyzes the synthesis of mevalonate**,** an intermediate necessary for cholesterol biosynthesis. The reaction it catalyzes is the rate-limiting step in cholesterol biosynthesis.

Statins are another family of drugs that treat **high cholesterol** by inhibiting **HMG-CoA reductase**. **Statins**, such as **Lovastatin**, are competitive inhibitors of **HMG-CoA reductase**, and they help patients reduce their cholesterol levels, thus **reducing the risk of heart attack and stroke** [1, 40, 69].

2.7.3 Hormones

Numerous different hormones are derived from cholesterol, and they also have a 4-ring structure. These include the three main sex hormones, estrogen, progesterone, and testosterone, as well as glucocorticoids and mineralocorticoids. Glucocorticoids (such as cortisol) are produced in the adrenal cortex and have important roles in a wide range of cellular functions, including metabolism. Mineralocorticoids (such as aldosterone) have a role in regulating sodium and water balance in the body [1].

2.7.4 Bile Acids

Bile acids, derived from cholesterol, are also steroids. They are amphipathic and can function as detergents that bind to and emulsify lipids [1, 72]. Due to their amphipathic nature, they can interact with both fat and water, enabling them to break up large droplets (emulsification) [73]. In the liver, bile acids are conjugated with amino acids such as glycine to form bile salts, which are their water-soluble forms. These are the major forms secreted into bile (digestive fluid) and are even more effective at emulsifying dietary fats.

2.8 Other Lipids

Other lipids exist aside from those discussed above. For example, prostaglandins, thromboxanes, and leukotrienes are members of a category of lipids known as eicosanoids, all of which are derived from fatty acids that contain 20 carbon atoms. They are derived from the metabolism of arachidonic acid, a type of essential fatty acid. They are involved in various biological processes, including the regulation of inflammation, blood flow, and ion transport. Some fat-soluble vitamins, such as vitamins A, D, E, and K, are also classified as lipids. Finally, gangliosides, a type of glycolipid, play important roles in some cell signaling pathways [1, 14, 74, 75].

2.9 Lipids in Drug Development

Lipids can be used for drug delivery. A novel mechanism for delivering certain drugs involves the use of nanoparticle technology. Nanoparticle delivery of drugs involves encasing drugs in lipid bilayers called liposomes (Fig. 9.19). Lipids on the surface of the liposome may, in some cases, also be modified by polyethylene glycol (PEG) to increase stability and efficiency. Liposomes sometimes also contain antibodies to aid in targeting them to the correct destination. This drug delivery method has various benefits. For example, it is beneficial for delivering hydrophobic drugs and provides a protective environment for them. Liposomes are typically delivered by IV, but they can sometimes be formulated for alternative methods, such as oral intake, inhalation,

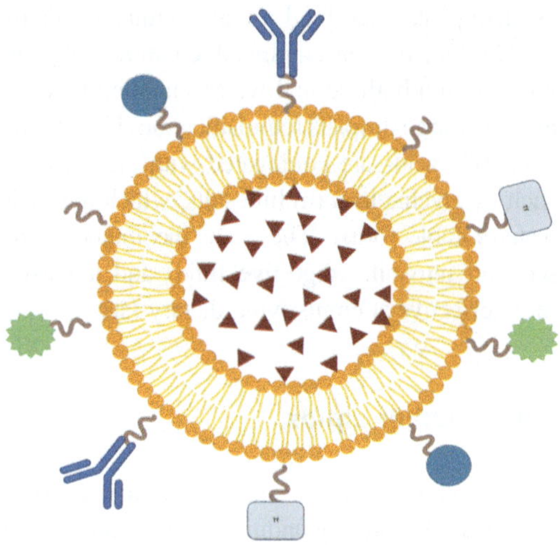

Fig. 9.19 Liposome-mediated drug delivery. In this depiction of a liposome, the liposome has antibodies on the outer surface, which have a role in targeting the liposome to the correct location. It also has other molecules on its outer surface, which can play various roles, such as providing stability, evading the immune system, and facilitating targeting. (*Source*: Created with BioRender.com)

injection, or even topical application. The anticancer drug **doxorubicin** is used to treat several types of **cancer** and was discussed in Chap. 6. Doxorubicin is an example of a drug approved for liposomal delivery. Doxorubicin can be delivered by other methods, but liposomal delivery has been shown to be the most effective. **Doxil** [76] is the brand name for the liposomal formulation of doxorubicin. Encapsulating doxorubicin in a liposome improves the targeting of the drug to the tumor and reduces side effects. It also helps prolong the circulation of the drug in the bloodstream, allowing it to accumulate at the site of the tumor [77–80]. This mechanism of drug delivery is discussed in more detail in Chap. 28.

2.10 Metabolic Diseases

As discussed previously with carbohydrates, many metabolic disorders are associated with disorders in fat metabolism. Obesity, for example, is associated with too much fat accumulation and can lead to other metabolic disorders such as type 2 Diabetes. Type 2 diabetes is characterized by insulin resistance, abnormalities in fat metabolism, and the accumulation of free fatty acids. Nonalcoholic fatty liver disease is also a disorder involving the inappropriate accumulation of fat tissue, in this case, in the liver [36]. These types of metabolic disorders will be discussed in Chap. 11.

2.11 Lipoprotein Particles and Lipid Transport

This chapter has covered various aspects of lipid structure and function. One final consideration is the transport of lipids throughout the body. Since lipids are hydrophobic, they cannot dissolve in water, which is the main component of blood. Therefore, they get transported through the bloodstream in the form of **lipoprotein particles**. We have already touched on some of these, such as LDLs, HDLs, and chylomicrons, but because of their importance as carriers, they will also be reviewed here. These carrier complexes are spherical structures. The main carriers are chylomicrons, VLDL, IDL, LDL, and HDL. These carriers not only carry lipids, but they also have other components. The main components that all of these particles carry are proteins, cholesterol, phospholipids, and TAGs. The proteins and phospholipids have a largely structural role in these complexes, and the main functions of the particles are the delivery of TAGs and cholesterol. The different particles vary in size, the proportion of their components, and the substances they specialize in carrying.

- Chylomicrons

Chylomicrons are the largest of these particles, and HDL is the smallest (see Fig. 9.18). Chylomicrons are formed in the intestine after fat absorption. They carry dietary TAGs and cholesterol to tissues. They have an especially high percentage of TAGs.

- VLDL, IDL, and LDL

Very low-density lipoproteins (VLDLs) **are produced in the liver and transport** endogenously synthesized (made in the body) TAGs to peripheral tissues. Approximately half of their components are TAGs, while the rest consists of cholesterol, phospholipids, and proteins. As VLDL particles release triacylglycerols, they become intermediate-density lipoproteins (IDLs) and eventually low-density lipoproteins (LDLs), which have progressively less TAG but more cholesterol. LDL particles have high levels of cholesterol, and they deliver cholesterol to cells throughout the body.

- HDL

Finally, high-density lipoproteins (HDLs) carry extra cholesterol from tissues and return it to the liver. This is called reverse cholesterol transport. In HDL particles, the concentration of protein is higher than the other components. Together, these carriers are important for packaging, delivery, and clearance of lipids throughout the body [14, 50].

- Apolipoprotein, Components of Lipoproteins

While lipoproteins are important for their lipid content, their function depends on the proteins they contain, specifically the apolipoproteins. Apolipoproteins are a family of proteins with multiple functions. They function as structural scaffolds, cofactors for enzymes, and ligands for receptors on target tissues. There are different Apo proteins with distinct functions, and some are localized to different particles. For example, ApoB-48 is required for the assembly of chylomicrons in the intestine, and ApoB-100 is in VLDL, IDL, and LDL. It forms a structural scaffold. ApoB-100 is the protein that binds to LDL receptors (LDLRs), which are mainly on liver cells. In other words, ApoB-100 serves as the ligand for LDLR, thereby mediating the uptake of LDL (and other particles) in the liver. Within the liver, particles generally get destroyed in the lysosome. ApoC-II is found mostly in VLDLs, HDLs, and chylomicrons. It is an essential activator of lipoprotein lipase (LPL), the enzyme that hydrolyzes TAGs into free fatty acids and glycerol. ApoE is found on chylomicron remnants, as well as on VLDL and IDL particles. It can also bind to the LDL receptor, leading to the clearance of its associated particles through lysosomal degradation. ApoA-I is the major structural protein of HDL. In addition to being a structural protein, it activates LCAT Lecithin-cholesterol acyltransferase), an enzyme in reverse cholesterol transport. This enzyme functions by converting free cholesterol to cholesterol esters, which are transported by HDL for collection in the liver [81, 82].

FDA-Approved Drugs Mentioned in This Chapter

Brand name	Generic name	First FDA approval year	Common conditions	Main target
Danyelza	*Naxitamab (naxitamab-gqgk)*	2020	Relapsed/refractory high-risk neuroblastoma	GD2 (Disialoganglioside 2)
Doxil	*Doxorubicin (liposomal)*	1995	Ovarian cancer, Kaposi's sarcoma, multiple myeloma	Intercalates DNA; delivered via liposome carrier
Kimmtrak	*Tebentafusp-tebn*	2022	Uveal melanoma	GP-100 (glycoprotein 100)
Repatha	*Evolocumab*	2015	High LDL cholesterol; cardiovascular risk; familial forms	PCSK9
Tryngolza	*Olezarsen*	2024	Familial chylomicronemia syndrome (FCS)	ApoC-III
Unituxin	*Dinutuximab*	2015	High-risk pediatric neuroblastoma	GD2 (Disialoganglioside 2)

Definitions

Acid A substance that donates hydrogen ions (H^+; often referred to as protons) in solution. They often have formulas that start with H, such as HCl (hydrochloric acid).

Aldehyde Organic functional group containing a carbonyl (C=O) bonded to a hydrogen at the terminal carbon (–CHO).

Aldose A monosaccharide with an aldehyde group (e.g., glucose).

Alpha (α) and beta (β) anomers Different forms that cyclic monosaccharides (sugars), differing in the orientation of the hydroxyl group attached to the anomeric carbon.

ApoC-III A protein that inhibits triglyceride clearance from the bloodstream.

Beta-oxidation The process by which fatty acids are broken down, mainly into acetyl-CoA units.

Cellulose A polysaccharide present in plants composed of β-1,4-linked glucose.

Chylomicrons Large lipoprotein particles that contain lipids, cholesterol, phospholipids, and proteins.

Condensation reaction Two molecules combine to form a larger molecule.

Dehydration reaction A type of condensation reaction where two molecules combine to form a larger molecule, with the elimination of a water molecule.

Disaccharide A carbohydrate made of two monosaccharides. Examples are lactose and sucrose.

Disialoganglioside 2 (GD2) A glycolipid (disialoganglioside) overexpressed in some cancers like neuroblastoma; target of the drug Unituxin.

Eicosanoids A category of lipids derived from the metabolism of arachidonic acid.

Ester group Functional group with the structure -COO- in which the carbonyl carbon (C=O) is bonded to an oxygen atom that is also bonded to another carbon atom. Commonly found in many fats and lipids.

Essential fatty acids Fatty acids that cannot be synthesized by the human body.

Ester bond A type of bond formed between a carboxylic acid and an alcohol, with the elimination of water (a condensation reaction). Connects the carbonyl carbon of the acid to the oxygen of the alcohol, creating an ester linkage.

Fatty acid The simplest form of lipids, A fatty acid consists of a hydrocarbon chain with a carboxyl group (COOH) at the alpha end and a methyl group at the other end, the omega end.

Fructose A monosaccharide found in fruits and honey.

Furanose A five-membered ring form of a monosaccharide. An example is ribose.

Glucose A monosaccharide and a main energy source for cells.

Glycogen A large polysaccharide composed of glucose molecules. The main storage form of glucose.

Glycosidic bond A covalent bond that links carbohydrate molecules to each other or to other types of molecules.

HDL (high-density lipoprotein) A lipoprotein that carries cholesterol. HDL particles transport excess cholesterol to the liver from various parts of the body, to be broken down and excreted.

Isoprene A five-carbon molecule (C_5H_8) that forms the basic unit of terpenes and terpenoids.

Ketone Functional group with a carbonyl (C=O) bonded to two carbons (R-CO-R'). The term ketone is also used to refer to a molecule that contains a ketone group.

Ketose A monosaccharide with a ketone functional group. An example is fructose.

Lactase The enzyme that hydrolyzes lactose, breaking it down into glucose and galactose. In individuals with lactose intolerance, this enzyme is deficient.

Lactose A disaccharide found in milk, made of glucose and galactose.

Low-density lipoprotein (LDL) A lipoprotein that delivers cholesterol to tissues; elevated levels are associated with atherosclerosis.

Lipase An enzyme that catalyzes the breakdown of TAGs into glycerol and free fatty acids.

Lipid Hydrophobic biomolecules primarily composed of carbon and hydrogen, involved in functions such as energy storage, signaling, and membrane structure.

Liposome A spherical vesicle made of a lipid bilayer, sometimes used for encasing drugs for delivery into patients.

Maltose A disaccharide made of two glucose units.

Monosaccharide The simplest form of carbohydrate, and the monomers that make up more complex carbohydrates.

Nucleophilic attack A reaction in which an electron-rich nucleophile donates an electron pair to a positive or partially positive charged atom (often a carbon) forming a new covalent bond.

Phospholipid A lipid containing two fatty acid tails, a glycerol backbone, and a phosphate group with a polar head. An important component of membranes.

Polysaccharide A carbohydrate, usually large, consisting of monosaccharide units linked by glycosidic bonds. Examples include starch, glycogen, and cellulose.

Proprotein convertase subtilisin/kexin type 9 (PCSK9) A protein that binds LDL receptors, promoting their internalization and degradation.

Pyranose A six-membered ring form of a monosaccharide.

Ribose A five-carbon (pentose) sugar. This sugar is found in ribonucleotides and deoxyribonucleotides that make up in RNA and DNA. It is also found in ATP.

Saturated fatty acid A fatty acid with no double bonds, allowing maximum hydrogen saturation.

Statins A class of drugs that inhibit HMG-CoA reductase, the rate-limiting enzyme in cholesterol biosynthesis.

Steroid A type of lipid with a four-ring carbon structure. Hormones like cortisol, sex hormones, and cholesterol are examples of steroids.

Sucrose A disaccharide composed of glucose and fructose. Sucrose is the term for table sugar.

Triacylglycerol (TAG) A molecule made of three fatty acids esterified to a glycerol backbone. It is the major form of energy storage in adipose tissue.

Unsaturated fatty acid A fatty acid that contains one or more double bonds.

Study Questions

Multiple-Choice Questions

1. Which of the following is a disaccharide composed of glucose and galactose?
   A. Sucrose
   B. Lactose
   C. Maltose
   D. Cellulose
2. What type of glycosidic bond links glucose units in the main chain of glycogen?
   A. β-1,4
   B. α-1,4
   C. α-1,6
   D. β-1,6
3. Which monosaccharide is most abundant and crosses the blood–brain barrier?
   A. Fructose
   B. Galactose
   C. Glucose
   D. Ribose
4. Which structural form is most common for glucose in the body?
   A. Linear
   B. D-glucose
   C. Furanose
   D. Pyranose
5. Which of the following best describes cellulose?
   A. A branched polysaccharide with ⍰-1,6 linkages
   B. A disaccharide used in milk
   C. A straight chain of glucose with ⍰-1,4 linkages
   D. A storage form of sugar in animals
6. Which is the chemical formula for glucose?
   A. C12H22O11
   B. C6H12O6

C. C2H6O
D. C10H2O6
E. None of the above

7. Which of the following is a monosaccharide?
A. Glucose
B. Sucrose
C. Cellulose
D. Starch
E. All of the above

8. Which polysaccharide is often found in plant cell walls and provides structural support?
A. Glycogen
B. Chitin
C. Cellulose
D. Starch

9. Which of the following lipids is the major component of cell membranes?
A. A, Triacylglycerols
B. Phospholipids
C. Steroids
D. Terpenoids

10. Which fatty acid nomenclature best describes linoleic acid?
A. C16:0
B. C18:2,Δ9,12
C. C12:0
D. C18:3,Δ9,12,15

11. Which of the following make trans-fats dangerous for our health?
A. Their ring structure increases the amount of LDL in the cell.
B. They are too short to be metabolized.
C. Due to their structures they are rigid and pack tightly together.
D. Their structure leads to inhibition of important enzymes.

12. What is the main mechanism of action for PCSK9 inhibitors like Repatha?
A. Enhance FA oxidation in mitochondria
B. Prevent PCSK9 from degrading LDL receptors
C. Inhibit HMG-CoA reductase
D. It functions by activating statins

13. Saturated fatty acids have
A. No double bonds between carbon atoms
B. One double bond between carbon atoms
C. Multiple double bonds between carbon atoms
D. No carbon atoms
E. They are saturated with double bonds between all the carbon atoms.

14. Phospholipids contain
A. glycerol
B. A phosphate group
C. Three hydro-carbon side chains
D. A and B
E. All of the above

True/False Questions

15. Humans can digest cellulose due to the presence of β-1,4 glycosidic bond-cleaving enzymes.

Fill-in-the-Blank Questions

16. The monosaccharide commonly found in ATP and RNA is _________.
17. The enzyme lacking in people with lactose intolerance is _________.
18. A polysaccharide used for energy storage in animals is called _________.
19. The ________ group is located at the alpha end of a fatty acid.
20. The bond that links two monosaccharides together is a _________ bond.

Short-Answer Questions

21. Explain why unsaturated fatty acids are usually liquid at room temperature.
22. Why are essential fatty acids termed "essential" in human nutrition?

Matching Questions

23. **Match the disaccharides with their monosaccharide components:**

Disaccharide	

1. Sucrose
2. Lactose
3. Maltose

Components:

A. Glucose + Glucose
B. Glucose + Fructose
C. Glucose + Galactose

24. **Match each lipid with its function**

Lipid	Function

A. Triacylglycerols
B. Phospholipids
C. Cholesterol

Function:

1. Energy storage
2. Cell membrane structure
3. Precursor to bile acids and steroid hormones

Answer Key

Multiple-Choice Questions

1. **Answer**: B
2. **Answer**: B
3. **Answer**: C
4. **Answer**: D
5. **Answer**: C
6. **Answer**: B
7. **Answer**: A
8. **Answer**: C
9. **Answer**: B
10. **Answer**: B
11. **Answer**: C.
12. **Answer**: B.
13. **Answer**: A
14. **Answer**: D

True False Questions

15. **Answer**: False

Fill-in-the-Blank Questions

16. **Answer**: Ribose
17. **Answer**: Lactase
18. **Answer**: Glycogen
19. **Answer**: Carboxyl
20. **Answer**: Glycosidic

Short-Answer Questions

21. **Answer**: Because of their double bonds, they usually bend and have kinks rather than having a linear structure. This prevents them from packing together tightly, and they are therefore more likely to be liquid.
22. **Answer**: Because humans lack the enzymes to introduce double bonds beyond the 9th carbon from the carboxyl end and therefore must obtain these fatty acids from the diet.

Matching Questions

23. **Answer**: 1–B, 2–C, 3–A
24. **Match each lipid with its function**
    Answer:
    A – 1, B – 2, C – 3

References

1. Berg JM, Tymoczko JL, Gatto GJ Jr, Stryer L. Biochemistry. 9th ed. New York: MacMillan Learning; 2019.
2. Cleveland Clinic. Carbohydrates. Available from: https://my.clevelandclinic.org/health/articles/15416-carbohydrates.
3. Campbell MK, Farrell SO. Biochemistry. Pacific Grove: Brooks/Cole; 2007.
4. Ghazarian H, Idoni B, Oppenheimer SB. A glycobiology review: carbohydrates, lectins and implications in cancer therapeutics. Acta Histochem. 2011;113(3):236–47. https://doi.org/10.1016/j.acthis.2010.02.004.
5. Fares A, Fares G. The chemistry of carbohydrates and their effects on the human body. Biomed J Sci Techn Res. 2025;61:54053–63. https://doi.org/10.26717/BJSTR.2025.61.009662.
6. Alberts B, Heald R, Johnson A, Morgan D, Raff M, Roberts K, Walter P, Wilson J, Hunt T. Molecular

biologyof the cell. New York, NY. W.W. Norton and Company; 2022.
7. Neoh GKS, Tan X, Chen S, Roura E, Dong X, Gilbert RG. Glycogen metabolism and structure: a review. Carbohydr Polym. 2024;346:122631. https://doi.org/10.1016/j.carbpol.2024.122631.
8. Zabala V, CCamilo C. An overview on starch structure and chemical nature. Starch-based nanomaterials. Cham: Springer International Publishing; 2020. p. 3–9.
9. Lebrilla CBLJ, Widmalm G, Prestegard JH. Oligosaccharides and polysaccharides. In: Varki ACR, Esko JD, Esko JD, Stanley P, editors. Essentials of Glycobiology. 4th ed. Cold Spring Harbor, NY. Cold Spring Harbor Laboratory Press; 2022.
10. Healthline. Is cellulose fiber safe to eat? Available from: https://www.healthline.com/nutrition/cellulose-fiber.
11. Mergenthaler P, Lindauer U, Dienel GA, Meisel A. Sugar for the brain: the role of glucose in physiological and pathological brain function. Trends Neurosci. 2013;36(10):587–97. https://doi.org/10.1016/j.tins.2013.07.001.
12. International Food Information Council. What is fructose 2020. Available from:https://foodinsight.org/what-is-fructose/.
13. Braunberger TL, Nahhas AF, Katz LM, Sadrieh N, Lim HW. Dihydroxyacetone: a review. J Drugs Dermatol. 2018;17(4):387–91.
14. Nelson DL, Cox MM, Hoskins AA. Lehninger principles of biochemistry. 8th ed. New York, NY Macmillan Learning; 2021.
15. Mellado-Mojica E, Seeram NP, López MG. Comparative analysis of maple syrups and natural sweeteners: carbohydrates composition and classification (differentiation) by HPAEC-PAD and FTIR spectroscopy-chemometrics. J Food Compos Anal. 2016;52:1–8. https://doi.org/10.1016/j.jfca.2016.07.001.
16. Theocharis AD, Skandalis SS, Gialeli C, Karamanos NK. Extracellular matrix structure. Adv Drug Deliv Rev. 2016;97:4–27. https://doi.org/10.1016/j.addr.2015.11.001.
17. Zhang X, Zhu X, Bi X, Huang J, Zhou L. The insulin receptor: an important target for the development of novel medicines and pesticides. Int J Mol Sci. 2022;23(14):7793.
18. Stowell SR, Ju T, Cummings RD. Protein glycosylation in cancer. Annu Rev Pathol. 2015;10:473–510. https://doi.org/10.1146/annurev-pathol-012414-040438.
19. Berois N, Pittini A, Osinaga E. Targeting tumor Glycans for cancer therapy: successes, limitations, and perspectives. Cancers (Basel). 2022;14(3) https://doi.org/10.3390/cancers14030645.
20. Voeller J, Sondel PM. Advances in anti-GD2 immunotherapy for treatment of high-risk Neuroblastoma. J Pediatr Hematol Oncol. 2019;41(3):163–9. https://doi.org/10.1097/mph.0000000000001369.
21. (dinutuximab) U. Part of the standard-of-care regimen for high-risk neuroblastoma for over 10 years. Available from: https://www.unituxin.com/hcp/.
22. United Therapeutics Corporation. Unituxin website. Unituxin. Available from:https://www.unither.com/research-and-medicine/unituxin#:~:text=Unituxin%20can%20cause%20serious%20and,may%20need%20to%20stop%20treatment.
23. FDA. Danyelza (Naxitamab (naxitamab gqgk) Approval Overview 2020. Available from:https://www.accessdata.fda.gov/scripts/cder/daf/index.cfm?event=overview.process&ApplNo=761171.
24. ecancer. FDA grants accelerated approval to naxitamab for high-risk neuroblastoma in bone or bone marrow. 2020. Available from: https://ecancer.org/en/news/19168-fda-grants-accelerated-approval-to-naxitamab-for-high-risk-neuroblastoma-in-bone-or-bone-marrow
25. Gupta S, McDonald JD, Ayabe RI, Khan TM, Gamble LA, Rossi AJ, et al. Perioperative MVT-5873, a fully human monoclonal antibody against a CA 19-9 epitope, for operable CA 19-9 producing pancreatic cancers, cholangiocarcinomas, and metastatic colorectal cancers. J Clin Oncol. 2020;38(15_suppl):TPS4655-TPS. https://doi.org/10.1200/JCO.2020.38.15_suppl.TPS4655.
26. O'Reilly EM, Borazanci EH, Yu KH, Varghese AM, Estrella H, Kamins D, et al. HuMab-5B1 (MVT-5873), a mAb targeting sLe^a, in combination with first-line gemcitabine plus nab-paclitaxel (gem/nab-P) for patients with pancreatic cancer (PDAC) and other CA19-9 positive malignancies. J Clin Oncol. 2018;36(15_suppl):e16235-e. https://doi.org/10.1200/JCO.2018.36.15_suppl.e16235.
27. Martinez-Perez D, Viñal D, Solares I, Espinosa E, Feliu J. Gp-100 as a novel therapeutic target in Uveal Melanoma. Cancers (Basel). 2021;13(23) https://doi.org/10.3390/cancers13235968.
28. The Antibody Society. FDA approves KIMMTRAK® (tebentafusp-tebn) for unresectable or metastatic uveal melanoma. 2022. Available from:https://www.antibodysociety.org/food-and-drug-administration/fda-approves-kimmtrak-tebentafusp-tebn-for-unresectable-or-metastatic-uveal-melanoma/#:~:text=On%20January%2026%2C%202022%2C%20Immunocore%20Holdings%20plc,with%20unresectable%20or%20metastatic%20uveal%20melanoma%20(mUM).
29. Cleveland Clinic. Lactose Intolerance 2023. Available from:https://my.clevelandclinic.org/health/diseases/7317-lactose-intolerance.
30. Deng Y, Misselwitz B, Dai N, Fox M. Lactose intolerance in adults: biological mechanism and dietary management. Nutrients. 2015;7(9):8020–35. https://doi.org/10.3390/nu7095380.
31. Schaefer, EC. What is Lactose? Available from: https://www.verywellhealth.com/what-is-lactose-5076054

32. Mayo Clinic. Lactose intolerance 2022. Available from: https://www.mayoclinic.org/diseases-conditions/lactose-intolerance/symptoms-causes/syc-20374232.
33. Di X, Liang X, Shen C, Pei Y, Wu B, He Z. Carbohydrates used in polymeric systems for drug delivery: from structures to applications. Pharmaceutics. 2022;14(4) https://doi.org/10.3390/pharmaceutics14040739.
34. Chiang AW, Li S, Spahn PN, Richelle A, Kuo CC, Samoudi M, et al. Modulating carbohydrate-protein interactions through glycoengineering of monoclonal antibodies to impact cancer physiology. Curr Opin Struct Biol. 2016;40{Chiang, 2016 #2686}:104–11. https://doi.org/10.1016/j.sbi.2016.08.008.
35. Alberts B, Bray D, Hopkin K, Johnson A, Lewis J, Raff M, Roberts K, Walter P. Essential Cell Biology. New York, NY. Norton and Company; 2023.
36. Bender DA. Introduction to nutrition and metabolism. Boca Raton: CRC Press; 2021.
37. de Carvalho C, Caramujo MJ. The various roles of fatty acids. Molecules. 2018;23(10) https://doi.org/10.3390/molecules23102583.
38. Yoon H, Shaw JL, Haigis MC, Greka A. Lipid metabolism in sickness and in health: emerging regulators of lipotoxicity. Mol Cell. 2021;81(18):3708–30. https://doi.org/10.1016/j.molcel.2021.08.027.
39. Marks F, Klingmüller U, Müller-Decker K. Cellular signal processing: an introduction to the molecular mechanisms of signal transduction. 2nd ed. New York, NY. Garland Science; 2017.
40. Hames BD, Hooper NM. BIOS Instant Notes in Biochemistry. New York, NY. Taylor and Francis; 2011.
41. MedlinePlus. LDL: The "Bad" Cholesterol 2024. Available from: https://medlineplus.gov/ldlthebadcholesterol.html.
42. Zivkovic AM, Telis N, German JB, Hammock BD. Dietary omega-3 fatty acids aid in the modulation of inflammation and metabolic health. Calif Agric (Berkeley). 2011;65(3):106–11. https://doi.org/10.3733/ca.v065n03p106.
43. Liu AG, Ford NA, Hu FB, Zelman KM, Mozaffarian D, Kris-Etherton PM. A healthy approach to dietary fats: understanding the science and taking action to reduce consumer confusion. Nutr J. 2017;16(1):53. https://doi.org/10.1186/s12937-017-0271-4.
44. Temkov M, Mureşan V. Tailoring the structure of lipids, Oleogels and fat replacers by different approaches for solving the trans-fat issue—a review. Foods. 2021;10(6):1376.
45. Ge Y, Liu W, Tao H, Zhang Y, Liu L, Liu Z, et al. Effect of industrial trans-fatty acids-enriched diet on gut microbiota of C57BL/6 mice. Eur J Nutr. 2019;58(7):2625–38. https://doi.org/10.1007/s00394-018-1810-2.
46. Mayo Clinic. Trans fat is double trouble for heart health. Available from: https://www.mayoclinic.org/diseases-conditions/high-blood-cholesterol/in-depth/trans-fat/art-20046114#:~:text=Trans%20fat%20is%20considered%20the,the%20leading%20killer%20of%20adults.
47. National Institutes of Health. Omega-3 Fatty Acids. Available from: https://ods.od.nih.gov/factsheets/Omega3FattyAcids-Consumer/#:~:text=for%20ALA%20alone.-,What%20foods%20provide%20omega%2D3s?,soy%20beverages%2C%20and%20infant%20formulas).
48. Rui L. Energy metabolism in the liver. Compr Physiol. 2014;4(1):177–97. https://doi.org/10.1002/cphy.c130024.
49. Kumari A, Kristensen KK, Ploug M, Winther AL. The importance of lipoprotein lipase regulation in atherosclerosis. Biomedicine. 2021;9(7) https://doi.org/10.3390/biomedicines9070782.
50. Feingold KR. Introduction to lipids and lipoproteins. In: Feingold KR, Ahmed SF, Anawalt B, Blackman MR, Boyce A, Ghrousos G, Corpas E, Dhatariya K, Dungan K, Hofland J, Kalra S, Kaltsas G, Kapoor N, editors. Endotext [Internet]. South Dartmouth: MDText.com; 2024.
51. Javed F, Hegele RA, Garg A, Patni N, Gaudet D, Williams L, et al. Familial chylomicronemia syndrome: an expert clinical review from the National Lipid Association. J Clin Lipidol. 2025;19(3):382–403. https://doi.org/10.1016/j.jacl.2025.03.013.
52. FDA. Tryngolza(Olezarsen) Approval overview 2024. Available from: https://www.accessdata.fda.gov/scripts/cder/daf/index.cfm?event=overview.process&ApplNo=218614.
53. Borén J, Packard CJ, Taskinen M-R. The roles of ApoC-III on the metabolism of triglyceride-rich lipoproteins in humans. Front Endocrinol. 2020;11 https://doi.org/10.3389/fendo.2020.00474.
54. Tryngolza. Introducing Tryngolza 2024. Available from: https://tryngolzahcp.com/.
55. Hillier SG, Lathe R. Terpenes, hormones and life: isoprene rule revisited. J Endocrinol. 2019;242(2):R9–R22. https://doi.org/10.1530/joe-19-0084.
56. Masyita A, Mustika Sari R, Dwi Astuti A, Yasir B, Rahma Rumata N, Emran TB, et al. Terpenes and terpenoids as main bioactive compounds of essential oils, their roles in human health and potential application as natural food preservatives. Food Chem X. 2022;13:100217. https://doi.org/10.1016/j.fochx.2022.100217.
57. Staels B, Fonseca VA. Bile acids and metabolic regulation: mechanisms and clinical responses to bile acid sequestration. Diabetes Care. 2009;32 Suppl 2(Suppl 2):S237–45. https://doi.org/10.2337/dc09-S355.
58. Medical News Today. What to know about lipoproteins, cholesterol, and diet. Available from: https://www.medicalnewstoday.com/articles/318712.
59. Hofmann AF. Biliary secretion and excretion in health and disease: current concepts. Ann Hepatol. 2007;6(1):15–27. https://doi.org/10.1016/S1665-2681(19)31949-0.
60. National Heart Lung, and Blood Institute. What is Atherosclerosis? 2024. Available from: https://www.nhlbi.nih.gov/health/atherosclerosis#:~:text=Email%20Print,blood%20flow%20to%20the%20brain.
61. Yu X-H, Fu Y-C, Zhang D-W, Yin K, Tang C-K. Foam cells in atherosclerosis. Clin Chim Acta.

2013;424:245–52. https://doi.org/10.1016/j.cca.2013.06.006.
62. Anderson RG, Brown MS, Goldstein JL. Role of the coated endocytic vesicle in the uptake of receptor-bound low density lipoprotein in human fibroblasts. Cell. 1977;10(3):351–64. https://doi.org/10.1016/0092-8674(77)90022-8.
63. Goldstein JL, Anderson RG, Brown MS. Coated pits, coated vesicles, and receptor-mediated endocytosis. Nature. 1979;279(5715):679–85. https://doi.org/10.1038/279679a0.
64. Islam MM, Hlushchenko I, Pfisterer SG. Low-density lipoprotein internalization, degradation and receptor recycling along membrane contact sites. Front Cell Dev Biol. 2022;10 https://doi.org/10.3389/fcell.2022.826379.
65. McKenney JM. Understanding PCSK9 and anti-PCSK9 therapies. J Clin Lipidol. 2015;9(2):170–86. https://doi.org/10.1016/j.jacl.2015.01.001.
66. FDA. Repatha(Evolocumab) Approval overview 2015. Available from: https://www.accessdata.fda.gov/scripts/cder/daf/index.cfm?event=overview.process&ApplNo=125522.
67. Hummelgaard S, Vilstrup JP, Gustafsen C, Glerup S, Weyer K. Targeting PCSK9 to tackle cardiovascular disease. Pharmacol Ther. 2023;249:108480. https://doi.org/10.1016/j.pharmthera.2023.108480.
68. Horton JD, Cohen JC, Hobbs HH. Molecular biology of PCSK9: its role in LDL metabolism. Trends Biochem Sci. 2007;32(2):71–7. https://doi.org/10.1016/j.tibs.2006.12.008.
69. Ritter JM, Flower RJ, Henderson G, Loke YK, MacEwan DJ, Rang HP. Rang and Dale's pharmacology. 10thed; London, England, Elsevier. 2023.
70. Amgen. Repatha wetbiste. Repatha. Available from: https://www.repatha.com/.
71. Ryman JT, Meibohm B. Pharmacokinetics of monoclonal antibodies. CPT: pharmacometrics & systems. Pharmacology. 2017;6(9):576–88. https://doi.org/10.1002/psp4.12224.
72. Lin S, Wang S, Wang P, Tang C, Wang Z, Chen L, et al. Bile acids and their receptors in regulation of gut health and diseases. Prog Lipid Res. 2023;89:101210. https://doi.org/10.1016/j.plipres.2022.101210.
73. di Gregorio MC, Cautela J, Galantini L. Physiology and physical chemistry of bile acids. Int J Mol Sci. 2021;22(4) https://doi.org/10.3390/ijms22041780.
74. Ricciotti E, FitzGerald GA. Prostaglandins and inflammation. Arterioscler Thromb Vasc Biol. 2011;31(5):986–1000. https://doi.org/10.1161/atvbaha.110.207449.
75. Stevens SL. Fat-Soluble Vitamins. Nurs Clin North Am. 2021;56(1):33–45. https://doi.org/10.1016/j.cnur.2020.10.003.
76. FDA. Doxil (Pegylated liposomal doxorubicin) Approval Overview 1995. Available from: https://www.accessdata.fda.gov/scripts/cder/daf/index.cfm?event=overview.process&ApplNo=050718.
77. Solomon R, Gabizon AA. Clinical pharmacology of liposomal Anthracyclines: focus on Pegylated liposomal doxorubicin. Clin Lymphoma Myeloma. 2008;8(1):21–32. https://doi.org/10.3816/CLM.2008.n.001.
78. Baxter. Doxil (doxorubicin HCl liposome injection) for Hospital Care 2025. Available from: https://www.baxter.com/doxil-doxorubicin-hcl-liposome-injection-hospital-care.
79. Abu Lila AS, Ishida T. Liposomal delivery systems: design optimization and current applications. Biol Pharm Bull. 2017;40(1):1–10. https://doi.org/10.1248/bpb.b16-00624.
80. Makwana V, Karanjia J, Haselhorst T, Anoopkumar-Dukie S, Rudrawar S. Liposomal doxorubicin as targeted delivery platform: current trends in surface functionalization. Int J Pharm. 2021;593:120117. https://doi.org/10.1016/j.ijpharm.2020.120117.
81. Ma Z, Zhong J, Tu W, Li S, Chen J. The functions of apolipoproteins and lipoproteins in health and disease. Mol Biomed. 2024;5(1):53. https://doi.org/10.1186/s43556-024-00218-7.
82. Sorci-Thomas MG, Bhat S, Thomas MJ. Activation of lecithin:cholesterol acyltransferase by HDL ApoA-I central helices. Clin Lipidol. 2009;4(1):113–24. https://doi.org/10.2217/17584299.4.1.113.

Carbohydrate Metabolism and Carbohydrates as an Energy Source

10

Abstract

Metabolism is a set of biochemical pathways that convert nutrients into energy and building blocks of molecules. Carbohydrate metabolism is especially important, since carbohydrates, particularly glucose, are the main energy source for cells. This chapter focuses on how glucose, the most abundant dietary carbohydrate, is broken down and used as fuel or for other metabolic processes. The metabolism of glucose begins with its digestion in the food we eat and culminates in the production of ATP, the main energy currency. ATP, as well as electron carriers and redox reactions, are covered here. The major pathways, glycolysis, the Krebs cycle, and the electron transport chain, followed by oxidative phosphorylation, are also discussed. The metabolism of carbohydrates is important for human health, but improper regulation of metabolic pathways can lead to disease. Metabolic disorders are also introduced in this chapter.

Keywords

ATP · Glycolysis · Krebs cycle · Oxidative phosphorylation · Electron transport chain · Metabolic disorders

Abbreviations

ADP	Adenosine diphosphate
AMP	Adenosine monophosphate
ATP	Adenosine triphosphate
DHAP	Dihydroxyacetone phosphate
ETC	Electron transport chain (mentioned in later context)
FAD	Flavin adenine dinucleotide (oxidized form)
$FADH_2$	Flavin adenine dinucleotide (reduced form)
G1P	Glucose-1-phosphate
G1P	Glucose-1-phosphate
G6P	Glucose-6-phosphate
GBE	1,4-α-glucan branching enzyme
GDP	Guanosine diphosphate
GK	glucokinase
GP	glycogen phosphorylase
GPI	Glucose-6-phosphate isomerase
GS	glycogen synthase
GTP	Guanosine triphosphate
HK	hexokinase
IEMs	Inborn errors of metabolism
LDH	Lactate dehydrogenase *(noted indirectly in lactate production)*
NAD^+	Nicotinamide adenine dinucleotide (oxidized form)
NADH	Nicotinamide adenine dinucleotide (reduced form)

A. Minden, *Biochemistry, Cell and Molecular Biology for Pharmacists and Health Professionals*,
https://doi.org/10.1007/978-3-032-20529-2_10

OAA	Oxaloacetate
PDC	Pyruvate dehydrogenase complex
PDH	Pyruvate dehydrogenase complex (both PDC and PDH are sometimes used)
PEP	Phosphoenolpyruvate
PFK-1	Phosphofructokinase-1
PGM	phosphoglucomutase
pG6Pase	glycose 6 phosphatase
PGI	phosphoglucse-isomerase
PK	pyruvate kinase
TCA cycle	Tricarboxylic acid cycle (also known as the Krebs cycle)
TCA cycle	Tricarboxylic acid cycle. Also referred to as the Krebs cycle
UDP	Uridine diphosphate
UTP	Uridine triphosphate
UGPase	UDP glucose, Phosphorylase

What You Will Learn

- Carbohydrate metabolism.
- What is metabolism and a basic introduction to metabolic disorders.
- Key concepts in metabolism, such as ATP as the main energy source, electron carriers, redox pairs, and redox reactions.
- How carbohydrates are first metabolized by the digestive system.
- The basic carbon metabolism pathways, including glycolysis, conversion of pyruvate to acetyl-CoA, the Krebs cycle, the electron transport chain, and oxidative phosphorylation.
- Shuttle systems such as the glycerol 3 phosphate shuttle and the malate aspartate shuttle, for transporting NADH to the mitochondria.
- Metabolic pathways that occur when oxygen is low and the generation of lactate.
- The Cori cycle.
- Introduction to the pentose phosphate pathway.
- How improper regulation of the central metabolic pathways relates to disease.

Introduction to the Chapter

This chapter explores carbohydrate metabolism. Metabolism is a set of biochemical pathways that convert nutrients into energy and building blocks of molecules. Carbohydrates are the main energy source for cells, and this chapter focuses on how glucose, the most abundant dietary carbohydrate, is broken down and used as fuel or for other metabolic processes. Glucose metabolism is an ideal place to start because, in many ways, the utilization of glucose for fuel is central to all types of metabolism. Metabolism can initially seem complex, so this chapter is broken down into easy-to-follow sections: Section 1 provides an overview of the field of metabolism, covering metabolic disorders and introducing key concepts such as energy currency, electron carriers, and redox reactions. These concepts are important for understanding all metabolic pathways. Section 2 introduces glucose metabolism, discussing how glucose is obtained through digestion and distributed to tissues and organs. Section 3 discusses the primary cellular respiration pathways through which glucose is metabolized, beginning with glycolysis in the cytosol and concluding with the electron transport chain in the mitochondria. This section also includes a brief introduction to the pentose phosphate pathway, which is important for nucleotide synthesis. Section 4 reviews the main balanced equations that are associated with the metabolic pathways discussed here.

1 Some Basic Concepts in Metabolism

1.1 What Is Metabolism?

Metabolism in biochemistry refers to the chemical reactions that occur in cells within the body as they process biomolecules to sustain life. These reactions usually involve molecules derived from the food we ingest, and the energy or storage products produced when these nutrients are digested. Metabolic reactions include catabolic reactions and anabolic reactions. In catabolic reactions, molecules are broken down. This leads to the generation of energy. Catabolic reactions will occur, for example, after a meal, when molecules such as glucose are broken down to produce energy. Anabolic reactions are the reactions that build biomolecules. These reactions do not provide energy; rather, they utilize energy to synthesize necessary biomolecules. For example, anabolic reactions can result in the production of more glucose after a period of fasting, such as after a night of sleep [1–5]. The field of metabolism encompasses the study of pathways involving glucose, fats, proteins, and amino acids. An important concept is homeostasis, which refers to the body's ability to maintain a balance between nutrient intake and energy expenditure.

Metabolism centers around several key pathways, often referred to as the **central metabolic pathways**. These pathways play roles in both the breakdown of macromolecules for energy and the biosynthesis of biomolecules when energy levels are sufficient and the necessary building blocks are available. At first glance, metabolic pathways can seem daunting due to the many complex steps involved. Over time, however, certain patterns make these pathways easier to follow. For example, tracking the number of carbons in molecules can be helpful (like seeing glucose, a 6-carbon molecule, being split into pyruvate, a 3-carbon molecule). Sometimes the terminology for molecules can include an indication of the number of carbons the molecule has. For example, glucose-1-phosphate, which has 6 carbons, can be referred to as glucose-1-phosphate (6C). It can also be referred to by its abbreviation, G1P (6C), or by its full name along with its abbreviation: glucose-1-phosphate (G1P, 6C).

Terminology also provides other helpful clues. Words ending in "-olyis" mean breakdown, such as glycolysis, the breakdown of glucose, and words ending in "-genesis" mean the production of new molecules (such as gluconeogenesis, the biosynthesis of glucose).

When studying these pathways, some steps are referred to as reversible (or bidirectional), while others are considered irreversible. A reversible step means that the same enzyme can catalyze both the forward reaction and the reverse reaction, depending on the cellular conditions. In contrast, an irreversible step means that the enzyme can only catalyze one direction, such as the forward direction. If needed, a different enzyme, or sometimes multiple enzymes, may catalyze the reverse reaction [6]. This section of the chapter covers basic metabolic processes. Much of this material draws from [1–3] and other references cited in the text.

1.2 Introduction to Metabolic Disorders

Studying metabolic pathways is important for pharmacy and healthcare students, not only for understanding how nutrients are metabolized and biomolecules are synthesized, but also because many human diseases are rooted in metabolic disorders. Some metabolic disorders are considered inborn errors in metabolism (IEMs).r42" These types of disorders are inherited genetic disorders in which a deficiency or malfunction of a specific enzyme or cofactor disrupts a metabolic pathway. As a result, the body may be unable to properly convert nutrients into energy. This can lead to toxic buildup of metabolic intermediates or deficiencies in certain essential compounds. Other metabolic disorders are not classified as IEMs and may be caused by factors such as diet, lifestyle, lack of exercise, combinations of genetic mutations, resistance to hormones like insulin, or autoimmune disorders [7, 8].

Some metabolic disorders that involve carbohydrate metabolism include type 2 diabetes, obesity, metabolic syndrome, and glycogen storage diseases. Other diseases are associated with the metabolism of other biomolecules, such as urea cycle disorders, which are linked to abnormalities in amino acid metabolism. Multiple pharmaceuticals have been developed in an effort to treat these conditions, but they remain a serious problem. Understanding normal metabolism is critical for treating metabolic disorders and developing novel therapeutic approaches.

This chapter will focus on glucose metabolism. Subsequent chapters in this section will cover lipid metabolism, followed by a discussion of metabolic disorders, specifically those involving carbohydrates and lipids. The chapters in this section will also address some current treatment strategies for these disorders and potential avenues for developing new treatments. Chapters in the next section will cover metabolism of other biomolecules, such as proteins, and relevant disorders.

1.3 Key Concepts in Metabolism

A key aspect of metabolic pathways is the generation, storage, and use of energy. An important outcome of catabolic reactions is the net generation of energy. Conversely, energy is used in anabolic reactions. To make sense of how glucose metabolism contributes to the cell's energy needs, we first need to discuss the major molecules that serve as the energy currency molecules: ATP and electron carriers. ATP is a direct energy currency. It donates energy to power cellular processes. The electron carriers are indirect forms of energy currency. They store and transfer high-energy electrons, which can be used to generate ATP.

1.3.1 ATP

The major energy currency in metabolism is **ATP**. Why is ATP such a good energy source? ATP has three phosphate groups (Fig. 10.1). Phosphate groups are negatively charged, and in ATP, these three strong negative charges repel each other. This repulsion contributes to potential energy in the molecule. The hydrolysis of ATP to ADP (adenosine triphosphate (three phosphates) to adenosine diphosphate (two phosphates)) and inorganic phosphate (consisting of a single phosphate) releases energy because the products are more stable and have a lower free energy than ATP. Thus, hydrolysis of ATP to ADP releases a significant amount of energy that can be used for energy-consuming reactions. The phosphate groups in ATP are the α, β, and γ phosphate groups, with the terminal phosphate group being the γ phosphate group. This terminal phosphate group is removed during reactions that use energy. The bonds between the phosphate groups are called phosphoanhydride bonds, which are high-energy covalent bonds. In a typical reaction, the phosphoanhydride bond between the β and γ phosphate groups on ATP is broken. Hydrolysis of this bond releases energy that can be coupled to other reactions [1, 3, 9].

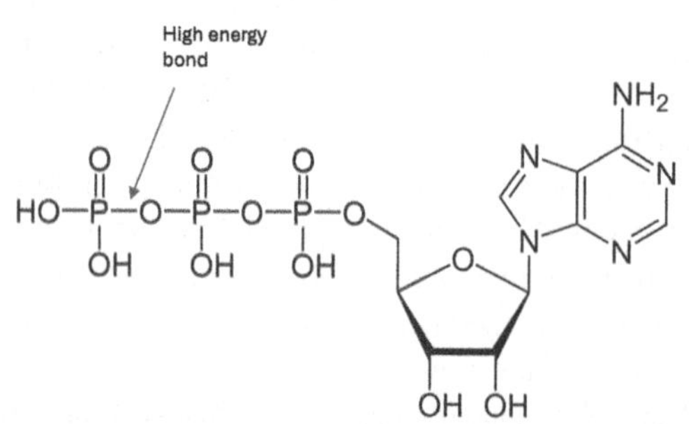

Fig. 10.1 The structure of ATP. (Source: Image adapted from NEUROtiker, via Wikimedia Commons, public domain. File:Adenosintriphosphat protoniert.svg)

1.3.2 Electron Carriers, Redox Pairs, and Redox Reactions

Electron carriers play essential roles in metabolism and are necessary for ATP production. They carry high-energy electrons, which can be donated to the electron transport chain, where energy is released to establish a proton gradient that powers ATP synthesis. This is especially important in organisms that can use oxygen, as the electron carriers release their electrons to generate ATP in the electron transport chain, which functions at the final stage of the major aerobic respiration pathways. The major electron carriers in metabolism include NAD^+/NADH and FAD/$FADH_2$, and in some cases $NADP^+$ and NADPH. They are often referred to as redox coenzymes or redox pairs. They are involved in redox reactions, reactions that involve the transfer of electrons between molecules.

Electron carriers harvest high-energy electrons from carbohydrates or the byproducts of carbohydrates, through redox reactions. In these reactions, one molecule loses electrons (it is oxidized) and the other, usually the carrier, gains electrons (it is reduced). This harvesting involves the removal and transfer of hydrogen (H) atoms (which contain an electron and a proton) from these carbon/hydrogen-rich molecules. Although both electrons and protons are moved in these

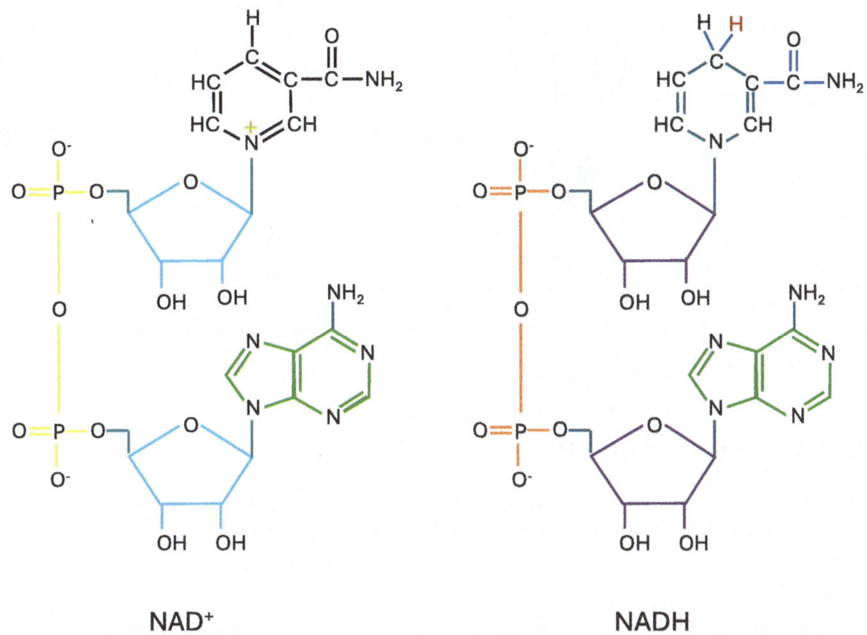

Fig. 10.2 Structures of NAD^+/NADH. NAD consists of a nicotinamide group, a ribose sugar, two phosphate groups linked to another ribose sugar, and an adenine. Nicotinamide is a pyridine ring (a 6-membered ring with 1 nitrogen plus a CONH2 group)

reactions, we usually describe the reactions in terms of electron transfer because it is the electrons that carry the energy being harvested. (The protons are involved more in establishing gradients or maintaining a charge balance.)

Let's discuss NAD^+ (nicotinamide adenine dinucleotide) as an example of how electron carriers operate. NAD^+ consists of a nicotinamide group (a pyridine ring with a $CONH_2$ attached), attached to a ribose sugar. This sugar is connected via two phosphates to another ribose, which is linked to adenine. NAD^+ has a positive charge; it is in its oxidized form. When it becomes reduced, NAD^+ accepts electrons that have been removed from another molecule, typically a carbon/hydrogen-rich molecule (such as a carbohydrate or its metabolite). This converts NAD^+ to NADH, and the nicotinamide ring loses its positive charge, becoming neutral (see Fig. 10.2 for the structures of NAD^+ and NADH).

In the reaction in which NAD^+ becomes reduced, it accepts the equivalent of two hydrogen atoms. These are transferred as a hydride ion (one proton plus two electrons, often depicted as H-) to the nicotinamide ring, while an additional proton is released into the surrounding solution. NAD^+ is now reduced to NADH, and NADH now carries the high-energy electrons it gained. The overall reaction is (see Fig. 10.3).

$$NAD^+ + 2H(2H^+ + 2e^-) \rightarrow NADH + H^+$$

This type of reaction is often written in shorthand as follows, with [H] representing the proton and electron coming from hydrogen:

$$NAD^+ + 2[H] \rightarrow NADH + H^+$$

In many metabolic reactions, the two hydrogen atoms in the above reaction come from carbon/hydrogen-rich molecules, such as carbohydrates or their metabolites. These carbon-rich molecules are thus oxidized, as the NAD^+ is reduced. This type of reaction happens often in metabolism. For example, it occurs when malate is oxidized to become oxaloacetate (a reaction that will be discussed in more detail later in this chapter, when the Krebs cycle is covered). In this reaction, malate loses two hydrogen atoms (thus it loses two electrons), and NAD^+ is reduced to NADH. NAD^+ is thus considered the oxidizing agent for this reaction. (The enzyme that catalyzes this reaction is **malate dehydrogenase.**) See Fig. 10.4.

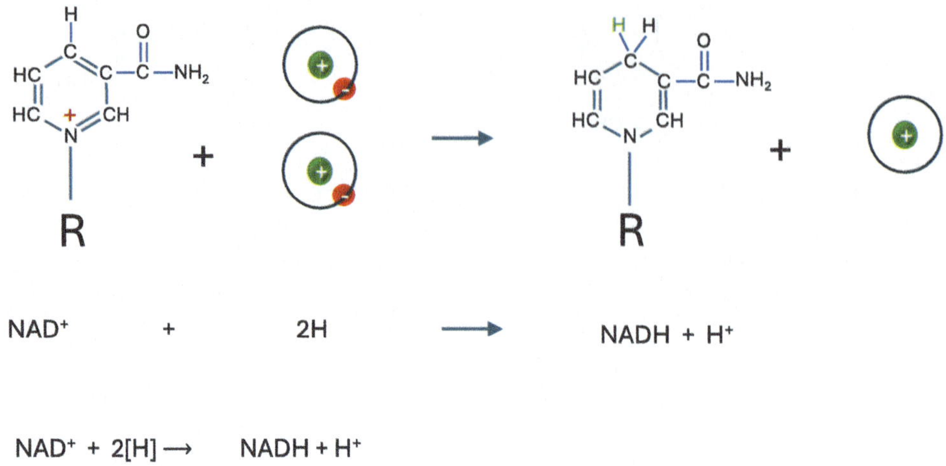

Fig. 10.3 Reduction reaction. A reaction in which NAD^+ is reduced to NADH is shown. In this figure, only the nicotinamide portion of NAD is shown. The rest of the molecule is represented by R

Fig. 10.4 Example of a metabolic redox reaction. In this reaction, reduction of NAD^+ to NADH is coupled to the oxidation of malate to oxaloacetate. This type of reaction is representative of many similar redox reactions that occur throughout metabolism

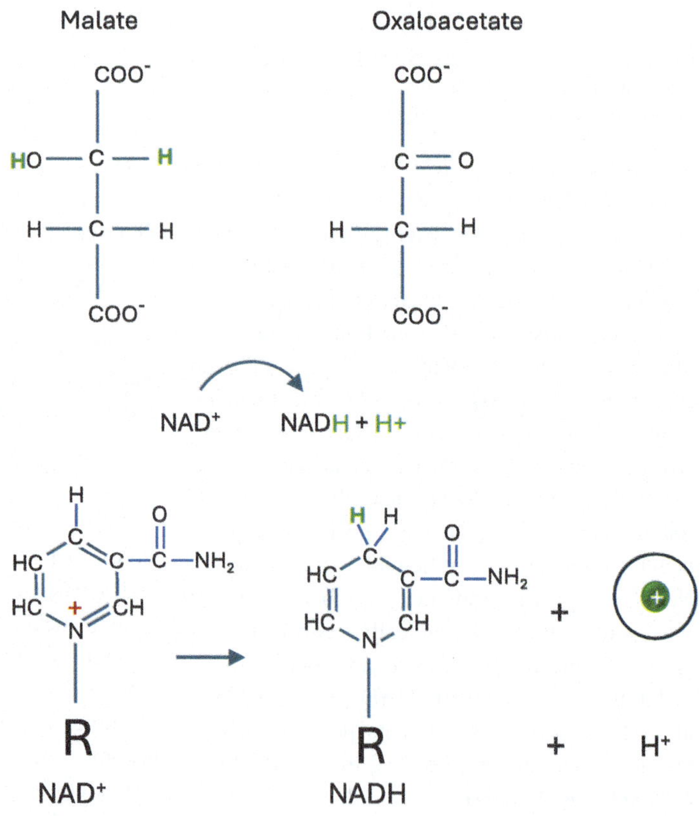

This type of reaction occurs at multiple steps in carbohydrate metabolism. The electrons transferred to carriers like NAD^+ are later released during the electron transport chain, yielding energy in the form of ATP. Electron carriers and redox reactions are also covered in [1, 3, 9].

Other electron carriers function similarly with slight differences in structure. $NADP^+$ (nicotinamide adenine dinucleotide phosphate) behaves similarly to NAD^+, while FAD (flavin adenine dinucleotide) also accepts two electrons, but it takes both protons too, forming $FADH_2$ without releasing any protons, in the following reaction:

$$\mathrm{FAD} + 2[\mathrm{H}] \rightarrow \mathrm{FADH_2}$$

2 Introduction to Glucose Metabolism, Digestion, and Nutrient Breakdown

2.1 Introduction to Glucose Metabolism

Having introduced some of the basic molecules important for metabolism, we now turn to some of the main steps in glucose metabolism, starting with its digestion. Most of the food we eat contains carbohydrates. Some foods, such as bread, are especially rich in carbohydrates. The carbohydrates we ingest are broken down into **glucose.** Glucose metabolism is often the focus for understanding how metabolism works. There are several reasons for this. (a) Glucose is a primary energy source for most cells, and its breakdown provides the energy needed to fuel many cellular functions. It is a highly efficient energy source, and certain tissues, such as the brain, rely heavily on glucose and cannot directly utilize lipids as an energy source [10]. (b) Regulating blood glucose levels is critical for health, and blood glucose levels that are too high or too low can lead to serious health problems such as diabetes. (c) Glucose metabolic pathways are tightly linked to the metabolism of other biomolecules, such as fats and proteins, and therefore understanding glucose metabolism is key to understanding the metabolism of other molecules. For these reasons, this chapter begins by focusing on glucose metabolism, specifically the breakdown of glucose (catabolic pathways) to produce energy. Before discussing how glucose is metabolized, this section gives an overview of how the glucose we consume is processed through the digestive system for distribution to tissues [1–4, 9].

2.2 Digestion and Nutrient Breakdown

The glucose that gets broken down to generate energy comes from the nutrients that we consume. A typical meal consists primarily of macronutrients, including carbohydrates, fats, and proteins, which are broken down into component molecules and integrated into metabolic pathways [10]. Before discussing the metabolic pathways, it is helpful to first briefly review how the food we eat is digested in the gastrointestinal (GI) tract and broken down into various molecules. The focus here is on the breakdown of nutrients into glucose and the distribution of glucose throughout the body. The major organs of the digestive system are shown in Fig. 10.5, and some of the organs that absorb glucose are shown in Fig. 10.6.

2.2.1 The Mouth to the Small Intestine

The first destination of carbohydrates in all the foods we eat is the gastrointestinal (GI) tract. The process of digestion begins in the mouth, where saliva contains the enzyme **salivary amylase**, which starts breaking down starch (a complex carbohydrate) into smaller polysaccharides and some maltose (a disaccharide consisting of two glucose molecules). The partially broken-down carbohydrates make their way to the intestine, where pancreatic amylase in the small intestine continues the breakdown into maltose. Other disaccharides like sucrose (glucose + fructose; from fruits and sugar) and lactose (glucose + galactose; from dairy) enter the intestine directly from the diet. Free fructose (a monosaccharide in

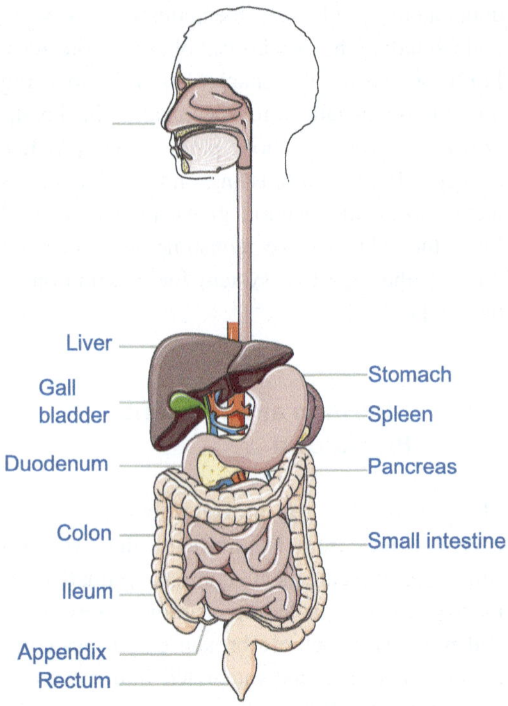

Fig. 10.5 Major organs of the digestive system. (**Source**: Image adapted from Servier Medical Art (https://smart.servier.com/), licensed under the Creative Commons Attribution 4.0 (CC BY 4.0) license)

fruits and added sweeteners) can also come directly from the diet.

Still within the small intestine, enzymes on the brush border of the **enterocytes** (the cells lining the small intestine) such as **lactase**, **sucrase**, and **maltase** catalyze the complete breakdown of these disaccharides (including maltose, sucrose, and lactose) to the monosaccharides glucose, galactose, and fructose, which are the main absorbable forms of dietary carbohydrates [11, 12].

Next, these monosaccharides are absorbed by enterocytes lining the intestine and then exit the small intestine to enter the bloodstream. Glucose, galactose, and fructose use different mechanisms for this. Glucose and galactose enter the enterocytes via the SGLT1 transporter, a sodium/glucose transporter, using active transport driven by the sodium gradient. The SGLT1 transporter is on the apical (luminal) side of the enterocytes. Now the molecules need to exit the enterocytes through the basolateral side. They do so via the **GLUT2** transporter. This transporter uses facilitated diffusion because glucose and galactose are now moving down their concentration gradients. Fructose enters the enterocytes

Fig. 10.6 Some important organs in glucose metabolism. (**Source**: Image adapted from Servier Medical Art (https://smart.servier.com/), licensed under the Creative Commons Attribution 4.0 (CC BY 4.0) license)

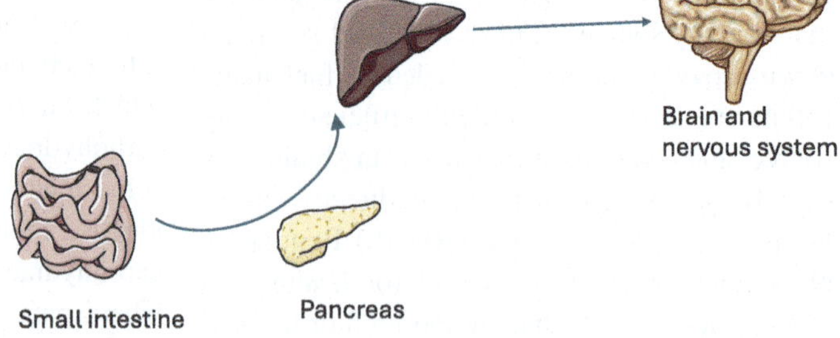

differently from glucose and galactose. Fructose enters the enterocytes via the **GLUT5** transporter. This transporter uses facilitated diffusion because the levels of fructose are higher in the intestinal lumen than in the enterocytes after a meal. (For this reason, fructose absorption might slow down if the concentration gradient is not favorable.) Like glucose and galactose, fructose also exits the enterocytes via the **GLUT2** transporter. Among these different monosaccharides, glucose is the most abundant and is the primary form that can be stored for later use. Glucose will therefore be the focus of the rest of the discussion on carbohydrate digestion, although in later sections, we will also discuss how the other sugars can also integrate into the glucose catabolic pathways, including glycolysis [13].

Once glucose exits the enterocytes through the GLUT2 transporters, it enters the interstitial fluid (the fluid between cells). From there, glucose diffuses passively into capillaries and thereby enters the bloodstream. The capillaries drain into the hepatic portal vein (via capillary villi, venules, and mesenteric veins) [3, 12, 14].

2.2.2 The Liver

The hepatic portal vein passes into the liver [15]. At this point, some of the glucose from the hepatic portal vein enters the hepatocytes (liver cells) via their GLUT2 transporters. This represents the "first pass" through the liver, a process where substances absorbed from the gut first pass through the liver before reaching the rest of the body [12]. The glucose in the liver can be stored in large amounts in the form of glycogen. The glucose that is not taken up by hepatocytes continues through the circulation to other tissues such as the pancreas, muscle, and adipose tissue. This happens because the hepatic vein drains into the inferior vena cava, leading to systemic circulation and allowing the blood to reach other organs [1, 3].

2.2.3 The Pancreas

Glucose that enters the systemic circulation also reaches the pancreatic β cells. Glucose enters these cells via **GLUT2** transporters. This glucose enters the cellular respiration pathways that lead to the generation of ATP (these pathways are discussed in the next section). This ATP in the pancreas plays an essential role in the secretion of the hormone insulin [1, 3]. Insulin plays a crucial role in glucose metabolism and will be discussed in more detail in Chap. 12.

2.2.4 Transport to Other Tissues

In addition to the liver and pancreas, glucose is also delivered to other tissues. Glucose enters muscle via the **GLUT4** transporter, brain and neurons, via the **GLUT1** and **GLUT3** transporter, respectively, and adipose tissue, via the **GLUT4** transporter. In muscle, this glucose can be stored as glycogen or used for energy, given that muscle has an especially high demand for it. Some glucose also enters the heart via the GLUT1 and GLUT4 transporters, although the heart primarily relies on fatty acids for energy at rest. Glucose is more important for the heart during exercise or stress. In the brain and neurons, glucose is the main energy source, and in adipose tissue, it can be converted into fat for long-term storage or broken down to produce energy. The liver primarily stores glucose (in the form of glycogen), although it uses some glucose for energy. Glucose can also exit the liver (particularly in the fasting state) via the GLUT2 transporter to enter circulation and be distributed to other tissues [3, 16, 17].

2.2.5 Glycogen Is the Storage Form of Glucose, Stored Mostly in the Liver

- Glycogen Storage in the Liver

The liver plays a central role in glucose metabolism. Glucose that is not taken up by the liver passes through and is transported to other tissues through the circulation. However, if circulating blood glucose levels are already sufficient and if the body's energy needs are met, then excess glucose gets stored in the liver. This glucose is stored in the form of **glycogen,** and it is stored in this form until the body needs more glucose (either due to a drop in glucose levels or an increased energy demand or both). The liver has a key role in both glucose storage (as glycogen) and its release (as glucose) into the bloodstream.

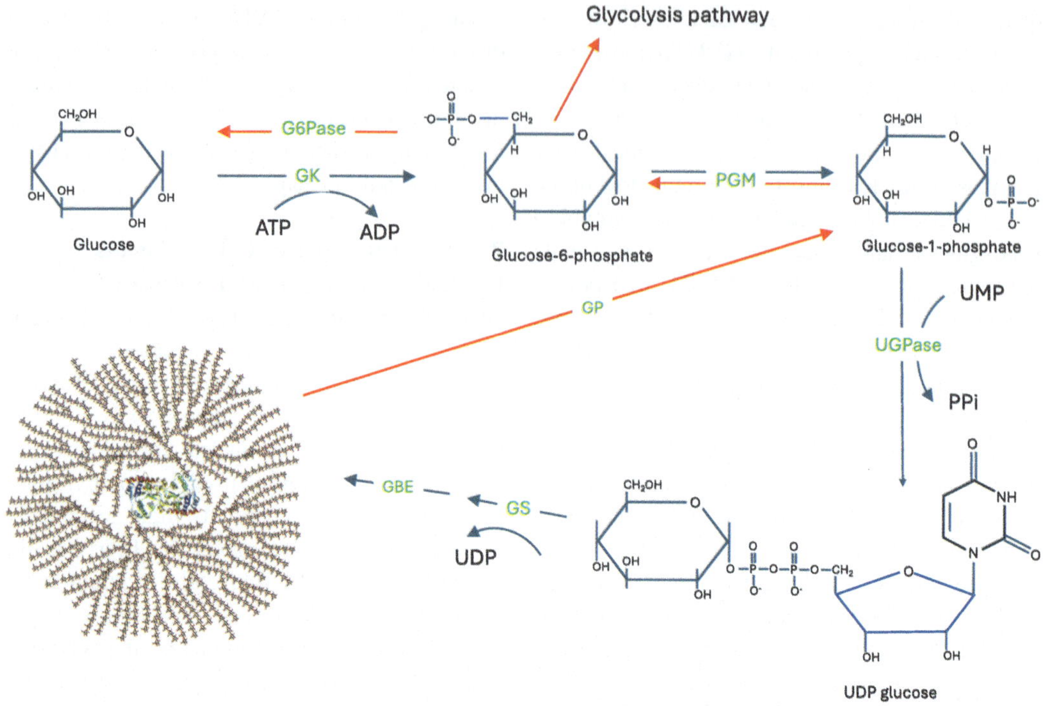

Fig. 10.7 Conversion of glucose to glycogen and back. Synthesis of glycogen from glucose is indicated by blue arrows. Catabolism of glycogen to glucose is indicated by red arrows. Some debranching enzymes, discussed in the text, are not shown in the figure. Abbreviations: G6Pase, glycose 6 phosphatase; GK, glucokinase; PGM, phosphoglucomutase; UGPase, UDP glucose Phosphorylase; GS, glycogen synthase; GBE, 1,4-α-glucan branching enzyme; GP, glycogen phosphorylase. (Source: Glycogen from Mikael Häggström. Public domain (Wikimedia Commons))

It is therefore a key player in regulating blood glucose levels and energy balance. While the liver is a major site of glycogen storage, skeletal muscle cells can also store some glycogen, but in the case of muscle, this glycogen is used by the muscle itself and not exported as glucose [1–4].

- Glycogenesis: The Synthesis of Glycogen from Glucose

As discussed above, glucose can be made into glycogen and stored in the liver. The formation of glycogen from glucose is called **glycogenesis**. The pathway from glucose to glycogen synthesis is shown in Fig. 10.7. It involves converting glucose to glucose-6-phosphate (G6P) by the enzyme **glucokinase** (GK; in the liver). This is an irreversible step that uses one ATP to convert glucose into a phosphorylated form that cannot exit the cell, thus trapping it inside the cell [18]. This reaction is followed by the conversion of G6P to glucose-1-phosphate (G1P) by the enzyme **phosphoglucomutase (PGM).** This is a reversible reaction that rearranges the phosphate. The next reaction activates glucose by attaching it to a UDP (uridine diphosphate). In this reaction, G1P combines with UTP (uridine triphosphate) to yield UDP-glucose and the release of two phosphates in the form of pyrophosphate. This reaction is catalyzed by the enzyme **UDP-glucose-pyrophosphorylase** (UGPase). UDP-glucose is then added onto a growing chain of glucose molecules, and in the process, UDP is released. This is an irreversible reaction catalyzed by **glycogen synthase** (GS). This reaction catalyzes the formation of α-1,4 glycosidic bonds between glucose molecules, leading to a chain of glucose molecules. Branching also occurs, and this is catalyzed by a glycogen branching enzyme called **1,4-α-glucan branching enzyme** (GBE). The end result is glycogen, a branched polymer of glucose molecules [1, 3]. These polymers can be quite large and can contain many thousands of glucose molecules. These major steps of glycogenesis are summarized in Sect. 4 (balanced equations) at the end of the chapter.

3 The Cellular Respiration Pathways

3.1 Introduction to Cellular Respiration Pathways

Having laid the groundwork, we can now delve into the main pathways of glucose metabolism. If the body needs energy, then instead of being stored, the glucose can enter the **cellular respiration pathways,** also known as the **central metabolic pathways**, leading to the generation of energy in the form of ATP. While this can occur in the liver, it occurs primarily in the organs that receive glucose from the liver. In cellular respiration, glucose gets broken down (oxidized) by a series of pathways that culminate in the production of energy in the form of ATP. This process usually occurs in the presence of oxygen (aerobic respiration), although under certain conditions, cellular respiration can occur even when oxygen levels are low. In aerobic respiration, glucose gets fully oxidized to produce a high yield of ATP, along with the release of carbon dioxide (CO_2) and water. When oxygen levels are low, less ATP is produced. In vertebrates, glucose is not completely broken down under these conditions, resulting in the production of lactic acid as a by-product. For this reason, cellular respiration under low-oxygen conditions is referred to as lactic acid fermentation, also known as anaerobic glycolysis. Here, we will focus on aerobic respiration, although we will also discuss situations where oxygen levels are low, later in the chapter. If all the cellular respiration pathways are completed, glucose will be completely broken down (a catabolic reaction) into CO_2 and ATP. These metabolic pathways operate in nearly every cell type, and they are especially active in cells with high energy needs, such as muscle, brain, and kidney [1–4, 17].

When studying these metabolic pathways, it should be noted that metabolic pathways are dynamic and adapt to the body's physiological state. For example, during the fed state, nutrients tend to be stored, and during fasting, nutrients are broken down to generate energy. When needed, biomolecules can also be synthesized in the body. These shifts in metabolic activity relate to how the body maintains homeostasis. While this chapter will focus on the breakdown of glucose, the balance between storage, breakdown, and synthesis of glucose and other biomolecules [11] will be addressed in upcoming chapters.

3.2 Glycogenolysis: The Breakdown of Glycogen to Form Glucose

Before discussing the central metabolic pathways, we must first discuss the source of the glucose that enters these pathways. Free glucose can enter the central metabolic pathways, but if not enough is available, glucose can be obtained from its stored form, glycogen. Glycogen is present in the liver and muscle. In the liver, glycogen can be broken down into glucose, which can be distributed to other tissues for use in metabolic pathways that produce ATP, but muscle does not export glucose to other cells.

The breakdown of glycogen to release glucose is called **glycogenolysis** and is summarized in Fig. 10.7. In this process, individual glucose-1-phosphate (G1P) molecules are released from glycogen in reactions catalyzed by the enzyme **glycogen phosphorylase** (GP). This enzyme uses an inorganic phosphate (Pi) to cleave glucose residues from the nonreducing ends (the ends where the terminal glucose has a free 4'-OH group) of glycogen chains. G1P is released, one residue at a time, from the ends of glycogen branches. When the enzyme encounters branch points (α-1,6 glycosidic linkages), it cannot proceed further without the help of another enzyme. This enzyme is a **glycogen debranching enzyme (GDE). GDE** is an enzyme complex with two catalytic activities, each carried out on a separate active site: **amylo-1,6-glucosidase** and **4-α-glucanotransferase**. These two activities work together to allow **glycogen phosphorylase** (GP) to continue releasing individual G1P molecules. As each G1P is released, it is converted to glucose-6-phosphate (G6P) by the reversible reaction catalyzed by **phosphoglucomutase** (PGM). G6P can then enter the glycolysis pathway as described below. In the liver (but not in muscle), the G6P can also be further broken down into glucose, in a reaction catalyzed by **glucose-6-phosphatase (G6Pase),** which can then be exported and distributed to other tissues. [1, 3] The total reaction from glycogen to glucose in the liver can be summarized as

$$\text{glycogen} \rightarrow \text{G1P} \rightarrow \text{G6P} \rightarrow \text{glucose}$$

3.3 The Main Stages of Cellular Respiration

In the sections below, we discuss the four main steps of **glucose metabolism: glycolysis, pyruvate decarboxylation** (which links glycolysis to the Krebs cycle), **the Krebs cycle,** (also called the TCA (triboxylic acid) cycle or the citric acid cycle), and **oxidative phosphorylation.** Together, these are referred to as the **cellular respiration pathways**, through which the cells break down glucose and obtain energy. They are commonly referred to as the **central metabolic pathways** or the **central carbon metabolism.**

3.3.1 Glycolysis

The breakdown of glucose begins with the process of glycolysis. Glycolysis is the metabolic pathway consisting of 10 enzymatic steps that breaks down glucose (6C) to two pyruvates (3C), also discussed in the following resources [1, 3, 19, 20] (see Fig. 10.8a). In the process, glucose is converted into a series of intermediates, which can be referred to as glycolytic intermediates.

- The 10 Major Steps in Glycolysis

The 10 steps of glycolysis can be divided into two halves, or phases. The **first half** is the **energy investment phase**. This phase uses energy in the form of **ATP**. The **second half** is the **energy payoff phase**, where energy is gained in the form of **ATP** as well as electron carriers.

– The First Half of the Pathway

Step 1: In the first step, **glucose** is **phosphorylated** and becomes **glucose-6-phosphate (G6P)**. This reaction is catalyzed by **glucokinase (GCK)** in the liver and **hexokinase** in most other cells, and one ATP is used. This reaction is irreversible, and it helps trap glucose in the cell. In the liver and muscle, G6P can also be generated from glycogen breakdown, as discussed above.

Step 2: **G6P** is converted to **fructose-6-phosphate (F6P)** in a reaction catalyzed by **phosphoglucose isomerase**, which rearranges the structure of the molecule.

Step 3: This is followed by a second phosphorylation step where **fructose-1,6-biphosphate (F1,6BP)** is formed. This molecule now has two phosphate groups. The reaction is catalyzed by **phosphofructokinase-1 (PFK-1)** and uses one ATP molecule. This is another irreversible step. It is the rate-limiting step of glycolysis and, therefore, its regulation is important. Regulation of **PFK-1** is complex and also involves another enzyme, **PFK-2**. The regulation of **PFK-1** is discussed here.

It should be noted that step 3, F6P to F1,6BP, catalyzed by PFK-1) is a rate limiting and therefore highly regulated step in glycolysis. **PFK-1**, activity is inhibited by high levels of ATP and citrate in the cytoplasm, which indicate the availability of sufficient energy, while AMP and ADP activate PFK-1 under low energy conditions. [20]. Another metabolite, **fructose-2,6-bisphosphate (F2,6BP)** is also a strong activator of PFK-1. The cellular concentration of F2,6BP is regulated by the bifunctional enzyme phosphofructokinase 2/fructose-2,6-biphosphatase-2 (PFK-2/FBPase-2), which links glycolysis with hormonal regulation, to be

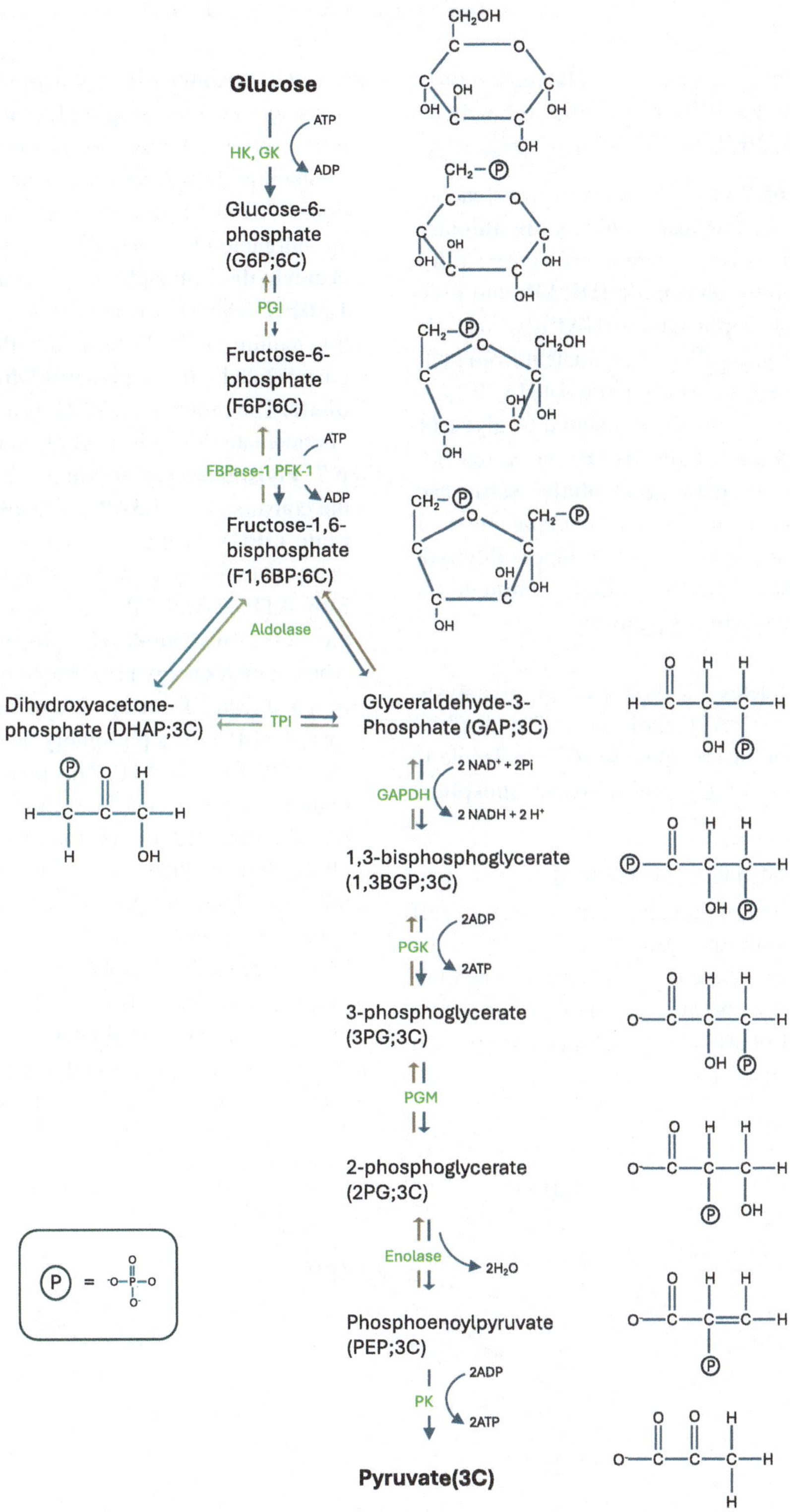

Fig. 10.8 **Top**. Glycolysis. The ten steps mediating the conversion of one glucose molecule to two molecules of pyruvate are shown. Phosphate groups are represened by a P in a circle. Abbreviations: HK, hexokinase; GK, glucokinase; PGI, phosphoglucose-isomerase (also sometimes referred to as glucose-6-phosphate isomerase (GPI); PFK-1, phospho-fructokinase-1; FBPase-1, fructose-1,6-biphosphatase; TPI, triose phosphate isomerase; GAPDH, glyceraldehyde-3-phosphate dehydrogenase; PGK, phosphoglycerate kinase; PGM, phosphoglycerate mutase; PK, pyruvate kinase (GK functions in the liver, HK in most other tissues). **Bottom**. Regulation of PFK-1.

discussed further in Chap. 12. The regulation of **PFK-1** and **F1,6BPase** by the bifunctional enzyme **PFK-2/FBPase-2** is shown in Fig. 10.8b.

Step 4: **F1,6BP** is now cleaved in a reaction catalyzed by **fructose biphosphate aldolase**. The result is two 3-carbon molecules: **dihydroxyacetone phosphate (DHAP)** and **glyceraldehyde 3-phosphate (GAP)** [we use the abbreviation GAP to differentiate it from glycerol-3-phosphate (often abbreviated G3P)].

Step 5: **DHAP** is rapidly converted to **glyceraldehyde 3-phosphate (GAP)** in a reaction catalyzed by **triose phosphate isomerase**. This reaction involves a rearrangement of bonds in the molecule. Only **glyceraldehyde 3-phosphate (GAP)** continues through the rest of the glycolysis pathway.

This completes the first part of glycolysis, which utilizes 2 ATP molecules. The result is the conversion of **one glucose (6C)** molecule to **two** molecules of **glyceraldehyde-3-phosphate (3C)**.

– The Second Half of the Pathway

The second half of glycolysis, the energy payoff phase, will result in a gain of ATP. Each of the following steps happens in duplicate for one molecule of glucose, because two molecules of glyceraldehyde-3-phosphate are obtained from each molecule of glucose.

Step 6: **Glyceraldehyde-3-phosphate (GAP)** is oxidized to **1,3-biphosphoglyerate (1,3BPG)**, in a reaction catalyzed by **glyceraldehyde-3-phosphate dehydrogenase**. During this reaction, NAD^+ is reduced to NADH. In addition, an inorganic phosphate (Pi) is added to **glyceraldehyde-3-phosphate,** resulting in **1,3BPG.** This step is notable because it does not require ATP. Instead, the Pi is directly incorporated into **glyceraldehyde-3-phosphate**. The product, **1,3BPG,** is a high-energy intermediate due to its acyl phosphate bond.

Step 7: **Phosphoglycerate kinase** then catalyzes the conversion of **1,3BPG** to **3-phosphoglycerate (3PG)**. In this reaction, one phosphate group is transferred from 1,3BPG to ADP, to form ATP, leaving 3PG. This step is an example of **substrate-level phosphorylation**, where a high-energy phosphate group is transferred directly from an intermediate (in this case, 1,3BPG) to a phosphorylated intermediate, ADP, to form ATP. This process does not require oxygen or the involvement of mitochondria and differs from oxidative phosphorylation, which occurs in the electron transport chain and depends on a protein gradient across the inner mitochondrial membrane. Oxidative phosphorylation generates a much larger amount of ATP and will be discussed in more detail later.

Step 8: Next, **phosphoglycerate mutase** catalyzes the conversion of **3PG** to **2-**

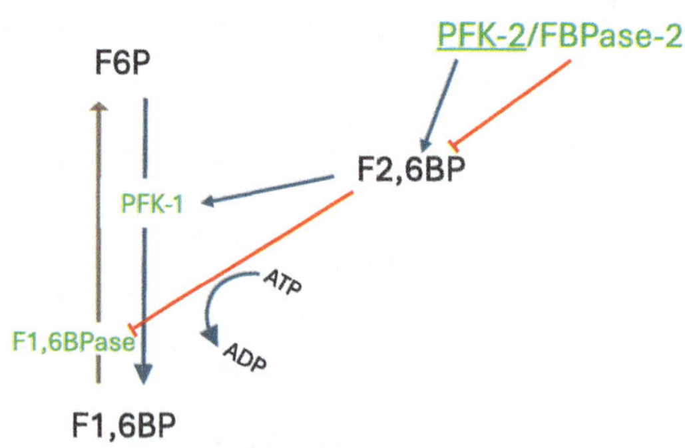

Fig. 10.8 (continued) PFK-1, the enzyme that converts F6P to F1,6BP, is an important regulatory point of glycolysis. It is regulated by the enzyme, PFK-2/FBPase-2 (phosphofructokinase-2/fructose-2,6-bisphosphatase), the bifunctional enzyme that regulates the level of F2,6BP (fructose-2,6-biphosphate) an allosteric activator of PFK-1 (and inhibitor of FBPase-1). F1,6BP, fructose 1,6 biphosphate; F6P, fructose-6-phosphate

phosphoglycerate (2PG). This reaction involves a rearrangement of the molecule.

Step 9: **Enolase** then catalyzes the conversion of **2PG** to **phosphoenolpyruvate (PEP)**, with the release of one water molecule.

Step 10: Finally, in the last step of glycolysis, **PEP** is converted to **pyruvate** in a reaction catalyzed by **pyruvate kinase (PK)**. In this reaction, one phosphate is removed from **PEP** and transferred to ADP, forming ATP; this is the second substrate-level phosphorylation step. As a result, one ATP molecule is produced for every PEP molecule converted.

Step 10 is irreversible and highly regulated; it is a major point of regulation in glycolysis. **Pyruvate kinase** is activated by **F1,6BP**, a product of an earlier step in glycolysis (see step 3 above). This process is known as feed-forward activation. The presence of **F1,6BP** indicates that glycolysis is proceeding efficiently, and it helps ensure that the final step keeps pace with earlier steps. **F1,6BP** binds to an allosteric site on **pyruvate kinase**, stabilizing its active conformation and increasing its catalytic activity.

In contrast, pyruvate kinase is inhibited by high levels of ATP, which indicates that the cell has enough energy. ATP is an allosteric regulator of pyruvate kinase, binding to its regulatory site and converting it to a less active form. Thus, glycolysis slows down when energy is abundant.

In some tissues, **pyruvate kinase** is also inhibited by alanine and inactivated in response to glucagon, a hormone that will be discussed in a later chapter.

- Net Results from Glycolysis

Glycolysis concludes with two molecules of **pyruvate** (**3C**) produced for every one molecule of glucose (**6C**). Two ATP molecules are consumed per glucose molecule in the first half of the reaction, but four ATP molecules are gained per glucose molecule in the second half. Thus, glycolysis results in a **net gain of 2 ATPs**. The ATPs that are gained are obtained via substrate-level phosphorylation. Glycolysis also concludes with the reduction of two NAD^+ molecules to two NADH molecules, which are high-energy electron carriers that will play important roles later, in the electron transport chain. Balanced equations corresponding to the major steps of glycolysis and other pathways discussed in this chapter are summarized in Sect. 4 at the end of the chapter.

- Other Sugars Can Feed into the Glycolysis Pathway

Glucose is the primary sugar in the diet and the main substrate processed by glycolysis. It is found in many fruits, grains, vegetables, starches, and honey. However, we also ingest other sugars. These include sucrose, found in table sugar, some fruits and processed foods, lactose, found in dairy products, and mannose, which is the least common but is found in some types of fruits and vegetables [20, 21]. These sugars interact with the glycolysis pathway [3, 22, 23], as follows.

- Lactose: Glucose and Galactose, and the Leloir Pathway

Lactose consists of glucose + galactose. It can be broken down by the enzyme **lactase** into glucose and galactose. Glucose can enter glycolysis directly, but galactose gets processed to lead to glycolysis more indirectly, via a pathway called the Leloir pathway [1, 24–26], as described below.

First, galactose is phosphorylated, becoming galactose-1-phosphate (Gal1P), catalyzed by **galactokinase (GALK)**, using 1 ATP. This traps the galactose inside the cell.

Next, Gal1P combines with UDP-glucose, resulting in UDP-galactose + G1P, catalyzed by **galactose-1-phosphate uridyl transferase (GALT)**. (The UDP-glucose used in this pathway is synthesized from G1P + UTP in a reaction that is common to several pathways, including glycogenesis discussed above.)

G1P, the product of this reaction, is now converted to G6P, catalyzed by **phosphoglucomutase**, and this G6P enters glycolysis.

The above pathway is the **Leloir pathway**. The UDP-galactose that results from this pathway can have several possible fates. It can be converted back to UDP glucose in a reaction catalyzed by **UDP-glalactose-4-epimerase (GALE)** to sustain the pathway, or it can be used for the synthesis of various other biomolecules.

- Sucrose

Sucrose is a disaccharide consisting of glucose and fructose. In order to enter glycolysis, it is first broken down into glucose + fructose, usually in a reaction catalyzed by **sucrase** in the small intestine. Glucose can then enter glycolysis.

Fructose is broken down largely in the liver, where it is first converted to fructose-1-phosphate (F1P) in a reaction catalyzed by **fructokinase**.

It is then cleaved into DHAP and glyceraldehyde, in a reaction catalyzed by **aldolase B**. The glyceraldehyde is phosphorylated by **triose kinase**, resulting in glyceraldehyde-3-phosphate (GAP), which can enter glycolysis.

- Mannose

Mannose is converted to mannose-6-phosphate (M6P) in a reaction catalyzed by **hexokinase** that uses 1 ATP.

M6P is then isomerized to F6P in a reaction catalyzed by **phosphomannose isomerase (PMI)**. F6P is a glycolysis intermediate and enters the glycolysis pathway.

3.3.2 Decarboxylation of Pyruvate

After glycolysis, **pyruvate** (3C) plays a central role in the next steps of cellular respiration, described below and in the following resources [1, 3, 22, 23]. While this step is often referred to as decarboxylation of pyruvate, it actually involves two events: decarboxylation of pyruvate and the addition of coenzyme A (CoA) to form acetyl-CoA. This process is described below and illustrated in Fig. 10.9a. (CoA is used in many of the reactions discussed in this section. The structure of CoA is shown in Fig. 10.9b).

Glycolysis occurs in the cytosol, and therefore, pyruvate is generated in the cytosol. The next set of reactions, however, which will make up the Krebs cycle, occurs in the mitochondria. Pyruvate, therefore, must enter the mitochondria. Pyruvate first crosses the outer mitochondrial membrane through porin channels (Voltate Dependent Anion Channel: VDAC), reaching the intermembrane space. It then enters the mitochondrial matrix by passing through a transporter called the mitochondrial pyruvate carrier (MPC) in the inner mitochondrial membrane [22].

In the mitochondrial matrix, a large multi-enzyme complex called the **pyruvate dehydrogenase (PDH) complex** [27] converts **pyruvate** (3C) to **acetyl-CoA** (2C). **PDH** is an enzyme complex that consists of three enzymes: **E1** (pyruvate dehydrogenase), **E2** (dihydrolipoyl transacetylase), and **E3** (dihydrolipoyl dehydrogenase). Conversion of pyruvate to acetyl-CoA is a multistep process. It involves the decarboxylation of pyruvate (removal of one carbon in the form of CO_2) and the addition of Coenzyme A (CoA) to form acetyl-CoA. The reaction involves the reduction of NAD^+ to yield NADH via the transfer of a hydride ion from pyruvate to NAD^+. This is an irreversible set of reactions. The overall reaction is as follows:

$$\begin{aligned}&\text{Pyruvate} + \text{NAD}^+ + \text{CoA}\\&\quad\rightarrow \text{Acetyl-CoA} + \text{NADH} + \text{CO}_2.\end{aligned}$$

One pyruvate molecule leads to the production of 1 acetyl-CoA, 1 NADH, and 1 CO_2. Since one glucose molecule leads to two pyruvates, these numbers are doubled for each glucose molecule.

This process of transporting pyruvate to the mitochondria and converting it to acetyl-CoA is the main link between glycolysis and the next step, the Krebs cycle.

As a side note, it is worth noting that under fasting conditions, pyruvate can be carboxylated rather than decarboxylated by the enzyme **pyruvate carboxylase (PC)**, resulting in the production of oxaloacetate (OAA), an intermediate in the Krebs cycle, which is discussed below. This reaction, catalyzed by **PC,** occurs primarily during fasting. It replenishes OAA, and it supports gluconeogenesis (the formation of more glucose). This pathway mediated by **PC** is not discussed in

A

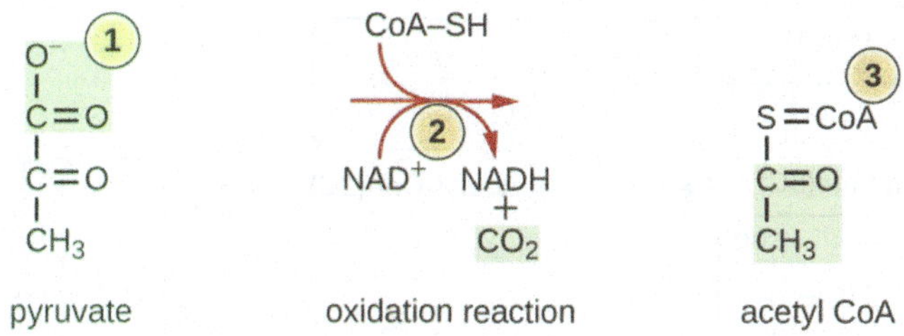

Pyruvate + CoA + NAD$^+$ → Acetyl-CoA + CO2 + NADH + H$^+$

B

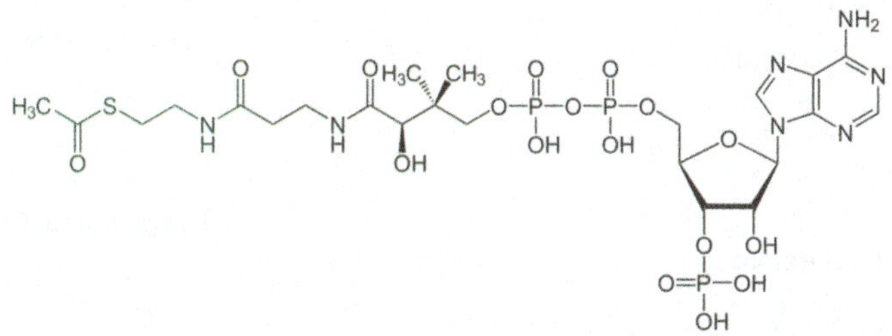

Fig. 10.9 (**a**) The reaction from pyruvate to acetyl-CoA, catalyzed by the pyruvate dehydrogenase complex (PDH, also sometimes called PDC). In this reaction, a carboxyl group is removed from pyruvate (3C), releasing CO_2 and leaving an acetyl group (2C). This acetyl group is transferred to coenzyme A (CoA) to form acetyl-CoA. In the process, NAD$^+$ is reduced to NADH. (**Source**: Adapted from *OpenStax Microbiology*, licensed under Creative Commons Attribution 4.0 (CC BY 4.0) license). (**b**) Structure of coenzyme A (CoA). CoA is a complex. Its thiol group (–SH) allows it to form thioester linkages with acyl groups, generating derivatives such as acetyl-CoA, succinyl-CoA, or fatty acyl-CoA. (**Source**: Public domain image by NEUROtiker, Wikimedia Commons. File: Acetyl-CoA2.svg)

this chapter, but **PC** and gluconeogenesis will be discussed in Chap. 12.

3.3.3 The Krebs Cycle (Also Called Citric Acid Cycle or the TCA Cycle)

The Krebs cycle takes place in the mitochondrial matrix. The steps of the Krebs cycle, also summarized in [1, 3, 22, 23, 28, 29] are outlined in Fig. 10.10.

Step 1: The Krebs cycle begins when **acetyl-CoA** (a 2-carbon molecule attached to CoA) interacts with **oxaloacetate (OAA, 4C)** (a 4-carbon molecule) to form **citrate** (6C). The enzyme that catalyzes this reaction is **citrate synthase (CS)**. As part of this reaction, **coenzyme A (CoA)** is released from acetyl-CoA as a free molecule in the form of **SH-CoA** (the thiol group (SH) is part of the CoA's structure and can serve as an attachment site for other molecules). Breaking the thioester bond between CoA and the 2-carbon acetyl group releases energy because the thioester bond is a high-energy bond, and this helps in the formation of citrate.

Step 2: **Citrate (6C)** is converted to **isocitrate (6C)**, in a reaction catalyzed by **aconitase (ACO)**. This involves a rearrangement of the molecule's functional groups.

Step 3: **Isocitrate (6C)** is then converted to **α-ketoglutarate (αKG, 5C)** in a reaction catalyzed by **isocitrate dehydrogenase (EDH)**. In this reaction, NAD$^+$ is reduced to **NADH**, and one molecule of CO_2 is released.

Step 4: **αKG (5C)** is oxidized to form **succinyl-CoA** (a 4-carbon molecule attached to CoA).

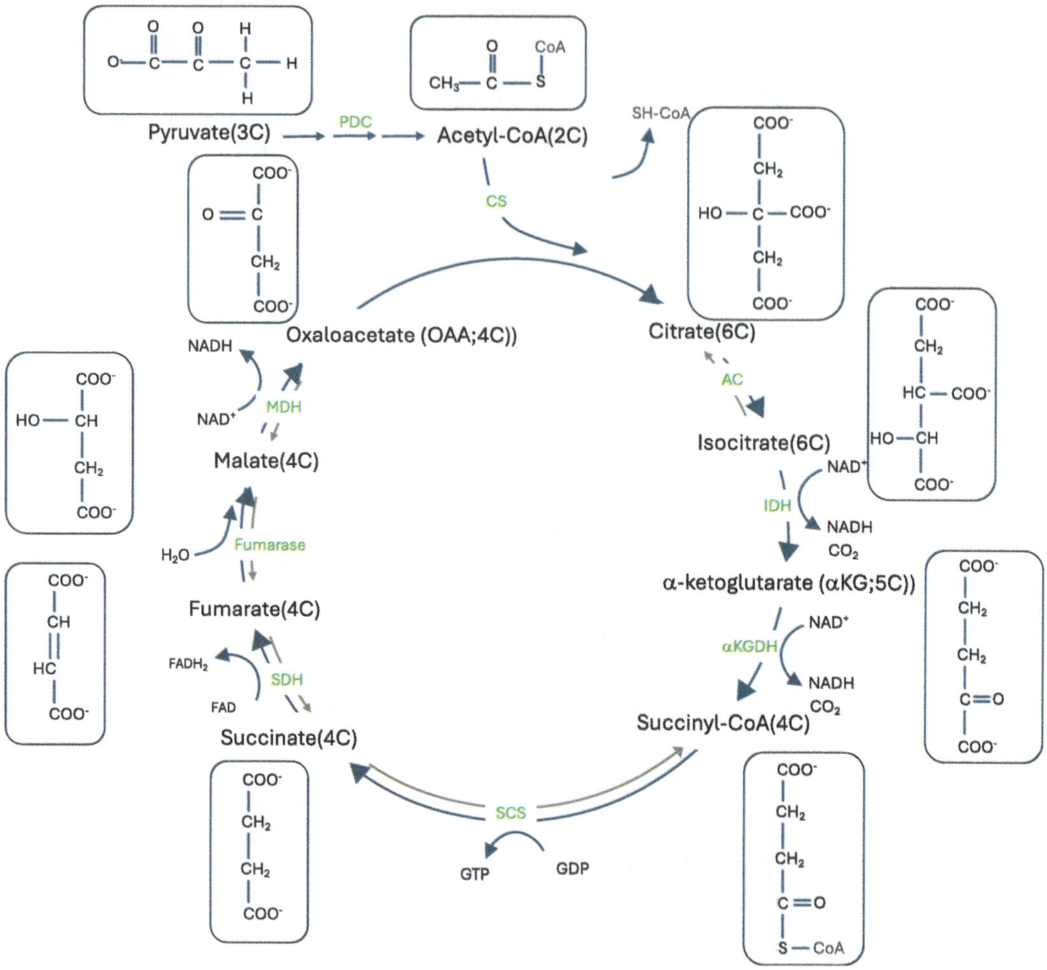

Fig. 10.10 Krebs cycle. The Krebs cycle, also called the citric acid cycle or TCA cycle, takes place in the mitochondria. PDH, pyruvate dehydrogenase complex; CS, citrate synthase; AC, aconitase; IDH, isocitrate dehydrogenase; aKGDH, a-ketoglutarate dehydrogenase complex; SCS, succinate-CoA synthetase; SDH, succinate dehydrogenase malate; MDH, malate dehydrogenase

The enzyme that catalyzes this reaction is **α-ketoglutarate dehydrogenase (α-KGDH)**. The reaction involves the addition of **CoA** to **αKG** via a thioester bond, the reduction of 1 **NAD**$^+$ to **NADH,** and the release of one CO_2. The high-energy thioester bond helps drive the next step in the cycle, which involves the phosphorylation of GDP to form GTP .

Step 5: Next, **succinyl-CoA** is converted to **succinate (4C)** in a reaction catalyzed by **succinyl-CoA synthetase (SCS)**. This involves the release of the **SH-CoA** molecule and the phosphorylation of one molecule of GDP to form **GTP.** In some cells, this GTP is converted to ATP by other enzymes.

Step 6: Next, **succinate (4C)** is oxidized to form **fumarate (4C)**. During this reaction, one FAD is reduced to $FADH_2$. The reaction is catalyzed by **succinate dehydrogenase (SDH)**.

Step 7: **Fumarate (4C)** is hydrated to **malate (4C)**. This is catalyzed by **fumarase (FH)**

and involves the addition of one water molecule to the complex.

Step 8: Finally, **malate (4C)** is oxidized to **oxaloacetate (OAA, 4C)**, in a reaction catalyzed by **malate dehydrogenase (MDH)**, completing the cycle.

The last step in the Krebs cycle is the production of **OAA** so that the entire cycle can begin again (where OAA interacts with acetyl-CoA, which was generated from pyruvate, to make citrate again). The result of one cycle of the Krebs cycle is 2 GTP (which can be converted to 2 ATP), 6 NADH, 2 $FADH_2$, and 4 CO_2. An important function of the Krebs cycle is harvesting high-energy electrons from the carbon molecules that enter the cycle. This is what leads to the production of NADH and $FADH_2$. The enzymes involved in this process are the **dehydrogenases**, which remove hydrogens from their substrates. The NADH and $FADH_2$ molecules are especially important here because they have been reduced (gained electrons) and can, therefore, function in the next step, which is the electron transport chain portion of oxidative phosphorylation [1, 3, 16].

3.3.4 Summary of the Glucose Catabolic Pathways, and the Importance of Bidirectionality

The pathways discussed until now can be considered glucose catabolic pathways, as they are involved in completely oxidizing (breaking down) glucose (6C) into CO_2. These pathways are important because they generate ATP for energy, as well as NADH and $FADH_2$, which will lead to even more ATP during oxidative phosphorylation, a process that will be discussed next.

The glucose catabolic pathways discussed until now are summarized in Fig. 10.11. These pathways are part of the central metabolic pathways of the cell, which include both catabolic and anabolic reactions (and when combined with the oxidative phosphorylation, discussed next, are constitute the major cellular respiration pathways). While the pathways we have discussed involve the breakdown of glucose, it is worth noting that cells sometimes need to synthesize glucose instead of breaking it down, especially during fasting or low-energy conditions. This can be achieved because many of the metabolic steps described above are reversible. Some steps, however, are irreversible, such as the conversion of pyruvate to acetyl-CoA, the conversion of acetyl-CoA to citrate, several key steps in glycolysis, and three steps in the Krebs cycle. When enzymes are irreversible, different enzymes may be required to catalyze the forward reactions and the reverse reactions. The enzymes in these pathways, both forward and reverse, are regulated by various hormones, including glucose and glucagon. These hormones have critical roles in determining whether glucose is broken down (glycolysis) or synthesized (gluconeogenesis). Gluconeogenesis and some of these hormones will be discussed further in Chaps. 12 and 13, but this chapter focuses on the breakdown of glucose (catabolic pathway).

The next section covers the steps that occur after the Krebs cycle is complete, specifically the oxidative phosphorylation step that leads to the generation of large amounts of ATP.

3.3.5 Oxidative Phosphorylation

After the Krebs cycle, the next step is **oxidative phosphorylation** (see Fig. 10.12). This is the step that generates the most ATP. This step includes the **electron transport chain (ETC)** + **chemiosmosis**, discussed below and also covered in [1, 3, 9, 22, 23, 28–30].

An important outcome of the Krebs cycle is the production of $FADH_2$ and NADH, which are high-energy electron carriers. These molecules will now function in the ETC. They will release their hydrogen atoms, which contain electrons (e^-) and protons (H^+). The protons will be released in the mitochondrial matrix, and the electrons will be passed through a series, or chain, of complexes that contain redox centers (molecules or ions that undergo reduction and oxidation as they transfer electrons). The transfer of electrons through the chain releases energy that powers the pumping of protons from the mitochondrial matrix into the intermembrane space. At the end of the chain, four electrons and four

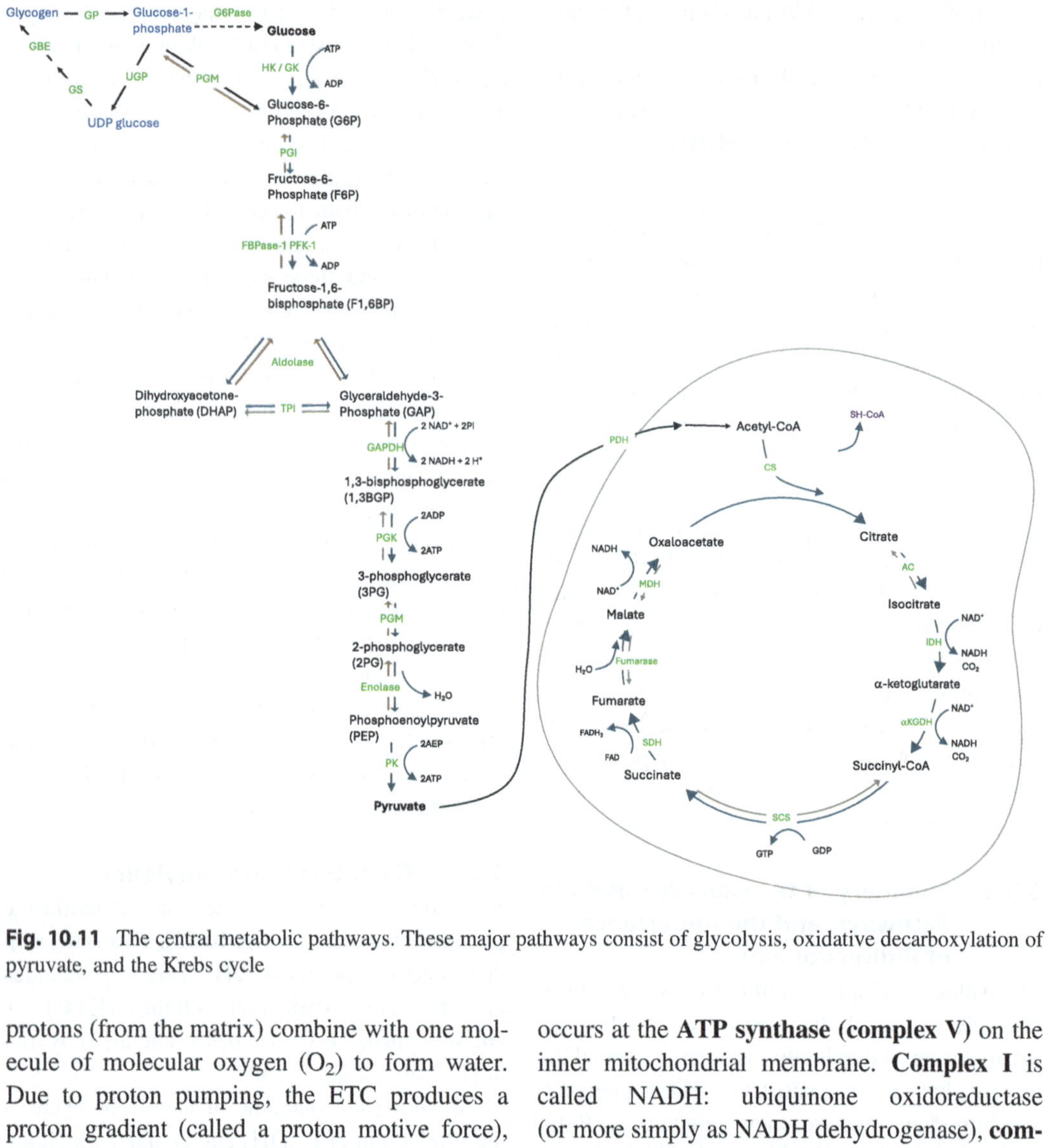

Fig. 10.11 The central metabolic pathways. These major pathways consist of glycolysis, oxidative decarboxylation of pyruvate, and the Krebs cycle

protons (from the matrix) combine with one molecule of molecular oxygen (O_2) to form water. Due to proton pumping, the ETC produces a proton gradient (called a proton motive force), where the level of protons in the inner membrane space is higher than in the mitochondrial matrix. These protons move down their electrochemical gradient through a complex (complex V) called an **ATP synthase**, which drives the production of ATP. This last step is called **chemiosmosis**. This produces large amounts of ATP, which provides energy for nearly all the activities of our cells. This generation of energy is the culmination of all the steps leading up to this point.

The ETC consists of four major enzyme complexes through which electrons pass: **complexes I, II, III, and IV**. Chemiosmosis occurs at the **ATP synthase (complex V)** on the inner mitochondrial membrane. **Complex I** is called NADH: ubiquinone oxidoreductase (or more simply as NADH dehydrogenase), **complex II** is succinate dehydrogenase (which also has a role in the Krebs cycle), **complex III** is ubiquinol: cytochrome c oxidoreductase (also called cytochrome c reductase or the cytochrome bc_1 complex), and **complex IV** is cytochrome c oxidase. **Complexes I, III,** and **IV** are integral membrane proteins that span the inner mitochondrial membrane; **complex II** is a peripheral membrane protein on the matrix side of the inner mitochondrial membrane, and **complex V,** the **ATP synthase**, spans the inner mitochondrial membrane. The electron transport chain also

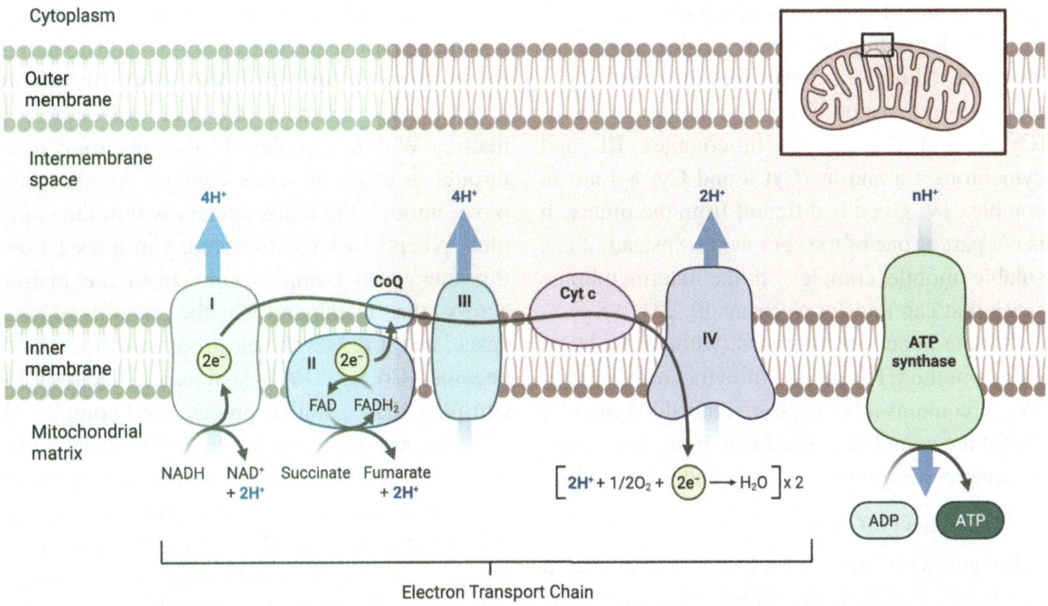

Fig. 10.12 Oxidative phosphorylation. Oxidative phosphorylation takes place at the inner mitochondrial membrane. It consists of the electron transport chain (ETC) and chemiosmosis (ATP synthesis)

consists of two mobile electron carriers: **coenzyme Q (CoQ)** and **cytochrome C (Cyt c)**.

Complexes I, II, and III have iron–sulfur clusters, which are critical for electron transfer. Complex IV also contains copper centers that are essential for electron transfer and oxygen reduction.

Before diving into how the ETC works, we will review the functions of some important components of the ETC, including flavin mononucleotide (FMN), iron–sulfur clusters, cytochromes, coenzyme Q, and copper centers. These components are key to the ETC because they contain redox centers through which electrons are passed along the chain.

- Molecules Containing Key Redox Centers Used in the Electron Transport Chain

– Flavin Mononucleotide (FMN)

Flavin mononucleotide (FMN) is a prosthetic group (which is a nonprotein component that binds to an enzyme) located in complex I. FMN serves as the **initial electron acceptor.** It can cycle between FMN (oxidized) and $FMNH_2$ (reduced). $FMNH_2$ has an additional two protons and two electrons, compared with FMN.

– Iron–Sulfur Clusters

Iron–sulfur clusters are prosthetic groups composed of inorganic sulfur atoms and iron. These clusters transfer electrons by cycling them between two oxidation states of iron: Fe^{3+} (oxidized) to Fe^{2+} (reduced; the form that carries an electron). Iron–sulfur clusters are present in complexes I, II, and III. Most iron–sulfur clusters, such as the [2Fe-2S], [3Fe-4S], and [4Fe-4S], are coordinated (held in place by bonds) by cysteine. The Rieske [2Fe-2S] cluster (in complex III) is coordinated by histidine instead of cysteine. [9, 23, 31]

– Cytochromes

Cytochromes are proteins that contain a prosthetic group, specifically a heme group, and they also play critical roles in electron transfer. Heme groups are ring-shaped molecules (made of carbon, hydrogen, and nitrogen), with an iron atom at their center. This iron also cycles between Fe^{3+} to Fe^{2+}, depending on whether it loses or gains electrons. Multiple cytochromes are involved in the electron transport chain. Cytochrome b and c_1 (Cyt b and Cyt c_1) are in complex III, and cytochromes a and a_3 (Cyt a and Cyt a_3) are in complex IV. Cyt c is different from the others. It is not part of one of the complexes. Instead, it is a soluble mobile complex in the intermembrane space that can transfer electrons [9, 23].

A note about heme groups: while most heme groups in the ETC are part of cytochromes, complex II contains a heme group embedded directly within the complex, called heme b (the two forms of heme b are called heme b_L and heme b_H).

– Coenzyme Q (CoQ)

Coenzyme Q (CoQ) is also called ubiquinone. It is a small, lipid-soluble, mobile electron carrier within the inner mitochondrial membrane that shuttles electrons between certain complexes. CoQ can exist in both oxidized (CoQ) and reduced forms. In its reduced form, it is $CoQH_2$, also called ubiquinol. In its reduced form, it picks up two protons and two electrons. It can also exist as a semiquinone intermediate (CoQH·), which carries one proton and an electron.

– Copper Centers

Copper centers are metal cofactors that also participate in electron transfer reactions. There are two main types: CuA and CuB. Copper atoms cycle between Cu^{2+} (oxidized) and Cu^{+} (reduced) as they donate or accept an electron. They are found only in complex IV.

- A Brief Overview of the ETC and Chemiosmosis

Before describing the ETC and chemiosmosis in some more detail, this section briefly summarizes the process.

The ETC involves the transfer of electrons through a series of complexes, which drives the pumping of protons across the inner mitochondrial membrane, from the matrix to the intermembrane space. This results in a proton gradient that drives the synthesis of ATP.

NADH starts this process by releasing a hydride (a proton and two electrons) and donating two electrons to complex I (via FMN). The proton is ultimately released into the mitochondrial matrix. Within complex 1, the electrons pass through a chain of redox centers. As electrons move through the redox centers within the complex, potential energy is stored. **Complex I** uses this energy to pump protons from the matrix across the membrane to the intermembrane space, and it passes the electrons to CoQ, which becomes $CoQH_2$. $FADH_2$ donates electrons to **complex II** in a similar process, and complex II also donates its electrons to CoQ, which also becomes $CoQH_2$. Unlike complex I, complex II does not also pump protons. CoQ then donates its electrons to **complex III**. Electrons from complex III are transferred to cytochrome c (Cyt c), which then carries the electrons to **complex IV**. In complex IV, oxygen is the final electron acceptor. In the process, four electrons (from the ETC) and four protons (from the mitochondrial matrix) combine with one molecule of O_2 to make two molecules of H_2O. **Complex IV** is the final destination of the electrons in this process, and it is the end of the electron transport chain. To summarize, the path these electrons take through in the electron transport chain is

$$\text{NADH} \rightarrow \text{complex I} \rightarrow \text{CoQ}, \text{FADH}_2 \rightarrow \text{complex II} \rightarrow \text{CoQ},$$

$$\text{CoQ} \rightarrow \text{CoQH}_2 \rightarrow \text{complex III} \rightarrow \text{cyt C} \rightarrow \text{complex IV} \rightarrow \text{H}_2\text{O}$$

The proton gradient that builds up as a result of the electron transport chain now powers **complex V**, the ATP synthase, which drives the synthesis of ATP, a process called chemiosmosis. See also [22] for more details.

- The Steps in the ETC Followed by Chemiosmosis

The ETC and chemiosmosis can be broken down into steps, as follows:

Step 1: **NADH passes electrons into complex I, which are passed to CoQ, which becomes $CoQH_2$.**

In this step, a series of redox reactions occur, which transfer electrons, which move from NADH through Complex I to CoQ, summarized as follows:

NADH → FMN → iron sulfur clusters → CoQ → $CoQH_2$

(CoQ becomes reduced to $CoQH_2$ after protons are taken up from the matrix.)

The main steps in step 1 are outlined below.

- NADH, located in the mitochondrial matrix, releases a hydride (two electrons (e^-) and a proton (H^+)), and as a result is oxidized to NAD^+.
- The hydride from NADH is now donated to flavin mononucleotide (FMN), which is tightly bound to **complex I** as a prosthetic group. FMN is considered to be the first electron acceptor in this complex. FMN also picks up another hydrogen (H^+) from the matrix, and as a consequence, it is reduced to $FMNH_2$. FMN is important because it can accept two electrons from NADH but it passes them one at a time to the next carriers.
- The electrons from $FMNH_2$ next move through a chain of electron acceptors within complex I, which are iron–sulfur clusters (including [2Fe-2S], [Fe-2S], [4Fe-4S], and [3Fe-4S]), each transferring a single electron stepwise to the next cluster until they reach **CoQ**, which is an electron carrier within the inner mitochondrial membrane.
- CoQ also picks up protons from the matrix and becomes $CoQH_2$.
- The redox reactions in complex I provide energy to drive conformational changes in the complex that pump four protons (4 H^+) from the matrix into the intermembrane space, per NADH oxidized.

Step 2: **$FADH_2$ donates electrons to complex II, which are also passed to CoQ, resulting in $CoQH_2$:**

This step involves complex II. It also includes a series of redox reactions, passing electrons from $FADH_2$ to CoQ as follows:

$FADH_2$ → iron sulfur clusters → heme b → CoQ → $CoQH_2$

(CoQ becomes reduced to $CoQH_2$ after protons are taken up from the matrix.)

This step involves $FADH_2$, which is also in the mitochondrial matrix. FAD is a permanent prosthetic group bound to complex II (which is also succinate dehydrogenase), and it has been reduced to $FADH_2$ during the oxidation of succinate to fumarate during the Krebs cycle. The main steps in step 2 are as follows:

- The $FADH_2$ bound to complex II releases its two electrons (e^-) and two protons (H^+) and, as a result, becomes oxidized to FAD.
- The electrons from $FADH_2$ are next transferred through a series of iron–sulfur clusters within complex II (including [2Fe-2S], [4Fe-4S], [3Fe-4S]), which include the first electron acceptors.
- The electrons then pass the heme b group within complex II.
- Next, the electrons are transferred to CoQ.
- CoQ also picks up protons from the matrix and becomes $CoQH_2$.

Complex II functions differently from complex I. It passes the electrons, but it does not pump protons across the membrane.

Step 3: **Electrons are transferred from $CoQH_2$ to complex III, then transferred to cytochrome c (Cyt c) or used to regenerate $CoQH_2$.**

This step is a multistep process. It involves the transfer of electrons from $CoQH_2$ to complex III, from which they will be transferred to their next destinations.

- The $CoQH_2$ from the previous steps (each $CoQH_2$ contains two electrons from steps 1 and 2) is passed to **complex III**. Complex III contains two CoQ binding sites: The Q_0 site on the side of the complex facing the intermembrane space and the Q_i complex on the matrix side.
- $CoQH_2$ from steps 1 and 2 binds to the Q_0 site, where $CoQH_2$ is oxidized to CoQ, releasing

two protons into the intermembrane space. Two electrons are also released, and they take separate paths:

(a) Path 1—one electron passes to **Cyt c**, as follows:

$$CoQH_2(Q_0) \rightarrow [2Fe-2S] \rightarrow \text{Cyt } c_1 \rightarrow \text{Cyt c}$$

- This electron passes through the Rieske [2Fe-2S] cluster (a sulfur cluster).
- Next, it goes through Cyt c_1 within complex III.
- Next, it is passed directly to Cyt c in the intermembrane space.
- The heme group within Cyt c cycles between Fe^{3+} and Fe^{2+} as the iron group accepts the donated electron.

(b) Path 2—the other electron: The other electron regenerates $CoQH_2$, via this path:

$$CoQH_2(Q_0) \rightarrow \text{Cyt b\_L} \rightarrow \text{Cyt b\_H} \rightarrow CoQ(Q_i) \rightarrow CoQH_2$$

- This electron passes from $CoQH_2$ through cytochrome b in complex III, moving sequentially through its two b-type hemes (b_L and b_H).
- It then reaches the Q_i site, on the matrix side of complex III. There, it reduces a bound CoQ to a semiquinone radical (this is a CoQ that has received one electron); CoQH·.
- In a second pass of an electron through the chain (called a cycle; see below), this semiquinone receives a second electron (from another $CoQH_2$ at the Q_0 site), and picks up two protons from the mitochondrial matrix, forming $CoQH_2$.

The above sequence of events is known as the Q cycle, which involves the regeneration of CoQH2. As a result of this entire step, a total of four protons are translocated to the intermembrane space for every $CoQH_2$ oxidized (two from the matrix and two from $CoQH_2$ oxidation).

Step 4: **Cytochrome c delivers electrons to complex IV, where oxygen is the electron acceptor.**

This is a key step in the ETC because it involves oxygen as the final electron acceptor. It is also the last step in the electron transport chain. (The next step will be the chemiosmosis step.) **Complex IV** has two cytochromes, **cytochrome a and a3**, which contain heme a and heme a3, respectively. It also contains two copper centers (CuA and CuB) that transfer electrons and participate in splitting the O_2 bond. The main steps in step 4 are as follows, where we will see electrons transfer in this sequence:

$$\text{Cyt c} \rightarrow \text{complex IV(CuA)} \rightarrow \text{Cyt a} \rightarrow \text{cyt a3 and CuB} \rightarrow O_2 \rightarrow H_2O$$

- Electrons arrive at complex IV from Cyt c.
- These electrons pass through CuA, within complex IV.
- Next, these electrons pass through Cyt a (which has heme a) within the complex.
- Next, electrons move to cytochrome a3 (heme a3) and CuB, which forms a binuclear center. Both heme a3 and CuB receive electrons from cytochrome a. The Iron (Fe) in heme a3 cycles between Fe^{3+} and Fe^{2+} as it accepts an electron, while CuB cycles between Cu^{2+} and Cu^{+} and also accepts an electron.
- In the binuclear center, heme a3 and CuB both now carry electrons, and both work together to bind and reduce O_2, which diffuses from the outside of the cell.
- O_2 binds between Fe (heme a3) and CuB, and the O_2 accepts electrons from both heme a (and in the process the Fe within heme a shuttles from Fe^{2+} to Fe^{3+}), and from CuB (which shuttles from Cu^{+} to Cu^{2+}), making O_2 the final electron acceptor. As a result, O_2, which accepts four electrons from this process and also accepts four protons, is reduced to two H_2O molecules.
- An important role for CuB in this process is binding oxygen and assisting in splitting its O–O bond.

- Energy is released when O_2 is reduced to H_2O, and this energy is used to drive two additional protons across the membrane into the intermediate space.
- This marks the end of the ETC. The next step is chemiosmosis, the ATP synthase step.

Complex IV plays a critical role due to its interaction with oxygen. Because O_2 is the final electron acceptor, the entire electron transport chain stops without oxygen, thus halting oxidative phosphorylation. This is why the entire process cannot work without oxygen [3, 32].

Step 5: **ATP synthase harnesses the proton gradient**.

This step involves **complex V**, which is the **ATP synthase**. This complex has a structure that is different than the other complexes we discussed.

Complex V is a large **rotary enzyme** (an enzyme that rotates part of its structure as part of its catalytic activity) embedded in the inner mitochondrial membrane. It uses the energy stored in the proton gradient to produce ATP. It has two main parts: the F_0 unit, which functions as a channel through which protons pass, and which rotates, and the F_1 unit, which has the ATP synthase activity. The structure and function of complex V are described below.

Embedded in the inner mitochondrial membrane is the F_0 region. This region spans the inner mitochondrial membrane. It contains subunit a and a rotating c-ring, which is made of c subunits, and it acts as a proton channel. One side faces the intermembrane space, and the other side faces the mitochondrial matrix.

By the time complex V functions, a significant proton-motive force (PMF) has been established. The PMF consists of stored energy in the form of an electrochemical gradient. This gradient consists of a concentration gradient (ΔpH) and an electrical gradient (Δψ), across the inner mitochondrial membrane. The protons will flow down their electrochemical gradient through **Complex V**, as follows:

Protons enter the complex through a half-channel in the a subunit (a half channel is only open on one side, in this case, the side facing the intermembrane space), and then bind to c subunits, and this causes the c-ring to rotate. After a full rotation, each proton exits the complex via another half channel. This half channel is open at the mitochondrial matrix side. This is called chemiosmosis.

The other part of complex V is the F_1 catalytic region. This is the part of the enzyme that has ATP synthase catalytic activity. This part of the complex is in the mitochondrial matrix. It has three catalytic sites for ATP synthesis, which are β subunits, discussed below. The F_1 unit connects to the F_0 unit via a central stock, called the γ subunit. Surrounding the central stock is a hexamer containing three α subunits and three β subunits (the $\alpha_3\beta_3$ hexamer). The central stock rotates along with the F_0 region, and this rotation is caused by the flow of protons through the F_0 unit as described above, but the hexamer is held stationary by the peripheral part of the stock (the part that extends into the matrix). The rotation of the central stalk against the stationary β subunits causes the β subunits to change their conformation. This conformational change drives the joining of ADP with Pi to form ATP, which is then released into the mitochondrial matrix. Each rotation of the γ subunit produces 3 ATP molecules. In mammalian mitochondria, this requires the translocation of about 8 protons. This whole process is driven by the rotational energy from the F_0 unit.

Large amounts of ATP are generated in this process. On average, 2.5 ATP are generated per NADH, and 1.5 ATP per $FADH_2$ [3, 29].

Protons, now back in the mitochondrial matrix, can be passed back across to the intermembrane space by complexes I, III, and IV during continued electron transport, thereby maintaining the proton gradient.

- Shuttle Systems for the Transport of Cytoplasmic NADH to the Mitochondria

As discussed above, the electron transport chain oxidizes NADH and $FADH_2$, which was produced from the Krebs cycle. These molecules are already present in the mitochondria, as this

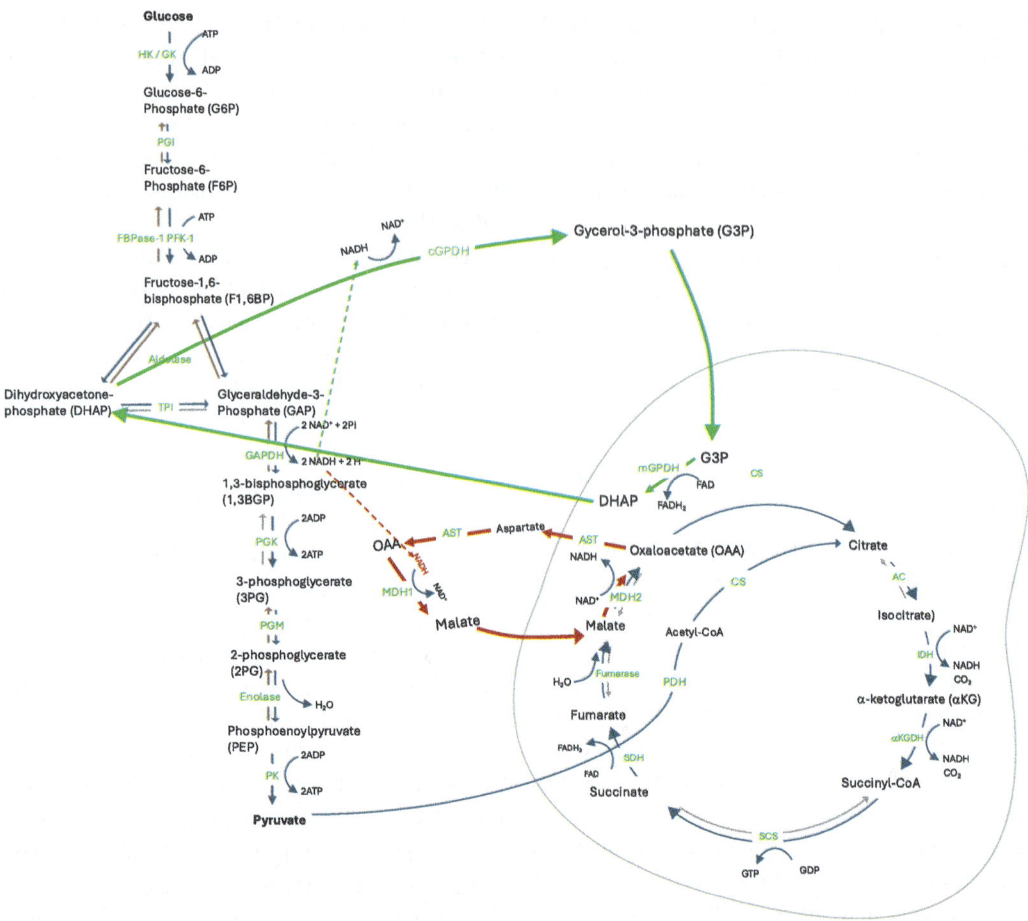

Fig. 10.13 The glycerol phosphate shuttle (green) and the malate aspartate shuttle (red) cGPDH, cytosolic glycerol 3 phosphate dehydrogenase; mGPDH, mitochondrial glycerol 3 phosphate dehydrogenase; MDH2, mitochondrial malate dehydrogenase; MDH1, cytoplasmic malate dehydrogenase; AST, aspartate aminotransferase

is where the Krebs cycle takes place. However, glycolysis also produces two NADHs, but these are in the cytosol (since glycolysis takes place in the cytosol). The electrons from these NADHs (which each have two electrons to donate) need to enter the mitochondria in order to contribute to the electron transport chain, but they cannot cross the inner mitochondrial membrane. Instead, their electrons must be transferred into the mitochondria via **shuttle systems**. There are two main shuttle systems used for this. One is the **glycerol-3-phosphate shuttle** found mostly in the **brain and muscle**, and the other is the **malate-aspartate shuttle**, which is more efficient and which is found in the **liver, kidney**, and **heart cells**. These shuttle systems are outlined in Fig. 10.13. Other tissues may use either shuttle depending on their metabolic needs. Both shuttles are discussed below.

– The Glycerol-3-Phosphate Shuttle

The **glycerol-3-phosphate shuttle** [1, 22] consists of the following steps:

Step 1: In the cytosol, cytosolic **DHAP** (see the glycolysis) can be reduced to **glycerol 3 phosphate (G3P)** (not to be confused with glyceraldehyde 3 phosphate (GAP)) by **cytosolic glycerol-3-phosphate dehydrogenase (cGPDH)** (encoded by the **GPD1 gene**). In the process, NADH is oxidized to NAD^+.

Step 2: **Glycerol-3-phosphate (G3P)** crosses through the outer mitochondrial membrane through porins, to reach the intermembrane space [33].

Step 3: In the mitochondrial intermembrane space, **G3P is oxidized back to DHAP**. This is catalyzed by **mitochondrial glycerol-3-phosphate dehydrogenase (mGPDH;** encoded by the **GPD2** gene) located on the outer face of the inner mitochondrial membrane, facing the intermembrane space. This reaction uses FAD, which is bound to GPD2 as a prosthetic group. This FAD is reduced to $FADH_2$, which then transfers its electrons directly to CoQ (ubiquinone), reducing it to $CoQH_2$. CoQH2 can now contribute to the ETC by donating two electrons to complex III. In this way, the two available electrons from NADH are salvaged and fed into the ETC.

Step 4: **DHAP** returns to the cytosol, also through porins in the outer mitochondrial membrane, completing the cycle. [34].

– The Malate-Aspartate Shuttle

The **malate-aspartate shuttle** [1, 23] relies on **cytosolic oxaloacetate (OAA)** to transfer electrons from cytosolic NADH to the mitochondria. (Although OAA is typically generated in the mitochondria, the malate-aspartate shuttle also regenerates OAA in the cytosol, which will be discussed in steps 4–6 below.) In this shuttle system, the following occurs:

Step 1: Electrons are transferred from cytosolic NADH to reduce cytoplasmic **OAA** to **malate** in a reaction catalyzed by **cytosolic malate dehydrogenase (MDH1)**. In this process **NADH → NAD$^+$**, NADH loses a hydride, which consists of two electrons and one proton.

Step 2: Although OAA cannot cross the mitochondrial membrane, Malate crosses into the mitochondria via a specific malate transporter, which is a **malate-α-ketoglutarate antiporter**. Malate enters the mitochondria through this antiporter, while αKG is transported out of the mitochondria.

Step 3: In the mitochondria, **malate** is re-oxidized to **OAA** in a reaction catalyzed by **mitochondrial malate dehydrogenase (MDH2)**, and in the process, new **NADH** is generated in the mitochondria, which can enter the ETC by donating its electrons to complex I.

Step 4: The oxaloacetate cannot cross the membrane to return to the cytosol. It is therefore converted to **aspartate** via a transamination reaction, catalyzed by **aspartate aminotransferase (AST)** (discussed in Chap. 17).

Step 5: Aspartate can exit the mitochondria via a glutamate-aspartate antiporter (glutamate will enter the mitochondria as aspartate exits).

Step 6: In the cytosol, **aspartate** can be transaminated back to **OAA**, completing the cycle within this shuttle [35]. Transamination is discussed in Chap. 14. The balanced reactions associated with these two shuttle systems are shown in Sect. 4 at the end of this chapter.

- Cyanide Is a Toxin that Works by Blocking the Electron Transport Chain (ETC)

The ETC, along with chemiosmosis, generates ATP and uses oxygen. Blocking the Electron Transport Chain (ETC) can have serious consequences because cells depend on ATP for nearly all of their functions. Cyanide is a highly deadly toxin. It is found in low levels in some natural sources, such as apple seeds, but most of these sources are poorly absorbed and not likely to be dangerous in normal quantities. It can be made industrially, however, in dangerous amounts. Cyanide is deadly because it blocks **complex IV** in the ETC. Cyanide works by binding tightly to the Fe^{3+} within heme a3 in the binuclear center. It binds at the site where O_2 normally binds, and therefore, O_2 cannot be reduced to H_2O. As a consequence, no electrons can be passed through to oxygen, and no water can be formed. The whole ETC then gets backed up and cannot be completed. ATP synthase thus gets shut down, and an energy crisis results. This leads to rapid cell death and respiratory failure and is often fatal [23].

3.4 Summary of Cellular Respiration Pathways

3.4.1 Recap of the Main Pathways

In summary, glucose breakdown involves four main stages: [1] glycolysis (the breakdown of glucose into two pyruvates), [2] decarboxylation of pyruvate; conversion of pyruvate to acetyl-CoA, [3] the Krebs cycle, and [4] oxidative phosphorylation. These are the main steps of cellular respiration (see Fig. 10.11). In addition to carbohydrates and sugars, other biomolecules can feed into the cellular respiration pathways discussed here. These include fats and proteins. The metabolism of fats will be discussed in the next chapters.

3.4.2 Total Energy Outcome from the Breakdown of Glucose during Cellular Respiration

The net energy result from glycolysis is 2 ATPs and 2 NADHs from one glucose molecule. The next step, converting pyruvate to acetyl-CoA, yields 2 more NADH from each glucose molecule. The Krebs cycle yields 2 ATPs, 6 NADH, and 2 $FADH_2$ from 1 glucose molecule.

Oxidative phosphorylation then uses the NADHs and $FADH_2$s to produce more ATP. The NADHs and $FADH_2$s from the Krebs cycle produce approximately 18 ATPs (15 from NADH and 3 from $FADH_2$). The NADHs from glycolysis and pyruvate decarboxylation produce about 10 ATPs (5 from glycolysis and 5 from pyruvate decarboxylation). In addition, 4 ATPs are generated from glycolysis and directly from the Krebs cycle, resulting in a total of approximately 32 ATPs per glucose molecule, although the exact number can vary depending on the cell's conditions. These numbers are based on the assumption that each NADH produces approximately 2.5 ATP and each $FADH_2$ produces approximately 1.5 ATP [1].

3.4.3 In What Tissues Does Cellular Respiration Occur?

The cellular respiration pathways, starting with the breakdown of glucose and culminating in ATP generation via the oxidative phosphorylation, can occur in any cell that has mitochondria. However, some tissues rely on ATP generation from oxidative phosphorylation more than others. Skeletal muscle and cardiac muscle have numerous mitochondria and rely heavily on oxidative phosphorylation, which generates ATP to meet their high energy demands. The brain is also highly dependent on ATP generated through oxidative phosphorylation to meet its energy needs. The liver carries out cellular metabolism pathways, but mostly for fat production (discussed in the next chapter), rather than for ATP production. Red blood cells lack mitochondria and therefore, unlike most cells, do not carry out oxidative phosphorylation at all [36, 37].

3.5 Electron Transport Chain (ETC) and Heat Generation

The electron transport chain (ETC) is critical for ATP generation, but it also indirectly contributes to another outcome, which is heat generation. While most of the energy from the proton gradient is used to generate ATP, the system is not entirely efficient. Some of the potential energy is instead lost as heat. This is partly due to some protons leaking back across their gradient without engaging the **ATP synthase**. Thus, the release of energy is dissipated as heat instead of being stored as ATP. In addition to some natural inefficiencies in the ETC system, which can lead to heat generation, uncouplers exist that can disrupt the protein gradient even more. As a result, even more energy is dissipated as heat rather than being converted into ATP. These uncouplers are generally tissue-specific and not abundant in most cells. An important example of an uncoupler is thermogenin (UCP-1). UCP-1 creates a channel in the inner mitochondrial membrane, allowing protons to flow down their gradient. UCP-1 is found in brown adipose tissue (BAT). BAT is a type of adipose tissue that is rich in mitochondria. It is more prevalent in babies than in adults. UCP-1 is especially important for maintaining heat in newborns. Adults can maintain heat, at least in part, by shivering, because the contraction of muscles uses up ATP, speeds up oxidative phosphorylation, and generates both heat. However, newborns cannot shiver; therefore, they rely entirely on UCP-1, which generates heat via ETC uncoupling. The UCP-1 mechanism is also important in adults under very cold conditions [3, 38, 39].

Dinitrophenol (DNP) is a drug that was used as a **weight loss** drug in the 1930s. Scientists found that this compound could cause weight loss, but they did not understand the mechanism of the drug until action later. We now know that DNP functions as a molecule that can shuttle protons across the inner mitochondrial membrane. It thus functions as an uncoupler by carrying protons across the mitochondrial membrane without the use of ATP synthase, uncoupling the ETC from ATP production. As a result, the ETC must work even harder to sustain the proton gradient and produce more ATP, thereby burning more fuel from glucose (and fatty acids) to try to restore the gradient. Meanwhile, the energy that should have been used for ATP production is instead released as heat. The drug led to increased metabolic rate and rapid fat loss, but at a serious cost because it turned out to be highly toxic. It led to excessive heat production due to the uncoupling, increased oxygen consumption, sweating, dehydration, and, in some cases, death due to overheating. The drug was subsequently banned for its use as a weight loss drug [3, 39, 40].

3.6 Metabolic Pathways when Oxygen Levels Are Low

The generation of a large percentage of ATP from glucose depends on the oxidative phosphorylation step, which requires oxygen as the final electron acceptor in the electron transport chain. Without oxygen, or when oxygen levels are low, the electron transport chain slows down or stops [1, 3, 22, 23].

Oxygen may become low or even depleted in certain situations. An important example is muscle. During vigorous exercise, it may not be possible for the muscle cells to take in oxygen at a high enough rate for the ETC to function effectively. When this happens, the muscle cells may need to supplement ATP production with lactic acid fermentation, a type of anaerobic metabolism (also called anaerobic glycolysis), described below.

As a consequence of low oxygen levels, such as during intense muscular activity, the ETC will decrease or terminate, and ATP levels in the cell begin to decline. Another consequence is that the levels of NADH increase and the levels of NAD^+ in the cell decrease. This is because NADH is normally oxidized to NAD^+ at complex I of the ETC, but the ETC is slowed down in low oxygen conditions. Thus, in low oxygen conditions, both ATP and NAD^+ levels are **low**. NAD^+ is required for glycolysis and the Krebs cycle; therefore, these two pathways also slow down. The Krebs cycle slows down not only because of low NAD^+ levels but also because it is not needed when the ETC is not operating.

3.6.1 The Generation of Lactate

This combination of factors alters the fate of the pyruvate produced during glycolysis. Under normal (aerobic) oxygen conditions, pyruvate is transported into the mitochondria, where **pyruvate dehydrogenase complex (PDH)** catalyzes pyruvate's conversion to **acetyl-CoA**, reducing NAD^+ to NADH. This process links glycolysis to the Krebs cycle. **PDH** is inhibited, however, by the low NAD^+ levels that result from **low oxygen** conditions. Additionally, another enzyme, **pyruvate dehydrogenase kinase (PDK)**, is activated in low-oxygen conditions. **PDK**, now active, phosphorylates **PDH,** and this further inhibits its activity. Consequently, instead of entering the mitochondria and being converted to acetyl-CoA, **pyruvate** stays in the cytoplasm and takes part in another reaction. Instead of being converted to acetyl-CoA, it is converted to **lactate** in a reaction catalyzed by **lactate**

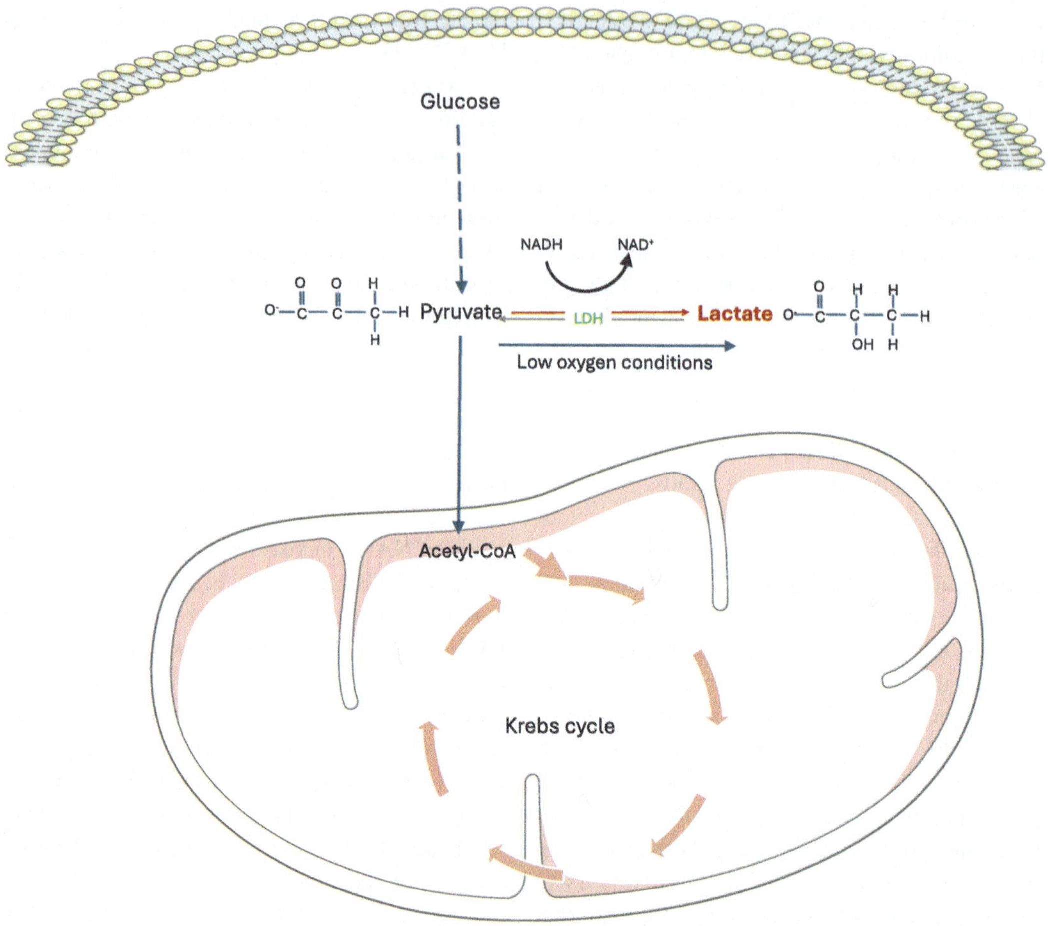

Fig. 10.14 Pyruvate to lactate. Under low oxygen conditions, pyruvate (produced as a result of glycoysis) can be converted to lactate instead of being converted to acetyl-CoA to enter the Krebs cycle. LDH, lactate dehydrogenase. (**Source**: Mitochondria adapted from Servier Medical Art (https://smart.servier.com/), licensed under the Creative Commons Attribution 4.0 (CC BY 4.0) license)

dehydrogenase (LDH) [1, 16]. In this process, NADH donates electrons and a proton to **pyruvate**, thus changing the pyruvate to **lactate** and generating NAD^+ (see Fig. 10.14) in the following reaction: Pyruvate + NADH + H^+ → Lactate + NAD^+.

This additional NAD^+ allows glycolysis to continue to some extent, and consequently, some ATP is still generated. The lactate produced becomes part of the **Cori cycle**, which helps manage low oxygen levels and the rise in lactate, as discussed below.

3.6.2 The Cori Cycle

The Cori cycle transports lactate, produced by anaerobic glycolysis in muscle, to the liver, where it is used to produce glucose (gluconeogenesis). This glucose can then be supplied back to the muscle for energy [1, 3, 22, 23] (see Fig. 10.15). The cycle works as follows.

- Muscle Produces Lactate

Muscle produces **lactate** from pyruvate during low oxygen conditions, as described above in a reaction catalyzed by **lactate dehydrogenase (LDH).**

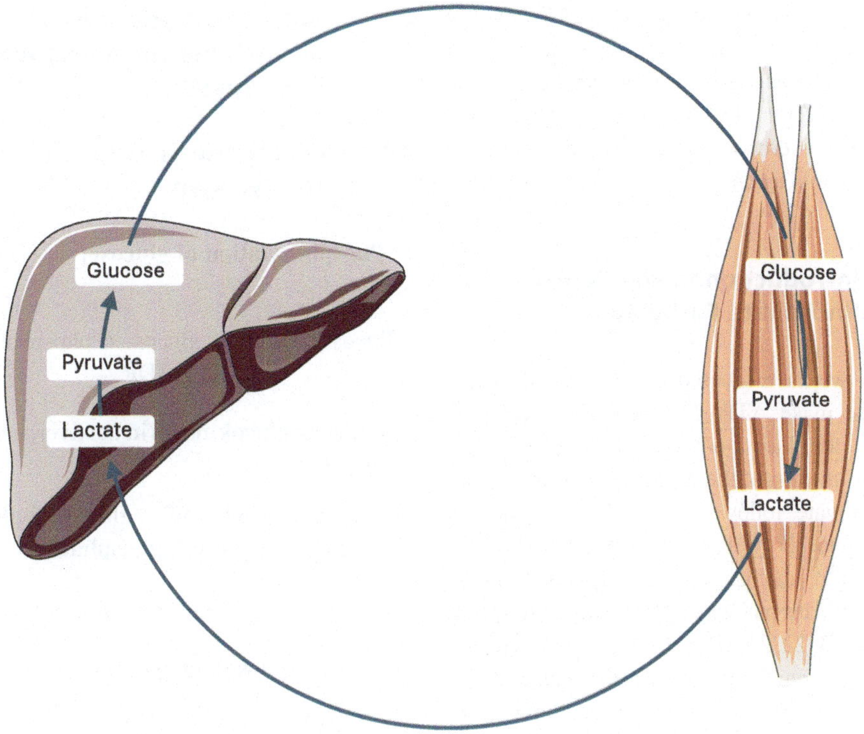

Fig. 10.15 The Cori cycle. The Cori cycle links muscle and liver metabolism. Lactate is produced in the muscle during low oxygen conditions. The lactate flows to the liver through the bloodstream, where it is converted to pyruvate and leads to glucose formation via glycolysis. This glucose then returns to muscle for ATP production and the production of more lactate

- Lactate Is Transported to Liver

The lactate is then transported to the **liver** via the bloodstream. In the liver, **lactate** is converted back to **pyruvate** in a reaction that is also catalyzed by **LDH** in the following reaction: Lactate + $NAD^+ \rightarrow$ pyruvate + NADH + H^+.

This pyruvate is then converted to **glucose** during the process of **gluconeogenesis**. Gluconeogenesis is a series of steps, discussed in more detail in Chap. 12, where some of the glycolysis pathway reactions are reversed, and some irreversible pathways are bypassed. In this process, **glucose** is produced, and some ATP is consumed. This cannot happen in muscle, because muscle lacks the necessary enzymes for gluconeogenesis. Instead, gluconeogenesis happens mostly in the liver.

- Glucose Is Sent Back to Muscle

The glucose produced in the liver is sent back to muscle via the bloodstream to replenish its supply. This is important because muscle cannot carry out gluconeogenesis. In muscle, this glucose undergoes **glycolysis**, producing a small amount of ATP (two ATPs per glucose molecule), which is especially valuable under low oxygen conditions.

In this way, the liver helps the muscle produce more ATP because it is in short supply due to low oxygen conditions. Glycolysis will also produce pyruvate, which is then converted to lactate and can be returned to the liver.

- Summary of the Cori Cycle

The Cori cycle forms a loop: the liver uses lactate from the muscle to produce more glucose, which is then sent back to the muscle, where it can generate more ATP and lactate. This cycle can continue temporarily while the muscle is in a state of low oxygen conditions [3, 4, 16] (see Fig. 10.13). This cycle helps provide energy to muscle under low oxygen conditions. However, it causes the liver to use a significant amount of energy (for gluconeogenesis to occur), and lactate accumulation in the liver can become a problem, so it is a short-term solution until sufficient oxygen is available again.

3.7 Introduction to the Pentose Phosphate Pathway

Thus far, we have focused on the breakdown of glucose to make ATP. In glycolysis, glucose is converted to glucose-6-phosphate (G6P), which then enters the rest of the glycolysis pathway, leading to pyruvate, followed by the Krebs cycle and oxidative phosphorylation. However, G6P can also be diverted to another pathway, depending on the cell's needs. This pathway is the pentose phosphate pathway [1, 3, 22, 23]. This pathway can be considered a shunt from the glycolysis pathway. A major outcome of this pathway is the production of pentose phosphates, which are five-carbon sugars (unlike glucose, which is a six-carbon sugar). Five-carbon sugars are essential because they are the basis of nucleotides, which are necessary for DNA, RNA, and ATP. Another important product of this pathway is NADPH, a reducing agent utilized in various pathways, including the one leading to fatty acid biosynthesis (discussed in Chap. 11).

Thus, after the formation of G6P, the flow of molecules can proceed through glycolysis or the pentose phosphate pathway, depending on the cell's metabolic needs. Low ATP levels will favor the glycolysis pathway. Low NADPH levels, or the need for more pentose sugars, will favor the pentose phosphate pathway. In some cases, both pathways operate. The pentose phosphate pathway is tightly linked to glycolysis, and many of its intermediates can also feed back into the glycolysis pathway as needed [1, 3, 4, 16, 17, 41]. (The pentose phosphate pathway is discussed in more depth in Chap. 15.)

4 Balanced Equations in Carbohydrate Metabolic Pathways

4.1 Glycogenesis: Glucose to Glycogen

1. **Phosphorylation of glucose:**

$$\text{Glucose} + \text{ATP} \rightarrow \text{glucose-6-phosphate(G6P)} + \text{ADP}$$

Enzyme: **glucokinase** (in the liver, hexokinase in other tissues).

2. **Isomerization of glucose-6-phosphate (G6P) to glucose-1-phosphate (G1P).**

$$\text{G6P} \rightleftharpoons \text{G1P}$$

Enzyme: **phosphoglucomutase.**

3. **Activation of glucose:**

$$\text{G1P} + \text{UTP} \rightarrow \textbf{UDP-glucose} + \text{PPi}$$

Enzyme: **UDP-glucose pyrophosphorylase.**

4. **Addition of UDP-glucose to the glycogen chain:**

$$\text{UDP-glucose} + \text{glycogen} \rightarrow \text{glycogen} + 1 + \text{UDP}$$

Enzyme: **glycogen synthase.**

And branching enzyme (also called: **1,4-α-glucan branching enzyme** or **amylo-(1,4 → 1,6)-transglycosylase**).

4.2 Glycogenolysis: Glycogen to Glucose

1. **Phosphorylation of glycogen:**

$$(\text{Glycogen})n + \text{Pi} \rightarrow (\text{Glycogen})n - 1 + \text{glucose-1-phosphate(G1P)}$$

Enzyme: **glycogen phosphorylase** and **glycogen debranching enzyme (GDE).**
(the above is a multi-step process)

2. **Isomerization of Glucose-1-phosphate (G1P) to glucose-6-phosphate (G6P):**

$$G1P \rightleftharpoons G6P$$

Enzyme: **phosphoglucomutase.**

3. **Dephosphorylation of glucose-6-phosphate (G6P) to glucose.**

$$G6P + H_2O \rightarrow Glucose + Pi$$

Enzyme: **glucose-6-phosphatase.**
(This step does not always occur, as the Glucose-6-phosphate can enter directly into glycolysis)

4.3 Glycolysis

4.3.1 First Half of the Pathway

1. **Phosphorylation of glucose:**

$$Glucose + ATP \rightarrow glucose\text{-}6\text{-}phosphate(G6P) + ADP$$

Enzyme: **hexokinase** (or **glucokinase** in the liver and pancreas).

2. **Isomerization of Glucose-6-phosphate (G6P) to fructose-6-phosphate (F6P):**

$$G6P <-> F6P$$

Enzyme: **phosphoglucose isomerase.**

3. **Phosphorylation of fructose-6-phosphate (F6P):**

$$F6P + ATP \rightarrow Fructose\text{-}1, 6\text{-}bisphosphate (F\text{-}1, 6BP)$$

Enzyme: **phosphofructokinase-1 (PFK-1).**

4. **Cleavage of fructose-1,6-bisphosphate (F1,6BP):**

$$F\text{-}1, 6BP \rightarrow Dihydroxyacetope\ phosphate(DHAP) + glyceraldehyde\ 3\text{-}phosphate(G\text{-}3P)$$

Enzyme: **fructose bisphosphate aldolase (aldolase).**

5. **Isomerization of Dihydroxyacetone phosphate (DHAP) to glyceraldehyde 3-phosphate (GAP):**

$$DHAP \rightleftharpoons G3P$$

Enzyme: **triose phosphate isomerase.**

4.3.2 Second Half of the Pathway

6. **glyceraldehyde-3-phosphate (GAP) to 1,3-bisphosphoglyerate (1,3-BPG):**

$$G3P + NAD^+ + Pi \rightarrow 1, 3\text{-}BPG + NADH + H^+$$

Enzyme: **glyceraldehyde-3-phosphate dehydrogenase.**

7. **1,3-biphosphoglycerate (1,3-BPG)** to **3-phosphoglycerate (3-PG)**:

$$1, 3\text{-}BPG + ADP \rightarrow 3PG + ATP$$

Enzyme: **phosphoglycerate kinase.**

8. **3-phosphoglycerate (3PG)** to **2-phosphoglucerate (2PG):**

$$3PG \rightleftharpoons 2PG$$

Enzyme: **phosphoglycerate mutase.**

9. **2-phosphoglycerate (2PG)** to **phosphoenolpyruvate (PEP):**

$$\mathbf{2PG \rightarrow PEP + H2O}$$

Enzyme: **enolase.**

10. **phosphoenolpyruvate (PEP)** to **pyruvate.**

$$PEP + ADP \rightarrow pyruvate + ATP$$

Enzyme: **pyruvate kinase.**

4.4 Decarboxylation of Pyruvate

$$\text{Pyruvate} + NAD^+ + \text{CoA} \rightarrow \text{Acetyl-CoA} + \text{NADH} + CO_2.$$

Enzyme: pyruvate dehydrogenase complex (PDH).

4.5 The Leloir Pathway: (Galactose to Glycolysis)

1. **Phosphorylation of galactose:**

$$\text{galactose} + \text{ATP} \rightarrow \text{galactose-1-phosphate} + \text{ADP}$$

Enzyme: **galactokinase.**

2. **Uridylyl transfer:**

$$\text{galactose-1-phosphate} + \text{UDP-glucose} \rightarrow \text{UDP-galactose} + \text{glucose-1-phosphate}$$

Enzyme: **galactose-1-phosphate uridylyl-transferase (GALT).**

3. **Epimerization:**

$$\text{UDP-galactose} \rightleftharpoons \text{UDP glucose}$$

Enzyme: **UDP-galactose 4′-epimerase (GALE)**

4. **Isomerization to glucose-6-phosphate.**

$$\text{glucose-1-phosphate} \rightleftharpoons \text{glucose-6-phosphate}$$

Enzyme: **Phosphoglucomutase.**

4.6 Fructose to Glycolysis

1. **Fructose to Fructose-1-phosphate (F1P).**

$$\text{Fructose} + \text{ATP} \rightarrow \text{F1P} + \text{ADP}$$

Enzyme: **fructokinase.**

2. **Fructose-1-phosphate (F1P) to DHAP and glyceraldehyde.**

$$\text{F1P} \rightarrow \text{DHAP} + \text{glyceraldehyde}$$

Enzyme: **aldolase B.**

3. **Glyceraldehyde → glyceraldehyde-3-phosphate (G3P).**

$$\text{Glyceraldehyde} + \text{ATP} \rightarrow \text{G3P} + \text{ADP}$$

Enzyme: **triose kinase.**

4. **G3P** and **DHAP** are **glycolysis** intermediates.

4.7 Mannose to Glycolysis

1. **Mannose to Mannose-6-phosphate (M6P)**

$$\text{Mannose} + \text{ATP} \rightarrow \text{M6P} + \text{ADP}$$

Enzyme: **hexokinase.**

2. **Mannose-6-phosphate (M6P) to Fructose-6-phosphate (F6P)** in an isomerase reaction:

$$\text{M6P} \rightarrow \text{F6P}$$

Enzyme: **phosphomannose isomerase.**

3. **Fructose-6-phosphate (F6P) is a glycolysis intermediate.**

4.8 Krebs Cycle Reactions

1. **Formation of citrate (6C):**

$$\text{Acetyl-CoA} + \text{Oxaloacetate(OAA)} + H_2O \rightarrow \text{Citrate} + \text{CoA-SH}$$

Enzyme: **citrate synthase.**

2. **Citrate to isocitrate (6C).**

$$\text{Citrate} \rightleftharpoons \text{isocitrate}$$

Enzyme: **aconitase.**

3. **Isocitrate to a-ketoglutarate (αKG) (5C).**

$$\text{Isocitrate} + NAD^+ \rightarrow \alpha KG + NADH + CO_2 + H^+$$

Enzyme: **isocitrate dehydrogenase.**

4. **α-ketoglutarate (αKG) to succinyl-CoA (4C).**

$$\alpha KG + NAD^+ + \text{CoA-SH} \rightarrow \text{Succinyl-CoA} + NADH + CO_2$$

Enzyme: **α-ketoglutarate dehydrogenase complex**.

5. **Succinyl-CoA to succinate (4C).**

$$\text{Succinyl-CoA} + GDP + Pi \rightarrow \text{Succinate} + \text{CoA-SH} + GTP$$

Enzyme: **succinyl-CoA synthetase.**
(in some cells GTP is converted to ATP by diphosphate kinase)

6. **Succinate to fumarate (4C).**

$$\text{Succinate} + FAD \rightarrow \text{Fumarate} + FADH_2$$

Enzyme: **succinate dehydrogenase.**

7. **Fumarate to malate (4C).**

$$\text{Fumarate} + H_2O \rightarrow \text{Malate}$$

Enzyme: **fumarase.**

8. **Malate — > Oxaloacetate (OA)(4C).**

$$\text{Malate} + NAD^+ \rightarrow OA + NADH + H^+$$

Enzyme: **malate dehydrogenase.**

9. **OA** can now combine with another **acetyl-CoA** to enter another cycle.

Outcome:
3 NADH, 1 $FADH_2$, 1 GDP = 1 ADP, 2 CO_2.
Times Two for One Glucose

4.9 The Glycerol 3 Phosphate Shuttle

1. DHAP is reduced to G3P:

$$DHAP + NADH + H^+ \rightarrow G3P + NAD^+$$

Enzyme: **cytoplasmic glycerol-3-phosphate dehydrogenase.**

2. G3P moves to the intermediate space.
3. G3P is oxidized back to DHAP:

This occurs on the mitochondrial inner membrane facing the intermembrane space.

$$G3P + FAD \rightarrow FADH2 + DHAP$$

Enzyme: **mitochondrial glycerol-3-phosphate dehydrogenase.**

4. **DHAP returns to the cytosol.**

4.10 The Malate Aspartate Shuttle

1. Reduction of OAA to malate in the cytosol:

$$OAA + NADH + H^+ \rightarrow NAD^+ + \text{malate}$$

Enzyme: **cytosolic malate dehydrogenase.**

2. **Transport of malate into the mitochondria.**

This uses a malate-α-ketoglutarate antiporter.

3. **Oxidation of malate to OAA in the mitochondria:**

$$\text{Malate} + NAD^+ \rightarrow OAA + NADH + H^+$$

Enzyme: **mitochondrial malate dehydrogenase.**

4. **Transamination of OAA to aspartate in the mitochondria:**

$$OAA + GLU \rightarrow \text{aspartate} + \alpha KG$$

Enzyme: **mitochondrial aspartate aminotransferase.**

5. **Aspartate is shuttled out of the mitochondria.**

   This uses a glutamate-aspartate antiporter.

6. **Transamination of aspartate back to OAA in the cytoplasm:**

$$\text{Aspartate} + \alpha\text{KG} \rightarrow \text{OAA} + \textbf{GLU}$$

Enzyme: **cytosolic aspartate aminotransferase.**

Definitions

Acetyl-CoA An important metabolic intermediate primarily from the oxidative decarboxylation of pyruvate by the pyruvate dehydrogenase complex. It enters the Krebs cycle, leading to energy production.

Allosteric regulation See Chap. 5; enzymes.

Anabolic reactions Metabolic pathways that build complex molecules from simpler ones, requiring energy input. An example is gluconeogenesis, the synthesis of glucose from precursors.

Catabolic reactions Metabolic pathways that break down molecules into smaller units, releasing energy. An example is glycolysis.

Central metabolic pathways A term used to describe the major metabolic pathways that make up cellular respiration pathways, including glycolysis, oxidative decarboxylation of pyruvate, the Krebs cycle, and oxidative phosphorylation.

Cori cycle A cycle that transports lactate, produced by anaerobic glycolysis in muscle, to the liver, to produce glucose (gluconeogenesis), which can then be supplied back to the muscle for energy.

Electron carriers Molecules that carry high-energy electrons that drive ATP synthesis, especially in organisms that can use oxygen. The major electron carriers in metabolism include NAD+/NADH and FAD/$FADH_2$ and in some cases NADP+ and NADPH. They are often referred to as **redox coenzymes** because they are involved in **redox reactions.**

Electron transport chain (ETC) A series of reactions in the mitochondria that oxidize NADH and $FADH_2$, mostly produced in the Krebs cycle, to generate ATP through oxidative phosphorylation.

FAD/$FADH_2$ A redox coenzyme pair involved in the Krebs cycle and electron transport chain.

Gluconeogenesis The synthesis of glucose from non-carbohydrate sources, such as amino acids and lactate.

Glucose-6-phosphate An intermediate in glycolysis and gluconeogenesis, also a branch point for the pentose phosphate pathway.

Glycogen The major storage form of glucose. It is a polysaccharide made of repeated units of glucose. It is found in the liver and muscle.

Glycogenesis The process of producing glycogen from glucose.

Glycogenolysis The breakdown of glycogen into glucose for energy production.

Glycolysis A cytoplasmic pathway that breaks down glucose into pyruvate, producing ATP and NADH.

Glycolytic intermediate Intermediate in the glycolysis pathway.

Hexokinase The enzyme that catalyzes the first step of glycolysis in most tissues, phosphorylating glucose to glucose-6-phosphate.

Homeostasis Refers to how the body maintains a stable environment by balancing the breakdown and biosynthesis of biomolecules as needed, depending on its needs.

Hydride ion A hydrogen atom with an extra electron. It has two electrons and one proton, giving it a negative charge.

Inborn errors of metabolism (IEMs) Inherited genetic disorders, in which a deficiency or malfunction of a specific enzyme or cofactor disrupts a metabolic pathway.

Insulin Resistance A condition in which cells in the body respond poorly to insulin, leading to elevated blood glucose levels.

Isomerize To convert a molecule into another molecule with the same chemical formula but a different structural organization.

Krebs cycle (TCA/citric acid cycle) A cyclic series of reactions that occurs in the mitochondria that generate high-energy electron carriers (NADH, $FADH_2$) and CO_2 from acetyl-CoA.

Lactate Product of anaerobic glycolysis when pyruvate is reduced to lactate.

Malate-aspartate shuttle A mechanism that transfers electrons from NADH into the mitochondria from the cytoplasm. It works through the interconversion and transport of malate, aspartate, and oxaloacetate.

Mitochondrial matrix The space inside mitochondria, within the inner membrane.

NAD^+/NADH Electron carrier molecules involved in redox reactions in glycolysis and the Krebs cycle.

Nicotinamide group The functional part of NAD^+ and $NADP^+$ involved in redox reactions. It contains a pyridine ring derived from niacin (vitamin B3).

Oxidative phosphorylation A mitochondrial process that generates large amounts of ATP, using energy from the electron transport chain.

Pentose phosphate pathway A metabolic pathway that forms a shunt from glycolysis that generates NADPH and ribose-5-phosphate. The ribose-5-phosphates are important for nucleotide synthesis.

Phosphofructokinase-1 (PFK-1) A key regulatory enzyme of glycolysis that catalyzes an irreversible step.

Pyridine ring A six-membered heterocycle (a ring made of different atoms) made of five carbons and one nitrogen.

Pyruvate The end product of glycolysis, it can be further metabolized to acetyl-CoA, which may enter the Krebs cycle, or it can be reduced to form lactate under low oxygen conditions.

Pyruvate decarboxylation The conversion of pyruvate into acetyl-CoA, linking glycolysis to the Krebs cycle.

Pyruvate dehydrogenase complex (PDH) The enzyme complex that catalyzes oxidative decarboxylation of pyruvate to form acetyl-CoA.

Redox reactions in metabolism Reactions that involve the transfer of electrons. The transfer occurs via **oxidation** (loss of electrons) of one molecule and **reduction** (gain of electrons) of another member of the redox pair.

Substrate-level phosphorylation The direct generation of ATP (or GTP) from a phosphorylated intermediate (ADP or GDP). This occurs in glycolysis and in one step of the Krebs cycle.

Study Questions

Multiple-Choice Questions

1. A characteristic of reversible reactions in metabolic pathways is
   A. The product and the substrate are the same
   B. They use different enzymes for forward and reverse reactions
   C. The same enzyme can catalyze both directions
   D. They only occur in the mitochondria
2. Type 1 diabetes mellitus is primarily caused by.
   A. Resistance to insulin
   B. Lifestyle and obesity
   C. Autoimmune destruction of pancreatic β-cells
   D. An abnormality in glycogen storage
   E. C and D
3. In the presence of oxygen (aerobic conditions), what is the main product of glycolysis in the cytosol?
   A. Lactic acid
   B. Pyruvate
   C. Acetyl-CoA
   D. Glucose-6-phosphate
4. Which of the following enzymes catalyzes an irreversible step in glycolysis?
   A. Phosphoglucomutase
   B. Phosphofructokinase-1
   C. Enolase
   D. Aldolase
5. Which shuttle system transfers NADH equivalents into the mitochondria?
   A. Urea cycle
   B. Beta-oxidation

C. Malate-aspartate shuttle
D. Glucose-alanine shuttle

6. In the electron transport chain, the final electron acceptor is.
   A. NAD^+
   B. FAD
   C. Oxygen
   D. Cytochrome c
7. Which of the following is the main purpose of catabolic reactions in metabolism?
   A. Building complex biomolecules
   B. Breaking down molecules to generate energy
   C. Storing energy as fat
   D. Producing oxygen for respiration
8. In anaerobic conditions, what is pyruvate primarily converted into in muscle cells?
   A. Acetyl-CoA
   B. Ethanol
   C. Lactate
   D. Oxaloacetate
9. Which of the following is NOT a product of glycolysis?
   A. ATP
   B. NADH
   C. $FADH_2$
   D. Pyruvate
10. What happens to excess glucose that is not immediately used for energy?
   A. It is excreted in the urine
   B. It is converted into glycogen for storage
   C. It is broken down into amino acids
   D. It is immediately oxidized in the Krebs cycle
11. Which of the following tissues primarily depends on glucose for energy and cannot use fatty acids as an alternative fuel source?
   A. Liver
   B. Skeletal muscle
   C. Brain
   D. Heart

Fill-in-the-Blank Questions

12. The metabolic process that breaks down molecules and releases energy is called___________.
13. ___________ refers to the body's ability to maintain balance between nutrient intake and energy needs.
14. _________is the major storage form of glucose in the liver and muscle.
15. _____________is the enzyme that converts pyruvate to acetyl-CoA before entering the Krebs cycle.
16. The major allosteric activator of phosphofructokinase-1 is __________.
17. Glycolysis occurs in the _________ of the cell.

Short-Answer Questions

18. Explain the difference between catabolic and anabolic reactions.
19. Why are mitochondrial disorders particularly damaging to tissues like the brain and heart?

Matching Question

20. Match the molecule to its metabolic function:

Molecule

A. NADH.
B. Pyruvate.
C. Acetyl-CoA.
D. ATP.
E. Glucose.

Function

1. Electron carrier.
2. Product of glycolysis.
3. Substrate for the Krebs cycle.
4. Energy currency.
5. Primary fuel source.

Answer Key

Multiple-Choice Questions

1. **Answer**: C
2. **Answer**: C
3. **Answer**: B
4. **Answer**: B
5. **Answer**: C
6. **Answer**: C
7. **Answer**: B
8. **Answer**: C
9. **Answer**: C
10. **Answer**: B
11. **Answer**: C

Fill-in-the-Blank Questions

12. **Answer**: Catabolism
13. **Answer**: Homeostasis
14. **Answer**: Glycogen
15. **Answer**: Pyruvate dehydrogenase
16. **Answer**: AMP
17. **Answer**: Cytoplasm

Short-Answer Questions

18. **Answer**: Catabolic reactions break down complex molecules to release energy, and anabolic reactions build complex molecules and require energy.
19. **Answer**: These tissues have high energy demands, and the mitochondria is the powerhouse of cells, producing large amounts of ATP.

Matching Question

20. **Answer**:
    A → 1, B → 2, C → 3, D → 4, E → 5

References

1. Chandel NS. Navigating metabolism. 2nd ed. Cold Spring Harbor: Cold Spring Harbor Laboratory Press; 2020.
2. Chandel NS. Carbohydrate metabolism. Cold Spring Harb Perspect Biol. 2021;13(1):a040568. https://doi.org/10.1101/cshperspect.a040568.
3. Berg JM, Tymoczko JL, Gatto GJ Jr, Stryer L. Biochemistry. 9th ed. New York, NY: MacMillan Learning; 2019.
4. Salway JG. Metabolism at a glance. Hoboken: Wiley-Blackwell; 2017.
5. Rigoulet M, Bouchez CL, Paumard P, Ransac S, Cuvellier S, Duvezin-Caubet S, et al. Cell energy metabolism: an update. Biochimica et Biophysica Acta (BBA) – Bioenergetics. 2020;1861(11):148276. https://doi.org/10.1016/j.bbabio.2020.148276.
6. Smith RL, Soeters MR, Wüst RCI, Houtkooper RH. Metabolic flexibility as an adaptation to energy resources and requirements in health and disease. Endocr Rev. 2018;39(4):489–517. https://doi.org/10.1210/er.2017-00211.
7. National Library of Medicine. Nutritional and metabolic disease. In: Genes and disease [Internet]. Bethesda: National Center for Biotechnology Information (US); 1998. Available from: https://www.ncbi.nlm.nih.gov/books/NBK22259/.
8. Saudubray J-M, Garcia-Cazorla À. Inborn errors of metabolism overview: pathophysiology, manifestations, evaluation, and management. Pediatr Clin N Am. 2018;65(2):179–208. https://doi.org/10.1016/j.pcl.2017.11.002.
9. Alberts B, Heald R, Johnson A, Morgan D, Raff M, Roberts K, Walter P, Wilson J, Hunt T. Molecular biology of the cell. W.W. Norton and Company; 2022.
10. Roche HM, Macdonald IA, Schols AMWJ, Lanham-New SA, editors. Nutrition and metabolism. Wiley-Blackwell; 2024.
11. Bender DA. Introduction to nutrition and metabolism. Boca Raton: CRC Press; 2021.
12. Hall JE, Ha.. ME. Guyton and Hall textbook of medical physiology. 14th ed. Elsevier; 2020.
13. Gromova LV, Fetissov SO, Gruzdkov AA. Mechanisms of glucose absorption in the small intestine in health and metabolic diseases and their role in appetite regulation. Nutrients. 2021;13(7):2474. https://doi.org/10.3390/nu13072474.
14. Bacharach SZ, Tordoff MG, Alhadeff AL. Glucose sensing in the hepatic portal vein and its role in food intake and reward. Cell Mol Gastroenterol Hepatol. 2023;16(2):189–99. https://doi.org/10.1016/j.jcmgh.2023.03.012.
15. Elsevier. Hepatic portal circulation. Available from: https://www.elsevier.com/resources/anatomy/cardiovascular-system/veins/hepatic-portal-circulation/16977?_gl=1*1kjtpr0*_gcl_au*ODQwOTA0MjMuMTc1MDczMjg5OC4yMTIxNj

A0MDIuMTc1MDc3NjA2NS4xNzUwNzc2MDY5*_ga*MTQzODg0NDY4NC4xNzUwNzMyODk5*_ga_MBTGG7KX5Y*czE3NTM0MzU0MzckbzI4JGcxJHQxNzUzNDM1NTUwJGoyOSRsMCRoMTQwMjA4MTMz.
16. Hames BD, Hooper NM. BIOS instant notes in biochemistry. Taylor and Francis; 2011.
17. Dashty M. A quick look at biochemistry: carbohydrate metabolism. Clin Biochem. 2013;46(15):1339–52. https://doi.org/10.1016/j.clinbiochem.2013.04.027.
18. Rajas F, Gautier-Stein A, Mithieux G. Glucose-6 phosphate, a central hub for liver carbohydrate metabolism. Meta. 2019;9(12):282. https://doi.org/10.3390/metabo9120282.
19. Kierans SJ, Taylor CT. Glycolysis: a multifaceted metabolic pathway and signaling hub. J Biol Chem. 2024;300(11):107906. https://doi.org/10.1016/j.jbc.2024.107906.
20. Chandel NS. Glycolysis. Cold Spring Harb Perspect Biol. 2021;13(5):a040535. https://doi.org/10.1101/cshperspect.a040535.
21. Sharma V, Ichikawa M, Freeze HH. Mannose metabolism: more than meets the eye. Biochem Biophys Res Commun. 2014;453(2):220–8. https://doi.org/10.1016/j.bbrc.2014.06.021.
22. Nelson DL, Cox MM, Hoskins A;A;. Lehninger principles of biochemistry. 8th ed. Macmillan Learning; 2021.
23. Rodwell VW, Bender DA, Kennely PJ, Weil PA, McGuinness OP. Harper's illustrated biochemistry. 32nd ed. New York: McGraw-Hill Education/Medical; 2022.
24. Holden HM, Rayment I, Thoden JB. Structure and function of enzymes of the Leloir pathway for galactose metabolism. J Biol Chem. 2003;278(45):43885–8. https://doi.org/10.1074/jbc.R300025200.
25. Coelho AI, Berry GT, Rubio-Gozalbo ME. Galactose metabolism and health. Curr Opin Clin Nutr Metab Care. 2015;18(4):422–7. https://doi.org/10.1097/mco.0000000000000189.
26. Conte F, van Buuringen N, Voermans NC, Lefeber DJ. Galactose in human metabolism, glycosylation and congenital metabolic diseases: time for a closer look. Biochim Biophys Acta Gen Subj. 2021;1865(8):129898. https://doi.org/10.1016/j.bbagen.2021.129898.
27. Kumar V, Greenberg ML. Emerging roles of pyruvate dehydrogenase phosphatase 1: a key player in metabolic health. Front Physiol. 2025;16:1596636. https://doi.org/10.3389/fphys.2025.1596636.
28. Arnold PK, Finley LWS. Regulation and function of the mammalian tricarboxylic acid cycle. J Biol Chem. 2023;299(2):102838. https://doi.org/10.1016/j.jbc.2022.102838.
29. Martínez-Reyes I, Chandel NS. Mitochondrial TCA cycle metabolites control physiology and disease. Nat Commun. 2020;11(1):102. https://doi.org/10.1038/s41467-019-13668-3.
30. Nolfi-Donegan D, Braganza A, Shiva S. Mitochondrial electron transport chain: oxidative phosphorylation, oxidant production, and methods of measurement. Redox Biol. 2020;37:101674. https://doi.org/10.1016/j.redox.2020.101674.
31. Read AD, Bentley RET, Archer SL, Dunham-Snary KJ. Mitochondrial iron–sulfur clusters: structure, function, and an emerging role in vascular biology. Redox Biol. 2021;47:102164. https://doi.org/10.1016/j.redox.2021.102164.
32. Online Mendelian Inheritance in Man O. COX6B1; cytochrome c oxidase subunit VIb polypeptide 1 (MIM #516030): Johns Hopkins University; Available from: https://omim.org/entry/516030.
33. AK Lectures. The Glycerol 3-phosphate shuttle 2015. https://aklectures.com/lecture/oxidative-phosphorylation/glycerol-3-phosphate-shuttle
34. Information NCfB. PubChem pathway summary for pathway PWY-6118, glycerol-3-phosphate shuttle, Source: BioCyc. 2025. Available from: https://pubchem.ncbi.nlm.nih.gov/pathway/BioCyc:META_PWY-6118.
35. Koch J, Broeks MH, Gautschi M, Jans J, Laemmle A. Inborn errors of the malate aspartate shuttle—update on patients and cellular models. Mol Genet Metab. 2024;142(4):108520. https://doi.org/10.1016/j.ymgme.2024.108520.
36. National Human Genome Research Institute. Mitochondria. 2025. Available from: https://www.genome.gov/genetics-glossary/Mitochondria#:~:text=Some%20different%20cells%20have%20different,the%20energy%20those%20mitochondria%20produce.
37. Anastasiadi AT, Arvaniti V-Z, Hudson KE, Kriebardis AG, Stathopoulos C, D'Alessandro A, et al. Exploring unconventional attributes of red blood cells and their potential applications in biomedicine. Protein Cell. 2024;15(5):315–30. https://doi.org/10.1093/procel/pwae001.
38. Bertholet AM, Kirichok Y. Mitochondrial H(+) leak and thermogenesis. Annu Rev Physiol. 2022;84:381–407. https://doi.org/10.1146/annurev-physiol-021119-034405.
39. Demine S, Renard P, Arnould T. Mitochondrial uncoupling: A key controller of biological processes in physiology and diseases. Cells. 2019;8(8):795. https://doi.org/10.3390/cells8080795.
40. Goldgof M, Xiao C, Chanturiya T, Jou W, Gavrilova O, Reitman ML. The chemical uncoupler 2,4-dinitrophenol (DNP) protects against diet-induced obesity and improves energy homeostasis in mice at thermoneutrality. J Biol Chem. 2014;289(28):19341–50. https://doi.org/10.1074/jbc.M114.568204.
41. TeSlaa T, Ralser M, Fan J, Rabinowitz JD. The pentose phosphate pathway in health and disease. Nat Metab. 2023;5(8):1275–89. https://doi.org/10.1038/s42255-023-00863-2.

Lipid Metabolism, Lipids in Diet, Health, and Disease 11

Abstract

Lipid metabolism includes the digestion, breakdown, synthesis, and regulation of fats. This chapter covers lipid metabolism. It discusses the digestion and breakdown (catabolism) of dietary fats. It also discusses the breakdown of stored fats, such as those stored largely in adipose tissue. The biosynthesis (anabolism) of fats is also covered. The interactions between lipid and carbohydrate metabolic pathways are discussed. The packaging of fats into particles, such as LDL particles, is covered. The regulation of fat metabolism via important hormones is also covered. Cholesterol metabolism is covered, along with how abnormalities in cholesterol metabolism can lead to health problems. Drugs that treat hypercholesterolemia are discussed. Disorders associated with lipid metabolism and treatment options are covered.

Keywords

Lipid metabolism · Beta oxidation · Cholesterol · Dietary cholesterol · HMG-CoA reductase

Abbreviations

ACC	Acetyl-CoA carboxylase
AMP	Adenosine monophosphate
ATP	Adenosine triphosphate
CoA	Coenzyme A
DAG	Diacylglycerol
DGAT	Diacylglycerol acyltransferase
FAS	Fatty acid synthase
FFA	Free fatty acid
GAP	In metabolism, stands for glycerol 3 phosphate (some textbooks also use G3P)
G3P	Glucose-3-phosphate (in some textbooks it is also used for glucose-3-phosphate)
GPAT	Glycerol-3-phosphate acyltransferase
HDL	High-density lipoprotein
HMG-CoA	3-Hydroxy-3-methylglutaryl-coenzyme A
LDL	Low-density lipoprotein
MAG	Monoacylglycerol
NAFLD	Non-alcoholic fatty liver disease
NASH	Non-alcoholic steatohepatitis
OAA	Oxaloacetate
TAG	Triacylglycerol (also known as triglyceride)
VLDL	Very low-density lipoprotein

What You Will Learn

- Lipid metabolism, both catabolism (breakdown) and anabolism (synthesis)
- How fats are broken down in the digestive system, and how they are transported to tissues via the bloodstream

A. Minden, *Biochemistry, Cell and Molecular Biology for Pharmacists and Health Professionals*,
https://doi.org/10.1007/978-3-032-20529-2_11

- Lipolysis and the breakdown of triacylglycerols (TAGs) from stored fats
- The fate of the glycerol and the fatty acids from both digested and stored fats
- The carnitine shuttle, which transports long-chain fatty acids to the mitochondria for beta oxidation
- Beta oxidation of fatty acids
- Fatty acid synthesis, and the role of the fatty acid synthase 1 (FAS-1) enzyme
- Regulation of fatty acid synthesis by hormones
- Cholesterol metabolism including dietary cholesterol and de novo cholesterol synthesis
- Packaging of cholesterol into lipoprotein particles
- The importance of the HMG-CoA reductase enzyme in cholesterol synthesis
- Cholesterol and human health, and drugs that are designed to treat high cholesterol
- Introduction to disorders involving lipid synthesis

In this chapter on lipid metabolism, we start with the digestion of dietary fats and then discuss both the breakdown (catabolism) and biosynthesis (anabolism) of fats in the body. Fat metabolism has multiple connections with the glucose metabolism pathways. For example, the oxidized fatty acids can lead to the formation of pyruvate or intermediates of the Krebs cycle. Conversely, some Krebs cycle intermediates play a role in the synthesis of fatty acids.

1 Digestion and Breakdown of Dietary Fats

1.1 Digestion of Fats and Breakdown of Digested Fats

Digestion of fats involves passing through the stomach, small intestine, and bloodstream, and then circulating to reach various tissues in the body. Throughout this process, fats are broken down into smaller components, which include glycerol and fatty acids, and packaged into particles such as chylomicrons and lipoproteins such as Low-density lipoprotein (LDL) and Very low-density lipoprotein (VLDL). These steps are outlined below and also outlined drawn from the following references, which also provide additional perspectives on the topic [1–8].

1.1.1 The Stomach

Dietary fats are primarily in the form of triacylglycerols (TAGs). After ingesting fats, they are partly broken down by **gastric lipases** in the stomach.

1.1.2 The Small Intestine

Next, in the duodenum, the first part of the small intestine, bile salts emulsify the fats into smaller droplets. Bile salts are produced in the liver; they are transported to the gallbladder, where they are stored and concentrated, and after a meal containing fat, they flow to the small intestine in response to hormonal stimulation [1, 5, 8]. Bile salts act as detergents; they can emulsify fats because they have hydrophilic and hydrophobic ends. The hydrophobic part binds to the fat, and the hydrophilic side interacts with the surrounding environment. This results in the fats being broken down into smaller droplets, which can be more readily accessible to **lipases. Pancreatic lipases**, together with the essential cofactor colipase, which travel from the pancreas to the small intestine, then break the resulting emulsified TAGs down into free fatty acids and 2-monoacylglycerols (2-MAGs; glycerol with one fatty acid esterified to it, on carbon 2; typically called **sn-2** for stereospecific numbering system). They work by hydrolyzing the ester bonds of the TAGs. This reaction proceeds through a DAG intermediate: TAG → DAG + FFA, followed by DAG + FFA, both catalyzed by **pancreatic lipases**.

Bile salts then have a second role, where they aggregate around these free fatty acids and 2-MAGs, along with other hydrophobic lipids such as cholesterol, to form **micelles**, which are small globules with hydrophobic fats on the inside and hydrophilic regions on the outside.

1.1.3 Enterocytes Lining the Intestine

These **micelles** travel to the brush border of **enterocytes**, which are the cells lining the **small intestine**. Here, the MAGs and fatty acids diffuse out of the micelles, and they are absorbed into the enterocytes at their apical membranes (the membrane facing the inside of the intestine). While some uptake into these enterocytes occurs via passive diffusion, long-chain fatty acids enter mostly through transporters. (Short- and medium-chain fatty acids are an exception; they bypass the chylomicrons and enter the portal blood directly, bound to albumin.) Once inside these cells, TAGs re-form (re-esterification) from the fatty acids and MAGs for chylomicron assembly. This occurs as part of a pathway called the monoacylglycerol pathway, which is catalyzed by **monoacylglycerol acyltransferase (MGAT)** and **diacylglycerol acyltransferase (DGAT)** in the smooth ER. Next, the TAGs are packaged into **chylomicrons**, which are large lipoprotein particles that also contain cholesterol, phospholipids, and proteins, including apolipoprotein B-48, which is necessary for chylomicron assembly.

1.1.4 The Bloodstream, Circulation, and Delivery to Other Tissues

These chylomicron particles are released from the enterocytes at their basolateral membranes via vesicle-mediated exocytosis. Unlike water-soluble nutrients, such as glucose, these lipid-rich particles are released into the lymphatic system and circulate through the peripheral tissues, rather than going directly to the liver. In the bloodstream, when chylomicrons reach capillaries of tissues such as muscle and adipose tissue, they encounter **lipoprotein lipases (LPLs)** bound to the luminal surface of the capillary endothelium, which hydrolyze the TAGs in the chylomicrons into fatty acids and 2-MAG. This reaction is extracellular; it occurs outside cells, at the chylomicron surface, and is often referred to as **lipoprotein lipase-mediated TAG hydrolysis**. The fatty acids are released from the chylomicrons and taken up by nearby tissues such as muscle and adipose tissue.

The remaining MAGs are taken up into tissues (such as muscle and fat) from the chylomicrons and further hydrolyzed within the cells by the enzyme **monoacylglycerol lipase (MGL)**, which is present in the cells, to yield glycerol and the remaining fatty acid.

- Fate of the Free Fatty Acids

Free fatty acids that result from this process, which are not needed for energy, are taken up by adipose tissue via transport proteins and re-esterified into TAGs. This involves the binding of free fatty acids and glycerol-3-phosphate (G3P). (The glycerol-3-phosphate can be generated in response to the glycolysis intermediate DHAP, a process that will be discussed later.) These TAGs are stored in lipid droplets. They serve as fat stores, which can be broken down, oxidized, and used as an energy source when needed. Insulin stimulates fatty acid uptake and TAG synthesis after a meal, a topic that will be covered in more detail in Chap. 12. Other free fatty acids may be taken up by muscle cells, transported to the mitochondria within those cells, oxidized, and used for energy.

- Fate of the Glycerol

The **glycerol** molecules from the chylomicrons do not get used by adipose tissue or muscle because these tissues lack significant **glycerol kinase (GK)** activity. Instead, the glycerol is released into the bloodstream and transported to the **liver**, where it will play other roles in metabolism. Specifically, in the liver, the glycerol can be phosphorylated by **glycerol kinase** to form **glycerol-3-phosphate**, which can enter glycolysis or gluconeogenesis. The chylomicrons now become chylomicron remnants and are also delivered back to the liver. This is the main mechanism by which the liver obtains glycerol from dietary sources, but it also receives glycerol from the breakdown of fat used for energy, which is discussed later [5, 9].

1.2 Breakdown (Catabolism) of Stored Fats

Most fat is stored in adipose tissue as TAGs (glycerol esterified to fatty acids). Here, we discuss what happens to this stored fat once it is needed for energy. The breakdown process differs slightly from that of digested fats and is referred to as **lipolysis**.

1.2.1 Lipolysis: Breakdown of TAGs from Stored Fats into Free Fatty Acids and Glycerol

When the body requires energy and there is not enough glucose available to supply it, hormones signal adipose cells to break down TAGs into free fatty acids and glycerol. These reactions occur inside adipocytes, at the surface of lipid droplets where TAGs are stored. This is the process of lipolysis. The reaction involves three steps (see Fig. 11.1):

- Step 1): Each TAG is broken down into diacylglycerol (DAG) and one free fatty acid, catalyzed by **adipose triglyceride lipase (ATGL)**. DAGs consist of glycerol esterified to two fatty acids.
- Step 2): The DAG is broken down to a MAG and another free fatty acid, catalyzed by **hormone-sensitive lipase (HSL)**. HSL activity is stimulated by hormonal signals and will be covered in more detail in Chap. 12.
- Step 3): This MAG is hydrolyzed to glycerol and the final free fatty acid, catalyzed by **monoglyceride lipase (MGL)**.

Once the free fatty acids and glycerol are released from the lipid droplet, they are then released into the bloodstream. In the bloodstream, free fatty acids bind to albumin and are transported to the liver, muscle, and other tissues, while glycerol is primarily taken to the liver [5, 10]. Note that this mechanism of TAG breakdown is different from the breakdown of digested TAGs, which, as discussed above, begins extracellularly on capillary endothelium and involves the hydrolysis of TAGs in chylomicrons.

The catabolism of digested and stored TAGs is summarized in Table 11.1.

The following sections describe how the **glycerol** and **free fatty acids** are metabolized further:

1.2.2 The Glycerol from the Breakdown of TAGs is Converted to Glycerol-3-Phosphate

The glycerol that results from the breakdown of TAGs is first converted to Glycerol-3-phosphate (G3P) in the liver, in a reaction catalyzed by **glycerol kinase (GK).** In this reaction, one phosphate group is added to the glycerol, using one ATP (Fig. 11.2a). Glycerol-3-phosphate then has

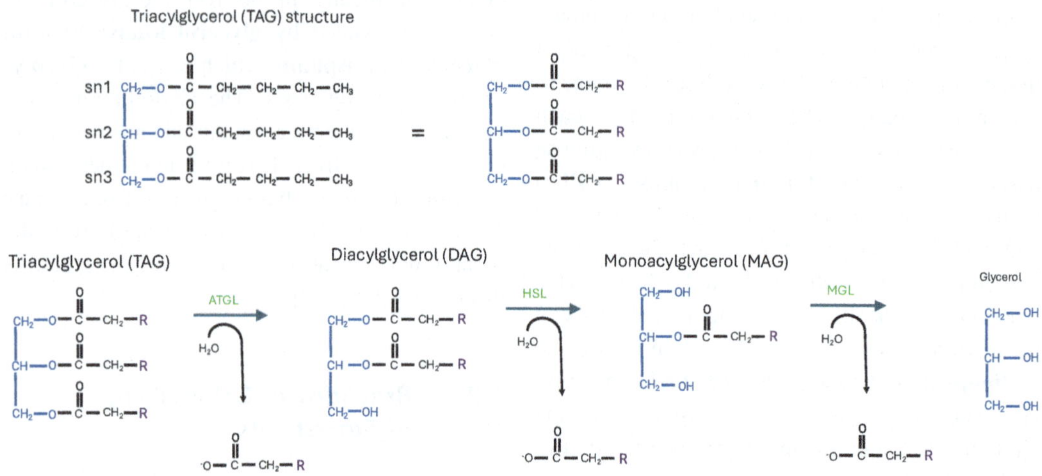

Fig. 11.1 Lipolysis: Catabolism of stored Triacylglycerols (TAGs). The top shows the basic structure of TAGs (in the structures on the right and bottom, the hydrocarbon chains are depected by R). The bottom shows the process of lipolysis in which stored TAGs are broken down into glycerol and fatty acids. This process uses the enzymes ATG, HSL, and MGL. In contrast, TAGs obtained from digestion are broken down by different enzymes, pancreatic lipases and lipoprotein lipases, although MGL catalyzes the hydrolysis of MAG to glycerol and one fatty acid in both contexts. Abbreviations used: *ATGL* adipose triglyceride lipase, *HSL* Hormone sensitive lipase, *MGL* monoacylglycerol lipase

Table 11.1 Summary of fat breakdown

Step	Location	Breakdown or re-esterification process	Breakdown or esterification product	Packaging/transport
Dietary fat				
TAG hydrolysis in the small intestine	Lumen of small intestine	TAG→DAG→MAG + FFAs Main enzymes: Pancreatic lipases	MAGs and FFAs	MAGs and FFAs are packaged into micelles
Uptake into enterocytes	Enterocyte brush border		MAGs and FFAs enter enterocytes	
Re-esterification	Enterocytes	TAGs are re-formed from MAGs and FFAs via the monoacylglycerol pathway	TAGs	TAGs are packaged into chylomicrons
Hydrolysis of TAGs in the bloodstream	Capillary endothelium	TAG→DAG→MAG + FFAs (extracellular. Enzyme: **lipoprotein lipase (LPL)** Followed by: MAG→glycerol + FFA (intracellular). Enzyme: **monoacylglycerol lipase (MGL)**	Glycerol and FFAs	FFAs are released and taken up by nearby tissues. FFAs that circulate bind to albumin. Glycerol is soluble and travels freely to the liver
Stored fat				
TAG hydrolysis	Adipocyte lipid droplet	TAG→DAG→MAG→ glycerol+FFAs Enzymes: TAG→DAG: ATGL. DAG→MAG: HSL. MAG→glycerol: MGL	Glycerol and FFAs	
Transport through bloodstream				FFAs bind to albumin. Glycerol is soluble in the plasma

several possible fates. First, it can be converted to dihydroxyacetone-phosphate (DHAP) in an oxidation reaction catalyzed by **glycerol-3-phosphate dehydrogenase (GPDH)** (Fig. 11.2b). (Glycerol 3 phosphate to DHAP is a reversible reaction, and glycerol can also be made from DHAP, also catalyzed by GPDH.) DHAP is a glycolytic intermediate, and it can be converted to glyceraldehyde-3-phosphate (GAP) in a reaction catalyzed by **triose phosphate isomerase (TPI)** (See Chap. 10, Figs. 10.9 and 10.13). From this stage, glyceraldehyde-3-phosphate, also a glycolytic intermediate, can lead to the breakdown of glucose into pyruvate, resulting in ATP production. This is important when energy levels are low, or gluconeogenesis, leading to the synthesis of more glucose during fasting. (Gluconeogenesis is discussed in more detail in Chap. 12). Alternatively, glycerol-3-phosphate can be used for the synthesis of more fatty acids. This is most likely to occur during the fed state [5, 6, 9].

1.2.3 Fate of Fatty Acids from the Breakdown of TAGs

Now that we've discussed the fate of the glycerol backbones, we'll turn to the fate of the free fatty acids that result from the breakdown of TAGs. These fatty acids are broken down further by the process of β oxidation, resulting in a significant amount of energy production. This occurs in various tissues, but mostly in muscle and liver. Muscle, in particular, relies on the β oxidation of free fatty acids for energy. The main steps are described below and and which draws from the following resources [5, 11, 12].

- Fatty Acid Activation in the Cytosol

In the cytosol, before they are broken down, long-chain fatty acids (containing 12 Cs or more)

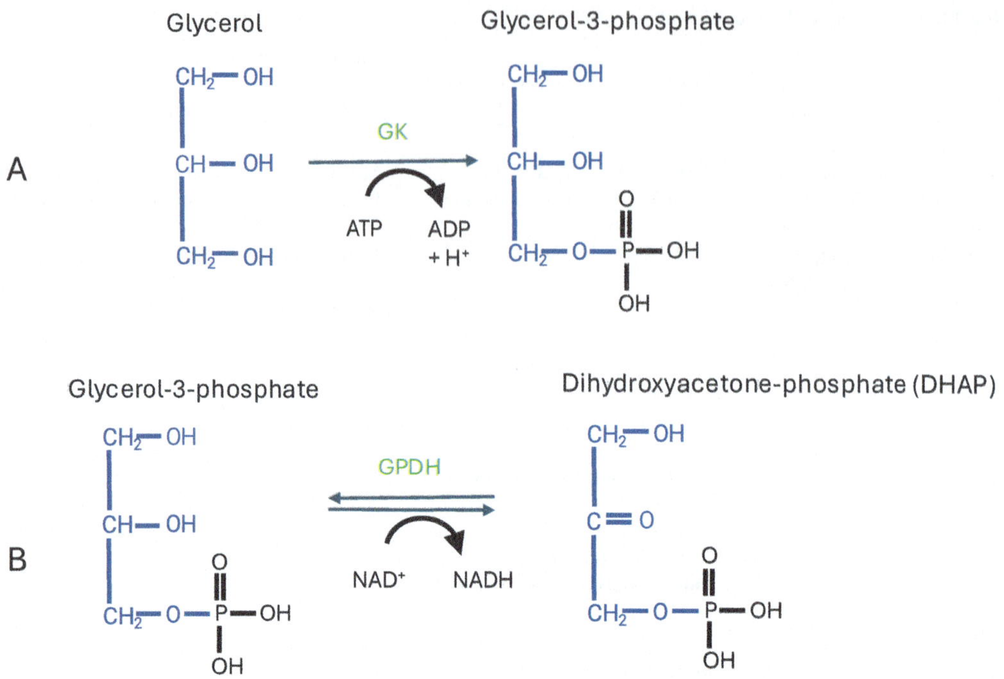

Fig. 11.2 The fate of the glycerol that results from triacylglycerol (TAG) catabolism. Glycerol is converted to glycerol 3-phosphate, and then to DHAP, which is a glycolysis intermediate

become bound to coenzyme A (CoA) to form fatty-acyl-CoA; long hydrocarbon chains attached to CoA via a thioester bond. The reaction in which fatty Acyl-CoA is formed is called fatty acid activation [13]. This involves two steps, described below.

Step 1): The fatty acid combines with ATP to form acyl-adenylate (acyl-AMP) [14], with the release of two pyrophosphates (PPi).

Step 2): This acyl-adenylate intermediate reacts with CoA to form fatty-acyl-CoA, with the release of AMP.

The overall final reaction is (see Fig. 11.3):

$$\text{Fatty acid} + \text{ATP} + \text{CoA} \rightarrow \text{Fatty acyl-CoA} + \text{AMP} + \text{PPi}$$

Both of these steps are catalyzed by **acyl-CoA synthetase (ACS)** (also called **fatty acid thiokinase**). There are different isoforms of the enzyme **ACS**. The one used here is **ACS long chain (ACSL)**. This enzyme is located on the cytoplasmic side of the outer mitochondrial membrane, which is where the reaction takes place.

Short-chain fatty acids (SCFAs) and medium-chain fatty acids (MCFAs) are generally considered to be those with approximately 1–6 carbons or 7–12 carbons, respectively (5-chain fatty acids are uncommon in humans) [15]. These SCFAs and MCFAs undergo the same reactions, where they have CoA attached. However, they first diffuse through the mitochondrial membrane, and the CoA is added within the mitochondrial matrix, with the same set of reactions occurring. However, they use a different form of ACS; medium (**ACSM**) and ACS short (**ACSS**), and these enzymes are in the mitochondrial matrix [15, 16].

This addition of the CoA makes the fatty acids more hydrophilic. Fatty acids are largely hydrophobic and, therefore, can potentially diffuse through the plasma membrane. By making them

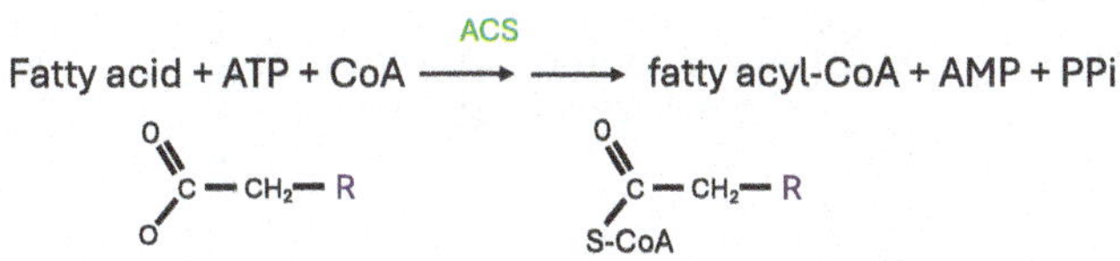

Fig. 11.3 Fatty acid activation in the cytosol. Fatty acid combines with ATP to form acyl-adenylate, and this intermediate then reacts with CoA to form fatty-acyl-CoA with the release of AMP. The overall reaction is shown in the figure. Abbreviation: *ACS* acyl-CoA-synthetase

more hydrophilic, the CoA helps to retain the fatty acids in the cell.

The fatty-acyl-CoAs will vary in size, and therefore, they will vary in full name, depending on the length of the fatty acid. For example, palmitoyl CoA is made from palmitate and contains a 16-carbon side chain, while stearoyl CoA contains an 18-carbon side chain [9].

Note that after activation, the carbon numbering of the fatty acyl-CoA remains unchanged. The carbon bound to CoA is still C1, the next carbon is C2, the α carbon, and the next is the β carbon, and so on.

- Carnitine Shuttle: Long-Chain Fatty-acyl-CoAs Are Transported into the Mitochondria via the Carnitine Shuttle

The next steps in fatty acid oxidation occur in the mitochondrial matrix. Long-chain fatty acids cannot readily pass from the cytoplasm through the mitochondrial membrane, however. Their transport across the mitochondrial membrane is instead facilitated by the **carnitine shuttle** (also called the carnitine transport system) (Fig. 11.4).

The **carnitine shuttle** depends on carnitine. Carnitine is a molecule concentrated in tissues like skeletal muscle, which use high amounts of fatty acids as an energy source. It is synthesized in the body, and vitamin C is a cofactor in its synthesis. Because of its chemical characteristics, carnitine can help bound substances cross the mitochondrial membrane, particularly the inner mitochondrial membrane, which is not very permeable.

Activated long-chain fatty acids are usually located at the cytoplasmic surface of the mitochondria, where the **ACSL** is located. While **ACSL** is at the cytosolic surface of the outer mitochondrial membrane, another enzyme, **carnitine-palmitoyl-transferase I (CPT-I)**, is embedded in the outer mitochondrial membrane. This enzyme catalyzes the transfer of the fatty-acyl group from fatty-acyl CoA to carnitine. In other words, it removes the CoA group and leaves behind a fatty acyl carnitine instead, on the cytoplasmic side of the mitochondria. The substrate for **CPT-1** is fatty acyl-CoA, and this helps explain why the fatty acid must be attached to CoA for this shuttle system to be utilized. Importantly, **CPT-1** is inhibited by malonyl-CoA, a key intermediate in fatty acid synthesis (described below). This inhibition is an important regulatory mechanism because it prevents fatty acid β oxidation when fatty acid synthesis is active. The resulting fatty-acyl carnitine can now cross the outer mitochondrial membrane and enter the intermembrane space [5, 6].

Now the fatty-acyl carnitine is in the intermembrane space of the mitochondria. The next step is the transfer of the fatty-acyl carnitine across the inner mitochondrial membrane into the mitochondrial matrix. This is catalyzed by **carnitine acylcarnitine translocase (CACT)**, which is embedded in the inner mitochondrial membrane.

Once inside the mitochondrial matrix, **carnitine palmitoyl-transferase (CPT-II)** catalyzes the transfer of the fatty acyl group from carnitine back to a molecule of CoA in the mitochondria, and releases the carnitine, thereby re-forming a fatty-acyl-CoA in the mitochondrial matrix. Carnitine is then transported back out of the mitochondrial membranes to the cytoplasmic side of the mitochondria, where it can participate in transporting another molecule [5, 9, 17–19].

- β Oxidation of fatty acids in the Mitochondria

Inside the mitochondria, β oxidation is the next step for the breakdown of the fatty acyl-CoA molecules. In this process, the hydrocarbon

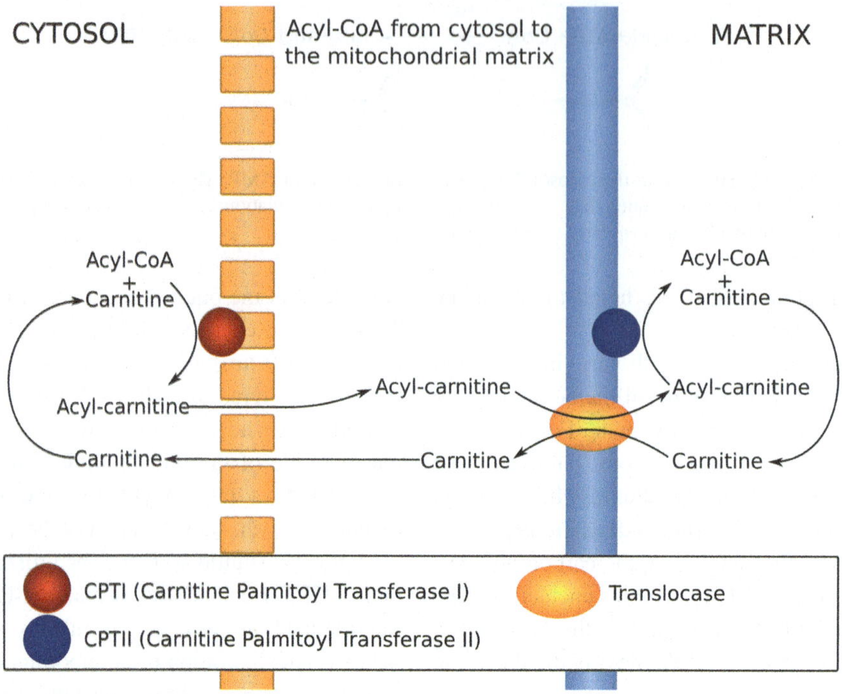

Fig. 11.4 The carnitine shuttle. The carnitine shuttle facilitates the transport of fatty acids across the mitochondrial membrane. See text for details. **Source**: Wikimedia Commons image, by Slagt. File: Acyl-CoA from cytosol to the mitochondrial matrix.svg. CC0 1.0

chains in the fatty-acyl-CoAs are broken down, and as a result, energy carriers $FADH_2$ and NADH are generated, which will later be used to generate ATP. The β-oxidation process consists of a repeated cycle of four reactions resulting in cleavage of bonds in which the fatty-acyl-CoA is shortened by 2 carbons in each round. The repeated reactions in β oxidation consist of oxidation, hydration, oxidation, and thiolysis reactions.

Before covering the steps in β oxidation, let us review the numbering of the carbons in a fatty acid, as discussed in Chap. 9: The carbon within the carbonyl carbon (C=H) is carbon 1 (C1). The next carbon (C2) is the alpha (α) carbon. The next carbon (C3) is the β carbon, and the next carbon (C4) is the γ carbon. This numbering continues in this way until the end of the fatty acid, the omega (ω) carbon. β oxidation of fatty acids gets its name because the oxidation happens at the β carbon (C3). The oxidation converts the β carbon to a carbonyl group, which allows the bond between the α carbon (C2) and the β carbon to be cleaved. This releases acetyl-CoA (C1–C2) and shortens the fatty acyl chain at each round.

The β oxidation steps are as follows (see Fig. 11.5)

Step 1): This first step is an **oxidation** step. Here, the enzyme **acyl-CoA dehydrogenase (ACAD)** catalyzes the removal of two hydrogen ions from the α and β carbons of the fatty acyl-CoA. (Multiple chain-length-specific ACAD isoforms exist, for different chain lengths.) One hydrogen is removed as a hydride (H^-) and the other as a proton (H^+), and both are transferred to FAD, which is thereby converted to $FADH_2$. This introduces a trans double bond between the α and β carbons, producing a **trans enoyl-Δ²-CoA**.

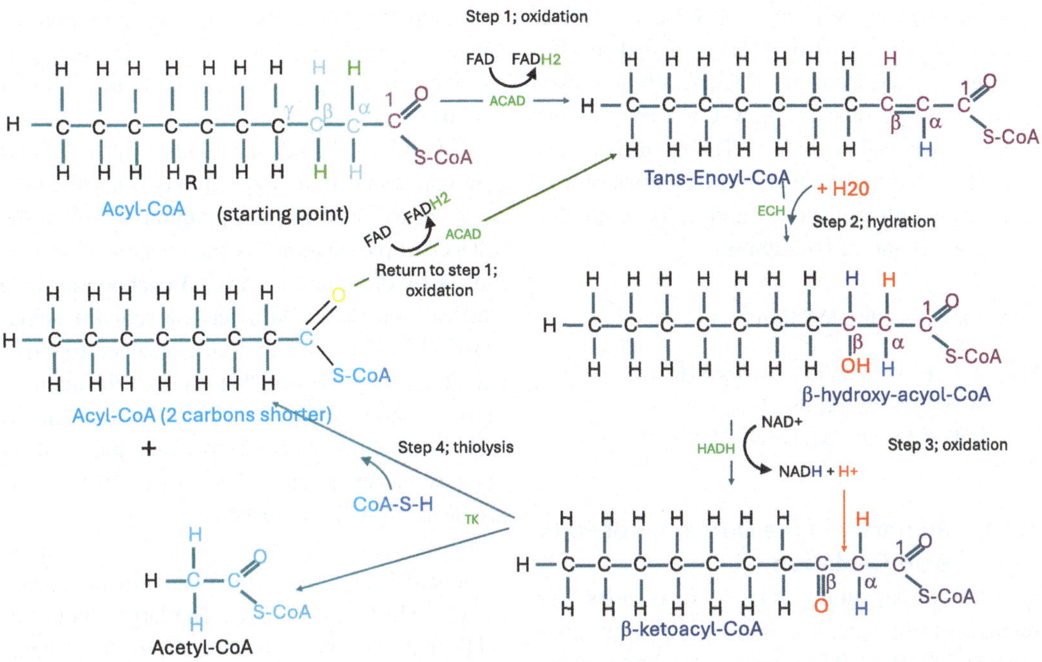

Fig. 11.5 Fatty acid oxidation steps. For each round of beta (β) oxidation, the fatty acid becomes shorter by two carbons. Abbreviations used in this figure: *ACAD* acyl-CoA dehydrogenase, *ECH* enoyl CoA hydratase, *HADH* beta-hydroxyacyl CoA dehydrogenase, *KT* beta-ketothiolase

Summary of the reaction:

$$R\text{–}CH_2\text{–}CH_2\text{–}CO\text{–}SCoA + FAD \rightarrow R\text{–}CH = CH\text{–}CO\text{–}SCoA(trans) + FADH_2$$

Step 2): This is a **hydration** step. The enzyme **enoyl CoA hydratase (ECH)** catalyzes the addition of an H_2O across the double bond between the α and β carbons of the newly formed enoyl-CoA. The β carbon (C3) gets the -OH group, and the α carbon (C2) gets the -H. The molecule now becomes a **β-hydroxy acyl-CoA.**

Summary of the reaction:

$$R\text{–}CH = CH\text{–}CO\text{–}SCoA + H_2O \rightarrow R\text{–}CH(OH)\text{–}CH_2\text{–}CO\text{–}SCoA$$

Step 3): The next step is another **oxidation** step. In a reaction catalyzed by **β-hydroxyacyl CoA dehydrogenase (HADH)**, a hydride is transferred from the β carbon of the **β-hydroxy acyl-CoA** to NAD^+, which then becomes NADH. (In the process, one proton from the β-hydroxy acyl-CoA is released into the mitochondrial matrix.) In this oxidation, the hydroxyl group (-OH) on the β-carbon is converted to a carbonyl (C=O), and the resulting molecule is **β-keto acyl-CoA.**

Summary of the reaction:

$$R\text{–}CH(OH)\text{–}CH_2\text{–}CO\text{–}SCoA + NAD^+ \rightarrow R\text{–}CO\text{–}CH_2\text{–}CO\text{–}SCoA + NADH + H^+$$

Step 4): Finally, this step is the **thiolysis** reaction. This reaction is catalyzed by the enzyme **β-ketothiolase** (or broadly called **thiolase** and abbreviated here as **KT**), which catalyzes the cleavage of the bond between the α carbon (C2) and β carbon (C3) of the **β-keto acyl-CoA** (see Fig. 11.5). This releases the first two carbons - C1 (the carbonyl carbon) and C2 (the α carbon) - in the form of acetyl-CoA. Now a new free CoA (CoASH) is added to the remaining chain at the C3 (β) carbon, which will now become the new C1 carbon of the new shortened acyl-CoA. The result is a new acyl-CoA molecule that is two carbons shorter than the previous one, and it is ready for another round of β oxidation.

Summary of the reaction:

$$R\text{–}CO\text{–}CH_2\text{–}CO\text{–}SCoA + CoASH \rightarrow R - CO - SCoA(\text{shortened by 2C}) + CH_3\text{–}CO\text{–}SCoA(\text{acetyl} - CoA)$$

1.2.4 Summary of the Outcomes of Fatty Acid Oxidation

In each β-oxidation cycle, two carbons are removed from the fatty acid, and the products include $FADH_2$, NADH, and acetyl-CoA. At the end of the process (after multiple rounds of β oxidation), for even-numbered FAs, which are the most common, only acetyl-CoA remains. Acetyl-CoA is a 2-carbon molecule attached to CoA. This acetyl-CoA, as well as each acetyl-CoA from each round of β oxidation, can then enter the Krebs cycle, leading to ATP generation. In addition, all the $FADH_2$s and NADHs generated in each step of β oxidation donate their electrons to the ETC and thereby lead to ATP generation. Fatty acids can be quite long hydrocarbon molecules. Longer chain fatty acids will produce more ATP because $FADH_2$, NADH, and acetyl-CoA are released due to more rounds of β oxidation; 16-carbon fatty acids (palmitate) are a common type of fatty acid that gets oxidized, but fatty acids can also be of other lengths.

For odd-numbered fatty acids, which are less common than even-numbered fatty acids, the fatty acids are oxidized to form **propionyl-CoA** in the last step of β oxidation, instead of acetyl-CoA (see Fig. 11.6). Propionyl-CoA is a **3-carbon** molecule attached to a CoA, and it gets converted to **succinyl-CoA** (a 4-carbon molecule attached to CoA) via a three-step reaction that involves the addition of one more carbon from CO_2 (carboxylation) (Fig. 11.6). Since succinyl CoA is an intermediate in the **Krebs cycle** (see Chap. 10, Fig. 10.10). This also leads to ATP production.

In summary, the β oxidation leads to large amounts of ATP due to the generation of acetyl-CoA, NADH, and $FADH_2$. The large amount of ATP that can be generated from fatty acids illustrates why fat is such an efficient source of energy. Furthermore, fat is usually abundant and, therefore, a readily available energy source. While the body usually has enough glycogen to last about 30 h before it needs to be replenished, it generally has enough fat to last about a month [5, 9, 11, 12].

2 Lipid Synthesis

Fats obtained in the diet are broken down (oxidized) in the body, but fatty acids can also be synthesized in the body. Fatty acids are

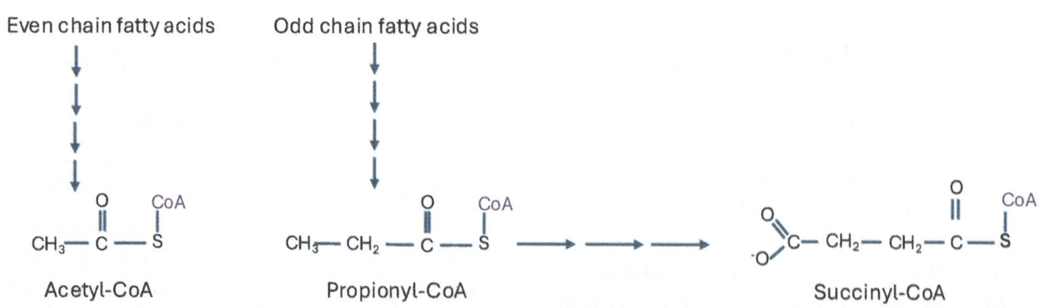

Fig. 11.6 β oxidation of odd-chain vs. even-chain fatty acids. Odd-chain fatty acids are oxidized to propionyl-CoA instead of acetyl-CoA. Propionyl-CoA can be converted to succinyl-CoA (structure on the bottom right), a Krebs cycle intermediate

synthesized from acetyl-CoA. Fatty acid synthesis occurs in the cytoplasm, predominantly in cells of the liver, adipose tissue, and lactating mammary glands. It occurs during a fed state when insulin and blood glucose levels are high. High ATP levels also positively regulate lipid synthesis because it leads to citrate accumulation, which in turn provides acetyl-CoA, an essential building block of fatty acids. This is because the Krebs cycle slows down when ATP levels are high, indicating that the cell already has sufficient energy. When this happens, citrate accumulates in the mitochondria and is then transported to the cytoplasm. There, it is converted to acetyl-CoA in a reaction catalyzed by **ATP-citrate lyase**.

In this section, we discuss **fatty acid** synthesis, the synthesis of **glycerol-3-phosphate,** which forms the backbone of the TAGs, and the esterification of fatty acids with glycerol-3-phosphate to form TAGs, which takes place in adipose tissue, as well as the liver. We also discuss some of the regulatory mechanisms involved in fat synthesis. The TAGs that are produced are transported, after being packed into VLDLs (very low-density lipoproteins) and sent to other tissues, largely to adipose tissue, where they can remain for long-term storage. We start with a discussion of fatty acid synthesis, outlined below and drawn in part from the following references [4, 5, 12].

2.1 Fatty Acid Synthesis

2.1.1 The Citrate Shuttle: Citrate Is Made into Acetyl-CoA, which Serves as the Main Building Block for Fatty Acids

Acetyl-CoA is the main building block of fatty acids. Therefore, before discussing the steps in fatty acid synthesis, it is important to discuss the source of the acetyl-CoA. Fatty acids are made from acetyl-CoA in the cytosol, and acetyl-CoA is made in response to the breakdown of glucose in glycolysis. However, much of the cell's acetyl-CoA is in the mitochondria, and acetyl-CoA cannot cross the mitochondrial membrane to enter the cytoplasm. To solve this, a shuttle system called the citrate shuttle is used (see also Appendix B). This shuttle uses energy in the form of ATP [5, 9, 20].

The basic premise of the **citrate shuttle** is that although acetyl-CoA cannot cross the mitochondrial membrane into the cytoplasm, citrate can do so. Once citrate crosses from the mitochondria to the cytoplasm, it can lead to the formation of acetyl-CoA, which is a starting point for fatty acid synthesis.

The first part of the citrate shuttle works as follows: Citrate is an intermediate in the Krebs cycle. Although it functions in the mitochondria during the Krebs cycle, it can also cross the mitochondrial membrane and enter the cytoplasm. It crosses using a transporter called the tricarboxylate carrier (TCC), also called the citrate carrier (CiC), which is a citrate malate antiporter. Once in the cytoplasm, citrate (C6) is undergoes CoA-dependent (thiolytic) cleavage in a reaction catalyzed by **ATP-citrate lyase (ACLY).** This cleavage leads to the formation of two molecules: (i) oxaloacetate (OAA, 4C) and (ii) acetyl-CoA (a 2-carbon molecule attached to CoA).

The reaction described above, catalyzed by **ACLY**, is in two parts, as follows: (a) A phosphate group from ATP is transferred to the enzyme (ACLY) citrate complex, forming a citryl-phosphate intermediate. (b) This citryl-phosphate intermediate is then cleaved into OAA (4C) and an acetyl-group, which is transferred to CoA. (c) The phosphate is released in the process. The result is cytoplasmic OAA and acetyl-CoA. The overall reaction, as a culmination of these steps, is as follows:

$$\text{Citrate} + \text{CoA} + \text{ATP} \rightarrow \text{acetyl} - \text{CoA} + \text{OAA} + \text{ADP} + \text{Pi}$$

Enzyme: **ACLY.**

An important outcome of this process is that cytoplasmic acetyl-CoA is now available, which is a key building block for fatty acid synthesis. Below, we discuss the role of acetyl-CoA as the starting point for fatty acid synthesis. The outcome of the OAA is important for completing the shuttle, however, and it is discussed later in Sect. 2.1.4. The entire citrate shuttle is outlined in Fig. 11.7.

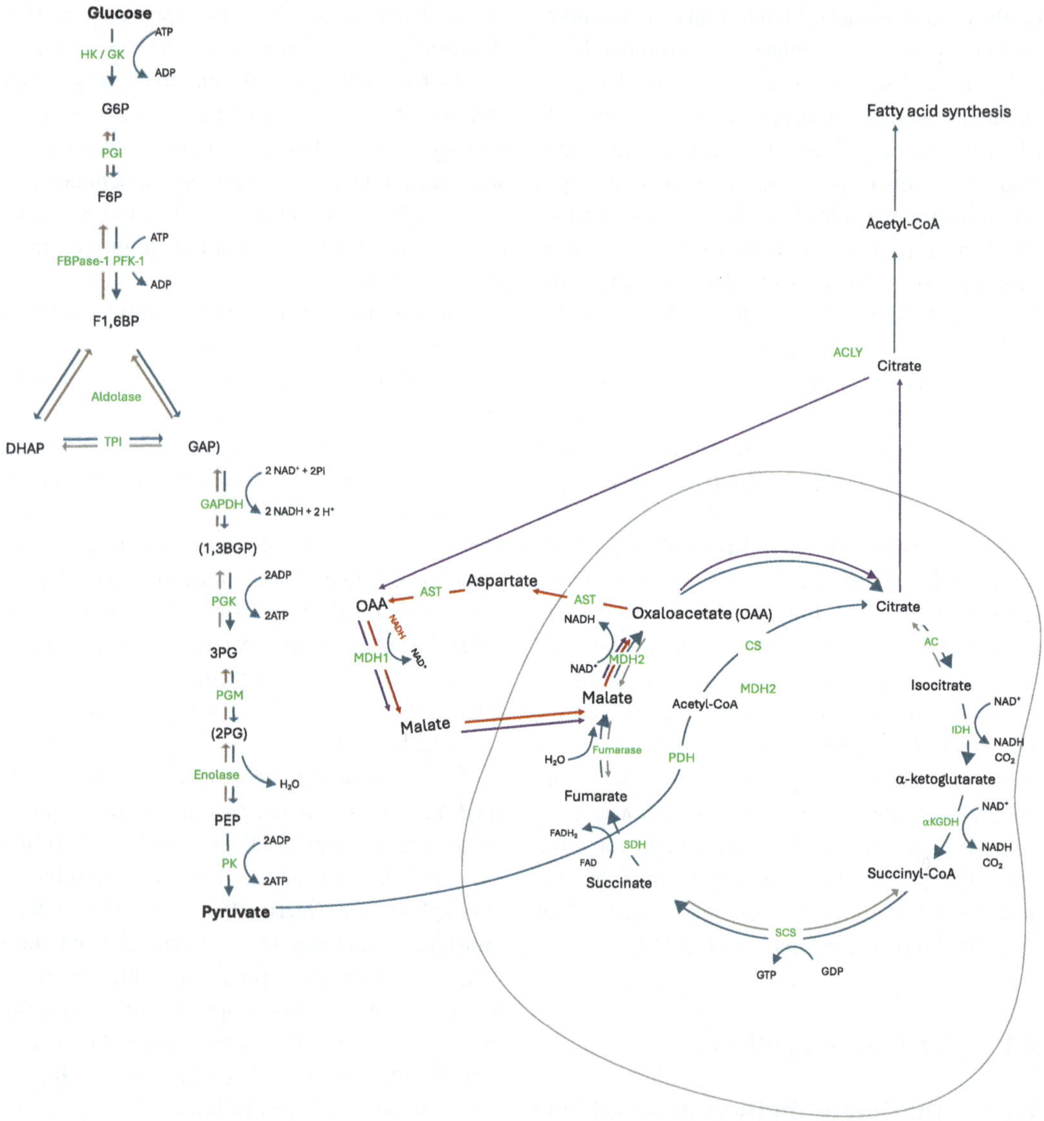

Fig. 11.7 The citrate shuttle. The citrate shuttle is indicated by the purple arrows. The malate aspartate shuttle (also described in Chap. 10, Fig 10.13) is also shown (red arrows), since both pathways use some of the same machinery. Abbreviation used in this figure: *ACLY* ATP-citrate lyase

2.1.2 The First Committed Step in Fatty Acid Synthesis: Synthesis of Malonyl-CoA from Acetyl-CoA, Catalyzed by Acetyl-CoA Carboxylase (ACC)

With acetyl-CoA now available as a starting factor in the cytoplasm, fatty acid synthesis can begin. The first step in fatty acid synthesis is the synthesis of malonyl-CoA from acetyl-CoA [5, 6, 9]. This step, the formation of malonyl CoA, is the committed step in fatty acid synthesis, and malonyl CoA is the first intermediate formed. In this reaction, acetyl-CoA (a 2-carbon molecule attached to CoA) is carboxylated to yield malonyl-CoA (a 3-carbon molecule attached to CoA).

This reaction (the formation of malonyl-CoA from acetyl-CoA) is catalyzed by the enzyme

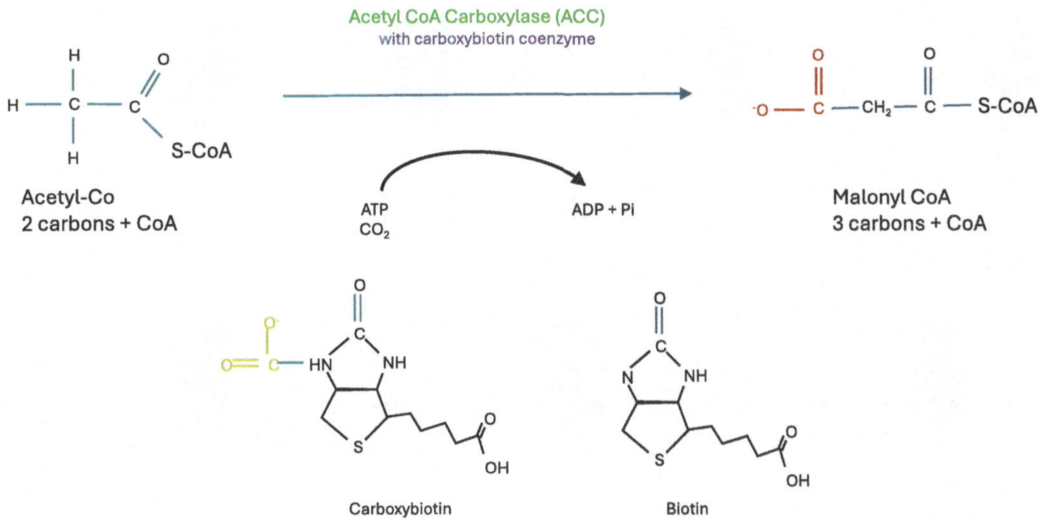

Fig. 11.8 Conversion of acetyl-CoA to malonyl-CoA. This is the first committed step in the process of fatty acid synthesis. It is catalyzed by acetyl-CoA carboxylase (ACC), which uses the coenzyme carboxybiotin

acetyl-CoA carboxylase (ACC) in a reaction that requires ATP. **ACC** requires carboxybiotin, a carboxylated form of Biotin (vitamin B7) as a prosthetic group. In the reaction, biotin temporarily binds a CO_2 group, becoming carboxybiotin, before transferring it to the acetyl-CoA to form malonyl-CoA. After this transfer, biotin is restored to its original form, and acetyl-CoA becomes malonyl-CoA (see Fig. 11.8). **ACC** is a very important enzyme because it catalyzes this first committed step in fatty acid synthesis. It is highly regulated, and its regulation will be discussed in more detail later.

2.1.3 Fatty Acid Elongation: Multiple Rounds of Reactions Lengthen the Fatty Acid Chain, Catalyzed by the Fatty Acid Synthase-1 (FAS-1) Complex

- The FAS-1 enzyme complex has a key role in FA biosynthesis

The next series of reactions is catalyzed by the **fatty acid synthase-1 (FAS-1) complex** (see Fig. 11.9). Both acetyl-CoA and malonyl-CoA can bind to **FAS-1**. **FAS-1** is a multi-enzyme complex that catalyzes the synthesis of fatty acids by allowing for the successive addition of two carbon units to a growing fatty acid chain. The **FAS-1** complex has two critical functional components:

The **first** is a **cysteine (Cys)** residue that has a thiol/sulfhydryl (-SH) group directly attached to it. This group is part of a **KS** (β-ketoacyl synthase) domain (a domain that catalyzes the condensation step in fatty acid synthesis). It is labeled Cys-S in Fig. 11.9.

The **second** is an **acyl-carrier protein (ACP),** which also has a thiol group on it, which is bound to phosphopantetheine, a derivative of vitamin B5. The **ACP** group acts as a swinging arm that can transfer intermediates between active sites on this complex. It is labeled ACP-S in Fig. 11.9.

FAS-1 is a complex, and it contains multiple distinct enzyme domains, each of which is responsible for a specific step in the fatty acid synthesis cycle. One of these domains is the enzyme **malonyl/acetyltransferase (MAT),** which is responsible for loading both **acetyl-CoA** and **malonyl-CoA** onto the ACP domain, reactions that will be described below.

The **FAS-1 complex** catalyzes multiple cycles of reactions as the fatty acid chain grows. In each cycle (except the first), two carbons are added to the fatty acid chain, using acyl-CoA and malonyl-CoA as substrates. Here is a breakdown of the

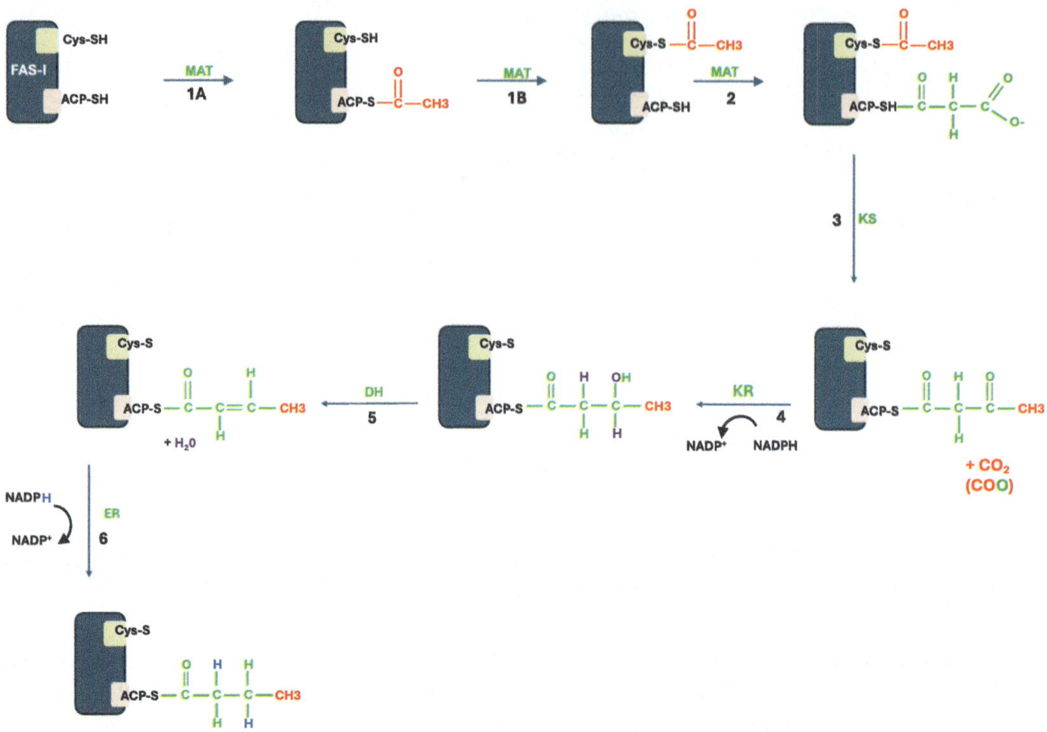

Fig. 11.9 Fatty acid elongation. Fatty acid elongation is catalyzed by the fatty acid synthase-1 (FAS-1) complex. See text for details. Abbreviations used in this figure: *FAS-1* fatty acid synthase-1, *MAT* malonyl/acetyltransferase, *KS* beta ketoacyl synthase, *KR* beta ketoacyl ACP reductase, *DH* 3-hydroxyacyl-ACP dehydratase, *Cys* cysteine, *ACP* acyl-carrier protein

steps, divided into initiation phase, elongation phase, and termination phase [5, 9, 21].

- Steps in Fatty Acid Biosynthesis Steps Catalyzed by the FAS-1 Complex
  - Initiation Phase.

Step 1): **Loading acetyl-CoA onto the ACP group on FAS-1**

The first two steps happen in quick succession, one after another, and are sometimes considered one step:

First, **two carbons** are transferred from an **acetyl-CoA** molecule to the **ACP domain** of the **FAS-1 complex**. **Second**, and in quick succession, these two carbons are transferred to the thiol group of the **KS domain** (via the **Cys** residue) of FAS-1. Both steps are catalyzed by **malonyl/acetyltransferase (MAT)**. The CoA is removed in this process.

Step 2): **Loading malonyl-CoA onto the ACP domain**

The **Malonyl group (3C)** from malonyl-CoA, generated as described above, now binds to FAS-1 at the **ACP** domain, which is now unoccupied. The CoA group of malonyl-CoA is now removed. This step is also catalyzed by the **MAT** activity of FAS-1.

- Elongation Phase (Repeated 7 Times).

Step 3): **Condensation reaction**

One of the carbons (the third carbon) from the **ACP** site (which contains the carbons from the malonyl group) is released as **CO_2**. The remaining malonyl-derived carbons now condense with the two carbons from the acetyl-group carbons on the **Cys** site. This leads to the formation of a new 4-carbon chain, which is transferred to the **ACP** group. This is a condensation reaction, catalyzed by the **KS domain (-β-ketoacyl-ACP synthase) of the FAS-1 enzyme (KS).**

Step 4): **Reduction of the β-keto group**

In this **reduction reaction,** a β-keto group (containing C=O) on the carbons linked to ACP is reduced to a hydroxyl (-OH) group. This reaction is catalyzed by the enzyme **β-ketoacyl ACP reductase (KR)**, and in the process, one NADPH is converted to $NADP^+$ [9].

Step 5): **Dehydration reaction**

Next is a **dehydration reaction.** H_2O is removed from the molecule, and a double bond is formed between two carbons. This is catalyzed by **3-hydroxyacyl-ACP dehydratase (DH).**

Step 6): **Reduction of the double bond**

The next reaction is a reduction reaction, which uses NADPH as the reducing agent. NADPH donates a hydride (H^-). This hydride (from the aqueous environment) and a proton are added to the growing fatty acid chain. This reaction is catalyzed by **enoyl ACP reductase (ER)** [9]. At the end of this process, a 4-carbon saturated fatty acyl chain is now attached to the **ACP**. This is called **butyryl-ACP** (because it is still bound to the ACP domain).

Step 7): **Repetition**

The remaining steps effectively repeat the above seven steps. **MAT** (see step 1) transfers the growing acyl-chain from the ACP domain to the Cys-SH site, and a new malonyl-CoA is added to the ACP domain. During the next condensation reaction, this new malonyl-CoA (also synthesized from acetyl-CoA, as above) will donate two carbons to the 4-carbon chain in another condensation reaction (and release one carbon as CO_2), extending the chain from four carbons to six, and so on. The series of reactions continues to repeat, and at each step, **two carbons** are added to the growing hydrocarbon chain. In mammals, this elongation process typically results in the **formation of palmitate**, a fatty acid with **16 carbons,** which is produced through 7 cycles of reactions.

– Termination phase

Step 8): **Release**

Palmitate is released from the enzyme in a reaction catalyzed by **thioesterase**. Later, **palmitate** can be shortened or elongated to form other fatty acids of different sizes, which require different enzymes.

• Summary of the fatty acid elongation steps

The reactions described above for synthesizing fatty acids require NADPH and ATP. ATP is provided by glucose metabolism. NADPH is supplied by the pentose phosphate pathway (discussed in Chap. 15) and by **malic enzyme (ME)**, as described in the next section.

2.1.4 The Fate of the Cytoplasmic Oxaloacetate (OAA) in the Citrate Shuttle

This ends the synthesis of fatty acids, but another point remains to be considered, which relates to OAA in the citrate shuttle. As discussed above, cytoplasmic acetyl-CoA, which is critical for the synthesis of fatty acids, accumulates due to the **citrate shuttle**. As a result of the citrate shuttle, citrate enters the cytoplasm and gets converted to acetyl-CoA and OAA. The cytosolic acetyl-CoA is used for fatty acid synthesis, but what happens to the **cytoplasmic OAA** that also results from the citrate shuttle? This cytoplasmic OAA must be processed, and OAA needs to be returned to the mitochondria to complete the cycle encompassing the citrate shuttle. There are several steps involved in the processing of OAA, as described below.

Cytoplasmic **OAA** is first reduced to cytosolic **malate** in a reaction catalyzed by **cytosolic malate dehydrogenase (MDH1)**, in the following reactions:

$$\text{OAA} + \text{NADH} + \text{H}^+ \rightarrow \textbf{malate} + \text{NAD}^+$$

The resulting **malate** has two possible fates:

1): Cytoplasmic **malate** can enter the **mitochondria** through a transporter, which is a malate-αKG antiporter. (OAA cannot cross the mitochondrial membrane.) Once there, malate can be converted to **OAA,** completing the citrate shuttle. This reaction is:

$$\text{Malate} + \text{NAD}^+ \rightarrow \textbf{OAA} + \text{NADH} + \text{H}^+$$

2): Another option is that cytosolic malate can be converted to pyruvate in the cytoplasm in a reaction catalyzed by **malic enzyme (ME)**. This reaction generates NADPH and releases one CO_2, as follows:

$$\text{Malate} + \text{NADP}^+ \rightarrow \text{Pyruvate} + \text{CO}_2 + \text{NADPH}$$

Pyruvate can then pass into the mitochondria via the **mitochondrial pyruvate carrier (MPC)**, and in order to complete the cycle, pyruvate can be converted to mitochondrial OAA, in a reaction catalyzed by **pyruvate carboxylase**: The reaction can be summarized as:

$$\text{Pyruvate} + \text{CO}_2 + \text{ATP} \rightarrow \textbf{OAA} + \text{ADP} + \text{Pi}$$

The complete citrate shuttle, including the recycling of OAA back to the mitochondria, is shown in Fig. 11.7.

The generation of NADPH (See 2 above) is an important outcome of the citrate shuttle. NADPH is essential for fatty acid synthesis [5, 9] (see step 4 in fatty acid elongation). In addition to supplying acetyl-CoA, the citrate shuttle also produces NADPH, which is needed for fatty acid synthesis. (Another major source of NADPH is the pentose phosphate pathway, which will be discussed in Chap. 15.)

2.2 Glycerol-3-Phosphate Synthesis and Assembly of Triacylglycerols (TAGs)

2.2.1 Triacylglycerol (TAG) Synthesis Starts with Glycerol-3-Phosphate

Once the **fatty acids** are made, they are combined with **glycerol-3-phosphate (G3P)** and assembled into **TAGs** in a series of esterification reactions. This occurs largely in the liver and adipose tissue.

Glycerol-3-phosphate is composed of a glycerol molecule with one phosphate residue, which will eventually be removed when the TAGs are formed. Glycerol-3-phosphate, rather than glycerol, is used as the starting backbone for fatty acids because its phosphate group activates it for acylation with fatty acids, and it is the required substrate for the enzymes that catalyze TAG assembly. Glycerol-3-phosphate synthesis can occur in most tissues, but its synthesis in the liver and fat is especially important for the synthesis of TAGs. Glycerol-3-phosphate can be synthesized differently in these tissues.

In **adipose tissue**, the glycerol-3-phosphate is all generated from the glycolysis intermediate DHAP, in a reaction catalyzed by **glycerol-3-phosphate-dehydrogenase (GDPH)** (see Fig. 11.2b, and Chap. 10, Fig. 10.13). In this reaction, DHAP is reduced to glycerol-3-phosphate, and in the process, NADH is oxidized to NAD^+:

$$\text{DHAP} + \text{NADH} + \text{H}^+ \rightleftharpoons \text{glycerol-3-phosphate} + \text{NAD}^+$$

In the **liver**, **glycerol-3-phosphate** can be made by two mechanisms. (**a**) It can also be made from DHAP, as above, also catalyzed by **GPD1**, or (b) in the liver, glycerol-3-phosphate can also be made by direct phosphorylation of glycerol. This is catalyzed by **glycerol kinase (GK)**, which is present in the liver but not in adipose tissue.

DHAP is important for fatty acid synthesis because it is an intermediate in glycerol-3-phosphate synthesis [5, 22]. It should be noted that DHAP can serve as a sort of branch point; it can (a) be made into glyceraldehyde-3-phosphate (GAP), a glycolytic intermediate, or (b) it can be made into glycerol-3-phosphate (G3P), a lipid synthesis (TAGs) intermediate. While glycolysis leads to the breakdown of glucose and the production of energy in the form of ATP, lipid synthesis leads to a storage form of energy.

2.2.2 Esterification of Fatty Acids and the Glycerol Backbone to Produce Triacylglycerols (TAGs)

The esterification (assembly) of TAGs from glycerol-3-phosphate happens in both the liver and adipose tissue. In **liver**, the newly synthesized

TAGs are packaged into VLDLs, which are exported, mostly to peripheral tissues such as adipose tissue and muscle. In **adipose tissue**, fatty acids released from VLDL hydrolysis are re-esterified to TAGs and stored. (The glycerol released from VLDL hydrolysis is not reused by adipose tissue for TAG synthesis because adipocytes lack glycerol kinase; this glycerol returns to the liver. Instead, adipose tissue obtains the glycerol-3-phosphate backbone from glycolysis intermediates.) In other tissues like muscle, these fatty acids are not stored as TAGs but are mostly oxidized to generate ATP for energy [6, 23]. In this section, we focus on the pathway by which fatty acids are re-esterified with glycerol-3-phosphate to form TAGs.

Fatty acids are combined with **glycerol-3-phosphate** in a multi-step process to form **TAGs**, as follows (Fig. 11.10): The **fatty acids** are first activated to **fatty acyl-CoA** in a reaction catalyzed by **acyl-CoA synthetase (ACS)** (also called **fatty acid thiokinase**). In this reaction, CoA is covalently attached to the fatty acid via a thioester bond, resulting in fatty acyl-CoA in the reaction: fatty acid + CoA + ATP → fatty acyl - CoA + AMP + PPi. (This is the same reaction that takes place when fatty acids are activated before their oxidation, although a different ACS isoform is used, with a different cellular location.)

Once activated, the fatty acyl-CoAs are then **esterified**, one at a time, to the hydroxyl groups of **glycerol-3-phosphate**. This process is catalyzed in several steps by different **acyltransferases**.

First, **lysophosphatidic acid (LPA)** is formed, where the **first fatty acyl-CoA** is added to the glycerol-3-phosphate (catalyzed by **glycerol-3-phosphate acyltransferase (GPAT)**).

Next, a **second fatty acyl-CoA** is added, resulting in **phosphatidic acid (PA)**. This reaction is catalyzed by **1-acyl-glycerol-3-phosphate acyltransferase (AGPAT)**. After this, **phosphatidic acid phosphatase (PAP)** catalyzes the removal of the phosphate group from the glycerol-3-phosphate backbone, leading to the formation of **DAG**.

Finally, the **third fatty acyl-CoA** is added to the complex, resulting in **the final TAG complex**. This reaction is catalyzed by **diacylglycerol acyltransferase (DGAT)**.

All of the above reactions involve the removal of the CoA from the fatty acyl-CoA and formation of an ester bond between the fatty acid and a hydroxyl group of the glycerol-3-phosphate [24–26].

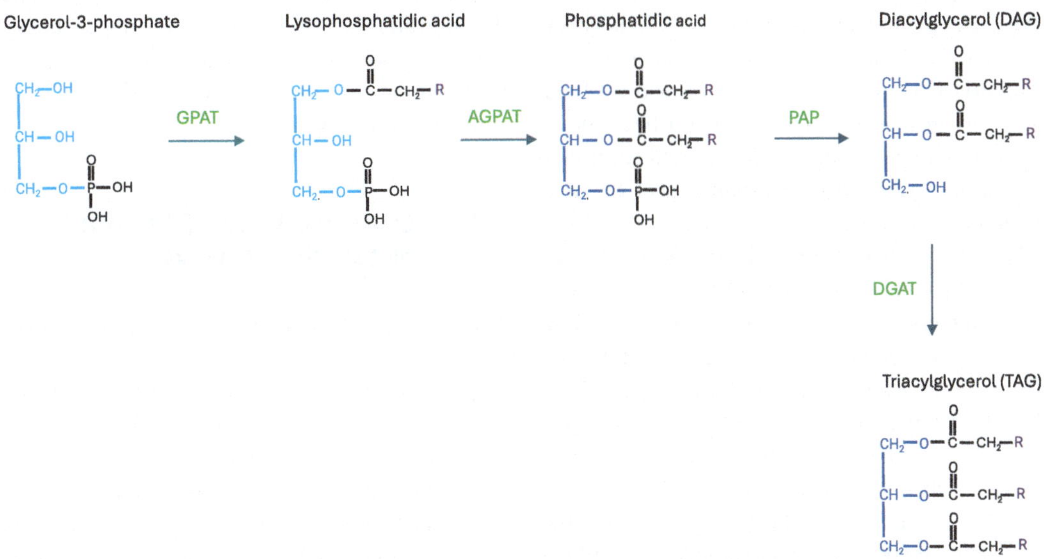

Fig. 11.10 Esterification of fatty acids and the glycerol backbone to produce triacylglycerol (TAG). See text for details. Abbreviations used are *GPAT* glycerol-3-phosphate acyltransferase, *AGPAT* 1-acyl-glycerol-3-phosphate acyltransferase, *PAP* phosphatidic acid phosphatase, *DGAT1* diacylglycerol acyltransferase

This pathway is called **the Kennedy pathway** of TAG biosynthesis. It was discovered in the 1950s by Kennedy and Weiss, who were working with rat liver extracts. They used radiolabeled glycerol and fatty acids to follow the fates of these molecules. From these studies, it was discovered that glycerol-3-phosphate serves as the starting backbone for lipid assembly and that fatty acyl-CoAs are added to this backbone sequentially. Before their study, lipid synthesis was thought to be a more random process. Their work led to the understanding that it is, in fact, an ordered process catalyzed by specific enzymes. This was an important foundation for understanding fat storage and lipid metabolism [25, 27, 28].

2.3 Fatty Acid Synthesis Is a Highly Regulated Process Due Largely to the ACC Enzyme

Synthesis of fats is a highly regulated process. It is important that the body does not synthesize fat unless it is needed, and that it can break down fat when energy levels are low. Much of this regulation is at the level of fatty acid synthesis. Fatty acid synthesis is tightly regulated and occurs only under specific conditions. A major source of regulation of fatty acid synthesis is the enzyme **acetyl-CoA carboxylase (ACC)** [5, 6, 9, 29]. This is the enzyme that catalyzes the synthesis of **malonyl-CoA** (Fig. 11.8), an early and essential step in fatty acid synthesis, described above. **ACC** is a tightly regulated enzyme and controlled at various levels. An important aspect of ACC regulation is its structural regulation. **ACC** can take on two main forms. It can form dimers, where two ACC proteins bind to each other. ACC dimers are inactive; they cannot catalyze the carboxylation of acetyl-CoA. However, these dimers can also self-polymerize and form larger enzyme complexes. These polymerized complexes, unlike the dimers, are catalytically active. The polymerization of the ACC dimers can be modulated by allosteric regulators. For example, **citrate** is a **positive** allosteric regulator of **ACC**, promoting its polymerization into its active form. Cytoplasmic citrate is an important precursor of fatty acid synthesis, as discussed above, because it can lead to acetyl-CoA formation. Thus, high levels of citrate signal an abundance of precursors for fatty acid synthesis and therefore encourage the activation of **ACC**. In contrast, LCFAs (long-chain fatty acids) inhibit **ACC** by promoting its depolymerization into inactive dimers. This prevents excess fatty acid synthesis under conditions where fatty acid levels are already high.

ACC is also regulated by **hormonal** signals [4, 5, 9, 30–32]. For example, **insulin** (the levels of which increase after glucose consumption) is a **positive** regulator of **ACC**. It stimulates the dephosphorylation of the enzyme by stimulating the phosphatase PP2A. This favors its polymerization and activation. In contrast, **glucagon** (whose levels rise during fasting) **inhibits ACC** and therefore inhibits fatty acid synthesis. **Glucagon** works by activating the enzyme **protein kinase A (PKA),** which phosphorylates and inhibits **PP2A**, thus preventing **ACC** activation. These hormonal signals ensure that **ACC** is active in the fed state but inactive during fasting. ATP levels also impact ACC activity. **Low ATP levels** (corresponding to **high AMP** levels) can result from fasting or heavy exercise. This leads to activation of **AMP-activated protein kinase (AMPK)**. **AMPK** phosphorylates **ACC**, and this **inhibits** its activity. Thus, fasting or heavy exercise also inhibits fatty acid synthesis [33]. Some of the regulators of ACC are summarized in Fig. 11.11.

2.4 Crosstalk Among Different Metabolic Pathways

The coverage of lipid metabolism in this chapter highlights the multiple points of crosstalk between different metabolic pathways. Some key fatty acid metabolism intermediates discussed so far have important roles in other metabolic pathways, such as glucose metabolism and amino acid metabolism. For example, as discussed above, DHAP is a precursor for glycerol-3-phosphate in lipid metabolism, but it is also an intermediate in glycolysis and

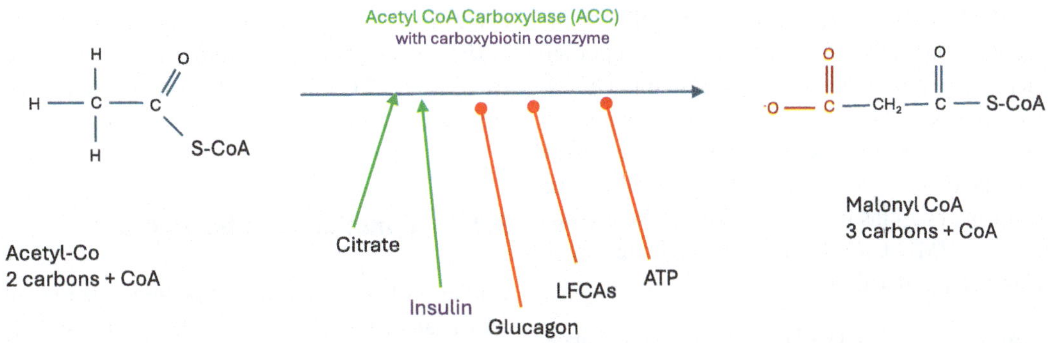

Fig. 11.11 Regulation of Acetyl CoA Carboxylase (ACC); ACC is a highly regulated enzyme. Making fatty acid synthesis highly regulated. Several factors that regulate ACC are illustrated here, including the hormones insulin and glucagon. (LFCA, Long chain fatty acids)

gluconeogenesis (synthesis of glucose). Likewise, acetyl-CoA is an important product of glycolysis and also serves as the key building block of lipids. Krebs cycle intermediates, such as citrate and oxaloacetate (OAA), link carbohydrate, lipid, and amino acid metabolism. Citrate exported to the cytosol, via the citrate shuttle, provides acetyl-CoA, which is needed for fatty acid synthesis, and the completion of the shuttle leads to the production of NADPH, which is also necessary for fatty acid synthesis. The citrate shuttle also leads to cytosolic OAA, which contributes to the generation of NADPH for fatty acid synthesis. OAA also has a key role in amino acid metabolism. In amino acid metabolism, OAA can be converted to the amino acid Aspartate (ASP) by **aspartate transaminase**, and ASP can be converted to the amino acid Asparagine (**ASN**). ASP and ASN can be used in protein synthesis, but ASP can also cross the mitochondrial membrane via transporters and be converted back to mitochondrial OAA by **mitochondrial aspartate transaminase**, thus contributing to the Krebs cycle. All of these connections, along with others, illustrate that the various metabolic pathways are intricately linked.

3 Cholesterol Metabolism

3.1 Dietary Cholesterol

Cholesterol, a sterol, is a lipid, and therefore its metabolism is briefly discussed here (although cholesterol and inhibitors of cholesterol synthesis are also discussed in other chapters throughout the book). Cholesterol is consumed in our diet, and when ingested, it is transported throughout the body in chylomicrons. These are large lipoprotein particles produced in the intestine after a meal to transport dietary fat and cholesterol.

3.2 De Novo Synthesis of Cholesterol

Cholesterol is also synthesized in the body de novo, mainly in the liver, with some synthesis occurring in other tissues such as the intestines [34]. The cholesterol biosynthesis pathway involves multiple steps. Similar to fatty acid synthesis, **acetyl-CoA** has a key role in cholesterol biosynthesis as a starting substrate (see Fig. 11.11). Cholesterol biosynthesis begins when there is excess **acetyl-CoA,** such as during the fed state after a meal. Two acetyl-CoA molecules bind together, forming **acetoacetyl-CoA**, in a reaction catalyzed by the enzyme **thiolase**. One CoA is released in the process. This is followed by the addition of a third acetyl-CoA molecule and the release of another CoA to lead to the formation of **HMG-CoA** (3-hydroxy-3-methylgutaryl-CoA), which is a 6-carbon molecule attached to CoA. This reaction is catalyzed by the enzyme **HMG-CoA synthase.**

HMG-CoA is the precursor of a multistep pathway that leads to cholesterol biosynthesis. The first step after **HMG-CoA** is formed is its reduction into **mevalonate,** a reaction catalyzed by **HMG-CoA reductase**, using NADPH as the reducing agent. This reaction is a very important step because it is the rate-limiting step in cholesterol biosynthesis, and it is tightly regulated. **HMG CoA reductase** is an integral membrane protein in the ER membrane. Its catalytic site faces the cytoplasm, and it catalyzes the conversion of HMG-CoA to mevalonate in the cytoplasm in the reaction:

$$\text{HMG-CoA} + 2\,\text{NADPH} + 2\,\text{H}^+ \rightarrow \text{Mevalonate} + 2\,\text{NADP}^+ + \text{CoA-SH}$$

This reaction uses 2 NADPHs and two H^+s and results in 2 $NADP^+$s.

After Mevalonate is formed, a series of enzymatic reactions finally leads to cholesterol synthesis. There are about 21 steps in total from HMG-CoA synthesis to cholesterol. The steps from acetyl-CoA to mevalonate, and the final structure of cholesterol, are outlined in Fig. 11.12.

3.3 Functions of Cholesterol

Cholesterol has multiple functions, some of which are covered in Chap. 10. It can become incorporated into cell membranes, where it regulates membrane fluidity and stability. It is also a precursor for the synthesis of other molecules, including bile salts and Vitamin D, as well as hormones such as estrogen and testosterone [8, 14].

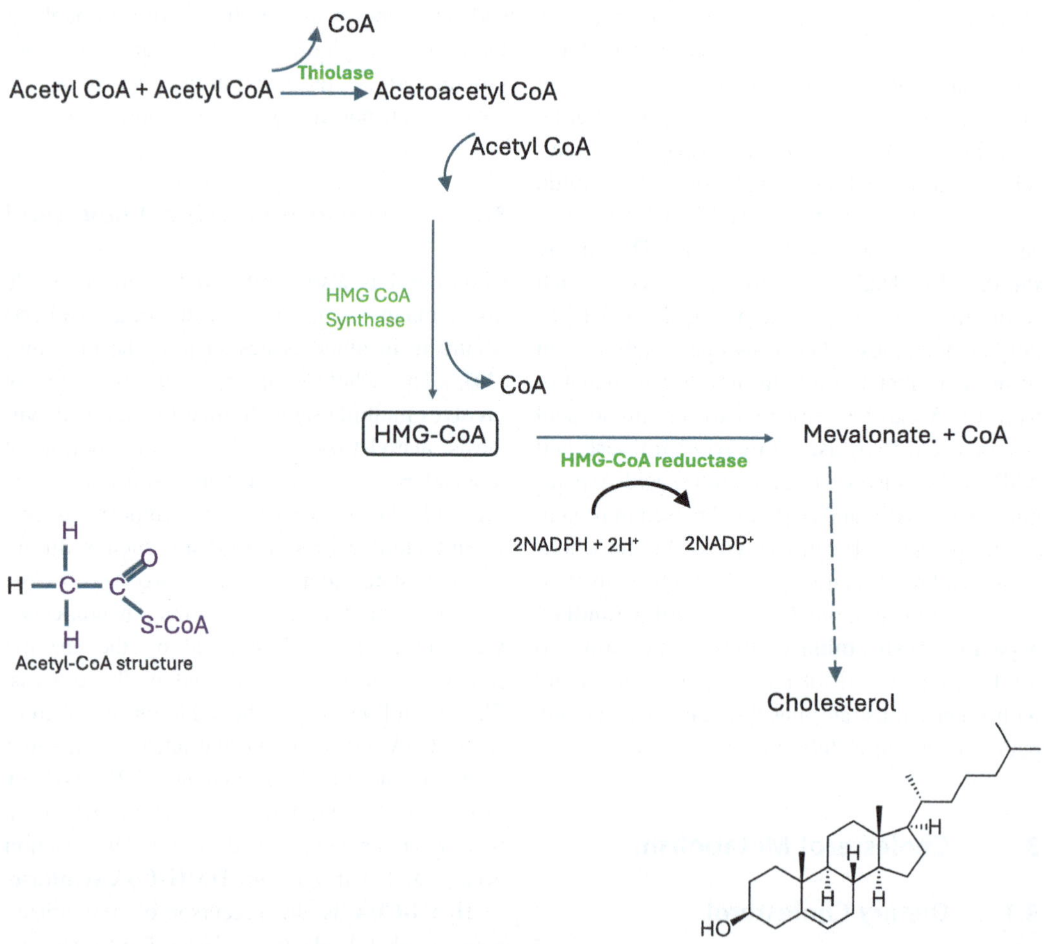

Fig. 11.12 Cholesterol synthesis. Cholesterol is synthesized via a multistep process from acetyl-CoA. Several of the steps are shown here. **Source**: Cholesterol structure from Vecteezy.com (Pro license)

3.4 Packaging of Cholesterol

Once it is made, cholesterol can be packaged into particles (low-density lipoprotein (LDL) particles), which transport cholesterol esters (cholesterol molecules with a fatty acid attached, making them more hydrophobic—their main cargo), and smaller amounts of TAGS, other lipids, and some protein. LDL delivers cholesterol from the liver to different parts of the body. Cholesterol can also be packaged into very low-density lipoprotein (**VLDL)** particles, which are also secreted by the liver. VLDL particles have a higher proportion of TAGs than LDL particles do. As they circulate, **lipoprotein lipase** removes TAGs, converting the LDLs first into intermediate density lipoprotein (IDL) and eventually into LDL particles [5, 6, 9].

3.5 The Importance of the HMG CoA Reductase Enzyme

HMG CoA reductase is an important enzyme because, as discussed above, it catalyzes the **rate-limiting step** in cholesterol biosynthesis [6]. Consequently, **HMG-CoA reductase** is an effective target for several cholesterol-lowering drugs. Specifically, **statins** are a family of drugs that act as competitive inhibitors of **HMG-CoA reductase** and are used to treat patients with **high cholesterol**.

HMG-CoA reductase is subject to normal regulation in the body by several different mechanisms. One of these is feedback regulation, where high levels of cholesterol will inhibit its enzymatic activity. For example, high cholesterol can lead to the activation of a family of proteins called **Insigs** (inhibitors of sterol regulatory element binding protein (SREBP)), which are located on the surface of the ER. **Insigs** can, in turn, regulate **HMG-CoA reductase** by different mechanisms. For example, Insigs can bind to **HMG-CoA reductase**, leading to its ubiquitination and degradation. This consequently reduces cholesterol synthesis.

Insigs also work by regulating the activity of the **SREBP transcription factors**. SREBPs are transcription factors that bind the promoters of various target genes. One of these, SREBP-2, regulates transcription of the gene encoding **HMG-CoA reductase**. SREBPs are synthesized as inactive precursors in the ER, and these inactive precursors are bound to SCAP (SREBP cleavage-activating protein). SCAP has a sterol-sensing domain (SSD) that can detect cholesterol levels in the cell. When cholesterol levels are **low**, SCAP, together with SREBP, leaves the ER, and SCAP transports SREBP to the Golgi, where it is cleaved by proteases. This cleavage changes SREBP to its active conformation. Active SREBP translocates to the nucleus and binds to the sterol response elements (SREs) in the promoters of target genes, including the HMG-CoA reductase gene. This increases transcription of the HMG-CoA reductase gene, thereby increasing the levels of HMG-CoA reductase protein. This, in turn, increases cholesterol biosynthesis. In contrast, when cholesterol levels are high, SCAP binds to Insigs, causing the SCAP-SREBP complex to be retained in an inactive state in the ER, where it is unable to activate the expression of HMG-CoA reductase [34, 35].

In addition to these regulatory mechanisms, high cholesterol can activate pathways that lead to the activation of kinases such as **AMP-activated protein kinase (AMPK)**. AMPK can phosphorylate **HMG-CoA reductase**, and this reduces its enzymatic activity. Consequently, when cholesterol levels are high, AMPK helps shut down the production of more cholesterol by inhibiting the activity of HMG-CoA reductase [5, 9, 12, 36].

4 Balanced Equations in Lipid Metabolism

4.1 Balanced Equations for the Conversion of Glycerol to Glycolytic Intermediates

1. Phosphorylation of glycerol

Glycerol + ATP → glycerol-3-phosphate + ADP

Enzyme: **glycerol kinase (GK).**

2. Oxidation of glycerol-3-phosphate

$$\text{Glycerol-3-phosphate} + NAD^{+} \rightarrow \text{DHAP} + NADH + H^{+}$$

Enzyme: **glycerol-3-phosphate dehydrogenase (GPD1).**

3. Isomerization of DHAP to glyceraldehyde-3-phosphate

$$\text{DHAP} \rightleftharpoons \text{Glyceraldehyde-3-phosphate}$$

Enzyme: **triose phosphate isomerase (TPI).**

Definitions

Acetyl-CoA carboxylase (ACC) An Enzyme that catalyzes the rate-limiting step in fatty acid synthesis that converts acetyl-CoA to malonyl-CoA.

Acyl carrier protein (ACP) A domain within the FAS-1 enzyme complex that holds the growing fatty acid chain and transfers it between catalytic sites.

Acyl-CoA aynthetase (ACS) Enzyme that catalyzes the activation of fatty acids by converting them to fatty acyl-CoA.

Acyl adenylate intermediate An activated fatty acid intermediate that is formed when a fatty acid reacts with ATP during fatty acid activation.

Adipose triglyceride lipase (ATGL) Enzyme that initiates lipolysis by cleaving TAGs into diacylglycerol (DAG) and free fatty acid (FFA).

Beta oxidation of fatty acids The breakdown of fatty acids, mostly to acetyl-CoA, with the generation of ATP, as well as NADH and $FADH_2$.

Bile salts Amphipathic molecules derived from cholesterol that emulsify fats in the small intestine for digestion.

Carnitine acylcarnitine translocase (CACT) Transports fatty acyl-carnitine across the inner mitochondrial membrane.

Carnitine Palmitoyltransferase I (CPT-I) Transfers the fatty acyl group from CoA to carnitine.

Carnitine palmitoyltransferase II (CPT-II) Transfers the fatty acyl group from carnitine back to CoA in the mitochondrial matrix.

Carnitine shuttle A system that transports long-chain fatty acids into the mitochondria for beta oxidation.

Citrate shuttle A transport system that moves citrate from the mitochondria to the cytoplasm to provide acetyl-CoA for fatty acid synthesis.

Citrate The deprotonated (anion) form of citric acid that predominates at physiological pH (around 7.4). Citrate is a Krebs cycle intermediate, and when acetyl-CoA combines with oxaloacetate, it forms citrate.

Citric acid The **protonated** form of citrate. Citric acid is a weak organic acid that has three carboxyl groups (-COOH). (In its pure form, it is usually found in citrus fruits like lemons and limes.)

De Novo lipogenesis The synthesis of fatty acids from non-lipid precursors like glucose, typically in the liver.

Diacylglycerols/diglycerides (DAG) Two fatty acids esterified to a glycerol backbone. The third hydroxyl group on the glycerol remains free (unesterified).

Diacylglycerol acyltransferase (DGAT) The enzyme that catalyzes the final step in triglyceride synthesis by attaching a third fatty acid to DAG.

Esterification A condensation reaction in which an alcohol and an acid react to form an ester bond, and water is released. Esterification is used to form triacylglycerols (TAGs) from glycerol and fatty acids.

Fatty acid synthase (FAS) A multi-enzyme complex responsible for synthesizing fatty acids. Palmitate is formed first, and other fatty acids are made from it.

Fatty acid activation The process in which fatty acids are converted to fatty acyl-CoA by attaching CoA, making them ready for β oxidation.

Fatty acyl-CoA Fatty acid covalently attached to coenzyme A; activated form of fatty acid for metabolism.

Glycerol-3-phosphate (G3P) A phosphorylated form of glycerol, and a precursor for triacylglycerol (TAG) synthesis.

HMG-CoA reductase The rate-limiting enzyme in cholesterol biosynthesis, and the target of statin drugs.

Hormone-sensitive lipase (HSL) Enzyme that breaks down DAG into MAG and FFA during lipolysis.

Insig proteins Regulatory proteins that inhibit cholesterol synthesis by promoting degradation of HMG-CoA reductase.

Kennedy pathway A pathway for TAG synthesis from glycerol-3-phosphate and fatty acyl-CoA involving acyltransferases.

Ketogenesis The metabolic pathway that produces ketone bodies from acetyl-CoA during periods of low carbohydrate availability.

Ketone bodies Alternative energy molecules produced in the liver from acetyl-CoA during fasting or starvation.

Low-density lipoprotein (LDL) See Chap. 13.

Lipolysis The breakdown of Triacylglycerols (TAGs) into free fatty acids and glycerol.

Lipoprotein lipase Enzyme located on capillary endothelium that hydrolyzes TAGs in chylomicrons into fatty acids and glycerol.

Malate aspartate shuttle A mechanism that transfers electrons from NADH in the cytoplasm into the mitochondria for oxidative phosphorylation.

Mevalonate A product formed from HMG-CoA during cholesterol synthesis, requiring NADPH as a reducing agent.

Malonyl-CoA A key intermediate in fatty acid synthesis that also inhibits the carnitine shuttle (thereby inhibiting fatty acid oxidation).

Micelles Small spherical structures formed by amphipathic molecules such as bile salts and fatty acids, with hydrophobic tails facing inward and hydrophilic regions on the outside. They help emulsify dietary fats for absorption.

Monoacylglycerols (MAGS) A Glycerol molecule with a single fatty acid chain attached via an ester bond.

Monoglyceride Lipase (MGL) Enzyme that catalyzes the final step of lipolysis, releasing glycerol and the last free fatty acid.

Non-alcoholic steatohepatitis (NASH) A severe form of NAFLD involving liver inflammation and damage.

Non-alcoholic fatty liver disease (NAFLD) A condition of fat accumulation in the liver not due to alcohol intake. It may progress to NASH.

Palmitate A 16-carbon saturated fatty acid, and the end product of the fatty acid synthesis pathway.

Propionyl-CoA A 3-carbon intermediate produced from β oxidation of odd-chain fatty acids; converted into succinyl-CoA.

Thiolysis A chemical reaction in which a thiol group (–SH) acts as a nucleophile to cleave a bond, typically an ester or thioester bond, resulting in the formation of a new thioester and the release of another product. This reaction is catalyzed by thiolase enzymes.

Triglyceride/triacylglycerol (TAG) The main form of stored fat in adipose tissue, composed of three fatty acids attached to a glycerol backbone.

Very low-density lipoprotein (VLDL) A lipoprotein synthesized in the liver that transports TAGs produced in the liver to tissues.

Study Questions

Multiple-Choice Questions

1. Which enzyme catalyzes the rate-limiting step in cholesterol biosynthesis?
   A. HMG-CoA synthase
   B. Acetyl-CoA carboxylase
   C. HMG-CoA reductase
   D. Fatty acid synthase
2. Which molecule is the starting point for fatty acid synthesis?
   A. Glycerol
   B. Palmitate
   C. Acetyl-CoA
   D. Cholesterol
3. What role does carnitine play in fat metabolism?
   A. Oxidizes fatty acids
   B. Stores fatty acids
   C. Transports fatty acids into mitochondria
   D. Synthesizes fatty acids

4. Which of the following is produced during each cycle of β oxidation?
   A. Glycerol
   B. NADPH
   C. Acetyl-CoA
   D. Cholesterol
5. Which enzyme catalyzes the conversion of glycerol to glycerol-3-phosphate?
   A. GPD1
   B. GK
   C. CPT-I
   D. ACLY
6. What is the function of CPT-I in fatty acid metabolism?
   A. Generates ATP from acetyl-CoA
   B. Catalyzes transfer of the fatty-acyl group from fatty-acyl CoA to carnitine
   C. Forms triacylglycerols from fatty acids
   D. Breaks down triglycerides to fatty acids
7. What is the major purpose of the citrate shuttle?
   A. Move fatty acids into mitochondria
   B. Move citrate across the ER membrane for proper folding
   C. Move citrate from the mitochondria into the cytoplasm for acetyl-CoA production
   D. To move citrate into the liver
8. Which of the following enzymes directly produces malonyl-CoA?
   A. ACLY
   B. ACC
   C. HMG-CoA Reductase
   D. CPT-II
9. Which lipoprotein particle is responsible for transporting dietary fats from the intestine?
   A. LDL
   B. VLDL
   C. Chylomicrons
   D. HDL
10. Which molecules are generated by each round of beta oxidation? (select all that apply).
   A. NADH
   B. $FADH_2$
   C. Acetyl-CoA
   D. ATP

Short-Answer Questions

11. Explain the role of the citrate shuttle in fatty acid synthesis.
12. In what way are odd-chain fatty acids metabolized differently in the final step of beta oxidation?

Matching Questions

13. Match the enzyme with its function:

Enzyme
   A. GPAT
   B. DGAT
   C. GPD1
   D. GK

Function:
   1. Converts glycerol to G3P
   2. Adds first fatty acid to G3P
   3. Adds final fatty acid to DAG to make TAG
   4. Converts DHAP to G3P

Answer Key

Multiple-Choice Questions

1. **Answer**: C
2. **Answer**: C
3. **Answer**: C
4. **Answer**: C
5. **Answer**: B
6. **Answer**: B
7. **Answer**: C
8. **Answer**: B
9. **Answer**: C
10. **Answer**: A, B, C

Short-Answer Questions

11. **Answer**: The citrate shuttle transports citrate from the mitochondria to the cytoplasm, where it is converted into acetyl-CoA and oxaloacetate. Acetyl-CoA is then used for fatty acid synthesis.

12. **Answer**: Because they yield propionyl-CoA (3-carbon, rather than acetyl-CoA (2-carbon)), which is then converted into succinyl-CoA, an intermediate of the Krebs cycle.

Matching Question

13. **Match the enzyme with its function:**
Answer: A → 2, B → 3, C → 4, D → 1

References

1. Iqbal J, Hussain MM. Intestinal lipid absorption. Am J Physiol Endocrinol Metab. 2009;296(6):E1183–94. https://doi.org/10.1152/ajpendo.90899.2008.
2. Omer E, Chiodi C. Fat digestion and absorption: Normal physiology and pathophysiology of malabsorption, including diagnostic testing. Nutr Clin Pract. 2024;39(S1):S6–S16. https://doi.org/10.1002/ncp.11130.
3. Xu E, Chen C, Fu J, Zhu L, Shu J, Jin M, et al. Dietary fatty acids in gut health: absorption, metabolism and function. Anim Nutr. 2021;7(4):1337–44. https://doi.org/10.1016/j.aninu.2021.09.010.
4. Marks D. Marks' essentials of medical biochemistry: a clinical approach. Lippincott Williams & Wilkins (LWW); 2017.
5. Chandel NS. Navigating metabolism. 2nd ed. Cold Spring Harbor: Cold Spring Harbor Laboratory Press; 2020.
6. Rodwell VW, Bender DA, Kennely PJ, Weil PA, McGuinness OP. Harper's illustrated biochemistry. 32nd ed. New York: McGraw-Hill Education / Medical; 2022.
7. Hall JE, Hall ME. Guyton and hall textbook of medical physiology. 14th ed. Elsevier; 2020.
8. Bender DA. Introduction to nutrition and metabolism. Boca Raton, FL: CRC Press; 2021.
9. Berg JM, Tymoczko JL, Gatto GJ Jr, Stryer L. Biochemistry. 9th ed. New York, NY: MacMillan Learning; 2019.
10. Nielsen TS, Jessen N, Jørgensen JOL, Møller N, Lund S. Dissecting adipose tissue lipolysis: molecular regulation and implications for metabolic disease. J Mol Endocrinol. 2014;52(3):R199–222. https://doi.org/10.1530/jme-13-0277.
11. Adeva-Andany MM, Carneiro-Freire N, Seco-Filgueira M, Fernández-Fernández C, Mouriño-Bayolo D. Mitochondrial β-oxidation of saturated fatty acids in humans. Mitochondrion. 2019;46:73–90. https://doi.org/10.1016/j.mito.2018.02.009.
12. Hames BD, Hooper NM. BIOS instant notes in biochemistry. Taylor and Francis; 2011.
13. Dikalov S, Panov A, Dikalova A. Critical role of mitochondrial fatty acid metabolism in Normal cell function and pathological conditions. Int J Mol Sci. 2024;25(12):6498.
14. Alberts B, Heald R, Johnson A, Morgan D, Raff M, Roberts K, Walter P, Wilson J, Hunt T. Molecular biology of the cell. W.W. Norton and Company; 2022.
15. Schönfeld P, Wojtczak L. Short- and medium-chain fatty acids in energy metabolism: the cellular perspective. J Lipid Res. 2016;57(6):943–54. https://doi.org/10.1194/jlr.R067629.
16. Yan S, Yang XF, Liu HL, Fu N, Ouyang Y, Qing K. Long-chain acyl-CoA synthetase in fatty acid metabolism involved in liver and other diseases: an update. World J Gastroenterol. 2015;21(12):3492–8. https://doi.org/10.3748/wjg.v21.i12.3492.
17. Laffel L. Ketone bodies: a review of physiology, pathophysiology and application of monitoring to diabetes. Diabetes Metab Res Rev. 1999;15(6):412–26.
18. Houten SM, Violante S, Ventura FV, Wanders RJA. The biochemistry and physiology of mitochondrial fatty acid β-oxidation and its genetic disorders. Annu Rev Physiol. 2016;78:23–44. https://doi.org/10.1146/annurev-physiol-021115-105045.
19. Longo N, Frigeni M, Pasquali M. Carnitine transport and fatty acid oxidation. Biochim Biophys Acta. 2016;1863(10):2422–35. https://doi.org/10.1016/j.bbamcr.2016.01.023.
20. Gnoni GV, Priore P, Geelen MJ, Siculella L. The mitochondrial citrate carrier: metabolic role and regulation of its activity and expression. IUBMB Life. 2009;61(10):987–94. https://doi.org/10.1002/iub.249.
21. Chan DI, Vogel HJ. Current understanding of fatty acid biosynthesis and the acyl carrier protein. Biochem J. 2010;430(1):1–19. https://doi.org/10.1042/bj20100462.
22. Salway JG. Metabolism at a glance. Hoboken: Wiley-Blackwell; 2017.
23. Saponaro C, Gaggini M, Carli F, Gastaldelli A. The subtle balance between lipolysis and lipogenesis: a critical point in metabolic homeostasis. Nutrients. 2015;7(11):9453–74. https://doi.org/10.3390/nu7115475.
24. Christie WWTL. Triacylglycerols: 2. Biosynthesis and metabolism. 2025. Available from: https://www.lipidmaps.org/resources/lipidweb/lipidweb_html/lipids/simple/tag2/index.htm.
25. Coleman RA, Mashek DG. Mammalian triacylglycerol metabolism: synthesis, lipolysis, and signaling. Chem Rev. 2011;111(10):6359–86. https://doi.org/10.1021/cr100404w.
26. Coleman RA, Lee DP. Enzymes of triacylglycerol synthesis and their regulation. Prog Lipid Res. 2004;43(2):134–76. https://doi.org/10.1016/S0163-7827(03)00051-1.
27. Kennedy EP, Weiss SB. The function of cytidine coenzymes in the biosynthesis of phospholipides. J Biol Chem. 1956;222(1):193–214.
28. Kennedy EP. Biosynthesis of complex lipids. Fed Proc. 1961;20:934–40.
29. Wang Y, Yu W, Li S, Guo D, He J, Wang Y. Acetyl-CoA carboxylases and diseases. Front Oncologia.

2022;12:836058. https://doi.org/10.3389/fonc.2022.836058.
30. Munday MR. Regulation of mammalian acetyl-CoA carboxylase. Biochem Soc Trans. 2002;30(6):1059–64. https://doi.org/10.1042/bst0301059.
31. Page TMB. Lipolysis and the oxidation of fatty acids. Available from: https://themedicalbiochemistrypage.org/lipolysis-and-the-oxidation-of-fatty-acids/.
32. Tong L. Acetyl-coenzyme a carboxylase: crucial metabolic enzyme and attractive target for drug discovery. Cell Mol Life Sci CMLS. 2005;62(16):1784–803. https://doi.org/10.1007/s00018-005-5121-4.
33. Garcia D, Shaw RJ. AMPK: mechanisms of cellular energy sensing and restoration of metabolic balance. Mol Cell. 2017;66(6):789–800. https://doi.org/10.1016/j.molcel.2017.05.032.
34. Dong X-Y, Tang S-Q, Chen J-D. Dual functions of Insig proteins in cholesterol homeostasis. Lipids Health Dis. 2012;11(1):173. https://doi.org/10.1186/1476-511X-11-173.
35. Xiaoping Z, Fajun Y. Regulation of SREBP-mediated gene expression. Sheng Wu Wu Li Hsueh Bao. 2012;28(4):287–94. https://doi.org/10.3724/sp.J.1260.2012.20034.
36. Burg JS, Espenshade PJ. Regulation of HMG-CoA reductase in mammals and yeast. Prog Lipid Res. 2011;50(4):403–10. https://doi.org/10.1016/j.plipres.2011.07.002.

12 Homeostasis and Regulation of Metabolism in Human Health

Abstract

Metabolic pathways are highly regulated to meet the body's constantly changing needs. This regulation is part of **homeostasis**, the process by which the body maintains a stable internal environment even as its metabolic needs and conditions fluctuate. Under some conditions, glucose and other biomolecules must be broken down to produce energy, while at other times, they must be synthesized and stored for future use. Maintaining a balance between energy production and storage is essential and depends on multiple factors, such as whether the body is in a fed or fasting state and the current energy demands. Hormones play a central role in sensing these needs and in turn regulating metabolic responses. This chapter explores how carbohydrate and lipid metabolic pathways are regulated and how hormones have a role in maintaining homeostasis. This chapter also introduces how disruption of metabolic homeostasis can lead to disease.

Keywords

Homeostasis · Fed state · Fasting state · Glucose levels · Ketone bodies

Abbreviations

AC	Adenylate cyclase
ACC	Acetyl CoA carboxylase
AMP	Adenosine monophosphate
AMPK	AMP activated protein kinase
ATGL	Adipose triglyceride lipase
cAMP	Cyclic adenosine monophosphate
CPT-1	Carnitine palmitoyltransferase 1
FBPase-2	fructose-2,6-biphosphatase
FFA	Free fatty acid
G3P	Glycerol-3-phosphate
G6P	Glucose-6-phosphate
G6Pase	Glucose-6-phosphatase
GAP	Glyceraldehyde-3-phosphate
GLP-1	Glucose-like peptide 1
GLUT	Glucose transporter
GLUT2	Glucose transporter type 2
GLUT4	Glucose transporter type 4
GP	Glycogen phosphorylase
GPCR	G protein coupled receptor
GS	Glycogen synthase
GSK3	Glycogen synthase kinase 3
HMG-CoA	3-Hydroxy-3-methylglutaryl coenzyme A
HSL	Hormone-sensitive lipase
LDH	Lactate dehydrogenase
mTOR	Mechanistic target of rapamycin
NAD^+	Nicotinamide adenine dinucleotide (oxidized form)
OAA	Oxaloacetate
PDH	Pyruvate dehydrogenase complex

A. Minden, *Biochemistry, Cell and Molecular Biology for Pharmacists and Health Professionals*, https://doi.org/10.1007/978-3-032-20529-2_12

PEPCK	Phosphoenolpyruvate carboxykinase
PFK-1	phosphofructokinase-1
PFK-2	phosphofructokinase-2
PGM	phosphoglucomutase
PK	Pyruvate kinase
PKA	Protein kinase A
TAG	Triacylglycerol

What You Will Learn

- The concept of homeostasis: the balance of energy intake and expenditure
- The fed and fasted states: the postprandial state and the postabsorptive state
- The central role of glucose and fats in maintaining homeostasis
- Important hormones in the fed and fasted state, especially the opposing roles of insulin and glucagon
- The roles of mTORC1 and AMPK in regulating anabolism and catabolism to maintain homeostasis
- The process of gluconeogenesis during the fasting state and its role in maintaining blood glucose levels
- The process of ketogenesis and how ketone bodies can serve as an alternative energy source during fasting
- Tissue-specific considerations, such as how the nervous system relies almost entirely on glucose for energy, while other tissues can use fats as an alternative energy source
- How disruption in metabolic homeostasis can lead to disease, which is discussed in more depth in the next chapter

1 Introduction to Homeostasis

1.1 Homeostasis: Balancing Energy Intake and Expenditure

Metabolic pathways are highly regulated to meet the body's constantly changing needs. This regulation is part of **homeostasis**, the process by which the body maintains a stable internal environment even as its metabolic needs and conditions fluctuate. Under some conditions, glucose and other biomolecules must be broken down to produce energy, while at other times, they need to be synthesized and stored for future use. This balance can be maintained because many reactions within metabolic pathways are reversible, meaning that the same enzyme can catalyze both forward and reverse reactions. In contrast, certain key regulatory steps are irreversible and require distinct enzymes and sometimes different intermediates to drive the reaction in each direction. Maintaining balance between energy production and storage is essential and depends on multiple factors, such as whether the body is in a fed or fasting state and the current energy demands. Hormones play a central role in sensing these needs and in turn regulating metabolic responses [1–5].

In this chapter, we will examine how metabolic pathways operate under various conditions and how hormonal signals contribute to regulating metabolism to maintain homeostasis. This involves a closer look at the metabolic pathways already covered in Chaps. 10 and 11 (carbohydrate/glucose metabolism and lipid metabolism), but here, we focus on how the pathways respond to different situations, such as an energy surplus or an energy deficit, and how hormonal signals influence the direction in which the pathways proceed. It may be useful, therefore, to refer back to those chapters when reading about specific pathways, steps, intermediates, and enzymes in the following sections. This chapter also refers to various signal transduction pathways and different types of receptors, including RTKs and GPCRs. While they are introduced briefly here, signal transduction pathways are discussed in more detail in Chap. 18. Referencing that chapter can provide more information about some of the signaling pathways discussed here.

1.2 Introduction to the Fed–Fast Cycle

A major factor in maintaining homeostasis is whether we are in a fed or fasting state, as well as how long it has been since our most recent meal. The biomolecule that is often used as a marker of these states is glucose. After a meal, glucose enters the bloodstream, but as hours pass, glucose levels may decrease. Maintaining stable blood glucose levels is vital for the proper functioning of our organs. To achieve this, the body has developed mechanisms to regulate blood glucose levels and energy stores, regardless of whether we are eating, exercising, resting, or fasting. This regulation is homeostasis. When we are in homeostasis, our body balances fuel production and consumption while meeting the body's energy demands. The goal is to ensure that energy production matches energy consumption. Caloric or energy homeostasis refers to the body's ability to maintain adequate, but not excessive, energy stores. When excess nutrients are consumed, such as after a large meal, they are often stored as glycogen or fat. Conversely, when energy is needed, the body breaks down stored resources, including glycogen and fats, into smaller molecules. These molecules can be further metabolized, resulting in the generation of ATP, the main energy currency [6].

Metabolic homeostasis can be observed throughout the **fed–fast cycle,** which typically spans a 24-h period. This cycle comprises periods of time when nutrients are being ingested, and periods that follow, when nutrients are not being consumed [6]. It can be broken down into phases or states. These phases are defined as approximate timeframes (in hours), but the timeframes can vary slightly according to different sources and may also vary somewhat among individuals.

The first state, **−1-,** is the **fed state**, also called the **postprandial state**. It begins immediately after a meal is eaten and lasts approximately 3–4 h. It is extended or reset if multiple meals are eaten in a day, but ends approximately 4 h after the last meal of the day. Next **−2-** is the **postabsorptive state**, which begins about 4–6 h after the last meal. This state can last up to about 12 h. Finally, **3**, a **fasting state** (also called the **basal state**) may begin, about 12 h after the last meal, and continues as long as food is not consumed. It should be noted that in the literature, the term postabsorptive state is sometimes used interchangeably with the fasting state, but they are actually separate states. Here we refer to the fasting state as beginning about 12 h after the last meal. This state may not be fully achieved in individuals who sleep for approximately 8 h, often eating shortly before bedtime and upon waking. The fasting state is of long duration, and depletion of glycogen stores progresses during this period [4, 7, 8].

This chapter discusses how the major metabolic pathways shift between the fed and fasting states in different tissues, and how key nutrients, such as glucose, lipids, and proteins, function in these tissues. The major changes that occur in liver, muscle, and adipose tissue are also summarized in Table 12.1.

1.3 Glucose and Fats: Central Biomolecules in Energy Homeostasis

The main biomolecules in the body are carbohydrates (which include glucose and other sugars), lipids, proteins, and nucleic acids. When discussing homeostasis, glucose and lipids are central, and this chapter focuses mostly on these two biomolecules. In particular, the regulation of their storage (an anabolic process) and breakdown (a catabolic process) in response to metabolic needs is critical. Both of these biomolecules have storage forms that can be saved for future use when energy is needed, but the types of storage forms differ.

Fat can be stored for a long time in the form of triacylglycerols (TAGs) in adipose tissue. It can be stored for several months if necessary, and it serves as a reliable energy source when needed. When broken down, the main components of fat, free fatty acids and glycerol, can lead to energy production or gluconeogenesis (gluconeogenesis will be covered later in this chapter). Fat yields approximately 9 kcal per gram, more than twice the energy yield of carbohydrates or proteins, making it an efficient source of energy for long-term storage [9].

The storage form of ***glucose is glycogen***, and its storage is less flexible than fat storage. Unlike fat, glycogen is a short-term storage. Glycogen storage is primarily in the liver, and some

Table 12.1 Fed and fasting states in different tissues

	Fed state	Fasting state
Liver		
Glucose uptake	Insulin independent (GLUT2) but increases due to flux→ (high blood glucose leads to liver uptake)	Decreases due to flux (net hepatic export of glucose)
Glucose to G6P	Increases	Decreases
Glycolysis (breakdown of glucose)	Increases, mainly to supply intermediates for lipid synthesis	Decreases
Glycogenesis (synthesis of glycogen for storage)	Increases	Decreases
Gluconeogenesis (synthesis of glucose)	Decreases	Increases
Lipid synthesis (including TAG synthesis)	Increases after glycogen storage capacity is full. The liver does not store much TAG. Instead, they are sent in VLDL to adipose tissue	Decreases
Lipid breakdown	Minimal	Increases (mostly beta-oxidation of fatty acids from adipose tissue lipolysis
Ketone body formation	Minimal	Increases during fasting, especially after a long fast
Skeletal muscle		
Glucose uptake	Increases. Insulin dependent (GLUT4)	Decreases except during exercise
Glycogenesis	Increases, especially in inactive muscle	Decreases
Glycolysis	Increases in active muscle for ATP production	Decreases in resting muscle while exercise uses glycogen derived muscle.
Protein synthesis	Increases	Decreases
Proteolysis	Decreases	Increases
Fatty acid oxidation	Decreases	Increases, becomes the major fuel at rest and in prolonged fasting
Ketone body use	Minimal	Ketone body use (not synthesis) increases during prolonged fasting
Adipose tissue		
Glucose uptake	Increases, insulin dependent (GLUT4)	Decreases
Glycerol-3-phosphate synthesis	Increases (from glycolysis intermediates). Kept for use locally	Decreases
TAG synthesis	Increases (esterifies free fatty acids from chylomicrons and VLDL)	Decreases
Lipolysis; the breakdown of TAGs to glycerol and fatty acids (adipocyte)	Inhibited by insulin (HSL is inhibited by insulin)	Increases (HSL and ATGL activated by catecholamines)
Fatty acid β-oxidation	Low	Increases locally but **mostfatty acids are exported to othertissues**
TAG lipolysis in capillaries; capillary lipoprotein lipase	Increases; hydrolysis of TAGs in chylomicrons and VLDL	Decreases

glycogen is present in muscle. Due to its structure and hydrophilic nature, glycogen is not stored as compactly as fat and is not stored in large amounts. The body's glycogen stores are typically depleted within 12–24 h. Glycogen is a polymer made of large amounts of glucose, and when glucose is needed for energy, glycogen is broken down into individual glucose units. When glucose is not stored as glycogen, it circulates in the bloodstream and is delivered to organs that need it. This is important because some organs, such as the brain, rely heavily on glucose for energy. The regulation of blood glucose levels is critical, however, and must be kept within a narrow range. Excessive or insufficient glucose levels can cause multiple problems very quickly [6, 10, 11].

1.4 Important Hormones and Glucose Sensors for Maintaining Homeostasis

The storage and breakdown of glucose, fat, and other biomolecules are mediated by hormones, which can respond to nutrient uptake or expenditure. Two of the most important hormones that help maintain homeostasis are **insulin** and **glucagon** [6, 10, 11]. Insulin and glucagon, which are secreted by the pancreas, are both peptide hormones, and they both use glucose levels as a signal of the body's energy status. They are introduced in this section, followed by a discussion of other glucose sensors.

1.4.1 The Pancreas as a Source of Metabolic Hormones

Insulin and glucagon, along with several other hormones involved in homeostasis, are secreted by the pancreas. Hormones are also released from other organs, but the pancreas is a critical organ in homeostasis and will therefore be briefly introduced here. The pancreas is both an endocrine and an exocrine organ located in the upper abdomen. Endocrine function refers to secretion of hormones, such as insulin and glucagon (endocrine hormones), into the bloodstream, while exocrine function involves secretion of enzymes, primarily digestive enzymes, into ducts leading to the digestive system. We have already discussed some of the digestive enzymes that are important in breaking down lipids and carbohydrates during digestion, which are produced in the exocrine portion of the pancreas.

The endocrine portion of the pancreas consists of islets of Langerhans, which are clusters of specialized cells. These cells include β cells, α cells, and δ cells. The β cells secrete insulin, the α cells secrete glucagon, and the δ cells produce somatostatin, a hormone that helps regulate the release of other pancreatic hormones. The islets of Langerhans also contain other cell types, such as PP (pancreatic polypeptide) cells and ε cells. The islets of Langerhans are also highly vascularized, which is important for the transport of hormones into the bloodstream [12, 13]. A diagram of the pancreas is shown in Fig. 12.1.

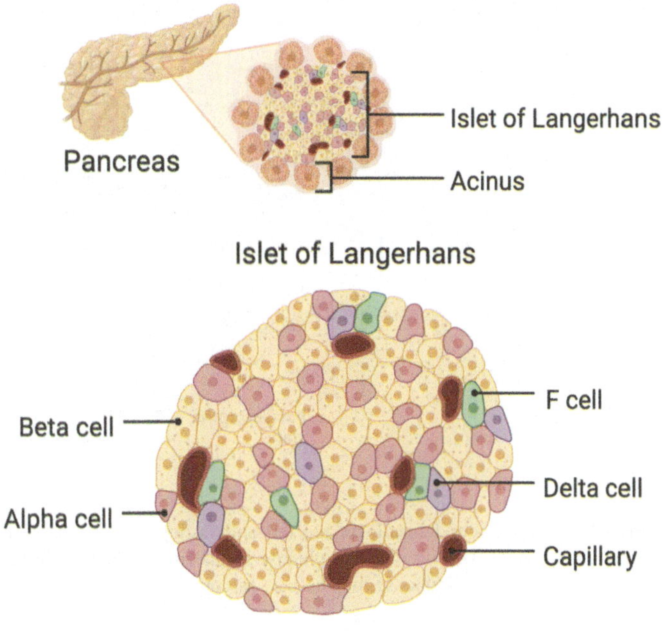

Fig. 12.1 The pancreas. The pancreas contains several specialized clusters of cells including the islets of Langerhans. The islets of Langerhans contain several cell types, including beta cells, which secrete insulin, and alpha cells, which secrete glucagon. (**Source**: Created with BioRender.com)

1.4.2 Insulin and Glucagon and Their Roles in Maintaining Homeostasis

Insulin is associated with **nutrient uptake**. It promotes nutrient storage and inhibition of energy production. Insulin is secreted by the pancreatic β cells in response to high blood glucose levels. It binds to insulin receptors on target cells, which are receptor tyrosine kinases (RTKs). It helps maintain and restore healthy blood glucose levels, even as they rise after nutrient intake. It does so by promoting the uptake of glucose into certain tissues. In some tissues, it also promotes storage. It promotes the storage of glucose (in the form of glycogen) in the liver and muscle, as well as the storage of fat in adipose tissue.

Glucagon is a hormone that is secreted by the α cells of the pancreas. As opposed to insulin, it is secreted in response to **low blood glucose** levels. Glucagon supports the release of energy rather than its uptake and storage. Glucagon stimulates the breakdown of glycogen to glucose in the liver, and this glucose can be released into the bloodstream to be used by various organs. It can also stimulate the synthesis of glucose (gluconeogenesis) from non-carbohydrate sources. Glucagon also binds to receptors on target cells, which are G protein-coupled receptors (GPCRs), specifically Gs-linked GPCRs. (RTKs and GPCRs are discussed in more detail in Chap. 18). The main metabolic effects of glucagon occur in the liver and, to a lesser extent, the kidney. Skeletal muscle does not respond directly to glucagon because it lacks glucagon receptors.

In summary, insulin is usually upregulated when blood glucose levels are high, while glucagon is upregulated when blood glucose levels are low. By their opposing actions, insulin and glucagon can maintain an appropriate level of blood glucose (see Fig. 12.2).

While insulin and glucagon are central hormones in maintaining homeostasis, several other hormones also have key roles, such as GLP-1 and GIP, which will be discussed later (see Sect. 2.1.2). Furthermore, nutrient sensors like mTORC1 and AMPK also have important roles in maintaining homeostasis, and are discussed below.

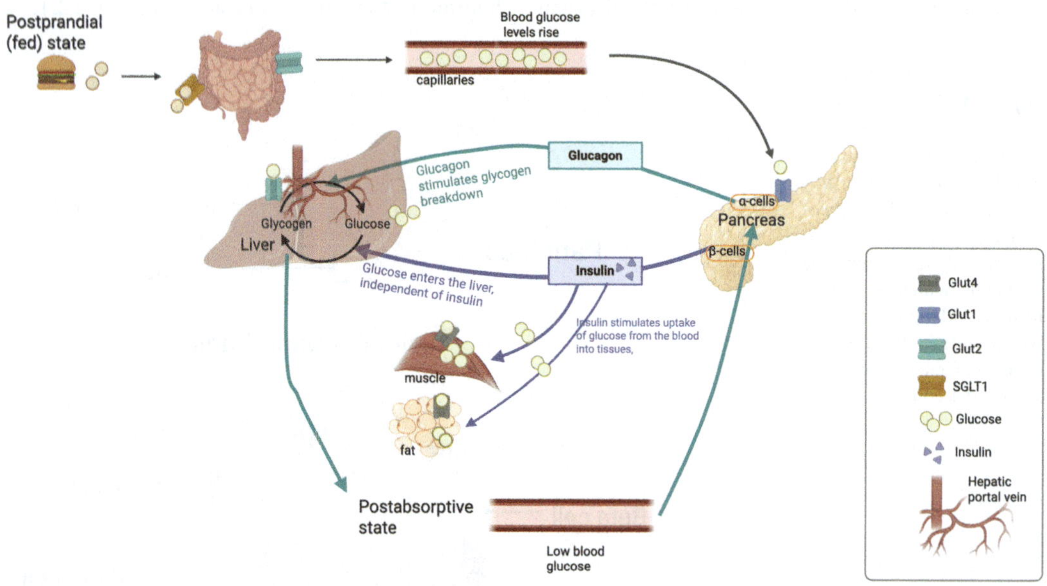

Fig. 12.2 Glucagon and insulin in the fed and fasting states. Insulin levels rise when blood glucose levels rise, leading to the uptake of glucose into tissues. In contrast, glucagon levels rise during the fasting state. See text for details. (**Source**: Created with BioRender.com)

1.4.3 Nutrient Sensors mTORC1 and AMPK Play Important Roles in Maintaining Homeostasis

mTORC1 (mechanistic target of rapamycin complex 1) and **AMPK** (AMP-activated protein kinase) are not hormones, but signaling molecules that are considered nutrient sensors, and they function in opposition to each other, and have important roles in the fed–fast cycle.

mTORC1 senses nutrient abundance. It is active in the fed state. It senses increased levels of amino acids, high levels of energy (ATP), and detects insulin and growth factors such as IGF-1. When active, it promotes anabolic pathways. It stimulates protein synthesis (via proteins like S6K and 4E-BP1), promotes lipid synthesis, and inhibits autophagy.

AMPK is active in the fasting state, or any other time when ATP levels are low. It senses low energy, specifically via a high AMP/ATP ratio or a high ADP/ATP ratio. It increases glucose uptake into cells. It inhibits anabolic pathways to conserve ATP. It activates autophagy, and it also inhibits mTORC1.

In summary, mTORC1 promotes anabolic processes such as synthesis and storage, while AMPK inhibits these processes to conserve energy when cellular energy levels are low [14–16]. While their roles in metabolic homeostasis are discussed in this chapter, the detailed signaling pathways mediated by mTORC1 and AMPK will be discussed in greater depth in Chap. 18.

1.5 Overview of the Next Sections

The next sections describe how homeostasis is achieved through the regulation of metabolic pathways in different tissues and in the major physiological states, organized as follows. **Section 12.2** focuses on the **postprandial (fed) state**, **Sect. 12.3** focuses on the **postabsorptive and fasting states**. **Section 12.4** briefly discusses the **return to the postprandial/fed state** after the postabsorptive and fasting states, and **Sect. 12.5** summarizes some of the important aspects of homeostasis and addresses some important tissue-specific considerations.

2 The Postprandial (Fed) State

This section focuses on the fed state, or postprandial state, the period just after a meal. We begin by discussing the hormones and nutrient sensors that are important in this state, and then examine the fate of biomolecules, particularly glucose and lipids.

2.1 Important Hormones and Nutrient Sensors in the Postprandial/Fed State

After a meal, glucose levels rise quickly. This leads to increased secretion of several hormones, as discussed below, and in [4, 6, 10, 17–20].

2.1.1 Insulin During the Postprandial/Fed State

- Secretion of Insulin from Pancreatic β Cells in Response to High Blood Glucose Levels

Insulin is a peptide hormone that is released into the bloodstream from pancreatic β cells in response to high blood glucose levels. It is the major hormone that responds to elevated glucose levels. Insulin is initially synthesized as pre-proinsulin, which is processed to proinsulin in the ER, and then cleaved in the Golgi to produce mature insulin and C-peptide, which are both stored in secretory granules, from which insulin is secreted upon glucose uptake [21]. Insulin secretion is initially triggered by glucose entering the pancreatic β cells. Glucose enters the pancreatic β cells mainly via GLUT1 transporters [22] (in human cells), and the GLUT transporters will be discussed more in the next section. Once glucose enters the pancreatic β cells, it is incorporated into the central metabolic pathways, leading to increased ATP production. The result is a rise in the ATP/ADP ratio. This, in turn, causes ATP-sensitive potassium channels to close, and this leads to membrane depolarization. This membrane depolarization leads to the opening of voltage-gated Ca^{2+} channels (VGCCs), and consequently, Ca^{2+} enters the pancreatic β cells. This increase in intracellular Ca^{2+} causes insulin-containing granules to fuse with the plasma membrane. As a consequence of the vesicle fusion with the membrane, insulin is released into the bloodstream, enters circulation, and binds to insulin receptors on target cells. This is an important example of vesicle transport. Once it is secreted, insulin can promote even more insulin secretion. This is because insulin binds to insulin receptors (IRs) on the pancreatic β cells. This leads to activation of pathways such as the PI3K/AKT pathway, which support pancreatic β cells' survival. This encourages the production of more insulin [23, 24].

- Insulin Signaling

Insulin binds to insulin receptors on target cells. The insulin receptor (IR) is a receptor tyrosine kinase (RTK). The IR is a covalent dimer, unlike some RTKs, which only dimerize upon ligand binding. Insulin binding to the IR activates the receptors. This results in tyrosine phosphorylation on the cytoplasmic side of the receptor, followed by the recruitment of insulin receptor substrate (IRS) to the phosphorylated receptors, which are also phosphorylated on multiple tyrosines by the insulin receptor. This, in turn, leads to activation of signaling pathways. One pathway that is commonly activated is the PI3 kinase (PI3K) pathway. In this pathway, PI3 kinase binds to the tyrosine-phosphorylated IRS. PI3 kinase then phosphorylates and activates the membrane phospholipid PIP2, which in turn becomes PIP3. PIP3 recruits AKT and PDK to the membrane. PDK1 phosphorylates AKT, activating it [25, 26]. AKT, in turn, activates several of the signaling pathways described in several sections below. The pathway from Insulin to AKT can be briefly outlined as (Insulin→IR → PI3K → PIP3 → AKT→) (For more detailed information about how these signaling pathways operate, refer to Chap. 18).

- Functions of Insulin

Once insulin is secreted from pancreatic β cells in response to glucose, insulin can circulate through the bloodstream, and it exerts its effects on various cell types by binding to insulin receptors on their cell surfaces and activating signaling pathways such as the PI3 kinase pathway and others in these cells, which can then carry out various responses.

Overall, increased circulating insulin is associated with nutrient storage and uptake into tissues, as well as the inhibition of the breakdown of storage molecules. Insulin stimulates the uptake of glucose from the bloodstream into muscle and adipose tissue. These are large tissues, and the absorption of glucose in these tissues can significantly help decrease circulating glucose. In addition to promoting glucose uptake into tissues, insulin also promotes the storage of glucose in the form of glycogen in the liver and muscle. It also promotes fat storage by enhancing fatty acid synthesis and promoting the uptake of fatty acids (from lipoproteins circulating in the bloodstream) into adipose tissue.

At the same time, insulin inhibits the breakdown of fat (lipolysis) and inhibits the breakdown of glycogen (glycogenolysis). Thus, in response to insulin, storage is encouraged, and breakdown of these stores is discouraged. Insulin also indirectly suppresses hepatic gluconeogenesis by downregulating the expression of key enzymes in this process. There is variability in this system, however, in different tissues and under different conditions, and this will be addressed later in the section.

2.1.2 Incretins: Glucagon-Like Peptide 1 (GLP-1) and Glucose-Dependent Insulinotropic Polypeptide (GIP) During the Postprandial/Fed State

Glucagon-like peptide 1 (GLP-1) and **glucose-dependent insulinotropic polypeptide (GIP)** are two other hormones that have important roles in the fed state. These two hormones are the main incretin hormones, and they participate in the incretin pathway. The incretin pathway is a gut-to-pancreas pathway that enhances glucose-dependent insulin secretion from pancreatic β cells in the fed state. In this pathway, the **incretins** are released from intestinal cells and they bind to receptors in the pancreas (specifically to G protein-coupled receptors (GPCRs) on pancreatic β cells). In response, they enhance glucose-stimulated insulin secretion, as follows.

GLP-1 is secreted from intestinal L cells (cells in the distal small intestine and colon) in response to increased nutrients, especially elevated levels of glucose and fat. The increase in GLP-1 in turn

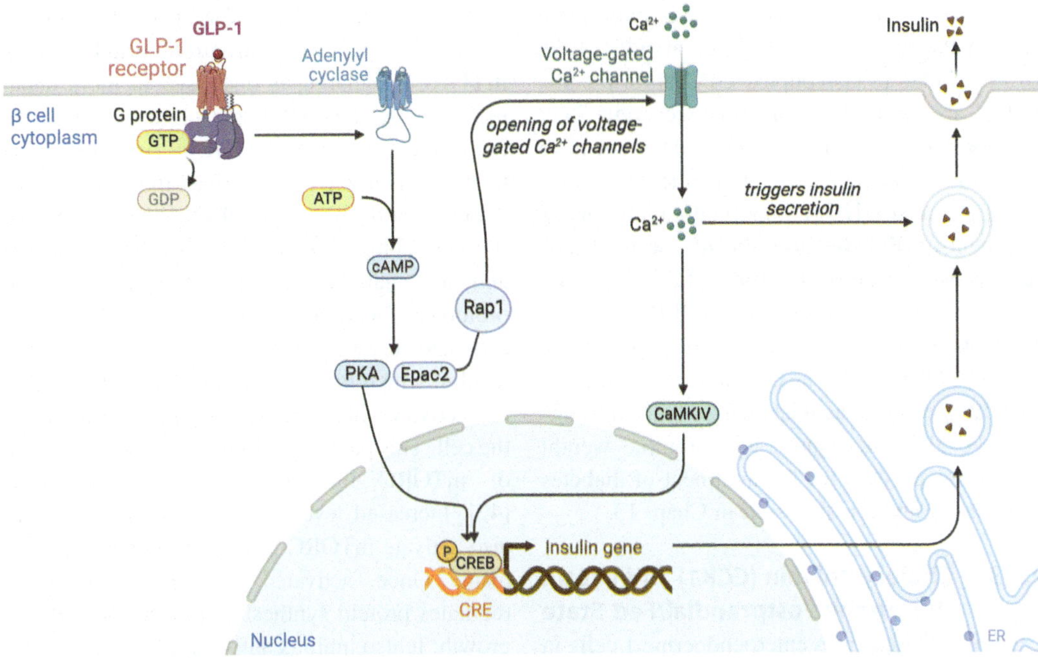

Fig. 12.3 GLP-1 signaling pathway. GLP-1 binds to a GPCR and activates a cAMP-dependent pathway that enhances insulin secretion by several mechanisms. See text for details. (**Source**: Created with BioRender.com)

enhances glucose-dependent insulin synthesis and secretion from the pancreatic β cells [20]. It occurs through several glucose-dependent mechanisms, summarized below and in Fig. 12.3.

GLP-1 binds to GLP-1 receptors (which are Gs-coupled GPCRs) on the pancreatic β cells, and activates a signaling pathway that increases insulin secretion. This pathway leads to increased levels of the second messenger cyclic AMP (cAMP), subsequent activation of protein kinase A (PKA), and activation of the protein **Epac2** (Exchange protein directly activated by cAMP 2). Epac2 is an exchange factor that activates the small GTPase Rap1. Rap1 affects membrane depolarization and enhances the opening of certain voltage-gated calcium channels. Together, PKA and Epac2 promote Ca^{2+}-dependent exocytosis of insulin granules [27]. This GLP-1 activity is glucose-dependent, and it increases insulin secretion only in the presence of elevated blood glucose levels.

The rise in intracellular Ca^{2+} also has another function, which is to stimulate **Ca^{2+}/calmodulin-dependent protein kinase IV (CaMKIV), one of the kinases that can phosphorylate** the transcription factor **CREB on Ser133. Phosphorylated** CREB, in turn, binds to its regulatory element in genes that are important for β-cell function, including those required for insulin production and cell survival. This Ca^{2+} and cAMP-dependent regulation replenishes insulin stores in β cells and supports long-term β-cell function [28–31].

GLP-1 also leads to increased insulin synthesis via its activation of **PKA.** PKA also phosphorylates CREB, which, as indicated above, binds to its regulatory element in the insulin gene promoter. This pathway also involves other transcription regulators such as PDX- and contributes to replenishing the insulin stores in the β cells [32].

At the same time**, GLP-1** also **inhibits** the secretion of **glucagon**. It does this both directly, by binding to GLP-1 receptors on pancreatic α cells, and indirectly by stimulating the secretion of **somatostatin**, a peptide hormone, from pancreatic δ cells [33]. By **upregulating insulin** and **downregulating glucagon**, **GLP-1** helps to maintain balanced blood glucose levels. An additional function of **GLP-1** is to slow down gastric emptying, thereby promoting satiety and helping maintain homeostasis [34].

GIP is secreted by K cells in the small intestine (in the duodenum and jejunum) [35]. Like GLP- it binds to Gs-coupled GPCRs on β cells, and it also **stimulates insulin secretion** in a glucose-dependent manner, but it does not have a big role in gastric emptying or satiety.

GLP-1 and GIP are important targets for the treatment of **diabetes and obesity**. Some drugs, such as **Mounjaro (tirzepatide)** (FDA approved 2022) [36], act as dual agonists of GLP-1 and GIP receptors, mimicking the effects of both of these incretins. Others, such as Ozempic (semaglutide), are GLP-1 receptor agonists. These types of drugs improve glycemic control and promote weight loss. They are used for the treatment of diabetes and obesity and are discussed in Chap. 13.

2.1.3 Cholecystokinin (CCK1) and Leptin During the Postprandial/Fed State

CCK is released from enteroendocrine I cells in the duodenum and jejunum of the small intestine in response to the intake of fats and proteins. This differs from GLP- which responds primarily to glucose and to a lesser extent, to fat [37]. CCK helps regulate digestion by various mechanisms. It promotes satiety by activating vagal afferent nerves, which are sensory nerves that transmit signals to the brainstem and hypothalamus [38, 39]. It stimulates pancreatic enzyme secretion, and it stimulates the gall bladder to release bile for fat digestion [10].

Leptin is produced by adipocytes (fat cells). Its release is not strictly associated with the fed state, unlike the other hormones we have discussed so far, as its levels correlate with fat mass rather than nutrient intake [40]. It sends signals to the brain (specifically the hypothalamus) to suppress appetite and increase energy expenditure, which helps regulate energy balance [41].

2.1.4 mTOR, A Nutrient Signaling Sensor, During the Postprandial/Fed State

In addition to hormonal regulation, cells also respond to nutrient availability through intracellular nutrient-sensing pathways, especially the mTORC1 (mechanistic Target of Rapamycin complex 1) pathway, introduced briefly above. **mTORC1** signaling is discussed in more detail in Chap. 18. **mTORC1** is a protein complex in the cell that includes mTOR, a serine/threonine kinase. It is involved in various intracellular signaling pathways. mTORC1 has multiple functions, but it is well known as a nutrient sensor and an anabolic regulator (meaning that it promotes anabolic processes). It becomes activated in response to high nutrient levels via insulin, which binds to the insulin receptor (IR), and activates the PI3K-AKT signaling pathway in the cell. This pathway inhibits negative regulators of mTORC thus leading to its activation [42]. Increased levels of some amino acids can also activate mTORC through a separate mechanism. Once activated, mTORC1 positively regulates protein synthesis, lipogenesis, and cell growth. It also inhibits autophagy, the process by which cells recycle internal components. This helps ensure that during the fed state, when nutrients are abundant, the body has the necessary components for growth, repair of damaged proteins and organelles, and storage of energy reserves (such as fats or protein). mTORC along with insulin, is an important mediator of the fed state, but overactivation of mTORC1 can lead to problems. Overactivation of mTORC1 is associated with conditions like obesity and insulin resistance [43], which will be discussed in the next chapter. mTORC1 functions in opposition to **AMPK**, which is a sensor for low ATP levels. AMPK is discussed later in the chapter.

2.2 Glucose Uptake into Different Tissues in the Postprandial/Fed State

In the fed state, after a meal, the biomolecules consumed, especially glucose, fat, and proteins, have specific fates. The overall trend in the fed state is the prioritization of storage over catabolism. However, there can be variations in this theme, and the fates of biomolecules differ across tissues and under various conditions. For example, while glucose is stored as glycogen in the liver, very little glucose is stored in neurons. Instead, it is broken down and used for energy in neurons, regardless of the metabolic state. In muscle, glucose storage can be prioritized, but this depends on the muscle's activity level.

This Sect. 12.2.2, discusses the fate of glucose during the postprandial/fed state. This will be followed by a discussion of the fate of fats and, to a lesser extent, protein, in the following Sect. 12.2.3.

2.2.1 Glucose Transporters

After a meal, the amount of glucose in the bloodstream rises. When homeostasis is functioning properly, this glucose will be absorbed into various tissues and, in some cases, stored in those tissues, thus quickly lowering blood glucose to healthy levels. In some tissues, this process of glucose absorption into tissues is regulated by insulin, which is secreted from pancreatic β cells into the bloodstream in response to elevated glucose levels. Glucose enters the cells of various tissues through glucose transporters, specifically GLUT transporters. There are several different GLUT transporters. The main ones are GLUT GLUT2, GLUT3, and GLUT4 (see Table 12.2), which are present on cells of different tissues. Of these transporters, GLUT4, in muscle and adipose cells, is insulin-dependent.

The absorption of glucose into different tissues in the postprandial/fed state, through various transporters, and its fate within these tissues are discussed in the following sections. The focus will be on the fate of glucose in the liver, pancreas, muscle, adipose tissue, and the nervous system.

2.2.2 Glucose Uptake and Its Fate in the Liver During the Postprandial/Fed State

- Glucose Uptake in the Liver Is Insulin-Independent

After digestion, glucose that is released from the enterocytes of the small intestine enters the hepatic portal vein, which leads to the liver. There, some of the glucose enters the hepatocytes (liver cells). Glucose enters the hepatocytes via **GLUT2**, which is an insulin-independent transporter. While glucose uptake in the liver is insulin independent, the amount of glucose entering the liver in the fed state is nonetheless elevated during the fed state because of the increased amount of insulin in the bloodstream.

- Phosphorylation of Glucose to Yield Glucose-6-Phosphate (G6P) Is Upregulated

Although glucose uptake into the hepatocytes is not insulin-dependent, insulin nonetheless has an important role. Specifically, insulin is critical not for glucose uptake, but for maintaining the glucose within the liver cells. This is because insulin regulates the phosphorylation of glucose to glucose-6-phosphate (G6P). Unphosphorylated glucose can move both into and out of the liver cells via Glut2. Phosphorylated glucose, however, cannot be transported through this transporter and is therefore trapped inside the cell.

Insulin leads to glucose phosphorylation in the liver mainly through its regulation of **glucokinase (GK)**, the enzyme that phosphorylates glucose to form **G6P**. **Insulin** binds to insulin receptors (**IR**) on the surface of hepatocytes (liver cells), and this activates signaling pathways in the cell, including the PI3K-AKT pathway. One outcome of **insulin-**

Table 12.2 Glucose transporters (GLUT)

GLUT	Transports	Main tissues expressed	Insulin responsiveness
GLUT1	Glucose	Widely expressed, including brain and RBCs and endothelial cells	No
GLUT2	Glucose	Liver, kidney, small intestine	No
GLUT3	Glucose (high affinity)	Neurons, especially the brain	No
GLUT4	Glucose	Skeletal muscle, adipose tissue, heart	Yes
GLUT5	Fructose	Small intestine, some in kidney, sperm	No

activated signaling is the increased expression of the gene encoding **GK**. This leads to increased levels of **GK** protein.

In addition, **insulin** also contributes to the release of **GK** from the nucleus, where it is sequestered during the fasting state [44, 45]. Thus, both by increasing **GK** levels and by releasing its inhibition, insulin leads to the phosphorylation of glucose to yield Glucose-6-phosphate (G6P). This G6P, now trapped in the cell, is prevented from returning to the bloodstream and a portion of the excess glucose will usually be stored in the liver in the form of glycogen [46] (see Fig. 12.4 for the roles of insulin and other hormones in glycolysis and gluconeogenesis).

- Increased Conversion of Glucose to Glycogen for Storage

Glucose that enters the liver in the fed state is primarily stored as glycogen because, in this state, the available energy supply exceeds the demand for energy. As described above, insulin plays a role in trapping glucose in the liver by promoting its phosphorylation to form glucose-6-phosphate (G6P). Much of this glucose is stored as glycogen.

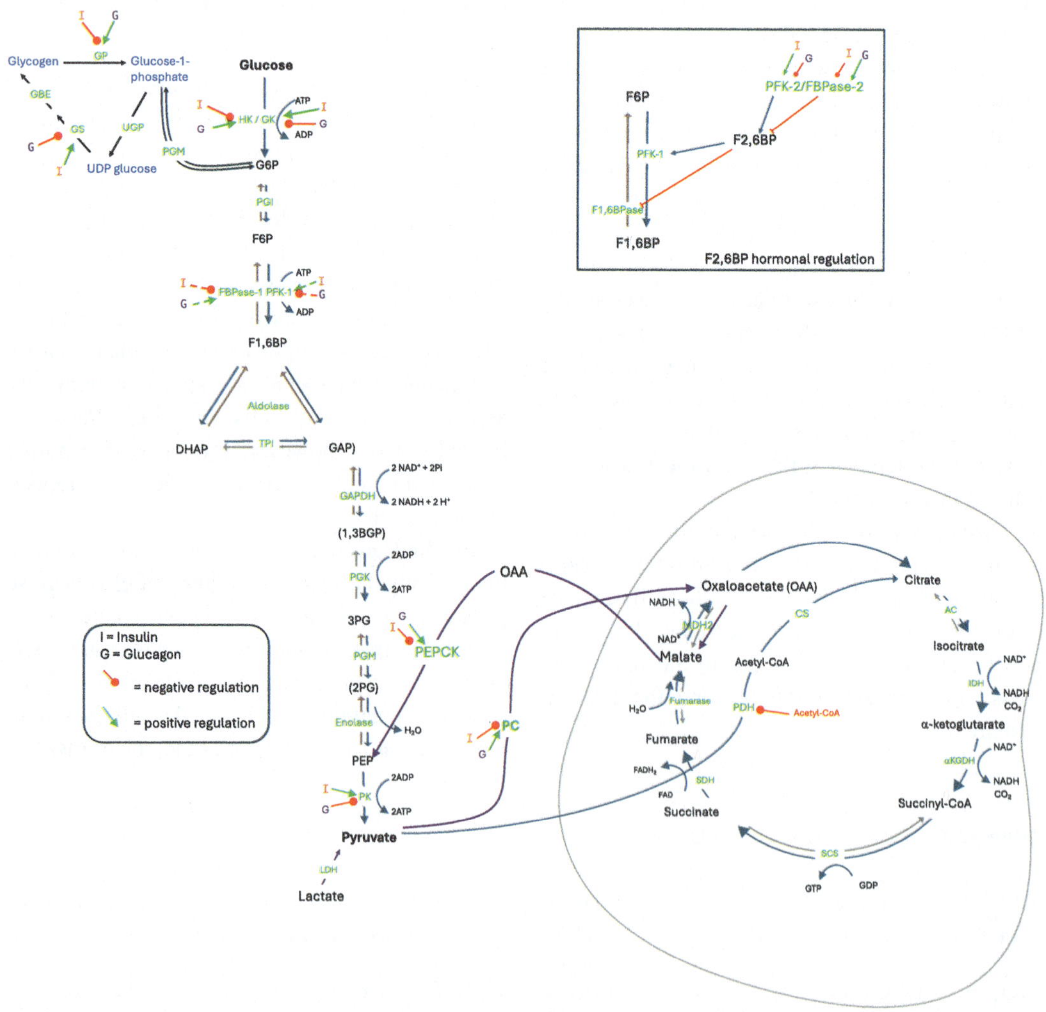

Fig. 12.4 Hormone regulation of the central carbon pathways and gluconeogenesis. Key steps in gluconeogenesis are indicated by purple arrows. Activation or inhibition of certain key enzymes by insulin (I) or glucagon (G) are illustrated by arrows (activation) or lines with a circle (inhibition)

Insulin has a role in both promoting glycogen synthesis (glycogenesis) and inhibiting the breakdown of glycogen into glucose (glycogenolysis). These effects are mediated by the **activation** of protein phosphatase 1 (PP1) and the **inhibition** of **glycogen synthase kinase (GSK-3)**. Insulin binding to the insulin receptor activates the PI3K-AKT signaling pathway in the cell as described above. **AKT** phosphorylates regulatory proteins that release **PP1** from inhibition, which increases **PP1's** activity [47]. (Insulin→IR → PI3K → PIP3 → AKT → PP1). **PP1** is a phosphatase (an enzyme that removes phosphate groups—dephosphorylation). **PP1** activation helps promote glycogen storage in several ways, as follows:

Once active, **PP1** dephosphorylates and thereby activates the enzyme **glycogen synthase (GS)** (this enzyme is active in its dephosphorylated state). **GS**, as discussed in Chap. 12, is a key enzyme in the synthesis of **glycogen**. It is the enzyme that can add UDP-glucose to a growing glycogen chain. This reaction can be summarized as:

Insulin → IR → PI3K → PIP3 → AKT → PP1 → GS → glycogen.

PP1 also has another function, which is to dephosphorylate and thereby **inhibit** the enzyme **glycogen phosphorylase (GP)**. **GP** is an enzyme required for glycogen breakdown into glucose by cleaving glycogen to release glucose-1-phosphate from the glycogen complex (see Chap. 10).This reaction can be summarized as:

Insulin→IR → PI3K → PIP3 → AKT → PP1--↓-GP-↓-glycogen-breakdown.

Thus, insulin both encourages glycogen synthesis and inhibits glycogen breakdown through its regulation of **PP1**.

In a separate branch of the same insulin-AKT pathway, AKT phosphorylates **glycogen synthase kinase 3 (GSK-3)** on a serine residue, thereby inhibiting its kinase activity. GSK-3 is an enzyme that can phosphorylate and inactivate **glycogen synthase (GS)**. Thus, the inhibition of **GSK-3** further encourages glycogen synthesis [8, 48], as follows:

Insulin→IR → PI3K → PIP3 → AKT-↓-GSK3-↓-GS → glycogen.

- Inhibition of Gluconeogenesis (the Synthesis of More Glucose)

Insulin also reduces the synthesis of new glucose by downregulating the transcription of the gene for the enzyme **PEPCK**, which has a key role in **gluconeogenesis**, the synthesis of glucose from non-carbohydrate sources, which is discussed later in the chapter.

- Glucose Uptake in the Liver Can Lead to Lipid Synthesis

One important outcome of glucose uptake in the liver, besides glycogen storage, is the synthesis of lipids, a process that occurs largely in the liver. It involves the use of glycolytic intermediates (intermediates in the glycolysis pathway), the assembly of TAGs, and their transport to adipose tissue. The steps are broken down as follows:

- From Excess Glucose to glycolytic Intermediates Required for Lipid Synthesis

While insulin promotes storage, hepatic (liver) storage of glycogen is limited. In the liver, most excess glucose beyond the glycogen storage capacity gets diverted to **fatty acid synthesis (lipogenesis)** and the formation of triacylglycerol (TAG). During the postprandial period, most of the incoming glucose in the liver is initially directed toward glycogen synthesis. Once glycogen stores approach capacity, additional glucose enters glycolysis, where the glycolytic intermediates become important for lipid synthesis.

- Regulation of Glycolysis by Insulin (PFK-PFK-2, and F2,6BP)

The rate of glycolysis, and consequently of lipogenesis, is strongly regulated by the enzyme **phosphofructokinase-1 (PFK1)**, which converts F6P to F6BP in glycolysis. Insulin regulates **PFK-1** by activating another enzyme, **PFK-2**, which is part of the bifunctional enzyme **PFK-2/FBPase-2**. (PFK-2/FBPase-2 and its regulation of PFK-1 are also described in Chap. 10). This regulatory mechanism applies to liver isoforms of PFK-2/FBPase-2, which is especially relevant here because **liver** is the main organ responsible for maintaining body glucose homeostasis.

Insulin activates PFK-2 via its regulation of protein **phosphatase-1 (PP1)**, mediated by the AKT signaling cascade, the same pathway described above for glycogen synthesis. Briefly, **PP1**, activated by insulin signaling, dephosphorylates and thereby activates **PFK-2.** Activated **PFK-2** leads to the formation of **F2,6BP**, which is a positive allosteric regulator of **PFK-1** [49]. Thus, activation of **PFK-1** in response to insulin occurs as follows: (Insulin→IR→PI3K→PIP3→AKT→PP1→PFK2 → F2,6BP → PFK1 → F6BP → glycolysis). Insulin also leads to the **inactivation** of the enzyme **F6BPase**, which is involved in gluconeogenesis and will be discussed in more detail in the next section. Inhibition of gluconeogenesis also causes the reaction to favor glycolysis.

– Glycolys Intermediates Provide Substrates for Lipogenesis

Two main substrates required for lipid synthesis are glycerol-3-phosphate (G3P) and acetyl-CoA. G3P forms the glycerol backbone of triacylglycerols (TAGs) and acetyl-CoA is a substrate for the fatty acid component of the TAGs. Both G3P and acetyl-CoA are derived from glycolytic intermediates. Acetyl-CoA is formed from pyruvate, which is the end product of glycolysis. Acetyl-CoA is in the mitochondria, and lipogenesis (fatty acid synthesis) occurs in the cytosol. To solve this, acetyl-CoA is converted to citrate, which can exit the mitochondria into the cytosol. This citrate can then be converted back to acetyl-CoA, to be used for lipogenesis (See Chap. 11 for the main steps in fatty acid synthesis from acetyl-CoA).

Glycerol-3-phosphate (G3P) is generated by two main routes in the liver. During the fed state, it is generated mostly from the glycolytic intermediate DHAP, which is reduced to form G3P (see Chap. 10, Fig. 10.13). G3P can also form from direct phosphorylation of glycerol in the liver via **glycerol kinase**, but this is more common in the fasting state when glycerol is released from adipose tissue.

– Assembly and Export of Triacylglycerols (TAGs)

Fatty acids and glycerol-3-phosphate (G3P; generated as described above) are esterified to form triacylglycerols (TAGs), the main storage form of lipids. The liver packages TAGs into very low-density lipoprotein (VLDL) particles. These particles are secreted into the bloodstream and delivered to other tissues, especially adipose tissue, where fats are stored [50].

In adipose tissue capillaries, the enzyme **lipoprotein lipase (LPL)** hydrolyzes the TAGs in the VLDLs, releasing the free fatty acids and glycerol. The free fatty acids enter adipocytes, where they are re-esterified with glycerol-3-phosphate, which is produced in the adipocytes (also from DHAP, as described above), and then used for storage. The released glycerol from the VLDLs returns to the liver, and it can be used for gluconeogenesis (discussed later) or for assembling new TAGs [51].

- Mild Increase in ATP Generation in the Liver Even During the Postprandial/Fed State

Overall, the liver is not a primary site for ATP generation in any metabolic state. However, even in the liver and in the fed state, some glucose does lead to ATP generation via the central metabolic pathways and oxidative phosphorylation.

2.2.3 Glucose Entry into Pancreatic β Cells During the Postprandial/Fed State Leads to Insulin Secretion

Glucose that enters the systemic circulation also reaches the pancreatic β cells. This is critical because it is the pancreatic β cells that produce and secrete insulin. Glucose enters these cells mainly via **GLUT1** transporters (in human cells [22]). GLUT1 is an insulin-independent transporter. Therefore, once glucose levels rise, it can enter the pancreatic beta cells regardless of the presence of insulin. Once glucose is transported into the pancreatic β cells, it leads to the secretion of insulin [23] (see Sect. 2.11).

2.2.4 Glucose Uptake and Fate in Muscle and Adipose Tissue in Response to Insulin in the Postprandial/Fed State

An important function of insulin in the fed state is to reduce blood glucose levels. A major way it does so is by promoting uptake of glucose into large tissues, especially skeletal muscle and adipose tissue. These tissues can take up a large amount of glucose, thereby lowering blood glucose levels effectively.

Unlike the liver and pancreas, uptake of glucose into muscle and adipose tissue cells is insulin-dependent and is mediated by the insulin-dependent GLUT4 transporters. **GLUT4** is insulin-dependent because insulin causes its translocation from the cytoplasm to the membranes of muscle and adipose cells, where it can transport glucose. It does so by binding to insulin receptors (IR) and activating the PI3 kinase/AKT pathway. AKT regulates targets involved in transporting the GLUT transporter to the cell surface via vesicle trafficking [52]. Specifically, it phosphorylates and inhibits AS160, which is a GAP for the small GTPase Rab. With AS160 inhibited, Rab is active and bound to GTP. Active Rab promotes the movement of GLUT4 storage vesicles (GSVs), which fuse with the membrane so that GLUT4 becomes membrane localized (see Fig. 12.5). This pathway can be summarized as: Insulin→IR → PI3K → PIP3 → AKT → →GLUT4 transport. This is a short-term response, but the signaling pathway activated by insulin can also positively regulate the gene encoding GLUT4, leading to a more long-term increase in GLUT4 levels. This signaling pathway is mediated by AKT indirectly and involves transcription factors such as MEF2 and GLUT4 enhancer factor (GEF) [53–55]. Thus, both the levels of GLUT4 and its translocation to the cell surface rise over time in response to insulin. As a consequence, in response to insulin, glucose can enter the cells of muscle and adipose tissue via the GLUT4 transporter [1, 4, 6].

- The Fate of Glucose in Muscle

In muscle, the fate of glucose depends on the tissue's activity level. In resting muscle, glucose is stored as glycogen. Besides the liver, muscle is one of the main tissues that stores glycogen. In active muscle, however, much of the glucose may enter the central metabolic pathways, including oxidative phosphorylation, leading to the generation of ATP.

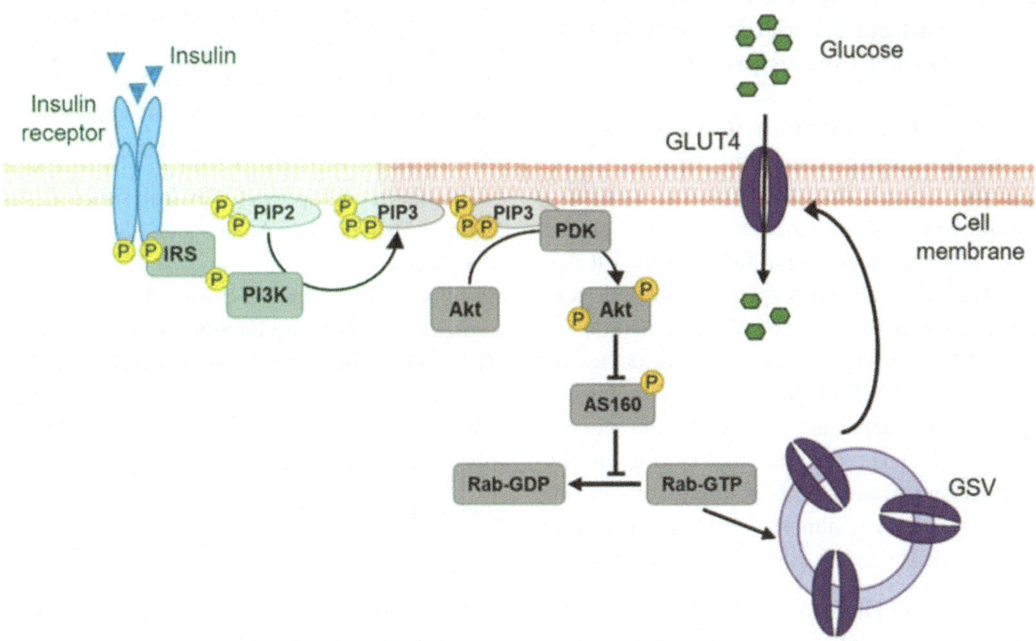

Fig. 12.5 Insulin signaling in muscle and adipose tissue. Insulin binds to the insulin receptor, an RTK which dimerizes and binds to the adapter protein IRS. IRS then leads to PI3K recruitment and activation of AKT. AKT then leads to the movement of GLUT4 to the cell surface as described in the text. (**Source**: Figure from Bauer et al. [106]. Licensed under the Creative Commons Attribution 4.0 (CC BY 4.0) license)

- The Fate of Glucose in Adipose Tissue

In adipose tissue, some of this glucose goes through the central metabolic pathways that lead to energy production. As the fed state progresses, however, energy needs are low. Instead of producing ATP, the main role of glucose in adipose tissue becomes the production of glycerol-3-phosphate, which will become the backbone for TAGs. Glucose is metabolized to DHAP (a glycolysis intermediate), which is then converted to glycerol-3-phosphate (G3P), catalyzed by **Glycerol-3-phosphate dehydrogenase (GPD1)** (In adipose tissue, glycerol-3-phosphate can only be made from DHAP, rather than by direct phosphorylation of glucose). The **glycerol-3-phosphate** that is produced is esterified with fatty acids to produce **TAGs**, While some of these fatty acids can be synthesized in adipose tissue, most of the fatty acids in adipose tissue are delivered from dietary fats (from chylomicrons) and from the liver, which are delivered in VLDLs [5, 56–58].

2.2.5 Glucose Uptake into the Brain and the Broader Nervous System During the Postprandial/Fed State

- Glucose Uptake in the Brain and Nervous System Is Insulin-Independent

Glucose uptake into the nervous system is insulin-independent. To enter the brain, glucose must pass the blood–brain barrier, which it does via **GLUT1** transporters on endothelial cells. Once in the brain, glucose enters the neurons via **GLUT3** transporters, and it enters astrocytes and glial cells via **GLUT1**. Both GLUT1 and GLUT3 are insulin-independent, allowing the nervous system to take up glucose in an insulin-independent manner. This is important since the nervous system is almost completely reliant on glucose as its primary energy source [1, 59–61].

- The Fate of Glucose in the Nervous System

In the brain, there is little glycogen storage. Instead, glucose enters glycolysis and the central metabolic pathways, also independent of insulin. Glucose is first phosphorylated to become G6P. In the brain and nervous system, this is catalyzed by **hexokinase**. Unlike glucokinase in the liver, hexokinase is not insulin-dependent. Glucose is continuously broken down in the brain and nervous system to generate ATP, although astrocytes can store some glucose as glycogen. Although insulin does not play a significant role in glucose storage in the brain, it does have other roles associated with the brain and nervous system. For example, it functions as a neuromodulator, affecting cognition and appetite regulation [62].

2.3 Fat and Protein Metabolism During the Postprandial/Fed State

Glucose metabolism is a major focus of homeostasis, but the metabolism of fats and other nutrients is also important. Here, we discuss fat metabolism in the fed state and include a brief introduction to protein metabolism.

2.3.1 Fat Metabolism During the Fed State

Most fat (both dietary fat and fat made by de novo lipogenesis) is stored during the fed state, and most of it is stored in adipose tissue. In adipose tissue, insulin strongly promotes fat storage while suppressing fat breakdown (lipolysis).

- Inhibition of Lipolysis (Hydrolysis of TAGs into Free Fatty Acids and Glycerol) in Adipocytes

An important way insulin positively regulates fat storage is by inhibiting **hormone-sensitive lipase (HSL)** and **adipose triglyceride lipase (ATGL)**, two enzymes in adipose tissue that break down TAGs into free fatty acids and glycerol [49], as follows (see also Chap. 11 Fig. 11.1):

Step 1: TAG + $H_2O \rightarrow$ DAG + free fatty acid. This step is catalyzed by the enzyme **ATGL.**

Step 2: DAG + $H_2O \rightarrow$ MAG + free fatty acid. This reaction is catalyzed by the enzyme **hormone-sensitive lipase (HSL)**. This is the main hormone-sensitive step, as discussed below.

Step 3: MAG + $H_2O \rightarrow$ glycerol + free fatty acid. This reaction is catalyzed by the enzyme

monoglyceride lipase (MGL). This step leads to the complete hydrolysis of TAGs into fatty acids and glycerol, but it is not a hormone-regulated step.

Insulin has an important role in step two. Insulin-activated signaling pathways mediated by **AKT** lead to a reduction in the levels of the second messenger **cAMP** (which is upregulated by glucagon, as discussed later in the chapter). The downregulation of **cAMP** by **insulin** is mediated by phosphorylation and activation of the **phosphodiesterase PDE3B,** by **AKT**, which degrades cAMP [63]. **cAMP** leads to **protein kinase A (PKA)** activation, and therefore, the reduction of **cAMP** leads to inactivation of **PKA** [64]. **PKA** phosphorylates and thereby activates HSL. By preventing **PKA** activity, insulin prevents this activation of HSL. This can be summarized as: Insulin→IR → PI3K → PIP3 → AKT → PDE3B-↓-cAMP→PKA → HSL.

Unlike **HSL**, **ATGL** is not directly phosphorylated by **PKA**, but **PKA** leads indirectly to its activation as well. **ATGL** requires a cofactor, called CGI-58. In the fed state, CGI-58 is sequestered by binding to the lipid droplet protein perilipin, keeping ATGL activity low. PKA, upregulated by glucagon, phosphorylates perilipin. This releases CGI-58, allowing ATGL to be active. Because insulin decreases **PKA** activity, it decreases the release of CGI-58 and thereby inhibits **ATGL** [65–67].

- Increased TAG Hydrolysis in Circulating Lipoproteins (Lipoprotein Lipase-Mediated Hydrolysis)

In addition to inhibiting the breakdown of fats in adipose tissue, insulin also promotes fat storage by increasing the activity of **lipoprotein lipase (LPL)** in vessels that supply adipose tissue. **LPL** is synthesized in adipocytes and transported to the luminal surface of capillary endothelial cells (blood vessel cells). In these cells, LPL hydrolyzes TAGs that are present in circulating lipoproteins (such as chylomicrons and VLDL, which come from digestion or from the liver, respectively). **LPL** releases two free fatty acids from the TAG, resulting in two free fatty acids plus MAG. The other free fatty acid is released by MGL. The resulting free fatty acids are then taken up by adipocytes and re-esterified into TAGs for storage. In the fed state, insulin increases **LPL** expression and activity in capillaries in adipose tissue [68]. This actually increases fat storage because the free fatty acids are now available for the synthesis of new TAGs for storage in the adipose tissue [8, 49, 67].

Importantly, in skeletal muscle, insulin has a different effect, where insulin does not increase LPL activity and even decreases it [69, 70]. In muscle, fat storage is not a priority, and most of the VLDL and chylomicrons are delivered to adipose tissue, where storage is prioritized.

Insulin regulates LPL activity at both the transcriptional level and post-transcriptional level, and at the level of translocation of LPL to the luminal surface of capillary endothelial cells [68]. Differential regulation relates to the presence of different combinations of transcription factors in various tissues.

- Insulin Stimulates Fatty Acid Synthesis

Insulin also stimulates fatty acid synthesis in the fed state, especially in the liver, by activating **acetyl-CoA carboxylase (ACC)**, the rate-limiting enzyme in fatty acid synthesis. ACC is regulated by phosphorylation on key serine residues [71]. It is inactive when phosphorylated and active when dephosphorylated. Insulin promotes ACC dephosphorylation through the PI3K–AKT signaling pathway, in part by increasing the activity of **protein phosphatase 2A (PP2A)**, and by antagonizing AMPK, thus increasing **ACC's** activity. **PP2A** is a phosphatase that dephosphorylates ACC, and this leads to its activation. **ACC** is also regulated by **AMPK** [16], as discussed in the next section, and **ACC** is regulated by metabolites such as citrate and long-chain fatty acids, as discussed in Chap. 11.

- Flexibility of Fat Metabolism

Even though fat storage is prioritized during the fed state, breakdown of fats (lipolysis) in adipose tissue can still occur if energy demands are especially high. This can occur, for example, during intense exercise. Under these conditions, catecholamines and AMPK (discussed more in the next section) can stimulate TAG hydrolysis in adipose tissue, where TAGs in adipose tissue are hydrolyzed into glycerol and free fatty acids. The resulting glycerol travels to the liver, where it is converted into glycerol-3-phosphate and then contributes to glycolysis or gluconeogenesis via the pathway: glycerol-3-phosphate → DHAP → glyceraldehyde-3-phosphate, a glycolytic intermediate. As a result, glycolysis and the other central metabolic pathways can be carried out, thus generating ATP. Meanwhile, the fatty acids can undergo β-oxidation, where they lead to acetyl-CoA, which enters the Krebs cycle, and this also leads to ATP generation [72] (See Chap. 11).

2.3.2 Protein Metabolism During the Postprandial/Fed State

During the fed state, insulin stimulates amino acid uptake into cells for protein synthesis, especially in muscle. During this state, protein breakdown (proteolysis) is minimal. Some amino acids may still be deaminated and used for energy if they are in excess and cannot be stored, but this is not a major energy source in the fed state [1]. Protein synthesis increases in response to insulin, and this is mediated by mTORC while proteolysis is inhibited (Protein metabolism, including deamination of amino acids, is discussed in more depth in Chap. 14).

2.4 Negative Regulation of Catabolic Pathways During the Postprandial/Fed State

While much of the glucose and fats are stored during the fed state, some are catabolized. Some glucose gets broken down and goes through the cellular respiration pathways to generate ATP. However, checks are in place to ensure that these catabolic pathways do not proceed for too long in most tissues in the fed state. As catabolic pathways proceed, intermediates, such as citrate and acetyl-CoA, eventually accumulate. These intermediates are strong negative regulators of catabolic pathways. Once their levels increase to a certain extent, this signals to the cell that energy levels are sufficient and that energy-producing pathways should stop. Citrate, for example, inhibits **phosphofructokinase (PFK-1)**. As described above, this enzyme regulates a key step in glycolysis: the conversion of F6P to F-6BP. Thus inhibition of PFK-1 slows down glycolysis due to the increasing the accumulation of citrate. Acetyl-CoA inhibits Pyruvate Dehydrogenase complex (**PDH**), the enzyme complex that converts pyruvate to acetyl-CoA. The accumulation of acetyl-CoA thus reduces the formation of more acetyl-CoA, thereby inhibiting the step between glycolysis and the Krebs cycle. As a result of these and other inhibitory factors, instead of continuing to be metabolized, glucose will primarily be stored as glycogen in most cells, although, as discussed above, this prioritization of storage can differ in different tissues.

Fatty acid synthesis is also subject to negative regulation. Malonyl-CoA, which builds up during fatty acid synthesis, is an allosteric inhibitor of **carnitine palmitoyltransferase I (CPT-1). CPT-1** is the enzyme that converts long-chain fatty acyl-CoA to acyl-carnitine (see Chap. 11), which is required for the transport of long-chain fatty acids across the mitochondria for subsequent oxidation. Thus, malonyl CoA slows down most fatty acid β-oxidation in many tissues [46]. In some tissues, however, such as skeletal muscle and heart, β-oxidation of fatty acids may continue during periods where it is inhibited elsewhere, either due to a lower amount of malonyl-CoA or due to an isoform of **CPT-1** that resists its inhibitory effects [49, 73].

3 The Postabsorptive and Fasting States

The **postabsorptive state** begins about 4–6 h after the last meal. This state can last up to about 12 h. A fasting state (also called the basal state) may begin about 12 h after the last meal and continues as long as food is not consumed. During the postabsorptive state (4–6 h after eating), glucose is no longer being absorbed from the digestive tract, and blood glucose levels gradually begin to decrease. This decrease in blood glucose levels signals a reduction in available energy and leads to decreased insulin secretion and an increase in **glucagon**.

3.1 Hormones and Nutrient Sensors in the Postabsorptive and Fasting States

3.1.1 Glucagon Has an Important Role in the Postabsorptive and Fasting States

- Glucagon Function

Glucagon is the major hormone secreted during the postabsorptive and fasting states. Glucagon is secreted from the α cells of the pancreas. It binds to its receptor (a GPCR) on target cells and it helps regulate biochemical pathways that increase glucose levels. When blood glucose levels decrease, glucagon signals the liver to break down the stored glycogen into glucose (glycogenolysis). This newly released glucose will exit the liver and enter the bloodstream, thus helping to replenish blood glucose levels. Glucagon also has other roles that are important in low glucose conditions. For example, it also plays a role in activating gluconeogenesis, which is the process of synthesizing glucose from non-carbohydrate sources. It is also involved in promoting fatty acid oxidation and ketogenesis [8, 46], processes that will be discussed later in the chapter. The metabolic changes that occur due to glucagon in the postabsorptive state are not due to glucagon alone, but to the balance between glucagon, which is increasing, and insulin, which is decreasing, in this state [4, 5].

- Glucagon Signaling

Unlike insulin, which binds to the RTK family of receptors, glucagon binds to G protein-coupled receptors (GPCRs). Specifically, it binds to GPCRs that bind to Gs (see Chap. 18 for details about GPCRs and Gs). GPCRs are seven-membrane-spanning proteins, and their activation by ligand leads to various signaling pathways. A pathway commonly activated by glucagon is the pathway leading to increased activity of **adenylate cyclase (AC)** followed by increased levels of the second messenger **cyclic AMP (cAMP)**, which in turn leads to activation of **protein kinase A (PKA)** (glucagon→ GPCR→AC → cAMP→PKA→). This pathway is similar to GLP- discussed above, but while GLP-1 binds to receptors mostly in the pancreatic βcells (as well as the CNS, and GUT), glucagon binds to receptors on liver and to a lesser extent kidney cells [74].

PKA, activated by the glucagon pathway, has several targets that shift metabolism toward glucose production and away from storage. For example, it phosphorylates and activates **phosphorylase kinase (PhK)**, which is important for the breakdown of glycogen. It also phosphorylates and inactivates **glycogen synthase (GS)**, thereby inhibiting glycogen synthesis, thus conserving glucose. It also leads to the inhibition of **PFK-1**, thereby inhibiting glycolysis, and it phosphorylates and inhibits **pyruvate kinase** (PK) in the liver, preventing the final step in glycolysis, the formation of pyruvate. PKA phosphorylates and activates CREB, which in turn activates the transcription of genes encoding several gluconeogenic enzymes, including **PEPCK, G6Pase, and F6BPase**, thereby increasing **gluconeogenesis** and decreasing glucose breakdown [75, 76]. Several of these steps impacted by glucagon are indicated by arrows in Fig. 12.4. They are also discussed in more detail later in the chapter.

3.1.2 Epinephrine in the Postabsorptive and Fasting States

In addition to **glucagon**, another important hormone in the fasting state is the catecholamine **epinephrine**, especially during stress or physical activity. Norepinephrine, a closely related catecholamine released from sympathetic nerve endings, has similar metabolic effects. These hormones also bind to GPCRs and can activate some of the same pathways activated by glucagon, and they enhance glucose breakdown in the liver and muscle.

3.1.3 AMP-Activated Protein Kinase (AMPK) in the Postabsorptive and Fasting States

AMP-activated protein kinase (AMPK), as described above, is a signaling molecule and energy sensor that has an important role in maintaining energy homeostasis [5]. **AMPK** is a kinase that functions as an **intracellular energy sensor** that responds to the rising **AMP/ATP ratio**. This ratio is high when ATP levels are low, and it is low when ATP levels are high. Thus, AMPK indicates to the cell when energy stores are being depleted (high AMP/ATP ratio). When activated, AMPK promotes catabolic pathways that generate ATP, such as fatty acid oxidation, and it inhibits anabolic pathways that consume ATP, thereby conserving energy under these conditions. AMPK acts in opposition to mTOR. While **mTOR promotes energy storage and growth** when nutrients are abundant, **AMPK inhibits growth processes and conserves energy** during fasting. This balance between AMPK and mTOR is crucial for maintaining metabolic homeostasis across different nutritional states [77].

3.2 Overview of the Adaptations to the Postabsorptive and Fasting State: Preserving Glucose and Shifting Energy Sources

Under conditions of low nutrients or fasting, the body maintains blood glucose levels through several key hormone (largely glucagon) regulated mechanisms [4]. Several of these mechanisms are introduced below, although there are some tissue-specific differences:

1. **Increased glycogenolysis**, especially in liver: **Glycogenolysis** is the breakdown of glycogen into glucose.
2. **Decreased glycogenesis**, also in liver: **Glycogenesis** is the conversion of glucose to glycogen for storage, in the liver.
3. **Inhibition of glycolysis**: Glycolysis is the breakdown of glucose. Inhibition of glycolysis in the liver is important because it prevents glucose from being broken down, allowing it to be exported from the liver and used by tissues that rely on it for energy. It also preserves intermediates that can be redirected toward gluconeogenesis, the formation of more glucose.
4. **Increased gluconeogenesis,** largely in the liver: Gluconeogenesis is the synthesis of glucose from noncarbohydrate precursors, such as lactate, glycerol, or amino acids. Gluconeogenesis is active during fasting and becomes the primary source of blood glucose once the liver's glycogen stores are depleted after a prolonged fast.
5. **Decreased fat synthesis** and **increased fat breakdown**: As cells reduce their reliance on glucose, they begin to rely instead on fatty acids as an energy source. Fat storage therefore decreases. By using fat as an energy source, glucose is saved for tissues like the brain and RBCs that rely primarily on glucose rather than fats for energy.
6. **Ketone body formation**: Finally, when energy sources get especially low, some tissues, including the brain, can also use ketone bodies (produced from the products of fatty acid oxidation in liver) as their energy source.

Each of these six basic responses outlined above is discussed in more detail in the following subsections, and Table 12.1 summarizes some of the metabolic processes in key tissues during both the fed and fasting states.

3.3 Increased Glycogenolysis (Glycogen to Glucose) in the Liver

The first step in **glycogenolysis** (glycogen breakdown into glucose) is catalyzed by the enzyme **glycogen phosphorylase (GP)**. As discussed in Chap. 11, *also see* Fig. 12.4, **GP** releases glucose-1-phosphate (G1P) monomers from glycogen. G1P is then converted to glucose-6-phosphate (G6P) in a reaction catalyzed by

phosphoglucomutase (PGM). This reaction involves transferring the position of the phosphate group from carbon-1 to carbon-6.

G6P can enter the glycolysis pathway, where it is broken down into metabolic intermediates for energy generation. However, in the postabsorptive and fasting states, it will instead be converted to glucose. The conversion of G6P to glucose in the liver is catalyzed by **glucose-6-phosphatase (G6Pase**; not by glucokinase, which catalyzes the irreversible conversion of glucose to G6P) (see Fig. 12.4). The resulting free glucose in the liver can then be released into the bloodstream and distributed to different tissues.

Glucagon plays a crucial role in **glycogenolysis** by activating a signaling pathway, mediated by **PKA**, that phosphorylates and activates **phosphorylase kinase (PhK). PhK** in turn phosphorylates and activates **glycogen-phosphorylase (GP)** (*see* Fig. 12.4) [74, 75, 78]. **GP,** in turn, as described above, breaks down glycogen into units of G1P. The pathway can be summarized as follows:

glucagon-glucagon→GPCR→AC→cAMP→ PKA → PhK → GP → glycogen→ glucose

3.4 Decreased Glycogenesis (Glucose to Glycogen) in Liver

As described in Chap. 10, the synthesis of **glycogen** from **glucose** in the **liver** requires the key regulatory reaction catalyzed by **glycogen synthase (GS)**. **Glucagon** functions by activating a signaling pathway that leads to the phosphorylation and, thereby, the inactivation of **GS**. This is also mediated by the signaling pathway leading to PKA. PKA activates other kinases, such as glycogen synthase kinase (GSK3), and it inhibits protein phosphatase-1 (PP1), which together keep GS in its phosphorylated, inactive form, summarized as: glucagon → GPCR → AC → cAMP → PKA-↓-GS (see Fig. 12.4). As a result, glucagon prevents glucose from being stored as glycogen in the liver, and instead, glucose is available for distribution in the bloodstream [1, 2, 79].

3.5 Inhibition of Glycolysis (the Breakdown of Glucose) in the Liver

Glucagon also **blocks glycolysis** in the liver, preventing glucose from being broken down. Instead, glucose can remain available to exit the liver and circulate in the bloodstream to organs that need it, especially the brain. This inhibition also occurs through glucagon's activation of the PKA-mediated signaling pathway, which targets important regulatory enzymes in glycolysis. Specifically, it leads to inhibition of the enzyme **PFK-1**, which converts **F6P** to **F6BP** (Fig. 12.4), and it leads to activation of the enzyme **F6BPase**, which converts F6BP to F6P. Glucagon regulates this step by modulating the activity of the bifunctional enzyme **PFK-2/ FBPase-2**. It does so as follows (these effects are specific to the liver isoform of PFK-2/ FBPase-2):

PKA, activated in response to **glucagon** signaling, phosphorylates both domains of **PFK-2/ FBPase-2**. Phosphorylation of the **PFK-2** domain (the kinase domain) **inhibits** its activity, and phosphorylation of the **FBPase-2** domain (the phosphatase domain) **stimulates** its activity. In response to inhibiting **PFK-2**, F2,6BP levels decrease, and therefore, PFK-1 activity decreases. At the same time, activation of **FBPas-2** leads to a decrease in the amount of F2,6BP. This results in an increase in the activity of **F6-BPase**, because F2,6BP is an allosteric inhibitor of F6-BPase. **FBPase-2** will be discussed below, in the context of gluconeogenesis. Insulin has an opposing effect on this enzyme. It activates PFK-2 and inhibits FBPase-2, thereby shifting the pathway toward glycolysis.

Glucagon also leads to inhibition of **pyruvate kinase (PK)**, the kinase that converts PEP to pyruvate, the last step in glycolysis. This is also mediated by **PKA**, which phosphorylates the liver form of **PK** and inhibits its activity. (glucagon→GPCR→AC→cAMP→PKA→PK)

This prevents pyruvate from being formed and instead diverts PEP toward gluconeogenesis, especially in the liver [3, 49].

3.6 Increased Gluconeogenesis

Glycogenesis is the use of non-carbohydrate energy sources for gluconeogenesis during the postabsorptive and fasting states. Glycogenesis involves multiple steps and enzymes, as discussed below.

3.6.1 When Glucose Levels Are Low, More Glucose Can Be Generated Via Gluconeogenesis

After a prolonged fast, glucose levels get very low, and over time, the body needs to make more glucose from non-carbohydrate sources; **gluconeogenesis** [4, 57]. A major non-carbohydrate source is **lactate.** Other sources include some amino acids (from protein breakdown) [80] and glycerol, from the breakdown of TAGs. These molecules integrate into the central metabolic pathways and are redirected toward glucose synthesis rather than glucose breakdown. **Glucagon** stimulates key enzymes in the gluconeogenesis pathway. The basic steps in these gluconeogenesis pathways are seen in Fig. 12.4. Gluconeogenesis using lactate as a substrate is discussed in some more detail below. An important aspect of gluconeogenesis is that it involves a reversal of many of the steps of glycolysis.

3.6.2 Lactate is an Important Substrate for Gluconeogenesis

An important substrate of **gluconeogenesis** is lactate. Lactate results from anaerobic glycolysis. Lactate can build up in heavily exercising muscles, as discussed in Chap. 10, when oxygen levels are low. This lactate can leave the muscle, go through the bloodstream, and enter the **liver**, as part of the Cori cycle. In the liver, lactate enters the gluconeogenesis pathway to generate glucose. Because lactate delivery to the liver can become very high during intense exercise, some gluconeogenesis can occur even in the fed state, although this is not normally a major contributor to blood glucose under resting conditions. Another more steady state of lactate comes from red blood cells (RBCs). RBCs, in fact, always carry out anaerobic glycolysis because they do not have mitochondria. RBC-derived lactate is continuously released into circulation and taken up by the liver. During fasting, rising glucagon levels upregulate gluconeogenic enzymes, and this pathway becomes a major source of glucose. Lactate from RBCs is delivered not only to the liver, as all tissues receive lactate from the blood, but only some tissues have the enzymes necessary for gluconeogenesis. Gluconeogenesis is primarily carried out in the liver [4, 5, 10, 57, 78, 81].

3.6.3 The Main Steps of Gluconeogenesis

Lactate from RBCs is transported to the liver. During the fasting state, this lactate serves as a substrate for the production of glucose through gluconeogenesis. Gluconeogenesis occurs as follows, and is shown in Fig. 12.4, and discussed in [8, 49, 57, 80, 82–84].

1. **Lactate to pyruvate**: First, lactate is converted to pyruvate in the cytoplasm of hepatocytes. This is an oxidation reaction catalyzed by **lactate dehydrogenase (LDH)**. This is a reversible reaction, and the same enzyme produces lactate from pyruvate (either under low oxygen conditions or in RBCs).

   Although pyruvate is made from PEP in glycolysis, pyruvate cannot be directly converted to PEP because the steps between PEP and pyruvate are irreversible. Instead, pyruvate leads to PEP formation through several steps, as follows.

2. **Transport of pyruvate into the mitochondria: Pyruvate** enters the mitochondria from the cytoplasm.
3. **Pyruvate to OAA**: In the mitochondrial matrix, pyruvate (3C) is carboxylated and converted to OAA (4C). This reaction is catalyzed by **pyruvate carboxylase (PC)**. This enzyme is positively allosterically regulated by acetyl-CoA, whose levels increase during fatty acid oxidation, a process stimulated by glucagon. **PC** also uses biotin as a cofactor. This step, pyruvate to OAA, is an important step as it is the first step that diverts pyruvate toward glucose production rather than energy production.

   The above reaction (pyruvate to OAA) must occur in the mitochondria because

pyruvate carboxylase (PC) is a **mitochondrial enzyme**. The newly formed OAA is therefore in the mitochondria, but the next reactions need to occur in the cytoplasm. OAA cannot cross the mitochondrial membrane because there is no OAA transporter in the mitochondrial membrane. Therefore, for OAA to enter the cytoplasm from the mitochondria, the following steps are employed.

4. **OAA to malate:** OAA is first converted to malate in a reaction catalyzed by **mitochondrial malate dehydrogenase (MDH2)**, which is a reversible reaction.
5. **Transport of malate to the cytosol**: This malate now exits the mitochondria through a specific **malate transporter** present in the mitochondrial membrane.
6. **Malate to cytoplasmic OAA**: Now, in the cytosol, malate is converted into cytoplasmic OAA in a reaction catalyzed by **cytoplasmic MDH (MDH1)**, resulting in cytoplasmic OAA (These steps 3–6 overlap with the malate–aspartate shuttle (see Chap. 10).
7. **OAA to PEP**: Finally, in the cytosol, OAA is converted to phosphoenolpyruvate (PEP) catalyzed by the enzyme **phosphoenolpyruvate-carboxykinase (PEPCK)**. The levels of **PEPCK** increase when **glucagon** is present (and blood glucose levels are low).
8. **The steps from PEP to glucose:** Now, the pathway can go 'backward' from PEP toward glucose. This involves employing enzymes along the glycolysis pathway. Most of these steps are reversible, but there are three steps that are irreversible (including the conversion of pyruvate to PEP, as covered above), and these steps require separate enzymes that are crucial for gluconeogenesis. The two other irreversible steps are discussed below:
   - **Fructose 6 biphosphate to fructose 6 phosphate:** One of the irreversible steps is the step between fructose 6 biphosphate and fructose 6 phosphate. To bypass this step, another glucagon-regulated enzyme is needed: **fructose-6-bisphosphatase (F6BPase)**. Glucagon regulation of this enzyme is described above and is mediated by glucagon regulation of the bifunctional enzyme PFK-2/FBPase-2 (see Fig. 12.4).

     After the formation of F6P, the pathway now goes all the way back to glucose-6-phosphate. This phosphorylated glucose is trapped in the cell and cannot be released into the bloodstream until the phosphate is removed in the next step.
   - **Glucose 6-phosphate to glucose**: The step between glucose and glucose-6-phosphate is the other irreversible step in glycolysis. In gluconeogenesis, the removal of this phosphate to generate glucose is catalyzed by **glucose-6-phosphatase (G6Pase)**, also a glucagon-regulated enzyme. Glucagon signaling leads to activation of PKA, as discussed above. PKA leads to phosphorylation of CREB, which binds to its binding site (a CRE site) within the promoter of the G6Pase gene. This increases the level of the G6Pase enzyme (In contrast, insulin, via AKT, leads to upregulation of transcription factors like FoxO which suppress transcription of both GCPase and PEPCK genes).

This is the final step in the synthesis of **glucose** (Importantly, **G6Pase** is present in the liver and kidney but not in muscle, and therefore this reaction, and consequently gluconeogenesis, cannot take place in muscle).

3.6.4 Overview of the Role of Glucagon in Gluconeogenesis

Glucagon regulates several key enzymes in the gluconeogenesis pathway through both post-translational mechanisms and gene expression.

For example, it indirectly leads to the activation of **pyruvate carboxylase (PC)**. It does so via its role in promoting fatty acid oxidation (discussed in Chap. 10), which increases mitochondrial acetyl-CoA levels. Acetyl-CoA in turn, is an allosteric activator of **PC**, thereby enhancing the conversion of pyruvate to Oxaloacetate (OAA).

Glucagon also activates expression of the gene-encoding **PEPCK**, the enzyme that converts OAA to PEP. Regulation of **PEPCK** gene expression by glucagon is due to the PKA → CREB pathway. As discussed above, glucagon leads to PKA activation via GPCR activation. PKA phosphorylates numerous targets. One important target is the transcription factor CREB. Phosphorylation of CREB activates it, and upon recruiting coactivators, CREB can bind to the promoter of the PEPCK gene, thereby increasing its transcription. (glucagon→ GPCR→AC → cAMP→PKA → CREB→ PEPCK).

Glucagon also **upregulates glucose-6-phosphatase (G6Pase)** at the transcriptional level, which is also mediated by CREB, similar to the PEPCK gene.

Glucagon also indirectly activates **F6BPase.** It does so by reducing levels of F2,6BP. F2,6BP is not only a positive allosteric regulator of **PFK-1**, as discussed above, but it is also an inhibitor of **F6BPase**. By reducing the level of F2,6BP, glucagon can promote the activity of **F6BPase**, thereby shifting the pathway toward gluconeogenesis (see Fig. 12.4 and also Chap. 11, Fig. 11.8b).

Together, these regulatory mechanisms promote gluconeogenesis in the liver during fasting (see Fig. 12.4).

3.6.5 Regulation of Gluconeogenesis by AMPK During the Postabsorptive and Fasting States

Gluconeogenesis is an important mechanism for the generation of glucose in the fasting state. However, it is subject to regulation. Glucogenesis may begin during the early stages of fasting. However, during severe fasting, gluconeogenesis may consume too many resources, causing ATP levels to become too low. When this happens, AMPK plays a role. AMPK senses low ATP levels and eventually stops the process of gluconeogenesis. It will do so by inhibiting the expression of certain gluconeogenic genes. For example, it can lead to downregulation of the gene for **PEPCK** and **G6Pase** [85].

3.6.6 AMPK in Skeletal Muscle During Fasting

During fasting, glucose levels are low and insulin levels are low. Although some glucose may still be present in the bloodstream, glucose uptake by skeletal muscle is reduced. However, in situations where energy demand increases, such as during intense exercise, AMPK becomes activated in response to an elevated AMP/ATP ratio. AMPK, in turn, promotes the translocation of GLUT4 transporters to the muscle cell membrane through an insulin-independent pathway. This allows for increased glucose uptake even in the absence of high insulin levels. In contrast to active muscle, resting skeletal muscle during fasting primarily relies on fatty acid oxidation and, in the early stages of fasting, also uses ketone bodies as an energy source. In prolonged fasting, however, skeletal muscle decreases its utilization of ketone bodies to conserve them for the brain, further increasing reliance on fatty acids for ATP production [4, 86, 87].

3.7 Decreased Fat Synthesis and Increased Fat Breakdown During the Postabsorptive and Fasting State, Leading to an Overall Reduction in Stored Fat

3.7.1 Inhibition of Fatty Acid Synthesis

During fasting, energy stores are mobilized rather than being replenished. Glucagon inhibits not only the storage of glycogen but also the synthesis of fats, another form of energy storage. The main mechanism by which it inhibits fatty acid synthesis is by inhibiting the activity of the enzyme **acetyl-CoA-carboxylase (ACC)** (see Chap. 10, Fig. 10.11). This is in contrast to insulin, which stimulates **ACC** and thereby stimulates the synthesis and storage of fatty acids [88]. Glucagon inhibits **ACC** via phosphorylation, which inactivates it. This can occur via PKA, which can

directly phosphorylate ACC, leading to its inactivation.
(glucagon→GPCR→AC→cAMP→PKA→ACC (inactivation)).

AMPK also has an important role in ACC regulation. **AMPK** is activated in this state due to sensing an increased AMP/ATP ratio. AMPK, in turn, also phosphorylates and inactivates ACC [5, 57, 89, 90].

3.7.2 Fat Breakdown (Lipolysis) in Adipocytes

While fat synthesis is inhibited, fat breakdown (lipolysis) is also regulated. Lipolysis is regulated by hormones, but in humans, glucagon's role in lipolysis is minimal or indirect compared to other hormones like catecholamines (epinephrine and norepinephrine). Lipolysis starts with the breakdown of TAGs into free fatty acids and glycerol. As discussed above, this involves three steps, catalyzed by ATGL, HSL, and MGL. Of these, ATGL and HSL are hormonally regulated; specifically, they are activated in the fasting state.

ATGL is indirectly activated by glucagon, epinephrine, and norepinephrine. As discussed above, ATGL requires a cofactor, CGI-58. This cofactor can be sequestered by perilipin. However, glucagon or the catecholamines, through a signaling pathway mediated by **PKA**, lead to phosphorylation of perilipin. This releases CGI-58, allowing ATGL to be active. (glucagon→GPCR→AC→cAMP→PKA→phosphorylated perilipin→active ATGL)

Epinephrine and norepinephrine function similarly to glucagon and are, in fact, the main hormones in this process.

HSL is regulated by direct phosphorylation by PKA in response to adrenaline (epinephrine) and noradrenalin (norepinephrine) as follows: (hormone→GPCR→AC → cAMP→PKA → HSL-phosphorylation)

3.7.3 The Fate of the Glycerol from Lipolysis: Gluconeogenesis

The glycerol that results from the breakdown of TAGs from fat stores is transported to the liver. Glycerol is hydrophilic, and it can enter the circulation without being bound to any carrier proteins. Once in the liver, glycerol in the postabsorptive state contributes primarily to gluconeogenesis. This process begins when glycerol is phosphorylated by **glycerol kinase** to form glycerol-3-phosphate (G3P), which is then oxidized to dihydroxyacetone phosphate (DHAP). DHAP can convert to glyceraldehyde 3 phosphate (GAP), and DHAP and glyceraldehyde 3 phosphate are condensed by aldolase and form fructose-6-bisphosphate (F6B). From here, the pathway works its way back toward glucose, as we have seen before with gluconeogenesis from lactate. Here again, certain enzymes in this process are hormone-sensitive. **Glucose-6-phosphatase** and **fructose-6-phosphatase** are positively regulated by glucagon signaling [8, 49] (see Fig. 12.4).

3.7.4 The Fate of the Fatty Acids from TAG Breakdown

In general, fatty acid oxidation in adipocytes is minimal, with most of it occurring in other tissues. The fatty acids released from the TAGs during lipolysis are mostly transported from adipose tissues to various tissues, with muscle being a primary destination. The liver also takes up some fatty acids for either oxidation or the formation of ketone bodies. Fatty acids are bound to a protein called serum albumin for transport. This is important because free fatty acids are hydrophobic on their own. Albumin can carry several fatty acid molecules at once. Once they reach their target tissue, the fatty acids are released from the albumin.

- β Oxidation of Even-Chain Fatty Acids During the Initial Stages of Fasting

In their target tissues, such as muscle, the fatty acids undergo β oxidation, a process described in more detail in Chap. 11. In the process of fatty acid β oxidation, the even-chain fatty acids, which are the most common, will lead to acetyl-CoA formation [91, 92]. This acetyl-CoA condenses with Oxaloacetate (OAA) and, in turn, leads to citrate production and thereby enters the Krebs cycle. The Krebs cycle results in the production of NADH and $FADH_2$, which will be utilized by the electron transport chain to generate ATP. Thus, even in a fasting state, there is a supply of ATP to be used for the body's functions.

AMPK has an important role in this process. As discussed above, AMPK senses the low ATP levels (via a high AMP/ATP ratio) and, in turn, phosphorylates ACC, inhibiting its activity. This, in turn, reduces malonyl-CoA levels. Malonyl-CoA is an initial building block of fatty acids, and it is a negative regulator of **carnitine palmitoyltransferase I (CPT1)**. By reducing malonyl-CoA, the inhibitory effect on **CPT1** is relieved. This is important for the transport of fatty acyl chains into the mitochondria for oxidation.

- Even-Chain Fatty Acids Under Prolonged Fasting: Ketogenesis

Under prolonged fasting conditions, acetyl-CoA can accumulate to an excessive level. This is because acetyl-CoA normally combines with OAA and enters the Krebs cycle, but under fasting conditions, most of the OAA is being used up for gluconeogenic pathways, as described above. Consequently, acetyl-CoA eventually accumulates. Under these conditions, the excess acetyl-CoA leads to the formation of ketone bodies [5, 93]. Ketone bodies (also called ketones) are important molecules in biochemistry and have important implications in homeostasis, diet, and nutrition. They are an important alternative energy source for the brain when glucose levels are low, and they are discussed in a separate section in more detail, following the discussion of odd-chain fatty acids below.

- β Oxidation of Odd-Chain Fatty Acids

The odd-chain fatty acids are less common than even-chain fatty acids, and they have a different outcome when oxidized. When odd-chain fatty acids are oxidized, they become propionyl CoA (three carbons plus CoA), which becomes succinyl CoA (four carbons plus CoA) and thus feeds into the Krebs cycle. As part of the Krebs cycle, succinyl-CoA is converted through the subsequent reactions of the cycle to fumarate, which is then converted to malate. Malate can then be transported across the mitochondrial membrane into the cytoplasm. There, it is converted to OAA by **cytosolic malate dehydrogenase**. The cytosolic OAA can then be used for gluconeogenesis, just as described above, when lactate is the substrate [49, 94, 95].

3.8 Ketone Bodies Are Formed and Are a Source of Energy Under Prolonged Fasting [6]

3.8.1 Acetyl CoA to Ketone Body Formation

If a fast goes on for a long time, even more energy sources are required. In some of the reactions discussed above, a large portion of the OAA gets shuttled from the mitochondria to the cytosol (via conversion to malate) and used up in gluconeogenesis reactions. Normally, acetyl-CoA interacts with OAA in the mitochondria to lead to citrate and entry into the Krebs cycle. When OAA levels are low, however, OAA is not available for interaction with acetyl-CoA. Not only this, but Acetyl-CoA levels increase in response to the oxidation of other molecules, especially fatty acids. This situation leads to the accumulation of acetyl-CoA, which, in turn, triggers the formation of ketone bodies in the liver. Ketone bodies are structures that form when acetyl-CoA molecules bind together. These ketone bodies are produced in the mitochondria but can diffuse across the mitochondrial membrane. They get transported to the bloodstream, from which they can reach many tissues that can oxidize ketones. The brain is an important destination because when glucose levels are low, the brain cannot use other energy sources, such as fatty acids, which cannot cross the blood-brain barrier. Ketones are also delivered to the heart and skeletal muscle. In these tissues, ketones can diffuse into the mitochondria and break down into acetyl-CoA. The brain still has plenty of mitochondrial OAA because it does not carry out gluconeogenesis. Thus, this acetyl-CoA can condense with OAA to form citrate, allowing it to enter the Krebs cycle and generate ATP for the brain. Ketone bodies are, therefore, a good energy

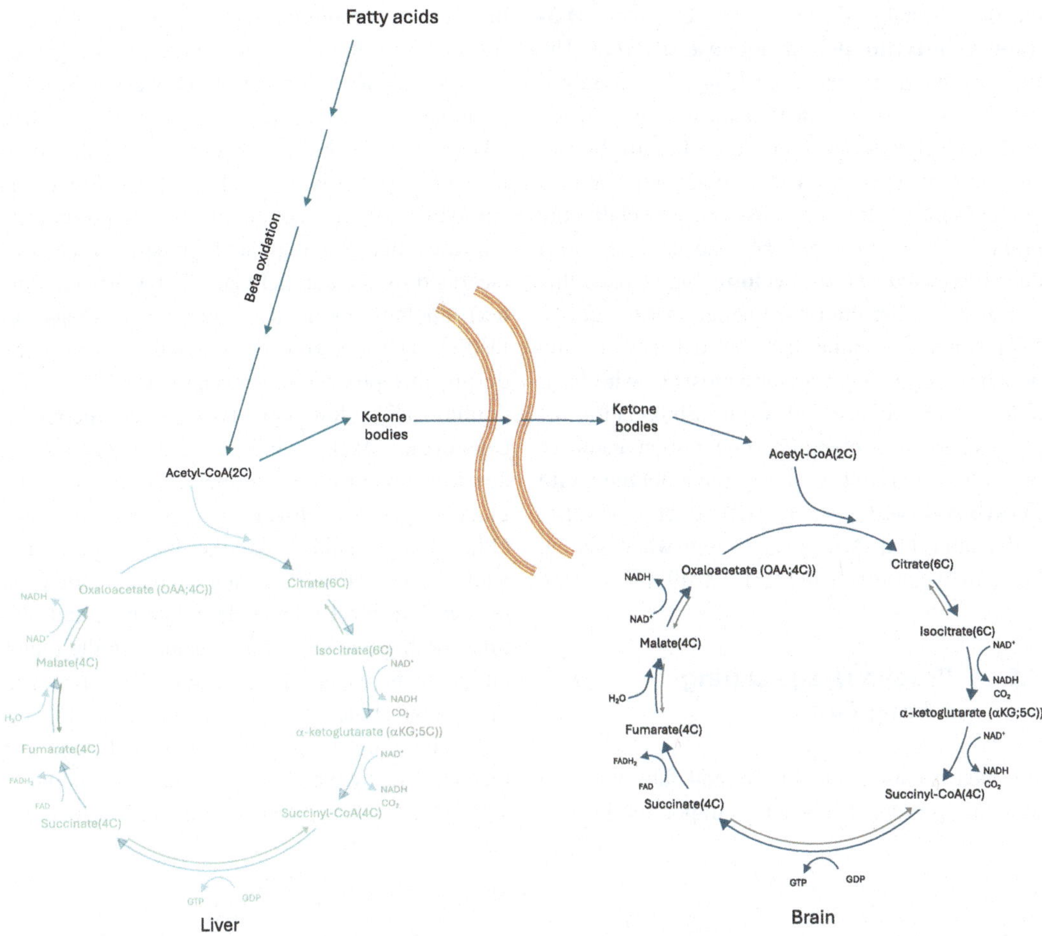

Fig. 12.6 Ketone bodies as a source of energy. During fasting, instead of entering the Krebs cycle, acetyl-CoA is diverted towards ketone body formation in the liver. These ketone bodies can be transported through the bloodstream to tissues such as the brain. In these tissues, including brain, ketone bodies are taken up by cells and converted into acetyl-CoA in the mitochondria, which can enter the Krebs cycle

source for the brain even under fasting conditions [5, 93, 96] (*see* Fig. 12.6).

3.8.2 What Are Ketone Bodies, and How Are They Formed?

Ketones are a broad class of organic compounds that have a carbonyl group bonded to two carbon atoms within a larger carbon-containing molecule. Ketone bodies are specific ketones derived primarily from acetyl-CoA, which is produced by fatty acid oxidation (although they can also come from other sources, such as ketogenic amino acids, discussed in Chap. 14) [5, 8, 93]. The formation of ketone bodies is a process known as ketogenesis, which involves the following steps. First, two molecules of acetyl-CoA are condensed to form acetoacetyl-CoA, catalyzed by the enzyme **thiolase**. Next, the acetoacetyl-CoA reacts with another molecule of acetyl-CoA to form **HMG-CoA**. This is catalyzed by the enzyme **HMG-CoA synthase (HMGCS)**. Next, HMG-CoA is cleaved to form the ketone body **acetoacetate** in a reaction catalyzed by the enzyme **HMG-CoA lyase (HMGCL)**. **Acetoacetate** can be further processed and reduced to form another ketone body, **D-3-hydroxybutyrate** (also called **beta-hydroxybutyrate (BHB)**), in a reversible reaction catalyzed by the enzyme **D-3-hydroxybutyrate dehydrogenase (BDH1)**. The forward reaction uses NADH, which is oxidized to NAD^+. Both **acetoacetate** and **D-3-hydroxybutyrate** are considered **ketone bodies** and are excreted from the **liver** and used as energy sources for other tissues, especially the brain. Some of the acetoacetate is also decarboxylated to form **acetone**, but most of this is exhaled rather than used as an energy source. Interestingly, acetone can be detected in the breath of people undergoing ketosis or who have diabetic ketoacidosis. The decarboxylation of acetoacetate to acetone can occur spontaneously or can be catalyzed by the enzyme **acetoacetate decarboxylase**. Ketone body production is especially important during starvation, when ketone bodies serve as the main fuel [5, 8, 49] (Fig. 12.7).

3.9 Protein Usage During a Long Fast

We have focused on glucose and fats, but in a long fast, protein also becomes important. Protein breakdown is minimal during a typical 24-h fed–fast cycle, as the body relies primarily on glycogen and fat stores for energy. However, during a prolonged fast (such as one lasting longer than 24 h), muscle protein begins to break down through proteolysis. This breakdown is stimulated by the action of several proteolytic enzymes that are regulated in part by cortisol (secreted by the adrenal gland in the adrenal cortex), another hormone produced in response to fasting. The amino acids released from proteins during this process are transported to the liver. Amino acids fall into two major metabolic categories. Many amino acids (glucogenic amino acids) provide intermediates that enter the gluconeogenic pathway to produce glucose. Other amino acids, like the ketogenic amino acids, can contribute to ketone body formation or feed into the Krebs cycle. Glutamine and alanine (both glucogenic amino acids) are the major amino acids released from muscle after a fast, and they contribute to gluconeogenesis [5, 6, 84]. These categories of amino acids and the mechanisms by which they can lead to gluconeogenesis are discussed in more detail in Chap. 14.

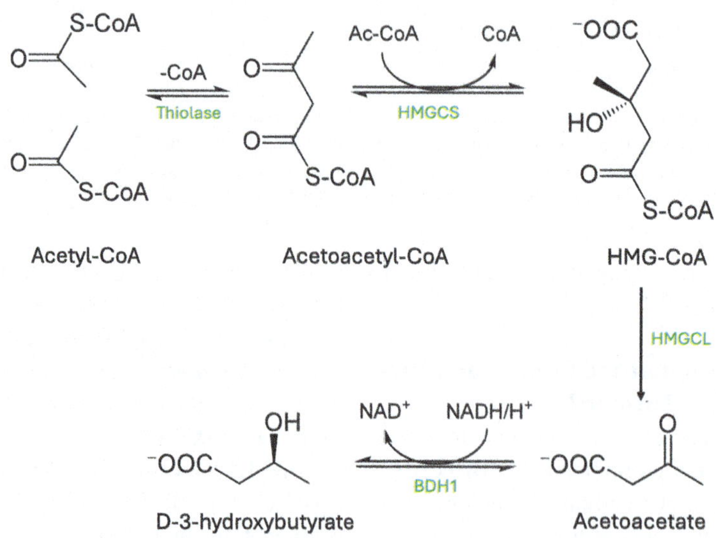

Fig. 12.7 Formation of ketone bodies from acetyl-CoA. The formation of ketone bodies involves the condensation of acetyl-CoA molecules. Acetoacetate and D-3-hydroxybutyrate are both ketone bodies. Acetone, which forms spontaneously from acetoacetate (and not shown here) is also a ketone body, but it is mostly exhaled. Abbreviations used here: HMGCS HMG-CoA synthase, HMBCL HMG-CoA-lyase, BDH1 D3-hydroxybutyrate dehydrogenase. (**Source:** Figure modified from 'Ketone bodies synthesis" by Yikrazuul, licensed under **CC0/public domain** via Wikimedia Commons https://commons.wikimedia.org/wiki/File:Ketone_bodies_synthesis.svg)

3.10 The Different Sources of Glucose Are Related to the Length of the Fast

We have discussed several possible sources of glucose available during the post-absorptive and fasting states. The sources used to produce more glucose can change over time and depend on the length of the fast. During the early part of a fast, the breakdown of glycogen (glycogenolysis) in the liver is a major source of glucose for the body. Over time, less glycogen becomes available, and alternative sources must be utilized. Therefore, as a fast continues once the liver's glycogen stores are depleted, fat breakdown (lipolysis) primarily in adipose tissue becomes more prominent. This involves the β oxidation of fatty acids, which becomes an important source of both energy (ATP) and glycerol, which contributes to gluconeogenesis. In the liver, βoxidation also produces large amounts of acetyl-CoA, which is converted into ketone bodies. These ketone bodies circulate to tissues such as the brain, muscle, and heart, where they serve as an alternative energy source to glucose, thereby reducing the need for glucose production and slowing down protein breakdown. Even more prolonged fasting can lead to protein breakdown (proteolysis) in muscle. This leads to the release and breakdown of amino acids, which are transported to the liver, where, during fasting, they are mostly used for gluconeogenesis. Regardless of the source, a key goal during fasting is to maintain an adequate level of glucose, which circulates in the bloodstream and serves as an energy source for various tissues. Circulating glucose is especially important for tissues that lack the ability to obtain energy from fatty acids, such as the brain and red blood cells. While RBCs still require glucose, much of the brain's ATP during prolonged fasting comes from ketone bodies. ATP production remains important; however, the major source of ATP during fasting is from fatty acid β oxidation [8, 10, 97].

4 Back to the Postprandial State: The Re-Fed State

Immediately after a fast, when nutrients are reintroduced, blood glucose levels rise again. Glucose is absorbed by tissues such as the brain, fat, muscle, and RBCs. These cells use the glucose for immediate energy (ATP) generation. The liver also takes up glucose, but instead of prioritizing ATP generation, the liver first replenishes its glycogen stores. Some of the liver's glucose does enter glycolysis, but the liver relies largely on fatty acids (obtained from adipose tissue or dietary fats) for energy. Once glycogen stores are replenished, glycolysis in the liver becomes more active, and excess glycolytic intermediates are converted to acetyl-CoA, which is used for fatty acid synthesis, and these fatty acids are transported to the adipose tissue [10, 46, 98]. This re-fed state maintains metabolic balance, ensures a balanced energy supply, and prepares the body for another cycle of fasting and feeding.

5 Summary and Tissue-Specific Considerations

This chapter has focused mainly on the liver, adipose tissue, skeletal muscle, and, to some extent, the nervous system. These tissues make the biggest contribution to maintaining homeostasis in the entire body [99]. Other tissues, however, are also important and must adapt to changes in nutrient availability during the fed–fast cycle. Cardiac muscle is continuously active and has a constant demand for ATP. It can oxidize glucose, fatty acids, and lactate in the fed state. The heart must avoid energy deficits, and it relies mostly on fatty acids and ketone bodies during fasting [100]. The brain depends almost entirely on glucose during the fed state, but during prolonged fasting, it can use ketone bodies, although a basal requirement for glucose remains [101]. Red blood cells lack mitochondria and rely entirely on anaerobic glycolysis during fasting. In fasting, the glucose they obtain is mostly from hepatic (liver) gluconeogenesis [8, 49]. Smooth muscle, in vascular and visceral organs, has a lower energy requirement than some other tissues, but its energy levels must be sustained. Smooth muscle can generate ATP from both glycolysis and oxidative metabolism, but unlike skeletal muscle, it only stores low amounts of glycogen, so it relies on glucose and other substrates such as fatty acids, supplied by the bloodstream. The kidney also contributes significantly to glucose production during fasting [102–105].

6 Balanced Equations

6.1 Gluconeogenesis (Lactate to Glucose)

These reactions happen twice because two pyruvates are required to make one glucose:

1. Lactate → Pyruvate
   Lactate + NAD^+ ⇌ Pyruvate + NADH + H^+
   Enzyme: Lactate dehydrogenase (LDH)
   Location: Cytosol
2. Pyruvate → Oxaloacetate (OAA)
   Enzyme: Pyruvate carboxylase (PC)
   Pyruvate + CO_2 + ATP → Oxaloacetate + ADP + Pi
   Location: Mitochondrial matrix
3. Oxaloacetate → Malate
   Oxaloacetate + NADH + H^+ → Malate + NAD^+
   Enzyme: Malate dehydrogenase (mMDH)
   Location: mitochondrial matrix, and then Malate is transported to the cytosol
4. Malate → Oxaloacetate
   Malate + NAD^+ → Oxaloacetate + NADH + H^+
   Enzyme: Malate dehydrogenase (cMDH)
   Location: cytosol, and all the subsequent reactions are also in the cytosol
5. Oxaloacetate → Phosphoenolpyruvate (PEP)
   Oxaloacetate + GTP → PEP + CO2 + GDP
   Enzyme: Phosphoenolpyruvate carboxykinase (PEPCK)
   Location: cytosol (in this version of the pathway)
   Note: Some tissues also contain mitochondrial PEPCK (PEPCK-M).
6. PEP → 2-Phosphoglycerate
   PEP + H_2O ⇌ 2-Phosphoglycerate
   Enzyme: Enolase
7. 2-Phosphoglycerate → 3-Phosphoglycerate
   2-Phosphoglycerate ⇌ 3-Phosphoglycerate
   Enzyme: Phosphoglycerate mutase
8. 3-Phosphoglycerate → 3-Bisphosphoglycerate
   3-Phosphoglycerate + ATP → 3-Bisphosphoglycerate + ADP
   Enzyme: Phosphoglycerate kinase
9. 3-Bisphosphoglycerate → Glyceraldehyde-3-phosphate (G3P)
   3-Bisphosphoglycerate + NADH + H^+ → G3P + Pi + NAD^+
   Enzyme: Glyceraldehyde-3-phosphate dehydrogenase (GAPDH)
10. G3P + DHAP → Fructose-6-bisphosphate
    (a) G3P ⇌ DHAP
    Enzyme: Triose phosphate isomerase
    (b) (G3P + DHAP → F6BP)
    Enzyme: Aldolase
11. Fructose-6-bisphosphate → Fructose-6-phosphate
    Fructose-6-bisphosphate + H_2O → Fructose-6-phosphate + Pi
    Enzyme: Fructose-6-bisphosphatase
12. Fructose-6-phosphate → Glucose-6-phosphate
    Fructose-6-phosphate ⇌ Glucose-6-phosphate
    Enzyme: Phosphoglucose isomerase
13. Glucose-6-phosphate → Glucose
    Glucose-6-phosphate + H_2O → Glucose + Pi
    Enzyme: Glucose-6-phosphatase
    Net Summary Reaction:
    2 Lactate +6 ATP + 2 GTP + → Glucose +6 ADP + 2 GDP + 6 Pi

6.2 Lipolysis: Breakdown of TAGs into Glycerol and FAs

Step 1. TAG + H_2O → DAG + FFA.
Enzyme: **ATGL.**
Step 2. DAG + H_2O → MAG + FFA. T
Enzyme: **hormone-sensitive Lipase (HSL)**.
Step 3: MAG + H_2O → glycerol + FFA.
Enzyme: **monoglyceride lipase (MGL)**.

FDA-Approved Drugs Covered in This Chapter

Commonly used brand name	Generic name	First FDA approval year	Main conditions	Main targets
Mounjaro	Tirzepatide	2022	Type 2 diabetes	GLP-1 and GIP
Ozempic	Semaglutide	2017	Type 2 diabetes	GLP-1

Definitions

Acetone A ketone body that is produced during ketogenesis that is primarily exhaled.

Adipocytes Fat cells.

Adipose tissue Tissue specialized in fat storage and endocrine signaling (e.g., leptin secretion).

AMP activated protein kinase (AMPK) An energy sensor that is activated when the AMP/ATP ratio is high.

Cyclic AMP (cAMP) Cyclic adenosine monophosphate; a second messenger that mediates glucagon signaling via protein kinase A (PKA).

De Novo lipogenesis The metabolic process by which excess carbohydrates (mostly glucose) are converted to fatty acids. These newly synthesized fatty acids can be esterified into triacylglycerols and stored in adipose tissue or the liver.

Glucagon A hormone secreted by pancreatic α cells when glucose levels are low. It stimulates glycogenolysis, gluconeogenesis, and lipolysis.

Glucagon-like peptide-1 (GLP-1) An incretin hormone that enhances insulin secretion in a glucose-dependent manner and slows down gastric emptying.

Gluconeogenesis The metabolic pathway that synthesizes glucose from non-carbohydrate sources like lactate, amino acids, and glycerol.

Glucose homeostasis The process by which the body maintains stable blood glucose levels through hormonal regulation, mainly involving insulin and glucagon.

Glucose transporter type 4 (GLUT4) An insulin-responsive glucose transporter found primarily in muscle and adipose tissue.

Glycogenesis The synthesis of glycogen from glucose, stimulated by insulin during the fed state.

Glycogenolysis The breakdown of glycogen into glucose. Glycogenolysis is stimulated by glucagon during fasting.

HMG-CoA synthase An enzyme that is important for the formation of ketone bodies.

Incretin hormones Gut-derived hormones that enhance insulin secretion in response to food intake. GLP-1 is an incretin hormone.

Insulin A hormone secreted by pancreatic β cells that promotes glucose uptake into tissues and promotes storage and anabolic processes.

Ketone bodies Energy-rich molecules synthesized in the liver from acetyl-CoA during prolonged fasting or low-carbohydrate diets.

Mechanistic target of rapamycin (mTOR) A protein kinase and signaling protein that functions as a nutrient and energy sensor. It promotes anabolic processes like protein and lipid synthesis when nutrient levels are high.

Phosphoenolpyruvate Carboxykinase (PEPCK) A key enzyme in gluconeogenesis that converts oxaloacetate to phosphoenolpyruvate (PEP).

Pyruvate carboxylase An enzyme in gluconeogenesis that converts pyruvate to oxaloacetate.

Study Questions

Multiple-Choice Questions

1. Which hormone promotes glucose uptake and storage in the fed state?
   A. Glucagon
   B. Cortisol
   C. Insulin
   D. Epinephrine
2. Which of the following glucose transporters is insulin-dependent?
   A. GLUT1
   B. GLUT2
   C. GLUT3
   D. GLUT4
3. What is the storage form of fat in adipose tissue?
   A. Glycogen
   B. Fatty acid
   C. Triacylglycerol (TAG)
   D. Glycerol-3-phosphate

4. During prolonged fasting, what is a major source of energy for the brain?
   A. Glucose
   B. Ketone bodies
   C. Glycerol
   D. Lactic acid
5. Which enzyme is inhibited by citrate to downregulate glycolysis?
   A. Phosphofructokinase-1 (PFK-1)
   B. Glucokinase
   C. Pyruvate kinase
   D. PEP carboxykinase

True/False Questions

6. The brain can store significant amounts of glycogen for energy use.
7. Glucagon promotes lipogenesis in adipose tissue during fasting.

Fill-in-the-Blank Questions

8. Insulin is secreted by the pancreatic _______ cells.
9. Lactate can lead to gluconeogenesis. The first step in this process is the conversion of lactate to _________.
10. Glucagon has an important role in glycogenolysis because it activates a signaling pathway, mediated by _____, that leads to activation of glycogen-phosphorylase.

Short-Answer Question

11. What is the role of GLP-1 in insulin regulation?

Matching Question

12. **Match each transporter with a tissue**
    Transporter
    GLUT2
    GLUT3
    GLUT4
    Tissue.
    A. Brain neurons
    B. Hepatocytes and pancreatic beta cells
    C. Skeletal muscle and adipose tissue

Answer Key

Multiple-Choice Questions

1. **Answer**: C
2. **Answer**: D
3. **Answer**: C
4. **Answer**: B
5. **Answer**: A

True/False Questions

6. **Answer**: False
7. **Answer**: False

Fill-in-the-Blank Questions

8. **Answer**: Beta
9. **Answer**: Pyruvate
10. **Answer**: cAMP

Short-Answer Question

11. **Answer**: GLP-1 is an incretin hormone released from intestinal L-cells that enhances insulin secretion from pancreatic beta cells in a glucose-dependent manner.

Matching Question

12. **Answers**:
    GLUT2 → B, GLUT3 → A, GLUT4→ C

References

1. Rahman MS, Hossain KS, Das S, Kundu S, Adegoke EO, Rahman MA, et al. Role of insulin in health and disease: an update. Int J Mol Sci. 2021;22(12):6403. https://doi.org/10.3390/ijms22126403.
2. Röder PV, Wu B, Liu Y, Han W. Pancreatic regulation of glucose homeostasis. Exp Mol Med. 2016;48 (3):e219. https://doi.org/10.1038/emm.2016.6.
3. Hall JE, Ha ME. Guyton and Hall textbook of medical physiology. 14th ed. Elsevier; 2020.
4. Bender DA. Introduction to nutrition and metabolism. Boca Raton: CRC Press; 2021.
5. Chandel NS. Navigating metabolism. 2nd ed. Cold Spring Harbor: Cold Spring Harbor Laboratory Press; 2020.
6. Roche HM, Macdonald IA, Schols AMWJ, Lanham-New SA, editors. Nutrition and metabolism. Wiley-Blackwell; 2024.
7. Devlin TM. Textbook of biochemistry with clinical correlations. 7th ed. Hoboken: John Wiley & Sons; 2010.
8. Berg JM, Tymoczko JL, Gatto GJ Jr, Stryer L. Biochemistry. 9th ed. New York: MacMillan Learning; 2019.
9. St. Jeor ST, Howard BV, Prewitt TE, Bovee V, Bazzarre T, Eckel RH, et al. Dietary protein and weight reduction. Circulation. 2001;104(15): 1869–74. https://doi.org/10.1161/hc4001.096152.
10. Rui L. Energy metabolism in the liver. Compr Physiol. 2014;4(1):177–97. https://doi.org/10.1002/cphy.c130024.
11. Meessen ECE, Warmbrunn MV, Nieuwdorp M, Soeters MR. Human postprandial nutrient metabolism and low-grade inflammation: a narrative review. Nutrients. 2019;11(12):3000. https://doi.org/10.3390/nu11123000.
12. Hu C, Chen Y, Yin X, Xu R, Yin C, Wang C, et al. Pancreatic endocrine and exocrine signaling and crosstalk in physiological and pathological status. Signal Transduct Target Ther. 2025;10(1):39. https://doi.org/10.1038/s41392-024-02098-3.
13. Da Silva XG. The cells of the islets of Langerhans. J Clin Med. 2018;7(3):54.
14. Yuan H-X, Xiong Y, Guan K-L. Nutrient sensing, metabolism, and cell growth control. Mol Cell. 2013;49(3):379–87. https://doi.org/10.1016/j.molcel.2013.01.019.
15. Ben-Sahra I, Manning BD. mTORC1 signaling and the metabolic control of cell growth. Curr Opin Cell Biol. 2017;45:72–82. https://doi.org/10.1016/j.ceb.2017.02.012.
16. Hardie DG. AMP-activated protein kinase: an energy sensor that regulates all aspects of cell function. Genes Dev. 2011;25(18):1895–908. https://doi.org/10.1101/gad.17420111.
17. Coll AP, Farooqi IS, O'Rahilly S. The hormonal control of food intake. Cell. 2007;129(2):251–62. https://doi.org/10.1016/j.cell.2007.04.001.
18. Dimitriadis GD, Maratou E, Kountouri A, Board M, Lambadiari V. Regulation of postabsorptive and postprandial glucose metabolism by insulin-dependent and insulin-independent mechanisms: an integrative approach. Nutrients. 2021;13(1):159. https://doi.org/10.3390/nu13010159.
19. Betts JG, Young KA, Wise JA, Johnson E, Poe B, Kruse DH, Korol O, Johnson JE, Womble M, DeSaix P. Anatomy and physiology. 2nd ed. Houston: OpenStax; 2023.
20. Nadkarni P, Chepurny OG, Holz GG. Regulation of glucose homeostasis by GLP-1. Prog Mol Biol Transl Sci. 2014;121:23–65. https://doi.org/10.1016/b978-0-12-800101-1.00002-8.
21. Vasiljević J, Torkko JM, Knoch KP, Solimena M. The making of insulin in health and disease. Diabetologia. 2020;63(10):1981–9. https://doi.org/10.1007/s00125-020-05192-7.
22. Caspi I, Tremmel DM, Pulecio J, Yang D, Liu D, Yan J, et al. Glucose transporters are key components of the human Glucostat. Diabetes. 2024;73(8): 1336–51. https://doi.org/10.2337/db23-0508.
23. Yunn N-O, Kim J, Ryu SH, Cho Y. A stepwise activation model for the insulin receptor. Exp Mol Med. 2023;55(10):2147–61. https://doi.org/10.1038/s12276-023-01101-1.
24. Boucher J, Kleinridders A, Kahn CR. Insulin receptor signaling in normal and insulin-resistant states. Cold Spring Harb Perspect Biol. 2014;6(1):a009191. https://doi.org/10.1101/cshperspect.a009191.
25. Burchfield JG, Diaz-Vegas A, James DE. The insulin signalling network. Nat Metab. 2025;7:1745–64. https://doi.org/10.1038/s42255-025-01349-z.
26. Hopkins BD, Goncalves MD, Cantley LC. Insulin–PI3K signalling: an evolutionarily insulated metabolic driver of cancer. Nat Rev Endocrinol. 2020;16 (5):276–83. https://doi.org/10.1038/s41574-020-0329-9.
27. Zaïmia N, Obeid J, Varrault A, Sabatier J, Broca C, Gilon P, et al. GLP-1 and GIP receptors signal through distinct β-arrestin 2-dependent pathways to regulate pancreatic β cell function. Cell Rep. 2023;42 (11):113326. https://doi.org/10.1016/j.celrep.2023.113326.
28. Sun P, Enslen H, Myung PS, Maurer RA. Differential activation of CREB by Ca2+/calmodulin-dependent protein kinases type II and type IV involves phosphorylation of a site that negatively regulates activity. Genes Dev. 1994;8(21):2527–39. https://doi.org/10.1101/gad.8.21.2527.
29. Van de Velde S, Wiater E, Tran M, Hwang Y, Cole PA, Montminy M. CREB promotes Beta cell gene expression by targeting its coactivators to tissue-specific enhancers. Mol Cell Biol. 2019;39(17). https://doi.org/10.1128/mcb.00200-19.
30. Liu B, Barbosa-Sampaio H, Jones PM, Persaud SJ, Muller DS. The CaMK4/CREB/IRS-2 cascade stimulates proliferation and inhibits apoptosis of β-cells. PLoS One. 2012;7(9):e45711.

31. Jhala US, Canettieri G, Screaton RA, Kulkarni RN, Krajewski S, Reed J, et al. cAMP promotes pancreatic beta-cell survival via CREB-mediated induction of IRS2. Genes Dev. 2003;17(13):1575–80. https://doi.org/10.1101/gad.1097103.
32. Buteau J. GLP-1 signaling and the regulation of pancreatic β-cells mass/function. Av Diabetol. 2011;27(1):3–8. https://doi.org/10.1016/S1134-3230(11)70002-3.
33. Seino Y, Fukushima M, Yabe D. GIP and GLP- the two incretin hormones: similarities and differences. J Diabetes Investig. 2010;1(1–2):8–23. https://doi.org/10.1111/j.2040-1124.2010.00022.x.
34. Holst JJ, Deacon CF, Vilsbøll T, Krarup T, Madsbad S. Glucagon-like peptide- glucose homeostasis and diabetes. Trends Mol Med. 2008;14(4):161–8. https://doi.org/10.1016/j.molmed.2008.01.003.
35. Reed J, Bain SC, Kanamarlapudi V. The regulation of metabolic homeostasis by Incretins and the metabolic hormones produced by pancreatic islets. Diabetes Metab Syndr Obes. 2024;17:2419–56. https://doi.org/10.2147/dmso.S415934.
36. FDA. Mounjaro (tirzepatide) Approval Overview 2022. Available from: https://www.accessdata.fda.gov/scripts/cder/daf/index.cfm?event=overview.process&ApplNo=215866.
37. Wang Y, Chandra R, Samsa LA, Gooch B, Fee BE, Cook JM, et al. Amino acids stimulate cholecystokinin release through the Ca2+-sensing receptor. Am J Physiol Gastrointest Liver Physiol. 2011;300(4): G528–37. https://doi.org/10.1152/ajpgi.00387.2010.
38. Cawthon CR, de La Serre CB. The critical role of CCK in the regulation of food intake and diet-induced obesity. Peptides. 2021;138:170492. https://doi.org/10.1016/j.peptides.2020.170492.
39. Raybould HE. Mechanisms of CCK signaling from gut to brain. Curr Opin Pharmacol. 2007;7(6):570–4. https://doi.org/10.1016/j.coph.2007.09.006.
40. Perakakis N, Farr OM, Mantzoros CS. Leptin in leanness and obesity: JACC state-of-the-art review. J Am Coll Cardiol. 2021;77(6):745–60. https://doi.org/10.1016/j.jacc.2020.11.069.
41. Kelesidis T, Kelesidis I, Chou S, Mantzoros CS. Narrative review: the role of leptin in human physiology: emerging clinical applications. Ann Intern Med. 2010;152(2):93–100. https://doi.org/10.7326/0003-4819-152-2-201001190-00008.
42. Dibble CC, Cantley LC. Regulation of mTORC1 by PI3K signaling. Trends Cell Biol. 2015;25(9): 545–55. https://doi.org/10.1016/j.tcb.2015.06.002.
43. Stanciu SM, Jinga M, Miricescu D, Stefani C, Nica RI, Stanescu S II, et al. mTOR dysregulation, insulin resistance, and hypertension. Biomedicine. 2024;12(8). https://doi.org/10.3390/biomedicines12081802.
44. Ashcroft FM, Lloyd M, Haythorne EA. Glucokinase activity in diabetes: too much of a good thing? Trends Endocrinol Metab. 2023;34(2):119–30. https://doi.org/10.1016/j.tem.2022.12.007.
45. Raimondo A, Rees MG, Gloyn AL. Glucokinase regulatory protein: complexity at the crossroads of triglyceride and glucose metabolism. Curr Opin Lipidol. 2015;26(2):88–95. https://doi.org/10.1097/mol.0000000000000155.
46. Rodwell VW, Bender DA, Kennely PJ, Weil PA, McGuinness OP. Harper's illustrated biochemistry. 32nd ed. New York: McGraw-Hill Education/Medical; 2022.
47. Li Q, Zhao Q, Zhang J, Zhou L, Zhang W, Chua B, et al. The protein phosphatase 1 complex is a direct target of AKT that links insulin signaling to hepatic glycogen deposition. Cell Rep. 2019;28(13): 3406–22.e7. https://doi.org/10.1016/j.celrep.2019.08.066.
48. Han H-S, Kang G, Kim JS, Choi BH, Koo S-H. Regulation of glucose metabolism from a liver-centric perspective. Exp Mol Med. 2016;48(3): e218-e. https://doi.org/10.1038/emm.2015.122.
49. Nelson DL, Cox MM, Hoskins AA. Lehninger principles of biochemistry. 8th ed. Macmillan Learning; 2021.
50. Alves-Bezerra M, Cohen DE. Triglyceride metabolism in the liver. Compr Physiol. 2017;8(1):1–8. https://doi.org/10.1002/cphy.c170012.
51. Feingold KR. Introduction to lipids and lipoproteins. In: Feingold KR, Ahmed SF, Anawalt B, Blackman MR, Boyce A, Ghrousos G, Corpas E, Dhatariya K, Dungan K, Hofland J, Kalra S, Kaltsas G, Kapoor N, editors. Endotext [Internet]. South Dartmouth: MDText.com; 2024.
52. Fazakerley DJ, Koumanov F, Holman GD. GLUT4 on the move. Biochem J. 2022;479(3):445–62. https://doi.org/10.1042/bcj20210073.
53. Karnieli E, Armoni M. Transcriptional regulation of the insulin-responsive glucose transporter GLUT4 gene: from physiology to pathology. Am J Physiol Endocrinol Metab. 2008;295(1):E38–45. https://doi.org/10.1152/ajpendo.90306.2008.
54. Knight JB, Eyster CA, Griesel BA, Olson AL. Regulation of the human GLUT4 gene promoter: interaction between a transcriptional activator and myocyte enhancer factor 2A. Proc Natl Acad Sci USA. 2003;100(25):14725–30. https://doi.org/10.1073/pnas.2432756100.
55. Wang T, Wang J, Hu X, Huang XJ, Chen GX. Current understanding of glucose transporter 4 expression and functional mechanisms. World J Biol Chem. 2020;11(3):76–98. https://doi.org/10.4331/wjbc.v11.i3.76.
56. Marks D. Marks' essentials of medical biochemistry: a clinical approach. Lippincott Williams & Wilkins (LWW); 2017.
57. Salway JG. Metabolism at a glance. Hoboken: Wiley-Blackwell; 2017.
58. Edwards M, Mohiuddin SS. Biochemistry: lipolysis. In: StatPearls. Treasure Island: StatPearls Publishing.

59. Mergenthaler P, Lindauer U, Dienel GA, Meisel A. Sugar for the brain: the role of glucose in physiological and pathological brain function. Trends Neurosci. 2013;36(10):587–97. https://doi.org/10.1016/j.tins.2013.07.001.
60. Agrawal R, Reno CM, Sharma S, Christensen C, Huang Y, Fisher SJ. Insulin action in the brain regulates both central and peripheral functions. Am J Physiol Endocrinol Metab. 2021;321(1):E156–e63. https://doi.org/10.1152/ajpendo.00642.2020.
61. Loh K, Zhang L, Brandon A, Wang Q, Begg D, Qi Y, et al. Insulin controls food intake and energy balance via NPY neurons. Mol Metab. 2017;6(6):574–84. https://doi.org/10.1016/j.molmet.2017.03.013.
62. Dienel GA. Brain glucose metabolism: integration of energetics with function. Physiol Rev. 2019;99(1):949–1045. https://doi.org/10.1152/physrev.00062.2017.
63. Kitamura T, Kitamura Y, Kuroda S, Hino Y, Ando M, Kotani K, et al. Insulin-induced phosphorylation and activation of cyclic nucleotide phosphodiesterase 3B by the serine-threonine kinase Akt. Mol Cell Biol. 1999;19(9):6286–96. https://doi.org/10.1128/mcb.19.9.6286.
64. Zhang J, Hupfeld CJ, Taylor SS, Olefsky JM, Tsien RY. Insulin disrupts β-adrenergic signalling to protein kinase A in adipocytes. Nature. 2005;437(7058):569–73. https://doi.org/10.1038/nature04140.
65. Brasaemle DL. Thematic review series: adipocyte biology. The perilipin family of structural lipid droplet proteins: stabilization of lipid droplets and control of lipolysis. J Lipid Res. 2007;48(12):2547–59. https://doi.org/10.1194/jlr.R700014-JLR200.
66. Yang A, Mottillo EP. Adipocyte lipolysis: from molecular mechanisms of regulation to disease and therapeutics. Biochem J. 2020;477(5):985–1008. https://doi.org/10.1042/bcj20190468.
67. Lass A, Zimmermann R, Oberer M, Zechner R. Lipolysis - a highly regulated multi-enzyme complex mediates the catabolism of cellular fat stores. Prog Lipid Res. 2011;50(1):14–27. https://doi.org/10.1016/j.plipres.2010.10.004.
68. Mead JR, Irvine SA, Ramji DP. Lipoprotein lipase: structure, function, regulation, and role in disease. J Mol Med (Berl). 2002;80(12):753–69. https://doi.org/10.1007/s00109-002-0384-9.
69. Sadur CN, Eckel RH. Insulin stimulation of adipose tissue lipoprotein lipase. Use of the euglycemic clamp technique. J Clin Invest. 1982;69(5):1119–25. https://doi.org/10.1172/jci110547.
70. Kiens B, Lithell H, Mikines KJ, Richter EA. Effects of insulin and exercise on muscle lipoprotein lipase activity in man and its relation to insulin action. J Clin Invest. 1989;84(4):1124–9. https://doi.org/10.1172/JCI114275.
71. Witters LA, Watts TD, Daniels DL, Evans JL. Insulin stimulates the dephosphorylation and activation of acetyl-CoA carboxylase. Proc Natl Acad Sci USA. 1988;85(15):5473–7. https://doi.org/10.1073/pnas.85.15.5473.
72. Vasim I, Majeed CN, DeBoer MD. Intermittent fasting and metabolic health. Nutrients. 2022;14:3. https://doi.org/10.3390/nu14030631.
73. McGarry JD, Brown NF. The mitochondrial carnitine palmitoyltransferase system — from concept to molecular analysis. Eur J Biochem. 1997;244(1):1–14. https://doi.org/10.1111/j.1432-1033.1997.00001.x.
74. Janah L, Kjeldsen S, Galsgaard KD, Winther-Sørensen M, Stojanovska E, Pedersen J, et al. Glucagon receptor signaling and Glucagon resistance. Int J Mol Sci. 2019;20(13). https://doi.org/10.3390/ijms20133314.
75. Jiang G, Zhang BB. Glucagon and regulation of glucose metabolism. Am J Physiol Endocrinol Metab. 2003;284(4):E671–E8. https://doi.org/10.1152/ajpendo.00492.2002.
76. Herzig S, Long F, Jhala US, Hedrick S, Quinn R, Bauer A, et al. CREB regulates hepatic gluconeogenesis through the coactivator PGC-1. Nature. 2001;413(6852):179–83. https://doi.org/10.1038/35093131.
77. Herzig S, Shaw RJ. AMPK: guardian of metabolism and mitochondrial homeostasis. Nat Rev Mol Cell Biol. 2018;19(2):121–35. https://doi.org/10.1038/nrm.2017.95.
78. Hames BD, Hooper NM. BIOS instant notes in biochemistry. Taylor and Francis; 2011.
79. Venugopal S, Sankar P, Jialal I. Physiology, glucagon. StatPearls. Treasure Island: StatPearls publishing; 2023.
80. Zhang X, Yang S, Chen J, Su Z. Unraveling the regulation of hepatic gluconeogenesis. Front Endocrinol (Lausanne). 2018;9:802. https://doi.org/10.3389/fendo.2018.00802.
81. van Wijk R, van Solinge WW. The energy-less red blood cell is lost: erythrocyte enzyme abnormalities of glycolysis. Blood. 2005;106(13):4034–42. https://doi.org/10.1182/blood-2005-04-1622.
82. Melkonian EA, Asuka E, Schury MP. Physiology, gluconeogenesis. In: StatPearls. Treasure Island: StatPearls Publishing; 2023.
83. Holeček M. Roles of malate and aspartate in gluconeogenesis in various physiological and pathological states. Metabolism. 2023;145:155614. https://doi.org/10.1016/j.metabol.2023.155614.
84. Shah A, Wondisford FE. Gluconeogenesis flux in metabolic disease. Annu Rev Nutr. 2023;43:153–77. https://doi.org/10.1146/annurev-nutr-061121-091507.
85. Srivastava RA, Pinkosky SL, Filippov S, Hanselman JC, Cramer CT, Newton RS. AMP-activated protein kinase: an emerging drug target to regulate imbalances in lipid and carbohydrate metabolism to treat cardio-metabolic diseases. J Lipid Res. 2012;53(12):2490–514. https://doi.org/10.1194/jlr.R025882.

86. Richter EA, Hargreaves M. Exercise, GLUT4, and skeletal muscle glucose uptake. Physiol Rev. 2013;93(3):993–1017. https://doi.org/10.1152/physrev.00038.2012.
87. Cahill GF Jr. Starvation in man. Clin Endocrinol Metab. 1976;5(2):397–415. https://doi.org/10.1016/s0300-595x(76)80028-x.
88. Galsgaard KD, Pedersen J, Knop FK, Holst JJ, Wewer Albrechtsen NJ. Glucagon receptor signaling and lipid metabolism. Front Physiol. 2019;10:413. https://doi.org/10.3389/fphys.2019.00413.
89. Kersten S. The impact of fasting on adipose tissue metabolism. Biochim Biophys Acta Mol Cell Biol Lipids. 2023;1868(3):159262. https://doi.org/10.1016/j.bbalip.2022.159262.
90. Duncan RE, Ahmadian M, Jaworski K, Sarkadi-Nagy E, Sul HS. Regulation of lipolysis in adipocytes. Annu Rev Nutr. 2007;27:79–101. https://doi.org/10.1146/annurev.nutr.27.061406.093734.
91. Houten SM, Violante S, Ventura FV, Wanders RJA. The biochemistry and physiology of mitochondrial fatty acid β-oxidation and its genetic disorders. Annu Rev Physiol. 2016;78:23–44. https://doi.org/10.1146/annurev-physiol-021115-105045.
92. Alberts B, Heald R, Johnson A, Morgan D, Raff M, Roberts K, Walter P, Wilson J, Hunt T. Molecular biology of the cell. W.W. Norton and Company; 2022.
93. Laffel L. Ketone bodies: a review of physiology, pathophysiology and application of monitoring to diabetes. Diabetes Metab Res Rev. 1999;15(6):412–26.
94. Gotoh N, Moroda K, Watanabe H, Yoshinaga K, Tanaka M, Mizobe H, et al. Metabolism of odd-numbered fatty acids and even-numbered fatty acids in mouse. J Oleo Sci. 2008;57(5):293–9. https://doi.org/10.5650/jos.57.293.
95. PharmaXChange. Oxidation of odd carbon chain length fatty acids 2013. Available from: https://pharmaxchange.info/2013/10/oxidation-of-odd-carbon-chain-length-fatty-acids/.
96. Newman JC, Verdin E. Ketone bodies as signaling metabolites. Trends Endocrinol Metab. 2014;25(1):42–52. https://doi.org/10.1016/j.tem.2013.09.002.
97. mindbodygreen. Stages of fasting: what to expect in the short & long term 2023. Available from: https://www.mindbodygreen.com/articles/stages-of-fasting?srsltid=AfmBOopr9ZoKxcW9CoZO0WSVn_TGvfPIuqfZiwG_ytjFaL8faEUd3og7.
98. OpenStax. 24.5 Metabolic states of the body. LibreTexts. Available from: https://med.libretexts.org/Bookshelves/Anatomy_and_Physiology/Anatomy_and_Physiology_1e_(OpenStax)/Unit_5%3A_Energy_Maintenance_and_Environmental_Exchange/24%3A_Metabolism_and_Nutrition/24.05%3A_Metabolic_States_of_the_Body.
99. Smith RL, Soeters MR, Wüst RCI, Houtkooper RH. Metabolic flexibility as an adaptation to energy resources and requirements in health and disease. Endocr Rev. 2018;39(4):489–517. https://doi.org/10.1210/er.2017-00211.
100. Shi X, Qiu H. New insights into energy substrate utilization and metabolic remodeling in cardiac physiological adaption. Front Physiol. 2022;13:831829. https://doi.org/10.3389/fphys.2022.831829.
101. Jensen NJ, Wodschow HZ, Nilsson M, Rungby J. Effects of ketone bodies on brain metabolism and function in neurodegenerative diseases. Int J Mol Sci. 2020;21(22). https://doi.org/10.3390/ijms21228767.
102. Lynch RM, Paul RJ. Energy metabolism and transduction in smooth muscle. Experientia. 1985;41(8):970–7. https://doi.org/10.1007/BF01952116.
103. Nakayama S, Clark JF. Smooth muscle and NMR review: an overview of smooth muscle metabolism. Mol Cell Biochem. 2003;244(1–2):17–30.
104. Kaneko K, Soty M, Zitoun C, Duchampt A, Silva M, Philippe E, et al. The role of kidney in the inter-organ coordination of endogenous glucose production during fasting. Mol Metab. 2018;16:203–12. https://doi.org/10.1016/j.molmet.2018.06.010.
105. Gerich JE, Meyer C, Woerle HJ, Stumvoll M. Renal gluconeogenesis: its importance in human glucose homeostasis. Diabetes Care. 2001;24(2):382–91. https://doi.org/10.2337/diacare.24.2.382.
106. Bauer, et al. A comprehensive in-vitro/in-vivo screening toolbox for the elucidation of glucose homeostasis modulating properties of plant extracts (from roots) and its bioactives. Front Pharmacol. 2024;15:1396292. https://doi.org/10.3389/fphar.2024.1396292.

Disorders of Carbohydrate and Lipid Metabolism, and Pharmacological Treatments

13

Abstract

Metabolic pathways are critical for controlling how the body generates energy, breaks down biomolecules, and maintains homeostasis. It is important for our health that these pathways function correctly. When they are disrupted, metabolic disorders can result. Metabolic disorders can range from mild to severe and life-threatening. They can lead to damage to organs and long-term health problems. This chapter discusses metabolic disorders, primarily those that affect carbohydrate and lipid metabolism and the body's ability to maintain homeostasis. Examples of metabolic diseases discussed here include diabetes, obesity, and non-alcoholic fatty liver disease. Treatments that affect metabolic pathways, such as GLP-1 agonists, are discussed. Excess cholesterol, metabolic syndrome, and obesity are also discussed, as well as current treatments and those in research stages.

Keywords

Metabolic disorders · Diabetes · GLP-1 agonists · Fatty liver disease · High cholesterol · Metabolic syndrome

Abbreviations

AMPK	AMP-activated protein kinase
CNS	Central nervous system
DAG	Diacylglycerol
DPP-4	Dipeptidyl peptidase-4
ETC	Electron transport chain
FFA	Free fatty acids
GLP-1	Glucagon-like peptide-1
GLUT4	Glucose transporter type 4
GIP	Glucose-dependent insulinotropic polypeptide
GI	Glycemic index
GL	Glycemic load
HbA1c	Hemoglobin A1c (glycated hemoglobin)
HDL	High-density lipoprotein
IR	Insulin receptor
JNK	c-Jun N-terminal kinase
LDL	Low-density lipoprotein
NAFLD	Non-alcoholic fatty liver disease
NASH	Non-alcoholic steatohepatitis
NADH	Nicotinamide adenine dinucleotide (reduced form)
NAD^+	Nicotinamide adenine dinucleotide (oxidized form)

A. Minden, *Biochemistry, Cell and Molecular Biology for Pharmacists and Health Professionals*,
https://doi.org/10.1007/978-3-032-20529-2_13

NFKB	Nuclear factor kappa-light-chain-enhancer of activated B cells
OAA	Oxaloacetate
PCSK9	Proprotein convertase subtilisin/kexin type 9
PKC	Protein kinase C
rbc	Red blood cell
ROS	Reactive oxygen species
SGLT1	Sodium-glucose linked transporter 1
SGLT2	Sodium-glucose linked transporter 2
TAG	Triacylglycerol
T3	Triiodothyronine
T4	Thyroxine
THRβ	Thyroid hormone receptor beta
VLDL	Very-low-density lipoprotein

What You Will Learn

- Different metabolic disorders are associated with abnormalities in the metabolic homeostasis of carbohydrates and fats.
- How understanding metabolic regulation aids in the development of drugs to treat disease.
- Diabetes, and the difference between type 1 and type 2 diabetes.
- Some causes of insulin resistance in diabetes.
- Some consequences of diabetes, such as ketoacidosis and other complications that can occur in response to high blood glucose levels.
- Methods for monitoring and diagnosing diabetes.
- Managing diabetes with diet and lifestyle changes.
- Medication for treating diabetes, including GLP-1 receptor agonists, drugs that target other hormone receptors, synthetic insulin, metformin, and others.
- Hepatic steatosis, fatty liver disease, including NAFLD and AFLD, and pharmacological interventions for these conditions.
- Nutrient excess and obesity, how obesity is treated by lifestyle changes, and in some cases by GLP-1 agonists.
- Excess cholesterol, health consequences of high cholesterol, and dietary and pharmacological treatments.
- Possible health benefits of a ketogenic diet.
- Metabolic syndrome is associated with several risk factors for metabolic diseases that occur together.
- Other metabolic disorders.
- How understanding metabolic regulation aids in the development of drugs to treat disease.

1 Overview Metabolic Disorders Associated with Homeostasis

The metabolic pathways discussed in this section so far are critical for controlling how the body generates energy, breaks down biomolecules, and maintains homeostasis. It is important for our health that these pathways function correctly. When they break down or do not function normally, metabolic disorders can result. Examples of metabolic disease linked to problems with homeostasis include diabetes, obesity, and non-alcoholic fatty liver disease. Metabolic disorders can range from mild to severe and life-threatening. They can lead to damage to organs and long-term health problems. Understanding how metabolic pathways work is crucial for understanding how their abnormal function can lead to disease. It is also important to develop ways to either treat or prevent metabolic disorders.

This chapter discusses metabolic disorders, primarily those that affect carbohydrate and lipid metabolism, as well as the body's ability to maintain homeostasis. The metabolic pathways discussed in the previous chapters in this section are closely linked to the disorders covered here, so it may be helpful to review those chapters before reading this one. Some sections below also discuss signal transduction. Signal transduction will be covered in more detail in Chap. 18, but it may also be helpful to revisit this chapter after reading that one.

2 Diabetes

2.1 Overview of Diabetes

A classic example of a metabolic disease associated with homeostatic abnormalities is diabetes mellitus, characterized by an impaired response to insulin. **Diabetes mellitus** is a major health concern, affecting millions of people throughout the world. It is a leading cause of health complications, including heart disease, kidney failure, limb abnormalities, and blindness. Diabetes is caused by either a lack of insulin or a failure to respond to insulin. Insulin is required for glucose to be absorbed into large tissues, such as muscle and adipose tissue, and for its retention in tissues like the liver. In diabetes, when insulin is lacking or the body fails to respond to it properly, muscle and adipose tissue are unable to efficiently absorb glucose. At the same time, the liver continues to release glucose into the bloodstream, and the storage of glucose as glycogen is reduced. Consequently, the clearance of glucose from the bloodstream after a meal is inhibited, leading to elevated circulating glucose levels, a condition known as hyperglycemia. In many individuals with diabetes, this leads to sharp increases in blood glucose after meals, known as postprandial glucose spikes, which contribute to hyperglycemia and tissue damage. These glucose spikes also trigger insulin spikes, as the pancreas releases more insulin to lower blood glucose. Glucose and insulin spikes can also occur in individuals without diabetes, especially after meals with a high glycemic index (discussed later), but frequent or large spikes may promote excessive fat storage and reduce fat breakdown. Over time, it can contribute to insulin resistance and weight gain, and diabetes. Additionally, large insulin spikes can sometimes overshoot, leading to a drop in blood glucose a few hours later (reactive hypoglycemia), which may increase hunger and promote overeating.

Persistent hyperglycemia is a major feature of diabetes and can lead to many health complications. There are two major types of diabetes: type 1 diabetes and type 2 diabetes, and a third type, gestational diabetes, which is less common [1–3]. Type 2 diabetes will be the main focus in this chapter, but type 1 diabetes and gestational diabetes are also introduced below.

2.2 Type 1 Diabetes

Insulin is normally produced by, and secreted from, the β cells of the pancreas in response to high blood glucose levels, such as after a meal. In type 1 diabetes (also known as insulin-dependent diabetes), these insulin-secreting β cells of the pancreas are destroyed over time. Consequently, patients with type 1 diabetes produce little to no insulin whatsoever. As a result, glucose cannot be cleared from the bloodstream, and blood glucose levels remain high (hyperglycemia). Type 1 diabetes is an autoimmune disorder that is associated with the destruction of pancreatic β cells, and it often begins in childhood or adolescence. This type of diabetes usually requires lifelong insulin replacement therapy. Insulin can be administered by injection or with an insulin pump. Careful monitoring of blood glucose levels is important for patients with type 1 diabetes [4, 5].

2.3 Type 2 Diabetes

2.3.1 Insulin Resistance

Type 2 diabetes, unlike type 1 diabetes, is not caused by an inability to produce insulin, at least in the early stages of the disease. Instead, it is caused by an inability to respond to insulin, known as insulin resistance. In this type of diabetes, insulin is produced, but the insulin receptors (IRs) on the surfaces of cells and tissues are not responsive to it. Under normal conditions, as described in Chap. 12, insulin is released from pancreatic β cells after a meal in response to high blood glucose levels, and this insulin subsequently binds to insulin receptors (IRs) on the surface of cells of many tissues. This activates a signaling pathway in the cells. In adipose tissue and muscle, this signaling causes the glucose transporter GLUT4 to move from the cytoplasm to the cell membrane, enabling glucose to be transported into the cells. In type 2 diabetes, insulin binding to its receptor (IR) is not necessarily impaired, at least initially, but IR-activated signaling is disrupted. As a consequence, GLUT4 translocation to the cell surface in muscle and adipose tissue is impaired, leading to reduced glucose uptake despite the presence of insulin. In the liver, impaired IR signaling can disrupt glucose storage as glycogen and can promote excessive glucose production and export into the bloodstream. These abnormalities in IR signaling contribute to insulin resistance, meaning that although insulin may be present, cells are not sensitive to its presence and therefore fail to respond to it.

Importantly, insulin resistance is not necessarily complete, and different tissues vary in their degree of responsiveness or resistance. Muscle and adipose tissue can become highly insulin-resistant in diabetes, impairing glucose uptake into these large tissues. On the other hand, the liver often remains partially sensitive to insulin, especially in responses involving lipid metabolism. As a result, insulin may still stimulate lipid synthesis in the liver, even as insulin resistance increases in certain tissues, leading to dyslipidemia, which is often observed in type 2 diabetes.

2.3.2 Disruption of Pancreatic β Cells Over Time

If insulin resistance persists over time, pancreatic β cells can also become dysfunctional. This is due to a combination of reasons that include increased metabolic demand on the pancreas (the need for the pancreas to secrete more insulin to compensate for increased insulin resistance throughout the body), chronic exposure to high blood glucose levels (glucotoxicity), elevated release of free fatty acids as a consequence of insulin resistance in adipose tissue (discussed more below), and inflammation [6, 7]. Over time, the reduction in functional pancreatic β cells leads to the additional problem of reduced insulin production. This exacerbates the condition because, in addition to insulin resistance, the pancreas is now unable to produce enough insulin to meet the body's increased demands. Although circulating insulin levels may not necessarily be lessened, they are no longer sufficient to maintain normal blood glucose levels. It is usually around this point, when blood glucose remains consistently elevated due to both insulin resistance and impaired insulin secretion, that the condition is clinically diagnosed as type 2 diabetes [8, 9].

2.4 Gestational Diabetes

Although type I and type 2 diabetes are the most common, there is also a third type of diabetes. The third type is gestational diabetes, and it occurs in some women during pregnancy. During pregnancy, the placenta produces hormones to support fetal growth. These hormones can reduce the mother's insulin sensitivity. This allows more glucose into the bloodstream, making it available to the fetus, but it also increases the mother's insulin demand. If her pancreatic β cells cannot keep up with this demand by producing enough insulin, blood glucose levels may rise too much, and this can result in gestational diabetes. All three types of diabetes are associated with elevated blood glucose levels, but in gestational diabetes, blood glucose levels usually return to normal after the birth of the baby [2, 10, 11].

2.5 Some Causes of Reduced Insulin Sensitivity in Type II Diabetes

2.5.1 Receptor Downregulation

Disruption in IR signaling, leading to insulin resistance, is a major problem in type 2 diabetes. This disruption can be due to various causes. For example, it may be due to downregulation of the insulin receptor in response to excessive sugar intake. The IR, like many receptors, can undergo ligand-induced internalization and degradation in lysosomes after repeated stimulation. This is eventually followed by the synthesis of new receptors. This is a normal regulatory function of receptors [12], but if insulin levels are chronically elevated, as in response to chronically high blood glucose levels, IR downregulation may overwhelm receptor synthesis. This can lead to a reduction in cell-surface receptors, contributing to decreased insulin sensitivity [13].

2.5.2 Cytokine-Activated Pathways and Inflammation

During obesity, adipose tissue enlarges, and adipocytes within it undergo hypertrophy (enlargement). Enlarged adipocytes undergo increased stress, such as hypoxia, due to their greater distance from capillaries. This stress can lead to the secretion of cytokines such as TNFα from the adipocytes. These cytokines can recruit immune cells, leading to inflammation (Inflammation generally refers to the recruitment and activation of immune cells, which secrete cytokines, which can lead to tissue changes such as vascular changes). These cytokines can also activate signaling pathways such as the JNK or NF-κB pathways, which may interfere with insulin signaling. One way they may do so is by leading to phosphorylation of the insulin receptor substrate (IRS) on serine residues on their cytoplasmic domains, which leads to insulin receptor inactivation [14–16].

2.5.3 Accumulation of Fatty Acids in Liver and Muscle, and an Increase in Fatty Acid Metabolites

Free fatty acids (FFAs) also impair insulin signaling. Adipose tissue normally stores TAGs. In obesity, the adipocytes enlarge (hypertrophy) and they can reach their maximum capacity [17]. At this point, they may not be able to take up additional fatty acids for esterification into TAGs. Normally, insulin inhibits lipolysis because it suppresses the activity of **HSL** (an enzyme involved in breaking down TAGs into free fatty acids). In insulin resistance, this inhibition is lost, and this, combined with adipocyte overcapacity, causes TAGs to be broken down into fatty acids. Fatty acids are therefore released from the adipocytes into the circulation. Liver and muscle take up these excess fatty acids. Liver and muscle are not specialized for long-term fat storage. Instead, these free fatty acids can be metabolized into lipid intermediates, such as DAG and ceramides (DAG is made from the combination of two fatty acids with glycerol, and ceramides consist of fatty acids that are modified by attachment to a sphingosine backbone). These intermediates can have various effects in cells. For example, DAG can lead to activation of the enzyme **protein kinase C (PKC). PKC** can also lead to phosphorylation on the insulin receptor substrate (**IRS**) on inhibitory sites, further disrupting IR signaling. Ceramides can also lead to mitochondrial dysfunction by inserting into mitochondrial membranes, thereby interfering with the electron transport chain [18]. Ceramides can also lead to pro-apoptotic signaling. One way they can do this is by inhibiting AKT signaling, an important survival protein [15, 19–23]

2.5.4 Nutrient-Sensing Pathways in Type 2 Diabetes: mTOR and AMPK

Chronic overnutrition can lead to persistent activation of the **mTOR** pathway, particularly mTORC1. Sustained mTOR1 signaling has been linked to **insulin resistance** by several mechanisms. For example, it activates **S6 kinase 1 (S6K1)**, a kinase that phosphorylates **IRS** at inhibitory serine residues. This reduces insulin signaling. Prolonged mTORC1 activation leads to further problems. mTORC1 normally activates anabolic pathways, which support cell growth. However, overactivation or prolonged activation of mTORC1 can lead to excessive protein synthesis, ER stress, lipid overload, and other issues, which can also contribute to impaired insulin sensitivity in muscle and liver.

In contrast to mTORC1, which is activated by nutrient abundance, **AMPK is activated when cellular energy levels are low.** Normally, AMPK activation promotes insulin sensitivity by enhancing glucose uptake in skeletal muscle (through enhancing GLUT4 translocation to the cell surface, inhibiting hepatic gluconeogenesis, and increasing fatty acid oxidation). AMPK also inhibits mTORC1 signaling, thereby putting the brakes on excessive anabolic activity. While mTORC1 may be overly active in type 2 diabetes, **AMPK activity is often reduced,** contributing to reduced insulin sensitivity. Furthermore, AMPK normally inhibits mTORC1, but when AMPK is deficient, mTORC1 activity can increase even further.

In summary, activation of **mTORC1**, along with inhibition of **AMPK**, can shift metabolism toward anabolic pathways and can interfere with the balance between storage and energy production. Some type 2 diabetes medications function by targeting these pathways.

2.6 Some Consequences of Diabetes

Diabetes can cause multiple complications. A few of these are discussed below, while others are mentioned throughout the rest of this chapter.

2.6.1 Ketoacidosis (Especially in Type 1 Diabetes)

- What is Ketoacidosis?

When insulin is functioning normally, its levels rise in response to elevated blood glucose. This promotes glucose uptake into muscle and fat, stimulates glycogen storage in the liver, and suppresses fat breakdown (lipolysis) in adipose tissue. In diabetes, however, insulin signaling is impaired. As a result, blood glucose levels become abnormally high (hyperglycemia), while glucose uptake into insulin-sensitive tissues, such as muscle and fat, is reduced. These tissues, therefore, will have low intracellular glucose levels even while blood glucose levels are high. This contributes to increased reliance on fat as an energy source for tissues. This leads to increased lipolysis (the breakdown of TAG). As a result, there is an increase in fatty acids entering circulation, and in muscle and liver, fatty acids may be oxidized to yield ATP.

A serious consequence of this metabolic shift is the increased production of ketone bodies by the liver, which can lead to a potentially serious condition called ketoacidosis. The liver plays the main role in this process. As insulin levels drop, and lipolysis increases in adipose tissue, the liver receives an influx of free fatty acids from adipose tissue. These fatty acids are oxidized in liver mitochondria, generating high levels of acetyl-CoA.

At the same time, low insulin levels prompt the liver to increase glucose production (gluconeogenesis). As a result, OAA is largely used up in gluconeogenesis (see Chap. 12) rather than interacting with acetyl-CoA. As a result, acetyl-CoA levels accumulate.

When acetyl-CoA is in excess, the liver diverts acetyl-CoA to the production of ketone bodies. These ketone bodies, produced by the liver, are secreted into the bloodstream. Blood pH drops because ketone bodies such as acetoacetate and β-hydroxybutyrate are acidic. This acidification of the blood has many consequences. For example, it disrupts the CNS because the brain requires a narrow pH range to operate effectively. Alterations in this pH can disrupt neurotransmission and other processes, many of which can lead to cognitive problems [1, 24, 25].

- Ketosis Compared with Ketoacidosis

When discussing ketoacidosis, it is important to remember that it differs from ketosis. **Ketosis** can occur in diabetes, but it is also a normal physiological response to low glucose availability. Ketosis is the accumulation of ketones. It can occur when the body uses fat as a major energy source. This can occur as a result of fasting or a low-carbohydrate diet. In both situations, the body utilizes fat as its primary energy source, resulting in ketone accumulation [26]. Ketosis can also occur due to low insulin levels or insulin resistance, as discussed above.

The ketones produced are important because they provide an energy source for the brain when glucose is unavailable [27, 28]. However, while ketosis is a normal controlled process which can reset as the diet changes, **ketoacidosis** involves the same basic metabolic pathways but involves a severe lack of insulin, very high blood sugar (but low glucose in the large tissues), and excessive production of ketones. The blood can become very acidic (acidosis) for a prolonged time, and this can severely impact the CNS. Unlike ketosis, ketoacidosis is a life-threatening condition that can happen in diabetes. It is most common in type 1 diabetes due to near or complete insulin deficiency [29].

2.6.2 High Blood Glucose Levels Can Lead to Other Complications

High blood glucose levels can cause damage over time, including to blood vessels, nerves, and organs. Blood vessel damage is particularly important, and these problems can occur in both type 1 and type 2 diabetes. It can occur because too much glucose enters endothelial cells, which line the vessels. Increased glucose metabolism through glycolysis and mitochondrial respiration in these cells leads to the overproduction of reactive oxygen species (ROS) due to excess electrons passing through the ETC. This results in oxidative stress, which can lead to inflammation, endothelial cell dysfunction, and cell death. Glucose can also react with proteins and lipids. Such proteins are considered glycated. Over time, these glycated products can undergo further modifications, forming irreversible structures called advanced glycation end-products (AGEs)

[30]. Proteins such as collagen can become glycated to form AGEs, for example, and this causes it to become stiffer. AGE-modified collagen in vessel walls can cause vessels to lose elasticity and stiffen. This stiffening can affect blood flow and lead to inflammation. Damage to the blood vessels can be especially problematic in the retina. It can cause retinal detachment, vision problems, and even vision loss [31].

High glucose can also damage the kidneys (diabetic nephropathy). Damage to vessels in the kidneys causes vessels to narrow and clog. This reduces the kidney's filtration efficiency, leading to protein leakage into the urine. This is called proteinuria and is an early sign of diabetic nephropathy (disease of the kidney) [32, 33]. High blood glucose can also damage certain nerves, particularly those in the hands and feet. This is called peripheral neuropathy. This is because excess glucose damages the small blood vessels that supply these nerves, reducing oxygen and nutrient delivery. Damage to these nerves can lead to numbness and injury [34]. People with diabetes are also at increased risk for cardiovascular disease, including heart failure and strokes. While this is largely due to the accumulation of LDL cholesterol, high glucose can also exacerbate this problem by damaging blood vessels and promoting inflammation [35, 36].

2.7 Monitoring and Diagnosing Diabetes

Diabetes is diagnosed and monitored by using two main types of blood sugar tests. One is the fasting blood glucose test. This measures the level of glucose in the blood after an overnight fast (In the United States, the units used are usually mg/dL (milligrams sugar per deciliter (dL) of blood) [37]. The other test is the HbA1c test (often called the A1c test) [38, 39]. The A1c test measures glycated hemoglobin, which is hemoglobin with glucose attached. Hemoglobin is present in red blood cells (RBCs), and glucose attaches to this hemoglobin in an irreversible reaction. This hemoglobin is now "glycated." This glycated hemoglobin is called HbA1C, and the HbA1C test measures the percentage of hemoglobin that is attached to glucose (glycated). This test is useful because it does not show day-to-day fluctuation. Once glucose binds to hemoglobin, it remains bound for the life of the RBCs, which have an average lifespan of about 120 days. In contrast, free glucose is transient. Therefore, the A1c number is a strong indicator of overall blood glucose levels [40]. In addition to these two important tests, other tests can also be used, such as continuous glucose monitoring [41].

2.8 Managing Diabetes

2.8.1 Diet and Lifestyle

Unlike type 1 diabetes, treatment with insulin will not always treat type 2 diabetes, especially at early stages. Lifestyle and dietary changes are important strategies for treating the disease. Dietary changes include reducing sugar intake and achieving a healthy weight. These changes have all been shown to increase insulin sensitivity and to lower circulating free fatty acids.

Individuals with diabetes or even those at risk for diabetes need to manage their blood sugar levels carefully. It is important to prevent **spikes** in blood sugar (glucose). Spikes in glucose can place excessive demand on pancreatic β cells and desensitize insulin signaling pathways due to persistent activation of insulin receptors. Therefore, in addition to moderating sugar and carbohydrate intake, two factors to consider are the glycemic index and glycemic load of foods. **Glycemic index (GI)** refers to how quickly a carbohydrate-containing food raises blood glucose levels, measured on a scale of 1 to 100. Some foods, such as white bread, have a high glycemic index and can cause a rapid rise in blood sugar. Others, such as some whole grains, will lead to a more gradual rise, which is more desirable. The other consideration is **glycemic load (GL)**. Glycemic load combines glycemic index (GI) of the food with the total amount of carbohydrates it contains. It is calculated as (GI x grams of carbohydrate per serving)/100. It provides a more accurate picture of how a food will affect blood glucose levels. For people with type 2 diabetes, foods with a low GI and a low GL are the most beneficial for managing the condition, although GL is particularly important [42, 43].

Other considerations are also important. For example, fiber, especially soluble fiber, is an important part of the diet. Soluble fiber is found in foods like oats, legumes, and fruits. Soluble fiber slows the absorption of glucose and prevents blood sugar spikes. Thus, a high-fiber diet helps improve insulin sensitivity and is beneficial for people with diabetes. Including protein in a meal can also help moderate the rise in blood sugar by slowing down digestion and promoting a feeling of fullness. It also helps preserve muscle mass. Preserving muscle mass is very important in diabetes. After a meal, glucose is taken up by muscle. If muscle mass is lost, there is less muscle to take up glucose, leading to higher blood glucose levels. Lean proteins are especially helpful. Healthy fats, such as those found in seeds, nuts, avocados, and olive oil, can help slow down the absorption of glucose. This is because they stimulate hormones like CCK and GLP-1, which slow stomach emptying into the small intestine; this, in turn, causes glucose to be absorbed more gradually into the bloodstream and reduces glucose spikes. Healthy fats also improve satiety and, on their own, do not raise blood sugar. Trans fats and excessive amounts of saturated fats, on the other hand, should be avoided due to the increased risk of cardiovascular problems. Balanced meals that combine beneficial factors are important [44–48].

2.8.2 Exercise and Weight Management

Exercise can help with type 2 diabetes because it activates insulin-independent synthesis of muscle GLUT4 and its translocation to the cell surface, thereby increasing insulin-independent glucose entry in the muscle [49]. There are various ways that exercise can help with this. For example, it increases the AMP/ATP ratio in cells because ATP consumption increases in active muscle. This activates **AMPK** (AMP-activated protein kinase), a key energy sensor. Activated **AMPK** initiates a signaling cascade that leads to the transport of GLUT4 via vesicular trafficking to the membrane [50]. Initiation of exercise also leads to a rise in intracellular Ca^{2+} levels, due to Ca^{2+} release from the sarcoplasmic reticulum. This intracellular Ca^{2+} promotes contraction by binding to troponin C, thereby allowing actin–myosin binding and contraction. Thus, the rise in intracellular Ca^{2+} has an important role in contraction. The Ca^{2+} also has another role, which is activating cellular signaling pathways, including a pathway that leads to transcription of the GLUT4 gene via transcription factors such as MEF2A, and possibly to its transport to the membrane. This appears to be mediated by **CAM Kinase** [51]. Thus, exercise contributes to glucose clearance by increasing GLUT4 expression, facilitating glucose import into muscle [52, 53].

Weight loss is also helpful for people with type 2 diabetes for several reasons. For example, extra free fatty acids in the bloodstream, often associated with excess body fat, can interfere with insulin signaling in the liver, because the liver, as discussed above, can metabolize excess fatty acids into metabolites that interfere with insulin signaling. Weight loss may therefore help restore insulin action [54]. Exercise and weight management are also important for type 1 diabetes, but people with type 1 diabetes need to be careful due to the risk of hypoglycemia, and they need to adjust their insulin doses accurately [55].

2.9 Medication and Insulin Therapy for Diabetes

Pharmacological intervention for the treatment of type 2 diabetes has come a long way. In the past two decades, numerous treatments have emerged. Their development was driven by an understanding of glucose metabolism and the regulation of homeostasis. New drugs target metabolic pathways involved in blood glucose control. Several of these will be discussed in this section.

2.9.1 GLP-1 Receptor Agonists for the Treatment of Diabetes and Obesity

GLP-1 receptor agonists have received considerable recent attention for regulating blood sugar and lipid storage. Although the first GLP-1 agonists were introduced in 2005, they did not gain widespread use until the approval of **Ozempic (semaglutide)** in 2017 [56]. This is largely due to Ozempic's less frequent dosing and stronger glucose-lowering and weight-loss effects than previous agonists [57].

GLP-1 agonists are biologics. They are peptide-based drugs that mimic the **GLP-1 hormone** and bind to and **activate the GLP-1 receptor**, which is a GPCR. GLP-1, as discussed in Chap. 12, is normally secreted from intestinal L cells in the fed state in response to increased nutrients, especially high glucose and fat. GLP-1 is an incretin hormone; it normally enhances insulin secretion from the pancreatic β cells in response to oral intake of glucose. It is glucose-dependent; it increases the body's insulin response, but only when blood glucose levels are elevated. Increased GLP-1 leads not only to insulin release from pancreatic β cells, but it also suppresses glucagon secretion from pancreatic α cells. As a reminder, insulin lowers blood glucose by promoting glucose uptake into tissues and encouraging its storage, while glucagon raises blood glucose through several mechanisms, including stimulating glucose release from the liver and promoting gluconeogenesis (see Chap. 12, Fig. 12.1).

GLP-1 and GLP-1 receptor agonists also play a role in slowing down gastric emptying (the movement of food from the stomach to the intestine) and increasing feelings of satiety. These effects of GLP-1 help improve glycemic control (blood glucose management) and aid in weight loss. This makes GLP-1 agonists valuable for managing both type 2 diabetes and obesity.

In summary, GLP-1 receptor agonists are effective in treating both **type 2 diabetes** and **obesity**. They enhance glucose-dependent insulin secretion and suppress glucagon, thereby improving blood glucose regulation. At the same time, since they reduce appetite and slow down gastric emptying, they also help with weight loss. These dual effects make GLP-1 agonists powerful drugs for treating diabetes and insulin, two closely linked conditions.

Some of the major GLP-1 receptor agonists are **Ozempic (semaglutide), Wegovy (semaglutide)** [58], and **Rybelsus (semaglutide)** [59]. These three drugs have the same active ingredients but different formulations (such as the excipients (inactive ingredients) and route of administration), dosage, and approved use. These medications were originally designed for the treatment of **type 2 diabetes** because of their roles in enhancing insulin secretion and inhibiting glucagon release. They also have the additional effect of promoting weight loss, which has led to their use in the treatment of **obesity**. The drugs are marketed for different purposes. In fact, some, such as **Wegovy,** are specifically approved and marketed for weight loss [60, 61], while **Ozempic** and **Rybelsus** are mainly indicated for diabetes. Other GLP-1 receptor agonists include **Trulicity (dulaglutide)** [62], **Saxenda (liraglutide)** [63], **Bydureon (exenatide extended-release)** [64], and **Victoza (liraglutide)** [65]. Like the drugs discussed above, they are also GLP-1 agonists, differing only slightly in formulation. They also differ in their mechanism and frequency of administration, as well as whether they are approved for the treatment of diabetes and obesity. All of these GLP-1 agonists except **Rybelsus** are administered by injection. **Rybelsus** is the only orally available GLP-1 agonist currently available. GLP-1 receptor agonists were originally developed for the treatment of diabetes, but they have now become very popular medications for obesity. In fact, they have become so popular that it has been challenging for companies to keep up with the demand [66].

2.9.2 Drugs That Target Multiple Receptors

Mounjaro (tirzepatide) [67] is a dual agonist of both the GLP-1 and GIP receptors, approved by the FDA in 2022 [68]. By activating both receptors, it enhances insulin secretion via the GLP-1 receptor, and it may increase insulin secretion further via the GIP receptor. It is used for type 2 diabetes [69]. **Tirzepatide** is also marketed as **Zepbound** for obesity.

Mazdutide is still in trials but is expected to be approved by the FDA for weight loss and glycemic control. It is a dual agonist of the **GLP-1 receptor** and of the **glucagon receptor**. This is an interesting combination because while

GLP-1 agonists stimulate insulin secretion, reduce appetite, and improve glycemic control, glucagon typically raises blood glucose levels and is suppressed in the fed state. This dual action for treating diabetes may at first seem counterintuitive, but when combined with GLP-1 activation, glucagon's actions are modulated. Its ability to increase blood glucose is offset by GLP-1-induced insulin secretion, but its beneficial effects, such as increased fat breakdown for energy, are preserved. This synergistic approach promotes greater weight loss than GLP-1 agonists alone [70, 71]. This drug is part of a new class of drugs that are designed to target multiple hormonal pathways.

Retatrutide is a drug intended for the treatment of obesity and other metabolic disorders. It is actually a triple agonist of GLP-1, GIP, and glucagon receptors. It is not yet FDA-approved and is currently in Phase II clinical trials, and appears to be effective for weight loss [72].

2.9.3 Other Medications for Diabetes

- Synthetic Insulin

GLP-1 agonists are an important family of medications for treating diabetes, but other treatments are also available. **Type 1** diabetes is treated differently from type 2 diabetes. Type 1 diabetes can be treated with insulin. **Lantus** [73]**, Toujeo** [74], and **Basaglar** [75] are all forms of **insulin glargine**, a synthetic form of insulin, classified as **biologics**. These drugs are administered by injection, usually by the patient. Insulin glargine has a structure that makes it last longer than natural insulin. It can be helpful for type 1 diabetes, where pancreatic β cells are destroyed and no longer produce insulin. Insulin therapy is also used in advanced type 2 diabetes because pancreatic β-cell dysfunction occurs as the disease progresses. In the case of type 2 diabetes, it is usually used only for patients whose blood sugar cannot be controlled by other medications and lifestyle changes [76, 77].

Some drugs for diabetes include a combination of synthetic insulin and GLP-1 agonists. Examples are **Xultophy (insulin degludec/liraglutide** [78]) and **Soliqua (insulin glargine/lixisenatide)** [79]. These combinations are meant to address multiple aspects of the disease and to improve both insulin deficiency and insulin resistance [60].

- Metformin

Glucophage (Metformin) [80, 81] (also has other brand names) is a commonly used drug to treat type 2 diabetes. Metformin has been used for decades and has been shown to lower blood sugar. Metformin improves the body's response to insulin. It inhibits hepatic gluconeogenesis (the production of glucose, largely in the liver). Its exact mechanism of action is still not completely understood. One proposed mechanism involves the inhibition of mitochondrial complex I in the electron transport chain, leading to decreased ATP production. This leads to an increase in the AMP:ATP and the ADP:ATP ratios, which activates **AMPK**. This, in turn, contributes to reduced gluconeogenesis in the liver because AMPK activates a signaling pathway that suppresses transcription of gluconeogenic enzymes such as **PEPCK** and **G6Pase**. Metformin also enhances GLUT4 translocation to the membrane in adipose and muscle cells, facilitating insulin entry into these cells. This is also due to AMPK signaling, which can promote an insulin-independent mechanism for GLUT4 translocation (see Chap. 14). AMPK can also increase transcription of the gene-encoding GLUT4, which also contributes to increased insulin transport. Metformin may also modulate signaling by hormones such as GLP-1 and glucagon. Although there are newer drugs that are increasingly being prescribed for diabetes, metformin is still being prescribed. Metformin is beneficial to many patients, although at later stages of the disease, when pancreatic β cells may be affected, insulin replacement therapy may also be necessary [80, 82–86].

- DPP-4 Inhibitors

DPP-4 (dipeptidyl peptidase-4) is an enzyme that breaks down the incretin hormones **GLP-1 and GIP**. It breaks them down shortly after they are secreted, so these hormones typically have short half-lives and remain in the circulation only briefly after nutrient ingestion. **DPP-4**

inhibitors, such as **Januvia (sitagliptin)** [87] and others, reduce the breakdown of GLP-1 and GIP1, thereby prolonging their activity. The result is enhanced glucose-dependent insulin secretion and improved blood glucose control. **DPP-4 inhibitors** do not cause weight loss, unlike GLP-1 agonists. These drugs are used for patients who do not tolerate some of the other available diabetes medications, and they are sometimes used in combination with other medications [88].

- Sulfonylureas

Sulfonylureas are among the oldest classes of drugs used to treat type 2 diabetes. Examples are **glipizide** [89] and **glyburide** [90]. They are still used to some extent but are largely replaced by newer drugs. These drugs act by blocking potassium channels on pancreatic beta cells. This leads to membrane depolarization and the opening of voltage-gated Ca^{2+} channels. This leads to Ca^{2+} influx, which triggers exocytosis of insulin from insulin granules. Consequently, more insulin is released into the bloodstream. These drugs come with some risks, such as the risk of hypoglycemia [80, 91, 92].

- Glucose Transport Inhibitors

Jardiance (empagliflozin) [93] (also discussed in Chap. 3) is a small-molecule drug used to treat diabetes. It has been used for some time, as it was first approved in 2014. This drug is an inhibitor of the sodium-glucose cotransporter 2 (SGLT2). This transporter is in the kidneys. It is on the apical (luminal) membrane of proximal tubule (a portion of the kidney) epithelial cells. It transports glucose from the tubular lumen into the epithelial cell. This glucose exits the cell through the basolateral membrane via the GLUT2 transporter, using passive transport, into the peritubular capillaries (small blood vessels surrounding the nephron, the kidney's functional unit), which are the filtering units of the kidney. The SGLT2 transporter transports glucose from the kidney filtrate into the bloodstream, preventing glucose from being lost in the urine. By inhibiting this transporter, **Jardiance** reduces the amount of glucose transported to the bloodstream, allowing some of it to be excreted in the urine instead. Jardiance also lowers blood pressure and appears to be good for the heart, and although newer drugs are available, Jardiance is still used today due to its well-known benefits and efficiency [94, 95].

Farxiga (dapagliflozin) [96] is also an inhibitor of the SGLT2 transporter and functions similarly to Jardiance. Likewise, **Brenzavvy (bexaglifozin)** [97], **Invokana (canagliflozin)** [98], and **Steglatro (ertugliflozin)** [99] function similarly [100].

Inpefa (sotagliflozin) [101], which was FDA approved in 2023, is a dual inhibitor of both SGLT2 and SGLT1. SGLT1 is also a sodium-glucose co-transporter and is the main glucose transporter in the cells of the small intestine, particularly in the enterocytes lining the brush border. It imports glucose from the inner part of the intestine (the intestinal lumen) into the intestinal epithelial cells that line the intestine. Once the glucose is transported into these intestinal lining cells, it is then transported out of the enterocytes via the GLUT2 glucose transporter into the bloodstream. Consequently, blocking SGLT1 slows down the absorption of glucose from food into enterocytes. Instead of passing through the enterocytes and being released into the bloodstream, glucose continues to transit through the GI tract. This lowers spikes in blood glucose levels after a meal. Furthermore, as more glucose reaches the lower intestine, it may lead to the release of GLP-1 due to the stimulation of L cells in the distal gut, which are responsive to glucose [102]. At the same time, **inpefa** inhibits SGL2, giving it a function similar to Jardiance and other SGL2 inhibitors, as it blocks the export of glucose from the kidney to the bloodstream.

3 Hepatic Steatosis

Hepatic steatosis is the term for fatty liver disease. There are two major types of hepatic steatosis: that caused by alcohol and that caused by metabolic problems. When caused by alcohol, the condition is called alcoholic fatty liver disease (AFLD). When it is not caused by alcohol, it is called non-alcoholic fatty liver disease (NAFLD). In both cases, steatosis occurs when excess fat accumulates in non-adipose tissues, especially the liver, and sometimes in skeletal muscle. NAFLD is closely linked to insulin resistance, and in advanced cases, it may contribute to pancreatic β-cell dysfunction. As a result, it can increase the risk of developing type 2 diabetes [103, 104]. This relationship is bidirectional, though, and type 2 diabetes can also increase the risk of fatty liver disease.

3.1 The Normal Role for the Liver in Fatty Acid Metabolism

The liver plays a central role in fatty acid metabolism. It is the main site of lipogenesis, the synthesis of fatty acids from excess carbohydrates (largely glucose). In addition, the liver obtains fatty acids from other tissues, mainly adipose tissue, which results from lipolysis that occurs in adipose tissue (the breakdown of TAGs into free fatty acids and glycerol, especially during fasting). The liver does not store significant amounts of fat. Instead, some of the fatty acids are oxidized and used as an energy source. Those that are not used are packaged into very-low-density lipoproteins (VLDLs) and transported to other tissues, especially adipose tissue and, to a lesser extent, to muscle [105–107].

3.2 Abnormalities in Fat Metabolism in Liver

In some metabolic conditions, TAG accumulation in the liver exceeds the liver's ability to oxidize the fatty acids or export them. As a result, fat builds up in the liver. One major cause of this is **excessive lipolysis** (increased breakdown of TAGs into fatty acids and glycerol) in adipose tissue. This excessive lipolysis can occur due to insulin resistance. Insulin resistance is associated with the lack of inhibition of **HSL** and **ATGL**, resulting in excessive TAG hydrolysis. The excess fatty acids that result increase the flow of fatty acids to the liver.

Another major cause of increased fat in the liver is increased de novo lipogenesis (DNL: the synthesis of fatty acids, which are subsequently esterified to form TAGs) in the liver. This is often stimulated by high-carbohydrate or high-fat diets. If fat production continues to exceed fat clearance, excess fatty acids accumulate in hepatocytes (liver cells). These fatty acids can be esterified with a glycerol backbone to form TAGs, which can result in steatosis (fatty liver disease) [108].

3.2.1 Alcoholic Fatty Liver Disease (AFLD)

One well-known cause of fatty liver that is not linked to abnormalities in metabolism is alcohol-induced fatty liver disease (AFLD). Alcohol metabolism disrupts hepatic function by altering the NADH/NAD^+ ratio because ethanol breakdown leads to NADH accumulation. High NADH levels inhibit the Krebs cycle and fatty acid oxidation in the mitochondria. Inhibition of the Krebs cycle results in the accumulation of acetyl-CoA. This acetyl-CoA can serve as a substrate for de novo fatty acid synthesis (see Chap. 11). Consequently, fatty acid synthesis increases while fat breakdown is simultaneously suppressed, promoting TAG accumulation in the liver [109].

3.2.2 Non-Alcoholic Fatty Liver Disease (NAFLD)

Unlike AFLD, nonalcoholic fatty liver disease (NAFLD) is strongly linked to problems with metabolic dysfunction, and it is unrelated to alcohol consumption. NAFLD develops due to excess TAG synthesis, impaired VLDL secretion, and chronic insulin resistance. A major component is increased de novo lipogenesis (DNL) in the liver, which can result from obesity, a high-carbohydrate diet, diabetes, or insulin resistance.

Impaired VLDL secretion can occur for several reasons. For example, excess fatty acids and ER stress can disrupt the machinery needed for VLDL assembly and lipid packaging. Insulin resistance may also play a role. This is a problem because VLDLs are responsible for exporting TAGs from the liver. When this process is disrupted, TAGs become trapped in the liver and accumulate there [110].

When fat accumulation persists, in some patients, the disease may progress from NAFLD to **nonalcoholic steatohepatitis (NASH)**, a more severe form of NAFLD. This is associated with severe liver inflammation, damage, and fibrosis. It increases the risk of cirrhosis, liver failure, and cancer (hepatocellular carcinoma) [111].

mTOR activation and **AMPK inactivation** may contribute to abnormal fat metabolism in the liver, and their roles have been more clearly established in non-alcoholic fatty liver disease [112, 113]. **mTOR activation in the liver promotes de novo lipogenesis (DNL),** stimulating the conversion of carbohydrates into fatty acids. Chronic overactivation of mTOR, especially during excess nutrient intake, enhances lipid synthesis and contributes to the accumulation of TAGs in the liver. This process plays a key role in NAFLD.

AMPK activation, as opposed to **mTOR, promotes fatty acid oxidation** and inhibits de novo lipogenesis (DNL). Thus, it has a protective function against hepatic fat accumulation. Excessive mTOR activity and reduced AMPK activity can collectively contribute to abnormal fat metabolism in the liver, leading to NAFLD.

Some therapies for treating hepatic steatosis focus on reducing mTOR activation and enhancing AMPK activity. Similar to the treatment of type 2 diabetes, exercise and caloric restriction are also beneficial because they activate AMPK signaling and reduce mTORC1 signaling [114].

3.2.3 Drugs Used to Treat Fatty Liver Disease

Rezdiffra (resmetirom) [115] is a new drug (FDA approved in 2024) that is used to treat some patients with **NASH**. Liver scarring can occur as a result of the inflammation that NASH causes as it progresses. This significantly increases the risk for liver dysfunction. **Rezdiffra** is used to treat liver scarring [116]. It works by acting on the important connection between thyroid function and the liver. The thyroid is a gland that produces two main hormones: thyroxine (T4) and triiodothyronine (**T3**) [117]. These thyroid hormones signal to the liver and stimulate several important functions. For example, **T3** binds to thyroid hormone receptor beta (THRβ) in liver cells after traveling from the thyroid to the liver. THRβ is a nuclear receptor. Upon binding to T3, it regulates the expression of genes, including genes that encode enzymes involved in β oxidation of fatty acids, thereby enhancing fatty acid oxidation in the liver. **Rezdiffra** is a small-molecule agonist of THRβ, which thereby increases the activation of the receptor. Activating this receptor increases fatty acid oxidation, thus reducing the amount of fatty tissue in the liver. Patients who have taken this drug have shown a decrease in liver scarring [118].

4 Nutrient Excess and Dyslipidemia (Abnormal Levels of Fats, Such as Cholesterol, in the Blood)

4.1 Caloric Excess and Obesity

When a large amount of calories is consumed, excess TAGs will be stored in adipose cells. Excess storage of TAGs is evolutionarily beneficial when food is scarce, but prolonged or chronic TAG storage can lead to metabolic disease such as diabetes. Obesity, defined as excessive fat accumulation of adipose tissue, is a consequence of long-term excess caloric intake. Below, we discuss several complications that are associated with excess caloric intake and obesity.

4.1.1 Excess Storage of TAGs in Adipose Tissue

When the amount of fat in the body becomes excessive, the storage capacity of the adipocytes can become limited or dysfunctional. In this case, the excess fat accumulates in non-adipose tissues, such as the liver, pancreas, skeletal muscle, and blood vessels (fatty acid spillover). This fat deposition in other tissues (ectopic fat deposition) contributes to metabolic disorders, including coronary heart disease, type 2 diabetes, some cancers, and hypertension [119, 120].

4.1.2 Abdominal Fat and Diabetes Risk

Body fat is distributed differently throughout the body. Its location makes a difference in metabolism. Subcutaneous fat, located beneath the skin, is the largest concentration of fat tissue, and it is usually the least harmful. In contrast, visceral fat, which surrounds the abdominal organs, is strongly associated with insulin resistance and type 2 diabetes. Visceral fat is often referred to as abdominal fat. This fat tissue is more metabolically active and releases free fatty acids directly into the portal vein, increasing fat delivery to the liver, which can contribute to hepatic insulin resistance [121]. Excess visceral fat can also secrete inflammatory cytokines (such as TNFα and IL-6), leading to inflammation. As storage capacity is exceeded, spillover occurs from these tissues, and this is the main source of fat that accumulates in non-adipose tissues such as liver, muscle, and pancreas (ectopic fat deposition). Increased abdominal fat is therefore a major risk factor for type 2 diabetes [122].

In contrast to abdominal fat, subcutaneous fat, located beneath the skin, is considered to be the least harmful. In fact, some studies suggest that subcutaneous fat in the lower body (gluteofemoral fat in the hips and thighs) may even be protective, as it acts as a safe storage site for fat and is less likely to release fatty acids into the circulation [123].

4.1.3 Abnormalities in LPL

Another problem that can lead to ectopic fat distribution is an abnormality in **LPL (lipoprotein lipase (LPL)). LPL** is located in the capillary endothelium of adipose tissue (as well as skeletal muscle and other tissues). It plays a central role in **lipid clearance from the bloodstream** by hydrolyzing circulating TAGs (in chylomicrons and VLDLs) into free fatty acids. Insulin stimulates **LPL** activity in the capillaries in adipose tissue. This leads to the breakdown of dietary and liver-derived TAGs in capillaries, making the components available for TAG storage in adipose tissue after meals. In conditions of **caloric excess and obesity**, increased adipose LPL activity initially enhances storage. However, over time, LPL can become dysregulated, and in some cases, this can contribute to the "spillover" effect of fats into non-fat tissue [124–126].

4.1.4 Altered Metabolism in Fat Tissue

Excessive adipose tissue leads to dysregulated metabolism within the adipose tissue itself. It can lead to increased lipolysis (breakdown of TAGs into fatty acids and glycerol), altered secretion of factors such as leptin and cytokines, and recruitment of immune cells that lead to inflammation. Excess fat that accumulates in muscle can also impact certain signaling pathways, including insulin signaling pathways, leading to insulin resistance.

Fatty acids that result from increased lipolysis can be metabolized into smaller components such as diacylglycerols (DAGs) and ceramide, and as discussed above, this can interfere with insulin signaling and promote insulin resistance. As a result, glucose uptake can be reduced, thus increasing blood glucose levels [3, 127].

4.2 Excess Cholesterol

4.2.1 Excess Cholesterol and LDL

Excessive cholesterol in the blood is a form of dyslipidemia (accumulation of too many lipids in the blood). Cholesterol is an essential component of cell membranes, and it is important for many biological functions. Some cholesterol in the body is essential. Excess cholesterol, however, especially in the form of LDL, can lead to multiple problems such as atherosclerosis, heart attack, and stroke. Excess cholesterol can accumulate in two ways. It can be caused by too much exogenous cholesterol (from the diet), or too much endogenous cholesterol, which is produced in the body, mostly in the liver. Excess dietary (exogenous) cholesterol can result from excessive consumption of foods that are high in saturated fat and trans-fat. Such foods include red meat, processed and fried foods, and foods made with saturated and trans-fats. High endogenous cholesterol can result from overproduction of cholesterol in the liver. Cholesterol is made normally in the body, especially by the liver, but excess production of cholesterol can be a result of genetic abnormalities such as **familial hypercholesterolemia**, or other metabolic disorders that lead to overproduction of cholesterol [128–131].

4.2.2 Health Consequences of High LDL Cholesterol

Cholesterol is packaged in the blood in lipoprotein particles, including low-density lipoprotein (LDL) particles, which transport cholesterol, other lipids, and protein to different parts of the body. When LDL cholesterol levels become too high, the particles become oxidized in the blood vessel walls, altering the overall structure of the LDL particles. This oxidation of LDL particles alters their structure, leading to an immune response where macrophages engulf the particles. These macrophages become engorged with the lipids they engulf, turning into foam cells. Foam cells accumulate in the arterial walls of blood vessels and contribute to the formation of atherosclerotic plaques along the cell walls of arteries [132]. Atherosclerotic plaques contribute to atherosclerosis, which is the accumulation of plaques that, over time, can result in the narrowing of arteries. This restricts blood flow and increases blood pressure. A possible outcome is a rupture of plaques, which can lead to a stroke or myocardial infarction (heart attack).

4.2.3 HDL and Protection Against Excess Cholesterol

In contrast to LDL (which is often called the "bad cholesterol," HDL particles are generally protective against excess cholesterol and are often called the "good cholesterol." HDL carries out reverse cholesterol transport, a process in which cholesterol is collected from peripheral tissues, including the walls of blood vessels, and transported to the liver for processing, redistribution, or eventual excretion. This function is largely mediated by the major HDL protein ApoA-I, which activates the enzyme **LCAT**. **LCAT** esterifies cholesterol into cholesterol esters, allowing it to be packed into HDL particles and delivered to the liver. In addition to its role in reverse transport, HDL also has anti-inflammatory and antioxidant functions that protect vessels from damage. Because of these functions, higher HDL levels are usually associated with lower cardiovascular risk, and to some extent, they can counterbalance high LDL levels. In contrast, reduced HDL levels can impair cholesterol clearance and increase the risk of atherosclerosis [133].

4.2.4 Treatment Options for High Cholesterol

Elevated LDL can be treated in different ways. An important approach is dietary changes, though this is not always sufficient to decrease cholesterol levels. Drugs are also used to control cholesterol, with statins being the main drugs used for this purpose. Two commonly used statins include **Lipitor (atorvastatin)** and **Crestor (rosuvastatin)**. Statins are inhibitors of the enzyme **HMG-CoA reductase**. By inhibiting HMG-CoA reductase, they reduce cholesterol synthesis in the liver (HMG-CoA reductase and its inhibition is also discussed in Chap. 5) [134].

Another option is PCSK9 inhibitors like **evolocumab** [135], which increases LDL receptor availability to enhance LDL clearance from the bloodstream, and which is discussed in Chap. 11. PCSK9 inhibitors are monoclonal antibodies that bind and inhibit PCSK9. PCSK9 is a soluble extracellular protein that binds to the LDL receptors on liver cells, and leads to degradation of LDL receptors in the lysosome. This reduces the number of LDL receptors on the cell surface. Too much PCSK9 can lead to too much degradation of LDL receptors, reducing the ability of the liver to clear cholesterol from the blood. PCSK9 inhibitors like evolocumab are monoclonal antibodies that block PCSK9 and prevent PCSK9 from interfering with LDLR's ability to bind LDL and clear it from the blood [136]. This treatment is more expensive and requires injection, and is less common than the use of statins [137].

4.3 Other Forms of Dyslipidemia

Excess cholesterol is a well-known type of dyslipidemia, but other forms exist. For example, impaired **LPL** function can contribute to hypertriglyceridemia (abnormally high TAGs in the blood). When LPL is impaired, TAGs in LPLs are not efficiently hydrolyzed. As a result, the lipoproteins fail to transition to their smaller remnant forms. Consequently, there is a decrease in the clearance of circulating TAG-rich lipoproteins, and TAG levels in the blood rise.

Dyslipidemia can also be caused by low levels of HDL cholesterol. Too little HDL can impair reverse cholesterol transport, reduce LDL clearance, and increase cardiovascular risk. Dyslipidemia is often present, associated with type 2 diabetes and metabolic syndrome. In addition, genetic disorders such as familial chylomicronemia (severely elevated TAGs) are inherited forms of dyslipidemia [138].

5 Possible Health Benefits of a High-Fat, Low-Carb Ketogenic Diet

A ketogenic diet involves restricting carbohydrate intake while maintaining moderate protein and high fat intake. This reduces the availability of glucose, which is normally the major energy source for most cells. It also reduces insulin levels, because glucose intake, rather than fat intake, increases insulin secretion from pancreatic β cells. Insulin is a hormone associated with storage, and when insulin levels are low, there is an increased breakdown of fat (TAGs) in the liver and adipose tissue. Therefore, a ketogenic diet is associated with an increased breakdown of TAGs. In a ketogenic diet, TAGs become the main energy source, and they are broken down into free fatty acids and glycerol. This leads in turn to ketone body formation. Ketone bodies form because there is an increase in acetyl-CoA due to the increased fatty acid oxidation (see Chap. 12), and a decrease in OAA due to the lack of carbohydrates to metabolize. Acetyl-CoA is the basis for the formation of ketone bodies. The cells in the body, especially the brain and muscle, adapt to using ketone bodies and fatty acids as their main energy source (the brain uses the ketone bodies, but it still relies on some glucose). This reduces reliance on glucose and glycogen, increasing reliance on fat stores for energy instead. Because of the reduced levels of insulin, the body continues to break down fat as an energy source. Thus, the ketogenic diet shifts the energy balance toward increased TAG degradation and provides health benefits associated with low glucose and the breakdown of fat [20, 139]. The ketogenic diet has multiple possible benefits, including contributing to weight loss and fat reduction, lower blood glucose levels, and reduced TAG levels. The lowering of insulin secretion may potentially also help in the prevention of diabetes. Long-term adherence to the diet may lead to problems, however, such as nutrient deficiency, fiber deficiency, and hypoglycemia [140, 141].

Ketogenic diets are also sometimes used as a treatment option for children with drug-resistant epilepsy. The ketogenic diet can provide an alternative fuel source for the brain (ketones), which may help stabilize neural activity. The ketones may also alter brain chemistry in a way that reduces seizure activity. For example, they have been shown to beneficially change the balance of inhibitory and excitatory neurotransmitters [15, 26, 142, 143].

6 Metabolic Syndrome

Many of the conditions discussed in this chapter share common underlying risk factors, such as insulin resistance or increased abdominal fat. When several risk factors occur together, this is referred to as metabolic syndrome. Metabolic syndrome is a cluster of conditions that together increase the risk of diseases, including heart disease, stroke, and type 2 diabetes. There are five official diagnostic criteria for having metabolic syndrome. They are (1) abdominal obesity, (2) elevated fasting glucose, (3) high blood pressure, (4) high TAGs, and (5) Low HDL cholesterol. Having three or more of these conditions increases the risk of serious health problems. Metabolic syndrome is often linked to obesity, physical inactivity, and insulin resistance, where the body cannot use insulin effectively. Lifestyle changes, such as improved diet, exercise, and weight loss, can help manage and prevent metabolic syndrome or prevent it from worsening [15, 144].

7 Fructose Malabsorption

Fructose malabsorption belongs to a different category than the other conditions discussed in this chapter because it is not directly related to problems in homeostasis. It is included here, however, because it also relates to the body's metabolism of sugar and can be managed by dietary changes. When dietary carbohydrates are digested into monosaccharides, those monosaccharides are absorbed from the intestinal lumen into the enterocytes, which are epithelial cells that line the intestine. The sugars enter the cells on the apical (luminal side) of the enterocytes. From within the cells, the sugars are transported across the basolateral membrane into the bloodstream, exiting the GI system to enter circulation.

From within the intestine, glucose and galactose enter the apical side of the enterocytes via the transporter **SGLT1**. This is a **sodium-dependent active transporter**. Because it is an active transporter, it allows the enterocytes to absorb glucose or galactose even when their concentrations inside the cell are high. In contrast, **fructose is absorbed via the GLUT5 transporter**. This transporter uses facilitated diffusion and thus does **not use energy**. As a consequence, GLUT5 does not concentrate fructose inside the cells. If fructose levels become too high in the enterocytes, then more fructose cannot enter these cells. While most individuals have enough GLUT5 transporters to handle this issue, some individuals have a lower level of GLUT5 expression. This, combined with the facilitated diffusion mechanism of transport, makes it difficult for these individuals to absorb fructose. People with this problem have **fructose malabsorption**. In these cases, not all the fructose is absorbed, and instead, it passes into the colon. There it is fermented by gut bacteria, leading to gas, bloating, and discomfort. Fructose malabsorption is not life-threatening, and people with fructose malabsorption handle the problem mainly with diet, especially by following a low-fructose diet. Foods that are high in fructose include some fruits, like apples and pears, honey, and high fructose corn syrup [145, 146].

FDA-Approved Drugs Covered in This Chapter

Brand name	Generic name	First FDA approval year	Main conditions	Main targets
Basaglar	*insulin glargine*	2015	Type 1 & type 2 diabetes	Insulin receptor
Brenzavvy	*bexagliflozin*	2023	Type 2 diabetes	SGLT2
Bydureon	*exenatide (extended-release)*	2017	Type 2 diabetes	GLP-1 receptor
Crestor	*Rosuvastatin*	2003	High cholesterol	HMG CoA reductase
Evolocumab	*evolocumab*	2015	High LDL cholesterol	PCSK9
Farxiga	*dapagliflozin*	2014	Type 2 diabetes	SGLT2
glipizide	*glipizide*	1978	Type 2 diabetes	K+ channels on beta cells
Glucophage	*metformin*	1995	Type 2 diabetes	AMPK
glyburide	*glyburide*	1984	Type 2 diabetes	K+ channels on beta cells
Inpefa	*sotagliflozin*	2023	Type 2 diabetes	SGLT1 and SGLT2
Invokana	*canagliflozin*	2013	Type 2 diabetes	SGLT2
Januvia	*sitagliptin*	2006	Type 2 diabetes	DPP-4
Jardiance	*empagliflozin*	2014	Type 2 diabetes	SGLT2
Lantus	*insulin glargine*	2000	Type 1 & type 2 diabetes	Insulin receptor
Mounjaro	*tirzepatide*	2022	Type 2 diabetes	GLP-1 and GIP receptors
Lipitor	*Atorvastatin*	1996	High cholesterol	HMG CoA Reductase
Ozempic	*semaglutide*	2017	Type 2 diabetes	GLP-1 receptor
Rezdiffra	*resmetirom*	2024	NASH	THRβ
Rybelsus	*semaglutide*	2019	Type 2 diabetes	GLP-1 receptor
Saxenda	*liraglutide*	2014	Obesity	GLP-1 receptor
Soliqua	*insulin glargine/ lixisenatide*	2016	Type 2 diabetes	Insulin and GLP-1 receptors
Steglatro	*ertugliflozin*	2017	Type 2 diabetes	SGLT2
Toujeo	*insulin glargine*	2015	Type 1 & type 2 diabetes	Insulin receptor
Trulicity	*dulaglutide*	2014	Type 2 diabetes	GLP-1 receptor
Victoza	*liraglutide*	2010	Type 2 diabetes	GLP-1 receptor
Wegovy	*semaglutide*	2021	Obesity	GLP-1 receptor
Xultophy	*insulin degludec/ liraglutide*	2016	Type 2 diabetes	Insulin and GLP-1 receptors

Definitions

Ceramides A class of lipids, specifically sphingolipids. They have a sphingosine backbone attached to a fatty acid.

Dyslipidemia Abnormal levels of lipids (fats), including cholesterol and triglycerides, in the blood.

HbA1c (glycated hemoglobin) A form of hemoglobin that has glucose molecules covalently attached. The HbA1c test is a measurement of the percentage of hemoglobin that is glycated.

Hepatic steatosis (fatty liver) Accumulation of triacylglycerides (TAGs) in liver cells.

Hyperglycemia Elevated levels of glucose in the blood, commonly associated with uncontrolled diabetes.

Insulin resistance A condition in which cells in muscles, fat, and liver do not respond well to insulin and therefore can't easily absorb glucose from the bloodstream or store glucose as glycogen.

Ketoacidosis A dangerous complication of diabetes, especially type 1, where ketone bodies accumulate excessively, leading to acidic blood pH.

Metabolic syndrome A cluster of conditions that together increase the risk of diseases including heart disease, stroke, and type 2 diabetes.

PCSK9 inhibitors Monoclonal antibodies that reduce LDL cholesterol by targeting PCSK9 and thereby preventing the degradation of LDL receptors on liver cells.

Reactive Oxygen Species (ROS) Chemically reactive molecules containing oxygen that can cause cellular damage and contribute to diabetic complications. Many ROS are free radicals, meaning they contain unpaired electrons, which makes them unstable and makes them seek out and interact with other molecules. Others are not free radicals but may have high redox potential, allowing them to oxidize other molecules due to their strong electron-accepting properties.

SGLT2 inhibitors Drugs that lower blood glucose by preventing glucose reabsorption in the kidney, leading to its excretion in urine.

Sulfonylureas One of the older classes of diabetes drugs. They stimulate insulin secretion by depolarizing pancreatic β-cell membranes.

Study Questions

Multiple-Choice Questions

1. Which of the following is a primary cause of Type 1 diabetes?
   A. Insulin resistance
   B. Excessive sugar intake
   C. Autoimmune destruction of β cells
   D. High-fat diet
   E. B and C
2. Which metabolic condition is most closely associated with the development of ketoacidosis?
   A. Type 2 diabetes
   B. Hepatic steatosis
   C. Fructose malabsorption
   D. Type 1 diabetes
3. What is the main mechanism of action of SGLT2 inhibitors like Jardiance?
   A. Increase insulin production
   B. Inhibit glucose absorption in the gut
   C. Block glucose reabsorption in the kidney
   D. Inhibit fat digestion
4. The energy sensor that is activated by a high AMP/ATP ratio is:
   A. AMPK
   B. mTORC1
   C. PEPCK
   D. Insulin
   E. GLP-1pt

True/False Questions

5. Metabolic syndrome requires the presence of all five diagnostic criteria.
6. The only types of diabetes are type 1 diabetes and type 2 diabetes.

Fill-in-the-Blank Questions

7. Type 2 diabetes is primarily caused by an inability of cells to respond to ________.
8. The HbA1c test measures the percentage of hemoglobin that is ________.
9. Sulfonylureas increase insulin secretion by blocking ________ channels in β cells.

Short-Answer Question

10. Why might patients with advanced type 2 diabetes require insulin therapy?

Answer Key

Multiple-Choice Questions

1. **Answer**: C
2. **Answer**: D
3. **Answer**: C
4. **Answer**: A

True/False Questions

5. **Answer**: False – It is diagnosed when three or more of the five criteria are present.
6. **Answer**: False – Gestational diabetes is a third type.

Fill-in-the-Blank Questions

7. **Answer**: Insulin
8. **Answer**: Glycated
9. **Answer**: Potassium

Short-Answer Question

10. **Answer**: Because long-term insulin resistance can disrupt pancreatic β-cell function over time, resulting in reduced production and secretion of insulin.

References

1. Rahman MS, Hossain KS, Das S, Kundu S, Adegoke EO, Rahman MA, et al. Role of insulin in health and disease: an update. Int J Mol Sci. 2021;22(12):6403. https://doi.org/10.3390/ijms22126403.
2. Plows JF, Stanley JL, Baker PN, Reynolds CM, Vickers MH. The pathophysiology of gestational diabetes mellitus. Int J Mol Sci. 2018;19(11):3342. https://doi.org/10.3390/ijms19113342.
3. Bender DA. Introduction to nutrition and metabolism. Boca Raton: CRC Press; 2021.
4. Boucher J, Kleinridders A, Kahn CR. Insulin receptor signaling in normal and insulin-resistant states. Cold Spring Harb Perspect Biol. 2014;6(1) https://doi.org/10.1101/cshperspect.a009191.
5. Galicia-Garcia U, Benito-Vicente A, Jebari S, Larrea-Sebal A, Siddiqi H, Uribe KB, et al. Pathophysiology of type 2 diabetes mellitus. Int J Mol Sci. 2020;21(17) https://doi.org/10.3390/ijms21176275.
6. Hudish LI, Reusch JE, Sussel L. β Cell dysfunction during progression of metabolic syndrome to type 2 diabetes. J Clin Invest. 2019;129(10):4001–8. https://doi.org/10.1172/jci129188.
7. Poitout V, Robertson RP. Glucolipotoxicity: fuel excess and β-cell dysfunction. Endocr Rev. 2008;29(3):351–66. https://doi.org/10.1210/er.2007-0023.
8. American Diabetes Association. Diagnosis and classification of diabetes mellitus. Diabetes Care. 2009;32(Suppl 1):S62–7. https://doi.org/10.2337/dc09-S062.
9. Fonseca VA. Defining and characterizing the progression of type 2 diabetes. Diabetes Care. 2009;32(Suppl 2):S151–6. https://doi.org/10.2337/dc09-S301.
10. Timberlake KC, Orgill MK. General, organic, and biological chemistry: structures of life. 6th ed. Pearson Education; 2018.
11. Mittal R, Prasad K, Lemos JRN, Arevalo G, Hirani K. Unveiling gestational diabetes: an overview of pathophysiology and management. Int J Mol Sci. 2025;26(5):2320. https://doi.org/10.3390/ijms26052320.
12. Marks F, Klingmüller U, Müller-Decker K. Cellular signal processing: an introduction to the molecular mechanisms of signal transduction. 2nd ed. Garland Science; 2017.
13. Lewis ST, Greenway F, Tucker TR, Alexander M, Jackson LK, Hepford SA, et al. A receptor story: insulin resistance pathophysiology and physiologic insulin resensitization's role as a treatment modality. Int J Mol Sci. 2023;24(13) https://doi.org/10.3390/ijms241310927.
14. Hotamisligil GS. Inflammation and metabolic disorders. Nature. 2006;444(7121):860–7. https://doi.org/10.1038/nature05485.
15. Berg JM, Tymoczko JL, Gatto GJ Jr, Stryer L. Biochemistry. 9th ed. New York: MacMillan Learning; 2019.

16. Nelson DL, Cox MM, Hoskins AA. Lehninger principles of biochemistry. 8th ed. Macmillan Learning; 2021.
17. Virtue S, Vidal-Puig A. Adipose tissue expandability, lipotoxicity and the Metabolic Syndrome – an allostatic perspective. Biochim Biophys Acta. 2010;1801(3):338–49. https://doi.org/10.1016/j.bbalip.2009.12.006.
18. Kogot-Levin A, Saada A. Ceramide and the mitochondrial respiratory chain. Biochimie. 2014;100: 88–94. https://doi.org/10.1016/j.biochi.2013.07.027.
19. Holland WL, Bikman BT, Wang LP, Yuguang G, Sargent KM, Bulchand S, et al. Lipid-induced insulin resistance mediated by the proinflammatory receptor TLR4 requires saturated fatty acid-induced ceramide biosynthesis in mice. J Clin Invest. 2011;121(5): 1858–70. https://doi.org/10.1172/jci43378.
20. Roche HM, Macdonald IA, Schols AMWJ, Lanham-New SA, editors. Nutrition and metabolism. Wiley-Blackwell; 2024.
21. Lee S-H, Park S-Y, Choi CS. Insulin resistance: from mechanisms to therapeutic strategies. Diabetes Metab J. 2022;46(1):15–37. https://doi.org/10.4093/dmj.2021.0280.
22. Moeschel K, Beck A, Weigert C, Lammers R, Kalbacher H, Voelter W, et al. Protein kinase C-ζ-induced phosphorylation of Ser318 in Insulin Receptor Substrate-1 (IRS-1) attenuates the interaction with the insulin receptor and the tyrosine phosphorylation of IRS-1*. J Biol Chem. 2004;279(24): 25157–63. https://doi.org/10.1074/jbc.M402477200.
23. Roszczyc-Owsiejczuk K, Zabielski P. Sphingolipids as a culprit of mitochondrial dysfunction in insulin resistance and Type 2 diabetes. Front Endocrinol (Lausanne). 2021;12:635175. https://doi.org/10.3389/fendo.2021.635175.
24. Calimag APP, Chlebek S, Lerma EV, Chaiban JT. Diabetic ketoacidosis. Dis Mon. 2023;69(3): 101418. https://doi.org/10.1016/j.disamonth.2022.101418.
25. Laffel L. Ketone bodies: a review of physiology, pathophysiology and application of monitoring to diabetes. Diabetes Metab Res Rev. 1999;15(6): 412–26. https://doi.org/10.1002/(SICI)1520-7560(199911/12)15:6<412::AID-DMRR72>3.0.CO;2-8.
26. Clinic C. Ketosis 2022. Available from: https://my.clevelandclinic.org/health/articles/24003-ketosis
27. Manninen AH. Metabolic effects of the very-low-carbohydrate diets: misunderstood "villains" of human metabolism. J Int Soc Sports Nutr. 2004;1 (2):7–11. https://doi.org/10.1186/1550-2783-1-2-7.
28. Rodwell VW, Bender DA, Kennely PJ, Weil PA, McGuinness OP. Harper's illustrated biochemistry. 32nd ed. New York: McGraw-Hill Education/ Medical; 2022.
29. Medline Plus. Diabetic ketoacidosis 2023. Available from: https://medlineplus.gov/ency/article/000320.htm#:~:text=The%20fat%20is%20broken%20down,a%20severe%20illness%20or%20infection
30. Giacco F, Brownlee M, Schmidt AM. Oxidative stress and diabetic complications. Circ Res. 2010;107(9):1058–70. https://doi.org/10.1161/CIRCRESAHA.110.223545.
31. National Eye Institute. Diabetic Retinopathy 2024. Available from: https://www.nei.nih.gov/learn-about-eye-health/eye-conditions-and-diseases/diabetic-retinopathy#:~:text=The%20damage%20to%20your%20eyes,that%20leak%20fluid%20or%20bleed
32. Volpe CMO, Villar-Delfino PH, Dos Anjos PMF, Nogueira-Machado JA. Cellular death, reactive oxygen species (ROS) and diabetic complications. Cell Death Dis. 2018;9(2):119. https://doi.org/10.1038/s41419-017-0135-z.
33. Natesan V, Kim SJ. Diabetic nephropathy – a review of risk factors, progression, mechanism, and dietary management. Biomol Ther (Seoul). 2021;29(4): 365–72. https://doi.org/10.4062/biomolther.2020.204.
34. CDC. Diabetes: Nerve Damage 2024. Available from: https://www.cdc.gov/diabetes/diabetes-complications/diabetes-and-nerve-damage.html#:~:text=Peripheral%20nerve%20damage%20is%20the,and%20bone%20and%20joint%20pain
35. Poznyak A, Grechko AV, Poggio P, Myasoedova VA, Alfieri V, Orekhov AN. The diabetes mellitus-atherosclerosis connection: the role of lipid and glucose metabolism and chronic inflammation. Int J Mol Sci. 2020;21(5) https://doi.org/10.3390/ijms21051835.
36. Low Wang CC, Hess CN, Hiatt WR, Goldfine AB. Clinical update: cardiovascular disease in diabetes mellitus: atherosclerotic cardiovascular disease and heart failure in Type 2 diabetes mellitus – mechanisms, management, and clinical considerations. Circulation. 2016;133(24): 2459–502. https://doi.org/10.1161/circulationaha.116.022194.
37. Mayo Clinic. Diabetes 2024. Available from: https://www.mayoclinic.org/diseases-conditions/diabetes/diagnosis-treatment/drc-20371451#:~:text=A%20fasting%20blood%20sugar%20level,separate%20tests%2C%20you%20have%20diabetes
38. American Diabetes Association. Classification and diagnosis of diabetes: standards of medical care in diabetes-2022. Diabetes Care. 2022;45(Suppl 1): S17–s38.https://doi.org/10.2337/dc22-S002.
39. National Institute of Diabetes and Digestive and Kidney Disease. Diabetes Tests & Diagnosis 2022. Available from: https://www.niddk.nih.gov/health-information/diabetes/overview/tests-diagnosis.
40. MedlinePlus. Hemoglobin A1C (HbA1c) Test. Available from: https://medlineplus.gov/lab-tests/hemoglobin-a1c-hba1c-test/#:~:text=An%20A1C%20test%20measures%20the,cells%20live%20about%20three%20months
41. National Institute of Diabetes and Digestive and Kidney Disease. Continuous Glucose Monitoring 2023. Available from: https://www.niddk.nih.gov/health-information/diabetes/overview/managing-diabetes/continuous-glucose-monitoring

42. Harvard Health Publishing. The lowdown on glycemic index and glycemic load 2023. Available from: https://www.health.harvard.edu/diseases-and-conditions/the-lowdown-on-glycemic-index-and-glycemic-load
43. Harvard T.H. Chan School of Public Health. Carbohydrates and Blood Sugar. Available from: https://nutritionsource.hsph.harvard.edu/carbohydrates/carbohydrates-and-blood-sugar/
44. Reynolds A, Mitri, J. Dietary advice for individuals with diabetes. 2024. South Dartmoutn (MA): MDText.com. Available from: https://www.ncbi.nlm.nih.gov/books/NBK279012/
45. Reynolds A, Mann J, Cummings J, Winter N, Mete E, Te Morenga L. Carbohydrate quality and human health: a series of systematic reviews and meta-analyses. Lancet. 2019;393(10170):434–45. https://doi.org/10.1016/s0140-6736(18)31809-9.
46. DeFronzo RA, Tripathy D. Skeletal muscle insulin resistance is the primary defect in type 2 diabetes. Diabetes Care. 2009;32(Suppl 2):S157–63. https://doi.org/10.2337/dc09-S302.
47. Little TJ, Horowitz M, Feinle-Bisset C. Modulation by high-fat diets of gastrointestinal function and hormones associated with the regulation of energy intake: implications for the pathophysiology of obesity2. Am J Clin Nutr. 2007;86(3):531–41. https://doi.org/10.1093/ajcn/86.3.531.
48. Dhillon J, Craig BA, Leidy HJ, Amankwaah AF, Osei-Boadi Anguah K, Jacobs A, et al. The effects of increased protein intake on fullness: a meta-analysis and its limitations. J Acad Nutr Diet. 2016;116(6):968–83. https://doi.org/10.1016/j.jand.2016.01.003.
49. Richter EA, Hargreaves M. Exercise, GLUT4, and skeletal muscle glucose uptake. Physiol Rev. 2013;93(3):993–1017. https://doi.org/10.1152/physrev.00038.2012.
50. Long YC, Zierath JR. AMP-activated protein kinase signaling in metabolic regulation. J Clin Invest. 2006;116(7):1776–83. https://doi.org/10.1172/jci29044.
51. Ojuka EO, Goyaram V, Smith JAH. The role of CaMKII in regulating GLUT4 expression in skeletal muscle. Am J Physiol Endocrinol Metab. 2012;303(3):E322–E31. https://doi.org/10.1152/ajpendo.00091.2012.
52. Hall JE, Ha ME. Guyton and Hall Textbook of medical physiology. 14th ed. Elsevier; 2020.
53. Ojuka EO, Goyaram V, Smith JA. The role of CaMKII in regulating GLUT4 expression in skeletal muscle. Am J Physiol Endocrinol Metab. 2012;303(3):E322–31. https://doi.org/10.1152/ajpendo.00091.2012.
54. Klein S, Gastaldelli A, Yki-Järvinen H, Scherer PE. Why does obesity cause diabetes? Cell Metab. 2022;34(1):11–20. https://doi.org/10.1016/j.cmet.2021.12.012.
55. McCall AL. Insulin therapy and hypoglycemia. Endocrinol Metab Clin N Am. 2012;41(1):57–87. https://doi.org/10.1016/j.ecl.2012.03.001.
56. FDA. Ozempic (semaglutide) Approval Overview 2017. Available from: https://www.accessdata.fda.gov/scripts/cder/daf/index.cfm?event=overview.process&ApplNo=209637
57. Zheng Z, Zong Y, Ma Y, Tian Y, Pang Y, Zhang C, et al. Glucagon-like peptide-1 receptor: mechanisms and advances in therapy. Signal Transduct Target Ther. 2024;9(1):234. https://doi.org/10.1038/s41392-024-01931-z.
58. FDA. Wegovy (semaglutide) Approval Overview 2021. Available from: https://www.accessdata.fda.gov/scripts/cder/daf/index.cfm?event=overview.process&ApplNo=215256
59. FDA. Rybelsus (semaglutide) Approval Overview 2019. Available from: https://www.accessdata.fda.gov/scripts/cder/daf/index.cfm?event=overview.process&ApplNo=213051
60. Nauck MA, Quast DR, Wefers J, Meier JJ. GLP-1 receptor agonists in the treatment of type 2 diabetes – state-of-the-art. Mol Metab. 2021;46:101102. https://doi.org/10.1016/j.molmet.2020.101102.
61. Singh G, Krauthamer M, Bjalme-Evans M. Wegovy (semaglutide): a new weight loss drug for chronic weight management. J Investig Med. 2022;70(1):5–13. https://doi.org/10.1136/jim-2021-001952.
62. FDA. Trulicity (dulaglutide) Approval Overview 2014. Available from: https://www.accessdata.fda.gov/scripts/cder/daf/index.cfm?event=overview.process&ApplNo=125469
63. FDA. Saxenda (liraglutide) Approval Overview 2014. Available from: https://www.accessdata.fda.gov/scripts/cder/daf/index.cfm?event=overview.process&ApplNo=206321
64. FDA. Bydureon [exenatide (extended-release)] Approval Overview. 2012. https://www.accessdata.fda.gov/scripts/cder/daf/index.cfm?event=overview.process&ApplNo=022200
65. FDA. Victoza (liraglutide) Approval Overview 2010. Available from: https://www.accessdata.fda.gov/scripts/cder/daf/index.cfm?event=overview.process&ApplNo=022341
66. Mattingly TJ II, Conti RM. Marketing and safety concerns for compounded GLP-1 receptor agonists. JAMA Health Forum. 2025;6(1):e245015-e. https://doi.org/10.1001/jamahealthforum.2024.5015.
67. FDA. Mounjaro (tirzepatide) Approval Overview 2022. Available from: https://www.accessdata.fda.gov/scripts/cder/daf/index.cfm?event=overview.process&ApplNo=215866
68. Syed YY. Tirzepatide: first approval. Drugs. 2022;82(11):1213–20. https://doi.org/10.1007/s40265-022-01746-8.
69. Ili Lily and Company. Mounjaro 2025. Available from: https://mounjaro.lilly.com/
70. Zhang B, Cheng Z, Chen J, Zhang X, Liu D, Jiang H, et al. Efficacy and safety of mazdutide in Chinese patients with Type 2 diabetes: a Randomized,

Double-Blind, Placebo-Controlled Phase 2 Trial. Diabetes Care. 2024;47(1):160–8. https://doi.org/10.2337/dc23-1287.
71. Bailey CJ, Flatt PR, Conlon JM. Multifunctional incretin peptides in therapies for type 2 diabetes, obesity and associated co-morbidities. Peptides. 2025;187:171380. https://doi.org/10.1016/j.peptides.2025.171380.
72. Jastreboff AM, Kaplan LM, Frías JP, Wu Q, Du Y, Gurbuz S, et al. Triple-hormone-receptor agonist retatrutide for obesity – a Phase 2 Trial. N Engl J Med. 2023;389(6):514–26. https://doi.org/10.1056/NEJMoa2301972.
73. FDA. Lantus (insulin glargine) Approval Overview. 2000. https://www.accessdata.fda.gov/scripts/cder/daf/index.cfm?event=overview.process&ApplNo=021081
74. FDA. Toujeo (insulin glargine) Approval Overview 2015. Available from: https://www.accessdata.fda.gov/scripts/cder/daf/index.cfm?event=overview.process&ApplNo=206538
75. FDA. Basaglar (insulin glargine) Approval Overview. 2015. https://www.accessdata.fda.gov/scripts/cder/daf/index.cfm?event=overview.process&ApplNo=021081
76. Lamb YN, Syed YY. LY2963016 insulin glargine: a review in Type 1 and 2 diabetes. BioDrugs. 2018;32(1):91–8. https://doi.org/10.1007/s40259-018-0259-4.
77. Toujeo. Toujeo: insuline glargine injection 300 Units/mL 2023. Available from: https://www.toujeo.com/.
78. FDA. Xultophy (insulin degludec; liraglutide) Approval Overview 2016. Available from: https://www.accessdata.fda.gov/scripts/cder/daf/index.cfm?event=overview.process&ApplNo=208583
79. FDA. Soliqua (insulin glargine; lixisenatide) Approval Overview 2016. Available from: https://www.accessdata.fda.gov/scripts/cder/daf/index.cfm?event=overview.process&ApplNo=208673
80. Whalen K, Lerchenfeldt S, Giordano C. Lippincott illustrated reviews. Pharmacology. Wolters Kluwer; 2021.
81. FDA. Glucophage (metformin) Approval Overview 1995. Available from: https://www.accessdata.fda.gov/scripts/cder/daf/index.cfm?event=overview.process&ApplNo=020357
82. Rena G, Hardie DG, Pearson ER. The mechanisms of action of metformin. Diabetologia. 2017;60(9):1577–85. https://doi.org/10.1007/s00125-017-4342-z.
83. Luengo A, Sullivan LB, Heiden MG. Understanding the complex-I-ty of metformin action: limiting mitochondrial respiration to improve cancer therapy. BMC Biol. 2014;12:82. https://doi.org/10.1186/s12915-014-0082-4.
84. Bahne E, Sun EWL, Young RL, Hansen M, Sonne DP, Hansen JS, et al. Metformin-induced glucagon-like peptide-1 secretion contributes to the actions of metformin in type 2 diabetes. JCI. Insight. 2018;3(23) https://doi.org/10.1172/jci.insight.93936.
85. Lee JO, Lee SK, Kim JH, Kim N, You GY, Moon JW, et al. Metformin regulates glucose transporter 4 (GLUT4) translocation through AMP-activated protein kinase (AMPK)-mediated Cbl/CAP signaling in 3T3-L1 preadipocyte cells. J Biol Chem. 2012;287(53):44121–9. https://doi.org/10.1074/jbc.M112.361386.
86. Dutta S, Shah RB, Singhal S, Dutta SB, Bansal S, Sinha S, et al. Metformin: a review of potential mechanism and therapeutic utility beyond diabetes. Drug Des Devel Ther. 2023;17:1907–32. https://doi.org/10.2147/dddt.S409373.
87. FDA. Januvia (sitagliptin) Approval Overview 2006. Available from: https://www.accessdata.fda.gov/scripts/cder/daf/index.cfm?event=overview.process&ApplNo=021995
88. Gilbert MP, Pratley RE. GLP-1 analogs and DPP-4 inhibitors in Type 2 diabetes therapy: review of head-to-head clinical trials. Front Endocrinol (Lausanne). 2020;11:178. https://doi.org/10.3389/fendo.2020.00178.
89. FDA. Glipizide (glipizide) Approval Overview 1984. Available from: https://www.accessdata.fda.gov/scripts/cder/daf/index.cfm?event=overview.process&ApplNo=017783
90. FDA. Glyburide (glyburide) Approval Overview 1984. Available from: https://www.accessdata.fda.gov/scripts/cder/daf/index.cfm?event=overview.process&ApplNo=017532
91. Sola D, Rossi L, Schianca GP, Maffioli P, Bigliocca M, Mella R, et al. Sulfonylureas and their use in clinical practice. Arch Med Sci. 2015;11(4):840–8. https://doi.org/10.5114/aoms.2015.53304.
92. Kourlas H, Papadopoulas J. Clinical Pharmacy Corner: sulfonylureas: clinical correlations; 2007. Available from: https://www.clinicalcorrelations.org/2007/10/17/clinical-pharmacy-corner-sulfonylureas/
93. FDA. Jardiance (empagliflozin) Approval Overview 2014. Available from: https://www.accessdata.fda.gov/scripts/cder/daf/index.cfm?event=overview.process&ApplNo=204629
94. Ndefo UA, Anidiobi NO, Basheer E, Eaton AT. Empagliflozin (Jardiance): a novel SGLT2 inhibitor for the treatment of Type-2 diabetes. P T. 2015;40(6):364–8.
95. Chilton R, Tikkanen I, Cannon CP, Crowe S, Woerle HJ, Broedl UC, et al. Effects of empagliflozin on blood pressure and markers of arterial stiffness and vascular resistance in patients with type 2 diabetes. Diabetes Obes Metab. 2015;17(12):1180–93. https://doi.org/10.1111/dom.12572.
96. FDA. Farxiga (dapagliflozin) Approval Overview. 2014. https://www.accessdata.fda.gov/scripts/cder/daf/index.cfm?event=overview.process&ApplNo=202293
97. FDA. Brenzavvy (bexagliflozin) Approval Overview 2023. Available from: https://www.accessdata.fda.gov/scripts/cder/daf/index.cfm?event=overview.process&ApplNo=214373

98. FDA. Invokana (canagliflozin) Approval Overview. 2013. https://www.accessdata.fda.gov/scripts/cder/daf/index.cfm?event=overview.process&ApplNo=204042
99. FDA. Steglatro (ertugliflozin) Approval Overview. 2017. https://www.accessdata.fda.gov/scripts/cder/daf/index.cfm?event=overview.process&ApplNo=209803
100. Tentolouris A, Vlachakis P, Tzeravini E, Eleftheriadou I, Tentolouris N. SGLT2 inhibitors: a review of their antidiabetic and cardioprotective effects. Int J Environ Res Public Health. 2019;16(16) https://doi.org/10.3390/ijerph16162965.
101. FDA. Inpefa (sotogliflozin) Approval Overview. 2023. https://www.accessdata.fda.gov/scripts/cder/daf/index.cfm?event=overview.process&ApplNo=216203
102. Azizogli AR, Vitti MR, Mishra R, Osorno L, Heffernan C, Kumar VA. Comparison of SGLT1, SGLT2, and dual inhibitor biological activity in treating Type 2 diabetes mellitus. Adv Ther (Weinh). 2023;6(12) https://doi.org/10.1002/adtp.202300143.
103. Jha BK, Sherpa ML, Imran M, Mohammed Y, Jha LA, Paudel KR, et al. Progress in understanding metabolic syndrome and knowledge of its complex pathophysiology. Diabetology. 2023;4(2):134–59.
104. Saklayen MG. The global epidemic of the metabolic syndrome. Curr Hypertens Rep. 2018;20(2):12. https://doi.org/10.1007/s11906-018-0812-z.
105. Alves-Bezerra M, Cohen DE. Triglyceride metabolism in the liver. Compr Physiol. 2017;8(1):1–8. https://doi.org/10.1002/cphy.c170012.
106. Ye J, DeBose-Boyd RA. Regulation of cholesterol and fatty acid synthesis, vol. 3. Cold Spring Harb Perspect Biol; 2011. p. 7. https://doi.org/10.1101/cshperspect.a004754.
107. Creative Proteomics. Triglyceride metabolism: structure, regulation, and role in metabolic diseases. Available from: https://www.creative-proteomics.com/resource/triglyceride-metabolism-structure-regulation-diseases.htm#:~:text=Triglycerides%20serve%20as%20the%20primary,utilized%20as%20an%20energy%20source
108. Saponaro C, Gaggini M, Carli F, Gastaldelli A. The subtle balance between lipolysis and lipogenesis: a critical point in metabolic homeostasis. Nutrients. 2015;7(11):9453–74. https://doi.org/10.3390/nu7115475.
109. Purohit V, Gao B, Song BJ. Molecular mechanisms of alcoholic fatty liver. Alcohol Clin Exp Res. 2009;33(2):191–205. https://doi.org/10.1111/j.1530-0277.2008.00827.x.
110. Cobbina E, Akhlaghi F. Non-alcoholic fatty liver disease (NAFLD) - pathogenesis, classification, and effect on drug metabolizing enzymes and transporters. Drug Metab Rev. 2017;49(2):197–211. https://doi.org/10.1080/03602532.2017.1293683.
111. NHS. Non-alcoholic fatty liver disease (NAFLD) 2022. Available from: https://www.nhs.uk/conditions/non-alcoholic-fatty-liver-disease/
112. Fang C, Pan J, Qu N, Lei Y, Han J, Zhang J, et al. The AMPK pathway in fatty liver disease. Front Physiol. 2022;13:970292. https://doi.org/10.3389/fphys.2022.970292.
113. Feng J, Qiu S, Zhou S, Tan Y, Bai Y, Cao H, et al. mTOR: a potential new target in nonalcoholic fatty liver disease. Int J Mol Sci. 2022;23(16) https://doi.org/10.3390/ijms23169196.
114. Marcondes-de-Castro IA, Reis-Barbosa PH, Marinho TS, Aguila MB, Mandarim-de-Lacerda CA. AMPK/mTOR pathway significance in healthy liver and non-alcoholic fatty liver disease and its progression. J Gastroenterol Hepatol. 2023;38(11):1868–76. https://doi.org/10.1111/jgh.16272.
115. FDA. Rezdiffra (resmetirom) Approval Overview. 2024. https://www.accessdata.fda.gov/scripts/cder/daf/index.cfm?event=overview.process&ApplNo=217785
116. Yale Medicine. FDA Approves Rezdiffa, the first drug for NASH, a form of fatty liver disease 2024. Available from: https://www.yalemedicine.org/news/rezdiffra-for-nash-nonalcoholic-steatohepatitis
117. Sinha RA, Singh BK, Yen PM. Direct effects of thyroid hormones on hepatic lipid metabolism. Nat Rev Endocrinol. 2018;14(5):259–69. https://doi.org/10.1038/nrendo.2018.10.
118. Karim G, Bansal MB. Resmetirom: an orally administered, smallmolecule, liver-directed, β-selective THR agonist for the treatment of non-alcoholic fatty liver disease and non-alcoholic steatohepatitis. touchREV Endocrinol. 2023;19(1):60–70. https://doi.org/10.17925/ee.2023.19.1.60.
119. Sakers A, De Siqueira MK, Seale P, Villanueva CJ. Adipose-tissue plasticity in health and disease. Cell. 2022;185(3):419–46. https://doi.org/10.1016/j.cell.2021.12.016.
120. Chait A, den Hartigh LJ. Adipose tissue distribution, inflammation and its metabolic consequences, including diabetes and cardiovascular disease. Front Cardiovasc Med. 2020;7:22. https://doi.org/10.3389/fcvm.2020.00022.
121. Lee WY. Inflammation in metabolic diseases and insulin resistance. Cardiovasc Prev Pharmacother. 2021;3(2):31–7.
122. Dhokte S, Czaja K. Visceral adipose tissue: the hidden culprit for Type 2 diabetes. Nutrients. 2024;16(7) https://doi.org/10.3390/nu16071015.
123. Alser M, Naja K, Elrayess MA. Mechanisms of body fat distribution and gluteal-femoral fat protection against metabolic disorders. Front Nutr. 2024;11 https://doi.org/10.3389/fnut.2024.1368966.
124. Frayn KN. Metabolic regulation: a human perspective. Wiley; 2022.
125. Wang H, Eckel RH. Lipoprotein lipase: from gene to obesity. Am J Physiol Endocrinol Metab. 2009;297

(2):E271–E88. https://doi.org/10.1152/ajpendo.90920.2008.

126. Nilsson-Ehle P. Impaired regulation of adipose tissue lipoprotein lipase in obesity. Int J Obes. 1981;5(6):695–9.
127. Samuel Varman T, Shulman GI. Mechanisms for insulin resistance: common threads and missing links. Cell. 2012;148(5):852–71. https://doi.org/10.1016/j.cell.2012.02.017.
128. MedlinePlus. Facts about trans fats 2024. Available from: https://medlineplus.gov/ency/patientinstructions/000786.htm
129. Duan Y, Gong K, Xu S, Zhang F, Meng X, Han J. Regulation of cholesterol homeostasis in health and diseases: from mechanisms to targeted therapeutics. Signal Transduct Target Ther. 2022;7(1):265. https://doi.org/10.1038/s41392-022-01125-5.
130. Nelson RH. Hyperlipidemia as a risk factor for cardiovascular disease. Prim Care. 2013;40(1):195–211. https://doi.org/10.1016/j.pop.2012.11.003.
131. Alberts B, Heald R, Johnson A, Morgan D, Raff M, Roberts K, Walter P, Wilson J, Hunt T. Molecular biology of the cell. W.W. Norton and Company; 2022.
132. Lusis AJ. Atherosclerosis. Nature. 2000;407(6801):233–41. https://doi.org/10.1038/35025203.
133. Rosenson RS, Brewer HB, Ansell BJ, Barter P, Chapman MJ, Heinecke JW, et al. Dysfunctional HDL and atherosclerotic cardiovascular disease. Nat Rev Cardiol. 2016;13(1):48–60. https://doi.org/10.1038/nrcardio.2015.124.
134. Katzung BG, Vanderah TW, Trevor AAJ. Katzung's basic and clinical pharmacology. 16th ed. McGraw-Hill Education; 2024.
135. FDA. Rapatha (evolocumab) Approval Overview 2015. Available from: https://www.accessdata.fda.gov/scripts/cder/daf/index.cfm?event=overview.process&ApplNo=125522
136. Roth EM, Davidson MH. PCSK9 inhibitors: mechanism of action, efficacy, and safety. RCM. 2018;19(S1):31–46. https://doi.org/10.3909/ricm19S1S0002.
137. Weintraub WS, Gidding SS. PCSK9 inhibitors: a technology worth paying for? PharmacoEconomics. 2016;34(3):217–20. https://doi.org/10.1007/s40273-015-0355-y.
138. Packard CJ, Boren J, Taskinen MR. Causes and consequences of hypertriglyceridemia. Front Endocrinol (Lausanne). 2020;11:252. https://doi.org/10.3389/fendo.2020.00252.
139. Margolis LM, Pasiakos SM, Howard EE. High-fat ketogenic diets and ketone monoester supplements differentially affect substrate metabolism during aerobic exercise. Am J Phys Cell Phys. 2023;325(4):C1144–C53. https://doi.org/10.1152/ajpcell.00359.2023.
140. Crosby L, Davis B, Joshi S, Jardine M, Paul J, Neola M, et al. Ketogenic diets and chronic disease: weighing the benefits against the risks. Front Nutr. 2021;8:702802. https://doi.org/10.3389/fnut.2021.702802.
141. Paoli A, Bianco A, Moro T, Mota JF, Coelho-Ravagnani CF. The effects of ketogenic diet on insulin sensitivity and weight loss, which came first: the chicken or the egg? Nutrients. 2023;15(14):3120.
142. Chandel NS. Navigating metabolism. 2nd ed. Cold Spring Harbor: Cold Spring Harbor Laboratory Press; 2020.
143. McGaugh E, Barthel B. A review of ketogenic diet and lifestyle. Mo Med. 2022;119(1):84–8.
144. Mayo Clinic. Metabolic syndrome 2025. Available from: https://www.mayoclinic.org/diseases-conditions/metabolic-syndrome/symptoms-causes/syc-20351916
145. Song A, Mao Y, Wei H. GLUT5: structure, functions, diseases and potential applications. Acta Biochim Biophys Sin Shanghai. 2023;55(10):1519–38. https://doi.org/10.3724/abbs.2023158.
146. Healthline. What is fructose malabsorption? 2018. Available from: https://www.healthline.com/health/fructose-malabsorption

Part V

Metabolism Beyond Carbohydrates and Lipids: Amino Acids, Nucleotides, One Carbon Metabolism, and Vitamins

This part continues the discussion of metabolism but moves beyond the core pathways involving carbohydrates and lipids. In this part, Chap. 14 discusses amino acid metabolism, which involves catabolic processes as well as the excretion of extra nitrogen units via the urea cycle. It also discusses synthesis of some amino acids that are not obtained from the diet. Chapter 15 covers nucleotide metabolism. It discusses the synthesis and breakdown of purines and pyrimidines. Chapter 16 covers one carbon metabolism, which includes the folate cycle, and the transfer of one carbon units. This pathway is also critical for the synthesis of nucleotides. Chapter 17 covers vitamins, a closely related topic because the folate pathway also involves the metabolism of vitamins, particularly the B vitamins. Disorders that occur when these metabolic pathways are improperly regulated are also addressed, such as urea cycle disorders or vitamin deficiencies, and the different treatment options will be discussed. Some enzymes in these pathways are important drug targets, for example, enzymes involved in nucleotide metabolism are important targets for blocking the growth of rapidly dividing cells, a topic that is also addressed here.

Amino Acid Metabolism, Associated Disorders, and Treatments

14

Abstract

Amino acid metabolism includes both the breakdown (catabolism) and synthesis (anabolism) of amino acids. This chapter starts with amino acid catabolism. An important part of amino acid catabolism is the processing of their amino groups, because these groups can be toxic to the cell once they are no longer part of the larger molecule. These amino groups must therefore be excreted from the body by a series of reactions that include transamination and deamination, followed by excretion via the urea cycle. Once the amino group is eliminated, only the carbon skeleton of the amino acid remains. Carbon skeletons of different amino acids have different outcomes, depending on whether they are ketogenic, glucogenic, or both. In addition to amino acid catabolism, amino acid synthesis (anabolism) is also discussed. While some amino acids are consumed in the body, some, the non-essential amino acids, can be synthesized de novo. Several disorders associated with amino acid metabolism are discussed, including urea cycle disorders. Treatment options are also addressed.

Keywords

Transamination reactions · Oxidative deamination · Urea cycle · Glucogenic amino acids · Ketogenic amino acids · Glucose–alanine cycle

Abbreviations

ALT	Alanine aminotransferase
ASL	Argininosuccinate lyase
ASP	Aspartate
ASS	Argininosuccinate synthetase
AST	Aspartate aminotransferase
BCKDC	Branched-chain ketoacid dehydrogenase complex
BH_4	Tetrahydrobiopterin
CPSI	Carbamoyl phosphate synthetase I
GDH	Glutamate dehydrogenase
GLU	Glutamate
GLN	Glycine
NH_3	Ammonia
NAG	N-acetylglutamate
NAGS	N-acetylglutamate synthase
OAA	Oxaloacetate
OTC	Ornithine transcarbamylase
PKU	Phenylketonuria
PLP	Pyridoxal phosphate
SAM	S-Adenosylmethionine
α-KG	Alpha-ketoglutarate

What You Will Learn

- How amino acids are metabolized.
- Transamination reactions and oxidative deamination of glutamate.
- The important role of glutamate as an amino group carrier.
- The urea cycle and the fate of the nitrogen group from amino acids.

A. Minden, *Biochemistry, Cell and Molecular Biology for Pharmacists and Health Professionals*,
https://doi.org/10.1007/978-3-032-20529-2_14

- Urea cycle disorders are mostly inborn errors of metabolism. Pharmacological treatments for urea cycle disorders.
- Glucogenic and ketogenic amino acids based on their carbon skeletons.
- Synthesis of amino acids.
- Amino acid carbon skeletons can participate in anaplerotic and cataplerotic reactions: those that replenish Krebs cycle intermediates and those that use Krebs cycle intermediates.
- The glucose–alanine cycle—an important link between protein metabolism and glucose metabolism.
- Diseases associated with amino acid metabolism, such as hepatic encephalopathy.
- Diagnosis of liver disorders by the presence of aminotransferase enzymes (AST and ALT) in the blood.

1 Introduction to Amino Acid Metabolism

This chapter covers the metabolism of amino acids. In the first part of this chapter, we explore amino acid breakdown (catabolism). While Chap. 8 covered how proteins are broken down into smaller parts, including amino acids, here, we discuss what happens to the resulting amino acids. An important part of the catabolism of amino acids is the processing of their amino groups because these groups can be toxic to the cell once they are no longer part of the larger molecule. These amino groups must therefore be excreted from the body by a series of reactions that include transamination and deamination, followed by excretion via the urea cycle. Once the amino group is eliminated, only the carbon skeleton of the amino acid remains. This chapter also discusses the different possible outcomes of the carbon skeletons, which can either be broken down for fuel or repurposed to synthesize other biomolecules. Before discussing the catabolism of amino acids, this chapter will briefly introduce the two main sources of these amino acids: dietary proteins and endogenous proteins.

The second part of this chapter discusses amino acid synthesis. While some amino acids are consumed in the body, some amino acids, the non-essential amino acids, can be made de novo in the body. The interconnection between amino acid metabolism pathways and carbohydrate metabolism pathways will also be addressed. Disorders associated with amino acid metabolism, such as urea cycle disorders and the inability to process certain amino acids, as well as their management, are also discussed throughout this chapter.

2 The Sources of Proteins That Are Broken Down into Amino Acids

2.1 Catabolism of Dietary Proteins

Dietary **proteins** that we eat are digested into fragments in the stomach. **Pepsin** is the main proteolytic enzyme of the stomach and is most active at the low acidic pH found there. The fragments that result then move to the duodenum, the first part of the small intestine. There, they are further digested by enzymes released from the pancreas, including **chymotrypsin** and **trypsin**, and even more by **aminopeptidases** in the brush border of the intestine. After this process, the proteins are broken down into smaller components, including amino acids, dipeptides, and tripeptides, which are then further broken down into amino acids. Finally, the resulting amino acids are absorbed into the intestinal cells (enterocytes) via the apical sides of these cells. They are transported out of the cells via transporters on the basolateral side and enter the hepatic portal vein, which carries them to the liver. There they are sorted. Some of these amino acids will be used to make proteins that the body needs, via the process of translation. These may either stay in the liver or enter the general circulation and travel to various parts of the body. Excess amino acids remaining in the liver, which may not be immediately needed for protein biosynthesis, may be catabolized (broken down) through a process involving transamination followed by deamination [1–4]. This is the process that will be described in more detail in this chapter. Some amino acids, specifically the branched chain amino acids (leucine, isoleucine, and valine), are instead metabolized mostly in muscle and other tissues, although the last step, of oxidative deamination, occurs happens in the liver [5].

2.2 Catabolism of Endogenous Proteins

As discussed in Chap. 8, even most endogenous proteins, which are made in the body for specific purposes, will eventually be broken down. Damaged or misfolded proteins, proteins that are no longer needed, or old proteins, for example, may be broken down. Chapter 8 covered a major pathway for protein degradation, the ubiquitin proteasome system. This system eventually breaks the protein down to amino acids. Some of these amino acids may remain in the cell and can be used in the synthesis of new proteins, but excess amino acids may be degraded further in the cell [4, 6, 7].

Amino acids from either situation, from dietary proteins or from degraded endogenous proteins, will be broken down (catabolized) if they are not used for the synthesis of other proteins. This amino acid degradation process includes transamination followed by deamination and will be discussed in the following sections.

3 Amino Acid Catabolism: Transamination (Amino Group Transfer) and Deamination (Amino Group Removal)

An essential part of the amino acid degradation process is the removal and subsequent excretion of the amino groups from amino acids. This is important because when amino groups (typically protonated as $-NH_3^+$ in amino acids are released from amino acids, they form ammonia (NH_3) or ammonium (NH_4^+), which are toxic. The body needs a way to manage or excrete these compounds. There are two major stages to the transfer and removal of amino groups from amino acids. **First** is the **transamination reaction**, where the amino group is transferred from the amino acid to another molecule, which then becomes the nitrogen carrier. **Second** is the **oxidative deamination reaction,** where the amino group is removed from this second molecule (the nitrogen carrier) so that it can enter the urea cycle and be excreted. Transamination reactions occur in many tissues, but oxidative deamination is carried out mostly in the liver (and to some extent in the kidney). The main basic mechanisms of transamination and deamination are described below and in the studies cited herein [8–13].

3.1 Transamination Reactions

3.1.1 Basics of Transamination Reactions

Amino ($-NH_2/-NH_3^+$) groups are removed from amino acids in a series of two main reactions that start with transamination reactions, which may occur in the cytoplasm or, in some cases, in the mitochondria. In transamination reactions, the amino group of an α-amino acid (the type of amino acid that makes up proteins) is reversibly transferred to an α-ketoacid (usually α-ketoglutarate (αKG)). As a result, the α-ketoacid accepts an amino group at its α-carbon, converting it into an amino acid (usually glutamate (Glu)). Meanwhile, the amino acid, having lost its amino group, forms a new carbonyl (=O) group at its α-carbon, converting it into an α-keto acid (see Fig. 14.1 for basic structures of

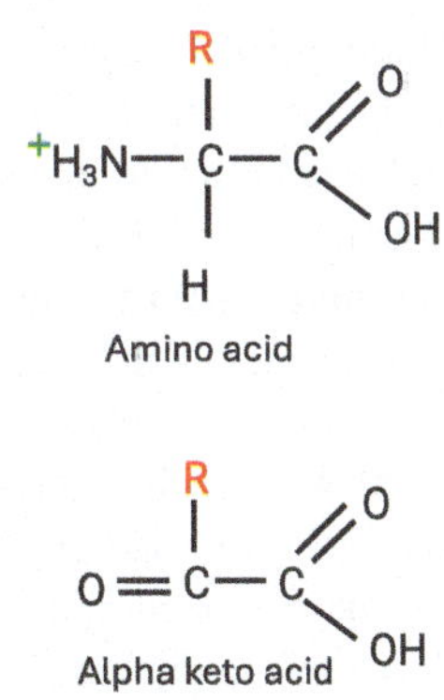

Fig. 14.1 Amino acid compared with an α-keto acid. An amino acid has an amino group at the α carbon, and an α-keto acid has a keto group at the α carbon

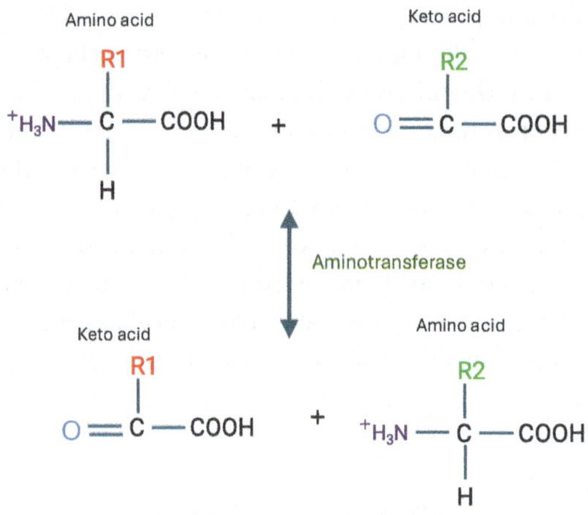

Fig. 14.2 Basic aminotransferase reaction. In these reactions, amino groups from an amino acid are transferred to an α-keto acid (usually α-ketoglutarate (αKG)), converting it to an amino acid, while the original amino acid becomes a keto acid

amino acids and ketoacids, and Fig. 14.2 for the basic transamination process).

3.1.2 Aminotransferases: The Enzymes That Catalyze Transamination Reactions

The family of enzymes that catalyze transamination reactions is called aminotransferases (also called transaminases). The **aminotransferase** enzyme is attached to a prosthetic group, pyridoxal phosphate (**PLP**), which is a derivative of pyridoxine (vitamin B6). It plays an important role in transferring the amino group from one molecule to another, which will be discussed later.

Different **aminotransferases** catalyze the transamination of different amino acids, and they are named according to the amino acids they act on. For example, below, we will talk about **alanine aminotransferase** (also called **alanine transaminase**; **ALT**) and **aspartate aminotransferase** (also called **aspartate transaminase**; **AST**), which catalyze the transamination of **alanine** (**Ala**) and **aspartate** (**Asp**), respectively. In, but in each case, the amino group is transferred to α-ketoglutarate (**αKG**) and results in **glutamate (Glu)** as the nitrogen carrier.

Some aminotransferases have dual isoforms, a cytoplasmic isoform and a mitochondrial isoform [6, 7].

3.1.3 Alanine Transamination

Overview of Alanine Transamination

Alanine aminotransferase (ALT) catalyzes the transfer of the amino group from **Ala** to **αKG** to form **Glu** and **pyruvate**, in the following reaction (also see Fig. 14.3a):

$$\text{Ala} + \alpha\text{KG} \rightleftharpoons \text{Glu} + \text{pyruvate}$$

This reaction, catalyzed by ALT, will be referred to throughout the text as the **alanine transamination reaction**. This is a reversible reaction, and the direction it takes depends on the tissue.

In muscle, abundant pyruvate (from glycolysis) drives the reaction in the reverse direction toward the formation of Ala. Ala then travels through the bloodstream to the liver as a nitrogen carrier. αKG is also generated in this reaction, and this αKG is a Krebs cycle intermediate, and it can thus enter the Krebs cycle.

In the liver, the reaction occurs mainly in the forward direction, which will be the major focus of this chapter. Ala that enters the liver from other tissues, including Ala from other transamination reactions, donates its amino group to αKG, forming Glu and pyruvate. The pyruvate serves as a substrate for gluconeogenesis. The glutamate (Glu) that results has two possible fates as follows:

(a) **Glu** is a nitrogen donor and in the liver, it can undergo oxidative deamination (catalyzed by glutamate dehydrogenase (GDH)), so that it releases the amino group, which can enter the urea cycle, for excretion. This reaction, oxidative deamination of glutamate is discussed more in Sect. 3.2 below.
(b) Alternatively, **Glu** can undergo transamination to form **Asp**, which releases a nitrogen group to the urea cycle.

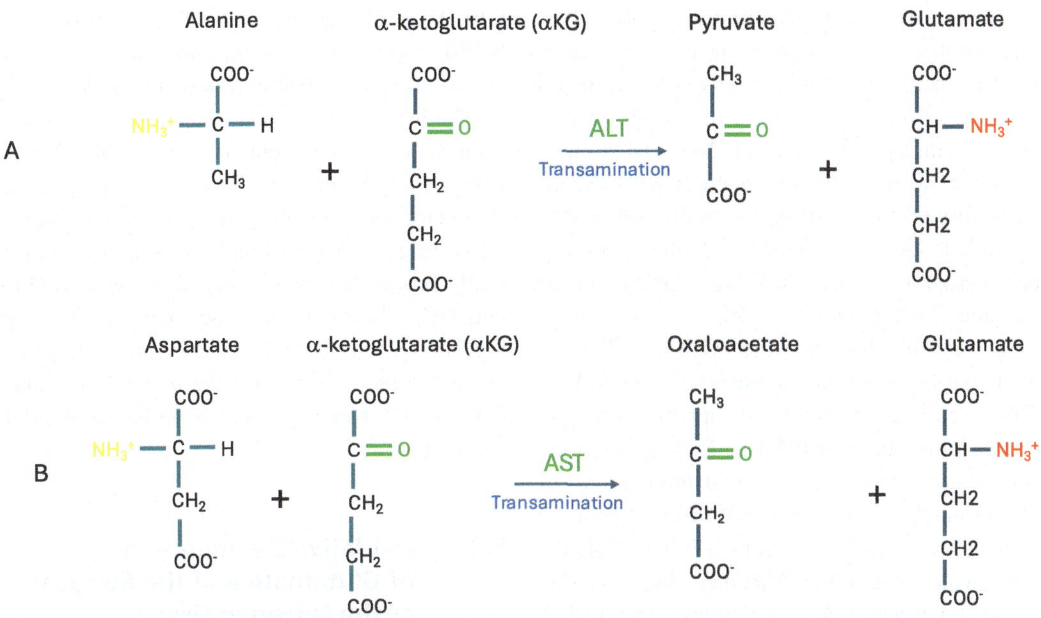

Fig. 14.3 Alanine transamination reaction (top) and aspartate transamination (bottom). In both cases, an amino group from the amino acid is transferred to αKG, which becomes glutamate. *ALT* Alanine aminotransferase, *AST* aspartate aminotransferase

3.1.4 Aspartate Transamination

The Aspartate transamination reaction occurs as follows, and is catalyzed by **aspartate aminotransferase (AST)** (see Fig. 14.3B):

$$\text{Asp} + \alpha\text{KG} \rightleftharpoons \text{OAA} + \text{Glu}$$

This reaction will be referred to throughout the text as the **aspartate transamination reaction**. As with the Ala transamination reaction, the direction of this reaction depends on tissue needs.

In the liver, especially in the cytosol, the reaction occurs largely in reverse. In this reaction, Glu, which may be obtained from the alanine transamination reaction described above, reacts with OAA to form Asp. This is important because Asp (like Glu) donates its amino group directly to the urea cycle (which is discussed later in Sect. 4).

The forward reaction can occur in many tissues whenever Asp is being catabolized. This can sometimes occur even in the liver. The Oxaloacetate (OAA) produced in this reaction can enter gluconeogenesis or the Krebs cycle, as it is a Krebs cycle intermediate.

3.1.5 Basic Mechanism of Action of the Aminotransferase Enzymes

Both reactions described above, as well as other transamination reactions, involve **transaminase enzymes**. The basic function of this family of enzymes is described here, with alanine aminotransferase (**ALT**) as a model.

The **ALT** enzyme that catalyzes the Ala transamination reaction (as seen in Fig. 14.3), like other **aminotransferases**, depends on **PLP (pyridoxal phosphate)**, a cofactor derived from vitamin B6. In this reaction, both Ala and the αKG bind **ALT**. **PLP** is also bound to **ALT**, covalently, via a **Schiff base linkage**. The PLP accepts the amino group from Ala, forming pyridoxamine phosphate (PMP), while Ala is converted to pyruvate. PMP then transfers the amino group to αKG, converting it to Glu. This restores PLP to its original form, allowing the enzyme to catalyze another cycle [11].

The Schiff base is an important part of this interaction. A Schiff base is a structure formed by the reaction an aldimine between a carbonyl group (C=O) group, typically on an aldehyde or ketone, with a primary amine (–NH_2), resulting in a carbon nitrogen double bond (C=N). Schiff bases formed from aldehydes, as discussed here, are called aldimines. The C=N double bond may be referred to as a Schiff base linkage or an aldimine linkage.

In the Schiff base between **ALT** and **PLP**, an amino group on a lysine in the active site of **ALT** forms a covalent bond with the aldehyde group in PLP, forming the **Schiff base**. This amino group is the ε amino group, the terminal amino group, of the lysine. This interaction stabilizes PLP in the enzyme's active site and this Schiff base is the internal aldimine. When **Ala** enters the active site, the amino group on **Ala** now interacts with **PLP**. This results in a new **Schiff base** forming between the **PLP** and the **Ala**. This forms the external aldimine. In the process, the Schiff base between **PLP** and **ALT** is replaced. The amino group transfer then occurs, where the **amino group** is transferred from **Ala** to **PLP**. As a result, **PLP** is converted to **PMP,** and **Ala** becomes deaminated and converted to **pyruvate**. Now **αKG** enters the active site and **PMP** donates this amino group to the **αKG**, which becomes **Glu**, **PLP** regenerates and reforms the original **Schiff base** (the internal aldimine) with the amino group on a **lysine** residue in the **ALT**.

The Schiff base is well suited for this function because it is stable but also reversible. It can ensure that the PLP binds to the enzyme while also being able to interact with substrates. Unlike a more rigid covalent bond, this type of bond is easily broken and re-formed, allowing it to shift between different intermediates [11]. See Fig. 14.4 for a diagram of an aminotransferase (A) and a the Schiff base (the internal aldimine) that forms between the active site lysine on ALT and PLP.

3.2 Oxidative Deamination of Glutamate and the Release of the Nitrogen Group

After most transamination reactions, glutamate (**Glu**) becomes the primary **nitrogen carrier**. However, this amino group still needs to be removed entirely and excreted from the body. This amino group is released from the **Glu** by **oxidative deamination of Glu**. This reaction happens primarily in the hepatocytes in the liver. As a result, the released amino group (NH_3) can readily enter the urea cycle, which also takes

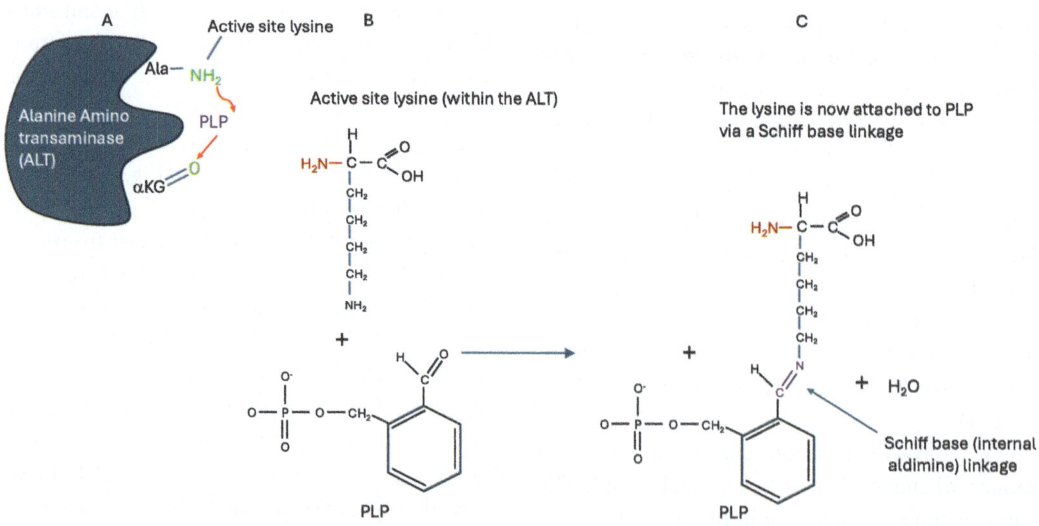

Fig. 14.4 Aminotransferase enzymes use the cofactor PLP (pyridoxal phosphate). In this example of ALT (alanine aminotransferase), (**A**) alanine, with its attached amino group (green) becomes bound to PLP, and it will transfer its nitrogen to PLP and which will then be transferred to αKG, which also interacts with the enzyme. In (**B**) and (**C**) thef formation of the Schiff base (internal aldimine) between the amino group (red) on the active site lysine and PLP is illustrated. See the text for details. (Amino groups are shown in their unprotonated forms)

place in the liver and will be discussed later in this chapter. The oxidative deamination reaction is catalyzed by the enzyme **glutamate dehydrogenase (GDH)** in the mitochondria, and during this reaction, NAD^+ or $NADP^+$ is reduced to NADH or NADPH. The reaction is as follows:

$$\text{Glu} + \text{NAD}^+ + \text{H}_2\text{O} \rightleftharpoons \alpha\text{KG} + \text{NH}_3 + \text{NADH} + \text{H}^+$$

This reaction proceeds through the formation of a ketimine intermediate, a transient compound formed by the oxidation of the amino group. This intermediate also has a Schiff base, and its full name is α-iminoglutarate. This intermediate is then hydrolyzed to release free ammonia (NH_3) and regenerate αKG. The reaction is summarized in Fig. 14.5. Once this reaction is complete, the **Glu** is now transformed back to **αKG**. **αKG**, which lacks an amino group, is now available to receive amino groups in more transamination reactions, or it can feed into the Krebs cycle.

The **Glu** that releases its amino group in these reactions may be in the cytosol or in the mitochondria, depending on where the transamination reaction took place. Oxidative deamination of Glu, however, occurs in the mitochondria. Glu, which is cytosolic, is transported into the mitochondria, usually through a glutamate-aspartate antiporter, which is also used in the malate–aspartate shuttle (the malate–aspartate shuttle is described in Chap. 10), and in some tissues, it can also be transported through a glutamate transporter.

An important outcome of the deamination reaction is the release of the amino group. This amino group will be processed by the urea cycle, where the excretion of nitrogen groups occurs.

3.3 The Important Role of Glutamate (Glu) in Amino group Transfer

3.3.1 Amino Group Transfer is the Main Strategy for Handling Nitrogen

Above, we covered some basic transamination reactions followed by oxidative deamination. Transamination reactions are key reactions in amino acid metabolism, but it is also important to note that not all amino acids undergo this straightforward "classic" transamination. Alanine, aspartate, glutamate, cysteine, and the branched-chain amino acids (valine, leucine, and isoleucine) undergo transamination reactions with αKG, and these are catalyzed by various aminotransferase enzymes. However, some amino acids undergo different first steps. (Appendix C summarizes some of the non-standard transamination routes by which certain amino acids remove their nitrogen groups.)

3.3.2 The Amino Groups of Most Amino Acids Are Transferred to α-Ketoglutarate (αKG) to yield Glutamate (Glu) as Part of Amino Acid Catabolism

While there is some variability, many acids ultimately funnel their nitrogen to **αKG**, producing **Glu**. **Glu** thus functions as a "collector" of amino groups in the cell. As discussed above, Glu can ultimately be deaminated, and its amino group is

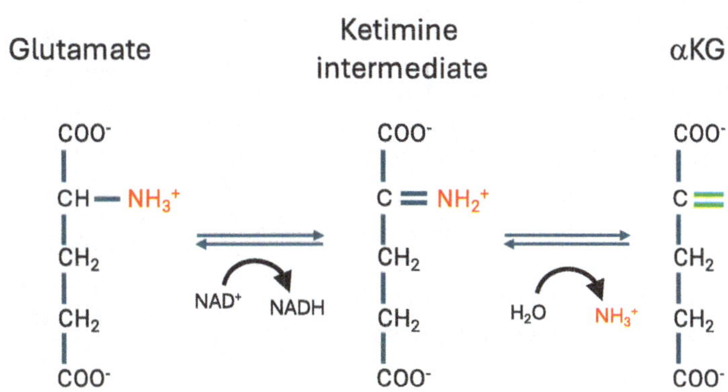

Fig. 14.5 Oxidative deamination of glutamate. In this reaction, glutamate is converted to αKG, as its amino group is released. (The full name of the ketimine intermediate is α-iminoglutarate)

passed into the urea cycle for excretion. One thing to note, however, is that although **Glu** is the main nitrogen carrier, it does not transport amino groups from various tissues to the liver. Instead, the amino group is transferred to another amino acid for transport, usually either Alanine (Ala) or Glutamine (Gln), depending on the tissue of origin, as follows.

3.3.3 Amino Groups Are Transferred from Glutamate to Other Groups for Transport to the Liver

For transport to the liver from muscle, the amino group on **Glu** is usually first transferred from **Glu** to pyruvate to form **Ala**, via the **alanine aminotransferase reaction** catalyzed by **ALT** (reverse direction: Glu + pyruvate ⇌ Ala + αKG).

In many other tissues, the amino group is transferred from Glu to form Gln, via a reaction catalyzed by glutamine synthetase (GS): Glu + NH_3 + ATP → Gln + ADP + Pi

(Gln carries two nitrogen atoms: one from Glu and one added as free ammonia via glutamine synthetase. This reaction is an ATP-dependent nitrogen-scavenging step that helps keep free NH_3 levels low and safe).

3.3.4 Once in the Liver, the Amino Groups Are Generally Transferred Back to Glutamate

Once in the liver, the amino groups on **Ala** or Gln are transferred back to **Glu**, as follows, so that **Glu** will be the main nitrogen carrier:

The amino group from Ala is transferred back to Glu via the alanine aminotransferase reaction catalyzed by ALT

Ala + αKG ⇌ Glu + pyruvate (catalyzed by ALT)

Gln is converted to Glu via a reaction catalyzed by mitochondrial glutaminase as follows: Gln + H_2O → Glu + NH_3 (catalyzed by mitochondrial glutaminase). The released NH_3 released in the Gln reaction can feed directly into the urea cycle (via the reaction catalyzed by mitochondrial **CPSI**, discussed later), and the other nitrogen group is now carried by Glu.

Glu is now the nitrogen carrier and one of the following reactions occurs:

(a) Glu can undergo oxidative deamination, catalyzed by glutamate dehydrogenase **(GDH)** in the following reaction: Glu → αKG + NH_3 The released NH_3 can exist in equilibrium with NH_4^+, which provides one of the nitrogens that contributes to the urea cycle, discussed next.

(b) Alternatively, Glu transfers the amino group to OAA to form Asp in the Aspartate aminotransferase reaction catalyzed by **AST**: OAA + GLU ⇌ Asp + αKG. The resulting Asp can release its amino group to the urea cycle.

4 The Urea Cycle and the Fate of the Nitrogen Group

The main function of the urea cycle is to rid the body of nitrogenous waste. The urea cycle occurs mainly in hepatocytes (liver cells). In the urea cycle, ammonia (NH_3) is converted to urea, transported via the bloodstream to the kidneys, and from there, it is exported from the body. The main steps of the urea cycle are discussed below and described in these resources [9–12].

4.1 The Main Steps of the Urea Cycle

The main steps of the urea cycle, which take place in the liver, are as follows, with balanced reactions after each step (See Fig. 14.6).

1. The cycle starts with **ammonia (NH_3)**. Ammonia comes from deaminated amino acids, released primarily by Glu or Gln. This ammonia, along with **bicarbonate** ($HC0_3^-$), enters the mitochondria of hepatocytes. The ammonia and bicarbonate form **carbamoyl phosphate** in a two-step reaction catalyzed by the enzyme **CPS I (carbamoyl phosphate synthetase 1)**

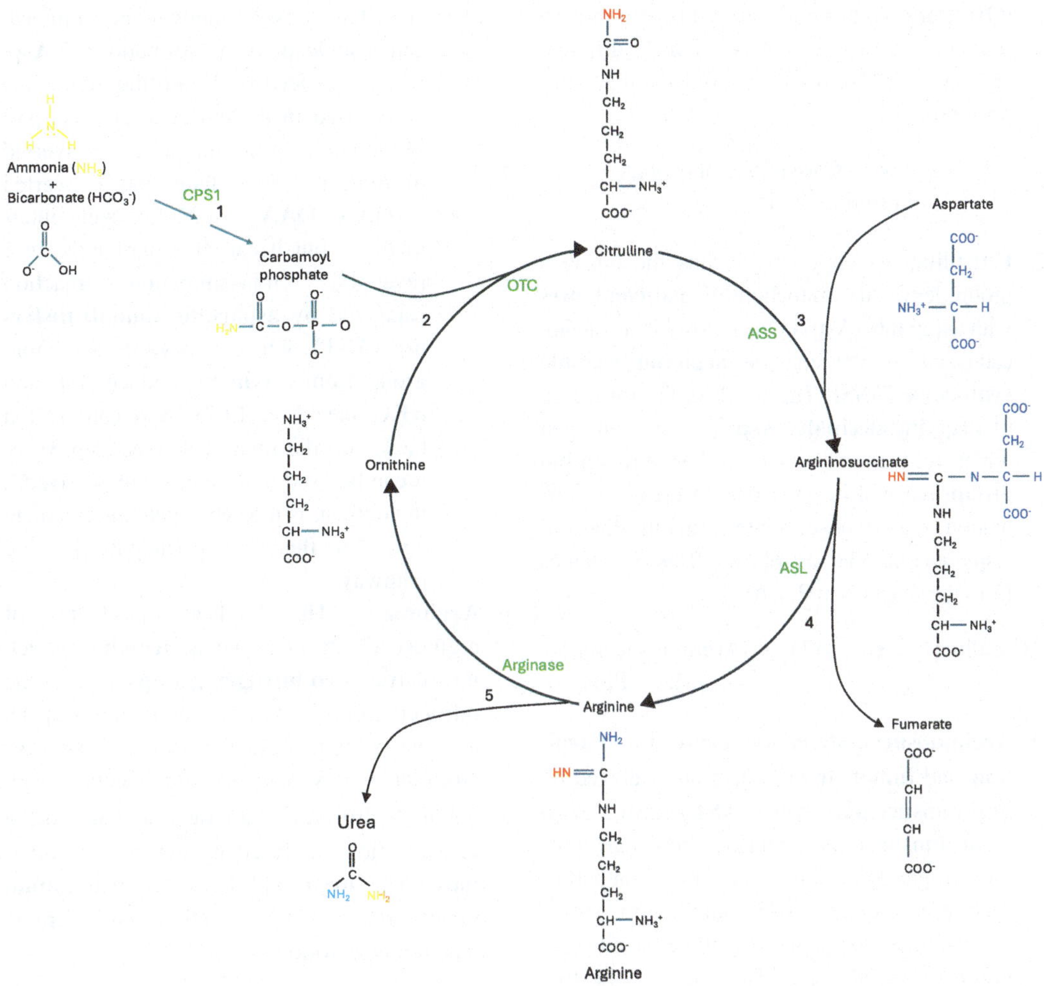

Fig. 14.6 The urea cycle. The urea cycle converts excess amino groups to urea. See text for details. Key intermediates and enzymes are shown. *OTC* ornithine transcarbamoylase, *ASS* argininosuccinate synthetase, *CPS I* carbamoyl phosphate synthetase 1

The two-step reaction is as follows, catalyzed by **CPS I**:

(a) Formation of carbamate

$$HCO_3^- + ATP \rightarrow \text{carboxyphosphate} + ADP$$

$$\text{Carboxyphosphate} + NH_3 \rightarrow \text{Carbamate} + Pi$$

(b) Formation of carbamoyl phosphate

$$\text{Carbamate} + ATP \rightarrow \text{carbamoyl phosphate} + ADP$$

The overall reaction is as follows:

$$NH_3 + HCO_3^- + 2\,ATP \rightarrow \text{carbamoyl phosphate} + 2\,ADP + Pi.$$

The carbamoyl phosphate that results now contains the nitrogen atom that came from the NH_3, and which will become part of **urea** at the end of the urea cycle. This reaction uses 2 ATPs.

2. **Ornithine** enters the mitochondria and reacts with the **Carbamoyl phosphate** to form **citrulline** in a reaction catalyzed by the enzyme **ornithine transcarbamylase (OTC)**.

(**Ornithine** is an amino acid, but it is not an α-amino acid and has some different characteristics from the amino acids that comprise proteins.)

$$\text{Ornithine} + \text{Carbamoyl phosphate} \rightarrow \text{citrulline} + \text{Pi}$$

3. **Citrulline**, which now contains the nitrogen group, leaves the mitochondria and **condenses** with **aspartate (Asp)** in the cytosol in a reaction catalyzed by the enzyme **argininosuccinate synthetase (ASS)**. The result is the formation of **argininosuccinate**. **Asp** is also a nitrogen donor to the urea cycle. It has **one amino group** that will become part of urea. The formation of argininosuccinate from citrulline and Asp uses one ATP, as follows (This Asp comes from the reaction in 3.3.4(b)):

$$\text{Citrulline} + \text{Asp} + \text{ATP} \rightarrow \text{Argininosuccinate} + \text{AMP} + \text{PPi}$$

4. **Argininosuccinate** is now **cleaved** into **arginine** and **fumarate** in a reaction catalyzed by **argininosuccinate lyase (ASL)** (also known as **arginosuccinase**). During these reactions, the nitrogen group from step 1 has been part of each molecule in the series and is now part of the arginine. Asp also contributes a nitrogen group to the arginine. The roles of the two products, arginine and **fumarate**, will be discussed individually below in steps 5 (for fumarate) and 6 (for arginine).

$$\text{Argininosuccinate} \rightarrow \text{arginine} + \text{fumarate}$$

5. **Fumarate: Fumarate** does not carry the nitrogenous waste. It can lead to various outcomes, depending on the needs of the cells, as described below in (a) and (b):
   (a) Fumarate is a Krebs cycle intermediate, and thus, it can feed into the **Krebs cycle**, from which it can be involved in either ATP generation or gluconeogenesis. (The relationship between the Krebs cycle and gluconeogenesis is discussed in Chap. 12)
   (b) As a Krebs cycle intermediate, fumarate can also lead to the formation of **Asp**, which can feed back into the urea cycle or be used in protein synthesis. As part of the Krebs cycle, fumarate is converted to **malate**, which then gets converted to **OAA**. **OAA** is a Krebs cycle intermediate, but it can also participate in a reversible transamination reaction catalyzed by **aspartate aminotransferase (AST)**, where it **accepts** an amino group from a **Glu** to produce **Asp** and **αKG** (see Fig. 14.7). **Asp** can be fed back into the urea cycle (see step 3), or it can be used in protein synthesis. **αKG**, in contrast, is a **Krebs cycle** intermediate and can therefore participate in that pathway.
6. **Arginine**: The other product of arginosuccinate cleavage is **arginine**, which now carries **two nitrogen groups**. One of the nitrogen groups is from the ammonia (step 1), and the other is from the aspartate (step 3). Arginine is key here because it gets broken down (hydrolyzed) into **urea** and **ornithine** in a reaction catalyzed by the enzyme **arginase**. The **urea** will have the **two amino groups** destined for excretion. The arginine reaction is as follows:

$$\text{Arginine} + H_2O \rightarrow \text{urea} + \text{ornithine}$$

Urea and ornithine are processed as follows:
   (a) **Ornithine** is now transported back to the mitochondria, and it can react with more **carbamoyl phosphate** to form **citrulline** and **restart the cycle.**
   (b) **Urea** is released into the bloodstream and transported to the **kidneys**. There, it is filtered out of the blood and excreted, thus ridding the body of two nitrogen atoms. One of these nitrogens_3 comes from the original ammonia that entered the urea cycle (for example, from the deamination of glutamate) and the other is from Aspartate.

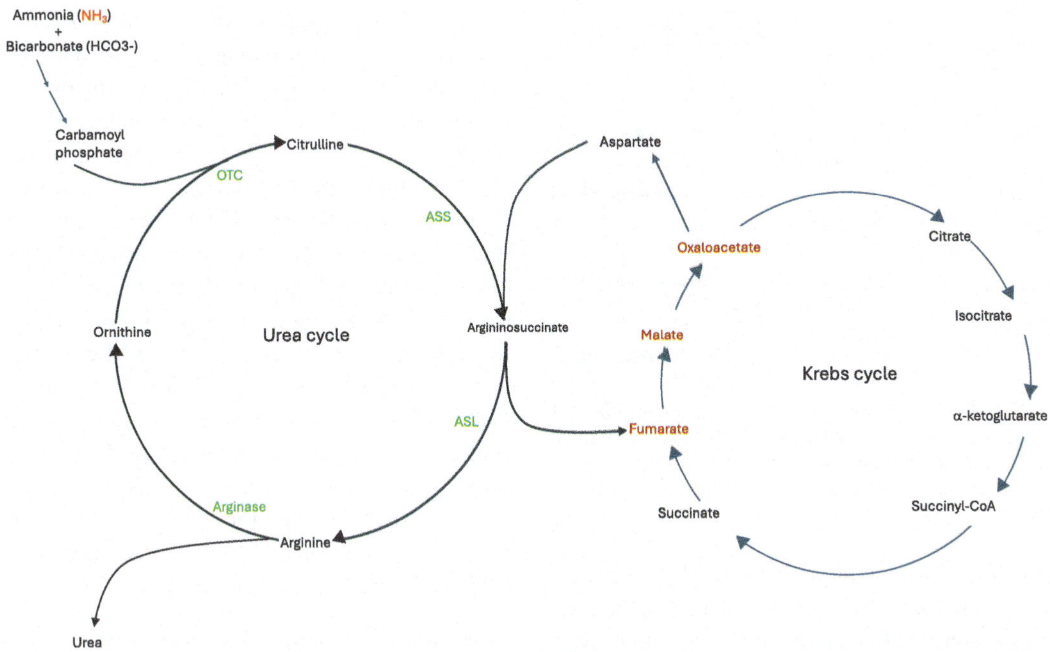

Fig. 14.7 The interconnection between the urea cycle and the Krebs cycle. The urea cycle and the Krebs cycle are interconnected and share common metabolites. Argininosuccinate is a key link

4.2 Outcomes of the Urea Cycle

The overall outcome of the urea cycle is:

$$CO_2 + NH_3 + 3\,ATP + \text{aspartate} + 2H_2O \rightarrow \text{urea} + 2\,ADP + 2Pi + AMP + PPi + \text{fumarate}$$

In this reaction, CO_2 is from the bicarbonate (HCO_3^-) and contributes to the formation of the carbamoyl phosphate; NH_4^+ is the ammonium, and 3 ATP molecules are used.

4.3 The Urea Cycle Is Linked to Other Metabolic Pathways

The urea cycle is tightly linked with the Krebs cycle and gluconeogenesis, and these pathways share some **intermediates** (see Fig. 14.7).

Arginosuccinate is an important link between the Krebs cycle and the urea cycle. **Arginosuccinate** is cleaved to **arginine** and **fumarate** (see step 4).

Arginine forms urea and ornithine, which can continue the urea cycle.

Fumarate is a Krebs cycle intermediate. As part of the Krebs cycle, it can be converted to **malate**, which can then become **OAA**.

OAA is also a **Krebs cycle** intermediate, and as discussed in Chap. 12, it is also linked to the gluconeogenic pathway. **OAA** can also lead to the formation of **Asp** via the **aspartate transamination reaction.**

Asp can feed back into the **urea cycle**, keeping it going.

Alternatively, as an amino acid, **Asp** can participate in the synthesis of **proteins**.

4.4 Regulation of the Urea Cycle

CPSI (Carbamoyl phosphate synthetase I), the mitochondrial enzyme that catalyzes the synthesis of carbamoyl phosphate, is central to the urea cycle (see step 1). The urea cycle can be regulated by whether or not **CPSI** is activated.

CPSI is regulated by an allosteric regulator, **N-acetylglutamate (NAG)**, which binds to it and

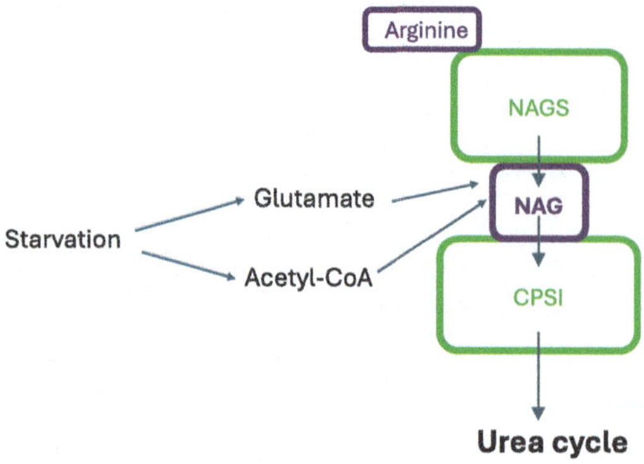

Fig. 14.8 Regulation of NAG (N-acetylglutamate). NAG is a key regulator of CPS I activity in the urea cycle, and its activity is tightly controlled by multiple factors. In this figure, enzymes are in green, and allosteric regulators are in purple. See text for details

activates its catalytic activity. N-acetylglutamate (NAG) is composed of a glutamate with an acetyl group attached to its α-amino group, and its regulation is important for the regulation of the urea cycle (see Fig. 14.8).

The synthesis of **N-acetylglutamate (NAG)** is catalyzed by the enzyme **N-acetylglutamate synthase (NAGS)** [14]. Specifically, **NAGS** catalyzes a reaction between acetyl-CoA and glutamate to form N-acetylglutamate (NAG) in the following reaction:

$$\text{Glutamate} + \text{Acetyl-CoA} \rightarrow \text{N-acetylglutamate (NAG)} + \text{CoA.Catalyzed by NAGS}$$

Therefore, **NAGS** is important for the synthesis of **NAG**, which is a positive regulator of **CPSI**, which is the rate-limiting enzyme in the urea cycle [6, 11]:

$$\text{NAGS} \rightarrow \text{NAG} \rightarrow \text{CPSI} \rightarrow \text{Urea cycle}$$

Due to its role in the regulation of CPSI, regulation of **NAGS** is important for the regulation of the urea cycle. NAGS is also regulated. An important positive regulator of **NAGS** is **arginine**. **Arginine** is an allosteric activator of **NAGS**. Arginine is a product of the urea cycle. Its presence, therefore, indicates an abundance of amino groups available for the production of urea, thus necessitating an active **NAGS** enzyme to produce **NAG**, an allosteric activator of **CPSI**.

Starvation also leads to increased **CPSI** activity. It does so by contributing to **NAG** production. Starvation leads to the breakdown of protein, and thus an increase in amino acids, including **glutamate**. Glutamate is the direct precursor of **N-acetylglutamate (NAG)** because it provides its glutamate backbone. Additionally, starvation increases acetyl-CoA levels due to increased fatty acid β-oxidation. Together, glutamate and acetyl-CoA lead to an increase in NAG production (in the reaction catalyzed by **NAGS**). The increase in **NAG** in turn positively regulates **CPSI** activity. This is important because, during starvation, proteins are broken down, and an active urea cycle is needed to clear out the resulting amino groups (See Fig. 14.8) [11, 14, 15]. The relationship between starvation and CPSI activity can be summarized as follows:

$$\text{Starvation} \rightarrow \text{protein and fat breakdown} \rightarrow \text{glutamate} + \text{acetyl-CoA} \rightarrow \text{NAG} \rightarrow \text{CPS I}$$

4.5 Urea Cycle Disorders

Urea cycle disorders (UCDs) can lead to elevated NH_4^+ levels in the blood (hyperammonemia). This may lead to disruption of the osmotic balance, especially in the brain, causing astrocyte swelling and altered neurotransmitter function. The CNS is especially sensitive to these changes, but problems also occur in other tissues. Other problems include changes in energy metabolism, altered levels of oxidative stress, altered nitric oxide synthesis, and abnormalities in various signal transduction pathways. As a compensation mechanism for the osmotic changes commonly seen in urea cycle disorders, cells activate a Na^+-K^+-$2Cl^-$ cotransporter (NKCC1), which is a membrane-bound transporter, to help restore the osmotic balance, but prolonged activation of this transporter can lead to more problems and ion imbalances [13, 16–19]. Deficiencies in several different enzymes in the urea cycle are known to cause Urea Cycle Disorders (UCDs). These are usually genetic disorders that affect the genes for important urea cycle enzymes. They are considered inborn errors of metabolism (IEMs). Some examples of UCDs, all of which lead to increased levels of nitrogen in the blood, are described below.

Argininosuccinate lyase deficiency (ASLD) is a genetic disorder caused by a mutation in the gene encoding the enzyme **ASL. ASL** catalyzes the cleavage of arginosuccinate to arginine and fumarate (see step (4) in the urea cycle), and this step of the urea cycle is therefore disrupted in the disorder [20].

In **carbamoyl phosphatase synthetase I deficiency (CPSID)**, the gene encoding the enzyme **CPSI** is disrupted. This leads to disruption of the very first step in the urea cycle, where NH_3 and bicarbonate form carbamoyl phosphate [21]. This UCD is of recent significance because in 2025, a baby with a disease was treated with a CRISPR-based gene therapy treatment [22]. This is a historic advancement in personalized medicine and will be discussed again in Chap. 25.

In **ornithine transcarbamoylase deficiency (OTCD)**, the gene for the enzyme **OTC** is disrupted (see step 2 of the urea cycle). Disruption of this enzyme disrupts the synthesis of **citrulline**, a key intermediate in the urea cycle. OTCD is an X-linked disorder. All UCDs are rare, but OTCD is the most common among them [23, 24].

Argininosuccinate synthetase (ASS) deficiency is also known as **citrullinemia**. In this disorder, the gene encoding **ASS** is disrupted. **ASS** is the enzyme that catalyzes the condensation of citrulline and aspartate, resulting in arginosuccinate. In this disorder, blood levels of citrulline are elevated.

UCDs can present at birth (neonatal), when there is a severe lack of enzyme function. A partial lack of enzyme function may show up later and be less severe. Overall, urea cycle disorders are rare, but have serious consequences because ammonia, which builds up in the bloodstream of patients, is toxic. UCD patients are at risk for neurological damage and premature death if steps are not taken to manage the disorder [25, 26]. Treatment for **UCDs** often involves supplementing the diet with **sodium benzoate** and **sodium phenylbutyrate**, which are FDA-approved nitrogen scavengers. These compounds facilitate the excretion of excess nitrogen. This is called nitrogen scavenger therapy.

Benzoate works by reacting with **glycine**, which contains **one nitrogen**. This combination leads to the formation of **hippurate**, which gets secreted, thus eliminating some nitrogen [27] (Fig. 14.9). The body synthesizes more glycine to replace the glycine that is removed.

Phenylacetate works in a similar way. **Phenylacetate** is activated by binding to CoA to form phenylacetyl-CoA. Phenylacetyl-CoA is then conjugated to glutamine (Gln, which contains two nitrogen atoms) to form **phenylacetylglutamine (PAGN)**. This compound is transported to the kidney and readily excreted in the urine, thus eliminating some nitrogen groups. In response, the body may increase **Gln** synthesis from **Glu** to replace the **Gln** that is used up. This synthesis consumes ammonia and is catalyzed by **glutamine synthetase**. This synthesis of Gln is an important part of this process, but it is not always sufficient to replace the Gln that is used up. In this case, ammonia toxicity can sometimes still occur [18, 20, 25]. The three reactions associated with phenylacetate are as follows:

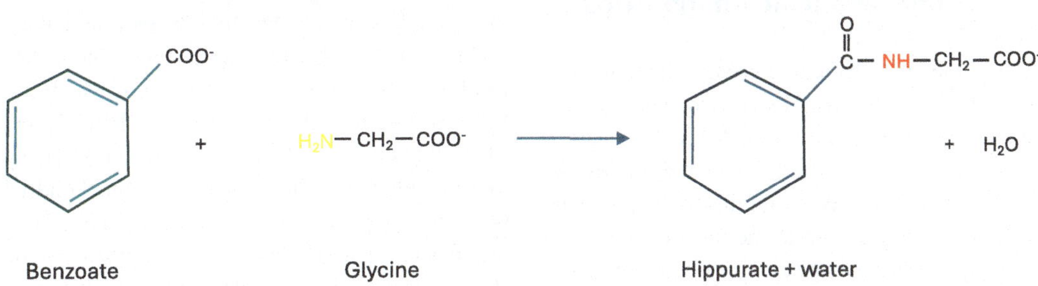

Fig. 14.9 Benzoate. Benzoate can help in eliminating some nitrogen, by interacting with glycine and forming hippurate

1. Phenylacetate + CoA + ATP → phenylacetyl-CoA + AMP + PPi
   Enzyme: **acyl-CoA synthetase**
2. Phenylacetyl-CoA + glutamine (2 nitrogens) → PAGN (2 nitrogens; excreted) + CoA-SH
   Enzyme: **PAGAT**
3. Glu (1 nitrogen) + NH_3 + ATP → Gln (has two nitrogens) + ADP + Pi
   Enzyme: **glutamine synthetase**

The glutamine formed in reaction 3 can then be used again in reaction 2.

In the above reactions, each molecule of phenylacetate ultimately removes two nitrogens as urinary PAGN. This is an important ammonia detoxification mechanism, especially for brain and liver function.

Several drugs are designed to provide the **benzoate** or **phenylacetate** supplementation. **Buphenyl (sodium phenylbutyrate)** [28] is an oral **small-molecule** drug and the first FDA-approved drug (1996) to treat UCDs. It contains **phenylbutyrate** and is converted to **phenylacetate** in the body [29]. A newer version, **Ravicti (glycerol phenylbutyrate)** [30], was approved in 2013 [31]. The newest phenylacetate drug is **Olpruva (sodium phenylbutyrate)** [32], FDA approved in 2022. It is designed to be more convenient to use than previous drugs, as it can be administered as a single dose per day [33]. **Ammonul (sodium benzoate + sodium phenylacetate)** [34] is a drug that contains both benzoate and phenylacetate [35], and is given to patients by IV.

5 The Carbon Skeletons of Amino Acids: Glucogenic and Ketogenic Amino Acids

After removal of the amino groups, the remaining part of the amino acid is a keto acid, usually referred to as the carbon skeleton. The carbon skeletons can have different metabolic fates. All of them can enter the central metabolic pathways at various points, often at the level of the Krebs cycle, where they can be completely broken down to CO_2, resulting in ATP generation. However, what distinguishes different amino acids is their potential anabolic outcomes: some carbon skeletons can be used in gluconeogenesis (glucose synthesis), others in ketogenesis (ketone body synthesis), and some in both. Based on these anabolic outcomes, amino acids are categorized as glucogenic, ketogenic, or both glucogenic and ketogenic (see Fig. 14.10 and Table 14.1) [8, 11]. The different categories are as follows:

5.1 Glucogenic Amino Acids

The largest percentage of amino acids are glucogenic amino acids. The carbon skeletons of these amino acids can be diverted to **gluconeogenesis** to **generate glucose**. Examples of glucogenic amino acids, including alanine and aspartate, are described below.

5.1.1 Alanine

Ala is transaminated to form pyruvate, which is the carbon skeleton that remains after removal of the amino group. This transamination reaction can occur in the cytosol or the mitochondria. Cytosolic pyruvate can enter the mitochondria via the mitochondrial pyruvate carrier (MPC). Once in the mitochondria, it can (a) be converted to acetyl-CoA via **pyruvate dehydrogenase** and combine with OAA, forming citrate, and enter the Krebs cycle for the generation of ATP. Alternatively (b), it can be converted to OAA in a reaction catalyzed by **pyruvate carboxylase**. This OAA can now lead to gluconeogenesis, but first, it must be exported from the mitochondria to the cytoplasm. It does so by first being converted to malate (via **malate dehydrogenase**), which can exit the mitochondria via a malate transporter (the malate-αKG antiporter). The resulting cytoplasmic malate can be converted to cytoplasmic OAA, which can be converted to PEP (catalyzed by **PEPCK**) and PEP continues through the gluconeogenic pathway to form glucose. (The transfer of mitochondrial OAA to cytoplasmic OAA overlaps with part of the malate aspartate shuttle and follows the same strategy described for lactate-induced gluconeogenesis described in Chap. 14.)

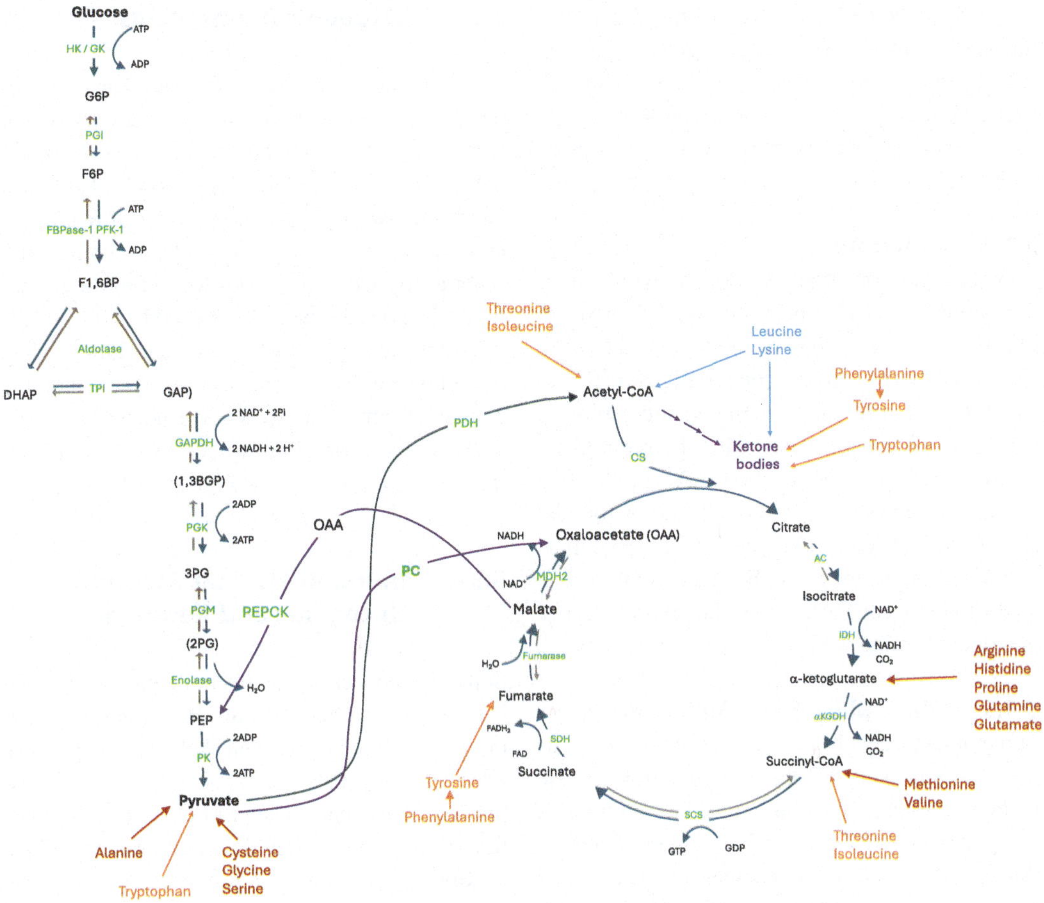

Fig. 14.10 Glucogenic and ketogenic amino acids. Glucogenic amino acids are in red, ketogenic amino acids are in blue, and amino acids that are both glucogenic and ketogenic are in orange font. The diagram illustrates how these amino acids feed into the central metabolic pathways

Table 14.1 Amino acid metabolism

Glucogenic	Ketogenic	Glucogenic and ketogenic
Alanine → pyruvate	Leucine → acetyl-CoA	Isoleucine → succinyl-CoA and acetyl-CoA
Cysteine → pyruvate	Lysine → acetyl CoA	Phenylalanine → fumarate and acetoacetate
Glycine → pyruvate		Threonine → succinyl CoA and acetyl CoA
Serine → pyruvate		Tryptophan → pyruvate and acetoacetate
Aspartate → oxaloacetate		Tyrosine → fumarate and acetoacetate
Arginine → alpha ketoglutarate		
Histidine → alpha ketoglutarate		
Proline → alpha ketoglutarate		
Asparagine → alpha ketoglutarate		
Glutamate → alpha ketoglutarate		
Methionine → succinyl CoA		
Valine → succinyl CoA		

Whether the pyruvate (the carbon skeleton) is funneled into (a) ATP generation or (b) gluconeogenesis depends on the metabolic needs of the cell, but because it can lead to gluconeogenesis, alanine is classified as a glucogenic amino acid.

5.1.2 Aspartate

Aspartate is converted to OAA when it is transaminated. OAA (the resulting carbon skeleton) may be cytoplasmic or mitochondrial OAA. Cytosolic OAA can be converted to PEP via **PEPCK** and enter the gluconeogenic pathway, leading to gluconeogenesis. Mitochondrial OAA can be transferred to cytosolic OAA, also leading to gluconeogenesis as described above for alanine, or it can combine with acetyl-CoA to initiate a round of the Krebs cycle. Because aspartate can lead to gluconeogenesis, it is also classified as a glucogenic amino acid.

5.1.3 Other Glucogenic Amino Acids

Some amino acids have αKG (mitochondrial) as their carbon skeleton after deamination. These include arginine, histidine, proline, glutamine, and glutamate. αKG is a Krebs cycle intermediate that can also lead to gluconeogenesis. It can lead to gluconeogenesis by proceeding through the Krebs cycle until malate (αKG → succinyl CoA → succinate → fumarate → malate), which then exits the mitochondria and proceeds through the gluconeogenic pathway described above. Some amino acids are broken down into succinyl-CoA (always mitochondrial), also a Krebs cycle intermediate, which can follow the same pathway. These include methionine and valine. Other amino acids besides alanine that are catabolized to pyruvate include cysteine, glycine, and serine. Some of these breakdown products occur after multiple steps. All of these amino acids are considered to be glucogenic amino acids because they can lead to gluconeogenesis as seen in Chap. 12. In summary, amino acids that are gluconeogenic are those that are converted to pyruvate, OAA, αKG, fumarate, or succinyl-CoA (either directly or after multiple steps).

5.2 Ketogenic Amino Acids

Some amino acids are ketogenic. These amino acids are converted into acetyl-CoA or acetoacetate (a ketone body) upon their transamination and subsequent steps. Acetyl-CoA can enter the Krebs cycle, but under low-glucose conditions, it is more likely to combine with other acetyl-CoA molecules to lead to the formation of ketone bodies. Ketone bodies are a very important energy source for the brain when nutrients are low. Leucine and lysine are the ketogenic amino acids. They lead, via multiple steps, to the formation of acetyl-CoA after their transamination [6, 7, 11] (Fig. 14.10).

5.3 Amino Acids That Are Both Glucogenic and Ketogenic

Some amino acids can be both glucogenic and ketogenic, meaning that after their transamination or deamination, their carbon skeletons can produce intermediates that enter gluconeogenesis and products that can form ketone bodies. For example, threonine can be broken down into succinyl-CoA, making it glucogenic, or it can be broken down into acetyl-CoA, which can form ketone bodies.

Tyrosine and phenylalanine are both ketogenic and glucogenic. Phenylalanine is broken down into tyrosine. Tyrosine is then converted to both fumarate and acetoacetate. Fumarate is a glucogenic intermediate that can lead to gluconeogenesis, while acetoacetate is a ketone body. Other amino acids in this category include isoleucine, tryptophan, and threonine. Isoleucine and threonine lead to both succinyl-CoA and acetyl-CoA, and tryptophan leads to both pyruvate and the ketone body acetoacetate.

6 Synthesis of Amino Acids (Anabolism)

While the sections above covered amino acid catabolism, this section explores different ways that amino acids are synthesized. Some amino acids are essential amino acids, and these need to be obtained in the diet as we cannot synthesize them. Others are nonessential amino acids, meaning we synthesize them in the body. The nonessential amino acids are Glycine (Gly), Alanine (Ala), Proline (Pro), Arginine (Arg), Serine (Ser), Asparagine (Asn), Glutamine (Gln), Aspartic acid (Asp), Glutamate (Glu), Cysteine (Cys), and Tyrosine (Tyr). Cys and Tyr are a bit different from the others because, although we synthesize them, they are synthesized from essential amino acids. Specifically, Tyr is synthesized from phenylalanine (Phe), and Cys is synthesized from methionine (Met). The synthesis of some amino acids, similar to their degradation, is closely linked to the glucose metabolism pathways, involving intermediates from the Krebs cycle and other pathways. Some amino acids can be synthesized by multiple mechanisms. Below is an overview of some mechanisms by which nonessential amino acids are synthesized, along with some links between central metabolic pathways and amino acids [9–11].

6.1 Synthesis of Serine

Glucose metabolism is a source for the synthesis of serine. Glucose is converted to glucose-6-phosphate (G6P), which can then enter glycolysis. One of the intermediates in the glycolysis pathway is 3-phosphoglyerate (3PG). In addition to leading to the formation of pyruvate in glycolysis, 3PG can also undergo a series of reactions that lead to Ser synthesis. This is called the phosphorylated pathway, and it is the main pathway leading to serine synthesis.

In this pathway, 3PG undergoes a series of reactions, beginning with its oxidation to 3-phospho-hydroxypyruvate, a reaction catalyzed by **phosphoglycerate dehydrogenase (PHDGH)**. 3-phospho-hydroxypyruvate is then transaminated, resulting in 3-phosphoserine in a reaction catalyzed by **phosphoserine aminotransferase (PSAT)**. In this reaction, an amino group is transferred from glutamate to the 3-phospho-hydroxypyruvate to form the 3-phosphoserine. A **phosphoserine phosphatase (PSP)** then catalyzes the removal of the phosphate group from 3-phosphoserine, leading to the production of Ser [9, 10] (Fig. 14.11).

The steps are summarized as follows:

1. Oxidation of 3PG

$$3PG + NAD^{+} \rightarrow 3 - \text{phosphohydroxypyruvate} + NADH$$

Enzyme: **PHDGH**

2. Transamination

$$3 - \text{phosphohydroxypyruvate} + \text{Glu} \rightarrow 3 - \text{phosphoserine} + \alpha\text{KG}$$

Enzyme: **PSAT**

3. Dephosphorylation

$$3 - \text{phosphoserine} + H_2O \rightarrow \text{Serine} + \text{Pi}$$

Enzyme: **phosphoserine phosphatase (PSP)**

6.2 Synthesis of Glycine and Cysteine

Ser can be converted to both Gly and Cys. Both of these mechanisms are linked to the folate and methionine cycles, which will be discussed in more detail in Chap. 16.

Ser can be converted to Gly in a reversible reaction where serine donates one carbon unit to tetrahydrofolate (THF), forming glycine and 5,10-methylene-THF (see Fig. 14.11). This reaction is catalyzed by the **enzyme serine hydroxymethyltransferase (SHMT)** [9, 10].

Ser can also lead to the formation of Cys (see Fig. 14.12). This reaction involves methionine (Met). Methionine is first converted to homocysteine (Hcy; a non-protein amino acid) through part of a cycle called the Activated Methionine cycle (also covered in Chap. 16). Homocysteine combines with serine in a condensation reaction catalyzed by **cystathionine β-synthase (CBS)** to form cystathionine (also a non-protein amino acid). Cystathionine is then cleaved into two molecules, one of which is Cys, and the other is α-keto-butyrate, in a reaction catalyzed by **cystathionine–γ-lyase (CGL)** [7, 11].

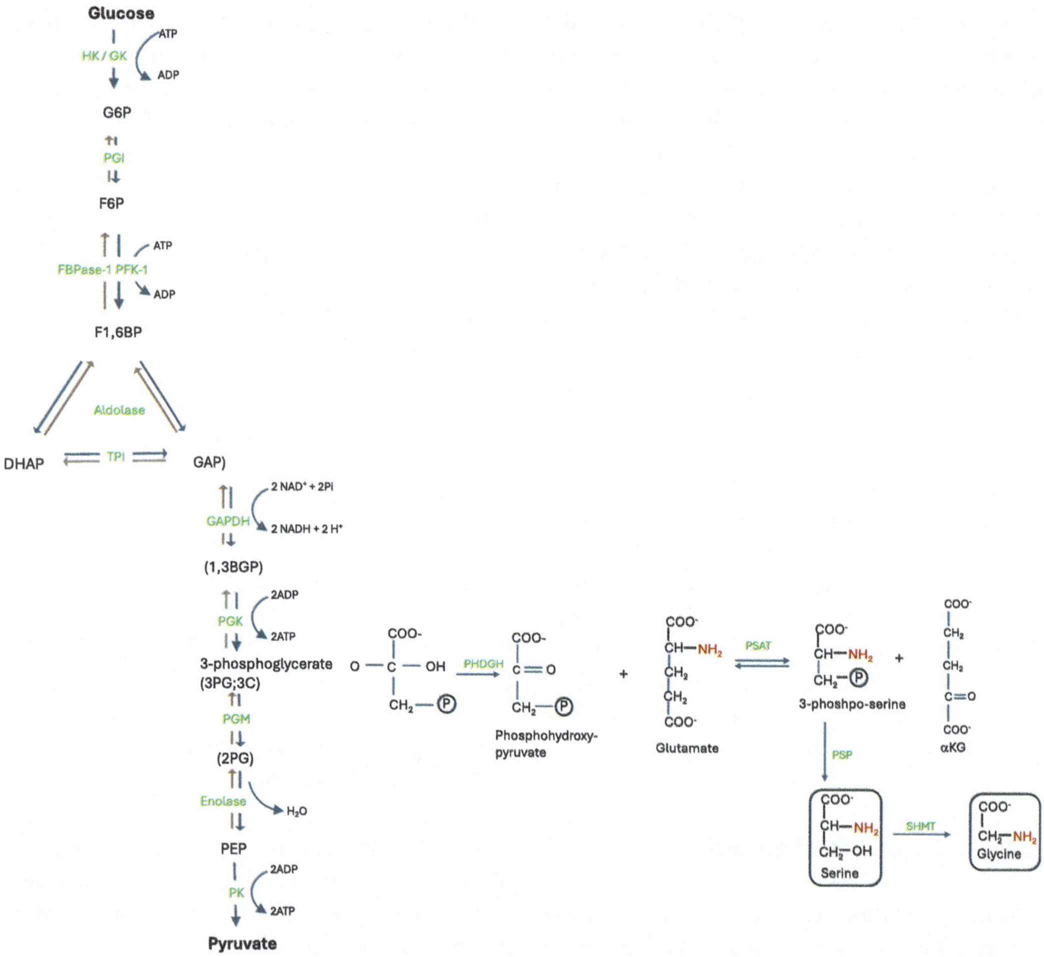

Fig. 14.11 Synthesis of serine and its connection to glycolysis. Serine can also be converted to glycine. See text for details. *PHDGH* phosphoglycerate dehydrogenase, *PSAT* phosphoserine aminotransferase, *PSP* phosphoserine phosphatase, *SHMT* serine hydroxymethyltransferase

6.3 Synthesis of Tyrosine

Tyr is synthesized from Phe, in a reaction catalyzed by **phenylalanine hydroxylase (PAH)** [11] (Fig. 14.13). **PAH** catalyzes the hydroxylation of Phe to form Tyr in a reaction that requires molecular oxygen. In this hydroxylation reaction, an -OH group is added to the aromatic ring of Phe. **Tetrahydrobiopterin (BH_4)** is a cofactor for **PAH** in this reaction. BH_4 functions as an electron donor, helping to reduce O_2 and enabling the insertion of a hydroxyl group onto Phe. As a result, BH_4 is oxidized to dihydrobiopterin (BH_2). To sustain the reaction, BH_2 gets recycled back to BH_4 in a reaction catalyzed by **dihydropterine reductase (DHPR)** using NADH as a reducing agent (Fig. 14.13).

The reaction from Phe to Tyr is as follows:

$$\text{Phe} + O_2 + BH_4 \rightarrow \text{Tyr} + H_2O + BH_2$$

This reaction is clinically important because abnormalities in the reaction can result in **phenylketonuria (PKU)**, an inborn error of metabolism where Phe accumulates because it cannot be

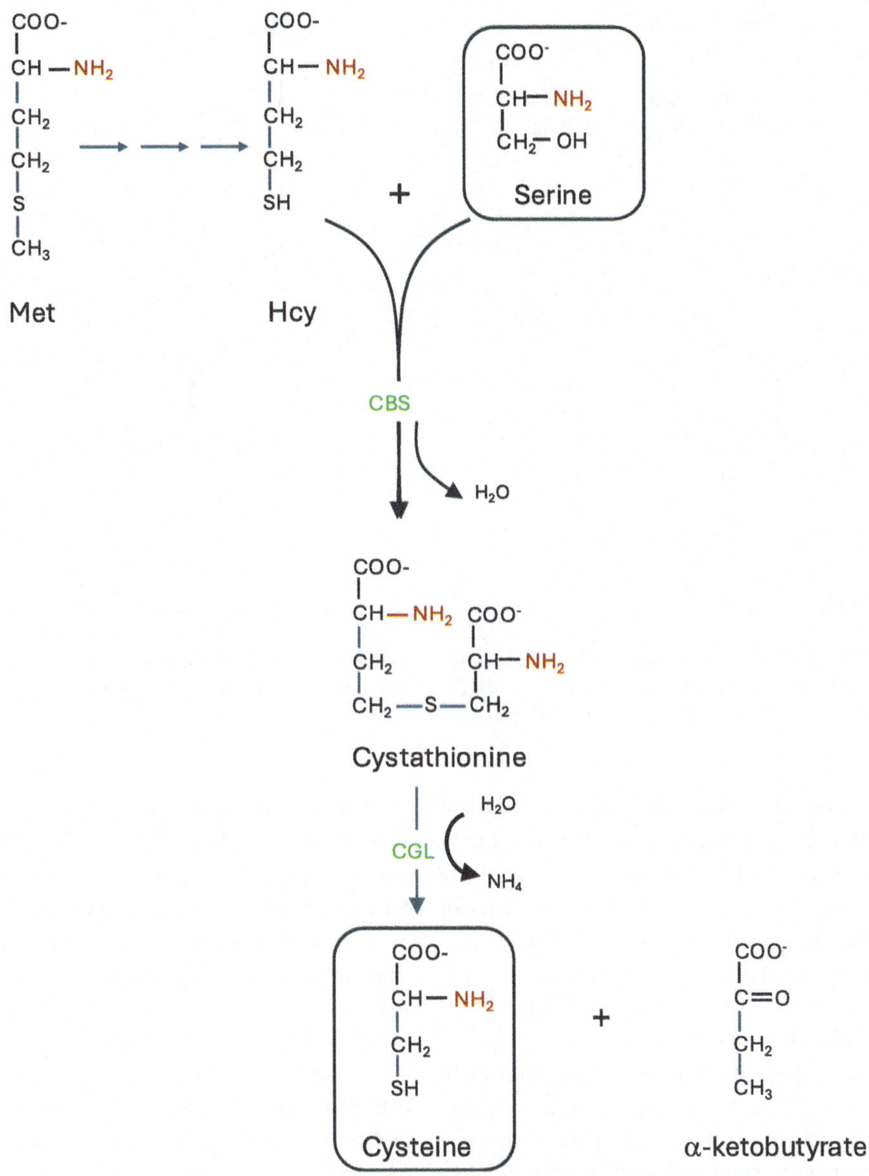

Fig. 14.12 Synthesis of cysteine from serine. See text for details. *CBS* cystathionine β-synthase, *CGL* Cystathionine lyase. (Amino groups are shown in their deprotonated forms)

converted to Tyr. This disorder is often caused by a genetic defect in the **phenylalanine hydroxylase (PAH)** enzyme. More rarely, it may also be caused by an abnormality in the metabolism of **BH_4**, specifically its synthesis or regeneration [36]. PKU can cause Phe to accumulate in the blood. This excess Phe gets converted to phenylketones, which lead to neurotoxicity [37]. The disorder also results in a deficit of Tyr, which can lead to multiple problems due to Tyr's key roles in the synthesis of neurotransmitters (dopamine, norepinephrine, and epinephrine), skin pigmentation, regulation of thyroid levels, and growth and metabolism. PKU is treated by a **low-Phe diet** and supplementation with Tyr [6, 38].

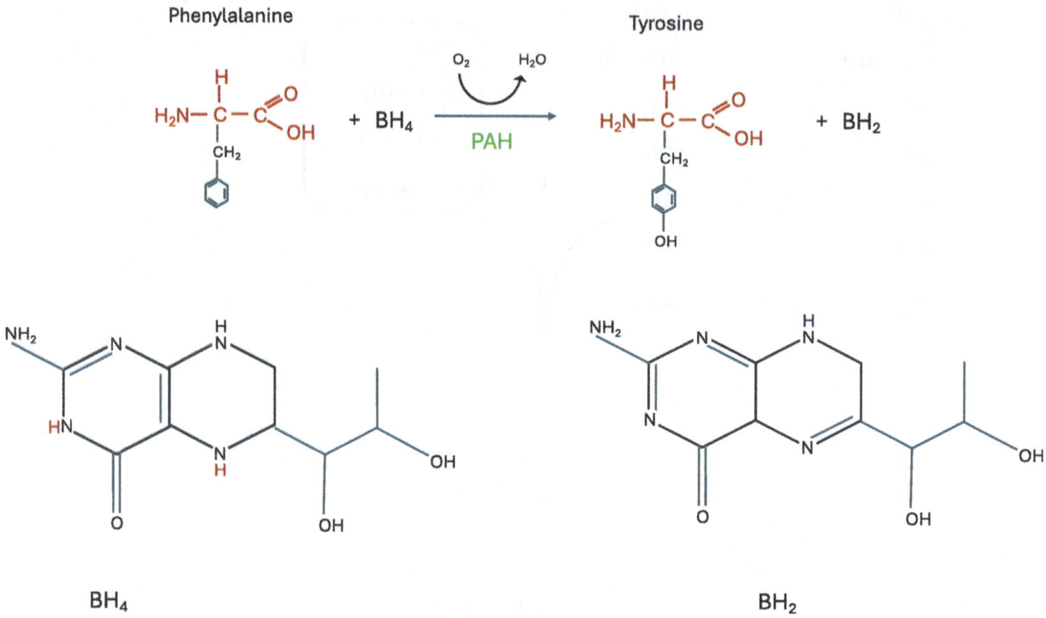

Fig. 14.13 Synthesis of tyrosine from phenylalanine. Structures of BH_4 and BH_2 are also indicated, as BH_4 is a cofactor in the reaction. *PAH* phenylalanine hydroxylase, *BH4* Tetrahydrobiopterin. (Amino groups are shown in their deprotonated forms)

Some individuals with PKU have some residual **PAH** activity. In these cases, it is possible to boost the activity of the enzyme by adding extra BH_4. The small-molecule drug **Kuvan (sapropterin dihydrochloride)** [39] is a synthetic form of BH_4. In some individuals, this drug can enhance the activity of **PAH**, but a **Phe-restricted diet** is still required [36, 40].

Palynziq (pegvaliase-pqpz) [41] is a **biologic** that uses a different approach. It is an enzyme substitution therapy. It is a pegylated form of a different enzyme, **phenylalanine ammonia lyase (PAL)**. Adding PEG (polyethylene glycol) to this enzyme protects it in the bloodstream and causes a reduced immune response [42]. **PAL** is a bacterial enzyme that, when administered to patients, converts Phe to trans-cinnamic acid and ammonia, which the body can excrete, thus preventing the toxic build-up of Phe. This treatment is sometimes used in patients who have responded poorly to Kuvan. Other ideas for treating this type of disorder involve using gene therapy to supplement or repair the PAH gene or other genes that may be mutated, such as those affecting BH_4 metabolism [36, 43].

Another abnormality that is associated with Tyr is **alkaptonuria**. This is a rare recessive disorder, and it is caused by a deficiency in the enzyme **homogentisate 1,2-dioxygenase (HGD)**. This enzyme has an important role in the Tyr degradation pathway. In the Tyr degradation pathway, Tyr is converted to p-Hydroxyphenylpyruvate (HPP), which is then converted to Homogentisate (HGA). This second reaction is catalyzed by 4-**hydroxyphenylpyruvate dioxygenase (HPPD)**. HGA is then converted to Maleylacetoacetate, in a reaction catalyzed by **Homogentisate 1,2-dioxygenase (HGD)**. Maleylacetoacetate is then converted to Fumarylacetoacetate, which is converted to Fumarate + Acetoacetate [7, 11].

$$\begin{aligned}\text{Tyr} &\rightarrow \text{p} - \text{Hydroxyphenylpyruvate}\\ &\rightarrow \mathbf{HGA}\ (\text{catalyzed by } \mathbf{HPPD})\\ &\rightarrow \text{Maleylacetoacetate (catalyzed by } \mathbf{HGD})\\ &\rightarrow \text{Fumarylacetoacetate}\\ &\rightarrow \text{Fumarate} + \text{acetoacetate}\end{aligned}$$

Alkaptonuria is caused by a **deficiency in HGD**. Its deficiency leads to the accumulation of **HGA,** an intermediate of Tyr degradation.

HGA accumulates and is excreted in the urine, which turns into a dark color. Skin pigmentation abnormalities and joint degeneration can also result due to the accumulation of **HGA** in connective tissues. The disorder is treated by dietary restrictions and **Harliku (Nitisinone NTBC)** [44, 45]. **Harliku** is a small-molecule inhibitor of HPPD, the enzyme that catalyzes the formation of HGA. By blocking HPPD, the drug reduces HGA production. As a consequence, less **HGA** accumulates. While this alleviates some symptoms, it also raises Tyr and Phe levels (Some Phe is normally converted to Tyr, but the decrease in Tyr slows this down) to levels that are too high, so patients require a low-protein diet that is Phe- and Tyr-restricted, and careful monitoring [46].

6.4 Synthesis of Alanine, Glutamate, Aspartate, and Asparagine

Ala, Glu, and Asp can be made from each other during transamination reactions.

6.4.1 Synthesis of Alanine (Ala) and Glutamate (Glu)

Above, we discussed transamination reactions, where amino groups are transferred between amino acids. Ala and Glu can be generated from the alanine transamination reaction (Ala + αKG ⇌ pyruvate + Glu), catalyzed by **ALT.** For example, Ala is generally made from pyruvate, when an amino group is transferred from Glu to pyruvate, to make Ala. Glu is made from αKG, where Ala or another amino acid transfers its amino group to αKG to make Glu.

6.4.2 Synthesis of Aspartate (Asp)

Asp can also be made from a transamination reaction, specifically the aspartate transamination reaction (Asp + αKG ⇌ Glu + OAA) catalyzed by **AST.** In this reaction, Asp is made from OAA, where OAA accepts an amino group from Glu to form Asp.

6.5 Synthesis of Glutamine (Gln) and Asparagine (Asp)

Gln can be made from **Glu.** This occurs in a reaction catalyzed by **glutamine synthetase (GS)**, where an ammonium ion (NH_4^+) (released from other amino acids) is transferred to Glu (see Fig. 14.14). This pathway is also important in nitrogen metabolism and ammonia detoxification, where Gln is a nitrogen carrier that releases nitrogen to the urea cycle for excretion. The reverse reaction, Gln to Glu, is catalyzed by **Glutaminase (GLS).** The forward and reverse reactions are as follows, respectively:

$$\mathrm{Glu} + \mathrm{NH_4}^+ + \mathrm{ATP} \rightarrow \mathrm{Gln} + \mathrm{ADP} + \mathrm{Pi}$$

$$\mathrm{Gln} + \mathrm{H_2O} \rightarrow \mathrm{Glu} + \mathrm{NH_4}^+$$

Ala and Ser can also be made from Gln, through a multistep process that involves first converting glutamine to glutamate. Asn can be

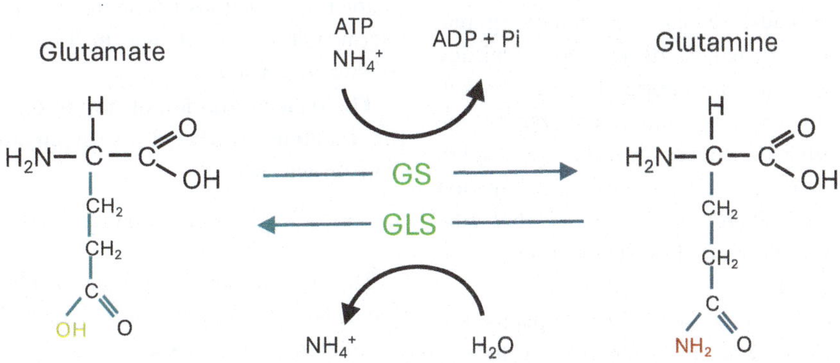

Fig. 14.14 Glutamate to glutamine. *GS* glutamine synthetase, *GLS* Glutaminase. (Amino groups are shown in their deprotonated forms)

made from Asp in two steps, involving **asparagine synthetase** [9, 11].

6.6 Synthesis of Proline (Pro) and Arginine (Arg)

Pro can be made from Glu through a series of multiple reactions [9, 10]. First, Glu becomes glutamate-γ-semialdehyde (in a reaction catalyzed by **pyrroline-5-carboxylate synthase (P5CS)**. Glutamate -γ-semialdehyde spontaneously cyclizes to form L-γ-pyrroline-5-carboxylate (L-P5C) and the two forms exist in an equilibrium. In the second reaction, L-P5C is reduced to Pro, catalyzed by **pyrroline-5-carboxylate reductase (P5CR)**. L-P5C can also be converted to ornithine in a reversible reaction catalyzed by the enzyme **ornithine aminotransferase (OAT)**, linking this reaction to the urea cycle [7].

Pro can also be synthesized from ornithine in an alternative pathway that can occur during metabolic stress or tissue repair. In this pathway, ornithine is transaminated to L-P5C, catalyzed by **OAT.** L-P5C is then reduced to form Pro, catalyzed by **P5CR** as described above.

OAT deficiency is associated with a disorder. When **OAT** is deficient, the levels of proline are reduced. In addition, there is an accumulation of excess ornithine because, without **OAT**, the ornithine cannot be converted to Pro. The accumulation of ornithine is more harmful than the lowered levels of Pro. Accumulation of ornithine is especially harmful in retinal pigment epithelial cells, where it can cause retinal degeneration (gyrate atrophy). It is treated by using a **low-arginine diet**, because arginine normally gets broken down into urea and ornithine, catalyzed by the enzyme **arginase** (see step 6 in the urea cycle). Thus, reducing Arg reduces the accumulation of ornithine. There is no cure, but the Arg-restricted diet can slow down disease progression [6, 7, 11].

Arg can be made by several mechanisms. For example, it can be made in the urea cycle as a result of cleavage of Argininosuccinate, or it can be made from citrulline, also a component of the urea cycle [10].

7 Anaplerotic and Cataplerotic Reactions

As discussed above, the carbon skeletons of many amino acids are Krebs cycle intermediates. Similarly, other biomolecules, such as fatty acids, can also be catabolized to Krebs cycle intermediates. The Krebs cycle has an important role in linking these various molecules. Specifically, it links anabolic and catabolic reactions, and therefore, it is amphibolic. This means it supplies intermediates that can be broken down for the production of energy (catabolism), and it can also supply intermediates that can be used for biosynthesis of larger molecules (anabolism). In other words, the Krebs cycle maintains a supply of intermediates that are used by two types of reactions: anaplerotic reactions and cataplerotic reactions. Anaplerotic reactions are reactions that replenish Krebs cycle intermediates so that the cycle can continue to function and support energy production. Cataplerotic reactions instead remove the intermediates for use in biosynthesis pathways, such as the synthesis of amino acids or nucleotides, or fatty acids. The balance of these different types of reactions helps to maintain a consistent amount of Krebs cycle intermediates so that the cycle is always able to function normally [4, 11, 47].

Some examples of anaplerotic reactions (those which replenish Krebs cycle intermediates) and cataplerotic reactions (those that use Krebs cycle intermediates) are shown in Fig. 14.15, and summarized below.

The transamination of Asp to form OAA in the reaction catalyzed by aspartate aminotransferase (AST):

$$\text{Asp} + \alpha\text{KG} \rightleftharpoons \mathbf{OAA} + \text{Glu}$$

The **forward** reaction is an anaplerotic reaction. It is anaplerotic because it results in a Krebs cycle intermediate, **OAA**.

The **reverse** reaction is a **cataplerotic** reaction.

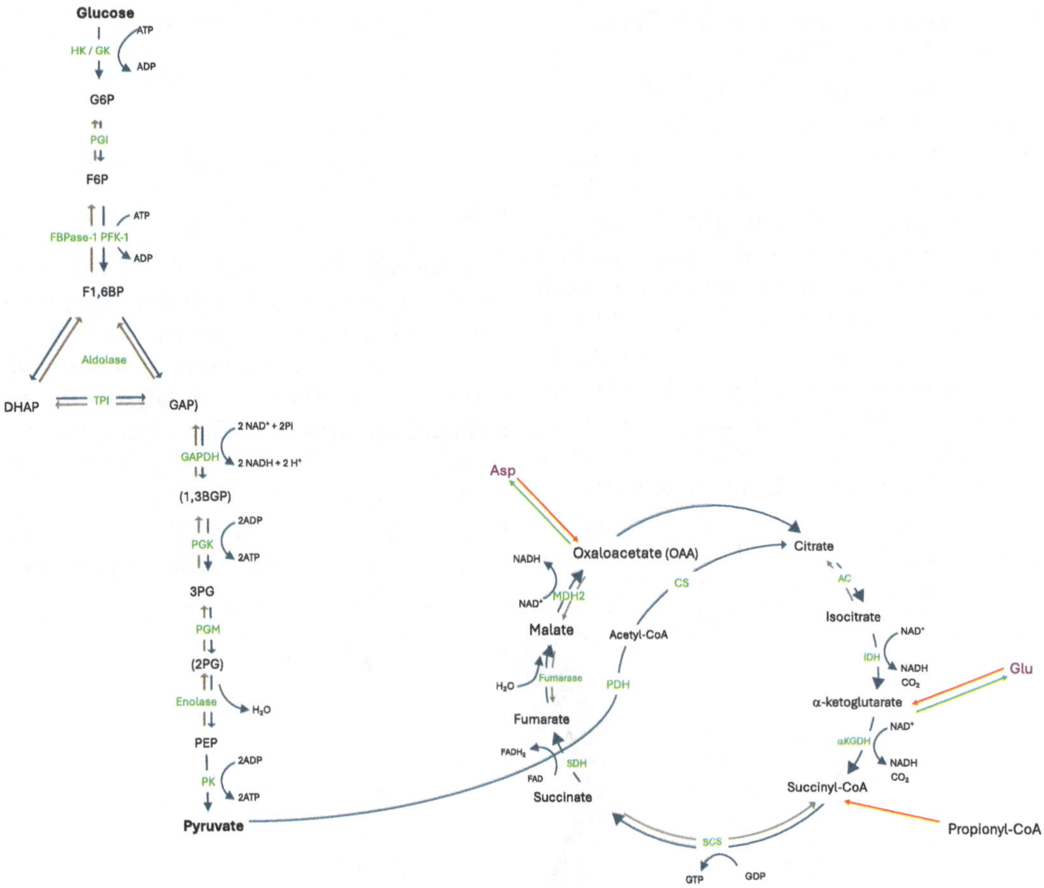

Fig. 14.15 Some examples of cataplerotic (red arrows) and anaplerotic (green arrows) reactions

In this case, **OAA**, a key Krebs cycle intermediate, is **used** to generate Asp, which can be used in nucleotide synthesis and protein synthesis, making this a cataplerotic reaction.

Oxidative deamination of Glu, catalyzed by **glutamate dehydrogenase (GDH)**:

$$\text{Glu} + \text{NAD}^+ + \text{H}_2\text{O} \rightleftharpoons \mathbf{\alpha KG}) + \text{NADH} + \text{H}^+ + \text{NH}_3$$

Here, in the **forward** reaction, **αKG** is generated. αKG is a Krebs cycle intermediate, and this is an **anaplerotic** reaction.

This reaction can occur in the **reverse** direction in some situations. This reaction uses αKG, a Krebs cycle intermediate, and it is a **cataplerotic** reaction.

Carboxylation of pyruvate to oxaloacetate, catalyzed by **pyruvate carboxylase (PC)**:

$$\text{pyruvate} + \text{CO}_2 + \text{ATP} \rightarrow \mathbf{OAA} + \text{ADP} + \text{Pi}.$$

This reaction does not involve amino acid metabolism, but it is also a good example of replenishment of Krebs cycle intermediates. OAA is produced in the **forward** reaction. **OAA** is a Krebs cycle intermediate, and therefore, this is also considered to be an **anaplerotic** reaction.

Conversion of propionyl-CoA to succinyl-CoA

Propionyl-CoA, which results from the β oxidation of odd-chain fatty acids, is converted in a multi-step process to **succinyl-CoA**, a Krebs cycle intermediate, making this an **anaplerotic** reaction.

8 The Glucose–Alanine Cycle

The glucose–alanine cycle (Fig. 14.16) is an important link between protein metabolism and glucose metabolism and between muscle and liver [48]. Above, we discussed how Glu is the main nitrogen carrier, but that the nitrogen group must be transferred to Ala for transport from muscle to liver. It must then be transferred back to Glu for transfer to the urea cycle. This set of events contributes to the glucose–alanine cycle, which helps provide glucose for ATP generation in muscle and, at the same time, helps the muscle clear harmful nitrogen groups. The cycle is largely based on the **alanine transamination reaction**, catalyzed by **ALT**, discussed earlier and shown again below:

Alanine transamination reaction:

$$\text{Ala} + \alpha\text{KG} \rightleftharpoons \text{Glu} + \text{pyruvate}$$

Enzyme: **ALT**

Here is how the glucose alanine cycle works:

- Muscle

In **muscle**, proteins are broken down under certain conditions, such as fasting or heavy exercise. When proteins are broken down, amino acids lose their amino groups through **transamination reactions catalyzed by aminotransferases**. For many amino acids, the amino group is transferred to αKG, and this leads to the formation of Glu and pyruvate. Glu now becomes the nitrogen carrier. This amino group (now on Glu) needs to be excreted from the body

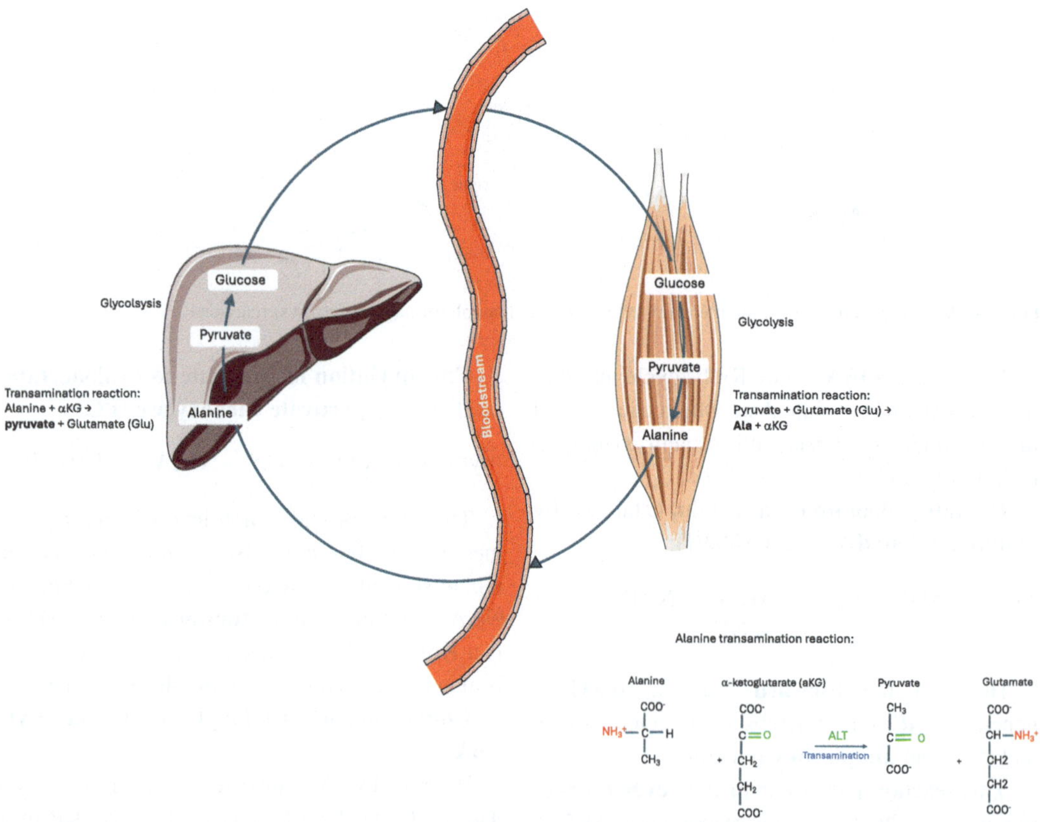

Fig. 14.16 The glucose–alanine cycle. The glucose–alanine cycle links muscle and liver metabolism. Muscle transfers amino groups to pyruvate to form alanine. Alanine travels to the liver to lead to glucose production. Glucose returns to muscle, where it is used for energy, completing the cycle and the amino group enters the urea cycle. (Source: Tissues were adapted from Servier Medical Art (https://smart.servier.com/), licensed under the Creative Commons Attribution 4.0 (CC BY 4.0) license)

via the urea cycle, which takes place in the liver. The amino group therefore needs to travel from the muscle to the liver. In order to be transported, this amino group needs to be transferred to pyruvate to form Ala in the alanine transamination reaction (reverse direction)

Ala is now the **nitrogen transporter**, and it will transport nitrogen to the **liver**, where it can efficiently cross the membrane to be used for nitrogen disposal, as described below.

- Liver

Once it reaches the liver, the **alanine transamination reaction** is involved again, this time in the **forward** direction. Now, **nitrogen** from **Ala** is then transferred to **αKG** to form **Glu**. Now, **Glu** is the nitrogen carrier. Meanwhile, **alanine** now becomes **pyruvate**.

The role of the **Glu** produced in the liver in this reaction is important. This **Glu** will become deaminated, and the **nitrogen group** will enter the **urea cycle**, to be eventually excreted in the urine (following transport to the kidney via the bloodstream).

The other product, **pyruvate**, is also an important part of this cycle. The pyruvate can lead to gluconeogenesis in the liver, thereby increasing glucose production. This glucose is released from the liver into the bloodstream, helping maintain blood glucose levels for the body as a whole. It can also travel to muscle, where it provides fuel.

- Back to Muscle

Back in the muscle, the **glucose** can enter **glycolysis** and the central metabolic pathways. As a result, more **pyruvate** is generated, and **ATP** is generated. **ATP** provides energy the muscle needs. **Pyruvate** can re-enter the **alanine transamination reaction** (the reverse reaction), thereby producing more **Ala** from Glu (which comes from transamination reactions in the muscle), and keeping the cycle going.

- Summary

In summary, the **glucose–alanine cycle** is important for nitrogen disposal, glucose homeostasis, and energy conservation in muscle, especially during fasting or intense muscle activity. It helps the muscle safely remove nitrogen by transporting amino groups, via Ala, to the liver, where the amino group enters the urea cycle for excretion, preventing toxic ammonia build-up in the muscle. The cycle helps maintain blood glucose levels even during fasting or intense exercise, providing glucose to the muscle for ATP production. It does so by providing the liver with Ala, which can then enter the alanine transamination reaction to form pyruvate. This pyruvate can be used for gluconeogenesis, producing glucose that can then be released into the bloodstream. This glucose contributes to maintaining blood glucose homeostasis, and it can also be taken up by muscle to generate ATP, supporting its activity. It also regenerates pyruvate so the glucose–alanine cycle can continue. Thus, the cycle both prevents nitrogen toxicity and supports glucose supply [6, 7, 11, 49].

9 Diseases Associated with Amino Acid Metabolism

9.1 Hepatic Encephalopathy (HE)

Hepatic Encephalopathy (HE) is a condition that is caused by damage to the liver. The liver can be damaged due to various causes, such as viral hepatitis, excess alcohol, non-alcoholic fatty liver disease (NAFLD), or reactions to some drugs. The liver is the major site of amino acid metabolism, and in HE, the damaged liver cannot properly process nitrogen waste from amino acid catabolism. Another consequence is that the urea cycle may become inefficient due to liver damage, which can lead to increased circulation of ammonia (NH_3). Ammonia can cause multiple problems. One problem is that ammonia crosses the blood–brain barrier, and this can lead to astrocyte swelling and cerebral edema. Ammonia also indirectly leads to alteration of amino acid transport in the brain, which alters neurotransmitter synthesis [50]. Both of these issues reduce the uptake of important amino acids into the brain. This creates problems because amino acids are essential for the production of certain neurotransmitters, and therefore, problems with cognition can occur. Some strategies for treating the disease include reducing ammonia production, enhancing its secretion, and supplementing with certain amino acids [51]. Dietary changes are also important, especially moderating protein intake. In advanced cases, a liver transplant may be required [52, 53].

9.2 Diagnosis of Liver Disorders

Aminotransferases, the enzymes that catalyze the transamination of amino acids, are normally located mostly in hepatocytes (liver cells), where the aminotransaminase reactions occur. However, when liver cells are damaged in liver disease, these enzymes may leak into the bloodstream. Therefore, liver disease can sometimes be diagnosed by detecting elevated levels of aminotransferases in the blood. In particular, the presence of alanine and aspartate aminotransferases (**ALT** and **AST**) is usually monitored in standard blood tests to assess liver function [54, 55].

Definitions

Aldimine The Schiff base that forms when the carbonyl is from an aldehyde.

Aldimine linkage The C=N bond within an Aldimine, analogous to the Schiff base linkage.

Alpha amino acids The types of amino acids that are the building blocks of proteins are the alpha-amino acids. In these amino acids, the amino group, carboxyl group, hydrogen, and side chain are all attached to the central (alpha) carbon atom.

Aminotransferase (transaminase) The type of enzyme that catalyzes transamination reactions. They catalyze the transfer of an amino group from an amino acid to an α-keto acid. Aminotransferases require PLP (active form of vitamin B6) as a coenzyme.

Argininosuccinate synthetase (ASS) An Enzyme that combines citrulline and aspartate to form argininosuccinate in the urea cycle.

Carbamoyl Phosphate Synthetase I (CPSI) A mitochondrial enzyme that catalyzes the first step of the urea cycle, using ammonia and bicarbonate.

Citrulline A non-standard amino acid formed in the urea cycle from ornithine and carbamoyl phosphate.

FDA-Approved Drugs Mentioned in This Chapter

Commonly used brand name	Generic name	First FDA approval year	Common conditions	Main targets
Ammonul	*Sodium benzoate + sodium phenylacetate*	2005	UCDs	Nitrogen disposal pathway
Buphenyl	*Sodium phenylbutyrate*	1996	Urea cycle disorders (UCDs)	Nitrogen disposal pathway
Harliku	*Nitisinone; NTBC*	2025	Alkaptonuria (AKU)	4-Hydroxyphenylpyruvate dioxygenase (HPPD)
Kuvan	*Sapropterin dihydrochloride (BH4 analog)*	2007	phenylketonuria (PKU)	PAH
Olpruva	*Sodium phenylbutyrate (suspension)*	2022	UCDs	Nitrogen disposal pathway
Palynziq	*Pegvaliase-pqpz*	2018	PKU	Phenylalanine
Ravicti	*Glycerol phenylbutyrate*	2013	UCDs	Nitrogen disposal pathway

Glucogenic amino acids After removal of the amino groups, the carbon skeletons of glucogenic amino acids can also contribute to glucose synthesis (gluconeogenesis).

Glucose–alanine cycle A metabolic pathway in which alanine transports nitrogen from muscle to liver. There, the alanine is converted to pyruvate for gluconeogenesis, and the amino group enters the urea cycle for excretion.

Glutamate Dehydrogenase (GDH) Enzyme that catalyzes oxidative deamination of glutamate to produce α-ketoglutarate and ammonia.

Hyperammonemia Elevated levels of ammonia in the blood, often due to defects in urea cycle enzymes.

Ketogenic amino acids After removal of the amino groups, the carbon skeletons of ketogenic amino acids give rise to acetyl-CoA or acetoacetate, contributing to ketone body formation.

Ornithine Transcarbamylase (OTC) An enzyme in the urea cycle that catalyzes the reaction between carbamoyl phosphate and ornithine to form citrulline.

Oxidative deamination A process in which the amino group is removed from an amino acid. This amino group will eventually excreted. An alpha-keto acid is formed in the process.

Phenylketonuria (PKU) A genetic disorder caused by a deficiency of phenylalanine hydroxylase, leading to the build-up of too much phenylalanine.

Pyridoxal phosphate (PLP) The active form of vitamin B6; a coenzyme essential for aminotransferase enzymes.

Schiff base A compound containing an imine (C=N) functional group formed by the reaction of a primary amine ($-NH_2$ with a carbonyl group (usually on an aldehyde or ketone) with the loss of water.

Schiff base linkage The actual C=N bond within the Schiff base.

Tetrahydrobiopterin (BH_4) A cofactor required for the activity of phenylalanine hydroxylase and other hydroxylases.

Urea cycle A cycle that rids the body of nitrogenous waste. The urea cycle occurs mainly in hepatocytes (liver cells).

Study Questions

Multiple-Choice Questions

1. Which enzyme catalyzes the conversion of alanine + αKG to glutamate + pyruvate?
   A. Alanine transaminase (ALT)
   B. Carbamoyl phosphate synthetase I (CPSI)
   C. Glutamate dehydrogenase (GDH)
   D. Argininosuccinate synthetase (ASS)
2. Which vitamin derivative acts as a coenzyme for aminotransferase enzymes?
   A. Biotin
   B. Pyridoxal phosphate (PLP)
   C. Folic acid
   D. NADH
3. In the urea cycle, which compound donates a second nitrogen to urea?
   A. Glutamate
   B. Alanine
   C. Aspartate
   D. Citrulline
4. Which amino acid is classified as both glucogenic and ketogenic?
   A. Leucine
   B. Lysine
   C. Alanine
   D. Threonine
5. Urea cycle disorders often involve which of the following:
   A. Genetic abnormalities in urea cycle enzymes
   B. Abnormalities in the ALT family of enzymes
   C. Too little ammonia
   D. Disorders in transamination

True/False Question

6. All amino acids are first transaminated and then undergo oxidative deamination via GDH.

Fill-in-the-Blank Questions

7. The enzyme ________ catalyzes the formation of carbamoyl phosphate in the urea cycle.

8. _________carries nitrogen from muscle to liver in the glucose–alanine cycle.
9. The removal of an amino group from glutamate is known as ________ deamination.
10. In hepatic encephalopathy, excess ________ crosses the blood–brain barrier.
11. ____________ reactions are reactions that replenish Krebs cycle intermediates so that the cycle can continue, leading to the generation of ATP

Short-Answer Question

12. What is the main function of the urea cycle?

Answer Key

Multiple-Choice Questions

1. **Answer**: A
2. **Answer**: B
3. **Answer**: C
4. **Answer**: D
5. **Answer**: A

True/False Question

6. **Answer**: False – some amino acids use alternative pathways.

Fill-in-the-Blank Questions

7. **Answer**: Carbamoyl phosphate synthetase I (CPSI)
8. **Answer**: Alanine
9. **Answer**: Oxidative
10. **Answer**: Ammonia (NH_3)
11. **Answer**: Anaplerotic

Short-Answer Question

12. **Answer**: To convert toxic ammonia (NH_3) into urea for excretion by the kidneys.

References

1. Kiela PR, Ghishan FK. Physiology of intestinal absorption and secretion. Best Pract Res Clin Gastroenterol. 2016;30(2):145–59. https://doi.org/10.1016/j.bpg.2016.02.007.
2. Hightower NC, Keeton WT, Encyclopaedia Britannica Editors. Human digestive system: Proteins 2024. Available from: https://www.britannica.com/science/human-digestive-system/Proteins
3. Roche HM, Macdonald IA, Schols AMWJ, Lanham-New SA, editors. Nutrition and metabolism. Wiley-Blackwell; 2024.
4. Chandel NS. Navigating metabolism. 2nd ed. Cold Spring Harbor: Cold Spring Harbor Laboratory Press; 2020.
5. Platell C, Kong SE, McCauley R, Hall JC. Branched-chain amino acids. J Gastroenterol Hepatol. 2000;15(7):706–17. https://doi.org/10.1046/j.1440-1746.2000.02205.x.
6. Rodwell VW, Bender DA, Kennely PJ, Weil PA, McGuinness OP. Harper's illustrated biochemistry. 32nd ed. New York: McGraw-Hill Education/Medical; 2022.
7. Nelson DL, Cox MM, Hoskins AA. Lehninger principles of biochemistry. 8th ed. Macmillan Learning; 2021.
8. Hames BD, Hooper NM. BIOS instant notes in biochemistry. Taylor and Francis; 2011.
9. Salway JG. Metabolism at a glance. Hoboken: Wiley-Blackwell; 2017.
10. Bender D. Amino acid metabolism. 3rd ed. Chichester: Wiley; 2012.
11. Berg JM, Tymoczko JL, Gatto GJ Jr, Stryer L. Biochemistry. 9th ed. New York, NY: MacMillan Learning; 2019.
12. Torres N, Tobón-Cornejo S, Velazquez-Villegas LA, Noriega LG, Alemán-Escondrillas G, Tovar AR. Amino acid catabolism: an overlooked area of metabolism. Nutrients. 2023;15(15) https://doi.org/10.3390/nu15153378.
13. Fry M. Essential biochemistry for medicine. 1st ed. Wiley; 2010.
14. Caldovic L, Ah Mew N, Shi D, Morizono H, Yudkoff M, Tuchman M. N-acetylglutamate synthase: structure, function and defects. Mol Genet Metab. 2010;100(Suppl 1):S13–9. https://doi.org/10.1016/j.ymgme.2010.02.018.
15. Gomez M, Jordá A, Cabo J, Grisolía S. Effect of starvation on the N-acetylglutamate system of rat liver. FEBS Lett. 1983;156(1):119–22. https://doi.org/10.1016/0014-5793(83)80260-9.
16. Jayakumar AR, Norenberg MD. The Na-K-Cl Co-transporter in astrocyte swelling. Metab Brain Dis. 2010;25(1):31–8. https://doi.org/10.1007/s11011-010-9180-3.
17. Jayakumar AR, Liu M, Moriyama M, Ramakrishnan R, Forbush B 3rd, Reddy PV, et al. Na-K-Cl Cotransporter-1 in the mechanism of

ammonia-induced astrocyte swelling. J Biol Chem. 2008;283(49):33874–82. https://doi.org/10.1074/jbc.M804016200.
18. Matsumoto S, Häberle J, Kido J, Mitsubuchi H, Endo F, Nakamura K. Urea cycle disorders – update. J Hum Genet. 2019;64(9):833–47. https://doi.org/10.1038/s10038-019-0614-4.
19. Ribas GS, Lopes FF, Deon M, Vargas CR. Hyperammonemia in inherited metabolic diseases. Cell Mol Neurobiol. 2022;42(8):2593–610. https://doi.org/10.1007/s10571-021-01156-6.
20. Nagamani SC, Erez A, Lee B. Argininosuccinate lyase deficiency. Genet Med. 2012;14(5):501–7. https://doi.org/10.1038/gim.2011.1.
21. National Organizations for Rare Disorders. Carbamoyl Phosphate Synthetase 1 Deficiency. Available from: https://rarediseases.org/rare-diseases/carbamoyl-phosphate-synthetase-i-deficiency/
22. Musunuru K, Grandinette SA, Wang X, Hudson TR, Briseno K, Berry AM, et al. Patient-specific in vivo gene editing to treat a rare genetic disease. N Engl J Med. 2025;392(22):2235–43. https://doi.org/10.1056/NEJMoa2504747.
23. Santos CD, Ratzlaff RA, Meder JC, Atwal PS, Joyce NE. Ornithine transcarbamylase deficiency: if at first you do not diagnose, try and try again. Case Rep Crit Care. 2017;2017:8724810. https://doi.org/10.1155/2017/8724810.
24. UCD In Mind. OTC Deficiency, the most common urea cycle disorder subtype. Available from: https://www.ucdinmind.com/urea-cycle-disorder-types/otc-deficiency#:~:text=Ornithine%20transcarbamylase%20(OTC)%20deficiency%2C%20or%20OTCD%2C%20is%20the,urea%20cycle%20disorder%20(UCD).&text=OTC%20deficiency%20may%20be%20life,of%20ammonia%20or%20hyperammonemic%20crises.&text=Prognosis%20for%20UCDs%20can%20vary,%25%20for%20late%2Donset%20hyperammonemia
25. Niemi A-K, Enns GM. Pharmacology review: sodium phenylacetate and sodium benzoate in the treatment of neonatal hyperammonemia. NeoReviews. 2006;7(9):e486–e95. https://doi.org/10.1542/neo.7-9-e486.
26. Häberle J, Boddaert N, Burlina A, Chakrapani A, Dixon M, Huemer M, et al. Suggested guidelines for the diagnosis and management of urea cycle disorders. Orphanet J Rare Dis. 2012;7(1):32. https://doi.org/10.1186/1750-1172-7-32.
27. Blanco A, Blanco G. Chapter 16 – Amino acid metabolism. In: Blanco A, Blanco G, editors. Medical biochemistry. Academic; 2017. p. 367–99.
28. FDA. Buphenyl (sodium phenylbutyrate) Approval Overview 2996. Available from: https://www.accessdata.fda.gov/scripts/cder/daf/index.cfm?event=overview.process&ApplNo=020572
29. Amgen. Buphenyl (sodium phenylbutyrate) prescribing information. https://www.pi.amgen.com/united_states/buphenyl/buphenyl_fpi_english.pdf
30. FDA. Ravicti (glycerol phenylbutyrate) Approval Overview 2013. Available from: https://www.accessdata.fda.gov/scripts/cder/daf/index.cfm?event=overview.process&ApplNo=203284
31. Amgen. Ravicti. What is UCD 2023. Available from: https://www.ravicti.com/ucd-specialist-finder?gad_source=1&gclid=CjwKCAiAqfe8BhBwEiwAsne6gc5YsexZgsDHZXi-kw0fkeR6RhgNzOooVf9IjjWcPwhzPxUGAhCmnhoC6CAQAvD_BwE&gclsrc=aw.ds
32. FDA. Olpruva [sodium phenylbutyrate (suspension)] Approval Overview 2022. Available from: https://www.accessdata.fda.gov/scripts/cder/daf/index.cfm?event=overview.process&ApplNo=214860
33. Acer Therapeutics Inc. OLPRUVA. Available from: https://olpruvahcp.com/?utm_medium=ppc&utm_source=Google&utm_content=About+Olpruva_Exact&utm_term=olpruva&utm_id=PS241029121101&gad_source=1&gclid=CjwKCAiAqfe8BhBwEiwAsne6gXbjzX5A7hgRu7IRF1xCKki8T2WJvLhsihb4ghqth7IrNrFRGBA78xoCAp8QAvD_BwE
34. FDA. Ammonul (sodium benzoate; sodium phenylacetate) Approval Overview 2005. Available from: https://www.accessdata.fda.gov/scripts/cder/daf/index.cfm?event=overview.process&ApplNo=020645
35. Drugs.com. Ammonul 2024. Available from: https://www.drugs.com/ammonul.html
36. Nulmans I, Lequeue S, Desmet L, Neuckermans J, De Kock J. Current state of the treatment landscape of phenylketonuria. Orphanet J Rare Dis. 2025;20(1):281. https://doi.org/10.1186/s13023-025-03840-y.
37. Rovelli V, Longo N. Phenylketonuria and the brain. Mol Genet Metab. 2023;139(1):107583. https://doi.org/10.1016/j.ymgme.2023.107583.
38. Farishian RA, Whittaker JR. Phenylalanine lowers melanin synthesis in mammalian melanocytes by reducing tyrosine uptake: implications for pigment reduction in phenylketonuria. J Invest Dermatol. 1980;74(2):85–9. https://doi.org/10.1111/1523-1747.ep12519975.
39. FDA. Kuvan [sapropterin dihydrochloride (BH4 analog)] Approval Overview 2007. Available from: https://www.accessdata.fda.gov/scripts/cder/daf/index.cfm?event=overview.process&ApplNo=022181
40. Williams RA, Bell DA, Hooper AJ, Burnett JR. Sepiapterin for the treatment of phenylketonuria. Expert Opin Pharmacother. 2025;26(8):933–8. https://doi.org/10.1080/14656566.2025.2498477.
41. FDA. Palynziq (pegvaliase-pqpz) Approval Overview 2018. Available from: https://www.accessdata.fda.gov/scripts/cder/daf/index.cfm?event=overview.process&ApplNo=761079
42. Pubchem. PubChem Compound Summary for CID 86278362, Pegvaliase: PubChem; 2025. Available from: https://pubchem.ncbi.nlm.nih.gov/compound/Pegvaliase
43. Harding CO, Longo N, Northrup H, Sacharow S, Singh R, Thomas JA, et al. Pegvaliase for the treatment

of phenylketonuria: final results of a long-term phase 3 clinical trial program. Mol Genet Metab Rep. 2024;39:101084. https://doi.org/10.1016/j.ymgmr.2024.101084.
44. FDA. Harliku (nitisinone; NTBC) Approval Overview 2017. Available from: https://www.accessdata.fda.gov/scripts/cder/daf/index.cfm?event=overview.process&ApplNo=209449
45. Mayrink FD, Dorneles G, da Silva IM, Areda CA. Efficacy and safety of Nitisinone for patients with alkaptonuria: a systematic review with metanalysis. Mol Genet Metab. 2025;145(1):109099. https://doi.org/10.1016/j.ymgme.2025.109099.
46. Thapa M, Bhatia M, Maurya VK. Alkaptonuria. Med J Armed Forces India. 2018;74(4):394–6. https://doi.org/10.1016/j.mjafi.2017.06.006.
47. Owen OE, Kalhan SC, Hanson RW. The key role of anaplerosis and cataplerosis for citric acid cycle function*. J Biol Chem. 2002;277(34):30409–12. https://doi.org/10.1074/jbc.R200006200.
48. ScienceFacts.net. Glucose-Alanine Cycle. Available from: https://www.sciencefacts.net/glucose-alanine-cycle.html
49. National Library of Medicine. PubChem Pathway Summary for Pathway SMP0087221, Glucose-Alanine Cycle, Source: PathBank 2025. Available from: https://pubchem.ncbi.nlm.nih.gov/pathway/PathBank:SMP0087221
50. Kumar G, Taneja A, Kandiah PA. Brain and the liver: cerebral edema, hepatic encephalopathy and beyond. Hepatic Crit Care. 2017:7. https://doi.org/10.1007/978-3-319-66432-3_8.
51. Hadjihambi A, Arias N, Sheikh M, Jalan R. Hepatic encephalopathy: a critical current review. Hepatol Int. 2018;12(Suppl 1):135–47. https://doi.org/10.1007/s12072-017-9812-3.
52. Butterworth RF. Pathophysiology of hepatic encephalopathy: a new look at ammonia. Metab Brain Dis. 2002;17(4):221–7. https://doi.org/10.1023/a:1021989230535.
53. Plauth M, Merli M, Kondrup J. Management of hepatic encephalopathy. N Engl J Med. 1997;337(26):1921–2. https://doi.org/10.1056/nejm199712253372614.
54. McGill MR. The past and present of serum aminotransferases and the future of liver injury biomarkers. EXCLI J. 2016;15:817–28. https://doi.org/10.17179/excli2016-800.
55. Adams LA. Biomarkers of liver fibrosis. J Gastroenterol Hepatol. 2011;26(5):802–9. https://doi.org/10.1111/j.1440-1746.2010.06612.x.

Nucleotide Metabolism and Therapeutic Targets 15

Abstract

Nucleotides serve as the building blocks of nucleic acids. They also have other functions. ATP, for example, is an essential energy source in cells. This chapter explores the metabolism of nucleotides (purines and pyrimidines), including both their synthesis and their catabolism. It covers the pentose phosphate pathway, which is crucial for generating 5-carbon sugars required for nucleotide synthesis. Different pathways involved in nucleotide metabolism are covered, including the de novo pathway, the interconversion pathway, and the salvage pathway. Some disorders associated with enzyme abnormalities important for purine and pyrimidine metabolism are discussed. Drugs that target enzymes involved in purine metabolism, such as those used to treat certain cancers, are discussed.

Keywords

Pentose phosphate pathway · Purine synthesis · Purine degradation · Pyrimidine metabolism · ADA deficiency

Abbreviations

ADA	Adenosine deaminase
AICAR	5-Aminoimidazole-4-carboxamide ribonucleotide
AMP	Adenosine monophosphate
CH_2-THF	$N5^5$,$N10^{10}$-Methylene tetrahydrofolate
CTP	Cytidine triphosphate
DHODH	Dihydroorotate dehydrogenase
DPYD	Dihydropyrimidine dehydrogenase
dTMP	Deoxythymidine monophosphate
dUMP	Deoxyuridine monophosphate
FAICAR	Formyl aminoimidazole carboxamide ribonucleotide
FGAR	Formylglycinamide ribonucleotide
GAR	Glycinamide ribonucleotide
GMP	Guanosine monophosphate
HGPRT	Hypoxanthine-guanine phosphoribosyltransferase
IMP	Inosine monophosphate
PRPP	Phosphoribosyl pyrophosphate
SAICAR	Succinylaminoimidazolecarboxamide Ribonucleotide
SCID	Severe combined immunodeficiency
THF	Tetrahydrofolate
UTP	Uridine triphosphate
XMP	Xanthosine monophosphate
R5P	Ribulose-5-phosphate
TS	Thymidylate synthase

What You Will Learn

- How nucleotides are metabolized.
- The pentose phosphate pathway is important for the generation of 5-carbon sugars that form the basis of nucleotides.

A. Minden, *Biochemistry, Cell and Molecular Biology for Pharmacists and Health Professionals*,
https://doi.org/10.1007/978-3-032-20529-2_15

- Purine synthesis by the de novo pathway, the interconversion pathway, or the salvage pathway.
- Purine degradation mechanisms.
- Some disorders associated with deficiencies in purine metabolism enzymes, such as ADA deficiency and HGPRT deficiency, and relevant treatments, including gene therapy and pharmacological intervention.
- Metabolism of pyrimidines by the de novo pathway or the salvage pathway.
- Pyrimidine catabolism.
- Pyrimidine metabolism enzymes as targets for drugs such as Aubagio, used to treat MS, or 5-FU, used to treat cancer.

1 Introduction

This chapter focuses on the metabolism of nucleotides (purines and pyrimidines) [1–3]. Nucleotides are critical because they are the building blocks of DNA and RNA, but they also have other functions. ATP, for example, is an essential energy source in cells. Nucleotide metabolism has links to the central metabolic pathways. Nucleotide synthesis, for example, is interconnected with carbohydrate metabolic pathways. In particular, the pentose phosphate pathway, a shunt from glycolysis, is central to nucleotide synthesis because it produces the 5-carbon sugars that are important components of nucleotide structures. This pathway is discussed below, followed by a discussion of nucleotide synthesis (anabolism) and breakdown (catabolism). (Most of the balanced reactions associated with the pathways discussed below are at the end of this chapter).

2 The Pentose Phosphate Pathway (PPP)

An important structure in the generation of nucleotides is ribose-5-phosphate (R5P). R5P is a metabolite that can be generated from glucose. However, glucose is a 6-carbon sugar, while R5P is a 5-carbon sugar, so several reactions need to occur to produce R5P. R5P is generated in the pentose phosphate pathway (PPP; also called the phosphogluconate pathway or the hexose monophosphate shunt), which is a shunt from glycolysis. This pathway is also the major source for generating NADPH, which is a cofactor required for the synthesis of various biomolecules, especially lipids.

Glycolysis starts with the conversion of glucose (6C) to glucose-6-phosphate (G6P, 6C) (catalyzed by **glucokinase** in the liver and **hexokinase** in most other tissues). G6P is a starting point for the rest of the glycolysis pathway, but if nucleotides are needed, or if more NADPH is needed, G6P can also feed into the PPP [3, 4] (see Fig. 15.1). The PPP can be divided into two phases, the oxidative stage and the nonoxidative stage, as described below and also covered in the following sources [1–4].

2.1 The Oxidative Stage of the PPP

The first step in this pathway is conversion of **G6P** to **6-phosphoglucono-δ-lactone (6C)**. This is an oxidation reaction, catalyzed by **Glucose-6-phosphate-dehydrogenase (G6PD)**, and involves the reduction of $NADP^+$ to NADPH + H^+.

Next, H_2O is added, and a hydrolysis reaction is carried out, resulting in **6-phosphogluconate (6C),** catalyzed by **6-phosphogluconolactonase (PGL).**

Now 6-**phosphogluconate (6C)** is converted to **ribulose-5-phosphate (Ru5P, 5C)**. In this step, one carbon is lost to CO_2 in a decarboxylation reaction, and $NADP^+$ is reduced to NADPH. This is a critical step because it involves the conversion of a 6-carbon molecule **(6-phosphogluconate)** to a 5-carbon molecule (**ribulose-5-phosphate**). The enzyme that catalyzes this reaction is **6-phosphogluconate-dehydrogenase (6PGD)**.

Ru5P can now become **ribose-5-phosphate (R5P)** (catalyzed by **ribose-5-phosphate-isomerase** (**RPI**)), or it can become **xylulose-5-**

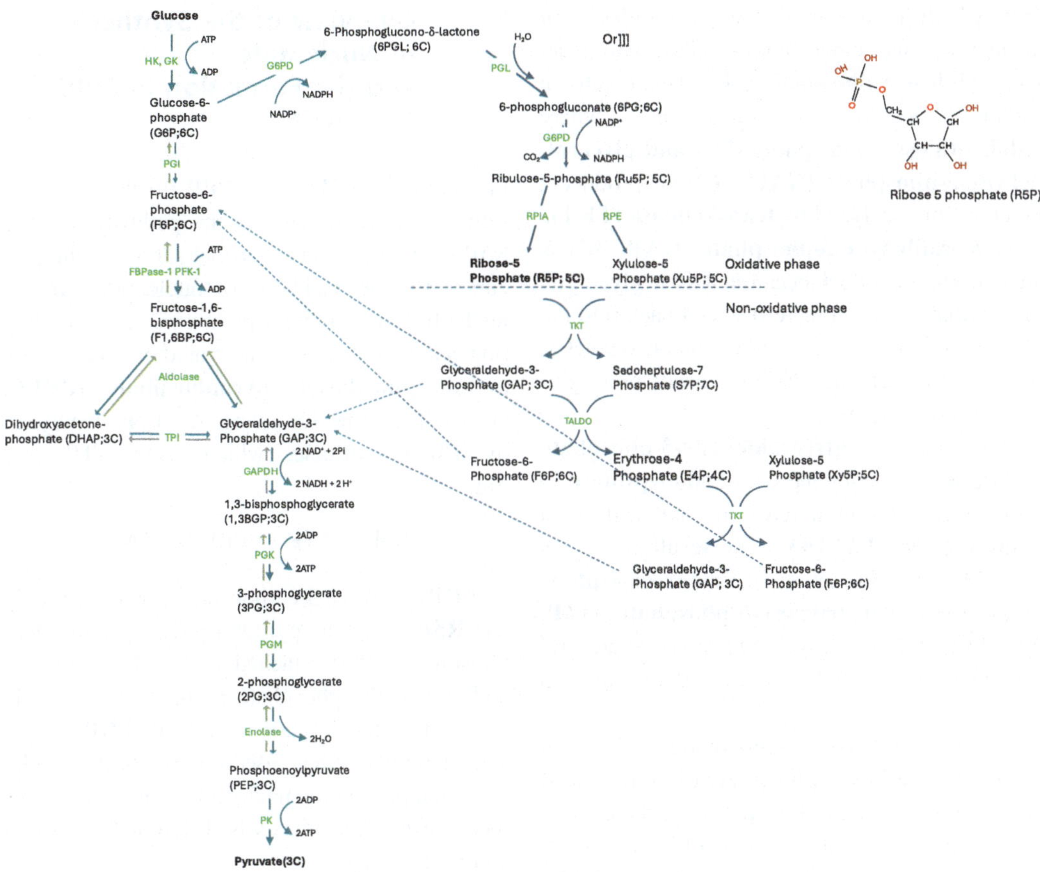

Fig. 15.1 The pentose phosphate pathway. This pathway is a shunt from glycolysis. Glycolysis and the pentose phosphate pathway are shown here. One important outcome is the formation of ribose 5-phosphate, a 5-carbon sugar, which is important for nucleotide synthesis. Aabbreviations: HK, hexokinase; GK, glucokinase; GPI, glucose-6-phosphate isomerase; FBPase-1, fructose-1,6-bisphosphatase 1; PFK-1, phosphofructokinase-1; TPI, triose phosphate isomerase; GAPDH, glyceraldehyde-3-phosphate dehydrogenase; PGK, phosphoglycerate kinase; PGM, phosphoglycerate mutase; PK, pyruvate kinase; G6PD, glucose-6-phosphate dehydrogenase; PGL, 6-phosphogluconolactonase; RPIA, ribose-5-phosphate isomerase A; RPE, ribulose-5-phosphate epimerase; TKT, transketolase; TALDO, transaldolase

phosphate (catalyzed by **ribulose-5-phosphate-3-epimerase (RPE)**). These reactions involve a rearrangement of bonds and atoms.

Ribose-5-phosphate (R5P) is an important product because it is a 5-carbon ring structure that forms the backbone of nucleotides. This part of the pathway is important for several reasons. It produces NADPH, which is important in various metabolic pathways and biosynthetic reactions, including the synthesis of fatty acids and cholesterol. R5P is important for nucleotide metabolism, which will be discussed below, and R5P is also important for the formation of Coenzyme A (CoA).

2.2 The Nonoxidative Stage of the PPP

The above marks the end of the oxidative stage of the PPP. Now in the second phase (the nonoxidative stage), **ribose-5-phosphate (R5P,**

5C), which is not used for the synthesis of nucleotides and other biomolecules, will react with **xylulose-5-phosphate (5C)**. This results in carbon skeleton rearrangement where **sedoheptulose-7-phosphate (7C) and glyceraldehyde-3-phosphate (GAP, 3C)** are produced, in a reaction catalyzed by **transketolase (TKT).**

Glyceraldehyde-3-phosphate (GAP, 3C) is an important product because it is a glycolytic intermediate and can therefore feed back into the glycolysis pathway or into gluconeogenesis, depending on whether the body needs more glucose or more energy.

Alternatively, **glyceraldehyde-3-phosphate (GAP, 3C)** can combine with **Sedoheptulose-7-phosphate (7C)** in a reaction catalyzed by a **transaldolase (TALDO)**. The result from this reaction is the formation of **erythrose-4-phosphate (4C)** and **fructose-6-phosphate (F6P, 6C)**. **F6P** is also a glycolytic intermediate and can also form a link between glycolysis and the PPP.

Finally, **erythrose-4-phosphate (4C)** can react with **xylulose-5-phosphate (5C)** (produced earlier in the first half of the pathway, see above), in a reaction also catalyzed by a **TKT**. This reaction produces more **F6P (6C) and glyceraldehyde-3-phosphate (GAP, 3C)**. Both of these products are glycolytic intermediates, thus allowing the carbons from the PPP to return to glycolysis.

All the steps in this stage are reversible, and thus R5P can also be made from glycolytic intermediates such as F6P or glyceraldehyde-3-phosphate (GAP), which can then feed into the PPP.

3 Overview of the Synthesis of Nucleotides and the Formation of PRPP from R5P

Formation of **R5P** in the **PPP** (in the oxidative stage) is a critical step in the pathway because R5P is the 5-carbon sugar that serves as the precursor for the backbone of nucleotides. In the nucleotide biosynthesis pathways, this 5-carbon ring sugar will first be activated by conversion to **phosphoribosyl-1-pyrophosphate (PRPP)** [1–3, 5] in the following reaction, catalyzed by **PRPP-synthase**, which uses ATP (see Fig. 15.2):

$$R5P + ATP \rightarrow PRPP + AMP$$

PRPP is also a 5-carbon sugar. It is essentially an **R5P** with a pyrophosphate group (two phosphates; PPi) attached to the 1′ carbon, in addition to the phosphate group that is already attached to the 5′ carbon of R5P. **PRPP** is an activated intermediate used for both purine nucleotide and pyrimidine nucleotide synthesis, but the mechanism of synthesis is different for purines and pyrimidines.

When **purine nucleotides** (those containing a purine base; adenine or guanine) are synthesized, the purine base will be built directly onto PRPP (at the 1′ carbon) in a step-by-step fashion.

In contrast, when **pyrimidines** (those containing a pyrimidine base; cytosine, thymine, or uracil) are synthesized, the pyrimidine bases are built first and then attached to the PRPP (also at the 1′ carbon).

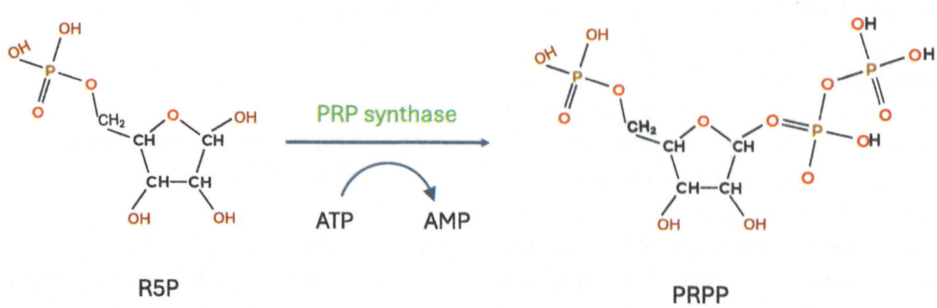

Fig. 15.2 Conversion of Ribose 5-phosphate (R5P) to PRPP. This is a key step in nucleotide synthesis. (**Source**: Structures rendered using MolViewPro)

Both purine metabolism (synthesis and catabolism) and pyrimidine metabolism are discussed below and also covered in [1–3, 5].

4 Purine Nucleotide Synthesis

Purine nucleotides are the nucleotides that contain a purine base. Purine bases are adenine or guanine. The purine nucleotides include adenosine monophosphate (AMP) and guanosine monophosphate (GMP). AMP and GMP can also be made into other forms, such as ATP and dATP. AMP and GMP can be synthesized (a) directly, via the de novo pathway. They can also be synthesized by two other mechanisms. These are (b) interconversion between different purines, and (c) recycling of nitrogenous bases via the salvage pathway. The **de novo** pathway will be discussed first.

4.1 Purine Synthesis by the De Novo Pathway

An important step in the de novo synthesis of purine nucleotides is the formation of **IMP (inosine monophosphate)** from **PRPP**. The process involves the replacement of the pyrophosphate on the 1′ carbon of PRPP with an amino group from Glycine (Gln), followed by the stepwise building of the purine ring onto this nitrogen. This ring will be hypoxanthine when it is completed. In the end, this yields IMP, which is a ribose with both a 5-phosphate and a hypoxanthine. Hypoxanthine is an important part of this structure. It is a purine base, but not one found in nucleic acids. IMP is thus a nucleotide; specifically, it is a purine nucleotide, though it is not one that is incorporated into DNA or RNA. Rather, it is the precursor of the major purine nucleotides. Once IMP is synthesized, it will be modified to form AMP and GMP, which can be converted to derivatives such as ATP, GTP, dATP, and dGTP [2, 3, 5–8]. The formation of IMP from PRPP and the synthesis of the purines from IMP are described in more detail below and in the studies cited herein [2, 3, 5–8].

4.1.1 PRPP to IMP

- Formation of IMP from PRPP Involves the Stepwise Addition of Atoms to PRPP

The steps between PRPP and IMP involve the sequential addition of carbon and nitrogen atoms onto PRPP (the 5-carbon ribose phosphate backbone), which ultimately form the nitrogenous purine base hypoxanthine. (The reactions between PRPP and IMP summarized in Fig. 15.3, and the chemical formulas for the various intermediates are illustrated in Fig. 15.4). Carbon transfer is an important part of this process. Some carbons that are added to PRPP come from molecules such as glycine (Gln), and others come from intermediates of the folate (vitamin B9) cycle, which is an essential metabolic pathway that facilitates the transfer of one-carbon units. The details of the folate cycle will be covered in Chap. 16. For now, it is important to know that this cycle supplies some of the carbons needed to make hypoxanthine, and thus to make IMP from PRPP.

- The Detailed Steps Between PRPP and IMP

The following steps are involved in building IMP from PRPP (see Fig. 15.3):

1. PRPP to 5PRA (5-phosphoribosylamine): First, **PRPP** reacts with **glutamine (Gln)**, where Gln donates an **amino group** to the 1 ‘carbon of PRPP. This replaces the pyrophosphate group (two phosphates) from that same carbon, which is now released. The resulting product is **5-phosphoribosylamine (5PRA)**.
2. 5PRA to GAR (glycinamide ribonucleotide): **5PRA** reacts with **glycine (Gly)**, forming **GAR**. This is an important reaction in which **Gly** is a **carbon donor**. In addition to donating 2 carbon atoms, glycine also donates a nitrogen atom. At this step, we begin to see a structure attached to the R5P, which will ultimately become the hypoxanthine nitrogenous base.
3. GAR to FGAR (formylglycinamide ribonucleotide): This step involves the **transfer** of a **one-carbon** group to the growing purine base structure by a mechanism that involves

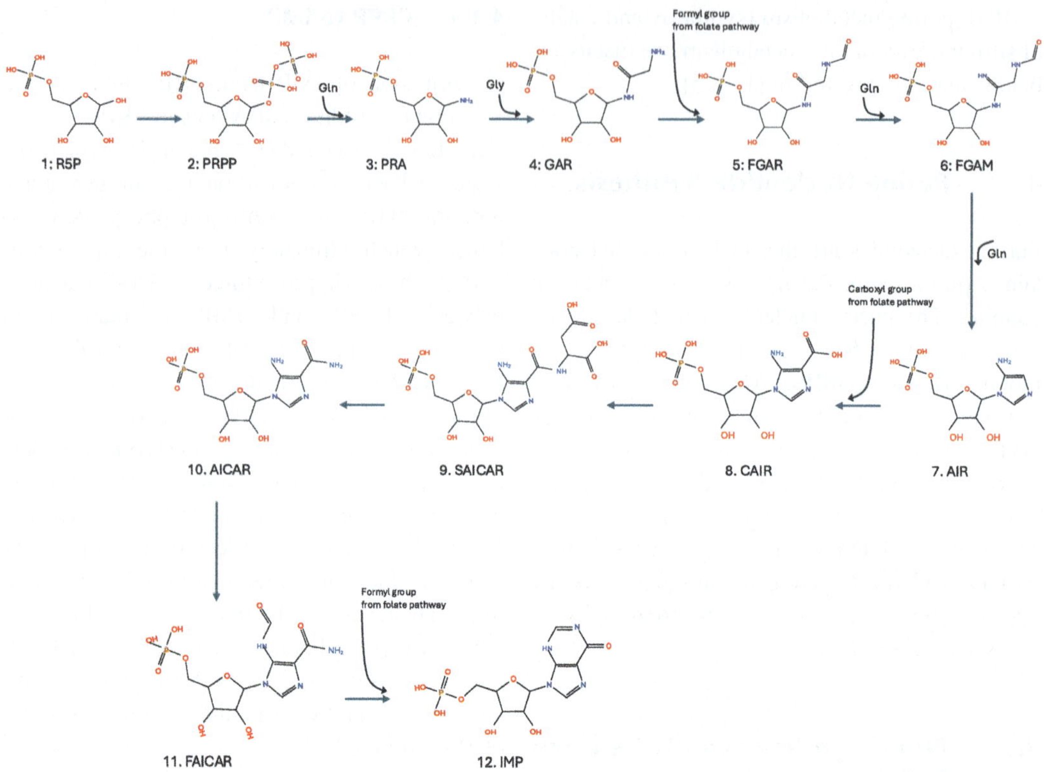

Fig. 15.3 R5P to PRPP to IMP. This series of steps is a key part of the de novo purine synthesis pathway. (**Source**: Structures rendered using MolViewPro)

R5P:	C5	H11		O8	P	
PRPP:	C5	H13		O14	P3	
5PRA:	C5	H12	N	O7	P	
GAR:	C7	H13	N2	O8	P	
FGAR:	C8	H15	N2	O9	P	*
FGAM:	C8	H16	N3	O8	P	
AIR:	C8	H14	N3	O7	P	
CAIR:	C9	H14	N3	O9	P	
SAICAR:	C13	H19	N4	O12	P	
AICAR:	C9	H15	N4	O8	P	
FAICAR:	C10	H15	N4	O9	P	*
IMP:	C10	H13	N4	O8	P	

Fig. 15.4 The chemical formulas for the intermediates between PRPP and IMP

the **folate cycle**. Specifically, it involves the transfer of a **formyl group** (containing a **carbon**, a hydrogen, and an oxygen) from **N10-formyl-THF** (an intermediate of the folate cycle) to **GAR**, resulting in **FGAR**. This step is significant, in that it involves the transfer of a one carbon unit from the folate cycle, to be discussed in Chap. 16.

4. FGAR to FGAM (formylglycinamidine ribonucleotide): In this step, an amino group is added to FGAR from Gln.
5. FGAM to AIR (aminoimidazole ribonucleotide). This step involves cyclization of the open-chain amidine group on FGAM to form the imidazole ring of AIR.
6. AIR to CAIR (carboxyaminoimidazole ribonucleotide). This step is a carboxylation reaction and involves the addition of a carboxyl group to AIR, from CO_2.
7. CAIR to SAICAR (succinylaminoimidazole carboxyamide ribonucleotide). In this reaction, Asp reacts with CAIR. In the process, Asp contributes a nitrogen and a four-carbon chain to form SAICAR.
8. SAICAR to AICAR (5-Aminoimidazole-4-carboxamide ribonucleotide). Here, a four-carbon group from the last step is removed. Fumarate is released as a result. Fumarate is a Krebs cycle intermediate and can thus be fed back into the central metabolic pathways.

9. AICAR to FAICAR (5-Formamidoimidazole -4-carboxamide ribonucleotide). Like step 3, this reaction also involves the folate cycle, where there is a transfer of **one carbon** to the growing structure in the form of a **formyl group** from **N10-formyl-THF**, a folate-derived carbon donor.
10. FAICAR to IMP. In this reaction, H_2O is released, and IMP is finally formed.

4.1.2 Conversion of IMP to the Purine Nucleotides AMP and GMP

In two separate two-step reactions, described below, **IMP** is converted to **AMP** (with the release of fumarate) and **GMP** (Figs. 15.5 *and* 15.6). AMP and GMP can then be converted to other forms that can be used in RNA, DNA, as an energy source, and in signaling.

- IMP to AMP

Conversion of **IMP to AMP** occurs in a two-step reaction: In **step 1**, **IMP** is converted to **adenylosuccinate** in a reaction catalyzed by **adenylosuccinate synthetase (ADSS)**. This reaction requires aspartate (Asp), and GTP is used as an energy source. In **step 2, adenylosuccinate** becomes **AMP**. This reaction is catalyzed by **adenylosuccinate lyase (ADSL)**. Fumarate is released from this reaction.

- IMP to GMP

Conversion of IMP to GMP also occurs in a two-step reaction: In **step 1**, IMP is oxidized to **xanthosine monophosphate (XMP)** in a reaction catalyzed by **IMP dehydrogenase (IMPDH)**. This reaction uses NAD^+, which is reduced to NADH. In **step 2**, an amination (addition of an

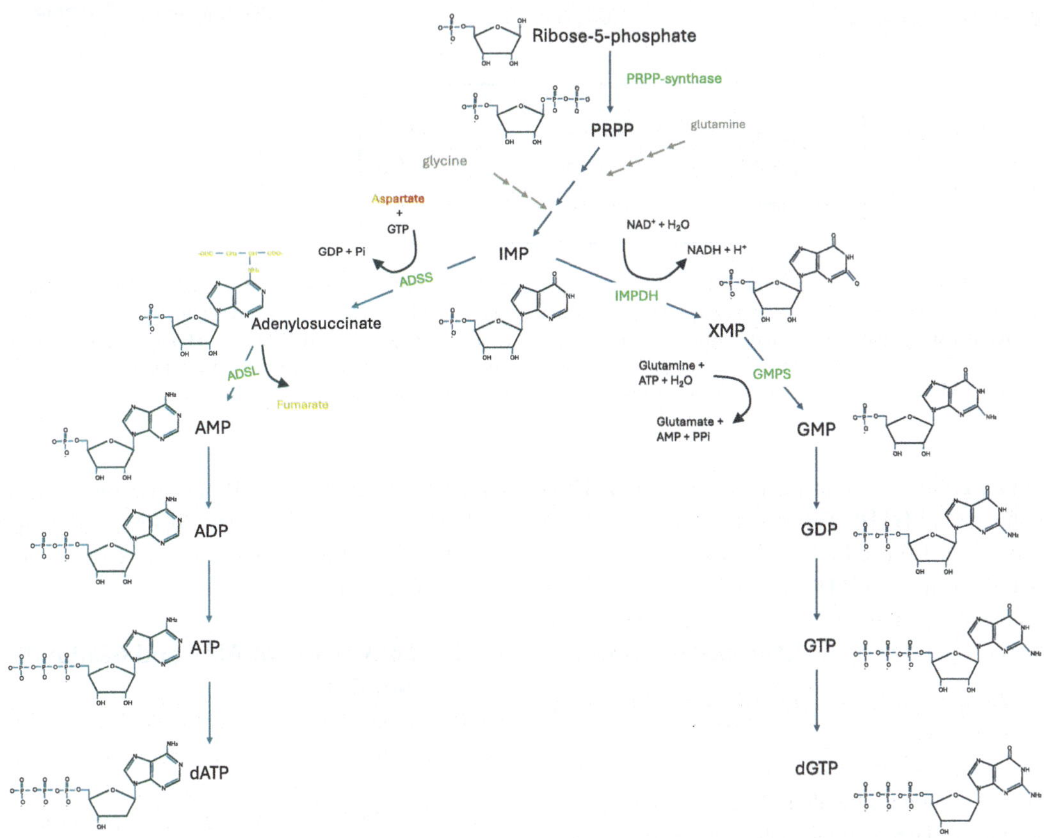

Fig. 15.5 De novo purine synthesis. This image shows the pathway from Ribose-5-phopphate (R5P) to IMP, ATP, GTP, dATP, and dGTP. Some enzymes are included in the figure: ADSS, adenylosuccinate synthetase; IMPDH, IMP dehydrogenase; ADSL adenylosuccinate lyase; GMPS, GMP synthetase

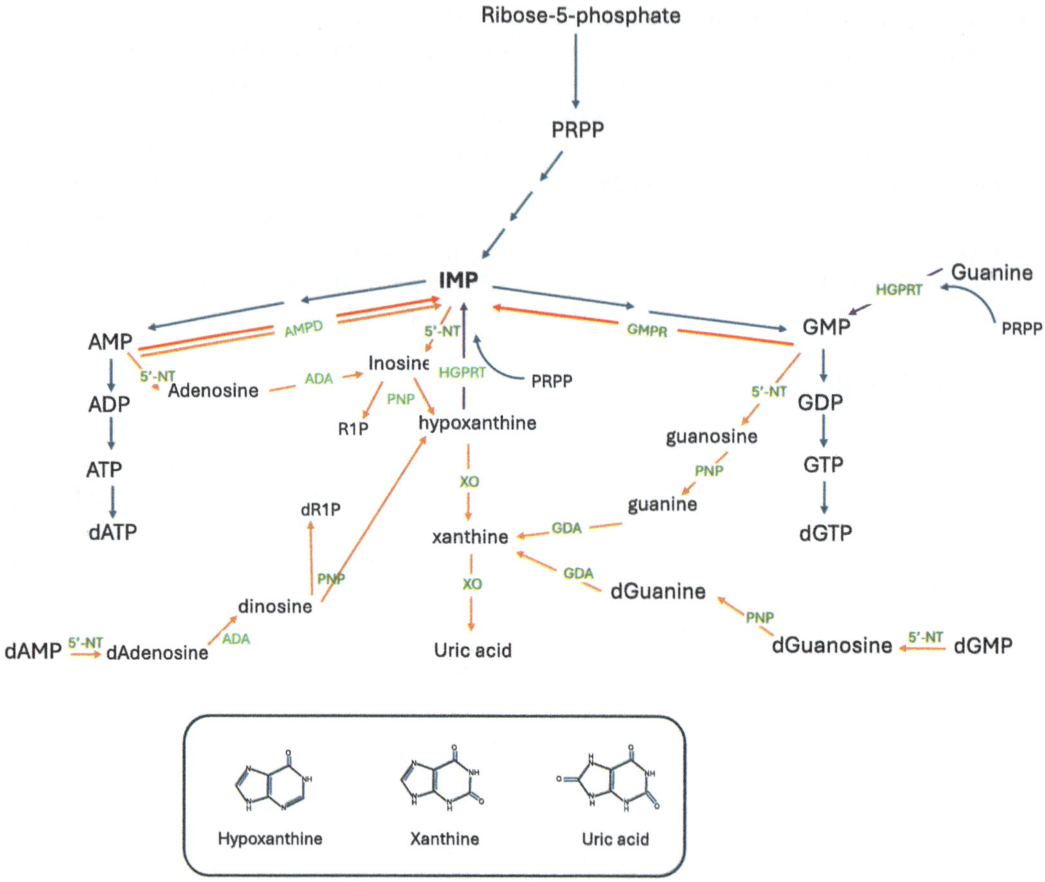

Fig. 15.6 Purine interconversion and salvage pathways, and purine catabolism. Salvage pathways; purple arrows, interconversion pathways: red arrows, catabolism: orange arrow. The de novo synthesis pathway (also see Fig. 15.5) is shown in blue arrows. Abbreviations include: AMPD, AMP deaminase; ADA Adenosine deaminase, PNP, Purine nucleoside phosphorylase; 5′-NT, 5′-nucleotidase; HGPRT, Hypoxanthine-guanine phosphoribosyltransferase; XO, Xanthine oxidase; GDA, Guanine deaminase; GMPR, GMP reductase

amino group) reaction occurs in which **XMP** is converted to **GMP**. This involves the donation of a nitrogen from **Gln** (which has 2 nitrogens) to **XMP**, forming **GMP**. In the process, **Gln** is converted to **Glu** (which has 1 nitrogen). This step is catalyzed by **GMP synthetase (GMPS)**.

- The End of the de Novo Pathway for Purine Synthesis

The formation of AMP and GMP, as described above, completes the de novo pathway for purine synthesis. The de novo pathway for purine synthesis starts with the synthesis of IMP from PRPP, as described above, followed by the synthesis of AMP and GMP. Even after their synthesis, AMP and GMP can undergo further modifications, as described below, allowing them to be used in nucleic acids, as energy currency, and in cell signaling.

4.1.3 Conversion of AMP and GMP ATP, and GTP

AMP and **GMP** are converted to **ATP** and **GTP** (adenosine triphosphate and guanosine triphosphate) via the addition of two more phosphate groups to AMP and GMP at the 5′ position of their sugar backbones. This occurs in two steps. The first step is catalyzed by **adenylate kinase (AK)** and **guanylate kinase (GK or GMPK)**, which produce **ADP** and **GDP**, respectively. (AK catalyzes the reaction AMP + ATP ⇌ 2 ADP, rather than using free phosphates.) This is followed by the second step, which is another phosphorylation reaction catalyzed by **nucleoside diphosphate kinase (NDPK)**. **NDPK** catalyzes the addition of the second phosphate group to the nucleotides. (NDPK transfers a phosphate from an existing nucleoside triphosphate, usually ATP, to ADP or GDP.) This involves transferring terminal phosphate groups from nucleoside triphosphates present in the cell, especially ATP, to the diphosphates (ADP and GDP) to make ATP and GTP.

4.1.4 Conversion of ADP and GDP to dADP and dGTP and to dATP and dGTP

GTP and **ATP** are purine nucleotides that serve as building blocks in RNA, and ATP also acts as an energy source. However, **dATP** and **dGTP** (deoxyATP and deoxyGTP) are the nucleotides used in DNA. Generation of dATP and dGTP involves the removal of a hydroxyl group on the 2′ carbon on the sugar portion of the molecule, in a reaction catalyzed by **ribonucleotide reductase (RNR)** (an enzyme that can convert ribonucleotides to deoxyribonucleotides). **RNR** only acts on diphosphates, so the diphosphates (ADP and GDP) are the starting point for these reactions. In these reactions, ADP and GDP (intermediates in the synthesis of ATP and GTP as described above) are converted to dADP and dGDP in reactions catalyzed by **RNR**. dADP and dGDP are then converted to dATP and dGTP (catalyzed by **NDPK**, as described above for ATP and GTP). The resulting **dATP** and **dGTP** are building blocks of DNA.

4.2 Purine Interconversion and Salvage Pathways

4.2.1 Interconversion of Purine Nucleotides

While AMP and GMP can be made individually in the de novo pathway, they can also be made from each other; in other words, they are **interconvertible**. In these conversions, AMP can be converted back to IMP, which can be converted to GMP. Similarly, GMP can be converted to IMP, which can be converted to AMP. These reactions are discussed below and can also be seen in Fig. 15.6.

- AMP Can be Converted to IMP, Which Can Then Be Converted to GMP

Conversion of AMP to GMP involves several reactions. **First, AMP** is converted to **IMP**. This reaction is catalyzed by **AMP deaminase (AMPD)**, which removes an amino group. Next, IMP is converted to GMP, exactly as described above in the de novo pathway.

- GMP Can Be Converted IMP, Which Can Then Be Converted to AMP

More rarely, **GMP** can also be converted to **AMP**. In this reaction, GMP is first converted to IMP in a reaction catalyzed by **GMP reductase (GMPR)**. IMP is then converted to AMP as above.

4.2.2 Purine Synthesis by the Salvage Pathway

Another mechanism by which purines can be created is by the **salvage pathway**. This pathway recycles existing purine bases rather than synthesizing them de novo. This pathway is more energy efficient than the de novo pathway because it allows the reuse of purine bases derived from nucleic acid catabolism or other metabolic processes. In the salvage pathway, the purine rings are not made from scratch, but the salvage pathway still relies on PRPP to provide the sugar phosphate backbone, to which purine bases are attached.

As an example, in the salvage pathway, **guanine** can be converted to **GMP** by combining with **PRPP** in a reaction catalyzed by **HGPRT** (hypoxanthine-guanine phosphoribosyltransferase). **Hypoxanthine** (which can come from the degradation of inosine or AMP) can also react with **PRPP** and be converted to **IMP.** More rarely, **adenine** can react with PRPP in a reaction catalyzed by **APRT** (adenine phosphoribosyltransferase) to be converted to **AMP** (not shown in the figure).

4.3 Summary of the De Novo Pathway, the Salvage Pathway, and the Interconversion Pathway

There are important differences among the de novo pathway, the salvage pathway, and the interconversion pathway. All of these pathways begin with PRPP. The de novo pathway builds a purine ring from scratch. This purine ring (hypoxanthine) is assembled stepwise onto PRPP, forming IMP, which then becomes AMP or GMP. In the salvage pathway, instead of building a new purine ring, free purine bases (such as adenine, guanine, and hypoxanthine) are recycled. These bases are attached to PRPP by specific transferases, such as **HGPRT** or **APRT**, to regenerate nucleotides. The interconversion pathway, instead of using free bases, utilizes existing nucleotides (such as AMP, GMP, IMP, or their phosphorylated forms) and converts them into other nucleotide forms. An example is the conversion of AMP to GMP via IMP [1–3].

5 Degradation (Catabolism) of Purines

Our cells undergo constant turnover, and many cells undergo apoptosis. This results in nucleic acids that are no longer needed. Nucleic acids in our cells that are no longer needed get broken down into nucleotides. This happens within all types of cells. Nucleic acids can also be obtained from the diet (from foods like meat and fish). These dietary nucleic acids are broken down into nucleotides, primarily in the small intestine, by **DNase** and **RNase** enzymes secreted from the pancreas, as well as by certain brush border enzymes. Regardless of the source, if they are not salvaged, nucleotides need to be broken down completely [1–3, 5]. Purine catabolism includes the reversal in some of the pathways that are used for purine synthesis (Fig. 15.6):

5.1 Catabolism of AMP

AMP can be broken down by two mechanisms, as follows.

5.1.1 The Main Pathway for AMP Catabolism

1. **AMP** to **IMP**: This reaction is catalyzed by **AMP deaminase (AMPD)**, and it involves the removal of an amino group from AMP.
2. **IMP** to **inosine**: This reaction is catalyzed by **5′-nuclotidase (5′-NT),** and it involves the removal of a phosphate group (dephosphorylation) from IMP.
3. **Inosine** to **hypoxanthine**: This reaction is catalyzed by **purine nucleoside phosphorylase (PNP)**. It breaks down inosine into **hypoxanthine** and ribose-1-phosphate.
4. **Hypoxanthine** to **xanthine**: This reaction is catalyzed by **xanthine oxidase (XO)** and it involves oxidation of the hypoxanthine to form xanthine.
5. **Xanthine** to **uric acid**: This reaction is also catalyzed by **xanthine oxidase (XO)**. Uric acid is then excreted in the urine.

[Alternative step 4]: Hypoxanthine, instead of being catabolized and converted to xanthine, followed by uric acid and excretion, can also be **salvaged** and converted back to **IMP** in a reaction catalyzed by **HGPRT** (see salvage pathway discussed above).

5.1.2 Alternative Pathway for AMP Catabolism (Involving Adenosine Deaminase (ADA)

1. **AMP** to **adenosine**, catalyzed by **5′ nucleoside (5′-NT).**
2. **Adenosine** to **inosine**, catalyzed by **adenosine deaminase (ADA).**
3. **Inosine** to **hypoxanthine** to **xanthine**: these steps are the same as 3–5 above.

5.2 Catabolism of dAMP

Catabolism of **deoxyAMP (dAMP)** is important because dAMP can come directly from the breakdown of DNA. It also involves the enzyme Adenosine Deaminase **(ADA)**. The steps are as follows:

1. Catabolism of **dAMP** to **dAdenosine**: This reaction removes a phosphate group and is catalyzed by **5′-nucleosidase (5′-NT)**.
2. **dAdenosine** to **dInosine**: Here, the amino group of dAdenosine is removed, in a reaction catalyzed by **ADA**.
3. **dInosine** to **hypoxanthine**: This reaction is catalyzed by **PNP**, where **dInosine** is cleaved into **hypoxanthine** and **deoxyribose-1-phosphate (dR1P)**.
4. **Hypoxanthine** to **xanthine** to **uric acid**: This step is catalyzed as in the steps above.

5.3 Catabolism of GMP

1. **GMP** to **guanosine**: In this reaction, a phosphate group is removed from GMP, in a reaction catalyzed by **5′-nucleotidase (5′-NT)**.
2. **Guanosine** to **guanine**: This reaction results in guanine and ribose-1-phosphate (R1P). The reaction is catalyzed by **PNP**.
3. **Guanine** is converted directly to **xanthine**: This reaction is catalyzed by **guanine deaminase (GDA)**.
4. **Xanthine** to **uric acid**: This is the same reaction taken by the AMP catabolism pathway, catalyzed by **xanthine oxidase (XO)**. Uric acid is excreted in the urine.

5.4 Catabolism of dGMP

1. **dGMP** to **dGuanosine**: This reaction involves the removal of a phosphate group, catalyzed by **5′-nucleotidase (5′-NT)**.
2. **dGuanosine** to **dGuanine**: In a reaction catalyzed by **PNP**, deoxyguanine (dGuanine) and deoxyribose-1-phosphate are formed from dGuanosine.
3. **dGuanine** to **xanthine**: This reaction is catalyzed by **guanine deaminase (GDA)**.
4. **Xanthine** to **uric acid**: This reaction is catalyzed as in the above reactions.

5.5 Catabolism of IMP

1. **IMP** is converted to **inosine**, catalyzed by **5′-nucleotidase**.
2. **Inosine** is converted to **hypoxanthine**, catalyzed by **PNP**.
3. **Hypoxanthine** is converted to **xanthine**, catalyzed by **xanthine oxidase**.
4. **Xanthine** is broken down to **uric acid**, also catalyzed by **xanthine oxidase**.

6 Some Key Enzymes in Purine Metabolism and Human Disease

6.1 Adenosine Deaminase (ADA) and ADA Deficiency

Adenosine Deaminase (ADA), one of the important enzymes described above, catalyzes the conversion of adenosine to inosine and dAdenosine to dInosine during purine catabolism. dAdenosine is a product of DNA breakdown, and its accumulation is especially harmful to cells. This becomes a serious problem when **ADA** is deficient, where adenosine can build up to harmful levels in cells. ADA deficiency is a rare disorder characterized by mutations in the gene encoding ADA, which cause its activity to be reduced or absent. This disorder results in **severe combined immunodeficiency (SCID)**, an inborn error of metabolism (IEM), which is associated with the accumulation of adenosine and dAdenosine. Without **ADA**, the high level of dAdenosine has toxic effects in cells. This is in part because dAdenosine can be phosphorylated and lead to the accumulation of dATP, which is an inhibitor of **RNR**, an enzyme discussed above and that has important roles in the synthesis of deoxynucleotides (it catalyzes the synthesis of dNTPs from NTPs). The cells of the immune system are especially sensitive to **ADA** deficiency because they divide frequently and thus replicate their DNA often. This leads to a build-up of dAdenosine due to dysfunctional **ADA**. The accumulation of dAdenosine can disrupt DNA synthesis and lead to apoptosis and death of the immune cells. Patients with ADA deficiency are therefore highly sensitive to infection as they are unable to carry out an adequate immune response. Neurological problems can also occur, as the disruption of metabolic pathways in the brain can also result from the accumulation of deoxyadenosine [9, 10].

Different treatment options are used to treat ADA deficiency. One approach is **hematopoietic stem cell transplantation (HSCT)** from a healthy donor. This can potentially restore a functioning immune system for a prolonged period of remission [11]. Another option is **enzyme replacement therapy (ERT)**, where the missing **ADA enzyme** is administered via injection (intramuscular or intravenous). This helps reduce symptoms, but the patient needs to be treated for life [12]. An important advancement for treating ADA deficiency is **gene therapy**. This involves removing the patient's hematopoietic stem cells, inserting a normal copy of the **ADA gene** into these cells, and then reinfusing the cells back into the patient. Gene therapy has been approved in some places (such as **Strimvelis** in Europe [13]) and continues to be studied in **clinical trials**. Gene therapy, including gene therapy for SCID, is discussed in more detail in Chap. 28. New delivery methods are being developed to optimize the treatment [14–18].

6.2 HGPRT and HGPRT Deficiency

HGPRT plays an essential role in the purine salvage pathway, recycling guanine to GMP and recycling hypoxanthine to IMP. A deficiency in **HGPRT** is an inborn error of metabolism. It causes hypoxanthine and guanine to accumulate. When hypoxanthine levels build up, it can be catabolized and converted to uric acid. Thus, a deficiency of HGPRT leads to two major problems: too much uric acid and deficient purine recycling. The accumulation of uric acid can lead to kidney stones and gout. Patients also have neurological symptoms, which may be due to the lack of GTP (due to the deficiency in guanine conversion to GMP). Symptoms can range in severity. The most severe form of **HGPRT deficiency** is known as **Lesch–Nyhan syndrome (LNS)**. Patients with the disorder have near near-complete deficiency in HGPRT and have severe neurological symptoms and cognitive problems. Patients can also have partial HGPRT deficiency and have milder symptoms.

Treatment for HGPRT deficiencies involves managing symptoms and using medications. **Zyloprim (allopurinol)** [19] is a small-molecule drug that inhibits **xanthine oxidase (XO)**. **XO** is an enzyme, also discussed above, that catalyzes the conversion of hypoxanthine to xanthine and then to uric acid. By blocking this enzyme, the drug lowers the production of uric acid, thus relieving some symptoms. **Allopurinol** works by acting as a substrate for XO. XO converts the allopurinol to oxypurinol in cells. Oxypurinol then functions as a long-acting inhibitor of XO. It binds tightly to XO's active site and prevents the enzyme from binding to its normal substrates, hypoxanthine and xanthine. Consequently, it reduces uric acid production [20, 21].

7 Metabolism of Pyrimidine Nucleotides

7.1 Pyrimidine Nucleotide Synthesis: De Novo Pathway

Pyrimidine nucleotides contain a pyrimidine base (cytosine, thymine, or uracil; they can be remembered by the acronym C.U.T.). The major pyrimidine nucleotides include **CTP, dCTP, dTTP, UMP,** and **UTP** (UTP is used only in RNA, and dTTP is used only in DNA). Two important precursors in pyrimidine synthesis are **PRPP** (which will form the sugar backbone) and **Orotic acid. Orotic acid** is a pyrimidine precursor with a ring structure. Pyrimidines are synthesized by a somewhat different reaction than purines. In pyrimidine synthesis, the ring structure, orotic acid, is made separately and is then added onto the sugar backbone (derived from PRPP), rather than being added in a stepwise manner, as described below and in the studies cited herein [2, 3, 5–7] (Fig. 15.7a).

7.1.1 The Formation of Orotic Acid

An important step in pyrimidine synthesis is the synthesis of orotic acid. **Orotic acid** is derived from a precursor: **carbamoyl phosphate.**

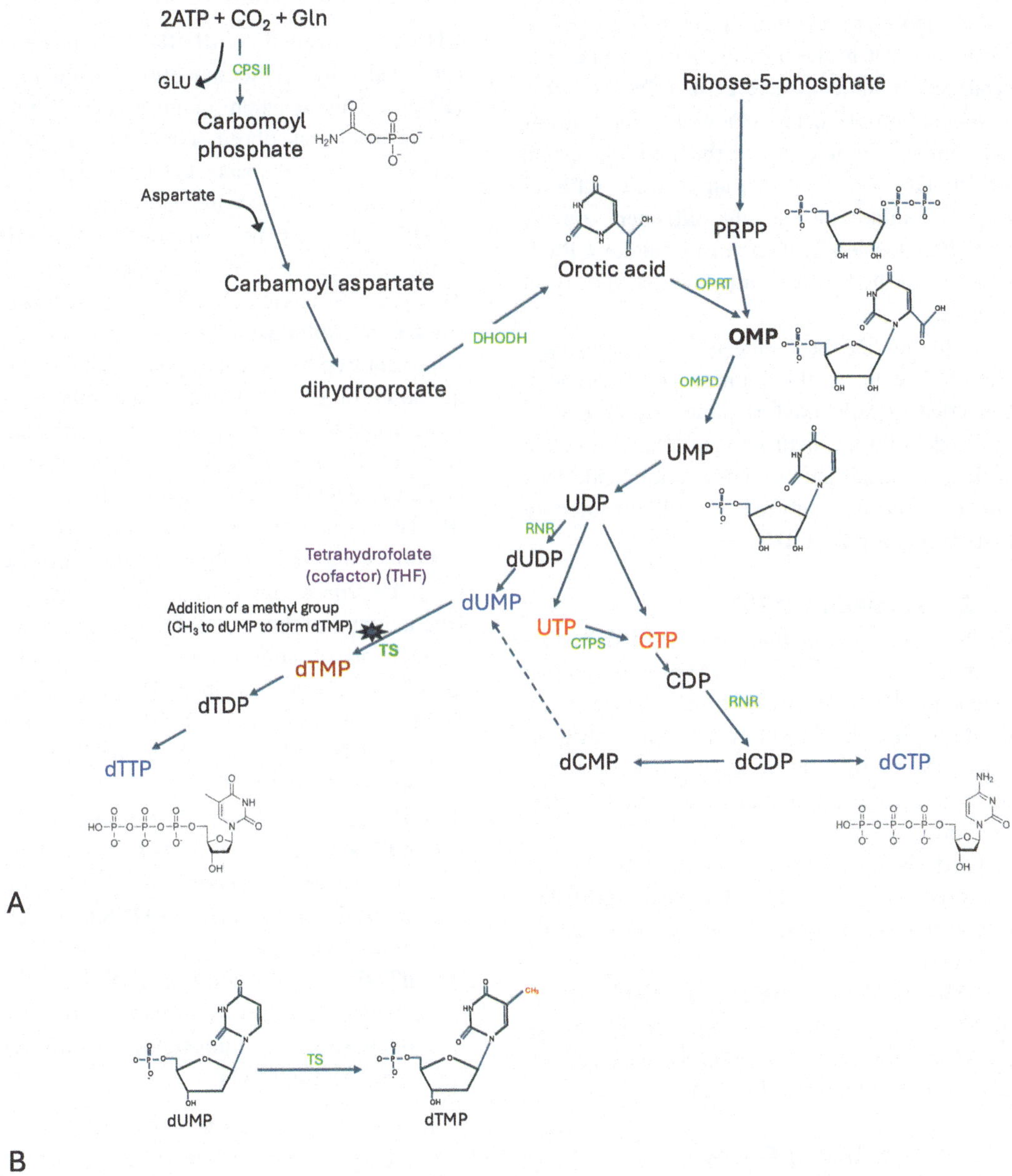

Fig. 15.7 (**a**) The de novo pyrimidine synthesis pathway. (**b**) The role for thymidylate synthase. This is a critical step, in which TS catalyzes the conversion of dUMP to dTMP. For this reason, TS is critical in pyrimidine synthesis. Abbreviations used in the figure: CPSII Carbamoyl phosphate synthetase II, OPRT Orotate phosphoribosyltransferase, DHODH Dihydroorotate dehydrogenase, OMPD Orotidine 5′-monophosphate decarboxylase, RNR Ribonucleotide reductase (RNR), CTPS CTP synthase, TS Thymidylate synthase

Carbamoyl phosphate is made from 2ATP and bicarbonate (derived from CO_2) and glutamine (Gln; as the nitrogen source, producing glutamate (Glu) as a byproduct), which are combined in a reaction catalyzed by **carbonyl phosphate synthase 2 (CPS II)**. Carbamoyl phosphate also has an important function in the urea cycle (see Chap. 14), but in the urea cycle, its synthesis is catalyzed by the enzyme **CPS I. CPS** uses free ammonia from the deamination of amino acids as the nitrogen source for carbamoyl phosphate (which has one amino group), while **CPS II** uses glutamine (gln) as the **nitrogen source. CPS II** is located in the cytosol, unlike **CPS I,** which is located in the mitochondria, mostly in liver cells.

In pyrimidine biosynthesis, once carbamoyl phosphate is made, it combines with Asp and is converted to carbamoyl-aspartate, which is then cyclized to form dihydroorotate. Dihydroorotate is then oxidized and becomes orotic acid. This last reaction is catalyzed by **dihydrootorate dehydrogenase DHODH**.

7.1.2 Formation of UDP

Orotic acid (a pyrimidine base) and **PRPP** now combine to form **OMP** (Orotidine monophosphate) in a reaction catalyzed by **Orotate phosphoribosyl transferase (OPRT).** Once OMP forms, the following reactions then occur to generate pyrimidines, based on this OMP structure:

OMP is converted to **UMP** in a reaction catalyzed by **OMP decarboxylase (OMPD).** This results in the loss of one carboxyl group from OMP.

UMP is then converted to **UDP**, which involves the addition of another phosphate group to UMP. This reaction is catalyzed by **UMP kinase (uridylate kinase).**

7.1.3 After UDP Is Formed, the Thymidine Pathway or the Cytidine Pathway Can Take Place

Next is a branch point. From here, **UDP** now has two choices: the **thymidine pathway** or the **cytidine pathway**.

- The Thymidine Pathway

In the thymidine pathway, UDP can be converted to dUDP, dUMP, dTMP, dTDP, and dTTP. Of these, **dTTP** is important for nucleic acids (DNA).

The thymidine pathway occurs as follows:

1. **UDP** is converted to **dUDP** in a reaction catalyzed by **ribonucleotide reductase (RNR)**. This reaction removes the 2′ OH group from the ribose, converting it to a deoxyribose. This marks the transition from ribonucleases to deoxyribonucleases.
2. **dUDP** can then be converted to **dUMP** through the removal of one phosphate group. This reaction is catalyzed by a **nucleotidase (dUDP phosphatase)**. This is a biologically important step because it discourages the accumulation of dUTP, which could otherwise form through phosphorylation of dUDP. Elevated dUTP would provide a risk because it could accidentally incorporate into DNA.
3. **dUMP** can now be converted to **dTMP** in a reaction catalyzed by **thymidylate synthase (TS). TS** requires the cofactor N^5, N^{10} methylene tetrahydrofolate (CH_2-THF), a derivative of THF, which comes from the folate cycle. CH_2-THF donates a methylene group to dUMP, converting it to dTMP. (During this reaction, CH_2-THF is oxidized and becomes dihydrofolate (DHF).)
4. **dTMP** is then converted to **dTTP** in two steps.
   (a) **dTMP** is converted to **dTDP** in a phosphorylation reaction catalyzed by **thymidylate kinase (TMPK),** which uses 1 ATP.
   (b) **dTDP** is converted to **dTTP**. This involves further phosphorylation by **nucleoside diphosphate kinase (NDPK)** using ATP.

Step 3 above, the conversion of **dUMP** to **dTMP** catalyzed by Thymidylate Synthase (**TS**), is important in pharmacy and medicine, and it is outlined in Fig. 15.7b. **TS** is essential for DNA synthesis since **dTTP** is a building block of DNA. **TS** has therefore become an important drug target for the treatment of several diseases because its inhibition can inhibit DNA replication. DNA replication is especially important in rapidly dividing cells, such as cancer cells [1]. The role of TS inhibition in disease treatment will be discussed more below and in other chapters.

- The Cytidine Pathway

The other branch point from UDP is the cytidine pathway.

In the cytidine pathway, UDP is converted to UTP, CTP, dCDP, and dCTP. Of these, UTP, CTP, and dCTP are important for nucleic acids (UTP and CTP for RNA and dCTP for DNA).

1. **UDP** is phosphorylated to become **UTP**. This process is catalyzed by **NDPK** and involves the transfer of a phosphate group from another nucleoside triphosphate, usually ATP, to UDP, forming **UTP**.
2. **UTP** is converted to **CTP** in a reaction catalyzed by **CTP synthase (CTPS)**. This reaction uses **glutamine (Gln)** as a **nitrogen donor**, and in the process, glutamine (Gln) is changed to **glutamate (Glu)**.
3. **CTP** is converted to **CDP** in a reaction catalyzed by **UTP diphosphatase.**
4. **CDP** can be converted to **dCTP** in two steps. First, **CDP** is converted to **dCDP** in a reaction catalyzed by **ribonucleotide reductase (RNR)**, which removes the 2′ OH group from CDP, converting it to dCDP, and in which NADPH, as the reducing agent, is oxidized to $NADP^+$. Next, dCDP is converted to dCTP in a phosphorylation reaction, catalyzed by **NDPK**.

   In summary, the above reactions create UTP, which can be used in RNA, and CTP and dCTP, which can be used in RNA and DNA, respectively.
5. Finally, **dCDP** can also become **dCMP** via dephosphorylation, and **dCMP** can become **dUMP** in a reaction catalyzed by **dCMP deaminase**. This facilitates crosstalk between the different pyrimidines because once dUMP is made, it can also form dTTP (via dTMP to dTDP to dTTP in the thymidine pathway).

7.2 Pyrimidine Salvage Pathway

Unlike in purine metabolism, in pyrimidine metabolism, the salvage pathway and interconversion pathways are generally grouped together and referred to as the salvage pathway. Salvage of the different nucleotides are described below.

7.2.1 Cytidine and Uridine Salvage

When RNA is broken down, cytidine (a nucleoside; base + sugar) may be released. This cytidine can be deaminated to uridine (also a nucleoside; base + sugar) by **cytidine deaminase (CDA)**. Uridine can then be phosphorylated to form **UMP** (a nucleotide; base plus sugar plus phosphate group) by **uridine–cytidine kinase (UCK)**. **UMP** can be processed through the pathways described above, where UMP is converted to UDP, leading to the synthesis of the pyrimidine nucleotides CTP and UTP, which can be used in RNA. Through further modification, UMP can also be made into dTTP, and CTP into dCTP, for use in DNA.

7.2.2 Deoxycytidine Salvage

When DNA is broken down, **deoxycytidine** (a nucleoside; base + sugar) may also be released. Deoxycytidine can be phosphorylated to yield **dCMP** (a nucleotide; base plus sugar plus phosphate group), in a reaction catalyzed by **deoxycytidine kinase (dCK)**. dCMP can then be further phosphorylated to **dCDP** and **dCTP**, or converted into dUMP, which leads to the synthesis of the main pyrimidine nucleotides in DNA: dCTP and dTTP. Through further reactions, these nucleotides can also be converted to CTP and UTP for use in RNA.

7.2.3 Salvage of Thymine

Thymine (free base) can also be released from DNA. Unlike cytidine or deoxycytidine, which are usually released as nucleosides, thymidine (the nucleoside) is rapidly interconverted with thymine (the free base) by **thymidine phosphorylase (TP)**. Thymine can be salvaged back into dTMP. It does so by combining with deoxyribose-1-phosphate to become thymidine (nucleoside; base + sugar) in the reverse reaction catalyzed by **thymidine phosphorylase (TP)**. This thymidine can be phosphorylated at the 5′ carbon of the deoxyribose sugar to become **dTMP** in a reaction catalyzed by **thymidine kinase (TK)**. As discussed above, dTMP can be further phosphorylated to dTDP and then dTTP for use in DNA synthesis.

8 Pyrimidine Catabolism

When DNA is degraded, cytosine and thymine can be released, and when RNA is degraded, cytosine and uracil can be released. If these products are not re-salvaged into other pyrimidine nucleotides, they can be broken down. In this process, cytosine is first deaminated to uracil. After this, the catabolism process starts, as follows, and is described in more detail in [2, 3, 5–7]:

1. **Uracil** gets broken down into dihydrouracil, and thymine gets broken down into dihydrothmine. Both reactions are catalyzed by **dihydropyrimidine dehydrogenase (DPYD)**, and they require NADPH as a reducing agent. This step, catalyzed by DPYD, is the rate-limiting step in pyrimidine catabolism.
2. Now, (2) the **dihydrouracil** gets processed to N-**carbamoyl-β-alanine** (catalyzed by **dihydropyrimidinase**) and **dihydrothrmine** gets processed to **N-carbamoyl-β-aminoisobutyrate** (also catalyzed by **dihydropyrimidinase**).
3. Next (3), **N-carbamoyl-β-alanine** gets processed to **β-alanine**, and **N-carbamoyl-β-aminoisobutyrate** gets processed to **β-aminoisobutyrate**. The enzyme that catalyzes both of these reactions is **beta-ureidopropionase (encoded by the UBP1 gene)**.

During the reaction catalyzed by **β-ureidopropionase**, ammonia (NH_3) and CO_2 are released [3, 5, 22].

9 Some Important Enzymes in Pyrimidine Metabolism and Human Disease Treatment

9.1 DHODH

The mitochondrial enzyme **DHODH** has an important role in the synthesis of **orotic acid** and thus is important for the synthesis of pyrimidines. Because pyrimidines are necessary for DNA synthesis, **DHODH** is considered a promising target for drugs designed to treat disorders characterized by excessive cell division (and hence DNA replication). Examples are cancer and automimmune disorders [23]. **Avara (leflunomide)** [24] is an example of a small-molecule DHODH inhibitor that is used to treat rheumatoid arthritis by suppressing T-cell proliferation. Preclinical studies have shown that it may also be promising for treating some kinds of cancer [25]. **Avara** is actually a prodrug, and it gets converted to its active metabolite in the liver. This metabolite is called **teriflunomide**. Teriflunomide inhibits **DHODH** by binding to it; it is a noncompetitive inhibitor.

Another FDA-approved DHODH inhibitor is **Aubagio (teriflunomide)** [26]. It was approved for the treatment of **multiple sclerosis (MS)**. MS is largely an autoimmune disease, and **Aubagio (teriflunomide)** relieves symptoms by reducing the proliferation of B and T cells [27]. **Aubagio (terifunomide)** is essentially the same drug as **Avara (leflunomide),** but it is not a prodrug. Instead, it is an active inhibitor that does not need to be metabolized further [28, 29].

9.2 Thymidylate Synthase (TS)

TS is an enzyme, discussed above, that is important for **dTTP** (thymidine triphosphate) synthesis and is therefore essential for DNA synthesis. It catalyzes the synthesis of dTMP from dUMP. dTMP, in turn, leads to dTTP synthesis through a series of sequential steps, as described above in the de novo pyrimidine synthesis pathway. Several different **TS inhibitors** are being used for cancer treatments by blocking DNA replication in rapidly dividing cells. These include **5-fluorouracil (5-FU),** which is also discussed in Chap. 6. **5-FU** is a pro-drug. It is converted in cells to its active metabolite **FdUMP** (5-fluoro-2-deoxyuridine monophosphate). FdUMP closely mimics dUMP, which is the natural substrate of TS. Thus, it is considered an analog of dUMP. When FdUMP binds to TS along with its cofactor (N^5,N^{10}-methylene-THF), it forms a stable complex that irreversibly inhibits **TS's** activity. **5-FU** thus functions as a **suicide inhibitor** of **TS**. It prevents TS from converting dUMP to dTMP, leading to a shortage of dTTP and thereby reducing or inhibiting DNA synthesis. **5-FU** is used to treat several types of cancer, especially those with high rates of cell proliferation and DNA synthesis [30].

9.3 DPYD and DPYD Deficiency

Dihydropyrimidine dehydrogenase (DPYD) is the enzyme necessary for the conversion of thymine to dihydrothymine and for the conversion of uracil to dihydrouracil, the rate-limiting steps in pyrimidine catabolism. DPYD deficiency is caused by mutations in the gene for **DPYD**. In this disorder, the lack of catabolism of uracil and thymidine leads to an excess of these molecules. As they accumulate, some of the uracil and thymidine are transported out of the cells. They can accumulate in places in the body, such as the blood, urine, and cerebrospinal fluid. While some patients have no symptoms, in some people, this build-up leads to symptoms such as neurological problems and delayed development, which can be severe. It is not completely understood how the build-up of these bases leads to the specific symptoms [31].

Individuals with **DPYD deficiency** also have abnormalities in the catabolism of anti-cancer drugs, particularly those of the **5-FU family. 5-FU** shares a similar pyrimidine ring structure to uracil. For this reason, **DPYD** catabolizes the 5-FU drug, and this prevents the drug from accumulating to levels that are too high. In people with **DPYD deficiency**, drugs such as 5-FU can accumulate to toxic levels, which is quite dangerous. Genetic testing before chemotherapy treatment can help determine whether an individual is deficient in DPYD and if so, alternative approaches to treatment need to be considered [31, 32].

10 Some Balanced Equations Associated with the Reactions Discussed in This Section

10.1 The Pentose Phosphate Pathway (PPP)

The Oxidative Stage of the PPP

1. Oxidation reaction:
   G6P (6C) + $NADP^+$ → 6-phosphoglucono-δ-lactone (6C) + NADPH + H^+
   Enzyme: **glucose-6-phosphate-dehydrogenase (G6PD)**
2. Hydrolysis:
   6-phosphoglucono-δ-lactone + H_2O → 6-phosphogluconate (6C) + H^+
   Enzyme: **lactonase**
3. Oxidative decarboxylation
   6-phosphogluconate (6C) + $NADP^+$ → ribulose-5-phosphate (Ru5P, 5C) + CO_2 + NADPH
   Enzyme: **6-phosphogluconate-dehydrogenase (6PGD)**
4. ribulose-5-phosphate (Ru5P, 5C) ⇌ ribose-5-phosphate (R5P, 5C)
   Enzyme: **ribose-5-phosphate-isomerase (RPI)**
   ribulose-5-phosphate (Ru5P, 5C) ⇌ xylulose-5-phosphate (5C)
   Enzyme: **ribulose-5-phosphate-3-epimerase (RPE)**
   The Nonoxidative Stage of the PPP
5. Ribose-5-phosphate (R5P, 5C) + xylulose-5-phosphate (5C) ⇌ sedoheptulose-7-phosphate (7C) + glyceraldehyde-3-phosphate (GAP, 3C)
   Enzyme: **transketolase (TKT)**
6. Glyceraldehyde-3-phosphate (GAP, 3C) + Sedoheptulose-7-phosphate (7C) ⇌ fructose-6-phosphate (F6P, 6C) + erythrose-4-phosphate (4C). Enzyme: **transaldolase (TALDO)**

7. Erythrose-4-phosphate (4C) + xylulose-5-phosphate (5C) ⇌ glyceraldehyde-3-phosphate (GAP, 3C) + F6P (6C) Enzyme: **transketolase (TKT)**
   Ribose-5-Phosphate to PRPP
   R5P + ATP → PRPP + AMP. Enzyme: **PRPP-synthase**

10.2 Purine Synthesis

De Novo Pathway

- **The Steps Between PRPP and IMP**
  1. PRPP to 5-PRA:
     PRPP + glutamine + H_2O → 5-PRA + glutamate (Glu) + PPi
     Enzyme: **amidophosphoribosyltransferase (GPAT)**
  2. 5-PRA to GAR:
     5-PRA + glycine + ATP → GAR + ADP + Pi. Enzyme: GAR synthetase
  3. GAR to FGAR:
     GAR + N10-formyl-THF → FGAR + THF. Enzyme: **GAR transformylase**
  4. FGAR to FGAM:
     FGAR + glutamine (Gln) + ATP + H_2O → FGAM + glutamate + ADP + Pi
     Enzyme: **FGAM synthetase**
  5. FGAM to AIR:
     FGAM + ATP → AIR + ADP + Pi. Enzyme: **AIR synthetase**
  6. AIR to CAIR:
     AIR + CO2 → CAIR. Enzyme: AIR carboxylase
  7. CAIR to SAICAR:
     CAIR + Aspartate + ATP → SAICAR + ADP + Pi. Enzyme: **SAICAR synthetase**
  8. SAICAR to AICAR:
     SAICAR → AICAR + Fumarate. Enzyme: **adenylsuccinate lyase**
  9. AICAR to FAICAR:
     AICAR + **N10-formyl-THF** → FAICAR + THF. Enzyme: **AICAR transformylase**
  10. FAICAR to IMP:
      FAICAR → IMP + H_2O. Enzyme: **IMP cyclohydrolase**
- **IMP to Purines**
  - IMP to AMP:
    1. IMP + Asp + GTP → adenylosuccinate + GDP + Pi
       Enzyme: **adenylosuccinate synthetase (ADSS)**
    2. Adenylosuccinate → AMP + fumarate
       Enzyme: **adenylosuccinate lyase (ADSL)**
  - IMP to GMP:
    1. IMP + NAD^+ + H_2O → XMP + NADH + H^+. Enzyme: **IMP dehydrogenase (IMPDH)**
    2. XMP + Gln + ATP + H_2O → GMP + Glu + AMP + PPi. Enzyme: **GMP synthetase (GMPS)**
  - Conversion of Monophosphates to Diphosphates
    1. AMP + ATP ⇌ 2ADP. Enzyme: **adenylate kinase**
    2. GMP + ATP ⇌ GDP + ADP Enzyme: **guanylate kinase**
  - Diphosphates to Triphosphates
    Formation of NTPs from NDPs
    $N_1DP + N_2TP \rightleftharpoons N_1TP + N_2DP$, Enzyme: **nucleoside diphosphate kinase (NDPK)** where the resulting N1TP is ATP or GTP, and the phosphate donor (N2TP) can be any nucleotide, but it is often ATP.
  - Ribonucleotides to Deoxyribonucleotides
    1. ADP + NADPH + H^+ → dADP + $NADP^+$ + H_2O. Enzyme: **ribonucleotide reductase**
       GDP + NADPH + H^+ → dGDP + $NADP^+$ + H_2O. Enzyme: **ribonucleotide reductase**
    2. **dADP + ATP ⇌ dATP + ADP.** Enzyme: **nucleoside diphosphate kinase (NDPK)**
       dGDP + ATP ⇌ dGTP + ADP. Enzyme: **nucleoside diphosphate kinase (NDPK)**

10.3 Purine Interconversion Pathways

- **AMP to IMP:**
  AMP + H_2O → IMP + NH_3. **Enzyme: AMP deaminase (AMPD)**
- **GMP to IMP:**
  GMP + NADPH + H^+ → IMP + NH_3 + $NADP^+$. **Enzyme: GMP reductase (GMPR)**

10.4 Purine Synthesis by the Salvage Pathway

- **Guanine to GMP**
  Guanine + PRPP → GMP + Ppi. Enzyme: **hypoxanthine-guanine phosphoribosyltransferase (HGPRT)**
- **Adenine to AMP:**
  Adenine + PRPP → AMP + PPi. Enzyme: Adenine phosphoribosyltransferase (APRT)

10.5 Catabolism of Purines

- **Catabolism of AMP**
  1. AMP to IMP:
     AMP + H_2O → IMP + NH_3 Enzyme: AMP deaminase
  2. IMP to inosine:
     IMP + H_2O → Inosine + Pi. Enzyme: 5′-nuclotidase
  3. Inosine to hypoxanthine:
     Inosine + Pi → hypoxanthine + ribose-1-phosphat.
     Enzyme: Purine nucleoside phosphorylase (PNP)
  4. Hypoxanthine to xanthine:
     Hypoxanthine + O_2 + H_2O → xanthine + H_2O_2 Enzyme. Xanthine oxidase (XO)
  5. Xanthine to uric acid:
     Xanthine + O_2 + H_2O → Uric acid + H_2O_2 Enzyme. Xanthine oxidase (XO)

  Alternative Step 4: Hypoxanthine to IMP (salvage)
  Hypoxanthine + PRPP → IMP + PPi. Enzyme. HGPRT
- **Catabolism of *AMP*: *Alternative* Pathway, Using ADA**
  1. AMP to adenosine:
     AMP + H_2O → Adenosine + Pi.
     Enzyme. Hypoxanthine-guanine phosphoribosyltransferase (HGPRT)
  2. Adenosine to inosine. Adenosine + H2O → Inosine + NH_3
     Enzyme. Adenosine deaminase (ADA)
  3. Inosine to hypoxanthine, Xanthine to uric acid (same as above)
     Inosine + Pi ⇌ hypoxanthine + ribose-1-phosphat.
     Enzyme. Purine nucleoside phosphorylase (PNP)
  4. Hypoxanthine to xanthine:
     Hypoxanthine + O_2 + H_2O → xanthine + H_2O_2 Enzyme. Xanthine oxidase (XO)
  5. Xanthine to uric acid:
     Xanthine + O_2 + H_2O → Uric acid + H_2O_2 Enzyme. Xanthine oxidase (XO)

  Alternative Step 4: Hypoxanthine to IMP (Salvage)
  Hypoxanthine + PRPP → IMP + PPi.
  Enzyme. Hypoxanthine-guanine phosphoribosyltransferase (HGPRT)
- **Catabolism of dAMP**
  1. dAMP to dAdenosine:
     dAMP + H_2O → dAdenosine + Pi. Enzyme. 5′-nucleosidase
  2. dAdenosine to dInosine.:
     dAdenosine + H_2O → dInosine + NH_3. Enzyme. Adenosine deaminase **(ADA)**
  3. dInosine to hypoxanthine.
     dInosine + Pi ⇌ hypoxanthine + Deoxyribose-1-phosphate.
     Enzyme. Purine nucleoside phosphorylase (PNP)
  4. Hypoxanthine to xanthine (4 and 5 are the same as above)
     Hypoxanthine + O_2 + H_2O → xanthine + H_2O_2 Enzyme. Xanthine oxidase (XO)
  5. Xanthine to uric acid:
     Xanthine + O_2 + H_2O → Uric acid + H_2O_2 Enzyme. Xanthine oxidase (XO)
- **Catabolism of GMP**
  1. GMP to guanosine:
     GMP + H_2O → guanosine + Pi. Enzyme. 5′-nucleotidase (5′-NT)
  2. Guanosine to guanine:
     Guanosine + Pi ⇌ Guanine + Ribose-1-phosphate.
     Enzyme. Purine nucleoside phosphorylase (PNP)
  3. Guanine to xanthine:
     Guanine + H_2O → Xanthine + NH_3. Enzyme. Guanine deaminase (GDA)
  4. Xanthine to uric acid:
     Xanthine + O_2 + H_2O → Uric acid + H_2O_2 Enzyme. Xanthine oxidase (XO)

- **Catabolism of dGMP**
  1. dGMP to dGuanosine:
     dGMP + H_2O → dGuanosine + Pi. Enzyme. 5′-nucleotidase
  2. dGuanosine to dGuanine:
     dGuanosine + Pi ⇌ dGuanine + deoxyribose-1-phosphate. Enzyme.
     Purine nucleoside phosphorylase (PNP)
  3. dGuanine to xanthine:
     dGuanine + H_2O → xanthine + NH_3 Enzyme. **guanine deaminase**
  4. Xanthine to uric acid, as catalyzed in the above reactions:
     Xanthine + O_2 + H_2O → Uric acid + H_2O_2 Enzyme. Xanthine oxidase (XO)
- **Catabolism of IMP**
  1. IMP to inosine:
     IMP + H_2O → Inosine + Pi. Enzyme. **5′-nucleotidase**
  2. Inosine to hypoxanthine:
     Inosine + Pi ⇌ hypoxanthine + ribose-1-phosphate.
     Enzyme. Purine nucleoside phosphorylase (PNP)
  3. hypoxanthine to xanthine:
     Hypoxanthine + O_2 + H_2O → xanthine + H_2O_2 Enzyme. Xanthine oxidase (XO)
  4. xanthine to uric acid:
     Xanthine + O_2 + H_2O → Uric acid + H_2O_2 Enzyme. Xanthine oxidase (XO)

10.6 Synthesis of Pyrimidines

- **Synthesis of Orotic Acid**
  1. Synthesis of carbamoyl phosphate:
     2ATP + HCO_3^- + Glutamine + H_2O → Carbamoyl phosphate +2ADP + glutamate
     Enzyme: **carbamoyl phosphate synthetase II (CPS II)**
  2. Formation of carbamoyl aspartate:
     Carbamoyl phosphate + Asp → carbamoyl aspartate + Pi.
     Enzyme: **aspartate transcarbamoylase (ATCase)**
  3. Cyclization to dihydroorotate:
     Carbamoyl aspartate → **dihydroorotate. Enzyme: Dihydroorotase (DHOase)**
  4. Oxidation to orotic acid (orotate):
     Dihydroorotate + CoQ (ubiquinone) → orotate (orotic acid) $CoQH_2$
     Enzyme: **dihydroorotate dehydrogenase (DHODH)**
- **Formation of UDP:**
  – Formation of OMP:
  1. Orotic acid to OMP:
     Orotic acid (orotate) + PRPP → OMP + Ppi
     Enzyme: **orotate phosphoribosyltransferase (OPRT)**
  2. OMP to UMP:
     OMP → UMP + CO_2 Enzyme: **orotidine 5′-phosphate decarboxylase** (OMPD)
  3. UMP to **UDP**:
     UMP + ATP → UDP + ADP Enzyme: **UMP kinase**

  – The thymidine pathway
  1. UDP to **dUDP:**
     UDP + NADPH + H+ → dUDP + NADP+ + H2O. Enzyme: **ribonucleotide reductase (RNR)**
  2. dUDP to dUTP to **dUMP** (important for the prevention of dUTP accumulation)
     (a) **dUDP + ATP → dUTP + ADP**
         Enzyme: **nucleoside diphosphate kinase (NDPK)**
     (b) **dUTP + H_2O → dUMP + PPi**
         Enzyme: **dUTPase**
  3. dUMP to **dTMP**
     dUMP + N^5, N^{10} methylene tetrahydrofolate-THF → dTMP + DHF (dihydrofolate)
     Enzyme: **thymidylate synthase (TS)**
  4. dTMP → **dTTP**
     (a) dTMP → **dTDP**:
         dTMP + ATP → dTDP + ADP. Enzyme: **thymidylate kinase (TMPK)**
     (b) dTDP → **dTTP**

dTDP + ATP → dTTP + ADP. Enzyme: **nucleoside diphosphate kinase (NDPK)**

- **The Cytidine Pathway**
  1. UDP to **UTP**:
     UDP + ATP → UTP + ADP. Enzyme: **nucleoside diphosphate kinase (NDPK)**
  2. UTP to **CTP**:
     UTP + glutamine + ATP + H_2O → CTP + glutamate (Glu) + ADP + Pi. Enzyme: **CTP synthase**
  3. CTP to CDP:
     CTP + ADP ⇌ CDP + ATP. Enzyme: **nucleoside diphosphate kinase (NDPK)**
  4. CDP to dCTP:
     (a) CDP to **dCDP**:
     CDP + NADPH + H^+ → **dCDP** + $NADP^+$ + H_2O. Enzyme: **ribonucleotide reductase (RNR)**
     (b) dCDP → **dCTP**:
     dCDP + ATP → **dCTP** + ADP. Enzyme: **nucleoside diphosphate kinase (NDPK)**
  5. **dCDP to dCMP to dUMP:**
     (a) Dephosphorylation of dCDP to **dCMP:**
     dCDP + H_2O → dCMP + Pi. Enzyme: **a nucleotidase**
     (b) dCMP to dUMP:
     dCMP + H_2O → dUMP + NH_3. Enzyme. **dCMP deaminase**

10.7 Pyrimidine Salvage Pathway

- **Cytidine and Uridine Salvage**
  1. Cytidine to uridine:
     Cytidine + H_2O → Uridine + NH_3. Enzyme: **cytidine deaminase**
  2. Uridine to form UMP:
     Uridine + ATP → UMP + ADP. Enzyme: **uridine-cytidine kinase (UCK)**
- **Deoxycytidine Salvage**
  1. Deoxycytidine to dCMP:
     Deoxycytidine + ATP → dCMP + ADP. Enzyme: **deoxycytidine kinase (dCK)**
  2. dCMP to dCDP and dCTP or converted into dUMP:
     (a) dCMP + ATP → **dCDP** + ADP. Enzyme: **dCMP kinase**
     (b) dCDP + ATP ⇌ **dCTP** + ADP. enzyme: **nucleoside diphosphate kinase (NDPK)**
     (c) dCMP + H_2O → **dUMP** + NH_3. **dCMP deaminase**
- **Salvage of Thymine**
  1. Thymine to thymidine:
     Thymine +2-deoxy-D-ribose-1-phosphate ⇌ **thymidine** + Pi.
     Enzyme: thymidine phosphorylase (TP)
  2. Thymidine to dTMP:
     Thymidine + ATP → dTMP + ADP. Enzyme: **thymidine kinase (TK)**

10.8 Pyrimidine Catabolism

Pre-step. Deamination of cytosine:
Cytosine + H_2O → uracil + NH_3. Enzyme: **cytosine deaminase**

1. Reduction of uracil and thymine:
   (a) Uracil to dihydrouracil:
   Uracil + NADPH + H^+ → dihydrouracil + $NADP^+$.
   Enzyme: **dihydropyrimidine dehydrogenase (DPYD)**
   (b) Thymine to Dihydrotymine:
   Thymine + NADPH + H^+ → Dihydrothymine + $NADP^+$.
   Enzyme: **dihydropyrimidine dehydrogenase (DPYD)**

2a. Dihydrouracil + H_2O → N-carbamoyl-β-alanine. Enzyme: **dihydropyrimidinase**

2b. Dihydrothymine + H_2O → N-carbamoyl-β-aminoisobutyrate.
Enzyme: **Dihydropyrimidinase**

3a. N-carbamoyl-β-alanine + H_2O → β-alanine + CO_2 + NH_3.
Enzyme: **β-Ureidopropionase (UPB1)**

3b. N-carbamoyl-β-aminoisobutyrate + H_2O → -β-aminoisobutyrate + CO_2 + NH_3
Enzyme: **β-Ureidopropionase (UPB1)**

FDA-Approved Drugs Mentioned in This Chapter

Brand name	Generic name	First FDA approval year	Main conditions	Main targets
Adrucil/ Efudex	5-fluorouracil	1962	Colorectal, breast, gastric, skin cancers	Thymidylate synthase
Aubagio	Teriflunomide	2012	Multiple sclerosis	DHODH
Arava	Leflunomide	1998	Rheumatoid arthritis	DHODH
Zyloprim	Allopurinol	1966	Gout, tumor lysis syndrome	Xanthine oxidase

Definitions

Adenosine deaminase (ADA) The enzyme that catalyzes the deamination of adenosine to inosine, and deoxyadenosine to deoxyinosine, during purine catabolism. Its deficiency leads to SCID (Severe Combined Immune Deficiency).

Allopurinol A drug used to treat gout. It inhibits xanthine oxidase, reducing uric acid production.

Dihydroorotate dehydrogenase (DHODH) Mitochondrial enzyme in pyrimidine synthesis that converts dihydroorotate to orotic acid.

Deoxythymidine monophosphate (dTMP) A DNA nucleotide synthesized from dUMP via thymidylate synthase.

Deoxyuridine monophosphate (dUMP) Precursor to dTMP in nucleotide synthesis.

Folate (folic acid) A B-vitamin necessary for one-carbon transfers, including those in nucleotide biosynthesis.

Glycinamide ribonucleotide (GAR) An intermediate in purine synthesis formed after glycine is added to PRPP.

Hypoxanthine-guanine phosphoribosyltransferase (HGPRT) An enzyme that is important in the purine salvage pathway.

Hypoxanthine A purine base that serves as an intermediate in nucleotide salvage pathways.

N^5,N^{10}-Methylene THF A derivative of THF, and part of the folate cycle. It donates a methyl group in thymidylate synthesis.

Nucleoside A base (purine or pyrimidine) attached to a sugar (ribose or deoxyribose), without a phosphate group.

Nucleotide A nucleoside with one or more phosphate groups, building blocks of DNA and RNA.

Orotic acid A pyrimidine precursor in the de novo synthesis pathway.

Phosphoribosyl pyrophosphate (PRPP) A 5-carbon sugar. It is a high-energy ribose phosphate donor used for both purine nucleotide and pyrimidine nucleotide synthesis.

Ribonucleotide reductase (RNR) An enzyme that catalyzes the reduction of ribonucleotides to deoxyribonucleotides, providing the nucleotides for DNA.

Salvage pathway A metabolic route that recycles free purines or pyrimidines into nucleotides.

Severe combined immunodeficiency (SCID) An inborn error of metabolism (IEM), which is associated with the accumulation of Adenosine and dAdenosine. It causes severely compromised immune function.

Thymidylate synthase Enzyme that catalyzes the conversion of dUMP to dTMP using methylated folate.

Uric acid The end product of purine metabolism in humans; excess levels of it can cause gout.

Xanthine oxidase An enzyme that converts hypoxanthine to xanthine and then to uric acid during purine degradation.

Study Questions

Multiple-Choice Questions

1. Which enzyme catalyzes the conversion of dUMP to dTMP?
   A. Ribonucleotide reductase
   B. Thymidine kinase
   C. Thymidylate synthase
   D. CTP synthase

2. In the salvage pathway for purine synthesis, what molecule is the main sugar-phosphate donor?
   A. ATP
   B. NADPH
   C. PRPP
   D. UMP
3. Which protein is inactive or deficient in severe combined immunodeficiency (SCID)?
   A. The SCID enzyme
   B. Immunoglobulin protein
   C. Adenosine deaminase
   D. xanthine oxidase
4. What molecule is produced by the DHODH catalyzed step in de novo pyrimidine synthesis?
   A. DNA repair enzyme
   B. Thymidine kinase
   C. Orotic acid
   D. dUMP

Fill-in-the-Blank Questions

5. PRPP stands for __________.
6. The rate-limiting enzyme in pyrimidine catabolism is __________.
7. In the thymidine synthesis pathway, the cofactor required by thymidylate synthase is __________.

Short-Answer Question

8. Briefly describe a major difference between purine and pyrimidine nucleotide synthesis.

Matching Question

9. Match each enzyme to its correct function:

   Enzyme
   A. Ribonucleotide reductase
   B. HGPRT
   C. Thymidine kinase
   D. Xanthine oxidase

   Function
   1. Converts ribonucleotides to deoxyribonucleotides
   2. Salvages guanine to GMP
   3. Converts thymidine to dTMP
   4. Converts hypoxanthine to uric acid

Answer Key

Multiple-Choice Questions

1. **Answer**: C
2. **Answer**: C
3. **Answer**: C
4. **Answer**: C

Fill-in-the-Blank Questions

5. **Answer**: Phosphoribosyl pyrophosphate
6. **Answer**: Dihydropyrimidine dehydrogenase (DPYD)
7. **Answer**: N^5,N^{10}-methylene tetrahydrofolate (CH_2-THF)

Short-Answer Question

8. **Answer**: In purine synthesis, the base is built stepwise directly onto PRPP. In pyrimidine synthesis, the base (orotic acid) is synthesized first, then attached to PRPP.

Matching Question

9. **Match each enzyme to its correct function:**
   Answer:
   A → 1, B → 2, C → 3, D → 4

References

1. Rodwell VW, Bender DA, Kennely PJ, Weil PA, McGuinness OP. Harper's illustrated biochemistry. 32nd ed. New York: McGraw-Hill Education/ Medical; 2022.
2. Berg JM, Tymoczko JL, Gatto GJ Jr, Stryer L. Biochemistry. 9th ed. New York: MacMillan Learning; 2019.
3. Chandel NS. Navigating metabolism. 2nd ed. Cold Spring Harbor: Cold Spring Harbor Laboratory Press; 2020.

4. TeSlaa T, Ralser M, Fan J, Rabinowitz JD. The pentose phosphate pathway in health and disease. Nat Metab. 2023;5(8):1275–89. https://doi.org/10.1038/s42255-023-00863-2.
5. Chandel NS. Nucleotide metabolism. Cold Spring Harb Perspect Biol. 2021;13(7). https://doi.org/10.1101/cshperspect.a040592.
6. Salway JG. Metabolism at a glance. Hoboken: Wiley-Blackwell; 2017.
7. Lane AN, Fan TW. Regulation of mammalian nucleotide metabolism and biosynthesis. Nucleic Acids Res. 2015;43(4):2466–85. https://doi.org/10.1093/nar/gkv047.
8. Pareek V, Pedley AM, Benkovic SJ. Human de novo purine biosynthesis. Crit Rev Biochem Mol Biol. 2021;56(1):1–16. https://doi.org/10.1080/10409238.2020.1832438.
9. Sauer AV, Brigida I, Carriglio N, Aiuti A. Autoimmune dysregulation and purine metabolism in adenosine deaminase deficiency. Front Immunol. 2012;3:265. https://doi.org/10.3389/fimmu.2012.00265.
10. Whitmore KV, Gaspar HB. Adenosine deaminase deficiency - more than just an immunodeficiency. Front Immunol. 2016;7:314. https://doi.org/10.3389/fimmu.2016.00314.
11. Haddad E, Hoenig M. Hematopoietic stem cell transplantation for severe combined immunodeficiency (SCID). Front Pediatr. 2019;7:481. https://doi.org/10.3389/fped.2019.00481.
12. Polmar SH, Stern RC, Schwartz AL, Wetzler EM, Chase PA, Hirschhorn R. Enzyme replacement therapy for adenosine deaminase deficiency and severe combined immunodeficiency. N Engl J Med. 1976;295(24):1337–43. https://doi.org/10.1056/nejm197612092952402.
13. European Medicines Agency. Strimvelis 2025. Available from: https://www.ema.europa.eu/en/medicines/human/EPAR/strimvelis.
14. Eshghi S, Mousakhan Bakhtiari M, Behfar M, Izadi E, Naji P, Jafari L, et al. Viral-based gene therapy clinical trials for immune deficiencies and blood disorders from 2013 until 2023 - an overview. Regen Ther. 2025;28:262–79. https://doi.org/10.1016/j.reth.2024.12.007.
15. Gaspar HB, Aiuti A, Porta F, Candotti F, Hershfield MS, Notarangelo LD. How I treat ADA deficiency. Blood. 2009;114(17):3524–32. https://doi.org/10.1182/blood-2009-06-189209.
16. Zheng X, Huang CH, Yan S, Rong MD. Advances and applications of genome-edited animal models for severe combined immunodeficiency. Zool Res. 2025;46(1):249–60. https://doi.org/10.24272/j.issn.2095-8137.2024.195.
17. Kohn DB. Gene therapy for adenosine deaminase severe combined immune deficiency-an unexpected journey of four decades. Immunol Rev. 2024;322(1):148–56. https://doi.org/10.1111/imr.13293.
18. Secord E, Hartog NL. Review of treatment for adenosine deaminase deficiency (ADA) severe combined immunodeficiency (SCID). Ther Clin Risk Manag. 2022;18:939–44. https://doi.org/10.2147/tcrm.S350762.
19. FDA. Zyloprim(Allopurinol) Approval Overview 1966. Available from: https://www.accessdata.fda.gov/scripts/cder/daf/index.cfm?event=overview.process&ApplNo=016084.
20. Torres RJ, Puig JG. Hypoxanthine-guanine phosophoribosyltransferase (HPRT) deficiency: Lesch-Nyhan syndrome. Orphanet J Rare Dis. 2007;2:48. https://doi.org/10.1186/1750-1172-2-48.
21. Nanagiri A, Shabbir N. Lesch-Nyhan syndrome. In: StatPearls. Treasure Island: StatPearls Publishing; 2025.
22. Jakubowski, H., and Flatt, P.M. (n.d.) Pathways of amino acid degradation. 18.5. Available from: https://bio.libretexts.org/Bookshelves/Biochemistry/Fundamentals_of_Biochemistry_(Jakubowski_and_Flatt)/02%3A_Unit_II-_Bioenergetics_and_Metabolism/18%3A_Nitrogen_-_Amino_Acid_Catabolism/18.05%3A_Pathways_of_Amino_Acid_Degradation
23. Sykes DB. The emergence of dihydroorotate dehydrogenase (DHODH) as a therapeutic target in acute myeloid leukemia. Expert Opin Ther Targets. 2018;22(11):893–8. https://doi.org/10.1080/14728222.2018.1536748.
24. FDA. Avara(Leflunomide) Approval Overview 1998. Available from: https://www.accessdata.fda.gov/scripts/cder/daf/index.cfm?event=overview.process&ApplNo=20905.
25. Niwata C, Nakagawa T, Naruse T, Sakuma M, Yamakado N, Akagi M, et al. Anticancer effect of the antirheumatic drug leflunomide on oral squamous cell carcinoma by the inhibition of tumor angiogenesis. Discov Oncol. 2025;16(1):53. https://doi.org/10.1007/s12672-025-01763-5.
26. FDA. Aubagio(Teriflunomide) approval overview 2012. Available from: https://www.accessdata.fda.gov/scripts/cder/daf/index.cfm?event=overview.process&ApplNo=202992.
27. Tilly G, Cadoux M, Garcia A, Morille J, Wiertlewski S, Pecqueur C, et al. Teriflunomide treatment of multiple sclerosis selectively modulates CD8 memory T cells. Front Immunol. 2021;12:730342. https://doi.org/10.3389/fimmu.2021.730342.
28. Bar-Or A, Pachner A, Menguy-Vacheron F, Kaplan J, Wiendl H. Teriflunomide and its mechanism of action in multiple sclerosis. Drugs. 2014;74(6):659–74. https://doi.org/10.1007/s40265-014-0212-x.
29. DrugBank. Teriflunomide 2025. Available from: https://go.drugbank.com/drugs/DB08880?utm.
30. Bijnsdorp IV, Comijn EM, Padron JM, Gmeiner WH, Peters GJ. Mechanisms of action of FdUMP[10]: metabolite activation and thymidylate synthase inhibition. Oncol Rep. 2007;18(1):287–91. https://doi.org/10.3892/or.18.1.287.

31. MedlinePlus. Dihydropyrimidine dehydrogenase deficiency 2015. Available from: https://medlineplus.gov/genetics/condition/dihydropyrimidine-dehydrogenase-deficiency/#causes.
32. Amstutz U, Henricks LM, Offer SM, Barbarino J, Schellens JHM, Swen JJ, et al. Clinical pharmacogenetics implementation consortium (CPIC) guideline for dihydropyrimidine dehydrogenase genotype and fluoropyrimidine dosing: 2017 update. Clin Pharmacol Ther. 2018;103(2):210–6. https://doi.org/10.1002/cpt.911.

Folate-Mediated One-Carbon Metabolism

16

Abstract

One-carbon metabolism is a network of pathways that involves the transfer of one-carbon units to various molecules in metabolic pathways. These one-carbon units are important for many metabolic pathways, such as those that involve nucleotide synthesis, amino acid metabolism, and methylation reactions. A major component of one-carbon metabolism and the main focus of this chapter is the folate pathway, a set of interconnected metabolic reactions that involve folate and its derivatives. Folate is the natural form of vitamin B9. The reactions in the folate pathway are involved in folate absorption, reduction to tetrahydrofolate (THF), one-carbon metabolism, and other processes. Within the folate pathway is a subset of reactions that form the folate cycle and other interconnected cycles. The core components of the folate cycle consist of a loop where THF (a folate derivative) is continuously interconverted and regenerated to sustain the various reactions within the folate pathway. An important cycle that is interconnected with the folate cycle is the activated methionine cycle (also called the methionine/SAM cycle). Together, the folate cycle and the activated methionine cycle have important roles in one-carbon transfer. Since folate is central to these pathways, folate deficiency can lead to serious problems. Likewise, vitamin B12 deficiency can lead to related health problems. These are discussed, as are drugs that target components of the folate pathway, such as certain chemotherapeutic agents.

Keywords

Folate · Folate cycle · One-carbon transfers · Vitamin B12 · Inhibitors of folate metabolism

Abbreviations

5-FU	5-Fluorouracil
B_{12}	Vitamin B12 (cobalamin)
DHFR	Dihydrofolate reductase
DM4	N2′-deacetyl-N2′-(4-methyl-N-acetyl-γ-mercapto-1-methylbutyryl)-maytansine
DNMT	DNA methyltransferase
dTMP	Deoxythymidine monophosphate
dUMP	Deoxyuridine monophosphate
FRα	Folate receptor alpha
MTHFR	Methylenetetrahydrofolate reductase
SHMT	Serine hydroxymethyltransferase
TS	Thymidylate synthase

What You Will Learn

- The importance of folate, the natural form of vitamin B9
- Dietary sources of folate

A. Minden, *Biochemistry, Cell and Molecular Biology for Pharmacists and Health Professionals*,
https://doi.org/10.1007/978-3-032-20529-2_16

- Folate metabolism, including the core folate cycle, the activated methionine cycle, and nucleotide synthesis pathways
- The role of these pathways in one-carbon transfer reactions
- Disorders related to the folate pathway, folate deficiency, vitamin B12 deficiency, and the folate trap
- Drugs that target components of the folate pathway, such as thymidylate synthase inhibitors, folate receptor antagonists, and other chemotherapeutic drugs

1 Introduction to the Folate Pathway, Interconnected Cycles, and One-Carbon Metabolism

This chapter covers the topic of **one-carbon metabolism** that is a system that involves the **transfer of one-carbon units** to various molecules involved in metabolic pathways [1–6]. These one-carbon units include **formyl** (–CHO), **methylene** (–CH_2–), **methenyl** (=CH–), or **methyl** (–CH_3) groups (see Fig. 16.1). They are important for many metabolic pathways, such as those that involve nucleotide synthesis, amino acid metabolism, and methylation reactions. A major component of one-carbon metabolism is the folate pathway. The folate pathway is a set of interconnected pathways that encompasses metabolic reactions involving folate and its derivatives. Folate is the natural form of vitamin B9. The reactions in the folate pathway have roles in the absorption of folate, reduction of folate to tetrahydrofolate (THF), one-carbon metabolism, and other processes. Within the folate pathway is a subset of reactions that form the folate cycle and other interconnected cycles. The core components of the folate cycle consist of a loop where THF (a folate derivative) is continuously interconverted and regenerated to sustain the various reactions within the folate pathway [4, 7, 8]. An important cycle that is interconnected with the folate cycle is the activated methionine cycle (also called the methionine/SAM cycle). Together, the folate cycle and the activated methionine cycle have important roles in one-carbon transfer. In addition to these cycles, folate derivatives also provide one-carbon units for biosynthetic processes. For example, N^{10}-formyl-THF, a derivative of THF, provides one-carbon units at two steps in the de novo purine synthesis pathway.

Many of the molecules discussed in this section function as substrates, products, enzymes, and coenzymes in various enzymatic reactions. When discussing enzymes, a substrate is a molecule that is chemically transformed during a reaction, a product is the molecule that results from that transformation, and a coenzyme is an organic molecule that has a role in assisting the enzyme in carrying out the reaction it catalyzes (see Chap. 5). In the folate pathway, the coenzymes often function by transferring chemical groups, specifically one-carbon groups. For example, we will discuss the molecule N^5,N^{10}-methylene-THF (CH_2-THF), a derivative of folate, below. This molecule acts as both a substrate and a coenzyme in a reaction catalyzed by the enzyme **thymidylate synthase (TS)**, which catalyzes the conversion of dUMP to dTMP. N^5,N^{10}-methylene-THF (CH_2-THF) is a substrate because its structure changes in the reaction (it is converted to dihydrofolate (DHF)), and it is a coenzyme because it donates a methylene group to dUMP to convert it to dTMP. This type of dual function is a common theme in the folate pathway.

There are strong links between the folate pathway and vitamins. Folate itself is a natural form of vitamin B9. Coenzymes derived from vitamins also have important roles in the pathway. Examples include FAD, derived from riboflavin (vitamin B2); the nicotinamide moiety of NADPH, derived from niacin (vitamin B3); PLP, derived from vitamin B6; and vitamin B12, a cofactor for the enzyme **methionine synthase**. The use of these vitamin derivatives highlights the nutritional relevance of this system.

The folate pathway consists of multiple interconnected pathways that will be discussed below. The balanced reactions associated with the folate pathway are also shown at the end of

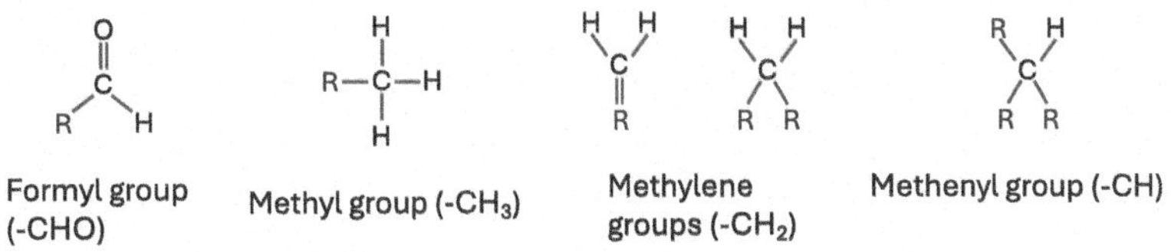

Fig. 16.1 One-carbon groups. Some important examples of one-carbon groups are shown here. One-carbon metabolism involves the transfer of such one-carbon groups

Table 16.1 Abbreviations and chemical formulas of folate and related metabolites

Abbreviation	Longer name	Full name	Formula
Folate			C19H19N7O6
THF	Tetrahydrofolate	5,6,7,8 tetrahydrofolate	C19H19N7O6
DHF	Dihydrofolate	7,8 dihydrofolate	C19H21N7O6
CH_2-THF	N^5,N^{10}-methylene-THF	N^5,N^{10} methylene-5,6,7,8-tetrahydrofolate	C20H23N7O6
5-MTHF	N^5-methyl-THF	5-methyltetrahydrofolate-THF	C20H25N7O6
CH^+ = THF	N^5,N^{10}-methenyl-THF		C20H21N7O6
CHO-THF	N^{10}-formyl-THF		C20H24N8O6
Hcy		Homocysteine	
Met		Methionine	
SAM		S-adenosyl methionine	
SAH		S-adenosyl homocysteine	

this chapter. Table 16.1 includes the names and abbreviations of key folate cycle intermediates. Many of these intermediates will be referred to by their abbreviations in the text. The table also includes chemical formulas of some of the intermediates. These can be helpful when tracking the fates of the various intermediates and one-carbon units transferred in these pathways.

Many metabolic pathways, including those covered in previous chapters in this textbook, are connected to the **folate pathway and interconnected cycles**. For example, purine synthesis is tightly linked to one-carbon metabolism, which supplies some of the one-carbon units that make up portions of the nitrogenous bases. Likewise, the generation of the pyrimidine dTMP also depends on one-carbon metabolism pathways. The levels of some amino acids are also linked to one-carbon metabolism, and methionine levels in the cell are regulated by the methionine cycle, which is linked to the folate cycle. The folate pathway and interconnected cycles discussed in this chapter are shown in Fig. 16.2. Structures of some components of the pathways, including the structure of folate, are shown in Fig. 16.3.

2 Folate Metabolism and Conversion of Folate to Its Active Derivatives

To begin the discussion of one-carbon metabolism and the folate and interconnected cycles, it is helpful to begin with a discussion of how folate is initially metabolized, starting with its ingestion from dietary sources. Once folate is ingested and absorbed, it is converted to DHF, which is then reduced to THF, its active form. THF in turn enters the folate cycle as explored below and also discussed by several resources including [1–6, 9].

2.1 Ingestion of Folate

Folate is obtained in our diet from sources such as dark leafy vegetables, some fruits, nuts, legumes, dairy, eggs, seafood, and meat. Many foods are also supplemented with folate because it plays such a vital role in metabolism. Folate is ingested mostly in its polyglutamate form, where

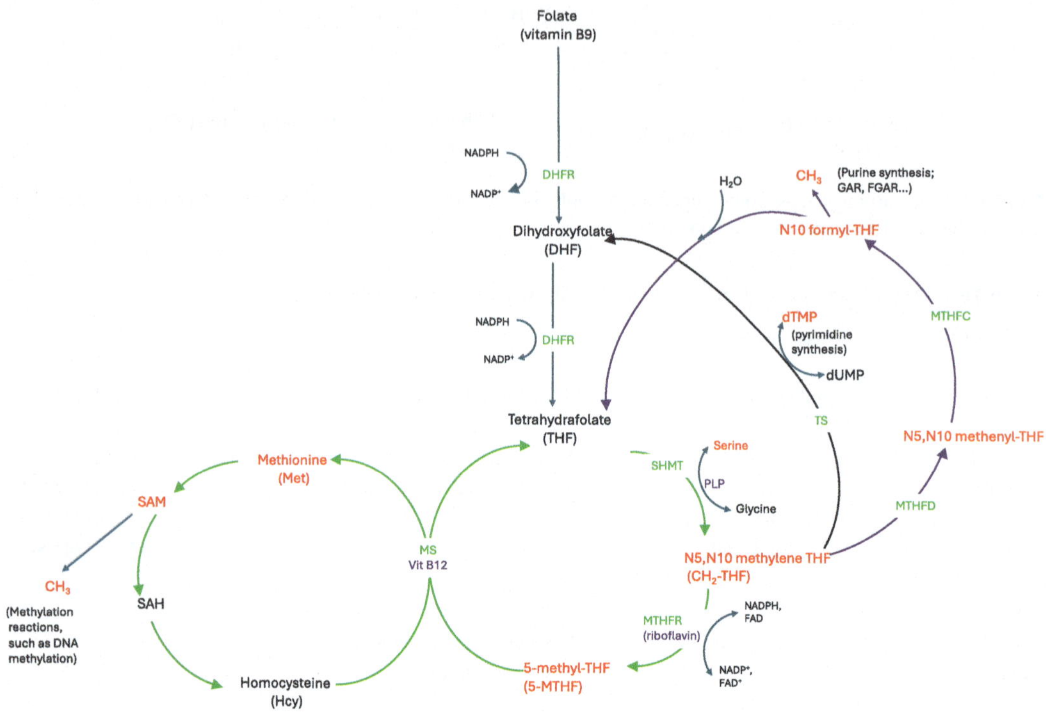

Fig. 16.2 Folate cycle and interconnected cycles. The folate pathway includes a set of interconnected reactions that involve folate and its derivatives and that involve the transfer of one-carbon units. See text for more details. Abbreviations of enzymes include the following: DHFR, dihydrofolate reductase; SHMT, serine hydroxymethyltransferase; MTHFR, methylenetetrahydrofolate reductase; MS, methionine synthase TS, thymidylate synthase; MTHFC, methenyltetrahydrofolate cyclohydrolase, MTHFD, methylenetetrahydrofolate dehydrogenase

it has multiple glutamic acid residues attached. At the brush border of the small intestine, it is deconjugated by **folate conjugase** to a monoglutamate form. (In this form, its full name is pteroylmonoglutamic acid). Folate is then absorbed in the duodenum and jejunum of the small intestine, via a proton-coupled folate transporter (PCFT) [10, 11].

2.2 Folate to DHF to THF

After absorption in the intestine, folate enters the portal circulation and reaches the liver. In the liver, **folate** is converted to **dihydrofolate (DHF)**, and **DHF** is converted to **tetrahydrofolate (THF)**. Both of these reactions are catalyzed by **dihydrofolate reductase (DHFR)** and require the oxidation of NADPH (the coenzyme in the reaction) to $NADP^+$. These reactions happen mostly in the cytosol of hepatocytes (liver cells).

2.3 THF to N^5,N^{10}-Methylene-THF (CH_2-THF)

Within the liver (and sometimes other tissues), **THF** initiates the **folate cycle**. The folate cycle consists of a series of interconversions of derivatives of **THF** that carry and transfer one-carbon units. The cycle allows cells to continuously regenerate THF and its derivatives, enabling the transfer of one-carbon units in biochemical reactions. The central components of the folate cycle are **THF** and its derivatives **N^5, N^{10}-methylene-THF (CH_2-THF)** and **5-methyl-THF (5-MTHF)** (from this point, **N^5,N^{10}-**

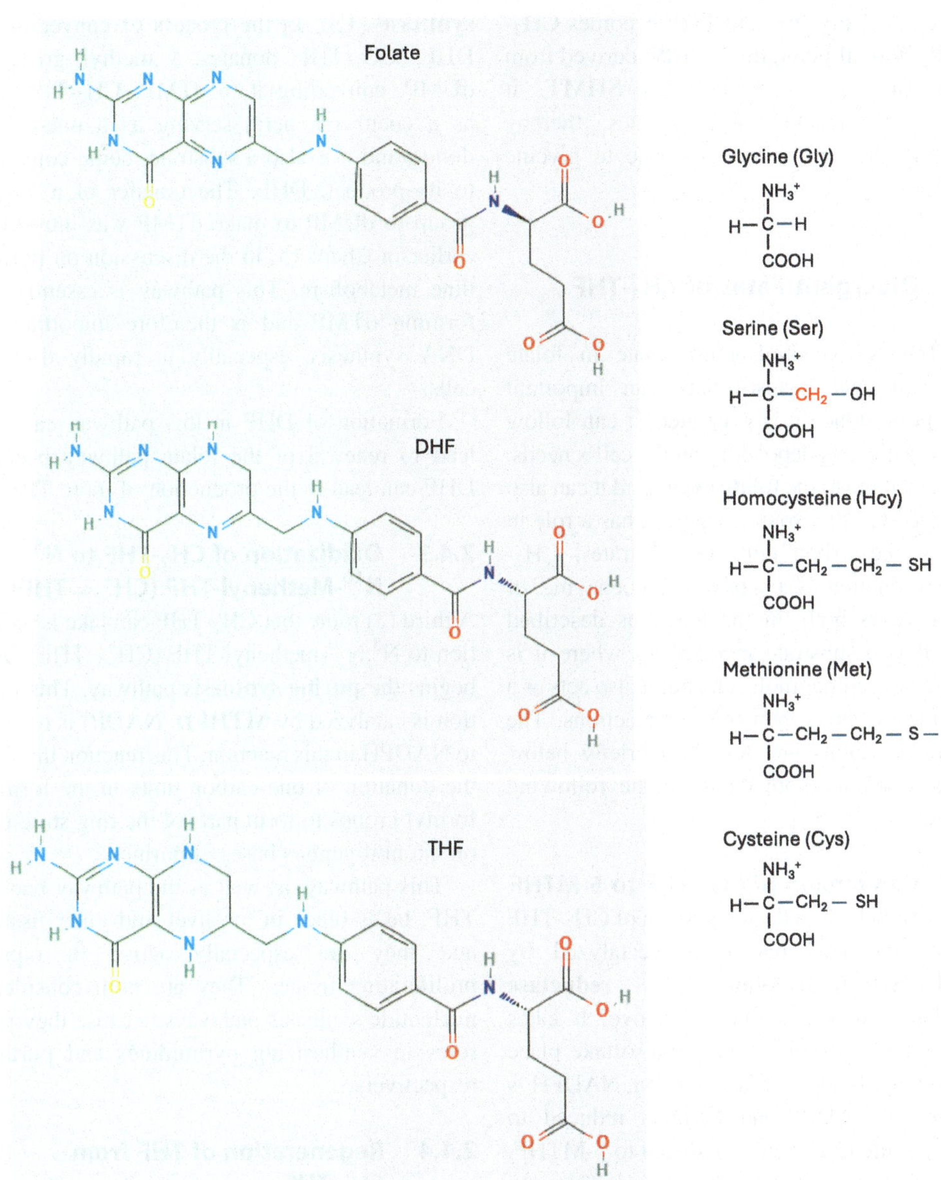

Fig. 16.3 The structures of some key intermediates in the folate pathway, including folate. Chemical formulas for several of the intermediates are shown in Table 16.1. (**Source**: Structures based on PubChem compound records: Folate (CID **135398658**), Dihydrofolate (CID **135398603**), and Tetrahydrofolate (CID **135444742**). PubChem data are released under CC0/public domain)

methylene-THF will be called **CH_2-THF** and **5-methyl-THF** will be called **5-MTHF**). These components also have branch points to other pathways.

After **THF** is formed, it can be converted to **CH_2-THF**. **CH_2-THF** differs from **THF** by the addition of a methylene (-CH_2-) bridge between the N^5 and N^{10} positions of the molecule; thus, it has an extra carbon and two extra hydrogen atoms. The conversion of **THF** to **CH_2-THF** is catalyzed by the enzyme **serine hydroxymethyltransferase (SHMT).** In this reaction, a **methylene (-CH_2-) group** is transferred from **serine** to **THF**, and as a result, **serine** is converted to **glycine**, and **THF** becomes **CH_2-THF**. **Pyridoxal phosphate (PLP)**, derived from vitamin B6, is a coenzyme for **SHMT**. It stabilizes the reaction intermediates, thereby facilitating the conversion of serine to glycine [2, 4, 12].

2.4 Divergent Fates of CH_2-THF

CH_2-THF is a central intermediate in folate metabolism, and it also marks an important branch point. Once it is generated, it can follow different pathways depending on the cell's needs. It can contribute to the **folate cycle**, and it can also be diverted to other routes, where it has a role in donating one-carbon units to substrates. CH_2-THF can function as a substrate, a coenzyme, or in some cases both, in the reactions described below. It is a substrate in reactions where it is converted to another molecule, but it also acts as a coenzyme in one-carbon transfer reactions. The four branch points are described briefly below and discussed in more detail in the following sections.

2.4.1 Conversion of CH_2-THF to 5-MTHF

The first branch (1) is the conversion of CH_2-THF to 5-MTHF. This reaction is catalyzed by **methylenetetrahydrofolate reductase (MTHFR),** already introduced above. It takes place mostly in the liver but can also take place in some other tissues. In this reaction, NADPH is oxidized to $NADP^+$ and FAD is reduced to $FADH_2$, while CH_2-THF is reduced to 5-MTHF. The reaction takes place in the cytosol. This step is important because THF, CH_2-THF, and 5-MTHF are the key components of the **folate cycle**. 5-MTHF also has a key role in the **activated methionine cycle**, which is directly connected with the folate cycle. 5-MTHF is considered the **active circulating form of folate** as it can be transported to various tissues through transporters.

2.4.2 Conversion of CH_2-THF Back to DHF

Second (2) CH_2-THF can also be converted back to DHF. This reaction is catalyzed by **thymidine synthase** (**TS**). In the process of conversion to DHF, CH_2-THF donates a methyl group to dUMP, converting it to dTMP. CH_2-THF acts as a coenzyme here, serving as a one-carbon donor, and it is also a substrate, being converted to its product, DHF. The transfer of a methyl group to dUMP to make dTMP was introduced earlier in Chap. 15, in the discussion on pyrimidine metabolism. This pathway is essential for forming dTMP and is therefore important for DNA synthesis, especially in rapidly dividing cells.

Formation of DHF in this pathway can also lead to renewal of the folate pathway because DHF can lead to the production of more THF.

2.4.3 Oxidization of CH_2-THF to N^5, N^{10}-Methenyl-THF (CH^+ = THF)

A third (3) route that CH_2-THF can take is oxidation to N^5,N^{10}-methenyl-THF (CH^+=THF). This begins the **purine synthesis** pathway. This reaction is catalyzed by **MTHFD**. $NADP^+$ is reduced to NADPH in this reaction. This reaction involves the donation of one carbon units in the form of formyl groups to form parts of the ring structures on the nitrogenous bases of purines.

This pathway, as well as the pathway back to THF, takes place in the liver and other tissues, and they are especially active in rapidly proliferating tissues. They are both considered nucleotide synthesis pathways because they play roles in synthesizing pyrimidines and purines, respectively.

2.4.4 Regeneration of THF from CH_2-THF

Fourth (4) CH_2-THF can regenerate THF via reversal of the reaction catalyzed by **SHMT**, and it uses **PLP** as the coenzyme. In this process, glycine is converted back to serine. This reaction helps balance glycine and serine levels in addition to restoring THF. It is important for amino acid metabolism and can occur in multiple tissues.

These four pathways that branch from CH_2-THF are described in more detail below. First, we discuss the reduction of CH_2-THF to 5-MTHF. This step is central to (a) the core folate cycle and (b) the activated methionine cycle, which are both tightly linked. After that, we discuss the roles of

CH_2-THF in nucleotide biosynthesis. These include its role as a carbon donor in dTMP synthesis, as it is converted back to DHF, and as a precursor for N^5,N^{10}-methenyl-THF ($CH^+ = THF$), which provides **formyl groups** for purine biosynthesis.

3 The Core Folate Cycle

The core component of the folate cycle consists of THF to CH_2-THF to 5-MTHF and back to THF. This cycle can continuously regenerate THF. This, in turn, maintains a supply of folate derivatives that can donate one-carbon units and can support interconnected cycles and pathways. The main components of the core folate cycle are described below. The first two steps (THF to CH_2-THF and CH_2-THF to 5-MTHF) were already introduced above but are described again here to solidify their roles in the folate cycle.

3.1 THF to CH_2-THF

The conversion of THF to CH_2-THF, already discussed above, can be considered the main starting point of the **folate cycle**. This step is catalyzed by **SHMT** and involves the conversion of serine to glycine. In this process, THF accepts a one-carbon group from serine, forming CH_2-THF, while serine is converted to glycine.

3.2 CH_2-THF to 5-MTHF

In the next step, **CH_2-THF** can become (is reduced to) **5-MTHF** in an irreversible reaction catalyzed by the enzyme **MTHFR.**

5-MTHF generated in this step is the main active circulating form of folate. It is generally generated in the liver, but it is considered the main circulating form because it can be transported to other tissues via the bloodstream and enter cells via transporters and receptors. The major 5-MTHF transporter is the reduced folate carrier (RFC, encoded by the SLC19A1 gene), which is present in most tissues. Other important transporter systems include the proton-coupled folate transporter (PCFT, encoded by the SLC46A1 gene), which is essential for intestinal absorption, and folate receptor α (FRα, encoded by the FOLR1 gene), which mediates uptake of 5-MTHF by receptor-mediated endocytosis in specialized tissues such as the brain and the placenta. FRα is also of medical importance because it is overproduced in certain cancer cells, as discussed more at the end of this chapter. Additional folate receptors, such as FRβ (encoded by the FOLR2 gene) contribute to tissue-specific transport, for example, in immune cells [13–15].

3.3 5-MTHF to THF

5-MTHF can donate a carbon unit in the form of a **methyl group** to homocysteine (Hcy). This generates methionine and regenerates THF. The enzyme that catalyzes this reaction is **methionine synthase (MS)**. THF can now re-enter the folate cycle, depending on the tissue type and enzyme availability. This step (5-MTHF to THF) provides an important link between the core folate cycle and the activated methionine cycle, which will be discussed in the next section.

4 The Activated Methionine Cycle

The **core folate cycle** is directly linked with the **activated methionine cycle** (also called the **methionine/SAM cycle)**. This cycle generates methionine from homocysteine and is also a source of methyl groups for other metabolic reactions. The cycle is described below.

4.1 Formation of Methionine from Homocysteine

The first step of the activated methionine cycle overlaps with the step in the core folate cycle in which 5-MTHF is converted to THF. As described above for the folate cycle, in this step, **5-MTHF** donates a **methyl group** to **homocysteine (Hcy;** a sulfur-containing amino acid**).** This converts **Hcy** to **methionine (MET)**, and in the process, **5-MTHF** is converted back to **THF**.

This reaction is catalyzed by **MS.**, which requires **vitamin B12 (cobalamin) as a cofactor**. If vitamin B12 is deficient, this reaction is impaired. This is sometimes called "folate trap," because although folate may be present in the body, in normal or even elevated levels, it is "trapped" in an unusable form (5-MTHF).

This step marks the beginning of the **activated methionine cycle**, which is directly linked to the folate cycle. Why is **Met** made in this cycle? Although **Met** is mainly consumed in the diet, this pathway is important for regulating the levels of **Met**. It also prevents the accumulation of Hcy, which is associated with cardiovascular disease and neurological functions if overproduced.

Another important outcome of this cycle is that the generation of Met leads to the formation of **S-adenosyl methionine (SAM)**. As discussed below, SAM is a universal **methyl donor** for various reactions.

4.2 Methionine Back to Homocysteine Via S-Adenosyl Methionine (SAM) and S-Adenosyl Homocysteine (SAH)

Hcy is an amino acid, though not one that is a building block of proteins (it is a non-proteinogenic amino acid). In the above reactions, **Hcy** is used to make **Met**, but **Hcy** is also **remade from Met**, thus completing a **cycle**. This starts with the formation of **S-adenosyl methionine (SAM)** from **Met** in a reaction where **Met** reacts with ATP to form **SAM**. This step is often called "activation" of methionine, and involves attaching an adenosyl group (derived from the adenosine on ATP) to Met, forming SAM. This reaction is catalyzed by **methionine adenosyltransferase (MAT)**.

SAM is a universal **methyl donor**. It releases its **methyl group**, donating it to a target molecule. This is an important step because this methyl group can be used for various reactions [1, 7]. The role of SAM as a methyl donor and the fate of its donated methyl group are discussed below. Once **SAM** donates its **methyl group**, it now becomes **S-adenosyl homocysteine (SAH)**. **SAH** is now hydrolyzed into two parts; one is adenosine, which is released, and the other that remains is **Hcy**, completing the cycle.

4.3 Completion of the Activated Methionine Cycle

This cycle, from **Hcy** to **Met** to **SAM** to **SAH** back to **Hcy,** completes the **active methionine cycle**. **Hcy** is now ready for another round of the cycle. This cycle ensures a continuous supply of **methyl groups** (via SAM) while maintaining Met and Hcy homeostasis. It should also be briefly noted here that although Hcy contributes to the methionine cycle, it also has another option, which is the transsulfuration pathway, where it is converted to cysteine (Cys)**.** Cys is a precursor for glutathione, and thus this pathway also leads to glutathione synthesis (this pathway is not covered in detail here).

4.4 SAM as a Universal Methyl Donor

Above, we discussed how **SAM** releases a **methyl group** as it is converted to **SAH**. This **methyl group**, released by **SAM**, can be used in various reactions. For example, this methyl group is the major source for the methylation of DNA, RNA, histones, and lipids. DNA methylation, for example, is important for the regulation of gene expression, and this methylation is highly dependent on the donation of methyl groups from **SAM**. DNA methylation is often on cytosines (Cs) in CpG groups, forming 5-methylcytosine. When DNA *is* methylated on cytosine, the methyl groups are transferred from **SAM** to the cytosine in the reaction (see Fig. 16.4):

$$\text{Cytosine} + \text{SAM} \rightarrow \text{5-methyl-cytosine} + \text{SAH}$$

The enzyme that catalyzes this reaction is **DNA methyltransferase (DNMT)**. The **DNMT** family has three members: **DNMT1**, **DNMT3a**,

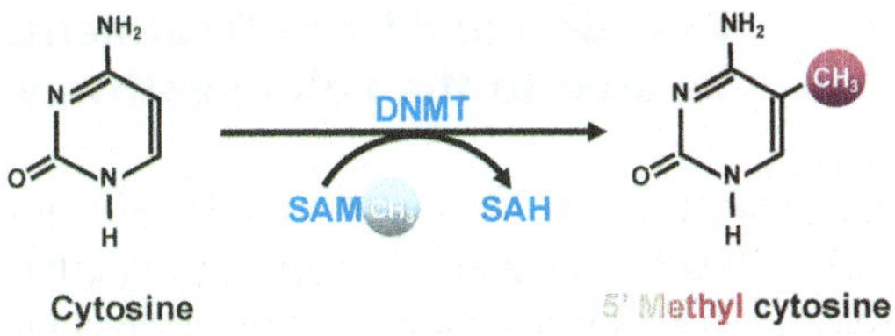

Fig. 16.4 Methylation of DNA. In DNA methylation, a methyl group is transferred from SAM to cytosine. (**Source:** Reproduced from Zakhari and Hoek [38]. Licensed under the Creative Commons Attribution 4.0 (CC BY 4.0) license)

and **DNMT3b**. **DNMT1** is used for maintenance methylation and requires the presence of hemimethylated DNA while the other two drive de novo methylation [7].

5 Nucleotide Synthesis Pathways That Branch from CH_2-THF

In Chap. 15, we discussed how one-carbon units are supplied to some of the nucleotides as they are being synthesized. A key source of these one-carbon units is the folate pathway, specifically the branches stemming from CH_2-THF. The two major pathways are those that contribute to pyrimidine synthesis (specifically dTMP) and purine synthesis (the donation of formyl groups to the purine ring). These are both discussed below.

5.1 CH_2-THF to DHF; Supply of a Methyl Group to Form dTMP

Pyridine synthesis involves a step where **dUMP** is converted to **dTMP**. This requires the addition of a **methyl group** to the pyrimidine ring of **dUMP**. This is because the pyrimidine base **thymine (T)** is identical to **uracil (U)**, except that **thymine** has a single extra **methyl group** on its ring structure. Now we can see where this extra methyl group comes from. It involves the conversion of **CH_2-THF** back to **DHF,** resulting in the release of a **one-carbon unit** from **CH_2-THF** in the form of a **methylene group (-CH_2-)**. This reaction is catalyzed by the enzyme **thymidylate synthase (TS)**. The **methylene group (-CH_2)** released in this reaction is added to **dUMP**, and it is reduced to a **methyl group (-CH_3)**, and **dUMP** is converted to **dTMP**, while **CH_2-THF** is converted to **DHF** [2, 4]. This pathway is referred to here as the thymidylate synthesis pathway.

5.2 CH_2-THF to N^5,N^{10}-Methenyl THF; Supply of Carbons for Purine Synthesis

Another route for **CH_2-THF** is the pyrimidine synthesis pathway. It starts with its conversion to **N^5,N^{10} methenyl-THF (CH^+=THF)**. This reaction is catalyzed by the enzyme **methylenetetrahydrofolate dehydrogenase (MTHFD),** and in the process, $NADP^+$ is reduced to NADPH. **CH^+=THF** then gets converted to **N^{10}-formyl-THF (CHO-THF)** in a hydrolysis reaction, catalyzed by **methenyltetrahydrofolate cyclohydrolase (MTHFC)**.

When discussing purine metabolism in Chap. 15, we discussed several steps between **PRPP** and **IMP** that involved transfers of one-carbon groups. In particular, the GAR to FGAR and AICAR to FAICAR reactions, which contribute to the growing purine ring structure, require the addition of one-carbon groups. These steps involve the donation of a **formyl group** from **CHO-THF** [9], which results in the addition of a carbon group to **GAR** or **AICAR**, the conversion of **CHO-THF** back to **THF**, and the generation of **FGAR** and **FAICAR**. **CHO-THF** is thus essential for the synthesis of purines as it provides the necessary one-carbon units [2, 4].

6 Recycling of CH_2-THF Back to THF

Finally, to maintain homeostasis, **CH_2-THF** can be recycled back to **THF** while glycine converts back to serine. This involves the reversal of the reaction catalyzed by **SHMT**. This can regulate the **balance of serine and glycine** in the cell and ensure that THF is available for ongoing folate metabolism pathways [2, 4].

7 Biotin (Vitamin B7) Is Another Source of Carbon Transfer, Playing a Role in Carboxylation Reactions

Biotin (vitamin B7) is not directly part of the folate-mediated one-carbon metabolism pathway, but it is briefly discussed here, because it also has a role of one-carbon transfer reactions. Unlike folate and its derivatives, which transfer methyl (-CH_3), methylene (-CH_2-), methenyl (-CH-), or formyl (-CHO) groups, biotin is specifically involved in carboxylation reactions, where it functions as a coenzyme that transfers carboxyl (-COOH) groups.

Biotin (vitamin B7) functions as a coenzyme for several carboxylase enzymes, where it serves as a carrier one one-carbon units. During catalysis, biotin is transiently carboxylated to form carboxybiotin, an enzyme-bound intermediate that binds the carboxyl group before it transfers it to the substrate. After the transfer, carboxybiotin is regenerated back to biotin, and another reaction can occur. An example of a biotin-dependent carboxylase enzyme is **acetyl-CoA carboxylase (ACC)**, involved in fatty acid synthesis. It catalyzes the addition of a carboxyl group (carboxylation) to acetyl-CoA, converting it to malonyl-CoA. Another example is **pyruvate carboxylase (PC)**, which is involved in gluconeogenesis and adds a carboxyl group to pyruvate to form oxaloacetate. Other examples are **propionyl-CoA-carboxylase (PCC)**, which is involved in amino acid and odd-chain fatty acid metabolism, and **3-methylcrotonyl-CoA carboxylase (MCC)**, which is involved in leucine catabolism. In all of these examples, **carboxybiotin** has a role in transferring the carboxyl group (COOH) to the substrate, and after this, the carboxybiotin is regenerated back to biotin and can then participate in more reactions [9, 16].

8 Disorders and Drug Treatments Related to the Folate Pathway

The folate pathway has important roles in many cellular functions. It plays an especially important role in DNA replication due to its involvement in the biosynthesis of nucleotides (purines and the pyrimidine dTTP). Cell replication depends on DNA replication, making the folate pathway critical for this process. It is especially important for cells that replicate rapidly and use a large supply of nucleotides. Because of this, deficiencies in folate are harmful to dividing cells. On the flip side, inhibition of the folate pathway is a promising mechanism for treating diseases characterized by excessive cell proliferation, such as cancer and autoimmune diseases.

8.1 Folate Deficiency

The folate pathway has crucial roles in DNA synthesis and, consequently, in cell division. Therefore, deficiencies in folate can be detrimental, especially to cells that divide rapidly. **Folate deficiency** is especially dangerous during **pregnancy** and can lead to **neural tube defects** in the baby. This is because cells in the neural tube divide especially rapidly and thus are highly dependent on DNA synthesis. If their proliferation is disrupted, this can lead to defects where the neural tube fails to close properly [17]. **Folic acid** is a synthetic form of **folate (vitamin B9)** that is used as a **supplement**. It is also added to many fortified foods. Folic acid in the form of supplements or fortified foods is recommended during pregnancy to help prevent neural tube defects [17].

8.2 Chemotherapy

8.2.1 DHFR Inhibitors

Antifolates are drugs that inhibit parts of the folate cycle. The first chemotherapeutic drugs ever developed were **antifolates**, which functioned by inhibiting **DHFR. DHFR** is the enzyme that catalyzes the conversion of **folate** to **DHF** and **DHF** to **THF**, thus initiating the folate pathway. The folate pathway is critical for nucleotide synthesis, and its inhibition consequently inhibits cell proliferation, especially in rapidly proliferating cells.

In 1947, Dr. Sidney Farber treated 16 children who had acute lymphoblastic leukemia (AML) with a drug called **aminopterin** [18]. **Aminopterin** is an amino derivative of **folic acid**. This means that it is a structural analog where a functional group is replaced with an amino (-NH_2) group. Specifically, a hydroxyl group in the folic acid structure is replaced with an amino group. Because it has a structure that is similar to folate, it functions as a competitive inhibitor of **DHFR**. Consequently, the drug inhibits the folate pathway, which is essential for DNA synthesis. The drug therefore kills rapidly dividing cells that depend on rapid DNA synthesis. Ten of the patients treated with the drug had temporary remission. This was the first ever successful remission of leukemia, and it laid the foundation for modern chemotherapy.

Methotrexate [19] was later developed as a less toxic analog of aminopterin. It is also an analog of folate, and the use of aminopterin has been replaced by methotrexate or other drugs in modern medicine [20]. **Methotrexate** also inhibits DHFR, and it is commonly used today in chemotherapy treatment. It is also used as an immunosuppressant to treat inflammatory conditions such as rheumatoid arthritis, psoriasis, and IBS, where it helps suppress the proliferation of immune cells in patients with an overactive immune system.

Aminopterin is the first antifolate to have been used. Today, there are multiple different antifolates, and they target different enzymes in the folate pathway. They have diverse structures and functions. Many of them are used to treat cancer and autoimmune diseases [1, 21–25].

8.2.2 Targeting a Folate Receptor

FRα is a receptor that utilizes receptor-mediated endocytosis to transport the circulating form of folate (5-MTHF) into cells, as discussed above. The FRα receptor is overexpressed on some cancer cells. Rapidly dividing cells, such as cancer cells, are particularly dependent on the folate pathway. Overexpression of the FRα receptor, therefore, helps these cells maintain a rapid level of proliferation. FRα has become a promising target for drug delivery for cancer [20]. **Elahere (mirvetuximab soravtansine)** [26] is an FDA-approved drug (2022) designed to target **cancer cells** that overexpress the **FRα receptor**. The drug does not actually target the FRα receptor. Rather, it takes advantage of the overexpressed receptor to gain entry into the cells by receptor-mediated endocytosis. **Elahere** is a biologic used to treat **ovarian cancers** that have resisted other treatments. Specifically, it is an antibody drug conjugate (ADC) that combines a targeting antibody with a toxic chemotherapeutic agent [27]. The antibody recognizes and binds to FRα. A linker combines the antibody with the chemotherapy agent, which is DM4 (mertansine derivative 4), a microtubule inhibitor. Microtubules are necessary for cell proliferation because of their important roles in chromosome segregation (see Chap. 20). Consequently, cells treated with this type of inhibitor stop replicating, and they undergo apoptosis. Normally, antibody-based drugs stay on the outside of the cell, but since FRα uses endocytosis to transport its cargo, the whole drug complex enters the cell via FRα. Once in the cell, the linker region of the drug is cleaved by lysosomal proteases. After cleavage, the DM4 is released into the cell and can disrupt the microtubules [28].

This approach is quite promising. Microtubule inhibitors are effective at killing rapidly growing cells but have serious side effects because, in addition to targeting cancer cells, they can also inhibit the proliferation of some healthy cells. **Elahere (mirvetuximab)** uses the power of microtubule inhibitors but directs them specifically to the targeted cells because the targeted cells overproduce FRα. Some side effects can still occur because some normal tissues also have FRα receptors. Nevertheless, side effects are reduced because FRα is not ubiquitously expressed on all cell types and its levels are particularly high in some types of cancer cells [28].

8.2.3 Thymidylate Synthase (TS) Inhibitors

Thymidylate Synthase (TS) catalyzes the methylation of dUMP to make dTMP (also called thymidylate). **TS** uses CH_2-THF as a coenzyme, which donates a methylene group (-CH_2) and a hydride ion ($H:^-$) to the dUMP as part of the reaction. CH_2-THF is also a substrate in this reaction and gets converted to DHF [9].

Cancer cells are rapidly dividing cells, and they need a large supply of thymidylate for their continued DNA synthesis. Blocking **TS** has therefore been a strategy for the development of some anticancer drugs. **Fluorouracil (5-FU)** is an anticancer drug that is a suicide inhibitor of **TS**. The active metabolite of **5-FU** is a pyrimidine analogue. It has a structure similar to dUMP. The drug binds **TS** along with CH_2-THF and irreversibly inhibits TS's activity [29–33]. **5-FU** is used to treat breast, colon, rectal, pancreatic, and stomach cancer [34].

8.3 Vitamin B12 Deficiency, the Folate Trap, and Megaloblastic Anemia

If vitamin **B12** is deficient, the conversion of **5-MTHF** to **THF** is impaired, because this reaction is catalyzed by the enzyme **MS**, which requires **vitamin B12** as a **cofactor**. As a result of this deficiency, **5-MTHF** accumulates and cannot be converted back to **THF**. This leads to what is referred to as **functional folate deficiency**; the body functions as if folate is deficient even if folate levels are normal. This is because the cells cannot carry out the normal folate pathway without vitamin B12. This functional folate deficiency is a problem because THF is required for nucleotide synthesis and other one-carbon transfer reactions. This condition is known as the "**folate trap**" (also called the methyl trap) [12]. This connection between vitamin B12 and the folate pathway explains why **vitamin B12 deficiency** can result in **megaloblastic anemia** that mimics folate deficiency [35]. Megaloblastic anemia is a type of anemia caused by impaired DNA synthesis [36], often due to either folate or vitamin B12 deficiency, resulting in the accumulation of large, immature red blood cells (megaloblasts) in the bone marrow. This is also called macrocytic anemia. Anemia refers to a lower-than-normal number of red blood cells or hemoglobin in the blood, and macrocytic anemia means that the blood cells are also larger than normal. They are large in this case because, although cell division is impaired, RNA and protein synthesis continue, allowing the cells to grow without dividing.

While folate deficiency and vitamin B12 deficiency can lead to the same type of problem, they are treated differently. Folate deficiency is treated with folate supplementation, or 5-MTHF supplementation, either orally or intravenously. Dietary improvements, such as eating leafy greens, are also helpful. This type of treatment is quite efficient.

In contrast to folate deficiency, vitamin B12 (cobalamin) deficiency can be more serious. Vitamin B12 also has other functions aside from its role in the folate pathway; specifically, it plays a role in odd-chain fatty acid metabolism, which contributes to the problem. Furthermore, while the most common cause of folate deficiency is poor nutrition, the most common cause of vitamin B12 deficiency is pernicious anemia, an autoimmune condition where antibodies attack intrinsic factor (IF), necessary for vitamin B12 absorption. Vitamin B12 deficiency is treated with vitamin B12 replacement, but high doses may be needed, and lifelong treatment may be required [37]. (Vitamin B12 deficiency and the folate trap are also discussed in Chap. 17).

9 Balanced Equations and Formulas

1. **Folate to DHF to THF**
   Folate + NADPH + H^+ → DHF + $NADP^+$
   DHF + NADPH + H^+ → THF + 2 $NADP^+$
   Enzyme for both: **Dihydrofolate reductase (DHFR)**
   Cofactor: NADPH

2. **THF to N^5,N^{10}-methylene-THF (CH_2-THF)**
THF + serine ⇌ N^5,N^{10}-methenyl-THF (CH_2-THF) + glycine + H_2O
Enzyme: **Serine hydroxymethyltransferase (SHMT)**
Cofactor: PLP
THF is a substrate and a coenzyme (because it accepts a 1 C unit)
3. **Outcomes for N^5,N^{10}-methylene-THF (CH_2-THF):**
   3A. **Reduction to N^5-methyl-THF (5-MTHF)**
   N^5,N^{10}-metheylene-THF (CH_2-THF) + NADPH + H^+ → N5-methyl-THF (5-MTHF) + $NADP^+$
   Enzyme: **Methylenetetrahydrofolate reductase (MTHFR)**
   NADPH is a coenzyme and a substrate.
   CH_2-THF is a coenzyme and a substrate.
   3B. **dTMP synthesis and conversion to DHF:**
   N^5,N^{10}-methylene-THF (CH_2-THF) + dUMP → dTMP + DHF
   Enzyme: **Thymidylate synthase (TS)**
   CH2-THF is a substrate and a coenzyme (a methyl donor).
   3C. **Oxidation to N^5,N^{10}-methenyl-THF**
   N^5,N^{10}-methylene-THF (CH_2-THF) + $NADP^+$ ⇌ N^5,N^{10}-methenyl-THF (CH^+ = THF) + NADPH + H^+
   Enzyme: **Methylenetetrahydrofolate dehydrogenase (MTHFD)**
   $NADP^+$ is a substrate and a coenzyme.
   3D. **Regeneration of THF**
   N^5,N^{10}-methylene-THF (CH_2-THF) + glycine + H_2O ⇌ serine + THF
   Enzyme: **Serine hydroxymethyltransferase (SHMT)**
   Coenzyme: PLP
   Activated methionine cycle reactions:
4. **The core folate cycle.**
   4A. **THF to N^5,N^{10}-methylene-THF (CH_2-THF)**
   See step 2 above.
   4B. **N^5,N^{10}-methylene-THF (CH_2-THF) to N^5-methyl-THF (5-MTHF)**
   See step 3A above.
   4C. **N^5-methyl-THF (5-MTHF)** to THF
   N^5-methyl-THF (5-MTHF) + Hcy → THF + Met
   Enzyme: **Methionine synthase (MS)**
   Cofactor: Vitamin B12 (methylcobalamin; accepts a methyl group and then donates it to Hcy)
5. **The activated methionine cycle**
   5A. **N^5-methyl-THF (5-MTHF) to THF**
   See 4C above.
   5B. **Met to SAM:**
   Met + ATP → SAM + Pi + PPi
   Enzyme: **Methionine adenosyltransferase (MAT)**
   5C. **SAM to SAH:**
   SAM + R-H (methyl-acceptor) → SAH + R-CH_3 (methylated product)
   Enzyme: Varies
   Example: Methylation of cytosine on DNA:
   SAM + cytosine → SAH + 5-methyl-cytosine
   Enzyme: **DNA methyltransferase (DNMT)**
   5D. **SAH to HCY:**
   SAH + H_2O → adenosine + HCY
   Enzyme: **S-adenosylhomocysteine hydrolase (SAHH)**
   5E. **HCY to MET:**
   HCY + 5-N^5-methyl-THF (5-MTHF) → MET + THF
   Enzyme: **Methionine synthase (MS)**

FDA-Approved Drugs Covered in This Chapter

Commonly used brand name	Generic name	First FDA approval year	Common conditions	Main targets
Elahere	*Mirvetuximab soravtansine*	2022	Cancer	Folate receptor-α
Trexall	*Methotrexate*	1953	Cancer, rheumatoid arthritis, psoriasis	DHFR
Adrucil	*Fluorouracil (5-FU)*	1962	Cancer	Thymidylate Synthase

Definitions

5-Fluorouracil (5-FU) Chemotherapy drug that inhibits thymidylate synthase (TS) by mimicking dUMP. It is a suicide inhibitor of TS.

5-Methyl-THF (5-MTHF) The major circulating form of folate. It donates a methyl group to homocysteine.

Aminopterin A folate analog and antifolate drug used in cancer treatment.

CH_2-THF (N5,N10-Methylene-THF) A folate derivative that serves as a one-carbon donor.

Cobalamin Also known as vitamin B_{12}, required for methionine synthase function.

Dihydrofolate reductase (DHFR) An enzyme that reduces folic acid to THF, in two steps. It is a target of the drug methotrexate.

DNA methyltransferase (DNMT) Enzyme that uses SAM to methylate cytosine residues.

Folate trap A type of functional folate deficiency due to B_{12} deficiency, it causes impaired nucleotide synthesis.

Folate receptor alpha (FRα) A receptor that is involved in the transport of 5-MTHF in some tissues. It is overexpressed in certain cancers and targeted by antifolate therapies.

Homocysteine A sulfur-containing amino acid that is methylated to form methionine or converted to cysteine.

Methionine synthase Enzyme that transfers a methyl group from 5-MTHF to homocysteine, forming methionine and regenerating THF.

Mirvetuximab soravtansine An antibody-drug conjugate targeting FRα used to treat cancer.

Methylenetetrahydrofolate reductase (MTHFR) Enzyme that converts CH_2-THF to 5-MTHF.

S-Adenosylhomocysteine (SAH) A product of methylation reactions in the methionine cycle.

S-Adenosylmethionine (SAM) A universal methyl donor in cellular methylation reactions.

Serine hydroxymethyltransferase (SHMT) Enzyme that transfers a one-carbon group from serine to THF, forming CH_2-THF.

Tetrahydrofolate (THF) A derivative of folate with a key role in one-carbon metabolism pathways.

Thymidylate synthase (TS) Enzyme that uses CH_2-THF to convert dUMP to dTMP, it is essential for DNA synthesis.

Study Questions

Multiple-Choice Questions

1. Which enzyme catalyzes the conversion of folate to DHF and then to THF?
   A. MTHFR
   B. SHMT
   C. DHFR
   D. TS
2. What coenzyme is required by thymidylate synthase (TS) to convert dUMP to dTMP?
   A. NAD^+
   B. FAD
   C. CH⍰-THF
   D. Biotin
3. Which of the following serves as a universal methyl donor in cells?
   A. SAH
   B. SAM
   C. THF
   D. ATP
4. Which drug targets folate receptor alpha (FRα) in ovarian cancer treatment?
   A. Methotrexate
   B. Aminopterin
   C. 5-Fluorouracil
   D. Mirvetuximab soravtansine

Fill-in-the-Blank Question

5. The conversion of THF to CH⍰-THF requires the enzyme __________.
6. The enzyme that reduces CH_2-THF to 5-MTHF is __________.
7. __________is a synthetic analog of folate used in chemotherapy and as an immunosuppressant to treat inflammatory disorders.
8. SAM is synthesized from methionine and __________.
9. The enzyme that catalyzes the methylation of cytosine using SAM is __________.

Short-Answer Questions

10. Explain how 5-FU (5-Fluorouracil) inhibits thymidylate synthase (TS).

Matching Question

11. Match Enzyme to Function
    Enzyme
    A. DHFR
    B. SHMT
    C. MTHFR
    D. DNMT
    Function
    1. Converts DHF to THF
    2. Transfers methylene from serine to THF
    3. Reduces CH_2-THF to 5-MTHF
    4. Catalyzes DNA methylation

Answer Key

Multiple-Choice Questions

1. **Answer**: C
2. **Answer**: C
3. **Answer**: B
4. **Answer**: D

Fill-in-the-Blank Question

5. **Answer**: Serine hydroxymethyltransferase (SHMT)
6. **Answer**: Methylenetetrahydrofolate reductase (MTHFR)
7. **Answer**: Methotrexate
8. **Answer**: ATP
9. **Answer**: DNA methyltransferase (DNMT)

Short-Answer Question

10. **Answer**: 5-FU is a suicide inhibitor of TS. The active metabolite of**5-FU**is a pyrimidine analogue. It mimics dUMP, binds to TS along with N5-N10-methylene-THF, and irreversibly inhibits TS's activity, halting DNA replication.

Matching Question

11. **Answers**: A → 1, B → 2, C → 3, D → 4

References

1. Ducker GS, Rabinowitz JD. One-carbon metabolism in health and disease. Cell Metab. 2017;25(1):27–42. https://doi.org/10.1016/j.cmet.2016.08.009.
2. Fox JT, Stover PJ. Chapter 1 Folate-mediated one-carbon metabolism. In: Vitamins & hormones, vol. 79. Academic Press; 2008. p. 1–44.
3. Shulpekova Y, Nechaev V, Kardasheva S, Sedova A, Kurbatova A, Bueverova E, et al. The concept of folic acid in health and disease. Molecules. 2021;26(12). https://doi.org/10.3390/molecules26123731.
4. Chandel NS. Navigating metabolism. 2nd ed. Cold Spring Harbor: Cold Spring Harbor Laboratory Press; 2020.
5. Salway JG. Metabolism at a glance. Hoboken: Wiley-Blackwell; 2017.
6. Holeček M. Serine metabolism in health and disease and as a conditionally essential amino acid. Nutrients. 2022;14(9). https://doi.org/10.3390/nu14091987.
7. Crider KS, Yang TP, Berry RJ, Bailey LB. Folate and DNA methylation: a review of molecular mechanisms and the evidence for folate's role. Adv Nutr. 2012;3(1):21–38. https://doi.org/10.3945/an.111.000992.
8. Zarou MM, Vazquez A, Vignir HG. Folate metabolism: a re-emerging therapeutic target in haematological cancers. Leukemia. 2021;35(6):1539–51. https://doi.org/10.1038/s41375-021-01189-2.
9. Berg JM, Tymoczko JL, Gatto GJ Jr, Stryer L. Biochemistry. 9th ed. New York: MacMillan Learning; 2019.
10. Zhao R, Matherly LH, Goldman ID. Membrane transporters and folate homeostasis: intestinal absorption and transport into systemic compartments and tissues. Expert Rev Mol Med. 2009;11:e4. https://doi.org/10.1017/s1462399409000969.
11. Reisenauer AM, Halsted CH, Jacobs LR, Wolfe BM. Human intestinal folate conjugase: adaptation after jejunoileal bypass. Am J Clin Nutr. 1985;42(4):660–5. https://doi.org/10.1093/ajcn/42.4.660.
12. Stover PJ. One-carbon metabolism-genome interactions in folate-associated pathologies. J Nutr. 2009;139(12):2402–5. https://doi.org/10.3945/jn.109.113670.
13. O'Connor C, Wallace-Povirk A, Ning C, Frühauf J, Tong N, Gangjee A, et al. Folate transporter dynamics and therapy with classic and tumor-targeted antifolates. Sci Rep. 2021;11(1):6389. https://doi.org/10.1038/s41598-021-85818-x.
14. Matherly LH, Wilson MR, Hou Z. The major facilitative folate transporters solute carrier 19A1 and solute carrier 46A1: biology and role in antifolate chemotherapy of cancer. Drug Metab Dispos. 2014;42(4):632–49. https://doi.org/10.1124/dmd.113.055723.

15. Nelson DL, Cox MM, Hoskins AA. Lehninger principles of biochemistry. 8th ed. Macmillan Learning; 2021.
16. Tong L. Structure and function of biotin-dependent carboxylases. Cell Mol Life Sci. 2013;70(5):863–91. https://doi.org/10.1007/s00018-012-1096-0.
17. Imbard A, Benoist J-F, Blom HJ. Neural tube defects, folic acid and methylation. Int J Environ Res Public Health. 2013;10(9):4352–89.
18. Spain PD, Kadan-Lottick N. Observations of unprecedented remissions following novel treatment for acute leukemia in children in 1948. J R Soc Med. 2012;105 (4):177–81. https://doi.org/10.1258/jrsm.2012.12k013.
19. FDA. Trexall (methotrexate) 1953. Available from: https://www.accessdata.fda.gov/scripts/cder/daf/index.cfm?event=overview.process&ApplNo=008085.
20. Visentin M, Zhao R, Goldman ID. The antifolates. Hematol Oncol Clin North Am. 2012;26(3):629–48, ix. https://doi.org/10.1016/j.hoc.2012.02.002.
21. Zwart NRK, Franken MD, Tissing WJE, Lubberman FJE, McKay JA, Kampman E, et al. Folate, folic acid, and chemotherapy-induced toxicities: a systematic literature review. Crit Rev Oncol Hematol. 2023;188: 104061. https://doi.org/10.1016/j.critrevonc.2023.104061.
22. Whalen K, Lerchenfeldt S, Giordano C. Lippincott illustrated reviews. Pharmacology. Wolters Kluwer; 2021.
23. Fowler B. The folate cycle and disease in humans. Kidney Int Suppl. 2001;78:S221–9. https://doi.org/10.1046/j.1523-1755.2001.59780221.x.
24. Lele AC, Mishra DA, Kamil TK, Bhakta S, Degani MS. Repositioning of DHFR inhibitors. Curr Top Med Chem. 2016;16(19):2125–43. https://doi.org/10.2174/1568026616666160216152540.
25. Bertino JR. Cancer research: from folate antagonism to molecular targets. Best Pract Res Clin Haematol. 2009;22(4):577–82. https://doi.org/10.1016/j.beha.2009.09.004.
26. FDA. Elahere (mirvetuximab) approval overview 2022. Available from: https://www.accessdata.fda.gov/scripts/cder/daf/index.cfm?event=overview.process&ApplNo=761310.
27. Heo YA. Mirvetuximab Soravtansine: first approval. Drugs. 2023;83(3):265–73. https://doi.org/10.1007/s40265-023-01834-3.
28. Mai J, Wu L, Yang L, Sun T, Liu X, Yin R, et al. Therapeutic strategies targeting folate receptor α for ovarian cancer. Front Immunol. 2023;14:1254532. https://doi.org/10.3389/fimmu.2023.1254532.
29. Ritter JM, Flower RJ, Henderson G, Loke YK, MacEwan DJ, Rang HP. Rang and Dale's pharmacology. 10th ed. Elsevier Health Sciences; 2023.
30. Peng M, Darko KO, Tao T, Huang Y, Su Q, He C, et al. Combination of metformin with chemotherapeutic drugs via different molecular mechanisms. Cancer Treat Rev. 2017;54:24–33. https://doi.org/10.1016/j.ctrv.2017.01.005.
31. Zhang N, Yin Y, Xu SJ, Chen WS. 5-Fluorouracil: mechanisms of resistance and reversal strategies. Molecules. 2008;13(8):1551–69. https://doi.org/10.3390/molecules13081551.
32. Sobich J, Prokopowicz M, Maj P, Wilk P, Zieliński Z, Frączyk T, et al. Thymidylate synthase-catalyzed, tetrahydrofolate-dependent self-inactivation by 5-FdUMP. Arch Biochem Biophys. 2019;674: 108106. https://doi.org/10.1016/j.abb.2019.108106.
33. Longley DB, Harkin DP, Johnston PG. 5-Fluorouracil: mechanisms of action and clinical strategies. Nat Rev Cancer. 2003;3(5):330–8. https://doi.org/10.1038/nrc1074.
34. Bijnsdorp IV, Comijn EM, Padron JM, Gmeiner WH, Peters GJ. Mechanisms of action of FdUMP[10]: metabolite activation and thymidylate synthase inhibition. Oncol Rep. 2007;18(1):287–91. https://doi.org/10.3892/or.18.1.287.
35. Abali EE, Cline SD, Franklin DA, Viselli SM. Lippincott illustrated reviews: biochemistry. LWW; 2025.
36. Torrez M, Chabot-Richards D, Babu D, Lockhart E, Foucar K. How I investigate acquired megaloblastic anemia. Int J Lab Hematol. 2022;44(2):236–47. https://doi.org/10.1111/ijlh.13789.
37. Castillo LF, Pelletier CM, Heyden KE, Field MS. New Insights into folate–vitamin B12 interactions. Annu Rev Nutr. 2025;45:23–39. https://doi.org/10.1146/annurev-nutr-120524-043056.
38. Zakhari S, Hoek JB. Epidemiology of moderate alcohol consumption and breast cancer: association or causation? Cancers. 2018;10(10):349. https://doi.org/10.3390/cancers10100349.

The Roles of Vitamins in Human Health and the Effects of Vitamin Deficiencies 17

Abstract

Vitamins and minerals are essential dietary components that support numerous biochemical processes, often functioning as cofactors for enzymes. This chapter explores vitamins and minerals, with special attention to those that function as enzyme cofactors, especially those that play key roles in metabolism. Vitamins are compounds that cannot be produced in sufficient amounts in the body. They are diverse in structure and function, but all vitamins must be ingested in small amounts nearly every day through our diet or supplements. Many vitamins are ingested in inactive or less active forms and then converted to their biologically active forms in the body. Vitamins can be water soluble or fat soluble. Both are described here. This chapter will discuss different categories of vitamins, including the B vitamins, vitamin C, vitamin D, vitamin A, vitamin K, and vitamin E, as well as essential minerals. Health problems caused by vitamin deficiencies will be discussed, as well as vitamin metabolism.

Keywords

B vitamins · Vitamin C · Vitamin D · Essential minerals · Vitamins in human health

Abbreviations

1,3-BPG	1,3-biphosphoglycerate
CoA	Coenzyme A
CYP2R1	25 hydroxylase
CYP27B1	1-α-hydroxylase
DHAP	Dihydroxyacetone phosphate
E1	Pyruvate dehydrogenase
E2	Dihydrolipoyl transacetylase
E3	Dihydrolipoyl dehydrogenase
FAD/ $FADH_2$	Flavin adenine dinucleotide (oxidized/reduced form)
Fe^{2+}, Fe^{3+}	Ferrous, ferric forms of iron
FMN	Flavin mononucleotide
G3P	Glyceraldehyde-3-phosphate
GADPH	Glyceraldehyde-3-phosphate-dehydrogenase
Glu	Glutamate (amino acid abbreviation)
GPD	Glycerol-3 phosphate dehydrogenase
IF	Intrinsic factor
IPG	Glucose-6-phosphate isomerase
LEO	Loss of electrons is oxidation
MCM	Methylmalonyl-CoA mutase
NAD^+/ NADH	Nicotinamide adenine dinucleotide (oxidized/ reduced form)
$NADP^+$/ NADPH	Nicotinamide adenine dinucleotide phosphate (oxidized/ reduced form)

A. Minden, *Biochemistry, Cell and Molecular Biology for Pharmacists and Health Professionals*,
https://doi.org/10.1007/978-3-032-20529-2_17

PGI	Phosphoglucose isomerase
PDC	Pyruvate dehydrogenase complex
PDH	Pyruvate dehydrogenase (PDH) complex
PLP	Pyridoxal 5′-phosphate (active form of vitamin B6)
PTH	Parathyroid hormone
RAR, RXR	Retinoic acid receptor, retinoid x receptor
RBC	Red blood cell
THF	Tetrahydrofolate
TKT	Transketolase
TPP	Thiamine pyrophosphate (active form of vitamin B1)
α-KGDH	Alpha-ketoglutarate dehydrogenase

What You Will Learn

- What are vitamins
- The difference between water-soluble and fat-soluble vitamins
- The B vitamins and their roles as cofactors in metabolic reactions
- Disorders involving the B vitamins, including vitamin B1 (thiamine) deficiency (beriberi), vitamin B12 (cobalamin) deficiency (anemia), vitamin B12-IF complex deficiency (pernicious anemia), and vitamin B7 (biotin) deficiency
- Vitamin C and its importance in collagen synthesis, catecholamine synthesis, and carnitine biosynthesis. The overall importance of vitamin C in health
- Vitamin D and its metabolism, including its synthesis in response to UV light
- Vitamin D deficiency and vitamin D supplements
- Vitamin A (retinoids) and their functions, including their important role in vision. Golden rice as which is supplemented with beta-carotene, a precursor of vitamin A
- Other vitamins, such as vitamin K and vitamin E
- Essential minerals, and some enzymes that require essential minerals, such as DNA polymerase

In this chapter, we discuss the vitamins and minerals we obtain from our diet that are essential for human nutrition. Many vitamins and minerals are important for the regulation of enzymes by serving as cofactors (which include coenzymes and prosthetic groups). Here, we will discuss different vitamins and minerals, focusing on those that are cofactors for enzymes, especially those that play key roles in metabolism.

1 Introduction to Vitamins

Vitamins are compounds that cannot be produced in sufficient amounts in the body. Vitamins are diverse in structure and function, but what they have in common is that we need to ingest them in small amounts nearly every day from our diet or from supplements [1]. Many (but not all) vitamins get ingested in inactive or less active forms and then get converted to their biologically active forms in the body. The active forms of many vitamins function as cofactors. Not all vitamins are cofactors, however, and some of them have various other essential roles in the body [1–3]. Vitamins can be water soluble or fat soluble. Both are described below.

1.1 Water-Soluble Vitamins

Water-soluble vitamins have a large percentage of polar groups that can form hydrogen bonds with water. Groups such as carboxyl groups (-COOH), hydroxyl groups (-OH), amines ($-NH_2$, NHR, $-NR_2$), and amides ($-CONH_2$, -COONHR, $-CONR_2$) are common in these vitamins. It is difficult to overdose on water-soluble vitamins because they are not easily stored and are excreted readily through the urine. Because they are easily excreted, they also need to be replenished in the diet every day to keep their levels constant. Examples of water-soluble vitamins are all the B vitamins and vitamin C [1, 2].

1.2 Fat-Soluble Vitamins

Some vitamins are fat soluble and thus stored in fat rather than water. These vitamins have a higher percentage of nonpolar groups than polar groups, and they include aliphatic groups and a high percentage of hydrocarbons. They may have some polar groups, such as hydroxyl (-OH) groups, but the nonpolar groups greatly outnumber these. These types of vitamins do not necessarily have to be obtained daily because excess vitamins can be stored in the body. However, it is possible to overdose on the fat-soluble vitamins because they are soluble in fat, and therefore, they can accumulate, such that they are not readily transported out of the body. The major fat-soluble vitamins are vitamins D, E, K, and A [1, 2, 4].

2 The B Vitamins

The B vitamins are a large and important category of water-soluble vitamins that have important roles in energy regulation and metabolism [1, 3, 5]. We have already discussed several B vitamins and their important roles in the folate pathways in Chap. 16.

2.1 What Are the B Vitamins

The B vitamins [1, 3] are a group of water-soluble vitamins that have important roles in metabolism. These vitamins can be referred to in different ways. They can be referred to as the vitamin type, such as **vitamin B1**, or by their actual name, in this case **thiamine**. When discussing biochemical reactions, they can also be referred to as their active forms, which is the form that may function as a cofactor (often specifically as a coenzyme). In this case, for vitamin B1, the active form is **thiamine pyrophosphate (TPP)**. Activation of vitamins can involve various chemical modifications, such as phosphorylation or combination with other molecules.

The B vitamins are summarized in Table 17.1, which includes the vitamin type, the actual name, the name of the active form, and information about some of their functions.

2.2 The Roles for Some of the B Vitamins in Regulating Enzymes That Are Important for Metabolism

B vitamins have multiple important roles in the central metabolic pathways, where many of them serve as precursors of coenzymes. The roles of some of the coenzymes derived from B vitamins involved in glucose metabolism and fat metabolism are diagrammed in Fig. 17.1. Below, we discuss some of their key roles in metabolic pathways, including the metabolism of carbohydrates, lipids, and proteins.

2.2.1 Glycolysis

In glycolysis, **NAD^+** is a coenzyme derived from **vitamin B3**, playing a crucial role in metabolic pathways. It acts primarily as an oxidizing agent by accepting electrons and becoming NADH. It is a coenzyme for the enzyme **glyceraldehyde 3-phosphate dehydrogenase (GAPDH)**. This enzyme catalyzes the oxidation of glyceraldehyde-3-phosphate (GAP) to 1,3-bisphosphoglycerate (1,3-BPG) [6–8].

2.2.2 Decarboxylation of Pyruvate; Pyruvate to Acetyl CoA

After glycolysis, two pyruvates are formed from the breakdown of one glucose. The pyruvates are then converted to acetyl-CoA through a process involving multiple enzymes and cofactors. The enzyme that catalyzes this reaction is actually a complex of enzymes called the **pyruvate dehydrogenase complex (PDH)**. This complex consists of three enzymes: **E1, E2,** and **E3**. E1 is pyruvate dehydrogenase, E2 is dihydrolipoyl transacetylase, and E3 is dihydrolipoyl dehydrogenase. This enzyme complex requires four coenzymes derived from members of the vitamin

Table 17.1 B vitamins

Vitamin	Common name	Active form	How the active form is made	Some functions	Dietary sources
B1	Thiamine	Thiamine pyrophosphate (TPP)	Phosphorylation of thiamine by **thiamine phosphokinase** (adds two phosphate groups)	Carbohydrate metabolism, converting pyruvate to acetyl CoA, pentose phosphate pathway, nervous system function	Whole grains, pork, legumes, nuts, seeds, brown rice, potatoes, fortified cereals
B2	Riboflavin	FAD, FMN	Phosphorylation of riboflavin is by **riboflavin kinase** to form FMN. FMN is converted to FAD by **FAD synthetase** using ATP.	Important for redox reactions. Electron acceptor in the Krebs cycle	Dairy, eggs, liver, meats, leafy greens, almonds, mushrooms
B3	Niacin	NAD^+, $NADP^+$	Niacin (nicotinic acid) and nicotinamide are converted into NMN or NR, which are precursors to NAD^+. NAD^+ is phosphorylated to form $NADP^+$.	Important in redox reactions, important in the Krebs cycle. Roles in DNA repair, cell signaling	Chicken, beef, turkey, fish, peanuts, whole grains, mushrooms, fortified cereals
B5	Pantothenic acid	Coenzyme A (CoA)	Pantothenic acid is converted to 4′-phosphopantothenate by **pantothenate kinase**, then to phosphopantetheine, which is finally combined with adenosine triphosphate (ATP) and cysteine to form CoA.	Addition of CoA to molecules, such as acetyl-CoA. Important in the Krebs cycle, important for synthesis of fatty acids, neurotransmitters, steroid hormones	Egg yolks, liver, chicken, whole grains, avocados, dairy, sunflower seeds, mushrooms
B6	Pyridoxine group	Pyridoxal phosphate (PLP)	Phosphorylated by **pyridoxal kinase**	Transamination, reactions in amino acid metabolism, deamination, decarboxylation. Roles in neurotransmitter synthesis, hemoglobin production, glycogen breakdown	Poultry, fish, bananas, potatoes, nuts, chickpeas, fortified cereals
B7	Biotin	Biotin (as enzyme cofactor0	Activated by **holocarboxylase synthetase**, which attaches biotin to apo-carboxylase	Coenzyme for carboxylase enzymes like pyruvate, acetyl-CoA, and propionyl-CoA carboxylase. Important for gluconeogenesis and fatty acid biosynthesis	Egg yolks, almonds, peanuts, whole grains, liver, sweet potatoes, salmon
B9	**Folate**: The name for the naturally occurring form found	Tetrahydrofolate (THF)	Folate is reduced by **dihydrofolate reductase (DHFR)** to dihydrofolate (DHF)	Important in the folate pathway. One carbon donor in purine, thymidine, and methionine synthesis	Leafy greens (spinach, kale), legumes, citrus fruits, avocado, fortified grains

(continued)

Table 17.1 (continued)

Vitamin	Common name	Active form	How the active form is made	Some functions	Dietary sources
	in food. Folic acid is the name for the synthetic form in fortified foods		and then to tetrahydrofolate (THF).		
B12	**Cobalamin**: The umbrella term for multiple vitamin B12 compounds	Methylcobalamin and adenosylcobalamin	Cyanocobalamin and hydroxocobalamin are converted by **methyltransferases** and **adenosyltransferases**.	Also important in one carbon transfer Methionine synthesis Works with folate Roles in blood cell formation, neurological function	Meat, fish, dairy, eggs, fortified cereals, nutritional yeast

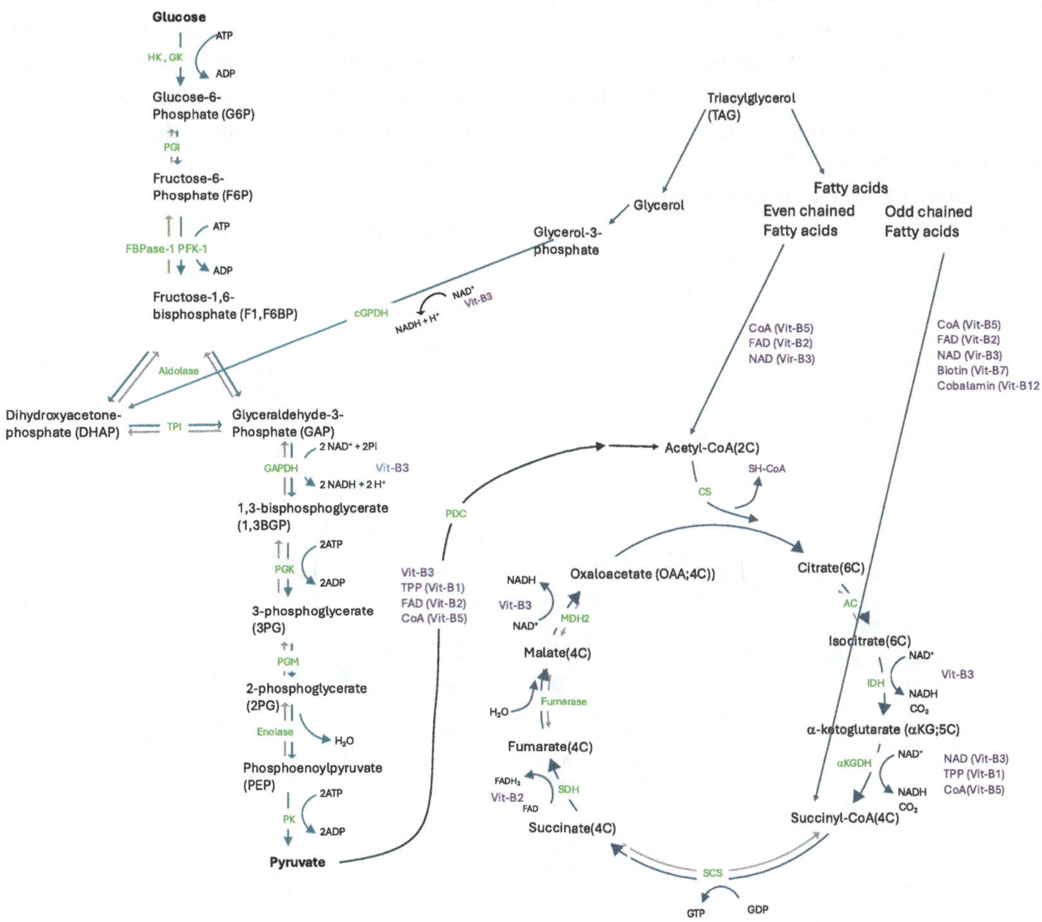

Fig. 17.1 Some of the B vitamins with roles in the central metabolic pathways. This figure highlights key steps in glycolysis and the Krebs cycle that require B vitamin-derived cofactors

B family and one prosthetic group. The coenzymes are thiamine pyrophosphate (TPP; derived from vitamin B1), FAD (derived from vitamin B2), NAD^+ (derived from vitamin B3), and coenzyme A (CoA; derived from vitamin B5). The prosthetic group is lipoyllysine, a derivative of lipoic acid. (Lipoyllysine is not a true coenzyme because it binds covalently to the enzyme. It is also not a vitamin, but it is an essential cofactor synthesized in small amounts in the body.) See Fig. 17.2 for an overview of **PDH** function, also described below [1, 3, 10,11].

The enzyme complex is organized as follows: **E1**: TPP is a coenzyme that binds tightly to E1. **E2**: Lipoyllysine (a lysine residue on the enzyme covalently attached to lipoic acid) is a prosthetic group bound to E2. Specifically, a lipoyllysine is a type of lipoamide. CoA also functions with E2. **E3**: FAD and NAD^+ are associated with E3. Together, they function in the conversion of **pyruvate** to **acetyl-CoA** (Figs. 17.1 and 17.2), catalyzed by **PDH**. This is a multistep process, as follows:

1. The first reaction is catalyzed by the **E1** component of **PDH**. In this reaction, pyruvate is decarboxylated in a reaction where carbon 1 (C1) is released from pyruvate as CO_2. The remaining two carbon molecules bind to the coenzyme TPP (derived from vitamin B1, thiamine), which then forms a complex called hydroxyethyl-TPP.
2. Next, the **hydroxyethyl group** (a two-carbon molecule originally from pyruvate) is transferred from the TPP to the lipoyllysine in the **E2** component. In its oxidized form, lipoyllysine contains a disulfide bond between two sulfur atoms (Fig. 17.2). During this transfer, the disulfide bond is reduced, forming two free sulfhydryl (-SH) groups (also called thiol groups). Simultaneously, the hydroxyethyl group is oxidized to an acetyl group. This oxidation involves the loss of two electrons and two protons, converting the terminal hydroxyl (-OH) group into a carbonyl (C=O), forming the acetyl group. The resulting product is acetyl-dihydrolipoyllysine.

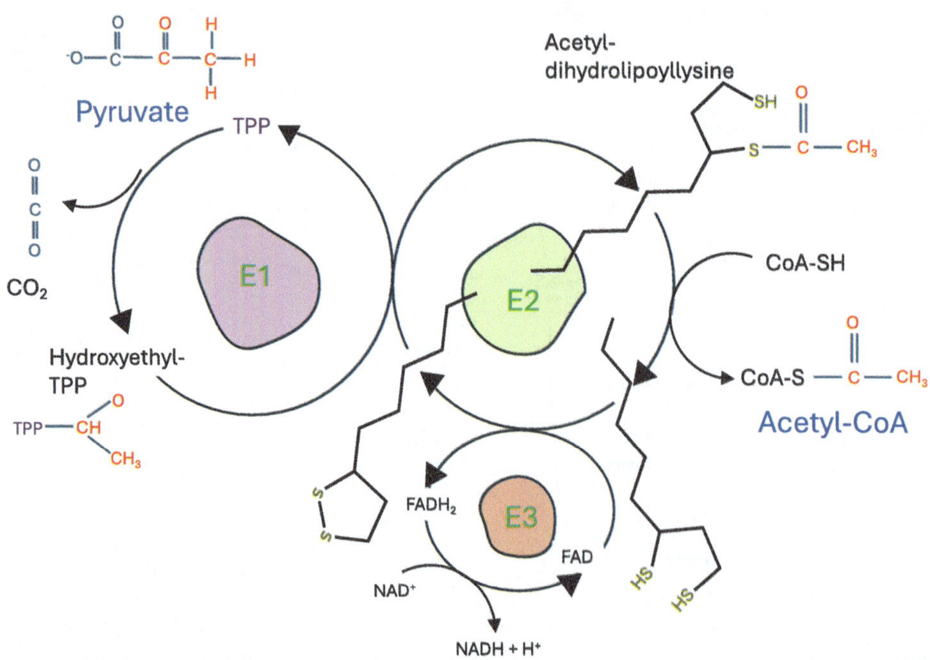

Fig. 17.2 The basic structure and function of the pyruvate dehydrogenase complex (PDH). PDH is a complex, and it contains three enzymes: E1, E2, and E3. See text for details. (*Source*: Lipoic acid was rendered using MolviewPro)

3. The acetyl group (CH_3-CO-) from acetyl-dihydrolipoyllysine is now transferred to coenzyme A (CoA; derived from vitamin B5, pantothenic acid), which contains a reactive thiol (-SH) group. This reaction forms a thioester bond. The result is the formation of acetyl-CoA, the final product of the **PDH** reaction, and leaves dihydrolipoyllysine.
4. The **E3** component of **PDH** then catalyzes the reoxidation of dihydrolipoyllysine, restoring its disulfide bond. In the process, electrons are transferred to FAD (derived from vitamin B2, riboflavin), reducing it to $FADH_2$. Subsequently, $FADH_2$ transfers electrons to **NAD^+ (derived from vitamin B3, niacin)**, resulting in NADH and H^+. This regenerates FAD, preventing its accumulation. The enzyme complex is now ready to catalyze another reaction.

2.2.3 Krebs Cycle

NAD^+ (from **vitamin B3**) has key roles in the Krebs cycle. NAD^+ functions as a coenzyme for three enzymes in this cycle, where it is reduced to NADH [7–9], as indicated in Fig. 17.1.

FAD (from **vitamin B2)** also has key roles in the Krebs cycle. It is a coenzyme for the enzyme **succinate dehydrogenase,** which leads to the conversion of succinate to fumarate [3, 11, 12].

TPP (from **vitamin B1**) is a coenzyme for **α-ketoglutarate dehydrogenase (α-KGDH),** which catalyzes the conversion of α-ketoglutarate (αKG) to succinyl CoA. **α-KGDH** functions similarly to **PDH**, and **TPP** has a similar role in this enzyme. Like **PDH**, **α-KGDH** is a multienzyme complex made up of three component enzymes. In **α-KGDH**, TPP facilitates the decarboxylation of αKG at **E1** and the formation of a hydroxyethyl intermediate, which is then transferred to lipoamide on **E2** [13, 14].

CoA (from **vitamin B5)** has a role every time a CoA molecule is added to a carbon intermediate. Thus, vitamin B5 is important not only for the conversion of pyruvate to acetyl CoA but also for other reactions such as the conversion of αKG to succinyl CoA in the Krebs cycle, via its role as a coenzyme for the enzyme complex **α-KGDH** [3, 10] (see Fig. 17.1).

2.2.4 Oxidative Phosphorylation

Oxidative phosphorylation functions a bit differently from other metabolic pathways, because it is driven by the electron transport chain, using redox carriers. Several B vitamins do have a role in the process, however, as follows and as described in [3, 15].

NADH (derived from **vitamin B3**) is a coenzyme for complex I. It donates electrons to the electron transport chain via this complex.

FMN and **FAD** (both derived from **vitamin B2**) function as prosthetic groups for complex I and complex II, respectively. They remain tightly bound to these complexes while transferring electrons to ubiquinone.

2.2.5 Metabolism of Fats

B vitamins are also important in the metabolism of fats. For example, the reaction from glycerol-3-phosphate to DHAP is catalyzed by **glycerol-3-phosphate-dehydrogenase-1 (GPD1)**. This is an important reaction in the breakdown of TAGs, and it uses **NAD^+** (vitamin B3) as a cofactor.

Prior to and during fatty acid breakdown by β-oxidation, **CoA (from vitamin B5)** gets added to fatty acid chains, and the resulting fatty acids form fatty-acyl-CoA. **CoA (from vitamin B5)** also plays a role in the carnitine shuttle, which facilitates fatty acid transport into the mitochondria (see Chap. 11), which is critical for fatty acid β-oxidation. β-oxidation of fatty acids also involves other B vitamins. It involves **FAD** (from **vitamin B2),** which is a cofactor for **acyl-CoA dehydrogenase**; **NAD** (from **vitamin B3)**, a cofactor for **β-keto-acyl-CoA dehydrogenase**; and **CoA** (from **vitamin B5**), a cofactor for **thiolase**, which has a role in the cleavage of the fatty acid chains.

While even-chain fatty acids are oxidized into acetyl-CoA, odd-chain fatty acids are oxidized to propionyl-CoA, which then enter the Krebs cycle as succinyl-CoA [16, 17]. In this reaction, propionyl-CoA is first carboxylated to form methylmalonyl-CoA. This carboxylation reaction is catalyzed by **propionyl-CoA carboxylase**, which uses the cofactor **biotin** (from **vitamin B7**). Methylmalonyl-CoA is converted to succinyl-CoA by the enzyme **methylmalonyl-CoA mutase**, which uses the cofactor **cobalamin** (from **vitamin B12**). This causes a rearrangement of the carbon skeleton. Succinyl-CoA then proceeds through the Krebs cycle. Fatty acid metabolism is discussed in Chap. 11, and the links to vitamin derivatives are outlined in Fig. 17.1 [3, 15, 17, 18].

2.2.6 Deamination of Amino Acids

During the breakdown of amino acids, transamination reactions occur. During these reactions, amino acids such as Ala or Asp lose an amino group (-NH_2), which is transferred to αKG, resulting in the formation of Glu and another molecule such as pyruvate or OAA. The transamination process (removal of the amino group and its addition to another molecule) is catalyzed by an enzyme in the **aminotransferase** family, and **PLP** (from **vitamin B6**) is an essential coenzyme in this process. **PLP** was discussed in Chap. 14. Briefly, **PLP** accepts the amino group and transfers it to its substrate. Pyruvate, oxaloacetate, and glutamate, products of deamination reactions, can feed back into cellular respiration pathways.

Glu, which can result from many of the transamination reactions, can also be converted to GABA, an inhibitory neurotransmitter. **PLP** has a key role in this process as well [3]. The role of **PLP** in converting glutamate to GABA can help explain why a deficiency in vitamin B6 is associated with seizures. The lack of sufficient levels of this inhibitory neurotransmitter leads to increased excitability in the nervous system, potentially resulting in seizures. Deficiency in vitamin B6 may be caused by factors such as alcoholism, impaired B6 metabolism, or metabolic disorders. It can be treated with lifelong supplementation with vitamin B6 or PLP [19].

Some amino acids (specifically Met, Val, isoleucine (Ile), and threonine (Thr)) can feed into the Krebs cycle by conversion to succinyl CoA. They do so via several steps and enzymes, where the amino acids are broken down into propionyl-CoA. Propionyl-CoA is then converted to succinyl-CoA in a multistep process as described above for even-chain fatty acids, which involves **propionyl-CoA carboxylase**, **methylmalonyl-CoA mutase (MCM), biotin (Vitamin B7)**, and **cobalamin (vitamin B12)** [3, 20].

2.2.7 The Pentose Phosphate Pathway and Nucleotide Synthesis

The pentose phosphate pathway (PPP), as discussed in Chap. 15, is a shunt from glycolysis and is important for the formation of purines, pyrimidines, and various biosynthetic intermediates. Several vitamin B-derived coenzymes are important in this pathway. **NAD^+** (from **vitamin B3**) is important for an early step in this pathway when glucose-6-phosphate (G6P) is converted to 6-phospho-gluconolactone. In this step, NAD^+ is reduced to NADH in a reaction catalyzed by **glucose-6-phosphate-dehydrogenase (G6PD)** [18]. **NAD^+** (from **vitamin B3**) is also necessary for the conversion of 6-phosphogluconate to ribulose-5-phosphate (Ru5P), catalyzed by **6PGD**. **TPP** (from **vitamin B1**) is a coenzyme for the enzyme **transketolase (TKT)**, which has a role in regulating the levels of ribose-5-phosphate (R5P), a sugar that forms the sugar phosphate backbone for purines and pyrimidines [5, 10].

B vitamins are essential for purine metabolism. In addition to regulating the levels of 5-carbon ring precursors in the pentose phosphate pathway, these compounds are also important for subsequent steps. The conversion of PRPP to IMP in purine metabolism requires one-carbon unit from the folate pathway. Likewise, the synthesis of dTMP from dUMP also requires the transfer of one-carbon unit from the folate pathway. The **folate pathway** is initiated from **folate**, a derivative of **vitamin B9.** The folate pathway also involves multiple other coenzymes derived from vitamin B, as described in Chap. 16. This includes **cobalamin**, derived from **vitamin B12**. The vitamins involved in the folate pathway are especially important in rapidly dividing cells, such as erythroblasts, the precursors of RBCs, because rapid cell division requires large amounts of DNA replication.

2.3 Disorders Involving B Vitamins

B vitamins are important, and although we only need them in small amounts, deficiencies in B vitamins can cause multiple serious health problems. In some cases, these problems can be treated with dietary supplements, and in other cases, the problem is more complex. A few examples are described below.

2.3.1 Vitamin B1 (Thiamine) Deficiency Causes Beriberi

Beriberi is a disorder that can result from too little **vitamin B1 (thiamine)**. It is rare in people with a varied diet, but it can occur in people with dietary deficiencies. People with alcohol addiction are also at increased risk for this deficiency because alcohol affects the body's ability to absorb **thiamine** [23]. Symptoms include weakness, mental confusion, and encephalopathy, which is associated with damage to the brain, pain, and damage to the heart.

The symptoms of this disorder are caused by a lack of **TPP**, the active form of **vitamin B1**. **TPP**, as discussed above, is an important coenzyme for the enzymes **PDH** and **αKGDH**. Reduced function of these enzymes due to **TPP** deficiency has important implications in the disorder, as follows:

The **PDH complex** plays an essential role in carbohydrate metabolism by catalyzing the generation of acetyl-CoA from pyruvate. This acetyl-CoA can enter the Krebs cycle, where it is further metabolized, leading to ATP generation. Without a fully functional **PDH,** ATP generation is reduced, and pyruvate and lactate accumulate (lactate forms from the excess pyruvate; see Chap. 10). The result is decreased energy production, and this is especially problematic for tissues with strong energy needs such as the brain [1, 3, 5, 9, 10].

α-KGDH catalyzes an important step in the Krebs cycle: the conversion of αKG to succinyl CoA. Therefore, its deficiency impairs the Krebs cycle, thus also leading to a decrease in ATP, as well as a rise in αKG (see Fig. 17.1).

Beriberi is treatable, especially if treatment starts before severe symptoms begin. Treatment involves dietary supplementation with **thiamine (vitamin B1)** and consumption of thiamine-rich foods and thiamine-fortified foods. In advanced cases, thiamine infusions may be used [13, 21, 22].

Interestingly, the symptoms of mercury and arsenic poisoning are similar, though not identical, to beriberi. This is because these toxins also inhibit **PDC** and **α-KGDH**, though they inhibit them in a slightly different way than thiamine deficiency. **Thiamine deficiency** mainly affects the **E1** enzyme of **PDC** and **α-KGDH**, due to the need for **TPP** by these enzymes. Arsenic and mercury, rather than affecting E1, specifically bind to the sulfhydryl groups on the reduced lipoyllysine attached to E2. Arsenic and mercury poisoning have wider effects than thiamine deficiency because they also act on other enzymes that have sulfhydryl groups, and some of these are involved in important metabolic functions such as regulation of mitochondrial function and ATP production [3, 23, 24].

Both arsenic and mercury are heavy metals that have a strong affinity for sulfhydryl groups. Therefore, sulfhydryl compounds can bind to these toxins and compete with them for binding to enzymes. These compounds act as chelators, forming stable complexes with arsenic or mercury, and they are often used as a treatment for arsenic or mercury poisoning. These treatments can be successful if started promptly, but they are not perfect, can lead to other complications, and cannot necessarily reverse damage that has already occurred [3, 15, 24, 25].

2.3.2 Vitamin B12 (Cobalamin) Deficiency Causes Anemia

Vitamin B12 (cobalamin) is especially important in cells that have high levels of proliferation because of its role in the biosynthesis of nucleotides (see Chap. 17). In the folate pathway, **vitamin B12** is important for producing methionine (Met) and for regenerating **THF** from **5-MTHF**, which continues the folate pathway and enables nucleotide synthesis (see Chap. 17). When vitamin B12 is deficient, there is therefore a shortage of nucleotides needed for DNA replication. This is especially important in rapidly proliferating cells like erythroblasts, the precursors of RBC. When vitamin B12 is deficient, these immature blood cells cannot divide properly, and instead they become enlarged cells called megaloblasts. These enlarged blood cells lead to ineffective erythropoiesis and cause a type of anemia called megaloblastic anemia. Neutrophils are also rapidly dividing cells, and their growth is also affected by a deficiency in vitamin B12. In these cells, cytoplasmic division continues even though the nuclear division is impaired, resulting in hypersegmented cells with multiple segmented nuclear lobes [26–28]. This increases the risk of infection in patients because these cells are important for defense. Vitamin B12 deficiency, also called cobalamin deficiency, is also called the "folate trap" and is discussed in Chap. 17.

Vitamin B12 (cobalamin) deficiency has various causes, including poor diet, a strict vegetarian diet, malabsorption, anemia, gastrointestinal problems, and some medications. The problem can be treated by injections with cobalamin, oral supplementation, and addressing the symptoms [6, 29].

2.3.3 Vitamin B12, Intrinsic Factor (IF), and Pernicious Anemia

In addition to getting enough **vitamin B12 (cobalamin)** in the diet, its proper absorption is also essential. Vitamin B12 comes from foods such as meat, dairy, and eggs. It is released from the food in the stomach, where it binds to a protein called haptocorrin (R-protein), which protects it from stomach acid. In the small intestine, pancreatic enzymes free vitamin B12 from haptocorrin. A glycoprotein called **intrinsic factor (IF)** then binds to vitamin B12, protecting it from digestion by enzymes and ensuring its absorption. After passing through the ileum of the small intestine, the **vitamin B12-IF complex** binds to receptors on intestinal cells, allowing vitamin B12 absorption into the bloodstream [30, 31]. **IF** is a critical part of the process. **IF** is produced by cells in the stomach, and it is essential for the absorption of vitamin B12. If **IF** is not present, vitamin B12 absorption is severely impaired, leading to a type of anemia called pernicious anemia. Lack of IF can therefore cause symptoms that are similar to vitamin B12 deficiency, but it is more difficult to treat. Deficiency in **IF** can occur for different reasons. For example, it may occur because IF is not produced correctly. Pernicious anemia, however, is defined by an autoimmune reaction in which IF is destroyed, and this is the most common mechanism for IF deficiency [32, 33].

Treatment for IF deficiency is more complex than the deficiency of vitamin B12 because it cannot be resolved simply by supplementing the diet with vitamin B12. Treatment must bypass the digestive system. This can be achieved by injecting vitamin B12. Another option is to give the patient very high oral doses that allow small amounts of vitamin B12 to be absorbed passively. Injections with vitamin B12 are the most common treatment, and these injections are given periodically, throughout life [5, 32–34].

2.3.4 Vitamin B7 (Biotin) Deficiency

Vitamin B7 (biotin) deficiency is very rare in people with a normal diet. In rare cases when biotin deficiency occurs, it causes symptoms due to its roles as a coenzyme for important carboxylases that are involved in metabolism. Examples are **pyruvate carboxylase, acetyl-CoA carboxylase,** and **propionyl-CoA carboxylase**. These enzymes are important in carbohydrate metabolism, gluconeogenesis, fatty acid synthesis, and amino acid metabolism. Deficiency in **vitamin B7** can cause neurological symptoms such as numbness, depression, and learning disorders because biotin is important for cellular metabolism in the nervous system. It can also cause skin and hair problems and rashes due to its role in fatty acid metabolism. Impaired fatty acid synthesis, for example, affects the health of hair follicles, which need fatty acids for proper structure and function. Biotin is also necessary for the production of keratin, which is needed for both skin and hair. Biotin deficiency can also interfere with the immune system's function and cause various metabolic complications.

Vitamin B7 deficiency can be caused, rarely, by a poor diet. It can also be caused by a deficiency in **biotinidase**, an enzyme that is important for the recycling and release of biotin from dietary proteins. It can also be caused by certain other medical conditions, alcohol use, and some medications. It is generally treated by supplementation, by dietary modifications, or by making changes in medications [35].

3 Vitamin C (Ascorbate)

Vitamin C is also a water-soluble vitamin. It is important for collagen synthesis, carnitine biosynthesis, catecholamine synthesis, and the synthesis of some neuropeptides. Vitamin C is a cofactor in several enzymatic reactions, primarily due to its function as a strong reducing agent, which can donate electrons to other molecules. Vitamin C is **ascorbic acid** in its protonated form (often found in supplements), while **ascorbate** is the main physiological form. Ascorbate can be reversibly oxidized to dehydroascorbic acid (DHA), which can be transported into cells and reduced back to ascorbate as part of its antioxidant function [36–39]. The structures of ascorbic acid and ascorbate are shown in Fig. 17.3.

3.1 Collagen Synthesis

Vitamin C has an important role in the production and maintenance of collagen. Collagen is a fibrous extracellular matrix protein that provides structure to many parts of the body, including skin, bone, tendon, cartilage, and various connective tissues. It is the most abundant protein in animals. There are several steps involved in the complete formation of mature collagen. An important step is hydroxylation (addition of an -OH group) of lysine and proline residues on collagen. Hydroxylation of proline is necessary for the hydrogen bonds that form in collagen, giving it its triple helix structure. Hydroxylation of lysine is also important because it allows for glycosylation, which is important for collagen crosslinking and helps provide strength and stability (see Fig. 17.3) [40]. The hydroxylation of proline and lysine is catalyzed by **prolyl hydroxylase** and **lysyl hydroxylase**, and **vitamin C (ascorbate)** is a **cofactor** in these reactions. This hydroxylation occurs in the ER [20].

Lysyl hydroxylase and **prolyl hydroxylase** are αKG-dependent enzymes. These types of enzymes require αKG, iron (Fe^{2+}), and oxygen (O_2). They use αKG as a co-substrate, which subsequently undergoes oxidative carboxylation, which produces succinate and CO_2. During the hydroxylation process, the ferrous form of iron, Fe^{2+}, is oxidized to the ferric form, Fe^{3+}. In this process, iron loses one electron. This electron is donated to molecular oxygen (O_2), which allows one of the oxygens to be used for hydroxylation. The other oxygen is used to oxidatively decarboxylate αKG. Thus, as a final result, CO_2 is released, αKG is converted to succinate, and Fe^{2+} is changed to Fe^{3+} (see Fig. 17.4).

After the hydroxylation reaction, **vitamin C** acts as a reducing agent, donating an electron back to iron. This regenerates iron from its oxidized form (Fe^{3+}) to its reduced form (Fe^{2+}), restoring the enzymes for future use. In the process, vitamin C (ascorbate) is oxidized by donating its electrons. Without vitamin C, Fe^{3+} would accumulate, and the enzyme would eventually

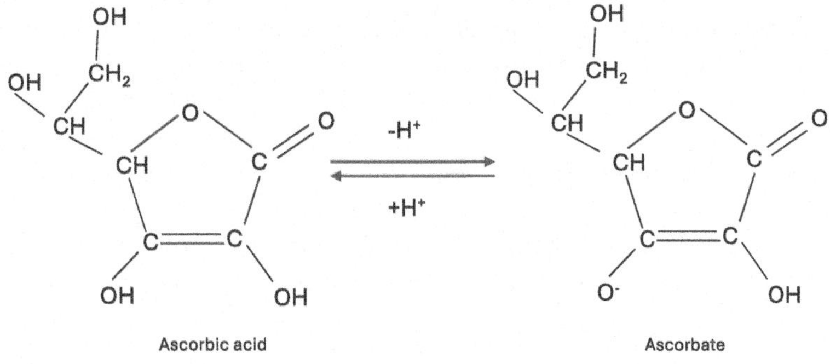

Fig. 17.3 The structure of vitamin C. Ascorbic acid and ascorbate forms are shown

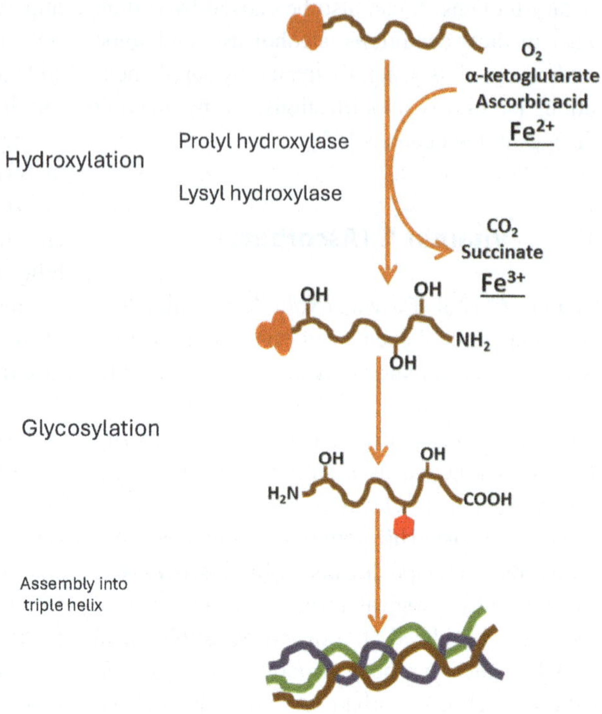

Fig. 17.4 Hydroxylation of collagen. Collagen hydroxylation depends on vitamin C. See text for details. (*Source*: Figure modified from Toxqui and Vaquero [74]. cc-by MDPI. Licensed under the Creative Commons Attribution 4.0 (CC BY 4.0) license)

become inactive. After hydroxylation, some of the hydroxylysine residues are also glycosylated, which involves the addition of sugars. After hydroxylation and glycosylation, three alpha chains of collagen form a triple helix, which is the secreted form of collagen that will ultimately assemble into collagen fibrils [3, 15, 41, 42].

3.2 Catecholamine Synthesis

Vitamin C is also a cofactor for the enzyme **dopamine hydroxylase.** This enzyme catalyzes the synthesis of the neurotransmitter norepinephrine from dopamine. Norepinephrine is part of the body's fight-or-flight response [36].

3.3 Carnitine Biosynthesis

Vitamin C is also a **cofactor** for the enzymes **trimethyllysine hydroxylase** and **gamma butyrobetaine hydroxylase**. These enzymes have important roles in **carnitine** biosynthesis [43]. Carnitine synthesis is important for the **transport of fatty acids** into the mitochondria for β-oxidation, following which they can be metabolized completely for ATP generation (see Chap. 11).

3.4 Vitamin C and Health

Vitamin C can be obtained in the diet from citrus fruits and various vegetables. Too little vitamin C can lead to **scurvy**, which is associated with an abnormality in connective tissue, which is made up of collagen. Scurvy is rare today, but it is famous for having been common among sailors in the nineteenth century. The symptoms of scurvy occur because a deficiency in vitamin C prevents hydroxylation of proline and lysine residues of **collagen**. Lack of proline hydroxylation is especially problematic because it has a severe effect on the structure and stability of collagen. In healthy tissues, there is a constant turnover of collagen, where it is continually degraded and replaced. A deficiency in vitamin C will lead to decreased stability of newly synthesized collagen because it will not be properly hydroxylated. As collagen is replaced, the new collagen will be unstable, and tissues will lose structure and rigidity. Over time, more and more defective collagen accumulates in tissues, leading to more severe scurvy symptoms. These symptoms include loose teeth, swollen and bleeding gums, fatigue, joint and muscle pain, and poor wound healing. The symptoms worsen over time [36, 38, 44].

While a severe deficit in vitamin C is associated with scurvy, even a moderate deficiency can lead to health problems. These can include impaired wound healing, which can be associated with ulcers and other wounds. Vitamin C deficiency is also associated with problems with the immune system, weakness, fatigue, cardiovascular disease, and increased risk for cancer [45].

4 Vitamin D

Another important vitamin is **vitamin D**. As a fat-soluble vitamin, it can last longer in the body than water-soluble vitamins like the B vitamins. Like most other vitamins, the physiological functions of vitamin D are not actually carried out by vitamin D but by its active metabolites. The two types of vitamin D are vitamin D3 and vitamin D2. Vitamin D3 is more abundant, more potent, and more stable. The metabolism of both forms will be discussed below. Vitamin D metabolites are not coenzymes, but they have roles in regulating calcium and phosphate homeostasis in the body. They also have roles in regulating gene expression and regulating the synthesis of some proteins [1, 46–48].

Unlike the B vitamins and most other vitamins we get from our diet, we mainly get vitamin D from the sun. UV light converts vitamin D from its precursor form in the skin to its active form. Some vitamin D precursors, however, can also come from the diet. The steps leading to the production of active vitamin D metabolites are as follows (Fig. 17.5) [1, 47, 49, 50]:

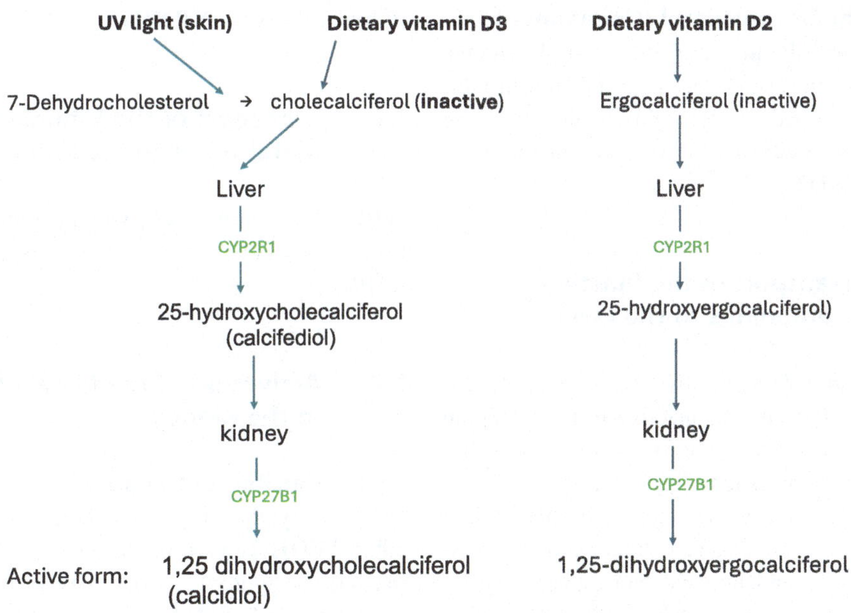

Fig. 17.5 Vitamin D. The main routes of vitamin D synthesis are shown. See text for details

4.1 The First Step in Vitamin D Activation Can Occur in Three Main, Alternative Ways

4.1.1 Formation of Cholecalciferol in Response to Skin Exposure to UV Light

This is the first step in the most common mechanism of activation of vitamin D3. UV-B light triggers the conversion of **7-dehydrocholesterol** (also called **provitamin D3**; a **precursor of vitamin D3** found naturally in the skin) to **cholecalciferol**, which is the **inactive form of vitamin D3**.

4.1.2 Dietary Vitamin D3, Usually in the Form of Cholecalciferol

Although sunlight is a major source of vitamin D, we can also ingest some vitamin D, primarily from foods like fish or liver. This ingested vitamin D is usually in the form of **cholecalciferol** (this is **the same inactive form of vitamin D3** generated from UV light, as described above).

4.1.3 Dietary Vitamin D2 in the Form of Ergocalciferol (Uncommon)

Another **occasional** source of ingestible vitamin D is from some plants, such as certain mushrooms that are exposed to UV light. In this case, we ingest **ergocalciferol**, which is an inactive form of **vitamin D2**.

4.2 Transport of the Inactive Metabolites to the Liver

Cholecalciferol (the inactive form of vitamin D3) and **ergocalciferol** (the inactive form of vitamin D2) are transported to the liver, as follows. After absorption or ingestion, they enter intestinal epithelial cells. There, they are incorporated into chylomicrons, which carry them into the lymphatic system and into the bloodstream. There, the chylomicrons are metabolized, and the vitamin D binds to **vitamin D binding protein (DBP)** until it reaches the liver, either as DBP-bound protein or in some cases within chylomicron remnants [52].

4.3 First Step of Activation of Vitamin D Metabolites in the Liver

4.3.1 First Step of Activation of the Vitamin D3 Metabolite Cholecalciferol in the Liver

In the liver, the enzyme **25-hydroxylase** (gene name; **CYP2R1)** adds a hydroxyl group to carbon-25 (C25) on **cholecalciferol.** This results in **25-hydroxycholecalciferol** (also known as **calcifediol, calcidiol**, or **25-hydroxyvitamin D** and abbreviated (**25-OH-D$_3$**)), which is still an **inactive** metabolite of **vitamin D3**. This is the major form of **circulating vitamin D**. It is the form that is usually tested for when measuring vitamin D levels in a blood test although it is actually only the precursor of the active vitamin D metabolite.

4.3.2 The Ergocalciferol Activation Pathway (for Vitamin D2)

For the less common **ergocalciferol**, the product made in the liver is **25-hydroxyergocalciferol** (also called **25-hydroxyvitamin D2** or **25 (OH)D$_2$**). This is also an **inactive** metabolite of **vitamin D2**. This reaction uses the **same enzyme**: **25-hydroxylase (CYP2R1)**.

4.4 Transport of the Vitamin D Metabolites to the Kidney

Calcifediol and **25-hydroxyergocalciferol** travel to the kidney via the bloodstream, bound to DBP.

4.5 Activation of the Metabolites in the Kidney

4.5.1 Calcifediol to Calcitriol

In the kidney, the enzyme **1-α-hydroxylase** (also **called 25(OH-1a-OHase**, gene name **CYP27B1**) transfers a hydroxyl group (-OH) to carbon 1 (C1) of **calcifediol**. The result is the formation of **1,25 dihydroxycholecalciferol** (also called **calcitriol, 1,25-dihydroxyvitamin D3,** or **1,25 (OH)$_2$D$_3$**). This is the **active metabolite of vitamin D3**.

4.5.2 The Ergocalciferol Pathway

Also in the kidney, **25-hydroxyergocalciferol** is converted to **1,25-dihydroxyergocalciferol (also called 1,25 dihydroxyvitamin D2** or **$1,25(OH)_2$, D_2)**, which is also an active metabolite, the active form of **vitamin D2**. (The same enzyme, **1-α-hydroxylase (CYP27B1),** catalyzes this reaction).

4.6 Regulation of the Enzymes That Catalyze the Production of Active Vitamin D Metabolites

The enzymes **25-hydroxylase (CYP2R1)** (in the liver) and **1-α-hydroxylase (CYP27B1)** (in the kidney) are crucial for the two steps that lead to the production of the active forms of both types of vitamin D, **calcitriol** and **1,25-dihydroxyergocalciferol**, from inactive **vitamin D3** and inactive **vitamin D2**, respectively. We need these enzymes to make active vitamin D metabolites every time we are exposed to sunlight (or ingest certain foods). The activity of **1-α-hydroxylase (CYP27B1)** is tightly regulated. Its regulation is complex, but two important components of 1-α-hydroxylase regulation are the levels of phosphate and calcium in the blood. 1-α-hydroxylase levels are upregulated by the presence of low phosphate levels in the blood, which itself responds to low calcium levels in the blood. When calcium levels are low, the parathyroid gland detects this and then produces and releases **parathyroid hormone (PTH)**. **PTH,** in turn, travels to the kidney and activates signaling pathways that lead to upregulation of the gene encoding **1-α-hydroxylase (CYP27B1)**. This increases active vitamin D metabolite levels [51, 53].

4.7 What Does Calcitriol (the Active Metabolite of Vitamin D3) Do?

Calcitriol's primary role is to increase calcium (Ca^{2+}) and phosphate (PO_4^{3-}) levels in the blood. Its production is mainly stimulated by parathyroid hormone (PTH), which increases the levels of **1-α-hydroxylase (CYP27B1)** in the kidney when blood calcium is low. Low phosphate levels also enhance calcitriol synthesis. **Calcitriol**, in turn, increases the absorption of calcium and phosphate in the intestine. It also increases reabsorption of calcium at the kidneys (renal reabsorption) to reduce its loss in the urine [47, 53–55].

Calcitriol promotes bone resorption, which is the process of osteoclasts breaking down bone tissue. Bone is made up largely of connective tissue, and it maintains its hardness due to minerals, largely hydroxyapatite ($Ca_{10}(PO_4)_6(OH)_2$), the crystal form of calcium phosphate. Bones are dynamic and are constantly broken down via osteoclasts and rebuilt via osteoblasts. The periodic breakdown of bones via osteoclasts is essential for maintaining blood mineral levels. Calcitriol promotes bone resorption by stimulating osteoblasts to express the cytokine RANKL. By producing RANKL, osteoblasts have a role in controlling osteoclasts, the cells that break down bone. When RANKL is produced, it activates osteoclasts and breaks down bone. An important mechanism by which osteoclasts do so is by dissolving the hydroxyapatite and releasing Ca^{2+} and PO_4^{3-} into the bloodstream.

Despite this function of calcitriol, calcitriol maintains bone health overall because it assures Ca^{2+} and PO_4^{3-} availability by multiple mechanisms. For example, as discussed above, calcitriol increases intestinal absorption of Ca^{2+} and PO_4^{3-} and reduces renal loss of Ca^{2+}, which ensures the availability of Ca^{2+} and PO_4^{3-} for bone [47, 56].

4.8 Vitamin D Deficiency

Vitamin D deficiency can have many causes, but a lack of UV light (due to lack of sunlight exposure) is a main cause. Furthermore, skin pigmentation can also make a difference, where light skin can produce more vitamin D than darker skin. There can also be problems with the absorption of vitamin D. Vitamin D is a fat-soluble vitamin, and it needs dietary fat in order to be absorbed from the intestine (in the case of ingested vitamin D) to subsequently reach the lymphatic system and enter the bloodstream and tissues. Fat absorption problems can be caused by problems with bile and with problems with the intestinal tract absorbing fat.

Diseases associated with vitamin D deficiency include rickets (in children) and osteomalacia (in adults). In rickets, the bones cannot develop properly due to low vitamin D. This leads to "bendy" bones because the bones in rickets consist primarily of collagen rather than mineralized hard bone. In osteomalacia (also known as soft bone disease), bones have already formed and hardened before the deficiency begins, but they become brittle due to vitamin D deficiency. Normally, calcium and phosphate are continuously removed from bone, but they are also redeposited in bone during mineralization. In vitamin D deficiency, however, calcium and phosphate absorption from the gut is reduced, so there is not enough available to replace what is lost. As a result, bones lose mineral content ($Ca_3(PO_4)_2$) and become soft and fragile [1, 47, 49].

4.9 Vitamin D Supplements

The most reliable way to treat a vitamin D deficiency is to take supplements. Dietary supplements can come in different forms. The most common is **cholecalciferol (vitamin D3)**. This type is naturally produced from **7-dehydrocholesterol** in response to UV light or is found in some dietary sources. The supplement is metabolized in the body to produce active vitamin D3 metabolites, just like the natural form of this compound. While this is the most common, some supplements are made from other sources of vitamin D, such as ergocalciferol, which is vitamin D2. Because vitamin D is fat soluble, there is a possibility of overdose, so taking the correct dose is important. Overdosing on vitamin D can lead to kidney stones due to alterations in calcium metabolism and increased calcium excretion, and abdominal problems due to the contraction of smooth muscles [57].

5 Vitamin A (Retinoids)

5.1 Introduction to Vitamin A

Vitamin A is a fat-soluble vitamin. It is obtained from plants and animals in the diet, mostly in the forms of retinol, retinal, and retinoic acid. **Retinol** is the major circulating metabolite in the blood, and it can reversibly form **retinal. Retinal**, in turn, can be irreversibly oxidized to **retinoic acid** [58]. Retinol has roles in transport and storage, and as the precursor to the other two active derivatives.

Vitamin A that comes from plants is generally in the form of **β-carotene**, which is considered a provitamin A carotenoid. **β-carotene** can be cleaved in the intestine to produce two molecules of retinal. Retinal, in turn, can be converted into **retinol** or **retinoic acid** when needed [58]. The structures of all β-carotene and its products are shown in Fig. 17.6.

Good animal sources of vitamin A are liver and kidney. They may come in any of the three primary forms although when they come from animal sources, they may be esterified with the long-chain fatty acid palmitate, such as retinyl palmitate. Once they are digested, they are hydrolyzed, and **retinol** is released and used [1, 59].

5.2 Retinal Is Important for Low-Light Vision

Vitamin A derivatives are not coenzymes, but one of the metabolites, **retinal**, functions similarly to a coenzyme because it interacts with a protein and regulates its function.

In the regulation of vision, retinal exists in two isomeric forms, the all-trans retinal (which we often simply call retinal) and the 11-cis retinal. The 11-cis form is a bent form of the molecule where the bond at carbon 11 is in the cis position. **In the all-trans form, the conformation of the geometry around the double bond is in trans so that the molecule takes on a straighter conformation.**

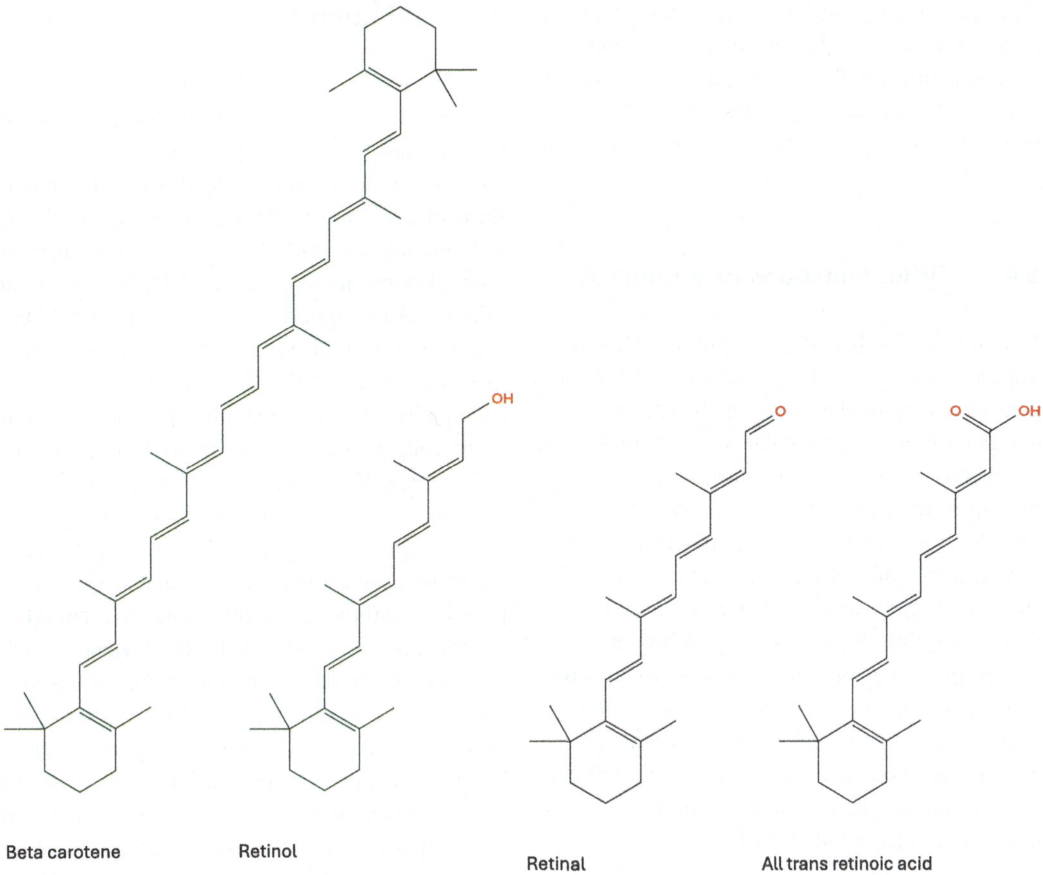

Fig. 17.6 Vitamin A and its precursors and metabolites. The structures of beta carotene, retinol, retinal, and all-trans retinoic acid are shown. (*Source*: Structures rendered using MolViewPro)

Retinal is a crucial component of a receptor protein called opsin, which is a G protein-coupled receptor (GPCR), in the rod cells of the retina. In the dark, retinal is bound to opsin in the 11-cis configuration, allowing it to fit into the opsin binding site. Opsin bound to 11-cis-retinal is called rhodopsin. When light strikes rhodopsin, this 11-cis-retinal isomerizes to all-trans retinal. This forces a conformational change in opsin. This activates opsin, which is now called metarhodopsin II. This activation activates the G protein transducin. This activates a signaling pathway that changes how light is converted into a neural signal. Because of this function, vitamin A is needed for night vision, and a lack of vitamin A can lead to blindness. Vitamin A deficiency is, in fact, a cause of blindness in children, and for this reason, many foods today are fortified with vitamin A [60–62].

5.3 Retinoic Acid Functions as a Hormone

Retinoic acid functions differently from retinal. It functions like a hormone and crosses the plasma membrane and nuclear membrane. Inside the cell, it functions as a ligand and binds to the nuclear receptors RAR and RXR. These receptors can then function as transcription factors and regulate the expression of various target genes, including those that have important roles in embryonic development, skin health, vision, and cell proliferation and differentiation [60, 63]. (Hormone receptors are discussed in more detail in Chap. 18.)

5.4 Other Functions of Vitamin A

Vitamin A also has other functions. Retinol is important for male fertility, supporting spermatogenesis and testicular function. In female reproduction, retinoic acid regulates the expression of genes that have roles in ovarian function and placental development. In the immune system, retinoic acid helps maintain mucosal barriers and regulates the activity of some immune cells, such as T cells and dendritic cells. Too little vitamin A thus leads not only to blindness or vision problems but also reproductive issues, skin damage, and immune system problems. Exogenous retinoic acid is often used as a treatment for acne because it plays a role in regulating skin turnover, maintaining skin health, and controlling inflammation [60].

5.5 Golden Rice Is a Type of Rice That Contains β-Carotene

Golden rice is a genetically altered form of rice that has genes from other organisms (specifically daffodil, a flowering plant, and a soil bacterium) that enable the rice to synthesize β-carotene. This genetically modified rice (a GMO: genetically modified organism) produces β-carotene, which can then be broken down into retinal, making this rice an excellent source of vitamin A. It is designed to help combat vitamin A deficiency in areas of the world where rice is a staple food and where vitamin A levels tend to be low in the population. The goal is to prevent blindness and vision loss caused by low levels of vitamin A. The rice is used to a limited extent in some countries, but it is controversial because of the use of GMOs. Nevertheless, it has the potential to help improve vision for many people [64–66].

6 Vitamin K

Vitamin K is a fat-soluble vitamin. The active metabolite of vitamin K is important for blood **coagulation** (clotting). Specifically, it is important for the modification, specifically **carboxylation**, of certain coagulation factors in the liver. Coagulation factors are proteins that function with platelets to stop further bleeding after an injury and to support tissue regeneration. Many coagulation factors are enzymes, and they are also discussed in Chap. 5. There are many different coagulation factors, but the primary vitamin K-dependent ones are factors II (prothrombin), VII, IX, and X, as well as proteins C and S.

These coagulation factors get carboxylated, where carboxyl groups (COO^-) are added to specific Glu residues on their side chains. Vitamin K promotes carboxylation by acting as a **cofactor** for the enzyme **γ-glutamyl carboxylase**, which carboxylates these coagulation factors. Carboxylation of the coagulation factors is important because it gives the factors a negative charge. This gives them a high affinity for calcium (Ca^{2+}), which has a positive charge. This, in turn, allows them to bind negatively charged membranes in platelets and other cells. These interactions between cells, especially platelets, and the coagulation factors, are important for clotting to occur [1, 3, 67]. (Coagulation factors are also discussed in Chap. 5.)

7 Vitamin E (α-Tocopherol)

Vitamin E is actually a group of fat-soluble compounds that include tocopherols and tocotrienols. Among these, α-tocopherol is the most biologically active form, and the main form in human tissues. Vitamin E is not a coenzyme. Instead, it is a strong antioxidant for peroxyl free radicals, those that affect fatty acids, such as those found on membranes. Vitamin E resides in the membrane. As an antioxidant, vitamin E plays an important role in protecting cells from oxidative damage caused by free radicals and reactive oxygen species (ROS). It has an essential role in maintaining the integrity of cell and organelle membranes [68]. By scavenging free radicals and neutralizing them, vitamin E helps prevent lipid peroxidation, where free radicals remove electrons from lipids. If left unchecked, this process would otherwise lead to damaged membranes and other detrimental effects in cells. Vitamin E is obtained from a variety of plant-based foods, especially vegetable oils, nuts, and seeds. Some sources are rich in α-tocopherol. Examples are almonds, sunflower seeds, and hazelnuts. α-tocopherol is the most biologically active form in humans [69]. Soybean and corn oils have higher amounts of γ-tocopherol, which is less efficiently retained in human tissues. The liver preferentially incorporates α-tocopherol into circulating lipoproteins, making it the predominant form of vitamin E in the body [1, 70].

8 Essential Minerals

8.1 What Are Essential Minerals?

In addition to vitamins, several different minerals are also essential for human nutrition. These include 13 essential minerals. Of these, six are called trace minerals, because although essential, we need them in very small amounts. See Table 17.2 for a complete list. A few examples

Table 17.2 Essential and trace minerals

Mineral	Main functions	Dietary sources	Enzymes (and some nonenzymes) they serve as cofactors for
Calcium (Ca)	Bones and teeth health, muscle contraction, nerve signaling	Dairy, leafy greens, tofu, almonds	
Phosphorous (P)	Bond structure, ATP production, DNA synthesis	Meat, diary, nuts beans, whole grains	
Magnesium (Mg)	ATP activation, nerve and muscle function	Nuts, seeds, whole grains, spinach	ATPases, DNA polymerase, alkaline phosphatase
Sodium (Na)	Fluid balance, nerve transmission, muscle contractions	Salt, seafood	
Potassium (K)	Heart function, nerve signaling, muscle contraction	Bananas, potatoes, legumes, dairy	
Chloride (Cl)	Fluid balance, stomach acid production	Table salt, seaweed, tomatoes	
Sulfur (S)	Components of some amino acids, protein structure	Meat, eggs garlic, onion	
Trace mineral			
Iron (Fe)	Oxygen transport	Red meat, spinach, legumes, fortified grains	Cytochromes, catalase
Zinc (Zn)	Immune function, DNA synthesis, wound healing	Meat, shellfish, nuts, dairy	Carbonic anhydrase, alcohol dehydrogenase, alkaline phosphatase, carboxypeptidases
Copper (Cu)	Iron metabolism, redox reactions, energy production	Shellfish, nuts, seeds, whole grains	Cytochrome c oxidase, superoxide dismutase
Manganese (Mn)	Bone formation, antioxidant activity, metabolism	Nuts, whole grains, tea, leafy vegetables	Superoxide dismutase, arginase, polymerases
Iodine (I)	Thyroid hormone production	Iodized salt, seafood, dairy	
Selenium (Se)	Antioxidant protection, thyroid function	Brazil nuts, seafood, meat, eggs	Glutathione peroxidase
Fluoride (F)	Tooth and bone health	Tea, fish, supplemented water	
Chromium (Cr)	Insulin action and glucose metabolism	Whole grains, broccoli, meat	
Molybdenum (Mo)	Detoxification and metabolism	Legumes, whole grains, nuts	Xanthine oxidase, sulfite oxidase

of enzymes that require essential minerals are described below; some of these have also been discussed in Chap. 5.

8.2 Some Enzymes That Require Essential Minerals

- Carbonic Anhydrase
  Carbonic anhydrase is an enzyme that catalyzes the conversion of carbon dioxide and water to bicarbonate and protons. It is involved in various metabolic reactions, including gluconeogenesis, fatty acid synthesis, and the urea cycle. Zinc (Zn^{2+}) is a cofactor for this enzyme [3].
- Cytochrome C Oxidase
  This enzyme is part of the electron transport chain and plays a role in the reduction of oxygen to water. Copper (Cu^{2+}) and iron (Fe^{2+}/Fe^{3+}) are cofactors for this enzyme [3, 18].
- DNA Polymerase
  DNA polymerase catalyzes the synthesis of DNA. It requires magnesium (Mg^{2+}) as a cofactor. Some polymerases also use manganese (Mn^{2+}) [71, 72].
- Alkaline Phosphatase
  This enzyme removes phosphate groups from proteins and other molecules. It uses zinc (Zn^{2+}) and magnesium (Mg^{2+}) as cofactors [73].

Definitions

Anemia Inability of the blood to carry sufficient oxygen to the body's tissues, either due to too few healthy red blood cells, too little hemoglobin in the blood cells, or both.

Beriberi A disorder due to a deficiency of vitamin B1 (thiamine). Symptoms are caused by a lack of the active form TPP.

Biotinidase An enzyme important for the recycling and release of biotin from dietary proteins. If deficient, biotin itself (vitamin B7) will also be deficient.

Coenzyme A small organic molecule that binds to an enzyme and is required for its activity. Often derived from vitamins (e.g., NAD^+ from niacin).

Decarboxylation The removal of a carboxyl group (–COOH) from a molecule, releasing CO_2.

Free radicals Free radicals are unstable molecules that have at least one unpaired electron. They are thus highly reactive and unstable. They oxidize other molecules by taking electrons. They can cause damage to lipids, proteins, and DNA.

Lipoamide The active form of lipoic acid that is covalently attached to a lysine in the PDC enzyme complex, forming lipoyllysine.

Lipoic acid A sulfur containing cofactor that participates in oxidative decarboxylation reactions.

Lipoyllysine A derivative of a lipoic acid that is attached to lysine in enzyme complexes. It cycles between oxidized (disulfide) and reduced (dithiol) states.

Megaloblastic anemia A blood disorder caused by impaired DNA synthesis, often due to folate or B12 deficiency, leading to large, immature red blood cells.

Methylmalonyl-CoA mutase A vitamin B12-dependent enzyme that converts methylmalonyl-CoA to succinyl-CoA in odd-chain fatty acid and amino acid metabolism.

Osteoclasts Cells that break down bone, releasing calcium into the bloodstream. Vitamin D increases osteoclast activity.

Pernicious anemia A specific type of anemia caused by a lack of intrinsic factor (IF), which protects vitamin B12 from digestive enzymes and is necessary for its absorption.

Peroxyl free radicals Free radicals that affect fatty acids, such as those found on membranes. Vitamin E is a strong antioxidant for these radicals.

Prosthetic group A nonprotein molecule that is tightly bound to an enzyme and essential for its activity.

Pyridoxal phosphate (PLP) The active form of vitamin B6; a coenzyme involved in amino acid metabolism, particularly transamination and decarboxylation.

Redox reaction A chemical reaction involving transfer of electrons; one molecule is oxidized, another is reduced.

Retinoids A collective term for the forms of vitamin A

Rickets Disease associated with "bendy bones" in children, caused by a vitamin D deficiency

Scurvy Disease associated with an abnormality in connective tissue. Caused by a vitamin C deficiency.

Thiamine pyrophosphate (TPP) The active coenzyme form of vitamin B1, required for decarboxylation of α-ketoacids.

Trace mineral Minerals that are essential, but that we need in only small amounts.

Transamination A reversible reaction that transfers an amino group from an amino acid to a ketoacid, commonly catalyzed by aminotransferases.

Vitamin An organic compound that cannot be produced in sufficient amounts in the human body. Needs to be ingested in small amounts nearly every day. Often gets converted into an active form to function as a coenzyme.

Study Questions

Multiple-Choice Questions

1. Which vitamin is a coenzyme for transketolase in the pentose phosphate pathway?
   A. Vitamin B2
   B. Vitamin B3
   C. Vitamin B1
   D. Vitamin B9
2. A deficiency of which vitamin can cause megaloblastic anemia due to impaired DNA synthesis?
   A. Vitamin B6
   B. Vitamin B12
   C. Vitamin A
   D. Vitamin C
3. Which vitamin is required for collagen hydroxylation and its deficiency leads to scurvy?
   A. Vitamin K
   B. Vitamin A
   C. Vitamin D
   D. Vitamin C
4. What is the active form of vitamin D3 produced in the kidney?
   A. Cholecalciferol
   B. 25-hydroxycholecalciferol
   C. 1,25-dihydroxycholecalciferol
   D. Ergocalciferol
5. If a healthy person eats food high in this vitamin (shown in picture), where will the majority of this vitamin be 24 h later?

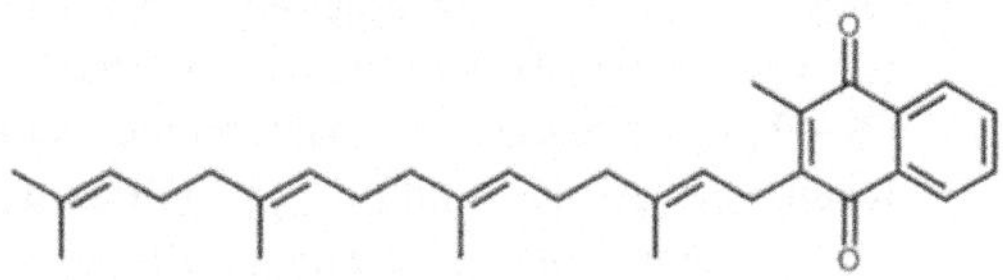

   A. Circulation
   B. Urine
   C. Adipose (fat) tissue and liver
   D. Bone
6. In a blood sample from a patient with beriberi, what changes would you predict compared to a healthy person?
   A. Elevated acetyl-CoA and succinyl-CoA
   B. Elevated pyruvate and ⍰-ketoglutarate
   C. Elevated NAD+ and CoA
   D. Elevated Mg2+
7. Which of the following can treat arsenic and mercury poisoning?
   A. Thiamine supplementation
   B. Sulfhydryl compounds
   C. Golden rice
   D. Cobalamin supplementation
   E. Exogenous retinoic acid
8. Vitamin C is involved in the synthesis of all of the following EXCEPT:
   A. Collagen
   B. Norepinephrine
   C. Nucleotides
   D. Carnitine

True/False Question

9. Fat-soluble vitamins cannot be stored in the body and must be replenished daily.

Fill-in-the-Blank Questions

10. ___________ is required for blood coagulation, specifically for carboxylation of certain coagulation factors
11. __________ is the active form of vitamin B1
12. Vitamin A deficiency is a major cause of __________ in children.
13. The forms of vitamin B are as follows: Biotin is vitamin __; cobalamin is vitamin __; folic acid (THF) is vitamin __; niacin (NAD+) is vitamin __; pantothenic acid (coenzyme A) is vitamin __; pyridoxine (PLP) is vitamin __, riboflavin (FAD) is vitamin __, and thiamine (TPP) is vitamin __.
14. In terms of the issues caused by certain vitamin deficiencies, thiamine deficiency causes ________, cobalamin deficiency causes ________, pyridoxine deficiency causes ________, vitamin A deficiency causes ________, vitamin C deficiency causes ________, and vitamin D deficiency causes ________ and ________.

Short-Answer Question

15. Explain why vitamin B12 deficiency can mimic folate deficiency symptoms.

Matching Question

16. **Match the vitamin with its function**:
    Vitamin
    A. Vitamin B6
    B. Vitamin C
    C. Vitamin D
    D. Vitamin E
    Function
    1. Hydroxylation of collagen
    2. Transamination reactions
    3. Calcium and phosphate homeostasis
    4. Antioxidant in membranes

Answer Key

Multiple-Choice Questions

1. **Answer**: C
2. **Answer**: B
3. **Answer**: D
4. **Answer**: C
5. **Answer**: C
6. **Answer**: B
7. **Answer**: B
8. **Answer**: C

True/False Question

9. **Answer**: False

Fill-in-the-Blank Questions

10. **Answer**: Vitamin K
11. **Answer**: Thiamine pyrophosphate (TPP)
12. **Answer**: Blindness
13. **Answer**:
    Biotin is vitamin $\mathbf{B_7}$
    Cobalamin is vitamin $\mathbf{B_{12}}$
    Folic acid (THF) is vitamin $\mathbf{B_9}$
    Niacin (NAD+) is vitamin $\mathbf{B_3}$
    Pantothenic acid (coenzyme A) is vitamin $\mathbf{B_5}$
    Pyridoxine (PLP) is vitamin $\mathbf{B_6}$
    Riboflavin (FAD) is vitamin $\mathbf{B_2}$
    Thiamine (TPP) is vitamin $\mathbf{B_1}$

14. **Answer**:
Thiamine deficiency causes beriberi
Cobalamin deficiency causes **anemia** (megaloblastic)
Pyridoxine deficiency causes **seizures**
Vitamin A deficiency causes **blindness**
Vitamin C deficiency causes **scurvy**
Vitamin D deficiency causes **rickets** (children) and **osteomalacia** (adults)

Short-Answer Question

15. **Answer**: Because B12 is required to regenerate THF from 5-MTHF; without it, THF is not regenerated, causing a functional folate deficiency that impairs DNA synthesis.

Matching Question

16. **Answers**:
A-2, B-1, C-3, D-4

References

1. Bender DA. Introduction to nutrition and metabolism. Boca Raton: CRC Press; 2021.
2. Medical News Today. What are vitamins, and how do they work? Available from: https://www.medicalnewstoday.com/articles/195878#what_are_vitamins.
3. Berg JM, Tymoczko JL, Gatto GJ Jr, Stryer L. Biochemistry. 9th ed. New York: MacMillan Learning; 2019.
4. Kubala JH. Can you overdose on vitamins? 2020. Available from: https://www.healthline.com/nutrition/can-you-overdose-on-vitamins#risks-of-high-doses.
5. Hanna M, Jaqua E, Nguyen V, Clay J. B Vitamins: functions and uses in medicine. Perm J. 2022;26(2):89–97. https://doi.org/10.7812/tpp/21.204.
6. Chandel NS. Carbohydrate metabolism. Cold Spring Harb Perspect Biol. 2021;13(1). https://doi.org/10.1101/cshperspect.a040568.
7. Peechakara BV, Gupta M. Vitamin B3. In: StatPearls. Treasure Island (FL): StatPearls Publishing; 2024.
8. Zapata-Pérez R, Wanders RJA, van Karnebeek CDM, Houtkooper RH. NAD(+) homeostasis in human health and disease. EMBO Mol Med. 2021;13(7):e13943. https://doi.org/10.15252/emmm.202113943.
9. Patel MS, Nemeria NS, Furey W, Jordan F. The pyruvate dehydrogenase complexes: structure-based function and regulation. J Biol Chem. 2014;289(24):16615–23. https://doi.org/10.1074/jbc.R114.563148.
10. Depeint F, Bruce WR, Shangari N, Mehta R, O'Brien PJ. Mitochondrial function and toxicity: role of the B vitamin family on mitochondrial energy metabolism. Chem Biol Interact. 2006;163(1):94–112. https://doi.org/10.1016/j.cbi.2006.04.014.
11. Kim HJ, Winge DR. Emerging concepts in the flavinylation of succinate dehydrogenase. Biochim Biophys Acta. 2013;1827(5):627–36. https://doi.org/10.1016/j.bbabio.2013.01.012.
12. Judge A, Dodd MS. Metabolism. Essays Biochem. 2020;64(4):607–47. https://doi.org/10.1042/ebc20190041.
13. Kerns JC, Gutierrez JL. Thiamin. Adv Nutr. 2017;8(2):395–7. https://doi.org/10.3945/an.116.013979.
14. Tretter L, Adam-Vizi V. Alpha-ketoglutarate dehydrogenase: a target and generator of oxidative stress. Philos Trans R Soc Lond Ser B Biol Sci. 2005;360(1464):2335–45. https://doi.org/10.1098/rstb.2005.1764.
15. Nelson DL, Cox MM, Hoskins AA. Lehninger principles of biochemistry. 8th ed. Austin: Macmillan Learning; 2021.
16. Salway JG. Metabolism at a glance. Hoboken: Wiley-Blackwell; 2017.
17. Wongkittichote P, Ah Mew N, Chapman KA. Propionyl-CoA carboxylase – a review. Mol Genet Metab. 2017;122(4):145–52. https://doi.org/10.1016/j.ymgme.2017.10.002.
18. Chandel NS. Navigating metabolism. 2nd ed. Cold Spring Harbor: Cold Spring Harbor Laboratory Press; 2020.
19. Kim JE, Cho KO. Functional nutrients for epilepsy. Nutrients. 2019;11:6. https://doi.org/10.3390/nu11061309.
20. Rodwell VW, Bender DA, Kennely PJ, Weil PA, McGuinness OP. Harper's illustrated biochemistry. 32nd ed. New York: McGraw-Hill Education/Medical; 2022.
21. WebMD. What is Beriberi. 2023. Available from: https://www.webmd.com/brain/what-is-beriberi.
22. Smith TJ, Johnson CR, Koshy R, Hess SY, Qureshi UA, Mynak ML, et al. Thiamine deficiency disorders: a clinical perspective. Ann N Y Acad Sci. 2021;1498(1):9–28. https://doi.org/10.1111/nyas.14536.
23. Ajsuvakova OP, Tinkov AA, Aschner M, Rocha JBT, Michalke B, Skalnaya MG, et al. Sulfhydryl groups as targets of mercury toxicity. Coord Chem Rev. 2020;417. https://doi.org/10.1016/j.ccr.2020.213343.
24. Balali-Mood M, Naseri K, Tahergorabi Z, Khazdair MR, Sadeghi M. Toxic mechanisms of five heavy metals: mercury, lead, chromium, cadmium, and arsenic. Front Pharmacol. 2021;12:643972. https://doi.org/10.3389/fphar.2021.643972.
25. Flora SJ, Pachauri V. Chelation in metal intoxication. Int J Environ Res Public Health. 2010;7(7):2745–88. https://doi.org/10.3390/ijerph7072745.
26. Stover PJ. One-carbon metabolism-genome interactions in folate-associated pathologies. J Nutr.

2009;139(12):2402–5. https://doi.org/10.3945/jn.109.113670.
27. Abali EE, Cline SD, Franklin DA, Viselli SM. Lippincott illustrated reviews: biochemistry. LWW; 2025.
28. Torrez M, Chabot-Richards D, Babu D, Lockhart E, Foucar K. How I investigate acquired megaloblastic anemia. Int J Lab Hematol. 2022;44(2):236–47. https://doi.org/10.1111/ijlh.13789.
29. Briani C, Dalla Torre C, Citton V, Manara R, Pompanin S, Binotto G, et al. Cobalamin deficiency: clinical picture and radiological findings. Nutrients. 2013;5(11):4521–39. https://doi.org/10.3390/nu5114521.
30. Herbert V. Absorption of vitamin B12 and folic acid. Gastroenterology. 1968;54(1):110–5.
31. Wuerges J, Garau G, Geremia S, Fedosov SN, Petersen TE, Randaccio L. Structural basis for mammalian vitamin B12 transport by transcobalamin. Proc Natl Acad Sci. 2006;103(12):4386–91. https://doi.org/10.1073/pnas.0509099103.
32. Guéant J-L, Guéant-Rodriguez R-M, Alpers DH. Chapter Nine – Vitamin B12 absorption and malabsorption. In: Litwack G, editor. Vitamins and hormones, vol. 119. Academic Press; 2022. p. 241–74.
33. Cleveland Clinic. Pernicious anemia 2022. Available from: https://my.clevelandclinic.org/health/diseases/22377-pernicious-anemia.
34. O'Leary F, Samman S. Vitamin B12 in health and disease. Nutrients. 2010;2(3):299–316. https://doi.org/10.3390/nu2030299.
35. Saleem F, Soos MP. Biotin deficiency. In: StatPearls. Treasure Island (FL): StatPearls Publishing. Copyright © 2025, StatPearls Publishing LLC; 2025.
36. Padayatty S, Levine M. Vitamin C: the known and the unknown and Goldilocks. Oral Dis. 2016;22(6):463–93. https://doi.org/10.1111/odi.12446.
37. Figueroa-Méndez R, Rivas-Arancibia S. Vitamin C in health and disease: its role in the metabolism of cells and redox state in the brain. Front Physiol. 2015;6:397. https://doi.org/10.3389/fphys.2015.00397.
38. Dresen E, Lee Z-Y, Hill A, Notz Q, Patel JJ, Stoppe C. History of scurvy and use of vitamin C in critical illness: a narrative review. Nutr Clin Pract. 2023;38(1):46–54. https://doi.org/10.1002/ncp.10914.
39. Du J, Cullen JJ, Buettner GR. Ascorbic acid: chemistry, biology and the treatment of cancer. Biochim Biophys Acta. 2012;1826(2):443–57. https://doi.org/10.1016/j.bbcan.2012.06.003.
40. Terajima M, Perdivara I, Sricholpech M, Deguchi Y, Pleshko N, Tomer KB, et al. Glycosylation and cross-linking in bone type I collagen*. J Biol Chem. 2014;289(33):22636–47. https://doi.org/10.1074/jbc.M113.528513.
41. Deshmukh SN, Dive AM, Moharil R, Munde P. Enigmatic insight into collagen. J Oral Maxillofac Pathol. 2016;20(2):276–83. https://doi.org/10.4103/0973-029x.185932.
42. Timoshnikov VA, Kobzeva TV, Polyakov NE, Kontoghiorghes GJ. Redox interactions of vitamin C and iron: inhibition of the pro-oxidant activity by deferiprone. Int J Mol Sci. 2020;21(11). https://doi.org/10.3390/ijms21113967.
43. Rebouche CJ. Ascorbic acid and carnitine biosynthesis. Am J Clin Nutr. 1991;54(6 Suppl):1147s–52s. https://doi.org/10.1093/ajcn/54.6.1147s.
44. Cleveland Clinic. Scurvy. 2022. Available from: https://my.clevelandclinic.org/health/diseases/24318-scurvy.
45. Oregon State University, Linus Pauling Institute. Vitamin C. Linus Pauling Institute; Micronutrient Information Center; 2025. Available from: https://lpi.oregonstate.edu/mic/vitamins/vitamin-C.
46. Courbebaisse M, Cavalier E. Vitamin D in 2020: an old pro-hormone with potential effects beyond mineral metabolism. Nutrients. 2020;12(11). https://doi.org/10.3390/nu12113378.
47. Ross AC, Taylor CL, Yaktine AL, Del Valle HB. In: IoMUCtRDRIfVDa, editor. Calcium. Institute of Medicine (US) Committee to Review Dietary Reference Intakes for Vitamin D and Calcium; Dietary Reference Intakes for Calcium and Vitamin D. Washington, DC: National Academies Press (US); 2011.
48. Jeon S-M, Shin E-A. Exploring vitamin D metabolism and function in cancer. Exp Mol Med. 2018;50(4):1–14. https://doi.org/10.1038/s12276-018-0038-9.
49. Dominguez LJ, Farruggia M, Veronese N, Barbagallo M. Vitamin D sources, metabolism, and deficiency: available compounds and guidelines for its treatment. Metabolites. 2021;11(4):255.
50. Bikle DD. Vitamin D metabolism, mechanism of action, and clinical applications. Chem Biol. 2014;21(3):319–29. https://doi.org/10.1016/j.chembiol.2013.12.016.
51. Burns-Whitmore B, Froyen EB, Isom KA. Vitamin D and calcium—an overview, review of metabolism, and the importance of co-supplementation. Dietetics. 2024;3(4):588–608.
52. Voltan G, Cannito M, Ferrarese M, Ceccato F, Camozzi V. Vitamin D: an overview of gene regulation, ranging from metabolism to genomic effects. Genes. 2023;14(9):1691.
53. National Institutes of Health. Vitamin D. 2024. Available from: https://ods.od.nih.gov/factsheets/VitaminD-HealthProfessional/#:~:text=Vitamin%20D%20promotes%20calcium%20absorption,leading%20to%20cramps%20and%20spasms.
54. DeLuca HF. The metabolism and functions of vitamin D. Adv Exp Med Biol. 1986;196:361–75. https://doi.org/10.1007/978-1-4684-5101-6_24.
55. Blaine J, Chonchol M, Levi M. Renal control of calcium, phosphate, and magnesium homeostasis. Clin J Am Soc Nephrol. 2015;10(7):1257–72. https://doi.org/10.2215/cjn.09750913.
56. Murshed M. Mechanism of bone mineralization. Cold Spring Harb Perspect Med. 2018;8(12). https://doi.org/10.1101/cshperspect.a031229.

57. Alshahrani F, Aljohani N. Vitamin D: deficiency, sufficiency and toxicity. Nutrients. 2013;5(9):3605–16. https://doi.org/10.3390/nu5093605.
58. Kedishvili NY. Retinoic acid synthesis and degradation. Subcell Biochem. 2016;81:127–61. https://doi.org/10.1007/978-94-024-0945-1_5.
59. Vilhais-Neto GC, Pourquié O. Retinoic acid. Curr Biol. 2008;18(5):R191–2. https://doi.org/10.1016/j.cub.2007.12.042.
60. Carazo A, Macáková K, Matoušová K, Krčmová LK, Protti M, Mladěnka P. Vitamin A update: forms, sources, kinetics, detection, function, deficiency, therapeutic use and toxicity. Nutrients. 2021;13(5). https://doi.org/10.3390/nu13051703.
61. Choi EH, Daruwalla A, Suh S, Leinonen H, Palczewski K. Retinoids in the visual cycle: role of the retinal G protein-coupled receptor. J Lipid Res. 2021;62:100040. https://doi.org/10.1194/jlr.TR120000850.
62. Kono M, Goletz PW, Crouch RK. 11-cis- and all-trans-retinols can activate rod opsin: rational design of the visual cycle. Biochemistry. 2008;47(28):7567–71. https://doi.org/10.1021/bi800357b.
63. Petkovich M, Chambon P. Retinoic acid receptors at 35 years. J Mol Endocrinol. 2022;69(4):T13–t24. https://doi.org/10.1530/jme-22-0097.
64. Goodman RE. Twenty-eight years of GM Food and feed without harm: why not accept them? GM Crops Food. 2024;15(1):40–50. https://doi.org/10.1080/21645698.2024.2305944.
65. Paine JA, Shipton CA, Chaggar S, Howells RM, Kennedy MJ, Vernon G, et al. Improving the nutritional value of Golden Rice through increased pro-vitamin A content. Nat Biotechnol. 2005;23(4):482–7. https://doi.org/10.1038/nbt1082.
66. Ye X, Al-Babili S, Klöti A, Zhang J, Lucca P, Beyer P, et al. Engineering the provitamin A (beta-carotene) biosynthetic pathway into (carotenoid-free) rice endosperm. Science. 2000;287(5451):303–5. https://doi.org/10.1126/science.287.5451.303.
67. Berkner KL, Runge KW. Vitamin K-dependent protein activation: normal gamma-glutamyl carboxylation and disruption in disease. Int J Mol Sci. 2022;23(10). https://doi.org/10.3390/ijms23105759.
68. Creative Proteomics. Water-soluble vitamins: importance, characteristics, and functions 2024. Available from: https://www.creative-proteomics.com/blog/index.php/water-soluble-vitamins-importance-characteristics-and-functions/.
69. Meena VK, Chand S, Shekhawat HVS, Choudhary K, Sharma JK, Sheera A, et al. Advances in plant tocopherol biosynthesis: from pathway elucidation to crop biofortification strategies. Discov Plants. 2025;2(1):9. https://doi.org/10.1007/s44372-025-00093-8.
70. Shahidi F, Pinaffi-Langley ACC, Fuentes J, Speisky H, de Camargo AC. Vitamin E as an essential micronutrient for human health: common, novel, and unexplored dietary sources. Free Radic Biol Med. 2021;176:312–21. https://doi.org/10.1016/j.freeradbiomed.2021.09.025.
71. Alberts B, Heald R, Johnson A, Morgan D, Raff M, Roberts K, Walter P, Wilson J, Hunt T. Molecular biology of the cell. New York: W.W. Norton and Company; 2022.
72. Balint E, Unk I. For the better or for the worse? The effect of manganese on the activity of eukaryotic DNA polymerases. Int J Mol Sci. 2024;25(1):363. https://doi.org/10.3390/ijms25010363.
73. Sharma U, Pal D, Prasad R. Alkaline phosphatase: an overview. Indian J Clin Biochem. 2014;29(3):269–78. https://doi.org/10.1007/s12291-013-0408-y.
74. Toxqui L, Vaquero MP. Chronic iron deficiency as an emerging risk factor for osteoporosis: a hypothesis. Nutrients. 2015;7(4):2324–44. https://doi.org/10.3390/nu7042324.

Part VI
Advanced Topics

The chapters in this part cover some more advanced topics in cell and molecular biology. They cover topics that are closely linked to cell function. In particular, they deal with cellular reactions that are critical for normal cell function but that, when improperly regulated, can lead to disease. This part starts with Chap. 18, which explores *signal transduction*. Signal transduction, or cell signaling, refers to the mechanisms by which cells respond to signals from other cells and from the environment. These may be signals that tell cells to enter the cell cycle and divide, differentiate, or even die. In some cases, they may also tell cells how to respond to metabolic changes such as rising glucose levels. Responding appropriately to signals is critical for normal cell function, but when genes involved in these signaling mechanisms are improperly regulated, this may result in diseases such as inflammatory diseases, metabolic disorders, or cancer. Understanding this pathway has been instrumental in developing treatments for multiple types of disorders. Closely related to cell signaling is the topic of *cancer biology*, particularly proto-oncogenes and oncogenes, and cancer cell metabolism. These topics are discussed in Chap. 19. Chapter 19 introduces basic features of cancer cells, followed by a discussion of proto-oncogenes. Proto-oncogenes are genes that have normal functions in cells but when improperly activated may become oncogenes and lead to cancer. Many proto-oncogenes also have roles in cell signaling, and this overlap will be explored. Metabolism is often altered in cancer cells, as exemplified by the Warburg effect, and this is also discussed in this chapter. Some of the treatment options for cancer are also discussed.

Chapter 20 covers the *cell cycle*, which is the ordered series of events that a cell goes through every time it divides into two daughter cells. These events are divided into distinct phases. The genes encoding proteins that regulate the cell cycle are also important for human health, and some of them have become targets for various drugs. Some genes that have roles in the cell cycle are tumor suppressor genes. Tumor suppressor genes are genes whose loss or inactivation can lead to unregulated cell growth and cancer, and they are covered in Chap. 21. Chapter 22 covers topics related to cell death or recycling of damaged cell parts: *apoptosis*, *necrosis*, and *autophagy*. Apoptosis is a form of programmed cell death, while necrosis is a less regulated form of cell death that can occur in response to injury or infection. Autophagy can lead to cell survival rather than cell death, and it involves the destruction and recycling of damaged organelles and other cellular contents. This is important for cell survival, but, in some situations, too much autophagy can also be associated with cell death. Chapter 23 covers *DNA damage and repair* mechanisms. DNA can be damaged either in response to mutagens or as part of the normal errors that can be incorporated during DNA synthesis. In many cases, DNA damage can be repaired, and it can sometimes be avoided altogether due to the proofreading mechanisms of DNA polymerase. Abnormalities in DNA repair mechanisms can lead to disease, and therefore a thorough understanding of these mechanisms is critical for their treatment.

Chapter 24 covers a brief discussion of the immune system and virology. Many diseases are associated with abnormalities in the immune system, and viral infections are a major global health concern. Immunology and virology are vast topics and are beyond the main scope of this textbook. However, since many drugs are involved in modulating the immune system, many biologics consist of monoclonal antibodies, and several important drugs deal with viral infections, it is helpful to have at least a basic understanding of these topics.

Signal Transduction and Drugs that Target Signaling Pathways

18

Abstract

The study of signal transduction, also called cell signaling, focuses on how cells transmit, receive, and respond to various signals. Eukaryotic cells often receive signals in the form of extracellular factors such as growth factors, hormones, and cytokines, referred to as **ligands**. While many ligands are extracellular, some ligands enter the cell and bind to intracellular receptors. Once ligands bind to receptors, they activate a series of events in the cell, called signaling pathways, which in turn lead to cellular responses, often in the form of changes in gene expression. This chapter explores several major signaling pathways in eukaryotic cells, including key receptors, protein kinases, second messengers, and nuclear factors. Examples of classic signaling pathways covered in this chapter include the ERK MAP Kinase pathway, the NF-κB pathway, the PI3K/AKT pathway, estrogen-activated signaling pathways, and GPCR-activated pathways that lead to PLC activation or calcium influx. Signaling pathways are critical for normal cell function, but their improper activation can lead to diseases such as cancer and inflammatory disorders. For this reason, numerous drugs have been developed that interfere with specific signaling pathways to treat a variety of different disorders. Examples of such drugs are discussed here.

Keywords

G protein-coupled receptors (GPCRs) · Receptor tyrosine kinases (RTKs) · Transmembrane receptors · Intracellular receptors · Hormones · Ligands

Abbreviations

AH	Aqueous humor
AhR	Aryl hydrocarbon receptor
ALK	Anaplastic leukemia kinase
AP-1	Activator protein 1 (a transcription factor complex of c-Jun and c-Fos)
Arp2/3	Actin-related protein 2/3 complex
BAD	Bcl2-associated death promoter
BMP	Bone morphogenic protein
cAMP	cyclic AMP
Cdc42	Cell division cycle 42 (a small GTPase)
CREB	cAMP response element binding protein
DAG	Diacylglycerol
DLLs	Delta-like ligands
DP1	D-type cyclic-dependent kinase partner 1
EGF	Epidermal growth factor
EGFR	Epidermal growth factor receptor
ER	Estrogen receptor
ERK	Extracellular Signal-Regulated Kinase

A. Minden, *Biochemistry, Cell and Molecular Biology for Pharmacists and Health Professionals*,
https://doi.org/10.1007/978-3-032-20529-2_18

FDA	Food and Drug Administration
FERM domain	Band 4.1/Ezrin/Radixin/Moesin domain
FSHD	Facioscapulohumeral muscular dystrophy
GAP	GTPase-activating protein
GEF	Guanine nucleotide exchange factor
GM-CSF	Granulocyte-macrophage colony-stimulating factor
GPCR	G protein-coupled receptor
GDF	Growth differentiation factor
GDP	Guanosine diphosphate
GRs	Glucocorticoid receptors
Gs	G stimulatory protein
Gi	G inhibitory protein
GTP	Guanosine triphosphate
Her2	Human epidermal growth factor 2
IFN-γ	Interferon gamma
IP_3	Inositol 1,4,5 triphosphate
IKK	IkB kinase
IL-6	Interleukin 6
IOP	Intraocular pressure
JAK	Janus Kinase
JNK	c-Jun N-terminal Kinase
MAPK	Mitogen-activated protein kinase
MAPKK	MAP Kinase Kinase
MAPKKK	MAP Kinase Kinase Kinase
MEK	Mitogen-activated protein kinase kinase
MET	Mesenchymal epithelial transition factor
MLC	Myosin light chain
MLCP	Myosin light chain phosphatase
mTOR	Mammalian target of rapamycin
mTORC1	mTOR complex 1
mTORC2	mTOR complex 2
Nemo	NFKB Essential Modulator
NFkB	Nuclear factor kappa light chain enhancer of activated B cells
PAK	p21-activated kinase
PDK1	Phosphoinositide-dependent kinase 1
PH domain	Pleckstrin Homology Domain
PI3K	Phosphoinositide 3-kinase
PIP_2	Phosphatidylinositol 4,5-bisphosphate
PIP3	Phosphatidylinositol 3,4,5-trisphosphate
PLC	Phospholipase C
PTH	Parathyroid hormone
Rac	Ras-related C3 botulinum toxin substrate (a small GTPase)
Rho	Ras homolog (a family of small GTPases)
RTK	Receptor tyrosine kinase
RTT	Retrotransposon-derived transcript
SERDs	Selective estrogen receptor downregulators
SERMs	Selective estrogen receptor modulators
SH2 domain	Src homology 2 domain
SH3 domain	Src homology 3 domain
Sos	Son of sevenless (an exchange factor)
STAT	Signal Transducer and Activator of Transcription
TLR	Toll-like receptor
TM	Trabecular meshwork
TGF-β	Transforming growth factor beta
TNF	Tumor necrosis factor
TNFR	Tumor necrosis factor receptor
TSC2	Tuberous sclerosis complex 2
VEGF	Vascular endothelial growth factor
VEGFR	Vascular endothelial growth factor receptor
wAMD	Wet age-related macular degeneration
WAVE	WASP family Verprolin-homologous Protein

What You Will Learn

- The basic mechanisms of signal transduction and how cells respond to signals
- The key components of signaling pathways, such as receptors, protein kinases, second messengers, and nuclear factors
- The major classes of signal transduction pathways, those mediated by intracellular receptors, single-pass transmembrane receptors, and G protein-coupled receptors (GPCRs)

- Examples of classic signaling pathways, such as the ERK MAP Kinase pathway, the NF-KB pathway, the PI3K/AKT pathway, estrogen-activated signaling pathways, and GPCR-activated pathways that lead to PLC activation or to calcium influx
- How abnormal regulation of signaling pathways can lead to diseases such as cancer or inflammatory disorders
- Drugs that interfere with specific signaling pathways, such as aromatase inhibitors that block estrogen signaling in breast cancer, biologics that block IL-23 signaling in inflammatory disorders, and beta-blockers that block GPCR-mediated signaling and are used for conditions such as high blood pressure

1 Introduction to Signal Transduction

1.1 Why Is the Study of Cell Signaling Important?

The study of signal transduction, also called cell signaling, focuses on how cells transmit, receive, and respond to various signals (see Fig. 18.1). In a multicellular organism, cells do not live in isolation but must communicate with other tissues and cells throughout the body. They continually receive signals that tell them to grow, divide, multiply, carry out specific functions, secrete various substances, or even stop growing, undergo apoptosis, and die. Cells receive these signals from surrounding cells and tissues. Even unicellular organisms need to respond to cues in their environments, such as signals that alert them to the level or type of nutrients in their environment. In general, eukaryotic cells receive signals in the form of extracellular factors such as growth factors, hormones, and cytokines, referred to as **ligands**. Many categories of ligands cannot diffuse across the plasma membrane and therefore cannot enter the cell. Instead, they bind to cell surface **receptors**. This can activate the receptors, which usually leads to a cascade of biochemical events in the cell. Often, this can lead to nuclear changes, such as the regulation of a transcription factor, resulting in the regulated expression (transcription) of specific genes that respond to the stimulus. The series of events that occur after receptor activation is referred to as a signaling pathway or signaling cascade. While many ligands function by binding to cell surface receptors, some small hydrophobic ligands may diffuse across the membrane and bind to receptors inside the cell. This can lead to a more direct response to the stimulus. This chapter focuses on how cells respond to signals that in turn alter cell function. This work is drawn largely from the following sources as well as sources listed in the tect [1–4].

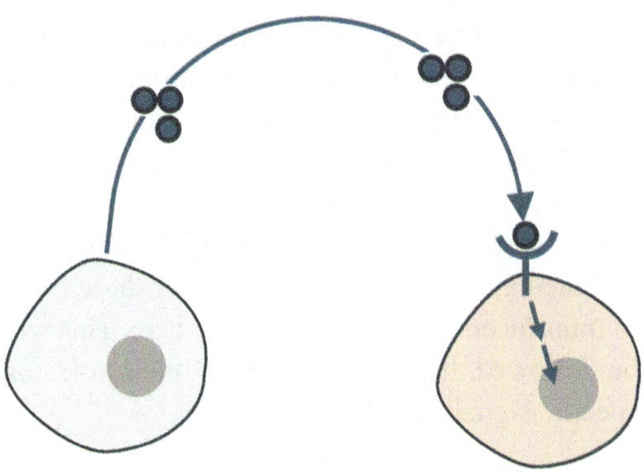

Fig. 18.1 Cell signaling. In the diagram, a signaling molecule is released from one cell and binds to a receptor on a target cell, activating a signaling pathway and eliciting a response

Cells can signal to each other in different ways. Cells that are distant from each other signal via **endocrine** signaling, where signaling molecules (ligands) may be from glands and subsequently travel through blood vessels in order to contact distant cells. Cells that are closer to each other may signal via **paracrine** signaling, in which the ligand only travels a short distance. In this case, one cell may produce and secrete a ligand, which then diffuses a short distance to a nearby cell. **Contact-dependent** signaling may occur if stimulation of one cell is caused by direct contact with a ligand on the surface of a nearby cell. In **autocrine signaling**, a cell may produce a ligand, which is then secreted and binds to a receptor on the same cell. Finally, **synaptic** signaling involves signaling between neurons or between neurons and muscle cells. Regardless of which mechanisms are used, receptor activation of the recipient cell may lead to the activation of a signaling pathway; a cascade of events in the cell culminating in a specific response to the stimulus [3, 4] (Fig. 18.2).

Understanding cell signaling is important for many reasons. Many disorders are caused by signaling pathways that are not functioning correctly. For example, in cancer, signals that regulate growth control may be improperly regulated, or signaling pathways may even be constitutively activated independently of a stimulus. In other cases, specific cell functions may be abnormal in disease, and this is also often associated with disruptions in signaling. For example, in type 2 diabetes, the signal transduction pathway normally activated by insulin may be improperly regulated. Dysregulation of some signaling pathways are also associated with inflammatory disorders such as arthritis.

Many different drugs target signal transduction pathways. For example, in Her2-positive breast cancer, the cell surface receptor Her2 is overproduced and overly active. This leads to increased activation of signaling pathways downstream to this receptor. Understanding this pathway paved the way to the development of **Herceptin (trastuzumab)** [5], a drug used to treat HER2-positive breast cancer. Herceptin is a monoclonal antibody that targets Her2 and inhibits the associated signaling pathway, thereby reducing the growth of the cancer cells. It is critical for pharmacists to have a strong understanding of cell signaling to understand the mechanisms of action of drugs that target signaling pathways. It is also important for researchers to understand cell signaling to develop new drugs targeting key signaling proteins associated with disease [6, 7].

Signal transduction is a broad field, and this chapter is organized into four sections. Section 1 covers an **overview** of what cell signaling is and why it is important. It also covers the basic components of signaling pathways, such as ligands, receptors, and intracellular signaling molecules. There are many different types of signaling pathways in cells. One way to categorize different signaling pathways is by the types of receptors to which ligands bind. The next three sections of this chapter discuss signaling pathways mediated by ligands that bind different types of receptors. Section 2 covers signaling pathways that are initiated when ligands bind to **single-pass transmembrane receptors**, such as receptor tyrosine kinases and cytokine receptors. Section 3 discusses signaling pathways that are mediated by **intracellular receptors**. Examples are signaling pathways that are regulated by hormones such as estrogen. In these types of pathways, the ligand must diffuse through the plasma membrane and enter the cell to bind to its receptor. Section 4 discusses one of the largest classes of receptors, which are the **G protein-coupled receptors (GPCRs)**. These are seven-transmembrane spanning receptors that activate trimeric G proteins in response to ligand binding. Throughout this chapter, various disorders involving dysregulated signaling pathways are discussed, along with drugs targeting components of these pathways, such as cell surface receptors.

1.2 Some of the Basic Signal Transducers and Mediators of Cell Signaling Pathways

When eukaryotic cells respond to a stimulus, they can activate signaling pathways that lead to a response. Eukaryotic signaling pathways involve the regulation of various proteins and second messengers (see Fig. 18.2). Some of these include ligands (such as growth factors, hormones, and cytokines), receptors, adapter proteins, protein kinases and phosphatases, GTP-binding proteins, transcription factors, second messengers, and ions. These can be considered signal transducers and mediators of a pathway. Some of these types of transducers are described briefly here. This will be followed by a discussion of their roles in specific signaling pathways [2, 3].

1.2.1 Ligands

Many ligands are small proteins or peptides that initiate signaling pathways by binding to receptors. Many ligands cannot diffuse across

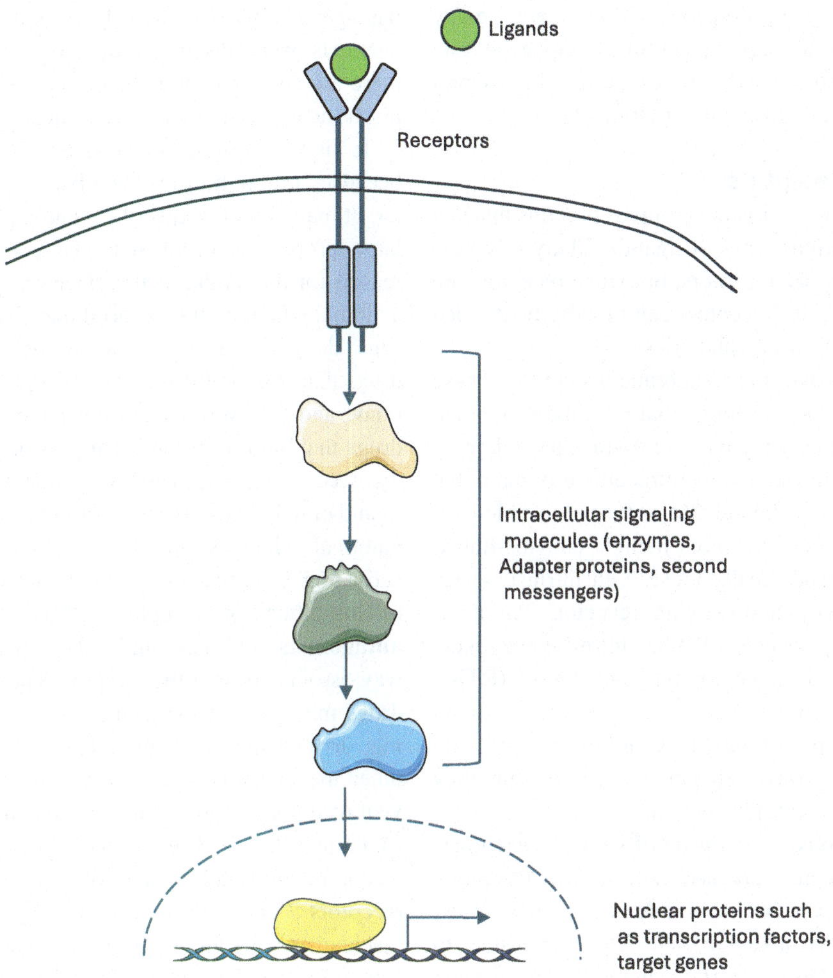

Fig. 18.2 The main components of a signaling pathway. Signaling pathways comprise various molecules, including ligands, receptors, intracellular signaling proteins, and nuclear proteins such as transcription factors. (*Source*: Created with BioRender.com)

the plasma membrane and therefore cannot enter the cell. Instead, these ligands bind to the extracellular side of cell surface receptors. Examples of such ligands include growth factors (such as EGF), cytokines (such as TNFα), and peptide hormones (such as insulin).

Unlike many growth factors and peptide hormones, some ligands are lipophilic (fat soluble), and they can diffuse across the plasma membrane and enter the cell. These are generally not peptides. Rather, they are usually small molecules that can cross the membrane. Once in the cell, these ligands bind to intracellular receptors that may be in the cytoplasm or the nucleus. Examples of this type of ligand are steroid hormones such as estrogen.

Ligands are sometimes categorized based on the types of functions they have. In general, growth factors regulate cell proliferation and differentiation, cytokines and interferons mediate immune responses, and hormones are secreted by endocrine glands. Hormones usually travel through the bloodstream, functioning at a distance from where they were secreted, and may regulate metabolic pathways. Many but not all ligands are mitogens, which are molecules that stimulate cell division. There are many exceptions to these generalities, however, and some ligands have different functions in different cells or under different conditions [3].

1.2.2 Receptors

Receptors are a diverse group of proteins that can bind to various types of ligands. Many receptors are cell surface receptors, but some receptors are inside the cell. Receptors can broadly be divided into the following categories:

Single-pass transmembrane receptors. These are cell surface receptors that span the membrane a single time. They have an extracellular domain that binds ligand and an intracellular domain that responds to the ligand. Major types of single-pass transmembrane receptors include enzyme-linked receptors (specifically those with intrinsic enzymatic activity) and cytokine receptors. An abundant type of receptor that has intrinsic enzymatic activity is the receptor tyrosine kinase (RTK). Receptor serine/threonine kinases also respond to some types of signals. Cytokine receptors do not have intrinsic enzymatic activity, but they may bind to cytoplasmic kinases.

Seven-pass transmembrane receptors. These receptors are also cell surface receptors, but they span the membrane seven times. They are generally linked on their cytoplasmic sides to a trimeric G protein (G proteins that have three subunits). They bind ligand on their extracellular sides. These receptors are called G protein-coupled receptors (**GPCRs**). They are abundant and have diverse functions.

Intracellular receptors. These receptors are inside the cell, either in the cytoplasm or the nucleus. Their ligands must be able to diffuse through the plasma membrane to bind to the receptor. Once bound to the receptor, these receptors can often lead directly to regulation of gene expression.

The above are the main types of receptors covered here, but there are also other types of receptors in cells. For example, the Toll-like receptors (TLRs) have important roles in immunity. In addition, ligand-gated ion channels are considered a type of receptor since they can mediate a response to a stimulus by allowing the passage of ions into the cell. (Ligand-gated ion channels were discussed in Chap. 2). Finally, there are some receptors that are atypical and do not fit neatly into any of these categories.

It is important to understand receptors because, although drugs have been developed to target many components of signaling pathways, a large proportion of them target receptors. One reason for this is that cell surface receptors have a domain facing the extracellular environment. This domain can be an accessible target for drugs that cannot diffuse across the plasma membrane and therefore cannot enter the cell. Some drugs that bind receptors are **agonists**. They bind the receptor in the same way that the natural ligands bind. These types of drugs cause conformational changes in the receptors, thereby activating signaling pathways. Other drugs lead to inhibition of receptors. These drugs are **antagonists**, and they inhibit the signaling pathway associated with the receptor. These types of drugs may prevent the natural ligand from binding, or they may prevent receptor activation by other mechanisms. For example, they may prevent receptor dimerization which can inhibit their catalytic activity. The extracellular domains of receptors are readily accessible to drugs, but receptors can be targeted by multiple mechanisms, including by drugs that enter the cell and target their intracellular domains [2, 8].

1.2.3 Adapter Proteins

Once receptors are bound to ligand, they are usually considered activated. Activated receptors can initiate a cascade of events in the cell. This can take many forms. One possibility is that the receptors can bind to adapter proteins. Adapter proteins are proteins whose primary function is to bind to other proteins, thereby bringing various proteins into proximity with each other. They have different domains that enable them to interact with receptors or other cellular proteins.

1.2.4 GTP-Binding Proteins

GTP-binding proteins are common in signaling pathways, and they usually play a role early in the pathway. This family of proteins can bind to GTP or GDP. They are often considered as a type of on/off switch. GTP-binding proteins are "on" when they are GTP bound and "off" when they are GDP bound. GTP-binding proteins have a low level of intrinsic GTPase activity, where they can hydrolyze their bound GTP to GDP, causing them to take an "off" position. This intrinsic GTPase activity is weak, however, and it requires the presence of a GTPase activation protein (GAP). When GTP-binding proteins are in the GTP-bound state, they can bind to other proteins, such as protein kinases, which leads to their activation. GTP-binding proteins can be heterotrimeric (having three subunits), or they can have a single subunit. Those with the single subunit will be introduced first in this chapter, and they are often called small GTP-binding proteins. They are also sometimes referred to as GTPases or small GTPases due to their enzymatic activity [2].

1.2.5 Enzymes

Enzymes within the cell have important roles in cell signaling pathways. **Protein kinases** are common and often abundant in signaling pathways. Protein kinases, also discussed in Chap. 5, are enzymes that catalyze the phosphorylation (the addition of a phosphate group) of other proteins. There are three main types of protein kinases. These are the tyrosine kinases, which phosphorylate their substrates on tyrosine residues; serine/threonine kinases, which phosphorylate their substrates on serine and/or threonine residues; and dual-specific kinases, which can phosphorylate their substrates on tyrosine, as well as serine/threonine.

In addition to protein kinases, other types of enzymes also have roles in signaling pathways. These include lipid kinases, phosphatases, proteases, and others, some of which will be discussed in subsequent sections.

1.2.6 Second Messengers

Second messengers are signaling molecules that are not proteins. Examples of second messengers are cyclic AMP (cAMP), ions such as calcium, and diacylglycerol (a lipid-based molecule). Second messengers can play crucial roles in propagating a signal.

1.2.7 Signaling Cascades and Nuclear Signals

Activation of receptors leads to a cascade of events called a signaling pathway or signaling cascade. These pathways may be mediated by the types of signaling molecules described above, like adapter proteins, second messengers, and enzymes. A signaling cascade can lead to different outcomes, but often, it results in the regulation of a nuclear signal, such as transcription factors, leading to the expression of certain target genes. For example, a signaling cascade may lead to activation of a kinase that phosphorylates a transcription factor, thereby stimulating its activity. The active transcription factor may, in turn, lead to the regulation of specific genes. In this way, the binding of a ligand to a cell surface receptor, through a series of events, can ultimately lead to the regulation of gene expression.

1.2.8 Signaling Pathway Organization

While there are many possible components of signaling pathways, the activation of a signaling pathway usually starts with the binding of ligand to a receptor, and many (but not all) signaling pathways end with the regulation of nuclear proteins such as transcription factors. When discussing a signaling pathway, the events that happen first, such as the binding of ligand to a receptor, are considered the upstream part of the pathway, and the events that happen at the end, such as activation of transcription factors and regulation of target genes, are the downstream part of the pathway [2] (see Fig. 18.2).

There are many different types of signaling pathways in cells, and in the next section, we will discuss some important signaling pathways. The next section of this chapter, Sect. 2, covers signaling pathways that are associated with **single-pass transmembrane receptors**, including enzyme-linked receptors. Section 3 covers signaling pathways that are activated by **intracellular receptors**, and Sect. 4 covers signaling pathways that are regulated by **GPCRs**. It is important to note, though, that although this chapter is organized according to receptor class, there is also a significant amount of overlap and crosstalk between these different types of signaling pathways.

2 Single-Pass Transmembrane Receptors and Associated Signaling Pathways

2.1 Enzyme-Linked Receptors (Receptors with Intrinsic Enzymatic Activity)

A major type of single-pass transmembrane receptor is the enzyme-linked receptor. This is an abundant and important family of cell surface receptors that activate numerous different signaling pathways. Enzyme-linked receptors span the plasma membrane one time. They have a cytoplasmic domain, an extracellular domain, and a transmembrane domain that consists of a hydrophobic region anchored in the membrane. Enzyme-linked receptors generally bind their ligand on the outside of the cell and have intrinsic enzymatic activity on the cytoplasmic domain [2].

2.2 Receptor Tyrosine Kinases (RTKs) and RTK-Activated Signaling Pathways

One of the most abundant types of enzyme-linked receptors is the receptor tyrosine kinases or RTKs (Fig. 18.3). RTKs are members of a large family of cell surface receptors. There are 20 different classes of RTKs, but the basic mechanisms by which they are regulated are conserved. RTKs have an extracellular ligand-binding domain and an intracellular tyrosine kinase domain, which also has several tyrosine residues. In most cases, ligand binding induces receptor dimerization (although in some cases, receptors are already dimers and instead undergo a conformational change). Dimerization or conformational change of RTKs in response to ligand binding brings their cytoplasmic kinase domains into proximity. This causes them to phosphorylate each other on specific tyrosine residues. This is often referred to as trans-autophosphorylation or sometimes as trans-phosphorylation. The phosphorylated tyrosine residues on the cytoplasmic side of the receptors serve as docking sites for proteins that have SH2 domains, which are domains that specifically bind to phospho-tyrosine residues. In many cases, these SH2-containing proteins are adaptor proteins that recruit other proteins. This initiates a signaling pathway inside the cell, which can lead to the regulation of cell function and gene expression.

One well-studied family of RTKs is **the EGF receptor (EGFR)** family. The EGFR contains a

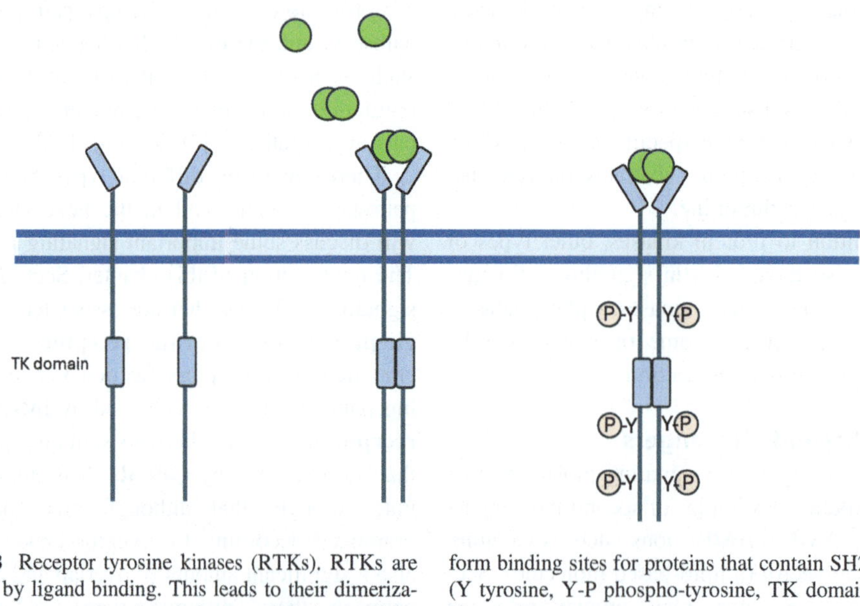

Fig. 18.3 Receptor tyrosine kinases (RTKs). RTKs are activated by ligand binding. This leads to their dimerization and trans-autophosphorylation on their intracellular tyrosine residues. These phospho-tyrosine residues can form binding sites for proteins that contain SH2 domains. (Y tyrosine, Y-P phospho-tyrosine, TK domain, tyrosine kinase domain

Table 18.1 EGFR family

Name	Also called	Also called	Major ligands
EGFR1 (EGFR)	ErbB1	Her1	EGF, TGFα
EGFR2	ErbB2	Her2	No known ligands
EGFR3	ErbB3	Her3	HRGs
EGFR4	ErbB4	Her4	HRGs, BTC, EREG

HRG Heregulin, *BTC* Betacellulin, *EREG* Epiregulin

ligand-binding site on its extracellular domain and a tyrosine kinase domain along with multiple tyrosines on its intracellular side. It dimerizes upon ligand binding and trans-autophosphorylates. As described above, this leads to activation of intracellular signaling pathways. There are four major members of this family (EGFR, EGFR2, EGFR3, and EGFR4), which also have different names, as illustrated in Table 18.1. Three of these four RTKs bind to ligands. Only one of them (EGFR2) does not have any known ligand but forms dimers with the other receptors in the family, thereby contributing to the transmission of a signal [2, 9].

The EGFR family is a classic example of RTKs. Often, EGFRs respond to signals such as growth factors, which encourage cell division. For example, they may regulate progression from G1 to S phase of the cell cycle (see Chap. 20), thus leading to cell division. Another important RTK is the **insulin receptor (IR)**. The IR is a bit different from some other RTKs because it preexists as a dimer (actually a heterotetramer because two dimers interact). The IR contains two α-subunits that are extracellular and two β-subunits that are intracellular. The α-subunits bind ligand (insulin) and the β-subunits have tyrosine kinase activity and tyrosine residues. Ligand (insulin) binding to the α-subunits leads to a conformational change that results in activation of the β-domains. These domains have kinase activity. They can trans-phosphorylate each other on tyrosine residues. Unlike many other RTKs, they also recruit and phosphorylate another protein called insulin receptor substrates (IRS). IRS binds to the IR (one IRS can bind to each β-domain) and acts as scaffolds that recruit other signaling proteins. In return, this leads to activation of the signaling pathway, similar to other RTKs [2, 10].

2.3 The MAP Kinase Pathway

2.3.1 Overview of the MAP Kinase Pathway

Different types of signaling pathways can be activated by RTKs. The MAP Kinase pathway is a classic example. MAP Kinase pathways are a family of well-known signaling pathways that can be activated by ligands that bind to RTKs (although they can also be activated by some other receptors). The MAP Kinase pathway is a good starting point for understanding how signaling pathways work as many other signaling pathways utilize some similar mechanisms.

In mammalian cells, several major related pathways are categorized as **MAP Kinase (MAPK)** pathways: the **ERK** (**E**xtracellular Signal-**R**egulated **K**inase), **p38**, and **JNK** pathways (c-**J**un **N** Terminal **K**inase) pathways.

The **ERK MAP Kinase pathway** is considered the **classical MAP Kinase pathway** (see Fig. 18.4). It is sometimes referred to simply as the MAP Kinase pathway even though it is one of several related pathways. MAP Kinase pathways begin with receptor activation via ligand binding, followed by other steps, which include activation of a **small GTP-binding protein**, followed by sequential activation of a series of three protein kinases (broadly classified as a MAP Kinase-Kinase-Kinase (**MAPKKK**), a MAP Kinase-Kinase (**MAPKK**), and a MAP Kinase (**MAPK**)). This sequential activation of kinases is called a **kinase cascade**, where each kinase phosphorylates and thereby activates the next kinase in the cascade. Once activated, MAPK, the final kinase in the cascade, translocates to the nucleus where it phosphorylates transcription factors, regulating their activity, and thereby leading to regulation of gene expression. The ERK MAP Kinase pathway is commonly activated by growth factors such as EGF, which can lead to cell cycle progression. The MAP Kinase pathway is described in more detail below.

2.3.2 The ERK MAP Kinase Pathway

The ERK MAP Kinase pathway (sometimes called simply the ERK or the MAP Kinase pathway) is considered the classic MAP Kinase pathway and serves as a good model for understanding how these pathways function. It is described in more detail below.

- Ligand Binding to an RTK Leads to Ras Activation

The ERK pathway (see Fig. 18.4) is initiated when a **ligand** binds to a receptor on its extracellular domain. Different types of receptors can lead to activation of MAP Kinase pathways, but they are commonly activated by RTKs, which will be discussed here. Commonly, the ligand is a growth factor such as **EGF** (epidermal growth factor), which is a small peptide that binds to its receptor, such as the **EGFR** (EGF receptor), which is an **RTK**, although numerous other ligand-receptor combinations can activate the ERK pathway. Ligand binding to the RTK leads to the dimerization of the receptor. Dimerization is a key step in the activation of RTKs. RTKs have tyrosine kinase activity on their cytoplasmic domains, and they also have tyrosine residues on their cytoplasmic domains. When the receptors dimerize, they can trans-phosphorylate each other on multiple tyrosine residues. This is considered receptor activation, and it results in a receptor with phosphorylated tyrosine (phospho-tyrosine) residues on its cytoplasmic sides (in the case of the IR, the receptor can also phosphorylate and recruit IRS). Phospho-tyrosine residues are critical in signaling because there is a type of protein that can bind specifically to phospho-tyrosine residues. These are proteins that have a domain called an SH2 domain (Src homology domain), which binds specifically to phospho-tyrosine

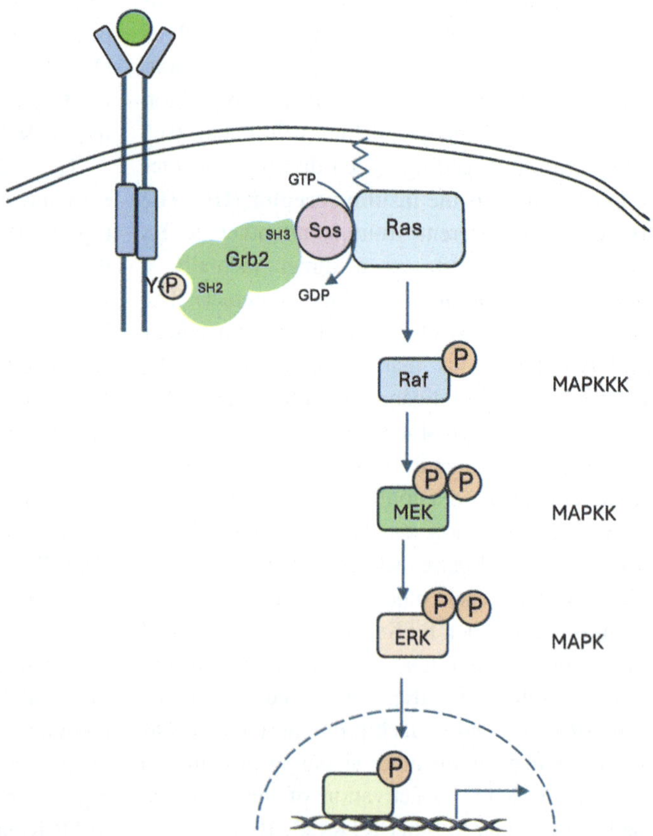

Fig. 18.4 The MAP Kinase pathway. Activated RTKs (as well as some other receptors) can lead to activation of the MAP Kinase pathway and to subsequent regulation of transcription factors, as described in the text. The pathway includes adapter proteins, small GTPases, kinases, other enzymes, and transcription factors. GRB2 is an important example of an adapter protein. It links the RTK to Sos, which in turn activates Ras

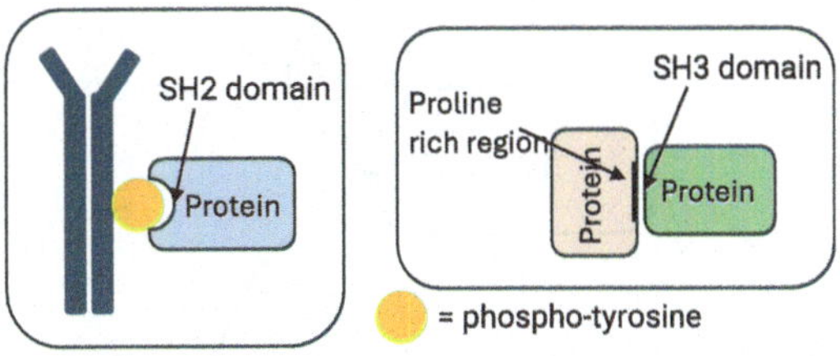

Fig. 18.5 Protein domains. Protein domains play a crucial role in cell signaling. SH2 domains bind phospho-tyrosine, while SH3 domains bind proline-rich regions. Other protein domains interact with other targets

residues (see Fig. 18.5). In the classical MAP Kinase pathway, the SH2 domain-containing protein that binds to the phospho-tyrosines on the RTK is GRB2. GRB2 contains not only an SH2 domain but also two SH3 domains. SH3 domains are also protein-interacting domains. They bind to proteins that are rich in proline residues (see Fig. 18.5a). GRB2 can therefore bind to more than one protein at once, via its SH2 and SH3 domains. Here, GRB2 binds to the RTK via its SH2 domain and to another protein, called Sos (son of sevenless), via one of its SH3 domains. GRB2 is an adapter protein. It does not contain catalytic activity, but by binding to two different proteins, it can bring different proteins into close proximity (see Fig. 18.4).

Sos becomes localized near the plasma membrane due to its interaction with GRB2 (which is near the cell surface due to binding to the RTK). This brings Sos into proximity to **Ras**, which is a membrane-bound **small GTP-binding protein** (Fig. 18.6), anchored to the membrane group by a farnesyl group. **Sos** is a guanine nucleotide exchange factor (**GEF**). The primary function of Sos is to catalyze the exchange of GDP for GTP on Ras, thus turning Ras to its GTP-bound, "on," state. This switch causes a conformational change in two flexible regions of Ras, called switch I and switch II. When Ras is GTP bound, these switch regions adopt an "on" conformation that creates the surface used to bind and activate downstream effectors such as Raf. Ras is a central protein in many signaling pathways. It is part of a larger family of **small GTP-binding proteins**. Ras functions as a molecular switch. It is "off" when bound to GDP (guanosine diphosphate; having two phosphates), but "on" when bound to GTP (guanosine triphosphate; having three phosphates). When it is in its "on" position, it can lead to activation of downstream signaling pathways, which are discussed below. Thus, the localization of Sos to the proximity of Ras switches Ras to its "on" state, due to GTP binding.

Later, after Ras has led to activation of the downstream pathways described below, the GTP on Ras will be hydrolyzed (it will lose one phosphate group) and become GDP. This will be facilitated by a **GAP** protein. GAP stands for **GTPase GTPase-activating protein**, and as the name indicates, GAP enhances **Ras's intrinsic GTPase activity**, which is not very efficient on its own. The GTPase activity of Ras catalyzes the removal of a phosphate from GTP, leaving Ras bound to in GDP, and as a result, Ras is now in its "off" position (Fig. 18.6). This cycle can repeat each time an RTK is activated by ligand. (Ras also has two important regions called the switch 1 and switch 2 regions, which are discussed in Chap. 19).

- Activation of Intracellular Signaling Proteins and Nuclear Signals

– Activation of Raf by Ras-GTP

Once Ras is GTP bound and thus in the "on" position, and before it becomes inactivated due to GAP, it contributes to activating the next step in the pathway. A key target of Ras is the protein kinase **Raf** (see Fig. 18.4). There are multiple members of the Raf family, which will all be referred to here as Raf. When Ras is activated, it binds to **Raf** and causes Raf to be recruited to the membrane and to be activated.

Activation of Raf occurs as follows: Raf has a kinase domain and an amino-terminal Ras binding domain (RBD). In unstimulated cells, Raf is in a closed conformation, in an inactive state. It is kept in this state partly through phosphorylation of one of its serine residues, Ser259. This phosphorylation allows it to bind to a protein called 14-3-3, which stabilizes Raf in its autoinhibited conformation. However, when Ras is GTP bound, its switch I region undergoes a conformational change that creates a binding surface for effectors.

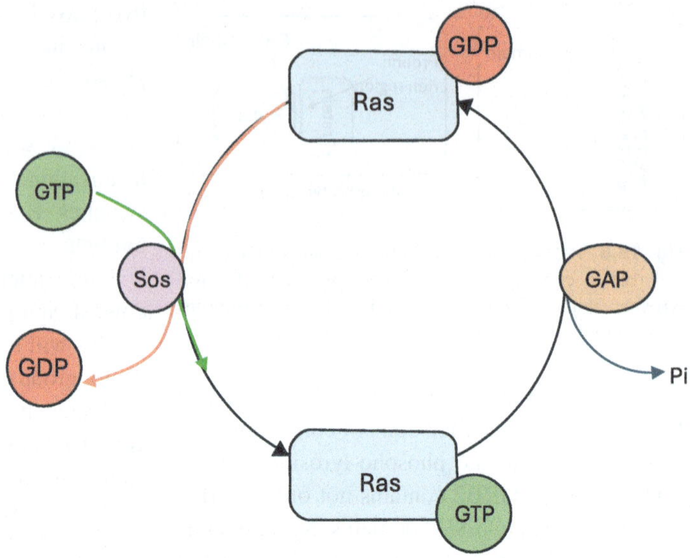

Fig. 18.6 Ras. Ras functions as a molecular switch. It is activated when bound to GTP and is inactive when bound to GDP. Sos and GAP play key roles in its regulation

Ras-GTP uses this surface to bind to Raf. This interaction recruits Raf to the plasma membrane. This promotes dephosphorylation of Ser259, which relieves the inhibitory 14-3-3 interaction. This, in turn, allows Raf to be phosphorylated on two other residues, which are positive regulatory phosphorylation sites: Ser338 and Tyr341. These sites are phosphorylated by other kinases such as Pak or Src family members. These events also lead to Raf dimerization, along with phosphorylation of two other positive regulatory sites in its activation loop, Thr491 and Ser494, possibly by trans-autophosphorylation. This further stabilizes the active conformation. These events cause Raf to take on an active conformation. Once it is active, Raf can phosphorylate its substrates, primarily MEK, described below [2, 11].

– Raf activation of MEK and downstream signaling proteins

Raf is the **MAPKKK** in the pathway, and it is a **serine/threonine kinase**. Once activated, Raf can phosphorylate its substrate. The main substrate of Raf is MEK. **MEK** is the **MAPKK** in the pathway, and it is a dual-specificity kinase. Once it is activated in response to phosphorylation by Raf, MEK can phosphorylate its substrate, **ERK**, which is the MAP Kinase (**MAPK**). ERK is phosphorylated on both threonine residues and tyrosine residues by **MEK**. The signaling cascade up to this point has occurred in the cytoplasm. Once ERK is phosphorylated, however, it undergoes a conformational change that allows it to be translocated into the nucleus through nuclear pores, due to exposure of a nuclear localization signal (NLS).

– Phosphorylation of the ERK MAP Kinase and subsequent regulation of transcription factors

Phosphorylated **ERK** is an active kinase, and once in the nucleus, phospho-ERK can phosphorylate transcription factors, such as TCFs (ternary complex factors). TCFs are a family of transcription factors that form a complex with other transcription factors, particularly SRF (serum response factor). Phosphorylation positively regulates TCFs' activity. Phosphorylated TCF, in cooperation with SRF, binds to regulatory elements of target genes and activates their expression (transcription). Importantly, these target genes include a family of genes known as immediate early genes, which encode other transcription factors (such as c-Fos and c-Myc), some of which drive the expression of genes that promote cell cycle progression. Importantly, they drive the expression of the gene encoding cyclin D1, which promotes G1 to S phase transition. In this way, the ERK-MAK Kinase pathway can mediate the mitogenic effect—the stimulation of cell proliferation via cell cycle entry—in response to growth factors [12, 13].

- Summary of the ERK MAP Kinase Pathway

To summarize, if we look at the pathway from start to finish (see Fig. 18.4), a growth factor binds to a cell surface receptor (in this case, an RTK), and this leads to a signal transduction pathway that leads to the nucleus, where genes that respond to the stimulus become expressed.

An important function of the MAP Kinase pathway is to respond to signals such as growth factors that regulate cell cycle progression and thus lead to division. The MAP Kinase pathway often responds to signals such as growth factors like EGF, which lead to increased expression (transcription) of genes that encourage cell cycle progression, cell growth, and proliferation [2, 3, 14]. Thus, EGF is considered to be a mitogen, and it can exert its effects via the MAP Kinase pathway. While growth factors such as EGF are classic examples of ligands that can activate the MAP Kinase pathway, another important example of a ligand that can activate the pathway is insulin. Insulin can bind to the IR, which is an RTK. Although the major insulin-activated pathway in metabolism is the PI3K-AKT pathway (discussed later), insulin can also activate the MAP Kinase pathway and regulate cell growth and survival. This is especially important in pancreatic β-cells, where insulin signaling through the MAP Kinase pathway has an important role in the growth and survival of those cells [15].

- Why Does the MAP Kinase Pathway Have Multiple Steps?

It is interesting that, like many signaling pathways, the ERK/MAP Kinase pathway involves multiple steps. There are several possible reasons for this. One reason may simply relate to the distance between the cell surface and the nucleus. Crossing this distance may be facilitated by the utilization of multiple enzymes located in different areas of the cytoplasm. Another reason may be related to amplification. With multiple steps involved, each enzyme (particularly the kinases) activates the next one in the cascade, and this allows the signal to be amplified at each step. Third, having multiple steps allows for crosstalk with other pathways. As we will discuss soon, multiple different pathways can be activated in a cell, and some of these pathways can have crosstalk with each other. The existence of multiple steps can increase the opportunity for crosstalk between different pathways.

- Negative Regulation of Pathways
  - Negative feedback mechanisms

After a signaling pathway has completed its job, it is important that it is terminated. This is critical because if the signaling pathway stays active for too long, it can lead to undesired effects such as uncontrolled cell growth. One mechanism for shutting down a signaling pathway is negative feedback. This may occur, for example, due to inhibitory phosphorylation of one or more of the mediators in the pathway [2, 3, 16]. For example, in some cells, after the sustained activation of the ERK pathway, once a large amount of phosphorylated ERK has accumulated, ERK can phosphorylate Sos or Raf, inhibiting their activity, thus shutting down the pathway.

- Receptor endocytosis

Another important way that signaling pathways, such as the MAP Kinase pathway, are terminated is by downregulation of the cell surface receptors. This involves receptor-mediated endocytosis (see Chap. 2), where the receptors cluster in clathrin-coated pits. These pits pinch off the membrane by endocytosis, forming endocytic vesicles. The vesicles they form fuse with endosomes, and from there, the receptor can recycle back to the membrane or be sent to the lysosome, where it is degraded. In this case, the signal is terminated and cannot occur again until more receptors are synthesized. This type of system can turn off a signaling pathway, or it can lead to a more subtle form of regulation and modulate how strongly a signaling pathway is activated [2].

2.3.3 RTK Signaling in Dysregulation in Disease and Therapy

- Cancer

Signaling pathways activated by RTKs have important roles in cell cycle progression and cell proliferation. Their uncontrolled activation, however, can be linked to cancer. These pathways are therefore of significant interest to cancer researchers. Many mediators of the MAP Kinase pathway have been linked to cancer. Ras is perhaps the best example as it is a well-known oncogene when it is improperly activated. Other signaling proteins, such as Raf, are known to be involved in some types of cancer, such as melanoma, when they are constitutively active. RTKs are also involved in cancer, especially when they are truncated, which can lead to constitutive ligand-independent activation of signaling pathways. The roles of signaling proteins in these and other pathways in cancer will be discussed in more detail in Chap. 19. In this chapter, we will focus on noncancer diseases.

- Wet Age-related Macular Degeneration (wAMD), Angiogenesis, and the VEGFR

MAP Kinase pathways are activated by various stimuli (ligands), often via binding to **RTKs**. Ligands and their associated RTKs can have important roles in human health, and their dysregulation can lead to some human health conditions. One important example is the Vascular Endothelial Growth Factor (VEGF), a ligand that binds the **Vascular Endothelial Growth Factor Receptor** (**VEGFR**). The binding of **VEGF** to **VEGFR** on the surface of endothelial cells (the cells that line the blood vessels) activates signaling pathways that lead to angiogenesis, which is the growth of blood vessels. This is because activation of VEGFR by VEGF on endothelial cells activates signaling pathways, including MAP Kinase pathways and other pathways that lead to their increased cell division and proliferation, and also pathways that regulate their migration, survival, and tube formation [17]. In this way, VEGF signaling promotes blood vessel formation, angiogenesis. Although angiogenesis is a normal and important process, improper regulation of angiogenesis can lead to health problems [2, 18].

An example of a disorder associated with abnormal angiogenesis is neovascular (wet) age-related macular degeneration (wAMD). wAMD is a progressive vision disorder characterized by abnormal growth of blood vessels at the macula, an area at the back of the retina. These blood vessels extend into the subretinal space, which is the area between the retina and the outer layers of the eye (see Fig. 18.7). The result is a gradual loss of central vision. wAMD is usually associated with increased expression of **VEGF**, which then binds to the **VEGFR** and activates signaling pathways leading to the formation of the blood vessels [19]. The finding that VEGF has an important role in this disorder has led researchers to develop drugs that target **VEGF** and VEGFR signaling for the treatment of wAMD.

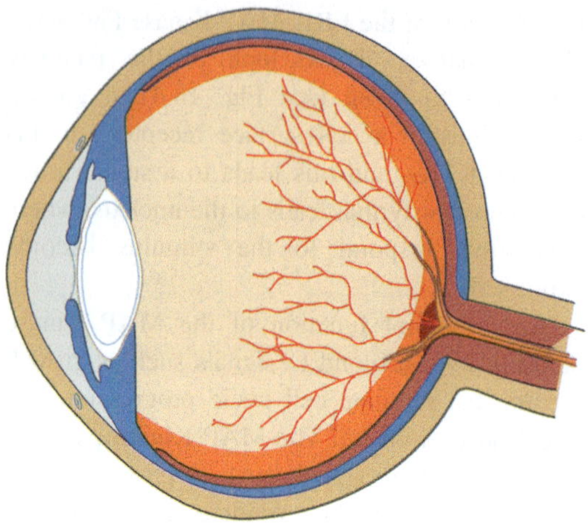

Fig. 18.7 Diagram of macular degeneration. Macular degeneration is associated with abnormal growth of blood vessels in the macula of the eye. Growth of blood vessels is regulated by VEGF. (*Source*: Illustration from Vecteezy.com (Pro license))

Monoclonal antibodies are one type of drug that is promising for targeting ligands and their associated RTKs. Monoclonal antibodies are antibodies generated in the lab, designed to bind a specific target (see Chap. 24). By binding to their targets, monoclonal antibodies can either mark them for destruction or block their activity [20, 21]. Monoclonal antibodies cannot penetrate the cell membrane to enter the cell, and therefore, they generally cannot target intracellular signaling proteins (such as Raf). RTKs and ligands are promising targets for monoclonal antibodies, however, because monoclonal antibodies can bind to their extracellular domains.

Vabysmo (faricimab) [22] is a biologic drug in the form of a monoclonal antibody that inhibits both VEGF and Ang-2 (FDA approved in 2022). VEGF binds the VEGFR, and Ang-2 is a ligand that binds to another RTK called Tie-2 that regulates blood vessel stability. This monoclonal antibody does not trigger an immune response, but it prevents the ligands from binding their receptors (VEGFR and Tie-2). As a result, angiogenesis is inhibited, and new blood vessel formation is reduced. This drug is injected into the eye and used to treat wAMD. Although it does not cure the disease, it can stabilize the disorder and delay the worsening of symptoms. This drug is also used to treat **diabetic macular edema**, which is associated with blood vessels in the eye. Because it binds to two targets (VEGF and Ang-2), **Vabysmo** is considered to be a bispecific monoclonal antibody [23].

Eylea (aflibercept) [24] is a slightly older drug (FDA approved 2011) that targets **VEGF**. It is a popular drug for the treatment of **wAMD**. Eylea is not an antibody, but it is a fusion protein that contains a part of the VEGF receptor (the part that binds to VEGF) and a portion of an IgG antibody. It acts as a decoy and prevents VEGF from binding to VEGFR on target cells in the eye. Both Eylea and Vabysmo block VEGF, although by different mechanisms. They differ in their dosing and the frequency with which they are used. Vabysmo may have advantages over Eylea because it has two important targets (VEGF and Ang-2) while Eylea specifically blocks VEGF [25].

2.4 The JNK and p38 Pathways Are Also MAP Kinase Pathways

2.4.1 Overview of the JNK and p38 Pathway

While the classic ERK-MAP Kinase pathway was the first to be identified in mammalian cells, two other major MAP Kinase pathways were identified later. These are the JNK (Jun N-terminal Kinase) and p38 pathways. Their activation follows the same overall pattern as other MAP Kinase pathways. While the proteins in these pathways differ from those described above for the ERK pathway, they belong to the same families of enzymes. Thus, a MAPKKK phosphorylates and activates a MAPKK (such as MKK4, 7, 3, or 6), which phosphorylates and activates the MAPK (JNK or p38) (Fig. 18.8). The classic ERK/MAP Kinase pathway is often activated by stimuli such as growth factors. In contrast, the JNK and ERK pathways are more likely to be activated by cytokines such as TNFα or even by stress signals that do not operate through a receptor, such as UV irradiation. There are exceptions to this, and there is often overlap as to which stimuli activate which pathway [2].

In the JNK and p38 pathways, Ras is not always the upstream activator. Instead, other small GTPases may activate the pathways under certain conditions, such as the **Rho GTPases Cdc42 and Rac**, which will be discussed later. Like Ras, these small GTP-binding proteins are switched "on" in their GTP-bound states. Once GTP bound, Cdc42 and Rac bind other proteins, leading to their activation. For example, they bind the serine/threonine kinases, leading to their activation, and these kinases may in turn activate kinase cascades that lead to activation of JNK and p38. JNK is a protein kinase that phosphorylates the transcription factor c-Jun. c-Jun usually forms a heterodimer with another protein, c-Fos. Together, **c-Jun** and **c-Fos** make up the transcription factor AP-1, which binds to target genes and regulates their expression [26]. p38 phosphorylates different transcription factors, such as ATF-2 and ELK-1. The response to both the JNK and p38 pathways is the regulation of gene expression. Depending on the stimulus and cell type, different genes may be regulated in response to these pathways, leading to different outcomes. For example, genes that are important for the regulation of cell division and proliferation may be activated in some cells. In contrast, in other cells, genes that are involved in promoting cell death by apoptosis may be regulated. In some cases, genes encoding cytokines may be regulated, potentially leading to a stress response [27, 28]. For this reason, these MAP Kinase pathways, particularly the p38 MAP Kinase pathway, are sometimes referred to as stress response pathways [2, 29].

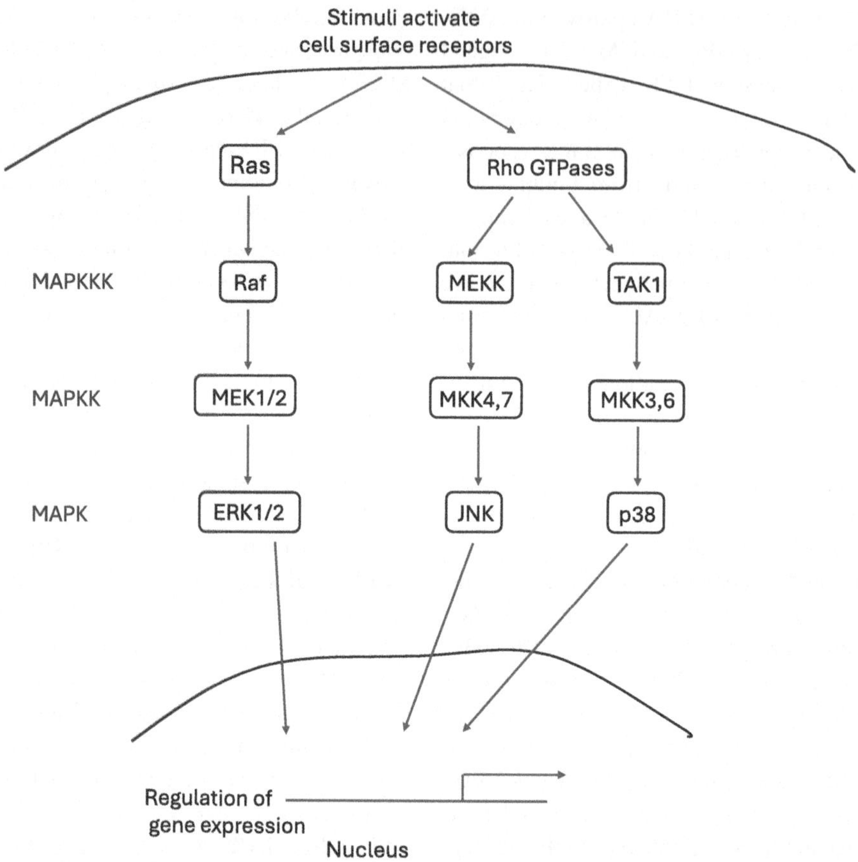

Fig. 18.8 MAP Kinase pathways are a family of related pathways. They include the ERK, JNK, and p38 pathways, described in the text

2.4.2 The JNK and p38 Pathways in Disease and as Drug Targets

Like the ERK pathway, dysregulation of the JNK or p38 pathways is sometimes associated with disease, and there is an interest in developing drugs that can target them. Developing such drugs has been challenging however. This may be due to nonspecific effects of some inhibitors and the crosstalk between different pathways. So far, there are no FDA-approved p38 or JNK inhibitors for disease treatment, but several experimental inhibitors are under investigation. **Dilmapimod** is a small-molecule **p38 inhibitor** that has been studied for its **anti-inflammatory** effects and may be promising for treating disorders such as arthritis and other inflammatory conditions. **Losmapimod** is also a p38 inhibitor, and it has been studied for its role in treating **facioscapulohumeral muscular dystrophy (FSHD), inflammatory diseases, and cancer** [30, 31].

2.5 Rho GTPases

2.5.1 Introduction to the Rho GTPases

Ras was discussed above as a small GTP-binding protein that is activated (GTP bound) after RTK activation. Ras is part of a larger family of related GTP-binding proteins. The Ras family includes H-Ras, N-Ras, and K-Ras, all of which function as molecular switches using similar mechanisms. In addition to these Ras family members, Ras is a member of an even larger superfamily of small GTP-binding proteins. Included in this superfamily are the Rho GTPases, which include Cdc42,

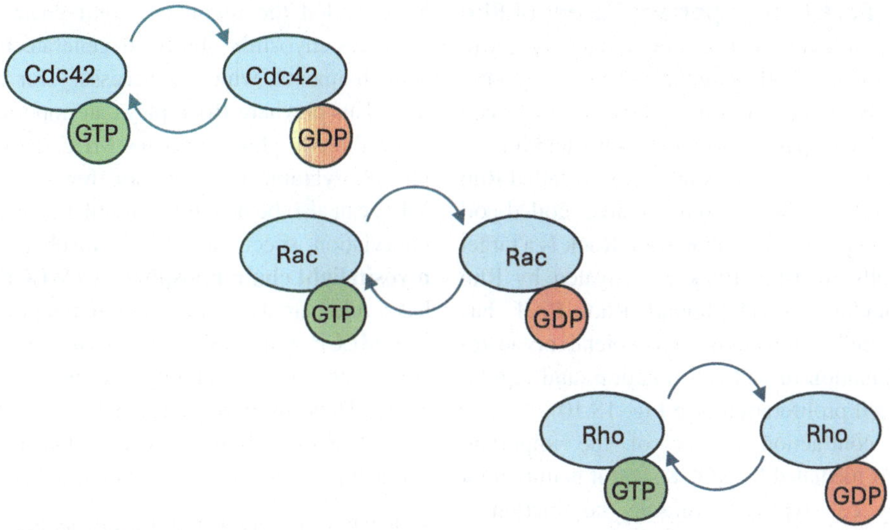

Fig. 18.9 Rho GTPases. Rho GTPases are members of the Rho family of GTP-binding proteins, which are part of the Ras superfamily. Major members of this family are Cdc42, Rac, and Rho. They can cycle between active GTP bound forms and inactive GDP bound forms

Rac, and Rho (see Fig. 18.9), along with multiple subfamily members. Two of these, Cdc42 and Rac, were alluded to above in the discussion of the JNK and p38 pathways. Like Ras, these GTP-binding proteins are also activated (GTP bound) by GEFs in response to various stimuli, such as RTK. Likewise, GAPs have a role in the hydrolysis of the bound GTP to GDP.

2.5.2 The Rho GTPases and the Actin Cytoskeleton

While we have focused on signaling pathways that lead to the nucleus and thereby regulate gene expression, signaling proteins can also function in different ways. The Rho GTPases can lead to nuclear responses, for example, by activating the JNK and p38 pathways, but they are especially well known for their roles in regulating the organization of the actin cytoskeleton. This is independent of MAP Kinases pathways and pathways leading to the nucleus. Their roles in cytoskeletal organization impact cell structure, shape, and motility. In fact, although we now know that they have multiple functions, the Rho GTPases were first known largely for their roles in regulating the actin cytoskeleton. For example, they lead to the formation of actin-rich structures such as filopodia, lamellipodia, stress fibers, and focal adhesions. Lamellipodia are actin-rich structures at the leading edges of migrating cells, and filopodia are actin-containing protrusions that project from some cells. Stress fibers are polymerized actin structures that can span the cell, and focal adhesions are complexes that can span the plasma membrane, connecting the stress fibers inside the cell to the extracellular matrix outside the cell. Specific Rho GTPases promote the formation of the different actin structures. Cdc42 induces filopodia, Rac leads to lamellipodia formation, and Rho leads to the formation of stress fibers and focal adhesion assembly [32–35].

The cytoskeletal structures regulated by the Rho GTPases are mediated by various Rho GTPase-interacting proteins [2, 36]. These changes are important for regulating cell shape, migration, and adhesion, all of which are critical for normal cell function. Improper control of these structures can be associated with disease. For example, in cancer, they play a role in metastasis, where cell migration and adhesion are not properly regulated [2, 37].

2.5.3 Rock Is an Important Target of Rho

The Rho proteins, like Ras, function by binding to other proteins and stimulating their activity. Numerous targets for Rho GTPases have been identified. One particularly interesting and important target is a serine/threonine kinase called Rho Kinase (also called Rho-associated coiled-coil containing protein kinase: **Rock**). **Rock** is a target specifically by Rho. Rock is activated by Rho upon binding to GTP-bound Rho. Rock has roles in cell contraction, cytoskeletal organization, regulation of cell-cell junction, and regulation of cell proliferation (see Fig. 18.10).

Cell contraction is one of the important functions mediated by ROCK, and it is important in many cell types. In muscle, contraction is important for promoting force and movement. In epithelial cells, it is important for maintaining cell shape, migration, wound healing, and other functions. In vascular smooth cells, the regulation of contraction is important for the regulation of blood pressure.

Rock phosphorylates two substrates that are important in contraction. Contraction involves the interaction between myosin, which is a motor protein, and actin. While there are multiple myosins, the classic myosin that is involved in contraction is myosin II. Myosin contains multiple domains. Among these, myosin light chain (MLC) is regulated by phosphorylation. Phosphorylation of MLC (specifically the part of the MLC called the regulatory light chain (RLC)) increases myosin's ability to generate force on actin filaments, which is necessary for contraction. This is where Rock plays an important role. Rock directly phosphorylates MLC (RLC). This phosphorylation is necessary for contractility. MLC can also be negatively regulated by dephosphorylation, specifically by a phosphatase called **myosin light chain phosphatase (MLCP)**. Rock has a role here as well because it can phosphorylate MLCP and inhibit its activity, which also contributes to maintaining the phosphorylated MLC. Thus, by phosphorylating both MLC and MCLP, RhoK allows contact between myosin and actin and encourages cell contraction [2, 38].

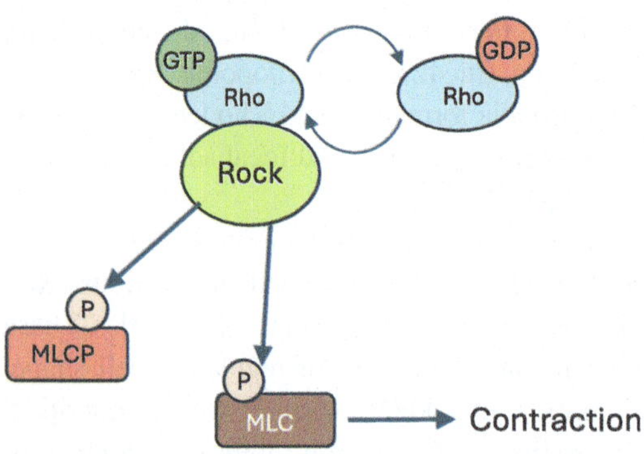

Fig. 18.10 Rho has a role in regulating contraction. When active, Rho binds to Rock. Rock is a protein kinase that phosphorylates targets that lead to contractility, including MLC (myosin light chain; activating) and MLCP (myosin light chain phosphatase; inhibiting)

- Rock as a Drug Target for Glaucoma

Improper regulation of Rock may be associated with numerous conditions, and Rock inhibitors have been tested for the treatment of various disorders [39]. One interesting finding is that Rock inhibitors can improve symptoms of **glaucoma**. Glaucoma is a leading cause of blindness. A major symptom of glaucoma is increased intraocular pressure (IOP). This is caused by blockage of the outflow of the aqueous humor (AH; a clear fluid in the area between the cornea and the lens [40]). A structure in the eye that plays an important part in the regulation of the IOP is the trabecular meshwork (TM), which contains a mixture of cells and extracellular material. The TM has a major role in regulating the outflow of AH. Rock can phosphorylate some components of the TM and increase its stiffness. This regulates permeability and has a role in decreasing the outflow of AH. **Inhibition of Rock,** in contrast, relaxes the TM, reduces its stiffness, and increases fluid drainage. This has **been shown to reduce IOP,** and for this reason, Rock has become a target for drugs that treat glaucoma.

Rocklatan (netarsudil and latanoprost ophthalmic solution) [41] is an FDA-approved Rock inhibitor eye drop used to treat **glaucoma**. This drug contains both **netarsudil** and **latanoprost** (both small molecules). **Natarsudil** is the **Rock inhibitor**. It inhibits the kinase activity of Rock by binding to its ATP-binding site so that it cannot phosphorylate its substrates. Rock inhibition helps restore the outflow of the AH through the TM due to its role in relaxing the TM. **Latanoprost** is a prostaglandin analog. Prostaglandins bind to receptors on ciliary muscle cells of the eye and induce the expression of matrix metalloproteinases, which are extracellular matrix (ECM) proteins. It thus remodels the ECM, making it easier for aqueous humor to drain from the eye. Another FDA-approved drug for glaucoma is **Rhopressa (nearside ophthalmic solution)** [42], which consists of **netarsudil** alone [43, 44].

- The PAK Kinases Are Targets of Rho GTPases

Members of the Ras superfamily lead to activation of target proteins by binding to them when they are in their active GTP-bound state. For example, when Ras is GTP bound, it binds to Raf and leads to its activation. Raf is thus considered to be a Ras target. What are the targets of other Ras superfamily members, specifically the Rho GTPases? One way to identify target proteins is by looking at evolutionary conservation. Proteins that have vital functions are likely to have been conserved throughout many years of evolution, and thus homologues of such proteins (or the genes that code for them) are present in different species and cell types. Yeast is a useful eukaryotic model for identifying evolutionarily conserved genes. This is because, although yeast is a relatively simple eukaryote, it performs many essential functions similar to those of more complex cells, such as cell replication and responding to cell signals. Yeast also has signal transduction pathways, which share many features with those in higher eukaryotic cells. For example, yeast also has a Cdc42 protein. In fact, yeast Cdc42 was identified before mammalian Cdc42 was discovered. In yeast, Cdc42 binds to and activates a protein called STE20, which is a serine/threonine kinase.

Interestingly, in mammalian cells, a homologous kinase with an amino acid sequence similar to STE20 was identified. It turns out that this protein binds to and is activated by mammalian Cdc42. To a lesser extent, it is also activated by Rac. The mammalian protein was named PAK for p21-activated kinase. It gets this name because the Rho GTPases are sometimes referred to as p21 due to their size of 21 KD (not to be confused with the p21 protein involved in cyclin/CCDK regulation in the cell cycle). In mammalian cells, Pak was found to be a member of a family of six related kinases named PAKs 1–6 [45, 46]. The PAK proteins all have a similar domain structure, with a regulatory domain that can bind to GTP-bound Cdc42 or Rac, and a catalytic domain with its kinase activity (see Fig. 18.11). The PAKs are serine/threonine kinases, and they phosphorylate various substrates when they are active. Their catalytic (kinase) activity is activated after they bind to Cdc42 or Rac through a different PAK domain called the GTPase binding domain (GBD), but they may also be activated by other mechanisms. Activation of the PAKs leads to the regulation of cytoskeletal organization and activation of various cellular signaling pathways. The PAKs have roles in several of the signaling pathways discussed in this text, such as the JNK and p38 pathways, as well as other pathways, such as the Wnt pathway, and pathways that regulate cell survival and apoptosis via phosphorylation of the pro-apoptotic protein BAD (Bcl2-associated death promoter). Some of the PAK kinases can also phosphorylate Raf, which in turn has an important role in the MAP Kinase pathways [47]. The PAKs are often dysregulated or upregulated in cancer and are therefore considered to be promising targets for the treatment of some types of cancer.

2.6 The PI3K/AKT Pathway

The AKT pathway can be activated by RTKs as well as other types of receptors. It has multiple functions. It is especially well known for its role in regulating cell survival and also plays important roles in metabolism. The pathway is mediated by the serine-threonine kinase AKT (also known as protein kinase B or PKB). The lipid kinase called PI3-kinase (PI3K;

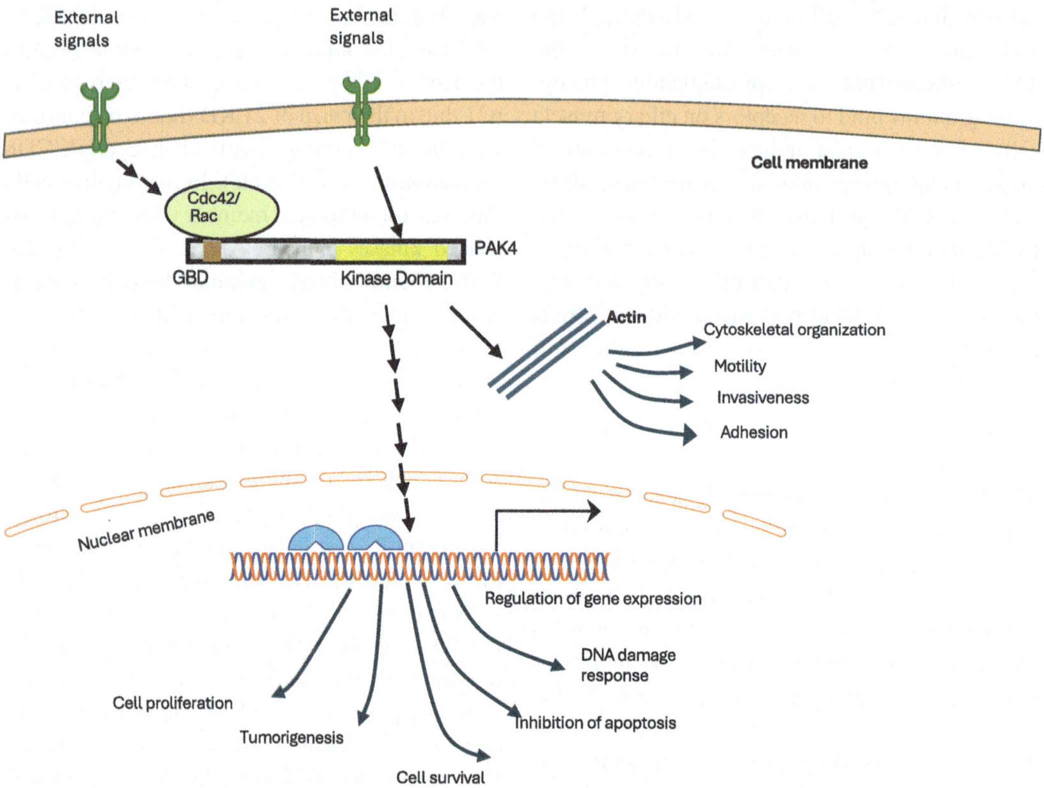

Fig. 18.11 PAK Kinases (PAKs). The PAKs bind to and are activated by the activated Rho GTPases, Cdc42 and Rac. PAKs are serine/threonine kinases that regulate signaling pathways that impact cell functions such as cell growth, survival, and cytoskeletal organization. (*Source*: Figure from Cordover E & Minden A. J Cancer Metastasis Treat 2020;6:45. https://doi.org/10.20517/2394-4722.2020.101. Licensed under the Creative Commons Attribution 4.0 (CC BY 4.0) license)

phosphoinositide-3-kinase) has a crucial role in activating AKT, and the AKT pathway, and is sometimes referred to as the PI3K/AKT pathway. Below is a discussion about how the AKT pathway is regulated, followed by a discussion of several of its functions, as well as the closely connected mTORC1 pathway [48].

2.6.1 The PI3K/AKT Pathway

- Recruitment of PI3K to the membrane

The AKT pathway relies on **PI3-Kinase (PI3K)**. Unlike many of the other kinases we have discussed so far, which are protein kinases, **PI3K is a lipid kinase**. It phosphorylates lipids rather than proteins. Specifically, **PI3K** is a **phosphoinositide kinase**. **Phosphoinositides** (also called phosphatidylinositides) are phosphorylated forms of **phosphatidylinositol** (PI), one of the main components of membranes. There are multiple forms of phosphoinositides. They are found naturally in membranes and can have important roles in cell signaling. The main phosphoinositide that PI3K phosphorylates is PIP_2 (phosphatidylinositol 4,5-bisphosphate), to form PIP3 (phosphatidylinositol 3,4,5-trisphosphate). The details of PI3K recruitment to the membrane are discussed below.

A receptor, often an **RTK**, gets activated by ligand binding. This leads to receptor dimerization and phosphorylation on cytoplasmic tyrosine residues of the dimerized receptor. Instead of Grb2 and Sos binding to the receptor as described above for the MAP Kinase pathway, in this pathway, **PI3K** binds to the **RTK**. This interaction occurs via the **SH2 domain** present in **p85**, which is the regulatory subunit of **PI3K**. This is a key step because it causes PI3K to be translocated to the plasma membrane, bringing it close to its substrate, PIP_2, a membrane lipid. PI3K phosphorylates PIP_2, converting it to PIP3.

- Recruitment of PDK1 and AKT to PIP3

Now another kinase, called **PDK1** (phosphoinositide-dependent kinase), gets involved. PDK1 binds to the newly formed PIP3 via its PH domain. A PH (Plextrin homology) domain is a domain that specifically binds to phosphoinositides. Likewise, **AKT** also binds to a PIP3 at the cell surface, via its own PH domain (both PDK1 and AKT have a higher affinity for PIP3 compared to PIP_2). PDK1 is a protein kinase, and AKT is its substrate. The outcome of this entire process so far is that PDK1 is brought into proximity to its substrate, AKT. PDK1 now phosphorylates AKT, leading to AKT's activation. AKT is also a kinase, and once it is activated, it phosphorylates various substrates.

2.6.2 Some Functions of the PI3K/AKT Pathway

- PI3K/AKT promotes cell survival

One important AKT substrate is **BAD (Bcl2-associated death promoter)**. **BAD** is a pro-apoptotic protein of the Bcl-2 family. In its unphosphorylated form, BAD leads to inhibition of the anti-apoptotic proteins Bcl-2 and Bcl-xL, thereby promoting apoptosis. When AKT phosphorylates BAD, phosphorylated BAD binds to 14-3-3 proteins instead of Bcl-2/Bcl-XL. This sequesters BAD so that it is inactive, and it releases Bck-2/Bcl-XL, thus blocking apoptosis. In this way, phosphorylation of BAD by AKT promotes cell survival. This is a major mechanism by which AKT induces cell survival, often in response to growth factors such as EGF and related ligands. The roles for the Bcl proteins discussed here are also discussed in Chap. 22 [2, 3, 49] (Fig. 18.12).

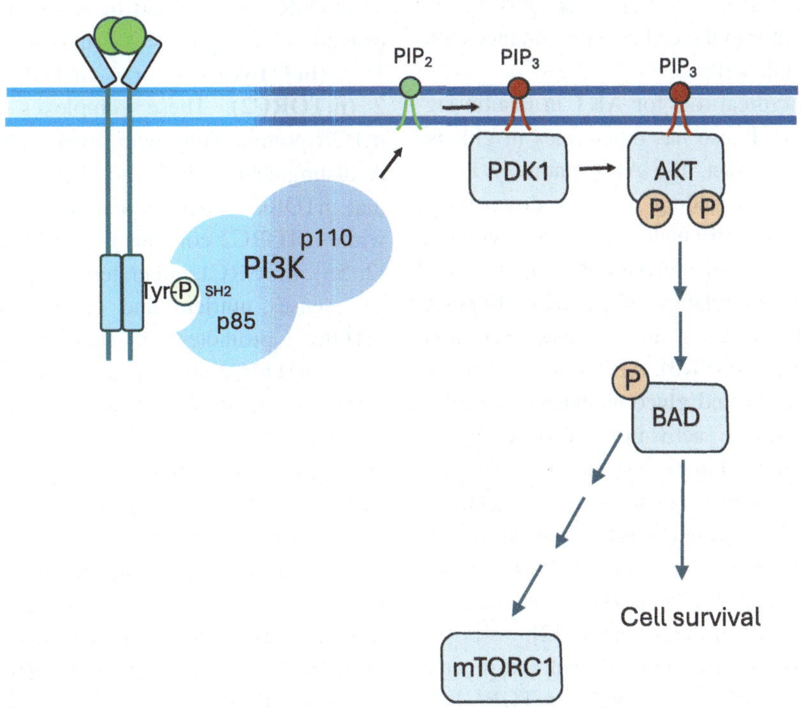

Fig. 18.12 The AKT pathway. AKT is a protein kinase that has a role in cell survival. This pathway involves the recruitment of PI3K (via its p85 subunit) to the activated receptor. Details are described in the text

- Role of the PI3K/AKT pathway in insulin signaling

In addition, another important activator of the PI3K/AKT pathway is insulin, via binding to the insulin receptor (IR). Insulin can also lead to cell survival pathways via Bad phosphorylation, but in addition to this, it also has roles in regulating metabolism. Insulin binds to the IR, especially in cells of tissues such as muscle and adipose tissue. This leads to auto-transphosphorylation and recruitment of IRS proteins to the receptor. Phosphorylated IRS proteins, in turn, recruit PI3K, which activates the PI3K/AKT pathway as described above for other activators. The AKT pathway is important in metabolism by phosphorylating a protein that allows GLUT4 to translocate from the cytoplasm to the cell surface. Specifically, it phosphorylates a protein called AS160, which is a GAP that keeps the GTP-binding protein Rab in its inactive GDP-bound form. Phosphorylation of AS160 by AKT relieves this inhibition so that Rab can become active. Rab in turn has an important role in transporting GLUT4-containing vesicles to the plasma membrane [50, 51]. Transport of the GLUT4 receptor to the cell surface enhances glucose uptake into cells.

This is a critical role for AKT in insulin signaling, but AKT also has other roles in glucose homeostasis. In fact, the AKT pathway is the main metabolic signaling pathway downstream of insulin. AKT stimulates glycogen synthesis (by inhibiting glycogenolysis). This is mediated by AKT phosphorylation of **protein phosphatase 1 (PP1)**, which in turn regulates **glycogen synthase kinase (GSK3)**. AKT also has roles in lipid metabolism and gluconeogenesis. In addition, Akt leads to activation of mTORC1, a nutrient-sensing kinase complex. Through mTORC1, Akt signaling can upregulate anabolic pathways such as protein synthesis and lipogenesis. Several of these functions of AKT are also discussed in Chap. 22, and mTORC1 is discussed in more detail later in the chapter [52].

In addition to regulating survival and glucose homeostasis, Akt also activates mTORC1, a nutrient-sensing kinase complex that regulates anabolic metabolism in response to nutrients, growth factors, and energy signals. The next section focuses on mTOR signaling, with a focus on its role in the mTORC1 complex in response to AKT signaling.

2.7 mTOR Signaling Pathways

2.7.1 mTORC1 Regulation and Activation

One of the most important downstream effects of **Akt** is the activation of **mTORC1.** mTORC1 is a complex containing the protein mTOR. mTOR is a serine/threonine kinase that regulates a wide range of cellular processes, including cell growth, proliferation, survival, and metabolism (see Fig. 18.12). **mTOR** stands for Mammalian Target of Rapamycin. Rapamycin is a bacterial toxin, and it is sometimes used as an immunosuppressant and an anticancer drug. mTOR got its name because it is a target of this toxin, but it is now known for many different functions, especially its role as a nutrient sensor and mediator of metabolism.

mTOR exists as part of one of two different protein complexes: mTOR complex 1 (**mTORC1**) or mTOR complex 2 (**mTORC2**). These complexes contain the mTOR protein along with several other proteins. A distinguishing difference between the two is that mTORC1 includes a protein called Raptor, while mTORC2 contains a protein called Rictor [2, 53]. mTORC1 integrates cell growth signals and senses nutrient and energy status, while mTORC2 promotes cell survival by activating AKT. mTORC2 also has a role in regulating the cytoskeleton. mTORC1 is the focus of this section.

mTORC1 regulation is mostly at the cytosolic surface of the lysosomal membrane, where mTORC1 is recruited by **Rag** proteins, which are **small GTP-binding proteins**, that are bound to the lysosomal membrane via a protein complex called the **ragulator complex**. Here, **mTORC1** can interact with **Rheb**, another **small GTP-binding protein**. **Rheb** is also found on the lysosomal membrane, and it directly **activates mTORC1**. **Rheb** is regulated by a complex called the **TSC1/TSC2** complex. In this complex, **TSC2** is a **GAP**. As a **GAP, TSC2** keeps **Rheb** in its **inactive** GDP-bound state. This is where the role of **AKT** comes into play. When active, **AKT** can directly **phosphorylates TSC2** on multiple sites. This inhibits **TSC2's** activity as a **GAP**. Now, **TSC2** can no longer inhibit the activity of **Rheb** (i.e., it can no longer keep it in its GDP-bound state). Now, **Rheb** is **active** (GTP bound) and can activate **mTORC1**. (mTORC1 regulation involves a complex combination of positive and negative regulatory interactions (see Fig. 18.13).

Full mTORC1 activation also requires nutrient inputs. The presence of amino acids in the cell indicates an abundance of nutrients. This can be detected by nutrient sensors in the cytoplasm, especially those that recognize leucine and arginine. These sensors in the cell activate **Rag** in response to leucine and arginine. When **Rag** is active and GTP bound in response to nutrient sensors, it recruits mTORC1 to the lysosomal membrane, where it can be activated by **Rheb**. Thus, **Rag** is also important for mTORC1 activation.

Full activation of mTORC1 requires both growth factor signaling through Akt and Rheb, nutrient availability through Rag. Once active, mTORC1 promotes anabolic processes such as protein, lipid, and nucleotide synthesis. In addition to these key mechanisms of mTORC1 activation, it should also be mentioned that other pathways can also lead to mTORC1 activation. For example, the ERK MAP Kinase pathway can also negatively regulate TSC2, leading to activation of mTORC1 [53–55].

These mechanisms of mTORC1 activation ensure that mTORC1 is only fully active when both growth factors, especially insulin (via AKT signaling), and nutrients are sufficient. Once activated, mTORC1 has multiple functions. Relevant to metabolism, activated mTORC1 promotes protein synthesis by phosphorylating S6K (S6 kinase), which enhances translation efficiency, and 4E-BP1, which releases the translation initiation factor eIF4E to start cap-dependent translation. mTORC also leads to lipid and nucleotide biosynthesis, and inhibits autophagy, supporting overall anabolic metabolism and cell growth. When mTORC1 signaling is very high, it can also lead to downregulation of insulin receptor signaling, via S6K-mediated phosphorylation of IRS on inhibitory sites [53, 54].

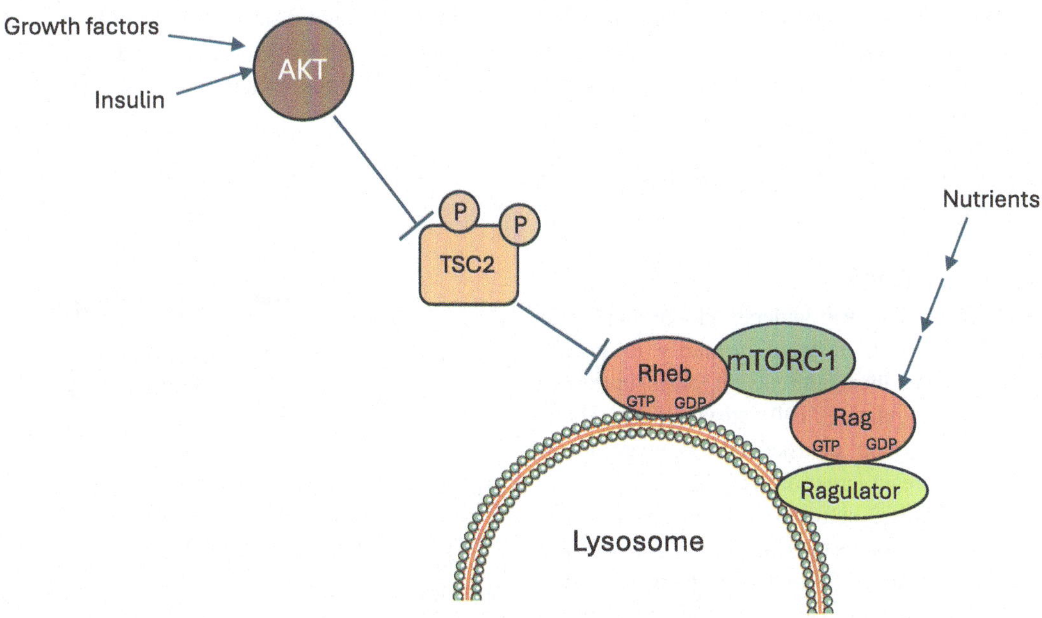

Fig. 18.13 Regulation of mTORC1 at the surface of the lysosome. mTORC1 is activated upon AKT activation. It is activated mainly at the surface of the lysosome, as described in the text

2.7.2 AMPK and Downregulation of mTORC1 in Low-Nutrient Conditions

In contrast to high-energy conditions, mTORC1 is inhibited under low-energy conditions. This is mediated by the kinase **AMP-activated protein kinase (AMPK)**, which is activated by an increased AMP/ATP ratio. An increased AMP/ATP ratio indicates low energy levels in the cell, leading to the elevation of AMPK, unlike mTORC1, which is activated when energy levels are high. When active, AMPK phosphorylates TSC2. While AKT phosphorylates TSC2 on inhibitory sites, AMPK phosphorylates TSC2 on stimulatory sites, thereby increasing TSC2's inhibitory effect on Rheb and consequently inhibiting mTORC1 activity. This process is the opposite of AKT, which inhibits TSC2's activity and activates mTORC1 activity. AMPK also phosphorylates Raptor, a component of the mTORC1 complex, and this also suppresses mTORC1 activity.

Thus, AMPK and mTORC1 have opposing roles in metabolism. AMPK inhibits mTORC1 activity, and it shifts metabolism toward catabolic pathways, which break down molecules and produce energy. In contrast, the activation of mTORC1 (by AKT and nutrients) shifts metabolism toward anabolic pathways. AMPK is upregulated when energy levels and nutrient levels are low (as in the fasting state), while mTORC1 is upregulated when nutrients and energy levels are plentiful (the fed state) [53, 54,56].

2.7.3 mTORC2

mTORC2 is less well understood than mTORC1 but is known to respond primarily to growth factors rather than nutrients. It plays important roles in regulating cell survival, metabolic homeostasis, and cytoskeletal organization. A key downstream target of mTORC2 is Akt itself, which mTORC2 activates through phosphorylation at a specific serine (Ser473). mTORC2 also regulates small GTP-binding proteins that influence cytoskeletal organization, thereby affecting cell shape and migration [2, 53].

2.8 PI3 Kinase

Above, we covered AKT signaling. PI3K signaling is an important component of AKT signaling. Therefore, before moving on to other signaling pathways, it is important to provide a basic overview of PI3K and discuss some of the drugs associated with PI3K/AKT signaling.

2.8.1 The Basic Structure of PI3K

PI3K is central to this pathway as it initiates the activation of AKT when it phosphorylates PIP_2 to produce PIP3. PI3K is regulated by a mechanism that depends on intramolecular interactions between its subunits. PI3 Kinase consists of two main subunits. These are a catalytic subunit (p110α, encoded by PI3KCA) and a regulatory subunit (p85α, encoded by PI3KR1) (see Figs. 18.12 and 18.14). The p110 subunit has the kinase activity and phosphorylates its substrate, and the p85 subunit is a regulatory subunit that can regulate P110's activity. The p85 subunit has two major SH2 domains. In p85, they are called the nSH2 (N-terminal SH2) and cSH2 (C-terminal SH2) (it also contains a third region called the iSH2, which is structurally distinct from SH2 domains and forms a coiled coil that stabilizes p110, not discussed in detail here). p85 can bind to and stabilize p110 and inhibit its activity under basal conditions. This interaction

Fig. 18.14 PI3 Kinase (PI3K). The basic structure of PI3K is shown here. PI3K contains two major domains: p85 and p110. p110 is the catalytic domain and p85 is the regulatory domain. (*Source*: Adapted from Elshaer et al., Antioxidants, 2018, 7(4), 47. https://doi.org/10.3390/antiox7040047. Licensed under the Creative Commons Attribution 4.0 (CC BY 4.0) license)

is mediated primarily by the SH2 domains in p85. Under basal conditions, nSH2 binds to a domain in p110 called the helical domain, and this suppresses its activity. (The cSH2 domain plays less of a role and mainly contributes to binding to adapter proteins). When RTKs become phosphorylated on tyrosines, these receptor phospho-tyrosines bind to the SH2 (especially the nSH2) of p85. This disrupts the nSH2-p110 interaction, allowing the p110 structure to take an open and active conformation. (It should be noted that although SH2 domains are best known for their roles in recognizing phospho-tyrosine residues, the nSH2 domain of p85 can also interact with p110 in a phospho-tyrosine-independent manner) [57, 58].

2.8.2 The Different Classes of PI3K

PI3K is important for the activation of AKT, promoting cell survival and regulating metabolism, but it is actually part of a family of related proteins. They are all lipid kinases, but they have a range of functions. There are three different classes of PI3K in mammals, which encompass eight different PI3Ks in total. These are Class I (containing Class IA and Class IB), Class II, and Class III. Class I is largely involved in signal transduction, cell growth, and survival, and it is the type discussed above. Class II is involved in membrane trafficking and endocytosis, and Class III is known for its role in autophagy and endosome signaling.

Class I PI3Ks phosphorylate PIP_2 and lead to AKT activation, as described above. This class has two subclasses, *class IA* and *class IB*. Class IA has three members, which contain the p110α, p110β, and p110δ isoforms, as well as the p85 regulatory subunit or one of its splice variants. They are correspondingly called PI3Kα, PI3Kβ, and PI3Kδ. They can interact with RTKs, including insulin receptors and other types of receptors. They differ from each other mostly in their tissue-specific expression patterns. PI3Kα is expressed in most tissues. PI3Kβ is also ubiquitously expressed, but it has some specialized immune functions, and PI3Kδ is mostly found in leukocytes. These different PI3K family members all have roles in cell growth, survival, metabolism, and proliferation. *Class IB* has one member, PI3Kγ, which contains p110γ. It is found mostly in leukocytes, and they interact mainly with GPCRs.

Class II contains three members: PI3K-C2α, PI3K-C2β, and PI3K-C2γ. They are known for their catalytic domains that differ somewhat from the other PI3Ks. They are involved in membrane trafficking. **Class III** has one member, Vps34, which is also called PI3K class III. It is involved in autophagy and vesicle trafficking [2, 3, 59, 60].

2.8.3 AKT and PI3K Inhibitors for Disease Treatment

The AKT pathway can promote cell survival. It is therefore a promising target for treating cancer cells since cancer cells often exhibit elevated levels of cell survival [61]. Some of the AKT and PI3K inhibitors in the market for cancer treatment are discussed in Chap. 19. In addition to cancer, the PI3K/AKT signaling pathway is implicated in other disorders as well. Some relevant inhibitors are described below:

JOENJA® (leniolisib) [62] is a **small-molecule** PI3K inhibitor (FDA approved 2023), which is used to treat activated **PI3K delta syndrome (APDS)**. APDS is an immunodeficiency disorder which is associated with abnormal growth of white blood cells caused by having hyperactive PI3Kδ. Although this activation leads to increased proliferation, this disorder leads to an imbalance between mature and functional immune cells, thus making immune responses inefficient [61, 63].

Vijoice (alpelisib) is a **small-molecule** PI3K inhibitor FDA approved in 2022. It is used to treat **PI3KCA-related overgrowth spectrum (PROS)**, which is caused by a gain-of-function mutation of the PIK3CA gene, which encodes the p110 subunit of PI3Kα. PROS is associated with a group of vascular and lymphatic conditions. It encompasses a wide range of disorders associated with PI3KCA mutation, largely characterized by abnormal cell growth, vascular and lymphatic abnormalities, and skin lesions [61].

2.9 Type I and Type II Cytokine Receptors and the JAK/STAT Pathway

Type I and type II cytokine receptors are single-pass transmembrane receptors. Ligands for this type of receptor include some cytokines, interleukins, interferons, and sometimes growth factors. Unlike enzyme-linked receptors, these receptors do not have intrinsic kinase activity and therefore they do not phosphorylate each other upon ligand binding. They may, however, be linked to tyrosine kinases, such as the JAKs, which can phosphorylate each other and the receptors.

2.9.1 JAK/STAT Pathway

- Overview of the Pathway

There are several examples of cytokine receptors that operate through the JAK/STAT pathway. These include the interferon-γ (IFN-γ) receptor; the growth hormone (GH) receptor; the interleukin receptors, including interleukin-6 (IL-6), IL-4, and IL-13 receptors; the leptin receptor; and the granulocyte colony-stimulating factor (G-CSF) receptor. These receptors have in common a signaling mechanism that involves JAK and STAT. JAK stands for Janus Kinase, while STAT stands for Signal Transducers and Activators of Transcription. Cytokine receptors that operate via JAK/STAT stably associate with JAKs, which are cytoplasmic tyrosine kinases (Fig. 18.15). There are four known JAKs in humans: JAK1, JAK2, JAK3, and Tyk2 (tyrosine kinase 2), and there are seven different STATs: STAT1, STAT2, STAT3, STAT4, STAT5A, STAT5B, and STAT6.

Cytokine binding to their receptors leads to a conformational change in the receptor, which brings two JAKs close to each other so that they phosphorylate each other. This phosphorylation increases the tyrosine kinase activity of the JAKs further. The JAKs, in turn, phosphorylate the tyrosines on the cytoplasmic tails of the cytokine

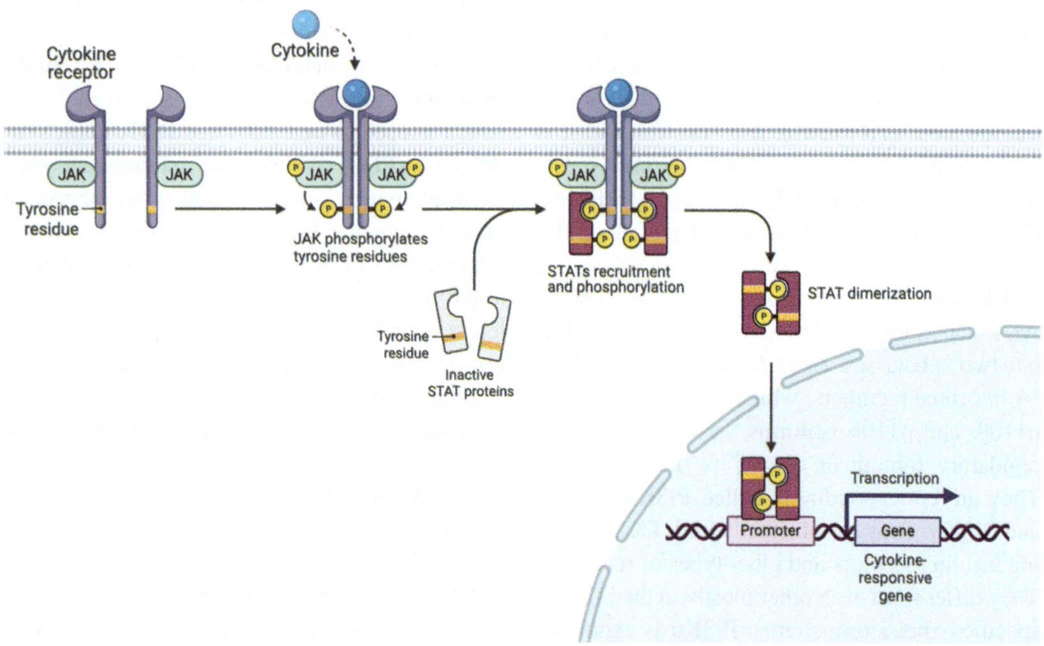

Fig. 18.15 The JAK/STAT pathway. JAKs are associated with cytokine receptors and become activated when the receptors are ligand bound. Activated JAKs phosphorylate tyrosine residues on the receptor, creating docking sites for STATs. STATs are then recruited and phosphorylated, and they dimerize. This results in the activation of STAT-regulated target genes. (*Source*: Created with BioRender.com)

receptors, creating phospho-tyrosine sites on the receptors. STATs can now bind to these phospho-tyrosines via the STAT SH2 domains. Once the STATs are brought into proximity to the JAKs via this interaction, the JAKs phosphorylate the STATs on tyrosine residues. This causes the STATs to dissociate from the receptors.

The SH2 domain on the released STAT now binds to a phospho-tyrosine on another STAT molecule, resulting in STAT dimers. The STAT dimers then translocate to the nucleus, where they function as transcription factors. In the nucleus, they bind, in combination with other transcription factors, to regulatory sequences in regulatory regions of various genes, thus regulating gene expression in response to the stimulus. For example, they may regulate genes that are important for immune responses upon stimulation with interferon or that regulate growth and metabolism in response to growth hormone. They also activate genes that can help turn off the pathway so that it does not stay on too long. In this way, the pathway is tightly controlled [2, 3, 64] (Fig. 18.15). The STATs are referred to as a type of **latent transcription factor**. This term describes transcription factors that are cytoplasmic but translocate to the nucleus to regulate transcription upon receptor stimulation.

- The JAK Proteins

The JAKs are considered non-receptor tyrosine kinases. They have several domains, including two main domains that are especially important for their catalytic activity: **JAK homology 1 (JH1)** and **JAK homology 2 (JH2)**. These domains are found in all four of the JAKs. **JH1** has the **catalytic activity**, and it phosphorylates its substrate, STAT. **JH2** is called a pseudokinase domain, and it is a **regulatory domain** for JH1. JH2 domains have structures somewhat similar to actual kinase domains. They even contain an activation loop and ATP-binding site, similar to a kinase, but they are not identical to the activation loops on functional kinases. For example, they are partly truncated and have some sequence differences compared with JH1 activation loops. The JH2 domain does not have any catalytic activity [65].

The JH2 domain has an inhibitory effect on the JH1 domain under basal conditions, but upon stimulation, it releases this inhibition and allows the JH1 domain to be active. In addition to these two domains, the JAKs also have an SH2 domain that interacts with other proteins and links them to cytokine receptors. They also have a FERM domain, which binds to other proteins and can target JAK to the membrane [66].

- JAKs and JAK Signaling Pathways as Drug Targets

Types I and II receptor signaling pathways, especially those involving JAKs, are frequently targets for drugs that treat inflammatory disorders. This is because many cytokines, interleukins, and interferons that signal through these receptors are linked to inflammation, especially when they are improperly regulated. JAK/STAT signaling often leads to upregulation of genes that encode inflammatory cytokines and chemokines. This can lead to recruitment of immune cells such as B and T cells, macrophages, and neutrophils, which lead to inflammation. Upregulation of cytokines can also lead to an autoregulatory loop, which can worsen inflammation. For this reason, blocking these pathways is often a therapeutic strategy for treating inflammatory disorders [67]. Several examples of these types of treatments are described here.

- Inhibition of TYK2

TYK2, one of the four **JAKs**, has been shown to have a role in psoriasis, an inflammatory skin condition. This is thought to be due to its role in the **IL-23** signaling pathway. **IL-23** is a pro-inflammatory cytokine that is secreted by certain immune cells, such as dendritic cells and activated macrophages. It binds to receptors on various T cells and activates JAK/STAT signaling, often through the JAKs JAK2 and TYK2. This can lead to activation of pathways that lead to the production of more inflammatory cytokines and granulocyte-macrophage colony-stimulating factor (GM-CSF). Excessive IL-23 signaling and subsequent improper regulation of the pro-inflammatory cytokine production can contribute to inflammation and abnormal skin cell growth in psoriasis. It also contributes to other inflammatory disorders such as Crohn's disease, rheumatoid arthritis (RA), and ulcerative colitis (UC) [68]. Some patients have a genetic mutation in the TYK2-JH2 gene. This mutation disrupts the ability of the JH2 to inhibit JH1, leading to increased activation of TYK2. This genetic mutation is only found in some patients and is probably not the main cause of the disease, but increased activity of IL-23 signaling via TYK2 is a common feature in patients with psoriasis [68].

Sotyktu (deucravacitinib) [69] is a drug that **inhibits Tyk2**. It is a small molecule that enters the cell. Unlike some kinase inhibitors that bind the ATP-binding site of a kinase, this compound binds allosterically to the JH2 domain of Tyk2. By stabilizing JH2 in its inhibitory conformation, the drug prevents activation of the JH1 domain of Tyk2. Importantly, this inhibitor is quite selective for TYK2 over other JAK family members (such as JAK1, JAK2, JAK3). This targeted approach helps alleviate adverse side effects. This drug was FDA approved for the treatment of moderate to severe **plaque psoriasis** in 2022. It is administered orally, and it may have broader potential in the treatment of other autoimmune and inflammatory disorders [65, 66, 70–72].

– Inhibition of IL-23

Increased **IL-23** signaling is also associated with other inflammatory disorders [73]. **Skyrizi (risankizumab-rzaa)** [74] is a drug that directly inhibits IL-23. It is used to alleviate symptoms of Crohn's disease, as well as plaque psoriasis [75, 76]. This biologic drug is a monoclonal antibody that specifically binds IL-23, thereby preventing it from binding to its receptor. This prevents it from activating downstream signaling pathways. **Skyrizi** is used to treat **plaque psoriasis, psoriatic arthritis (PsA), and Crohn's disease**.

Other IL-23 inhibitors that are also used to treat inflammatory diseases include **Omvoh (mirikizumab-mrkz)** (FDA: 2023) [77], **Ilumya (tildrakizumab)** [78], and **Tremfya (guselkumab)** [79]. These drugs are all **monoclonal antibodies** and are categorized as biologics. IL-23 is a cytokine made up of two subunits, p40 and p19. While the p40 subunit is found on another cytokine (IL-12), p19 is specific to IL-23 [80]. These drugs target the P19 subunit of IL-23 [76]. This is helpful for the specificity of the drug. They are used for **ulcerative colitis (UC), plaque psoriasis, and psoriatic arthritis**.

– Inhibition of IL-4 and IL-13 signaling

Dupixent (dupilumab) [81] is a biologic composed of a monoclonal antibody that binds to the IL-4-receptor alpha subunit (**IL-4Rα**), which is a subunit that is shared by the **receptors for both IL-4 and IL-13**. IL-4 and IL-13 are cytokines that have important roles in activating type 2 helper cells (Th2) in the immune response. This activation is overactive in certain conditions, such as **atopic dermatitis and asthma**. By blocking ligand binding to IL-4R (the shared subunit on both the IL-4 and IL-13 receptors), **Dupixent** prevents both IL-4 and IL-13 from binding to their receptors. This prevents activation of the JAK/STAT pathway by these receptors, which primarily act through STAT6. This inhibition, in turn, reduces TH2-mediated inflammation. The pathways activated by these receptors also have other functions that directly affect the skin, such as reducing skin barrier proteins like filaggrin, which influence skin permeability and susceptibility to irritants and allergens. Treatment with Dupixent can reduce inflammation, restore barrier function, and help improve symptoms like redness, swelling, and itchiness. The drug can also help with allergy symptoms such as airway hyperreactivity [8, 82].

– Inhibition of JAK1

Rinvoq (upadacitinib) [83] is a small-molecule drug for the treatment of **rheumatoid arthritis** and several other inflammatory diseases [84]. It is an oral medication, and it functions by inhibiting JAK1. JAK1 mediates signaling pathways that involve the STAT transcription factors and that lead to the production of inflammatory cytokines such as IL-6, IL-12, IL-23, and IFN-gamma. These cytokines have important roles in inflammation and reducing them is beneficial for the treatment of arthritis, as well as other inflammatory disorders such as psoriatic arthritis, atopic dermatitis (eczema), UC, and Crohn's disease. Rinvoq enters the cell and binds to the ATP-binding site of JAK1's kinase domain. This inhibits its kinase activity and thereby inhibits downstream signaling.

– Other JAK inhibitors

Myelofibrosis (MF) is a disorder linked to improper activation of JAK2. Myelofibrosis is a complex disease involving increased collagen production by fibroblasts in the bone marrow and some other organs [85]. (Myelo refers to bone marrow, and fibrosis refers to the development of fibrous tissue or scar tissue). In MF, scar tissue builds up within the bone marrow, which impairs its ability to produce blood cells normally. Some symptoms include fatigue, enlarged spleen, abnormal bleeding, bone pain, and anemia. In many patients, **MF** has been linked to a mutation in **JAK2**, which leads to its improper activation in hematopoietic stem cells in the bone marrow. This mutation is associated with uncontrolled growth of hematopoietic stem cells, and as a result, megakaryocytes are overproduced. These megakaryocytes and inflammatory cytokines can stimulate fibroblasts in the bone marrow to produce excess collagen, which is a major feature of the disease [86].

The drug **Vonjo (pacritinib)** is an oral kinase inhibitor. It is a small molecule that blocks the activity of both wild-type **JAK2** and the mutant **Jak2 (JAK2V617F)** that is often found in **MF**. It is a competitive inhibitor that binds directly to JAK. It is highly specific to JAK2 as opposed to the other JAKs, which is important for reducing side effects. It has other targets, though, in addition to JAK2. It also binds to and inhibits the activity of **IRAK** (interleukin 1 receptor-associated kinase). IRAK is a cellular kinase that mediates other pathways, such as those activated by the IL-1 receptor, leading to pathways such as the MAP Kinase pathway and the NFKB pathway. IRAK-mediated pathways also have roles in the disease, most likely due to their roles in inflammation. In addition, Vonjo also inhibits **FTL3 (Fms-like tyrosine kinase 3)** as an off-target effect. FTL3 is an RTK that is activated by the cytokine FTL3L. It is found mostly on hematopoietic cell progenitors, and it activates signaling pathways that are important for hematopoietic cell survival and differentiation. **Vonjo** was FDA approved for the treatment of **myelofibrosis (MF)** in 2022 [87, 88].

Ojjaara (momelotinib) [89] is also a small-molecule **JAK2** inhibitor designed to treat **MF**. It was FDA approved in 2023. It blocks JAK2 and AVCR1 (Activin A Receptor 1; activin is discussed in more detail later in the TGFβ section), a receptor in liver cells, which helps reduce the fibrosis that can occur in the disease [90]. **Jakafi (ruxolitinib)** [91] is also used to treat **myelofibrosis** by targeting the **Jak/Stat pathway**. It is a kinase inhibitor that reduces the activity of **JAK1 and JAK2**, which can be improperly activated in MF [92, 93].

Xeljanz (tofacitinib) [94] and **Olumiant (baricitinib)** [95] are used to treat certain **inflammatory conditions**, especially **rheumatoid arthritis (RA)**. Both are small-molecule **JAK** inhibitors and, by targeting the JAK-STAT pathway, they disrupt the production of cytokines [96–98]. Xeljanz targets JAK1 and JAK3, and Olumiant targets JAK1 and JAK2.

JAK signaling pathways have been implicated in hair loss. **Alopecia areata (AA)** is an autoimmune disease in which the immune system attacks the hair follicles. It is associated with the overexpression of pro-inflammatory cytokines that signal through the JAK/STAT pathway. Some studies have shown that inhibiting **JAK3** can reverse some symptoms of **AA**. **Litfulo (ritlecitinib)** (FDA approved in 2023) is a **JAK3** inhibitor that is used to treat AA [99]. It binds to an amino acid in the ATP-binding site of the **JH1 domain of JAK3**. **Litfulo** inhibits not only **JAK3** but also a tyrosine kinase called **Tec**. Tec is an intracellular tyrosine kinase involved in signaling pathways that are associated with T-cell activation [100]. Other JAK inhibitors that are being used to treat AA are **Olumiant (baricitinib)**, discussed above, and **Leqselvi (deuruxolitinib)** (FDA approved in 2024) [101], which target **JAK1 and JAK2** [102].

2.9.2 TNF Receptor Signaling

- TNF and the TNF Receptor

TNF is an important cytokine that can activate multiple different pathways in the cell. There are two main TNFs: TNFα and TNFβ, but TNFα is more broadly expressed in different tissues. **TNFα** binds to the TNF receptor (**TNFR**). This type of receptor is different than the type I and II receptors discussed above. TNF binding to the TNFR causes three TNFRs to form a trimer. This type of receptor does not have intrinsic enzymatic activity. Instead, it interacts with various intracellular proteins that often function as adapter proteins. Once it is bound to ligand and activated, TNFRs can activate a number of pathways in the cell. These include the JNK and p38 pathways, the ERK pathway, AKT, the JAK/STAT pathway, and the NF-κB pathway, and others [2, 3, 103]. In addition to these signaling pathways, TNF can also trigger apoptosis through different mechanisms, which will be discussed in Chap. 22. The NFKB pathway will be the focus in this section, and it is summarized in Fig. 18.16.

TNF signaling is complex because it can lead to different responses in different situations. It can lead to cell death in some cells, and it can lead to inflammatory responses and cell survival in other cells. The **NF-κB** pathway is one of the most well-known signaling pathways activated by TNF. NF-κB has important roles in inflammation, and it also has other roles such as immunity, cell

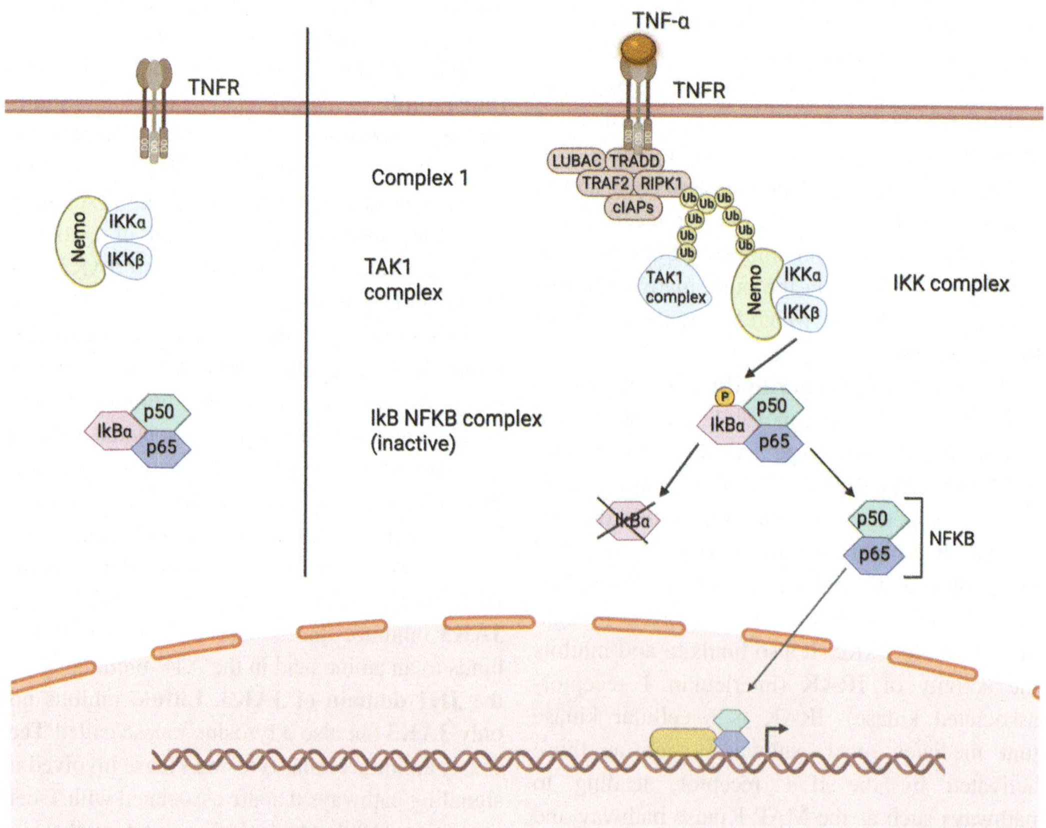

Fig. 18.16 NFKB pathway. The left side shows the TNF receptor (TNFR) in the absence of TNF. The right side illustrates the activation of the NF-κB pathway upon TNF binding to its receptor. **Source**: Created with Biorender.com

survival, cell proliferation and differentiation, and metabolism [2, 3]. When improperly regulated, the NF-κB pathway can also be linked to disease, including inflammatory disorders and tumorigenesis [2].

- NF-κB (Nuclear Factor Kappa-B)

– Introduction to the NF-kB pathway

The NF-κB pathway is a somewhat complex and tightly regulated signaling system. It is complicated because it encompasses multiple intracellular signaling molecules. Not only this, but several proteins in the pathway are regulated through inhibitory molecules that must be inactivated before NF-κB is finally activated. The pathway will therefore be broken down into steps, below. The following steps are involved in the activation of NF-κB in response to the binding of TNFα to the TNF receptors (TNFRs) [104, 105].

– The major signaling complexes in canonical TNF-induced NF-kB activation

Before discussing the NF-kB pathway, let's discuss three main signaling complexes that drive canonical NF-kB activation. (Another important complex, complex II, is involved in apoptosis and not discussed in this section).

(a) **TNFR1 complex I**. The complex includes the trimerized TNF receptor, bound to other proteins including TRADD, RIPK1 (also called RIP1), TRAF2, cIAP1/3, and LUBAC. (TRADD is an adapter protein, RIPK1 is a serine/threonine kinase, TRAF2 is an adapter protein that recruits CiAP1/2, cIAP1/2 are the main ubiquitin ligases in the complex, and LUBAC is also a linear ubiquitin ligase.

(b) **The TAK1 complex**. This complex consists of the kinase TAK1 (transforming growth factor β-activated kinase 1) bound to the adaptor proteins TAB2, TAB3, and TAB1 (TAK1-binding proteins 2, 3, and 1). TAB2 and TAB3 are adapter proteins that contain ubiquitin-binding domains and that allow docking to ubiquitin chains, while TAB1 binds directly to TAK1 and promotes its activation, via ubiquitin-induced oligomerization and autophosphorylation.

(c) **The IKK complex**. This complex consists of IKKα (IkB Kinase α), IKKβ (IkB kinase β), and NEMO (NF-kB essential modulator, also called IKKγ or IkB kinase γ). IKKα and IKKβ are serine/threonine kinases, where IKKβ is the main active kinase in the canonical NF-kB pathway, and NEMO has a scaffolding and regulatory role.

– Ligand binding to the TNF receptors and recruitment of **complex 1**

The two main TNFRs are TNFR1 and TNFR2. TNFR1 is more commonly involved in NFKB signaling and is the main one discussed here. The binding of **TNF** to **TNFR1** leads to receptor trimerization. Following trimerization, the death domain (DD) on the TNFR recruits the protein TRADD. TRADD, in turn, serves as a scaffold to recruit RIPK1 (also called RIP1), TRAF2, cIAP1/2, and LUBAC (linear ubiquitin assembly complex). This is **complex 1**.

cIAP1/2 (recruited by TRAF2) catalyze K63-linked ubiquitination of RIPK1, and LUBAC extends these chains with M1-linked ubiquitin. LUBAC can also add M1 chains to NEMO, discussed below, and to other complex members. These ubiquitin units are not for targeting the complex for degradation, but instead for serving as scaffolds.

– Recruitment and activation of the **IKK complex**

The ubiquitin chains on components of **complex 1** are recognized by the ubiquitin binding domains in TAB2/3, allowing recruitment of the TAK1 complex. **NEMO** in the **IKK complex** also binds the ubiquitin chains. This brings the IKK complex into the vicinity of complex I at the TNF receptor. Upon recruitment, **TAK1** becomes activated, and it **phosphorylates the IKKβ** on specific serine residues, **thus activating the IKK complex**.

– NF-kB is held inactive in the cytoplasm by IkBα

In resting cells, NF-kB dimers (most commonly p65/p50, as discussed below) are sequestered in the cytoplasm by binding to the inhibitory protein IkBα. IKbα masks the nuclear localization sequence of NF-kB, preventing its entry into the nucleus until upstream signaling triggers IkBα phosphorylation and degradation.

- The active IKK complex has a role in releasing NFKB from IkB

At this point, the active **IKK complex is active**. Now, the active IKK complex phosphorylates IkBα on specific serine residues within a region called the destruction box. This marks IkBα for ubiquitination and proteasomal degradation. As IKBa is degraded, the NFKB dimers (usually p65/p50) are released and can then translocate to the nucleus.

- NF-kB, now released, translocates to the nucleus

Once released, NF-κB is now free to translocate into the nucleus. There, it functions as a transcription factor, regulating the expression of genes involved in immune responses, inflammation, cell proliferation, and survival [2, 106].

- Subunits of NF-kB

NF-kB is not actually a single protein but a family of transcription factors that can form homo- and heterodimers. The dimers are made of two subunits. There are different combinations of dimers that can make up different NFKB family members, but the **p65/p50** heterodimer is the main dimer that is activated by the canonical pathways.

- Summary of the pathway and gene expression regulation

In summary, NF-κB is initially sequestered in the cytoplasm due to its interaction with **IKB**. After TNF stimulation, NF-κB is released and translocates to the nucleus, and as a transcription factor, it regulates gene expression. NF-κB regulates the expression of a number of different genes. These include genes for certain cytokines and chemokines, which are important for an immune response; genes encoding anti-apoptotic proteins, which promote cell survival; genes encoding some growth factors; and other growth-promoting genes. Improperly regulated NF-κB is linked to some types of cancer, inflammation, and other disorders [2, 3, 107]. It should be noted that NF-kB is not only activated by the TNF receptor but also by other receptors, most notably the Toll-like receptor (TLR). TLR plays important roles in the innate immune system. They are expressed on immune cells, and they respond to pathogens such as bacteria, parasites, and damaged cells.

- Drugs That Target TNF Pathways

Because of its roles in activating pathways that are involved in inflammation, the TNF pathway is considered a good target for the treatment of inflammatory disorders. **Humira (adalimumab)** [108] is an example of an anti-inflammatory drug. It is a biologic that consists of a monoclonal antibody directed against TNF. By binding to TNF, it prevents TNF from binding to the TNF receptor. It is a widely used drug that was FDA approved in 2002. It is used to treat disorders that are associated with inflammation or autoimmunity. Examples of disorders it is used to treat include **rheumatoid arthritis, psoriasis, Crohn's disease, ulcerative colitis, and psoriasis** (see Fig. 18.17). It is given to patients as a subcutaneous injection every other week although the dose and timing can vary for different conditions and different individuals [20, 109, 110]. By blocking TNF signaling, **Humira** reduces NFKB signaling, and this reduces transcription of pro-inflammatory genes.

Enbrel (etanercept) [111] is also a **biologic** that works by attaching to and blocking TNF. It is used to treat several **autoimmune disorders, including rheumatoid arthritis (RA), psoriatic arthritis (PsA), ankylosing spondylitis (AS), and polyarticular juvenile idiopathic arthritis (JIA)**. **Enbrel** is an **engineered fusion protein** rather than a straightforward monoclonal antibody. It contains the ligand-binding domain of the TNF receptor (TNFR2) linked to part of an IgG region of an antibody. The antibody helps stabilize the drug. This molecule acts as a sort of **decoy receptor** that will bind to soluble TNF in cells and prevent it from binding to its cell surface receptor, thereby blocking TNF-activated signaling in cells [110, 112].

MLN120B is a **small-molecule IKK inhibitor** that specifically blocks IKK (IkB kinase). It does so because it is an ATP-competitive

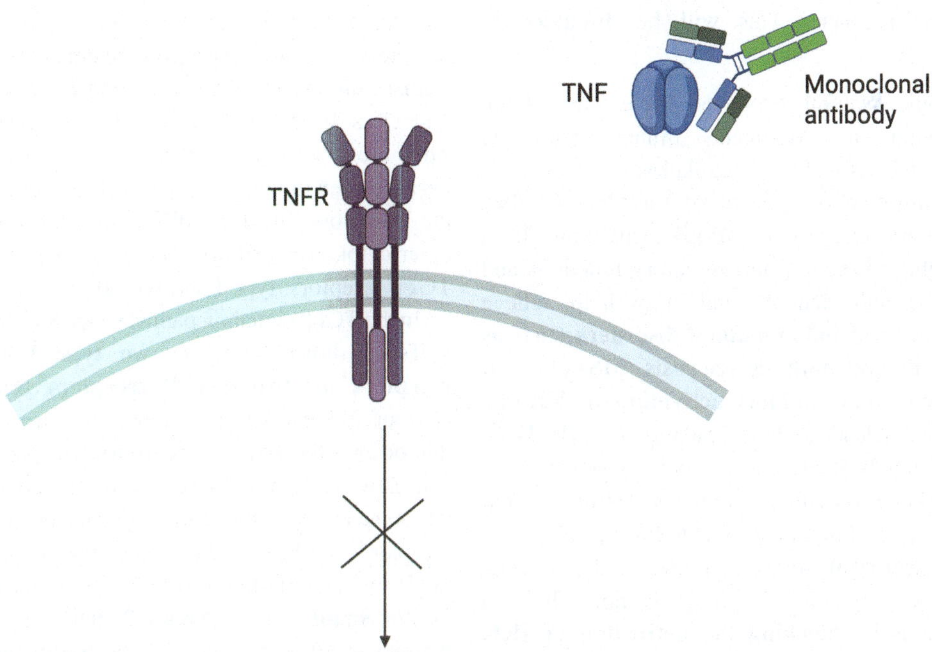

Fig. 18.17 Monoclonal antibodies that inhibit TNF. Some biologics inhibit TNF or TNFR by binding to them and preventing TNF-activated signaling. (*Source*: Created with BioRender.com)

inhibitor of IKKβ. It binds to its kinase domain and prevents it from phosphorylating IkBα. By inhibiting the activity of IKK, it blocks the phosphorylation of IκB and subsequent NF-κB activation. It has been investigated in preclinical studies for its potential anti-inflammatory effects. It has been tested on experimental animal models of several **inflammatory disorders** such as rheumatoid arthritis, cancer, and neuroinflammation [113–115].

RANK-L (Receptor activator of Nuclear Factor kappa B ligand) is a **cytokine** in the **TNF family**. It has an important role in bone remodeling. It activates signaling pathways that promote the formation, maturation, differentiation, and activation of osteoclasts, cells that resorb (break down) bone. RANKL works by binding to its receptor, **RANK**, on osteoclast precursors. It then triggers a signaling cascade that activates NFKB and other transcription factors. These pathways are essential for osteoclastogenesis. Osteoclastogenesis is the process by which osteoclasts form, and this includes the maturation of osteoclast precursors. RANK is found in various cells, but osteoclasts and their precursors are a major site of RANK expression. Osteoclasts are important for the breakdown of bone. Bone breakdown is a normal physiological function because there is a balance between the synthesis of new bone and bone breakdown. However, in diseases like **osteoporosis**, where bone breakdown predominates, it may be beneficial to slow down osteoclast function. Inhibition of RANKL signaling is one way to achieve this [116]. **Prolia (denosumab)** [117] is a **monoclonal antibody** drug that specifically targets RANKL. It prevents RANKL from interacting with RANK on osteoclast precursors. This prevents signaling pathways that allow osteoclasts to mature and differentiate. Blocking the production of osteoclasts leads to a decrease in bone breakdown and, consequently, to an increase in bone density and a reduced risk of fractures. This drug is given to patients with **osteoporosis** to protect them from further bone breakdown and fractures [118].

Many of the drugs discussed in this section are used for inflammatory conditions such as arthritis and psoriasis. Blocking the TNF receptor and NFKB signaling is also used in some types of cancer treatment. This will be discussed in Chap. 19.

- Some Natural Products Have Also Been Found to Modulate Inflammation via the NFKB Pathway Regulation

Curcumin, a component of turmeric, exhibits inhibitory effects on NF-κB activation. It is thought to have anti-inflammatory functions and other health benefits and may help relieve symptoms of **inflammatory disorders such as arthritis and multiple sclerosis (MS)** [119]. It has been shown to **block activation of IKK** (the kinase that leads to the degradation of IkB). Thus, inhibiting IKK inhibits nuclear translocation of NFKB and subsequently reduces the transcription of some pro-inflammatory cytokines [120].

Resveratrol, found in grapes and red wine, can also inhibit NF-κB activation. It also functions by **blocking the activation of IKK** and inhibiting NF-kB activation. Resveratrol is also thought to have **anti-inflammatory effects** [121]. In addition to blocking IKK, resveratrol may also regulate NFKB by a different mechanism. It can activate a protein called Sirtuin 1 (SIRT1). SIRT1 leads to deacetylation of the p65 subunit of NFKB, and this reduces NFKB's activity [122].

2.10 The TGF-β Pathway

2.10.1 Overview of the Pathway

TGFβ proteins are extracellular factors that can act as hormones or as local mediators and have a variety of biological functions. Upon binding to cell surface receptors, they can influence a wide range of biological processes, depending on the cell type. These responses include regulation of proliferation, differentiation, extracellular matrix (ECM) production, cell death, tissue repair, and immune regulation. Two important members of this family of ligands are the **TGFβ/activin** family and the **bone morphogenic protein (BMP)** family. These proteins all act through single-pass transmembrane receptors with a serine/threonine kinase domain on the cytosolic side and a ligand-binding domain on the extracellular side. These receptors are thus **receptor serine/threonine kinases** but are usually considered to be a separate class from other enzyme-linked receptors. They are a somewhat unique class of receptors because even though TGF-β is a type of cytokine, the receptors function differently from many other cytokine receptors. There are two types of TGFb receptors: type I and type II.

In the TGFβ signaling pathway (see Fig. 18.18), a TGFβ dimer binds to **two type I TGFβ receptors** and **two type II receptors**, bringing all four of their kinase domains together. Once this occurs, the **type II receptors** phosphorylate and activate the **type I receptors** on serine and threonine residues, thus forming active **tetrameric** receptor complexes. The active type I receptor binds two cytoplasmic proteins that are members of the **Smad** family (**Smad 2** and/or **3**). This binding is often mediated by the **Smad anchor for receptor activation (SARA)**. Upon binding to the receptor, Smad gets phosphorylated by the receptor on serine residues. These phosphorylated Smads dissociate from the receptor and bind in the cytoplasm to another Smad, **Smad4**, resulting in a heterotrimeric complex with three Smads (SMAD2/3 and SMAD4). This Smad complex then translocates to the **nucleus**. In the nucleus, it interacts with other **transcription factors** and binds to regulatory elements of target genes, thereby regulating their expression. Like STAT, the Smad complex is considered a latent transcription factor [3, 123].

The functions of the TGFβ family are variable and depend on the cell type they act on and the context. In some situations, TGFβ can suppress cell growth or lead to cell death, such as in some early cancer cells, but in other situations, such as in later stages of cancer, it may lead to increased cell growth and proliferation. TGFβ also has roles in immune regulation, where it can have several functions. For example, it can cause immune suppression by inhibiting the functions of some cells in the immune system, and it can help regulate inflammation due to its impact on the growth of immune cells [2, 3].

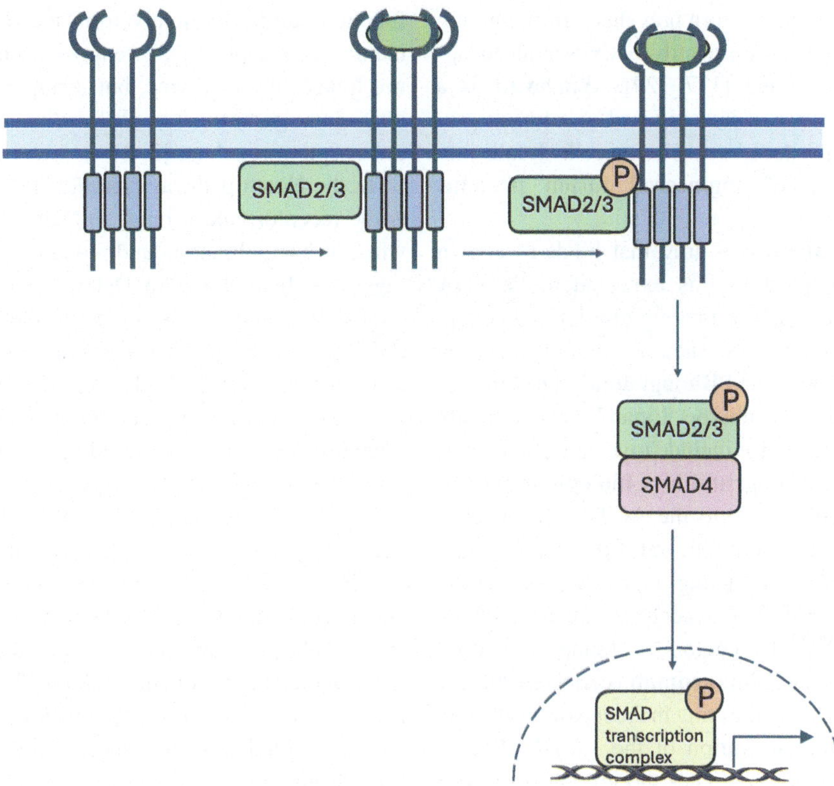

Fig. 18.18 The TGF-beta pathway. TGF-beta binds to TGF-beta receptors and leads to SMAD translocation to the nucleus, followed by regulation of gene expression

2.10.2 TGF-β in Disease and Medicine

While the TGFβ pathway would appear to be an attractive target for drugs due to its role in cell proliferation and inflammatory responses, the development of anti-TGF-β drugs has been challenging. Most of the drugs that have been developed to target TGF-β have eventually been discontinued. This may be due to ineffectiveness, side effects, or both. Inhibiting TGFβ is complicated by the fact that TGFβ can have different functions in different situations [124]. Nevertheless, recently, some drugs targeting TGFβ/SMAD signaling have been approved or are in the pipeline.

While targeting TGF-β has been challenging, a promising drug that inhibits a member of the TGF-β superfamily has been approved by the FDA recently (**2024**). This drug, **Winrevair (sotatercept-csrk)** [125], is designed to treat **pulmonary arterial hypertension (PAH)** [126]. TGF-β encompasses a large superfamily of ligands as well as a family of receptors through which they signal. Ligands include the TGF-β subfamily and BMP subfamily, as well as the GDF subfamily and the activins and inhibins. Each subfamily includes multiple members, and most of them signal through SMAD proteins although some of them use other mechanisms. These ligands can bind to different combinations of type I and type II receptors within the TGF-β receptor superfamily. In **PAH**, BMP signaling, which has antiproliferative functions, is reduced, but activin levels and activin signaling (which have proliferative functions) are increased. This leads to abnormal proliferation of vascular cells, which in turn leads to vascular remodeling. **Winrevair** acts by trapping and thereby inhibiting **activin**. The goal is to restore regulated vascular cell proliferation. It is a **fusion protein** that has the Fc domain of an antibody (IgG) linked to the activin-binding domain of the activin receptor. This causes the drug to bind to the activin, which in turn prevents it from binding to its receptor. Consequently, activin-induced signaling is inhibited [127–129]. **Winrevair** is a **biologic** because it is composed of a fusion protein that is produced using living cells. It falls into the category of **engineered fusion proteins** [123, 130].

Bimagrumab is a drug that binds to activin type II receptors and is expected to be FDA approved in 2025 for treating **obesity** and certain **metabolic disorders**. There are two main activin type II receptors, and **Bimagrumab** binds mostly to the ActR11B subtype. Ligands that normally bind to this receptor include myostatin, activin A, and others. **Bimagrumab**, a **biologic monoclonal antibody**, binds to the ACTR11B receptor, inhibiting the natural ligand from binding and preventing the signaling pathway from being activated [131]. This receptor, a member of the TGFβ family of receptors, is found on several tissue types, but **Bimagrumab** is effective largely due to its role in inhibiting the receptor on skeletal muscle cells. Activation of the ACTR11B normally negatively regulates the growth of muscle cells. This prevents excessive muscle growth, but reduced muscle mass is sometimes associated with increased fat mass, especially in some metabolic disorders. By inhibiting this receptor, **Bimagrumab** has been shown in clinical trials to reduce fat accumulation and increase muscle [132, 133].

2.11 The Notch Pathway

2.11.1 Overview of the Notch Pathway

The Notch pathway has important roles in animal development. It has roles in controlling cell-fate choices and regulating pattern formation during the development of most tissues. It is also involved in the continual cell renewal in tissues such as the gut, and it has been shown to regulate the production of neural cells in Drosophila. Notch signaling usually involves contact-dependent signaling. Specifically, this type of contact-dependent signaling is called juxtacrine signaling, where the ligand and receptor remain bound to the cell membranes of adjacent cells. Notch is a single-pass transmembrane protein. It functions quite differently from the other transmembrane receptors that we have discussed. It is not linked to an enzyme, but it requires proteolytic cleavage to function. It functions as a latent transcription regulator [2, 3].

In the Notch pathway (see Fig. 18.19), Notch is the receptor, and it binds to Delta, the ligand, which is a membrane-bound protein on neighboring cells. In addition to Delta, there are other related ligands called **delta-like ligands (DLLs)**, which are also membrane bound, that can bind to Notch. Following the interaction between Notch and Delta, Notch is cleaved by **proteases**, which are activated upon this interaction. The cleavage of Notch occurs in three steps called S1, S2, and S3 [2, 134, 135]. S1 cleavage occurs before contact is made between Delta and Notch; S2, followed by S3, occurs in response to Delta-Notch binding [136]. Following cleavage, the cytoplasmic part of Notch (the tail) is called notch intracellular domain (NICD). This domain becomes released from the membrane-bound part. This NICD now translocates to the nucleus. Here, it functions as a DNA-binding protein, and in the nucleus, it binds to another DNA-binding protein, Rbpsuh (Rbp-j), forming a complex. In this complex, Rbpsuh is converted from a transcriptional repressor to a transcriptional activator due to being associated with coactivators instead of corepressors. Once activated, Rbpsuh binds DNA and regulates the expression of target genes [137]. Some of its target genes include genes that have a role in proliferation and differentiation, cell cycle control, cell fate determination, and others [136].

2.11.2 Possible Relationship Between Notch Pathway Mediators and Alzheimer's Disease

γ-secretase is one of the protease complexes that cleaves Notch in response to Delta ligation [2, 136]. In addition to the Notch pathway, Y-secretase is also responsible for the cleavage of other single-pass membrane proteins. **Presenilin** is the catalytic subunit of γ-secretase. Mutations in the presenilin gene are frequently associated with early-onset familial **Alzheimer's disease**. This is thought to be due to disruption of the Notch pathway and also to disruption of the

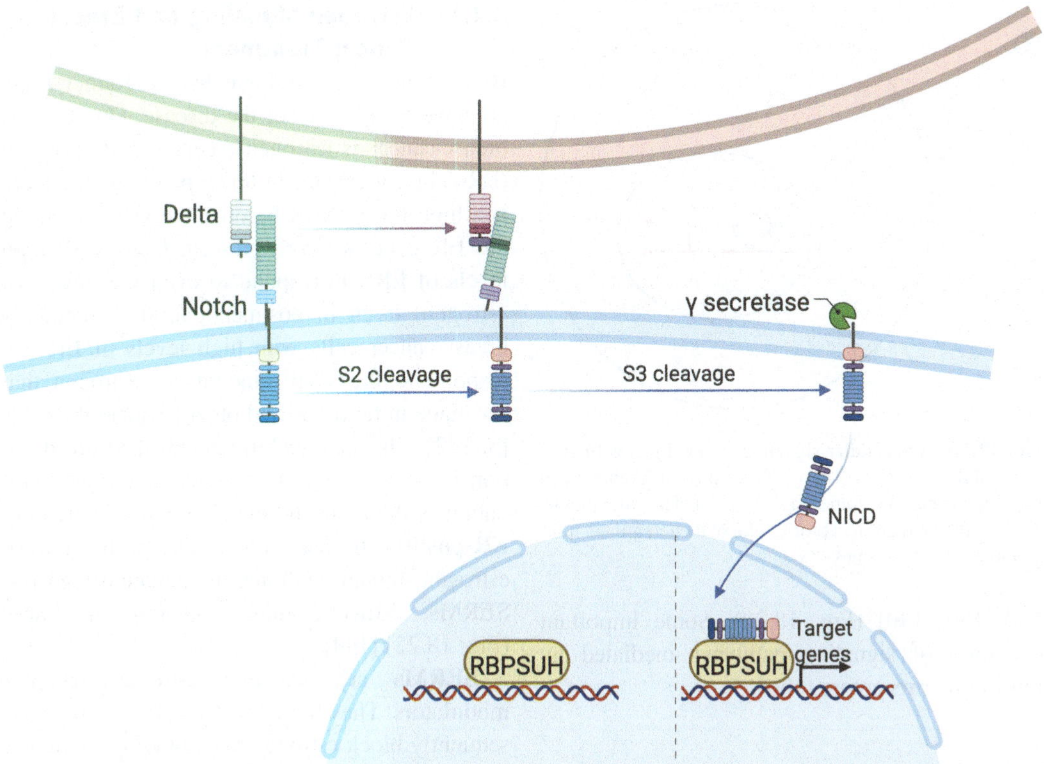

Fig. 18.19 Notch signaling. Notch signaling is an example of contact-dependent signaling, also called juxtacrine signaling. The ligand for the Notch receptor is Delta, on an adjacent cell. The pathway is described in more detail in the text. (*Source*: Created with Biorender.com)

normal cleavage of the amyloid precursor protein (APP), which is also cleaved by presenilin. The accumulation of abnormally cleaved APP in the brain may contribute to Alzheimer's. Targeting presenilin has been explored as a treatment for Alzheimer's disease as well as other neurodegenerative diseases [138].

3 Intracellular Receptor Pathways

3.1 Introduction to Signaling via Intracellular Receptors

Intracellular receptors are receptors that are located inside the cell, in the cytoplasm or the nucleus. The ligands that bind to these receptors have a lipophilic structure so that they can diffuse across the plasma membrane and enter the cell. Unlike growth factors such as EGF, this type of ligand is typically small, hydrophobic, and uncharged, allowing it to diffuse through the plasma membrane. These types of ligands are often hormones that are secreted from endocrine glands and then travel some distance to their target cells. They are transported through the bloodstream, where they are usually bound to carrier proteins that protect them from the aqueous environment. Once they get to their target cell, they dissociate from the carriers and enter the cell. Inside the cell, either in the cytoplasm or the nucleus, this type of ligand binds and activates the internal receptor. Activated intracellular receptors usually function as transcription factors and regulate the expression of target genes [2, 3,

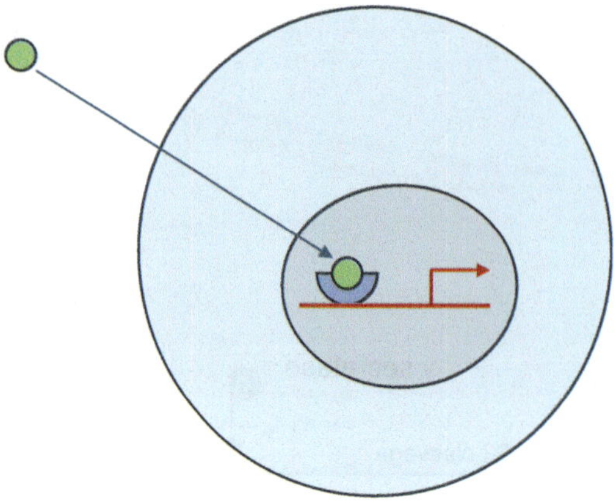

Fig. 18.20 Intracellular signaling. Some ligands bind to intracellular receptors. These ligands pass through the cell membrane and, in some cases, the nuclear membrane. They activate their receptors, which then regulate transcription of target genes

109, 139, 140] (Fig. 18.20). Some important examples of signaling pathways mediated by intracellular receptors follow.

3.2 Estrogen Signaling

Estrogen signaling is an important example of intracellular signaling. Estrogens actually encompass a group of related hormones, the most prominent of which is estradiol. Estrogens are produced mostly in the ovaries and in smaller amounts in other organs and glands, and in fat tissue. It is transported to target cells via the bloodstream, and it is bound to carriers during its transit. Once it reaches target cells, estrogen disengages from the carrier protein and can diffuse through the plasma membrane. In the cytosol, estrogen can bind to **estrogen receptors (ERs)**. The **estrogen/ER complex** can then cross the nuclear membrane and enter the nucleus, where it can function as a transcription factor and bind to ER Response Elements (EREs) in the regulatory elements of target genes. In this way, it regulates the expression of various genes that mediate a response to estrogen. These include genes that are associated with cell growth, cell cycle control, cell survival, and angiogenesis (Fig. 18.21) [141]. This type of receptor is considered a **nuclear receptor**.

3.2.1 Estrogen Signaling and Breast Cancer Treatment

Breast cancer is not a single disease. Rather, there are several subtypes of the disease. One type of breast cancer is estrogen receptor (ER) positive (ER+) breast cancer. In this type of cancer, there is a high level of ER because the gene encoding the ER gene is overexpressed. Cells with high levels of ER can respond to even low levels of estrogen. Even in postmenopausal women, the breast cancer cells with high levels of ER can respond to the small amounts of estrogen that are made in fat tissue and other peripheral tissues [8, 142]. This can lead to uncontrolled proliferation in these cells [143]. Blocking estrogen signaling is therefore the main strategy for treating **ER-positive breast cancer**. Drugs that reduce estrogen signaling fall into three main categories: SERMs, SERDs, and aromatase inhibitors (Fig. 18.22) [144].

SERMs are selective estrogen receptor modulators. They bind directly to ERs. They consequently block estrogen from binding to the ER so that downstream signaling pathways cannot be activated.

SERDs are selective estrogen receptor downregulators. SERDs are a newer class of drugs. They are used to treat advanced ER+ breast cancer. SERDs not only bind to the ERs, but they also promote the degradation of the ERs via ubiquitination, reducing their total levels in the cells. SERDs are in a category of drugs called degraders [145].

Aromatase inhibitors are the third type of inhibitor. Instead of targeting the ER, they block the production of estrogen. They do so by blocking **aromatase**, an enzyme that plays an important role in estrogen synthesis. These three types of drugs all represent various mechanisms for blocking estrogen signaling.

3.2.2 Some Examples of SERMs, SERDs, and Aromatase Inhibitors

- SERMs

The drugs **Nolvadex (tamoxifen)** and **Fareston (toremifene)** [146] are **SERMs**. They are **small molecules** that bind to the **estrogen receptor (ER)**, preventing estrogen from binding to

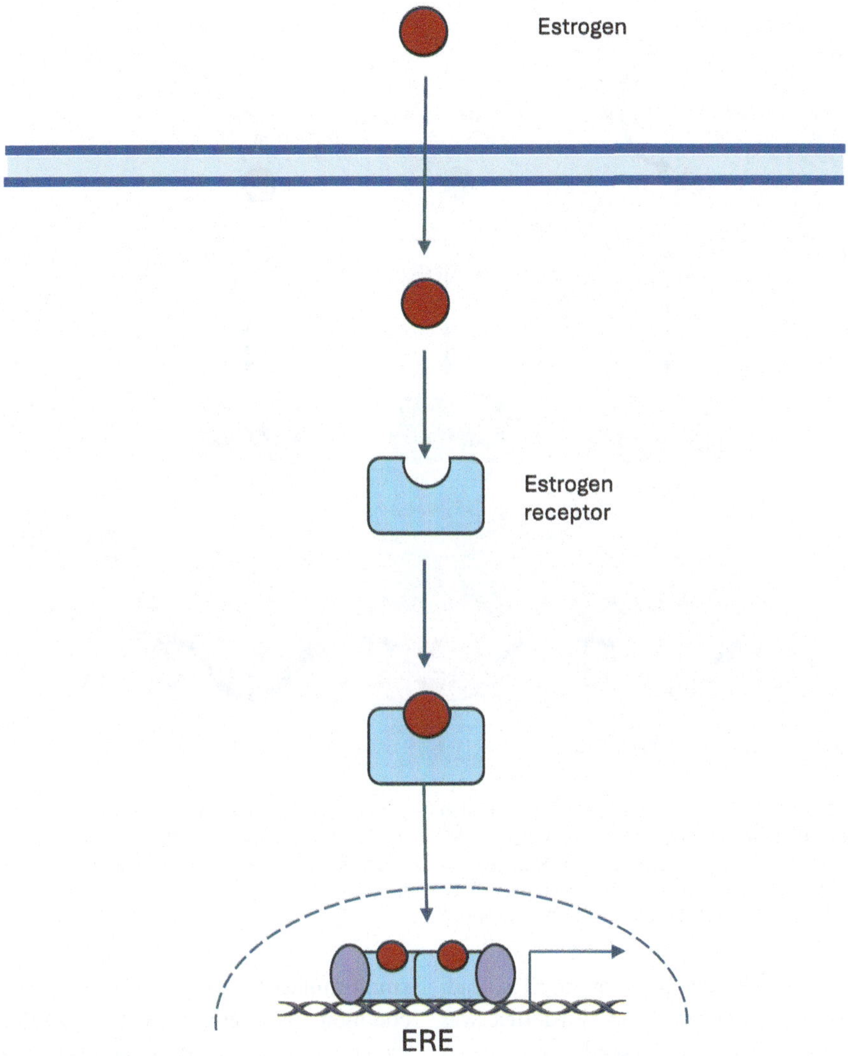

Fig. 18.21 Estrogen signaling. Estrogen binds to intracellular receptors. This leads to receptor dimerization followed by activation of transcription of target genes

it. **Tamoxifen** [147] was FDA approved in **1977**, and **Fareston** was approved in **1997. Tamoxifen** has a structure that is somewhat similar to estrogen. It binds to the ER at the site where estrogen normally binds and prevents estrogen from binding to it [148]. **Fareston** is only slightly different from tamoxifen in structure and functions similarly [144]. These drugs are given orally. The development of tamoxifen was a groundbreaking step in cancer treatment. It was the first SERM developed, but more importantly, it represented a movement toward targeted cancer treatments rather than relying primarily on nonspecific treatments such as chemotherapy and radiation. The drug significantly increased the prognosis of women with **ER+ breast cancer** and paved the way for the development of more targeted treatments for cancer.

- SERDs

Orserdu (elacestrant) [149] and **Faslodex (fulvestrant)** [150] are **SERDs**. Like the SERMs they also bind the ER, but they can be even more powerful because they have the additional benefit of causing the degradation of the

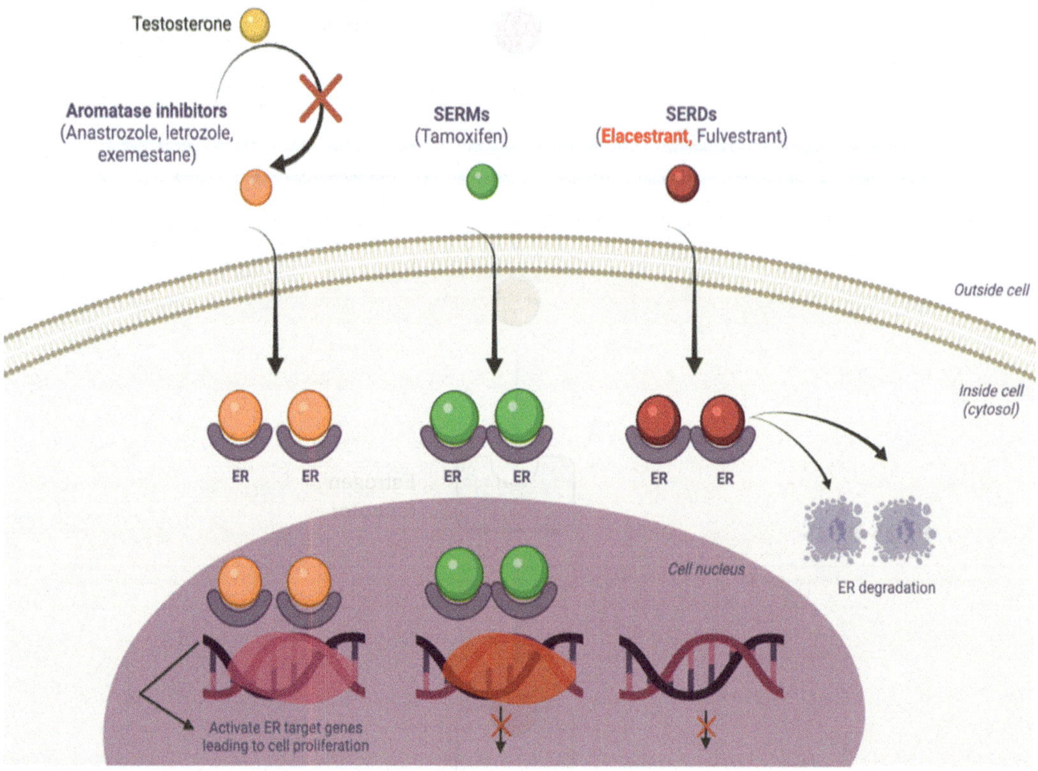

Fig. 18.22 Types of estrogen signaling inhibitors used in cancer treatment. Estrogen signaling inhibitors can work in different ways, including inhibition of estrogen production (Aromatase inhibitors) or inhibition at the estrogen receptor (ER) level (SERMs and SERDs). (*Source*: Figure from Hernando et al., Int. J. Mol. Sci. 2021, 22, 7812. https://doi.org/10.3390/ijms22157812, licensed under the Creative Commons Attribution 4.0 (CC BY 4.0) license)

ERs. They act by ubiquitination of the ERs, leading to proteasomal degradation. **Faslodex** was approved in 2002, but **Orserdu** was FDA approved more recently, in **2023**, and has led to more promising outcomes.

Faslodex (fulvestrant) works by binding to the ER at the ligand-binding domain and preventing estrogen from binding to it. It also causes the ER to misfold. This misfolding targets it for ubiquitination followed by degradation by the ubiquitin proteasome pathway [151]. **Faslodex** is larger than a typical small molecule and can be considered a **synthetic steroidal antiestrogen.** It is administered by intramuscular injection.

Orserdu (elacestrant) was approved for ER+/Her2- metastatic breast cancer with a specific mutation in the estrogen receptor. **Orserdu** is a **small-molecule** drug used for patients with a common mutation in the ER, the ESR1 mutation, which can cause ligand-independent ER activation. It is usually used for patients who have already been treated with another therapy, such as SERMs. Orserdu also binds to the ER and causes misfolding, leading to ubiquitination and proteasomal degradation. It is more bioavailable than Faslodex and is given orally, which is an advantage for patients [152–154].

- Aromatase Inhibitors

Examples of **aromatase inhibitors (Ais)** are **Arimidex (anastrozole)** [155], **Aromasin (exemestane)** [156], and **Femara (letrozole)** [157]. These **small-molecule** drugs are enzyme inhibitors, approved between **1995 and 2005**. **AIs** prevent estrogen production by inhibiting **aromatase**.

Aromatase is an enzyme that catalyzes the synthesis of **estrogen from androgens**, such as testosterone and androstenedione. While estrogen is made in the ovaries of premenopausal women, postmenopausal women no longer produce estrogen although low levels may remain present. The main source of this estrogen is fat tissue. This is because **aromatase** in fat tissue (and to a lower amount in some other tissues) can convert circulating androgens (from the adrenal glands) to estrogen. Aromatase inhibitors are therefore particularly effective in postmenopausal women [144, 158, 159].

The aromatase inhibitors **Arimidex (anastrozole)** and **Femara (letrozole)** are competitive inhibitors of **aromatase**. They bind to the active site of the enzyme, but they do not structurally resemble the androgen hormone. They are therefore considered **nonsteroidal inhibitors** of aromatase. They bind to the enzyme reversibly [160, 161].

Aromasin (exemestane) also inhibits **aromatase**, but it is a **steroidal inhibitor** with a structure similar to androstenedione. It is also a suicide inhibitor. It binds to aromatase and, due to a covalent modification, it converts it into a permanently inactive enzyme. It is therefore an irreversible inhibitor [162].

3.2.3 Estrogen in Hormone Replacement Therapy (HRT)

In ER+ breast cancer, the goal is to reduce estrogen and estrogen signaling. However, there are other situations where it is desirable to increase estrogen signaling. Estrogen is administered as part of hormone replacement therapy (HRT) to relieve **symptoms of menopause** in women, where estrogen levels have become low. The resulting increase in estrogen levels activates estrogen-regulated signaling pathways, thus alleviating some symptoms caused by low estrogen levels [163]. In HRT, estrogen that is administered can be synthetic, which is identical to human estrogen, or it can come from animal or plant sources.

HRT can relieve some symptoms in menopausal women, such as hot flashes and bone loss, but it also may come with some risks. Studies have shown that treatment with HRT may lead to a slightly increased risk of breast cancer, as well as other health risks. Therefore, patients must balance the risks and benefits before using this type of treatment [164].

3.3 Glucocorticoids and Mineralocorticoids

3.3.1 Glucocorticoids and Glucocorticoid Receptors

Glucocorticoids are **steroid hormones** produced in the adrenal glands. The primary glucocorticoid in humans is **cortisol**, also called **hydrocortisone**. Cortisol passes through the membrane and binds to **glucocorticoid receptors (GRs)**. Like the estrogen receptor, the bound GR is also a **nuclear receptor**. Upon binding with cortisol, the **hormone-receptor complex translocates** from the **cytoplasm** to the **nucleus**, where it acts as a **transcription factor**. There, it binds to the GRE elements (glucocorticoid response elements) on the regulatory elements of target genes, leading to the regulation of gene expression.

Glucocorticoid signaling has several important functions. Importantly, glucocorticoids are well known for their roles in the stress response. Glucocorticoids are released from the adrenal glands in response to stress and help the body respond. They respond by regulating the expression of genes that mobilize energy stores, increase blood pressure, and modulate the immune response. The products of these genes help prepare the body for "fight-or-flight" situations. Some of the genes regulated by glucocorticoid receptors are those involved in gluconeogenesis (glucose production), metabolism of lipids and proteins, suppression of inflammatory cytokine production, cell cycle control, cell survival, and the genes encoding neurotransmitter receptors. In addition to their roles in stress responses, glucocorticoids also have **anti-inflammatory** and **immunosuppressive effects**. They can suppress the immune system and inhibit the production of some cytokines, thus reducing inflammation [3, 109].

The GR is expressed in most tissues. Some tissues and cells in which they are especially important include immune cells, where they have roles in immune suppression; the liver, where they are involved in gluconeogenesis; adipose tissue, where they are important for fat metabolism; muscle, where they are important for protein catabolism; and the brain, where they are important for stress responses [165].

3.3.2 Mineralocorticoids and Mineralocorticoid Receptors

Mineralocorticoids are also hormones produced in the adrenal gland. They bind the mineralocorticoid receptors (MRs), which are also nuclear receptors. The major mineralocorticoid is **aldosterone**. The binding of mineralocorticoids to the MR causes it to translocate to the nucleus and regulate gene expression. Activation of MRs by mineralocorticoid binding leads to different results depending on the tissue. In epithelial cells, it leads to the expression of genes including those that encode proteins that regulate the transport of water and ions, such as Na/K+ pumps, and sodium channels [166]. Expression of MRs is limited compared with the expression of GR. MR levels are highest in the kidney, where it has an important role in regulating blood pressure and sodium reabsorption, and it is also found in several other tissues. In contrast, GR is expressed in nearly all tissues. Some glucocorticoids can also bind to mineralocorticoid receptors [167].

3.3.3 Drugs Targeting Glucocorticoid and Mineralocorticoid Receptors

- Deltasone (Prednisone), Decadron (Dexamethasone), and Cortef (Hydrocortisone)

Deltasone (prednisone), Decadron (dexamethasone), and **Cortef (hydrocortisone)** [168] are **small molecules** that are synthetic **glucocorticoids**. They are **agonists** that mimic glucocorticoids, especially cortisol. They are **anti-inflammatory and immunosuppressive** drugs. They bind to glucocorticoid receptors and activate glucocorticoid signaling pathways similarly to glucocorticoids themselves. They therefore activate the expression of genes that are typically activated by glucocorticoids. The brand names Deltasone and **Decadron** are discontinued, but their generic counterparts (prednisone and dexamethasone) are still widely used. **Cortef** is still available as a brand name and generic (hydrocortisone) [8, 169]. These drugs are all glucocorticoid agonists, but they all function slightly differently:

Deltasone (prednisone) [170] is a synthetic prodrug. It is a strong anti-inflammatory drug. It gets converted from its prodrug form to its active form in the liver due to a liver enzyme that catalyzes a modification which reduces one ketone (=O) to a hydroxyl group (-OH) on the molecule. The active drug binds to GR and activates a GR-mediated response. This leads to **upregulation of anti-inflammatory genes** and **downregulation of some pro-inflammatory genes**. It leads to a reduction in immune cell activation, suppression of cytokines, and decreased inflammation, pain, and swelling. In contrast to the **GR**, it is not very effective at binding the **MR** [167]. Prednisone is given orally. It is used for multiple purposes such as the treatment of autoimmune diseases, allergic reactions, respiratory problems like asthma, skin inflammation, and others [8, 20].

Decadron (dexamethasone) is also a synthetic **glucocorticoid**, but it has a longer half-life than natural glucocorticoids. It is not a prodrug, and it binds with high affinity to the GR. Like Deltasone, it regulates inflammatory and anti-inflammatory genes, but it is a more potent drug. It does not effectively bind MR. It is used as a **mild anti-inflammatory** and also for conditions where the adrenal glands do not produce sufficient hormones. It is usually given orally, by IV or IM injection, injected into a joint, or sometimes intranasally or as eye drops [8, 20].

Cortef (hydrocortisone) is a **synthetic cortisol**. It is identical to biological cortisol, and it is shorter acting than Decadron. By binding to GR, it leads to the **regulation of inflammation** although less potently than the above drugs. It also has a role in maintaining glucose metabolism, regulating stress, and immune balance. Cortef (hydrocortisone) is commonly used topically in the forms of creams and ointments. Sometimes, it is also given as tablets, or injection, or as suppositories. It is normally used to treat skin irritations like eczema, rashes, and itching [8, 20].

- Agamree (Vamorolone) and Treatment of Duchenne Muscular Dystrophy (DMD)

Duchenne muscular dystrophy (DMD) is caused by the loss or misfunction of the dystrophin protein, which is a protein that helps stabilize muscle fibers. Consequently, DMD leads to muscle weakness. This, in turn, leads to inflammation, which also contributes to the symptoms of the disease. Some types of treatment for DMD involve reducing this inflammation [171].

Agamree (vamorolone) (FDA: 2023) is a drug that can bind both **GRs** and **MRs** though it is an agonist of GR and an antagonist of MR [172]. This drug is used for the treatment of **Duchenne muscular dystrophy (DMD)**. Like other GR agonists, it binds GR and translocates to the nucleus. It leads to downregulation of pro-inflammatory genes, similar to other GR agonists. This is important because inflammation in DMD is associated with damaged muscle and disease progression.

One problem with other, more traditional GR agonists is that they can have side effects. For example, they can be associated with sodium retention, fluid retention, hypertension, bone loss, and tissue fibrosis. Some of these side effects are due to high levels of GR signaling, and others to off-target effects due to interaction with the MR. **Agamree** solves some of these problems because it is "selectively dissociative." It binds to the GR and translocates to the nucleus just like the other glucocorticoid agonists discussed above. However, while it encourages repression of pro-inflammatory genes, it is designed in a way that it reduces activation of genes whose products cause some of the side effects.

Another important advantage of this drug is that it also acts as an **antagonist of the MR**. This is different from the drugs discussed above, some of which can weakly activate MR. This helps to reduce side effects associated with sodium and water retention, edema, hypertension, and weight gain [173, 174].

- IL-6 Is an Important Target for Glucocorticoid Agonist Anti-Inflammatories

Glucocorticoid receptors (GRs) have anti-inflammatory effects due to the genes they regulate. There are a number of different genes regulated by **GRs** that mediate their effects. One important one is the gene for IL-6. GRs can have positive effects on the expression of some genes and negative effects on others. Glucocorticoid signaling has been shown to lead to negative regulation of the interleukin-6 **(IL-6) gene**. This suppression is often indirect, through GR's interference with various pro-inflammatory transcription factors. In some cases, GRs may also bind to regulatory sequences that regulate IL-6 gene expression. The repression of transcription may also occur through competition for shared transcription cofactors, which are redirected by GR toward anti-inflammatory gene targets. This limits their ability to regulate pro-inflammatory genes like IL-6 [175]. IL-6 plays a complex role in immune responses, primarily acting as a pro-inflammatory cytokine. It activates various signaling pathways by binding to its receptor (a type I cytokine receptor), which can trigger downstream effects such as JAK-mediated signaling and promote inflammation. Excessive IL-6 production is associated with chronic inflammation and other disorders such as immune diseases [176–178]. Glucocorticoid agonist drugs are used to treat inflammatory disorders such as allergies, rheumatoid arthritis, asthma, and inflammatory bowel disease. Although multiple target genes are involved, the suppression of IL-6 expression is an important example of how GR agonists lead to anti-inflammatory effects. In addition to treating inflammation, GR agonists can also be used to suppress the immune system after organ transplantation, and they are often used as immunosuppressive drugs [109, 175].

3.4 AhR (Aryl Hydrocarbon Receptor) Signaling

3.4.1 Overview of AhR Signaling

AhR is an intracellular receptor that has roles in the regulation of the maintenance of normal healthy skin. It leads to the regulation of the expression of skin barrier proteins as well as to the regulation of immune cells. **AhR** responds to various **ligands**, including environmental chemicals, which cross the membrane and bind to AhR in the cell. AhR is bound to other proteins in the cytoplasm when it is inactive, which include cytoplasmic chaperones [2]. These proteins keep AhR inactive. However, upon binding to ligands, AhR disengages from some of its binding proteins and translocates into the nucleus. There, it heterodimerizes with another protein (ARNT), forming an active transcription factor complex. The AhR-ARNT complex then binds to XRE elements (xenobiotic response elements) in the regulatory elements of target genes, thus regulating gene expression [2]. Some of the genes that the AHR pathway regulates are those that are involved in detoxification, metabolism, and immune responses. In some cases, the AHR pathway leads to the elimination of the very ligands that bind to it [179, 180].

3.4.2 AhR Signaling and Psoriasis

Improper regulation of AhR signaling has been implicated in the skin disorder psoriasis, which involves a thickening of patches of skin and is associated with inflammation. While there are other approaches for treating psoriasis, activation of AhR is also a promising psoriasis. **Vtama (tapinarof)** [181] is an **AhR agonist** used to treat psoriasis, and it leads to activation of the AhR pathway. **Tapinarof** is a **small molecule** naturally produced by pathogenic bacteria carried by nematodes. It can diffuse across the plasma membrane and enter the cell. In the cell, it binds directly to the ligand-binding domain of AHR, thereby activating it. **Tapinarof** has been shown to induce the expression of certain skin barrier genes that tend to be downregulated in psoriasis. These are genes that have roles in keratinocyte differentiation. **It also downregulates certain inflammatory cytokines** [180]. **Tapinarof** is applied topically to the skin. It was FDA approved in 2022 [182]. Although the use of tapinarof as an agonist is promising, it should also be noted that the role of AhR in psoriasis is complex because both too much downregulation and too much upregulation of AhR have been associated with the disease. This makes treatment and dosage complex [183].

3.5 PPAR (Peroxisome Proliferator-Activated Receptor) Signaling

3.5.1 PPAR Signaling and Expression

PPARs are members of the **nuclear hormone receptor family**. They have diverse roles, such as regulation of differentiation, and metabolic functions, such as the regulation of **lipid and glucose homeostasis**. There are three major PPARs (PPARα, PPAR β/δ, and PPARγ). They have four main domains. These include an amino-terminal domain that interacts with co-regulators, a DNA-binding domain that binds DNA, a hinge region, and a carboxyl-terminal domain that binds ligand. PPARs form heterodimers with retinoid X receptor (RXR; on their own, these receptors bind to retinoids, derived from vitamin A, and they are also nuclear receptors). Together, this complex binds target DNA on specific elements within promoter and enhancer regions called peroxisome proliferator response elements (PPREs). PPAR activity and the effects they have on transcription can be modified by various coactivators or corepressors.

The different PPARs have different tissue-specific expression patterns. **PPARα** is found in metabolically active tissues, including liver, heart, and skeletal muscle, where it has important roles in lipid metabolism and fatty acid oxidation. **PPAR β/δ** is found in all tissues, but it is high in the skeletal muscle, skin, and brain. It has roles in fatty acid oxidation and generation of mitochondria, especially in skeletal muscle. **PPARγ** is highest in adipose tissue and macrophages.

The PPAR/RXR complexes are regulated mostly by PPAR ligands, and not by retinoids, which are RXR ligands. Ligands that bind the PPAR/RXR complexes include fatty acids, leukotrienes, and prostaglandins, and others. All of them can diffuse across the membrane to enter the cell. They can also be activated by some synthetic molecules [184, 185]. The PPARs have various functions, largely related to **fatty acid and glucose metabolism**. For example, PPARα regulates genes in the liver that are associated with fatty acid oxidation and maintaining lipid homeostasis. PPARα is upregulated during fasting, and this helps cells mobilize and catabolize fatty acids [185]. PPARβ/δ are involved in glucose homeostasis, insulin sensitivity, and mitochondrial function. They also increase fatty acid oxidation. PPARγ is also important for glucose homeostasis as well as lipid storage, and it improves insulin sensitivity. Overall, the PPARs are critical regulators of metabolic pathways [184, 185].

3.5.2 Drugs That Target PPARs

PPARs are targets for various drugs, including those involved in treating diabetes, cardiovascular disease, cancer, and others. Such drugs are limited, however, due to the side effects they have, and some of the drugs that target PPAR have been replaced by other treatments.

Fibrates, such as **Tricor (fenofibrate)** [186] (also goes by several other brand names), are examples of PPARα agonists. Tricor is a **small molecule** that activates PPARα by binding to it. It is used to treat **dyslipidemia**, which involves abnormal levels, usually too high, of lipids in the blood. It works because PPAR leads to increased transcription of genes encoding enzymes involved in fatty acid oxidation and the breakdown of TAGs [187].

Thiazolidinediones (TZDs) are drugs that are agonists of PPARγ. These drugs include **Actos (pioglitazone)** [188] and **Avandia (rosiglitazone)** [189]. They have been used to treat **type 2 diabetes** by improving insulin sensitivity, especially in adipose tissue, muscle, and liver [190]. These drugs work via several mechanisms. One important mechanism is the regulation of genes involved in adipocyte differentiation. PPARγ activation leads to the formation of new small insulin-sensitive adipocytes. This increases the capacity of the adipose tissue to store lipids and reduces lipid accumulation in circulation and in other organs. These drugs are helpful for treating diabetes, but they are largely replaced by other drugs due to the adverse effects they have [191].

While some PPAR drugs have been of limited success, there is still interest in PPAR as drug targets due to their important roles in metabolism. A newer PPAR drug that has shown some promise is **Iqirvo (elafibranor)** [192]. It is a **small-molecule agonist** for **PPARα** and PPARβ/δ. It activates the PPARs by binding to their ligand-binding domains. It is used to treat **primary biliary cholangitis (PBC)** which is a rare liver disease. PBC is an autoimmune disease, and it causes destruction of bile ducts in the liver, leading to bile acid buildup, which is toxic, and leading to inflammation, and eventually cirrhosis. PPARα increases expression of genes involved in detoxification and bile acid transport, and it reduces bile acid synthesis. This helps reduce liver cell damage. PPARδ reduces the production of pro-inflammatory cytokines in the liver and helps modulate the immune response, which is abnormally regulated in the disease. This drug was given accelerated approval by the FDA in 2024. **Livdelzi (seladelpar)** [193] is a similar drug used to treat PBC; it was FDA approved in 2024 and it is a **PPARδ agonist** [194–196].

4 G Protein-Coupled Receptor Pathways (GPCRs)

4.1 Basic Mechanisms of Operation of GPCRs and Heterotrimeric G Proteins

4.1.1 Basic Mechanisms

G protein-coupled receptors (GPCRs) are the largest class of cell receptors and are also targets of numerous drugs. These receptors are seven membrane-spanning receptors. On their extracellular sides, they bind ligand, and when bound to ligand, they can interact with trimeric G proteins on their cytosolic sides. GPCR-mediated pathways have many different roles, including regulation of cell proliferation, differentiation, and migration, ion transport, sight, smell, and taste, among others. There are hundreds of different GPCRs in humans and a wide array of different ligands. Almost half of all known drugs target GPCR-regulated signaling pathways. The G proteins that bind to GPCRs are varied, but all have similar structures and operate by similar basic mechanisms. These G proteins are considered trimeric G proteins as they have three subunits: the α, β, and γ subunits (referred to as Gα, Gβ, and Gγ; these **trimeric G proteins** are sometimes referred to as **heterotrimeric G proteins** or sometimes simply as **G proteins**) [3].

In resting cells, the three subunits of trimeric G proteins (the heterotrimer) are generally bound together, and the Gα subunit is bound to GDP. In this state, the G proteins are inactive. The entire heterotrimer is anchored to the membrane via a lipid modification, and it is loosely associated with the GPCR. The main steps involved in activating trimeric G protein are highly conserved among the different GPCRs, although some exceptions exist. Generally, when the GPCR is activated by ligand binding to its extracellular domain, it undergoes a conformational change, and this increases its affinity for the heterotrimeric G protein. The GPCR then acts as a **guanine nucleotide exchange factor (GEF)** and promotes the exchange of bound GDP to bound GTP on the Gα subunit. This GTP binding activates the subunit and leads to its release from the βγ subunits. This active Gα subunit can now interact with and activate various target proteins to begin a signaling pathway. In some cases, the free βγ subunits can also interact with and activate target proteins (Fig. 18.23)

The next two sections below introduce basic categories of G proteins and GPCRs, and in the sections that follow, some of the major GPCR-activated signaling pathways will be described in more depth [2, 3, 197, 198].

4.1.2 The Categories of Heterotrimeric G Proteins

There are many different G proteins, and they can be categorized in different ways, but they are generally separated into four major groups based on their Gα subunits. The Gα subunits are categorized as **Gα12/13, Gαi, Gαs,** and **Gαq**. G proteins with these different Gα subunits are often

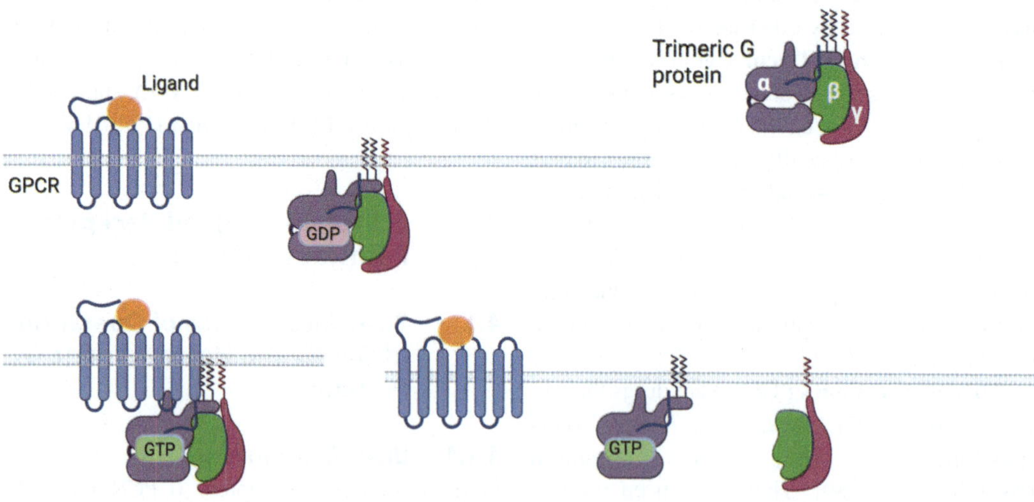

Fig. 18.23 Trimeric G proteins. α, β, and γ subunits are shown. Upon binding to the G Protein Coupled Receptor (the GPCR), GDP is exchanged for GTP on the alpha subunit, and the α subunit and βγ/ subunits activate downstream pathways. **Source**: Created with biorender.com

referred to simply as **G12/13, Gi, Gs,** and **Gq**. The G proteins associated with these different Gα subunits can activate different, although sometimes overlapping, pathways, but they also have certain key features in common [2, 3, 197, 199].

4.1.3 The Different Families of G Protein-Coupled Receptors (GPCRs)

There are different families of GPCRs that differ in which G proteins they bind, where they are expressed, and what functions they have. The different families of GPCRs also have subfamilies. The different GPCRs are known for activating different signaling pathways. Two important pathways are the adenylate cyclase and phospholipase C (PLC) pathways. Below is a summary of some of the different categories of GPCRs, along with a brief introduction to which intermediates and signaling pathways they activate. A more detailed and inclusive list of the main GPCRs is shown in Table 18.2.

- The α-Adrenergic Receptors

There are two categories of alpha-adrenergic receptors. These are **the alpha-1-adrenergic**

Table 18.2 The major categories of G Protein Coupled Receptors (GPCRs)

Receptor type	Subtypes	G protein
β-adrenergic receptors	β 1, β 2, β 3	Gs
α-adrenergic receptors	α-1-adrenergic receptor	Gq
	α-2-adrenergic receptor	Gi
Muscarinic acetylcholine receptors	M1, M3, M5	Gq
	M2, M4	Gi
Dopamine receptors	D1, D5	Gs
	D2, D3, D4	Gi
Serotonin (5-HT) receptors	5-HT_1	Gi
	5-HT_2	Gq
	5-HT_3	Not a GPCR
	5-HT_4	Gs
Histamine receptors	H1	
	H2	Gs
	H3, H4	Gi
Opioid receptors	Mu, delta, kappa	Gi
Glutamate receptors	Group I	Gq
	Group II, III	Gi
Angiotensin (AT) receptors	AT1	Gq
	AT2	Gi (?)
Adenosine receptors	A1	Gi
	A2A, A2B	Gs
	A3	Gq
Chemokine receptors	CCR1-10, CXCR1-6, XCR1, CX3CR	Gi
Thromboxane receptors	TP-α and TP-β (isoforms)	Gq, G12/13, Gi
Endothelin receptors	ET-A, ET-B	Gq, Gi
Glucagon receptor		Gs (sometimes Gq)
Parathyroid hormone receptor (PTH)	PTH1R	Gs, Gq
	PTH2R	Gs
Rhodopsin		Gt (transducin)
S1P receptors	S1P1R1, 4	Gi
	S1PR2, 3	Gi, Gq, G12/13
	S1PR5	Gi, G12/13
C5aR	C5aR1 (CD88)	Gi
	C5aR2 (C5L2)	

receptors and the **alpha-2-adrenergic receptors**.

The **alpha-1-adrenergic receptor** mainly binds to **Gq**. It primarily leads to the **phospholipase-C (PLC)** pathway, and this leads to increased intracellular **calcium** levels. These receptors are on several cell types, but they are mainly found on **vascular smooth muscle cells,** and their activation in these cells can cause vasoconstriction.

The **alpha-2-adrenergic receptor** couples with **Gi**. It primarily **inhibits the adenylate cyclase pathway**. It is also present on several cell types, but mainly on **presynaptic neurons**, where its activation inhibits neurotransmitter release [2, 200].

- The β-Adrenergic Receptors

These include the β-1, β-2, and β-3 adrenergic receptors, which primarily couple to **Gs**. The **β-1-adrenergic receptor** is mostly in the heart and kidneys. In the heart, it has a role in increasing heart rate and contraction force. The β **-2-adrenergic receptor** is found in smooth muscle of various tissues. In the lungs, it is involved in muscle relaxation and bronchodilation. The β**-3-adrenergic receptor** is found in adipose tissue and has a role in regulating the breakdown of TAGs and thermogenesis. The major pathway all the β-adrenergic receptors activate is the **adenylate cyclase pathway** [2, 200].

- Muscarinic Acetylcholine Receptors

These include the M1, M2, M3, M4, and M5 muscarinic receptors. M1, M3, and M5 muscarinic receptors bind mainly to **Gq**, and they primarily activate the PLC pathway. **M2** and **M4** bind mostly to **Gi**, and they inhibit the adenylate cyclase pathway. The **M1 muscarinic receptor** functions primarily in the CNS and is involved in cognitive functions. **M2** is mainly in the heart, and its activation decreases heart rate and contractility. **M3** is found in smooth muscle and glands, where it induces contraction and secretion. **M4** is mostly in the CNS, and it regulates neurotransmitter release. **M5** is also found in the CNS, where it may be involved in dopamine regulation [201].

- Dopamine Receptors

The subtypes of these receptors include the **D1-like receptors (consisting of D1 and D5)**, and the **D2-like receptors (consisting of D2, D3, and D4). The D1-like** receptors couple with **Gs** and stimulate the adenylate cyclase pathway. The **D2-like** receptors couple with **Gi** and inhibit the adenylate cyclase pathway. These receptors are involved in the regulation of mood, motivation, and motor control [202].

- Serotonin Receptors (5-HT)

There are several families of these receptors. These include 5-HT_1, 5-HT_2, 5-HT_3, 5-HT_4, and others. **5-HT_1** couples with **Gi**, **5-HT_2** is coupled with **Gq**, and **5-HT_4** is coupled with **Gs**. They are involved in the regulation of mood, appetite, and gastrointestinal mobility (5-HT_3 is an ion channel and not a GPCR) [203].

- Histamine Receptors

Subtypes of these receptors include H1, H2, H3, and H4 histamine receptors. **H1** is coupled with **Gq, H2** is coupled with **Gs**, and **H3 and H4** are coupled with **Gi**. They mediate allergic reactions, gastric acid secretions, and neurotransmission [204].

- Opioid Receptors

Subtypes include the **mu, delta, and kappa opioid receptors**. They couple with **Gi** and inhibit the **adenylate cyclase pathway**. They are involved in pain regulation and reward and addictive activity [205].

- The Glucagon Receptor

The glucagon receptor is a GPCR that is linked mostly to **Gs**. The binding of glucagon to its receptor activates the adenylate cyclase pathway. Glucagon is released during low glucose states (fasting), and it plays an important role in metabolism (see Chap. 12). In hepatocytes, the pathway activated by glucagon binding to its GPCR leads to increased **glycogenolysis**, **inhibition of glycogen synthesis**, and stimulates **gluconeogenesis**. In adipose tissue, glucagon signaling also supports **lipolysis**. The glucagon receptor plays an important role in maintaining blood glucose levels during fasting, acting as a functional counterpart to insulin signaling.

4.2 Some of the Important Signaling Pathways Activated by GPCRs

Although certain of the basic mechanisms by which GPCRs operate are similar, the signaling pathways that they activate also have important differences. Here, we discuss some of the main signaling pathways that are mediated by GPCRs. Two of the major pathways are the adenylate cyclase pathway and the PLC pathway [2, 3, 206].

4.2.1 Gs and the Adenylate Cyclase Pathway

The **adenylate cyclase pathway** (see Fig. 18.24) is an important GPCR-activated pathway. It involves several second messengers, including cyclic AMP (**cAMP**), and various intracellular signaling proteins, such as protein kinase A (**PKA**) and the transcription factor **CREB**. The activation of this pathway is primarily mediated by **Gs** and can be downregulated by **Gi** [206]. Activation of the pathway, mediated by **Gs**, is discussed in this section.

- Overview of the Adenylate Cyclase Pathway
  – Summary of the major steps in the adenylate cyclase pathway

The main steps in the adenylate cyclase pathway, which will be discussed in more detail below, are as follows:

GPCR activation → Gs → adenylate cyclase → cAMP → PKA → CREB → regulation of transcription.

This is the classical pathway, but PKA also has other targets leading to diverse responses in addition to transcription regulation.

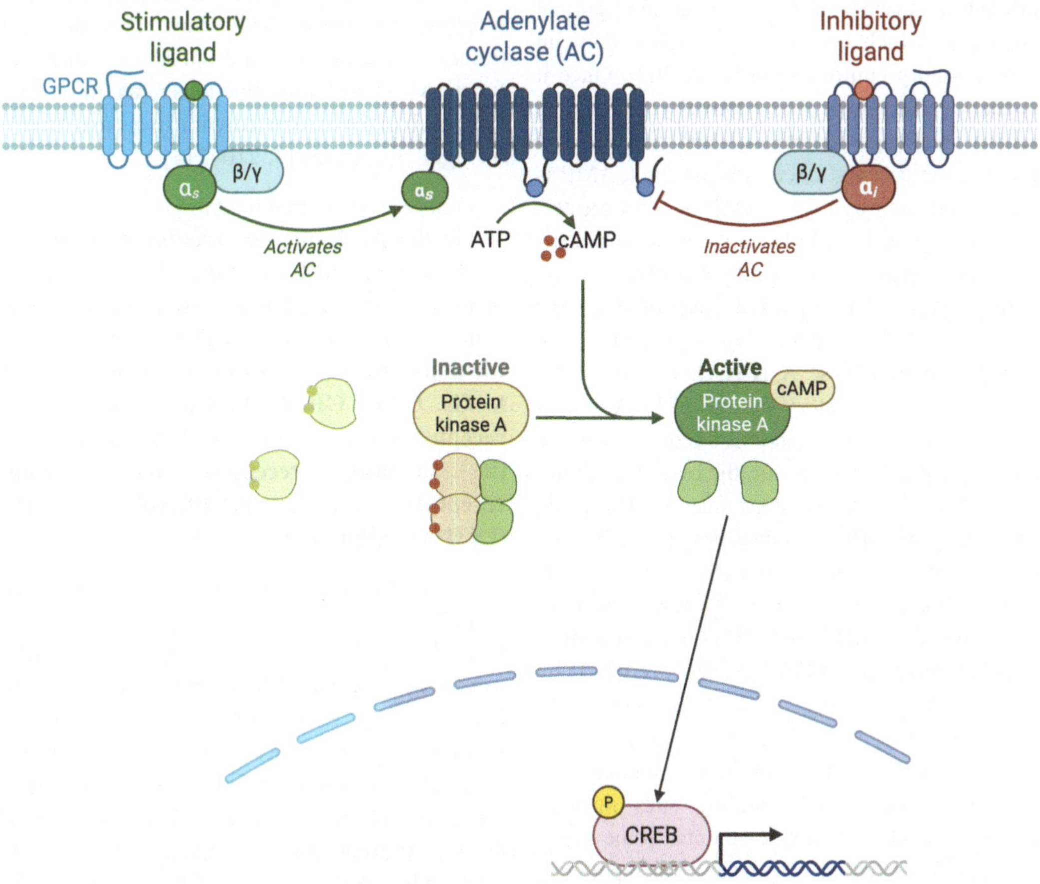

Fig. 18.24 The cAMP pathway. G proteins linked to GαS lead to activation of adenylate cyclase, while Gαi can inhibit it. Adenylate cyclase leads to increased levels of cyclic AMP, which in turn triggers downstream signaling

– Description of the adenylate cyclase pathway
β-adrenergic receptors, commonly linked to **Gs**, are one of the main types of GPCRs that activate the **adenylate cyclase pathway** (although other GPCRs and other G proteins can also activate the pathway). When the β-adrenergic receptor is activated by a ligand (such as epinephrine), it undergoes a conformational change and interacts with the trimeric G protein. The Gα subunit, originally bound to GDP, associates with the GPCR. The GPCR then functions as a guanine nucleotide exchange factor (GEF), promoting the exchange of GDP for GTP on the Gα subunit. After this, the Gα subunit dissociates from the Gβγ subunits.

The GTP-bound Gα subunit now binds directly to the membrane-bound enzyme **adenylate cyclase** (also sometimes called **adenylate cyclase** and abbreviated **AC**). This direct interaction leads to the **activation of the adenylate cyclase** due to a conformational change in which the enzyme is relieved of an autoinhibitory conformation [2, 3, 207]. Once it is active, adenylate cyclase converts **ATP** to the second messenger **cAMP**. cAMP then diffuses through the cell and elicits responses. (cAMP is later inactivated by cyclic AMP phosphodiesterase (PDE), which hydrolyzes its cyclic structure, opening its ring structure and terminating its signaling function). An important function of cAMP is to lead to the activation of a protein kinase called **protein kinase A (PKA)**. It does so by binding to the regulatory subunit of PKA, which leads to the regulatory subunit's release from the catalytic subunit. As a result, the catalytic subunit of PKA is now active. Once active, PKA can phosphorylate different substrates. These include other cellular enzymes, as well as transcription factors. One important target of PKA is the transcription factor **CREB (cAMP response element binding protein)**. PKA regulates **CREB** by translocating into the nucleus and phosphorylating it, leading to its activation. By regulating CREB, PKA can thereby regulate target genes with CREB binding sites in their promoters and regulatory elements. Some genes that are regulated in response to this pathway have important roles in responses to stress, as well as cell growth, survival, immune responses, and regulation of some neurotransmitters [208].

- Types of Receptors and Ligands That Are Linked to the Adenylate Cyclase Pathway via Gs

– β-adrenergic receptors
The major type of **GPCR** that is known for binding **Gs** and activating the **adenylate cyclase pathway** is the **β-adrenergic receptor**. The major ligands that bind to them are **epinephrine** and **norepinephrine** (**adrenaline** and **noradrenaline**). Epinephrine and norepinephrine are **catecholamines**, which are small molecules derived from tyrosine, that can function as both hormones and, in some cases, neurotransmitters. They are secreted from the adrenal medulla (within the adrenal glands), and some norepinephrine is also secreted from sympathetic nerve terminals [3, 209]. The three major subcategories of β-adrenergic receptors, **β-1, β-2,** and **β-3** receptors, described briefly above, have different tissue distributions and mediate different functions. Their major functions relate to physiological changes associated with stress and "fight-or-flight" responses [3, 210–214].

– Other receptors that bind to Gs
While the **β-adrenergic receptor** is the major receptor linked to **Gs** and the adenylate cyclase pathway, other receptors can also be linked to this pathway. **For example, the D-1-like receptor of the dopamine receptor family is mainly linked to Gs. Other GPRCs that are linked to the receptor are the A2A and A2B members of the adenosine receptors, the glucagon receptors, and the parathyroid hormone (PTH) receptors** [202, 215–217].

- Some Functions of the Adenylate Cyclase Pathway

In cardiac tissue, β-adrenergic receptor signaling can lead to **increased heart rate**. This is due to PKA's ability to phosphorylate two different channel proteins, which in turn leads to faster cardiac action potentials (brief changes in voltage across the cardiac cells) and heart rate. It also leads to **increased contractility**. This is due to PKA phosphorylating proteins involved in contraction, such as troponin I and myosin-binding protein. The ligands that bind to the β-adrenergic receptors in these cells are mostly catecholamines (epinephrine and norepinephrine) [218–220].

In **adipose tissue**, the adenylate cyclase pathway can lead to the **breakdown of stored fat**. Stored fat serves as a source of energy as it can be broken down to produce ATP during stress. PKA phosphorylates and thereby activates **HSL (hormone-sensitive lipase)**, which has a role in breaking down TAGs stored in adipose cells. In addition, **PKA** phosphorylates other targets such as perilipin. Phosphorylation of perilipin makes the lipid droplets accessible to lipases like HSL, which break down TAGs [221, 222].

In the **central nervous system**, norepinephrine (and to a lesser extent epinephrine) functions as a neurotransmitter that binds to the **β-adrenergic** receptor on neurons, leading to Gs activation. This leads to positive regulation of **neuronal activity**, which leads to increased **alertness** and **arousal**. This is mediated in part by **PKA** phosphorylation of **CREB**, which leads to the regulation of expression of genes that are associated with synaptic plasticity (briefly described as ability of synapses to strengthen or weaken due to experience, which is important in learning and memory), neuronal excitability (how likely a neuron is to fire in response to generate an action potential—the signal that travels down an axon), and neurotransmitter release. PKA also phosphorylates certain ion channels in neurons, leading to increased neurotransmission (the entire process involving action potential, neurotransmitter release, binding of neurotransmitters to the postsynaptic neuron, and postsynaptic response) [223, 224]. Norepinephrine and epinephrine can also bind to other receptors in the nervous system [225].

In the **liver,** the adenylate cyclase pathway can lead to the **release of glucose into the bloodstream**. This helps to provide energy for cells. This is due to **PKA** phosphorylation of **CREB**, leading to the expression of genes involved in the regulation and synthesis of glucose in these cells. **PKA** also phosphorylates proteins involved in glucose metabolism, such as **glycogen phosphorylase**, which releases glucose from glycogen stores [226]. The liver is a good example of a tissue where the adenylate cyclase pathway is activated by a GPCR that is not the β-adrenergic receptor. In the liver, the main GPCR that activates the pathway is the glucagon receptor, which also links to Gs. The main ligand that binds this receptor is glucagon, released from pancreatic alpha cells [227].

An especially important example of **adenylate cyclase** signaling is the pathway activated by glucagon binding to the **glucagon receptor**. The glucagon receptor is a Gs-coupled GPCR found in hepatocytes. Glucagon is secreted from pancreatic α-cells during **fasting**. It then activates signaling pathways that are important for maintaining glucose homeostasis during fasting. This starts by glucagon binding to its receptor and activating the adenylate cyclase pathway. This leads to PKA activation, which in turn phosphorylates multiple targets to regulate glucose metabolism. For example, it stimulates **glycogen breakdown** by phosphorylating **phosphorylase kinase**, which in turn activates glycogen phosphorylase, an important enzyme in the breakdown of glycogen (see Chap. 12). PKA also **inhibits glycogen synthesis** by phosphorylating and thereby inactivating glycogen synthase. As discussed above, PKA also phosphorylates and activates the transcription factor CREB. CREB increases the expression of genes encoding gluconeogenic enzymes. For example, along with other transcription factors, CREB can increase transcription of the PEPCK gene. Likewise, along with coactivators, CREB also leads to upregulation of the gene encoding glucose-6-phosphatase. Both of these enzymes play important roles in gluconeogenesis and thus help maintain blood glucose levels during fasting.

- Drugs Targeting the β-Adrenergic Receptors (the β-Receptor)

Beta-blockers are well-known drugs that **block** the action of **β-adrenergic receptors**. They are used to treat **hypertension, angina, and arrhythmias**. Some beta-blockers include **Inderal (propranolol)**, **Lopressor (metoprolol)**,

Tenormin (atenolol), and **Zebeta (bisoprolol)**. These **small-molecule** drugs primarily act in cardiac tissue, which, as described above, is an important site of β-adrenergic receptor activation. These drugs primarily block the β-1 adrenergic receptor although **Inderal** can also block the β-2 receptor with equal effectiveness. They function as competitive inhibitors (antagonists) by binding directly to the β-receptor and thereby preventing the natural ligands (primarily epinephrine and norepinephrine) from binding. By blocking these GPCRs, the beta blockers **reduce heart rate and contractility, lower blood pressure, and regulate heart rhythm** [8, 20, 228, 229].

The other major type of drugs is the **beta-2-agonists**. The agonists **activate the β-2-adrenergic receptor** rather than blocking it. An important example is **Ventolin (albuterol)** [230]. This type of **small-molecule** drug **activates** certain **β-adrenergic receptors on smooth muscle cells in the airways**. This leads to relaxation of the bronchial smooth muscles. Specifically, it relaxes the bronchioles, which are small airways in the lungs. By relaxing these muscles, these drugs cause bronchodilation, which is the widening of the bronchioles. This can lead to improved airflow into the lungs. These drugs are used to treat asthma and **chronic obstructive pulmonary disease (COPD)**. Beta-2-agonists activate the receptors by binding directly to them, acting as ligands, and stimulating the downstream signaling pathway [109, 231, 232].

4.2.2 Gi and Inhibition of the Adenylate Cyclase Pathway

- Inhibition of the Adenylate Cyclase Pathway

Gi, like Gs, acts on **adenylate cyclase**. However, unlike **Gs**, which **activates adenylate cyclase activity**, **Gi inhibits adenylate cyclase activity** (see Fig. 18.24). **Gi** is usually linked to **different receptors**, such as the **alpha-2-adrenergic receptors** and others. G proteins containing Gαi have mechanistic features in common with other G proteins. Specifically, their associated GPCRs are activated by ligand, the Gαi subunit becomes GTP bound and therefore active. Similar to Gαs, GTP-bound Gαi also binds to adenylate cyclase. However, rather than activating adenylate cyclase, it changes its conformation into an inactive state. The result is inhibition of the adenylate cyclase pathway. Gi can, in turn, lead to decreased heart rate and contractility, reduced calcium influx into cells, regulation of cell proliferation, anti-inflammatory effects, reduction of sensory perception, and other functions [3].

The Gβγ subunits released from the Gαi also have important functions. Together, they have a role in modulating potassium channels in some cells, specifically inwardly rectifying potassium (GIRK) channels (Kir3 channels). They directly bind to the channels, which leads to a conformational change that causes them to open. This leads to an efflux of K+ ions out of the cell, causing hyperpolarization of the membrane potential. This has various effects depending on the cell type. It can make neurons less likely to fire an action potential, or in cardiac cells, it can slow the heart rate. These outcomes contribute to the inhibitory effects of Gi [233].

- Drugs That Inhibit Adenylate Cyclase via Gi

Opioids are drugs that **activate opioid receptors,** which are found on nerve cells and are coupled to **Gi**. Opioids **reduce the activity of nerve cells**, which involves reduced excitability and reduced neurotransmitter release. They are used **to treat pain**. Most opioids are small molecules, and examples are **morphine** and **fentanyl**. When these drugs bind the receptors, the receptors become activated, leading to Gi activation. Gi, in turn, leads to downregulation of the adenylate cyclase pathway. This, in turn, reduces neuronal activity [234]. As described above, activation of the Gαi pathway also leads to the opening of inward rectifying potassium channels (due to Gβγ), causing potassium ions to flow out of the neuron. This makes the neuron less likely to fire an action potential, and thus it becomes less excitable. Together, these two effects reduce pain signals [235]. These drugs are used to treat pain but can have some detrimental effects on the central nervous system and can also become highly addictive [2, 20, 236].

Delta-9-tetrahydrocannabinol (THC) is the active component in **cannabis**. **THC** is a **small**

molecule, and it acts as an agonist on **cannabinoid receptors**, including **CB1**, which are linked to **Gi**. As with the opiates, THC **reduces the activity of neurons** due to activation of Gα, and it also regulates the activity of inward rectifying calcium channels via the associated Gβγ subunits. In addition, there is also some evidence that THC can bind directly to the inward rectifying potassium channels and regulate their activity. THC is used for many reasons, including the **treatment of pain, nausea, and stress** [2, 20, 237–240].

4.2.3 Gq and the PLC Pathway

- Overview of the PLC Pathway and Gq
  - Summary of the pathway

The main steps in the PLC pathway can be summarized as follows and summarized in Fig. 18.25:

GPCR → Gq → PLC → DAG + IP_3 → Ca^{2+} influx from the ER → Ca^{2+}-dependent responses including activation of PKC and other responses

- Detailed description of the pathway

The **alpha-1-adrenergic receptor** is an example of a receptor that binds to **Gq**. The main ligands for this type of receptor are **norepinephrine** and **epinephrine**. When receptors are activated, Gαq becomes GTP bound and separates from the βγ subunits. After this, the Gα binds to **phospholipase C (PLC)**. **PLC** is a membrane-bound enzyme, and when activated, it cleaves the **PIP_2** (PI-4,5 bisphosphate; a membrane lipid described previously in this chapter) into **diacylglycerol (DAG)** and **IP_3**. Upon its cleavage, **DAG** (a diglyceride made of a glycerol attached to two fatty acids) remains membrane bound. In contrast, **IP_3**, which is a second messenger, is released from the membrane upon the cleavage of PIP_2 [2, 3, 206].

IP_3 diffuses through the cell and binds to calcium channels on the ER membrane, called **IP_3 receptors**. The ER is rich in calcium, while the cytoplasm is generally low in calcium. By binding to these IP_3 receptors, IP_3 causes the channels to open, and calcium can flow out of the ER and

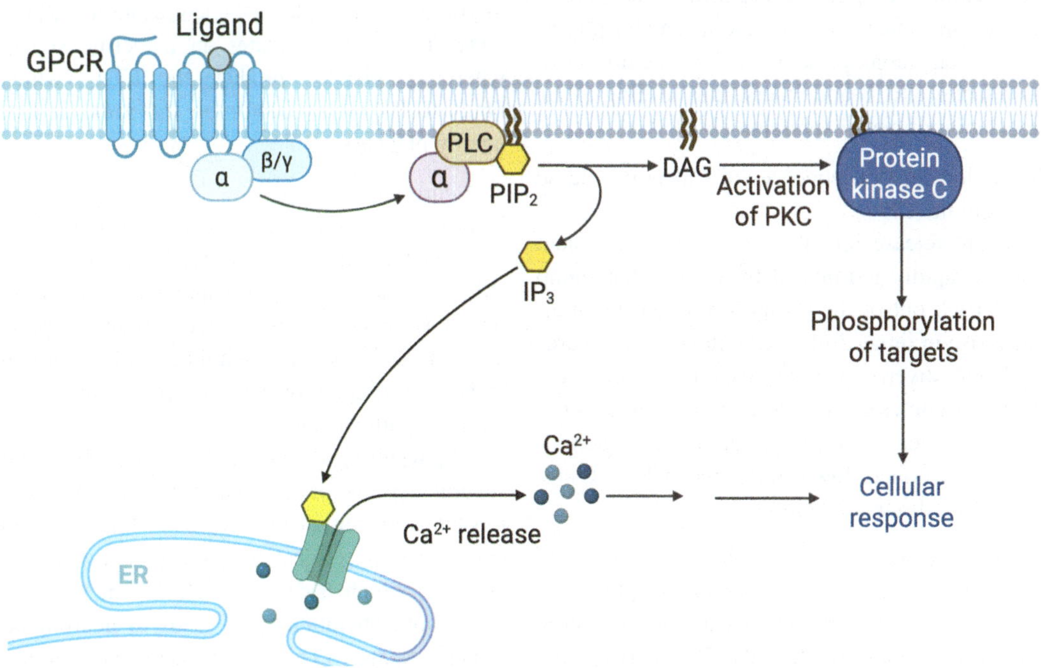

Fig. 18.25 The Phospholipase C (PLC) pathway. In the PLC pathway, the GPCR leads to cleavage of PIP_2 to yield DAG and IP_3, leading to calcium signaling and PKC activation. See the text for details (*Source*: Created with Biorender.com)

into the cytoplasm. Thus, the levels of **intracellular calcium rise**. This increased level of intracellular calcium has many functions.

One important function of calcium is the activation of **protein kinase C (PKC)**. PKC activation requires both calcium and DAG. These are important for relieving PKC's autoinhibitory conformation. PKC, in turn, phosphorylates various targets. These include ion channels, receptors, cytoskeletal proteins, transcription factors, and other kinases [2, 3].

Another important mechanism by which calcium can function is via calmodulin, which is a calcium sensor. **Calmodulin** is a protein that has four binding sites for calcium. When calcium levels are high in the cell, calcium can bind to these binding sites on calmodulin. This leads to a conformation change of calmodulin. While **calmodulin** has no catalytic activity, this conformational change causes it to interact with other proteins and thereby regulate their activity. For example, the kinase **CaM-kinase** (**CAMK**; calcium calmodulin-dependent kinase) becomes activated due to its interaction with the calcium-calmodulin complex. **CaM-kinase** is thus a calcium-responsive kinase, and it can phosphorylate various target proteins, such as the transcription factor CREB and others, thus leading to the regulation of gene expression [241].

- Biological Consequences of Calcium Release from the ER

Calcium release into the cytoplasm is transient and is rapidly terminated by pumps that pump Ca^{2+} back out of the cytoplasm. Even this transient rise in intracellular calcium can lead to crucial and diverse cellular responses. There are many examples of how different cells respond to calcium release. A few examples of calcium-dependent effects downstream to GPCR-PLC signaling are discussed below.

– Smooth Muscle Contraction

Binding of ligands such as norepinephrine to alpha-1-adrenergic receptors leads to activation of Gq in smooth muscle cells. This activates the PLC pathway and leads to calcium release from the ER.

This calcium binds to calmodulin, and the Ca^{2+}-calmodulin complex interacts with **myosin light chain kinase (MLCK)**, which leads to its activation due to a conformational change. Activated MLCK phosphorylates myosin light chain (MLC), and this allows myosin to interact with actin, leading to smooth muscle contraction [3, 20].

– Platelet Aggregation

Platelet aggregation is an early step in forming blood clots. The process is initiated when thromboxane-2 (TXA_2) binds its GPCRs on the surface of platelets. This activates Gq and the PLC pathway, leading to a rise in intracellular Ca^{2+} [242].

Increased Ca^{2+} activates several Ca^{2+}-dependent enzymes involved in platelet activation and cell shape changes. For example, Ca2+-calmodulin leads to **MLCK** activation in platelets, leading to MLC phosphorylation, and this has an important role in cytoskeletal organization that leads to cell shape changes in these cells [242]. **PKC** activation in response to Ca^{2+} and DAG also mediates signaling events that are important for platelet aggregation [242–244]. TXA_2 also initiates an autocrine signal, where the platelets produce the TXA_2, which binds to its own receptor on the same cell or nearby platelets.

– Neurotransmission

Neurotransmission can be modulated by GPCRs such as the muscarinic acetylcholine receptors (M1, M3, and M5) [245] and some serotonin receptors, which couple to Gq. These receptors are activated by acetylcholine and serotonin (5-HT). Activation of these receptors can lead to the PLC pathway and release of Ca^{2+}.

The resulting rise in Ca^{2+} levels can regulate several aspects of synaptic function. One key mechanism involves Ca2+ binding to synaptotagmin, a calcium sensor on synaptic vesicles, which triggers vesicle fusion with the presynaptic membrane and leads to neurotransmitter release. (This mechanism was briefly described in Chap. 3). Ca^{2+} can also lead to activation of PKC, which also phosphorylates various synaptic proteins, increasing neurotransmitter release [246, 247].

– Immune Responses

GPCRs, including **muscarinic acetylcholine receptors** on immune cells, can be activated by acetylcholine. This can lead to activation of the PLC pathway and Ca^{2+} release.

The resulting Ca^{2+} has important roles in immune cell activation, proliferation, and regulation of cytokine production. For example, **Ca^{2+}-calmodulin** can bind to a phosphatase called **calcineurin** and activate it. Calcineurin dephosphorylates a calcium-regulated transcription factor called **NFAT (nuclear factor of activated T cells)**. Dephosphorylated NFAT enters the nucleus and promotes transcription of genes such as the gene for IL-2, which is critical for T-cell activation. Ca^{2+}-calmodulin also leads to activation of **CAMK** (specifically **CAMKII**). CAMKII phosphorylates transcription factors that are important for B- and T-cell activation. **PKC** also contributes to immune responses by regulating cytokine production, cell survival, and adhesion [248]. Some immune responses involve Gq although other G proteins and pathways are also involved in immune responses [249, 250].

– Hormone Secretion

Hormone secretion from cells such as pancreatic cells and pituitary cells can be triggered by GPCRs, such as muscarinic acetylcholine receptors, which are activated by **acetylcholine** and which signal through Gq. This leads to PLC signaling and increased intracellular Ca^{2+}.

This increased Ca^{2+} can stimulate the release of hormones from vesicles. For example, in pancreatic β-cells, it can lead to insulin secretion due to binding to synaptotagmin-like proteins. Although the M3 muscarinic receptors are major contributors to this process, other receptor types and signaling pathways can also regulate hormone release [251, 252].

- Some Drugs That Target Receptors that Signal via Gq and the PLC Pathway

Drugs that block the alpha-adrenergic receptors are called **alpha blockers**. Alpha blockers bind the alpha-1 adrenergic receptors at the site where ligand normally binds. Consequently, they prevent ligand binding to the receptor. As a result, the receptor does not get activated, and downstream signaling is prevented. **Cardura (doxazosin)** and **Minipress (prazosin)** are examples of **alpha-1-adrenergic receptor** antagonists. They are **small molecules** that **block** the alpha-adrenergic receptors, thereby blocking Gq signaling in smooth muscle cells. This prevents contraction and instead leads to vasodilation and relaxation of the smooth muscles. Consequently, these drugs lower blood pressure and are used to treat disorders such as **hypertension** and **benign prostatic hyperplasia (BPH)** [109, 253, 254].

Aspirin (acetylsalicylic acid) has multiple purposes. It is used as a **pain reliever**, which is discussed in Chap. 5. In some patients, however, it is also used to prevent heart attacks and strokes. Aspirin works by inhibiting the enzyme **cyclooxygenase (COX)**. It does so by acetylating a serine residue on COX. This permanently prevents COX from catalyzing the synthesis of thromboxanes. Therefore, one outcome of taking aspirin is reduced production of **thromboxane A2 (TXA2)** [255]. The decrease in TXA2 leads to a decrease in activation of the **thromboxane receptor (found on platelets)**, which is a **Gq**-linked receptor. Inhibition of this receptor reduces platelet aggregation (see above), making the blood less prone to forming clots. This can be beneficial in preventing heart attacks and strokes [109].

Sofdra (sofpironium) (FDA approved in 2024) is a prescription topical gel designed to **decrease sweating** in **primary axillary hyperhidrosis**, a disorder that is characterized by excessive sweating. **Sofdra** is a **small-molecule competitive inhibitor** of muscarinic acetylcholine receptors, especially the **M3** type, in sweat glands. The drug binds to the receptors and **inhibits Gq signaling**. By blocking the activity of these receptors in the sweat glands, **Sofdra** reduces sweating [256].

Urecholine (bethanechol) is a **small molecule** that mimics acetylcholine and is an **agonist** of **muscarinic acetylcholine receptors**, particularly the **M3 receptors** located in the bladder and

GI tract. It is used **to stimulate bladder contractions**, and this is linked to **Gq** signaling and calcium mobilization. It is used to treat **urinary retention** and some other GI conditions. It causes the detrusor muscle in the bladder to contract, which helps alleviate urinary retention. It can also be used to treat GI disorders. In the GI system, it can facilitate peristalsis, contractions that move food through the GI tract. This can relieve conditions associated with delayed gastric emptying. This drug can have side effects, and sometimes other drugs can be preferable [20, 257, 258].

Salagen (pilocarpine) is also an agonist of **muscarinic acetylcholine receptors (M3)**, linked to **Gq**. It is used to treat **xerostomia (dry mouth)**. Xerostomia can be linked to dental caries and other health problems. This drug promotes salivation by binding to the M3 receptors (though it can also bind the other muscarinic receptors). Activation of the M3 receptor in the salivary gland leads to a rise in intracellular calcium. Increased intracellular calcium has been shown to lead to saliva release from acinar epithelial cells (secretory cells in the salivary gland) [259, 260].

The drug is largely known for treating **xerostomia**, but it is also sometimes used to treat glaucoma. This is because it causes the sphincter muscle on the iris to contract, leading to pupil constriction. The resulting small pupil helps open the trabecular network, and this helps improve drainage from the eye to lower intraocular pressure [261].

4.3 Downregulation of GPCR Pathways

Like any signaling pathway, once the GPCR pathways are turned on, there must be a mechanism to turn them back off. One mechanism for this is desensitization. Prolonged exposure to a stimulus can lead to desensitization. This depends on a family of kinases called GRKs (G protein-coupled receptor kinases), which phosphorylate GPCRs. The activity of GRKs can be upregulated in response to G protein signaling. In response, GRKs phosphorylate multiple serines and threonines in activated GPCRs. Phosphorylated GPCRs then bind to a protein called **arrestin (Arr)**. As a result, the receptor can no longer bind G proteins, thereby preventing further stimulation of the signaling pathway. In addition, arrestin also has a role in stimulating endocytosis of the receptor so that no new signaling can occur until new receptor is produced [3].

4.4 More Drugs That Act on GPCR Pathways

GPCRs are common targets for drugs. Several additional examples that were not covered above are described here.

4.4.1 Drugs That Regulate CXCR4 (WHIM Syndrome and Multiple Myeloma)

CXCR4 is a **Gi**-linked **GPCR** that has an important role in inflammation. It is also linked to immune responses and cancer [262]. **Activating mutations** in the gene encoding this receptor are associated with a rare disorder called **WHIM syndrome** (WHIM stands for warts, hypogammaglobulinemia). WHIM syndrome is associated with myelokathexis. Myelokathexis is characterized by neutropenia (low levels of neutrophils in the blood) due to abnormal retention of neutrophils in the bone marrow [263]. CXCR4 signaling plays an important role in regulating neutrophil migration from the bone marrow to the peripheral bloodstream. Specifically, when it is activated by its ligand (stromal-derived-factor-1 (**SDF-1 or CXCL12**)), it helps **in retaining cells in the bone marrow** [262]. In WHIM, CXCR4 (the receptor) is improperly activated. This improper activation disrupts the normal regulation of neutrophil migration. **Xolremdi (mavorixafor)** is a drug that was FDA approved in **2024** to treat WHIM. It is an inhibitor of the CXCR4 receptor. It is a **small molecule** that binds the extracellular domain of CXCR4 and thereby blocks its binding to ligand. The result is inhibition of CXCR4 signaling. Inhibition of this pathway leads to increased mobilization of neutrophils and lymphocytes. The drug has been shown to improve neutrophil and lymphocyte count, which are typically low in patients with the disease [264, 265].

Aphexda (motixafortide) is another drug that affects **CXCR4**. It was first FDA approved in 2023. It is used together with **filgrastim (G-CSF)**, a synthetic form of **granulocyte colony-stimulating factor (G-CSF)** [266]. **Filgrastim (G-CSF)** stimulates the mobilization of hematopoietic stem cells in the bone marrow, which give rise to the white blood cells (**WBCs**) that are important for the immune system. It does so because CSF, a cytokine which binds to cytokine receptors, activates pathways such as the JAK/STAT, PI3K, and MAP Kinase pathways that lead to proliferation, survival, and differentiation of WBCS [267]. Treatment with **Aphexda**, together with **filgrastim,** is used for the purpose of **collecting blood cells from patients undergoing medical treatments**. This combination of drugs increases the production of WBCs from the bone marrow stem cells and encourages them to migrate to the peripheral blood. The result is increased availability of WBCS, which can be collected from patients. This treatment is used for patients who have **multiple myeloma** and who will receive autologous transplantations of the WBCs collected from their own blood. For this treatment, the WBCs are collected from the patients, who are then given chemotherapeutic treatments. The WBCS are protected from the chemotherapeutic treatment because they have been collected and saved. Once the treatment is finished, the WBCs are infused back into the patients [268]. **Filgrastim** is a **biologic** because it is a synthetic form of CSF. **Aphexda** is a cyclic peptide that is larger than a typical small molecule, so it does not fit perfectly into the small-molecule category. It can be considered a **peptide therapy** [269].

4.4.2 High Blood Pressure

Endothelin is a peptide ligand secreted from endothelial cells, which binds to GPCRs on smooth muscle cells, such as those lining the vessels [270, 271]. Endothelin is known to lead to vasoconstriction, and in some individuals, it is a strong inducer of **high blood pressure** [272]. Endothelin can bind to two receptors (both GPCRs), ET-A (linked to Gq) and ET-B. ET-A (linked to G1 and other G proteins) activation leads to vasoconstriction, which is associated with high blood pressure. ET-B activation is more complex; it can contribute to vasoconstriction but can also counteract the effects of ET-B. Furthermore, while ET-A receptors are on smooth muscle cells, ET-B receptors are also found on endothelial cells [273]. They are both considered to be potential targets for the treatment of high blood pressure. **Tryvio (aprocitentan)** is a new medication (FDA approved in 2024). It is a **small molecule** that is an antagonist of both ET-A and ET-B receptors. It functions by directly binding to the receptors, thereby preventing their natural ligands from binding. It is an oral medication, and the first new **blood pressure medication** in 40 years [274]. It is designed to treat **high blood pressure** that has been refractory to other treatments [275].

4.4.3 Ulcerative Colitis (UC)

Velsipity (etrasimod) (FDA approved in 2023) is a **small-molecule** drug that antagonizes the **sphingosine-1-phosphate (S1P)** receptor by binding to it. The S1P receptors (S1PRs) are a class of GPCRs found on lymphocytes and other immune cells. **Velsipity** causes the receptors to be internalized and degraded. GPCRs in this class have roles in controlling the migration of lymphocytes from the lymph nodes to various sites, including the intestine. **Ulcerative colitis (UC)** is associated with inflammation in the digestive system, and blocking the SLPRs (especially S1PR1) relieves inflammation by preventing movement of the immune cells to the digestive system [276, 277].

4.4.4 Antidepressant

Exxua (gepirone) (FDA approved in 2023) is an antidepressant. Rather than targeting neurotransmitters like some other antidepressants, it targets and activates the receptor that serotonin normally binds to. It is a small molecule and an **agonist** of the GPCR **5-HT$_1$A**, which is one of the **serotonin** receptors that interacts with Gi [278]. This receptor has a role in **regulating mood and anxiety**, and the drug **improves depressive symptoms** [279].

4.4.5 Veozah (Fezolinetant) for Vasomotor Symptoms (VMS) During Menopause

Veozah (fezolinetant) (FDA approved in 2023) blocks NK3R, a GPCR that is found on the membranes of some neurons. NK3R is linked to Gq and binds to the neuropeptide neurokinin B (NKB) [280]. It is found in the CNS and some other tissues. It is involved in regulating temperature, mood, stress, reproduction, and vasodilation. Certain neurons in the hypothalamus, known as KNDy neurons, are inhibited by estrogen but stimulated by NKB, both of which operate via different receptors. This balance is important for the regulation of body temperature. During menopause, the estrogen signal decreases, leaving NKB signaling unopposed, which results in increased KNDy activity and affects temperature regulation. This leads to VMS, which includes vasodilation, heat, and sweating. **Veozah** is a **small-molecule** drug that binds **NK3R** directly and blocks its ability to bind to NKD. This **reduces heat signaling** [281–283]. Importantly, this is a hormone-independent mechanism, and it may thus avoid some of the safety concerns of other treatments, particularly HRT.

4.4.6 Migraines

Migraines are a chronic and painful condition characterized by severe headaches. Although the origin of migraines is incompletely understood, they are thought to involve a series of complex interactions in the brain. An important player in this process is **CGRP** (calcitonin gene-related peptide), which is released from various sources, including trigeminal nerve endings. Trigeminal nerves are cranial nerves that have sensory and motor functions. The nerve endings are the terminal parts of the nerve that innervate tissue such as skin or, in this case, the dura mater—the protective layer of the brain [284]. **CGRP** binds to its receptor, called **CALCRL** (calcitonin receptor-like receptor), which is a GPCR on target neurons. Binding to **CALCRL** requires that GALCRL bind a co-receptor, which is a single-pass transmembrane receptor called **RAMP1**. Activation of this receptor complex leads to activation of a pathway mediated by Gs, which leads to vasodilation and other changes that are thought to lead to the pain in migraines. Because this mechanism is at least partly understood, the proteins involved are considered to be potential targets for drug treatment [285]. **Zavzpret (zavegepant)** is a new (FDA approved in **2023**) treatment for migraines. It is an **antagonist of CGRP**. It is a **small molecule** that binds to **the CALCRL-RAMP1 complex** and thereby inhibits the receptor's interaction with ligand (**CGRP**), which reduces the symptoms of **migraines** [286].

4.5 Drugs That Target Second Messengers

Following GPCR activation, second messengers play important roles in transmitting signals [206]. Second messengers can therefore also be drug targets. Gs activation, for example, leads to upregulation of cAMP. **cAMP** is formed from ATP by a reaction catalyzed by **adenylate cyclase**, which involves the removal of two phosphate groups and cyclization of the remaining molecule to form cAMP. cAMP is only present transiently. Soon after it is generated, it is inactivated by **cyclic AMP phosphodiesterase (PDEs)**. **PDEs** are enzymes that hydrolyze cAMP, breaking its cyclic structure and converting it to AMP (specifically 5'-AMP) [3] (Fig. 18.26). Two important PDEs that target cAMP are **PDE-3 and PDE-4.**

Inhibitors of **PDE-3 and PDE-4** are used as drugs to **treat chronic obstructive pulmonary disease (COPD)**. By targeting PDEs, cAMP can remain active longer, which may be an advantage in COPD. COPD is a progressive lung disease that is associated with reduced airflow. There are several reasons that inhibiting **PDEs** is promising for treating the disease. An important reason is that cAMP mediates various responses, including relaxation of smooth muscle cells. This relaxation can lead to bronchodilation, which eases airflow and thus reduces the symptoms of COPD. cAMP can also lead to inhibition of inflammatory

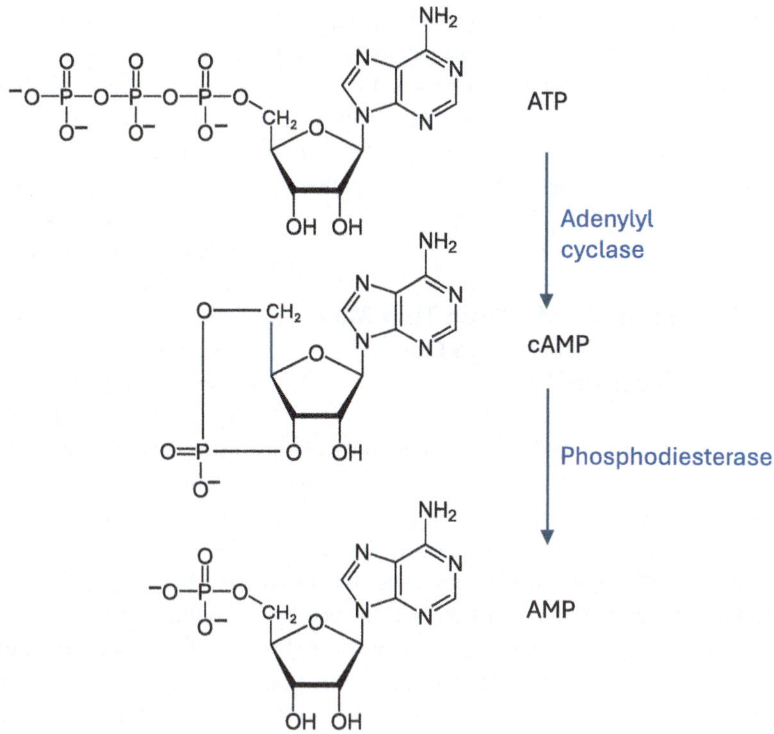

Fig. 18.26 Formation of cAMP by adenylate cyclase and its inactivation by phosphodiesterase

responses by regulating the expression of inflammatory cytokines. Maintaining cAMP for longer is thus a potential advantage [287].

Daliresp (roflumilast) [288] (also known by the brand name **Zoryve**) is a **PDE4 inhibitor** used to treat **COPD**. It reduces inflammation and increases bronchodilation. It is also used to treat other disorders such as plaque psoriasis and atopic dermatitis [289, 290]. It is a **small molecule** and a **competitive inhibitor** of **PDE4**. It binds to **PDE4** at the site where **PDE4** normally binds its substrate, **cAMP**.

Ohtuvayre (ensifentrine) (FDA approved 2024) [291] is also a **small molecule**, but this drug is a dual inhibitor of both **PDE-3** and **PDE-4** [292]. It is used alone or together with other treatments to treat **COPD**. Inhibition of both PDEs was shown to have a synergistic effect and to be more effective than inhibiting either PDE alone.

4.6 Some Pathways Can Be Regulated by More Than One Receptor Type

We have discussed RTKs and GPCRs as some of the major types of receptors that lead to the activation of intracellular signaling pathways. Although we have described these as separate units, such as MAP Kinase activation by RTKs and PLC activation by GPCRs, there is actually significant overlap between pathways activated by different receptors. For example, PLC can be activated not only by GPCRs but, in some cases, by RTKs. Likewise, in some cases, GPCRs can activate MAP Kinase pathways. Signal transduction is thus complex, characterized by crosstalk and overlap between different pathways. In some cases, multiple pathways can converge to produce similar outcomes, while in others, they may act antagonistically, with one pathway inhibiting another. This interconnectedness adds complexity to the development and use of pharmaceuticals for manipulating signaling pathways. Because of the potential for crosstalk between pathways, designing drugs requires careful consideration of the relationships between different pathways [3, 293].

4.7 Example of a Drug That May Target More Than One Receptor Type

Thyroid eye disease (TED) is a complication of Graves' disease. It is also called **Graves' orbitopathy**. It can cause vision problems such as double vision. Graves' disease is an autoimmune disorder where the immune system makes antibodies (autoantibodies) that bind a receptor called the thyroid-stimulating hormone receptor (TSHR). **TSHR** is a **GPCR**, and it is present on thyroid follicular cells. Unlike some other autoimmune diseases, the antibodies **activate** rather than inactivate their target, in this case the TSHR. Activation of this receptor leads to increased thyroid hormone production (hyperthyroidism).

Thyroid eye disease (TED) is a complication of Graves' disease. It is also caused by the autoantibodies binding to **TSHR**, but in this case, the antibodies bind to TSHR on **orbital fibroblasts**, which are connective tissue cells around the eyes. Here as well, the antibodies mimic thyroid hormone (TSH) and activate the receptor abnormally. Evidence has shown that in TED, TSHR forms a complex with another receptor, IGF-1R. IGF-1R (insulin-growth-factor-1 receptor) is an RTK, making this an unusual complex because it combines these two different types of receptors. Despite the two kinds of receptors, the autoantibodies activate IGF-1R signaling. This co-stimulation of two receptors on the orbital fibroblasts leads to inflammation and other problems. It causes tissue to expand behind the eyes, and this can lead to eye problems like double vision.

Tepezza (teprotumumab) [294] is a **biologic** that consists of a **monoclonal antibody** that targets **IGF-1R** and **inhibits IGF-1R** signaling. It was FDA approved in 2020 and used to treat **thyroid eye disease (TED)**. It reduces swelling proptosis (eye bulging), improves vision, and relieves the double vision symptoms.

FDA-Approved Drugs Discussed in This Chapter

Commonly used brand name	Generic name	First FDA approval year	Major conditions treated	Target
Actos	*Pioglitazone*	1999	Diabetes	PPARγ
Agamree	*Vamorolone*	2023	Duchenne muscular dystrophy	GR, MR
Aphexda	*Motixafortide*	2023	Multiple myeloma	CXCR4
Arimidex	*Anastrozole*	1995	ER+ breast cancer	Aromatase
Aromasin	*Exemestane*	1999	ER+ breast cancer	Aromatase
Avandia	*Rosiglitazone*	1999	Diabetes	PPARγ
Cortef	*Hydrocortisone*	1952	Inflammation, adrenal insufficiency	GR
Daliresp	*Roflumilast*	2011	COPD	PDE4
Deltasone	*Prednisone*	1955	Inflammation, autoimmune disorders	
Dupixent	*Dupilumab*	2017	Atopic dermatitis, asthma	IL-4Ra
Enbrel	*Etanercept*	1998	RA, psoriasis	TNF
Eylea	*Aflibercept*	2011	wAMD	VEGF
Fareston	*Toremifene*	1997	ER+ breast cancer	ER
Faslodex	*Fulvestrant*	2002	ER+ breast cancer	ER
Femara	*Letrozole*	2001	ER+ breast cancer	Aromatase
Herceptin	*Trastuzumab*	1998	HER2+ breast cancer	HER2
Humira	*Adalimumab*	2002	RA, UC, psoriasis	TNF

(continued)

Commonly used brand name	Generic name	First FDA approval year	Major conditions treated	Target
Ilumya	*Tildrakizumab*	2018	Plaque psoriasis	IL-23
Iqirvo	*Elafibranor*	2024	Primary biliary cholangitis	PPARa
Jakafi	*Ruxolitinib*	2011	Myelofibrosis	JAK1, JAK2
JOENJA	*Leniolisib*	2023	Activated PI3K delta syndrome	PI3K-delta
Leqselvi	*Deuruxolitinib*	2024	Alopecia areata	JAK1, JAK2
Litfulo	*Ritlecitinib*	2023	Alopecia areata	JAK3, Tec
Livdelzi	*Seladelpar*	2024	Primary biliary cholangitis	PPARd
Nolvadex	*Tamoxifen*	1977	ER+ breast cancer	ER
Ohtuvayre	*Ensifentrine*	2024	COPD	PDE3, PDE4
Ojjaara	*Momelotinib*	2023	Myelofibrosis	JAK2
Olumiant	*Baricitinib*	2018	RA	JAK1, JAK2
Omvoh	*Mirikizumab*	2023	Ulcerative colitis	IL-23
Orserdu	*Elacestrant*	2023	ER+ breast cancer	ER
Prolia	*Denosumab*	2010	Osteoporosis	RANKL
Rhopressa	*Netarsudil*	2017	Glaucoma	ROCK
Rinvoq	*Upadacitinib*	2019	RA, UC, eczema	JAK1
Rocklatan	*Netarsudil + latanoprost*	2019	Glaucoma	ROCK
Skyrizi	*Risankizumab*	2019	Psoriasis, Crohn's disease	IL-23
Sotyktu	*Deucravacitinib*	2022	Psoriasis	TYK2
Tepezza	*Teprotumumab*	2020	Thyroid eye disease	IGF-1R
Tremfya	*Guselkumab*	2017	Psoriasis, psoriatic arthritis	IL-23
Tricor	*Fenofibrate*	1993	Dyslipidemia	PPARα
Vabysmo	*Faricimab*	2022	wAMD, diabetic macular edema	VEGF, Ang-2
Ventolin	*Albuterol*	1981	Asthma	β-adrenergic receptor
Vijoice	*Alpelisib*	2022	PIK3CA-related overgrowth	PI3K
Vonjo	*Pacritinib*	2022	Myelofibrosis	JAK2
Vtama	*Tapinarof*	2022	Psoriasis	AhR
Winrevair	*Sotatercept*	2024	Pulmonary arterial hypertension	Activin
Xeljanz	*Tofacitinib*	2012	RA	JAK1, JAK3

Definitions

Adapter protein Non-enzymatic proteins like GRB2 that link two or more proteins together.

Agonists An agonist is an activator. In signaling it is usually a molecule that binds to a receptor and mimics the effect of the normal ligand for that receptor. An example is a growth factor agonist, one that mimics the growth factor, binds to the receptor that the natural ligand binds to, and activates the receptor.

Antagonists An antagonist is an inhibitor. In signaling it is usually a molecule that binds to a receptor and blocks or inhibits its activation.

Autocrine signaling Refers to a type of cell signaling where a cell releases a signal (usually a hormone or cytokine) that binds to receptors on its own surface, influencing its own activity.

AKT A serine/threonine kinase that promotes cell survival and growth; activated by PI3K signaling. Also known as protein kinase B.

Bispecific antibody A monoclonal antibody engineered to bind two different targets simultaneously.

Bronchodilation The widening of the air passages (bronchi) in the lungs, which helps improve airflow and is typically induced by medications like bronchodilators in the treatment of conditions like asthma and chronic obstructive pulmonary disease (COPD).

Catalyze To accelerate a chemical reaction. Enzymes are biological catalysts.

Cell migration The movement of cells from one location to another within the body. This is a crucial process in many biological functions, including wound healing, immune responses, and development.

Cytokine A type of small signaling protein that binds to cell surface receptors, often involved in inflammatory responses, immune responses, and cell signaling and communication.

Dimerization The process by which two protein molecules (monomers) bind together to form a dimer.

Endocytosis A process by which cells internalize molecules, including receptors and ligands.

FERM Domain A protein domain found in proteins like JAK that can mediate interactions with other proteins and anchor them to the membrane.

Growth factor A type of small signaling protein that binds to receptors. They often regulate cell division and bind to transmembrane receptors such as RTKs.

GTPase Enzymes that bind and hydrolyze GTP. They act as molecular switches in signaling pathways. A classic example is Ras.

Guanine nucleotide exchange factor (GEF) Proteins that exchange GDP for GTP on GTPases such as Ras. An example is Sos.

Glucocorticoids Steroid hormones produced in the adrenal gland and that regulate metabolism, inflammation, and stress responses.

Hormone Hormones are signaling molecules released by glands that travel via the bloodstream to their target cells, which can be distant from the hormone's origin.

Janus Kinase (JAK) A family of tyrosine kinases associated with cytokine receptors that phosphorylate and activate STAT proteins.

Kinase A type of enzyme that catalyzes the transfer of phosphate groups to their substrates. The substrate is often a protein, but it can also be a lipid or other molecule.

Latent transcription factor A transcription factor that is present in an inactive form within the cytoplasm of a cell but can be activated by signaling events. STATs are examples.

Ligand In signaling a ligand is a molecule that binds to a receptor to initiate a cellular response. It can be a protein, peptide, hormone, or small molecule.

MAP kinase pathway A signaling cascade involving three sequential kinases (MAPKKK → MAPKK → MAPK) that ultimately regulate gene expression.

Mitogen Usually a ligand that can stimulate cell division. Its name comes from the stimulation of mitosis.

Monoclonal antibody A lab-engineered antibody that binds a specific target (antigen), used in therapies like cancer or inflammation treatment.

mTOR (Mammalian Target of Rapamycin) A kinase involved in regulating cell growth, metabolism, and survival.

Myosin light chain (MLC) A regulatory component of myosin that must be phosphorylated to promote actin-myosin contraction in cells.

Neurotransmission Communicate between neurons or between neurons and other cells via the release and reception of neurotransmitters.

Paracrine A form of cell signaling in which the ligand travels only a short distance to its receptor on its target cell.

Pleckstrin homology domain (PH domain) Protein domain that binds specifically to phosphoinositide lipids like PIP_3.

Phosphorylation Addition of a phosphate group (PO_4^{3-}) to a molecule, often to serine, threonine, or tyrosine residues on proteins.

Phosphoinositide 3-Kinase (PI3K) A lipid kinase that phosphorylates PIP_2 to form PIP_3, initiating AKT signaling.

PIP_2/PIP_3 Membrane lipids that act as signaling molecules; PIP_3 recruits proteins like AKT to the plasma membrane.

Proteases Enzymes that catalyze the breakdown of proteins by hydrolyzing peptide bonds.

Ras A small GTPase that plays a key role in growth factor signaling and is frequently mutated in cancer.

Receptor tyrosine kinase (RTK) A type of cell surface receptor that spans the membrane once and has tyrosine kinase activity on its cytoplasmic domain.

Rho GTPases A subfamily of GTPases that regulate the actin cytoskeleton and cell motility. Includes Cdc42, Rac, and Rho.

Rho-associated protein kinase (Rock) A serine/threonine kinase regulated by Rho GTPase. It has important roles in cytoskeletal dynamics and smooth muscle contraction.

Second messenger Small, nonprotein molecules like Cyclic AMP (cAMP), calcium ions, or Diacylglycerol (DAG) that propagate signals inside the cell.

Src homology 2 domain (SH2 Domain) Protein domain that binds specifically to phosphorylated tyrosines.

Signal transducer and activator of transcription (STAT) A family of transcription factors activated by JAKs.

Target proteins Proteins that are directly influenced or modified by other proteins, such as the substrate of an enzyme.

Toll-like receptors (TLRs) A family of innate immune receptors that recognize microbial components and trigger immune responses.

Transcription factor A protein that binds DNA at promoters or other cis-regulatory elements and regulates gene expression.

Vascular endothelial growth factor (VEGF) A protein that stimulates angiogenesis by activating VEGF receptors on endothelial cells.

Study Questions

Multiple-Choice Questions

1. Which of the following ligands can diffuse across the plasma membrane to bind intracellular receptors?
   A. Insulin
   B. Epidermal growth factor (EGF)
   C. Estrogen
   D. TNF-alpha
2. Which signaling molecule directly phosphorylates ERK in the MAPK pathway?
   A. Ras
   B. MEK
   C. Raf
   D. GRB2
3. Which protein kinase does the drug Rocklatan target to treat glaucoma?
   A. Raf
   B. ERK
   C. Rock
   D. PAK
4. What type of protein is GRB2 in the MAP Kinase pathway?
   A. Receptor
   B. GTPase
   C. Adapter protein
   D. Kinase
5. Which of the molecules below as an example of a second messenger is involved in cell signaling?
   A. GRB2
   B. Ras
   C. Calcium
   D. PAK
6. In cell signaling, when Ras is GTP bound, it usually:
   A. Inhibits kinase activation
   B. Binds to transcription factors
   C. Activates a signaling pathway
   D. Transports ligands into the nucleus
7. What disease is associated with abnormal VEGF-induced blood vessel growth in the eye?
   A. Glaucoma
   B. Rheumatoid arthritis
   C. Wet age-related macular degeneration (wAMD)
   D. Psoriasis
8. Dupixent works by inhibiting:
   A. VEGF directly
   B. Rock kinase
   C. IL-4 and IL-13 receptor signaling
   D. JAK2 kinase
9. Which of the following best describes the primary signaling effect of Gs protein

activation by its associated GPCR in the adenylate cyclase pathway?
  A. Activation of phospholipase C, leading to increased IP_3 and DAG production
  B. Inhibition of adenylate cyclase, resulting in decreased cAMP levels
  C. Activation of adenylate cyclase, leading to increased cAMP production and activation of protein kinase A (PKA)
  D. Direct opening of ion channels, causing membrane depolarization
10. Adenylate cyclase is a:
  A. Membrane-bound enzyme
  B. Cytoplasmic enzyme
  C. cAMP-activated enzyme
  D. A receptor tyrosine kinase
  E. A type of trimeric G protein

True/False Question

11. Activated ERK translocates to the nucleus. ______

Fill-in-the-Blank Questions

12. ______ is a monoclonal antibody that targets the HER2 receptor in Her2 positive breast cancer.
13. Rock phosphorylates ______ to promote cellular contraction.
14. ______ and ______ are the two main Rho GTPases involved in lamellipodia and filopodia formation.
15. The drug ______ inhibits the JAK protein TYK2 and is used to treat psoriasis.
16. PIP_2 is converted to PIP_3 by the enzyme ______.
17. ______ is the small GTPase that is activated by Sos in the ERK signaling pathway.

Short-Answer Questions

18. Explain the role of adaptor proteins in signal transduction pathways.
19. How does the structure of GPCRs differ from RTKs?

Answer Key

Multiple-Choice Questions

1. **Answer**: C
2. **Answer**: B
3. **Answer**: C
4. **Answer**: C
5. **Answer**: C
6. **Answer**: C
7. **Answer**: C
8. **Answer**: C
9. **Answer**: C
10. **Answer**: A

True/False Question

11. **Answer**: True

Fill-in-the-Blank Questions

12. **Answer**: Herceptin (trastuzumab)
13. **Answer**: Myosin light chain (MLC)
14. **Answer**: Rac and Cdc42
15. **Answer**: Sotyktu (deucravacitinib)
16. **Answer**: PI3K
17. **Answer**: Ras

Short Answer Questions

18. **Answer**: Adaptor proteins link proteins together. For example, they may link substrates near the kinase that phosphorylates them. Adapter proteins, however, do not have any catalytic activity themselves.
19. **Answer**: GPCRs span the membrane seven times and signal through G-proteins, while RTKs span once and have intrinsic kinase activity.

References

1. Iqbal J, Sun L, Zaidi M. Complexity in signal transduction. Ann N Y Acad Sci. 2010;1192:238–44. https://doi.org/10.1111/j.1749-6632.2010.05388.x.
2. Marks F, Klingmüller U, Müller-Decker K. Cellular signal processing: an introduction to the molecular mechanisms of signal transduction. 2nd ed. New York, NY. Garland Science; 2017.
3. Alberts B, Heald R, Johnson A, Morgan D, Raff M, Roberts K, Walter P, Wilson J, Hunt T. Molecular biology of the cell. New York, NY. W.W. Norton and Company; 2022.
4. Cooper G, Adams K. The cell: a molecular approach. New York, NY. Sinauer Associates; 2022.
5. FDA. HERCEPTIN (trastuzumab) Approval Overview 1998. Available from: https://www.accessdata.fda.gov/scripts/cder/daf/index.cfm?event=overview.process&ApplNo=103792.
6. Yip HYK, Papa A. Signaling pathways in cancer: therapeutic targets, combinatorial treatments, and new developments. Cells. 2021;10(3):659.
7. McCubrey JA, Steelman LS, Chappell WH, Sun L, Davis NM, Abrams SL, et al. Advances in targeting signal transduction pathways. Oncotarget. 2012;3(12):1505–21. https://doi.org/10.18632/oncotarget.802.
8. Katzung BG, Vanderah TW, Trevor AAJ. Katzung's basic and clinical pharmacology. 16th ed. ed. New York, NY. McGraw-Hill Education; 2024.
9. Lemmon MA, Schlessinger J. Cell signaling by receptor tyrosine kinases. Cell. 2010;141(7):1117–34. https://doi.org/10.1016/j.cell.2010.06.011.
10. Chemmalar S, Intan Shameha AR, Che Abdullah CA, Ab Razak NA, Yusof LM, Ajat M, et al. Busting the breast cancer with AstraZeneca's Gefitinib. Adv Pharmacol Pharm Sci. 2023;2023:8127695. https://doi.org/10.1155/2023/8127695.
11. Spencer-Smith R, Morrison DK. Regulation of RAF family kinases: new insights from recent structural and biochemical studies. Biochem Soc Trans. 2024;52(3):1061–9. https://doi.org/10.1042/bst20230552.
12. Lavoie H, Gagnon J, Therrien M. ERK signalling: a master regulator of cell behaviour, life and fate. Nat Rev Mol Cell Biol. 2020;21(10):607–32. https://doi.org/10.1038/s41580-020-0255-7.
13. Fowler T, Sen R, Roy AL. Regulation of primary response genes. Mol Cell. 2011;44(3):348–60. https://doi.org/10.1016/j.molcel.2011.09.014.
14. Braicu C, Buse M, Busuioc C, Drula R, Gulei D, Raduly L, et al. A comprehensive review on MAPK: a promising therapeutic target in cancer. Cancers (Basel). 2019;11(10) https://doi.org/10.3390/cancers11101618.
15. De Meyts P. The insulin receptor and its signal transduction network. In: Feingold KRASF, Anawalt B, et al., editors. Endotext [Internet]. South Dartmoutn: MDText.com, Inc.; 2016.
16. Morrison DK. MAP kinase pathways. Cold Spring Harb Perspect Biol. 2012;4(11) https://doi.org/10.1101/cshperspect.a011254.
17. Nilsson M, Heymach JV. Vascular endothelial growth factor (VEGF) pathway. J Thorac Oncol. 2006;1(8):768–70.
18. Apte RS, Chen DS, Ferrara N. VEGF in Signaling and disease: beyond discovery and development. Cell. 2019;176(6):1248–64. https://doi.org/10.1016/j.cell.2019.01.021.
19. Fleckenstein M, Schmitz-Valckenberg S, Chakravarthy U. Age-related macular degeneration: a review. JAMA. 2024;331(2):147–57. https://doi.org/10.1001/jama.2023.26074.
20. Ritter JM, Flower RJ, Henderson G, Loke YK, MacEwan DJ, Rang HP. Rang and Dale's pharmacology. 10th ed; London, England, Elsevier. 2023.
21. Fauvel B, Yasri A. Antibodies directed against receptor tyrosine kinases: current and future strategies to fight cancer. MAbs. 2014;6(4):838–51. https://doi.org/10.4161/mabs.29089.
22. USFaDAF. Vabysmo(Faricimab) Approval Overview 2022. Available from: https://www.accessdata.fda.gov/scripts/cder/daf/index.cfm?event=overview.process&ApplNo=761235.
23. Panos GD, Lakshmanan A, Dadoukis P, Ripa M, Motta L, Amoaku WM. Faricimab: transforming the future of macular diseases treatment - a comprehensive review of clinical studies. Drug Des Devel Ther. 2023;17:2861–73. https://doi.org/10.2147/dddt.S427416.
24. FDA. Eylea(Aflibercept) Approval Overview 2011. Available from: https://www.accessdata.fda.gov/scripts/cder/daf/index.cfm?event=overview.process&ApplNo=125387.
25. Liberski S, Wichrowska M, Kocięcki J. Aflibercept versus Faricimab in the treatment of neovascular age-related macular degeneration and diabetic Macular Edema: a review. Int J Mol Sci. 2022;23(16) https://doi.org/10.3390/ijms23169424.
26. Shaulian E, Karin M. AP-1 as a regulator of cell life and death. Nat Cell Biol. 2002;4(5):E131–E6. https://doi.org/10.1038/ncb0502-e131.
27. Zeke A, Misheva M, Reményi A, Bogoyevitch MA. JNK signaling: regulation and functions based on complex protein-protein partnerships. Microbiol Mol Biol Rev. 2016;80(3):793–835. https://doi.org/10.1128/mmbr.00043-14.
28. Zarubin T, Han J. Activation and signaling of the p38 MAP kinase pathway. Cell Res. 2005;15(1):11–8. https://doi.org/10.1038/sj.cr.7290257.
29. Wagner EF, Nebreda ÁR. Signal integration by JNK and p38 MAPK pathways in cancer development. Nat Rev Cancer. 2009;9(8):537–49. https://doi.org/10.1038/nrc2694.
30. DrugBank. Dilmapimod. Available from: https://go.drugbank.com/drugs/DB12140.
31. Yeung YT, Yin S, Lu B, Fan S, Yang R, Bai R, et al. Losmapimod overcomes Gefitinib resistance in

non-small cell lung cancer by preventing Tetraploidization. EBioMedicine. 2018;28:51–61. https://doi.org/10.1016/j.ebiom.2018.01.017.

32. Nobes CD, Hall A. Rho, rac, and cdc42 GTPases regulate the assembly of multimolecular focal complexes associated with actin stress fibers, lamellipodia, and filopodia. Cell. 1995;81(1):53–62. https://doi.org/10.1016/0092-8674(95)90370-4.
33. Hall A. Rho GTPases and the actin cytoskeleton. Science. 1998;279(5350):509–14. https://doi.org/10.1126/science.279.5350.509.
34. Cherfils J, Zeghouf M. Regulation of small GTPases by GEFs, GAPs, and GDIs. Physiol Rev. 2013;93(1): 269–309. https://doi.org/10.1152/physrev.00003.2012.
35. Nobes CD, Hall A. Rho GTPases control polarity, protrusion, and adhesion during cell movement. J Cell Biol. 1999;144(6):1235–44. https://doi.org/10.1083/jcb.144.6.1235.
36. Stradal TEB, Scita G. Protein complexes regulating Arp2/3-mediated actin assembly. Curr Opin Cell Biol. 2006;18(1):4–10. https://doi.org/10.1016/j.ceb.2005.12.003.
37. Haga RB, Ridley AJ. Rho GTPases: regulation and roles in cancer cell biology. Small GTPases. 2016;7(4):207–21. https://doi.org/10.1080/21541248.2016.1232583.
38. Liao JK, Seto M, Noma K. Rho kinase (ROCK) inhibitors. J Cardiovasc Pharmacol. 2007;50(1): 17–24. https://doi.org/10.1097/FJC.0b013e318070d1bd.
39. Amin E, Dubey BN, Zhang SC, Gremer L, Dvorsky R, Moll JM, et al. Rho-kinase: regulation, (dys)function, and inhibition. Biol Chem. 2013;394(11):1399–410. https://doi.org/10.1515/hsz-2013-0181.
40. Hall JE, Ha ME. Guyton and Hall textbook of medical physiology. 14th ed. Philadelphia, PA, Elsevier; 2020.
41. FDA. Rocklatan(Netarsudil + Latanoprost) Approval Overview 2019. Available from: https://www.accessdata.fda.gov/scripts/cder/daf/index.cfm?event=overview.process&ApplNo=208259.
42. FDA. Rhopressa(Netarsudil) Approval Overview 2017. Available from: https://www.accessdata.fda.gov/scripts/cder/daf/index.cfm?event=overview.process&ApplNo=208254.
43. Pagano L, Lee JW, Posarelli M, Giannaccare G, Kaye S, Borgia A. ROCK inhibitors in corneal diseases and glaucoma-a comprehensive review of these emerging drugs. J Clin Med. 2023;12(21) https://doi.org/10.3390/jcm12216736.
44. Winkler NS, Fautsch MP. Effects of prostaglandin analogues on aqueous humor outflow pathways. J Ocul Pharmacol Ther. 2014;30(2–3):102–9. https://doi.org/10.1089/jop.2013.0179.
45. Kumar R, Sanawar R, Li X, Li F. Structure, biochemistry, and biology of PAK kinases. Gene. 2017;605: 20–31. https://doi.org/10.1016/j.gene.2016.12.014.
46. Rudolph J, Crawford JJ, Hoeflich KP, Chernoff J. p21-activated kinase inhibitors. Enzyme. 2013;34 Pt. B:157–80. https://doi.org/10.1016/b978-0-12-420146-0.00007-x.
47. Chetty AK, Ha BH, Boggon TJ. Rho family GTPase signaling through type II p21-activated kinases. Cell Mol Life Sci. 2022;79(12):598. https://doi.org/10.1007/s00018-022-04618-2.
48. Marks D. Marks' essentials of medical biochemistry: a clinical approach. Philadelphia, PA. Lippincott Williams & Wilkins (LWW); 2017.
49. Glaviano A, Foo ASC, Lam HY, Yap KCH, Jacot W, Jones RH, et al. PI3K/AKT/mTOR signaling transduction pathway and targeted therapies in cancer. Mol Cancer. 2023;22(1):138. https://doi.org/10.1186/s12943-023-01827-6.
50. Sakamoto K, Holman GD. Emerging role for AS160/TBC1D4 and TBC1D1 in the regulation of GLUT4 traffic. Am J Physiol Endocrinol Metab. 2008;295(1): E29–37. https://doi.org/10.1152/ajpendo.90331.2008.
51. Mîinea CP, Sano H, Kane S, Sano E, Fukuda M, Peränen J, et al. AS160, the Akt substrate regulating GLUT4 translocation, has a functional Rab GTPase-activating protein domain. Biochem J. 2005;391(Pt 1):87–93. https://doi.org/10.1042/bj20050887.
52. Saltiel AR. Insulin signaling in health and disease. J Clin Invest. 2021;131(1) https://doi.org/10.1172/jci142241.
53. Saxton RA, Sabatini DM. mTOR signaling in growth, metabolism, and disease. Cell. 2017;168(6): 960–76. https://doi.org/10.1016/j.cell.2017.02.004.
54. Durán RV, Hall MN. Regulation of TOR by small GTPases. EMBO Rep. 2012;13(2):121–8. https://doi.org/10.1038/embor.2011.257.
55. Nguyen TP, Frank AR, Jewell JL. Amino acid and small GTPase regulation of mTORC1. Cell Logist. 2017;7(4):e1378794. https://doi.org/10.1080/21592799.2017.1378794.
56. Hindupur SK, González A, Hall MN. The opposing actions of target of rapamycin and AMP-activated protein kinase in cell growth control. Cold Spring Harb Perspect Biol. 2015;7(8):a019141. https://doi.org/10.1101/cshperspect.a019141.
57. Vogt PK, Hart JR, Gymnopoulos M, Jiang H, Kang S, Bader AG, et al. Phosphatidylinositol 3-kinase: the oncoprotein. Curr Top Microbiol Immunol. 2010;347:79–104. https://doi.org/10.1007/82_2010_80.
58. Zhang X, Vadas O, Perisic O, Anderson KE, Clark J, Hawkins PT, et al. Structure of lipid kinase p110β/p85β elucidates an unusual SH2-domain-mediated inhibitory mechanism. Mol Cell. 2011;41(5): 567–78. https://doi.org/10.1016/j.molcel.2011.01.026.
59. Doepfner KT, Boller D, De Laurentiis A, Guerreiro AS, Marinov M, Arcaro A. Recent patents of gene sequences relative to the phosphatidylinositol 3-kinase/Akt pathway and their relevance to drug

discovery. Recent Pat DNA Gene Seq. 2007;1(1): 9–23. https://doi.org/10.2174/187221507779814461.
60. Jean S, Kiger AA. Classes of phosphoinositide 3-kinases at a glance. J Cell Sci. 2014;127(5): 923–8. https://doi.org/10.1242/jcs.093773.
61. Vanhaesebroeck B, Perry MWD, Brown JR, André F, Okkenhaug K. PI3K inhibitors are finally coming of age. Nat Rev Drug Discov. 2021;20(10): 741–69. https://doi.org/10.1038/s41573-021-00209-1.
62. FDA. JOENJA(Leniolisib) Approval Overview 2023. Available from: https://www.accessdata.fda.gov/scripts/cder/daf/index.cfm?event=overview.process&ApplNo=217759.
63. Pharming Healthcare Inc. Joenja website. Available from: https://joenja.com/.
64. Schwartz D, Bonelli M, Gadina. Type I/II cytokines, JAKs, and new strategies for treating autoimmune diseases. Nat Rev Rheumatol. 2016;12(25-36) https://doi.org/10.1038/nrrheum.2015.167.
65. Henry SP, Jorgensen WL. Progress on the pharmacological targeting of Janus Pseudokinases. J Med Chem. 2023;66(16):10959–90. https://doi.org/10.1021/acs.jmedchem.3c00926.
66. Roskoski R Jr. Deucravacitinib is an allosteric TYK2 protein kinase inhibitor FDA-approved for the treatment of psoriasis. Pharmacol Res. 2023;189:106642. https://doi.org/10.1016/j.phrs.2022.106642.
67. Schwartz DM, Kanno Y, Villarino A, Ward M, Gadina M, O'Shea JJ. JAK inhibition as a therapeutic strategy for immune and inflammatory diseases. Nat Rev Drug Discov. 2017;17(1):78. https://doi.org/10.1038/nrd.2017.267.
68. Shang L, Cao J, Zhao S, Zhang J, He Y. TYK2 in immune responses and treatment of Psoriasis. J Inflamm Res. 2022;15:5373–85. https://doi.org/10.2147/jir.S380686.
69. FDA. Sotyktu(Deucravacitinib) Approval Overview 2022. Available from: https://www.accessdata.fda.gov/scripts/cder/daf/index.cfm?event=overview.process&ApplNo=214958.
70. Catlett IM, Aras U, Hansen L, Liu Y, Bei D, Girgis IG, et al. First-in-human study of deucravacitinib: a selective, potent, allosteric small-molecule inhibitor of tyrosine kinase 2. Clin Transl Sci. 2023;16(1): 151–64. https://doi.org/10.1111/cts.13435.
71. Hoy SM. Deucravacitinib: first approval. Drugs. 2022;82(17):1671–9. https://doi.org/10.1007/s40265-022-01796-y.
72. Granau AM, Boye TL, Jensen KB, Nielsen OH. Deucravacitinib (Sotyktu™) for plaque psoriasis. Trends Pharmacol Sci. 2023;44(4):252–3. https://doi.org/10.1016/j.tips.2023.01.004.
73. Ghoreschi K, Balato A, Enerbäck C, Sabat R. Therapeutics targeting the IL-23 and IL-17 pathway in psoriasis. Lancet. 2021;397(10275):754–66. https://doi.org/10.1016/s0140-6736(21)00184-7.
74. FDA. Skyrizi(Risankizumab) Approval Overview 2019. Available from: https://www.accessdata.fda.gov/scripts/cder/daf/index.cfm?event=overview.process&ApplNo=761262.
75. D'Haens G, Panaccione R, Baert F, Bossuyt P, Colombel JF, Danese S, et al. Risankizumab as induction therapy for Crohn's disease: results from the phase 3 ADVANCE and MOTIVATE induction trials. Lancet. 2022;399(10340):2015–30. https://doi.org/10.1016/s0140-6736(22)00467-6.
76. Menter A, Strober BE, Kaplan DH, Kivelevitch D, Prater EF, Stoff B, et al. Joint AAD-NPF guidelines of care for the management and treatment of psoriasis with biologics. J Am Acad Dermatol. 2019;80(4): 1029–72. https://doi.org/10.1016/j.jaad.2018.11.057.
77. FDA. Omvoh(Mirikizumab) Approval Overview 2023. Available from: https://www.accessdata.fda.gov/scripts/cder/daf/index.cfm?event=overview.process&ApplNo=761279.
78. FDA. Ilumya(Tildrakizumab) Approval Overview 2018. Available from: https://www.accessdata.fda.gov/scripts/cder/daf/index.cfm?event=overview.process&ApplNo=761067.
79. FDA. Tremfya(Guselkumab) Approval Overview 2017. Available from: https://www.accessdata.fda.gov/scripts/cder/daf/index.cfm?event=overview.process&ApplNo=761061.
80. Krueger JG, Eyerich K, Kuchroo VK, Ritchlin CT, Abreu MT, Elloso MM, et al. IL-23 past, present, and future: a roadmap to advancing IL-23 science and therapy. Front Immunol. 2024;15:1331217. https://doi.org/10.3389/fimmu.2024.1331217.
81. FDA. Dupixent(Dupilumab) Approval overview 2017. Available from: https://www.accessdata.fda.gov/scripts/cder/daf/index.cfm?event=overview.process&ApplNo=761055.
82. Harb H, Chatila TA. Mechanisms of Dupilumab. Clin Exp Allergy. 2020;50(1):5–14. https://doi.org/10.1111/cea.13491.
83. FDA. Rinvoq(Upadacitinib) Approval Overview 2019. Available from: https://www.accessdata.fda.gov/scripts/cder/daf/index.cfm?event=overview.process&ApplNo=211675.
84. Mohamed MF, Bhatnagar S, Parmentier JM, Nakasato P, Wung P. Upadacitinib: mechanism of action, clinical, and translational science. Clin Transl Sci. 2024;17(1):e13688. https://doi.org/10.1111/cts.13688.
85. Tefferi A. Primary myelofibrosis: 2023 update on diagnosis, risk-stratification, and management. Am J Hematol. 2023;98(5):801–21. https://doi.org/10.1002/ajh.26857.
86. Perner F, Perner C, Ernst T, Heidel FH. Roles of JAK2 in aging, inflammation, hematopoiesis and malignant transformation. Cells. 2019;8(8) https://doi.org/10.3390/cells8080854.
87. Mascarenhas J. Pacritinib for the treatment of patients with myelofibrosis and thrombocytopenia. Expert

Rev Hematol. 2022;15(8):671–84. https://doi.org/10.1080/17474086.2022.2112565.
88. Hart S, Goh KC, Novotny-Diermayr V, Tan YC, Madan B, Amalini C, et al. Pacritinib (SB1518), a JAK2/FLT3 inhibitor for the treatment of acute myeloid leukemia. Blood Cancer J. 2011;1(11):e44. https://doi.org/10.1038/bcj.2011.43.
89. FDA. Ojjaara(Momelotinib) Approval Overview 2023. Available from: https://www.accessdata.fda.gov/scripts/cder/daf/index.cfm?event=overview.process&ApplNo=216873.
90. Keam SJ. Momelotinib: First Approval. Drugs. 2023;83(18):1709–15. https://doi.org/10.1007/s40265-023-01964-8.
91. FDA. Jakafi(Ruxolitinib) Approval Overview 2011. Available from: https://www.accessdata.fda.gov/scripts/cder/daf/index.cfm?event=overview.process&ApplNo=202192.
92. Incyte Corporation. Jakafi® (ruxolitinib) Available from: https://www.jakafi.com/.
93. Wolfe L. Ruxolitinib in Myelofibrosis and Polycythemia Vera. J Adv Pract Oncol. 2016;7(4):436–44.
94. FDA. Xeljanz(Tofacitinib) Approval Overview 2012. Available from: https://www.accessdata.fda.gov/scripts/cder/daf/index.cfm?event=overview.process&ApplNo=203214.
95. FDA. Olumiant(Baricitinib) Approval Overview 2018. Available from: https://www.accessdata.fda.gov/scripts/cder/daf/index.cfm?event=overview.process&ApplNo=207924.
96. Berekmeri A, Mahmood F, Wittmann M, Helliwell P. Tofacitinib for the treatment of psoriasis and psoriatic arthritis. Expert Rev Clin Immunol. 2018;14(9):719–30. https://doi.org/10.1080/1744666x.2018.1512404.
97. King B, Ohyama M, Kwon O, Zlotogorski A, Ko J, Mesinkovska NA, et al. Two phase 3 trials of Baricitinib for Alopecia Areata. N Engl J Med. 2022;386(18):1687–99. https://doi.org/10.1056/NEJMoa2110343.
98. Kubo S, Nakayamada S, Tanaka Y. JAK inhibitors for rheumatoid arthritis. Expert Opin Investig Drugs. 2023;32(4):333–44. https://doi.org/10.1080/13543784.2023.2199919.
99. Ritlecitinib (Litfulo) for severe alopecia areata. Med Lett Drugs Ther. 2023;65(1690):185–6. https://doi.org/10.58347/tml.2023.1690a.
100. Robinson MF, Damjanov N, Stamenkovic B, Radunovic G, Kivitz A, Cox L, et al. Efficacy and safety of PF-06651600 (Ritlecitinib), a novel JAK3/TEC inhibitor, in patients with moderate-to-severe rheumatoid arthritis and an inadequate response to methotrexate. Arthritis Rheumatol. 2020;72(10):1621–31. https://doi.org/10.1002/art.41316.
101. FDA. Leqselvi(Deuruxolitinib) Approval Overview 2024. Available from: https://www.accessdata.fda.gov/scripts/cder/daf/index.cfm?event=overview.process&ApplNo=217900.
102. King BA, Craiglow BG. Janus kinase inhibitors for alopecia areata. J Am Acad Dermatol. 2023;89(2s):S29–s32. https://doi.org/10.1016/j.jaad.2023.05.049.
103. Webster JD, Vucic D. The balance of TNF mediated pathways regulates inflammatory cell death signaling in healthy and diseased tissues. Front Cell Dev Biol. 2020;8:365. https://doi.org/10.3389/fcell.2020.00365.
104. Hayden MS, Ghosh S. Regulation of NF-κB by TNF family cytokines. Semin Immunol. 2014;26(3):253–66. https://doi.org/10.1016/j.smim.2014.05.004.
105. Hayden MS, Ghosh S. NF-κB, the first quarter-century: remarkable progress and outstanding questions. Genes Dev. 2012;26(3):203–34. https://doi.org/10.1101/gad.183434.111.
106. Lim KH, Staudt LM. Toll-like receptor signaling. Cold Spring Harb Perspect Biol. 2013;5(1):a011247. https://doi.org/10.1101/cshperspect.a011247.
107. Mitchell S, Vargas J, Hoffmann A. Signaling via the NFκB system. Wiley Interdiscip Rev Syst Biol Med. 2016;8(3):227–41. https://doi.org/10.1002/wsbm.1331.
108. FDA. HUMIRA (adalimumab) Approval Overview 1001. Available from: https://www.accessdata.fda.gov/scripts/cder/daf/index.cfm?event=overview.process&ApplNo=125057.
109. Whalen K, Lerchenfeldt S, Giordano C. Lippincott illustrated reviews. Philadelphia, PA. Pharmacology Kluwer; 2021.
110. Horiuchi T, Mitoma H, Harashima S, Tsukamoto H, Shimoda T. Transmembrane TNF-alpha: structure, function and interaction with anti-TNF agents. Rheumatology (Oxford). 2010;49(7):1215–28. https://doi.org/10.1093/rheumatology/keq031.
111. FDA. Enbrel (etanercept): Drugs@FDA Approval Overview. Silver Spring: U.S. Food and Drug Administration; 1998. Available from: https://www.accessdata.fda.gov/scripts/cder/daf/index.cfm?event=overview.process&ApplNo=103795.
112. Amgen, Inc. Enbrel Mechanism of action: https://www.enbrelpro.com/mechanism-of-action
113. Ansaldi D, Hod EA, Stellari F, Kim JB, Lim E, Roskey M, et al. Imaging pulmonary NF-kappaB activation and therapeutic effects of MLN120B and TDZD-8. PLoS One. 2011;6(9):e25093. https://doi.org/10.1371/journal.pone.0025093.
114. Hideshima T, Chauhan D, Kiziltepe T, Ikeda H, Okawa Y, Podar K, et al. Biologic sequelae of I {kappa}B kinase (IKK) inhibition in multiple myeloma: therapeutic implications. Blood. 2009;113(21):5228–36. https://doi.org/10.1182/blood-2008-06-161505.
115. Wen D, Nong Y, Morgan JG, Gangurde P, Bielecki A, DaSilva J, et al. A selective small molecule IκB Kinase β inhibitor blocks nuclear factor κB-mediated inflammatory responses in human fibroblast-like Synoviocytes, chondrocytes, and

mast cells. J Pharmacol Exp Ther. 2006;317(3): 989–1001. https://doi.org/10.1124/jpet.105.097584.
116. Zhao W, Qian J, Li J, Su T, Deng X, Fu Y, et al. From death to birth: how osteocyte death promotes osteoclast formation. Front Immunol. 2025;16:1551542. https://doi.org/10.3389/fimmu.2025.1551542.
117. FDA. Prolia(Denosumab) Approval Overview 2010. Available from: https://www.accessdata.fda.gov/scripts/cder/daf/index.cfm?event=overview.process&ApplNo=125320.
118. Chi Y, Li Y, Cheng Q, Cai H, Zeng Y, Sheng H, et al. Efficacy and safety of narlumosbart, an anti-RANKL monoclonal antibody, in postmenopausal women with osteoporosis: a multi-center, randomized, double-blind, placebo- and active-controlled, phased II study. EClinicalMedicine. 2025;85:103329. https://doi.org/10.1016/j.eclinm.2025.103329.
119. Gachpazan M, Habbibirad S, Kashani H, Jamialahmadi T, Rahimi HR, Sahebkar A. Targeting nuclear factor-Kappa B signaling pathway by Curcumin: implications for the treatment of multiple sclerosis. Adv Exp Med Biol. 2021;1291: 41–53. https://doi.org/10.1007/978-3-030-56153-6_3.
120. Buhrmann C, Mobasheri A, Busch F, Aldinger C, Stahlmann R, Montaseri A, et al. Curcumin modulates nuclear factor kappaB (NF-kappaB)-mediated inflammation in human tenocytes in vitro: role of the phosphatidylinositol 3-kinase/Akt pathway. J Biol Chem. 2011;286(32):28556–66. https://doi.org/10.1074/jbc.M111.256180.
121. Manna SK, Mukhopadhyay A, Aggarwal BB. Resveratrol suppresses TNF-induced activation of nuclear transcription factors NF-κB, activator protein-1, and apoptosis: potential role of reactive oxygen intermediates and lipid Peroxidation1. J Immunol. 2000;164(12):6509–19. https://doi.org/10.4049/jimmunol.164.12.6509.
122. Yeung F, Hoberg JE, Ramsey CS, Keller MD, Jones DR, Frye RA, et al. Modulation of NF-kappaB-dependent transcription and cell survival by the SIRT1 deacetylase. EMBO J. 2004;23(12):2369–80. https://doi.org/10.1038/sj.emboj.7600244.
123. Wrana JL. Signaling by the TGFβ superfamily. Cold Spring Harb Perspect Biol. 2013;5(10):a011197. https://doi.org/10.1101/cshperspect.a011197.
124. Tschernia NP, Gulley JL. Tumor in the crossfire: inhibiting TGF-β to enhance cancer immunotherapy. BioDrugs. 2022;36(2):153–80. https://doi.org/10.1007/s40259-022-00521-1.
125. FDA. Winrevair(Sotatercept) Approval Overview 2024. Available from: https://www.accessdata.fda.gov/scripts/cder/daf/index.cfm?event=overview.process&ApplNo=761363.
126. Kang C. Sotatercept: first approval. Drugs. 2024;84(7):857–62. https://doi.org/10.1007/s40265-024-02058-9.
127. Yung LM, Yang P, Joshi S, Augur ZM, Kim SSJ, Bocobo GA, et al. ACTRIIA-Fc rebalances activin/GDF versus BMP signaling in pulmonary hypertension. Sci Transl Med. 2020;12(543) https://doi.org/10.1126/scitranslmed.aaz5660.
128. Madonna R, Biondi F. Sotatercept: new drug on the horizon of pulmonary hypertension. Vasc Pharmacol. 2024;157:107442. https://doi.org/10.1016/j.vph.2024.107442.
129. Hoeper MM, Badesch DB, Ghofrani HA, Gibbs JSR, Gomberg-Maitland M, McLaughlin VV, et al. Phase 3 trial of Sotatercept for treatment of pulmonary arterial hypertension. N Engl J Med. 2023;388(16): 1478–90. https://doi.org/10.1056/NEJMoa2213558.
130. Ryanto GRT, Musthafa A, Hara T, Emoto N. Inactivating the uninhibited: the tale of activins and Inhibins in pulmonary arterial hypertension. Int J Mol Sci. 2023;24(4) https://doi.org/10.3390/ijms24043332.
131. Kaur M, Misra S. Bimagrumab: an investigational human monoclonal antibody against activin type II receptors for treating obesity. J Basic Clin Physiol Pharmacol. 2024;35(6):325–34. https://doi.org/10.1515/jbcpp-2024-0065.
132. Heymsfield SB, Coleman LA, Miller R, Rooks DS, Laurent D, Petricoul O, et al. Effect of Bimagrumab vs placebo on body fat mass among adults with type 2 diabetes and obesity: a phase 2 randomized clinical trial. JAMA Netw Open. 2021;4(1):e2033457. https://doi.org/10.1001/jamanetworkopen.2020.33457.
133. Kanbay M, Siriopol D, Copur S, Hasbal NB, Güldan M, Kalantar-Zadeh K, et al. Effect of Bimagrumab on body composition: a systematic review and meta-analysis. Aging Clin Exp Res. 2024;36(1):185. https://doi.org/10.1007/s40520-024-02825-4.
134. Kopan R, Ilagan MX. The canonical Notch signaling pathway: unfolding the activation mechanism. Cell. 2009;137(2):216–33. https://doi.org/10.1016/j.cell.2009.03.045.
135. Bray SJ, Gomez-Lamarca M. Notch after cleavage. Curr Opin Cell Biol. 2018;51:103–9. https://doi.org/10.1016/j.ceb.2017.12.008.
136. Zhou B, Lin W, Long Y, Yang Y, Zhang H, Wu K, et al. Notch signaling pathway: architecture, disease, and therapeutics. Signal Transduct Target Ther. 2022;7(1):95. https://doi.org/10.1038/s41392-022-00934-y.
137. Vasyutina E, Lenhard DC, Wende H, Erdmann B, Epstein JA, Birchmeier C. RBP-J (Rbpsuh) is essential to maintain muscle progenitor cells and to generate satellite cells. Proc Natl Acad Sci USA. 2007;104(11):4443–8. https://doi.org/10.1073/pnas.0610647104.
138. Sun Y, Islam S, Michikawa M, Zou K. Presenilin: a multi-functional molecule in the pathogenesis of Alzheimer's disease and other neurodegenerative diseases. Int J Mol Sci. 2024;25(3) https://doi.org/10.3390/ijms25031757.

139. Yen PM. Classical nuclear hormone receptor activity as a mediator of complex biological responses: a look at health and disease. Best Pract Res Clin Endocrinol Metab. 2015;29(4):517–28. https://doi.org/10.1016/j.beem.2015.07.005.
140. Meier CA. Regulation of gene expression by nuclear hormone receptors. J Recept Signal Transduct Res. 1997;17(1–3):319–35. https://doi.org/10.3109/10799899709036612.
141. Fuentes N, Silveyra P. Estrogen receptor signaling mechanisms. Adv Protein Chem Struct Biol. 2019;116:135–70. https://doi.org/10.1016/bs.apcsb.2019.01.001.
142. Qureshi R, Picon-Ruiz M, Aurrekoetxea-Rodriguez I, Nunes de Paiva V, D'Amico M, Yoon H, et al. The major pre- and postmenopausal Estrogens play opposing roles in obesity-driven mammary inflammation and breast cancer development. Cell Metab. 2020;31(6):1154–72.e9. https://doi.org/10.1016/j.cmet.2020.05.008.
143. Clusan L, Ferrière F, Flouriot G, Pakdel F. A basic review on Estrogen receptor signaling pathways in breast cancer. Int J Mol Sci. 2023;24(7) https://doi.org/10.3390/ijms24076834.
144. Patel HK, Bihani T. Selective estrogen receptor modulators (SERMs) and selective estrogen receptor degraders (SERDs) in cancer treatment. Pharmacol Ther. 2018;186:1–24. https://doi.org/10.1016/j.pharmthera.2017.12.012.
145. Wang L, Guillen VS, Sharma N, Flessa K, Min J, Carlson KE, et al. New class of Selective Estrogen Receptor Degraders (SERDs): expanding the toolbox of PROTAC Degrons. ACS Med Chem Lett. 2018;9(8):803–8. https://doi.org/10.1021/acsmedchemlett.8b00106.
146. FDA. Fareston(Toremifene) Approval Overview 1997. Available from: https://www.accessdata.fda.gov/scripts/cder/daf/index.cfm?event=overview.process&ApplNo=020497.
147. FDA. NOLVADEX (tamoxifen) Approval Overview 1977. Available from: https://www.accessdata.fda.gov/scripts/cder/daf/index.cfm?event=overview.process&ApplNo=17970.
148. Goodsell DS. The molecular perspective: tamoxifen and the estrogen receptor. Stem Cells. 2002;20(3):267–8. https://doi.org/10.1634/stemcells.20-3-267.
149. FDA. Oserdu(Elacestrant) Approval Overview 2023. Available from: https://www.accessdata.fda.gov/scripts/cder/daf/index.cfm?event=overview.process&ApplNo=217639.
150. FDA. Faslodex(Fulvestrant) Approval Overview 2002. Available from: https://www.accessdata.fda.gov/scripts/cder/daf/index.cfm?event=overview.process&ApplNo=021344.
151. Hosford SR, Shee K, Wells JD, Traphagen NA, Fields JL, Hampsch RA, et al. Estrogen therapy induces an unfolded protein response to drive cell death in ER+ breast cancer. Mol Oncol. 2019;13(8):1778–94. https://doi.org/10.1002/1878-0261.12528.
152. Ferro A, Generali D, Caffo O, Caldara A, De Lisi D, Dipasquale M, et al. Oral selective estrogen receptor degraders (SERDs): the new emperors in breast cancer clinical practice? Semin Oncol. 2023;50(3):90–101. https://doi.org/10.1053/j.seminoncol.2023.08.002.
153. FDA. FDA approves elacestrant for ER-positive, HER2-negative, ESR1-mutated advanced or metastatic breast cancer 2023. Available from: https://www.fda.gov/drugs/resources-information-approved-drugs/fda-approves-elacestrant-er-positive-her2-negative-esr1-mutated-advanced-or-metastatic-breast-cancer#:~:text=On%20January%2027%2C%202023%2C%20the,inhibitor%20(n=73).
154. Stemline Therapeutics, Inc. Oserdu. Available from: https://www.orserduhcp.com/.
155. FDA. Arimidex(Anastrozole) Approval overview 1995. Available from: https://www.accessdata.fda.gov/scripts/cder/daf/index.cfm?event=overview.process&ApplNo=020541.
156. FDA. Aromasin(Exemestane) Approval overview 1999. Available from: https://www.accessdata.fda.gov/scripts/cder/daf/index.cfm?event=overview.process&ApplNo=020753.
157. FDA. Femara(Letrozole) Approval Overview. 2001. https://www.accessdata.fda.gov/scripts/cder/daf/index.cfm?event=overview.process&ApplNo=020726
158. Wendler, R. Diagnosed with breast cancer after menopause? Aromatase inhibitors can help 2022. Available from: https://www.mdanderson.org/cancerwise/diagnosed-with-breast-cancer-after-menopause%2D%2Daromatase-inhibitors-can-help.h00-159542112.html.
159. Brueggemeier RW. Aromatase, aromatase inhibitors, and breast cancer. Am J Ther. 2001;8(5):333–44. https://doi.org/10.1097/00045391-200109000-00007.
160. DrugBank. Anastrozole. Available from: https://go.drugbank.com/drugs/DB01217?utm.
161. Drugbank. Letrozole. Available from: https://go.drugbank.com/drugs/DB01006.
162. Lombardi P. Exemestane, a new steroidal aromatase inhibitor of clinical relevance. Biochim Biophys Acta (BBA) – Mol Basis Dis. 2002;1587(2):326–37. https://doi.org/10.1016/S0925-4439(02)00096-0.
163. Bush NJ, Griffin-Sobel JP. Hormone replacement therapy in postmenopausal women. Oncol Nurs Forum. 2003;30(5):747–9. https://doi.org/10.1188/03.Onf.747-749.
164. Sourouni M, Kiesel L. Menopausal hormone therapy and the breast: a review of clinical studies. Breast Care (Basel). 2023;18(3):164–71. https://doi.org/10.1159/000530205.
165. Lockett J, Inder WJ, Clifton VL. The glucocorticoid receptor: isoforms, functions, and contribution to glucocorticoid sensitivity. Endocr Rev. 2024;45(4):593–624. https://doi.org/10.1210/endrev/bnae008.

166. Agarwal MK, Mirshahi M. General overview of mineralocorticoid hormone action. Pharmacol Ther. 1999;84(3):273–326. https://doi.org/10.1016/S0163-7258(99)00038-8.
167. Sevilla LM, Pérez P. Roles of the glucocorticoid and mineralocorticoid receptors in skin pathophysiology. Int J Mol Sci. 2018;19(7):1906. https://doi.org/10.3390/ijms19071906.
168. FDA. Cortef(Hydrocortisone) Approval overview 1952. Available from: https://www.accessdata.fda.gov/scripts/cder/daf/index.cfm?event=overview.process&ApplNo=008697.
169. Drugs.com. Generic Cortef Availability 2025. Available from: https://www.drugs.com/availability/generic-cortef.html.
170. FDA. Deltasone (prednisone) Approval Overview. 1955. https://www.accessdata.fda.gov/drugsatfda_docs/nda/2012/202020Orig1s000ClinPharmR.pdf?utm
171. Duan D, Goemans N, Takeda S, Mercuri E, Aartsma-Rus A. Duchenne muscular dystrophy. Nat Rev Dis Primers. 2021;7(1):13. https://doi.org/10.1038/s41572-021-00248-3.
172. Keam SJ. Vamorolone: first approval. Drugs. 2024;84(1):111–7. https://doi.org/10.1007/s40265-023-01986-2.
173. Kourakis S, Timpani CA, Campelj DG, Hafner P, Gueven N, Fischer D, et al. Standard of care versus new-wave corticosteroids in the treatment of Duchenne muscular dystrophy: can we do better? Orphanet J Rare Dis. 2021;16(1):117. https://doi.org/10.1186/s13023-021-01758-9.
174. Heier CR, Yu Q, Fiorillo AA, Tully CB, Tucker A, Mazala DA, et al. Vamorolone targets dual nuclear receptors to treat inflammation and dystrophic cardiomyopathy. Life Sci Alliance. 2019;2(1) https://doi.org/10.26508/lsa.201800186.
175. Gerber AN, Newton R, Sasse SK. Repression of transcription by the glucocorticoid receptor: a parsimonious model for the genomics era. J Biol Chem. 2021:296. https://doi.org/10.1016/j.jbc.2021.100687.
176. Kaur S, Bansal Y, Kumar R, Bansal G. A panoramic review of IL-6: structure, pathophysiological roles and inhibitors. Bioorg Med Chem. 2020;28(5):115327. https://doi.org/10.1016/j.bmc.2020.115327.
177. Chang WT, Hong MY, Chen CL, Hwang CY, Tsai CC, Chuang CC. Mutant glucocorticoid receptor binding elements on the interleukin-6 promoter regulate dexamethasone effects. BMC Immunol. 2021;22(1):24. https://doi.org/10.1186/s12865-021-00413-z.
178. Fuster JJ, Walsh K. The good, the bad, and the ugly of interleukin-6 signaling. EMBO J. 2014;33(13):1425-7-7. https://doi.org/10.15252/embj.201488856.
179. Larigot L, Juricek L, Dairou J, Coumoul X. AhR signaling pathways and regulatory functions. Biochimie Open. 2018;7(1–9) https://doi.org/10.1016/j.biopen.2018.05.001.
180. Fernández-Gallego N, Sánchez-Madrid F, Cibrian D. Role of AHR ligands in skin homeostasis and cutaneous inflammation. Cells. 2021;10(11) https://doi.org/10.3390/cells10113176.
181. FDA. Vtama(Tapinarof) Approval Overview 2022. Available from: https://www.accessdata.fda.gov/scripts/cder/daf/index.cfm?event=overview.process&ApplNo=215272.
182. Keam SJ. Tapinarof cream 1%: first approval. Drugs. 2022;82(11):1221–8. https://doi.org/10.1007/s40265-022-01748-6.
183. Furue M, Hashimoto-Hachiya A, Tsuji G. Aryl hydrocarbon receptor in Atopic Dermatitis and Psoriasis. Int J Mol Sci. 2019;20(21) https://doi.org/10.3390/ijms20215424.
184. Grygiel-Górniak B. Peroxisome proliferator-activated receptors and their ligands: nutritional and clinical implications--a review. Nutr J. 2014;13:17. https://doi.org/10.1186/1475-2891-13-17.
185. Li Y, Pan Y, Zhao X, Wu S, Li F, Wang Y, et al. Peroxisome proliferator-activated receptors: a key link between lipid metabolism and cancer progression. Clin Nutr. 2024;43(2):332–45. https://doi.org/10.1016/j.clnu.2023.12.005.
186. FDA. Tricor(Fenofibrate) Approval Overview 1993. Available from: https://www.accessdata.fda.gov/scripts/cder/daf/index.cfm?event=overview.process&ApplNo=021656.
187. Rosenson RS. Fenofibrate: treatment of hyperlipidemia and beyond. Expert Rev Cardiovasc Ther. 2008;6(10):1319–30. https://doi.org/10.1586/14779072.6.10.1319.
188. FDA. Actos(Pioglitazone) Approval overview 1999. Available from: https://www.accessdata.fda.gov/scripts/cder/daf/index.cfm?event=overview.process&ApplNo=021073.
189. FDA. Avandia(Rosiglitazone) Approval overview 1999. Available from: https://www.accessdata.fda.gov/scripts/cder/daf/index.cfm?event=overview.process&ApplNo=021071.
190. Chiarelli F, Di Marzio D. Peroxisome proliferator-activated receptor-gamma agonists and diabetes: current evidence and future perspectives. Vasc Health Risk Manag. 2008;4(2):297–304. https://doi.org/10.2147/vhrm.s993.
191. Hong F, Xu P, Zhai Y. The opportunities and challenges of peroxisome proliferator-activated receptors ligands in clinical drug discovery and development. Int J Mol Sci. 2018;19(8) https://doi.org/10.3390/ijms19082189.
192. FDA. Iqirvo(Elafibranor) Approval Overview 2024. Available from: https://www.accessdata.fda.gov/scripts/cder/daf/index.cfm?event=overview.process&ApplNo=218860.
193. FDA. Livdelzi (Seladelpar) Approval Overview 2024. Available from: https://www.accessdata.fda.gov/scripts/cder/daf/index.cfm?event=overview.process&ApplNo=217899.

194. IPSEN. Ipsen's Iqirvo® receives U.S. FDA accelerated approval as a first-in-class PPAR treatment for primary biliary cholangitis 2024. Available from: https://www.ipsen.com/press-release/ipsens-iqirvo-receives-u-s-fda-accelerated-approval-as-a-first-in-class-ppar-treatment-for-primary-biliary-cholangitis/
195. Lildelzi. Lower is better with Livdelzi 2024. Available from: https://www.livdelzihcp.com/.
196. Kamata S, Honda A, Ishikawa R, Akahane M, Fujita A, Kaneko C, et al. Functional and structural insights into the human PPARα/δ/γ targeting preferences of anti-NASH investigational drugs, Lanifibranor, Seladelpar, and Elafibranor. Antioxidants (Basel). 2023;12(8) https://doi.org/10.3390/antiox12081523.
197. Hilger D, Masureel M, Kobilka BK. Structure and dynamics of GPCR signaling complexes. Nat Struct Mol Biol. 2018;25(1):4–12. https://doi.org/10.1038/s41594-017-0011-7.
198. Alhosaini K, Azhar A, Alonazi A, Al-Zoghaibi F. GPCRs: the most promiscuous druggable receptor of the mankind. Saudi Pharm J. 2021;29(6):539–51. https://doi.org/10.1016/j.jsps.2021.04.015.
199. Hamm HE. The many faces of G protein signaling *. J Biol Chem. 1998;273(2):669–72. https://doi.org/10.1074/jbc.273.2.669.
200. Insel PA. Structure and function of Alpha-Adrenergic receptors. Am J Med. 1989;87(2):S12–S8. https://doi.org/10.1016/0002-9343(89)90108-3.
201. Nathanson NM. A multiplicity of muscarinic mechanisms: enough signaling pathways to take your breath away. Proc Natl Acad Sci. 2000;97(12):6245–7. https://doi.org/10.1073/pnas.97.12.6245.
202. Beaulieu JM, Gainetdinov RR. The physiology, signaling, and pharmacology of dopamine receptors. Pharmacol Rev. 2011;63(1):182–217. https://doi.org/10.1124/pr.110.002642.
203. Pithadia AB, Jain SM. 5-Hydroxytryptamine receptor subtypes and their modulators with therapeutic potentials. J Clin Med Res. 2009;1(2):72–80. https://doi.org/10.4021/jocmr2009.05.1237.
204. Haas HL, Sergeeva OA, Selbach O. Histamine in the nervous system. Physiol Rev. 2008;88(3):1183–241. https://doi.org/10.1152/physrev.00043.2007.
205. Stein C. Opioid Receptors. Annu Rev Med. 2016;67:433–51. https://doi.org/10.1146/annurev-med-062613-093100.
206. Newton AC, Bootman MD, Scott JD. Second messengers. Cold Spring Harb Perspect Biol. 2016;8(8) https://doi.org/10.1101/cshperspect.a005926.
207. Weis WI, Kobilka BK. The molecular basis of G protein-coupled receptor activation. Annu Rev Biochem. 2018;87:897–919. https://doi.org/10.1146/annurev-biochem-060614-033910.
208. Levitzki A, Marbach I, Bar-Sinai A. The signal transduction between β-receptors and adenylyl cyclase. Life Sci. 1993;52(26):2093–100. https://doi.org/10.1016/0024-3205(93)90724-H.
209. Tank AW, Lee WD. Peripheral and central effects of circulating catecholamines. Compr Physiol. 2015;5(1):1–15. https://doi.org/10.1002/cphy.c140007.
210. Wallukat G. The beta-adrenergic receptors. Herz. 2002;27(7):683–90. https://doi.org/10.1007/s00059-002-2434-z.
211. Siegel GJ, Agranoff BW, Albers RW, Molinoff PB, editors. Basic neurochemistry: molecular, cellular, and medical aspects. 6th ed. Philadelphia, PA. Lippincott-Raven Publishers; 1999.
212. Schena G, Caplan MJ. Everything you always wanted to know about β(3)-AR * (* but were afraid to ask). Cells. 2019;8(4) https://doi.org/10.3390/cells8040357.
213. Velmurugan BK, Baskaran R, Huang C-Y. Detailed insight on β-adrenoceptors as therapeutic targets. Biomed Pharmacother. 2019;117:109039. https://doi.org/10.1016/j.biopha.2019.109039.
214. Dongdem JT, Etornam AE, Beletaa S, Alidu I, Kotey H, Wezena CA. The β(3)-Adrenergic receptor: structure, physiopathology of disease, and emerging therapeutic potential. Adv Pharmacol Pharm Sci. 2024;2024:2005589. https://doi.org/10.1155/2024/2005589.
215. Borea PA, Gessi S, Merighi S, Vincenzi F, Varani K. Pharmacology of Adenosine receptors: the state of the art. Physiol Rev. 2018;98(3):1591–625. https://doi.org/10.1152/physrev.00049.2017.
216. Authier F, Desbuquois B. Glucagon receptors. Cell Mol Life Sci. 2008;65(12):1880–99. https://doi.org/10.1007/s00018-008-7479-6.
217. Pavlos NJ, Friedman PA. GPCR Signaling and trafficking: the long and short of it. Trends Endocrinol Metab. 2017;28(3):213–26. https://doi.org/10.1016/j.tem.2016.10.007.
218. Baldwin TA, Dessauer CW. Function of adenylyl cyclase in heart: the AKAP connection. J Cardiovasc Dev Dis. 2018;5(1):2. https://doi.org/10.3390/jcdd5010002.
219. Wu Y, Zeng L, Zhao S. Ligands of adrenergic receptors: a structural point of view. Biomolecules. 2021;11(7) https://doi.org/10.3390/biom11070936.
220. Morano I. The contractile machines of the heart. Adv Exp Med Biol. 2024;1441:417–33. https://doi.org/10.1007/978-3-031-44087-8_21.
221. Carmen G-Y, Víctor S-M. Signalling mechanisms regulating lipolysis. Cell Signal. 2006;18(4):401–8. https://doi.org/10.1016/j.cellsig.2005.08.009.
222. Miyoshi H, Souza SC, Zhang H-H, Strissel KJ, Christoffolete MA, Kovsan J, et al. Perilipin promotes hormone-sensitive lipase-mediated adipocyte lipolysis via phosphorylation-dependent and -independent mechanisms *. J Biol Chem. 2006;281(23):15837–44. https://doi.org/10.1074/jbc.M601097200.

223. Hussain LS, Reddy V, Maani C. Physiology, noradrenergic synapse. Treasure Island, FL. StatPearls Publishing; 2023.
224. Cleveland Clinic. Epinephrine (adrenaline): What it is, function, deficiency & side effects. Cleveland Clinic 2022. Available from: https://my.clevelandclinic.org/health/articles/22611-epinephrine-adrenaline.
225. Keil MF, Briassoulis G, Stratakis CA, Wu TJ. Protein kinase A and anxiety-related behaviors: a mini-review. Front Endocrinol (Lausanne). 2016;7 (83) https://doi.org/10.3389/fendo.2016.00083.
226. Benchoula K, Parhar IS, Madhavan P, Hwa WE. CREB nuclear transcription activity as a targeting factor in the treatment of diabetes and diabetes complications. Biochem Pharmacol. 2021;188: 114531. https://doi.org/10.1016/j.bcp.2021.114531.
227. Kimura T, Pydi SP, Pham J, Tanaka N. Metabolic functions of G protein-coupled receptors in hepatocytes-potential applications for diabetes and NAFLD. Biomolecules. 2020;10(10) https://doi.org/10.3390/biom10101445.
228. do Vale GT, Ceron CS, Gonzaga NA, Simplicio JA, Padovan JC. Three generations of β-blockers: history, class differences and clinical applicability. Curr Hypertens Rev. 2019;15(1):22–31. https://doi.org/10.2174/1573402114666180918102735.
229. Klabunde RCPC. Beta-Adrenoceptor Antagonists (Beta-Blockers) 2023. Available from: https://cvpharmacology.com/cardioinhibitory/beta-blockers#:~:text=Beta%2Dblockers%20are%20drugs%20that,intrinsic%20sympathomimetic%20activity%20(ISA).
230. FDA. Ventolin(Albuterol) Approval Overview 1981. Available from: https://www.accessdata.fda.gov/scripts/cder/daf/index.cfm?event=overview.process&ApplNo=020983.
231. Billington CK, Penn RB, Hall IP. β(2) Agonists. Handb Exp Pharmacol. 2017;237:23–40. https://doi.org/10.1007/164_2016_64.
232. Marques L, Vale N. Salbutamol in the management of asthma: a review. Int J Mol Sci. 2022;23 (22) https://doi.org/10.3390/ijms232214207.
233. Nguyen H, Glaaser IW, Slesinger PA. Direct modulation of G protein-gated inwardly rectifying potassium (GIRK) channels. Front Physiol. 2024;15: 1386645. https://doi.org/10.3389/fphys.2024.1386645.
234. Pathan H, Williams J. Basic opioid pharmacology: an update. Br J Pain. 2012;6(1):11–6. https://doi.org/10.1177/2049463712438493.
235. Reeves KC, Shah N, Muñoz B, Atwood BK. Opioid receptor-mediated regulation of neurotransmission in the brain. Front Mol Neurosci. 2022;15:919773. https://doi.org/10.3389/fnmol.2022.919773.
236. Lambert DG. Opioids and opioid receptors; understanding pharmacological mechanisms as a key to therapeutic advances and mitigation of the misuse crisis. BJA Open. 2023;6:100141. https://doi.org/10.1016/j.bjao.2023.100141.
237. Peyravian N, Deo S, Daunert S, Jimenez JJ. Cannabidiol as a novel therapeutic for immune modulation. Immunotargets Ther. 2020;9:131–40. https://doi.org/10.2147/itt.S263690.
238. Zou S, Kumar U. Cannabinoid receptors and the endocannabinoid system: signaling and function in the central nervous system. Int J Mol Sci. 2018;19 (3) https://doi.org/10.3390/ijms19030833.
239. Krishna Kumar K, Shalev-Benami M, Robertson MJ, Hu H, Banister SD, Hollingsworth SA, et al. Structure of a signaling Cannabinoid receptor 1-G protein complex. Cell. 2019;176(3):448–58.e12. https://doi.org/10.1016/j.cell.2018.11.040.
240. Lin YF. Potassium channels as molecular targets of endocannabinoids. Channels (Austin). 2021;15(1): 408–23. https://doi.org/10.1080/19336950.2021.1910461.
241. Bhattacharyya M, Karandur D, Kuriyan J. Structural insights into the regulation of Ca(2+)/Calmodulin-dependent protein Kinase II (CaMKII). Cold Spring Harb Perspect Biol. 2020;12(6):10.1101/cshperspect.a035147.
242. Bye AP, Unsworth AJ, Gibbins JM. Platelet signaling: a complex interplay between inhibitory and activatory networks. J Thromb Haemost. 2016;14 (5):918–30. https://doi.org/10.1111/jth.13302.
243. Nakahata N. Thromboxane A2: physiology/pathophysiology, cellular signal transduction and pharmacology. Pharmacol Ther. 2008;118(1):18–35. https://doi.org/10.1016/j.pharmthera.2008.01.001.
244. Offermanns S. Activation of platelet function through G protein-coupled receptors. Circ Res. 2006;99(12): 1293–304. https://doi.org/10.1161/01.Res.0000251742.71301.16.
245. Habibi M. Acetylcholine☆. Reference module in neuroscience and biobehavioral psychology. Amsterdam, Netherlands. Elsevier; 2017.
246. Levitan IB, Kaczmarek LK. The neuron: cell and molecular biology. 4th ed. New York, NY. Oxford University Press; 2015.
247. Shen J-x, Yakel JL. Nicotinic acetylcholine receptor-mediated calcium signaling in the nervous system. Acta Pharmacol Sin. 2009;30(6):673–80. https://doi.org/10.1038/aps.2009.64.
248. Isakov N, Altman A. Regulation of immune system cell functions by protein Kinase C. Front Immunol. 2013;4 https://doi.org/10.3389/fimmu.2013.00384.
249. Vig M, Kinet JP. Calcium signaling in immune cells. Nat Immunol. 2009;10(1):21–7. https://doi.org/10.1038/ni.f.220.
250. Mashimo M, Moriwaki Y, Misawa H, Kawashima K, Fujii T. Regulation of immune functions by non-neuronal acetylcholine (ACh) via muscarinic and nicotinic ACh receptors. Int J Mol Sci. 2021;22 (13) https://doi.org/10.3390/ijms22136818.
251. Molina J, Rodriguez-Diaz R, Fachado A, Jacques-Silva MC, Berggren PO, Caicedo A. Control of insulin secretion by cholinergic signaling in the human pancreatic islet. Diabetes. 2014;63(8):2714–26. https://doi.org/10.2337/db13-1371.

252. Zhu L, Rossi M, Doliba NM, Wess J. Beta-cell M3 muscarinic acetylcholine receptors as potential targets for novel antidiabetic drugs. Int Immunopharmacol. 2020;81:106267. https://doi.org/10.1016/j.intimp.2020.106267.
253. Clinic C. Alpha-blockers 2022. Available from: https://my.clevelandclinic.org/health/treatments/22321-alpha-blockers.
254. Clinic M. Alpha blockers 2023. Available from: https://www.mayoclinic.org/diseases-conditions/high-blood-pressure/in-depth/alpha-blockers/art-20044214#:~:text=Alpha%20blockers%20are%20a%20type,flow%20and%20lowers%20blood%20pressure.
255. Undas A, Brummel-Ziedins KE, Mann KG. Antithrombotic properties of aspirin and resistance to aspirin: beyond strictly antiplatelet actions. Blood. 2007;109(6):2285–92. https://doi.org/10.1182/blood-2006-01-010645.
256. Inc BS. Sofdra™ (sofpironium) topical gel. 2024.
257. patsnap Sb. What is the mechanism of Bethanechol Chloride? 2024. Available from: https://synapse.patsnap.com/article/what-is-the-mechanism-of-bethanechol-chloride.
258. online D. Bethanechol. Available from: https://go.drugbank.com/drugs/DB01019.
259. Kapourani A, Kontogiannopoulos KN, Barmpalexis P. A review on the role of pilocarpine on the management of Xerostomia and the importance of the topical administration systems development. Pharmaceuticals (Basel). 2022;15(6) https://doi.org/10.3390/ph15060762.
260. Turner RJ, Sugiya H. Understanding salivary fluid and protein secretion. Oral Dis. 2002;8(1):3–11. https://doi.org/10.1034/j.1601-0825.2002.10815.x.
261. Yang MC, Lin KY. Drug-induced acute angle-closure glaucoma: a review. J Curr Glaucoma Pract. 2019;13(3):104–9. https://doi.org/10.5005/jp-journals-10078-1261.
262. Bianchi ME, Mezzapelle R. The Chemokine receptor CXCR4 in cell proliferation and tissue regeneration. Front Immunol. 2020;11:2109. https://doi.org/10.3389/fimmu.2020.02109.
263. Foundation ID. WHIM Syndrome. Available from: https://primaryimmune.org/understanding-primary-immunodeficiency/types-of-pi/whim-syndrome.
264. Milanesi S, Locati M, Borroni EM. Aberrant CXCR4 Signaling at crossroad of WHIM syndrome and Waldenstrom's macroglobulinemia. Int J Mol Sci. 2020;21(16) https://doi.org/10.3390/ijms21165696.
265. X4 Pharmaceuticals I. About Xolremdi 2024. Available from: https://xolremdihcp.com/about-xolremdi.
266. Aphexda. A+pheresis: Achieve stem cell collection targets in multiple myeloma (MM) in fewer apheresis sessions 2024. Available from: https://www.aphexda.com/.
267. Zhang F, Xia Y, Su J, Quan F, Zhou H, Li Q, et al. Neutrophil diversity and function in health and disease. Signal Transduct Target Ther. 2024;9(1):343. https://doi.org/10.1038/s41392-024-02049-y.
268. (n.d.). A. FDA-approved APHEXDA® (motixafortide). 2024. Available from: https://www.aphexda.com/.
269. Hoy SM. Motixafortide: first approval. Drugs. 2023;83(17):1635–43. https://doi.org/10.1007/s40265-023-01962-w.
270. Davenport AP, Hyndman KA, Dhaun N, Southan C, Kohan DE, Pollock JS, et al. Endothelin. Pharmacol Rev. 2016;68(2):357–418. https://doi.org/10.1124/pr.115.011833.
271. Speck D, Kleinau G, Szczepek M, Kwiatkowski D, Catar R, Philippe A, et al. Angiotensin and endothelin receptor structures with implications for signaling regulation and pharmacological targeting. Front Endocrinol (Lausanne). 2022;13:880002. https://doi.org/10.3389/fendo.2022.880002.
272. Clinic C. Endothelin 2022. Available from: https://my.clevelandclinic.org/health/articles/23139-endothelin.
273. Schneider MP, Boesen EI, Pollock DM. Contrasting actions of endothelin ET(A) and ET(B) receptors in cardiovascular disease. Annu Rev Pharmacol Toxicol. 2007;47:731–59. https://doi.org/10.1146/annurev.pharmtox.47.120505.105134.
274. (AAFP) AFP. New Kid on the Block: Aprocitentan for the Treatment of Difficult-to-Treat Hypertension 2024. Available from: https://www.aafp.org/pubs/afp/afp-community-blog/entry/new-kid-on-the-block-aprocitentan-for-the-treatment-of-difficult-to-treat-hypertension.html#:~:text=In%20March%202024%2C%20the%20U.S.,Pugh%20class%20B%20and%20C.
275. Dhillon S. Aprocitentan: first approval. Drugs. 2024;84(7):841–7. https://doi.org/10.1007/s40265-024-02053-0.
276. Shirley M. Etrasimod: first approval. Drugs. 2024;84(2):247–54. https://doi.org/10.1007/s40265-024-01997-7.
277. Bravo G, Cedeño RR, Casadevall MP, Ramió-Torrentà L. Sphingosine-1-Phosphate (S1P) and S1P signaling pathway modulators, from current insights to future perspectives. Cells. 2022;11(13) https://doi.org/10.3390/cells11132058.
278. Garcia-Garcia AL, Newman-Tancredi A, Leonardo ED. 5-HT(1A) [corrected] receptors in mood and anxiety: recent insights into autoreceptor versus heteroreceptor function. Psychopharmacology. 2014;231(4):623–36. https://doi.org/10.1007/s00213-013-3389-x.
279. Keam SJ. Gepirone extended-release: first approval. Drugs. 2023;83(18):1723–8. https://doi.org/10.1007/s40265-023-01975-5.
280. Sun W, Yang F, Zhang H, Yuan Q, Ling S, Wang Y, et al. Structural insights into neurokinin 3 receptor activation by endogenous and analogue peptide agonists. Cell Discov. 2023;9(1):66. https://doi.org/10.1038/s41421-023-00564-w.

281. Administration. USFaD. FDA approves novel drug to treat moderate to severe hot flashes caused by menopause. 2023. Available from: https://www.fda.gov/news-events/press-announcements/fda-approves-novel-drug-treat-moderate-severe-hot-flashes-caused-menopause.
282. Medicine U. New treatments for hot flashes target neurons. UW Medicine Newsroom. 2020. Available from: https://newsroom.uw.edu/blog/new-treatments-hot-flashes-target-neurons.
283. tablets Vf. For the treatment of moderate to severe Vasomotor Symptoms (VMS)—commonly referred to as hot flashes and night sweats—due to menopause. Available from: https://www.veozahhcp.com/mechanism-of-action.
284. Levy D, Moskowitz MA. Meningeal mechanisms and the migraine connection. Annu Rev Neurosci. 2023;46:39–58. https://doi.org/10.1146/annurev-neuro-080422-105509.
285. Fila M, Chojnacki J, Sobczuk P, Chojnacki C, Blasiak J. Nutrition and Calcitonin Gene Related Peptide (CGRP) in Migraine. Nutrients. 2023;15(2):289.
286. Dhillon S. Zavegepant: First Approval. Drugs. 2023;83(9):825–31. https://doi.org/10.1007/s40265-023-01885-6.
287. Nourian YH, Salimian J, Ahmadi A, Salehi Z, Karimi M, Emamvirdizadeh A, et al. cAMP-PDE signaling in COPD: review of cellular, molecular and clinical features. Biochem Biophys Rep. 2023;34:101438. https://doi.org/10.1016/j.bbrep.2023.101438.
288. FDA. Daliresp(Roflumilast) Approval Overview 2011. Available from: https://www.accessdata.fda.gov/scripts/cder/daf/index.cfm?event=overview.process&ApplNo=022522.
289. Bodkhe S, Nikam M, Sherje AP, Khan T, Suvarna V, Patel K. Current insights on clinical efficacy of roflumilast for treatment of COPD, asthma and ACOS. Int Immunopharmacol. 2020;88:106906. https://doi.org/10.1016/j.intimp.2020.106906.
290. Online D. Roflumilast 2023. Available from: https://go.drugbank.com/drugs/DB01656.
291. FDA. Ohtuvayre(Ensifentrine) Approval Overview 2024. Available from: https://www.accessdata.fda.gov/scripts/cder/daf/index.cfm?event=overview.process&ApplNo=217389.
292. Keam SJ. Ensifentrine: first approval. Drugs. 2024;84(9):1157–63. https://doi.org/10.1007/s40265-024-02081-w.
293. Natarajan K, Berk BC. Crosstalk coregulation mechanisms of G protein-coupled receptors and receptor tyrosine kinases. Methods Mol Biol. 2006;332:51–77. https://doi.org/10.1385/1-59745-048-0:51.
294. USFaDAF. Tepezza(Teprotumumab) Approval Overview 2020. Available from: https://www.accessdata.fda.gov/scripts/cder/daf/index.cfm?event=overview.process&ApplNo=761143.

Cancer Cells, Oncogenes, and Therapeutic Interventions

19

Abstract

Cancer is a complex disease that is fundamentally a disease of the cell. Cancer cells have characteristics that distinguish them from normal cells. These often include uncontrolled cell division, increased cell survival, loss of differentiation, and dysregulated cell migration. These changes are often accompanied by alterations in metabolism, which can enable cancer cells to divide rapidly and to survive in challenging environments. Cancer is mainly caused by normal cellular genes that function improperly, such as proto-oncogenes that, when improperly activated, cause uncontrolled cell growth. This chapter introduces cancer cell biology, focusing on oncogenes and cancer cell metabolism. Characteristics that distinguish cancer cells from normal cells are discussed, followed by an introduction to proto-oncogenes and oncogenes. Proto-oncogenes are normal cellular genes that, when improperly activated, can lead to cancer. V-Src was the first oncogene discovered, and this chapter covers some history of its discovery and that of other oncogenes. Additional oncogenes discussed include oncogenic Ras, Raf, and Myc. Therapeutic strategies for inhibiting oncogenic activity are discussed, along with the challenges of regulating uncontrolled growth in cancer cells. Cancer metabolism is also addressed in this chapter, including the Warburg effect, where cancer cells preferentially use aerobic glycolysis. Understanding how cancer cells grow and divide, along with identifying oncogenes that cause cancer, has led to the development of numerous targeted cancer therapies. This chapter discusses drugs designed to treat cancer, which vary in their efficacy.

Keywords

Cancer cells · Oncogenes · Activating mutations · Src · Ras · Signaling proteins

Abbreviations

ABL	Abelson Murine Leukemia Viral Oncogene Homolog
ADC	Antibody drug conjugate
AKT	Also known as protein kinase B (PKB), a key signaling molecule
ALV	Avian leukosis virus
AML	Acute myeloid leukemia
ATP	Adenosine triphosphate (often mentioned in kinase activity)
BCR	Breakpoint cluster region gene
BCR-ABL	Fusion gene from translocation of BCR and ABL genes (Philadelphia chromosome)
Car-T cell	Chimeric antigen receptor T cell (check)
CDK	Cyclin-dependent kinase

A. Minden, *Biochemistry, Cell and Molecular Biology for Pharmacists and Health Professionals*,
https://doi.org/10.1007/978-3-032-20529-2_19

CML	Chronic myeloid leukemia
DNA	Deoxyribonucleic acid
ECM	Extracellular matrix (often referenced in invasion/metastasis discussions)
EGFR	Epidermal growth factor receptor
EMT	Epithelial-mesenchymal transition
FGFR	Fibroblast growth factor receptor
GDP	Guanine diphosphate
GEF	Guanine nucleotide exchange factor
GTP	Guanine triphosphate
HER2	Human epidermal growth factor receptor 2
HNSCC	Head and neck squamous cell carcinoma
iCCA	Intrahepatic cholangiocarcinoma
NSCLC	Non-small cell lung cancer
IgH	Immunoglobulin heavy chain
mTOR	Mechanistic (or mammalian) target of rapamycin
p53	Protein 53 (a tumor suppressor involved in DNA repair and apoptosis)
PDGF	Platelet-derived growth factor
PI3K	Phosphoinositide 3-kinase
PI3KCA	PI3 kinase catalytic domain
RB (or pRB)	Retinoblastoma protein
RSV	Rous sarcoma virus
RTK	Receptor tyrosine kinase
TME	Tumor microenvironment
V600E	Valine to glutamate at amino acid 600
VEGF	Vascular endothelial growth factor

What You Will Learn

- Some basic characteristics of cancer cells, such as increased cell proliferation and survival
- The concepts of metastasis, angiogenesis, and the tumor microenvironment
- How cancers are classified
- Benign and malignant tumors
- The steps in cancer development and progression
- Oncogenes and proto-oncogenes, and some history about their discovery, including the discovery of Src and Ras
- Proto-oncogenes become oncogenic if there is an activating mutation on one allele.
- Signaling pathways discussed in the previous chapters that are associated with cancer, when they are improperly activated
- Drugs that target oncogenes such as oncogenic Ras and Raf
- Drugs that target other signaling molecules involved in cancer, such as those in the FGF pathway, the PI3Kinase pathway, and AKT
- Gleevec as the first kinase inhibitor that was used as a targeted cancer drug
- Cancer cell metabolism and the Warburg effect
- The clonal origins of cancer

1 Introduction to the Chapter

Cancer is a complex disease that is fundamentally a disease of the cell. Cancer cells have characteristics that distinguish them from normal cells. These usually include factors such as uncontrolled cell division, increased cell survival, loss of differentiation, and dysregulated cell migration. These changes are often accompanied by alterations in metabolism, which can enable cancer cells to divide rapidly and to survive in challenging environments. Unlike some disorders, which are caused mainly by external factors, such as a bacterial infection, an injury, or a toxin, cancer is mostly caused by normal cellular genes that are functioning improperly. These genes can include both proto-oncogenes (which are called oncogenes when they are improperly activated) and tumor suppressor genes. While this chapter will focus on oncogenes and cancer metabolism, tumor suppressor genes will be discussed later, in Chap. 21.

This chapter is designed to introduce readers to cancer cell biology, focusing on oncogenes and cancer cell metabolism. Section 2 is an overview of the distinguishing characteristics of **cancer cells** that differentiate them from normal cells. Sections 3 and 4 discuss **proto-oncogenes and oncogenes**. Proto-oncogenes are normal cellular genes that, when improperly activated, can lead to cancer. Section 3 provides some history about the first oncogene to be discovered, v-Src, as well as oncogenic Ras, the first human oncogene to be discovered. Section 4 expands on some additional oncogenes identified after the early discovery of these genes. These include oncogenic Raf, Myc, and multiple others. The functions of the proteins encoded by these oncogenes, and their relationship to cellular signaling pathways, are discussed. Therapeutic strategies for inhibiting oncogenic activity are discussed, along with the challenges of regulating uncontrolled growth in cancer cells. Section 5 shifts focus to the metabolism of cancer cells. Cancer cells are often rapidly dividing cells that sometimes cluster densely in tumors where nutrients can be low. Because of this, cancer cells often reprogram their metabolism. An important example is the Warburg effect, where cancer cells preferentially use **aerobic glycolysis**. This type of altered metabolism is introduced in this final section.

Understanding how cancer cells grow and divide, along with identifying oncogenes that cause cancer, has led to the development of numerous drugs used for targeted cancer treatments. This chapter includes a discussion of drugs designed to treat cancer, which vary in their efficiency. Studies of cell signaling and the identification of oncogenes have allowed researchers to pinpoint the basic causes of some types of cancers. Thus, while early cancer drugs tended to be nonspecific and directed generally toward rapidly dividing cells, the identification of specific genes involved in cancer has allowed researchers to develop more targeted drugs. This approach has been beneficial in targeting individual tumor types, reducing side effects, and saving or prolonging many lives. Nevertheless, cancer is complicated, and cancer cells tend to develop multiple mutations over time; therefore, in general, single-drug approaches are often not successful in the long term. In most cases, a combination of drugs is more effective than a single drug. The types of drugs used may also need to change over time as cells develop more mutations and develop resistance to treatments. Furthermore, it is also important to note that although newer drugs have many advantages due to specificity, older and more general drugs are still used in many cases, either alone or more often in combination with other drugs. These are important concepts to keep in mind when reading about and discussing various drugs that inhibit cancer cells.

2 Overview of Cancer Cell Biology

2.1 Some Basic Characteristics of Cancer Cells: Disruption of Cellular Homeostasis

Cancer is associated largely with uncontrolled cell proliferation. Cancer biology is a complex and ever-changing field, and the references at the end of the chapter offer valuable sources for understanding the disease's complexity in more detail. For the purpose of this chapter, however, a brief introduction to the key factors that differentiate cancer cells from normal cells will be discussed before delving into the discussion of oncogenes and cancer cell metabolism. Some of the key changes that differentiate cancer cells from normal cells are discussed below.

Cancer cells are also often referred to as **transformed cells**, and the term **oncogenic transformation** is sometimes used to describe the process of a cell becoming a cancer cell, either in the body or in culture. Cancer cells differ from normal cells in several important ways, leading to their uncontrolled proliferation and their ability to migrate and invade various tissues. Overall, these differences contribute to the **disruption of cellular homeostasis**. In normal cells, **cell proliferation, growth inhibition,** and **cell death** are closely balanced and regulated according to the body's needs. This balance can be considered cellular **homeostasis**. In cancer cells, this regulation is disrupted. This loss of regulation leads to several problems, including persistent proliferation, resistance to growth inhibitory signals, evasion of apoptosis, replicative immortality, sustained angiogenesis, and activation of invasion and metastasis. These are often referred to as the six **hallmarks of cancer** (a term formalized by

[1]), although there are also other characteristics that are common in cancer cells. Some of the important differences commonly seen between cancer cells and normal cells [2–4] are as follows.

2.1.1 Increased Cell Proliferation and Survival

Increased proliferation is a critical feature of cancer cells, and it is the major topic discussed in this chapter on oncogenes. In normal cells, cell proliferation is tightly controlled. Cells proliferate when more cells are needed, and cells stop growing when they do not receive signals encouraging cell proliferation. This can be linked to cell signaling pathways covered in the previous chapter. Normal cells may proliferate when they respond to a growth factor that binds to a cell surface receptor but stop proliferating when the signal is terminated and the growth factor is no longer present. Cancer cells may lose this tight regulation, allowing them to proliferate even when the proper signals are not present. This may result, for example, from improper regulation of a signaling molecule such as a growth factor receptor.

Another way that cancer cells can proliferate uncontrollably is by **evading apoptosis, the process of cell death**. Normal cells undergo a controlled level of cell death when cells are damaged or no longer needed. Cancer cells, however, often evade apoptosis and therefore may survive too long.

Normal cells also undergo a process called replicative senescence, which is based on a limit as to how many times a cell can divide. Cancer cells may bypass this in a process referred to as **replicative immortality**. One important mechanism that contributes to replicative immortality involves **telomeres** and the enzyme **telomerase**. Telomeres are protective sequences at the ends of chromosomes. In normal somatic cells, telomeres shorten with each cell division, eventually triggering cellular senescence when they become too short. Cancer cells, however, often activate the enzyme telomerase, which prevents telomere shortening. By maintaining telomere length, telomerase enables cancer cells to divide indefinitely, bypassing normal limits on cellular lifespan. Telomerase is active in most human cancers, making it a potential target for cancer treatment. (Telomeres and telomerase are also discussed in Chap. 6).

All of these factors contribute to increased proliferation of the cancer cells, also discussed in [2–4] and in other references throughout this chapter.

2.1.2 Changes in Cell Behavior and Morphology

Certain changes in cell growth characteristics are also often altered in cancer cells, and these changes can support the increased rate of cell proliferation. An example is the **loss of contact inhibition**. Normal cells stop proliferating when they get too close to their neighboring cells; this is called contact inhibition. Cancer cells often lose this inhibition, allowing them to continue proliferating and crowding together even when in close proximity to other cells. Another change is the loss of anchorage independence. Normal cells, with few exceptions, such as blood cells, need to be in contact with a surface in order to survive. This surface may be extracellular matrix (ECM) material on the surface of an organ (or on the surface of a culture dish). Cancer cells lose this inhibition, and they may exhibit what is known as **anchorage-independent growth**. Thus, they can proliferate without being attached to a physical surface. This can be demonstrated in culture by growing cells on a soft surface such as a layer of agar rather than a rigid culture dish [3].

2.1.3 Decrease in Differentiation

Lack of differentiation is another key characteristic of cancer cells. Normal cells differentiate into specific cell types (such as bone, muscle, neurons, and others). In contrast, many cancer cells lose their ability to differentiate, and they may appear undifferentiated or poorly differentiated. This is tightly linked to increased cell proliferation as differentiated cells often stop proliferating while undifferentiated cells may maintain the ability to proliferate [1].

2.1.4 Increased Angiogenesis

Increased angiogenesis, the formation of blood vessels, is also a change that is often seen in tumors. Cancer cells need a source of nutrients and oxygen, and as tumors grow, they often develop mechanisms to stimulate angiogenesis, allowing vessels to infiltrate the tumor and provide nutrients. For this reason, the proliferation of endothelial cells, the cells that line the blood vessels, is sometimes targeted in cancer treatment. Likewise, VEGF signaling, which promotes angiogenesis, is also a common target, and tumors may secrete VEGF in order to promote angiogenesis [5].

2.1.5 Alterations in the Tumor Microenvironment

A tumor does not exist in a vacuum but is an integral part of its environment, called the tumor microenvironment (TME). This environment contains many types of cells besides cancer cells, such as stromal and immune cells, and it also includes the extracellular matrix. The contents and shape of this environment are influenced by the cancer cells, which can produce signals that manipulate their environment. Cancer cells can also develop ways to evade detection by cells of the immune system by producing immune checkpoint proteins like PD-L1, which suppress the activity of local immune cells [6].

2.1.6 Metastasis

Metastasis, the invasion of surrounding tissues, is an important feature of cancer. Metastasis involves the cells entering the bloodstream or lymphatic system and establishing secondary tumors in distant organs. Only a small proportion of cancer cells successfully metastasize, but those that do can cause the disease to spread.

An important process that often contributes to metastasis is the **epithelial-mesenchymal transition (EMT)**. A high percentage of cancer cells arise from epithelial cells. EMT is a process where epithelial cells, which are normally stationary and adhere tightly to each other or to an extracellular matrix, lose their cell-cell adhesion and polarity and acquire some characteristics of mesenchymal cells, a type of cell found in connective tissues. In embryonic mesoderm or in adult connective tissues, these cells are usually loosely organized and migratory. The cells that obtain these mesenchymal characteristics become more motile and invasive. During EMT, cancer cells downregulate epithelial markers such as E-cadherin and upregulate mesenchymal markers such as N-cadherin and vimentin. This transition allows cancer cells to detach from the primary tumor, invade surrounding tissues, and migrate through the extracellular matrix [3].

EMT is often reversible; once metastatic cancer cells reach a new site, they may undergo a mesenchymal-epithelial transition (MET) to reestablish epithelial characteristics and form secondary tumors [7].

While many cancer cells are epithelial in origin, some are not [4]. These cells may also metastasize although they use different mechanisms that also involve changes in cell migration, invasiveness, and survival. In either case, metastatic cells need to continue proliferating, even in their new environments, even if their environments are harsh and have low levels of growth factors. Therefore, cells in their new environments may acquire an increased number of growth factor receptors that are improperly active, allowing them to survive in conditions where they would not normally survive. Genes encoding proteins important for cell migration or survival in harsh environments are putative targets for cancer therapy [8].

2.1.7 Other Changes Often Seen in Cancer Cells

Metabolism may also change in cancer cells due to their rapid growth, often in environments where they are densely packed together. Common metabolic changes involve a heavy reliance on glucose. **Genetic instability** is also common in cancer cells. Cancer cells often have impaired DNA repair mechanisms, and this can lead to the accumulation of genetic mutations. Sometimes, this can be associated with aneuploidy, the presence of an abnormal number of chromosomes, which is common in some cancers [3].

2.2 Classification of Cancers

Not all cancer cells are the same, and cancers are classified based on their tissue of origin. Carcinomas arise from epithelial cells and are the most common type of cancer, likely due to the high rate of cell turnover in these tissues. Sarcomas arise from connective or muscle tissue, while leukemias, lymphomas, and myelomas originate from blood or immune cells. Cancers of the nervous system include gliomas and medulloblastomas [2, 3].

2.3 Benign and Malignant Tumors

Another important consideration is whether tumors are benign or malignant. Benign tumors exhibit some of the features described above, such as an increase in proliferation, but they do not metastasize or invade other tissues. Even with their increased capacity for growth, their growth usually has some limitations. Malignant tumors, on the other hand, have the ability to invade other tissues and metastasize. This ability to metastasize is the major distinguishing characteristic of malignant cells [2].

2.4 Cancer Development and Progression

Cancer development is a gradual process. Many cancers are clonal in origin, meaning that the cancer begins with a single mutated cell that has some growth advantages. Over time, this cell may divide during clonal expansion, producing a population of genetically identical daughter cells. As these cells divide, some daughter cells may accumulate additional mutations, which provide even more growth advantages. Over time, even if the tumor started with a single mutation in a single cell, it develops into a genetically diverse group of cells. Tumor progression may be slow at first, but it can accelerate as cancer cells develop more mutations and become genetically unstable, leading to rapid genetic changes [2, 3].

The accumulation of new mutations as the cancer cells divide is an important aspect of tumor development. Mutations that cancer cells accumulate can be broadly classified into two types: driver mutations and passenger mutations. **Driver mutations** directly contribute to cancer development by conferring a growth, survival, or metastatic advantage to the cell (these can potentially be mutations in either pro-oncogenes or tumor suppressor genes, which will be discussed in more detail below and in upcoming chapters). In contrast, **passenger mutations** do not contribute to cancer progression but arise simply because of the cancer cell's increased mutation rate and genomic instability. Distinguishing between these two types of mutations can be challenging, but it is crucial in order to develop treatments that precisely target the correct genes [3].

3 Oncogenes and the First Oncogenes Discovered

3.1 Introduction to Oncogenes

The main types of cellular genes that are known to cause cancer when mutated are proto-oncogenes and tumor suppressors. These genes code for proteins that either encourage or inhibit cell replication, respectively. Understanding how these genes work normally and how their mutation can lead to cancer is essential to understanding cancer and developing new treatments. Here, we will discuss cellular genes involved in cancer, with an emphasis on proto-oncogenes. **Proto-oncogenes** are genes often associated with cell proliferation that can become **oncogenes** when they obtain activating mutations. It is common for proto-oncogenes to encode signaling proteins. This is not surprising because many signaling pathways are involved in regulating cell proliferation, and cancer is usually associated with uncontrolled cell proliferation. Therefore, improper activation of some signaling pathways may lead to unchecked cell proliferation. Likewise, many cancer drugs specifically target signaling pathways. This chapter, therefore, builds on the previous chapter, which covered signal transduction. While Chap. 18 covered signaling pathways and signaling proteins, this chapter covers some of these same proteins but discusses their links to cancer and unregulated cell proliferation.

3.2 Oncogenes Compared to Tumor Suppressor Genes

When mutations in cellular genes cause cancer, the genes are usually proto-oncogenes or tumor suppressor genes. A proto-oncogene encodes a protein whose normal function is usually to positively regulate cell proliferation. Certain types of mutations in a proto-oncogene can cause it to be an oncogene. These mutations usually cause the encoded protein to have an increased level of activity or to be constitutively active, often independently of external or internal signals. This type of mutation can result in excessive and unregulated cell growth, which can lead to cancer or make cells more susceptible to becoming cancer cells. This unregulated cell proliferation can occur even when the mutation only occurs on one allele of the gene. In other words, the mutations are dominant. This is an important feature that differentiates oncogenes from tumor suppressor genes.

In contrast to oncogenes, the products of tumor suppressor genes (such as retinoblastoma (Rb)) are negative regulators of cell proliferation. Tumor suppressor genes can cause uncontrolled cell growth when the genes are inactivated, such as by an inactivating mutation, loss of expression, or loss of part or all of the gene. This is unlike oncogenes, where activating mutations are the culprit. Unlike oncogenes, tumor suppressor genes must be inactivated on both alleles to cause unregulated cell growth. Thus, the mutations are recessive [3, 4, 9].

In this chapter, proto-oncogenes and oncogenes are the main focus, while tumor suppressors are covered in Chap. 21. There are different ways that proto-oncogenes can be mutated to become oncogenes. In general, the mutations must be activating mutations. Activating mutations can include any of the following: a point mutation that leads to constitutive activation of the gene product (the protein), mutations that enhance the stability of the encoded protein, deletion of a gene encoding a regulatory domain of a protein, overexpression of the gene (enhanced transcription), or gene amplification (multiple copies of the gene). Some genes that encode proteins that positively regulate cell growth do not fit the classic definition of oncogenes and are also associated with cancer, and these will be covered here as well.

3.3 Viruses and the Discovery of Src, the First Oncogene to Be Identified

Now we return to the discussion of oncogenes, which were found to be an important cause of malignant transformation. First, however, is some history about how oncogenes were first discovered.

3.3.1 The Discovery of v-Src

Although viruses cause only a small minority of human cancers, tumor viruses led to the discovery of oncogenes, which was instrumental in understanding cancer. This type of virus will be discussed here to provide a historical perspective on the discovery of oncogenes, particularly the identification of **Src**, the first oncogene discovered. In **1911**, **Peyton Rous** published the results from a groundbreaking study of a sarcoma, a type of tumor, in chickens [3, 10–12]. He removed a sarcoma from a chicken, broke it up into small pieces, and passed it through a filter. The filtrate was then injected into other chickens, and the exciting result was that more tumors formed in these chickens. This study indicated that something in the sarcomas was sufficient to cause more tumors to form. Furthermore, because of the fine filtration, the filtrates appeared to be a **virus**.

Decades later, more experiments were done with the filtrates from sarcomas. By this time, biochemical studies indicated that the virus obtained from the tumors appeared to contain RNA rather than DNA. Studies conducted by **Howard Temin in 1958** further assessed the ability of this virus to transform cells [13]. The tumors from chickens infected with the filtrate, as described above, were isolated again, ground up, and filtered. This filtrate was used to inoculate other chickens and, importantly, was also

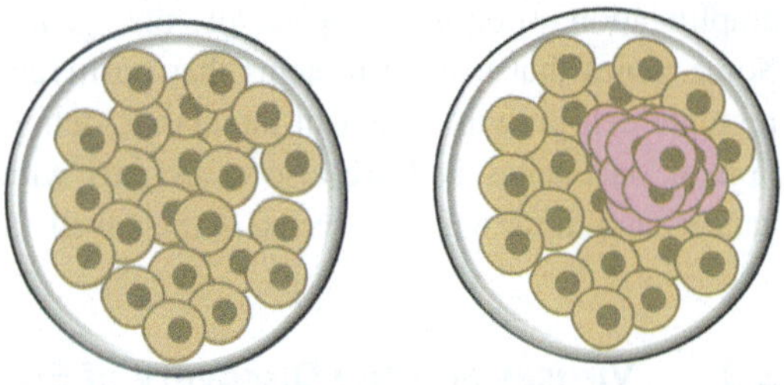

Fig. 19.1 Focus formation in a culture dish. In culture, transformed cells will cluster together and form foci. In the diagram, one focus is seen in the dish of cells on the right side

introduced to cultured chicken fibroblasts. Remarkably, a portion of the cultured cells became **transformed**. This was detected by observing that foci developed in some of the culture dishes of cells. Foci are abnormal clusters of cells representing cells that have lost their normal growth controls (Fig. 19.1). While normal cells grow in a monolayer and then stop growing once they reach confluence, foci can pile up on each other and grow without their normal controls, thus resulting in clumps or aggregates of cells.

These results in cultured cells were pivotal because they provided strong evidence that the filtrate contained genetic material, presumably from a virus, that could transform normal cells, even in vitro. In **1962**, **Howard Temin** proposed the **provirus hypothesis**. This was part of the RNA tumor virus model developed at this time, suggesting that RNA viruses could cause cancer. The provirus hypothesis proposes that RNA viruses can convert their RNA into DNA although the exact mechanism is not yet understood. Later, in **1970**, **Temin** and **Baltimore** independently discovered **reverse transcriptase**. This enzyme can synthesize DNA using RNA as the template. This supported the idea that some RNA viruses, namely, retroviruses, could replicate in the host cell by integrating a DNA copy of their RNA genome into the host cell. The virus derived from the sarcoma was determined to be a retrovirus. It was named the **Rous sarcoma virus (RSV)** after the scientist who first discovered it. Prior to this, it was called simply the "sarcoma virus" [14, 15].

In the **1970s and 1980s**, **Harold Varmus and Michael Bishop** investigated RSV in more detail. They determined that one gene in the RSV's genome was responsible for the transformation of cells; this gene was v-Src (viral Src). The scientists studied v-Src and found that the v-Src gene had the key role in transforming cells and causing tumors to form. Furthermore, their work led to the discovery that v-Src is not originally a viral gene but that it was acquired from a **host cell** during the viral infection. This gene is **c-Src** (cellular Src). **V-Src** was similar but not identical to this normal cellular gene. This was a key finding that led to the **discovery of cellular oncogenes**, which are normal genes that could cause cancer if mutated or dysregulated [16]. In **1989, Varmus and Bishop** received a **Nobel Prize** in Physiology and Medicine for their research on **oncogenes** [17].

To better understand how v-Src was picked up by the RSV, it is helpful to learn a bit about the life cycle of a typical retrovirus [3, 18–20]. A retrovirus is a specific type of RNA virus. It infects the cell and releases its RNA into the cytoplasm. This RNA gets reverse-transcribed into DNA in the host cell, catalyzed by **reverse transcriptase**. This DNA then integrates into the host cell genome via the viral enzyme **integrase**. Now, every time the host cell's machinery replicates its DNA, the viral DNA will also be replicated along with it. Likewise, it will be transcribed and translated. The retroviral sequences can thereby be translated to produce more virus particles, thereby increasing the number of retroviruses. These viruses can bud from the membrane and infect other cells (Fig. 19.2).

In the case of the chicken sarcoma study, ALV (Avian leukosis virus) is now known to have been the original virus that infected the chicken cells. This was discovered in the 1970s. During the infection, the ALV genome happened to integrate into the host genome in a region close to the c-Src gene. During viral replication, an RNA recombination event caused a copy of the Src sequence to be incorporated into the viral genome. As a result, when the region of the genome containing the inserted viral sequence was transcribed and

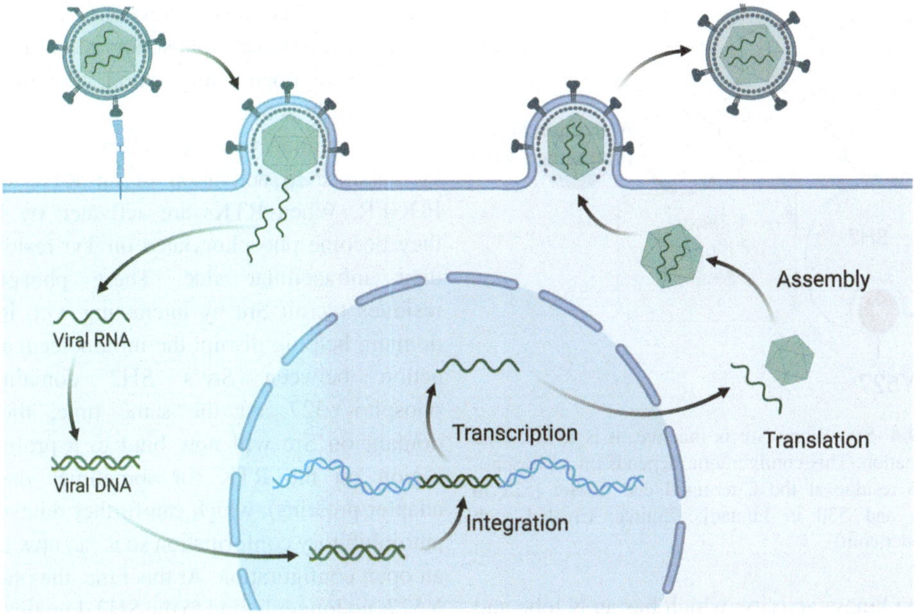

Fig. 19.2 Retroviruses. Retroviruses are a type of RNA virus. The RNA is inserted into the cell and reverse transcriptase catalyzes the formation of DNA from the RNA, which can incorporate into the host genome. See the text for details (Source: Created with BioRender.com)

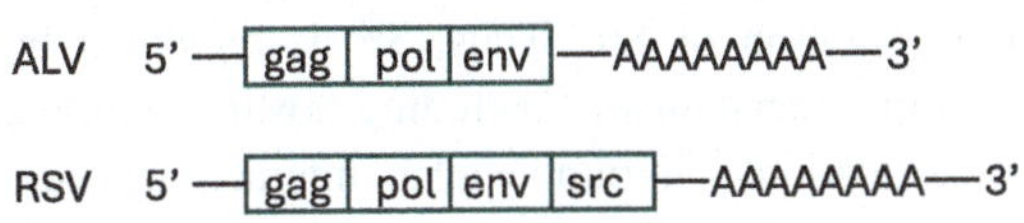

Fig. 19.3 ALV and RSV viruses. V-Src is part of the RSV

translated, the inserted Src sequence was transcribed and translated and packaged into new virions. Therefore, the viral RNA acquired a large portion of the Src gene. This gave rise to the new virus, now named **RSV**, which had the host's c-Src integrated into its genome. However, the Src gene incorporated by this new virus slightly differed from the original c-Src because the acquired gene lacked one small part of the cellular gene. Thus, it was slightly truncated when integrated into the new virus. This new, slightly smaller sequence was now called **v-Src** and was constitutively active, which enabled RSV to transform cells [12] (Fig. 19.3).

3.3.2 Structure and Function of Src

Understanding the normal function of c-Src is important for determining how **v-Src** causes transformation. C-Src, which is the normal cellular Src, can be considered a **proto-oncogene**. It does not cause cancer unless it is mutated or truncated. C-Src is a **non-receptor tyrosine kinase** [21]. This means that unlike some tyrosine kinases, it is not part of a cell surface receptor. C-Src has various roles in signaling pathways within the cell, including regulation of cell proliferation, differentiation, adhesion, and migration. c-Src mediates its effects by phosphorylating other signaling molecules in signal transduction pathways. c-Src can be activated by extracellular stimuli. These may include growth factors, cytokines, and extracellular matrix (ECM) proteins.

c-Src is subject to complex modes of regulation. At the N-terminal end of Src is a site that becomes myristoylated. Myristoylation is the covalent linkage of myristic acid, which has a role in anchoring Src to the plasma membrane. This site is followed by an SH3 domain and then an SH2 domain. SH3 domains bind to proline-rich sequences on proteins, and SH2 domains bind to phospho-tyrosine residues. Following these domains is the catalytic domain, the domain

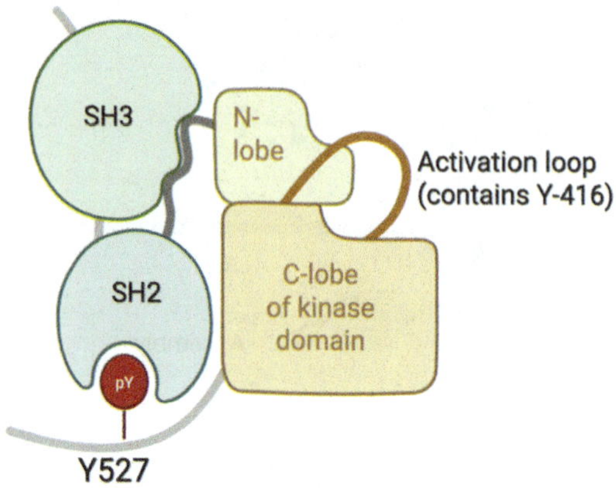

Fig. 19.4 Src. When Src is inactive, it is in a closed conformation. This configuration depends on a phosphotyrosine residue at the C-terminal end of Src [527 in chicken and 530 in human]. (Source: Created with BioRender.com)

with the kinase activity, which has an N lobe and a C lobe. Near the C-terminal end of the c-Src protein is an important Tyr residue (Y527 in chicken and Y530 in humans), which is often phosphorylated. Within the catalytic domain is the activation loop, which controls whether the kinase is turned on. Within this activation loop is another important tyrosine residue (Y416) (see Fig. 19.4).

Physical interactions within the Src enzyme (intramolecular interactions) keep it in a closed, inactive conformation when it is not active or needed. These interactions are between the SH2 domain and the phospho-Tyr-527 residue and between the SH3 domain and proline-rich sequences within Src.

For Src activity to be turned on, it needs to take on an open configuration. This requires three changes need to occur sequentially. First, removing the C-terminal phosphate (phospho-Y527) by a phosphatase is required to relieve the interaction between the Src SH2 domain and this phospho-Tyr. Second, the SH3 domain must bind to an external proline-rich sequence, breaking the intramolecular interaction with the proline-rich sequence on Src. Finally, the Tyr residue in the activation loop (Y416) must be phosphorylated. This is often catalyzed by another kinase or sometimes by Src autophosphorylation. When this residue is unphosphorylated, the activation loop obstructs the active site, preventing access to the substrate. However, phosphorylation of Y416 stabilizes the open configuration, allowing Src to become fully active [21].

One common way these changes can occur is in response to activation of an RTK such as PDGFR. When RTKs are activated by ligand, they become phosphorylated on Tyr residues on their intracellular side. These phospho-Tyr residues recruit Src by interacting with its SH2 domain, helping disrupt the intramolecular interaction between Src's SH2 domain and phospho-Y527. At the same time, the SH3 domain on Src will now bind to a proline-rich region on the RTK (or sometimes on other adapter proteins), which can further relieve Src's autoinhibitory conformation so it can now take on an open configuration. At this time, the phospho-Y527, no longer bound to the SH2 domain on Src, becomes dephosphorylated by a phosphatase, removing its inhibitory effect. Next, Tyr in the activation loop (Y-416) gets phosphorylated, usually by autophosphorylation, due to the now open configuration of Src. Thus, Src is regulated by multiple mechanisms, including posttranslational modifications, intramolecular interactions, and protein-protein **interactions** [4, 21]. Once activated, Src can activate pathways such as the MAP kinase pathway. Depending on the cell context, activation of Src can also lead to activation of other signaling proteins such as STAT3 and STAT5, c-Myc, Cyclin-D, Ras, Raf, PI3K, AKT, mTOR, or PDK1 [3, 4, 21, 22].

3.3.3 Differences Between v-Src and c-Src

In contrast to c-Src, v-Src, an oncogene, is truncated. Specifically, it is missing a stretch of DNA with the C-terminal tyrosine (Y527) residue (corresponding to Y530 in the human protein). That tyrosine is critical to the intramolecular interaction that can keep Src in a closed conformation. Without this tyrosine residue, Src always remains in an open, active conformation. Hence, the truncated Src that is part of the RSV virus is in a constitutively active conformation. This active Src can constitutively activate pathways that regulate cell growth and proliferation, and it led to sarcomas in the infected chicken cells. While Src was identified originally in chicken sarcoma, dysregulation of Src is now known to be associated with several types of cancer in humans [3, 23, 24].

Src was the first oncogene identified and also the first to be identified in a retrovirus. Later, other oncogenes were found in other retroviruses. Some oncogenes were also found in DNA viruses. Most oncogenes, however, are not found in viruses. The work on Src was important, not necessarily because of the relationship between Src and retroviral infection in chickens but because it led to the understanding that oncogenes could develop from normal cellular genes. Abnormal activation or expression of Src in human cancers is generally independent of any retroviral infection. Scientists have now identified many different oncogenes, thanks to this earlier work, and have found that, in general, oncogenes are normal cellular genes that have become improperly activated. This improper activation is not usually due to a virus even though oncogenes were discovered in viruses. Instead, causes for improper regulation of cellular genes include mutation, overexpression, chromosomal translocations, or gene amplification [3].

3.4 Discovery of New Oncogenes

Work on v-Src generated excitement about identifying new proto-oncogenes in human cells, with the hope that a better understanding of the causes of human cancers can lead to improved treatments. An assay was developed to identify human oncogenes. In this assay, DNA was isolated from human tumors. This DNA was broken into fragments, introduced into normal, non-tumor mouse cells (fibroblasts), and grown in culture dishes. Most of the cultured cells grew like normal cells, forming a single layer, or monolayer, on the surface of the culture dish. Occasionally, however, foci would grow and appear in the monolayer of cells. Foci, as described above, are aggregates of cells that are beginning to grow uncontrollably. They are now often used as an assay to determine whether cells are transformed. Scientists reasoned that these foci contained genes from the cancer cells that could cause the normal cells to take on characteristics of cancer cells. Scientists isolated the foci that had formed and extracted DNA from them. They could distinguish the tumor DNA (from the cancer cells) from the mouse DNA (from normal tissues) because human DNA contains unique sequences called ALU sequences. Once they isolated the human DNA, they analyzed the genes and were able to identify those that specifically led to the formation of foci (transformation). These genes were analyzed further, with the expectation that some of them would be oncogenes. This method, as well as other studies, led to the identification of multiple new human **oncogenes** and their corresponding **proto-oncogenes**. The first oncogene to be discovered using this method was Ras [3].

3.5 Ras

The types of studies described above contributed to the discovery of the **Ras oncogene**. The discovery of the Ras oncogene was very important in the cancer field, and Ras is still one of the most well-known oncogenes. Studies with Ras showed that a single mutation can cause normal cellular Ras (the **Ras proto-oncogene**) to become the Ras oncogene. Ras is part of a large superfamily of related proteins, and three key members of the family are H-Ras (Harvey-Ras), N-Ras (neuroblastoma-Ras), and K-Ras (Kirsten-Ras). Ras genes are important to understand because they are mutated in a high percentage of human cancers [25].

3.5.1 Normal Ras Compared with Oncogenic Ras

- Normal Ras

The basic function of normal Ras, the proto-oncogene, was already covered in Chap. 18. To briefly summarize, Ras functions as a molecular switch. It is active when it is bound to GTP and inactive when bound to GDP. Ras is regulated by two main proteins, a **GAP** (GTPase-activating protein) and a **GEF** (guanine nucleotide exchange factor). **GAP** is a **GTPase-activating protein**. It can bind Ras and stimulate Ras's intrinsic GTPase activity, causing GTP to hydrolyze to GDP, and thus Ras is inactivated. The GEF, in contrast, can catalyze the change of GDP to GTP on Ras so that Ras becomes active [3].

Ras has two important regions in it called **Switch 1** and **Switch 2**. In its active state, the two switch regions change their conformation to allow Ras to activate effector proteins such as Raf [26]. When GAP binds to Ras, it binds to a region that includes the Switch II region. It creates a conformational change that stabilizes the transition state for GTP hydrolysis and thus increases the rate at which Ras can hydrolyze GTP. Switch 2 must be in the correct conformation in order for GAP to bind [27]. GTP does not bind the switch regions, but it binds to a pocket region that contains a loop called a P-loop, which is adjacent to the switch regions [28] (see Chap. 18, Fig. 18.6).

- Oncogenic Ras and Cancers with Ras Mutations

The oncogenic form of H-Ras (the H-Ras oncogene) has a single point mutation, compared with wild-type Ras (the proto-oncogene). This single point mutation (Ras(G12V), originally found in bladder cancer, leads to a single amino acid change at codon 12, where glycine is changed to valine. This region is in the Switch II region, and as described above, it is an important region for GAP binding, which promotes GTP hydrolysis. Thus, this mutation causes Ras to remain in the GTP-bound (active) state by interfering with GAP binding to Ras and thereby interfering with Ras's intrinsic GTPase activity. While this mutation only changes a single amino acid, it has devastating consequences, causing Ras to remain GTP-bound and active in cells, which can lead to uncontrolled cell growth. Similar point mutations have been found in the other major Ras family members. They all cause Ras to be constitutively active (GTP-bound), most often via disruption of the intrinsic GTPase activity of Ras [3].

Ras is frequently activated in human cancers. Among the different Ras family members, K-Ras mutations are the most common Ras mutations in cancer. For example, K-Ras mutations are often found in pancreatic ductal adenocarcinoma, colorectal adenocarcinoma, and multiple myeloma. N-Ras mutations are often found in melanoma. H-Ras mutations are less common. They are sometimes found in head and neck squamous cell carcinoma [29].

3.5.2 Targeting Ras in Human Cancers

Targeting Ras has been challenging. One reason for this is that Ras has many important functions in normal cells, and inhibiting Ras can thus disrupt too many normal cell functions in addition to destroying the cancer cells [30, 31]. Nevertheless, there have been some recent promising examples of Ras inhibitors. Two of these inhibitors target a mutant K-Ras called K-Ras(G12C), a mutation that is found in about 13% of non-small cell lung cancer (NSCLC) tumors. K-Ras(G12C) has a single point mutation that changes a glycine to a cysteine. This single amino acid change is near the Switch 2 region, disrupting the interaction between K-Ras and GAP, and therefore, Ras remains in an active GTP-bound state.

Lumakras (sotorasib) [32] is a **small-molecule** inhibitor that targets **mutant K-RAS** (G12C) in **non-small cell lung carcinoma (NSCLC)**. This drug binds covalently and irreversibly to the cysteine residue that is unique to the G12C mutant and causes Ras to accumulate in a permanently GDP-bound form. The region that the drug binds to in the mutant Ras is only accessible when Ras is GDP-bound. Although the G12C mutation impairs GAP-mediated GTP hydrolysis, a small proportion of mutant Ras proteins still spontaneously hydrolyze GTP to GDP. This transient GDP-bound state exposes the binding pocket, allowing the drug to covalently attach. Since the drug binds covalently, it then locks Ras in the inactive GDP conformation, and gradually, this "locked" Ras accumulates. The fact that the drug specifically targets the mutant Ras is important because it prevents the drug from binding to normal cellular Ras. Promising results were obtained in clinical trials, and the drug received accelerated FDA approval in 2021 [33].

Another example of a drug that targets **Ras** is **Krazati (adagrasib)** [34]. This drug was FDA approved for **NSCLC** in 2022 and for **colorectal carcinoma** (CRC) in 2024. It is also a **small-molecule** drug that inhibits tumor growth in cancers that have a mutant K-Ras (G12C). Like **Lumakras**, **Krazati** also binds to the mutant KRAS and locks it in its inactive GDP-bound state [35, 36]. Both Lumakras and Krazati are given orally.

3.6 Oncogene Addiction

Cancer often starts because of a single genetic mutation, such as an activating mutation in a proto-oncogene. However, once the uncontrolled cell proliferation begins, this tends to lead to genomic instability and the accumulation of many additional mutations over time. Thus, while the cancer cells might have started as a homogeneous group, as the tumor grows, the cells may become quite heterogeneous genetically, with various cells carrying multiple genetic mutations. Interestingly, however, many cancer cells remain highly dependent on the continued activity of one specific oncogene. This is often called **oncogene addiction**. In these cases, the initiating oncogene continues to play a central role in maintaining the malignant phenotype. This is important because it impacts both research and disease treatment. While many different mutant genes may be identified in cancer cells, it is often the initiating oncogene that must be targeted to stop their growth. This is one reason why the idea of targeted inhibition of cancer cells is a growing and important field [37].

4 More Oncogenes, and Their Connections with Signaling Pathways

Src and Ras, described above, were the first two oncogenes discovered. Since their discovery, numerous oncogenes have been identified, along with other genes that, although not fitting the classical definition of oncogenes, are also aberrantly activated in cancer cells. Many of these genes encode signaling proteins, including those we talked about in Chap. 18. The next section will cover a selection of these genes, organized where possible by their roles in cell signaling.

4.1 Receptor Tyrosine Kinases (RTKs)

In normal cells, RTKs are only activated in response to ligand binding. However, some RTKs are proto-oncogenes, and they can become oncogenes. This can be caused by mutations that make them constrictively active, usually independently of ligand stimulation. RTKs that are well known for becoming oncogenes include members of the EGFR family. Several RTK family members that are important in cancer are discussed here.

4.1.1 EGFRvIII

- Functions of EGFRvIII

EGFRvIII (EGFR variant III) is a **truncated** form of the **EGF receptor,** specifically **EGFR1**. This truncated form remains in an activated state even without binding to ligand. This constitutive activation of the receptor results in continuous activation of the downstream pathways, regardless of whether ligand is present. This variant is a frequent mutation in glioblastoma (a type of brain cancer) as well as some other cancers.

Although EGFRvIII does not require a ligand for activation, it can lead to cancer cell growth by enhancing autocrine signaling loops. **Glioblastoma** cells often express both EGFRvIII and wild-type EGFR, and they produce and secrete ligands such as EGF and TGF-α. These ligands bind to receptors on the same or neighboring cells. This leads to autocrine activation of the growth-promoting signals, resulting in aggressive tumor cell growth, and it is associated with poor clinical outcomes [38, 39].

This type of cancer is difficult to treat. EGFRvIII is considered a potential drug target, but so far, there are no FDA-approved drugs to target it. There are several studies in progress, but success has been difficult due in part to the blood-brain barrier (BBB) and other complications [39, 40].

- Possible Treatments Under Investigation

Antibody-drug conjugates (ADCs) and CAR-T cell therapy are under investigation for targeting EGFRvIII. **ADCs** work by linking an antibody to a cytotoxic drug. For targeting EGFRvIII, the antibody component recognizes EGFRvIII on **tumor cells** and delivers the cytotoxic drug directly to the cancer cells. Several EGFRvIII-targeted ADCs are currently under investigation [41].

Another promising treatment under investigation is CAR-T cell therapy. In **CAR-T Cell Therapy,** the patient's T cells are isolated and then genetically modified in the lab to express a receptor that binds to EGFRvIII. These modified T cells are then transfused back into patients. The goal is to generate T cells that will bind and attack the EGFRvIII-expressing cancer cells. Some early studies using this methodology have been promising [42].

4.1.2 HER2/Neu and Breast Cancer

EGFR2 is also called HER2/Neu, and it has an important role in cancer, especially breast cancer. HER2 is somewhat different from the other EGFR family members because it does not bind to ligands. Instead, Her2 is usually activated by forming dimers with other members of the EGFR family. Once it dimerizes, its binding partner can bind the ligand, leading to the activation of signaling pathways such as the MAP kinase pathway, as well as other pathways that can lead to cell proliferation (see Chap. 18).

The HER2/Neu gene functions as an oncogene when overexpressed or if its gene is amplified. This can sometimes lead to ligand-independent activation, and it is frequently associated with a subset of breast cancers called **HER2-positive breast cancer**. Several drugs have been developed to block HER2 activity as a treatment for HER2-positive breast cancer. Some of these drugs also block other members of the EGFR family and may be beneficial for the treatment of other cancers as well [3, 43].

Herceptin (trastuzumab) [44] is a drug that is now commonly used to treat **HER2-positive breast cancer**. Herceptin is a **monoclonal antibody** that was FDA approved in 1998. It is used alone or in combination with other drugs. It specifically binds to **Her2**. By binding to Her2, **Herceptin** prevents the receptor from dimerizing and thus prevents its activation of downstream pathways [45]. Herceptin can also lead to an immune response (ADCC: antibody-dependent cellular toxicity), which specifically targets the cancer cells it binds to. Herceptin was a real breakthrough in treating HER2-positive breast cancer. Development of this highly specific drug significantly improved the prognosis for this type of cancer. Several **biosimilars** of Herceptin have also been made. These include **Ogivri** and **Herzuma,** which both target Her2 [46].

Tykerb (lapatinib) [47] is another drug for the treatment of **HER2-positive breast cancer**. It is used in combination with other treatments, such as chemotherapy or hormonal therapy. It is a **small-molecule** dual kinase inhibitor that targets **two RTKs**: **HER2** and **EGFR1 (Her1)**, both of which are overactive in some types of breast cancer, and which can form heterodimers with each other [48, 49]. Rather than binding to the ligand domain of the receptors, this drug **enters the cell by diffusing across the membrane** and binds to the **ATP-binding site** of the kinase domains of these two receptors on their cytoplasmic domains. By inhibiting their tyrosine kinase activity, the drug shuts down signaling from the receptors. **Perjeta (pertuzumab)** [50] is also a monoclonal antibody that recognizes Her2, but it binds to a different part of Her2 than Herceptin. It is used in combination with Herceptin [51].

Kadcyla (ado-trastuzumab emtansine) [52] was introduced in Chap. 1. It is an **antibody drug conjugate (ADC),** and it was FDA approved to treat **HER2-positive breast cancer** [53]. It is a biologic that contains a monoclonal antibody binding to Her2, linked to a highly toxic chemotherapeutic substance. This is an example of targeted chemotherapy delivery of a toxic agent while minimizing harm to the normal cells. The antibody part of the molecule is **trastuzumab (Herceptin)**. The drug is attached to **DM1**

(emtansine), a microtubule inhibitor that is effective at killing rapidly dividing cells. Although trastuzumab binds to the extracellular part of the receptor, receptors periodically get internalized by endocytosis. Thus, some of this drug also gets into the cell. Once in the cell, the DM1 is trafficked to the lysosome, and the DM1 is released from the rest of the molecule. Once released, it enters the cytoplasm where it does its job of interfering with microtubules and blocking cell proliferation [54]. **Enhertu (fam-trastuzumab)** [55] is also an antibody drug conjugate (ADC). It was FDA approved in 2019. It also targets a chemotherapy drug to the **HER2-positive cancer cells**, but in this case, the chemotherapy drug is a topoisomerase inhibitor [56].

HER2 and HER3 are overexpressed not only in breast cancer but also in other types of cancer. **Zenocutuzumab (Bizengri)** [57] is a **bispecific antibody** that targets both **Her2** and Her3. It was recently approved, having received accelerated approval in 2024. It is approved for tumors that contain an NRG1 fusion gene and is to be used for treating advanced unresectable or metastatic **non-small cell lung cancer and some types of pancreatic cancer** with this fusion [58]. This antibody uses a "dock and block" system. It docks to HER2, which positions itself so that it can efficiently prevent HER3 from binding to its ligand (neuregulin 1 (NRG1)). This prevents HER2/HER3 heterodimerization and consequently blocks oncogenic signaling by **HER2-HER3** heterodimers. The drug also leads to antibody-dependent cell-mediated cytotoxicity (ADCC) and which leads to cancer cell killing [59].

4.1.3 FGF Receptor (FGFR)

Another important RTK in human health is the fibroblast growth factor receptor (FGFR). There are four main FGFRs in humans, FGFR1–FGF4. They play roles in cell growth, development, wound healing, and tissue repair [60]. Improper regulation of FGFR2 is implicated in some cancers, particularly **bile duct cancer**. One type of bile duct cancer that has an especially poor prognosis is **intrahepatic iCCA (cholangiocarcinoma)**. In this type of bile duct cancer, the tumor cells originate in the liver. FGFR2 appears to play an important role in this cancer when it is either constitutively activated, overexpressed, or adopts a conformation that allows it to constitutively form a dimer. Overactivation of this receptor leads to activation of downstream signaling pathways that are independent of ligand stimulation [3, 61].

Lytgobi (futibatinib) [62] is a drug that inhibits FGFR2, FDA approved in 2022. It is being used to treat some patients with iCCA. It inhibits FGFR2 by binding directly to it [63] (Fig. 19.5). There are also other anti-FGFR drugs, such as **Truseltiq (infigratinib)** and **Pemazyre (pemigatinib)** [64], that are being used to treat **iCCA, as well as other cancers**, including bladder, gastric, and breast cancers [65]. **Truseltiq** was FDA approved in 2021, but it was removed in 2024 due to difficulties with the clinical trials [66]. **Pemazyre (pemigatinib)** was first FDA approved in 2020 and is being used for **metastatic cholangiocarcinoma** and other cancers associated with FGFR [67]. Some examples of drugs that inhibit FGFR and FGFR-activated signaling pathways are illustrated in Fig. 19.5.

4.1.4 FLT3 (Fms-Like Tyrosine Kinase 3) and Acute Myeloid Leukemia (AML)

FLT3 is an **RTK** that is normally expressed in hematopoietic stem cells. The ligand that binds to it is FLT3L. The binding of FLT3L to FLT3 activates downstream pathways such as the MAP kinase pathway, PI3 kinase pathway, and STAT pathways [68]. FLT3 signaling has an important role in the development of cells in the immune system [69]. It is activated by ligand binding, which leads to the activation of signal transduction pathways that regulate cell fate. FLT3 is frequently mutated in **acute myeloid leukemia (AML)**. These mutations lead to its constitutive activity and, thus, constitutive activation of downstream signaling pathways. Two major types of mutations in FLT3 exist. These are internal tandem duplication (ITD) and mutations in the tyrosine kinase domain (TKD).

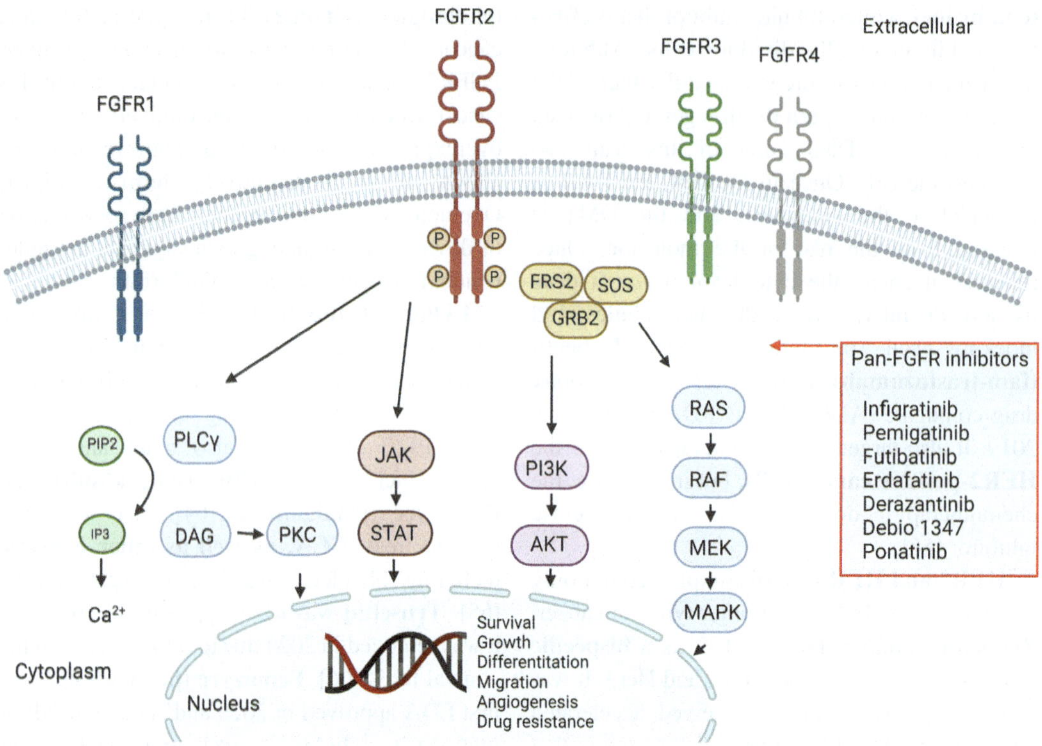

Fig. 19.5 Some anticancer drugs are inhibitors of FGFR signaling. FGFR is a family of RTKs, and its inhibition can inhibit multiple signaling pathways that are involved in the regulation of cell growth and function. FGF signaling pathways, their functions, and some FGF pathway inhibitors are shown above. (*Source*: Reproduced from Laffi, A., et al. Frontiers Oncol. (2022), 12: 860453. doi:10.3389/fonc.2022.860453. Licensed under the Creative Commons Attribution 4.0 (CC BY 4.0) license)

The ITD mutation is the most common. It is an in-frame duplication of a DNA sequence that disrupts a part of the juxtamembrane domain, which is the cytoplasmic region adjacent to the transmembrane domain. This domain usually keeps FLT3 in an inactive conformation until it is activated. The ITD mutation, therefore, makes FLT3 constitutively active.

The TKD mutation is within the kinase domain. It changes the conformation of the kinase domain, also making it constitutively active even in the absence of ligand. Both types of mutations in FLT3 are associated with poor prognosis.

Drugs that target FLT3's kinase activity have been beneficial in some patients with **AML.** These drugs include **Rydapt (midostaurin)** [70] and **Xospata (gilteritinib)** [71], which can block the activity of FLT3 with both types of mutations. **Rydapt (midostaurin)** and **Xospata (gilteritinib)** are **small molecules** that diffuse across the plasma membrane. They are **competitive inhibitors** that compete with ATP for binding to the ATP-binding site on FLT3. **Xospata** is more specific toward FLT3, while **Rydapt** can also inhibit some other RTKs [72].

4.1.5 Other Inhibitors of RTKs or Their Ligands

RTKs or ligands of RTKs are common targets for cancer therapy. Below are a few more examples:

Tagrisso (osimertinib) [73] is a **small-molecule tyrosine kinase inhibitor** that inhibits **EGFR1.** It is used for the treatment of **NSCLC** in which there are activating mutations in EGFR1 [74].

Lenvima (lenvatinib) [75] is a **small-molecule multi-kinase inhibitor** that treats advanced **thyroid cancer**. It inhibits members of the

VEGFR, FGFR, and PDGFR families as well as RET and KIT. These RTKs may be activated or overexpressed in these cancers [76].

Alecensa (alectinib) [77] is effective in treating certain types of **NSCLC**. It targets the **RTK ALK (anaplastic lymphoma kinase)**, which is overactive in some types of NSCLC [78].

Sutent (sunitinib) [79] is used to treat various kinds of cancer, including renal cell carcinoma (RCC; **kidney cancer**), some **gastrointestinal stromal tumors** (GISTs), and **pancreatic neuroendocrine tumors**. It targets several **RTKs,** including **VEGFR**, **PDGFR**, and **KIT** [80].

Fotivda (tivozanib) [81] is a VEGFR inhibitor that was FDA approved in 2021 to treat **RCC (kidney cancer)**. By inhibiting the VEGFR, it inhibits angiogenesis, which is an important mechanism for treating RCC [82].

Tecentriq (atezolizumab) + Avastin (bevacizumab) is an immunotherapy combination drug. It is a **biologic** used to treat **hepatocellular carcinoma (HCC)**. **Avastin** uses a monoclonal antibody to target VEGF-A. VEGF-A is the ligand that binds to the **VEGFR** and has an important role in angiogenesis. The other component of the drug, **Tecentriq,** is an immune checkpoint inhibitor. This type of inhibitor, discussed in more detail in Chap. 24, helps the immune system target the cancer cells. This drug combination prevents angiogenesis, cutting off the blood supply to the cell, and at the same time helps kill the cells due to an immune reaction [83].

4.2 Raf

4.2.1 Raf in Human Cancers

Signaling proteins that function downstream to RTKs and other receptors can also be proto-oncogenes and can become oncogenes when they are improperly activated. Ras, as discussed above, is an important example. Raf, which normally gets activated by Ras, is also constitutively activated in certain types of cancer. Human cells have three main Raf proteins: A-Raf, B-Raf, and C-Raf (Raf-1). They are all serine/threonine kinases with important roles in the MAP kinase pathway and other pathways (see Chap. 18). B-RAF is the most well studied of the different Raf proteins and the most commonly associated with cancer. It is often mutated in cancer, especially in melanoma, a cancer that develops from the pigment cells of the skin. The most common BRAF mutation in cancer is V600E (valine to glutamate at amino acid 600). This point mutation is found in approximately 50% of melanomas and some other cancers. This point mutation causes B-RAF to be constitutively active, regardless of the presence or absence of extracellular signals [84].

4.2.2 Drugs That Target Raf

Activated Raf is considered a promising target for cancers associated with inappropriate Raf activation. Drugs have been developed that bind directly to the Raf kinase domain, targeting the ATP binding site. These are competitive inhibitors because they prevent ATP from binding to Raf and thereby inhibit substrate phosphorylation [85, 86].

Zelboraf (vemurafenib) [87] is a **small-molecule** drug that targets **B-RAF-V600E.** It targets the ATP-binding domain of mutant Raf but not the normal wild-type Raf, thus reducing side effects. It is effective in treating some melanoma patients who have a BRAF V600E mutation. **Tafinlar (dabrafenib)** [88] and **Braftovi (encorafenib)** [89] are small-molecule anti-**B-Raf** inhibitors used to treat **melanoma and other cancers**, targeting the V600E mutant. **Ojemda (tovorafenib)** [90] is a small-molecule Raf inhibitor that was FDA approved in 2024. It inhibits the B-Raf-V600E mutant as well as other mutations and wild-type Raf. It is designed to treat patients with **low-grade glioma (LGG)** who have a rearrangement or mutation in the B-Raf gene [86, 91–93].

4.3 PI3 Kinase (PI3K)

4.3.1 Roles for PI3K in Cancer

As discussed in Chap. 18, PI3K is a lipid kinase that is activated by some RTKs and other receptors [94]. It phosphorylates PIP2 and activates downstream pathways, particularly those that regulate cell survival. PI3K has a catalytic domain (P110) and a regulatory domain (p85), both encoded by separate genes. Mutations in **PI3KCA (p110)** have been found in several human cancers, including breast, colorectal, ovarian, brain, and other cancers. PIK3CA mutations have been found in about 25% of breast cancers and are associated with a poorer prognosis compared with some other breast cancers. These mutations are often point mutations that lead to constitutive activation of the kinase. Various mutations have been identified in PI3K in cancer. The most common PI3K mutations are in two "hotspots" in the PI3KCA (p110) gene, specifically in exons 9 and 20. Mutations in these regions account for about 80–90% of all PIK3CA mutations in human cancers. These mutations include **H1047R** (substitution of histidine for arginine) and **E545K** (substitution of glutamic acid for lysine). The E545K mutant may disrupt the inhibition of p110 by p85, thus leading to constitutive activation of the kinase. The exact mechanism for the other mutations in PI3KCA may differ and may not be thoroughly understood. Mutations in PI3K lead to dysregulation of signaling pathways, such as the PI3K/AKT/mTOR pathway, which can lead to uncontrolled cell growth and survival. Some drugs that target the PI3K/AKT/mTOR pathway may be promising for treating these cancers.

4.3.2 PI3K Inhibitors

Zydelig (idelalisib) [95] is an example of a drug that targets the catalytic subunit of an isoform of **PI3K (PI3Kδ)**. It is approved for some types of blood cancer patients who have previously received other treatments. For example, it is approved for the treatment of **chronic lymphocytic leukemia (CLL)** [96–98].

Aliqopa (copanlisib) [99] also **inhibits the PI3K catalytic domain**, but it inhibits the isoforms PI3Kα and PI3Kδ [100]. It is used to treat **follicular lymphoma** (a white blood cell cancer). Several other PI3K inhibitors have been FDA approved more recently, some of which are used in combination with other treatments [3].

Copiktra (duvelisib) [101] is a **small-molecule PI3K inhibitor**. It inhibits the isoforms PI3Kδ and PI3Kγ. It binds specifically to the ATP-binding region, which is similar in the different PI3K family members. It is used to treat **chronic lymphocytic leukemia (CLL)** and follicular lymphoma, a cancer that affects B cells [97, 102]. Overall, some of the PI3 kinase inhibitors have some serious side effects, which limit their use.

4.4 AKT and AKT Inhibitors

PI3K signaling is closely linked to AKT signaling, which is associated with cell survival. AKT itself is not necessarily activated by mutation, but it is often activated due to upstream enzymes, particularly PI3K. When activated, AKT leads to increased cell survival, cell growth, proliferation, metabolism, and other processes that provide a growth advantage. It therefore also has potential as a target for cancer treatment.

Truqap (capivasertib) [103] is the first FDA-approved (2023) **AKT inhibitor** (developed by AstraZeneca (keep?)). It is a **small-molecule** and competitive inhibitor that binds to the ATP-binding site of AKT, thereby inhibiting its kinase activity. It is used to treat **hormone-positive breast cancer** [104, 105].

4.5 mTOR and mTOR Inhibitors

mTOR signaling is also closely linked to AKT and has critical roles in cell survival and proliferation. mTOR is not an oncogene, but it is tightly connected to oncogenic cell growth. Rapamycin is an mTOR inhibitor, and it is used as an immunosuppressant. It is also used to treat certain types of cancer and, however, often used in combination with other treatments. By inhibiting the **mTOR** pathway, which plays a role in cell growth and proliferation, **rapamycin** can slow or inhibit **cancer cell** growth [106].

Afinitor (everolimus) [107], also called **Zortress**, is an analogue of rapamycin [108]. It functions similarly but has improved viability. It inhibits the growth of some cancer cells via targeting **mTOR**. Both Afinitor and rapamycin can have side effects due to their immunosuppressive functions but are nonetheless considered beneficial due to their ability to **slow the growth of cancer cells** [109]. **Torisel (temsirolimus)** [110] is another mTOR inhibitor. It is used largely for the treatment of **renal cancer**. It also binds to FKB12, an mTOR binding protein, and inhibits mTOR [111].

4.6 TNF and NFKB

Cytokine signaling pathways are also important in cancer. Several components of the TNF/NFKB pathway, for example, have oncogenic roles, even though some of them not classically oncogenes. NFKB is sometimes constitutively active in cancers, as well as several members of the NFKB pathway. Several inhibitors have been made, targeting the NFKB pathway.

Adcetris (brentuximab vedotin) [112] is a drug that targets **CD30**, a **TNF receptor family member** also known as TNFRSF8 (TNFR Superfamily Member 8). It is found on activated B and T cells and is overexpressed in some cancers, such as Hodgkin lymphoma, ALCL, and certain T-cell lymphomas. **Adcetris (brentuximab vedotin)** is an **antibody drug conjugate (ADC)**. It contains a monoclonal antibody against CD30 and is linked to the cytotoxic drug MMAE (monomethyl auristatin E), a microtubule disruptor. It is designed to direct the drug to tumors that overexpress CD30. It is used to treat several types of lymphomas. The drug binds to the receptors and gets internalized by endocytosis. Once in the cell, it can stop the growth of rapidly dividing cells by interfering with microtubules, leading to cell death [113, 114].

Velcade (bortezomib) [115] is a **small-molecule** drug that prevents the degradation of IκB, thereby inhibiting NF-κB activation. IkB is normally targeted for proteasomal degradation after it is ubiquitinated. **Velcade is a proteasome inhibitor**. It was, in fact, **the first proteosome inhibitor** to be approved by the FDA. By inhibiting proteasomes, it prevents IkB degradation and consequently inhibits the activation of NF-kB. It has been approved for treating **multiple myeloma** and **mantle cell lymphoma**. It was FDA approved in 2003 [116].

4.7 Myc

Myc is a transcription factor. Transcription factors function as terminal effectors of signaling pathways, leading to the regulated transcription of target genes in response to a stimulus. An important example is **Myc**. Myc is a transcription factor that gets upregulated in response to several signaling pathways, such as the ERK pathway and the PI3K pathway. Myc has important roles in cell signaling and regulating cell growth, survival, and differentiation. It forms a heterodimer with another protein, **Max**. The Myc/Max heterodimer binds to DNA sequences called E-boxes and regulates transcription of various genes that have E-boxes in their regulatory elements (Fig. 19.6). This can lead to the regulation of cell function and proliferation [117, 118]. The reason for this is that Myc can regulate the expression of target genes that drive cell cycle progression and cell division. For example, Myc can regulate Cyclin D2 and Cdk4 [119], which are important cell cycle regulatory proteins. It can also regulate genes crucial for metabolism [8] and those involved in ribosome biogenesis, which is essential for protein production [120]. Thus, it regulates genes that are important for increasing cell size and for supporting cell metabolism. It can also repress certain genes that have roles in cell cycle arrest, such as p21 and p15, which are both CKIs (cyclin-dependent kinase inhibitors) [121].

Myc is a proto-oncogene, and improper activation of Myc can cause it to function as an **oncogene**. Generally, improper activation of the Myc gene involves its overproduction. This is often due to **DNA amplification**, **point mutations**, or **changes in regulatory elements**

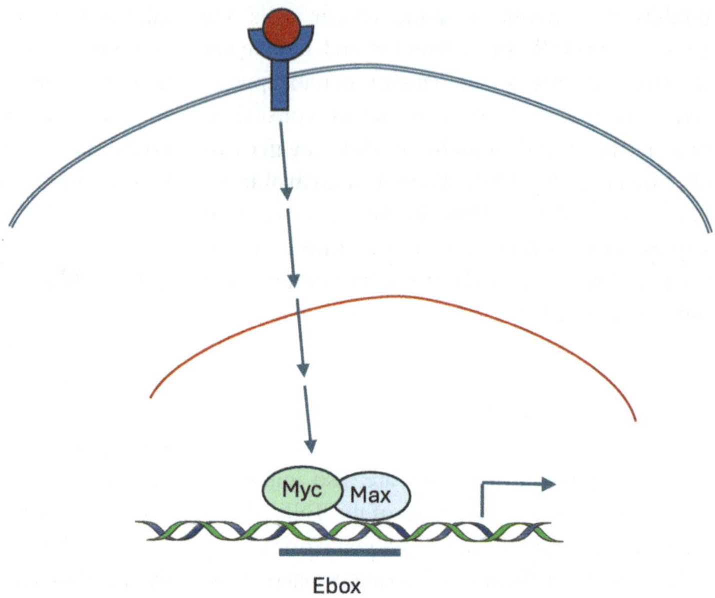

Fig. 19.6 Myc. Myc is a transcription factor. It responds to stimuli and binds to E-box regions in target genes in combination with the transcription factor Max

that control Myc gene expression. **DNA amplification** can result from errors in DNA replication. **Point mutations** in the gene are small changes that may lead to improper activation of the protein. For example, such mutations can lead to the production of a protein that is more stable than normal, in contrast to the rapid turnover that is typical of the Myc protein. **Changes in regulatory elements** that control Myc gene expression can occur due to a chromosomal translocation, which places a strong regulatory element next to the Myc gene. Other mutations may affect Myc's interaction with Max [3, 4, 118].

Burkitt lymphoma is an example of a cancer that involves a chromosomal translocation involving Myc. This translocation causes the Myc gene to be placed under the transcriptional control of a strong enhancer that normally regulates an immunoglobulin heavy chain (IgH) gene. The translocation is between chromosome 14 (containing the heavy chain IgH gene) and 8 (containing the Myc proto-oncogene). The result is abnormally high levels of Myc gene expression. The Myc oncogene is also implicated in other types of cancer, such as lymphoma, leukemia, breast cancer, and lung cancer. In these cases, Myc is activated by mutation, DNA amplification, or overexpression. While Myc plays a key role in several types of cancer, unfortunately, directly targeting Myc with drugs remains challenging [3].

4.8 Notch Signaling

Abnormalities in notch signaling have links to multiple types of cancer [4, 122]. Notch signaling was covered in Chap. 18. As a brief review, Notch is a single-pass transmembrane receptor that binds to a membrane-bound ligand called Delta on the surface of neighboring cells. Other ligands that are related to delta are called DLLs. Notch activation by Delta or DLLs involves proteolytic cleavage of NOTCH in three steps called S1, S2, and S3. It is cleaved by proteases, which include γ-secretase, which cleaves at the S2 stage. Once it is cleaved, the cytoplasmic domain of Notch travels to the nucleus and, along with other nuclear proteins, plays a role in the regulation of target genes.

Desmoid tumors are rare soft tissue tumors. They are often benign, but they can be aggressive, cause pain, and have deleterious effects on nearby organs and tissues [123]. Desmoid tumors are associated with fibroblastic overgrowth (an increased number of fibroblasts and increased fibroblast activity, which includes ECM production). Notch signaling is important in fibroblast growth, and components of the notch pathway have been shown to be upregulated in some desmoid tumors [124]. For this reason, targeting notch is an approach for treating desmoid tumors. A good way to inactivate Notch is to prevent its cleavage, which is essential for its activity.

Ogsiveo (nirogacestat) [125] is an FDA-approved (2023) Notch inhibitor that targets **γ-secretase** and is used to treat **desmoid tumors**. It is a **small-molecule** drug that functions as an **enzyme inhibitor**. It binds the active site of γ-secretase and prevents it from cleaving its substrate, Notch [126].

A high percentage of **small cell lung cancer (SCLC)** cells overproduce one of the delta-like ligands, DLL3, on their surface [127, 128]. **Imdelltra (tarlatamab-dlle)** [129] is a drug that targets **DLL3**, and it was FDA approved in 2024 to treat SCLC. This drug is called a **T-cell engager**. It is a bispecific antibody that binds **DLL3** on the lung cancer cells and **CD3**, which is found on **T cells**. In this way, it brings the cancer cells into proximity to the T cells, thus encouraging the T cells to attack the cancer cells. This type of drug is a **biologic** as it is antibody based. It functions differently from most other monoclonal antibodies, however, as it is genetically engineered in the lab to bind to two targets. It is considered a **monoclonal antibody-based biologic** [128].

Other approaches have been investigated for targeting the Notch pathway to treat various types of cancer. These include other approaches such as inhibiting DLLs, inhibiting other proteases that cleave Notch, blocking both the extracellular and intracellular domains of Notch, and inhibiting the Notch transcription complex [130–133].

4.9 Gleevec (Imatinib)

Gleevec (imatinib) [134] is a very important cancer drug. Although it does not target a classical oncogene, it targets a cellular protein that is highly linked to dysregulation of cell growth. It is an excellent example of how drugs can target proteins that disrupt normal growth control with a high success rate. Gleevec was a breakthrough drug for cancer treatment (FDA approved in 2001). It was developed to treat **chronic myelogenous leukemia (CML)**. In most cases of CML, there is a chromosomal translocation in which the gene encoding the **tyrosine kinase c-abl** (on **chromosome 9**) is randomly inserted next to the bcr (breakpoint cluster region) gene on **chromosome 22**. The chromosome resulting from this fusion is called the **Philadelphia chromosome**. The result of this translocation is the overexpression of a fusion gene encoding a protein called **Bcr-Abl**. Abl is a tyrosine kinase. Bcr has a low level of kinase activity, but it is known mostly for its role as a scaffold protein. The Bcr-Abl fusion protein, however, functions as a constitutively active tyrosine kinase. It phosphorylates a number of different targets that, in turn, lead to uncontrolled proliferation in CML (Fig. 19.7).

Identifying the Philadelphia chromosome had important implications. For example, it helped solidify the idea that cancer is **clonal in origin**. This is because within a single CML patient, all their CML cells had the same fusion with the

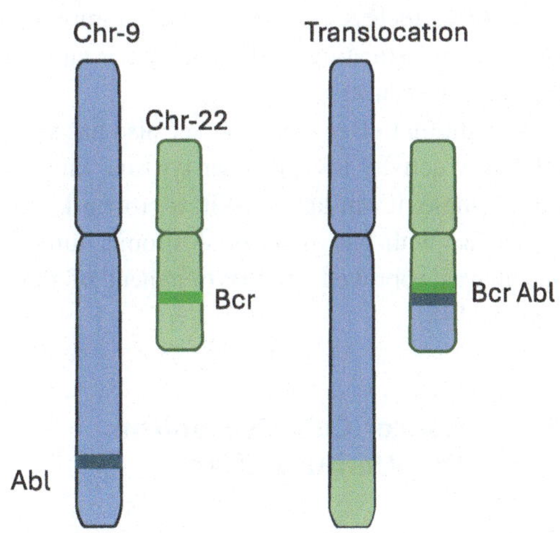

Fig. 19.7 Philadelphia chromosome. The Philadelphia chromosome is a translocation between chromosomes 9 and 22, found in Chronic Myelogenous Leukemia (CML). The translocation leads to a fusion of the BCR and Abl gene, and the synthesis of the BCR/ABL protein

exact same breakpoint in the DNA. This means that the fusion occurs at the same exact location within the BCR and Abl genes, indicating that the cancer cells are all clones. Between different patients, translocation still occurs between BCR and Abl, but the breakpoints may differ slightly. An even more important reason that discovering BCR-ABL was groundbreaking is that it provided a clear target for researchers to identify a drug that could treat the disease [4, 135].

Gleevec (imatinib) was developed for this purpose. It is a highly competitive inhibitor of BCR-Abl. It binds directly to the ATP-binding site on BCR-Abl and inhibits ATP from binding, thereby inhibiting its kinase activity. Gleevec is used successfully to treat CML. Since Bcr-Abl is found only in tumor cells, this drug targets tumor cells specifically, minimizing side effects. The drug has an over 80% 10-year overall survival rate. This led to a great deal of promise that similarly effective cancer drugs could be made quickly [136, 137]. This level of success has been more elusive for other cancers, however, unfortunately. CML had a very clear chromosomal abnormality found in most patients. Most other cancers have turned out to be more complex. They can be driven by multiple mutations, there is heterogeneity between different patients, and cells have mechanisms for developing resistance. Nevertheless, the finding was an important step for investigating and developing targeted therapies for cancer.

In addition to Bcr-Abl, Gleevec also has some inhibitory activity toward other tyrosine kinases. For example, it can target **c-kit**, a protein kinase associated with gastrointestinal stromal tumors, and it was approved for the treatment of these tumors [135].

5 Cancer Cell Metabolism, the Warburg Effect

Cancer cells do not always carry out metabolism in the same way that normal cells do. Instead, many cancer cells alter their metabolism to support their rapid cell division and environmental conditions. An important example is the Warburg effect, in which many cancer cells metabolize glucose through aerobic glycolysis.

5.1 The Warburg Effect and Aerobic Glycolysis

The most efficient way for normal cells to generate ATP from glucose in the presence of oxygen is via cellular respiration pathways, which include glycolysis, decarboxylation of pyruvate, the Krebs cycle, and finally, oxidative phosphorylation in the mitochondria and the electron transport chain.

Under conditions of low oxygen (hypoxia), however, the electron transport chain cannot operate due to the lack of oxygen. As a result, the Krebs cycle slows down or halts, and pyruvate is converted to lactate instead of proceeding through the Krebs cycle. This regenerates NAD^+, which is necessary to keep glycolysis running. This is important since NAD^+ cannot be generated when the ETC is not functioning. This type of metabolism sometimes occurs temporarily in highly active muscle cells. It is referred to as lactic acid fermentation and is a type of anaerobic metabolism. It is described in more detail in Chap. 10.

Since the metabolism of glucose to lactate tends to occur in low oxygen conditions, it was a surprise when Otto Warburg found in 1920 that many cancer cells preferentially metabolize glucose to lactate, even when oxygen is available. This is called aerobic glycolysis because it functions like anaerobic metabolism, but it occurs even in the presence of oxygen. This metabolic behavior is now known as the **Warburg effect**. Aerobic glycolysis is a much less efficient system for generating ATP compared to full aerobic respiration. Aerobic glycolysis generates 2 ATP molecules from one glucose molecule, while full aerobic respiration generates 30–32 ATPs. Nevertheless, many cancer cells prioritize aerobic glycolysis even though the output of ATP is much lower than oxidative phosphorylation. Surprisingly, aerobic glycolysis has advantages and supports the growth of cancer cells [8, 138–140]. Some aspects of aerobic glycolysis and its advantages for rapidly dividing cancer cells are discussed below.

5.2 Increased Glycolysis Is a Consequence of the Warburg Effect

One interesting outcome of the Warburg effect is that it encourages an increase in glycolysis. There are several reasons for this. One reason is that the lower yield of ATP from glycolysis (as opposed to oxidative phosphorylation) reduces feedback inhibition that would typically result from high ATP levels that would be obtained from aerobic respiration. Another reason is because of NAD+. In the Warburg effect, cells regenerate NAD^+ rapidly by converting pyruvate to lactate, bypassing the need to rely on the slower electron transport chain for NAD^+ regeneration. This rapid accumulation of NAD^+ supports increased levels of glycolysis. Cancer cells also have an increased level of glucose transporters in their membranes, which leads to a greater level of glucose influx. This also contributes to the high rate of glycolysis [139, 141].

Another factor that increases the rate of glycolysis in cancer cells undergoing aerobic glycolysis is the regulation of the transcription factor **Hif-1** (Hif-1, hypoxia-inducible factor-1, is also covered in Chap. 21). Normally, Hif-1 has a role in increasing the rate of glycolysis during hypoxia (low oxygen). Hif-1 increases the expression of several genes. Among these are genes encoding several enzymes important for glycolysis, as well as genes encoding glucose transporters. This helps ensure sufficient ATP generation from glycolysis alone, which is important in hypoxic conditions. In normal cells with normal oxygen levels, Hif-1 is hydroxylated by **prolyl hydroxylase,** which causes it to be ubiquitinated and destroyed by proteasomes. Thus, under these normal oxygen conditions, Hif-1 is continuously degraded, thus preventing its accumulation, and therefore, it has no opportunity to be functional. **Prolyl hydroxylase** depends on **oxygen** (as well as α-ketoglutarate and vitamin C). Thus, in hypoxic (low oxygen) conditions, **decreased O2** prevents the **prolyl hydroxylase** from hydroxylating **Hif-1**. Consequently, **Hif-1 is not degraded** and the result is increased glycolysis [142, 143].

Although Hif-1 helps cells deal with occasional hypoxic conditions, cancer cells use Hif-1 for their own benefit, even in normoxia (normal oxygen) conditions. In cancer cells undergoing aerobic glycolysis, the Krebs cycle is not functioning, even if oxygen is present. Consequently, some Krebs cycle intermediates, such as succinate and fumarate, accumulate and can move from the mitochondria to the cytosol. This is important because these intermediates function as **competitive inhibitors** of **prolyl hydroxylase**. Thus, these **intermediates**, which accumulate because the cycle is inactive, **prevent the degradation of Hif-1**. The increased Hif-1 activity, in turn, contributes to the increased rate of glycolysis in cancer cells since glycolysis intermediates are among the proteins whose genes are regulated by Hif-1 [144].

5.3 Some Additional Growth Advantages Due to Oxidative Glycolysis in Cancer Cells

5.3.1 The Increased Production of Metabolic Intermediates

The increase in glycolysis in cancer cells offers several advantages beyond the generation of ATP. Importantly, it ensures a supply of biosynthetic precursors from the glycolysis pathway, which are needed for rapid cell proliferation [145]. These include precursors for some amino acids, fatty acids, as well as nucleotides when the pentose phosphate shunt is engaged. The increase in glycolysis can also lead to an increase in NADPH, which can be used for FA synthesis.

5.3.2 Advantages of Lactic Acid Production

Producing lactic acid may also have certain advantages for cancer cells. Lactic acid can exit the cell and lead to increased acidity in the surrounding tissues. This can suppress the functions of some cells, such as immune cells in the tumor microenvironment (TME). This may weaken the ability of these immune cells to target and destroy the cancer cells [146–148].

5.3.3 Adaptation to Environmental Conditions

Cancer cells do not necessarily begin in a low-oxygen environment, but as the tumor grows, its size can restrict the flow of oxygen to some of the tumor cells. Thus, over time, parts of the tumor can become hypoxic. Therefore, another reason that cells may switch to aerobic glycolysis is to prepare the cells for hypoxic conditions that are likely to develop as the tumor progresses [149].

5.4 Mutations or Altered Expression of Some Genes Affects Cancer Cell Metabolism

There are numerous genes whose mutation or activation can lead to alterations in metabolism and encourage the Warburg effect in cancer cells. A few of these are described below.

5.4.1 Hexokinase

The enzyme **hexokinase** catalyzes the conversion of glucose to glucose-6-phosphate, the first step in glycolysis [150]. This step is essential for trapping glucose in the cell and committing it to further processing. In many cancers, **hexokinase-2 (HK2)** is dysregulated. This can occur at multiple levels, including the level of gene transcription, protein stability, and post-translational modifications. Overexpression or hyperactivation of **hexokinase** can enhance glycolysis. This provides a supply of ATP and various metabolic intermediates, helping the tumor to grow and helping support the Warburg effect. Hexokinase also has other functions in addition to its role in glucose metabolism, such as promoting immune evasion, stemness, resistance to apoptosis, regulation of autophagy, and increased tumor invasion and migration [150, 151].

5.4.2 c-Myc

c-Myc was discussed above as an important oncogene. It regulates the expression of various genes that are important for cell proliferation, and too much Myc can lead to oncogenesis. C-Myc is a transcription factor and, in addition to some of the target genes discussed above, can also lead to upregulated expression of some glycolytic genes, including enzymes like hexokinase 2, enolase, and lactate dehydrogenase. It can also regulate the expression of other genes that are important for metabolism. For example, it regulates the gene for the glucose transporter GLUT1. By increasing the transcription of these genes, c-Myc enhances glycolytic flux (the processing of glucose through metabolic pathways) and contributes to the Warburg effect [120].

5.4.3 K-Ras

K-Ras is a well-known oncogene, also discussed above. When it is active, Kras and other Ras family members can activate signaling pathways that lead to the regulation of genes, or in some cases, increased production or stability of proteins, that lead to cell growth regulation. For example, Ras can lead to increased transcription of the gene for Cyclin D1 [152, 153], c-Myc [154], and Bcl-2 [155], which are involved in cell cycle progression, cell proliferation, and cell survival. K-Ras mutations are common in cancers that exhibit metabolic changes. Activated K-Ras can regulate the expression of glycolytic enzymes and glucose transporters, which enhance glycolytic flux and support the Warburg effect. For example, it regulates the transcription of the gene for GLUT1, a glucose transporter, hexokinase 2, and phosphofructokinase 1 (PFK1), the rate-limiting enzyme in glycolysis [156, 157].

FDA-Approved Drugs Discussed in This Chapter

Commonly used brand name	Generic name	First FDA approval year	Major conditions treated	Main target
Adcetris	*Brentuximab vedotin*	2011	Hodgkin lymphoma	CD30
Afinitor	*Everolimus*	2009	Various cancers	mTOR
Alecensa	*Alectinib*	2015	NSCLC (ALK+)	ALK
Aliqopa	*Copanlisib*	2017	Follicular lymphoma	PI3K
Braftovi	*Encorafenib*	2018	Melanoma (BRAF V600E)	BRAF
Copiktra	*Duvelisib*	2018	CLL, follicular lymphoma	PI3K
Enhertu	*Fam-trastuzumab deruxtecan*	2019	HER2+ breast cancer	HER2
Fotivda	*Tivozanib*	2021	RCC	VEGFR
Gleevec	*Imatinib*	2001	CML, GIST	BCR-Abl
Herceptin	*Trastuzumab*	1998	HER2+ breast cancer	HER2
Imdelltra	*Tarlatamab-dlle*	2024	SCLC	DLL3
Herzuma	*Trastuzumab-pkrb*	2018	Her2+ breast cancer	Her2
Kadcyla	*Ado-trastuzumab emtansine*	2013	HER2+ breast cancer	HER2
Krazati	*Adagrasib*	2022	NSCLC, colorectal cancer	KRAS
Lenvima	*Lenvatinib*	2015	Thyroid cancer	VEGFR, FGFR, PDGFR, RET, KIT
Lumakras	*Sotorasib*	2021	NSCLC (KRAS G12C)	KRAS
Lytgobi	*Futibatinib*	2022	iCCA	FGFR2
Ogivri	*Trastuzumab-dkst*	2017	Her2 _ breast cancer	Her2
Ogsiveo	*Nirogacestat*	2023	Desmoid tumors	Gamma-secretase (Notch signaling
Ojemda	*Tovorafenib*	2024	Low-grade glioma	BRAF
Pemazyre	*Pemigatinib*	2020	Cholangiocarcinoma	FGFR
Perjeta	*Pertuzumab*	2012	HER2+ breast cancer	HER2
Rydapt	*Midostaurin*	2017	AML (FLT3+)	FLT3
Sutent	*Sunitinib*	2006	RCC, GIST, pNET	VEGFR, PDGFR, KIT
Tafinlar	*Dabrafenib*	2013	Melanoma (BRAF V600E)	BRAF
Tagrisso	*Osimertinib*	2015	NSCLC (EGFR+)	EGFR
Tecentriq + Avastin	**Atezolizumab +bevacizumab**	**2020**	HCC	PD-L1, VEGF-A
Torisel	Temsirolimus	2007	Renal cancer	mTOR
Truqap	**Capivasertib**	**2023**	Breast cancer	AKT
Tykerb	Lapatinib	2007	HER2+ breast cancer	HER2, EGFR
Velcade	Bortezomib	2003	Multiple myeloma	Proteasome
Xospata	Gilteritinib	2018	AML (FLT3+)	FLT3
Zelboraf	Vemurafenib	2011	Melanoma (BRAF V600E)	BRAF
Zenocutuzumab	**Bizengri**	**2024**	NSCLC, pancreatic cancer	HER2, HER3
Zydelig	Idelalisib	2014	CLL	PI3Kdelta

Definitions

Angiogenesis The formation of new blood vessels from preexisting vessels.

Apoptosis Programmed cell death, a tightly regulated process that eliminates damaged or unnecessary cells.

Bcr-Abl A fusion protein with constitutive tyrosine kinase activity, associated with chronic myelogenous leukemia (CML). Targeted by the drug Gleevec (imatinib).

Cyclin-dependent kinases (CDKs) Enzymes that regulate the cell cycle. CDKs can be dysregulated in some cancers.

Cholangiocarcinoma A type of bile duct cancer.

Clonal expansion The process by which a single cell with mutations proliferates to form a tumor mass.

Confluence In cell culture, cells reach confluence when they have covered the available surface in a monolayer.

Constitutively active The state in which an enzyme remains active, often independent of both external and internal signals.

Cyclins Proteins with an important role in the cell cycle. Cyclins are necessary for the activation of CDKs.

Driver mutation A genetic alteration that directly contributes to cancer development by conferring a growth advantage.

Foci Abnormal clusters of cells that have lost their normal growth controls.

Genomic instability Increased mutation rates due to defective DNA repair, allowing cancer cells to evolve rapidly.

Glioblastoma A type of brain cancer.

Human epidermal growth factor receptor 2 (HER2) A receptor tyrosine kinase (RTK) that promotes cell proliferation. Overexpressed in some breast cancers.

Imatinib (Gleevec) A tyrosine kinase inhibitor used to treat cancers like CML by targeting BCR-ABL.

Invasion In cancer, invasion refers to the ability of cancer cells to break through surrounding tissues and basement membranes and invade other tissues.

Ligand-independent activation The constant activity of a receptor, without the need for ligand binding.

Melanoma A type of skin cancer.

Metastasis The spread of cancer from its original site to distant organs or tissues.

Oncogene/proto-oncogene A proto-oncogene is a normal cellular gene that promotes cell growth and survival under controlled conditions. When it acquires an activation mutation or is overexpressed, it can become an oncogene, leading to uncontrolled cell proliferation and cancer. Activating mutations on one allele are sufficient to lead to uncontrolled cell growth.

Passenger mutation A mutation that occurs during cancer progression, but unlike a driver mutation, does not directly contribute to malignancy.

Philadelphia chromosome A chromosomal abnormality caused by translocation between chromosomes 9 and 22, producing the BCR-ABL oncogene, found in CML.

PI3K-AKT pathway A signaling pathway that promotes cell survival. It can be overactive in some cancers.

Replicative immortality A hallmark of cancer where cells acquire the ability to divide indefinitely.

Retrovirus A type of RNA virus. It infects the cell and inserts its entire genetic content (in the form of RNA) into the cell. This RNA gets reverse-transcribed into DNA in the host cell, catalyzed by **reverse transcriptase**. This DNA then integrates into the host cell genome.

Sarcoma A type of connective tissue cancer.

Transformation (oncogenic transformation) In the field of cancer biology, transformation refers to the process by which cells acquire growth characteristics associated with cancer, such as uncontrolled cell growth. The process that cells undergo to reach this state is called **transformation** or oncogenic **transformation.**

Translocation A rearrangement of parts between nonhomologous chromosomes.

Trastuzumab (Herceptin) A monoclonal antibody drug that targets HER2 receptors in HER2+ breast cancer.

Tumor suppressor gene A gene that normally prevents uncontrolled cell growth. Loss or inactivation of such a gene can lead to cancer. Mutations that inhibit activity must occur on both alleles in order to cause uncontrolled cell growth and cancer.

Warburg effect A phenomenon where cancer cells favor glycolysis over oxidative phosphorylation, even in oxygen-rich environments. This is called aerobic glycolysis.

Study Questions

Multiple-Choice Questions

1. Which class of genes, when they have an activating mutation, may contribute to uncontrolled cell division and tumor formation?
   A. Tumor suppressor genes
   B. Proto-oncogenes
   C. DNA repair genes
   D. Checkpoint genes
2. The Philadelphia chromosome results from a translocation between which two chromosomes?
   A. 9 and 21
   B. 8 and 14
   C. 9 and 22
   D. 13 and 17
3. HER2 is an example of which type of protein?
   A. Tumor suppressor
   B. GTPase
   C. Receptor tyrosine kinase
   D. DNA repair enzyme
4. Imatinib (Gleevec) is used to treat cancers involving which oncogenic fusion protein?
   A. BCL2
   B. HER2
   C. BCR-ABL
   D. EGFR
5. Which hallmark of cancer is associated with the formation of new blood vessels to supply nutrients to tumors?
   A. Invasion and metastasis
   B. Inducing angiogenesis
   C. Deregulating cellular energetics
   D. Avoiding immune destruction
6. Which of the following could be most likely to be a proto-oncogene?
   A. Gene that encodes a protein that promotes repair of damaged DNA
   B. Gene that encodes a protein that promotes cell differentiation
   C. A gene that encodes a protein that promotes cell proliferation
   D. A gene that encodes a protein that puts a brake on the cell cycle
   E. A gene that encodes the Rb protein
7. Which of the following accurately describes the role of a GAP in Ras signaling?
   A. GAP leads to activation of Ras by promoting GTP binding
   B. GAP leads to inactivation of Ras by promoting GTP hydrolysis
   C. GAP stabilizes Ras in its active GTP-bound form
   D. GAP prevents Ras from binding to effector proteins
   E. GAP leads to translocation of Ras to the nucleus
8. When bound to ___, Ras is ______. (Select<u>all</u>that apply)
   A. GTP; active
   B. GDP; active
   C. GTP; inactive
   D. GDP; inactive
9. Which of the following best describes the Warburg effect observed in cancer cells?
   A. Increased reliance on oxidative phosphorylation even under low oxygen conditions
   B. Preferential use of glycolysis to generate energy even in the presence of oxygen
   C. Exclusive use of fatty acid oxidation for energy production

   D. Complete suppression of glucose uptake by cancer cells
10. What is a major consequence of the Warburg effect in cancer cell metabolism?
   A. Decreased production of lactate and increased ATP yield
   B. Increased lactate production and acidification of the tumor microenvironment
   C. Reduced glucose uptake and increased mitochondrial activity
   D. Enhanced efficiency of energy production with minimal metabolic waste

Fill-in-the-Blank Questions

11. The type of cancer that develops from the pigment cells of the skin is called: ______________.
12. _____ is a monoclonal antibody that targets HER2 in breast cancer therapy.
13. Tumors may secrete _____ to promote the growth of new blood vessels.
14. The _____ chromosome is commonly found in chronic myeloid leukemia and is a result of a translocation.
15. The important residue that is present in c-Src but deleted in v-Src is: ______________.
16. The protein that catalyzes the exchange of GDP for GTP on Ras, thus activating Ras, is called a ______________.

Short-Answer Questions

17. Describe the role of telomerase in cancer cell immortality.
18. How do cancer cells induce angiogenesis and why is it important?

Answer Key

Multiple-Choice Questions

1. **Answer**: B
2. **Answer**: C
3. **Answer**: C
4. **Answer**: C
5. **Answer**: B
6. **Answer**: C
7. **Answer**: B
8. **Answer**: A, D
9. **Answer**: B
10. **Answer**: B

Fill-in-the-Blank Questions

11. **Answer**: Melanoma
12. **Answer**: Herceptin (Trastuzumab)
13. **Answer**: VEGF
14. **Answer**: Philadelphia
15. **Answer**: A tyrosine residue
16. **Answer**: Guanine nucleotide exchange factor (GEF)

Short-Answer Questions

17. **Answer**: Telomerase extends telomeres, preventing senescence and allowing cancer cells to divide indefinitely.
18. **Answer**: They induce angiogenesis by releasing VEGF, a factor that promotes angiogenesis. It is important, because blood vessels supply the tumor with necessary nutrients and oxygen.

References

1. Hanahan D, Weinberg RA. Hallmarks of cancer: the next generation. Cell. 2011;144(5):646–74. https://doi.org/10.1016/j.cell.2011.02.013.
2. Cooper G, Adams K. The cell: a molecular approach. New York, NY. Sinauer Associates; 2022.
3. Weinberg RA. The biology of cancer. 3rd ed. New York, NY. Norton and company; 2023.
4. Alberts B, Heald R, Johnson A, Morgan D, Raff M, Roberts K, Walter P, Wilson J, Hunt T. Molecular biology of the cell. W.W. Norton and Company; 2022.
5. Nishida N, Yano H, Nishida T, Kamura T, Kojiro M. Angiogenesis in cancer. Vasc Health Risk Manag. 2006;2(3):213–9. https://doi.org/10.2147/vhrm.2006.2.3.213.

6. Baghban R, Roshangar L, Jahanban-Esfahlan R, Seidi K, Ebrahimi-Kalan A, Jaymand M, et al. Tumor microenvironment complexity and therapeutic implications at a glance. Cell Commun Signal. 2020;18(1):59. https://doi.org/10.1186/s12964-020-0530-4.
7. Pei D, Shu X, Gassama-Diagne A, Thiery JP. Mesenchymal–epithelial transition in development and reprogramming. Nat Cell Biol. 2019;21(1):44–53. https://doi.org/10.1038/s41556-018-0195-z.
8. Chandel NS. Navigating Metabolism. 2nd ed. Cold Spring Harbor: Cold Spring Harbor Laboratory Press; 2020.
9. Torry DS, Cooper GM. Proto-oncogenes in development and cancer. Am J Reprod Immunol. 1991;25(3):129–32. https://doi.org/10.1111/j.1600-0897.1991.tb01080.x.
10. Rous P. A Sarcoma of the fowl transmissible by an agent separable from the tumor cells. J Exp Med. 1911;13(4):397–411. https://doi.org/10.1084/jem.13.4.397.
11. Vogt PK. A humble chicken virus that changed biology and medicine. Lancet Oncol. 2009;10(1):96. https://doi.org/10.1016/S1470-2045(08)70338-3.
12. Weiss RA, Vogt PK. 100 years of Rous sarcoma virus. J Exp Med. 2011;208(12):2351–5. https://doi.org/10.1084/jem.20112160.
13. Temin HM. The control of cellular morphology in embryonic cells infected with rous sarcoma virus in vitro. Virology. 1960;10:182–97. https://doi.org/10.1016/0042-6822(60)90038-6.
14. Coffin JM, Fan H. The discovery of reverse transcriptase. Ann Rev Virol. 2016;3:29–51. https://doi.org/10.1146/annurev-virology-110615-035556.
15. Vogt PK. Peyton Rous: homage and appraisal. FASEB J. 1996;10(13):1559–62. https://doi.org/10.1096/fasebj.10.13.8940303.
16. Stehelin D, Varmus HE, Bishop JM, Vogt PK. DNA related to the transforming gene(s) of avian sarcoma viruses is present in normal avian DNA. Nature. 1976;260(5547):170–3. https://doi.org/10.1038/260170a0.
17. Newmark P. Nobel for oncogenes. Nature. 1989;341(6242):475. https://doi.org/10.1038/341475b0.
18. Potlakayala S, Miles A, Patel S, Patel A, Wolbrette G, Kane S, Lookfoon N, Noble A, Shen X, Rudrabhatia S. Retroviruses: reversing the dogma of life - a review. J Clin Microbiol Biochem Technol. 2022; https://doi.org/10.32474/CTBM.2022.03.000160.
19. Vogt PK. Retroviral oncogenes: a historical primer. Nat Rev Cancer. 2012;12(9):639–48. https://doi.org/10.1038/nrc3320.
20. National Human Genome Research Institute. Retroviruses 2024. Available from: https://www.genome.gov/genetics-glossary/Retrovirus.
21. Marks F, Klingmüller U, Müller-Decker K. Cellular signal processing: An introduction to the molecular mechanisms of signal transduction. 2nd ed. New York, NY. Garland Science; 2017.
22. Roskoski R. Src protein-tyrosine kinase structure, mechanism, and small molecule inhibitors. Pharmacol Res. 2015;94:9–25. https://doi.org/10.1016/j.phrs.2015.01.003.
23. Roskoski R. Src protein–tyrosine kinase structure and regulation. Biochem Biophys Res Commun. 2004;324(4):1155–64. https://doi.org/10.1016/j.bbrc.2004.09.171.
24. Krishnan H, Miller WT, Goldberg GS. SRC points the way to biomarkers and chemotherapeutic targets. Genes Cancer. 2012;3(5–6):426–35. https://doi.org/10.1177/1947601912458583.
25. Weiss RA. A perspective on the early days of RAS research. Cancer Metastasis Rev. 2020;39(4):1023–8. https://doi.org/10.1007/s10555-020-09919-1.
26. Kratz CP, Niemeyer CM, Zenker M. An unexpected new role of mutant Ras: perturbation of human embryonic development. J Mol Med. 2007;85(3):227–35. https://doi.org/10.1007/s00109-006-0135-4.
27. Kim S-H. Conformational switch of Ras proteins. Mol Cells. 1993;3(3):229–32. https://doi.org/10.1016/S1016-8478(23)17154-8.
28. Wang Y, Ji D, Lei C, Chen Y, Qiu Y, Li X, et al. Mechanistic insights into the effect of phosphorylation on Ras conformational dynamics and its interactions with cell signaling proteins. Comput Struct Biotechnol J. 2021;19:1184–99. https://doi.org/10.1016/j.csbj.2021.01.044.
29. Hobbs GA, Der CJ, Rossman KL. RAS isoforms and mutations in cancer at a glance. J Cell Sci. 2016;129(7):1287–92. https://doi.org/10.1242/jcs.182873.
30. Cherfils J, Zeghouf M. Regulation of small GTPases by GEFs, GAPs, and GDIs. Physiol Rev. 2013;93(1):269–309. https://doi.org/10.1152/physrev.00003.2012.
31. Simanshu DK, Nissley DV, McCormick F. RAS proteins and their regulators in human disease. Cell. 2017;170(1):17–33. https://doi.org/10.1016/j.cell.2017.06.009.
32. FDA. Lumakras(Sotorasib) Approval Overview 2021. Available from: https://www.accessdata.fda.gov/scripts/cder/daf/index.cfm?event=overview.process&ApplNo=214665
33. Blair HA. Sotorasib: first approval. Drugs. 2021;81(13):1573–9. https://doi.org/10.1007/s40265-021-01574-2.
34. FDA. Krazati(Adagrasib) Approval Overview 2022. Available from: https://www.accessdata.fda.gov/scripts/cder/daf/index.cfm?event=overview.process&ApplNo=216340
35. Mausey N, Halford Z. Targeted therapies for previously "Undruggable" KRAS-mutated non-small cell lung cancer: a review of Sotorasib and Adagrasib. Ann Pharmacother. 2024;58(6):622–35. https://doi.org/10.1177/10600280231197459.

36. Dhillon S. Adagrasib: First Approval. Drugs. 2023;83(3):275–85. https://doi.org/10.1007/s40265-023-01839-y.
37. Weinstein IB, Joe A. Oncogene addiction. Cancer Res. 2008;68(9):3077–80; discussion 80. https://doi.org/10.1158/0008-5472.Can-07-3293.
38. Hatanpaa KJ, Burma S, Zhao D, Habib AA. Epidermal growth factor receptor in glioma: signal transduction, neuropathology, imaging, and radioresistance. Neoplasia. 2010;12(9):675–84. https://doi.org/10.1593/neo.10688.
39. An Z, Aksoy O, Zheng T, Fan QW, Weiss WA. Epidermal growth factor receptor and EGFRvIII in glioblastoma: signaling pathways and targeted therapies. Oncogene. 2018;37(12):1561–75. https://doi.org/10.1038/s41388-017-0045-7.
40. Gan HK, Cvrljevic AN, Johns TG. The epidermal growth factor receptor variant III (EGFRvIII): where wild things are altered. FEBS J. 2013;280(21): 5350–70. https://doi.org/10.1111/febs.12393.
41. Ho ECH, Qiu R, Miller E, Bilotta MT, FitzGerald D, Antignani A. Antibody drug conjugates, targeting cancer-expressed EGFR, exhibit potent and specific antitumor activity. Biomed Pharmacother. 2023;157: 114047. https://doi.org/10.1016/j.biopha.2022.114047.
42. Lin YJ, Mashouf LA, Lim M. CAR T cell therapy in primary brain tumors: current investigations and the future. Front Immunol. 2022;13:817296. https://doi.org/10.3389/fimmu.2022.817296.
43. Swain SM, Shastry M, Hamilton E. Targeting HER2-positive breast cancer: advances and future directions. Nat Rev Drug Discov. 2023;22(2): 101–26. https://doi.org/10.1038/s41573-022-00579-0.
44. FDA. HERCEPTIN (trastuzumab) Approval Overview 1998. Available from: https://www.accessdata.fda.gov/scripts/cder/daf/index.cfm?event=overview.process&ApplNo=103792.
45. Gemmete JJ, Mukherji SK. Trastuzumab (herceptin). AJNR Am J Neuroradiol. 2011;32(8):1373–4. https://doi.org/10.3174/ajnr.A2619.
46. Miller EM, Schwartzberg LS. Biosimilars for breast cancer: a review of HER2-targeted antibodies in the United States. Ther Adv Med Oncol. 2019;11: 1758835919887044. https://doi.org/10.1177/1758835919887044.
47. FDA. Tykerb(Lapatinib) Approval Overview 2007. Available from: https://www.accessdata.fda.gov/scripts/cder/daf/index.cfm?event=overview.process&ApplNo=022059.
48. Asif HM, Sultana S, Ahmed S, Akhtar N, Tariq M. HER-2 positive breast cancer - a mini-review. Asian Pac J Cancer Prev. 2016;17(4): 1609–15. https://pubmed.ncbi.nlm.nih.gov/27221828/
49. Waks AG, Winer EP. Breast cancer treatment: a review. JAMA. 2019;321(3):288–300. https://doi.org/10.1001/jama.2018.19323.
50. FDA. Perjeta(Pertuzumab) Approval Overview 2012. Available from: https://www.accessdata.fda.gov/scripts/cder/daf/index.cfm?event=overview.process&ApplNo=125409.
51. Tóth G, Szöőr Á, Simon L, Yarden Y, Szöllősi J, Vereb G. The combination of trastuzumab and pertuzumab administered at approved doses may delay development of trastuzumab resistance by additively enhancing antibody-dependent cell-mediated cytotoxicity. MAbs. 2016;8(7):1361–70. https://doi.org/10.1080/19420862.2016.1204503.
52. FDA. Kadcyla (ado trastuzumab emtansine) Appproval Overview 2013. Available from: https://www.accessdata.fda.gov/scripts/cder/daf/index.cfm?event=overview.process&ApplNo=125427.
53. Genentech, Inc. Kadcyla. Proposed mechanism of action (MOA) for KADCYLA. Available from: https://www.kadcyla-hcp.com/early-breast-cancer/about/how-kadcyla-works.html
54. Liang K, Mei S, Gao X, Peng S, Zhan J. Dynamics of endocytosis and degradation of antibody-drug conjugate T-DM1 in HER2 positive cancer cells. Drug Des Devel Ther. 2021;15:5135–50. https://doi.org/10.2147/dddt.S344052.
55. FDA. Enhertu(Fam-trastuzumab deruxtecan) Approval Overview 2019. Available from: https://www.accessdata.fda.gov/scripts/cder/daf/index.cfm?event=overview.process&ApplNo=761139.
56. Martín M, Pandiella A, Vargas-Castrillón E, Díaz-Rodríguez E, Iglesias-Hernangómez T, Martínez Cano C, et al. Trastuzumab deruxtecan in breast cancer. Crit Rev Oncol Hematol. 2024;198:104355. https://doi.org/10.1016/j.critrevonc.2024.104355.
57. FDA. Zenocutuzumab(Bizengril) Approval Overview 2024. Available from: https://www.accessdata.fda.gov/scripts/cder/daf/index.cfm?event=overview.process&ApplNo=761352.
58. FDA. FDA grants accelerated approval to zenocutuzumab-zbco for non-small cell lung cancer and pancreatic adenocarcinoma 2024. Available from: https://www.fda.gov/drugs/resources-information-approved-drugs/fda-grants-accelerated-approval-zenocutuzumab-zbco-non-small-cell-lung-cancer-and-pancreatic?utm.
59. Gandullo-Sánchez L, Ocaña A, Pandiella A. HER3 in cancer: from the bench to the bedside. J Exp Clin Cancer Res. 2022;41(1):310. https://doi.org/10.1186/s13046-022-02515-x.
60. Gallo LH, Nelson KN, Meyer AN, Donoghue DJ. Functions of Fibroblast Growth Factor Receptors in cancer defined by novel translocations and mutations. Cytokine Growth Factor Rev. 2015;26 (4):425–49. https://doi.org/10.1016/j.cytogfr.2015.03.003.
61. Ilyas SI, Borad MJ. The rise of the FGFR inhibitor in advanced biliary cancer: the next cover of time magazine? J Gastrointest Oncol. 2016;7(5):789–96. https://doi.org/10.21037/jgo.2016.08.12.

62. FDA. Lytgobi(Futibatinib) Approval Overview 2022. Available from: https://www.accessdata.fda.gov/scripts/cder/daf/index.cfm?event=overview.process&ApplNo=214801
63. Hoy SM. Futibatinib: a review in locally advanced and metastatic Cholangiocarcinoma. Target Oncol. 2024;19(3):473–80. https://doi.org/10.1007/s11523-024-01059-8.
64. FDA. Pemazyre(Pemigatinib) Approval Overview 2020. Available from: https://www.accessdata.fda.gov/scripts/cder/daf/index.cfm?event=overview.process&ApplNo=213736.
65. Katoh M, Loriot Y, Brandi G, Tavolari S, Wainberg ZA, Katoh M. FGFR-targeted therapeutics: clinical activity, mechanisms of resistance and new directions. Nat Rev Clin Oncol. 2024;21(4):312–29. https://doi.org/10.1038/s41571-024-00869-z.
66. Administration USFaD. WITHDRAWN: FDA grants accelerated approval to infigratinib for metastatic cholangiocarcinoma 2024. Available from: https://www.fda.gov/drugs/resources-information-approved-drugs/withdrawn-fda-grants-accelerated-approval-infigratinib-metastatic-cholangiocarcinoma.
67. Patel TH, Marcus L, Horiba MN, Donoghue M, Chatterjee S, Mishra-Kalyani PS, et al. FDA approval summary: Pemigatinib for previously treated, unresectable locally advanced or metastatic Cholangiocarcinoma with FGFR2 fusion or other rearrangement. Clin Cancer Res. 2023;29(5):838–42. https://doi.org/10.1158/1078-0432.Ccr-22-2036.
68. Takahashi S. Downstream molecular pathways of FLT3 in the pathogenesis of acute myeloid leukemia: biology and therapeutic implications. J Hematol Oncol. 2011;4:13. https://doi.org/10.1186/1756-8722-4-13.
69. Daver N, Cortes J. Molecular targeted therapy in acute myeloid leukemia. Hematology. 2012;17(supl):s59–62. https://doi.org/10.1179/102453312X13336169155619.
70. FDA. Rydapt(Midostaurin) Approval Overview 2017. Available from: https://www.accessdata.fda.gov/scripts/cder/daf/index.cfm?event=overview.process&ApplNo=207997.
71. FDA. Xospata(Gilteritinib) Approval Overview 2018. Available from: https://www.accessdata.fda.gov/scripts/cder/daf/index.cfm?event=overview.process&ApplNo=211349.
72. Novatcheva ED, Anouty Y, Saunders I, Mangan JK, Goodman AM. FMS-Like Tyrosine Kinase 3 inhibitors for the treatment of acute myeloid Leukemia. Clin Lymphoma Myeloma Leuk. 2022;22(3):e161–e84. https://doi.org/10.1016/j.clml.2021.09.002.
73. FDA. Tagrisso(Osimertinib) Approval Overview 2015. Available from: https://www.accessdata.fda.gov/scripts/cder/daf/index.cfm?event=overview.process&ApplNo=208065.
74. Tian Z, Cen L, Wei F, Dong J, Huang Y, Han Y, et al. EGFR mutations in non-small cell lung cancer: classification, characteristics and resistance to third-generation EGFR-tyrosine kinase inhibitors (review). Oncol Lett. 2025;30(2):375. https://doi.org/10.3892/ol.2025.15121.
75. FDA. Lenvima(Lenvatinib) Approval Overview 2015. Available from: https://www.accessdata.fda.gov/scripts/cder/daf/index.cfm?event=overview.process&ApplNo=206947.
76. Chen Y, Dai S, Cheng C-s, Chen L. Lenvatinib and immune-checkpoint inhibitors in hepatocellular carcinoma: mechanistic insights, clinical efficacy, and future perspectives. J Hematol Oncol. 2024;17(1):130. https://doi.org/10.1186/s13045-024-01647-1.
77. FDA. Alecensa(Alectinib) Approval Overview 2015. Available from: https://www.accessdata.fda.gov/scripts/cder/daf/index.cfm?event=overview.process&ApplNo=208434.
78. Kumari S, Akhter M, Gupta GD, Sharma K. Progression and expansion of ALK inhibitors against NSCLC: a dual target approach. Eur J Med Chem. 2025;293:117722. https://doi.org/10.1016/j.ejmech.2025.117722.
79. FDA. Sutent(Sunitinib) Approval Overview 2006. Available from: https://www.accessdata.fda.gov/scripts/cder/daf/index.cfm?event=overview.process&ApplNo=021938.
80. Choi HY, Chang J-E. Targeted therapy for cancers: from ongoing clinical trials to FDA-approved drugs. Int J Mol Sci. 2023;24(17):13618. https://doi.org/10.3390/ijms241713618.
81. FDA. Fotivda(Tivozanib) Approval Overview 2021. Available from: https://www.accessdata.fda.gov/scripts/cder/daf/index.cfm?event=overview.process&ApplNo=212904
82. Passi I, Billowria K, Kumar B, Chawla PA. Tivozanib: a new hope for treating renal cell carcinoma. Anti Cancer Agents Med Chem. 2023;23(5):562–70. https://doi.org/10.2174/1871520622666220617103126.
83. Ventura I, Sanchiz L, Legidos-García ME, Murillo-Llorente MT, Pérez-Bermejo M. Atezolizumab and Bevacizumab combination therapy in the treatment of advanced hepatocellular cancer. Cancers (Basel). 2023;16(1) https://doi.org/10.3390/cancers16010197.
84. Holderfield M, Deuker MM, McCormick F, McMahon M. Targeting RAF kinases for cancer therapy: BRAF-mutated melanoma and beyond. Nat Rev Cancer. 2014;14(7):455–67. https://doi.org/10.1038/nrc3760.
85. Ullah R, Yin Q, Snell AH, Wan L. RAF-MEK-ERK pathway in cancer evolution and treatment. Semin Cancer Biol. 2022;85:123–54. https://doi.org/10.1016/j.semcancer.2021.05.010.
86. Huang T, Karsy M, Zhuge J, Zhong M, Liu D. B-Raf and the inhibitors: from bench to bedside. J Hematol

Oncol. 2013;6(1):30. https://doi.org/10.1186/1756-8722-6-30.
87. Khaddour K, Kurn H, Zito PM. Vemurafenib. StatPearls [Internet]. StatPearls Publishing LLC; 2025.
88. FDA. Tafinlar(Dabrafenib) Approval Overview 2013. Available from: https://www.accessdata.fda.gov/scripts/cder/daf/index.cfm?event=overview.process&ApplNo=202806.
89. FDA. Braftovi(Encorafenib) Approval Overview 2018. Available from: https://www.accessdata.fda.gov/scripts/cder/daf/index.cfm?event=overview.process&ApplNo=210496
90. FDA. Ojemda(Tovorafenib) Approval Overview 2024. Available from: https://www.accessdata.fda.gov/scripts/cder/daf/index.cfm?event=overview.process&ApplNo=218033
91. Vemurafenib (Zelboraf) for metastatic melanoma. Med Lett Drugs Ther. 2011;53(1374):77–8.
92. Day One Biopharmaceuticals, Inc. OJEMDA (tovorafenib) Available from: https://www.ojemda.com/.
93. Dhillon S. Tovorafenib: First Approval. Drugs. 2024;84(8):985–93. https://doi.org/10.1007/s40265-024-02069-6.
94. Marks D. Marks' essentials of medical biochemistry: a clinical approach. Philadelphia, PA. Lippincott Williams & Wilkins (LWW); 2017.
95. FDA. Zydelig(Idelalisib) Approval Overview 2014. Available from: https://www.accessdata.fda.gov/scripts/cder/daf/index.cfm?event=overview.process&ApplNo=205858.
96. Cheah CY, Fowler NH. Idelalisib in the management of lymphoma. Blood. 2016;128(3):331–6. https://doi.org/10.1182/blood-2016-02-702761.
97. Vanhaesebroeck B, Perry MWD, Brown JR, André F, Okkenhaug K. PI3K inhibitors are finally coming of age. Nat Rev Drug Discov. 2021;20(10):741–69. https://doi.org/10.1038/s41573-021-00209-1.
98. Rai K, Jain P. Chronic lymphocytic leukemia (CLL)-then and now. Am J Hematol. 2016;91(3):330–40. https://doi.org/10.1002/ajh.24282.
99. FDA. Aliqopa(Copanlisib) Approval Overview 2017. Available from: https://www.accessdata.fda.gov/scripts/cder/daf/index.cfm?event=overview.process&ApplNo=209936.
100. Mensah FA, Blaize JP, Bryan LJ. Spotlight on copanlisib and its potential in the treatment of relapsed/refractory follicular lymphoma: evidence to date. Onco Targets Ther. 2018;11:4817–27. https://doi.org/10.2147/ott.S142264.
101. FDA. Copiktra(Duvelisib) Approval Overview 2018. Available from: https://www.accessdata.fda.gov/scripts/cder/daf/index.cfm?event=overview.process&ApplNo=211155.
102. Rodrigues DA, Sagrillo FS, Fraga CAM. Duvelisib: a 2018 novel FDA-approved small molecule inhibiting phosphoinositide 3-Kinases. Pharmaceuticals (Basel). 2019;12(2) https://doi.org/10.3390/ph12020069.
103. FDA. Truqap(Capivasertib) Approval Overview 2023. Available from: https://www.accessdata.fda.gov/scripts/cder/daf/index.cfm?event=overview.process&ApplNo=218197
104. Shirley M. Capivasertib: first approval. Drugs. 2024;84(3):337–46. https://doi.org/10.1007/s40265-024-01998-6.
105. AstraZenica. Truqap (capivasertib). Available from: https://www.truqap.com/
106. Abraham RT, Wiederrecht GJ. Immunopharmacology of rapamycin. Annu Rev Immunol. 1996;14:483–510. https://doi.org/10.1146/annurev.immunol.14.1.483.
107. FDA. Afinitor(Everolimus) Approval Overview 2009. Available from: https://www.accessdata.fda.gov/scripts/cder/daf/index.cfm?event=overview.process&ApplNo=022334.
108. Hasskarl J. Everolimus. Recent Results Cancer Res. 2018;211:101–23. https://doi.org/10.1007/978-3-319-91442-8_8.
109. Yu L, Wei J, Liu P. Attacking the PI3K/Akt/mTOR signaling pathway for targeted therapeutic treatment in human cancer. Semin Cancer Biol. 2022;85:69–94. https://doi.org/10.1016/j.semcancer.2021.06.019.
110. FDA. Torisel(Temsirolimus) Approval Overview 2007. Available from: https://www.accessdata.fda.gov/scripts/cder/daf/index.cfm?event=overview.process&ApplNo=022088.
111. Pfizer. Torisel (temsirolimus) 2024. Available from: https://www.pfizermedicalinformation.com/torisel.
112. FDA. Adcetris(Brentuximab vedotin) Approval Overview 2011. Available from: https://www.accessdata.fda.gov/scripts/cder/daf/index.cfm?event=overview.process&ApplNo=125388.
113. Dumitru AV, Țăpoi DA, Halcu G, Munteanu O, Dumitrascu DI, Ceaușu MC, et al. The polyvalent role of CD30 for cancer diagnosis and treatment. Cells. 2023;12(13) https://doi.org/10.3390/cells12131783.
114. Heiser RA, Cao AT, Zeng W, Ulrich M, Younan P, Anderson ME, et al. Brentuximab Vedotin-driven microtubule disruption results in endoplasmic reticulum stress leading to immunogenic cell death and antitumor immunity. Mol Cancer Ther. 2024;23(1):68–83. https://doi.org/10.1158/1535-7163.Mct-23-0118.
115. FDA. Vasotec(Enalapril) Approval Overview. 1985. https://www.accessdata.fda.gov/drugsatfda_docs/label/2014/018998s079lbl.pdf

116. Fricker LD. Proteasome inhibitor drugs. Annu Rev Pharmacol Toxicol. 2020;60:457–76. https://doi.org/10.1146/annurev-pharmtox-010919-023603.
117. Eilers M, Eisenman RN. Myc's broad reach. Genes Dev. 2008;22(20):2755–66. https://doi.org/10.1101/gad.1712408.
118. Dhanasekaran R, Deutzmann A, Mahauad-Fernandez WD, Hansen AS, Gouw AM, Felsher DW. The MYC oncogene - the grand orchestrator of cancer growth and immune evasion. Nat Rev Clin Oncol. 2022;19(1):23–36. https://doi.org/10.1038/s41571-021-00549-2.
119. Bouchard C, Dittrich O, Kiermaier A, Dohmann K, Menkel A, Eilers M, et al. Regulation of cyclin D2 gene expression by the Myc/Max/Mad network: Myc-dependent TRRAP recruitment and histone acetylation at the cyclin D2 promoter. Genes Dev. 2001;15(16):2042–7. https://doi.org/10.1101/gad.907901.
120. Miller DM, Thomas SD, Islam A, Muench D, Sedoris K. c-Myc and cancer metabolism. Clin Cancer Res. 2012;18(20):5546–53. https://doi.org/10.1158/1078-0432.Ccr-12-0977.
121. García-Gutiérrez L, Delgado MD, León J. MYC oncogene contributions to release of cell cycle brakes. Genes (Basel). 2019;10(3) https://doi.org/10.3390/genes10030244.
122. Shi Q, Xue C, Zeng Y, Yuan X, Chu Q, Jiang S, et al. Notch signaling pathway in cancer: from mechanistic insights to targeted therapies. Signal Transduct Target Ther. 2024;9(1):128. https://doi.org/10.1038/s41392-024-01828-x.
123. Kingwell K. γ-secretase inhibitor notches first approval. Nat Rev Drug Discov. 2024;23(1):9. https://doi.org/10.1038/d41573-023-00204-8.
124. Timbergen MJM, Smits R, Grünhagen DJ, Verhoef C, Sleijfer S, Wiemer EAC. Activated signaling pathways and targeted therapies in desmoid-type fibromatosis: a literature review. Front Oncol. 2019;9:397. https://doi.org/10.3389/fonc.2019.00397.
125. FDA. Ogsiveo(Nirogacestat) Approval Overview 2023. Available from: https://www.accessdata.fda.gov/scripts/cder/daf/index.cfm?event=overview.process&ApplNo=217677
126. Kummar S, Bui N, Messersmith WA, Whiting J, Portnoy M, Lim A, et al. Nirogacestat-the pathway to approval of the first treatment for desmoid tumors, a rare disease. Ther Adv Rare Dis. 2025;6:26330040251317546. https://doi.org/10.1177/26330040251317546.
127. Dhillon S. Tarlatamab: First Approval. Drugs. 2024;84(8):995–1003. https://doi.org/10.1007/s40265-024-02070-z.
128. Health and Pharma. FDA approves IMDELLTRA for lung cancer after a 40% response rate 2024. Available from: https://healthandpharma.net/fda-imdelltra-tarlatamab-lung-cancer-amgen.
129. FDA. Imdelltra(Tarlatamab-dlle) Approval Overview 2024. Available from: https://www.accessdata.fda.gov/scripts/cder/daf/index.cfm?event=overview.process&ApplNo=761344
130. Allen F, Maillard I. Therapeutic targeting of notch signaling: from cancer to inflammatory disorders. Front Cell Dev Biol. 2021;9:649205. https://doi.org/10.3389/fcell.2021.649205.
131. Li D, Masiero M, Banham AH, Harris AL. The notch ligand JAGGED1 as a target for anti-tumor therapy. Front Oncol. 2014;4:254. https://doi.org/10.3389/fonc.2014.00254.
132. Yang J, Sun Q, Liu X, Yang Y, Rong R, Yan P, et al. Targeting Notch signaling pathways with natural bioactive compounds: a promising approach against cancer. Front Pharmacol. 2024;15:1412669. https://doi.org/10.3389/fphar.2024.1412669.
133. Moore G, Annett S, McClements L, Robson T. Top notch targeting strategies in cancer: a detailed overview of recent insights and current perspectives. Cells. 2020;9(6):1503. https://doi.org/10.3390/cells9061503.
134. FDA. Gleevec(imatinib): drugs@FDA approval overview silver Spring. U.S. Food and Drug Administration; 2001. Available from: https://www.accessdata.fda.gov/scripts/cder/daf/index.cfm?event=overview.process&ApplNo=021588
135. Cohen P, Cross D, Jänne PA. Kinase drug discovery 20 years after imatinib: progress and future directions. Nat Rev Drug Discov. 2021;20(7):551–69. https://doi.org/10.1038/s41573-021-00195-4.
136. Hochhaus A, Larson RA, Guilhot F, Radich JP, Branford S, Hughes TP, et al. Long-term outcomes of Imatinib treatment for chronic myeloid leukemia. N Engl J Med. 2017;376(10):917–27. https://doi.org/10.1056/NEJMoa1609324.
137. Reddy EP, Aggarwal AK. The ins and outs of bcr-abl inhibition. Genes Cancer. 2012;3(5–6):447–54. https://doi.org/10.1177/1947601912462126.
138. Bouillaud F, Hammad N, Schwartz L. Warburg effect, glutamine, succinate, alanine, when oxygen matters. Biology (Basel). 2021;10(10) https://doi.org/10.3390/biology10101000.
139. Vander Heiden MG, Cantley LC, Thompson CB. Understanding the Warburg effect: the metabolic requirements of cell proliferation. Science. 2009;324(5930):1029–33. https://doi.org/10.1126/science.1160809.
140. Liberti MV, Locasale JW. The Warburg effect: how does it benefit cancer cells? Trends Biochem Sci. 2016;41(3):211–8. https://doi.org/10.1016/j.tibs.2015.12.001.
141. Lu J, Tan M, Cai Q. The Warburg effect in tumor progression: mitochondrial oxidative metabolism as an anti-metastasis mechanism. Cancer Lett. 2015;356(2 Pt A):156–64. https://doi.org/10.1016/j.canlet.2014.04.001.
142. Masoud GN, Li W. HIF-1α pathway: role, regulation and intervention for cancer therapy. Acta Pharm Sin

B. 2015;5(5):378–89. https://doi.org/10.1016/j.apsb.2015.05.007.
143. Sharma A, Sinha S, Shrivastava N. Therapeutic targeting hypoxia-inducible factor (HIF-1) in cancer: cutting Gordian Knot of cancer cell metabolism. Front Genet. 2022;13:849040. https://doi.org/10.3389/fgene.2022.849040.
144. Martínez-Reyes I, Chandel NS. Mitochondrial TCA cycle metabolites control physiology and disease. Nat Commun. 2020;11(1):102. https://doi.org/10.1038/s41467-019-13668-3.
145. Shiratori R, Furuichi K, Yamaguchi M, Miyazaki N, Aoki H, Chibana H, et al. Glycolytic suppression dramatically changes the intracellular metabolic profile of multiple cancer cell lines in a mitochondrial metabolism-dependent manner. Sci Rep. 2019;9(1):18699. https://doi.org/10.1038/s41598-019-55296-3.
146. Luengo A, Li, Z, Vander Heiden M. National Cancer Institute. New Clarity on the Warburg Effect 2021. Available from: https://www.cancer.gov/research/key-initiatives/ras/news-events/dialogue-blog/2021/vander-heiden-warburg-effect#:~:text=This%20phenotype%20is%20referred%20to%20as%20%E2%80%9Caerobic,as%20the%20Warburg%20effect%20in%20this%20context.
147. Luengo A, Li Z, Gui DY, Sullivan LB, Zagorulya M, Do BT, et al. Increased demand for NAD(+) relative to ATP drives aerobic glycolysis. Mol Cell. 2021;81(4):691–707.e6. https://doi.org/10.1016/j.molcel.2020.12.012.
148. Choi SY, Collins CC, Gout PW, Wang Y. Cancer-generated lactic acid: a regulatory, immunosuppressive metabolite? J Pathol. 2013;230(4):350–5. https://doi.org/10.1002/path.4218.
149. Al Tameemi W, Dale TP, Al-Jumaily RMK, Forsyth NR. Hypoxia-modified cancer cell metabolism. Front Cell Dev Biol. 2019;7:4. https://doi.org/10.3389/fcell.2019.00004.
150. Bender DA. Introduction to nutrition and metabolism. Boca Raton: CRC Press; 2021.
151. Guo D, Meng Y, Jiang X, Lu Z. Hexokinases in cancer and other pathologies. Cell Insight. 2023;2(1):100077. https://doi.org/10.1016/j.cellin.2023.100077.
152. Hitomi M, Stacey DW. Cyclin D1 production in cycling cells depends on Ras in a cell-cycle-specific manner. Curr Biol. 1999;9(19):1075–S2. https://doi.org/10.1016/S0960-9822(99)80476-X.
153. Filmus J, Robles AI, Shi W, Wong MJ, Colombo LL, Conti CJ. Induction of cyclin D1 overexpression by activated ras. Oncogene. 1994;9(12):3627–33.
154. Sears R, Leone G, DeGregori J, Nevins JR. Ras enhances Myc protein stability. Mol Cell. 1999;3(2):169–79. https://doi.org/10.1016/S1097-2765(00)80308-1.
155. Kinoshita T, Yokota T, Arai K, Miyajima A. Regulation of Bcl-2 expression by oncogenic Ras protein in hematopoietic cells. Oncogene. 1995;10(11):2207–12.
156. Kerk SA, Papagiannakopoulos T, Shah YM, Lyssiotis CA. Metabolic networks in mutant KRAS-driven tumours: tissue specificities and the microenvironment. Nat Rev Cancer. 2021;21(8):510–25. https://doi.org/10.1038/s41568-021-00375-9.
157. Wu R, Zhu H, He Q, Yuan T, Yang B. Metabolic reprogramming in KRAS-mutant cancers: proven targetable vulnerabilities and potential therapeutic strategies. Drug Discov Today. 2024;29(12):104220. https://doi.org/10.1016/j.drudis.2024.104220.

20 Regulation of the Cell Cycle and Abnormalities in Disease

Abstract

The cell cycle is a tightly regulated process that controls when cells divide, ensuring that cell division occurs completely and without errors. Each time the cell divides into two daughter cells, it goes through the cell cycle. Several important events must occur during each cell cycle. The cell needs to grow in size; chromosomes containing the DNA need to replicate; the chromosomes then need to segregate; and finally, the cell needs to divide into two daughter cells. Since the cell cycle is fundamental to every cell type, the process is highly conserved throughout evolution, and the basic mechanisms of the cell cycle apply to nearly all eukaryotic cells. The basic mechanisms regulating the cell cycle are discussed here, including the main phases: M, G1, S, and G2. The molecules involved in controlling the cell cycle, especially cyclins and cyclin-dependent kinases, are also discussed. Some history about the identification of these mechanisms is also explored. Serious health problems can result when the cell cycle is not working correctly, leading to excessive or insufficient cell proliferation, or proliferation at the wrong times or under the wrong conditions. An important example is cancer. For this reason, numerous cancer therapeutics have been developed that target cell cycle regulatory proteins and different phases of the cell cycle. Examples of such types of drugs are discussed throughout the chapter.

Keywords

Cell cycle phases · Cyclins · Cyclin-dependent kinases (CDKs) · MPF · Restriction points · Checkpoints

Abbreviations

APC	Anaphase-promoting complex
APC/C	Anaphase-promoting complex/ cyclosome
ATM	Ataxia-telangiectasia mutated (a kinase involved in DNA damage response; included here as it's often referenced in checkpoint context)
ATR	ATM and Rad3-telated (another kinase in DNA damage sensing pathways)
CDC	Cell division cycle (used in names like CDC25 phosphatase)
CDK	Cyclin-dependent kinase
CKD	Cyclin-dependent kinase (alternate abbreviation sometimes used)
CKI	CDK inhibitor
DNA	Deoxyribonucleic acid
E2F	E2 factor (a family of transcription factors)
G0	Gap 0 phase (resting/quiescent phase)

A. Minden, *Biochemistry, Cell and Molecular Biology for Pharmacists and Health Professionals*,
https://doi.org/10.1007/978-3-032-20529-2_20

G1	Gap 1 phase
G2	Gap 2 phase
M phase	Mitosis
p53	Tumor suppressor protein 53
Rb	Retinoblastoma protein
RNA	Ribonucleic acid (commonly referenced alongside DNA processes)
S phase	Synthesis phase (DNA replication)

What You Will Learn

- The basics of the cell cycle and the cell cycle phases: G1, S, G2, and M phase.
- Nuclear division, and the stages of M phase.
- The importance of microtubules in the cell cycle, the basic structure of microtubules, and microtubule-disrupting drugs that can inhibit growth of rapidly dividing cells by blocking the cell cycle at M phase.
- The changes in DNA throughout the cell cycle, DNA replication, and DNA condensation.
- The concept of ploidy, diploid and haploid cells, and cells that have abnormal chromosomal numbers.
- Restriction points and checkpoints regulate the cell cycle.
- Some historical background about cell cycle regulation in Xenopus, yeast, sea urchin, and mammalian cells, the discovery of MPF, and how it led to an understanding of cyclins and CDKs.
- The molecular basis of cell cycle regulation, including the roles of cyclins and cyclin-dependent kinases (CDKs) that regulate different phases of the cell cycle.
- The different cyclins that function at different phases of the cell cycle including Cyclins B, D, E, and A.
- Cyclin D and Rb phosphorylation, and progression through S phase.
- P53 inhibits the cell cycle in response to DNA damage, so that damaged DNA does not replicate. The challenge of p53 inactivation in human cancers.
- Links between the cell cycle and cell metabolism.
- Drugs that target the cell cycle, such as anticancer drugs that target DNA synthesis in S phase.
- Drugs that inhibit CDK4/6 inhibit cell cycle progression and are used for cancer treatment.

1 Fundamentals of the Cell Cycle

1.1 Cell Cycle Basics

All living things are made up of cells. Every time the cell divides into two daughter cells, it goes through the cell cycle (Fig. 20.1). Several important events must occur each time the cell divides. The cell needs to grow in size, chromosomes containing the DNA need to replicate, the DNA, organized into chromosomes, then needs to segregate, and finally, the cell needs to divide into two daughter cells. When groups of cells divide, we often refer to this as cell proliferation. The cell cycle is a tightly regulated process that controls when cells divide, ensuring that cell division occurs completely and without errors. Since the cell cycle is fundamental to every cell type, the process is highly conserved throughout evolution. Therefore, scientists have been able to study the

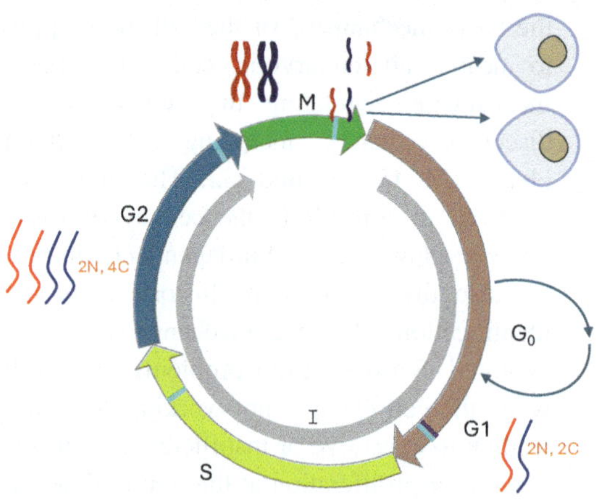

Fig. 20.1 Phases of the cell cycle. G1, S, G2, and M phases are shown. The G1 restriction point is illustrated by a purple line, and checkpoints are illustrated by light blue lines. A pair of representative chromosomes is illustrated, as well as an illustration of their level of condensation at different phases

cell cycle in various types of cells and have found that the basic mechanisms apply to nearly all types of eukaryotic cells. The primary cell types that have been used to study the cell cycle are yeast (*Saccharomyces cerevisiae* (budding yeast), and *Saccharomyces pombe* (fission yeast)), oocytes (especially Xenopus and sea urchin oocytes), and various mammalian cells. Understanding how the cell cycle works and how it is controlled is important for many reasons. Cells make up our tissues and organs. Serious problems can result when the cell cycle is not working correctly, leading to excessive or insufficient cell proliferation, or proliferation at the wrong times or under the wrong conditions. An important example is cancer. In many types of cancer, the cell cycle may be improperly regulated. Instead of proceeding through the cell cycle when more cells are needed and exiting the cell cycle at other times, cancer cells may proceed through the cell cycle and divide without the normal restrictions. The result can be uncontrolled cell proliferation and invasion of other tissues. In fact, many anticancer drugs inhibit the cell cycle, and understanding how the cell cycle works is helping to make this possible. The cell cycle is the main focus of this chapter, also discussed in [1–5] and other references throughout the text.

1.2 Phases of the Cell Cycle

The cell cycle can be divided into phases. The two major phases are interphase (I) and M phase (M). Interphase is divided further into three phases, starting with G1, then S, and then G2 phase. During G1 (Gap1) phase, the cell grows in size and carries out its normal functions. Some proteins and organelles are synthesized during this time in preparation for S phase. During S phase, DNA is replicated. (S phase stands for synthesis phase, and the terms DNA replication and DNA synthesis can be used interchangeably.) During this phase, the entire genome is duplicated. This is followed by G2 (Gap2) phase. During this phase, the cell continues growing and performing its functions. Once G2 phase ends, interphase is over, and M phase begins. M phase consists of mitosis, where the nuclear material splits in two, and cytokinesis, where the entire cell, including the cytoplasm, divides into two daughter cells. Following M phase, a new G1 phase starts, and the cell cycle can begin again for each daughter cell. In many cells, there is also a G0 phase, which occurs at the end of the G1 phase. This is a stage at which a cell is metabolically active but not advancing through the cell cycle. Some cells are in G0 temporarily, but others stay in that phase for the individual's entire life. For example, most neurons generally do not duplicate once an individual reaches adulthood, but the cells carry on their functions normally throughout life. This can be considered a long-term or permanent G0 phase, where cells perform their normal functions but do not duplicate [1–6] (Fig. 20.1).

1.3 M Phase and Mitosis

M phase is important to understand because it is the phase of the cell cycle where the chromosomes segregate, and the cell divides into two daughter cells. M phase starts after G2 phase is completed (Fig. 20.1). It can be divided into the following subphases as follows and in [2–4], and as illustrated in Fig. 20.2.

- Prophase

During prophase, the chromatin, which consists of DNA and associated proteins, becomes highly condensed. These condensed structures take on the characteristic form of chromosomes, which can be seen under the microscope. Each chromosome has two sister chromatids joined together at the **centromere**, a structure near the center of the chromosome. Microtubules present throughout the cell disassemble during prophase as they prepare to take their essential roles in mitosis. The mitotic spindle (also called the spindle apparatus), an array of microtubules that will pull the sister chromatids apart by the end of M phase, begins to form during this phase.

- Prometaphase

During this phase, disruption of the nuclear envelope occurs. With the nuclear envelope out of the

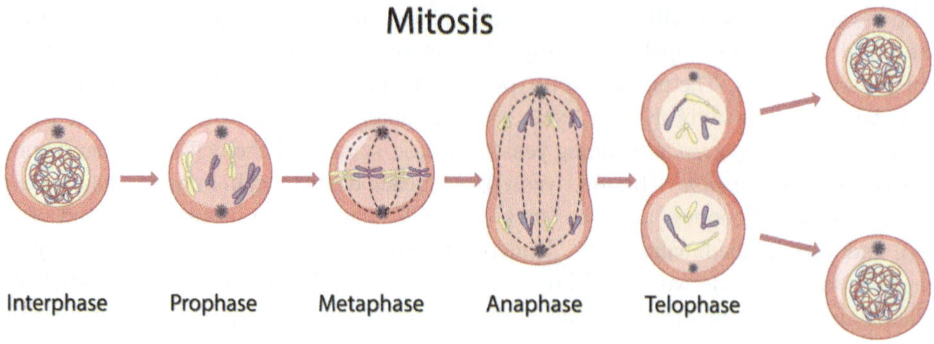

Fig. 20.2 The stages of mitosis. Following interphase, the stages of mitosis include prophase, metaphase, anaphase, and telophase are illustrated. Telophase is followed by cytokinesis. After cytokinesis, interphase starts again (prometaphase is not shown). (*Source*: Illustration from Vecteezy.com (Pro license))

way, microtubules can now enter the nuclear region. A subset of microtubules, known as **kinetochore microtubules**, binds to the chromosomes via the **kinetochores**, which are structures located at the centromere region. These kinetochore microtubules exert tension on the chromosomes. The other microtubules in the mitotic spindle are the polar and astral microtubules.

- Metaphase

During this phase, the kinetochore microtubules align the chromosomes in a central plane in the middle of the cell, called the **metaphase plate**. (Sometimes prometaphase and metaphase are considered as one phase.)

- Anaphase

In this phase, the chromosomes separate. Each sister chromatid separates from its duplicate to form two separate new chromosomes. The kinetochore microtubules shorten. In doing so, they pull the chromosomes apart. Polar microtubules elongate during this time, sliding past the chromosomes.

- Telophase

The separated chromosomes arrive at opposite poles of the cell, and the kinetochore microtubules depolymerize and thus disappear. At the end of telophase, new nuclear envelopes form around each set of chromosomes, and the condensed chromatin expands. The nucleus and nuclear envelope then re-form.

- Cytokinesis

During cytokinesis, the cytoplasm divides via cleavage, and at the end of this stage, there are two separate daughter cells.

1.4 Microtubules

1.4.1 Functions of Microtubules in the Cell Cycle

Microtubules have critical roles in mitosis. They are, in fact, absolutely essential for the process due largely to their crucial roles in separating the chromosomes. Microtubules are important components of the cytoskeleton, the other two being actin and intermediate filaments. During interphase and during times when cells are not actively dividing, the microtubules are located throughout the cell. During this time, they play roles in maintaining cell structure, facilitating cell movement, attaching to surfaces, supporting organelles, and transporting intracellular vesicles. Once the cells prepare to go through M phase, microtubules throughout the cell disassemble and take on their new roles in the mitotic spindle, where they control the separation of chromosomes. Since microtubules are essential for cell division, they are potential targets for drugs that are designed to slow down or stop the growth of rapidly dividing cells, such as cancer cells. It is, therefore, critical to have an understanding of their basic structures.

1.4.2 Basic Structure of Microtubules

Microtubules are made of repeating units of smaller proteins: these are the α-tubulin and β-tubulin subunits. One α-tubulin subunit and one β-tubulin subunit form a tubulin dimer. These dimers polymerize together to form a protofilament. Protofilaments, in turn, interact together to create a hollow tube. There are usually 13 protofilaments in a tube, and this tube is the basic structure of the microtubule [2] (Fig. 20.3).

1.4.3 The Centrosome, Nucleation, and the Assembly of Microtubules

Microtubules are assembled by a process called **nucleation**. Microtubules have polarity, having a plus end and a minus end, and they grow from the plus end. Nucleation involves the sequential addition of tubulin dimers to a growing chain at the plus end. The dimers that are added are GTP-bound, which is discussed below.

Nucleation often occurs at **microtubule-organizing centers (MTOCs)**. There are two main types of MTOCs: the basal bodies and the centrosomes. The basal bodies form the nucleation site for microtubules in cilia and flagella, and the **centrosomes** are important during mitosis. In this chapter, the focus is only on the **centrosome** since it is important in the cell cycle. A centrosome contains two centrioles comprising triplets of microtubules, surrounded by pericentriolar material (PCM) (also called the pericentriolar matrix) (Fig. 20.4). The PCM is rich in a third tubulin subunit, which is used primarily for nucleation and is called γ-tubulin. γ-Tubulin and accessory proteins form a ring structure, which α-tubulin and β-tubulin subunits polymerize onto. It thus has an important role in getting the microtubule started and forming the minus end of the structure, but it does not remain a part of the final microtubule [2, 7].

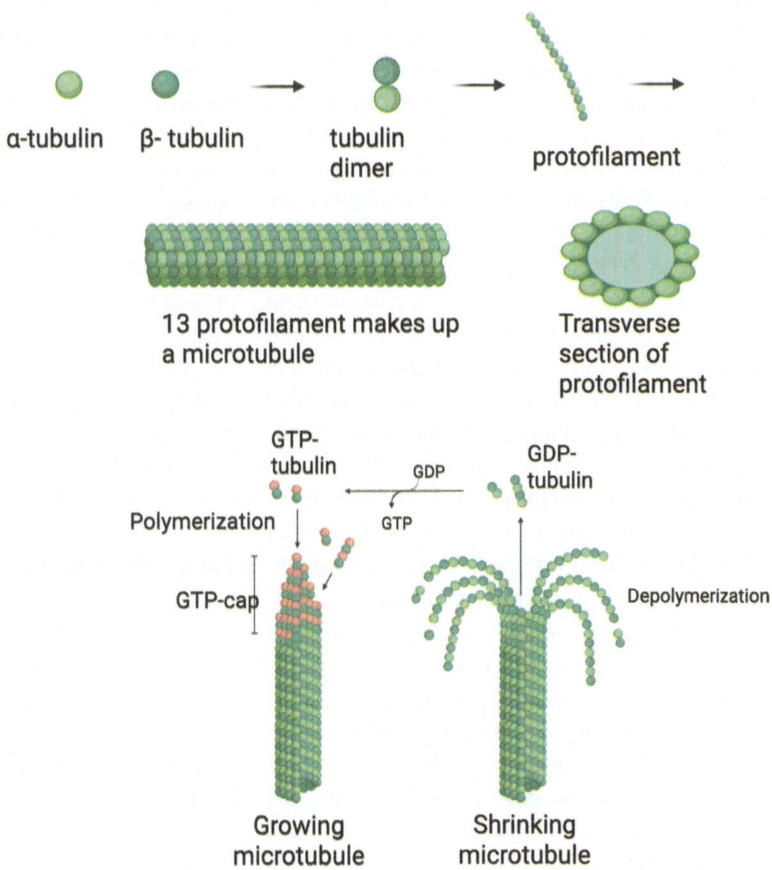

Fig. 20.3 Microtubules. Microtubules are dynamic structures made of protofilaments that contain alpha and beta tubulin subunits. (*Source*: Created with BioRender.com)

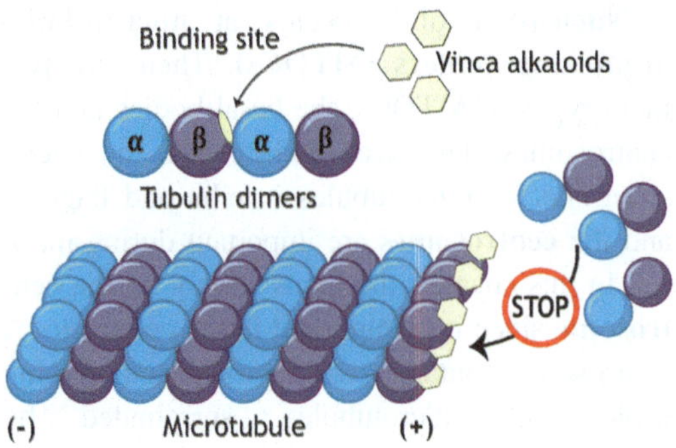

Fig. 20.4 Vinca alkaloids. Vinca alkaloids are derived from the periwinkle plant, and they inhibit microtubule polymerization. This type of drug is used to treat some types of cancer. (*Source*: Figure from Škubník J, et al. Biology (Basel). 2021 Aug 31;10(9):849. DOI: https://doi.org/10.3390/biology10090849. Licensed under the Creative Commons Attribution 4.0 (CC BY 4.0) license)

1.4.4 Dynamic Instability of Microtubules

Microtubules are dynamic structures that regularly grow and shrink. This dynamic nature is called dynamic instability. Specifically, they grow and shrink from the positive end. The dynamic instability is due to the α-subunits and β-subunits and their interactions with GTP or GDP. α-Tubulin and β-tubulin can both bind to GTP. α-Subunits are always **GTP**-bound, but the **β-subunits** are **GTPases**, which can hydrolyze their bound GTP, to form GDP. Tubulin dimers with GTP bound to both the α-subunits and β-subunits are considered GTP-bound tubulin dimers. Only these GTP-bound dimers bind to the plus ends of microtubules, and they bind very stably. GTP tubulin favors a straight conformation and fits in well with the microtubule structure. After being added onto the plus end of the microtubule, the GTP is eventually hydrolyzed to GDP due to the GTPase activity of β-tubulin subunits in the microtubule. GDP tubulin tends to take on a curved structure, and if too much of it builds up, it can bend and destabilize the microtubule structure. Therefore, if the concentration of GTP-bound subunits in the cell decreases, then the GTP-bound dimers will be added more slowly, and the process of GTP hydrolysis approaches the plus end of the microtubule. When GTP hydrolysis catches up to the growing end, then the GTP-rich end of the microtubule, called the GTP cap, is lost. The resulting GDP-bound dimers at the plus end obtain a curved conformation, causing strain on the structure. The subunits now bind more weakly to other tubulin units, resulting in the depolymerization of the tubulin dimers and leading to shrinkage. Whether the microtubule grows or shrinks depends on the balance between the rate of tubulin addition and the rate of GTP hydrolysis. When the cellular concentration of GTP-tubulin dimers is high, the addition of dimers can occur at a high enough rate to maintain the GTP cap, and microtubule growth is prioritized over microtubule shrinking (Fig. 20.3) [2, 8].

1.4.5 The Mitotic Spindle

The mitotic spindle, the structure involved in segregating the chromosomes, is anchored to two centrosomes in regions called the spindle poles. Microtubules radiate out from these poles during mitosis to play their roles in chromosome separation (see Chap. 2; Fig. 2.6). The microtubules in this array include the **kinetochore** microtubules, the **polar** microtubules, and the **astral** microtubules, all anchored at their minus ends at the spindle pole regions. The kinetochore microtubules are the microtubules that bind to the chromosomes. As they shorten, the sister chromatids pull apart. The polar microtubules slide past the chromosomes and lengthen. As these polar microtubules lengthen, the spindle poles move apart, and the cell widens to prepare for division. The **astral** microtubules face away from the chromosomes toward the membrane and have roles in positioning the mitotic spindle [2, 3, 9].

1.4.6 Microtubule-Disrupting Drugs

Cancer is often characterized by rapidly proliferating cells. Slowing down or stopping the growth of these cells is therefore a promising strategy for treating the disease. Microtubules are considered to be promising targets for anticancer drugs due to their essential roles in the cell cycle. Inhibition of microtubule polymerization can block cell cycle progression and thereby disrupt tumor cell growth. Microtubule inhibitors have the advantage of blocking rapidly dividing cells, but they can also lead to severe side effects by impacting some cells that divide normally. Despite their side effects, these types of drugs can be effective in treating certain types of cancer [10]. Some examples are described here.

Velban (vinblastine sulfate) (also known by the brand name Alkaban-AQ) [11] and **Oncovin (vincristine)** [12] (also known as the brand names Marqibo and Vincasar PFS) are **small-molecule** chemotherapy drugs that **disrupt microtubule** dynamics during **M phase**. These drugs have been used since the 1960s, but are still used for the treatment of some cancers.

Vinblastine binds the **β-subunit** of **tubulin dimers** at a domain called the vinca binding domain [10]. This binding site is at the boundary between the α- and β-dimers (see Fig. 20.4). The binding of vinblastine to the vinca domain results in the formation of tubulin-vinblastine complexes. Due to its altered conformation, this complex cannot align properly with other tubulin dimers and consequently cannot properly assemble onto growing microtubule chains. This inhibits polymerization of new tubulin units and thus impacts mostly the plus end of microtubules. The vinblastine can also bind to tubulin units already in the microtubules, causing them to change their shape and leading to depolymerization. By disrupting polymerization and depolymerization, vinblastine affects the dynamic instability of microtubules. This disruption of microtubule dynamics is detrimental to dividing cells because cell division requires microtubule dynamics that lead to the correct segregation of chromosomes during M phase. Vincristine works similarly to vinblastine [13–16].

Both vincristine and vinblastine are **vinca alkaloids**, natural plant products, isolated from the leaves of the periwinkle plant (*Catharanthus roseus*). The periwinkle plant has been used in traditional medicine for centuries, and vincristine and vinblastine were first isolated and used as cancer chemotherapeutic drugs in the 1950s. Because they disrupt microtubule dynamics, they work largely by blocking the M phase of the cell cycle. They were initially used to treat childhood leukemias and subsequently used to treat several types of cancer [10]. Now, vincristine is most commonly used to treat leukemias, lymphomas, and neuroblastoma, while vinblastine is used to treat Hodgkin's lymphoma, testicular cancer, lymphomas, neuroblastoma, and other malignancies [17, 18]. Both vincristine and vinblastine are still used, although they are increasingly being replaced by more specific and targeted treatments.

Taxol (paclitaxel) [19] is a **small-molecule** chemotherapy drug that also disrupts microtubule dynamics. Taxol was also originally isolated from a natural product, specifically the bark of the Pacific yew tree. It is part of a family of drugs called the **taxanes**. Instead of preventing microtubule polymerization, Taxol mostly prevents microtubule disassembly; thus, it stabilizes, or "freezes," the microtubules. Taxol also binds to β-tubulin subunits of microtubules, but while vincristine and vinblastine bind to tubulin at the vinca region on free α-β-tubulin-dimers or on β-subunits near the plus end of the microtubule, Taxol binds to a different region on β-tubulin, in a region that is on the inside of the already assembled microtubule (the luminal side). This region is the taxane-binding site. Because microtubule dynamics are critical for cell replication, Taxol also inhibits the cell cycle at M phase, preventing proper separation of chromosomes during cell division and leading to cell death. It is used to treat some cancers, including **breast** and **ovarian cancers** [10, 15, 16].

The microtubule-disrupting drugs were among the first to successfully treat any type of cancer. They are effective tools for inhibiting the growth of cancer cells because they block the cell cycle, and thus they target rapidly growing cells, which is typical of many cancer cells. Microtubule-inhibiting drugs, specifically the vinca alkaloids, were the first agents to successfully treat childhood leukemias, changing the disease from being mostly incurable to highly treatable. However, the drugs are also problematic because of side effects. Some of these side effects are due to the death of normal cells, in addition to the cancer cells, and can lead to digestive problems and neurotoxicity. Other problems relate to cells developing resistance to treatment. Microtubule-disrupting drugs are not specific to cancer cells, and more recent development of targeted therapies may have benefits over these more broadly acting drugs, but in general, as with most cancer drugs, optimizing the dose for each patient and considering various combination therapies are essential [16].

1.5 DNA Changes in Quantity and Chromatin Structure Throughout the Cell Cycle

1.5.1 DNA Quantity Changes Throughout the Cell Cycle

DNA changes in quantity throughout the cell cycle. Using C to represent DNA quantity DNA goes from 2C in G1 to 4C by the end of S phase, and back to 2C again after M phase, when a new round of the cell cycle begins in each daughter cell (Fig. 20.1). When the DNA quantity is at 2C, this means there are two copies of every chromatid. At 4C, there are four copies of every chromatid [7]. It is important to remember that 2C and 4C in this case refer to the number of **chromatids**, described below.

1.5.2 Chromosomes and Chromatids

Two essential concepts relating to DNA throughout the cell cycle are the concepts of **chromosomes** and **chromatids**: **after S phase**, a **chromosome** consists of **two identical sister chromatids** joined together at the centromere [7, 20]. At this point, the DNA content is 4C. This is because there are two homologous chromosomes for each chromosome in the cell, one maternal and one paternal, and since DNA replication has taken place in S phase, each of these chromosomes consists of **two sister chromatids for a total of four chromatids.** Once the chromosomes are separated at the **end of M phase,** the sister **chromatids** separate, and now these chromatids constitute the two daughter **chromosomes** (Fig. 20.1). At this point, the DNA content becomes 4C for one maternal chromosome and one paternal chromosome, and neither has been replicated yet. Once the cells go through S phase, each chromatid will replicate, and the cell returns to the state where each chromosome consists of two identical sister chromatids. Humans have **23 pairs** of chromosomes, one from the mother and one from the father. Thus, they have 46 chromosomes total (22 pairs of autosomes and 1 pair of sex chromosomes (X and Y)). There are 46 chromosomes before DNA replication (S phase), and there are still 46 chromosomes after DNA replication. After DNA replication, however, each chromosome consists of a pair of identical sister chromatids.

1.5.3 Chromatin Structure Changes Throughout the Cell Cycle

The overall morphology of the DNA also changes throughout the cell cycle. During interphase, the DNA is not highly condensed and cannot be readily visualized under a microscope. DNA becomes highly condensed during M phase, at which point the individual chromosomes are visible through a microscope. The DNA becomes decondensed again after M phase. Chromosome condensation involves DNA-binding proteins, especially histones. Histones (described in more detail in Chap. 6) comprise a family of proteins that bind to DNA. DNA wraps around the histones and forms a coil. These coils become supercoiled, creating the highly condensed chromosome structures present in M phase [1–4].

1.6 Ploidy

An important concept in cell biology and genetics is ploidy. Ploidy relates to the number of complete sets of chromosomes in a cell, as discussed below and in [2, 7].

1.6.1 Diploid Cells

Most somatic cells in mammals are diploid. Our somatic cells have two sets of each chromosome, one maternal set and one paternal set. This is referred to as 2N.

It is important to distinguish between the use of 2N to describe diploid cells and the use of C to describe DNA content during the cell cycle (as in Fig. 20.1). When referring to **ploidy**, 2N represents the number of **chromosome sets** in a cell (diploid cells are 2N at G1 phase before chromosome replication, but they remain 2N even after S phase). In the context of the **cell cycle**, however, C refers to the total DNA content in the cell. A diploid human cell in G1 phase is 2N 2C and has one chromatid per chromosome, but once the DNA replicates in S phase, the cell remains 2N (a diploid cell) but becomes 4C because each chromosome now consists of two chromatids. These are distinct concepts and should not be confused. One refers to ploidy (N) and the other to DNA content (C) in different phases of the cell cycle.

1.6.2 Haploid Cells

One type of cell in our body is a haploid cell. That is the **germ cell**: egg or sperm. These cells have one set of each chromosome; they are 1N. Germ cell precursors undergo meiosis, and at the end of meiosis, the chromosome number is reduced to half, resulting in one set of each chromosome. The egg remains a haploid cell until it is fertilized by sperm, also a haploid cell. After the egg is formed by sperm, it becomes a diploid cell (a zygote), which will develop into an embryo.

1.6.3 Polyploidy

Animal cells are mostly diploid, with the exception of germ cells, which are haploid. Polyploidy is seen in some contexts, particularly in plants. Polyploidy refers to cells that have more than two sets of chromosomes, greater than 2N. For example, some plant cells might be triploid or tetraploid, meaning that they have three or four sets of chromosomes (3N or 4N). While polyploidy is rare in animal cells, it can sometimes be seen in cancer cells. Some cancer cells may acquire additional sets of chromosomes, leading to abnormal ploidy states such as 3N or 4N, or even higher. These cells are considered to be polyploid.

1.6.4 Aneuploidy

In aneuploid cells, there is an abnormality in chromosome number, but not in the complete set of chromosomes. For example, in trisomy 21, associated with the condition Down syndrome, there is an extra copy of chromosome 21, but the other chromosomes are normal in number. Cells with trisomy 21 are aneuploid. Aneuploids can be trisomy, such as trisomy 21 and trisomy 18, or monosomy (where one chromosome of a pair is missing), such as monosomy 7. Sometimes aneuploidy can occur in a mosaic form. For example, in some patients with mosaic Down syndrome, some cells have trisomy 21, and other cells have the normal set of chromosomes.

1.6.5 Euploidy

Euploidy is a broad term used to describe some of the above situations. Euploidy means that there is a balanced set of complete chromosomes. This may be the case in normal cells, which are 2N, or abnormal cells, such as cancer cells, which may be 3N, 4N, or more. Any cell with a complete set of chromosomes is euploid. Thus, cells that are haploid, diploid, or triploid (1N, 2N, 3N) and so on are all euploid. In contrast, cells with abnormal chromosome numbers, such as in trisomy 21, are aneuploid and are not considered euploid. Aneuploid cells have an incomplete or unbalanced chromosome set.

1.7 Regulation of the Cell Cycle by Restriction Points and Checkpoints

The different phases of the cell cycle must be precisely regulated every time the cell divides. Each phase follows the previous phase at a precise time and when conditions are right. For example, M phase can only occur after S phase has been completed in its entirety, and S phase cannot start if the conditions are not suitable for DNA replication. Furthermore, cell duplication usually does not occur at all if the cell is not in an environment containing nutrients or growth factors. The regulation of the cell cycle thus involves the coordination of various external and internal cues. Restriction points and checkpoints are the two primary mechanisms of cell cycle regulation. **Restriction points** usually monitor the presence of nutrients or growth factors, and cell size. Cells typically have one restriction point. Once cells pass a restriction point, they commit to another complete round of the cell cycle without relying on external cues, as long as all the individual phases are completed correctly. **Checkpoints** ensure that the events that take place during different phases of the cell cycle are coordinated with one another. They ensure that the various phases occur in the appropriate order and that the cell does not enter one phase until the previous phase is correctly completed. They do not respond to external cues. There are multiple checkpoints throughout the cell cycle, discussed below and also covered in [2–4, 21–25].

1.7.1 Restriction Points

One restriction point, the **G1 restriction point** (also called **the R point)**, controls the cell cycle in most animal cells and in *Saccharomyces cerevisiae* (budding yeast). This restriction point controls the transition from G1 phase to S phase. Cells do not pass this restriction point unless there are sufficient nutrients or growth factors in their environment. If cells do not pass this restriction point, they may stay in G1 or G0 phase, either indefinitely or until the conditions are right to pass the G1 restriction points. Some cells are controlled via a **G2 restriction point** instead of a G1 restriction point. The G2 restriction point controls the transition from G2 to M phase. Cells with a G2 restriction point include oocytes of some species, and *Saccharomyces pombe* (fission yeast) (Fig. 20.1).

1.7.2 Checkpoints

Checkpoints are different from restriction points in that there are multiple checkpoints throughout the cell cycle, and they do not depend on external cues such as growth factors or nutrients. Checkpoints instead coordinate the different phases of the cell cycle with each other. They ensure that one phase is completed correctly before the subsequent phase begins. Some examples of checkpoints include the M phase checkpoint, the G1 checkpoint, the G2 checkpoint, and the S phase checkpoint [22].

The **M phase checkpoint** (also called the spindle assembly checkpoint, SAC) monitors chromosome alignment. Cells cannot pass this checkpoint unless the chromosomes are perfectly aligned in the center plane of the cell. This checkpoint ensures that all kinetochores are properly attached to spindle microtubules before progression to anaphase. This is important so that each daughter cell receives precisely half the genetic material. If it fails, daughter cells can obtain an abnormal number of chromosomes, which may result in aneuploidy.

The **G1 checkpoint** is closely linked to the G1 restriction point. It monitors several things, including the presence of damaged DNA. Due to this checkpoint, the cell cycle will stop at G1 phase if damaged DNA is present. Thanks to this checkpoint, damaged DNA does not get replicated during S phase and thus does not get passed down to future generations of cells. Cells have DNA repair mechanisms, and once the DNA is repaired, the cell may pass through the G1 checkpoint and enter S phase, where DNA gets replicated.

The **G2 checkpoint** (not shown here) controls the transition from G2 to M phase. This phase also monitors for damaged DNA as well as unreplicated DNA. This checkpoint ensures that M phase does not start until S phase is complete. It also ensures that any damaged DNA will not be passed on to subsequent generations of cells.

The **S phase checkpoint** monitors for damaged DNA during replication, as well as stalled replication forks or incomplete replication. It can pause S phase to allow for DNA repair.

If the checkpoints fail, the consequences can be severe. Failure of the G1 checkpoint can lead to mutations occurring because damaged DNA can be replicated. Failure of the S phase checkpoint can lead to genomic instability due to replication of DNA even if replication is stalled or if it has mutations. Failure of the G2 checkpoint can lead to chromosome fragmentation or DNA that does not segregate properly. Failure of the M phase checkpoint can lead to aneuploidy, and this in turn can lead to cancer.

1.8 Drugs That Target Various Phases of the Cell Cycle

Often in cancer, the cell cycle is not properly controlled. Cells may divide at the wrong rate, fail to respond properly to external cues, or may incorrectly bypass checkpoints. In addition to targeting microtubules, as described above, anti-cancer drugs have been developed to target different phases of the cell cycle, with the goal of slowing down or stopping the growth of rapidly growing cells. A common target for cancer therapies is DNA replication, and therefore, a number of cancer drugs target S phase. A few examples are here.

Cytosar-U (cytarabine) [26] is a small-**molecule chemotherapy drug** that interferes with DNA synthesis during **S phase**. It is used to treat leukemia and lymphoma. Sometimes it is used in combination with other drugs. Cytarabine is an analog of deoxycytidine (dCTP). It is similar to dCTP but has an arabinose sugar instead of deoxyribose. In the cell it gets phosphorylated to form cytarabine triphosphate (Ara-CTP). Ara-CTP competes with dCTP and gets incorporated into DNA as DNA replicates. This interferes with normal DNA synthesis and leads to chain termination and inhibition of DNA replication during S phase. This leads to an accumulation of damaged DNA, which can lead to cell death [27, 28]. This drug was developed in the 1960s but remains an important treatment for leukemia [29].

Rheumatrex (methotrexate) [30] is also a **small-molecule** chemotherapy drug targeting cells in **S phase**. It works by **inhibiting folic acid synthesis**, which is required for **DNA synthesis**. This process is described in more detail in Chap. 16. Methotrexate inhibits folic acid synthesis by **competitively binding to the enzyme dihydrofolate reductase (DHFR)**, which converts folate to its active form. This active form of folic acid, detrahydrofolate, is necessary for DNA and RNA synthesis. In addition to **cancer treatment**, this drug is used to treat **inflammation-related diseases, such as rheumatoid arthritis, psoriasis, and Crohn's disease**. This is because inflammatory disorders involve improperly controlled growth of multiple types of cells, such as immune system cells, blood vessel cells, and some epithelial cells [31–34].

Adriamycin (doxorubicin) [35] is also a **small-molecule chemotherapy** drug. It targets cells in **S and G2 phases**. Adriamycin was previously covered in Chap. 6. Adriamycin inserts itself into the DNA helix. This distorts the DNA and interferes with DNA replication. This is due in part to its interference with the ability of the enzyme **topoisomerase II** to function. Adriamycin may also work by other mechanisms, such as the generation of free radicals leading to damaged DNA. The drug is used to treat several cancers, including breast, lung, and bladder cancers [36–38].

VePesid (etoposide) [39, 40] also inhibits **topoisomerase II** and targets cells in **S and G2 phases**, but it does so by a different mechanism than Adriamycin. It is a **small-molecule** chemotherapy drug used to treat several cancers, including **lung cancer, testicular cancer, and lymphomas** [41–44]. Etoposide is also an older drug, having been developed in the 1980s, which is still regularly used for some cancers, often in combination with other treatments.

Gemzar (gemcitabine) [45] is a **small-molecule** chemotherapy drug that interferes with **DNA synthesis**, thus targeting cells in **S phase**. **Gemzar** is a prodrug. It gets phosphorylated in the cell to become its active form. Its active form is **gemcitabine triphosphate (dFdCTP)**, which competes for dCTP for incorporation into DNA as it is being replicated. This leads to chain termination, halting DNA synthesis. The drug also interferes with the enzyme **ribonucleotide reductase**. Ribonucleotide reductase is an enzyme that catalyzes the conversion of ribonucleotides to deoxyribonucleotides and is therefore important for maintaining a pool of deoxyribonucleotides for DNA synthesis. Cells treated with the drug cannot replicate their DNA and thus cannot pass through the cell cycle, and they ultimately die. The drug is used to treat **pancreatic, lung, breast, and ovarian cancers**, among others [33, 46, 47].

2 Molecular Basis of Cell Cycle Regulation

Cell cycle regulation is complex and involves many steps. Complex molecular reactions are required to regulate and coordinate these steps. To understand the molecular basis of cell cycle regulation, it is helpful to review some history of cell cycle research. In particular, the discovery of a protein complex called **MPF** led to the identification of **cyclins and cyclin-dependent kinases (CDKs)**, which have essential roles in coordinating the different phases of the cell cycle. Early studies on the molecular basis of cell cycle regulation involved studying Xenopus (frog) oocytes, yeast, sea urchin eggs, and mammalian cells [1, 48], and will be discussed here.

2.1 History of the Discovery of MPF

2.1.1 Studies Involving Xenopus Oocytes

Key studies on the cell cycle regulation were done in **1971** by Masui and Markert [49]. This involved research studying meiosis in germ cells, specifically, in Xenopus oocytes. While somatic cells and germ cells both go through the cell cycle and share many conserved features, somatic cells divide by mitosis to produce two identical diploid daughter cells, and germ cells undergo meiosis and produce gametes, which are haploid, having half the starting number of chromosomes. Although the topic of this chapter is the cell cycle in somatic cells, it is helpful to have a basic understanding of meiosis because early studies of cell cycle regulation were carried out by studying oocytes undergoing meiosis [2, 50].

- Basic Review of Oogenesis and Meiosis
  - Overview of Oogenesis and Meiosis

Briefly, meiosis is a process of cell division that is unique to germ cells. It reduces the number of chromosome sets by half, resulting in four haploid cells for each diploid cell that started the process. In females, this occurs as part of oogenesis, the process by which oogonia give rise to mature oocytes. Oogonia (singular is oogonium) are diploid precursor cells that enter meiosis and give rise to four haploid cells.

Before meiosis begins, the oogonium undergoes the G1, S, and G2 phases of the cell cycle (although the G2 phase is usually very short). After S phase, the cell is called a primary oocyte. It is still diploid, but it has duplicated chromosomes. After the brief G2, the cell enters meiosis. Meiosis consists of meiosis I and meiosis II. Meiosis I begins with prophase I, where chromosomes condense and homologous chromosomes pair up. At this stage, crossing over occurs, where segments of genetic material are exchanged between homologous chromosomes. Metaphase I is next, where the chromosomes align along the metaphase plate at the center of the cell. At anaphase I, the homologous chromosomes are separated to opposite poles of the cell. Then, during telophase I and cytokinesis, the cell divides into two cells. These cells are haploid. They have one copy of each chromosome, but each chromosome contains two chromatids. The chromosomes partially decondense at this stage.

This split differs from mitosis. In mitosis, sister chromatids are pulled apart at the centromere, so that each daughter cell gets an identical set of chromosomes consisting of a single chromatid (now called a chromosome again). In meiosis I, homologous chromosomes pair up as tetrads, and it is these pairs, rather than the sister chromatids, that are separated. Each daughter cell gets one chromosome from each tetrad, either maternal or paternal, chosen at random, and the sister chromatids remain joined.

During meiosis II, the chromosomes condense again at prophase II. In metaphase II, the individual chromosomes align at the metaphase plate. During anaphase II the sister chromatids of each chromosome separate and move to opposite poles, in a manner similar to what is seen in mitosis. After telophase II and cytokinesis, four haploid cells result from one oogonium.

It should be noted that during oogenesis, the cells do not actually proceed past metaphase II unless fertilization occurs. Once it does so, the process continues until a mature, fertilized ovum results. Furthermore, unlike male gamete production, oogenesis only results in one functional haploid cell. The remaining cells are extruded as polar bodies and do not develop into ova.

Meiosis is distinct from mitosis, with specific differences. For example, during metaphase I in meiosis, crossing over occurs between homologous chromosomes, leading to the exchange of genetic material. Genetic exchange is rarer in mitosis. Certain basic mechanisms, however, are conserved in cells undergoing mitosis and those undergoing meiosis.

– The Two Arrests in Xenopus Oocyte Development

During the process of oogenesis, oocytes usually undergo two arrests. Because the cell studies we will be describing involve Xenopus (frog) oocytes, the following will cover these two arrests in Xenopus oocytes. Oocytes in other animals also generally have two arrests, but there are some key differences, not covered here.

During oogenesis, and after the cell has gone through S phase, the cell, now referred to as a primary oocyte, transitions through a very brief G2 phase and enters prophase of meiosis I (specifically the diplotene stage of prophase). The primary oocyte now arrests at this prophase I of meiosis. This phase, though technically already part of meiosis, resembles G2 phase because DNA replication is complete and the cell is growing. This is the **first arrest**, often referred to as the **G2 arrest**, even though cells have already technically entered the first part of meiosis. This is a prolonged arrest while the cell grows in size. This first arrest is released once the cell is stimulated by progesterone, which is typically released by the female when stimulated by a male frog.

The primary oocyte now resumes meiosis I, and after homologous chromosomes are separated, it becomes a secondary oocyte and enters meiosis II. In meiosis II, this secondary oocyte undergoes a second arrest, which is referred to as the M phase arrest, and it occurs during metaphase II. The cell remains in this second arrest until fertilization by sperm. Following fertilization by sperm, the secondary oocyte completes meiosis II, a process that separates sister chromatids and results in a mature ovum [50, 51].

– Summary of Meiosis and the Two Arrests in Xenopus Oocytes

Oogonia are diploid precursor cells that enter meiosis and become primary oocytes. The primary oocyte initiates meiosis I but arrests in prophase I. Upon hormonal stimulation, it completes meiosis I to become a secondary oocyte, a haploid cell that has chromosomes with sister chromatids. The secondary oocyte arrests again at metaphase II, and at this stage it is referred to as a mature oocyte. Completion of meiosis II occurs only upon fertilization, at which point the cell becomes a mature ovum.

• Identification of MPF Activity in Xenopus Oocytes

Now that we have reviewed the basic steps of meiosis, we can discuss the discovery of MPF in frog oocytes. To understand the mechanism behind the release from the **G2 phase arrest** in the frog oocytes, studies were conducted in which a small amount of cytoplasm was extracted from frog oocytes that had been hormonally stimulated to mature and reach the **metaphase II arrest**. This cytoplasm was then transferred into the cytoplasm of G2 phase-arrested oocytes using a microinjection technique. Interestingly, this transfer or cytoplasm was sufficient to cause the G2-arrested oocytes to **resume meiosis**, complete meiosis I, and arrest at metaphase II. These studies were groundbreaking because they demonstrated that some internal cytoplasmic factor is sufficient for triggering the cells to mature (i.e., proceed through meiosis until metaphase II), rather than requiring external stimulation. The cell's own machinery appeared to be sufficient. While the researchers did not yet know the nature of this cellular factor, they named it **MPF** for **maturation-promoting factor** [1, 50]. Similar types of activity were later found in other organisms such as starfish, mouse, and surf clam [48].

2.1.2 Genetic Studies in Yeast

Several years later, in 1976, Lee Hartwell and Paul Nurse also studied the cell cycle, but their model systems were yeast [52–54]. They studied the two major types of yeast, *Saccharomyces cerevisiae* (budding yeast) and *Saccharomyces pombe* (fission yeast). In this study, the scientists used yeast genetics approaches. Yeast has a relatively small genome compared to many other eukaryotes, and it is thus an ideal system for carrying out genetic studies. In this system, the researchers treated the yeast with mutagens that lead to random DNA mutations. They used conditions that are well established by yeast geneticists that lead to approximately one mutation per yeast cell or less. They also used conditions where the mutations were temperature sensitive (temperature-sensitive mutants); these types of mutations could be turned on or off at specific temperatures. They used this method to screen for mutant yeast that had abnormalities in progressing through the cell cycle. After identifying these cell cycle-deficient strains, yeast genetics techniques were used to determine which specific genes were mutated. By doing this type of yeast genetics, the researchers identified a gene that they named **Cdc2** in *Saccharomyces pombe*, and a homologous gene that they named **Cdc28** in *Saccharomyces cerevisiae*. **Cdc** stands for cell division cycle. Yeasts with mutations in these genes were unable to progress from G1 to S phase. These yeast cells were unable to synthesize the proteins normally encoded by these genes. While this study differs from the Xenopus study described above, it had in common the finding that a factor within the cell is essential for progression through the cell cycle. Later, the protein product of Cdc was found to have an important connection to MPF [1, 55].

2.1.3 Studies in Mammalian Cells

Cytoplasm injection experiments were conducted to determine whether MPF was also present in mammalian cells [48]. These experiments were done as early as 1979 [56]. Mammalian cells can be arrested at M phase under certain conditions. Cytoplasm from M phase-arrested mammalian cells was injected into Xenopus oocytes that were at the G2-arrested state. Strikingly, this caused the Xenopus oocytes to mature, i.e., to exit the arrest and progress through meiosis. The conclusions from these studies were that mammalian cells also contain a cytoplasmic factor that can cause Xenopus oocytes to progress through meiosis. This factor appeared to be highly conserved because it could function across different species (the factor from mammalian cells was introduced into Xenopus eggs). Studies carried out in mammalian cells did not involve the same types of cytoplasmic transfer experiments, but rather they used various biochemical analyses and showed that G2 to M transition in mammalian cells is driven by a similar mechanism. This factor was also referred to as MPF, and it was later renamed **M phase-promoting factor** to emphasize its role in the transition to mitosis (check for clarity).

2.1.4 Studies in Sea Urchin Eggs

The cell cycle was studied in sea urchin eggs in 1982 by Tim Hunt and colleagues [57]. Sea urchin eggs can be a valuable model for studying cell duplication because they grow relatively quickly and efficiently in the lab. **Sea urchin eggs** were **fertilized** and synchronized, then grown in culture, and monitored as they progressed through different cell cycle phases. The cells were grown in media that contained radioactively labeled methionine (35S Methionine). This is an amino acid that and thus gets incorporated into newly made proteins. This gave scientists a mechanism for identifying newly produced proteins in the cell. The egg cells were harvested at different cell cycle phases and lysed, and the proteins were isolated. The proteins were then electrophoresed on an SDS polyacrylamide gel, which separates proteins according to size. The proteins on the gel were then transferred to a membrane, and the membrane was analyzed by autoradiography so that bands would appear representing any radioactively labeled, and hence newly synthesized, proteins. What the researchers found was that the levels of certain proteins cycled throughout the cell cycle. They would be present in certain stages in the cycle, disappear, and be detectable again later. They named these proteins **cyclins** because their presence cycled throughout the cell cycle. They identified two major cyclins that differed by size, which they named **cyclin A** and **cyclin B**. The conclusion was that certain proteins in the sea urchin eggs show a cyclic pattern throughout the cell cycle. We will see later that these were also subsequently found to be components of **MPF** [1, 55].

2.2 Purification of MPF

Experiments described above indicate that MPF is a cytoplasmic factor. Scientists next wanted to understand what this factor consisted of. In order to do so, the cytoplasm from the meiosis-arrested eggs from the Xenopus experiment was isolated, with the aim of purifying MPF from the cytoplasm [48]. The goal was to isolate the smallest components of the cytoplasm that were sufficient to release the G2 arrest of oocytes and cause them to progress through meiosis [58]. This involved carrying out fractionation and chromatography of the cytoplasm, followed by functional assays to pinpoint the smallest component that could drive release of the G2 arrest. Once the purification process was complete, the scientists identified a complex of two proteins: a 32-kd protein and a 45-kd protein. See Fig. 20.5.

The **45-kd protein** turned out to be a homolog of **Cyclin B**, initially found in the sea urchin egg study, and the **32-kd** protein was a homolog of **cdc2**, which is encoded by the gene initially identified in the yeast study. This complex of proteins, although initially identified in a study of gametes and meiosis, was found to be a key complex for regulating the cell cycle in somatic cells, at the G2 to M phase transition. These two proteins turned out to be highly conserved across species, and analogous proteins were found in all other eukaryotic cells. In some cases, the homologs in other species have been given different names. For example, the Cdc2 protein, also identified in yeast, stands for cell division cycle protein, but a homologous protein identified in mammalian cells is referred to instead as CDK1 for cyclin-dependent kinase 1. The activity of MPF (now known to be a complex of cyclin B and CDK1) oscillates throughout the cell cycle.

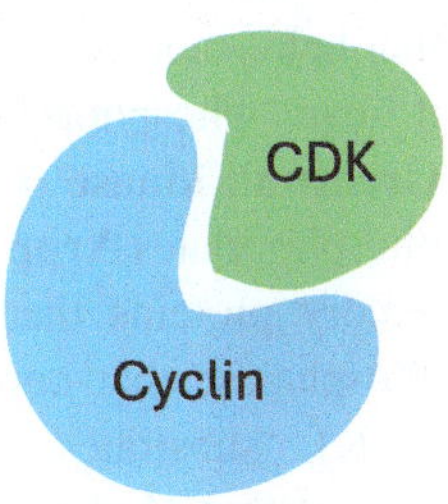

Fig. 20.5 Cyclin CDK complexes. Cyclin CDK complexes regulate different phases of the cell cycle. The CDK complex is a kinase that depends on cyclin for its activation

This is due in large part to the presence or absence of the cyclin component. Furthermore, Cyclin B is part of a family of proteins called cyclins, and CDK1 is also a part of a family of CDKs. The rest of this chapter will focus on the functions of the cyclin/CDK complexes in regulating the cell cycle in mammalian eukaryotic somatic cells.

2.3 The Functions of MPF in Regulating the Progression from G2 to M Phase

2.3.1 The Main Components of MPF

In mammalian somatic cells, the **MPF** complex contains **Cyclin B** and **CDK1** (the mammalian homolog of Cdc2), and it promotes progression from G2 phase to M phase. Cyclin B is a cyclin whose levels vary throughout the cell cycle (see Fig. 20.6). The amount of CDK1 does not cycle, although its activity (rather than its amount) does cycle. CDK1 is a protein kinase, specifically a serine/threonine kinase. As a protein kinase, it has the role of phosphorylating its substrates and thereby regulating its substrate's activity. Specifically, CDK is a cyclin-dependent kinase in that it must be bound to a cyclin in order for its kinase activity to be active [4, 22, 50].

2.3.2 Key Functions of MPF and Its *Substrates* That Promote, Maintain, and Complete M Phase

In somatic cells, the main function of MPF is to control the progression from G2 to M phase. By the end of G2 phase, CDK1 and Cyclin B are bound together, and CDK1 is active. CDK1 has a crucial role in what happens next. As a part of MPF, **CDK1** phosphorylates and thereby regulates several substrates that are important for progression to M phase. CDK1 has to be bound to Cyclin B in order to phosphorylate its substrate. Some important CDK1 substrates are as follows.

- Nuclear Lamins MAPs and Motor Proteins are Phosphorylated by CDK1

Nuclear lamins [59] are an important example of a CDK1 substrate. Nuclear laminas are a type of protein similar to intermediate filaments but

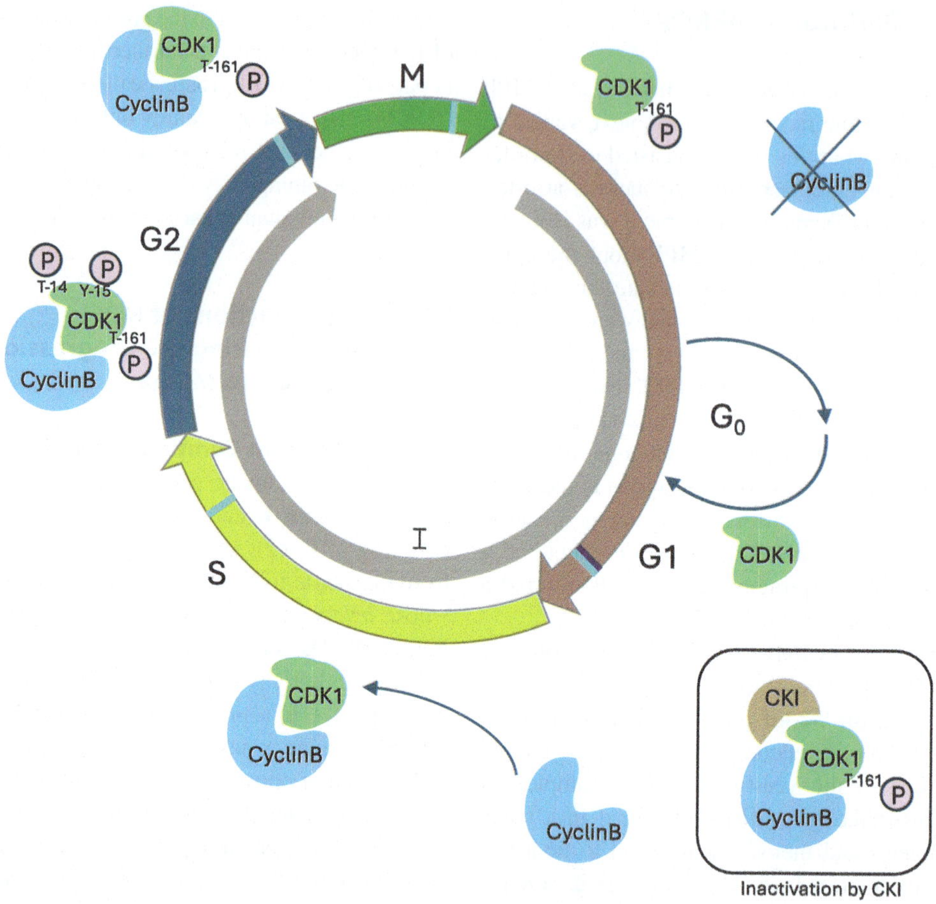

Fig. 20.6 Cyclin B/CDK1. The cyclin B/CDK1 complex plays a crucial role in regulating the G2-M phase transition. The complex is also regulated by phosphorylation and dephosphorylation and it can be inactivated by CKIs. See text for details.

found in the nucleus. In that sense, they are analogous to a type of cytoskeletal protein in the nucleus. They are major components of a structure called the **nuclear lamina**. The nuclear lamina is a structure found on the inner surface of the nuclear membrane. It provides structural support to the nucleus and helps maintain its shape. Phosphorylation of nuclear lamins changes their structure within the nuclear lamina. Phosphorylation is the major event that causes the nuclear lamina to break down during mitosis, and since they provide structural support, this leads to nuclear membrane breakdown.

Another CDK1 substrate is **MAP**. MAPs are **microtubule-associated proteins (MAPs)**: they play key roles in the reorganization of microtubules, which is essential for chromosome segregation in M phase [2, 60]. CDK1 also phosphorylates **motor proteins** that have roles in separating mitotic chromosomes, and it phosphorylates substrates that are important for chromosome condensation.

- Greatwall kinase (Gwl is also phosphorylated by CDK1

CDK1 also phosphorylates and activates proteins that help stabilize the mitotic state. For example, it can phosphorylate and thereby activate **Greatwall kinase (Gwl)**. Activated Gwl in turn phosphorylates **Arpp19** and **Ensa**, which are regulatory proteins that inhibit **PP2A**. PP2A is a phosphatase that dephosphorylates some of the CDK1 substrates. Therefore, its inhibition ensures that M phase is maintained [2, 61].

- APCC is phosphorylated by CDK1

Finally, at metaphase, **CDK1 phosphorylates APC/C (anaphase-promoting complex/ cyclosome), an E3 ubiquitin ligase complex**. Phosphorylation of **APC/C** facilitates its binding to, and activation by, its coactivator protein, **Cdc20**. The resulting **APC/C ^{Cdc20} complex**, in turn, promotes ubiquitination and degradation of certain proteins. First, it ubiquitinates a protein called **securin**, targeting it for degradation. This is an important step in triggering the separation of the chromatids and the beginning of anaphase. The other important target that it ubiquitinates is **Cyclin B**. This leads to the degradation of cyclin B. Once cyclin B is degraded, CDK1 is inactivated and can no longer phosphorylate its targets. At the same time, inhibition of **PP2A** by **Greatwall kinase** is relieved. Active PP2A now dephosphorylates CDK1 substrates, and **M phase is complete** [2–4, 50, 62–65].

2.3.3 Regulation of MPF

MPF, consisting of cyclin B and CDK1, regulates G2 to M phase entry, but MPF itself is also tightly regulated. Due to its precise regulation, it is only active at the correct phase, when cells are exiting G2 and entering M phase (Fig. 20.6). MPF is regulated at multiple levels, as follows:

(a) **Cyclin B synthesis and MPF formation**. Cyclin B synthesis begins in S phase [66, 67]. Cyclin B synthesis is required for CDK1 to function, so this is an essential step, but it is still insufficient alone. Next, Cyclin B forms a complex with CDK1. The complex with Cyclin B is essential for CDK1's activity, but activation of CDK1 is still not complete; several phosphorylation and dephosphorylation steps are still required.

(b) **CDK1 is phosphorylated and dephosphorylated**: CDK1 is phosphorylated on three regulatory sites: Thr-14, Tyr-15, and Thr-161. Several kinases carry out this phosphorylation. WEE1 and MYT1 phosphorylate Tyr-15 and Thr-14, respectively, and CAK (Cdk-activating kinase) phosphorylates thr-161. Thr-14 and Tyr-15 are negative regulatory phosphorylation sites, meaning that the CDK cannot be active when these two sites are phosphorylated. In contrast, thr-161 is a positive regulatory site.
At this stage, where all three of these sites are phosphorylated (including the two negative regulatory sites), the CDK1 remains inactive. Full activation requires the CDC25 phosphatase, which removes the inhibitory phosphate groups on Thr-14 and Tyr-15, leaving only Thr-161 phosphorylated. The result is an active MPF complex.

(c) Now that **CDK1 is active**, it can phosphorylate the substrates that lead to M phase, such as nuclear laminas, MAPs and others, as discussed above. This results in **M phase progression**.

(d) **Positive feedback loops in MPF activation:** MPF activation is reinforced by positive feedback mechanisms. Active **CDK1** phosphorylates and **activates CDC25**, thus amplifying its own activation. CDK1 also phosphorylates and thereby **inhibits** the activity of WEE1, preventing its inhibitory activity. These feedback loops can help reinforce the M phase signal until CDK1 triggers the mechanisms that lead to the end of M phase.

2.4 Other Cyclins and CDKs in the Cell Cycle

MPF is made up of Cyclin B and CDK1, and together these proteins regulate the transition from G2 phase to M phase. However, cyclins and CDKs are members of families of proteins that operate at different stages of the cell cycle. In each case, the different Cyclin/CDK complexes operate by a similar mechanism as described above for MPF (Figs. 20.7 and 20.8), but have different substrates, regulate different phases of the cell cycle, and can be regulated by different regulatory phosphorylation sites. There are multiple cyclins in eukaryotic cells, and they are active at different phases of the cell cycle. Cyclin D in

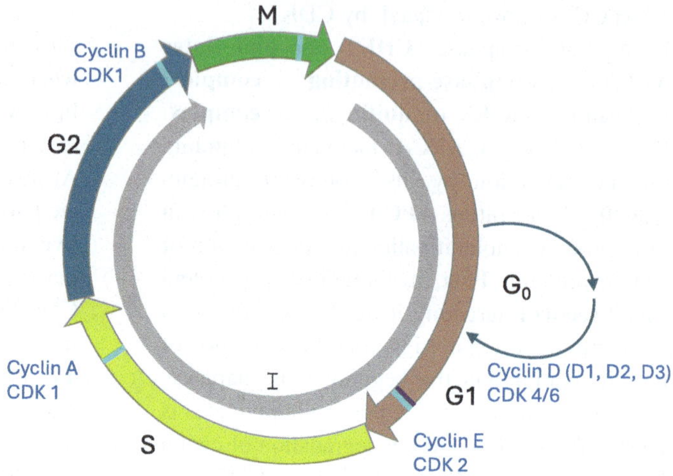

Fig. 20.7 Cyclin/CDK complexes throughout the cell cycle. Different cyclin/CDK complexes have different roles during different phases of the cell cycle

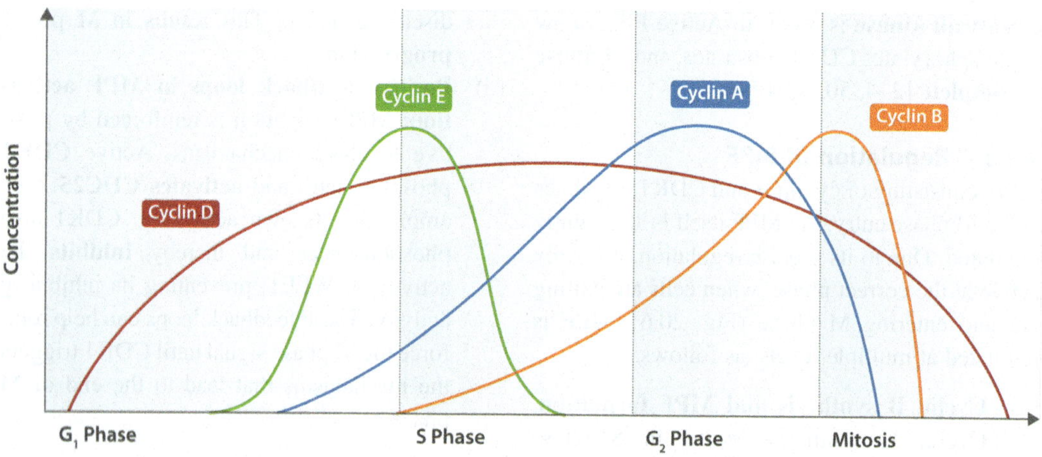

Fig. 20.8 The levels of different cyclins in the cell during different phases. Cyclin levels fluctuate at various stages of the cell cycle. The levels of Cyclin D, Cyclin E, Cyclin A, and Cyclin B throughout the cell cycle are shown here. (*Source*: Figure adapted from *Cyclin Expression.svg*, public domain image by WikiMiMa via Wikimedia Commons (https://commons.wikimedia.org/wiki/File:Cyclin_Expression.svg))

complex with CDK4 or CDK6 is important for regulating the G1 to S transition. Cyclin E and CDK2 play a role later in the G1 to S transition. Cyclin A in complex with CDK2 plays a role in the initiation and progression of DNA replication in S phase, and later, when it complexes with CDK1, it plays a role in the transition from G2 to M, along with the Cyclin B/CDK1 complex. The Cyclin B/CDK1 complex, as we discussed, has a role in the transition from G2 phase to M phase. As with all cyclins, the levels of the different cyclins change throughout the cell cycle, where Cyclin D levels rise during G1 and S phase, Cyclin E levels rise later in G1 phase, Cyclin A levels rise at the end of S phase, and in G2 phase, and cyclin B levels are highest at the end of G2 phase and in M phase [4, 50, 68]. These different cyclin/CDK complexes and the proteins they interact with are discussed individually, in more detail below.

2.4.1 Cyclin D, CDK4/6, Rb, and G1 Phase

• Cyclin D/CDK4/6 Complexes in G1 Phase

Cyclin D plays a key role in promoting progression through G1 phase. Cyclin D is unique among the cyclins. Other cyclins regulate checkpoints, ensuring proper progression through the different phases of the cell cycle. Cyclin D, however, has a role in the G1 restriction point (the R point) as well as in the G1 checkpoint. As a mediator of the restriction point, unlike other cyclins, Cyclin D is regulated by extracellular stimuli such as growth factors, which is a key feature differentiating restriction points from checkpoints.

Cyclin D's activity is followed by Cyclin E, which drives the transition from G1 phase to S phase. There are three cyclin D family members: Cyclin D1, Cyclin D2, and Cyclin D3. They function similarly but have different expression patterns and can thus be found in different cells. Cyclin D1 often receives the most attention as a primary player in the cell cycle. It is also important in cancer because the cyclin D1 gene is frequently overexpressed in cancer. Cyclin D can interact with two different CDKs: CDK4 and CDK6. Both CDKs work in the same way, and which one Cyclin D binds to may depend on which one is abundant in the cell [4, 69].

Cyclin D, as discussed above, differs from most of the other cyclins because it responds to extracellular signals. Remember that when cells pass from G1 phase into S phase, at the G1 restriction point, they depend on external cues such as growth factors. In cultured cells, these clues can be nutrients in the media. The presence of growth factors can activate signaling pathways that, in turn, lead to activation of expression (transcription) of the cyclin D gene. For example, its gene is upregulated in response to pathways that have been covered in Chap. 18, such as the MAP Kinase pathway, in response to stimuli such as growth factors. In fact, cyclin D is a critical target of the MAP Kinase pathway. This is a good example, therefore, of how a signaling pathway can lead to transcriptional changes that impact cell function, in this case, cell cycle progression [4, 70, 71].

Once Cyclin D is upregulated in the cell, it can bind to CDK4 or CDK6. Which one it binds to can depend on the cell type. In many cells, it binds preferentially to CDK4, and in other cells, it may bind to CDK6. The two have the same basic functions and can also compensate for each other. Here, we will use CDK6 to represent both.

• Rb Is a Key Cyclin D Substrate

CDK6, like other CDKs, is a protein kinase that phosphorylates substrates. A key substrate that it phosphorylates is **Rb (Retinoblastoma protein)** (Fig. 20.9). Rb has a role in putting a brake on the cell cycle. An important way that it does so is by binding to the transcription factor **E2F**. E2F, along with the transcription factor DP1, binds to the promoters of target genes and stimulates their transcription. Some of the target genes it activates are those that are important for entering S phase and carrying out the process of DNA synthesis/replication. Rb inhibits E2F by binding to it, and this prevents E2F from promoting the expression of genes that lead to S phase entry (Fig. 20.9).

The ability of Rb to inhibit the E2F transcription factor is regulated by phosphorylation. When Rb is hyperphosphorylated, i.e., phosphorylated on multiple sites, its ability to bind to and thereby inhibit E2F is blocked. Thus, E2F-regulated genes are effectively "derepressed" in response to Rb phosphorylation. Rb phosphorylation occurs in a sequence of steps that start when Cyclin D/ CDK6 phosphorylates Rb on multiple serine and threonine residues. This phosphorylation weakens the interaction between E2F and Rb and partly inhibits the ability of Rb to repress E2F. Rb gets further phosphorylated—hyperphosphorylated—by Cyclin E/CDK2 late in G1 phase, discussed below. This will lead to complete release (derepression) of E2F and allows cells to progress from G1 phase to S phase, where DNA is replicated [22, 72].

• FOXM1

Rb is the main substrate of CDK6 and plays the key role in G1 phase progression and subsequently to S phase transition, but another CDK6 substrate that plays a role in the cell cycle is the transcription factor **Forkhead box M1**

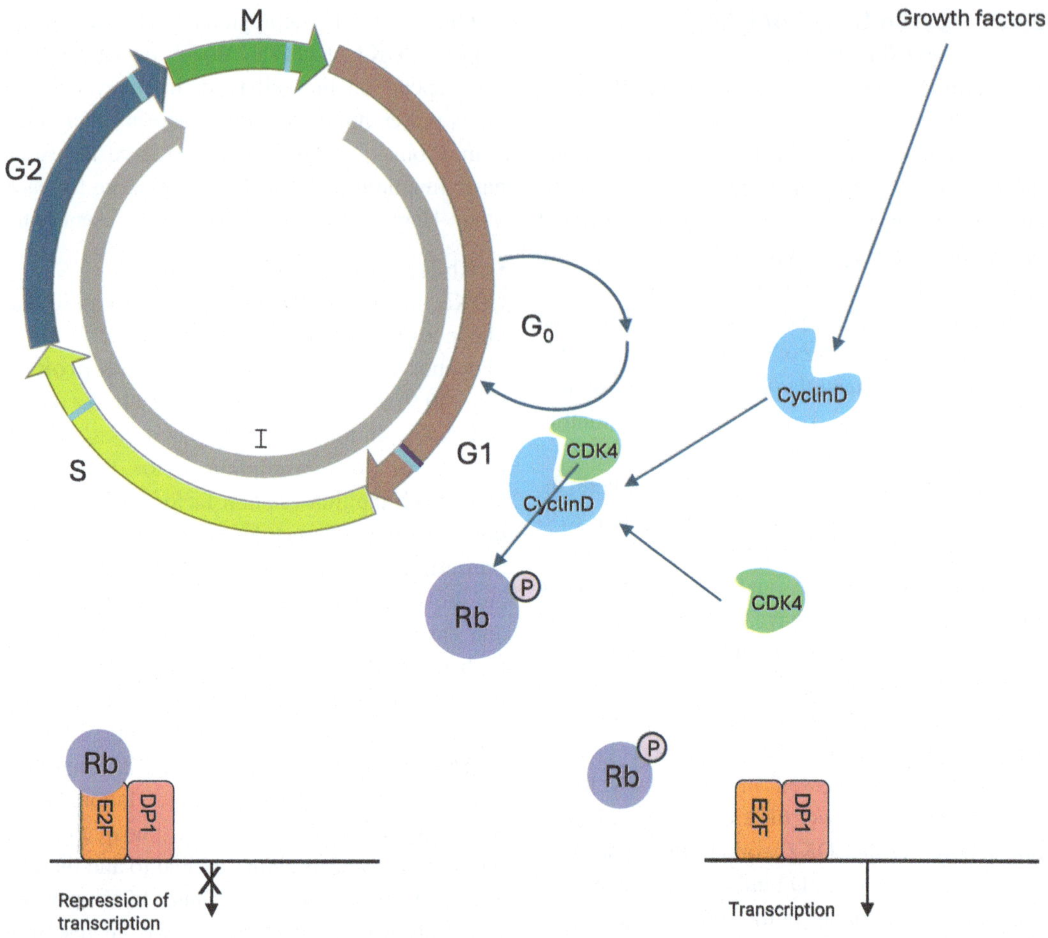

Fig. 20.9 The role for Rb at the G1 to S phase transition. Rb is regulated by CyclinD/CDK4; CDK4 phosphorylates Rb. Phosphorylated Rb leads to derepression of E2F/DP1 and enables the transition to S phase

(FOXM1) [73]. FOXM1 is mostly involved in G2/M phase, but it also has a role in G1 to S transition [74]. Forkhead is regulated by CDK6 in two ways. First, E2F (released from inhibition by Rb in response to CDK6 phosphorylation) leads to transcription of the FOXM1 gene. Second, after translation, the FOXM1 protein becomes phosphorylated by CDK4/6, and this enhances FOXM1's activity. FOXM1 now drives the expression of some of the genes that are important for DNA replication. Later, FOXM1 also becomes phosphorylated by CDK2, and at that point it will play important roles in progression to mitosis [75, 76].

- The Decline of Cyclin D

At the end of G1 phase, prior to the start of S phase, cyclin D levels begin to decline as transcription of the Cyclin D gene decreases due to a reduction in growth factors. In addition, under some conditions, such as stress or metabolic changes, Cyclin D also gets phosphorylated by **GSK3-β** [77]. This phosphorylated Cyclin D gets targeted for nuclear export and degradation via the ubiquitin proteasome pathway. Once another round of the cell cycle proceeds, new Cyclin D needs to be made for another round of G1 to S phase progression to occur [2, 3, 62, 63, 67, 69, 78].

- Cyclin D and CDK4/6 and Cancer

Since the CyclinD/CDK6 complex is critical for G1-S phase entry, it has a key role in controlling cell division. This complex is also highly important in cancer. Cyclin D is overexpressed in some types of cancer. Rb, the target of the CyclinD/CDK complex, is a tumor suppressor, meaning that its inactivation can cause cancer [1–4]. Rb will be discussed more in Chap. 21.

2.4.2 Cyclin E, CDK2, and S Phase

Rb is partly phosphorylated by CyclinD/CDK6, and this leads to partial inactivation of E2F and this in turn increases transcription of some E2F target genes. One of the genes that the E2F transcription factors activate (in response to this partial phosphorylation of Rb) is **Cyclin E**. Cyclin E also has a critical role in S phase entry. Cyclin E, in combination with CDK2, phosphorylates various substrates. One of the important substrates of CDK2 is also Rb. The initial phosphorylation of Rb by cyclin D/CDK6 weakens the interaction between E2F and Rb and partly activates E2F. It also primes Rb for further phosphorylation on additional serine and threonine sites, specifically by Cyclin E/CDK2. This further phosphorylation is considered hyperphosphorylation, and it leads to Rb's complete inactivation, leading to its complete release of E2F. E2F-regulated genes are consequently completely derepressed, promoting entry to S phase [67].

Once Rb is hyperphosphorylated and E2F is active, genes important for entering S phase and beginning DNA replication can be transcribed. Rb stays phosphorylated during this process, but the phosphate groups are removed later in the cell cycle, and cells cannot enter S phase again until Rb is phosphorylated again, first in response to Cyclin D/CDK6, and then in response to Cyclin E/CDK2.

In addition to hyperphosphorylation of Rb, Cyclin E/CDK2 has other substrates. For example, it has a key role in the assembly of the **pre-initiation complex** at replication origins. By phosphorylating proteins like Treslin [79], Cyclin E/CDK2 facilitates the recruitment of factors like CDC45 and GINS, and this enables the formation of the CMG complex, the active DNA helicase. As a result, Cyclin E/CDK2 activity has a role in the formation of the replication bubbles, which are critical steps at the beginning of DNA replication [2, 4, 80, 81].

2.4.3 Cyclin A/CDK1,2

Cyclin A levels rise during S phase and reach their highest levels in G2 phase. When Cyclin A levels become high during the S phase, it eventually replaces Cyclin E in the complex with CDK2. This new complex (Cyclin A/CDK2) has a role in DNA replication as well as completion of DNA replication and the end of S phase. Cyclin A is somewhat different from the other cyclins because it has a role in the regulation of two phases of the cell cycle by binding to two distinct CDKs. In S phase, cyclin A binds to CDK2, and as the cell progresses into G2 phase, Cyclin A also begins to bind to CDK1, helping the cell get ready for mitosis. Over time, Cyclin B replaces Cyclin A as the main binding partner of CDK1, thus leading to the formation of MPF (Cyclin B/CDK1). Cyclin A appears to have a role in stabilizing the Cyclin B/CDK1 complex, thus beginning M phase. Cyclin A is ultimately degraded, and the Cyclin B/CDK1 complex plays the major role in M phase entry, as described above. Now, another round of the cell cycle can take place [50, 82].

2.5 CDK Inhibitors (CKIs)

Another type of regulatory protein not discussed yet is the **cyclin-dependent kinase inhibitor (CKI)**. CKIs are proteins that can bind to Cyclin/CDK complexes and **block** their activity, specifically the kinase activity of the CDKs. CKIs may not always be present at high levels, but they can by upregulated in response to various extracellular stimuli or stress conditions. CKIs fall into two major categories: the CIP/KIP family and the INK4 family. CIP stands for CDK-interacting protein, KIP stands for kinase-inhibitory protein, and INK stands for inhibitors of CDK4. The **CIP/KIP** family inhibits a wide range of cyclin/CDK complexes, including those that function at G1/S phase and those that function at G2/M phase. This family includes p21 (CDKN1A) and p27 (CDKN1B). The **INK4** families are CDK4/CDK6 inhibitors; they are important in early G1 phase. Examples of the INK family are p16 (CDKN2A) and p15 (CDKN2B).

Any of the Cyclin/CDK complexes can be inhibited by CKIs, but here, we will discuss several pathways that can lead to the upregulation of CKIs that, in turn, inhibit the CyclinD/CDK6 complex. Such CKIs play critical roles in the cell cycle, and their improper regulation can be associated with uncontrolled cell growth and cancer [2, 83, 84].

p53 is an example of a protein that can lead to the upregulation of a CKI that can inhibit the CyclinD/CDK6 complex, and consequently, it has a negative impact on the G1 to S transition. P53 levels in the cell are usually low, but when DNA is damaged, the cell can sense the damaged DNA, and this leads to increased expression of the p53 gene and subsequent increased production of the p53 protein. This sensing of DNA damage involves proteins such as ATM and ATR, which help initiate a response to DNA damage. These proteins activate p53 as part of the DNA damage checkpoint pathway. They will be discussed in more detail in Chap. 21. p53 is a transcription factor, and when its levels rise, it binds to the regulatory elements of genes and stimulates their transcription. One of its important target genes is **the gene for p21**. This is significant because **p21** is a **CKI** that can bind to CyclinD/CDK6 complexes and inhibit their activity (see Fig. 20.10). Consequently, once p21 levels rise, the cell cycle is halted at the G1 to S transition so that damaged DNA does not get replicated. The cell has repair mechanisms, and the DNA may be repaired during the pause introduced by p53 and p21. Once the DNA is repaired, p53 levels go down, and as a consequence, p21 levels also go down. The cell can now pass through S phase and repair DNA. By upregulating p21, p53 has a vital role in responding to damaged DNA and also to other cellular stresses. It helps ensure that cells with damaged DNA do not divide. p53 is a tumor

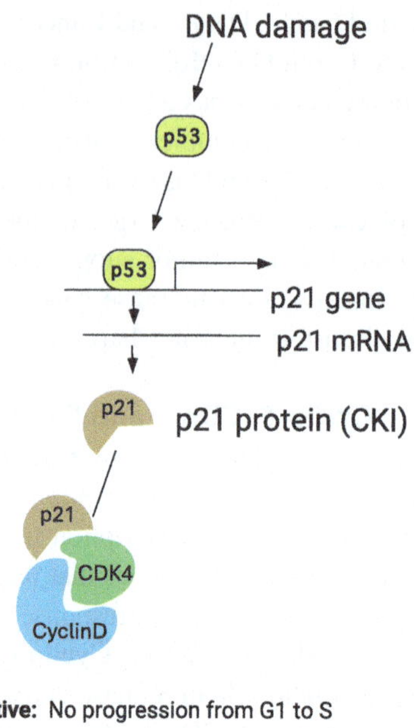

Fig. 20.10 p53 inhibits the cell cycle. P53 is activated by cellular stress such as DNA damage. It is a transcription factor that upregulates expression of the p21 gene and synthesis of p21. p21 is a CKI that inhibits cell cycle progression by binding to Cyclin/CDK complexes

suppressor protein. Its absence or inactivation is dangerous because without p53, the cell can replicate damaged DNA. The p53 gene is inactivated or mutated in many human cancers. In fact, up to 50% of human cancers have abnormalities in p53 [4, 85, 86].

TGF-β is another example of a protein that can lead to upregulation of **CKIs**. TGF-β is a ligand that binds to cell surface receptors. It is a complicated ligand because, in some cells and under some conditions, it can lead to increased cell division, but under other conditions, it can inhibit cell division. The mechanism by which it can inhibit division is the upregulation of CKIs, including **p15** and **p27**. These CKIs can also block CyclinD/CDK complexes and thus block G1 to S transition [4].

2.6 Links Between Cell Cycle and Metabolism

Finally, the cell cycle has important links to the central glucose metabolic pathways. In addition to growth factors, the cell cycle is sensitive to the availability of glucose and energy. Adequate glucose levels encourage cell cycle progression, but when glucose levels are low, the cell cycle may stop or pause until glucose becomes available again. Likewise, adequate energy (ATP) has a positive effect on cell cycle progression. There are many points of connection between glucose metabolism and the cell cycle, and the G1 to S phase transition involving cyclin D is a major interaction point. A few important examples of links between glucose metabolism and the cell cycle are described below.

2.6.1 GSK3β Responds to Glucose Levels

Glycogen synthase 3 beta (GSK-3β) is the key enzyme that links glucose metabolism to cyclin D degradation. GSK-3β is the enzyme that **inhibits glycogen synthase**, the enzyme that synthesizes glycogen from glucose. GSK-3β is active, and therefore inactivates glycogen synthase, when glucose levels are **low**. When glucose levels are high, however, insulin signaling activates the PI3-AKT pathway, which inhibits GSK-3β activity so that glycogen storage can proceed. In contrast, when glucose levels are low, GSK3β is active. This, in turn, leads to inhibition of glycogen synthesis. GSK-3b also has another function besides the regulation of glycogen synthesis. When it is active, GSK3β also phosphorylates **Cyclin D1** at threonine 286. This marks Cyclin D for nuclear export and ubiquitination, and degradation [77]. Thus, low glucose levels lead to activation of GSK3β, which leads not only to inhibition of glycogen synthesis but also to degradation of cyclin D and inhibition of G1 to S progression. Consequently, low glucose levels discourage cell cycle progression [87].

2.6.2 AMPK Responds to Glucose and Energy Levels

AMPK (AMP-activated protein kinase) is an energy sensor. It responds to conditions where glucose levels are low and AMP levels are high (indicating **low ATP levels**). Under these conditions, AMPK phosphorylates and activates p53, and this leads to upregulation of p21 and thus to downregulation of the G1 to S transition as described above. AMPK may also lead to cyclin D degradation. Thus, while AMPK responds largely to low energy (ATP) levels, GSK3β responds mostly to low glucose levels, and both can shut down the cell cycle at the G1 to S phase transition under specific conditions [88, 89].

2.6.3 mTOR

mTOR also has an important role in linking glucose availability to cell cycle progression. mTOR is an important integrator of nutrient availability and cell growth signals. When glucose is abundant, mTORC1 becomes active through intracellular signaling pathways. One important function of active mTORC1 is to promote the translation of cyclin D1 mRNA. Consequently, the levels of cyclin D protein increase, which promotes G1 to S phase progression. Conversely, when glucose levels become very low, mTOR is inhibited, cyclin D1 levels decrease, and the cells arrest in G1 [90, 91].

2.6.4 p53 Responds to Glucose Levels

p53 responds to stress such as UV irradiation and DNA damage, but it also responds to metabolic stress, specifically stress caused by low levels of glucose. This can shut down the G1 to S transition due to the upregulation of p21 [92].

2.7 Cancer Therapies That Inhibit CDKs

Cyclin D and CDK4/6 have critical roles in cell cycle progression, particularly the progression from G1 to S phase. This regulatory checkpoint is frequently dysregulated in cancer.

In some cancers, such as some breast cancers, **Cyclin D levels are upregulated**, leading to increased CDK4/6 activity, elevated

Rb phosphorylation, and enhanced G1 to S progression.

In other cases, **CDK4/6 is overexpressed**, often due to gene amplification. This is most common in melanoma, some sarcomas, and some cases of breast, lung, and bladder cancers. While the most common mechanism of CDK4 upregulation in cancer is gene amplification, CDK4 can also sometimes be regulated by activating mutations. These mutations are in the kinase domain and can lead to its constitutive activation. These activating mutations are most commonly found in melanoma.

Cyclin D and CDK4/6 are promising therapeutic targets for treating cancer. In particular, inhibiting CDK4/6 is an important approach for restoring control of the cell cycle in cancer cells. Several different types of **CDK4/6 inhibitors** have been developed and are used to treat certain types of **cancer** [67]. These inhibitors inhibit or reduce Rb phosphorylation, and they also indirectly inhibit other CDK4/6 targets such as FOXM1. This, in turn, leads to inhibition of G1 to S phase transition and thus leads to cell death. These types of inhibitors are sometimes administered in combination with other treatments. For example, they may be used together with hormone therapies (those that block estrogen signaling, also called endocrine therapy) to treat patients with hormone receptor (HR)-positive HER2-negative advanced or metastatic breast cancer. Such inhibitors have been shown to improve progression-free survival in metastatic breast cancer [93–95]. Some examples are discussed below.

Ibrance (palbociclib) [96] is a **small-molecule CDK4/6** inhibitor that was FDA-approved in 2015, marketed by Pfizer. It is used for HR^+ (hormone receptor positive), HER2-metastatic **breast cancer**, but is currently being tested for other types of cancer. Ibrance inhibits both **CDK4** and **CDK6** by binding to them. It **binds the ATP-binding sites of these kinases**. Thus, it is a **competitive inhibitor,** and it blocks the ability of **CDK4** and **CDK6** to phosphorylate Rb and other substrates. This causes cell cycle arrest at G1 phase [95, 97].

Ibrance is one of Pfizer's top-selling drugs (as of 2023). The development of this drug has brought about a major and important shift in breast cancer treatment, particularly metastatic HER2+ breast cancer. It has, in fact, become the new standard of care for this type of cancer if it has metastasized. It is the first CDK4/6 inhibitor approved for this type of cancer, and it has significantly improved survival rates. It is generally used together with endocrine therapy. As a targeted therapy, it reduces side effects compared with nonspecific treatments such as chemotherapy, and it has proven to be much more effective than hormone therapy alone. It has also paved the way for the development of new CDK4/6 inhibitors.

Kisqali (ribociclib) [98] and **Verzenio (abemaciclib)** [99] are also competitive inhibitors of **CDK4/6**. They were FDA-approved in **2017** and are also used to treat **HR+ and HER2- breast cancer**. **Alvocidib, also known as flavopiridol**, is an experimental drug that inhibits multiple CDKs. It can induce cell cycle arrest and apoptosis in **cancer cells** [100, 101].

Another CDK4/6 inhibitor is **Cosela (trilaciclib)** [102] (G1 Therapeutics). This drug was FDA-approved in 2021. It is administered by IV injection. It has been approved for small-cell lung cancer and other cancers. It is also used as a **myelopreservation therapy**. This is a treatment that protects bone marrow cells (myelo refers to bone marrow), which are often damaged by chemotherapy. This drug works by temporarily halting the growth of hematopoietic stem cells during chemotherapy, thus protecting them from the toxic effects of the therapy. Other inhibitors are also at various stages of development [100, 103, 104].

FDA-Approved Drugs Covered in This Chapter

Brand name	Generic name	First FDA approval year	Major conditions treated	Target
Adriamycin	*Doxorubicin*	1974	Breast, lung, bladder cancers	Topoisomerase II
Blenoxane	*Bleomycin*	1973	Various cancers	DNA damage
Cosela	*Trilaciclib*	2021	SCLC, myelopreservation	CDK4/6
Cytosar-U	*Cytarabine*	1969	Leukemia, lymphoma	DNA synthesis
Gemzar	*Gemcitabine*	1996	Pancreatic, lung, breast, ovarian cancer	DNA synthesis
Ibrance	*Palbociclib*	2015	HR+/HER2- breast cancer	CDK4/6
Kisqali	*Ribociclib*	2017	HR+/HER2- breast cancer	CDK4/6
Oncovin	*Vincristine*	1963	Various cancers	Microtubules
Rheumatrex, Trexall, etc.	*Methotrexate*	1953	Cancer, RA, psoriasis	DHFR
Taxol	*Paclitaxel*	1992	Breast, ovarian cancers	Microtubules
Velban	*Vinblastine sulfate*	1965	Various cancers	Microtubules
VePesid	*Etoposide*	1983	Lung, testicular cancer, lymphoma	Topoisomerase II
Verzenio	*Abemaciclib*	2017	HR+/HER2- breast cancer	CDK4/6

Definitions

Anaphase-promoting complex (APC) A protein complex that triggers the separation of sister chromatids by targeting proteins like securin for degradation, ultimately leading to the completion of M phase.

Aneuploidy Abnormal number of chromosomes in a cell, often resulting from mitotic errors.

Apoptosis Programmed cell death

Cyclin-dependent kinase (CDK) A kinase that is active when bound to a cyclin, and which has important roles in cell cycle progression.

CDK inhibitor (CKI) A protein that binds CDKs and inhibits their activity, blocking cell cycle progression.

Cell cycle The ordered series of events that a cell goes through to grow and divide, including G1, S, G2, and M phases.

Centrosome A microtubule-organizing center that duplicates before mitosis and organizes the spindle apparatus.

Checkpoint Checkpoints coordinate the different phases of the cell cycle. They ensure that one phase is completed correctly before the subsequent phase begins.

Cyclins A family of proteins whose concentrations vary at different phases of the cell cycle. They are necessary for the activation of CDKs.

Cytokinesis The final step of cell division, where the cytoplasm splits, forming two daughter cells.

DNA replication The process of copying the entire genome, also called DNA synthesis, occurs during S phase.

E2 factor (E2F) A transcription factor needed for transcribing genes that are necessary for S phase gene expression.

Free radicals Molecules or atoms that have one or more unpaired electrons in their outer shell, making them unstable and reactive. They seek to pair with electrons on other molecules.

G_0 phase A quiescent state where cells exit the cycle and are not dividing but remain metabolically active.

G1 phase The first gap phase after mitosis, where the cell grows and monitors internal and external conditions before DNA synthesis in S phase.

G2 phase The second gap phase, where the cell prepares for mitosis by checking for DNA errors and accumulating resources.

Kinetochores Protein complexes on chromosomes where microtubules attach.

M phase The phase where cell division occurs, including nuclear division (mitosis) and cytoplasmic division (cytokinesis).

P53 A tumor suppressor protein that responds to DNA damage by inducing cell cycle arrest or apoptosis.

Retinoblastoma protein (Rb protein) The retinoblastoma protein; inhibits progression from G1 to S by inhibiting E2F transcription factors.

Restriction point The point in late G1 (in most cells) after which the cell is committed to enter the S phase regardless of external signals.

S phase The synthesis phase, in which DNA is replicated.

Spindle apparatus A structure made of microtubules that separates chromosomes during mitosis. Also called the mitotic spindle.

Spindle assembly checkpoint The checkpoint that ensures all chromosomes are properly attached to the spindle before proceeding to anaphase.

Study Questions

Multiple-Choice Questions

1. Which phase of the cell cycle involves DNA replication?
   A. G1 phase
   B. G2 phase
   C. M phase
   D. S phase
2. During mitosis, the categories of microtubules that bind directly to the chromosomes are the:
   A. Kinetochore microtubules.
   B. Polar microtubules.
   C. Astral microtubules.
   D. All of the microtubules contact the chromosomes directly.
   E. None of the above—only the centrioles contact the chromosomes.
3. What protein inhibits the cell cycle in response to DNA damage?
   A. Ras
   B. mTOR
   C. p53
   D. AKT
4. During which phase does the cell prepare for mitosis by checking for DNA damage and ensuring proper DNA replication?
   A. G1 phase
   B. S phase
   C. G2 phase
   D. Cytokinesis
5. Velban (vinblastine) is a chemotherapy drug that disrupts microtubule dynamics during M phase. What effects do you expect it to have in cells due to disrupting microtubules?
   A. Leads to cell death of rapidly dividing cells
   B. Causes cells to speed up division
   C. Increases cellular energy production
   D. Induces differentiation in dividing cells
   E. Stabilizes intermediate filaments
6. P53 leads to an increase in the levels of p21. This leads to cell cycle arrest because p21 is a:
   A. Transcription factor
   B. Kinase
   C. Cyclin-dependent kinase inhibitor (CKI)
   D. Cyclin-dependent kinase (CDK)
   E. M phase-promoting factor (MPF)

Fill-in-the-Blank Questions

7. The _____ checkpoint ensures that all chromosomes are properly attached to the spindle before anaphase.
8. _____ is the phase where cells exit the cycle and enter a nondividing state.

Short-Answer Questions

9. Describe how CDK activity is regulated in the cell cycle.
10. How does DNA damage affect progression through the cell cycle?
11. How does p53 put a brake on the cell cycle?

Answer Key

Multiple-Choice Questions

1. **Answer**: D
2. **Answer**: A
3. **Answer**: C
4. **Answer**: C
5. **Answer**: A
6. **Answer**: C

Fill-in-the-Blank Questions

7. **Answer**: Spindle assembly
8. **Answer**: G0

Short-Answer Questions

9. **Answer**: CDKs are regulated by binding to cyclins and by phosphorylation and dephosphorylation on specific regulatory sites. They can also be inhibited by CKIs.
10. **Answer**: DNA damage can activate checkpoint pathways. This often involves p53, which puts a halt on the cell cycle until the damage can be repaired.
11. **Answer**: p53 is a transcription factor. It regulates the expression of the gene for p21. P21 is a CKI and it can inhibit cyclin/CDK complexes, such as the cyclin/CDK4,6 complex, and thus put a break on the cell cycle.

References

1. Uzbekov R, Prigent C. A journey through time on the discovery of cell cycle regulation. Cells. 2022;11 (4) https://doi.org/10.3390/cells11040704.
2. Alberts B, Heald R, Johnson A, Morgan D, Raff M, Roberts K, Walter P, Wilson J, Hunt T. Molecular biology of the cell. New York, NY: W.W. Norton and Company; 2022.
3. Marks F, Klingmüller U, Müller-Decker K. Cellular signal processing: an introduction to the molecular mechanisms of signal transduction. 2nd ed. New York, NY: Garland Science; 2017.
4. Morgan DO. The cell cycle: principles of control. London: New Science Press; 2007.
5. Nurse P, Masui Y, Hartwell L. Understanding the cell cycle. Nat Med. 1998;4(10):1103–6. https://doi.org/10.1038/2594.
6. Hunt T. Getting in and out of mitosis. Rambam Maimonides Med J. 2011;2(3):e0051. https://doi.org/10.5041/rmmj.10051.
7. Cooper G, Adams K. The cell: a molecular approach. New York, NY: Sinauer Associates; 2022.
8. Horio T, Murata T. The role of dynamic instability in microtubule organization. Front Plant Sci. 2014;5 https://doi.org/10.3389/fpls.2014.00511.
9. Goodson HV, Jonasson EM. Microtubules and microtubule-associated proteins. Cold Spring Harb Perspect Biol. 2018;10(6) https://doi.org/10.1101/cshperspect.a022608.
10. Jordan MA, Wilson L. Microtubules as a target for anticancer drugs. Nat Rev Cancer. 2004;4(4):253–65. https://doi.org/10.1038/nrc1317.
11. U.S. Food and Drug Administration. Velban(Vinblastine sulfate) Approval Overview 1965. https://www.accessdata.fda.gov/drugsatfda_docs/label/2010/020449s044lbl.pdf
12. U.S. Food and Drug Administration. Oncovin (vincristine) Approval Overview 1963. Available from: https://www.accessdata.fda.gov/scripts/cder/daf/index.cfm?event=overview.process&ApplNo=75493.
13. Ritter JM, Flower RJ, Henderson G, Loke YK, MacEwan DJ, Rang HP. Rang and Dale's pharmacology. 10th ed. London, England; 2023.
14. Gigant B, Cormier A, Dorléans A, Ravelli RBG, Knossow M. Microtubule-destabilizing agents: structural and mechanistic insights from the interaction of Colchicine and Vinblastine with Tubulin. In: Carlomagno T, editor. Tubulin-binding agents: synthetic, structural and mechanistic insights. Berlin/Heidelberg: Springer Berlin Heidelberg; 2009. p. 259–78.
15. Whalen K, Lerchenfeldt S, Giordano C. Lippincott illustrated reviews. Pharmacology. Philadelphia, PA: Wolters Kluwer; 2021.
16. Mukhtar E, Adhami VM, Mukhtar H. Targeting microtubules by natural agents for cancer therapy. Mol Cancer Ther. 2014;13(2):275–84. https://doi.org/10.1158/1535-7163.Mct-13-0791.
17. John Ines Center. Vinblastine and vincristine: life-saving drugs from a periwinkle 2018. Available from: https://www.jic.ac.uk/blog/vinblastine-and-vincristine-life-saving-drugs-from-a-periwinkle/#:~:text=Since%20then%2C%20vincristine%2C%20which%20inhibits,produce%201%20g%20of%20vinblastine.
18. Taub JW, Buck SA, Xavier AC, Edwards H, Matherly LH, Ge Y. The evolution and history of Vinca alkaloids: from the Big Bang to the treatment of pediatric acute leukemia. Pediatr Blood Cancer. 2024;71(11):e31247. https://doi.org/10.1002/pbc.31247.

19. U.S. Food and Drug Administration. Taxol (paclitaxel) Approval Overview 1992. Available from: https://www.accessdata.fda.gov/scripts/cder/daf/index.cfm?event=overview.process&ApplNo=20262.
20. Alberts B, Bray D, Hopkin K, Johnson A, Lewis J, Raff M, Roberts K, Walter P. Essential cell biology. New York, NY: Norton and Company; 2023.
21. Barnum KJ, O'Connell MJ. Cell cycle regulation by checkpoints. Methods Mol Biol. 2014;1170:29–40. https://doi.org/10.1007/978-1-4939-0888-2_2.
22. Pellarin I, Dall'Acqua A, Favero A, Segatto I, Rossi V, Crestan N, et al. Cyclin-dependent protein kinases and cell cycle regulation in biology and disease. Signal Transduct Target Ther. 2025;10(1):11. https://doi.org/10.1038/s41392-024-02080-z.
23. Blagosklonny MV, Pardee AB. The restriction point of the cell cycle. Cell Cycle. 2002;1(2):103–10.
24. Nigg EA. Mitotic kinases as regulators of cell division and its checkpoints. Nat Rev Mol Cell Biol. 2001;2(1):21–32. https://doi.org/10.1038/35048096.
25. Musacchio A, Salmon ED. The spindle-assembly checkpoint in space and time. Nat Rev Mol Cell Biol. 2007;8(5):379–93. https://doi.org/10.1038/nrm2163.
26. U.S. Food and Drug Administration. Cytosar-U (Cytarabine) Approval Overview 1969. Available from: https://www.accessdata.fda.gov/scripts/cder/daf/index.cfm?event=overview.process&ApplNo=071868.
27. National Cancer Institute. Cytarabine. Available from: https://www.cancer.gov/publications/dictionaries/cancer-drug/def/cytarabine.
28. Li Z, Guo JR, Chen QQ, Wang CY, Zhang WJ, Yao MC, et al. Exploring the antitumor mechanism of high-dose Cytarabine through the metabolic perturbations of Ribonucleotide and Deoxyribonucleotide in human Promyelocytic Leukemia HL-60 cells. Molecules. 2017;22(3) https://doi.org/10.3390/molecules22030499.
29. Löwenberg B, Pabst T, Vellenga E, van Putten W, Schouten HC, Graux C, et al. Cytarabine dose for acute myeloid leukemia. N Engl J Med. 2011;364(11):1027–36. https://doi.org/10.1056/NEJMoa1010222.
30. U.S. Food and Drug Administration. Trexall (methotrexate) 1953. Available from: https://www.accessdata.fda.gov/scripts/cder/daf/index.cfm?event=overview.process&ApplNo=008085.
31. Koźmiński P, Halik PK, Chesori R, Gniazdowska E. Overview of dual-acting drug methotrexate in different neurological diseases, autoimmune pathologies and cancers. Int J Mol Sci. 2020;21(10) https://doi.org/10.3390/ijms21103483.
32. Huennekens FM. The methotrexate story: a paradigm for development of cancer chemotherapeutic agents. Adv Enzym Regul. 1994;34:397–419. https://doi.org/10.1016/0065-2571(94)90025-6.
33. Katzung BG, Vanderah TW, Trevor AA. J. Katzung's basic and clinical pharmacology. 16th ed. New York, NY: McGraw-Hill Education; 2024.
34. Beardsley GP, Moroson BA, Taylor EC, Moran RG. A new folate antimetabolite, 5,10-dideaza-5,6,7,8-tetrahydrofolate is a potent inhibitor of de novo purine synthesis. J Biol Chem. 1989;264(1):328–33.
35. U.S. Food and Drug Administration. Adriamycin (Doxorubicin) Approval overview 1974. Available from: https://www.accessdata.fda.gov/scripts/cder/daf/index.cfm?event=overview.process&ApplNo=050467.
36. Taymaz-Nikerel H, Karabekmez ME, Eraslan S, Kırdar B. Doxorubicin induces an extensive transcriptional and metabolic rewiring in yeast cells. Sci Rep. 2018;8(1):13672. https://doi.org/10.1038/s41598-018-31939-9.
37. National Cancer Institute. Adriamycin. Available from: https://www.cancer.gov/publications/dictionaries/cancer-terms/def/adriamycin.
38. Sritharan S, Sivalingam N. A comprehensive review on time-tested anticancer drug doxorubicin. Life Sci. 2021;278:119527. https://doi.org/10.1016/j.lfs.2021.119527.
39. U.S. Food and Drug Administration. VePesid (Etoposide) capsules. Approval Overview 1983. Available from: https://www.accessdata.fda.gov/scripts/cder/daf/index.cfm?event=overview.process&ApplNo=075635.
40. U.S. Food and Drug Administration. VePesid (Etoposide) injection. Approval Overview 1983. Available from: https://www.accessdata.fda.gov/scripts/cder/daf/index.cfm?event=overview.process&ApplNo=018768.
41. Kim HS, Lee YS, Kim DK. Doxorubicin exerts cytotoxic effects through cell cycle arrest and Fas-mediated cell death. Pharmacology. 2009;84(5):300–9. https://doi.org/10.1159/000245937.
42. Nitiss JL. Targeting DNA topoisomerase II in cancer chemotherapy. Nat Rev Cancer. 2009;9(5):338–50. https://doi.org/10.1038/nrc2607.
43. Hande KR. Etoposide: four decades of development of a topoisomerase II inhibitor. Eur J Cancer. 1998;34(10):1514–21. https://doi.org/10.1016/S0959-8049(98)00228-7.
44. Drug Bank. Etopiside 2025. Available from: https://go.drugbank.com/drugs/DB00773.
45. U.S. Food and Drug Administration. Gemzar (Gemcitabine) Approval Overview 1996. Available from: https://www.accessdata.fda.gov/scripts/cder/daf/index.cfm?event=overview.process&ApplNo=020509.
46. Miao X, Koch G, Ait-Oudhia S, Straubinger RM, Jusko WJ. Pharmacodynamic modeling of cell cycle effects for Gemcitabine and Trabectedin combinations in pancreatic cancer cells. Front

Pharmacol. 2016;7:421. https://doi.org/10.3389/fphar.2016.00421.
47. Ciccolini J, Serdjebi C, Peters GJ, Giovannetti E. Pharmacokinetics and pharmacogenetics of Gemcitabine as a mainstay in adult and pediatric oncology: an EORTC-PAMM perspective. Cancer Chemother Pharmacol. 2016;78(1):1–12. https://doi.org/10.1007/s00280-016-3003-0.
48. Kishimoto T. Entry into mitosis: a solution to the decades-long enigma of MPF. Chromosoma. 2015;124(4):417–28. https://doi.org/10.1007/s00412-015-0508-y.
49. Masui Y, Markert CL. Cytoplasmic control of nuclear behavior during meiotic maturation of frog oocytes. J Exp Zool. 1971;177(2):129–45. https://doi.org/10.1002/jez.1401770202.
50. Lodish H, Berk A, Kaiser CA, Krieger M, Bretscher A, Ploegh H, Martin KC, Yaffe MB, Amon A. Molecular cell biology. 9th ed. New York, NY: Macmillan Learning; 2020.
51. Jessus C, Munro C, Houliston E. Managing the oocyte meiotic arrest-lessons from frogs and jellyfish. Cells. 2020;9(5) https://doi.org/10.3390/cells9051150.
52. Hartwell LH, Culotti J, Pringle JR, Reid BJ. Genetic control of the cell division cycle in yeast. Science. 1974;183(4120):46–51. https://doi.org/10.1126/science.183.4120.46.
53. Hartwell LH, Unger MW. Unequal division in Saccharomyces cerevisiae and its implications for the control of cell division. J Cell Biol. 1977;75(2 Pt 1):422–35. https://doi.org/10.1083/jcb.75.2.422.
54. Nurse P, Thuriaux P, Nasmyth K. Genetic control of the cell division cycle in the fission yeast Schizosaccharomyces pombe. Mol Gen Genet. 1976;146(2):167–78. https://doi.org/10.1007/bf00268085.
55. Holloway SL. Getting control of the cell cycle literature. Cell. 1996;87(4):597–600. https://doi.org/10.1016/S0092-8674(00)81379-3.
56. Sunkara PS, Wright DA, Rao PN. Mitotic factors from mammalian cells induce germinal vesicle breakdown and chromosome condensation in amphibian oocytes. Proc Natl Acad Sci USA. 1979;76(6):2799–802. https://doi.org/10.1073/pnas.76.6.2799.
57. Evans T, Rosenthal ET, Youngblom J, Distel D, Hunt T. Cyclin: a protein specified by maternal mRNA in sea urchin eggs that is destroyed at each cleavage division. Cell. 1983;33(2):389–96. https://doi.org/10.1016/0092-8674(83)90420-8.
58. Lohka MJ, Hayes MK, Maller JL. Purification of maturation-promoting factor, an intracellular regulator of early mitotic events. Proc Natl Acad Sci USA. 1988;85(9):3009–13. https://doi.org/10.1073/pnas.85.9.3009.
59. Jeong S, Ahn J, Jo I, Kang S-M, Park B-J, Cho H-S, et al. Cyclin-dependent kinase 1 depolymerizes nuclear lamin filaments by disrupting the head-to-tail interaction of the lamin central rod domain. J Biol Chem. 2022;298(9):102256. https://doi.org/10.1016/j.jbc.2022.102256.
60. Fourest-Lieuvin A, Peris L, Gache V, Garcia-Saez I, Juillan-Binard C, Lantez V, et al. Microtubule regulation in mitosis: tubulin phosphorylation by the cyclin-dependent kinase Cdk1. Mol Biol Cell. 2006;17(3):1041–50. https://doi.org/10.1091/mbc.e05-07-0621.
61. Lorca T, Castro A. The Greatwall kinase: a new pathway in the control of the cell cycle. Oncogene. 2013;32(5):537–43. https://doi.org/10.1038/onc.2012.79.
62. Chow JP, Poon RY, Ma HT. Inhibitory phosphorylation of cyclin-dependent kinase 1 as a compensatory mechanism for mitosis exit. Mol Cell Biol. 2011;31(7):1478–91. https://doi.org/10.1128/mcb.00891-10.
63. Liu K, Zheng M, Lu R, Du J, Zhao Q, Li Z, et al. The role of CDC25C in cell cycle regulation and clinical cancer therapy: a systematic review. Cancer Cell Int. 2020;20:213. https://doi.org/10.1186/s12935-020-01304-w.
64. Hershko A. Mechanisms and regulation of the degradation of cyclin B. Philos Trans R Soc Lond Ser B Biol Sci. 1999;354(1389):1571–5; discussion 5-6. https://doi.org/10.1098/rstb.1999.0500.
65. Shevah-Sitry D, Miniowitz-Shemtov S, Teichner A, Kaisari S, Hershko A. Role of phosphorylation of Cdc20 in the regulation of the action of APC/C in mitosis. Proc Natl Acad Sci. 2022;119(35):e2210367119. https://doi.org/10.1073/pnas.2210367119.
66. Sherr CJ. Cancer cell cycles. Science. 1996;274(5293):1672–7. https://doi.org/10.1126/science.274.5293.1672.
67. Roskoski R. Cyclin-dependent protein kinase inhibitors including palbociclib as anticancer drugs. Pharmacol Res. 2016;107:249–75. https://doi.org/10.1016/j.phrs.2016.03.012.
68. Malumbres M, Barbacid M. Cell cycle, CDKs and cancer: a changing paradigm. Nat Rev Cancer. 2009;9(3):153–66. https://doi.org/10.1038/nrc2602.
69. Gao X, Leone GW, Wang H. Chapter Four - Cyclin D-CDK4/6 functions in cancer. In: Tew KD, Fisher PB, editors. Advances in cancer research. 148: Cambridge, MA: Academic Press; 2020. p. 147–69.
70. Terada Y, Inoshita S, Nakashima O, Kuwahara M, Sasaki S, Marumo F. Regulation of cyclin D1 expression and cell cycle progression by mitogen-activated protein kinase cascade. Kidney Int. 1999;56(4):1258–61. https://doi.org/10.1046/j.1523-1755.1999.00704.x.
71. Tchakarska G, Sola B. The double dealing of cyclin D1. Cell Cycle. 2020;19(2):163–78. https://doi.org/10.1080/15384101.2019.1706903.
72. Narasimha AM, Kaulich M, Shapiro GS, Choi YJ, Sicinski P, Dowdy SF. Cyclin D activates the Rb tumor suppressor by mono-phosphorylation. elife. 2014;3:e02872. https://doi.org/10.7554/eLife.02872.

73. Anders L, Ke N, Hydbring P, Choi Yoon J, Widlund Hans R, Chick Joel M, et al. A systematic screen for CDK4/6 substrates links FOXM1 phosphorylation to senescence suppression in cancer cells. Cancer Cell. 2011;20(5):620–34. https://doi.org/10.1016/j.ccr.2011.10.001.
74. Wierstra I. The transcription factor FOXM1 (Forkhead box M1): proliferation-specific expression, transcription factor function, target genes, mouse models, and normal biological roles. Adv Cancer Res. 2013;118:97–398. https://doi.org/10.1016/b978-0-12-407173-5.00004-2.
75. Katzenellenbogen BS, Guillen VS, Katzenellenbogen JA. Targeting the oncogenic transcription factor FOXM1 to improve outcomes in all subtypes of breast cancer. Breast Cancer Res. 2023;25(1):76. https://doi.org/10.1186/s13058-023-01675-8.
76. Liao GB, Li XZ, Zeng S, Liu C, Yang SM, Yang L, et al. Regulation of the master regulator FOXM1 in cancer. Cell Commun Signal. 2018;16(1):57. https://doi.org/10.1186/s12964-018-0266-6.
77. Diehl JA, Cheng M, Roussel MF, Sherr CJ. Glycogen synthase kinase-3beta regulates cyclin D1 proteolysis and subcellular localization. Genes Dev. 1998;12(22):3499–511. https://doi.org/10.1101/gad.12.22.3499.
78. Zou T, Lin Z. The involvement of ubiquitination machinery in cell cycle regulation and cancer progression. Int J Mol Sci. 2021;22(11) https://doi.org/10.3390/ijms22115754.
79. Fagundes R, Teixeira LK. Cyclin E/CDK2: DNA replication, replication stress and genomic instability. Front Cell Dev Biol. 2021;9:774845. https://doi.org/10.3389/fcell.2021.774845.
80. Chu C, Geng Y, Zhou Y, Sicinski P. Cyclin E in normal physiology and disease states. Trends Cell Biol. 2021;31(9):732–46. https://doi.org/10.1016/j.tcb.2021.05.001.
81. Bertoli C, Skotheim JM, de Bruin RA. Control of cell cycle transcription during G1 and S phases. Nat Rev Mol Cell Biol. 2013;14(8):518–28. https://doi.org/10.1038/nrm3629.
82. Yam CH, Fung TK, Poon RY. Cyclin A in cell cycle control and cancer. Cell Mol Life Sci. 2002;59(8):1317–26. https://doi.org/10.1007/s00018-002-8510-y.
83. Pérez-Hidalgo L, Moreno S. CDK inhibitors. In: Dubitzky W, Wolkenhauer O, Cho K-H, Yokota H, editors. Encyclopedia of systems biology. New York: Springer New York; 2013. p. 214–20.
84. Lim S, Kaldis P. Cdks, cyclins and CKIs: roles beyond cell cycle regulation. Development. 2013;140(15):3079–93. https://doi.org/10.1242/dev.091744.
85. Olivier M, Hollstein M, Hainaut P. TP53 mutations in human cancers: origins, consequences, and clinical use. Cold Spring Harb Perspect Biol. 2010;2(1):a001008. https://doi.org/10.1101/cshperspect.a001008.
86. Engeland K. Cell cycle regulation: p53-p21-RB signaling. Cell Death Differ. 2022;29(5):946–60. https://doi.org/10.1038/s41418-022-00988-z.
87. Wang L, Li J, Di LJ. Glycogen synthesis and beyond, a comprehensive review of GSK3 as a key regulator of metabolic pathways and a therapeutic target for treating metabolic diseases. Med Res Rev. 2022;42(2):946–82. https://doi.org/10.1002/med.21867.
88. Jones RG, Plas DR, Kubek S, Buzzai M, Mu J, Xu Y, et al. AMP-activated protein kinase induces a p53-dependent metabolic checkpoint. Mol Cell. 2005;18(3):283–93. https://doi.org/10.1016/j.molcel.2005.03.027.
89. Hardie DG. AMP-activated protein kinase: an energy sensor that regulates all aspects of cell function. Genes Dev. 2011;25(18):1895–908. https://doi.org/10.1101/gad.17420111.
90. Fingar DC, Blenis J. Target of rapamycin (TOR): an integrator of nutrient and growth factor signals and coordinator of cell growth and cell cycle progression. Oncogene. 2004;23(18):3151–71. https://doi.org/10.1038/sj.onc.1207542.
91. Fingar DC, Richardson CJ, Tee AR, Cheatham L, Tsou C, Blenis J. mTOR controls cell cycle progression through its cell growth effectors S6K1 and 4E-BP1/eukaryotic translation initiation factor 4E. Mol Cell Biol. 2004;24(1):200–16. https://doi.org/10.1128/mcb.24.1.200-216.2004.
92. Labuschagne CF, Zani F, Vousden KH. Control of metabolism by p53 - Cancer and beyond. Biochim Biophys Acta Rev Cancer. 2018;1870(1):32–42. https://doi.org/10.1016/j.bbcan.2018.06.001.
93. Ye F, Dewanjee S, Li Y, Jha NK, Chen ZS, Kumar A, et al. Advancements in clinical aspects of targeted therapy and immunotherapy in breast cancer. Mol Cancer. 2023;22(1):105. https://doi.org/10.1186/s12943-023-01805-y.
94. Goel S, DeCristo MJ, McAllister SS, Zhao JJ. CDK4/6 inhibition in cancer: beyond cell cycle arrest. Trends Cell Biol. 2018;28(11):911–25. https://doi.org/10.1016/j.tcb.2018.07.002.
95. Braal CL, Jongbloed EM, Wilting SM, Mathijssen RHJ, Koolen SLW, Jager A. Inhibiting CDK4/6 in breast cancer with Palbociclib, Ribociclib, and Abemaciclib: similarities and differences. Drugs. 2021;81(3):317–31. https://doi.org/10.1007/s40265-020-01461-2.
96. U.S. Food and Drug Administration. Ibrance (Palbociclib) Approval Overview 2015. Available from: https://www.accessdata.fda.gov/scripts/cder/daf/index.cfm?event=overview.process&ApplNo=207103.
97. Bonelli M, La Monica S, Fumarola C, Alfieri R. Multiple effects of CDK4/6 inhibition in cancer: from cell cycle arrest to immunomodulation. Biochem Pharmacol. 2019;170:113676. https://doi.org/10.1016/j.bcp.2019.113676.

98. U.S. Food and Drug Administration. Kisqali (Ribociclib) Approval Overview 2017. Available from: https://www.accessdata.fda.gov/scripts/cder/daf/index.cfm?event=overview.process&ApplNo=209092.
99. U.S. Food and Drug Administration. Verzenio (Abemaciclib) U.S. Food and Drug Administration (FDA) 2017. Available from: https://www.accessdata.fda.gov/scripts/cder/daf/index.cfm?event=overview.process&ApplNo=208716.
100. Morrison L, Loibl S, Turner NC. The CDK4/6 inhibitor revolution — a game-changing era for breast cancer treatment. Nat Rev Clin Oncol. 2024;21(2):89–105. https://doi.org/10.1038/s41571-023-00840-4.
101. Joshi H, Tuli HS, Ranjan A, Chauhan A, Haque S, Ramniwas S, et al. The pharmacological implications of Flavopiridol: an updated overview. Molecules. 2023;28(22):7530.
102. U.S. Food and Drug Administration. Cosela (Trilaciclib) Approval Overview 2021. Available from: https://www.accessdata.fda.gov/scripts/cder/daf/index.cfm?event=overview.process&ApplNo=214200.
103. Dhillon S. Trilaciclib: first approval. Drugs. 2021;81(7):867–74. https://doi.org/10.1007/s40265-021-01508-y.
104. Goel S, Bergholz JS, Zhao JJ. Targeting CDK4 and CDK6 in cancer. Nat Rev Cancer. 2022;22(6):356–72. https://doi.org/10.1038/s41568-022-00456-3.

21 Tumor Suppressor Genes and Cancer

Abstract

Tumor suppressor genes encode proteins that can put a brake on the cell cycle. Tumor suppressor genes can lead to cancer when they contain an inactivating mutation or deletion. Unlike oncogenes, mutations generally need to occur on both alleles, rather than only one allele, in order to cause cancer. This chapter covers tumor suppressor genes, including some history of the discovery of the retinoblastoma gene, the first tumor suppressor gene to be identified. Loss of heterozygosity (LOH), an important concept in the inactivation of tumor suppressor genes, is explored. Some important tumor suppressor genes, such as p53, NF1, and BRCA, are discussed, along with their associations with various cancers. Some tumor suppressor genes are tightly linked with cell cycle functions, and these connections will also be explored. While treatment of cancers caused by the inactivation of tumor suppressor genes is challenging, some treatment options are discussed. The concept of synthetic lethality for the treatment of cancer is also discussed.

Keywords

Retinoblastoma (Rb) · p53 · Breast cancer gene (BRCA) · Poly(ADP-ribose) polymerase (PARP) · Hypoxia-inducible factor 1 (HIF-1)

Abbreviations

APC	Adenomatous polyposis coli
ATM	Ataxia-telangiectasia mutated
BRCA	Breast cancer susceptibility gene
CDK	Cyclin-dependent kinase
CIN	Chromosomal instability
CKI	Cyclin-dependent kinase inhibitor
DSB	Double-stranded break
DSH	Disheveled
EPO	Erythropoietin
FZD	Frizzled
GAP	GTPase-activating protein
HIF-1α	Hypoxia-inducible factor 1-alpha
HR	Homologous recombination
LOH	Loss of heterozygosity
MEK	Mitogen-activated protein kinase kinase
NF1	Neurofibromin 1
p53	Tumor protein 53
PARP	Poly(ADP-ribose) polymerase
PDGF	Platelet-derived growth factor
PHD	Prolyl hydroxylase domain enzyme
PN	Plexiform neurofibroma
RAD51	Radiation-sensitive 51
RASSF1S	Ras association domain family one isoform A
Rb	Retinoblastoma protein
RBC	Red blood cell
SCLC	Small-cell lung cancer

A. Minden, *Biochemistry, Cell and Molecular Biology for Pharmacists and Health Professionals*,
https://doi.org/10.1007/978-3-032-20529-2_21

SSB Single-stranded break
TGF-α Transforming growth factor alpha
VEGF Vascular endothelial growth factor
VHL Von Hippel-Lindau

What You Will Learn

- The definition of tumor suppressors, and the two main categories of tumor suppressors.
- Tumor suppressor genes can cause cancer if there is an inactivating mutation on both alleles, as described by the "two-hit hypothesis."
- The concept of loss of heterozygosity (LOH) and its importance in the inactivation of tumor suppressor genes.
- Challenges with treating cancers caused by the inactivation of tumor suppressors.
- Rb was the first tumor suppressor identified.
- Some important tumor suppressor genes that are covered include Rb, p53, NF1, BRCA, and VHL.
- Promoter methylation (epigenetic modifications) can lead to the downregulation of gene expression without altering the gene sequence.
- PARP inhibitors for the treatment of BRCA mutant breast cancer.
- Drugs that target the HIF-1 pathway.
- Synthetic lethality as a mechanism for treating cancers caused by tumor suppressor gene inactivation.

1 Overview of Tumor Suppressor Genes

As discussed in Chap. 19, a single activating mutation on one allele of a proto-oncogene can cause it to become an oncogene and lead to uncontrolled cell proliferation. In contrast, **tumor suppressor genes** require **two inactivating mutations**, one on each allele, to promote the uncontrolled proliferation of cells (Fig. 21.1). Tumor suppressor genes are genes that encode proteins whose normal functions are to constrain or suppress cell division, unlike

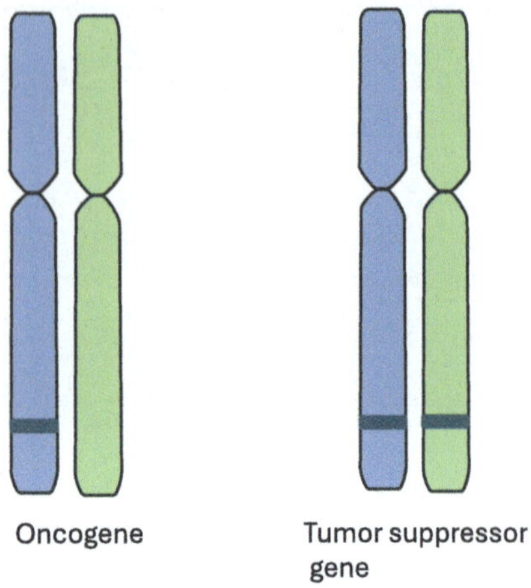

Fig. 21.1 Oncogenes compared to tumor suppressor genes. Oncogenes promote cancer when just one allele obtains an activating mutation (indicated by the shaded bar), while tumor suppressor genes require two inactivating mutations, one on each allele (also indicated by shaded bars)

proto-oncogenes, which usually code for genes that encourage cell division. In contrast to oncogenes, tumor suppressor genes are linked to cancer once the genes have mutations that disrupt their activity (inactivating mutations) or if the genes are deleted altogether. This chapter discusses the functions of tumor suppressor genes in normal and cancer cells. Much of the material in this chapter is drawn from the following [1, 3] and other references within the text.

2 The Major Categories of Tumor Suppressor Genes

Tumor suppressor genes can be broadly divided into two types:

A. Genes whose products suppress normal cell cycle progression and/or induce differentiation. These can be considered gatekeepers that regulate cell cycle control. A classic example is retinoblastoma (Rb).
B. Genes that suppress the cell cycle in response to certain stresses. These can be considered caretakers that maintain genomic stability. A classic example of this type of tumor

suppressor is p53, which inhibits cell cycle progression in response to stressful stimuli such as DNA damage (see Chap. 20).

3 The Retinoblastoma (Rb) Tumor Suppressor

3.1 Rb Was the First Tumor Suppressor to Be Identified

Rb was the first tumor suppressor gene to be discovered. It is the gene responsible for retinoblastoma, a rare childhood tumor of the retina. The first documented case of the disease was a child who had the disease in the 1800s. By the early twentieth century, physicians started noticing that there was a familial pattern to the disease, and in the 1900s, pedigree analysis showed an autosomal dominant inheritance pattern. In 1971, Dr. Alfred Knudson proposed the "two-hit hypothesis" based on retinoblastoma cases he had studied. He hypothesized that two mutations in the gene (one in each allele) were required for cancer to develop, and this was the first suggestion of the existence of what we now call tumor suppressor genes. Its mechanism of transmission supports Knudson's two-hit hypothesis. He noticed that there were both hereditary cases of the disease and sporadic cases. He proposed that in hereditary cases of Rb, one mutation is inherited (a germline mutation) and the disease would occur if another mutation occurred in a somatic cell (a somatic mutation). In contrast, in sporadic cases, both mutations must occur in a somatic cell. These sporadic cases were therefore rarer and also had a later onset. In the 1980s, researchers started mapping the genetic locus of the gene and found that it was located on chromosome 13. Over time, the gene for Rb was sequenced and studied. Rb has become a model for understanding other tumor suppressors and other cancers that are driven by tumor suppressor genes [1–3].

After the Rb gene was identified, scientists delved into studying its function. It was found to be a nuclear phosphoprotein, and we now know that it has a vital role in putting a brake on the cell cycle (see Chap. 20). Specifically, it suppresses the cell cycle by preventing G1 to S phase transition unless the appropriate growth factors and conditions are present, and its activity is tightly regulated by phosphorylation. The role of Rb as a cell cycle inhibitor explains why it is a tumor suppressor. Inactivation or deletion of Rb gives cells a strong growth advantage because they lack an important brake system that normally controls cell division [4].

3.2 Sporadic Versus Familial Rb

As indicated above, there are two mechanisms by which the Rb can be transmitted: sporadic retinoblastoma, in which there is no family history of the disease, and familial retinoblastoma, in which a parent also had the disease. Sporadic retinoblastoma is usually unilateral, occurring only in one eye. Following treatment or surgery, the risk of recurrence is low. In contrast, familial retinoblastoma is usually bilateral, occurring in both eyes. Patients with this type of retinoblastoma are at increased risk of developing other cancers later in life, especially osteosarcomas. In both cases, loss of function of both alleles of Rb is necessary to lead to cancer [1, 3].

3.3 Types of Cancers Caused by Rb

While the Rb gene is most commonly associated with the eye cancer **retinoblastoma**, Rb mutation is also associated with some other cancers. These include **osteosarcoma**, **soft tissue sarcoma**, **small-cell lung cancer** (SCLC), and **bladder cancer** [5, 6].

3.4 Challenges of Treating Cancers Caused by Rb Mutations

In general, the treatment of cancers caused by tumor suppressors is more challenging than treating those caused by oncogenes. While treating cancers caused by an oncogene may include the use of a specific inhibitor that targets the product of the oncogene, tumor suppressor genes present a more complicated problem. Tumor suppressors are already inhibited, and a mechanism for compensating for this inhibition is needed. Instead of direct inhibition of an activated protein, more general treatments such as chemotherapy, radiation, and surgery are often used to treat cancers in which inactivation of a tumor suppressor gene is the main cause [3, 7].

4 Loss of Heterozygosity (LOH) and Tumor Suppressors

Loss of heterozygosity (LOH) is a key concept in the tumor suppressor field. For a tumor suppressor gene such as Rb to lead to uncontrolled cell proliferation and cancer, there must be loss of heterozygosity (LOH) at the gene's locus. LOH means that there are abnormalities leading to loss of function in **both alleles** of the tumor suppressor gene, rather than only one (Fig. 21.2). This is a key characteristic of tumor suppressor genes [1].

Inactivating mutations in tumor suppressor genes are key to their loss of function. Different types of genetic abnormalities can lead to loss of function of an allele. These can include point mutations in regions essential for gene function, deletions in the gene, complete loss of the gene, absence of an entire chromosome, or epigenetic silencing, which involves epigenetic changes in the DNA, such as methylation. LOH depends not only on the first mutation but also on the development of a second mutation on the other allele. The first mutation may be an inherited mutation, or it may be a somatic cell mutation that occurred by chance. The second mutation, or second "hit," often happens by random chance, but certain mechanisms may increase the odds of a second hit slightly more than random chance. **Mitotic recombination** can occasionally occur in somatic cells, and this can lead to **LOH** by replacing a wild-type allele with a mutated one on the homologous chromosome. Likewise, **gene conversion** can copy a mutant sequence onto the functional allele during a DNA repair process. Thus, the possibility of getting the "second hit" may be a bit higher than random chance [1].

5 Some Other Tumor Suppressor Genes

After Rb was identified as a tumor suppressor, numerous other tumor suppressor genes have been found. Tumor suppressor genes are implicated in many different human cancers. Some of these will be discussed here.

5.1 p53

5.1.1 Functions of p53

The TP53 gene encodes the p53 protein. p53 puts a brake on the cell cycle in response to cellular stresses, especially DNA damage. P53 is often called the "guardian of the genome" [8] because it has such a crucial role in putting a brake on the cell cycle when DNA is damaged. This prevents the inheritance by daughter cells of damaged DNA and allows time for damaged DNA to be repaired. P53 is also discussed in Chap. 20. An important mechanism by which p53 puts a brake

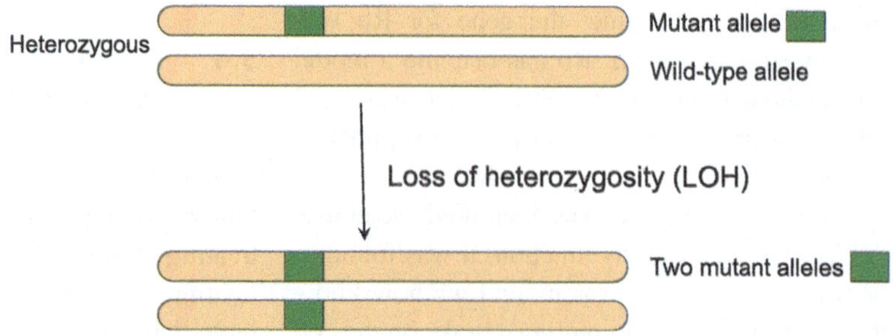

Fig. 21.2 Loss of heterozygosity. Loss of heterozygosity is important for tumor suppressor gene inactivation. It involves the loss of function or inactivation of one allele when the other allele is already inactivated

on the cell cycle is by leading to upregulation of p21, a CKI that inhibits cyclin/CDKs. P53 is thus highly protective against passing down damaged DNA to daughter cells, which could otherwise lead to cancer. Mutations in TP53 are one of the most common genetic alterations in cancer, found in about 50% of cancers. p53 mutations are frequent in ovarian, colorectal, lung, and esophageal cancers. They are less frequent in leukemia and retinoblastoma [3].

5.1.2 Treatment for Cancers Caused by Loss of Normal p53

As with all cancers involving tumor suppressor genes, developing drugs is challenging. There are currently no FDA-approved drugs that directly target TP53 mutations. Some research involves studying ways to restore p53 function indirectly, such as developing drugs that target pathways downstream of p53. For example, drugs that target CDK4/6 may inhibit the growth of some cancer cells in which p53 is mutated, because such cells may have high levels of CyclinD activity due to a lack of p53 activity (see Chap. 20). In some cases, p53 may not be absent in cancer but mutated or improperly folded. Restoring such a mutant p53 into its wild-type state is another area of investigation. Gene therapy is also an area of research. This would involve introducing the wild-type p53 gene into the cells of patients where both alleles of p53 are mutated [9, 10].

5.1.3 P53 and MDM2

Some cancers may have normal p53 but still function as p53 mutant cells because of overexpression of a protein called MDM2 (mouse double minute 2). MDM2 is a protein that binds to p53 [3]. It is an E3 ubiquitin ligase and, upon binding to p53, it targets it for degradation [11]. A related protein, MDM4 (also called MDMX), also binds to p53 and keeps it inactive [12]. These MDM proteins have the important roles of keeping p53 in check under normal growth conditions. However, if MDM2 is overproduced, the control of p53 may be too strong, and p53 can no longer respond effectively to stimuli such as DNA damage. In this case, the cell behaves like a p53 mutant cell, even though p53 is not actually mutated, and this can also lead to cancer. In these cases, blocking MDM2 is the goal for restoring p53 activity. MDM inhibitors are in research phases or clinical trials.

5.2 NF1 Neurofibromatosis (NF)

5.2.1 NF1 and Cancer

Neurofibromatosis (NF1) is a disease associated with mutations in both alleles of the NF1 gene [3]. This familial disorder is associated with an increased number of Schwann cells and other cells in the nervous system. Schwann cells are cells that form a myelin sheath, which is a protective layer that wraps around peripheral nerves. In NF, patients are prone to the development of benign tumors in these sheaths. While patients with NF frequently develop benign tumors, occasionally, these can become malignant, and these malignant tumors are called neurofibrosarcomas. Patients who have NF are also at increased risk for glioblastomas (a type of brain tumor), pheochromocytomas (tumors arising from the adrenal glands), and myelogenous leukemias [13].

5.2.2 The Function of NF1

The NF1 gene was identified and cloned in 1990. Researchers found that its genetic transmission resembles that of tumor suppressors, such as Rb, indicating that it was likely to be a tumor suppressor gene. The NF1 gene is expressed largely in neural tissues but also in some other tissues. By sequencing the gene, its function became better understood. The exciting finding after sequencing the NF1 gene is that it has sequence similarity to a yeast gene called IRA, which encodes a Ras GAP protein. The idea that NF1 could function as a GAP protein provided important clues about how it regulates cell proliferation. Like most proteins, GAP proteins are members of a family of proteins, with different GAPs having different tissue specificity. NF1 turns out to be an important GAP in some types of cells, such as neuroectodermal cells. As discussed in

Chap. 18, GAP proteins are GTPase-activating proteins. They have the role of switching Ras to its inactive state by increasing its GTPase activity so that GTP gets hydrolyzed to GDP. In tissues where NF1 is the main GAP, the absence of NF1 causes Ras to stay constitutively active. Thus, a lack of NF1 in some cells may lead to a similar outcome as a constitutively active Ras mutant (*See* Chap. 18). This, in turn, leads to uncontrolled cell proliferation and thereby increases the risk of the development of benign or malignant cancers [3].

Because of the link between NF1 and Ras activity, Ras pathway intermediates have been tested as possible drug targets for NF. One potential target is **MEK**, which functions downstream to Ras. **Koselugo (selumetinib)** [14] is a **small-molecule** drug for **NF** that targets **MEK**. It has been shown to reduce the formation of plexiform neurofibromas (PNs), which are benign tumors that grow along nerves in patients with NF. Selumetinib is a noncompetitive inhibitor of MEK. It binds to an allosteric site, not the ATP-binding site. Selumetinib was FDA-approved in 2020 [15, 16]. It is still in use, but a newer MEK inhibitor approved in 2025 is **Gomekli (mirdametinib)** [17], which is also a noncompetitive MEK inhibitor [18, 19].

5.3 APC (Adenomatous Polyposis Coli)

5.3.1 The Wnt, Frizzled, and APC/β-Catenin Pathway

The **APC** gene is a tumor suppressor gene that is frequently mutated in colon cancer. APC is an integral part of the **Wnt signaling pathway** (Fig. 21.3). Its role in regulating the protein **β-catenin** in this pathway is key to its role in regulating cell division. The **Wnt pathway** involves the ligand **Wnt**, a secreted glycoprotein that binds to the extracellular domain of a receptor called **Frizzled (FZD)** and a co-receptor LRP5/6. FZD is a seven-pass transmembrane receptor with features similar to a GPCR, but that is not linked to a G protein, and LRP is a low-density lipoprotein-related protein. When Wnt is not present in the local extracellular space, the FZD receptor remains unoccupied. Under these conditions, the cytoplasmic protein **Disheveled (DSH)** stays inactive. When DSH is inactive, a complex in the cell called the destruction complex binds to **β-catenin**. The **destruction complex** contains **Axin**, **CK1α**, **GSK-3β**, and **APC**. This complex leads to the phosphorylation of **β-catenin**. **CK1a** first phosphorylates β-catenin, allowing **GSK-3β**, to add additional phosphate groups, which target b-catenin for ubiquitylation and subsequent **destruction**. **β-Catenin** has a role in promoting cell proliferation, and therefore, its destruction inhibits cell proliferation [1, 3, 20–23].

In contrast, when **Wnt** is present, it binds to the **FZD receptor**. This interaction leads to the activation of **DSH** in the cell. Active DSH **inhibits GSK-3β**. Now, GSK-3β, along with the rest of the destruction complex, cannot lead to phosphorylation and subsequent destruction of **β-catenin**. β-Catenin can instead enter the nucleus, where it can interact with a transcription factor complex called the **TCF/Lef** complex. TCF/LEF binds to the regulatory elements of target genes and leads to increased transcription of genes important for **cell proliferation**, such as the gene for c-Myc and Cyclin D1. TCF/LEF bind to these regulatory elements even without β-catenin, but without β-catenin, they recruit co-repressors to keep target genes inactive. β-Catenin binding, however, changes the TCF/LEF complex into a transcriptional activator. Thus, β-catenin leads to activation of target genes that are involved in cell growth regulation. Thus, the major mechanism by which **APC** functions as a **tumor suppressor** is by being part of the **destruction complex** that leads to the **destruction of β-catenin**, thereby **inhibiting cell proliferation**.

5.3.2 The Wnt Pathway in the Colon

The Wnt pathway has an especially important role in regulating cell growth in the colon. The colon's lining is made up of large numbers of crypts, which are folds in the tissue that significantly increase the surface area of the colon. These crypts have many vital roles, including

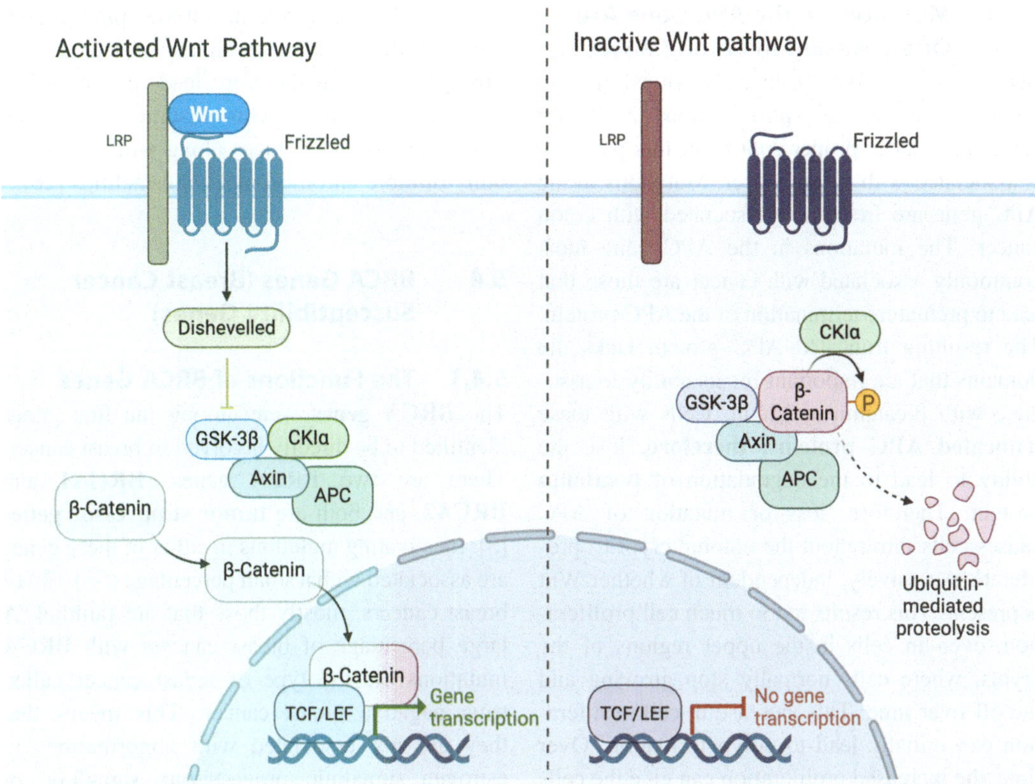

Fig. 21.3 The β-catenin pathway. The β-catenin pathway is activated in response to Wnt signaling. APC is an important signaling protein in the Wnt pathway, and its inactivation is associated with cancer, especially colorectal cancer. **Source**: Created using biorender.com

roles in absorption and secretion. Cells in the crypts include enterocytes, goblet cells, enteroendocrine cells, and some stem cells (as well as Paneth cells, which help support the stem cells). The cells within the crypt are responsive to the levels of Wnt in the surrounding environment. **Wnt** levels are generally high near the crypt bases (the bottom of the crypts), where Wnt ligands are produced mainly by Paneth cells and some mesenchymal cells, after being released by surrounding cells. Wnt binding to FZD receptors on these cells leads to the pathway described above, consequently keeping **β-catenin** levels **high**. The cells at the crypt bases, therefore, maintain the ability to divide, and some of them have stem cell-like characteristics in that they can divide and remain largely undifferentiated. As the progeny of these cells migrate upward, away from the crypt bases, toward the crypt top (also called the luminal surface or surface epithelium), and consequently they move away from the high Wnt levels. There is little Wnt in the regions surrounding the upper parts of the crypts. Therefore, in these cells at the crypt tops, β-catenin is destroyed, and these cells stop dividing and eventually die by apoptosis. The dead cells are shed into the lumen of the colon. This process of out-migration from the crypt bases to the crypt tops takes several days. This system is highly protective against cancer because cells die within such a short time before they have a chance to develop mutations. This is an important feature in the harsh environment of the digestive system [1, 3].

5.3.3 Mutations in the APC Gene Are Often Present in Colon Cancer

Because of the Wnt pathway's critical role in regulating colon cell proliferation, it is not surprising that improper control of this pathway is associated with colon cancer. Mutations in the APC gene are frequently associated with colon cancer. The mutations in the APC gene most commonly associated with cancer are those that lead to premature termination of the APC protein. The resulting truncated APC protein lacks the domains that are important for its ability to associate with β-catenin and axin. Cells with these **truncated APC proteins, therefore,** lose the ability to lead to the degradation of **β-catenin** protein. Therefore, loss or mutation of APC causes cells throughout the colonic crypt to proliferate excessively, independent of whether Wnt is present. This results in too much cell proliferation, even in cells in the upper regions of the crypts, where cells normally stop growing and die off over time. This increase in cell proliferation can initially lead to polyp formation. Over time, the increased proliferation can give the cells the time to develop other mutations. This can lead to cancer, and the development of multiple mutations can complicate the treatment [24, 25].

5.3.4 APC and Chromosome Instability (CIN)

While the role of APC in the Wnt pathway is often considered to be its classic role, like many proteins, it also has other functions. Another, more recently discovered function of APC appears to involve regulating chromosome stability. Cells lacking APC have been shown to have an increase in chromosomal instability (CIN), which is a high rate of changes in chromosome number during cell division. This can lead to the accumulation of too many chromosomes (such as tetraploidy). This can occur because chromosomes do not segregate properly during mitosis. Surprisingly, this may be due to APC. In addition to its function in the Wnt pathway as discussed above, APC proteins can also localize to centrosomes and kinetochores and interact with the microtubule arrays that form the mitotic spindle. APC has a role in mitotic spindle function and, therefore, in chromosome segregation. Loss of APC can therefore lead to aneuploidy (incorrect number of chromosomes). This can lead to growth advantages along with the possibility of more chromosomal abnormalities [26].

5.4 BRCA Genes (Breast Cancer Susceptibility Genes)

5.4.1 The Functions of BRCA Genes

The **BRCA genes** were among the first genes identified to be directly involved in breast cancer. There are two BRCA genes, **BRCA1** and **BRCA2**, and both are **tumor suppressor genes** [3]. Inactivating mutations in either of these genes are associated with a small percentage (5–10%) of breast cancers, mostly those that are familial. A large percentage of breast cancers with BRCA mutations are the type of breast cancer called triple-negative breast cancer. This means that they are not associated with abnormalities in estrogen signaling, progesterone signaling, or HER2 mutations. These are some of the most difficult breast cancers to treat. Mutations in the BRCA genes are largely associated with breast cancer but are also associated with some other cancers, particularly ovarian cancer [27].

The functions of the BRCA gene products are complex, but one of their most important functions is in **DNA repair**. BRCA1 and BRCA2 have important roles in the repair of double-stranded breaks (**DSBs**) in DNA due to their roles in **homologous recombination** (**HR**). The repair process they carry out is highly efficient and largely error-free. Multiple different proteins are involved in this repair process. A protein called **ATM** (which stands for ataxia-telangiectasia mutated) is a protein kinase that is recruited to damaged DNA, specifically at DSB sites. It binds indirectly, via a set of intermediary proteins called the MRN complex (Mre11, Rad50, and Nbs1). There, it becomes activated by autophosphorylation. Activated **ATM** then **phosphorylates** substrates, and these include **BRCA1**. Phosphorylated BRCA1 then localizes to DNA damage sites, where it plays a crucial role in DNA repair. BRCA1 also recruits another protein, PALB2, which connects BRCA1 to BRCA2. BRCA2 is thus also recruited to the damaged DNA. **BRCA2** facilitates the loading of another protein, **RAD51**, onto the DNA, which is also necessary for efficient DNA repair (This process is also discussed in Chap. 23). The entire DNA repair process can be inhibited when there are inactivating mutations in the BRCA1 or BRCA2 genes. As a result of such mutations, cells with damaged DNA can proliferate, and this can in turn lead to cancer [28].

While a key function of the BRCA proteins is in DNA repair, they also have some other functions. They have roles in cell cycle regulation, regulation of gene expression, regulation of protein turnover, and regulation of chromosome segregation [29]. The basic functions of BRCA and ATM are summarized in Fig. 21.4 also discussed in Chap. 23.

5.4.2 PARP Inhibitors, BRCA, and Cancer Treatment

Treatment of cancers with BRCA mutations is complex, as it is for all cancers linked to tumor suppressors. One idea involves treatment with **PARP inhibitors**. As described above, BRCA repairs double-stranded breaks (DSBs). PARP (poly(ADP-ribose) polymerase) is also a DNA repair protein. While BRCA repairs double-stranded breaks, PARP repairs single-stranded breaks (SSBs) through the BER pathway (see Chap. 23). Inhibition of both PARP and BRCA together has been shown to lead to **synthetic lethality**. Synthetic lethality is based on the concept that a combination of two defects can lead to loss of viability, while either defect alone does not. In cancer cells with a deficiency in **one** DNA repair pathway, inhibition of the **second** DNA repair pathway often creates synthetic lethality, and the cells die. In this way, the risk of passing down damaged DNA to daughter cells is prevented.

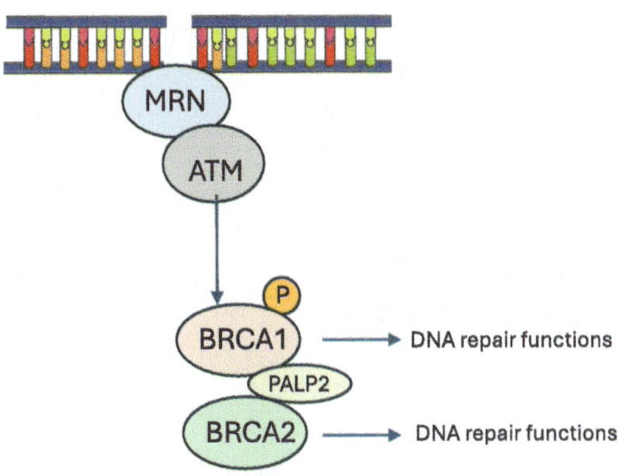

Fig. 21.4 BRCA: BRCA1 is recruited to double-stranded breaks and has DNA repair functions. (*Source*: Image adapted from Servier Medical Art (https://smart.servier.com/), licensed under the Creative Commons Attribution 4.0 (CC BY 4.0) license)

Synthetic lethality in response to PARP and BRCA occurs as follows: when PARP is inhibited, SSBs accumulate. During replication, these SSBs can, in some cases, lead to DSBs. Normally, these DSBs would be repaired by a process using BRCA proteins. However, when BRCA is also deficient, the DSBs accumulate. When they accumulate to the point of lethality, the cells die. This is synthetic lethality [30].

For these reasons, **PARP inhibitors** are considered to be promising for the treatment of **breast cancer with BRCA mutations**. Some PARP inhibitors are currently being used to treat breast or ovarian cancers, as well as some other cancers that are associated with BRCA1/BRCA2 mutations. These include **Lynparza (olaparib)** [31], **Rubraca (rucaparib)** [32], and **Zejula (niraparib)** [33]. These drugs are **competitive inhibitors** of **PARP**. They bind to its active site and prevent the binding of PARP's normal substrate, which is NAD^+ [34].

6 The VHL (Van Hippel-Lindau) Tumor Suppressor

VHL disease is a hereditary predisposition to the development of a variety of different types of tumors. These include clear-cell carcinomas of the kidney, pheochromocytomas (adrenal gland tumors), and hemangioblastomas (blood vessel tumors) of the central nervous system and the retina. These tumors are often associated with loss-of-function mutations in the gene encoding the **VHL tumor suppressor**. VHL is also inactivated in most **sporadic kidney carcinomas**. To understand the mechanism by which VHL inactivation leads to tumor formation, it is helpful to understand the main function of VHL in normal cells [3].

A major role of the **VHL** protein is to regulate the stability of HIF-1α in an oxygen-dependent manner. HIF-1α (hypoxia-inducible factor 1α) is a transcription factor that functions with another transcription factor, HIF-1β, and the two together are called HIF-1. VHL has the role of binding to the transcription factor HIF-1α and targeting it for destruction, but HIF-1α is only targeted for destruction under normoxia conditions, when oxygen levels are normal. The regulation of the VHL-HIF-1 pathway under normoxia and hypoxia (low oxygen) conditions is described below (see Fig. 21.5).

6.1 The VHL-HIF Pathway Normoxia

In normoxia, VHL prevents the transcription of genes that are necessary only under hypoxic conditions. It does so by promoting the destruction of the transcription factor HIF-1α, as follows: under normoxia conditions, **proline hydroxylases (PHDs)** add hydroxyl (OH) groups to proline residues on HIF-1α, thus converting them to hydroxyproline residues. These OH residues enable the binding of the **VHL** protein to **HIF-1α**. VHL, together with two other proteins, is a part of a **ubiquitin ligase complex**, leading to ubiquitination of HIF-1α. Ubiquitylation targets HIF-1α for **destruction**, and HIF-1α is degraded very quickly. Thus, under normoxia conditions, HIF-1α accumulates to only a very low level and is mostly inactive. This is important because, as we will discuss next,

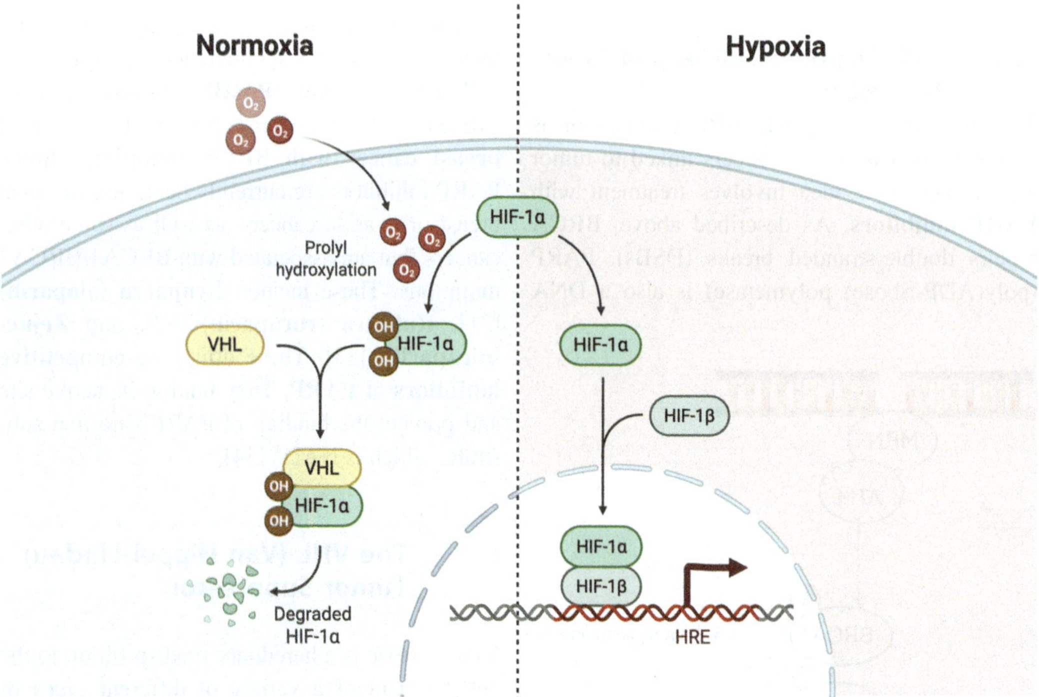

Fig. 21.5 VHL-HIF signaling. VHL binds to HIF-1alpha and leads to its destruction under normoxic conditions. Under hypoxic conditions, HIF-1α, along with HIF-1β, binds to promoter elements on target genes and regulates gene expression. **Source**: Created using biorender.com

when it is active, HIF-1a promotes the transcription of genes that are necessary for responding to hypoxia, but which could cause the cells that are in normoxia to have a dangerous growth advantage.

6.2 VHL During Hypoxia

Under **low-oxygen** conditions (**hypoxia**), **proline hydroxylase** does not add hydroxyl groups to the prolines on HIF-1α because the proline hydroxylase depends on normal oxygen levels to function. Consequently, **VHL** cannot bind to **HIF-1α** or target it for destruction. **HIF-1α** levels now rise quickly as it can no longer be destroyed. HIF-1α then translocates to the nucleus, where it dimerizes with **HIF-1β** (also called **ARNT**, aryl hydrocarbon receptor nuclear translocator) to form an active transcription factor referred to as HIF-1. **HIF-1** binds to the regulatory elements of many target genes (Hif regulatory elements, HREs) that are important for responding to hypoxic conditions and that allow cells to grow and survive under these stressful conditions. These include genes involved in angiogenesis (formation of new blood vessels), erythropoiesis (production of new red blood cells), energy metabolism, and glucose transport. As a result, the cells can survive the low-oxygen conditions until the conditions change. These target genes are discussed in more detail below.

6.3 VHL as a Tumor Suppressor

Cells are most often in **normoxia** conditions, and VHL is always present, keeping the levels of HIF-1α negligible and thereby preventing the transcription of HIF-1a target genes. If VHL is inactivated, however, the HIF-1α transcription factor can never be targeted for destruction, even in normoxia conditions. Consequently, the transcription of target genes can be upregulated, even under these normoxia conditions. Many of the target genes HIF-1α regulates (such as those involved in angiogenesis, erythropoiesis, energy metabolism, and glucose transport) provide a growth advantage for cells. This is why VHL is a tumor suppressor, as its absence can lead to a strong growth advantage to cells even though they are growing under normoxia conditions.

6.4 Some Target Genes of HIF-1α

Some of the **HIF-1 target genes** that give cells a growth advantage and that are important for cancer include **VEGF** (vascular endothelial growth factor), **PDGF** (platelet-derived growth factor), and erythropoietin (**EPO**). VEGF causes the growth of endothelial cells, which are blood vessel cells. This enables blood vessels to grow, which provides a source of oxygen to the cells. PDGF also stimulates the growth of cells that are important for blood vessels, including endothelial cells, mesenchymal cells, pericytes, and fibroblasts. **EPO** leads to erythropoiesis, the production of red blood cells (RBCs), which supply oxygen to the tissue [35, 36].

6.5 The Hydroxylation Reaction

The reaction for hydroxylation of HIF-1α is as follows:

$$\begin{aligned} &\mathrm{HIF} - 1\alpha - \mathrm{Pro} + \mathrm{O_2} + \alpha - \mathrm{ketoglutarate} \\ &\rightarrow \mathrm{HIF} - 1\alpha - \mathrm{Hyp} + \mathrm{CO_2} + \mathrm{succinate} \end{aligned}$$

Enzyme: **PHD2 (prolyl hydroxylase)**

Where -Pro is an unmodified proline residue and -Hyp is a hydroxylated proline residue.

In this reaction the **PHD**, requires iron (Fe^{2+}) as a cofactor and uses molecular oxygen (O_2) in oxidative decarboxylation of α-ketoglutarate. In this process, one oxygen atom of the O_2 is incorporated into the proline residue of HIF-1α, converting it to hydroxyproline (the hydrogen (H) will come from the surrounding solvent).

The other oxygen atom is used to decarboxylate the a-ketoglutarate. This produces succinate and CO_2 [37].

6.6 Drugs That Target the HIF-1α Pathway

Targeting the **HIF-1α** pathway in cancer has been challenging, as is the case for many tumor suppressor genes. However, the idea of manipulating the HIF-1α pathway is considered to be promising for treating other disorders besides cancer. As discussed above, the HIF-1 pathway is associated with the regulation of various target genes, including those that regulate erythropoiesis (formation of RBCs). This makes it an interesting target for disorders that are associated with anemia (lowered RBC levels). **Chronic kidney disease** is an example of a disease that is often associated with renal anemia [38]. A drug approved by the FDA in 2024, **Vafseo (vadadustat)** [39], is a **small-molecule** drug designed to improve this. This drug **inhibits the proline hydroxylases that inactivate HIF-1**. These are **(HIF)-prolyl-4-hydroxylases PHD1, PHD2, and PHD3**. The drug inactivates the PHs by binding to their active sites. It is a competitive inhibitor, and it mimics the natural substrate of the PHD enzymes, α-ketoglutarate. By inactivating these **PHDs**, HIF-1 stays active even in normoxia conditions, and can promote an increase in erythropoiesis, thus improving renal anemia. Another drug that functions similarly is **Jesduvroq (daprodustat)** (FDA-approved in 2023) [40], which also stabilizes HIF by inhibiting the **proline hydroxylase**, and is also used to **reduce renal anemia in kidney disease**.

7 Promoter Methylation Can Lead to Downregulation of Gene Expression

Tumor suppressors are often inactivated or inhibited by gene mutation, but they can also be regulated by the downregulation of the expression of the genes that code for them. Downregulation of gene expression can be regulated by upstream elements and transcription factors, but promoter methylation is another mechanism by which gene expression may be downregulated. Promoter methylation is an epigenetic change that can impact gene expression. It can lead to inhibition of gene expression without altering the nucleotide sequence of the gene and plays an important role in the inhibition of expression of some tumor suppressor genes [1, 3].

Repression of gene expression due to promoter methylation is notable in the BRCA1 gene. Early studies involving BRCA1 suggested that abnormalities in BRCA1 were only associated with familial breast cancer and that BRCA1 never contributes to sporadic breast cancer. This is because DNA mutations in the BRCA1 gene were never found in sporadic breast cancer. Later, however, it was found that the BRCA1 gene was inactive in a percentage of sporadic breast carcinomas. It was not detected in previous studies because, in sporadic carcinomas, the BRCA1 gene is inactivated by methylation, an epigenetic modification, rather than through DNA mutations that change the DNA sequence [41–43].

Promoter methylation is an important mechanism for inhibiting the expression (transcription) of tumor suppressor genes and other genes. The type of methylation that inhibits gene expression occurs in gene promoters on cytosines, in CpG regions, which are regions consisting of a cytosine followed by a guanine. Regions that are rich in these CpG sequences are sometimes referred to as CpG islands. Genes that are inactivated by methylation can be more difficult to detect compared with genes inactivated by DNA mutation because traditional DNA sequencing methods cannot detect changes in methylation. Methylation is considered to be an **epigenetic** modification of DNA. It is heritable and can be passed down to future generations of cells. Briefly, promoter methylation negatively regulates gene expression because it causes the DNA to take on a more closed conformation, which is less accessible to transcription factors [1]. This is discussed in more detail in Chap. 8. Several other oncogenes that are sometimes regulated by promoter methylation are discussed below.

Some tumor suppressor genes, such as **Runx3** (which is linked to stomach cancer), are almost always inactivated by methylation and not by mutation or deletion [44]. Others, such as **Rb**, are often inactivated by DNA mutation in cancer, although their genes can sometimes also be inactivated by promoter methylation [45]. **p16^{INK4A}** is a CKI that, like some other CKIs, is also a tumor suppressor because its normal function is to block cell cycle progression. In many cancers, the promoter of the p16 gene is **hypermethylated, leading to repression of its expression.** This has been seen in several types of cancers, including **colorectal** [46]**, breast** [47], **esophageal** [48], **and head and neck** [49] **cancers**. Methylation of the promoters of certain genes, such as the p16 promoter, can be used as a diagnostic tool for some types of cancer [50, 51].

The **RASSF1A** (Ras association domain family one isoform A) gene is a tumor suppressor that is downregulated in some solid tumors. It appears to function by interfering with various signaling pathways, as well as by modulating Ras activity. Some regions of this gene's promoter contain "CpG islands," which are methylated in many solid tumors [52]. The **APC** gene is also a tumor suppressor that is sometimes regulated by promoter methylation. This is seen in some colon cancers, and APC promoter methylation has also been detected in gastric cancer, lung cancer, and others [53].

Definitions

Adenomatous polyposis coli (APC) A tumor suppressor that forms part of the β-catenin destruction complex, promoting β-catenin degradation.

Anemia A blood disorder in which the body does not have enough healthy RBCs or hemoglobin to carry oxygen efficiently.

Angiogenesis Formation of new blood vessels.

BRCA1/2 Tumor suppressors in homologous recombination repair. Frequently mutated in familial breast cancer.

CpG islands Regions in which cytosine is followed by a guanine in DNA. The cytosines can sometimes be methylated, and when this occurs in promoters and enhancers, it can lead to suppression of transcription.

Disheveled (DSH) Intracellular Wnt effector that prevents β-catenin degradation.

Epigenetic DNA change A heritable change in the DNA that does not involve a change in DNA sequence. Instead, it involves a modification of the DNA or histones, such as methylation. (*Epi- in Greek means on/above.*)

EPO A hormone that stimulates red blood cell production.

Frizzled Wnt receptor on the cell membrane.

Glioblastomas Tumors of astrocyte lineage in the brain.

Hemangioblastomas Tumors of the blood vessels.

FDA-Approved Drugs Discussed in This Chapter

Commonly used brand name	Generic name	First FDA approval year	Major conditions treated	Main targets
Gomekli	*Mirdametinib*	2025	Neurofibromatosis type 1 (NF1)	MEK
Jesduvroq	*Daprodustat*	2023	Anemia from chronic kidney disease	PHD (prolyl hydroxylase)
Koselugo	*Selumetinib*	2020	Neurofibromatosis type 1 (NF1)	MEK
Lynparza	*Olaparib*	2014	Breast, ovarian, prostate cancers (BRCA+)	PARP
Rubraca	*Rucaparib*	2016	Ovarian, prostate cancers (BRCA+)	PARP
Vafseo	*Vadadustat*	2024	Anemia from chronic kidney disease	PHD (prolyl hydroxylase)
Zejula	*Niraparib*	2017	Ovarian cancer (BRCA+)	PARP

Homologous recombination An accurate DNA repair mechanism that uses a sister chromatid as a template to accurately repair double-stranded breaks.

Hypoxia Low-oxygen conditions.

Hypoxia-inducible factor 1 alpha (HIF-1α) Transcription factor activated in hypoxia.

Loss of heterozygosity (LOH) Loss of function of the remaining functional allele of a gene where the other allele is already inactivated. When it occurs with tumor suppressor genes it contributes to tumor development.

MEK Kinase in the MAPK pathway, target of several drugs including selumetinib.

Myelogenous leukemia (CML) A rare cancer of the blood and bone marrow, characterized by too many white blood cells.

Neurofibromin 1 (NF1) The NF1 gene encodes neurofibromin, a Ras-GAP family member. Loss or inactivation of the NF-1 gene leads to excessive Ras activity and causes the genetic disorder neurofibromatosis type 1 (NF1).

Neurofibrosarcomas Malignant tumors of the peripheral nerve sheath that occasionally develop in patients with neurofibromatosis type 1.

Normoxia Normal oxygen conditions.

Osteosarcoma Rare aggressive bone cancer, usually in children and young adults.

p16INK4A CDK inhibitor (CKI), sometimes silenced by methylation in cancer.

p21 A CDK (CKI) inhibitor upregulated by p53.

p53 A transcription factor and key tumor suppressor that regulates the cell cycle in response to DNA damage.

Pheochromocytomas Tumors of the adrenal glands.

Plexiform neurofibromas (PNs) Benign tumors that grow along nerves in patients with NF.

Poly(ADP-ribose) polymerase (PARP) Enzyme that repairs single-stranded breaks, a target in BRCA therapy.

Prolyl hydroxylase domain enzyme (PHD) Enzyme that hydroxylates HIF-1α to tag for degradation.

RASSF1A Tumor suppressor that is often epigenetically silenced.

Retinoblastoma protein (Rb) A protein with an important role in the cell cycle. When unphosphorylated, Rb blocks the cell cycle at G1/S. The Rb gene was the first tumor suppressor gene to be identified.

Synthetic lethality Occurs when the combination of two gene defects or inhibitions causes cell death, while each defect alone does not lead to cell death. It is sometimes an approach for targeted cancer therapies.

Tetraploidy A chromosomal abnormality in which a cell has four sets of chromosomes instead of the usual two.

Transforming growth factor alpha (TGF-α) Growth factor promoting proliferation of various cells, including epithelial cells.

Tumor suppressor gene A gene that inhibits cell division and prevents uncontrolled growth. Loss can lead to cancer when inactivated.

Von Hippel-Lindau (VHL) A tumor suppressor. Its normal function is to respond to oxygen levels and regulate HIF-1α degradation.

β-catenin A transcription factor regulated by the Wnt pathway. It can be improperly regulated in cancer, especially colon cancer.

Study Questions

Multiple-Choice Questions

1. Which of the following best illustrates loss of heterozygosity (LOH)?
   A. When a mutation occurs in only one copy of a gene
   B. When a person inherits one dysfunctional copy of an oncogene on one chromosome
   C. When a person inherits one dysfunctional copy of a tumor suppressor gene on the maternal chromosome, and the corresponding copy on the paternal chromosome becomes mutated later in life
   D. When a person inherits one dysfunctional copy of a tumor suppressor gene on the maternal chromosome, and later develops a second mutation in the same tumor suppressor gene on the same maternal chromosome
   E. Either C or D

2. NF1 encodes a GAP. This means that inactivation of NF1 mimics:
   A. Inactive Ras
   B. Active Ras
   C. Inactivation of Sos
   D. Inactivation of p53
   E. A and C
3. What is the mechanism by which BRCA1 contributes to DNA repair?
   A. Inhibiting CDK4/6
   B. Activating Ras
   C. Promoting homologous recombination
   D. Degrading β-catenin
4. What is the role of the APC gene in the Wnt pathway?
   A. Promotes Wnt ligand production
   B. Inhibits β-catenin degradation
   C. Part of the destruction complex that degrades β-catenin
   D. Stimulates Cyclin D1 directly
5. Which drug inhibits prolyl hydroxylase and stabilizes HIF-1α to treat renal anemia?
   A. Lynparza
   B. Vafseo
   C. Koselugo
   D. Rubraca
6. Which of the following could most likely be coded for by a tumor suppressor gene?
   A. A protein that helps prevent progression through the cell cycle
   B. A protein that helps prevent apoptosis
   C. A protein that codes for a DNA repair enzyme
   D. A protein that forms part of a growth factor signaling pathway
   E. A or B
7. The product of which gene prevents cell cycle progression in response to DNA damage?
   A. Rb
   B. p53
   C. Esterase D
   D. Ras
   E. A and C
8. In the intestinal tissue, Wnt concentration is highest:
   A. Near the base of the crypt
   B. Near the most differentiated cells
   C. Near the least differentiated cells
   D. A and B
   E. A and C
9. VHL causes __________ of HIFα via __________ in conditions of normoxia.
   A. activation; ubiquitylation
   B. activation; methylation
   C. degradation; methylation
   D. degradation; ubiquitylation

True/False Questions

10. Mutations in BRCA genes are responsible for most sporadic breast cancers.
11. Phosphorylation of β-catenin should be associated w/ increased proliferation.

Fill-in-the-Blank Questions

12. Loss of function in both alleles of a tumor suppressor gene is termed ________________.
13. The transcription factor targeted for degradation by VHL under normoxia conditions is _______.
14. Selumetinib is a MEK inhibitor used to treat tumors in patients with ________.
15. BRCA repairs ______-stranded DNA breaks, while PARP repairs ______-stranded DNA breaks. Inhibition of both together can lead to ____________.
16. _________ in the _________ can suppress the transcription of a tumor suppressor gene, and does so in the example of $p16^{INK4A}$, which is altered in a specific patient population (state which one).

17. When Wnt is present: disheveled is ______ (active or inactive?), the destruction complex is ______ (activated or inactivated?), β-catenin ______ (does or does not?) get degraded, and cellular proliferation ___________ (proceeds or halts?).

Short-Answer Question

18. Explain how VHL acts as a tumor suppressor under normoxic conditions.

Matching Question

19. Match each tumor suppressor gene to its associated function or cancer:
    Gene Function/Role
    A. Rb
    B. p53
    C. APC
    D. NF1
    E. BRCA1
    Function:
    1. Prevents G1 to S transition
    2. Induces p21 expression after DNA damage
    3. Regulates β-catenin in Wnt pathway
    4. Regulates Ras by acting as a GAP
    5. Involved in homologous recombination repair

Answer Key

Multiple-Choice Questions

1. **Answer**: C
2. **Answer**: B
3. **Answer**: C
4. **Answer**: C
5. **Answer**: B
6. **Answer**. A
7. **Answer**: B
8. **Answer**: E
9. **Answer:** D

True/False Questions

10. **Answer**: False
11. **Answer**: False: When β-catenin is phosphorylated, it is marked for degradation, which means less of it will be present to be able to stimulate an increase in cell proliferation, thus associating β-catenin phosphorylation with *decreased* proliferation.

Fill-in-the-Blank Questions

12. **Answer:** Loss of heterozygosity
13. **Answer:** HIF-1α
14. **Answer:** NF1
15. **Answer**: Double, single, synthetic lethality
16. **Answer**: *Methylation* in the *promoter*: $p16^{INK4A}$ promoter is methylated in 44% of normal bronchial epithelial cells of current/former smokers.
17. **Answer**: When Wnt is **present:** Disheveled is *active*, the destruction complex is *inactivated*, β-catenin *does not* get degraded, and cellular proliferation *proceeds*.
    (By the same logic, when Wnt is **absent:** disheveled is *inactive*, the destruction complex is *active*, β-catenin *does* get degraded, and cellular proliferation *halts*.)

Short-Answer Question

18. **Answer**: VHL promotes degradation of HIF-1α via ubiquitination under normoxic conditions, preventing transcription of genes that promote cell survival and proliferation.

Matching Question

19. **Answers**:
    A. 1. B. 2, C. 3, D. 4, E. 5

References

1. Alberts B, Heald R, Johnson A, Morgan D, Raff M, Roberts K, Walter P, Wilson J, Hunt T. Molecular biology of the cell. New York: W.W. Norton and Company; 2022.
2. Berry JL, Polski A, Cavenee WK, Dryja TP, Murphree AL, Gallie BL. The RB1 story: characterization and cloning of the first tumor suppressor gene. Genes (Basel). 2019;10(11). https://doi.org/10.3390/genes10110879.
3. Weinberg RA. The biology of cancer. 3rd ed. New York: Norton and Company; 2023.
4. Pellarin I, Dall'Acqua A, Favero A, Segatto I, Rossi V, Crestan N, et al. Cyclin-dependent protein kinases and cell cycle regulation in biology and disease. Signal Transduct Target Ther. 2025;10(1):11. https://doi.org/10.1038/s41392-024-02080-z.
5. Fletcher O, Easton D, Anderson K, Gilham C, Jay M, Peto J. Lifetime risks of common cancers among retinoblastoma survivors. JNCI J Natl Cancer Inst. 2004;96(5):357–63. https://doi.org/10.1093/jnci/djh058.
6. Linn P, Kohno S, Sheng J, Kulathunga N, Yu H, Zhang Z, et al. Targeting RB1 loss in cancers. Cancer. 2021;13(15):3737. https://doi.org/10.3390/cancers13153737.
7. Liu Y, Hu X, Han C, Wang L, Zhang X, He X, et al. Targeting tumor suppressor genes for cancer therapy. BioEssays. 2015;37(12):1277–86. https://doi.org/10.1002/bies.201500093.
8. Lane DP. p53, guardian of the genome. Nature. 1992;358(6381):15–6. https://doi.org/10.1038/358015a0.
9. Hassin O, Oren M. Drugging p53 in cancer: one protein, many targets. Nat Rev Drug Discov. 2023;22(2):127–44. https://doi.org/10.1038/s41573-022-00571-8.
10. Nishikawa S, Iwakuma T. Drugs targeting p53 mutations with FDA approval and in clinical trials. Cancer. 2023;15(2):429. https://doi.org/10.3390/cancers15020429.
11. Wu S, Xu Y. MDM2 degrader as a promising therapeutic strategy for cancer treatment. J Med Chem. 2025;68(13):13246–8. https://doi.org/10.1021/acs.jmedchem.5c01642.
12. Marine JC, Jochemsen AG. MDMX (MDM4), a promising target for p53 reactivation therapy and beyond. Cold Spring Harb Perspect Med. 2016;6(7). https://doi.org/10.1101/cshperspect.a026237.
13. Kresak JL, Walsh M. Neurofibromatosis: a review of NF1, NF2, and schwannomatosis. J Pediatr Genet. 2016;5(2):98–104. https://doi.org/10.1055/s-0036-1579766.
14. U.S. Food and Drug Administration. Koselugo (Selumetinib) approval overview. 2020. Available from: https://www.accessdata.fda.gov/scripts/cder/daf/index.cfm?event=overview.process&ApplNo=213756.
15. Gross AM, Dombi E, Widemann BC. Current status of MEK inhibitors in the treatment of plexiform neurofibromas. Childs Nerv Syst. 2020;36(10):2443–52. https://doi.org/10.1007/s00381-020-04731-2.
16. Campagne O, Yeo KK, Fangusaro J, Stewart CF. Clinical pharmacokinetics and pharmacodynamics of selumetinib. Clin Pharmacokinet. 2021;60(3):283–303. https://doi.org/10.1007/s40262-020-00967-y.
17. U.S. Food and Drug Administration. Gomekli (Mirdametinib) approval overview. 2025. Available from: https://www.fda.gov/drugs/resources-information-approved-drugs/fda-approves-mirdametinib-adult-and-pediatric-patients-neurofibromatosis-type-1-who-have-symptomatic?utm.
18. Roskoski R. Properties of FDA-approved small molecule protein kinase inhibitors: a 2025 update. Pharmacol Res. 2025;216:107723. https://doi.org/10.1016/j.phrs.2025.107723.
19. Hoy SM. Mirdametinib: first approval. Drugs. 2025;85(7):977–84. https://doi.org/10.1007/s40265-025-02190-0.
20. Marks F, Klingmüller U, Müller-Decker K. Cellular signal processing: an introduction to the molecular mechanisms of signal transduction. 2nd ed. New York: Garland Science; 2017.
21. MacDonald BT, Tamai K, He X. Wnt/beta-catenin signaling: components, mechanisms, and diseases. Dev Cell. 2009;17(1):9–26. https://doi.org/10.1016/j.devcel.2009.06.016.
22. Nusse R, Clevers H. Wnt/β-catenin signaling, disease, and emerging therapeutic modalities. Cell. 2017;169(6):985–99. https://doi.org/10.1016/j.cell.2017.05.016.
23. Clevers H, Nusse R. Wnt/β-catenin signaling and disease. Cell. 2012;149(6):1192–205. https://doi.org/10.1016/j.cell.2012.05.012.
24. Schatoff EM, Leach BI, Dow LE. Wnt signaling and colorectal cancer. Curr Colorectal Cancer Rep. 2017;13(2):101–10. https://doi.org/10.1007/s11888-017-0354-9.
25. Zhang L, Shay JW. Multiple roles of APC and its therapeutic implications in colorectal cancer. J Natl Cancer Inst. 2017;109(8). https://doi.org/10.1093/jnci/djw332.
26. Rusan NM, Peifer M. Original CIN: reviewing roles for APC in chromosome instability. J Cell Biol. 2008;181(5):719–26. https://doi.org/10.1083/jcb.200802107.
27. Godet I, Gilkes DM. BRCA1 and BRCA2 mutations and treatment strategies for breast cancer. Integr Cancer Sci Ther. 2017;4(1). https://doi.org/10.15761/icst.1000228.
28. Yoshida K, Miki Y. Role of BRCA1 and BRCA2 as regulators of DNA repair, transcription, and cell cycle in response to DNA damage. Cancer Sci. 2004;95(11):866–71. https://doi.org/10.1111/j.1349-7006.2004.tb02195.x.
29. Venkitaraman AR. Cancer susceptibility and the functions of BRCA1 and BRCA2. Cell. 2002;108(2):171–82. https://doi.org/10.1016/S0092-8674(02)00615-3.

30. Janysek DC, Kim J, Duijf PHG, Dray E. Clinical use and mechanisms of resistance for PARP inhibitors in homologous recombination-deficient cancers. Transl Oncol. 2021;14(3):101012. https://doi.org/10.1016/j.tranon.2021.101012.
31. U.S. Food and Drug Administration. Lynparza (Olaparib) approval overview. 2014. Available from: https://www.accessdata.fda.gov/scripts/cder/daf/index.cfm?event=overview.process&ApplNo=208558.
32. U.S. Food and Drug Administration. Rubraca (Rucaparib) approval overview. 2016. Available from: https://www.accessdata.fda.gov/scripts/cder/daf/index.cfm?event=overview.process&ApplNo=209115.
33. U.S. Food and Drug Administration. Zejula (Niraparib) approval overview. 2017. Available from: https://www.accessdata.fda.gov/scripts/cder/daf/index.cfm?event=overview.process&ApplNo=208447.
34. Rose M, Burgess JT, O'Byrne K, Richard DJ, Bolderson E. PARP inhibitors: clinical relevance, mechanisms of action and tumor resistance. Front Cell Dev Biol. 2020;8:564601. https://doi.org/10.3389/fcell.2020.564601.
35. Kim H, Shim BY, Lee SJ, Lee JY, Lee HJ, Kim IH. Loss of Von Hippel-Lindau (VHL) tumor suppressor gene function: VHL-HIF pathway and advances in treatments for metastatic renal cell carcinoma (RCC). Int J Mol Sci. 2021;22(18). https://doi.org/10.3390/ijms22189795.
36. Shaw P, Dwivedi SKD, Bhattacharya R, Mukherjee P, Rao G. VEGF signaling: role in angiogenesis and beyond. Biochim Biophys Acta Rev Cancer. 2024;1879(2):189079. https://doi.org/10.1016/j.bbcan.2024.189079.
37. Strowitzki MJ, Cummins EP, Taylor CT. Protein hydroxylation by hypoxia-inducible factor (HIF) hydroxylases: unique or ubiquitous? Cells. 2019;8(5). https://doi.org/10.3390/cells8050384.
38. Haase VH. Hypoxia-inducible factor-prolyl hydroxylase inhibitors in the treatment of anemia of chronic kidney disease. Kidney Int Suppl (2011). 2021;11(1):8–25. https://doi.org/10.1016/j.kisu.2020.12.002.
39. U.S. Food and Drug Administration. Vafseo (Vadadustat) approval overview. 2024. Available from: https://www.accessdata.fda.gov/scripts/cder/daf/index.cfm?event=overview.process&ApplNo=215192.
40. U.S. Food and Drug Administration. Jesduvroq (Daprodustat) approval overview. 2023. Available from: https://www.accessdata.fda.gov/scripts/cder/daf/index.cfm?event=overview.process&ApplNo=216951.
41. Dobrovic A, Simpfendorfer D. Methylation of the BRCA1 gene in sporadic breast cancer. Cancer Res. 1997;57(16):3347–50.
42. Xu X, Gammon MD, Zhang Y, Bestor TH, Zeisel SH, Wetmur JG, et al. BRCA1 promoter methylation is associated with increased mortality among women with breast cancer. Breast Cancer Res Treat. 2009;115(2):397–404. https://doi.org/10.1007/s10549-008-0075-5.
43. Panagopoulou M, Panou T, Gkountakos A, Tarapatzi G, Karaglani M, Tsamardinos I, et al. BRCA1 & BRCA2 methylation as a prognostic and predictive biomarker in cancer: implementation in liquid biopsy in the era of precision medicine. Clin Epigenetics. 2024;16(1):178. https://doi.org/10.1186/s13148-024-01787-8.
44. Tsang YH, Lamb A, Chen LF. New insights into the inactivation of gastric tumor suppressor RUNX3: the role of H. pylori infection. J Cell Biochem. 2011;112(2):381–6. https://doi.org/10.1002/jcb.22964.
45. Vélez-Cruz R, Johnson DG. The retinoblastoma (RB) tumor suppressor: pushing back against genome instability on multiple fronts. Int J Mol Sci. 2017;18(8). https://doi.org/10.3390/ijms18081776.
46. Shima K, Nosho K, Baba Y, Cantor M, Meyerhardt JA, Giovannucci EL, et al. Prognostic significance of CDKN2A (p16) promoter methylation and loss of expression in 902 colorectal cancers: cohort study and literature review. Int J Cancer. 2011;128(5):1080–94. https://doi.org/10.1002/ijc.25432.
47. Wang L, Tang L, Xie R, Nie W, Chen L, Guan X. p16 promoter hypermethylation is associated with increased breast cancer risk. Mol Med Rep. 2012;6(4):904–8. https://doi.org/10.3892/mmr.2012.1001.
48. Das M, Saikia BJ, Sharma SK, Sekhon GS, Mahanta J, Phukan RK. p16 hypermethylation: a biomarker for increased esophageal cancer susceptibility in high incidence region of North East India. Tumour Biol. 2015;36(3):1627–42. https://doi.org/10.1007/s13277-014-2762-7.
49. Wong TS, Man MW, Lam AK, Wei WI, Kwong YL, Yuen AP. The study of p16 and p15 gene methylation in head and neck squamous cell carcinoma and their quantitative evaluation in plasma by real-time PCR. Eur J Cancer. 2003;39(13):1881–7. https://doi.org/10.1016/s0959-8049(03)00428-3.
50. Brooks J, Cairns P, Zeleniuch-Jacquotte A. Promoter methylation and the detection of breast cancer. Cancer Causes Control. 2009;20(9):1539–50. https://doi.org/10.1007/s10552-009-9415-y.
51. Belinsky SA, Nikula KJ, Palmisano WA, Michels R, Saccomanno G, Gabrielson E, et al. Aberrant methylation of p16(INK4a) is an early event in lung cancer and a potential biomarker for early diagnosis. Proc Natl Acad Sci USA. 1998;95(20):11891–6. https://doi.org/10.1073/pnas.95.20.11891.
52. García-Gutiérrez L, McKenna S, Kolch W, Matallanas D. RASSF1A tumour suppressor: target the network for effective cancer therapy. Cancer. 2020;12(1):229.
53. Zhu L, Li X, Yuan Y, Dong C, Yang M. APC promoter methylation in gastrointestinal cancer. Front Oncol. 2021;11:653222. https://doi.org/10.3389/fonc.2021.653222.

Cell Death, Survival, and Recycling Mechanisms; Apoptosis, Necrosis, and Autophagy

22

Abstract

Normal development and homeostasis require a balance between cell growth and cell death. The major cell death mechanism in eukaryotic cells is apoptosis, a programmed series of events that irreversibly leads to cell death. There are other ways that cell death can occur, however. One of these is necrosis. Unlike apoptosis, necrosis is not a controlled process, and it leads to an inflammatory response. In addition to balancing cell survival and cell death, cells also need to recycle or destroy damaged organelles. This is the process of autophagy. Autophagy is important for cell survival, but at high levels, it can also lead to cell death. This chapter explores all of these topics. First, it covers apoptosis, including the proteins involved, such as caspases and Bcl proteins, and discusses the different pathways leading to apoptosis. This is followed by a brief overview of necrosis. Next, this chapter delves into autophagy. The different steps that lead to autophagy are covered, including packaging cellular components, such as damaged organelles, into vesicles, vesicular fusion, and fusion with lysosomes for destruction. The role of autophagy in cell survival is discussed, as well as its potential role in cell death. Throughout this chapter, diseases associated with abnormalities in these cellular processes are discussed, as are therapeutic strategies for their treatment.

Keywords

Apoptosis · Caspases · Intrinsic pathway · Extrinsic pathway · Autophagy · Necrosis

Abbreviations

Apaf-1	Apoptotic protease activating factor 1
Bcl-2	B-cell lymphoma 2
DSB	Double-stranded break
Fas	First apoptosis signal receptor
FasL	Fas ligand
LC3	Microtubule-associated protein 1A/1B-light chain 3
MLKL	Mixed lineage kinase domain-like protein
MOMP	Mitochondrial outer membrane permeabilization
PARP	Poly(ADP-ribose) polymerase
RIPK1	Receptor-interacting protein kinase 1
SSB	Single-stranded break
TNF	Tumor necrosis factor
TNFR	Tumor necrosis factor receptor
TUNEL	Terminal deoxynucleotidyl transferase dUTP nick end labeling

What You Will Learn

- The main characteristics of apoptosis and programmed cell death.

A. Minden, *Biochemistry, Cell and Molecular Biology for Pharmacists and Health Professionals*,
https://doi.org/10.1007/978-3-032-20529-2_22

- Types of proteins involved in apoptosis, including caspases and Bcl-2 proteins, as well as APAF1, IAP, and others.
- Mitochondrial proteins that are important in apoptosis including cytochrome C, AIF, and SMAC/Diablo.
- The intrinsic and extrinsic pathways to apoptosis.
- The role for MOMP: mitochondrial outer membrane permeabilization in the intrinsic pathway.
- Cytokines such as Fas and TNF can activate the extrinsic pathway.
- Crosstalk between the intrinsic and the extrinsic pathways of apoptosis.
- Assays for identifying apoptotic cells such as TUNEL assays to identify fragmented DNA, identification of cleaved PARP or caspases, or using flow cytometry to identify phosphatidylserine concentration on the outer membrane.
- Extracellular signals that regulate apoptosis and cell survival.
- Diseases associated with apoptosis such as cancer, neurodegenerative diseases, or developmental disorders.
- Drugs designed to restore apoptosis in cancer cells such as BH3 mimetics.
- Other forms of cell death such as necrosis, pyroptosis, and necroptosis.
- Autophagy, the process by which cells digest intracellular components and proteins. Autophagy does not necessarily lead to cell death, and can be necessary for cell survival.
- Autophagy can include macroautophagy, microautophagy, and chaperone-mediated autophagy.
- Some of the basic proteins involved in autophagy and the basic steps of macroautophagy.
- Diseases associated with autophagy include Parkinson's disease, Huntington's disease, and others.
- Drugs designed to treat lysosomal storage disease such as Pompe disease.

Normal development and homeostasis require a balance between cell growth and cell death. The major cell death mechanism in eukaryotic cells is apoptosis, a programmed series of events that irreversibly leads to cell death. There are other ways that cell death can occur, however. One of these is necrosis. Necrosis is different than apoptosis because it is not a controlled process, and it leads to an inflammatory response. In addition to balancing cell survival and cell death, cells also need to recycle or destroy damaged organelles. This is the process of autophagy. Autophagy is important for cell survival, but at high levels it can also lead to cell death.

These topics will all be covered in this chapter, which is organized into three sections. Section 1 explores apoptosis, including the proteins involved, such as caspases and Bcl proteins, and discusses the pathways leading to apoptosis: the extrinsic and intrinsic pathways. Section 2 gives an overview of necrosis, which is closely linked to an inflammatory response. Section 3 deviates from the discussion of cell death and covers autophagy. This section will cover the different steps involved in autophagy, including packaging cellular components such as damaged organelles into vesicles, fusion of the vesicles, and their fusion with lysosomes for destruction.

1 Apoptosis

1.1 Apoptosis: Balancing Cell Growth, Survival, and Death

1.1.1 Introduction to Apoptosis

Apoptosis is a type of **programmed cell death** that is tightly regulated. Apoptosis is the most common mechanism for cell death. It is vital in developing and maintaining the body's tissues and organs. During development, a balance between cell growth and cell death is essential. The correct formation of organs and tissues depends not only on cell growth but also on cell death. For example, for the digits (fingers and toes) to form correctly, the webbing that forms between them during embryonic development must be removed. This requires death by apoptosis of the cells that make up those tissues. The developing nervous system also requires apoptosis. During development, nerve cells are produced in excess. Many of the resulting excess nerve cells die by apoptosis so that the nervous system will ultimately become the correct size.

Cells in the adult also die by apoptosis, which is important for eliminating unwanted cells. For example, in some cases, damaged cells die by apoptosis, especially when affected by factors such as UV irradiation, which leads to DNA damage. Cells that recognize internal damage, such as damaged organelles, may also undergo apoptosis. Apoptosis is important in the normal function of the immune system. During an infection, the population of B and T cells expands. After the infection, most of these cells are no longer needed. At this point, the excess B and T cells are eliminated by apoptosis. In general, organs stay the same size in adults. Even this size maintenance depends on a balance between cell proliferation and apoptosis.

Cells have two main pathways leading to apoptosis: the extrinsic pathway and the intrinsic pathway. The extrinsic pathway is activated by cell surface receptors. These receptors, such as the Fas receptor, receive signals that trigger cell death, and they activate pathways in the cell, leading to apoptosis. The intrinsic pathway does not involve a cell surface receptor but involves triggers such as DNA damage. Both pathways will be discussed in this chapter. See Figure 22.1. The material in this chapter draws from the

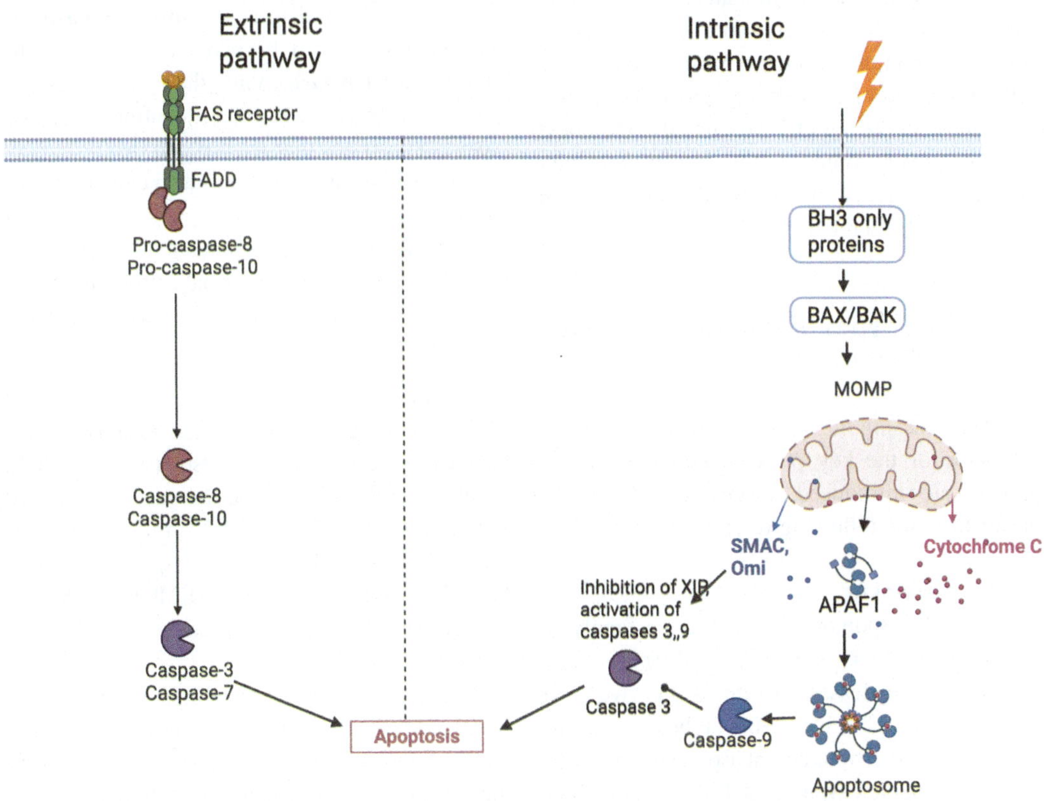

Fig. 22.1 Apoptosis: The extrinsic and intrinsic pathways. Some of the key steps in these pathways are shown here. More details are outlined in the text. (*Source*: Image adapted from Servier Medical Art (https://smart.servier.com/), licensed under the Creative Commons Attribution 4.0 (CC BY 4.0) license)

following references, which also provide additional perspective [1–7].

1.1.2 Characteristics of Apoptotic Cells

Apoptosis is characterized by chromatin degradation, proteolysis, apoptotic body formation, and changes in membrane permeability. During apoptosis, the cell shrinks, chromatin condenses and fragments, the nuclear envelope dissolves, and blebs form at the cell surface, forming protrusions. Apoptotic bodies can also form, which are membrane-enclosed pieces of cell. Macrophages engulf apoptotic cells, blebs, and apoptotic bodies. The apoptosis process happens quickly, and there is typically no inflammatory response.

While apoptosis is the most common and most well-characterized mechanism of cell death, other mechanisms also exist. Another mechanism is necrosis. Necrosis is not programmed or orderly, and unlike apoptosis, it can lead to an inflammatory response. Necrosis is more likely to occur after a severe injury or other type of damage to the cell. This chapter focuses on apoptosis, but necrosis will also be briefly discussed. The next sections cover some of the basic molecular mechanisms behind apoptosis, also discussed in the following references [1–4].

1.2 The Types of Proteins Involved in Apoptosis

Before discussing the molecular basis of apoptosis, some of the key proteins that play in the process are introduced, discussed below, and drawn from the following references [1–5, 7].

1.2.1 Caspases

Central to the apoptosis process is the family of proteins called **caspases** [2, 5, 8] (Fig. 22.2). Caspases are **proteases** (proteins that cleave other proteins). They have a **cysteine** at their active site, and they cleave their targets at aspartate. This is the basis of the name **caspase**, the full name of which is **cysteine aspartate protease** (**c** is for cysteine, **asp** is for aspartate, and **ase** is the ending for enzymes). Caspases are zymogens, which are

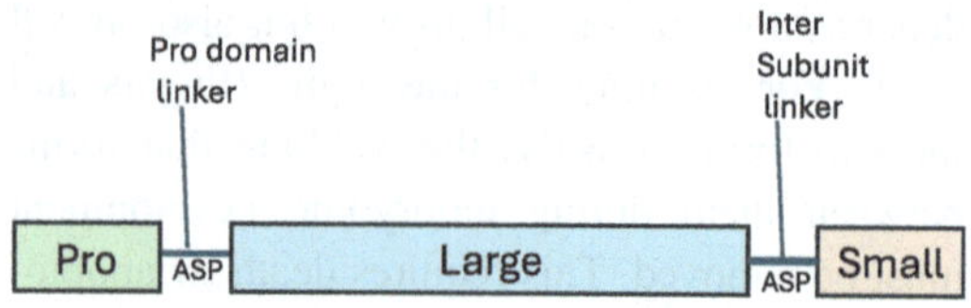

Fig. 22.2 Caspases. This image shows the basic structure of caspases. The pro-domain is cleaved when the caspase is activated

proteins that must be modified, usually by cleavage, to be activated. In cells not undergoing apoptosis, caspases exist as latent zymogens. These are caspases that are still in their inactive form. They can also be referred to as **procaspases**. Caspases are activated by their **cleavage**, in which a **pro-domain** is removed. Once caspases are activated by a stimulus that triggers apoptosis, a cascade of caspases can be activated, leading to the irreversible process of apoptosis.

In caspase cascades, the first caspase to be activated is referred to as the **initiator caspase**, and the last one to be activated is called the **executioner caspase**. While the initiator caspase cleaves other caspases, the executioner caspase cleaves substrates that lead to cell death. Executioner caspases are sometimes also called effector caspases.

It should also be noted that not all caspases are involved in apoptosis. Some have other functions, but here, we focus only on caspases that are involved in apoptosis.

- Initiator Caspases

The main initiator caspases are **caspase-8** and **caspase-9**. Caspase-8 is also called FLICE, Mach-1, and MCH5, and caspase-9 is called ICD-LAP6 and Mch6. Other initiator caspases include caspase-1 (ICE), caspase-2 (Ich-1, Nedd2), and caspase-10 (FLICE-2, Mch4) [1]. Caspase-8 and caspase-9 will be the focus here.

Initiator caspases are monomers when they are inactive but dimers when they are active. Dimerization happens when an apoptotic signal leads to the formation of a protein complex that recruits multiple initiator caspase monomers. In these complexes, the monomers dimerize. The dimers **cleave their dimer partner**, and as a result, the

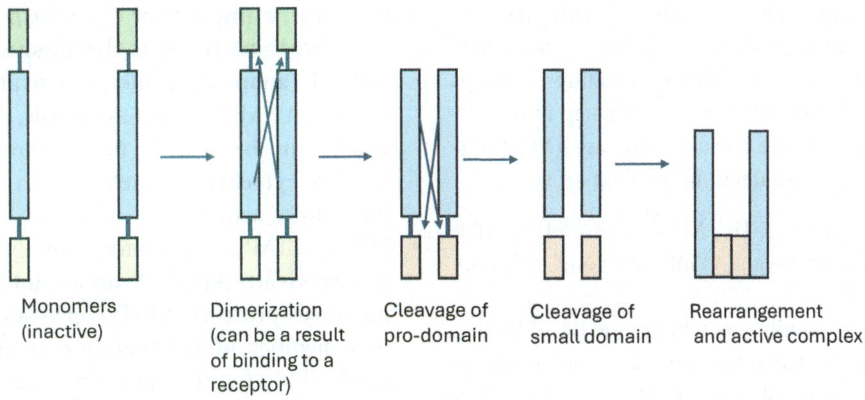

Fig. 22.3 Caspase activation. This figure illustrates one mechanism for caspase activation. Activation involves cleavage and dimerization

pro-domain is removed (Fig. 22.3). This is the **first cleavage**. A **second cleavage** can then occur between the small and large domains, followed by rearrangement and **complete activation** of the **initiator caspases**, which can now cleave and activate the executioner caspases.

- Executioner Caspases

The main executioner caspase is **caspase-3** (also called CPP32, Yama, Apopain, SCA-1, LICE). Other executioner caspases include caspase-6 (also called Mch-2) and caspase-7 (also called Mch-3, ICE-LAP-3, CMH-1).

Executioner caspases pre-exist as inactive dimers and are activated by cleavage, usually by initiator caspases. The system is amplified because initiator caspases can activate multiple executioner caspases. Active **executioner caspases** catalyze **many** protein cleavage events in the cell that are responsible for cell death and engulfment. The process they activate is irreversible; once it starts, there is no turning back.

1.2.2 Bcl-2 Proteins and the Balance Between Pro-apoptotic and Pro-survival Proteins

Another important family of proteins that are involved in apoptosis (particularly the intrinsic pathway) is the **Bcl-2 family** of proteins. The Bcl-2 family consists of **pro-survival proteins** as well as **pro-apoptosis proteins**. The balance between the two determines whether the cell will survive or undergo apoptosis. The Bcl-2 families all have in common a **BH3 domain**, which is a protein interaction domain. The Bcl-2 family proteins are highly conserved, and in addition to mammalian cells, they also regulate apoptosis in cells in other organisms. For example, homologs are found in *C. elegans*, an organism that was key to elucidating the mechanism of apoptosis in all organisms. The Bcl-2 family members are cytoplasmic proteins. They can exist either freely within the cytoplasm or on the surface of the mitochondrial outer membrane (MOM), depending on the conditions in the cell. They all have key roles in regulating the permeability of the mitochondria. This is important for the intrinsic apoptosis pathway because in this pathway, certain mitochondrial proteins exit the mitochondria and enter the cytoplasm, resulting in apoptosis. There are three main classes of Bcl-2 proteins. While it is a large family of proteins, some key members of each group are described below, and discussed in the following references [4, 9, 10].

- BAK and BAX Are Pro-apoptotic

BAK and BAX are **pro-apoptotic**. They are not expressed in non-apoptotic cells, but their expression is upregulated by some stimuli that trigger apoptosis. These cytosolic proteins can bind the mitochondrial outer membrane (MOM) and induce mitochondrial outer membrane permeabilization (MOMP) by creating openings (pores) in the MOM. This is a first step in triggering apoptosis in the intrinsic pathway. (MOMP is discussed in more detail in the next section.)

- Bcl-2 and Bcl-XL Are Examples of Pro-survival, Anti-apoptotic Family Members

Bcl-2 is the founding member of the Bcl-2 family. Bcl-2 and related family members are found in the cytoplasm of non-apoptotic cells. They **inhibit apoptosis** by inhibiting the functions of **Bak** and **Bax**.

- BH3-Only Proteins, Including Bad, Bim, Bid, Puma, and Noxa, Are a Second Class of Pro-apoptotic Proteins

BH3-only proteins help promote apoptosis by regulating the activities of **Bak** and **Bax**. They are not expressed in non-apoptotic cells but will be expressed upon some apoptotic stimuli. They are called **BH3-only proteins** because they have BH3 domains, a conserved domain found in all the BCL-2 proteins, but they lack other conserved domains.

1.2.3 Some Mitochondrial Proteins That Will Cause Apoptosis if Released from the Mitochondria

The following proteins are located inside the mitochondria. Their release from the mitochondria triggers apoptosis in the intrinsic pathway. They are discussed below, and in the following references [1–4, 7]

- **Cytochrome c (pro-apoptosis):** Cytochrome c is released from the mitochondria, and it promotes apoptosis. Its function in apoptosis is completely different from its other function in the electron transport chain.
- **AIF (apoptosis-inducing factor):** AIF also plays a role in inducing apoptosis if it is released from the mitochondria.
- **SMAC/Diablo** and **Omi (also pro-apoptosis):** These proteins can bind **XIAP** (see below) and prevent **XIAP** from inhibiting apoptosis.

1.2.4 Some Important *Cytoplasmic* Proteins Besides the Caspases

- **APAF1 (apoptotic protease activating factor 1): APAF1** is **pro-apoptosis,** but it is inactive in the absence of cytochrome c. It requires **cytochrome c** release into the cytoplasm to be active.
- IAP (XIAP): Inhibitor of apoptosis (pro-survival)**: XIAP (X-linked inhibitor of apoptosis protein): XIAP is pro-survival. It is a member of the inhibitor of apoptosis rotein (IAP) family and** has a role in **keeping caspases inactive** during cell survival states. In mammals, XIAP can bind to **caspase-9, caspase-3, and caspase-7** and **inactivate** them by marking them for destruction.

1.3 The Process of Apoptosis

1.3.1 Intrinsic and Extrinsic Pathways

As discussed above, two main pathways lead to apoptosis: the intrinsic and extrinsic pathways. The intrinsic pathway, also sometimes called the mitochondrial pathway, involves a response to stimuli, such as UV irradiation, which do not involve receptor activation. The extrinsic pathway involves a ligand binding to cell surface receptors. The two pathways converge at the executioner caspase step. Both are discussed below.

1.3.2 Intrinsic Pathway

- Overview of the Intrinsic Pathway

Most death by apoptosis in vertebrates occurs via the intrinsic pathway. The intrinsic pathway involves the mitochondria. Important components of this apoptotic pathway are contained in the mitochondria, and they only cause apoptosis if they are released from the mitochondria (specifically from the intermembrane space) into the cytoplasm. When they are released, apoptosis can occur. Signals that induce the intrinsic pathway may include stimuli like UV light and the DNA damage it causes. The intrinsic pathway can also be activated by other stimuli such as oxidative stress, hypoxia, ER stress, loss of growth factors, and others [1–4, 7].

• The Roles of the Bcl-2 Family Members in the Intrinsic Pathway

The **Bcl-2** family of proteins [11] plays key roles in triggering the intrinsic pathway because they control the permeabilization of the mitochondrial outer membrane (MOM). Specifically, they control "pores" that allow some contents to be released from the mitochondria. The process by which some Bcl-2 family members permeabilize the mitochondrial membrane is called mitochondrial outer membrane permeabilization (MOMP). MOMP is a fundamental concept in apoptosis (Fig. 22.4). The Bcl proteins have different roles in this process, as follows.

Bax and **Bak** are **pro-apoptotic** proteins that promote **MOMP** by creating pores in the mitochondrial outer membrane (MOM). Upon the cell receiving an apoptotic signal such as UV light, Bax and Bak become activated and translocate to the cytoplasmic side of the MOM. There, they aggregate into oligomers. These oligomers insert into the MOM, forming pores (MOMP). The formation of pores allows certain contents to flow out, such as cytochrome c, which starts the process of apoptosis and will be discussed below.

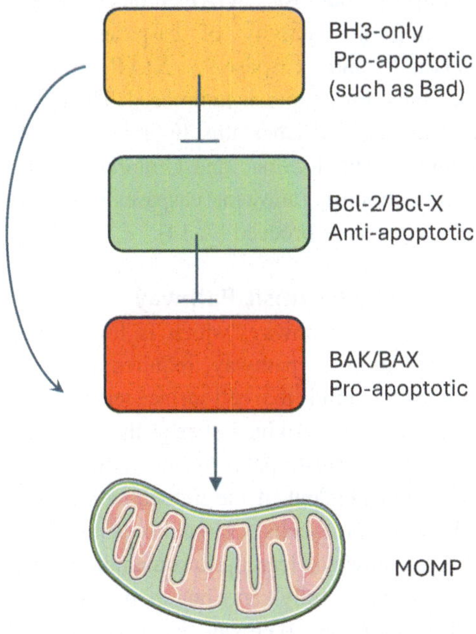

Fig. 22.4 BCL proteins. BH3, BCL-2/Bcl-X, and Bak/Bax all have important roles in the intrinsic pathway by controlling MOMP. Bak/Bax promote apoptosis. BCL-2/BCL-X are anti-apoptotic proteins that inhibit Bak/Bax. BH3 only proteins can promote apoptosis by inhibiting BCL-2/BCL-X or activating Bak/Bax. The the text for more details. (*Source*. Image of Mitochondria adapted from Servier Medical Art (https://smart.servier.com/), licensed under the Creative Commons Attribution 4.0 (CC BY 4.0) license)

Bcl-2 and **Bcl-XL** are also **Bcl-2 family members**, but these are **anti-apoptotic** proteins. These proteins bind the BH3 domains of **Bak** and **Bax**, preventing their oligomerization and, in turn, preventing pore formation. Thus, they **prevent MOMP**.

The **BH3-only proteins** (such as **Bad**, **Bim**, **Bid**, **Puma**, and **Noxa**) are **pro-apoptotic**. They also contribute to **promoting MOMP** when they are produced in response to apoptotic stimuli. These proteins contribute to MOMP in at least two ways. Some, like **Bad**, inhibit some anti-apoptotic **Bcl-2** family proteins, such as **Bcl-2**, by binding to them via their BH3 domains. Thus, they block the anti-apoptotic activity of the **Bcl-2 family** proteins. Doing so allows the pro-apoptotic Bcl-2 family members, like **Bax** or **Bak**, to oligomerize and trigger **MOMP**, as described above. Another way some BH3-only proteins, such as Bim and Bid, function is by **directly** binding to and **activating Bax** and **Bak**, thus directly stimulating their ability to trigger MOMP.

• Regulation of Expression of Bcl-2 Genes

The amounts of the different Bcl-2 proteins in the cell can be regulated at the gene expression level in response to apoptotic stimuli. For example, DNA damage can regulate the expression of some Bcl-2 genes. If DNA becomes damaged in a way that it cannot be repaired, this will lead to the accumulation of the transcription factor p53. The **p53** accumulation in the nucleus will activate transcription of the **BH3-only genes (Puma and Noxa)**. The products of these genes then encourage **MOMP**, leading to apoptosis and eliminating the cells. Mutations in DNA will also cause cells to stop producing **pro-survival proteins** like **Bcl2**, and instead, cells will express the pro-apoptosis proteins [2, 4, 9, 10, 12].

- The Steps That Occur After MOMP

Once the pores are open, as described above, **cytochrome c, AIF, Diablo (Smac), and Omi** are released from mitochondria into the cytosol [6]. Cytochrome c is especially important for promoting apoptosis. Once it is in the cytoplasm, **cytochrome c** binds to **APAF-1** (a cytoplasmic protein) in the presence of **ATP** or **dATP**. ATP/dATP serves as a cofactor that promotes a conformational change in APAF-1, enabling it to oligomerize into a heptameric structure. This recruits **procaspase-9**, and the resulting complex is called an **apoptosome**. Within the apoptosome, the **procaspase-9** proteins come into proximity with each other and subsequently dimerize within the complex. Procaspase-9 has a low level of protease activity even before it is cleaved. Within the dimers, the **procaspase-9** proteins cleave their binding partners, fully activating them. Thus, as a result, the **apoptosome** leads to the activation of procaspase-9, which now becomes **active caspase-9**. **Caspase-9** is the first caspase activated in this pathway and is thus considered the **initiator caspase**. Once active, **caspase-9** will activate an **executioner caspase,** usually **caspase-3**. It does so by cleaving **procaspase-3** in a process called proteolytic trimming. This is a gradual cleavage process that can involve sequential cleavage events, resulting in an active **caspase-3** [1–4, 7, 13]. The role of cytochrome c is essential here because it triggers apoptosome formation by binding to APAF-1 and facilitating its oligomerization in the presence of ATP/dATP (Fig. 22.1).

The major outcome of the intrinsic pathway is that the **initiator caspase** leads to activation of the **executioner caspase**, usually **caspase-3**. Interestingly, the extrinsic pathway also leads to activation of **caspase-3**, and thus, the two pathways converge upon **caspase-3** activation. The mechanisms by which caspase-3 can lead to apoptosis will be discussed after describing the first part of the extrinsic pathway below.

- AIF and Chromatin Degradation

Besides **caspase-3 activation**, another crucial function of the intrinsic pathway is the release of **AIF**. This is because one characteristic of apoptosis is chromatin degradation. **AIF**, released from the mitochondria, has an important role in this. AIF translocates to the nucleus and induces chromosome condensation and fragmentation, together with endonucleases. The resulting fragments and nucleotides are packaged in apoptotic bodies. Apoptotic bodies do not remain part of the cell, and they are eventually engulfed by phagocytic cells [6, 7].

- Release of Diablo/Smac and Omi-HtrA2

Other proteins released from the mitochondria include **Diablo/Smac** and **Omi/HtrA2**. These proteins can **inhibit XIAP** [7]. **XIAP** is a protein that has an important role in protecting any inappropriate activation of apoptosis. Cells always have some inactive procaspases, and it is important that they do not get activated at the wrong time, as this will trigger an irreversible process leading to apoptosis. **XIAP** protects cells by inhibiting the activity of caspases such as caspase-3 and caspase-9. **XIAP** must be inactivated for apoptosis to proceed. The intrinsic pathway accomplishes this by releasing XIAP inhibitors (Diablo/Smac and Omni/HtrA2) from the mitochondria, allowing caspases to be active and apoptosis to proceed [7, 14].

1.3.3 The Extrinsic Pathway

The extrinsic pathway, which is less common than the intrinsic pathway, requires an external stimulus that binds to a cell surface receptor. Such stimuli include cytokines, such as the **Fas ligand**. Unlike the intrinsic pathway, the extrinsic pathway is independent of the mitochondria. Nevertheless, in some cells, crosstalk occurs between the two pathways, leading to some mitochondrial involvement to fully carry out apoptosis.

The extrinsic pathway is activated when ligands bind to receptors. The types of receptors involved in this pathway are sometimes called "death receptors." They are single-membrane-spanning transmembrane receptors with an extracellular ligand-binding domain and an intracellular "death domain." The receptors form heterotrimers after binding to ligand. Two important examples of such receptors are the Fas receptor and the TNFα receptor. The ligands that bind to these receptors are members of the TNF family. They are trimers, and they trimerize the receptors [2, 7, 15, 16].

- The Fas Receptor
  – Fas Receptor Signaling

The Fas receptor (Fas, CD95) is a good example of a TNF family receptor that can lead to apoptosis when activated by ligand binding (see Figs. 22.1 and 22.5). Fas signaling has several functions, including regulating the numbers of B and T cells. The primary form of the Fas ligand (FasL or CD95L) is a nonsoluble form of the ligand. It is a transmembrane protein bound to the surface of cells, often on activated T cells. The binding of FasL to the receptors causes the receptors to trimerize [13]. The Fas receptor spans the membrane once. The binding site for the Fas ligand is on the extracellular side of the receptor. On the cytoplasmic tail of the receptor is the **death domain (DD)**. Once bound to its ligand (FasL), the trimerized Fas receptor recruits the **adapter** protein **FADD** (Fas-associated death domain) via interactions with the DDs on both FADD and the receptor. FADD also contains a death effector domain (DED). The DED on FADD recruits multiple **initiator caspases** (usually **procaspase-8**), which also have DEDs, through DED-DED interactions. These components assemble at the receptor to form the multiprotein **death-inducing signaling complex (DISC)** at the membrane. This allows the dimerization of procaspase-8. Dimerization, in turn, leads to cleavage of the dimer partners, resulting in active caspase-8 [17, 18]. (While caspase-9 is generally the initiator caspase for the intrinsic pathway, caspase-8 is generally the initiator caspase for the extrinsic pathway) [2, 7].

Caspase-8, an initiator caspase, is now active. **Caspase-8** now activates **procaspase-3** by cleaving it into its active form of **caspase-3**. This is a

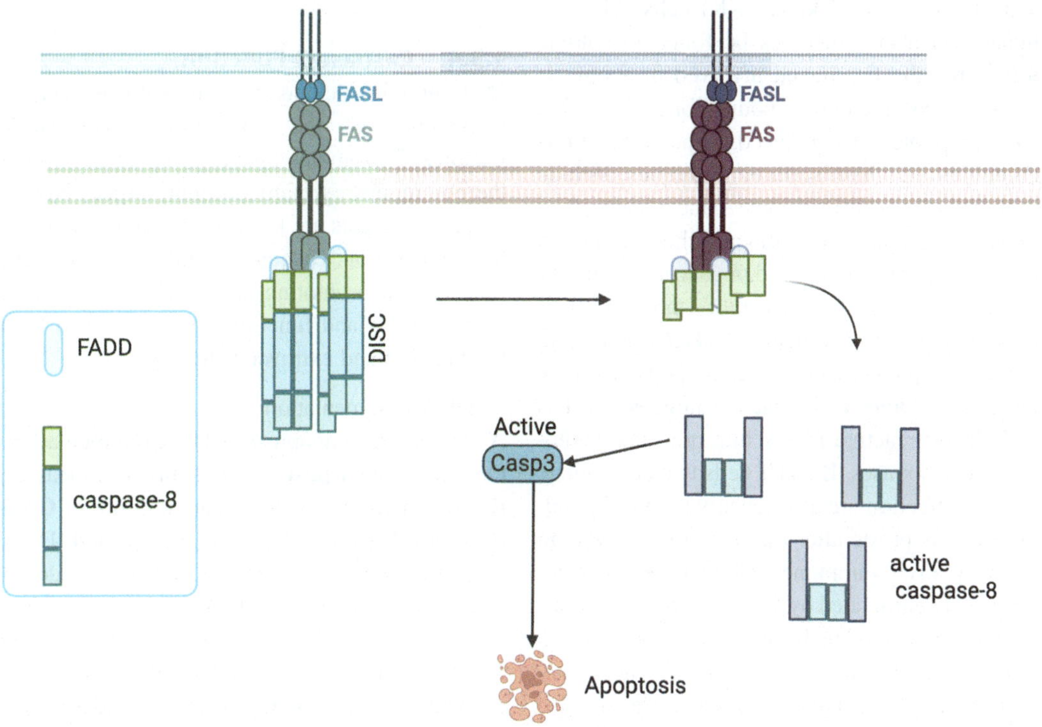

Fig. 22.5 The extrinsic pathway. The extrinsic pathway, shown here mediated by Fas and FasL. Caspase-8 binds to trimerized FAS, leading to the activation of caspase-8, which in turn activates caspase-3, ultimately resulting in apoptosis. (*Source*: Created with BioRender.com)

rapid activation step. **Caspase-3** functions as an **executioner caspase** by activating pathways leading to cell death (discussed more below). Caspase-3 can also amplify the apoptotic signal by cleaving and activating additional molecules of caspase-3. It can also cleave and activate other caspases, such as caspase-6 and caspase-7. In this way, although caspase-3 is an executioner caspase, it can also act as an **initiator caspase**, thus serving a dual role.

The Fas pathway is also subject to negative regulation. Some cells have inhibitory proteins that counteract the apoptosis pathway. For example, **FLIP** is a protein that is similar in structure to caspase-8 but does not have protease activity. It can dimerize with caspase-8 in the DISC on the Fas receptor and prevent it from activating apoptosis [18].

– Some Biological Functions of Fas Signaling

FasL (CD95L) is a **membrane-bound ligand** found on the surface of killer T lymphocytes and also on natural killer (NK) cells. The Fas ligand can also sometimes be found in soluble forms, but the Fas ligand involved in apoptosis is usually the membrane-bound form. Cells that are susceptible to Fas-induced apoptosis have **Fas receptors** on their surface. Examples include virus-infected cells or some cancer cells. The Fas-induced apoptosis pathways allow T cells to target and eliminate infected or cancerous cells by binding to the Fas receptor through their membrane-bound Fas ligands, thereby activating the cellular pathway that leads to apoptosis in the target cell. Other cells that are targeted by Fas include self-reactive T cells, thus preventing autoimmune responses. B and T cells that accumulate after an infection are also sensitive to Fas ligand-induced apoptosis after the infection is over. In this way, excess immune cells that are no longer needed are eliminated by Fas-mediated apoptosis. Overall, the Fas pathway helps maintain an immune cell balance, where only the necessary immune cells survive. The fas pathway is also important for immune-privileged cells. This is important for certain tissues like the eye, where Fas helps protect them from immune-mediated damage [2, 13, 19–22].

• TNFα Signaling

The TNFα receptor 1 (TNFR1), binding to tumor necrosis factor alpha (TNFα), can also lead to activation of apoptosis via the extrinsic pathway. The TNFα pathway is complex because it can lead to both cell survival and apoptosis, depending on the cell context. Under certain conditions, the activated (and trimerized) TNFα receptor activates the NF-κB pathway, and this pathway favors cell proliferation. In this case, the activated TNFα receptor initiates the formation of membrane-bound **complex-1**, which includes TRADD, TRAF2, and RIPK1. This complex activates the NFKB pathway, which has roles in cell survival and inflammation. However, under conditions where cell survival signals are blocked or insufficient, a cytosolic **complex II** can form. In this case, TRADD recruits FADD and the initiator **caspase-8**. This leads to activation of caspase-8, which then leads to activation of executioner **caspase-3** and apoptosis (Fig. 22.6) [15, 23].

1.3.4 Caspase-3 Function

Both the intrinsic and extrinsic pathways lead to activation of the executioner **caspase-3**. **Caspase-3** promotes apoptosis by cleaving and thereby regulating many different targets. Not all caspase-3 targets are known, but below, some of the targets of **caspase-3** and their relevant functions in promoting apoptosis are discussed. These targets and functions are relevant to both the intrinsic and extrinsic pathways [1].

• DNA Fragmentation

A key feature of apoptosis is DNA fragmentation. One way that **caspase-3** causes this is by cleaving **iCAD**. The **iCAD** protein is an inhibitor of **CAD** (Caspase-Activated DNase), also called DNA fragmentation factor (DFF). CAD is an endonuclease that fragments DNA. In non-apoptotic cells, iCAD binds to CAD, and this inhibits CAD's ability to cleave DNA. When iCAD is cleaved by **caspase-3**, however, it dissociates from CAD. Now, CAD can translocate to the nucleus, where it can bind and cleave DNA between the nucleosomes. DNA fragmentation is an integral part of the apoptotic process and

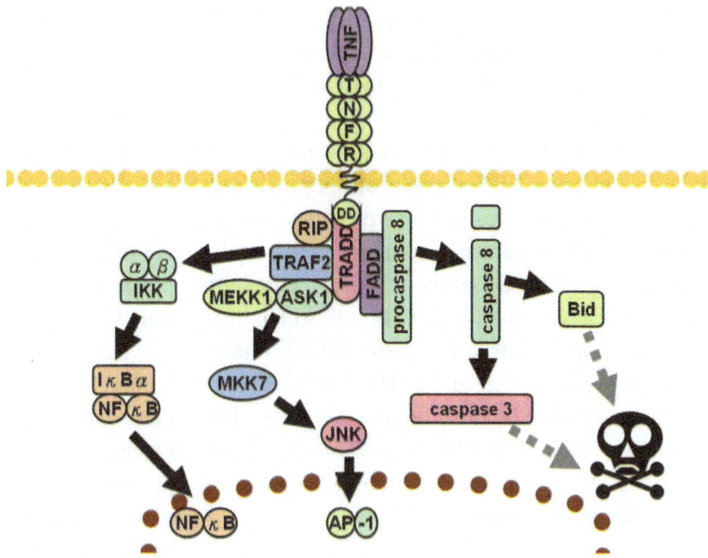

Fig. 22.6 TNF signaling. TNF signaling can lead to activation of multiple different pathways, both pro- and anti-apoptotic, depending on the cell context. See text for more details. (*Source*: Image by Subclavian, public domain, via Wikimedia Commons (https://commons.wikimedia.org/wiki/File:TNF_signaling.jpg))

serves as a good marker that scientists can visualize to determine whether cells are undergoing apoptosis. When scientists visualize the DNA fragments on a polyacrylamide gel, they see a "ladder" pattern because of the DNA fragmentation [24, 25].

Another widely used technique for assessing DNA fragmentation is called the TUNEL (terminal deoxynucleotidyl transferase dUTP Nick End Labeling) assay. In this assay, the free 3' OH ends of the DNA breaks are labeled with a fluorescent label so that they can be visualized in the lab. The DNA can be visualized by microscopy or flow cytometry, as the labeled DNA of apoptotic cells fluoresces more brightly than that of living cells due to the increased number of DNA breakpoints. This method allows scientists to detect the percentage of apoptotic cells in cultured cells or in tissue [24, 26].

- PARP Cleavage

PARP (Poly(ADP-ribose) polymerase) is another target that is cleaved and inactivated by executioner caspases such as **caspase-3** and **caspase-7**. PARP has roles in DNA repair and other vital cellular functions. Cleavage and subsequent inactivation of PARP by caspases is an important step in apoptosis because its inactivation inhibits DNA repair and other cellular processes, such as the regulation of morphological changes. Scientists also use PARP as a marker for apoptosis, and researchers often analyze the presence of cleaved PARP versus uncleaved PARP to assess apoptosis levels in cells [7, 27, 28].

- Flippase and Scramblase

Flippase and scramblase regulate the distribution of phospholipids on the plasma membrane. Apoptotic cells do not burst. Instead, the cell's contents remain membrane-bound, and everything is phagocytosed, usually by macrophages. The phagocytosis of the cells is triggered by membrane changes. Specifically, negatively charged phosphatidylserine (PS) has an important role. Normally, PS and phosphatidylethanolamine (PE) are mostly in the inner part of the phospholipid bilayer. They are kept there by an enzyme called **phospholipid flippase**. In apoptotic cells, however, they accumulate on the outside layer. This is because **caspase-3** cleaves and inactivates the flippase. **Caspase-3** also cleaves and **activates** an enzyme called **phospholipid scramblase**. The normal role of the **scramblase** is to transfer phospholipids nonspecifically between the two layers, but it has a net effect of shuffling PS outward. By inactivating the flippase and activating the scramblase, the executioner caspase can cause PS to concentrate at the outer membrane. The location of PS at the outer surface of the membrane is important because bridging proteins can link PS on the outer surfaces of cells to the phagocytic cells. Cells also have other markers ("eat me" signals) on their surface that encourage phagocytosis [29].

The accumulation of PS on the outer membrane is an important outcome of apoptosis and subsequent phagocytosis. It also provides another good tool for recognizing apoptotic cells in the lab. This can be done by staining cells with Annexin V conjugated with FITC, as Annexin V is a protein that binds to PS on the outer leaflet of the plasma membrane, and FITC is a fluorescent tag that allows visualization of the cells. The amount of staining by Annexin V in a group of cells can indicate to scientists the percentage of cells undergoing apoptosis [7].

- Some Other Caspase-3 Targets

Caspase-3 can cleave and inactivate **nuclear lamins**. This causes the breakdown of the nuclear lamina, which leads to nuclear envelope breakdown, one of the characteristics of apoptotic cells. **Caspase-3** also cleaves some **cytoskeleton regulators**, and this can contribute to the blebbing that occurs in apoptotic cells [30, 31].

1.3.5 Crosstalk Between Extrinsic and Intrinsic Pathways

In some cells, the extrinsic pathway requires the intrinsic pathway to complete the apoptosis process. This crosstalk is mediated largely by the protein **Bid** (one of the BH-only domain proteins). In non-apoptotic cells, **Bid** is inactive. However, when death receptors (such as the Fas receptor or TNFR) activate the **extrinsic pathway**, **caspase-8** (the **initiator caspase** activated in this pathway) cleaves **Bid**, thus activating it. In its active form, it is called **tBID**. Once active, **tBid** activates **Bax** and **Bak**. Now, **Bax** and **Bak** can oligomerize and activate **MOMP**. **tBid** also acts by blocking the **Bcl-2** proteins, thus preventing them from blocking **Bax** and **Bak**. This contributes to mitochondrial membrane permeabilization. In this way, **caspase-8**, activated by the extrinsic pathway, can also activate the intrinsic pathway [32, 33] (see Fig. 22.1).

1.3.6 Extracellular Signals That Regulate Cell Survival or Apoptosis

Certain extracellular signals can help determine whether cells undergo apoptosis or survive. Here, we covered signals that induce apoptosis, such as Fas and, in some cases, TNF. Survival factors can instead activate pathways that suppress apoptosis. An example of a pathway that can suppress apoptosis is any pathway that leads to AKT activation. Growth factors, for example, can bind to cell surface receptors such as Receptor Tyrosine Kinases (RTKs) and lead to activation of **AKT**. (The AKT pathway is discussed in Chap. 18). Once activated, AKT, a protein kinase, can phosphorylate **Bad** on serine residues. This phosphorylation causes **Bad** to associate with a protein called **14-3-3**. This interaction sequesters **Bad**, preventing it from binding and inhibiting **Bcl-2/Bcl-XL**, which are pro-survival proteins. Thanks to AKT activation and Bad phosphorylation, **Bcl-2/Bcl-XL** are now free to inhibit cell death [2, 34].

1.3.7 How Apoptotic Cells Are Identified in the Lab

Several techniques are available to identify apoptotic cells in culture in the lab or to determine what percentage of a group of cells is undergoing apoptosis. As discussed above, using Annexin V-FITC to stain the PS on the outer membranes of apoptotic cells is an important method for identifying them. The stained cells are usually subject to flow cytometry, which is a method for visualizing fluorescence in cells in suspension. Using staining and flow cytometry, it is possible to assess the percentage of cells in a population undergoing apoptosis. Usually in these experiments, the cells are stained not only with Annexin-V-FITC but also with PI (propidium iodide). PI will permeate cells in which membranes are damaged and bind to DNA. Thus, by staining with both, researchers can assess the percentage of cells that are viable (negative for both PI and Annexin-FITC), necrotic (a process to be discussed in the next section:

positive for PI but negative for FITC), early apoptotic (positive for Annexin-FITC but negative for PI), or late apoptotic (positive for both Annexin-FITC and PI). PI is useful for assessing late-stage apoptosis in cultured cells, even though loss of membrane integrity does not always occur in apoptotic cells in vivo. In the body, apoptotic cells are phagocytosed before any real membrane loss can be seen. However, in cultured cells, in late-stage apoptosis, some membrane integrity is eventually lost [7, 35].

Another mechanism for visualizing apoptotic cells, such as those in tissues, is to use microscopy, which can reveal phagocytes that have engulfed apoptotic remnants. Microscopy can also be used to visualize cytochrome c release, which occurs in apoptotic cells [36].

As discussed above, PARP is cleaved in apoptotic cells but not in live cells. Therefore, SDS gel electrophoresis and western blot analysis can be carried out to visualize cleaved and uncleaved PARP. Apoptotic cells are expected to have cleaved PARP. Finally, apoptosis is associated with the degradation of DNA. DNA in the cells can be analyzed by agarose gel electrophoresis. If the DNA is degraded, a smear of bands will be visualized on the gel rather than discrete bands [7, 37].

1.3.8 Apoptosis and Disease

- Abnormalities in Apoptosis Can Lead to Disease

Abnormalities in apoptosis can be associated with disorders such as cancer, neurodegenerative diseases, autoimmune diseases, ischemic injury, viral infections, and developmental disorders.

In cancer, resistance to apoptosis is a hallmark of the disease, allowing the cells to evade death and proliferate uncontrollably. Therapies that target apoptosis pathways are important strategies for treating the disease.

Too much apoptosis can also cause disorders. Neurodegenerative diseases like Alzheimer's disease and Parkinson's disease are associated with loss of neurons that occurs in part due to apoptosis. Conditions such as strokes can involve apoptosis of cells that lack oxygen or nutrients. Viruses can manipulate apoptotic pathways and, as a result, can enhance viral replication or evade immune responses. Certain developmental disorders are also associated with defects in apoptosis. An example is **syndactyly**, which presents as webbed fingers or toes. This is due to failure of apoptosis in the developing limb buds [38]. Neural tube defects, in contrast to too much apoptosis, may be caused by disruption of apoptosis during neural tube closure during development [1, 39].

- Drugs Designed to Restore Apoptosis in Cancer Cells

Certain drugs are designed in an effort to restore apoptosis in cancer cells. One example is a family of drugs called **BH3 mimetics** that mimic the function of **BH3-only proteins** to induce apoptosis in cancer cells, such as those found in leukemia and lymphoma.

An example of a BH3 mimetic is **Navitoclax**, which is an investigational drug being evaluated in clinical trials for the treatment of several types of **cancer**. As a **BH3 mimetic**, it has functional similarity to BH3-only proteins, although it does not mimic the structure of these proteins. It is a small-molecule drug designed to fit into the binding grooves of anti-apoptotic/pro-survival proteins like BCL-2, BCL-XL, and BCL-W (also a pro-survival protein, though not covered above). This displaces the pro-apoptotic proteins like BIM, which are otherwise sequestered by these anti-apoptotic proteins. Consequently, **Navitoclax** can promote apoptosis [40].

Another example of a BH3 mimetic is **Venclexta (venetoclax)** [41]. Venetoclax also binds to BCL-2, but it differs from **Navitoclax**. While **Navitoclax** binds to BCL-2, BCL-XL, and BCL-W, **venetoclax** is selective for BCL-2. Thus, although it has side effects, it may pose less risk due to its selectivity [42].

Venclexta (venetoclax) is FDA-approved (2016) for the treatment of some cancers (**CLL and AML**), and **Navitoclax** is still under investigation.

Dulanermin is a different kind of drug that is not a BH3 mimetic and is still under investigation for the treatment of several types of **cancer**.

Dulanermin activates the extrinsic apoptosis pathway rather than the intrinsic pathway. It is a biologic, consisting of a recombinant form of a ligand called the TNF-related apoptosis-inducing ligand (TRAIL). TRAIL binds to two receptors, DR4 (TRAIL-R1) and DR5 (TRAIL-R2), which are found on the surfaces of some cancer cells. The binding of ligand to these receptors leads to the recruitment of FADD, activation of caspase-8, and cell death. However, tumors often have ways to resist apoptosis. Clinical efficiency of the drug is limited, therefore, as it has been difficult to overcome the problem of resistance [43–45].

2 Necrosis and Other Forms of Cell Death

2.1 Necrosis

Some cells die by necrosis rather than apoptosis. In apoptosis, the cell initiates a programmed sequence of events leading to its own death. This is a tightly regulated process with specific purposes, such as removing damaged or unwanted cells, maintaining a balance of different cells, eliminating harmful cells, and regulating the development of various organs and tissues. Apoptosis is orderly and does not lead to inflammation or damage to surrounding cells.

In contrast to apoptosis, necrosis is a form of cell death characterized by cell swelling, uncontrolled cell lysis, and inflammation. It often occurs in response to significant injury, toxins, ischemia (lack of blood supply), infection, or, in some cases, autoimmune reactions. Necrosis is not a programmed process, and the cell does not initiate its own death. Necrotic cells release their cellular contents into the surrounding tissue, which can trigger an inflammatory response. Sometimes, the body tries to repair damaged tissue containing necrotic cells by releasing growth factors that promote cell proliferation and tissue repair [3, 4, 46].

Necrosis is not an orderly process, but there are some features that many necrotic cells share. Plasma membrane damage, for example, is a feature that differentiates apoptosis from necrosis. While the plasma membrane remains intact during apoptosis, it may become damaged in necrosis, allowing contents to leak out and cause damage to surrounding tissues. Organelles may become disrupted during necrosis. If the lysosome is damaged, its contents, including lysosomal enzymes, may leak out into the cytoplasm, destroying proteins and organelles. The mitochondria tend to swell and eventually become inactivated. This leads to a deficiency in ATP production, which in turn inhibits the activity of many enzymes and cellular processes. Many proteins, including enzymes, are denatured and degraded in cells undergoing necrosis. An inflammatory response, which does not result from apoptosis, is a common reaction to necrotic cells, allowing immune cells to attack the damaged cells and tissues [1, 47, 48].

2.2 Other Forms of Cell Death

While this chapter focuses on apoptosis and necrosis as major forms of cell death, it should be briefly mentioned that other forms can also occur. These are described here briefly.

Pyroptosis is a highly inflammatory type of programmed cell death. It generally occurs in immune cells in response to infection. **Caspase-1** is a key mediator of this type of cell death, and it is activated by inflammasomes, which are protein complexes in immune cells that can sense pathogenic signals. Activation of caspase-1 ultimately leads to the formation of pores in the membrane, resulting in membrane swelling, rupture, and the release of pro-inflammatory cytokines. The types of cells that are subject to this type of cell death are mostly immune cells [49].

Necroptosis is a necrosis-like cell death, but it is regulated. It occurs when apoptosis is blocked, particularly downstream of death receptors like **TNFR1**. When caspase-8 is absent or dysfunctional, **RIPK1** interacts with **RIPK3** and activates it; RIPK3 then phosphorylates **MLKL**. Phosphorylated MLKL then translocates to the membrane and forms pores, causing it to rupture and leading to inflammation. The type of cells that die by necroptosis are largely immune cells, but necroptosis can also occur in epithelial cells and neurons under some conditions [50].

Another specialized form of cell death is **ferroptosis**. This is an iron-dependent cell death associated with lipid peroxidation, which leads to oxidative damage to membranes. It can affect renal tubule cells, some neurons and hepatocytes, and some cancer cells [51].

Some cells can also die by **autophagic cell death**, where excessive autophagy leads to cell death [52]. Autophagy will be discussed in the next section.

3 Autophagy

3.1 Introduction to Autophagy

Autophagy is included in this section even though it does not necessarily lead to cell death and can, in fact, be important for cell survival. Autophagy is characterized by the internal degradation of cellular components, which can include proteins and organelles. These cellular components are destroyed by being packaged in a vesicle that fuses with the lysosome, which contains enzymes capable of digesting their contents. Autophagy can result in cell survival because cellular components can be recycled for raw material and energy, but under some conditions, it can also lead to cell death. The three major types of autophagy are microautophagy, macroautophagy, and chaperone-mediated autophagy. In this chapter, we will focus on **macroautophagy** (Fig. 22.7) and provide a brief introduction of the other two types. Macroautophagy is often referred to simply as autophagy and will be referred to that way here [4, 53–55].

Signals that lead to autophagy can include nutrient deprivation as well as other cellular stress signals, and sometimes autophagy can be triggered by signals that are involved in cell differentiation. In autophagy, proteins or whole organelles can be destroyed. These may be proteins or organelles that are nonfunctional or no longer needed, and they can become a source of energy or raw materials such as amino acids. Cellular components, proteins, or organelles to be destroyed are packaged in double-membraned vesicles called **autophagosomes (APs)**. The autophagosomes fuse with lysosomes, resulting in a structure called an **autophagolysosome** (also called **an autolysosome) (AL)**. Lysosomes contain hydrolytic enzymes that can break down many types of biomolecules. Therefore, the formation of the autophagolysosome leads to the destruction of the proteins or organelles engulfed in it (see Fig. 22.7) [4, 53, 54, 56].

An important trigger of autophagy is nutrient deprivation. Under conditions of low nutrients, cells may require alternative energy sources. One way to increase energy in the cell is to recycle cellular components and break them down to yield energy and raw material for the cell to use. There are a number of steps in autophagy, and several different gene products are involved.

3.2 Some of the Proteins Involved in Autophagy

There are a number of different proteins involved in autophagy. Some of the major ones are outlined here, and their functions are discussed in more detail in the next sections and also covered in [53, 54, 56]:

- The **ULK1 complex** (called the ATG1 complex in yeast.): This complex consists of the proteins ULK1, ATG13, FIP200, and ATG101. This complex is regulated by nutrient signals via mTORC1 and AMPK.
- The **nucleation complex**: This complex consists of class III PI3K, Beclin 1, ATG14L, and VPS15.
- The **ATG12-5-16L1 complex:** This complex consists of ATG12, ATG5, and ATG16L1.
- **LC3 processing and conjugation machinery**. This involves LC3, ATG7, ATG3, and ATG4.

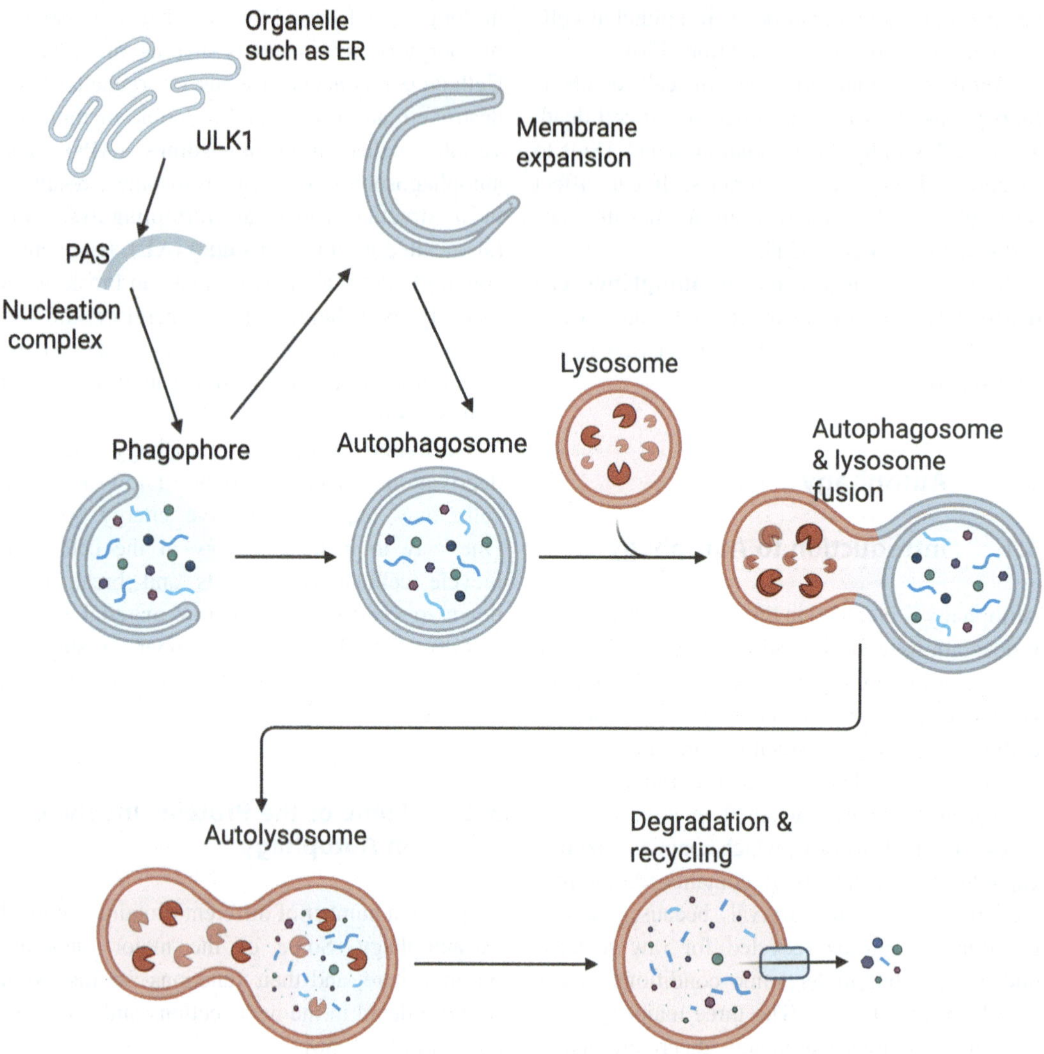

Fig. 22.7 Macroautophagy. Autophagy is a process in which cellular components, including proteins and organelles, are packaged into vesicles and subsequently destroyed. See text for details. (*Source*: Created with BioRender.com)

- **The autophagosome fusion, maturation, and degradation machinery**: This complex is involved in autosomephagolysosome fusion. It consists of SNARE proteins, LAMP1/2, Rab7, and HOPS.

Many of the proteins in this process are called **ATG** proteins. This stands for **autophagy-related gene** protein. These ATG proteins were identified in yeast when scientists were studying genes essential for autophagy [57, 58], but the names continued to be used in mammals.

3.3 Basic Steps in Autophagy (Macroautophagy) and Formation of the Autophagolysosome

The main goal in the steps described below is to create a type of membrane-bound vesicle called the **autophagosome (AP)**. The source of the membrane material for this vesicle is other organelles, usually the cell's ER. This AP fuses with a lysosome at which point it will be called an **autophagolysosome (AL)** (also called an autolysosome). At this point, the lysosomal enzymes digest the cellular components of the vesicle. The different steps involved in creating the autophagosome, followed by the autophagolysosome, are outlined below and discussed in [4, 53, 54].

3.3.1 Initiation of Autophagy: Formation of the ULK1 Complex and PAS

The initiation of autophagy is triggered by cellular stress, especially nutrient deprivation. This stress inhibits mTORC1, and this allows assembly and activation of a complex called **ULK1**. The ULK1 complex consists of several proteins, including ULK1 (a serine/threonine kinase), ATG13, FIP200, and ATG101. This complex is called an **initiation complex**. It phosphorylates substrates that allow the first steps in forming the autophagosome to begin. One of the main functions of **ULK1** is to **recruit membranes from organelles**, especially the ER, to form a structure called the **pre-autophagosomal structure (PAS)**. This is a small piece of membrane that is not enclosed into a vesicle. It is the starting site for developing the autophagosome; the starting steps for forming the next part of the structure are called the nucleation steps (Fig. 22.7).

3.3.2 Formation of the Nucleation Complex and the Phagophore

The **nucleation complex** is a complex that is formed on the PAS before the elongation step. It contains **PI3K** (class III) and several other proteins, including **Beclin1, VPS15, and ATG14L**. This complex is activated downstream of the ULK1 complex, and it is recruited to the **PAS**. This complex facilitates the next steps, in which the **PAS** is extended by the recruitment of lipids and proteins, making a larger segment of membrane. This larger structure is called the **phagophore**, also sometimes called the **isolation membrane**. It is a cup-shaped, unclosed membrane (Fig. 22.7).

The process above involves **class III PI3K.** PI3K is a lipid kinase. There are three classes of PI3K (classes I, II, and III). In Chap. 18, class I is discussed. In autophagy, the important PI3K is class III. The catalytic subunit of class III PI3K is also called **VPS34**. VPS34 primarily phosphorylates phosphatidylinositol (PI) in the membrane to produce phosphatidylinositol 3-phosphate (PI3P). Formation of PI3P is an important step because it recruits membrane-binding proteins that help shape and stabilize the phagophore membrane [59, 60].

3.3.3 Recruitment of the ATG12-5-16L1 Complex

During elongation, the membrane will expand and begin to capture cargo. During the early part of the elongation phase, however, the major change is the addition of multiple protein complexes onto the phagophore. The main complex is the **ATG12-ATG5-ATG16L1 complex.** This complex recruits a number of other proteins to prepare the membrane for the next step.

3.3.4 LC3 Processing and Lipidation

Now, a protein called **LC3 (microtubule-associated protein 1 light chain 3)** plays a central role in the membrane expansion and cargo recognition process. **LC3** is first synthesized as a precursor. It is then cleaved by the cysteine protease **ATG4**, resulting in **LC3-I**, which is cytosolic. LC3-I, through several steps, is conjugated to **phosphatidylethanolamine (PE)**, generating **LC3-II**. This is called the **lipidated form of LC3.** This lipidated LC3 stably associates with the membrane of the autophagosome and is involved in membrane remodeling and recruitment of cargo adaptor proteins [53, 54].

3.3.5 Elongation of the Phagophore and Formation of the Autophagosome

After LC3 lipidation the phagophore undergoes expansion and elongation. This involves the addition of new membrane material from other organelles. This expansion continues until the membrane fully encloses the cargo. This step involves further elongating the phagophore until it fuses to form a vesicle. This enclosed vesicle is the **autophagosome**. This is a key step because cellular components destined for destruction get trapped in this autophagosome structure.

What gets trapped in the autophagosome is a selective process. There are several mechanisms by which cellular components are selected. For example, cellular components can be ubiquitinated by a type of ubiquitin that is recognized by proteins on the inside of the autophagosome [61].

3.3.6 Fusion to Form the Autophagolysosome

The next step is fusion of the **autophagosome (AP) with** the **lysosome** to make the **autophagolysosome**. This is the final destruction stage because the lysosomal enzymes destroy the cellular components. In cases where the cell survives, these components can be broken down into their basic parts and reused by the cell [53].

This phase is also highly regulated. Rab7 is a small GTPase that helps guide the mature autophagosome to the lysosome. Once the vesicle is close to the membrane, the HOPS (homotypic fusion and protein sorting) complex brings the vesicle and the lysosome together. SNARE (soluble NSF attachment protein receptor) proteins then lead to the actual membrane fusion by forming complexes that bridge the two membranes. LAMP1 and LAMP2 (lysosome-associated membrane proteins) are also involved in the fusion process [62].

3.4 Outcomes of Autophagy

Autophagy is usually associated with cell survival. This is because the cellular components can be recycled and used for cells that are otherwise low on nutrients and raw material. Too much autophagy, however, can lead to cell death. This may occur, for example, when too many mitochondria are destroyed, leading to insufficient ATP energy. Cell death can also occur when there are mutations in some of the ATG genes, causing autophagy to occur incorrectly. This can lead to the accumulation of unhealthy cells that, and in some cases, this can lead to cell death [63].

3.5 Regulation of Autophagy

Autophagy occurs in response to a specific stimulus. A key trigger for autophagy is nutrient deprivation. The process is mainly regulated at the initiation step, at the level of ULK1. Specifically, it can be activated by AMP-activated protein kinase (AMPK), a kinase that is sensitive to increased AMP levels. This is important because AMP levels increase during nutrient deprivation (**low nutrient** levels), where ATP levels decrease relative to AMP levels. In this way, nutrient deprivation can start the entire autophagy process by activating **ULK1**. AMPK functions by inhibiting mTOR. In contrast to AMPK, mTOR is a protein that can put a brake on ULK1. mTOR is a nutrient-sensitive protein kinase that is activated in response to high nutrient levels. mTOR is a kinase that can phosphorylate and inhibit ULK1 in response to these high nutrient levels, so that autophagy does not occur when nutrients are plentiful. mTOR, however, can be indirectly inhibited by AMP Kinase under low nutrient conditions so that the process of autophagy can proceed. Autophagy can also be induced by hormonal stimuli like glucagon, pharmacological agents such as rapamycin and torin1, diseases including cancer, and genotoxic drugs like etoposide [64].

3.6 Other Forms of Autophagy

While the focus of this chapter is **macroautophagy**, two other types of autophagy, microautophagy and chaperone-mediated autophagy, are briefly introduced here.

3.6.1 Microautophagy

In this type of autophagy, cytosolic components are directly taken up by the lysosome itself. This is a result of invagination of the lysosome membrane. It may be selective or nonselective. The different types of microautophagy include constitutive, vesicle-mediated, and selective or nonselective targeting of proteins and organelles [54].

3.6.2 Chaperone-Mediated Autophagy

In this type of autophagy, targeted proteins are translocated across the lysosomal membrane in a complex with chaperone proteins (such as Hsc-70). These chaperone-protein complexes are recognized by the lysosomal membrane receptor, lysosome-associated membrane protein 2A (LAMP-2A). This results in the unfolding and degradation of the targeted proteins [54, 65].

3.7 Autophagy in Disease

Abnormalities in autophagy encompass various disorders, including Crohn's disease, neurodegenerative diseases like Huntington's, Parkinson's, and Alzheimer's, heart disease, age-related conditions, metabolic disorders, lysosomal disease, and cancer [54, 66]. Some disorders and diseases associated with autophagy are discussed below.

3.7.1 Parkinson's Disease

Parkinson's disease (PD) is associated with degeneration of dopamine neurons, and it is thought to be associated with **abnormalities in autophagy**. In particular, PD is thought to be linked to abnormalities in mitophagy, which refers specifically to the clearance of damaged mitochondria. In particular, the clearance of damaged mitochondria appears to be a problem in the disease, potentially leading to the death of dopamine neurons. The protein **Parkin** is an **E3 ligase** that plays an important role in targeting damaged mitochondria for degradation. Abnormalities in Parkin function or its recruitment to damaged mitochondria (often via the PINK1 pathway) may contribute to the accumulation of damaged mitochondria in the dopamine neurons, leading to their death. PINK1 is a kinase that accumulates on the outer mitochondrial membrane of damaged mitochondria and can, in turn, activate the activity of Parkin [67].

3.7.2 Huntington's Disease

Huntington's disease (HD) is a neurodegenerative disorder associated with a mutant Huntingtin protein (mHTT) that has an abnormally long polyglutamine sequence. This mutant protein forms toxic aggregates in neurons, leading to motor dysfunction and neurological symptoms. This **mutant Huntington protein (mHTT) interferes with normal recognition of cargo for autophagy**. This can lead to autophagosomes that do not function correctly, disrupting autophagic flux and causing the accumulation of toxic proteins. In contrast to the mutant HTT, the normal HTT helps facilitate autophagy and has a role in cargo recognition and trafficking [68–70].

Enhancing autophagy is a possible strategy for treating Huntington's disease. Felodipine (FELL-HD) is being tested in clinical trials for the treatment of HD due to its ability to upregulate autophagy [71].

3.7.3 Cancer

Autophagy can have tumor-suppressive effects by removing damaged organelles and protein aggregates that can lead to genomic instability and cancer. It can also regulate inflammation by removing cellular components that can trigger inflammatory responses, which are also associated with cancer. Autophagy also has other functions, such as suppressing oncogenic signaling pathways by degrading certain cell cycle regulatory proteins or signaling proteins. Some tumors are found to have lost the Beclin gene, which, as described above, codes for a protein that has an important role in autophagy. Furthermore, the loss of Beclin, ATG5, and ATG7, all important autophagy proteins, can lead to tumor susceptibility in mouse studies [72]. Autophagy in cancer is complex, however, because in advanced cancers, autophagy can have the opposing effect, where it can support tumor survival under stressful conditions such as hypoxia and lack of nutrients [5, 73].

3.7.4 Crohn's Disease

Some patients with Crohn's disease have a genetic variant in the **ATG16L1** gene, **T300A (a threonine to alanine substitution)**, which is associated with increased disease risk. ATG16L1, as discussed above, has an important role in the autophagy pathway. The T300A variant leads to a more unstable form of the ATG16L1 protein, and this reduces the efficiency of the autophagy process, especially in cells such as **Paneth cells**, which are specialized intestinal cells. These cells are important for gut barrier integrity and antimicrobial defense. The reduced efficiency of autophagy can lead to increased inflammation, abnormalities in microbial clearance, and problems with the intestinal barrier. These problems are common in Crohn's disease, suggesting that the ATG16L1(T300A) mutation may be one of the factors linked to the disease [74].

3.7.5 Lysosomal Storage Diseases

Lysosomal storage diseases (LSDs) are associated with impaired autophagy because of the important roles lysosomes have in degrading the cargo of the autophagosomes. When lysosomes are not functioning properly, as in LSDs, autophagic flux is disrupted, and this leads to the accumulation of undegraded material (autophagic flux refers to the entire process of autophagy, starting with the formation of the autophagosome to the degradation and recycling of its contents) [75]. In lysosomal storage diseases, undegraded material accumulates in the lysosomes. Most lysosomal storage diseases result from deficiencies of single lysosomal enzymes. **Gaucher's disease** is the most common lysosomal storage disease. A mutation in the gene for (Glucocerebrosid**ase**, GCase) is associated with **Gaucher's disease** type I. This gene encodes a lysosomal enzyme required for the breakdown of certain glycolipids. The disease is associated with easy bruising due to low platelets, enlarged spleen and liver, anemia, and bone damage. The deficiency of **GCase** leads to an accumulation of glucocerebroside, a glycolipid, in tissues including the liver. This impairs lysosome function and makes it so they cannot break down autophagocytic vesicles. This leads to toxicity in the cells [76].

Pompe disease is also a lysosomal storage disease. In Pompe disease, the gene for **alpha-glucosidase (GAA)** is mutated. GAA is necessary for the breakdown of glycogen in the lysosome (though other enzymes are necessary for its breakdown in the cytoplasm). It leads to muscle weakness and wasting and is fatal if not treated. One way to treat the disease is by enzyme replacement therapy, by injecting genetically engineered GAA, which can alleviate some symptoms. **Nexviazyme (avalglucosidase alfa-ngpt)** [77] is a **biologic** that consists of a **synthetic GAA**. It was FDA-approved for the treatment of **Pompe disease** in 2021. The combination of **Pombiliti (cipaglucosidase alfa-atga)** [78] and **Opfolda (miglustat)** [79] was approved for adults with **late-onset Pompe disease (LOPD)** in 2023. **Pombiliti** is a synthetic replacement form of **GAA**, and **Opfolda stabilizes Pombiliti** within the bloodstream so that it works more effectively [80–82].

Fabry disease is also a rare lysosomal disease. It is an inherited disease caused by a mutation in the gene for **alpha galactosidase (GLA)**, an enzyme that breaks down sphingolipids. This leads to the accumulation of **globotriaosylceramide (Gb3 or GL-3)** within lysosomes of the cells of some tissues. It is treated by enzyme replacement therapy. **Elfabrio (pegunigalsidase alfa-iwxj)** [83] is a **biologic**, which is a synthetic form of GLA (FDA-approved in 2023) that is used for the treatment of **Fabry disease** [84, 85].

3.7.6 Chloroquine Inhibits Autophagy and is Used for Disease Treatment and Research

Chloroquine (CQ) is a medicine first used to treat malaria [86] during WWII, but in the late twentieth century, researchers recognized that the drug also inhibits autophagy. This was discovered in cultured cells where GQ was found to lead to lysosomal swelling, impaired degradation activity, and increased pH (alkalinization) of lysosomes. The altered pH disrupted the function of lysosomal enzymes, which require an acidic environment to function properly. This disruption in lysosomal activity prevents the degradation of autophagosomes, leading to the accumulation of autophagic vesicles that contain undigested cellular components [87]. These findings led researchers to explore whether CQ may be effective at treating other disorders, including those that are associated with autophagy. CQ has now been widely used in experimental models to study autophagy [88] and has been investigated as a treatment for cancer [89] and neurodegenerative diseases [90].

CQ is under investigation for the treatment of several diseases. For example, although autophagy can have a tumor-suppressive role, some cancer cells also depend on autophagy to survive under stressful conditions, such as nutrient deprivation and hypoxia. CQ and its derivative, hydroxychloroquine, are being investigated in clinical trials as adjunct therapies in **cancer** treatment with the goal of inhibiting autophagy in the cancer cells [91]. Its potential use is complicated, however, by the fact that autophagy can have both cancer survival properties and anticancer properties [86].

CQ is also being used as a **tool to study neurodegenerative diseases** in relation to autophagy. For example, in diseases like Alzheimer's disease, autophagy can be dysregulated. Since CQ can inhibit autophagy, it can be used experimentally in animal studies and cell culture studies to understand how interfering with autophagy can affect disease progression or the accumulation of cellular markers of the disease [86, 92, 93].

CQ has also been explored for its **antiviral properties**, especially during the COVID-19 pandemic. Some viruses exploit the cell's autophagy machinery for their own benefit and replication. By interfering with autophagy, CQ may disrupt the viral life cycle, but results have been inconsistent and inconclusive [86, 92, 93].

Definitions

Apoptotic protease activating factor 1 (APAF-1) Apoptotic protease activating factor 1. It is pro-apoptosis, but it is inactive in the absence of cytochrome c. It requires cytochrome c release into the cytoplasm to be active.

Apoptosis A form of programmed cell death involving regulated cellular disassembly without provoking inflammation.

Apoptosome A protein complex formed by cytochrome c, APAF1, and dATP/ATP that activates caspase-9 in the intrinsic apoptotic pathway.

Autophagosome A double-membraned vesicle that engulfs cytoplasmic components.

Autophagy A catabolic process involving the degradation of cellular components via lysosomes, often for survival, but can also contribute to cell death.

Bax Pro-apoptotic protein that promotes cytochrome c release.

BCL-2 family A family of proteins that regulate mitochondrial membrane permeability and apoptosis, including pro- and anti-apoptotic members.

Beclin-1 A key autophagy regulator initiating vesicle nucleation.

BH3-only proteins A subclass of pro-apoptotic proteins within the Bcl-2 family that contain only the BH3 domain. Examples include Bid, Bad, Bim, Puma, and Noxa.

Caspases Cysteine-aspartic proteases involved in apoptosis, classified as initiator (e.g., caspase-8, caspase-9) or executioner (e.g., caspase-3, caspase-7).

FDA-Approved Drugs Discussed in This Chapter

Commonly used brand name	Generic name	First DA approval year	Major conditions	Main targets
Elfabrio	*Pegunigalsidase alfa-iwxj*	2023	Fabry disease	GLA
Nexviazyme	*Avalglucosidase alfa-ngpt*	2021	Pompe disease	GAA
Opfolda	*Miglustat*	2023	Late-onset Pompe disease	GAA stabilizer
Pombiliti	*Cipaglucosidase alfa-atga*	2023	Late-onset Pompe disease	GAA
Venclexta	*Venetoclax*	2016	CLL, AML	BCL-2

Cytochrome C A mitochondrial protein that activates caspase-9 when released.

FAS A receptor that initiates apoptosis via the extrinsic pathway.

FASL Ligand for Fas receptor, induces apoptosis.

Flow cytometry A technique used to analyze the characteristics of individual cells or particles as they flow in a fluid stream through a laser. It can be carried out along with fluorescent labeling of cells.

LC3 A protein that plays a central role in the membrane expansion and cargo recognition process during autophagy.

Lysosomal storage diseases Genetic disorders characterized by deficient lysosomal enzymes leading to accumulation of undegraded substances in lysosomes. This can disrupt lysosomal function and disrupt the cell.

Mitochondrial outer membrane permeabilization (MOMP) A key event in intrinsic apoptosis where proteins like Bax/Bak create pores in the outer mitochondrial membrane.

Necrosis A form of unregulated cell death due to injury, often involving membrane rupture and inflammation.

Poly(adp-ribose) polymerase (PARP) An enzyme involved in DNA repair. During apoptosis it is cleaved by caspase-3 to prevent DNA repair and promote cell death.

Phosphatidylserine A phospholipid commonly found mostly on the inner membrane leaflet, but found in the outer membrane leaflet during apoptosis.

tBID Truncated BID; a pro-apoptotic fragment generated when caspase-8 cleaves BID, allowing crosstalk between extrinsic and intrinsic apoptotic pathways.

Tunel assay Technique for detecting DNA fragmentation in apoptosis.

Zymogen An inactive precursor of an enzyme, which requires a biochemical change (often cleavage) to become active.

Study Questions

Multiple-Choice Questions

1. Felodipine is being tested as a treatment for Huntington's disease because it:
   A. Replaces mutant huntingtin with normal protein
   B. Inhibits mitochondrial fission
   C. Upregulates autophagy and may enhance clearance of toxic proteins
   D. Directly blocks dopamine neuron degeneration
2. The release of cytochrome c from mitochondria is a hallmark of which apoptotic pathway?
   A. Extrinsic
   B. Necroptotic
   C. Intrinsic
   D. Pyroptotic
3. Which death receptor initiates the extrinsic apoptotic pathway?
   A. Fas
   B. Caspase-9
   C. TNF-α
   D. Cytochrome c
4. Which process is typically characterized by membrane swelling and rupture?
   A. Apoptosis
   B. Autophagy
   C. Necrosis
   D. Senescence
5. Which caspase is considered an initiator in the intrinsic apoptotic pathway?
   A. Caspase-3
   B. Caspase-6
   C. Caspase-8
   D. Caspase-9
6. Which molecule is often used to detect apoptotic cells due to its exposure on the outer membrane?
   A. DNAse
   B. Phosphatidylserine
   C. Cytochrome c
   D. Bcl-xL

7. BH3-only proteins (e.g., Bid, Bim, Puma) promote apoptosis by:
   A. Directly cleaving caspase substrates
   B. Forming pores in the plasma membrane
   C. Inhibiting anti-apoptotic Bcl-2 family members and/or activating Bax/Bak
   D. Phosphorylating survival kinases
8. Which caspase is the primary executioner caspase in both intrinsic and extrinsic apoptotic pathways?
   A. Caspase-8
   B. Caspase-3
   C. Caspase-9
   D. Caspase-1
9. Which of the following is *not* a characteristic feature of apoptotic cells?
   A. Chromatin condensation
   B. Proteolytic cleavage of substrates
   C. Formation of membrane blebs
   D. Inflammatory response
10. Which "death receptor" adapter protein recruits and activates caspase-8 upon Fas ligand binding?
   A. TRADD
   B. FADD
   C. MyD88
   D. RIPK1

True/False Questions

11. Autophagy always leads to cell death.
12. Necrosis is a tightly regulated, noninflammatory form of cell death.
13. Macroautophagy under severe nutrient deprivation can sometimes lead to cell death rather than survival.

Fill-in-the-Blank Questions

14. Caspase-8 is primarily involved in the ______ apoptotic pathway.
15. Cytochrome c binds to ______ to form the apoptosome.
16. **Autophagy** is a cellular process in which damaged organelles and other cellular components are enclosed within a double-membraned structure called the __________, which later fuses with lysosomes for degradation and recycling.
17. Caspase-3 cleaves ______, a DNA repair enzyme, during apoptosis.
18. ______ is a marker for early apoptosis due to its externalization due to the enzyme scramblase.

Short-Answer Question

19. Explain how the intrinsic pathway of apoptosis is activated.

Answer Key

Multiple-Choice Questions

1. **Answer**: C
2. **Answer**: C
3. **Answer**: A
4. **Answer**: C
5. **Answer**: D
6. **Answer**: B
7. **Answer**: C
8. **Answer**: B
9. **Answer**: D
10. **Answer**: B

True/False Questions

11. **Answer**: False – Autophagy can be cytoprotective or cytotoxic depending on context.
12. **Answer**: False
13. **Answer**: True

Fill in the Blank Questions

14. **Answer**: Extrinsic
15. **Answer**: Apaf-1
16. **Answer**: Autophagosome

17. **Answer**: PARP
18. **Answer**: Phosphatidylserine

Short Answer Question

19. **Answer**: The intrinsic pathway is triggered by internal stress signals like DNA damage, leading to mitochondrial outer membrane permeabilization (MOMP) and cytochrome c release. This forms the apoptosome with Apaf-1 and caspase-9, initiating apoptosis.

References

1. Elmore S. Apoptosis: a review of programmed cell death. Toxicol Pathol. 2007;35(4):495–516. https://doi.org/10.1080/01926230701320337.
2. Marks F, Klingmüller U, Müller-Decker K. Cellular signal processing: an introduction to the molecular mechanisms of signal transduction. 2nd ed. Garland Science; 2017.
3. D'Arcy MS. Cell death: a review of the major forms of apoptosis, necrosis and autophagy. Cell Biol Int. 2019;43(6):582–92. https://doi.org/10.1002/cbin.11137.
4. Alberts B, Heald R, Johnson A, Morgan D, Raff M, Roberts K, Walter P, Wilson J, Hunt T. Molecular biology of the cell. W.W. Norton and Company; 2022.
5. Weinberg RA. The biology of cancer. 3rd ed. W. W. Norton and Company; 2023.
6. Mustafa M, Ahmad R, Tantry IQ, Ahmad W, Siddiqui S, Alam M, et al. Apoptosis: a comprehensive overview of signaling pathways, morphological changes, and physiological significance and therapeutic implications. Cells. 2024;13(22):1838. https://doi.org/10.3390/cells13221838.
7. Bose K. In: Bose K, editor. Proteases in apoptosis: pathways, protocols and translational advances. 1st ed. Springer; 2015.
8. Dho SH, Cho M, Woo W, Jeong S, Kim LK. Caspases as master regulators of programmed cell death: apoptosis, pyroptosis and beyond. Exp Mol Med. 2025;57 (6):1121–32. https://doi.org/10.1038/s12276-025-01470-9.
9. Kaloni D, Diepstraten ST, Strasser A, Kelly GL. BCL-2 protein family: attractive targets for cancer therapy. Apoptosis. 2023;28(1–2):20–38. https://doi.org/10.1007/s10495-022-01780-7.
10. Czabotar PE, Garcia-Saez AJ. Mechanisms of BCL-2 family proteins in mitochondrial apoptosis. Nat Rev Mol Cell Biol. 2023;24(10):732–48. https://doi.org/10.1038/s41580-023-00629-4.
11. Vogler M, Braun Y, Smith VM, Westhoff MA, Pereira RS, Pieper NM, et al. The BCL2 family: from apoptosis mechanisms to new advances in targeted therapy. Signal Transduct Target Ther. 2025;10(1):91. https://doi.org/10.1038/s41392-025-02176-0.
12. Chipuk JE, Bouchier-Hayes L, Green DR. Mitochondrial outer membrane permeabilization during apoptosis: the innocent bystander scenario. Cell Death Differ. 2006;13(8):1396–402. https://doi.org/10.1038/sj.cdd.4401963.
13. Murphy K, Weaver C, Berg LJ. Janeway's immunobiology. 10th ed. New York: W.W. Norton & Company; 2022.
14. Obexer P, Ausserlechner MJ. X-linked inhibitor of apoptosis protein – a critical death resistance regulator and therapeutic target for personalized cancer therapy. Front Oncol. 2014;4:197. https://doi.org/10.3389/fonc.2014.00197.
15. Gough P, Myles IA. Tumor necrosis Factor receptors: pleiotropic signaling complexes and their differential effects. Front Immunol. 2020;11:585880. https://doi.org/10.3389/fimmu.2020.585880.
16. Idriss HT, Naismith JH. TNF alpha and the TNF receptor superfamily: structure-function relationship (s). Microsc Res Tech. 2000;50(3):184–95. https://doi.org/10.1002/1097-0029(20000801)50:3<184::Aid-jemt2>3.0.Co;2-h.
17. Roufayel R. Regulation of stressed-induced cell death by the Bcl-2 family of apoptotic proteins. Mol Membr Biol. 2016;33(6–8):89–99. https://doi.org/10.1080/09687688.2017.1400600.
18. Green DR. The death receptor pathway of apoptosis. Cold Spring Harb Perspect Biol. 2022;14(2) https://doi.org/10.1101/cshperspect.a041053.
19. Strasser A, Jost PJ, Nagata S. The many roles of FAS receptor signaling in the immune system. Immunity. 2009;30(2):180–92. https://doi.org/10.1016/j.immuni.2009.01.001.
20. Peter ME, Hadji A, Murmann AE, Brockway S, Putzbach W, Pattanayak A, et al. The role of CD95 and CD95 ligand in cancer. Cell Death Differ. 2015;22 (4):549–59. https://doi.org/10.1038/cdd.2015.3.
21. Rathmell JC, Goodnow CC. Autoimmunity: the Fas track. Curr Biol. 1995;5(11):1218–21. https://doi.org/10.1016/S0960-9822(95)00241-7.
22. Green DR, Ware CF. Fas-ligand: privilege and peril. Proc Natl Acad Sci USA. 1997;94(12):5986–90. https://doi.org/10.1073/pnas.94.12.5986.
23. Marks D. Marks' essentials of medical biochemistry: a clinical approach. Lippincott Williams & Wilkins (LWW); 2017.
24. Martinez M, Reif R, Pappas D. Detection of apoptosis: a review of conventional and novel techniques. Anal Methods. 2010;2 https://doi.org/10.1039/c0ay00247j.
25. Eskandari E, Eaves CJ. Paradoxical roles of caspase-3 in regulating cell survival, proliferation, and tumorigenesis. J Cell Biol. 2022;221(6) https://doi.org/10.1083/jcb.202201159.

26. Guo M, Lu B, Gan J, Wang S, Jiang X, Li H. Apoptosis detection: a purpose-dependent approach selection. Cell Cycle. 2021;20(11):1033–40. https://doi.org/10.1080/15384101.2021.1919830.
27. Los M, Mozoluk M, Ferrari D, Stepczynska A, Stroh C, Renz A, et al. Activation and caspase-mediated inhibition of PARP: a molecular switch between fibroblast necrosis and apoptosis in death receptor signaling. Mol Biol Cell. 2002;13(3):978–88. https://doi.org/10.1091/mbc.01-05-0272.
28. Wang Y, Luo W, Wang Y. PARP-1 and its associated nucleases in DNA damage response. DNA Repair (Amst). 2019;81:102651. https://doi.org/10.1016/j.dnarep.2019.102651.
29. Nagata S, Sakuragi T, Segawa K. Flippase and scramblase for phosphatidylserine exposure. Curr Opin Immunol. 2020;62:31–8. https://doi.org/10.1016/j.coi.2019.11.009.
30. Faleiro L, Lazebnik Y. Caspases disrupt the nuclear-cytoplasmic barrier. J Cell Biol. 2000;151(5):951–9. https://doi.org/10.1083/jcb.151.5.951.
31. Green DR. Caspases and their substrates. Cold Spring Harb Perspect Biol. 2022;14(3) https://doi.org/10.1101/cshperspect.a041012.
32. Jan R, Chaudhry GE. Understanding apoptosis and apoptotic pathways targeted cancer therapeutics. Adv Pharm Bull. 2019;9(2):205–18. https://doi.org/10.15171/apb.2019.024.
33. Westphal D, Kluck RM, Dewson G. Building blocks of the apoptotic pore: how Bax and Bak are activated and oligomerize during apoptosis. Cell Death Differ. 2014;21(2):196–205. https://doi.org/10.1038/cdd.2013.139.
34. Downward J. PI 3-kinase, Akt and cell survival. Semin Cell Dev Biol. 2004;15(2):177–82. https://doi.org/10.1016/j.semcdb.2004.01.002.
35. Hingorani R, Deng J, Elia J, McIntyre C, Mittar D. BD biosciences. Detection of apoptosis using the BD Annexin V FITC assay on the BD FACSVerse (TM) system. 2011. Available from: https://www.bdbiosciences.com/content/dam/bdb/marketing-documents/BD_FACSVerse_Apoptosis_Detection_AppNote.pdf.
36. Waterhouse NJ, Trapani JA. A new quantitative assay for cytochrome c release in apoptotic cells. Cell Death Differ. 2003;10(7):853–5. https://doi.org/10.1038/sj.cdd.4401263.
37. McLennan ABA, Elcock L, Turner P. BIOS instant notes in molecular biology. 4th ed. New York: Garland Science; 2012.
38. Malik S. Syndactyly: phenotypes, genetics and current classification. Eur J Hum Genet. 2012;20(8):817–24. https://doi.org/10.1038/ejhg.2012.14.
39. Favaloro B, Allocati N, Graziano V, Di Ilio C, De Laurenzi V. Role of apoptosis in disease. Aging (Albany NY). 2012;4(5):330–49. https://doi.org/10.18632/aging.100459.
40. Mohamad Anuar NN, Nor Hisam NS, Liew SL, Ugusman A. Clinical review: Navitoclax as a pro-apoptotic and anti-fibrotic agent. Front Pharmacol. 2020;11:564108. https://doi.org/10.3389/fphar.2020.564108.
41. U.S. Food and Drug Administration. Venclexta (Venetoclax) approval overview. 2016. Available from: https://www.accessdata.fda.gov/scripts/cder/daf/index.cfm?event=overview.process&ApplNo=208573.
42. Li Q, Cheng L, Shen K, Jin H, Li H, Cheng Y, et al. Efficacy and safety of Bcl-2 inhibitor Venetoclax in hematological malignancy: a systematic review and meta-analysis of clinical trials. Front Pharmacol. 2019;10:697. https://doi.org/10.3389/fphar.2019.00697.
43. Ashkenazi A, Fairbrother WJ, Leverson JD, Souers AJ. From basic apoptosis discoveries to advanced selective BCL-2 family inhibitors. Nat Rev Drug Discov. 2017;16(4):273–84. https://doi.org/10.1038/nrd.2016.253.
44. Lampson BL, Davids MS. The development and current use of BCL-2 inhibitors for the treatment of chronic lymphocytic leukemia. Curr Hematol Malig Rep. 2017;12(1):11–9. https://doi.org/10.1007/s11899-017-0359-0.
45. Ouyang X, Shi M, Jie F, Bai Y, Shen P, Yu Z, et al. Phase III study of dulanermin (recombinant human tumor necrosis factor-related apoptosis-inducing ligand/Apo2 ligand) combined with vinorelbine and cisplatin in patients with advanced non-small-cell lung cancer. Investig New Drugs. 2018;36(2):315–22. https://doi.org/10.1007/s10637-017-0536-y.
46. Park W, Wei S, Kim B-S, Kim B, Bae S-J, Chae YC, et al. Diversity and complexity of cell death: a historical review. Exp Mol Med. 2023;55(8):1573–94. https://doi.org/10.1038/s12276-023-01078-x.
47. Karch J, Molkentin JD. Regulated necrotic cell death: the passive aggressive side of Bax and Bak. Circ Res. 2015;116(11):1800–9. https://doi.org/10.1161/circresaha.116.305421.
48. Akadeum Life Sciences. Necrosis andapoptosis are two main types of cell death in the body. 2024. Available from: https://www.akadeum.com/blog/necrosis-vs-apoptosis-processes-necrotic-cell-death-apoptosis-steps/?cn-reloaded=1.
49. Broz P. Pyroptosis: molecular mechanisms and roles in disease. Cell Res. 2025;35(5):334–44. https://doi.org/10.1038/s41422-025-01107-6.
50. Tummers B, Green DR. Mechanisms of TNF-independent RIPK3-mediated cell death. Biochem J. 2022;479(19):2049–62. https://doi.org/10.1042/bcj20210724.
51. Liang D, Minikes AM, Jiang X. Ferroptosis at the intersection of lipid metabolism and cellular signaling. Mol Cell. 2022;82(12):2215–27. https://doi.org/10.1016/j.molcel.2022.03.022.
52. Denton D, Kumar S. Autophagy-dependent cell death. Cell Death Differ. 2019;26(4):605–16. https://doi.org/10.1038/s41418-018-0252-y.

53. Wong EE. Autophagy and signaling. Boca Raton: CTC Press; 2017.
54. Glick D, Barth S, Macleod KF. Autophagy: cellular and molecular mechanisms. J Pathol. 2010;221(1): 3–12. https://doi.org/10.1002/path.2697.
55. Klionsky DJ, Eskelinen EL, Deretic V. Autophagosomes, phagosomes, autolysosomes, phagolysosomes, autophagolysosomes... wait, I'm confused. Autophagy. 2014;10(4):549–51. https://doi.org/10.4161/auto.28448.
56. Boya P, Codogno P, Rodriguez-Muela N. Autophagy in stem cells: repair, remodelling and metabolic reprogramming. Development. 2018;145(4) https://doi.org/10.1242/dev.146506.
57. Ohsumi Y. Molecular mechanism of autophagy in yeast, Saccharomyces cerevisiae. Philos Trans R Soc Lond Ser B Biol Sci. 1999;354(1389):1577–80; discussion 80-1. https://doi.org/10.1098/rstb.1999.0501.
58. Tsukada M, Ohsumi Y. Isolation and characterization of autophagy-defective mutants of Saccharomyces cerevisiae. FEBS Lett. 1993;333(1–2):169–74. https://doi.org/10.1016/0014-5793(93)80398-e.
59. Li H, Wen X, Ren Y, Fan Z, Zhang J, He G, et al. Targeting PI3K family with small-molecule inhibitors in cancer therapy: current clinical status and future directions. Mol Cancer. 2024;23(1):164. https://doi.org/10.1186/s12943-024-02072-1.
60. Jean S, Kiger AA. Classes of phosphoinositide 3-kinases at a glance. J Cell Sci. 2014;127(5):923–8. https://doi.org/10.1242/jcs.093773.
61. Shaid S, Brandts CH, Serve H, Dikic I. Ubiquitination and selective autophagy. Cell Death Differ. 2013;20 (1):21–30. https://doi.org/10.1038/cdd.2012.72.
62. Lőrincz P, Juhász G. Autophagosome-lysosome fusion. J Mol Biol. 2020;432(8):2462–82. https://doi.org/10.1016/j.jmb.2019.10.028.
63. Liu S, Yao S, Yang H, Liu S, Wang Y. Autophagy: regulator of cell death. Cell Death Dis. 2023;14(10): 648. https://doi.org/10.1038/s41419-023-06154-8.
64. Tamargo-Gómez I, Mariño G. AMPK: regulation of metabolic dynamics in the context of autophagy. Int J Mol Sci. 2018;19(12):3812. https://doi.org/10.3390/ijms19123812.
65. Medical News. What is autophagy. Available from: https://www.news-medical.net/life-sciences/What-is-Autophagy.aspx?showform.
66. Levine B, Kroemer G. Autophagy in the pathogenesis of disease. Cell. 2008;132(1):27–42. https://doi.org/10.1016/j.cell.2007.12.018.
67. Pickrell AM, Youle RJ. The roles of PINK1, parkin, and mitochondrial fidelity in Parkinson's disease. Neuron. 2015;85(2):257–73. https://doi.org/10.1016/j.neuron.2014.12.007.
68. Sarkar S, Rubinsztein DC. Huntington's disease: degradation of mutant huntingtin by autophagy. FEBS J. 2008;275(17):4263–70. https://doi.org/10.1111/j.1742-4658.2008.06562.x.
69. Cortes CJ, La Spada AR. The many faces of autophagy dysfunction in Huntington's disease: from mechanism to therapy. Drug Discov Today. 2014;19(7):963–71. https://doi.org/10.1016/j.drudis.2014.02.014.
70. Ehrnhoefer D. Unraveling the Cause of Dysfunctional Autophagy in HD. 2015. Available from: https://huntingtonstudygroup.org/hd-insights/unraveling-the-cause-of-dysfunctional-autophagy-in-hd/#:~:text=By:%20Dagmar%20Ehrnhoefer%2C%20PhD,degradation%20can%20take%20place%20efficiently.
71. Andresen K, Cutting E, Apostolopoulos D, Evans AH, Oakley L, Dayimu A, et al. Trial to assess the tolerability of using felodipine to upregulate autophagy as a treatment of Huntington's disease (FELL-HD): a phase II, single-centre, open-label, dose-finding trial protocol. BMJ Open. 2024;14(8):e087983. https://doi.org/10.1136/bmjopen-2024-087983.
72. Rakesh R, PriyaDharshini LC, Sakthivel KM, Rasmi RR. Role and regulation of autophagy in cancer. Biochim Biophys Acta (BBA) – Mol Basis Dis. 2022;1868(7):166400. https://doi.org/10.1016/j.bbadis.2022.166400.
73. Yang ZJ, Chee CE, Huang S, Sinicrope FA. The role of autophagy in cancer: therapeutic implications. Mol Cancer Ther. 2011;10(9):1533–41. https://doi.org/10.1158/1535-7163.Mct-11-0047.
74. Salem M, Ammitzboell M, Nys K, Seidelin JB, Nielsen OH. ATG16L1: a multifunctional susceptibility factor in Crohn disease. Autophagy. 2015;11(4): 585–94. https://doi.org/10.1080/15548627.2015.1017187.
75. Platt FM, d'Azzo A, Davidson BL, Neufeld EF, Tifft CJ. Lysosomal storage diseases. Nat Rev Dis Primers. 2018;4(1):27. https://doi.org/10.1038/s41572-018-0025-4.
76. Stirnemann J, Belmatoug N, Camou F, Serratrice C, Froissart R, Caillaud C, et al. A review of Gaucher disease pathophysiology, clinical presentation and treatments. Int J Mol Sci. 2017;18(2) https://doi.org/10.3390/ijms18020441.
77. U.S. Food and Drug Administration. Nexviazyme (Avalglucosidase alfa-ngpt) approval overview. 2021. Available from: https://www.accessdata.fda.gov/scripts/cder/daf/index.cfm?event=overview.process&ApplNo=761194.
78. U.S. Food and Drug Administration. Pombiliti (Cipaglucosidase alfa-atga) approval overview. 2023. Available from: https://www.accessdata.fda.gov/scripts/cder/daf/index.cfm?event=overview.process&ApplNo=761204.
79. U.S. Food and Drug Administration. Opfolda (Miglustat) approval overview. 2023. Available from: https://www.accessdata.fda.gov/scripts/cder/daf/index.cfm?event=overview.process&ApplNo=215211.
80. Amicus Therapeutics. Possible + Opfolda. Available from: https://pombilitiopfolda.com/.
81. Kohler L, Puertollano R, Raben N. Pompe disease: from basic science to therapy. Neurotherapeutics. 2018;15(4):928–42. https://doi.org/10.1007/s13311-018-0655-y.

82. Dhillon S. Avalglucosidase alfa: first approval. Drugs. 2021;81(15):1803–9. https://doi.org/10.1007/s40265-021-01600-3.
83. U.S. Food and Drug Administration. Elfabrio (Pegunigalsidase alfa-iwxj) approval overview. 2023. Available from: https://www.accessdata.fda.gov/scripts/cder/daf/index.cfm?event=overview.process&ApplNo=761161.
84. Turkmen K, Baloglu I. Fabry disease: where are we now? Int Urol Nephrol. 2020;52(11):2113–22. https://doi.org/10.1007/s11255-020-02546-3.
85. ELFABRIO. For your ERT-experienced and treatment-naïve patients: the PEGylated ERT for Fabry. Available from: https://hcp.elfabrio.com/.
86. Zhou W, Wang H, Yang Y, Chen ZS, Zou C, Zhang J. Chloroquine against malaria, cancers and viral diseases. Drug Discov Today. 2020;25(11):2012–22. https://doi.org/10.1016/j.drudis.2020.09.010.
87. Fedele Anthony O, Proud CG. Chloroquine and bafilomycin A mimic lysosomal storage disorders and impair mTORC1 signalling. Biosci Rep. 2020;40(4) https://doi.org/10.1042/bsr20200905.
88. Mauthe M, Orhon I, Rocchi C, Zhou X, Luhr M, Hijlkema KJ, et al. Chloroquine inhibits autophagic flux by decreasing autophagosome-lysosome fusion. Autophagy. 2018;14(8):1435–55. https://doi.org/10.1080/15548627.2018.1474314.
89. Varisli L, Cen O, Vlahopoulos S. Dissecting pharmacological effects of chloroquine in cancer treatment: interference with inflammatory signaling pathways. Immunology. 2020;159(3):257–78. https://doi.org/10.1111/imm.13160.
90. Pedrioli G, Patani R, Paganetti P. Chloroquine, the coronavirus crisis, and neurodegeneration: a perspective. Front Neurol. 2020;11:596528. https://doi.org/10.3389/fneur.2020.596528.
91. Morgan MJ, Gamez G, Menke C, Hernandez A, Thorburn J, Gidan F, et al. Regulation of autophagy and chloroquine sensitivity by oncogenic RAS in vitro is context-dependent. Autophagy. 2014;10(10):1814–26. https://doi.org/10.4161/auto.32135.
92. Hamano T, Enomoto S, Shirafuji N, Ikawa M, Yamamura O, Yen SH, et al. Autophagy and tau protein. Int J Mol Sci. 2021;22(14) https://doi.org/10.3390/ijms22147475.
93. Nelson MP, Shacka JJ. Autophagy modulation in disease therapy: where do we stand? Curr Pathobiol Rep. 2013;1(4):239–45. https://doi.org/10.1007/s40139-013-0032-9.

23 DNA Proofreading, Damage, and Repair: Implications in Health and Disease

Abstract

DNA encodes all the traits in an organism, and these traits are passed down from generation to generation. Errors in DNA can occur occasionally. Errors in DNA caused by incorrect nucleotides are called mutations. While some errors are inevitable, DNA polymerase has proofreading mechanisms that prevent most errors during DNA replication. In addition to prevention, cells also have DNA repair mechanisms to correct most errors that occur despite proofreading. These proofreading and repair systems are highly effective; without them, there would be many more errors. Nevertheless, a low level of errors still occurs despite these safeguards. Some mutations are harmful and can lead to disease, developmental disorders, fertility problems, and other adverse outcomes, but not all mutations are harmful. In fact, some mutations are necessary for the survival of a species. A low level of mutations is associated with evolutionary changes that allow species to adapt to a changing environment and, therefore, to evolve and survive. This chapter will cover different types of mutations that can occur in DNA and their different possible causes, including specific mutagens and toxins. It will also cover the proofreading and repair mechanisms in the cell that keep mutation levels low. It discusses disorders that arise from DNA mutations or faulty repair mechanisms. In addition to small mutations in DNA, large mutations can occur, such as chromosomal translocations or an abnormal chromosome number. These can cause serious diseases such as cancer and are difficult to treat.

Keywords

DNA mutations · Proofreading · DNA repair · Single-stranded breaks · Double-stranded breaks · Nonhomologous end joining (NHEJ)

Abbreviations

BER	Base excision repair
NER	Nucleotide excision repair
MMR	Mismatch repair
HDR	Homology-directed repair
NHEJ	Nonhomologous end joining
RER	Ribonucleotide excision repair
DSB	Double-stranded break
SSB	Single-stranded break
AP site	Apurinic site
APE1	AP endonuclease 1
XPB/XPD	Xeroderma Pigmentosum group B/D (helicase subunits in TFIIH)
TFIIH	Transcription factor IIH
EXO1	Exonuclease 1
PCNA	Proliferating cell nuclear antigen
ROS	Reactive oxygen species
UV	Ultraviolet (radiation)

A. Minden, *Biochemistry, Cell and Molecular Biology for Pharmacists and Health Professionals*,
https://doi.org/10.1007/978-3-032-20529-2_23

PAH	Polycyclic aromatic hydrocarbons
DSBR	Double-stranded break repair (general category)
Ku70/Ku80	Ku autoantigen p70/p80 (binds DNA ends in NHEJ)
DNA-PKcs	DNA-dependent protein kinase catalytic subunit
XRCC4	X-ray repair cross complementing protein 4
Ligase IV	DNA ligase IV (for sealing ends in NHEJ)
ATM	Ataxia-telangiectasia mutated (kinase)
TCR	Transcription-coupled repair
GGR	Global genomic repair

What You Will Learn

- Different types of DNA mutations
- Germline mutations compared to somatic mutations, large-scale and small mutations
- Some causes of DNA mutations, including those that occur during DNA synthesis, those that are caused by mutagens, and those that occur spontaneously
- Different types of mutations including base substitutions, depurination, bulky adducts, and single- and double-stranded breaks
- Missense mutations, nonsense mutations, silent mutations, and frameshift mutations
- Proofreading mechanisms of DNA polymerase that can prevent many mutations
- DNA repair mechanisms including homology-directed repair (HDR), nonhomologous end joining (NHEJ), mismatch repair (MMR), base excision repair (BER), and others
- Disorders that are associated with faulty DNA repair mechanisms, including some forms of breast cancer, severe combined immune deficiency (SCID), Lynch syndrome, and Xeroderma pigmentosa (XP) and some treatment options for these disorders
- DNA damage sensors such as ATR and ATM and regulation of the cell cycle
- Large chromosomal aberrations and their roles in some cancers such as Burkitt lymphoma
- Abnormalities in chromosome number: aneuploidy and polyploidy

1 Introduction

DNA contains the blueprint for all the traits in an organism, and these traits are passed down from generation to generation. It is, therefore, vital that the DNA remains mostly error-free. Errors associated with incorrect nucleotides in DNA are called mutations. There are different times in the cell cycle in which mutations can occur. Some mutations can result from errors that occur in S phase during the DNA replication process, while other mutations occur after DNA replication, in other phases of the cell cycle. While some errors are inevitable, there are proofreading mechanisms in place to prevent most errors from occurring during DNA replication. In addition to prevention, cells also have DNA repair mechanisms to repair most errors that occur despite the proofreading mechanisms. These proofreading and repair systems are highly effective; without them, there would be many more errors. Nevertheless, a low level of errors manages to occur despite these safeguards. Some mutations are harmful and can lead to disease, developmental disorders, fertility problems, and other outcomes, but not all mutations are a problem. In fact, some mutations are necessary for the survival of a species. A low level of mutations is associated with evolutionary changes that allow species to adapt to a changing environment and, therefore, to evolve and survive [1–5]. This chapter will cover different types of mutations that can occur in DNA and their different possible causes. It will also cover the proofreading and repair mechanisms that are in place in the cell to keep the level of mutations low.

2 Germline Versus Somatic Mutations

2.1 Germline Mutations

Some mutations are germline mutations; these mutations are present in the fertilized egg, and they are hereditary. Once this egg develops into an individual, every cell in that individual's body will have these mutations. Germline mutations often consist of errors that occur during DNA replication, but they can also occur independently of DNA replication [6].

2.2 Somatic Mutations

Other mutations are somatic mutations, also called acquired mutations. These mutations occur in somatic cells and are acquired during an individual's life. They may occur during DNA replication, or they may occur independently of DNA replication in response to mutagens or environmental stress. In either case, these somatic mutations are not found in every cell in the body and are not passed down to future generations [6].

3 The Scale of Different Types of DNA Damage

3.1 Large-Scale Mutations

Mutations in DNA can be large or small. Large-scale mutations are those that impact entire chromosomes or large portions of chromosomes. They can include deletions or duplications of large segments of DNA, insertions of large pieces of DNA into the wrong part of a chromosome, and chromosomal inversions, where a segment of the DNA is inverted within a chromosome. They can include chromosomal translocations, where sequences belonging to one chromosome are translocated onto another chromosome. They can also include changes in the number of chromosomes, such as duplication or loss of entire chromosomes.

3.2 Small-Scale Mutations

Small-scale mutations include mutations in a few nucleotides or even just one nucleotide. Mutations in a single nucleotide are called point mutations. Point mutations are small but can sometimes have a large impact. They can lead to changes in the genetic code, frameshifts, or inserted stop codons. Small-scale mutations can have a range of biological consequences. Some mutations have no effect at all. These may be mutations that occur in noncoding regions, or they may be in coding regions where the encoded amino acids either do not change or do not change in a way that impacts the function of the encoded protein. In other cases, even a single-base pair change can have a major impact, where it can change the entire function of the encoded protein [2, 3, 7]. The small-scale mutations will be the main focus of this chapter.

4 Some Causes of DNA Mutations

4.1 Mutations Can Occur During DNA Synthesis (DNA Replication)

During DNA replication, some errors can occur as the new DNA strand is synthesized. These are often the result of an incorrect nucleotide being added to the growing, newly synthesized DNA chain (the daughter strand). This can result in an error called a **base substitution**. Other errors that can occur during DNA synthesis include incorrect **insertions** or **deletions** of nucleotides and errors caused by **replication fork stalling** [2].

4.2 Mutations Can Occur in Response to Mutagens

While mutations can occur due to the incorporation of incorrect nucleotides during DNA synthesis, errors can also occur after DNA synthesis. In these cases, mutagens are sometimes the cause. Different types of mutagens can cause errors in DNA. These generally fall into the categories of chemical, physical, or biological mutagens.

Chemical mutagens are mutagens that can physically interact with DNA and cause chemical modifications in the DNA. Some examples are as follows. **Alkylating agents** add alkyl groups to nucleotides. These are generally human-made and not found in nature. **Deaminating**

compounds, such as nitrous acid, remove amine groups from nucleotides. They can be found in some dietary sources, through environmental exposure, and are formed in the body. **Polycyclic aromatic hydrocarbons (PAHs)**, such as **benzo [a]pyrene**, are found in coal, cigarette smoke, and grilled meats. They lead to several types of mutations, such as base substitutions. **Hydroxylating agents**, such as aflatoxins, add hydroxyl groups to DNA and subsequently lead to mutations. They are produced by certain molds that grow on food, such as peanuts and grains. **Benzene** is a mutagen found in gasoline and industrial solvents. It can cause mutations due to the metabolites it produces, which can alter some nucleotides and DNA sequences. Certain **heavy metals** like arsenic and cadmium, which can come from sources such as cigarette smoke, contaminated food and water, and certain foods, and **base analogs**, which may be found in some drugs or laboratories, have structures similar to natural nucleotides and are also chemical mutagens. They can cause mutations by different mechanisms, such as by being incorporated into the DNA [8].

Physical agents interact with DNA and cause damage due to their interactions. They can, for example, cause breaks in the DNA or changes in the DNA's structure. They include **ionizing radiation** (X-rays, gamma rays, alpha particles) and **nonionizing radiation** such as UV radiation [8].

Biological mutagens come from biological sources. In some cases, they may mutate DNA indirectly as a response to cellular processes that they trigger in the cell. They can include some viruses, transposable elements, *Helicobacter pylori*, and some parasites [9].

4.3 Some Mutations Can Occur Spontaneously

Sometimes, mutations can occur even in quiescent cells where the DNA is not replicating, and mutagens are absent. These can be due to factors such as spontaneous DNA breaks, exposure to reactive oxygen species (ROS) within the cell, and spontaneous intracellular chemical reactions that lead to DNA modification [10].

5 Some Types of Mutations in DNA

Different types of mutations can occur in DNA, either during replication, in response to mutagens, or spontaneously. If not repaired, some of these mutations may have severe effects and may also lead to the incorporation of more mutations [4]. Some of the major types of mutations that can occur are described below.

5.1 Base Substitutions

Base substitutions occur when one nucleotide is substituted for another. These types of mutations can occur **spontaneously** during DNA replication, and they can also occur in response to **mutagens** such as **alkylating agents, base analogs, hydroxylating agents,** and **ionizing irradiation** [2, 3]. The two major examples of base substitutions are transition mutations and transverse mutations.

5.1.1 Transition Mutations

These are mutations where a purine (A or G) is replaced by another purine, or a pyrimidine (C or T) is replaced by another pyrimidine [4, 11]. One way this can occur is in response to **deaminating enzymes. Deaminating enzymes** remove amine (-NH_2) groups from the bases of nucleotides, most often from cytosine. In this hydrolytic reaction, a water molecule participates in replacing the amino group with a carbonyl oxygen, converting the cytosine to uracil. This will cause the complementary strand to incorporate an A instead of a G, thus leading to a transition mutation, and on the next round of replication, this A will pair with a T (Fig. 23.1).

5.1.2 Transversion Mutations

These are mutations in which a purine is replaced by a pyrimidine or a pyrimidine is replaced by a purine. One possible cause is **irradiation**, which can lead to the formation of free radicals. An example of such a free radical is the superoxide radical, an oxygen molecule with an unpaired electron (O_2^-). Nucleotide bases such as guanine

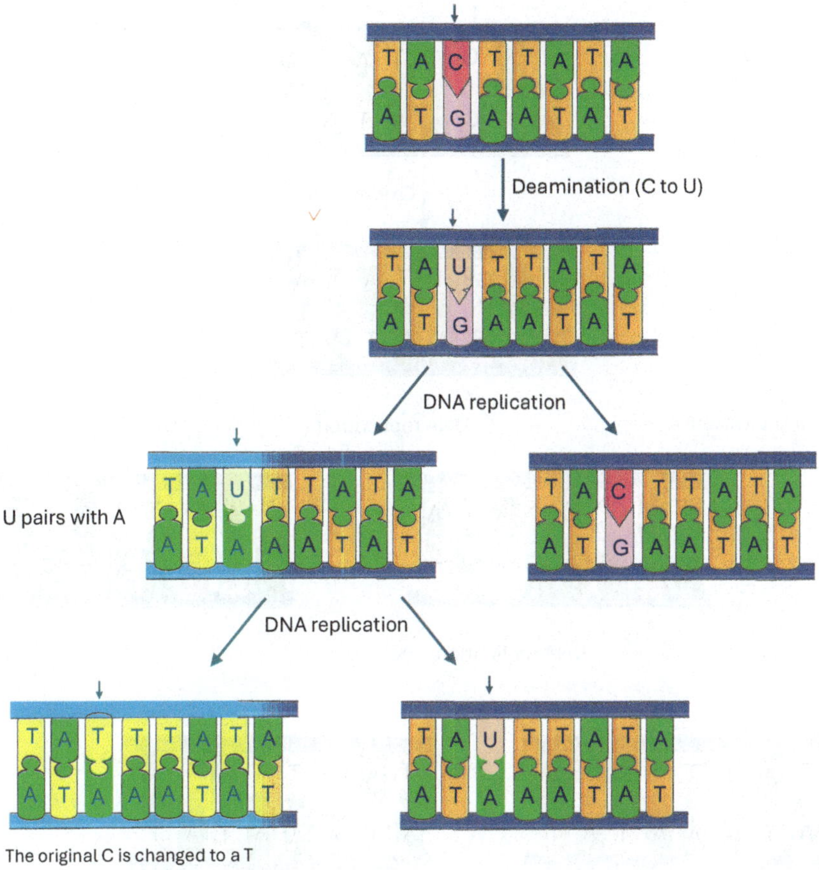

Fig. 23.1 Transition mutation. Transition mutations are mutations where a purine is replaced by another purine or a pyrimidine is replaced by another pyrimidine. This can be a response to deamination of a nucleotide. (*Source*: Image adapted from Servier Medical Art (https://smart.servier.com/), licensed under the Creative Commons Attribution 4.0 (CC BY 4.0) license)

can interact with this superoxide radical to form 8-OxyG (8-oxoguanine) (sometimes abbreviated as G^{O} in figures. 8-OxyG can mimic thymine in structure, and therefore it will base pair with A instead of C, causing a transversion mutation (Fig. 23.2), where, in the next round of replication, this A till pair with a T. Both transition mutations and transversion mutations can be broadly categorized as base substitutions [1, 12].

5.2 Depurination

Depurination involves the removal of certain bases from the nucleotide, leaving an apurinic (AP) site. This can result from the cleavage of the N-glycosidic bond between the base and the sugar-phosphate backbone, resulting in the loss of the purine base. Specifically, it involves the removal of guanine or adenine from the rest of the nucleotide. It can be caused by a **spontaneous event** during DNA replication, or after DNA replication, or by **mutagens** such as acid, or by heat [3] (Fig. 23.3).

5.3 Bulky Adducts

Bulky adducts refer to large modifications or attachments to DNA. These involve compounds that bind directly to DNA via covalent modification. They can cause structural distortions of the

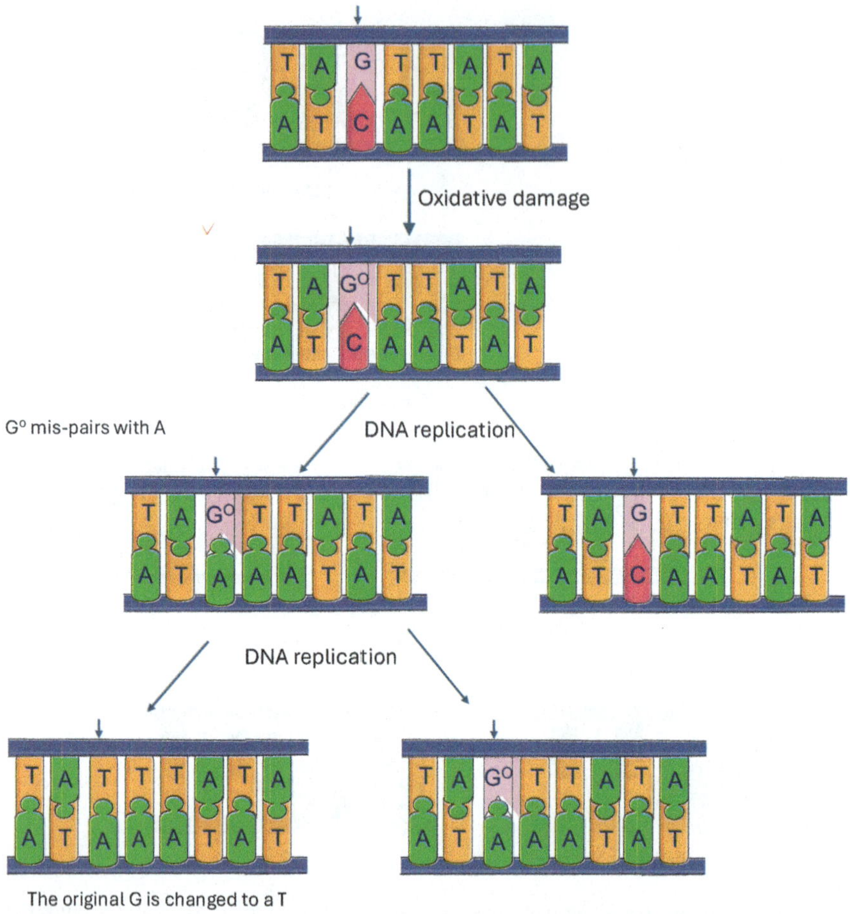

Fig. 23.2 Transversion mutation. Transversion mutations are mutations where a purine is replaced by a pyrimidine, or vice versa. This can result from DNA damage that alters base-pairing properties, such as irradiation of guanine to 8-OxyG (depicted here as G^{O}), which can mismatch with adenine and lead to a G:C to T:A transversion. (*Source*: Nucleotides adapted Servier Medical Art (https://smart.servier.com/), licensed under the Creative Commons Attribution 4.0 (CC BY 4.0) license)

double helix, thus changing the DNA's conformation. These mutations occur in response to mutagens such as **polycyclic aromatic hydrocarbons (PAHs)**, **aflatoxins** (produced by fungi and a contaminant in some foods), **alkylating agents**, and metabolic products of **benzene**, such as benzo[a]pyrene. These mutations can disrupt normal DNA structure and cause mutations such as base mutations, insertions, or deletions [4, 13].

One common and well-studied type of bulky adduct is the **thymidine dimer**. Thymidine dimers cannot occur spontaneously, but they can form in the DNA in response to physical mutagens such as **UV light**. Thymidine dimers involve dimerization between two adjacent thymidines via covalent bonds. This can disrupt the structure of the DNA and lead to more mutations [14] (see Fig. 23.4).

5.4 Double-Stranded Breaks and Single-Stranded Breaks

These are breaks in the backbone of the DNA (see Fig. 23.5). They can occur **spontaneously** during

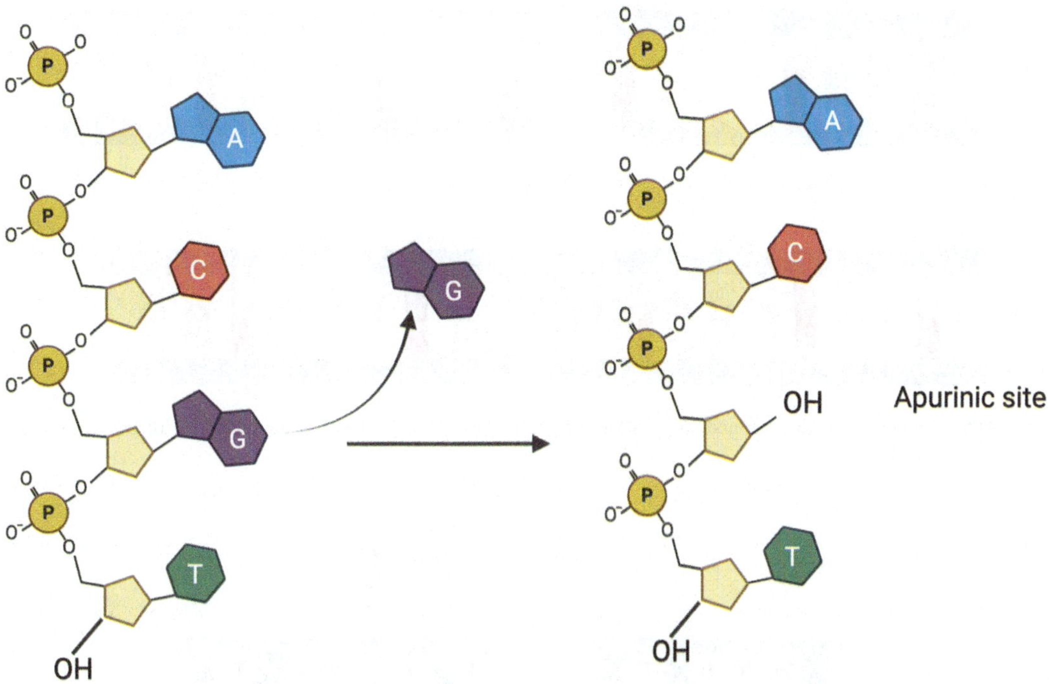

Fig. 23.3 Depurination. Depurination involves the removal of certain bases, leaving an apurinic (AP) site. (*Source*: Created with BioRender.com)

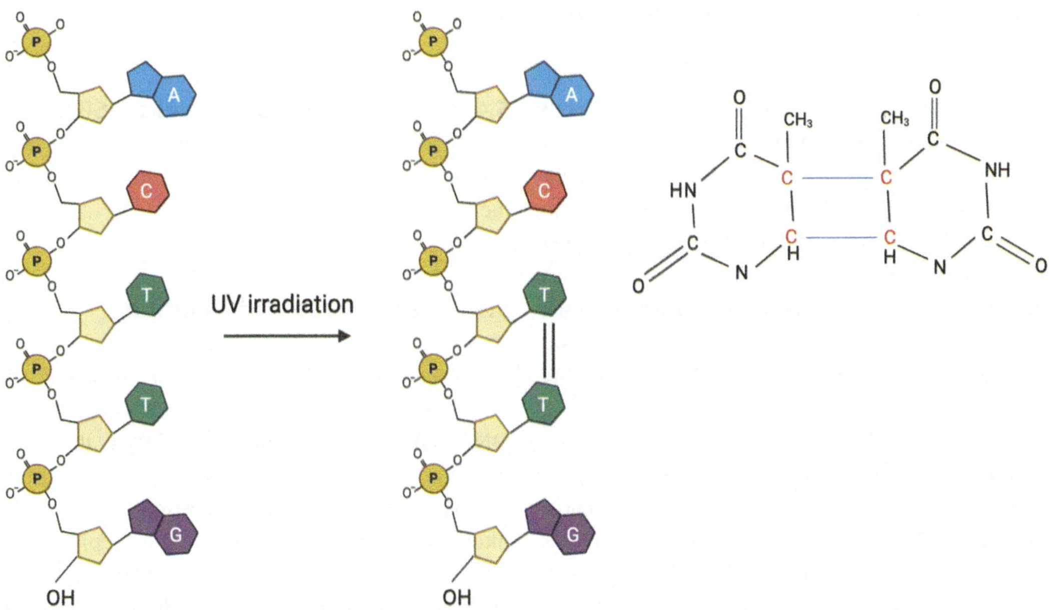

Fig. 23.4 Bulky adducts and thymidine dimers. These can occur in response to certain toxins or mutagens, which cause structural distortions. One common bulky adduct is the thymidine dimer. (*Source*: Created with BioRender.com. (combine with 3))

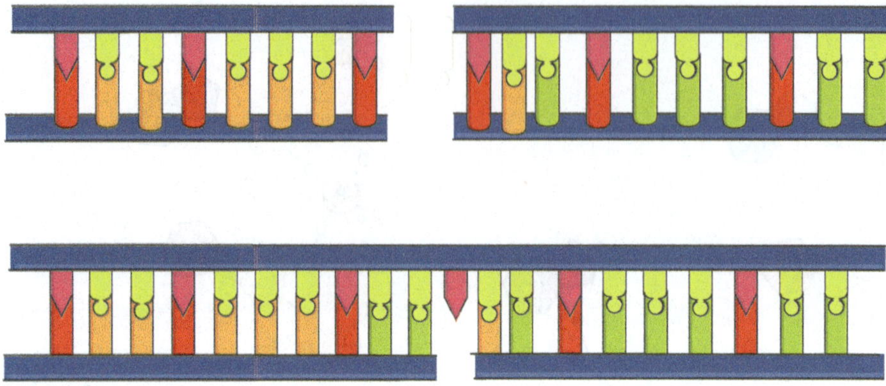

Fig. 23.5 Double-stranded break and single-stranded break. These breaks occur in the backbone of DNA. (*Source*: Created with BioRender.com)

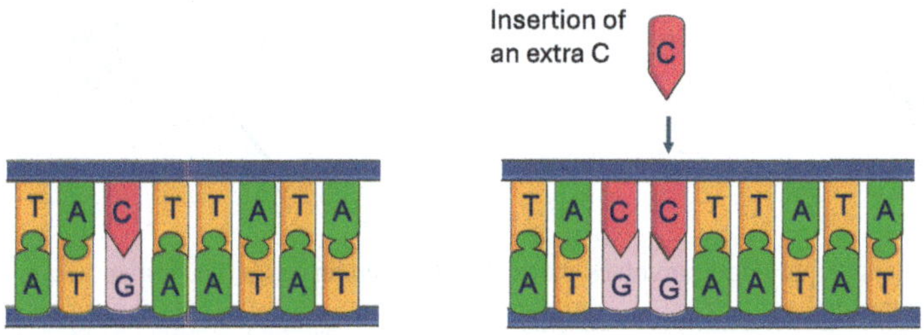

Fig. 23.6 Insertions. Insertions can result from the insertion of nucleotides or DNA fragments into the DNA. (*Source*: DNA and nucleotides adapted from Servier Medical Art (https://smart.servier.com/), licensed under the Creative Commons Attribution 4.0 (CC BY 4.0) license)

DNA replication. For example, they can result from problems with **topoisomerase** activity and errors in DNA unwinding. They can also occur spontaneously after DNA replication, or they can be caused by various stimuli. For example, they can result from **ionizing radiation**, such as **X-rays, reactive oxygen species, heavy metals**, and **various chemicals**. They can also be caused by **mechanical stress**, such as tension on the DNA [2, 3].

5.4.1 Double-Stranded Breaks (DSBs)

DSBs are breaks in the DNA where both strands, including the backbones, are broken. DSBs can be quite dangerous and can lead to more mutations and chromosomal rearrangements. Sometimes, DSBs can cause the DNA to break into multiple small pieces. These may be repaired, but sometimes, the repair process incorporates errors.

5.4.2 Single-Stranded Breaks (SSBs)

SSBs are breaks in only one strand of DNA. They can be less detrimental than DSBs but can nevertheless lead to further mutations.

5.5 Insertions and Deletions

Insertions and deletions are nucleotides or fragments of DNA that get inserted into the DNA chain, or fragments of DNA that get deleted, respectively. Insertions and deletions can occur **during DNA replication**, but are less likely to occur spontaneously after replication. They can occur in response to mutagens, however, such as **intercalating agents** or **alkylating agents**. They can also occur in other situations, such as when DNA **repair mechanisms** do not function properly, during homologous recombination, or due to errors that occur during mitosis and meiosis [2, 3, 8]. (See Fig. 23.5 for ss and ds breaks, and Fig. 23.6 for an example of an insertion).

6 Mutations Can Impact the Amino Acid Sequence of the Encoded Protein

Many of the types of mutations described above consist of point mutations (single-base pair mutations) or other mutations that are small in scale. Small mutations may have no impact on gene function if they occur in a noncoding region. If they occur in a coding region, they may have devastating effects on the genes that they mutate, but in other cases, they may have little or no impact on gene function. Below, mutations are categorized by what kind of impact they have on the DNA sequence and how they may impact the functions of the genes they encode [2, 3, 8, 15] (Fig. 23.7).

6.1 Missense Mutations

Missense mutations are single-nucleotide changes that result in the incorporation of a single **incorrect amino acid** once the gene is transcribed and translated. This may have a large or small effect on the protein's function, or it may have no effect at all. An important example of a missense mutation that is detrimental is one that can occasionally be found in the gene that encodes **hemoglobin**. This missense mutation is Glu (GAG) → Val (GTG). It is usually written as GAG → GTG or Glu-6-VAL (since the mutation affects amino acid 6). This missense mutation leads to an important change in the gene product hemoglobin, resulting in **sickle cell disease**,

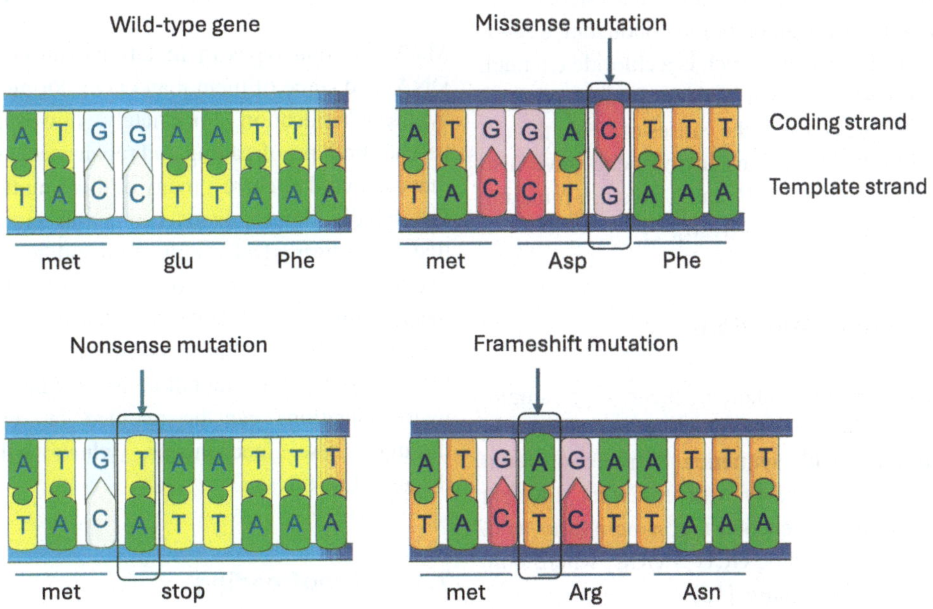

Fig. 23.7 Types of mutations based on their outcomes. A missense mutation changes an amino acid in the encoded sequence, and a nonsense mutation causes a termination sequence, resulting in a truncated protein. A frameshift mutation disrupts the reading frame and alters the entire sequence downstream from the mutation. (*Source*: Nucleotides adapted from Servier Medical Art (https://smart.servier.com/), licensed under CC BY 4.0)

severely affecting human health in individuals with the mutation [16, 17]. Understanding the function of this missense mutation has allowed scientists to develop treatments for the disease and has led to the development of the first CRISPR-based drug treatment, which is covered in Chap. 25.

6.2 Nonsense Mutations

This type of mutation is also a single-nucleotide change, but instead of changing a codon to a different amino acid, it inserts a **stop codon** into the code. As a result, a truncated protein will form. For example, Arg (CGA) → Stop (TGA) at amino acid 928 on the **dystrophin gene** causes the encoded protein to be truncated and nonfunctional. This is the cause of a small proportion of **muscular dystrophy** cases. **Cystic fibrosis (CF)** is another example of a disorder in which a nonsense mutation is often found. The nonsense mutation in CF is in the gene that encodes the cystic fibrosis transmembrane conductance regulator (CFTR) protein, which is a chloride channel. The nonsense mutation leads to a truncated form of CFTR, and is one of the causes of CF (although multiple other mutations in the gene have also been found to cause CF, and a deletion mutation is more common) [16, 18–20].

6.3 Silent Mutations

In this type of mutation, there is a single-nucleotide change in the DNA but **no change in the amino acid sequence**. This can occur because of redundancy, where several different DNA codons can encode the same amino acid. An example would be GCU > GCC, where both codons code for alanine [16].

6.4 Frameshift Mutations

In a frameshift mutation, an extra nucleotide (or nucleotides) is added or deleted. If the nucleotides added are in a coding region but are not in multiples of 3, then this disrupts the entire reading frame and thus leads to a complete change in the amino acid sequence of the resulting protein. The result is a protein that encodes an entirely incorrect sequence of amino acids after the frameshift. This can lead to a protein with incorrect amino acids, and in some cases, it can lead to an in-frame stop codon, resulting in premature termination of the protein. Here is an example of a frameshift mutation:

Wild type: ATG TTA GAC TGA
Mutation: ATG TT**C** **C**AG ACT GA

In this mutation, an extra **CC** is added between T5 and A6 of the wild-type gene. This changes the entire coding sequence of the gene downstream to the **CC** and will completely disrupt the amino acid sequence of the encoded protein [1, 16, 21, 22].

7 Proofreading and DNA Repair

Many different types of mutations can occur in DNA, and some of them may occur during DNA synthesis/replication. Fortunately, however, the cell has both proofreading and repair mechanisms that can avoid or correct many of these mutations. Proofreading mechanisms can avoid mutations that occur during DNA replication and, to a lesser extent, in response to mutagens. Repair mechanisms can fix mutations that occur either spontaneously or in response to mutagens [3]. Some level of mutation is inevitable, but many mutations can be avoided or repaired through these mechanisms, which will be discussed below.

7.1 Proofreading

When DNA is being replicated, some errors can occur on the new strand (the daughter strand). For example, an incorrect base may be incorporated. However, proofreading mechanisms ensure that such errors are infrequent. This is due mostly to the proofreading functions of **DNA polymerase**.

DNA polymerase is a multifunctional enzyme that not only adds new nucleotides to DNA during DNA synthesis but also corrects mistakes through its proofreading function. DNA polymerase has separate domains for DNA synthesis and for proofreading (exonuclease) activity. The proofreading activity of DNA polymerase is also known as its exonuclease activity, and it allows DNA polymerase to remove incorrectly added nucleotides. Because of this dual role of DNA polymerase, it ensures that the DNA synthesis process is efficient and largely error-free [23]. Two of DNA polymerase's major proofreading mechanisms are discussed below.

7.1.1 Nucleotide Selection

The DNA polymerase's first job in promoting error-free DNA replication is to ensure that the correct nucleotides are incorporated as the DNA is synthesized. Every time a nucleotide is added to the 3′ end of the growing DNA strand during DNA synthesis, the DNA polymerase can detect whether it binds to its complementary base pair, indicating that it is the correct binding partner. If the correct nucleotide is added, the DNA polymerase allows nucleotide incorporation to complete, the correct bonds are formed, and DNA synthesis proceeds. In contrast, if an incorrect nucleotide is added to the growing chain, the DNA polymerase has a mechanism to prevent it from being incorporated. Instead, the nucleotide is released, allowing the correct nucleotide to be added instead [1, 23].

7.1.2 Exonucleolytic Proofreading (3′ to 5′ Exonuclease Activity)

Despite the efficiency of the above system, errors can still occur, and an incorrect nucleotide may still be incorporated. Incorporation of an incorrect nucleotide results in a change in the conformation of the 3′ OH at the end of the growing DNA chain. This, in turn, makes it difficult for the DNA polymerase to add the next base. DNA polymerase instead stops and removes the wrong nucleotide from the 3′ end of the strand using its **exonuclease** activity, which is the activity that can remove nucleotides. After this happens, the correct nucleotide can be added instead, and DNA synthesis can continue.

This exonuclease activity of DNA polymerase can help explain why DNA synthesis requires a primer while RNA synthesis does not. DNA polymerase depends on the correct position and conformation of the 3′ OH before it can add the next nucleotide. This requirement prevents the incorporation of an incorrect nucleotide. This helps explain why DNA synthesis requires a primer. Without a primer, there would be no 3′ OH group for DNA polymerase to recognize when the first nucleotide is added, and consequently, nucleotide incorporation could not begin. It also helps explain why the initial primers that prime DNA synthesis are made from RNA rather than DNA. RNA polymerase does not have proofreading functions. Therefore, it also does not require a 3′ OH group for the addition of nucleotide. Instead, the RNA primer can form de novo without a preexisting nucleotide sequence.

Why does RNA polymerase not have proofreading activity? This may be because proofreading of RNA is not quite as important as DNA proofreading. While errors in RNA may lead to some problems, they are not as detrimental as errors in DNA because RNA sequences do not get replicated or passed down to future generations [23, 24].

7.2 Repair of DNA That Has Already Been Damaged

Despite proofreading mechanisms, DNA can still be mutated either during or after DNA replication, either spontaneously or in response to mutagens. Repair mechanisms exist to correct those errors. There are multiple different repair mechanisms, but the five major categories are described below. These are homology-directed repair (HDR) and nonhomologous end joining (NHEJ), which both repair double-stranded breaks in DNA, and base excision repair (BER), mismatch repair (MMR), and nucleotide excision repair (NER), which primarily repair single-stranded errors. In addition to these five major categories, ribonucleotide excision repair (RER) will also be discussed, as well as special consideration for **telomeres** [2, 3, 16].

7.2.1 Homology-Directed Repair (HDR)

- Overview of HDR

Homology-directed repair (HDR) is a mechanism that can be used to repair double-stranded breaks (DSBs) in DNA. HDR usually takes place **in G2 phase or late S phase** of the cell cycle, after DNA has been replicated in S phase. This is because in somatic cells, this method requires the presence of a sister chromatid.

There are two main HDR pathways, both of which repair double-stranded breaks. One is called the double-stranded break repair (DSBR) model, and the other is called the synthesis-dependent strand annealing (SDSA) pathway. The DSBR pathway is especially important in meiosis. The SDSA model is most prevalent in somatic cells. Both have many similarities and involve many of the same repair proteins and enzymes [25, 26]. The SDSA pathway is discussed here (see Fig. 23.8).

Briefly, in the DNA with the DSB, the 5′ ends of the breaks are resected by nucleases (meaning that the DNA on the 5′ ends is trimmed back) to create single-stranded 3′ overhangs. Each new single-stranded region then "searches" for a homologous "template." In somatic cells, this template will typically be the sister chromatid, which provides the correct sequence that can be used to repair the DSB. The ends with the 3′ overhangs invade this homologous template, causing partial unwinding and strand displacement of the template DNA, which causes a loop to form (a displacement (D) loop). These 3′ overhangs anneal to their complementary sequences in the **homologous template**, after

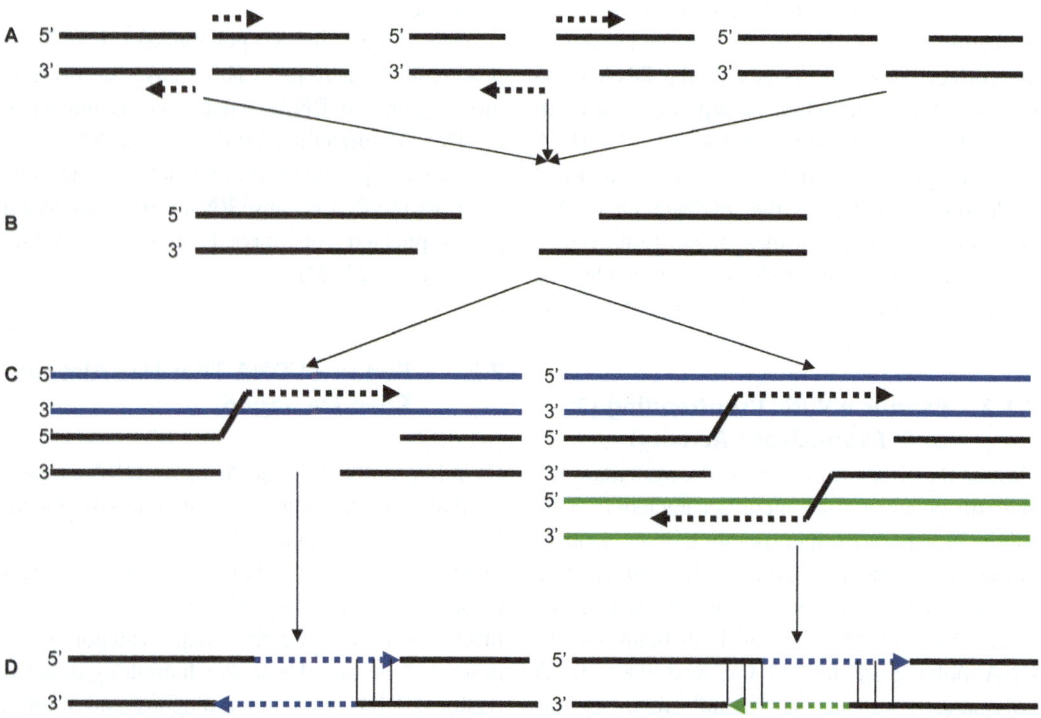

Fig. 23.8 HDR and the SDSA pathway. **A**. A double-stranded break occurs, which can lead to a blunt break or overhangs. **B**. Resection occurs, leading to 3′ overhangs. **C**. One 3′ overhang can invade the complementary DNA molecule, annealing to its complementary sequence. The other 3′ end may also invade a DNA molecule. DNA synthesis occurs. **D**. The correct sequence has now been copied into the break. (*Source*: Image reproduced from McGrath et al. [95]. Licensed under the Creative Commons Attribution 4.0 (CC BY 4.0) license)

which the newly synthesized strands dissociate from the template and anneal to the other resected end of the DSB. As a result, the original DNA with the DSB will incorporate specific sequences that are present in the homologous template, thus repairing the DSB [8, 26–29]. These are the basic steps in HDR, and they are described in more detail below.

- Detailed Mechanism of HDR and Steps Involved

1. DSB recognition and resection: After a **DSB** forms, the **MRN** complex (consisting of **Mre11-Rad50-Nbs1**) binds to the break site. This starts the **resection process**, creating 3′ overhangs. **Mre11** is especially important because it has **endonuclease activity**. A protein called **CtiP** (C-terminal binding protein interacting protein) collaborates with the **MRN** complex to initiate resection by promoting its endonuclease activity. The **BRCA1** protein also has an important role in this process. For example, it interacts with the **MRN** complex and helps recruit it to the **DSBs**.
2. ss stabilization: **Replication Protein A** (**RPA**) then binds and coats the ssDNA overhangs that result from the resection. This stabilizes the resulting ssDNA and prevents secondary structures from forming.
3. Rad51 filament formation: **Rad51** is a recombinase. (A recombinase is an enzyme that facilitates recombination, the process where DNA strands are broken and rejoined in new combinations.) Rad51 is recruited to the RPA-coated ssDNA. It displaces RPA and forms a Rad51-coated filament along the DNA. **BRCA1** also has a role in this step, where it helps stabilize the binding of **RAD51** to the DNA. Likewise, **BRCA2** also has a role in stabilizing the interaction between RAD51 and the DNA.
4. Strand invasion and D-loop formation: The next step is **strand invasion**. The **Rad51-ssDNA filament** recognizes its homologous sequence on a sister chromatid. The filament promotes strand invasion into the homologous sequence, forming a displacement loop (**D-loop**).
5. DNA synthesis: **DNA polymerase** δ extends the invading strand using the homologous DNA as a template, copying the correct sequence into the break. After synthesis, the newly extended strand dissociates from the D-loop and anneals to the other resected end of the break.

- Diseases Associated with the HDR Process

Breast and ovarian cancer can be associated with mutations in **BRCA1** and **BRCA2** genes, both of which have important roles in the HDR process. Cells with mutations in these genes cannot repair DSB breaks efficiently. As a result, the cells accumulate DNA damage. This can lead to genomic instability, which can lead to even more damaged DNA. The resulting mutations in DNA can sometimes include mutations in genes that are important for controlling cell growth. Treatment of **BRCA1 and BRCA2-associated cancer** can involve the use of PARP inhibitors such as **Lynparza (olaparib)** [30] and **Zejula (niraparib)** [31]. This mechanism involves the concept of synthetic lethality, to be discussed in Chap. 22 [32].

7.2.2 Nonhomologous End Joining (NHEJ)

- Overview of NHEJ

This is another mechanism that repairs **double-stranded breaks (DSBs)** and is also one of the double-stranded break repair (**DSBR**) pathways. This DSB repair mechanism is more likely to occur in **G1 phase** of the cell cycle. During G1 phase, the DNA has not yet replicated, and therefore, a homologous template sequence is not available, making the HDR method impractical. In nonhomologous end joining (NHEJ), the two ends of DNA on either side of a DSB are joined together with the help of a ligase enzyme and associated proteins. Thus, there is no need for a homologous template. This method repairs the break, but it is not perfect. There is usually a loss or gain of some nucleotides during this process due to the error-prone nature of NHEJ. This can potentially result in mutations, a frameshift, or an in-frame stop codon [2, 33], described as follows (Fig. 23.9).

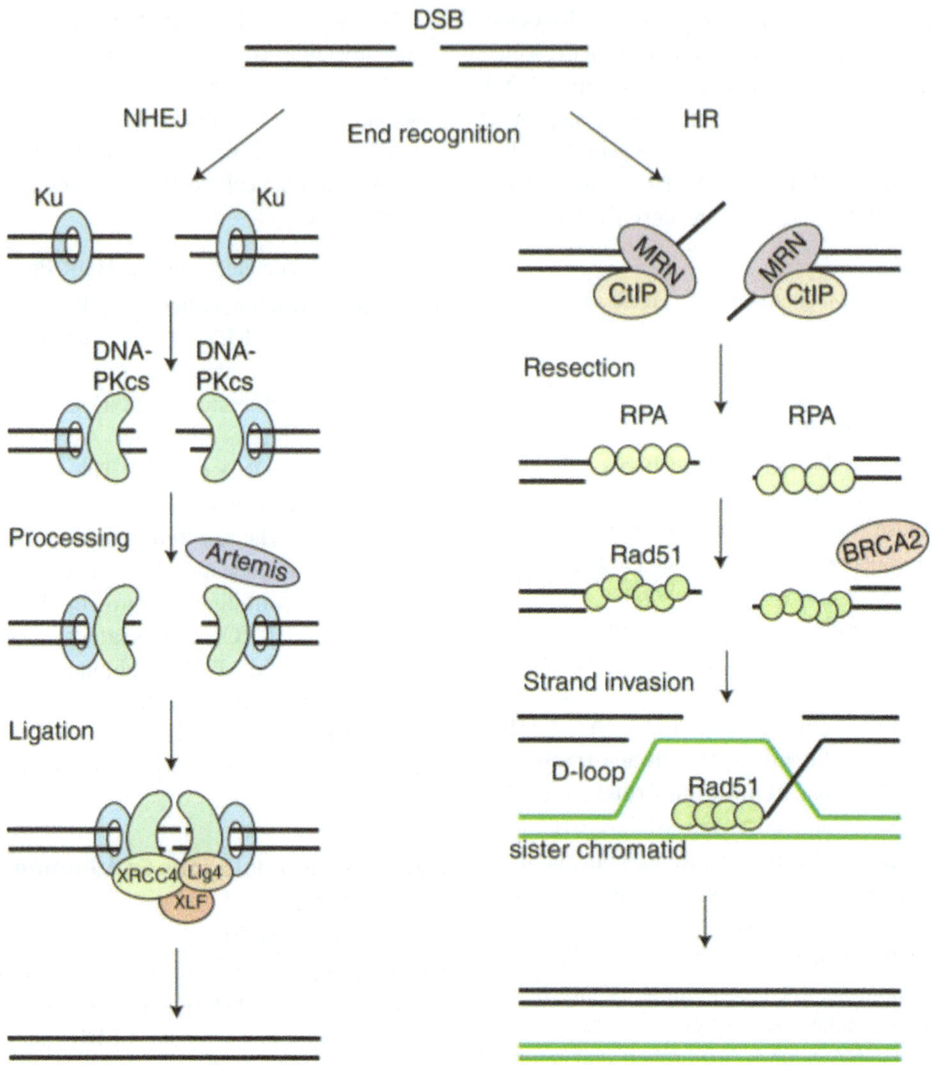

Fig. 23.9 Nonhomologous end joining (NHEJ). NHEJ is another mechanism for repairing double-stranded breaks, but it is more error-prone than HR. The basic steps of NHEJ (left) are compared with those of HR (right) in this figure. See text for details. (*Source*: Figure from Brandsma and Gent [96]. Licensed under the Creative Commons Attribution 2.0 (CC BY 2.0) license)

- Detailed Mechanism of NHEJ

The detailed sequence of events in NHEJ is as follows:

1. At the break site, minimal resection may sometimes occur, leaving behind a small section of ss 3′ overhangs, although blunt ends can also be repaired. **Ku70/80** are DNA-binding proteins that are rapidly recruited to these broken ends and form **heterodimers**. They stabilize and protect the DNA ends.
2. **Ku70/80 dimers** recruit **DNA-PKcs** (DNA protein kinase catalytic subunit) to the complex. **DNA-PKcs** interact with **Ku70/80**, forming a complex called **DNA-PK**, which is a **holoenzyme**.

3. The **DNA-PK** now becomes activated due to autophosphorylation. This loosens the DNA-PK complex and allows access for processing. It also facilitates recruitment of **Artemis** to the DNA. The **DNA-PK** phosphorylates **Artemis**, an **endonuclease** that can trim the ends of the overhangs and resolve hairpin structures that form, although Artemis is not always required.
4. Later, **XRCC4**, **XLF**, and **ligase IV** will also be recruited to the complex. **XRCC4** and **XLF** are scaffold proteins that have roles in stabilizing **ligase IV** and helping align the DNA ends. **Ligase IV**, in turn, repairs the DS break, completing the repair.

- NHEJ in Human Health and Disorders

NHEJ is not only involved in the repair of damaged DNA. It also has an important role in B and T cells, where it plays a role in V(D)J recombination (variable (diversity) joining recombination). V(D)J is a process that generates DNA sequences encoding diverse antigen receptors and antibodies, allowing the immune system to respond to a wide array of pathogens. The V(D)J recombination process occurs during the development of B and T cells. During this process, DNA breaks are induced by **Rag1/2** - (recombinase-activating gene) endonucleases. These breaks are produced in the T-cell receptor (TCR) genes of T cells and in the immunoglobulin receptor genes of B cells. They are then **repaired** by **NHEJ**. The repair process leads to a diversity of sequences. This leads to diversity in TCRs and antibodies. This is critical for immune responses to a variety of pathogens and to antibody diversity [34].

Because of its roles in V(D)J, **abnormalities in NHEJ** can lead to **immune deficiencies**. One of the most severe of these is **SCID (severe combined immune deficiency)**, which is a rare group of genetic disorders. Patients with SCID lack T cells. Patients, therefore, lack a functional immune system and are extremely vulnerable to infections. SCID can have many causes, not all of them involving DNA repair pathways. For example, common causes include a deficiency in the IL-2 receptor gene, which is critical for T-cell signaling, or a deficiency in ADA (adenosine deaminase), resulting in toxic metabolism accumulation. SCID can also result from abnormalities in V(D)J, however. For example, it can be caused by a deficiency in the Rag1/2 genes. Rag1/2 mutation disrupts the induction of DSBs necessary for V(D)J recombination in T cells. SCID can also be caused by mutations in NHEJ-specific genes such as **DNA-PKcs** and **Artemis**. These mutations impair the repair of the DSBs that are introduced during V(D)J recombination. Because V(D)J and NHEJ are essential for the development of B cells and T cells and the generation of diversity, these mutations severely disrupt the immune system [35–37].

Treatments for SCID involve stem cell replacement, where healthy hematopoietic stem cells can replace the abnormal cells. Other treatments involve gene therapy and enzyme replacement therapy for ADA deficiency. Other potential treatments under investigation include the use of CRISPR to repair genetic abnormalities in patients' cells [37, 38]. (Gene therapy and CRISPR technology will be discussed in Chap. 25).

7.2.3 Base Excision Repair (BER)

- Overview of BER

This is a mechanism for repairing small errors on a single strand that do not distort the shape of the double helix. These errors are caused by damaged bases, and they usually affect a single base. The base may be damaged by mechanisms such as deamination, alkylation (including methylation), and oxidative stress. An example of an error that may be repaired by BER is an error that involves the conversion of cytosine to uracil by deamination of the cytosine. During the BER process, the DNA strand with the error (in this case, the uracil) is nicked, creating a **single-stranded break**. Several enzymes remove the damaged base. The base is then replaced with the correct base (in this case, cytosine), and the backbone is repaired. This repair mechanism happens mostly in **G1 and G2 phases** of the cell cycle [2, 39]. The detailed mechanism of BER is described below.

- Detailed Mechanism of BER

BER is a multistep process. Below is a description of the main steps.

1. **Pre-repair step:** When the damage occurs, and before the repair process begins, **PARP1 (poly ADP-ribose polymerase 1)** acts as a sensor that detects the damage, particularly SSBs. PARP binds to the break. It then becomes active, and it uses NAD^+ as a substrate to synthesize and attach poly (ADP-ribose) (PAR) chains. It primarily attaches PAR chains to itself (auto-PARylation), and to some extent it also attaches PAR to some nearby proteins, such as histones and some of the proteins that will be involved in the repair process PARylation [40]. PARylation induces negative charges [41] that help relax the chromatin structure and loosen its packaging to allow recruitment of BER proteins such as XRCC1, DNA polymerase-β, and DNA ligase III, to the site of the damage. Once the BER repair machinery is successfully recruited to the DNA, PARP auto-PARylation leads to its own dissociation from the DNA, as the negative charges accumulate, and PAR chains are degraded by enzymes such as PARG (poly(ADP-ribose) glycohydrolase), and the repair can proceed [42].
2. **Recognition and removal of the damaged base** by a **DNA glycosylase:** A DNA glycosylase is a family of enzymes that have a role in DNA repair in the BER pathway. They are called glycosylases because they cleave glycosidic bonds, which are the type of bonds that connect a nitrogenous base to its sugar. A **glycosylase** recognizes a base that may be damaged, modified, or misplaced. There are specific glycosylases for different bases. For example, as indicated above, a common mutation that can be repaired by BER is the **deamination of cytosine to uracil**. In this case, a glycosylase specific for uracil called a **uracil-DNA glycosylase** (**UDG**; also called uracil-N-glycosylase or UNG) removes the uracil from the chain, leaving behind an **apurinic/apyrimidinic site**, which is a site in the DNA containing a sugar-phosphate residue lacking a base and also referred to as an abasic site [43].
3. **AP endonuclease 1 (APE1) action: AP endonuclease 1 (APE1)** is an enzyme that recognizes the **AP site** and cleaves the DNA backbone at this location. This creates a nick and removes the sugar-phosphate residue at the abasic site created in the step described above.
4. **Gap filling by DNA Polymerase β**: The enzyme **DNA Polymerase β** fills in the single-nucleotide gap by inserting the correct nucleotide complementary to the undamaged template strand,
5. **Sealing by DNA ligase:** Finally, **DNA ligase** seals the nick in the DNA backbone. This often involves a scaffolding protein called **XRCC1**. XRCC1 binds molecules together. It ensures that DNA ligase binds to the DNA, and it also stabilizes DNA polymerase in the previous step.

7.2.4 Mismatch Repair (MMR)

- Overview of MMR

MMR (Fig. 23.10) is a mechanism for repairing incorrectly paired bases called **base-base mismatches** that result from DNA replication errors. The errors it corrects are usually **single-stranded errors**. These errors usually cause some distortion of the helix and are often larger than the errors repaired by BER. Small loops (insertion-deletion loops) can form due to extra or skipped nucleotides. The MMR method repairs the incorrectly paired base matches and the loops that may result. The MMR process is most likely to take place during the **G2 phase** of the cell cycle, because it involves correcting errors that occurred in S phase [2, 44, 45].

- Detailed Mechanism of MMR

1. **Recognition of the mismatch or loop:** The process starts with the recognition of the mismatch or the small loop. The **MSH2/MSH6 complex (MutSα)** recognizes base-base mismatches and small loops. **MSH2/MSH3**

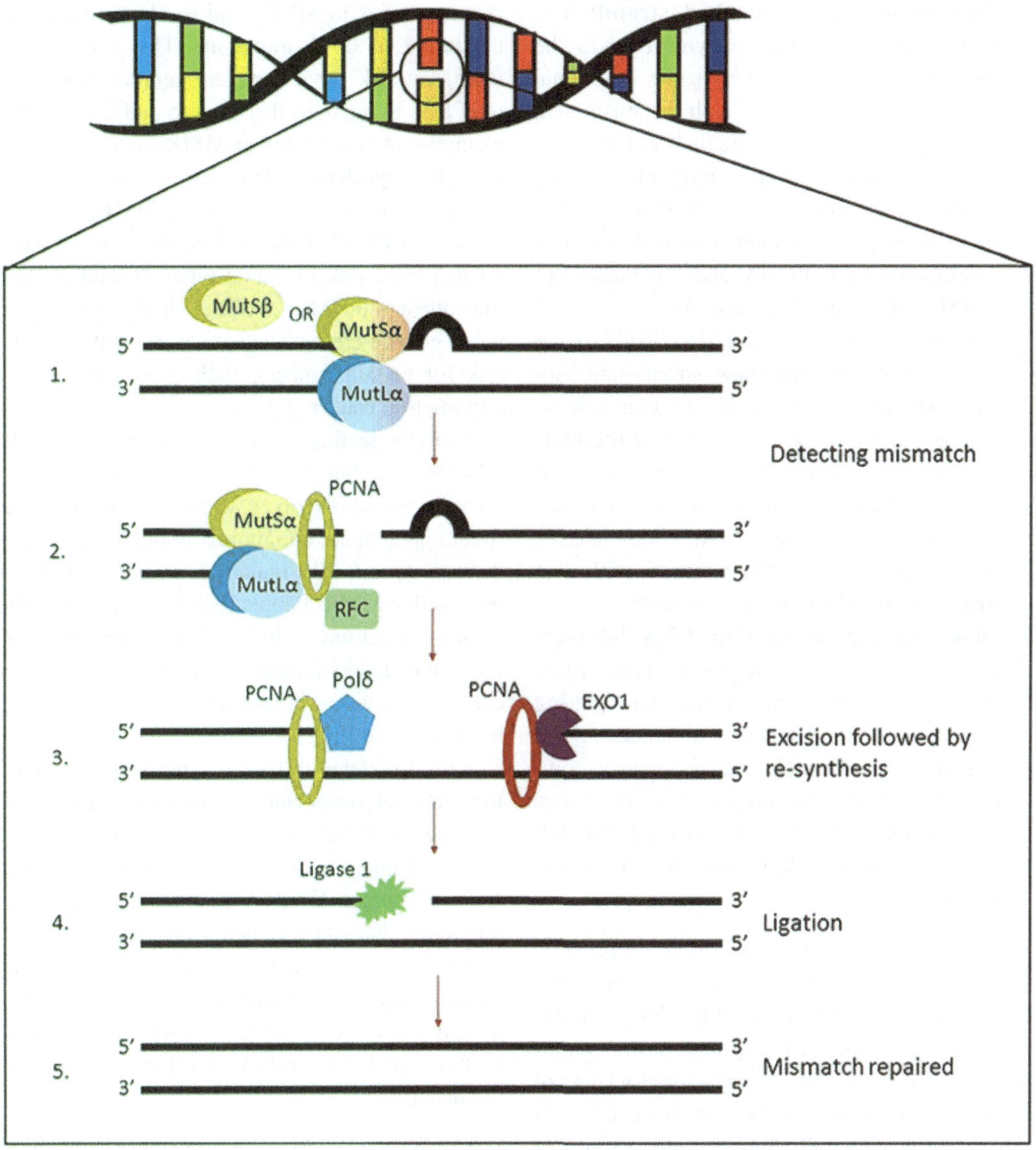

Fig. 23.10 Mismatch repair. Mismatch repair (MMR) is a mechanism for repairing incorrectly paired bases. The detailed process is described in the text. (*Source*: Image from Pecina-Slaus et al. [97]. Licensed under the Creative Commons Attribution 4.0 (CC BY 4.0) license)

complex (MutSβ) recognizes larger loops (more than 2 bps) [46].

2. **Recruitment of MutLα and nicking:** The **MLH1/PMS2 (MutLα)** complex is recruited to the mismatch site by the **MutSα complexes**. **MutLα** interacts with the **MutSα** complexes and other repair proteins. **PMS2**, the catalytic component of MutLα, has **endonuclease** activity, and it introduces a single-stranded nick in the area around the mismatch (in the newly synthesized strand). This nick defines the strand to be excised during the repair process. In this step, the sliding clamp protein **PCNA** (proliferating cell nuclear antigen) interacts with MutLα and enhances the endonuclease activity of PMS2.

3. **Excision of the mismatched strand:** The **exonuclease 1 (EXO1)** enzyme excises the mismatched sequences and some additional nucleotides on the strand with the mismatch. This excision starts from the nick and creates a single-stranded gap. The entire mismatched nucleotide is removed, rather than just the base. Some of the sequence around it is also usually removed. **PCNA** also interacts with **EXO1** and stimulates its activity.
   The entire repair process described so far involves both **endonuclease** activity and **exonuclease** activity. To clarify, an exonuclease cleaves the phosphodiester bonds **at the ends** of a nucleic acid strand. In this case, the exonuclease starts from a preexisting nick and proceeds along the strand. An endonuclease, in contrast, cleaves DNA **from within a strand**. It introduces a nick internally.
4. **Filling the gap and ligation: DNA Polymerase δ** fills in the excised region with the correct nucleotides using complementary base pairing with the complementary template. **DNA Ligase I** seals the nick, and the repair is complete. **PCNA** also has a role in this step, where it helps ensure the efficiency of the DNA Pol.
5. **Summary of the Role for PCNA in the MMR Process**

PCNA has important roles in this entire process. As a sliding clamp, it makes a circle around the double-stranded DNA, and it slides along it. It can serve as a base for recruiting and various repair proteins. It stimulates the endonuclease activity of MutLα and the exonuclease activity of Exo1. It also interacts with and stimulates the activity of DNA polymerase δ. For MutLα, PCNA allosterically activates the enzyme's activity. For Exo1, it interacts with the enzyme, stabilizes it, and helps align it properly on the DNA. It acts as a sort of sliding guide that keeps Exo1 stably on the DNA as it progresses. PCNA also binds to DNA Pol-δ and this binding helps stimulate its activity [3, 47–49].

- Disorders Associated with MMR
  - Lynch Syndrome and Hereditary Nonpolyposis Colorectal Cancer (HNPCC)

Abnormalities in MMR lead to abnormalities in the repair of DNA mutations. This can lead to increased risk for cancer, as genetic mutations may lead to uncontrolled growth and cancer. For example, abnormalities in MMR genes can lead to **Lynch syndrome**. This disorder is caused by mutations in the genes encoding MSH2, MSH6, MLH1, or PMS2. Patients with this disease have an increased risk for some types of cancer over the course of their life due to the inability to repair some genetic errors. They are at particularly high risk for certain cancers, such as colorectal and endometrial cancers [50].

Genetic testing is recommended for people who have a family member with the disease. Genetic testing involves screening for the gene mutations that are commonly associated with the disease, specifically mutations in genes that are involved in MMR. People found to be positive for Lynch syndrome by genetic testing are recommended to undergo early screening for cancer. In some cases, preventative surgery may also be carried out [51].

Closely related to Lynch syndrome is **hereditary nonpolyposis colorectal cancer (HNPCC)**. It is the most common cause of hereditary colon cancer, although patients are also at risk for other types of cancer. **HNPCC** results from the same mutations that cause Lynch syndrome. The main difference is that while Lynch syndrome can occasionally develop due to a random mutation in somatic cells, HNPCC is always a germline mutation and associated with a family history of the disease [52].

- Constitutional Mismatch Repair Deficiency (CMMRD)

Constitutional mismatch repair deficiency (CMMRD) is a rare disease, also caused by mutations in MMR genes. Like Lynch syndrome, it also leads to an increased risk for cancer, but in the case of CMMRD, the risk is for early-onset cancers in children or young adults [44, 53]. It is usually caused by germline mutations in the genes for **PMS2**, and sometimes in other genes that are associated with MMR, such as MLH1, MHL2, and MSH6. Consequently, genetic errors can accumulate in cells, which may sometimes lead to uncontrolled cell proliferation. Patients are not only at risk for cancer, but they may also develop benign growths. Patients may also have café au lait spots, which involve changes in skin pigmentation, similar to patients with neurofibromatosis (NF1). This may be due to genetic changes in the NF1 gene due to loss of MMR [54].

7.2.5 Nucleotide Excision Repair (NER)

- Overview of NER

The nucleotide excision repair mechanism is effective for repairing DNA in which **bulky adducts** have been attached to the DNA. As described above, bulky adducts are large modifications of DNA that can distort the structure of the double helix. For example, these types of mutations often result from UV-induced pyrimidine dimers, specifically thymine dimers. This results in a kink in the DNA. NER repair enzymes recognize these distortions. NER enzymes detect the distortions caused by these mutations, excise the damaged region, and synthesize a new, undamaged strand. This type of repair is most likely to occur during **G1 and G2** phases. This mechanism generally repairs **single-stranded errors** [2, 4, 55]. Many, but not all, of the repair proteins in this process are in the family of **XP** proteins, which stand for Xeroderma pigmentosum proteins. This naming is due to the finding that mutations in these genes are associated with the genetic disorder xeroderma pigmentosum (XP) discussed more later.

- Detailed Mechanism of NER

NER operates through two sub-pathways, **the global genome repair (GGR) pathway**, which is the major pathway, and **the transcription-coupled repair (TCR) pathway** [56]. Both are described below and some of the steps are summarized in Fig. 23.11.

- Global Genome Repair (GGR)

This mechanism removes lesions from anywhere within the entire genome, including non-transcribed regions:

1. **Damage recognition by the DDB complex: The DDB complex** (made of **DDB1** and **DDB2** (also called **XPE**) detects and binds the damaged DNA. **DDB2** recruits an E3 ubiquitin ligase complex (the **CUL4-DDB1-DDB2 E3 complex**) [57], which ubiquitinates histones and other proteins to make the DNA more accessible to the repair machinery. It also recruits the **XPC-RAD23B** complex. **XPC** (Xeroderma pigmentosum group C) is a protein that recognizes and binds damaged DNA, particularly at bulky adducts or distortions in the DNA helix. It often works together with **RAD23** protein which helps stabilize the complex. DDB2 is eventually ubiquitinated.
2. **Recruitment of TFIIH and DNA unwinding: TFIIH** is a protein complex with **helicase** activity (it includes subunits XPB and XPD). It is recruited to the DNA by the **XPC-RAD23B** complex. It unwinds the DNA around the lesion to create a localized single-stranded bubble that allows for recruitment of additional factors.
3. **Recruitment of XPA: XPA** then binds to the ssDNA and interacts with **RPA** (Replication Protein A), a ssDNA-binding protein that coats the DNA and prevents the ssDNA from reannealing. XPA functions as a type of damage verifier because it binds preferentially to the distorted DNA structures that form in response to the damage [58].
4. Recruitment of the **ERCC-XPF** and **XPG endonucleases: ERCC1-XPF** and **XPG endonucleases** are now recruited to the complex. ERCC-XPF makes an excision on the 5′ side of the lesion, and **XPG** makes an incision on the 3′ side. Together, these cuts lead to an excision of a short segment of ssDNA that contains the damage.
5. **DNA Polymerase δ/ε** fills in the gap: **DNA Polymerase δ/ε** fills in the gap with new nucleotides using the undamaged complementary DNA strand as the template. **Ligase I** seals the remaining nick, completing the repair [59, 60].

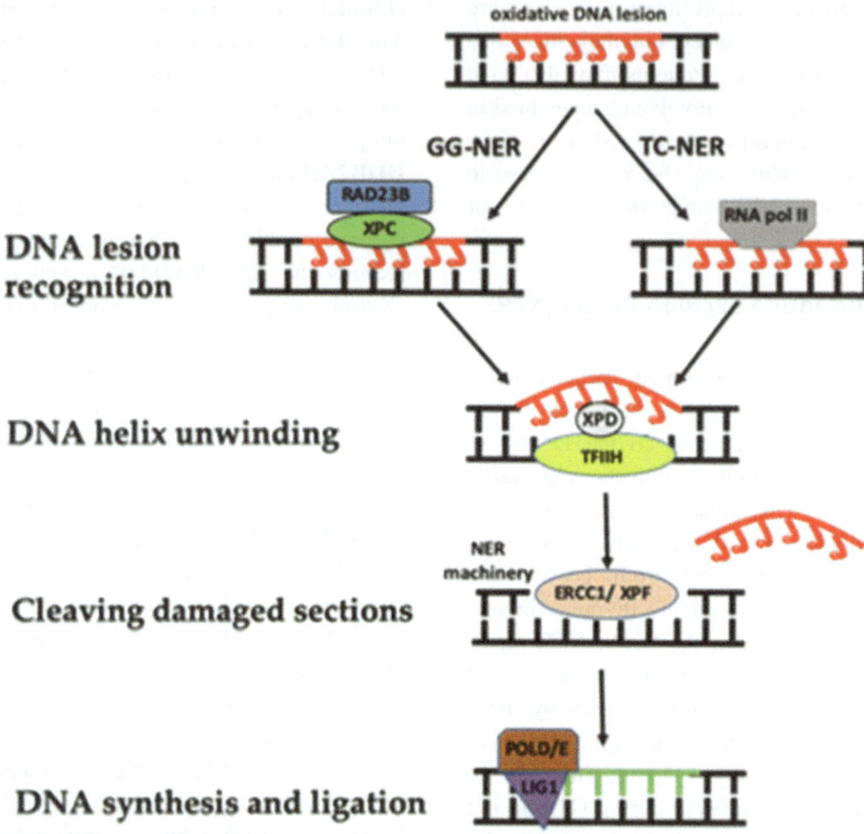

Fig. 23.11 Nucleotide excision repair (NER). GGR and TCR pathways are outlined here. This mechanism is effective for repairing DNA in which bulky adducts are attached to the DNA. The steps are described in more detail in the text. (Source: Adapted from Hu et al. [98]. Licensed under the Creative Commons Attribution (CC BY 4.0) license)

- Transcription-Coupled Repair (TCR)

This mechanism is a sub-pathway that focuses on the repair of DNA lesions on the transcribed strand of actively expressed genes [61, 62]. It is triggered when RNA polymerase stalls during transcription due to the bulky DNA error. It involves the following steps:

1. **RNA polymerase stalling**: **RNA Pol II** encounters a DNA lesion, often a bulky adduct, causing it to stall. This signals the need for repair and initiates TCR.
2. **CSB** (ERCC6; Cockayne syndrome B protein) is recruited: **CSB** is recruited to the stalled RNA Pol II, and it recruits **CSA** (ERCC8; Cockayne syndrome A protein), forming a complex. This complex displaces the **RNA Pol II** and initiates the repair process.
3. **CSA** is part of an E3 ligase complex called the **DDB1-CUL4 based E3 ubiquitin ligase complex**. This complex can ubiquitinate **RNA Pol II** or other associated factors, which causes it to pause or, in some cases, may lead to its degradation [63]. The rest of the steps are the same as the GGR method, starting with **TFIIH** recruitment.

- Diseases Associated with NER

Xeroderma pigmentosum (XP) is a group of diseases with mutations in the different XP genes, most commonly XPC. Patients with XP are especially sensitive to sunlight. This is because UV light from sunlight can cause mutations such as thymidine dimers, which cannot be repaired when XP genes are mutated. Patients have an increased rate of skin cancers due to a lack of DNA repair after exposure to sunlight [35, 64, 65].

Treatment and preventative measures for XP involve strict limitations on sunlight exposure for the skin and eyes, including protective clothing and high SPF sunscreen. Surgical removal of any lesions that develop is also important, as well as laser and photodynamic therapy (PDT) to destroy cancer cells. Gene therapy is under investigation, with the goal of reintroducing a functional copy of the XP genes that are mutated [66].

7.2.6 Ribonucleotide Excision Repair (RER)

This mechanism involves monitoring for the accidental incorporation of ribonucleotides (RNA) instead of deoxyribonucleotides (DNA). Repair involves cleavage of DNA by specific nucleases that recognize the incorporated ribonucleotide and replace the ribonucleotide with a deoxyribonucleotide [3, 67].

7.2.7 Special Considerations for Telomeres

The ends of chromosomes provide a unique problem because they could potentially be seen by the repair machinery as a double-stranded break and then "repaired." A specific structure is in place to prevent this. Telomeres, the structures at the ends of chromosomes, contain a protective cap called **Shelterin**. Shelterin is a complex of proteins (consisting of TRF1, TRF2, POT1, TPP1, TIN2, and Rap1) that protects telomeres, thereby protecting the ends of the chromosomes. One of its functions is to aid in the formation of T-loops. These are sequences of DNA that form loop structures and are stabilized by proteins. They play a critical role in protecting telomeres from the DNA repair machinery by preventing the formation of blunt ends at the ends of the chromosomes [3, 68–70] (Fig. 23.12).

8 DNA Damage Sensors and the Cell Cycle

DNA damage and repair are tightly connected to the regulation of the cell cycle. It is important that

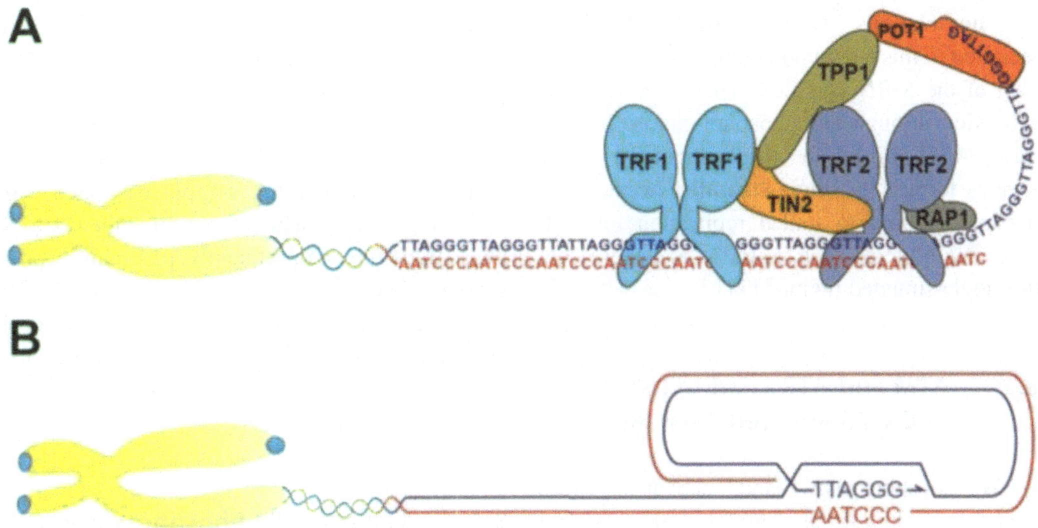

Fig. 23.12 Telomere structure and the shelterin complex. (**A**) shows the shelterin complex on the telomere. (**B**) shows the telomere with the T loop formation at the end. (*Source*: Reproduced from Doksani [99]. Licensed under the Creative Commons Attribution (CC BY 4.0) license)

damaged DNA is not replicated during S phase and is not passed on to daughter cells during M phase. Checkpoints in the cell cycle have important roles in ensuring that damaged DNA does not get maintained throughout the cell cycle or passed on to daughter cells. Regulators of cyclins and CDKs have important roles in pausing the cell cycle when DNA damage is present, until DNA repair can take place. DNA damage must first be recognized, however, so that the cell has a signal telling it to pause while the DNA damage machinery becomes engaged. **Some** of the DNA damage recognition machinery is discussed below.

8.1 ATM and ATR Are DNA Damage Sensors

DNA damage sensors recognize damaged DNA. They have a critical role in pausing the cell cycle so DNA repair can occur. **ATR** and **ATM** are the main DNA damage sensors. They are serine/threonine kinases, and they sense damaged DNA and bind to it. Their roles are discussed below.

ATM binds primarily to **double-stranded** DNA breaks. As discussed above, double-stranded breaks can be recognized by the **MRN complex**, which recruits the machinery required for **homologous recombination**. **ATM** is recruited to this site by interacting with a component of the MRN complex. This leads to ATM activation through autophosphorylation. In contrast, **ATR** binds to **single-stranded** DNA regions formed by **single-stranded breaks**, and it can also bind to stalled replication forks. It binds via proteins such as **RPA**, and these coat the single-stranded regions [71].

8.2 ATM and ATR Lead to Inhibition of Cdc25 and Cell Cycle Arrest

ATM phosphorylates and activates **Chk2** (Checkpoint kinase 2). Chk2 is a kinase, and one of the substrates is **Cdc25**. Likewise, **ATR** phosphorylates and activates **Chk1** (checkpoint kinase 1), and Chk1 also phosphorylates **Cdc25**. **Cdc25** is a phosphatase, and its activity is inhibited by phosphorylation. The normal role of Cdc25 is to enable the activation of the cyclin-dependent kinase Cdk1 (which mostly functions at the G2 to M phase transition) and Cdk2 (which mostly functions at the G1 to S phase transition). Cdc25, when active, activates these Cdks by removing their inhibitory phosphates (it dephosphorylates them). Thus, the net result is that activation of ATM and ATR in response to DNA damage can lead to cell cycle arrest [72], due to inhibition of Cdc25. This cell cycle arrest may be temporary, but it allows the DNA repair to take place before the cell cycle resumes. The role of Cdc25 in regulating Cyclin/CDK activity is also discussed in Chap. 21 (see Fig. 23.13).

8.3 ATM Also Leads to p53 Phosphorylation, Which Also Leads to Cell Cycle Arrest

In addition to Chk2, **ATM** also phosphorylates **p53** at Ser-15. This decreases p53's interaction with **MDM2**, which is a negative regulator of p53. **Chk2**, phosphorylated and activated by ATM, also phosphorylates p53, this time at serine 20. This stabilizes p53 even more and blocks MDM2 more effectively (unphosphorylated p53 binds MDM2 which is a ubiquitin ligase which leads to p53 degradation). This phosphorylation of p53 occurs mostly at the G1 checkpoint. p53 is a transcription factor, and an especially important p53 target gene is p21. P21 is a CKI, and it inhibits Cyclin/CDK complexes such as CyclinD/CDK4,6, which function at the G1 to S phase transition. Consequently, p21 inhibits progression from G1 through the S phase of the cell cycle, thus causing the cell cycle to arrest until DNA repair can take place [72, 73]. (The role for p53 and p21 in cell cycle regulation is also discussed in Chap. 20).

8.4 ATR and p53 Phosphorylation

ATR also phosphorylates p53 at Ser15, but is less efficient than ATM. Chk1 is the main substrate of ATR. Chk1 phosphorylates MDM2, which disrupts the MDM2-p53 interaction, and this

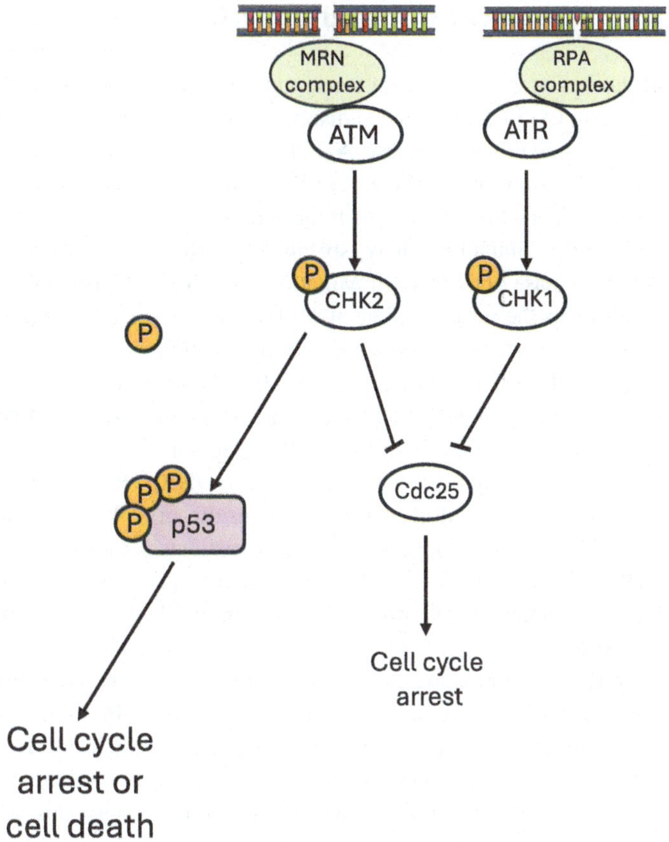

Fig. 23.13 ATM and ATR. ATM and ATR have roles in cell cycle arrest and DNA repair in response to double-stranded and single-stranded breaks in DNA. See text for details. (*Source*. DNA fragments from Servier Medical Art (https://smart.servier.com/), licensed under the Creative Commons Attribution 4.0 (CC BY 4.0) license)

thereby encourages p53's activity. Thus, ATR can also mediate cell cycle arrest or apoptosis through p53 [74, 75].

8.5 Under Some Conditions, p53 Can Lead to Apoptosis

p53 levels and activity increase in response to DNA damage, as indicated above. p53 levels and activity can be upregulated to high or low levels depending on the extent of DNA damage. Low levels of active p53 may respond to a level of DNA damage that is repairable, while high levels of p53 may result from a response to irreparable DNA damage. The low levels of p53 are associated with an arrest in the cell cycle that is then reversed after the DNA damage is repaired. In contrast, high levels of active p53 indicate to the cell that the DNA damage is too expansive or irreparable, and in this case, the cell may undergo apoptosis instead of cell cycle arrest. In this situation, the high levels of p53 will activate pro-apoptosis genes, such as Bad and Bax (described in Chap. 22) [76, 77].

8.6 ATM and ATR Phosphorylate BRCA1 and Other Targets, Leading to DNA Repair

In addition to promoting cell cycle arrest, ATM and ATR promote DNA repair. Another important substrate of both ATM and ATR is BRCA1. Phosphorylation of BRCA1 by these kinases is crucial for BRCA1's role in HR, as discussed above. ATM can also phosphorylate a histone variant in DSBs and a part of the MRN complex. By phosphorylating the MRN complex, ATM can initiate the homologous recombination process. ATR can also activate the nucleotide excision repair pathway. Thus together, ATM and ATR can pause the cell cycle and repair DNA so that it does not get passed down to future generations.

8.7 Diseases Associated with ATM

Mutations in the ATM gene cause **ataxia-telangiectasia (A-T)**, a rare genetic disorder. Different types of mutations can cause A-T. For example, some mutations may affect ATM's expression levels. Others may lead to a truncated protein, while other mutations may specifically affect ATM's kinase activity [78]. Mutations in ATM can lead to the accumulation of DSBs. This, in turn, can lead to genomic instability, which increases the risk of some cancers, such as breast cancer and leukemias. The increase in DSBs can also lead to other health problems. For example, the nervous system is especially sensitive to DNA damage and, therefore, A-T can result in neurodegeneration. In general, cells in patients with A-T can be particularly sensitive to any type of radiation exposure, which can lead to DSBs.

V(D)J recombination can be affected in A-T. The process of V(D)J starts with the formation of a ds break in the DNA of T cells and B cells. This is followed by repair by the NHEJ, which leads to diversity at the breakpoint. Before the repair occurs, though, ATM binds to the ds break and stabilizes the ends. Thus, the loss of functional ATM disrupts the V(D)J process. This leads to impaired immune function [79–81].

Treatments for A-T mostly involve managing symptoms and preventive measures, such as avoiding radiation and other risk factors, and early screening for disease. Some potential treatments in the research stage involve gene replacement to introduce wild-type copies of ATM into cells, and CRISPR gene editing to repair the ATM gene [80, 82].

9 Large Aberrations in DNA That Occur at the Chromosome Level

Up until now in this chapter, we have focused primarily on small mutations that occur in one or a few nucleotides of the DNA. Here, larger aberrations that occur at the chromosome level are briefly discussed. These types of large mutations can cause significant damage and are more challenging for the cell to repair compared to the smaller mutations. Large aberrations may impact chromosome structure, and some may be large enough to be visible in a karyotype by microscopy. Some types of large aberrations, considered chromosomal aberrations, are illustrated in Fig. 23.14a, and include the following.

9.1 Chromosomal Rearrangements

9.1.1 Deletions

When deletions occur, a small segment of a chromosome might be deleted, or in more extreme cases, an entire chromosome may be deleted. If a deletion spans the coding sequence of the DNA, it can have severe consequences. Certain diseases are associated with deletions. For example, in **cri du chat syndrome**, part of chromosome 5 is deleted. The name of this disease is French for cry of the cat, stemming from the fact that babies with this disorder have a high-pitched cry that sounds like a cat. The disease is characterized by delayed development and intellectual disability. Several genes on chromosome 5 are deleted in patients with the disease. The loss of CTNND2 (catenin delta 2) may be especially important. This is a signaling protein that has important roles in the nervous system [83, 84].

Other disorders associated with large chromosomal deletions include **Williams syndrome**, in which part of chromosome 7 is deleted, and **Prader-Willi syndrome**, in which there is a deletion on chromosome 15 [79, 85].

9.1.2 Duplications and Amplifications

Duplications and amplifications involve extra copies of DNA within a chromosome. If certain genes are duplicated, this can impact cell proliferation. For example, extra copies of genes that promote cell proliferation may lead to tumors.

If duplications occur in the germ cells, genetic disorders can result. For example, individuals with **Charcot-Marie-Tooth disease type 1A**

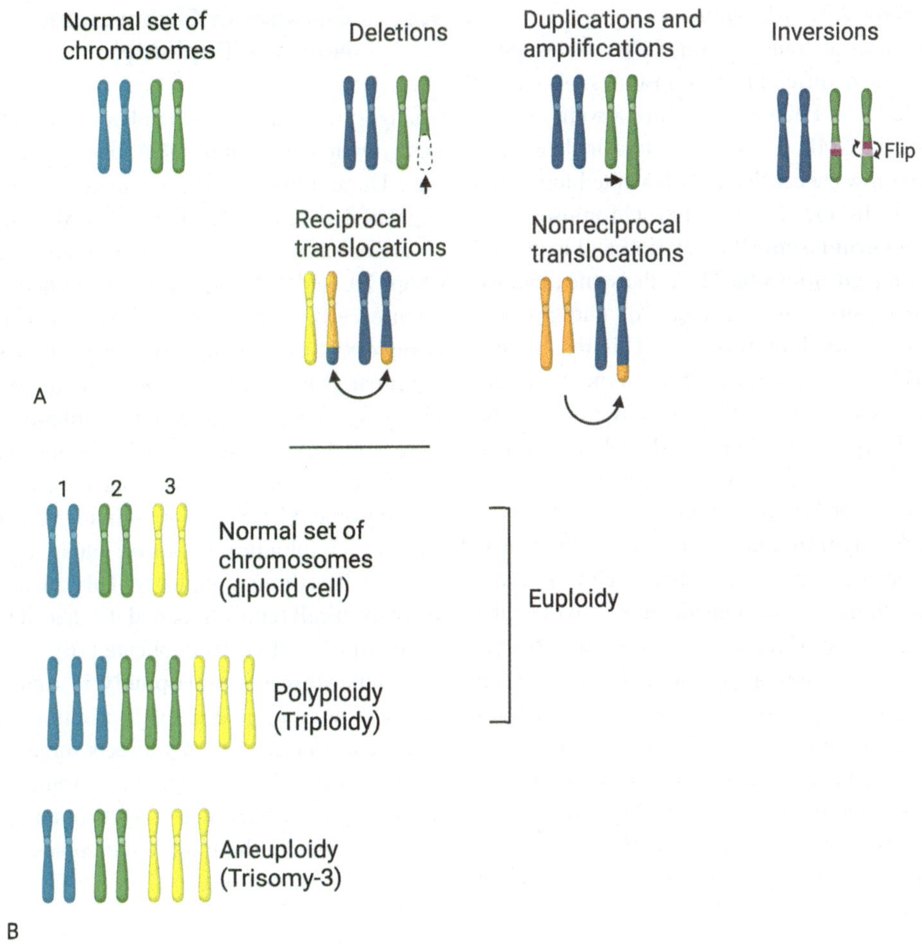

Fig. 23.14 Large chromosomal abnormalities. (**a**) shows abnormalities that can happen within chromosomes, such as deletions and translocations. (**b**) shows changes in ploidy, such as triploidy or aneuploidy. (*Source*: Created with BioRender.com)

usually have a duplication in **chromosome 17**. This disorder involves nervous system abnormalities, which can worsen over time. It is associated with damage to the myelin sheath covering nerves. Duplication of the gene for **PMP22** (Peripheral myelin protein 22) is one of the main causes. PMP22 encodes a protein that is important for the maintenance of the myelin sheath [86]. A duplication in **chromosome 15** can also lead to a genetic disorder, called **Dup15q syndrome**, which leads to a range of developmental disabilities [87, 88].

9.1.3 Inversions

In an inversion, a segment of DNA is flipped, and therefore is in a different orientation compared with the rest of the DNA in the chromosome. This can lead to disruption of genes and their regulatory elements [3].

9.1.4 Translocations

In a translocation, the material from one chromosome is transferred to a different chromosome. Translocations can be reciprocal or nonreciprocal [89].

- Reciprocal Translocations

In a reciprocal translocation, DNA is swapped between two different chromosomes. A classic example is the Philadelphia chromosome, which is found in CML (chronic myeloid leukemia), a bone marrow cancer that affects white blood cells (WBCs). In the Philadelphia chromosome, a piece of **chromosome 9** is missing and is instead fused with **chromosome 22**. At the same time, the reverse occurs, where a piece of chromosome 22 is added to chromosome 9. This results in a new DNA sequence encoding a new protein called BCR-ABL1. The discovery of the Philadelphia chromosome BCR-Abl was crucial for developing a new cancer treatment [3, 10, 90], and is discussed in more detail in Chap. 19.

Burkitt lymphoma is a B-cell cancer that is also caused by reciprocal chromosomal translocations. These translocations involve the proto-oncogene Myc, which is on chromosome 8. The most common translocation is between chromosomes 8 and 14, but translocations between chromosome 8 and other chromosomes are also found. The translocations lead to activation and improper regulation of Myc, which can lead to uncontrolled cell growth [10, 91]. (Burkitt lymphoma and Myc are discussed more in Chap. 19).

- Nonreciprocal Translocations

In a nonreciprocal translocation, genetic material is transferred from one chromosome to another, but there is no swap in genetic material between the two chromosomes [89].

9.1.5 Transpositions

In transpositions, genetic material is moved from one position in a chromosome to another in the same or a different chromosome. This type of change, as well as translocations, can have different outcomes. For example, it can cause genetic material to be separated from its normal regulatory elements and thus change how the expression of certain genes is controlled. It can also lead to genomic instability and can lead to disruption of genetic material [3, 92].

9.2 Changes in Chromosome Number (Euploidy)

Changes in chromosome number can lead to serious growth abnormalities that may not be reparable. These types of chromosome changes are expressed as euploidy (Fig. 23.14b). This type of abnormality is discussed in more detail in Chap. 20, and only briefly reviewed here. Some disease cells, such as cancer cells, can exhibit **polyploidy**. This means they have **more than the normal number of copies** of chromosomes. This can include, for example, triploidy, where there are three copies of each chromosome, or tetraploidy, where there are four copies of each chromosome. Generally, an individual with triploidy or tetraploidy will not be viable, but some cancer cells exhibit polyploidy. This can arise due to errors in cell replication, and it can lead to even more errors as the cells duplicate [16].

In other diseases, **aneuploidy** is seen. Aneuploidy is an abnormal number of chromosomes, where the abnormality only affects some (or one) chromosome [21]. An important example is trisomy 21, associated with the disorder Down's syndrome, where there are three copies of chromosome 21 instead of two [16].

9.3 Repair Mechanisms for Large Chromosomal Rearrangements

Polyploidy and aneuploidy are usually considered permanent in the cell, as mechanisms for their repair are either ineffective or absent. Translocations are also generally considered to be irreparable, although the cell's machinery may prevent some translocations from occurring in the first place. A major way the cellular machinery may prevent a translocation involves the repair of a DSB, because formation of a DSB is usually the first step in a chromosomal translocation. As we have covered above, DSBs can often be repaired, and this repair process may prevent the translocation from occurring. This repair process often involves NHEJ or HR at the site of the break [93, 94].

FDA-Approved Drugs Discussed in This Chapter

Commonly used brand name	Generic name	First FDA approval year	Common conditions	Major targets
Lynparza	*Olaparib*	2014	Several cancers	PARP inhibitor
Zejula	*Niraparib*	2017	Several cancers	PARP inhibitor

Definitions

Aflatoxins Fungal toxins that bind DNA and form adducts.

alkylating agents Chemicals that add alkyl groups to DNA bases, causing mispairing or strand breaks.

AP endonuclease 1 (APE) An endonuclease that cleaves the phosphodiester bond at an AP site during BER.

Apurinic/apyrimidinic (AP) site A site in DNA where a base has been removed, leaving just the sugar-phosphate backbone.

Base excision repair (BER) A mechanism for repairing small errors on a single strand of DNA that do not distort the shape of the double helix. These errors are caused by damaged bases, and they usually affect a single base. The base may be damaged by mechanisms such as deamination, alkylation, and oxidative stress.

D-Loop A displacement loop that occurs during homologous recombination.

Depurination Loss of a purine base (adenine or guanine), leaving an abasic site.

DNA glycosylase An enzyme that cleaves glycosidic bonds, the bonds that link the base to the sugar. Specific glycosylases remove specific bases, leaving an abasic site.

DNA-PKcs A protein kinase that stabilizes DNA ends in NHEJ.

Endonuclease An enzyme that cleaves the phosphodiester bonds within a DNA or RNA molecule, creating cuts.

Exonuclease An enzyme that removes nucleotides one at a time from the ends of a DNA or RNA strand, working either 3′ to 5′ or the other way around. It can also follow an endonuclease and start at a nick.

Exonuclease 1 (EXO1) An exonuclease that removes mismatched DNA during MMR.

Exonucleolytic proofreading The removal of mismatched nucleotides by DNA polymerase's 3'→5' activity during DNA replication.

Frameshift mutation Insertion or deletion of bases that shifts the reading frame of a gene.

Global genomic repair (GGR) A sub-pathway of NER that removes lesions from anywhere within the entire genome, including non-transcribed regions.

Homology-directed repair (HDR) Accurate repair of double-stranded breaks using a homologous DNA template. Active in S/G2 phase.

Ku70/Ku80 Proteins that recognize and bind to DNA ends in NHEJ.

Ligase IV A ligase that functions during NHEJ, which repairs double-stranded breaks during the NHEJ process.

Mismatch repair (MMR) A mechanism for repairing incorrectly paired bases called base-base mismatches that result from DNA replication errors.

MutSα/MutLα Protein complexes that recognize base-base mismatches and loops.

Nonhomologous end joining (NHEJ) DNA repair of double-stranded breaks without a template; this mechanism is error-prone.

Nucleotide excision repair (NER) A repair mechanism for bulky lesions like thymidine dimers. It involves a dual incision to remove a segment of DNA.

Polycyclic aromatic hydrocarbons (PAHs) Environmental carcinogens that form bulky DNA adducts.

Ribonucleotide excision repair (RER) A mechanism that removes misincorporated ribonucleotides from DNA.

Shelterin complex A protein complex that binds telomeres and protects them from being recognized as DNA damage.

Silent mutation A mutation that changes a nucleotide but does not change the encoded amino acid.

T-loop A loop structure that forms at the ends of telomeres.

Thymidine dimers Formed by dimerization between two adjacent thymidines via covalent bonds. This can disrupt the structure of the DNA and lead to more mutations.

Transcription-coupled repair (TCR) A sub-pathway of the NER pathway that focuses on the repair of DNA lesions on the transcribed strand of actively expressed genes. It is triggered when RNA polymerase stalls during transcription at a DNA lesion, such as a bulky DNA adduct.

Study Questions

Multiple-Choice Questions

1. Which repair mechanism uses a sister chromatid as a template for accurate repair of double-strand breaks?
   A. NHEJ
   B. BER
   C. Homology-directed repair (HDR)
   D. RER
2. Which enzyme removes damaged bases during base excision repair (BER)?
   A. DNA ligase
   B. DNA polymerase
   C. DNA glycosylase
   D. Ku70
3. What type of DNA damage does UV light most commonly cause?
   A. Oxidized guanine
   B. Abasic sites
   C. Thymidine dimers
   D. Ribonucleotide insertions
4. Which DNA repair process corrects bulky lesions such as thymidine dimers and other helix-distorting adducts?
   A. Base excision repair
   B. Mismatch repair
   C. Nucleotide excision repair
   D. Telomere maintenance

True/False Question

5. ATM is a damage sensor that binds mostly to double-stranded breaks.

Fill-in-the-Blank Questions

6. DNA polymerase's 3′ to 5′ activity that removes mismatched nucleotides is called ______.
7. Nonhomologous end joining (NHEJ) is more ______ than HDR because it lacks a template.
8. The complex that protects the ends of telomeres is the _________ complex

Short-Answer Questions

9. What are some differences between HDR from NHEJ?
10. Describe the main consequence of a frameshift mutation caused by DNA repair error.
11. ATM phosphorylates ________, which in turn can phosphorylate CDC25

Matching Question

12. Match the mutagen type to its damage mechanism.
    Mutagen Type
    A. UV radiation
    B. Alkylating agent
    C. Gamma radiation
    D. Aflatoxins
    E. ROS
    DNA effect
    1. Thymidine dimers
    2. Base modification
    3. DNA strand breaks
    4. Bulky adducts
    5. Oxidative base damage

Answer Key

Multiple-Choice Questions

1. **Answer**: C
2. **Answer**: C
3. **Answer**: C
4. **Answer**: C

True/False Question

5. **Answer**: True

Fill-in-the-Blank Questions

6. **Answer**: Exonucleolytic proofreading
7. **Answer**: Error-prone
8. **Answer**: Shelterin

Short-Answer Questions

9. **Answer**: HDR uses a sister chromatid as a template. It results in accurate repair. NHEJ does not require a template, but it is more error-prone.
10. **Answer**: Since it can affect the reading frame, it may lead to a change in the amino acid sequence and/or insertion of a stop codon in the coding region
11. **Answer**: Chk2

Matching Question

12. **Answer**:
A:1, B:2, C:3, D:4, E:5

References

1. Klug WS, Cummings MR, Spencer CA, Palladino MA, Killian DJ. Concepts of genetics. 12th ed. Pearson; 2019.
2. Carusillo A, Mussolino C. DNA damage: from threat to treatment. Cells. 2020;9(7) https://doi.org/10.3390/cells9071665.
3. Alberts B, Heald R, Johnson A, Morgan D, Raff M, Roberts K, Walter P, Wilson J, Hunt T. Molecular biology of the cell. W.W. Norton and Company; 2022.
4. Chatterjee N, Walker GC. Mechanisms of DNA damage, repair, and mutagenesis. Environ Mol Mutagen. 2017;58(5):235–63. https://doi.org/10.1002/em.22087.
5. Fry M. Essential biochemistry for medicine. 1st ed. Wiley; 2010.
6. Mohrenweiserf HW, Jones IM. Review of the molecular characteristics of gene mutations of the germline and somatic cells of the human. Mutat Res-Fund Mol Mech Mutagen. 1990;231(1):87–108. https://doi.org/10.1016/0027-5107(90)90179-8.
7. McLennan ABA, Elcock L, Turner P. BIOS instant notes in molecular biology. 4th ed. New York: Garland Science; 2012.
8. Brown TA. Genomes. 5th ed. CRC Press; 2023.
9. Kumari S, Sharma S, Advani D, Khosla A, Kumar P, Ambasta RK. Unboxing the molecular modalities of mutagens in cancer. Environ Sci Pollut Res Int. 2022;29(41):62111–59. https://doi.org/10.1007/s11356-021-16726-w.
10. Weinberg RA. The biology of cancer. 3rd ed. W.W. Norton and Company; 2023.
11. Nabel CS, Manning SA, Kohli RM. The curious chemical biology of cytosine: deamination, methylation, and oxidation as modulators of genomic potential. ACS Chem Biol. 2012;7(1):20–30. https://doi.org/10.1021/cb2002895.
12. Hahm JY, Park J, Jang E-S, Chi SW. 8-Oxoguanine: from oxidative damage to epigenetic and epitranscriptional modification. Exp Mol Med. 2022;54(10):1626–42. https://doi.org/10.1038/s12276-022-00822-z.
13. Hwa Yun B, Guo J, Bellamri M, Turesky RJ. DNA adducts: formation, biological effects, and new biospecimens for mass spectrometric measurements in humans. Mass Spectrom Rev. 2020;39(1–2):55–82. https://doi.org/10.1002/mas.21570.
14. Beukers R, Eker AP, Lohman PH. 50 years thymine dimer. DNA Repair (Amst). 2008;7(3):530–43. https://doi.org/10.1016/j.dnarep.2007.11.010.
15. Chen J, Potlapalli R, Quan H, Chen L, Xie Y, Pouriyeh S, et al. Exploring DNA damage and repair mechanisms: a review with computational insights. BioTech. 2024;13(1):3. https://doi.org/10.3390/biotech13010003.
16. Hartwell LH, Goldberg ML, Fischer JA, Hood L, Reynolds AE, Aquadro CF, Silver LM. Genetics: from genes to genomes. 6th ed. New York: McGraw Hill; 2017.
17. Thom CS, Dickson CF, Gell DA, Weiss MJ. Hemoglobin variants: biochemical properties and clinical correlates. Cold Spring Harb Perspect Med. 2013;3(3):a011858. https://doi.org/10.1101/cshperspect.a011858.
18. Gao QQ, McNally EM. The dystrophin complex: structure, function, and implications for therapy.

Compr Physiol. 2015;5(3):1223–39. https://doi.org/10.1002/cphy.c140048.
19. Sanderlin EJ, Keenan MM, Mense M, Revenko AS, Monia BP, Guo S, et al. CFTR mRNAs with nonsense codons are degraded by the SMG6-mediated endonucleolytic decay pathway. Nat Commun. 2022;13(1):2344. https://doi.org/10.1038/s41467-022-29935-9.
20. Cystic Fibrosis Foundation. Types of CFTR Mutations. Available from: https://www.cff.org/research-clinical-trials/types-cftr-mutations.
21. Strachan T, Read A. Human molecular genetics. 5th ed. New York: Garland Science; 2019.
22. Wang X, Dong Q, Chen G, Zhang J, Liu Y, Cai Y. Frameshift and wild-type proteins are often highly similar because the genetic code and genomes were optimized for frameshift tolerance. BMC Genomics. 2022;23(1):416. https://doi.org/10.1186/s12864-022-08435-6.
23. Reha-Krantz LJ. DNA polymerase proofreading: Multiple roles maintain genome stability. Biochim Biophys Acta (BBA) – Proteins Proteomics. 2010;1804(5):1049–63. https://doi.org/10.1016/j.bbapap.2009.06.012.
24. DePamphilis ML, Bell SD. Genome duplication. Garland Science; 2010.
25. Bhattacharjee S, Nandi S. Choices have consequences: the nexus between DNA repair pathways and genomic instability in cancer. Clin Transl Med. 2016;5(1):45. https://doi.org/10.1186/s40169-016-0128-z.
26. Elbakry A, Löbrich M. Homologous recombination subpathways: a tangle to resolve. Front Genet. 2021;12 https://doi.org/10.3389/fgene.2021.723847.
27. Jensen NM, Dalsgaard T, Jakobsen M, Nielsen RR, Sørensen CB, Bolund L, et al. An update on targeted gene repair in mammalian cells: methods and mechanisms. J Biomed Sci. 2011;18(1):10. https://doi.org/10.1186/1423-0127-18-10.
28. Chen CC, Feng W, Lim PX, Kass EM, Jasin M. Homology-directed repair and the role of BRCA1, BRCA2, and related proteins in genome integrity and cancer. Annu Rev Cancer Biol. 2018;2:313–36. https://doi.org/10.1146/annurev-cancerbio-030617-050502.
29. Prakash R, Zhang Y, Feng W, Jasin M. Homologous recombination and human health: the roles of BRCA1, BRCA2, and associated proteins. Cold Spring Harb Perspect Biol. 2015;7(4):a016600. https://doi.org/10.1101/cshperspect.a016600.
30. FDA. Lynparza(Olaparib) approval overview. 2014. Available from: https://www.accessdata.fda.gov/scripts/cder/daf/index.cfm?event=overview.process&ApplNo=208558.
31. FDA. Zejula(Niraparib) approval overview. 2017. Available from: https://www.accessdata.fda.gov/scripts/cder/daf/index.cfm?event=overview.process&ApplNo=208447.
32. Katzung BG, Vanderah TW, Trevor AAJ. Katzung's basic and clinical pharmacology. 16th ed. McGraw-Hill Education; 2024.
33. Stinson BM, Loparo JJ. Repair of DNA double-Strand breaks by the nonhomologous end joining pathway. Annu Rev Biochem. 2021;90:137–64. https://doi.org/10.1146/annurev-biochem-080320-110356.
34. Gullickson P, Xu YW, Niedernhofer LJ, Thompson EL, Yousefzadeh MJ. The role of DNA repair in immunological diversity: from Molecular mechanisms to clinical ramifications. Front Immunol. 2022;13:834889. https://doi.org/10.3389/fimmu.2022.834889.
35. O'Driscoll M. Diseases associated with defective responses to DNA damage. Cold Spring Harb Perspect Biol. 2012;4(12) https://doi.org/10.1101/cshperspect.a012773.
36. Weterings E, Chen DJ. The endless tale of non-homologous end-joining. Cell Res. 2008;18(1):114–24. https://doi.org/10.1038/cr.2008.3.
37. Immune Deficiency Foundation. Severe combined immuneodeficiency (SCID). 2019. Available from: https://primaryimmune.org/understanding-primary-immunodeficiency/types-of-pi/severe-combined-immunodeficiency-scid#:~:text=Severe%20combined%20immunodeficiency%20(SCID)%20is,past%20two%20years%20of%20age.
38. Iancu O, Allen D, Knop O, Zehavi Y, Breier D, Arbiv A, et al. Multiplex HDR for disease and correction modeling of SCID by CRISPR genome editing in human HSPCs. Mol Ther Nucleic Acids. 2023;31:105–21. https://doi.org/10.1016/j.omtn.2022.12.006.
39. Krokan HE, Bjørås M. Base excision repair. Cold Spring Harb Perspect Biol. 2013;5(4):a012583. https://doi.org/10.1101/cshperspect.a012583.
40. Wei H, Yu X. Functions of PARylation in DNA damage repair pathways. Genomics Proteomics Bioinformatics. 2016;14(3):131–9. https://doi.org/10.1016/j.gpb.2016.05.001.
41. Nosella ML, Kim TH, Huang SK, Harkness RW, Goncalves M, Pan A, et al. Poly(ADP-ribosyl)ation enhances nucleosome dynamics and organizes DNA damage repair components within biomolecular condensates. Mol Cell. 2024;84(3):429–46.e17. https://doi.org/10.1016/j.molcel.2023.12.019.
42. Morales J, Li L, Fattah FJ, Dong Y, Bey EA, Patel M, et al. Review of poly (ADP-ribose) polymerase (PARP) mechanisms of action and rationale for targeting in cancer and other diseases. Crit Rev Eukaryot Gene Expr. 2014;24(1):15–28. https://doi.org/10.1615/critreveukaryotgeneexpr.2013006875.
43. Jacobs AL, Schär P. DNA glycosylases: in DNA repair and beyond. Chromosoma. 2012;121(1):1–20. https://doi.org/10.1007/s00412-011-0347-4.
44. Fishel R. Mismatch repair. J Biol Chem. 2015;290(44):26395–403. https://doi.org/10.1074/jbc.R115.660142.

45. Pećina-Šlaus N, Kafka A, Salamon I, Bukovac A. Mismatch repair pathway, genome stability and cancer. Front Mol Biosci. 2020;7:122. https://doi.org/10.3389/fmolb.2020.00122.
46. Cline SD, Hanawalt PC. Who's on first in the cellular response to DNA damage? Nat Rev Mol Cell Biol. 2003;4(5):361–73. https://doi.org/10.1038/nrm1101.
47. Pluciennik A, Dzantiev L, Iyer RR, Constantin N, Kadyrov FA, Modrich P. PCNA function in the activation and strand direction of MutLα endonuclease in mismatch repair. Proc Natl Acad Sci. 2010;107(37):16066–71. https://doi.org/10.1073/pnas.1010662107.
48. Chen X, Paudyal SC, Chin R-I, You Z. PCNA promotes processive DNA end resection by Exo1. Nucleic Acids Res. 2013;41(20):9325–38. https://doi.org/10.1093/nar/gkt672.
49. Slade D. Maneuvers on PCNA rings during DNA replication and repair. Genes. 2018;9(8):416. https://doi.org/10.3390/genes9080416.
50. CDC. About lynch syndrome. 2024. Available from: https://www.cdc.gov/colorectal-cancer-hereditary/about/lynch-syndrome.html#:~:text=What%20Lynch%20syndrome%20is,Certain%20types%20of%20skin%20cancers.
51. Yale Medicine. Lynch syndrome. 2025. Available from: https://www.yalemedicine.org/conditions/lynch-syndrome#:~:text=Lynch%20syndrome%20is%20an%20inherited,the%20risk%20of%20developing%20cancer.
52. Cleveland Clinic. Lynch syndrome and HNPCC. 2022. Available from: https://my.clevelandclinic.org/health/diseases/17195-lynch-syndrome-and-hnpcc.
53. Wimmer K, Etzler J. Constitutional mismatch repair-deficiency syndrome: have we so far seen only the tip of an iceberg? Hum Genet. 2008;124(2):105–22. https://doi.org/10.1007/s00439-008-0542-4.
54. Medline Plus. Constitutional mismatch repair deficiency syndrome. 2020. Available from: https://medlineplus.gov/genetics/condition/constitutional-mismatch-repair-deficiency-syndrome/#causes.
55. Spivak G. Nucleotide excision repair in humans. DNA Repair (Amst). 2015;36:13–8. https://doi.org/10.1016/j.dnarep.2015.09.003.
56. Khobta A, Sarmini L. Transcription-coupled nucleotide excision repair: a faster solution or the only option? Biomolecules. 2025;15(7):1026. https://doi.org/10.3390/biom15071026.
57. Jackson S, Xiong Y. CRL4s: the CUL4-RING E3 ubiquitin ligases. Trends Biochem Sci. 2009;34(11):562–70. https://doi.org/10.1016/j.tibs.2009.07.002.
58. Sugitani N, Sivley RM, Perry KE, Capra JA, Chazin WJ. XPA: a key scaffold for human nucleotide excision repair. DNA Repair (Amst). 2016;44:123–35. https://doi.org/10.1016/j.dnarep.2016.05.018.
59. Shivji MK, Podust VN, Hübscher U, Wood RD. Nucleotide excision repair DNA synthesis by DNA polymerase epsilon in the presence of PCNA, RFC, and RPA. Biochemistry. 1995;34(15):5011–7. https://doi.org/10.1021/bi00015a012.
60. Shuck SC, Short EA, Turchi JJ. Eukaryotic nucleotide excision repair: from understanding mechanisms to influencing biology. Cell Res. 2008;18(1):64–72. https://doi.org/10.1038/cr.2008.2.
61. Kokic G, Wagner FR, Chernev A, Urlaub H, Cramer P. Structural basis of human transcription–DNA repair coupling. Nature. 2021;598(7880):368–72. https://doi.org/10.1038/s41586-021-03906-4.
62. Selby CP, Lindsey-Boltz LA, Li W, Sancar A. Molecular Mechanisms of Transcription-Coupled Repair. Annu Rev Biochem. 2023;92:115–44. https://doi.org/10.1146/annurev-biochem-041522-034232.
63. Fischer Eric S, Scrima A, Böhm K, Matsumoto S, Lingaraju Gondichatnahalli M, Faty M, et al. The molecular basis of CRL4DDB2/CSA ubiquitin ligase architecture, targeting, and activation. Cell. 2011;147(5):1024–39. https://doi.org/10.1016/j.cell.2011.10.035.
64. Sugasawa K. Xeroderma pigmentosum genes: functions inside and outside DNA repair. Carcinogenesis. 2008;29(3):455–65. https://doi.org/10.1093/carcin/bgm282.
65. Medline Plus. XPC gene. 2023. Available from: https://medlineplus.gov/genetics/gene/xpc/#:~:text=Xeroderma%20pigmentosum&text=Variants%20in%20the%20XPC%20gene,abnormalities%20and%20others%20do%20not.
66. Hossain M, Hasan A, Shawan M, Banik S, Jahan I. Current therapeutic strategies of Xeroderma Pigmentosum. Indian J Dermatol. 2021;66(6):660–7. https://doi.org/10.4103/ijd.ijd_329_21.
67. Kellner V, Luke B. Molecular and physiological consequences of faulty eukaryotic ribonucleotide excision repair. EMBO J. 2020;39(3):e102309. https://doi.org/10.15252/embj.2019102309.
68. de Lange T. Shelterin: the protein complex that shapes and safeguards human telomeres. Genes Dev. 2005;19(18):2100–10. https://doi.org/10.1101/gad.1346005.
69. Schmutz I, de Lange T. Shelterin. Curr Biol. 2016;26(10):R397–R9. https://doi.org/10.1016/j.cub.2016.01.056.
70. Doksani Y. The response to DNA damage at Telomeric repeats and its consequences for telomere function. Genes (Basel). 2019;10(4) https://doi.org/10.3390/genes10040318.
71. Maréchal A, Zou L. DNA damage sensing by the ATM and ATR kinases. Cold Spring Harb Perspect Biol. 2013;5(9) https://doi.org/10.1101/cshperspect.a012716.
72. Morgan DO. The cell cycle: principles of control. New Science Press; 2007.
73. Engeland K. Cell cycle regulation: p53-p21-RB signaling. Cell Death Differ. 2022;29(5):946–60. https://doi.org/10.1038/s41418-022-00988-z.
74. Tibbetts RS, Brumbaugh KM, Williams JM, Sarkaria JN, Cliby WA, Shieh SY, et al. A role for ATR in the DNA damage-induced phosphorylation of p53. Genes Dev. 1999;13(2):152–7. https://doi.org/10.1101/gad.13.2.152.

75. Gottifredi V, Karni-Schmidt O, Shieh SS, Prives C. p53 down-regulates CHK1 through p21 and the retinoblastoma protein. Mol Cell Biol. 2001;21(4): 1066–76. https://doi.org/10.1128/mcb.21.4.1066-1076.2001.
76. Karimian A, Ahmadi Y, Yousefi B. Multiple functions of p21 in cell cycle, apoptosis and transcriptional regulation after DNA damage. DNA Repair. 2016;42:63–71. https://doi.org/10.1016/j.dnarep.2016.04.008.
77. Chipuk JE, Kuwana T, Bouchier-Hayes L, Droin NM, Newmeyer DD, Schuler M, et al. Direct activation of Bax by p53 mediates mitochondrial membrane Permeabilization and apoptosis. Science. 2004;303 (5660):1010–4. https://doi.org/10.1126/science.1092734.
78. Putti S, Giovinazzo A, Merolle M, Falchetti ML, Pellegrini M. ATM kinase dead: from ataxia telangiectasia syndrome to cancer. Cancers (Basel). 2021;13 (21) https://doi.org/10.3390/cancers13215498.
79. Rothblum-Oviatt C, Wright J, Lefton-Greif MA, McGrath-Morrow SA, Crawford TO, Lederman HM. Ataxia telangiectasia: a review. Orphanet J Rare Dis. 2016;11(1):159. https://doi.org/10.1186/s13023-016-0543-7.
80. Choi M, Kipps T, Kurzrock R. ATM mutations in cancer: therapeutic implications. Mol Cancer Ther. 2016;15(8):1781–91. https://doi.org/10.1158/1535-7163.Mct-15-0945.
81. Lee J-H. ATM in immunobiology: from lymphocyte development to cancer immunotherapy. Transl Oncol. 2025;52:102268. https://doi.org/10.1016/j.tranon.2024.102268.
82. Mehri A, Toosi MB, Tavasoli AR, Saberi-Karimian M. The latest developments for the treatment of ataxia telangiectasia: a narrative review. Cerebellum. 2024;23(6):2607–15. https://doi.org/10.1007/s12311-024-01746-2.
83. Medline Plus. Cri-du-chat syndrome. 2022. Available from: https://medlineplus.gov/genetics/condition/cri-du-chat-syndrome/#causes.
84. Plus M. CTDNND2 gene; catenin delta 2. 2010. Available from: https://medlineplus.gov/genetics/gene/ctnnd2/.
85. Cerruti MP. Cri du Chat syndrome. Orphanet J Rare Dis. 2006;1:33. https://doi.org/10.1186/1750-1172-1-33.
86. Muscular Dystrophy Association. Charot-Marie-Tooth Disease (CMT). Available from: https://www.mda.org/disease/charcot-marie-tooth/types/cmt1#:~:text=What%20are%20the%20symptoms%20of,(stop%20breathing%20while%20sleeping).
87. Chance PF, Pleasure D. Charcot-Marie-tooth syndrome. Arch Neurol. 1993;50(11):1180–4. https://doi.org/10.1001/archneur.1993.00540110060006.
88. Lusk L, Vogel-Farley V, DiStefano C, Jeste S. 1993 Maternal 15q duplication syndrome. In: Adam MP, Feldman J, Mirzaa GM, Pagon RA, Wallace SE, Amemiya A, editors. GeneReviews(®). Seattle: University of Washington, Seattle. Copyright © 1993–2024, University of Washington, Seattle. GeneReviews is a registered trademark of the University of Washington, Seattle. All rights reserved.
89. Zhang Y, editor. Chromosomal translocation: advances in experimentsl medicine and biology. Springer; 2018.
90. Sacha T. Imatinib in chronic myeloid leukemia: an overview. Mediterr J Hematol Infect Dis. 2014;6(1): e2014007. https://doi.org/10.4084/mjhid.2014.007.
91. Fang H, Wang W, Medeiros LJ. Burkitt lymphoma. Hum Pathol. 2024:105703. https://doi.org/10.1016/j.humpath.2024.105703.
92. Torella A, Budillon A, Zanobio M, Del Vecchio Blanco F, Picillo E, Politano L, et al. Alu-mediated insertions in the DMD gene: a difficult puzzle to interpret clinically. Int J Mol Sci. 2023;24(11) https://doi.org/10.3390/ijms24119241.
93. Pfeiffer P, Goedecke W, Kuhfittig-Kulle S, Obe G. Pathways of DNA double-strand break repair and their impact on the prevention and formation of chromosomal aberrations. Cytogenet Genome Res. 2004;104(1–4):7–13. https://doi.org/10.1159/000077460.
94. Hromas R, Williamson E, Lee SH, Nickoloff J. Preventing the chromosomal translocations that cause cancer. Trans Am Clin Climatol Assoc. 2016;127:176–95.
95. Pace JK II, Sen SK, Batzer MA, Feschotte C. Repair-mediated duplication by capture of proximal chromosomal DNA has shaped vertebrate genome evolution. PLoS Genet. 2009;5(5):e1000469. https://doi.org/10.1371/journal.pgen.1000469
96. Brandsma I, Gent DC. Pathway choice in DNA double strand break repair: observations of a balancing act. Genome Integr. 2012;3(1):9. https://doi.org/10.1186/2041-9414-3-9.
97. Pecina-Slaus, et al. Mismatch repair pathway, genome stability and cancer. Front Mol Biosci. 2020;7:122. https://doi.org/10.3389/fmolb.2020.00122.
98. Hu Y, et al. Role of oxidative DNA damage and repair in atrial fibrillation and ischemic heart disease. Int J Mol Sci. 2021;22(8):3838. https://doi.org/10.3390/ijms22083838.
99. Doksani Y. The response to DNA damage at telomeric repeats and its consequences for telomere function. Genes. 2019;10(4):318. https://doi.org/10.3390/genes10040318.

24 Introduction to the Immune System and Virology for Pharmaceutical Sciences

Abstract

Understanding concepts such as metabolic pathways, enzyme function, and cell cycle regulation is central to understanding human health and identifying drug targets. However, to fully understand disease and drug development, some discussion of immunology and virology is also necessary. Many of today's therapies, including vaccines, antivirals, and immunotherapies, rely on our understanding of how the immune system functions and how it responds to pathogens such as viruses. A wide array of therapies involves modulating immune processes. Immunology and virology are vast fields, and this chapter provides only a concise overview of key concepts in these fields, which are essential for understanding many of today's therapeutic treatments. Some examples of drugs and disease treatment mechanisms that involve the immune system are also explored. Antiviral medications and vaccines are also discussed.

Keywords

Immune system · Cells of the immune system · Viral infections · DNA viruses · RNA viruses

Abbreviations

3TC	Lamivudine
ADCC	Antibody-dependent cellular cytotoxicity
AIDS	Acquired immunodeficiency syndrome
APC	Antigen-presenting cell
AZT	Zidovudine (a reverse transcriptase inhibitor)
BCR	B-cell receptor
C1–C9	Complement proteins
CTL	Cytotoxic T lymphocyte
CYP3A4	Cytochrome P450 3A4
Fab	Fragment antigen binding
Fc	Fragment crystallizable
HIV	Human immunodeficiency virus
ICI	Immune checkpoint inhibitor
Ig	Immunoglobulin (e.g., IgG, IgM)
IL	Interleukin (e.g., IL-1, IL-6)
MAC	Membrane attack complex
MASP	Mannose-binding lectin-associated serine protease
MHC	Major histocompatibility complex
NA	Neuraminidase
NK	Natural killer (cell)

A. Minden, *Biochemistry, Cell and Molecular Biology for Pharmacists and Health Professionals*,
https://doi.org/10.1007/978-3-032-20529-2_24

PAMPs	Pathogen-associated molecular patterns
PD-1	Programmed cell death protein 1
PD-L1	Programmed death ligand 1
PEG	Polyethylene glycol
PNH	Paroxysmal nocturnal hemoglobinuria
TCR	T-cell receptor
Th	T helper (cell)
TIL	Tumor-infiltrating lymphocyte
TNF	Tumor necrosis factor

What You Will Learn

- This chapter provides a short introduction to the immune system and viruses. It is meant to give students a basic understanding of these topics, rather than the depth that would be covered in an immunology or virology textbook.
- Overview of basic concepts in immunology, including the innate immune system, the adaptive immune response, cells of the immune system, and the complement system.
- Viral infections, including DNA viruses, RNA viruses, and retroviruses.
- Some examples of drugs or disease treatment methods that involve the immune system, including monoclonal antibodies, immune checkpoint inhibitors, tumor-infiltrating lymphocytes (TILs), and complement inhibitors.
- Antiviral drugs and vaccines, including HIV inhibitors, Paxlovid, Tamiflu, and disease-preventing vaccines.

This book is intended to give readers a solid foundation in how core concepts from biochemistry and cell biology can help us understand human disease and can drive the development of effective therapies. Understanding topics we have covered, such as metabolic pathways, enzyme function, and the cell cycle, is central to understanding human health and identifying drug targets. However, to fully understand disease and drug development, some discussion of immunology and virology is also necessary. Many of today's therapies, including vaccines, antivirals, and immunotherapies, rely on our understanding of how the immune system functions and how it responds to pathogens such as viruses. A wide array of therapies involves modulating immune processes. Immunology and virology are huge fields that require their own textbooks. In fact, many comprehensive immunology textbooks are available, and some references for such books are at the end of this chapter. This chapter provides a concise overview of key concepts in immunology and virology, along with examples of therapeutic treatments that involve the immune system, the complement system, and viral infections. Much of this discussion draws from the following reference, which also provides additional insites [1–3].

1 The Immune Response

The immune system protects us from foreign pathogens such as bacteria and viruses, and from our own cells when they develop abnormalities. The immune system is comprised of two major components, the **innate immune system** and the **adaptive immune system**. The innate immune system responds quickly and non-specifically to various threats, while the adaptive immune system responds more slowly but with higher specificity and immunological memory [1–3].

1.1 Key Terms to Know

Before discussing the immune system, it is helpful to understand a few important terms: a **pathogen** is an agent, often a microorganism, which can cause disease. An **antigen** is any molecule that can be recognized by the immune system, and it is usually recognized as foreign. **PAMPs** (pathogen-associated molecular patterns) are conserved structures found on groups of pathogens. **Cytokines** are small extracellular signaling proteins released by immune cells that have important roles in cell signaling, inflammation, and the regulation of immune responses. **Phagocytosis** is the mechanism that some immune cells use to engulf debris or pathogens. **Antibodies** are Y-shaped proteins that are made by B cells and secreted. They bind to antigens and are highly specific. The **T-cell receptor (TCR)** is a receptor on T cells that recognizes antigens presented by other cells. **MHC** (major histocompatibility complex) refers to a set of proteins on cell surfaces that display antigen fragments for recognition by T cells. **MHC class I** is found on almost all nucleated cells, and it displays endogenous antigens. **MHC class II** is on antigen-presenting cells and displays mostly exogenous antigens. An endogenous antigen is from inside the cell, such as a viral protein or an abnormal cellular protein. An exogenous antigen is from the outside of the cell, such as a bacterial protein or a toxin. **Degranulation** is the process by which some immune cells release the contents of their granules into the environment or to a target cell. These granules contain molecules such as toxic enzymes, inflammatory mediators, cytokines, perforin, and others [1, 3, 4].

1.2 The Innate Immune System

1.2.1 Physical and Chemical Barriers

Physical and chemical barriers are important components of the innate immune system, which is a nonspecific immune response and our first line of defense against invading pathogens. This aspect of the innate immune system consists of physical and chemical barriers that keep the pathogen out. Examples include the skin and mucous membranes, and environmental conditions. This type of defense is rapid and nonspecific. It does not respond to specific pathogens, nor does it have a memory of what pathogens it has encountered before [1, 4].

The presence of barriers is not always enough. Some pathogens may still get through these barriers. For example, they may pass through a cut in the skin. This is where some of the cells of the innate immune system play important roles. The cells of the innate immune system include macrophages and dendritic cells [5, 6].

1.2.2 The Roles of Macrophages

Macrophages are among the first immune cells recruited to the site of infection because they recognize certain structures on pathogens, called **PAMPs** (pathogen-associated molecular patterns). They can bind to these PAMPs through their own receptors, often the Toll-like receptors (TLRs).

Once macrophages detect the pathogen, they begin the process of phagocytosis. If the pathogen is not completely phagocytosed, the macrophages then secrete cytokines, such as TNFα, IL-1, and IL-6. This, in turn, leads to inflammation. This inflammatory response involves increased vascular permeability and the recruitment of additional immune cells. These include more macrophages, **monocytes**, **neutrophils**, and **natural killer (NK) cells**. These different cell types all have roles in eliminating the pathogen, leading to its destruction. The inflammatory response also leads to fever and other responses. This cytokine response is a normal and important response, but if cytokine signaling becomes too high, it can lead to too much inflammation and inflammatory disease. For this reason and others, cytokines such as TNF have become important drug targets [4, 7, 8].

1.2.3 Dendritic Cells

Dendritic cells are also phagocytic cells, but they have another role, in activating the adaptive immune response. After ingesting a pathogen, dendritic cells migrate to the lymph nodes. There, they "present" antigen fragments to T cells. Consequently, dendritic cells are important antigen-presenting cells. This will serve as a transition to the **adaptive immune response**, to be discussed below [6].

1.2.4 The Complement System

Another part of the innate immune system is the complement system [9, 10]. In the innate immune response, the complement system is commonly activated through either the alternative pathway or the lectin pathway. The complement system contributes to pathogen destruction through a cascade of reactions that lead to opsonization (tagging for destruction), recruitment of inflammatory cells, and cell lysis.

1.3 The Adaptive Immune Response

Not all pathogens are stopped by the innate immune response. Those that make it past this response are handled by the **adaptive immune response**. This response is highly specific. It recognizes specific pathogens and can distinguish between "self" and "nonself" antigens. This response also involves cell **memory**, resulting in a rapid response upon re-exposure to a pathogen.

The adaptive immune response includes two main arms: humoral immunity and cell-mediated immunity. Humoral immunity involves antibodies produced by B cells. Cell-mediated immunity involves T cells. Some T cells kill infected or abnormal cells ($CD8^+$ cytotoxic T cells), while others ($CD4^+$ helper T cells) help other immune cells. Broadly, within these two arms, the adaptive immune response involves two main types of cells, **T cells** and **B cells**, as well as the antibodies that B cells produce [1, 3].

1.3.1 T Cells

T cells originate from hematopoietic stem cells (a type of adult stem cell) in the bone marrow, and they mature in the thymus. T cells circulate throughout the body and are activated when they encounter antigen presented by an antigen-presenting cell (**APC**), such as **dendritic cells (DCs)**. The DCs display exogenous antigen fragments on **class II molecules** on their surface. These are recognized by a class of T cells called **$CD4^+$ T cells**. When these $CD4^+$ T cells have not yet encountered their specific antigen, they are naive T cells. These naive $CD4^+$ T cells recognize the presented antigen via their **T-cell receptors (TCRs)**, which is a key step in their activation. To become fully active, they also require a second signal called co-stimulation, which is provided through interaction with the dendritic cell. This two-signal requirement is known as the **co-stimulation model** of T-cell activation. Finally, $CD4^+$ T cells differentiate into active **T helper cells (Th cells)** [11].

Depending on the surrounding cytokines in the environment, the Th cells can differentiate into several subtypes, including Th1, Th2, Th17, Tfh, and Treg cells. T helper (Th) cells are not cytotoxic themselves (i.e., they do not directly target and kill abnormal cells), but they activate other cells that are part of the response. The different Th subtypes have different functions. Briefly, **Th1** promotes a cytotoxic response by stimulating macrophages to fight intracellular pathogens such as viruses or some bacteria. **Th2** cells are best at fighting parasites and are involved in allergic reactions, and **Th17** cells fight extracellular bacteria and fungi. **Tfh** are the main Th cells that activate B cells (although Th2 also activate B cells), which is important for **antibody production**, and **Treg** have roles in preventing autoimmunity [4, 12].

As an example, let's discuss Th1. Th1 cells secrete cytokines (especially IL-2 and IFN-γ) that lead to activation of another class of T cells, the **$CD8^+$ T** cells and macrophages. The CD8+ T cells recognize antigens presented on **MHC class I molecules**, which are found on all nucleated cells. The CD8+ cells then mature into cytotoxic T lymphocytes (**CTLs**), which are capable of killing infected or abnormal cells. This activation and maturation of the CD8+ T cells requires the following: (a) recognition of an antigen on a **MHC class I** molecule, (b) co-stimulation (provided by dendritic cells), and (c) help from the Th cells (especially the Th1 cells), which secrete cytokines as discussed above.

The mature **CTLs** can now find and destroy abnormal cells or cells that are infected with a virus or intracellular bacterium. They can do so by two main mechanisms. They can release substances that lead to apoptosis of the target cells. These include perforin, which form pores in the target cell, granzymes, which enter the pores and trigger apoptosis by activating caspases, and granulysin, which has microbial activity [13]. Alternatively, they can activate the FasL-Fas receptor pathway, which leads to activation of signaling pathways that trigger apoptosis [1, 14].

1.3.2 B Cells

B cells also derive from hematopoietic stem cells in the bone marrow. In the bone marrow, they undergo a maturation process that causes them to express a unique membrane-bound antibody, also called a B-cell receptor (BCR). Each B cell has a unique receptor that recognizes a unique antigen. When circulating B cells encounter their matching antigen, they internalize and process the antigen and present fragments on **MHC**

class II. Then, interaction with a Th cell leads to full activation of the B cell. This interaction involves cytokines and co-stimulatory signals.

Once activated, the B cells proliferate and differentiate. As a result, they become either **plasma cells** or **memory B cells**. **Plasma cells** secrete their unique antibody and make large quantities of it. **Memory B cells** do not secrete antibodies, but they maintain the genetic rearrangement that encodes the same antibody specificity. While the plasma cells are short-lived, the memory B cells persist for a long time. They can rapidly become plasma cells if the same antigen is encountered another time. These plasma cells can then secrete antibodies, which play a critical role in combating the infection. Memory B cells are an important outcome because they are ready to mount an immune response quickly if a second infection occurs. This is why second infections with a pathogen usually lead to a quicker response, which may even occur before significant symptoms are noticed. (Importantly, T cells can also form long-term memory cells after their activation, and the presence of these types of cells that can persist after an infection is important for how vaccines work [4]).

1.3.3 Antibody

The plasma B cells secrete large amounts of antibody. There are five major categories of antibodies: IgA, IgD, IgE, IgG, and IgM. They have different functions in responding to and neutralizing the pathogens they recognize. Antibodies are Y-shaped proteins that have two identical light chains and two identical heavy chains [15–17] (see Fig. 24.1). At one end are two **antigen-binding domains** (**Fabs**). This region is variable, which allows antibody specificity as different antibodies can recognize and bind to different pathogens via this variable region. The other end is the fragment crystallizable end (**Fc**). A hinge region holds the Fab and the Fc regions together. The Fab ends are identical, and they bind the **antigen**. By binding to the antigen, the antibody can accomplish several things:

(a) They can neutralize the pathogen, thereby inactivating it.

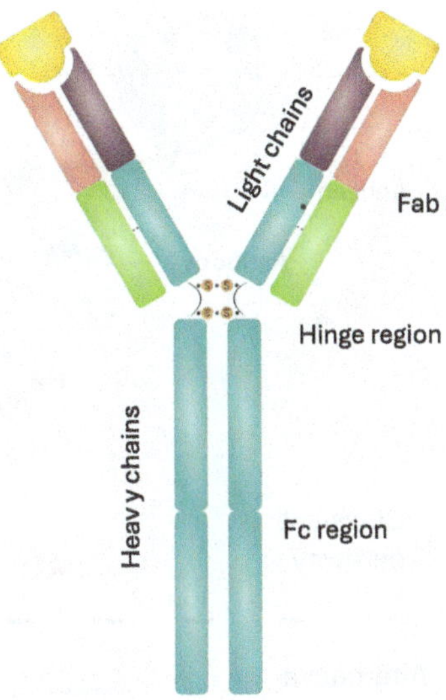

Fig. 24.1 Antibody structure. Structure of an antibody showing two heavy chains, two light chains, and Fc region, and other important components described in the text. (*Source*: Illustration from Freepik.com (premium license))

(b) The antibody can also lead to the recruitment of other immune cells, which can destroy the pathogen. This is mediated by the Fc region of the antibody, which can bind to receptors on various cells. For example, the Fc region can bind to receptors on macrophages, neutrophils, and mast cells, which can trigger responses such as phagocytosis, degranulation, and antibody-dependent cellular cytotoxicity (ADCC). In ADCC, natural killer (NK) cells recognize the Fc region on the antibody bound to the infected or abnormal cells, and then release cytotoxic granules to induce apoptosis in the targeted cell.

(c) Finally, the antibody can lead to activation of the complement system via the classical pathway, which is described next.

1.4 The Complement System

The complement system is also part of the immune system [18] (see Fig. 24.2). It involves a series of proteins in the blood and plays a role in both the innate and adaptive immune systems.

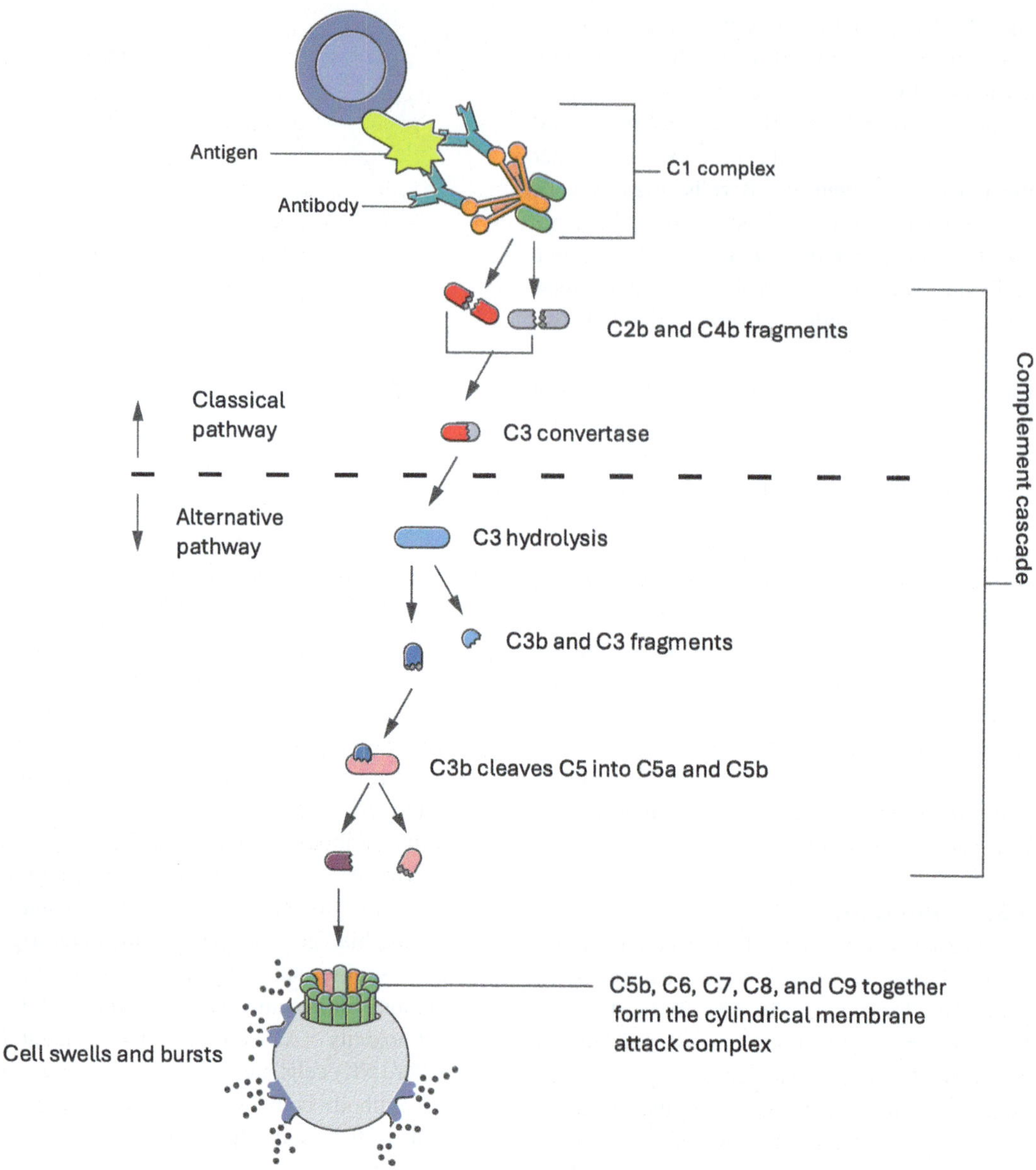

Fig. 24.2 The complement system: an overview of its key components. The complement system is part of the immune response. The classical, lectin, and alternative pathways are shown. (*Source*: Figure adapted from https://commons.wikimedia.org/wiki/File:Complement_pathway.svg)

The complement proteins include a series of nine core proteins called **C1–C9**, along with many regulatory and accessory proteins. These proteins are zymogens. Thus, they are inactive precursor proteins that become active when they are cleaved. The complement system can be activated by one of the three different pathways, the **classical pathway**, the **lectin pathway**, and the **alternative pathway**, summarized below and in [19].

1.4.1 The Classical Pathway

The classical pathway is the pathway that can be activated in response to antibody-antigen complexes. The **classical pathway** starts when **C1** binds to the antibody-antigen complex. C1 contains three subunits: C1q, C1r, and C1s. C1r and C1s are **proteases**. C1q binds to the antibody complex, and thus it operates in the **adaptive immune response**. When C1q binds the antibody, it leads to a conformational change, which results in activation of **C1r**, which then cleaves and thereby activates **C1s**. **C1s** now cleaves **C4** and **C2**. **C4** is cleaved into **C4a** and **C4b**, and **C2** is cleaved into **C2a** and **C2b**. **C2a** now binds to **C4b**. This new complex is called **C3 convertase** (also called C4b2a), and it is a key enzyme in the classical complement pathway.

C3 convertase then cleaves **C3** into two parts: **C3a** and **C3b**.

C3a is released into circulation, and it leads to inflammation and vasodilation. It also acts as an inflammatory mediator that helps attact immune cells, which enhances the immune response.

C3b binds directly to the pathogen, thereby coating it. This process, called opsonization, tags the pathogen or infected cell for destruction by phagocytes, which recognize **C3b**. The **C3b** can also bind to the **C3 convertase (C4b2a)**, forming a complex called a **C5 convertase**.

This **C5 convertase** splits **C5** into **C5a** and **C5b**. **C5a** is released into circulation, and it can recruit more phagocytes. **C5b** interacts with **C6**, **C7**, **C8**, and **C9** to form the **membrane attack complex** (**MAC**). **MAC** attaches to the membrane of the pathogen and makes pores in it, causing it to lyse.

1.4.2 The Lectin Pathway

Another complement pathway is the **lectin pathway**. This pathway can function in the **innate immune response** because it is not initiated by antibody-antigen complex. Lectins are proteins that can bind to certain carbohydrate patterns on pathogens. Some lectins can bind to sugar molecules on pathogens and, in turn, activate complement. When lectins bind to a microbe, they combine with proteins called **MASP1** and **MASP2**, which are **serine proteases**. **MASP2** gets activated upon this interaction, and it then cleaves **C4** and **C2**. The resulting **C4b** and **C2a** then combine to form **C3 convertase**. This **C3 convertase** then cleaves **C3** to form **C3a** and **C3b**. The rest of the pathway is the same as the classical pathway, with C3a and C3b playing their specific roles.

1.4.3 The Alternative Pathway

The alternative pathway can also take place in the absence of antibody-antigen complex and thus can also be part of the **innate immune response**. This pathway starts with **C3b**, which is usually generated from the spontaneous hydrolysis of C3. This **C3b** interacts directly with the pathogen. This bound **C3b** can bind to a protein called **Factor B**. Another factor, **Factor D**, then cleaves Factor B into two parts, Ba and Bb. This results in the formation of a complex called **C3bBb**, which is bound to the pathogen. **C3bBb** is an **alternative C3 convertase**. This alternative C3 convertase then splits **C3** into **C3a** and **C3b**. The rest of the steps follow the same series of events as above.

2 Viral Infection

Viral infections are common causes of human illnesses. The two main categories of viruses are DNA viruses and RNA viruses. An important subcategory of RNA viruses is the retrovirus [20, 21].

2.1 DNA Viruses

Some common DNA viruses are the herpesvirus, adenovirus, papillomavirus, and poxvirus [22]. DNA viruses have DNA as their genetic material, which can either be double-stranded or single-stranded. The DNA is contained within a capsid, a protective coat composed of several proteins encoded by the viral genome. Some viruses also have an additional layer, which is an envelope. Some viruses also have enzymes and proteins necessary for their replication, such as viral DNA polymerase.

The virus attaches to the host cell and enters it, usually through receptor-mediated endocytosis or membrane fusion. The viral DNA is then released into the host cell. A set of early genes is transcribed to produce proteins necessary for viral

DNA replication. Viral DNA replication then occurs, and another set of genes called the late genes is expressed. These genes encode structural proteins for the virus. Now, new viral particles (virions) are assembled containing the replicated DNA and the structural proteins. The new virions exit the host cell. Often, they cause the cell to lyse, and they infect other cells.

Sometimes, latency occurs. In this case, the virus remains dormant in the cell for some time. The DNA remains in an episome (a piece of DNA outside the host genome, usually circular), or in some cases, it may integrate into the host DNA. This viral DNA can remain dormant. This is called latency, and the virus may remain inactive in this way without causing symptoms. Certain conditions, such as stress and a weakened immune system, can cause a latent virus to reactivate, replicate, and form new viral particles [23].

2.2 RNA Viruses

RNA viruses contain RNA instead of DNA as their genetic material [22]. Some examples are the influenza virus, SARS-CoV-2, which causes COVID-19, and the hepatitis C virus. The RNA virus attaches to specific receptors on the cell surface, allowing it to enter the host. Inside the cell, the viral capsid is removed. What happens next depends on what type of RNA the virus is made of. Positive-sense RNA viruses have the sense strand RNA. This RNA can directly function as mRNA. This RNA gets translated by the host cell's machinery, and the translated proteins include proteins needed for viral replication as well as structural proteins that get packaged into viral particles. Negative-strand RNA viruses have the antisense strand. This viral RNA must be converted into positive-sense RNA by a viral-RNA dependent RNA polymerase (provided by the virus) before proceeding as above. Some RNA viruses are double-stranded. These strands are separated before transcription occurs. For all three mechanisms, once transcription occurs, the host cell's machinery carries out translation, including translation of the viral structural proteins. This causes new virions to be assembled. These new virions are released and can infect other cells, and in some cases, this can lead to cell lysis [24].

2.3 Retroviruses

Retroviruses are a subtype of RNA viruses [25]. Examples are **HIV** and human T lymphotropic virus (**HTLV**). This type of RNA virus also enters the host cell and inserts its RNA into the cell. However, rather than being translated directly, this RNA undergoes reverse transcription. This is catalyzed by a viral enzyme called **reverse transcriptase**, and this is a crucial part of the retrovirus system. The result is newly synthesized viral DNA. This DNA becomes integrated into the host cell's genome. The host cell then uses its own machinery for transcription and translation of the viral genes to produce more virus. New viral particles are made, and the virus is released from the cell, allowing it to infect other cells. Retroviruses are challenging to treat because they integrate into the host cell's genome. (Retroviruses were also discussed in Chap. 19.)

3 Some Examples of Drugs That Involve the Immune System, and Antiviral Treatments

3.1 Drugs That Function by Modulating the Immune System

A wide array of therapies involves the immune system. Such therapies may activate or modulate the immune system, or they may involve monoclonal antibodies that bind to targets. Some such drugs were already discussed in the previous chapter. Some of these drugs and a few other examples of drugs that involve the immune system are described below.

3.1.1 Monoclonal Antibodies (mAbs)

Monoclonal antibodies (mAbs) are antibodies developed in the lab that recognize a specific antigen, and drugs that utilize or are made of mAbs are considered **biologics**. Monoclonal antibodies are derived from a single clone of B cells and recognize one specific epitope of one antigen. They are usually IgG antibodies. Monoclonal antibodies are large and generally cannot readily enter the cell, so they need to be generated against extracellular targets. One common type of target for mAbs is cell surface proteins such as receptors. When mAbs bind to such targets, they may inactivate them. Another type of target is in the extracellular environment, such as a ligand, like a cytokine. By binding to ligand, the mAb prevents it from binding to its cell surface receptor [1, 26].

Herceptin (trastuzumab) [27, 28] and **Humira (adalimumab)** [26, 29, 30] are two examples of **monoclonal antibodies**, which were previously introduced in Chap. 18. Herceptin (trastuzumab) is a monoclonal antibody that targets Her2, a member of the EGF receptor family. It is used to treat **HER2-positive breast cancer**. Humira (adalimumab) targets the cytokine TNFα and prevents it from binding to its receptor. It is used to treat **inflammation in diseases such as arthritis**. **Leqembi (lecanemab-irmb)** is another example of a **monoclonal antibody** [26, 31]. It was developed to help treat **Alzheimer's disease** and was FDA-approved in 2023. It is an mAb that binds to amyloid beta, a component of the amyloid plaques found to accumulate in the disease. There is no cure for Alzheimer's disease, but the drug appears to slow down the decline in some patients.

Monoclonal antibodies can function in different ways. For example, coating a targeted cell, pathogen, or ligand with a monoclonal antibody can neutralize the target, preventing it from functioning. The mAb can neutralize a toxin or a virus, for example, preventing it from infecting cells or harming the body. It can also prevent a cell surface receptor from responding to a signal. In some cases, coating a cell or pathogen can also lead to activation of various immune cells, such as natural killer (NK) cells or macrophages, which in turn target and kill the coated cells. This is called antibody-dependent cytotoxicity (ADCC). Antibody-coated cells can lead to activation of complement, which can result in the destruction of the cells. This is called complement-dependent cytotoxicity **(CDC)** [3, 17, 32, 33].

3.1.2 Immune Checkpoint Inhibitors (ICIs)

Immune checkpoint inhibitors (ICIs) [34] are a promising type of treatment for some types of cancer. To understand their function, it is helpful to provide a brief overview of how immune cells may interact with tumor cells. A tumor cell displays a tumor antigen on its surface, using its MHC class I molecules. (A tumor antigen is a fragment of protein (a peptide) derived from abnormal proteins in the tumor.) This causes a T cell to recognize the tumor. It does so via its **TCR** (T-cell receptor), which binds to the tumor antigen-MHC complex. This interaction should activate the T cell so that it can attack the tumor. Tumor cells, however, have often developed ways to evade this activation of T cells. Some tumor cells have a cell surface protein called **PD-L1** (programmed death ligand 1). This protein is often amplified in tumor cells. PD-L1 interacts directly with a receptor on T cells called **PD-1** (programmed cell death protein 1) [35]. When these two proteins interact, T-cell activity is inhibited. Thus, when tumors overexpress PD-L1, the result can be evasion of the immune response by T cells. (It should be noted that the term programmed cell death in reference to PD-1 and PD-L1 is a bit misleading, as the receptors do not actually primarily lead to cell death but instead suppress T-cell activity.)

Immune checkpoint inhibitors (ICIs) are drugs (currently, they are all mAbs) designed to prevent this inhibition of T cells by tumor cells. ICIs can bind to PD-L1 on the tumor cells or PD-1 on the T cells (see Fig. 24.3). This prevents the interaction between PD-1 and PD-L1, and these drugs allow the T cells to kill the tumor cells. Thus, checkpoint inhibitors work by allowing the immune system to do its job and destroy cells that it recognizes as cancer cells.

A well-known example of an immune checkpoint inhibitor is **Keytruda (pembrolizumab)** [36], which binds to **PD-1**. It is approved for the treatment of **melanoma**, **non-small cell lung cancer, head and neck squamous cell**

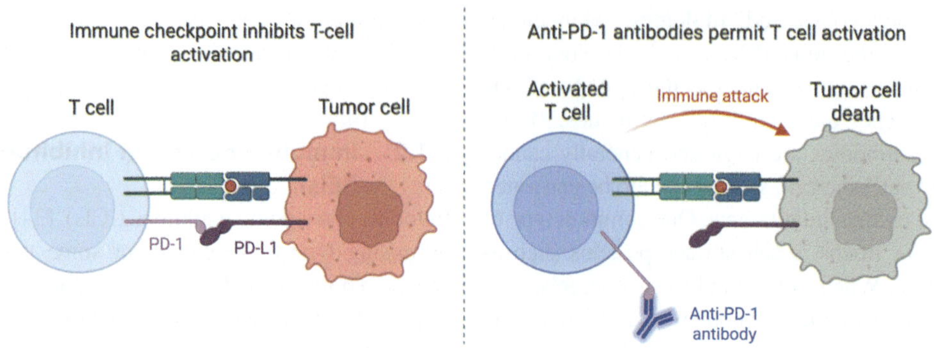

Fig. 24.3 Immune checkpoint inhibitors. Immune checkpoint inhibitors can bind to PD-1 or PD-L1, allowing T cells to mount a response against cancer cells. (*Source*: Created with BioRender.com)

carcinoma, classical Hodgkin lymphoma, and others. It is used either alone or in combination with other treatments [35].

3.1.3 Tumor-Infiltrating Lymphocyte (TIL) Therapy

Amtagvi (lifileucel) [37] is a **TIL-based therapy** FDA-approved in 2024 for the treatment of **melanoma**. Tumor-infiltrating lymphocytes (TILs) are T cells that migrate into tumors. They are also part of the immune system's response to eliminate cancer cells. TIL therapy [38] involves harvesting TILs from the patient and expanding them in the lab. Cytokines are used to encourage their expansion. They are then reinfused into the patient to mount an attack on the cancer cells that is stronger than the body's natural immune response [39, 40].

3.1.4 Complement Inhibitors

Pegcetacoplan (Empaveli) is a **complement inhibitor**, FDA-approved in 2021. Specifically, it is a PEGylated peptide that binds to C3. The PEG (polyethylene glycol, a synthetic polymer) increases the peptide's half-life. This drug inhibits the cleavage of C3 by binding to it, thus halting the amplification of the complement cascade. It is used to treat **paroxysmal nocturnal hemoglobinuria (PNH)**, which is a life-threatening disease where red blood cells (RBCs) are destroyed by unregulated complement activity [41, 42]. It is also being studied for other disorders such as age-related macular degeneration (AMD) [43].

3.2 Antiviral Drugs and Vaccines

3.2.1 HIV Inhibitors

HIV is the virus responsible for causing AIDS. Before a treatment was found, AIDS was nearly always fatal. HIV is a retrovirus and it infects cells, especially T cells. HIV binds to CD4, a cell surface protein on T cells, and this allows it to bind its coreceptor CCR5. Once it does so, the viral contents—specifically the RNA—enter the cells. The viral RNA is reverse-transcribed into DNA in a reaction catalyzed by the virus's **reverse transcriptase enzyme**. This DNA can then integrate into the host genome and new viral particles are made, as described in Chap. 19. Several enzymes play key roles in the HIV life cycle. Inhibitors targeting these enzymes form the basis of today's highly effective HIV treatments, which have saved many lives [26, 44] (see Fig. 24.4).

Three major classes of enzyme inhibitors are used to combat HIV infection. These are **reverse transcriptase inhibitors**, **protease inhibitors**, and **integrase inhibitors**. Drugs using these types of inhibitors are given to patients who already have an **HIV infection**, to suppress viral replication and slow disease progression.

- Reverse Transcriptase Inhibitors

Reverse transcriptase inhibitors are drugs used to treat **HIV/AIDS** by blocking the reverse transcription of viral RNA into DNA. Examples of HIV reverse transcriptase inhibitors include

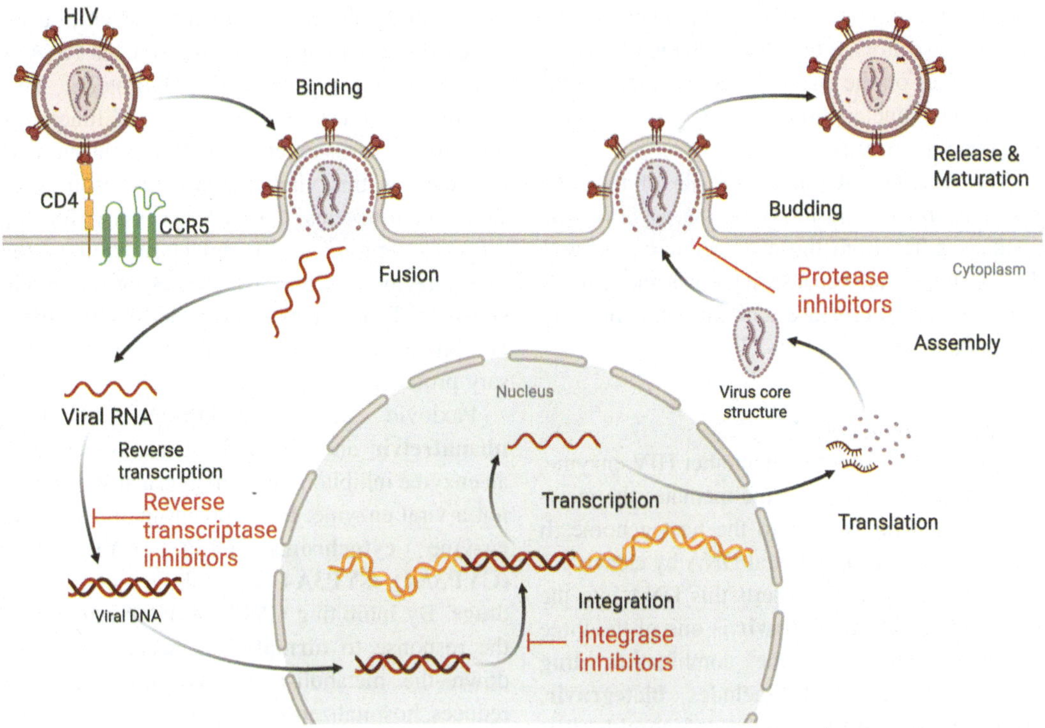

Fig. 24.4 HIV infection and HIV inhibitors. The HIV virus is a retrovirus and its mode of infection is shown here and discussed in the text. HIV inhibitors can function at several levels, including reverse transcription inhibitors, integrase inhibitors, and protease inhibitors. (*Source*: Created with BioRender.com)

Retrovir (zidovudine), AZT, **Epivir (lamivudine, 3TC)** [45], **Emtriva (emtricitabine)** [46], and **Vemlidy (tenofovir)** [47] (also present in formulations like **Viread**) [48]. These drugs are nucleoside reverse transcriptase inhibitors (NRTIs). They are analogs of the nucleotides normally found in cells, and they get phosphorylated in the host cell into their triphosphate forms. However, they lack a 3′ OH group, so when they are incorporated into the viral DNA in the reverse transcription reaction, phosphodiester bonds cannot form and chain elongation stops [49].

- Protease Inhibitors

HIV protease inhibitors can inhibit HIV infection by blocking the activity of the **HIV protease**. When HIV infects cells, its RNA is reverse-transcribed, integrated, transcribed, and translated into large polyproteins. These polyproteins must be cleaved into individual functional HIV proteases in order to produce functional individual viral particles. This viral protease is an aspartyl protease. The cleavage occurs during and after the budding process (the process where the viral particles are packaged into vesicles and exported). By inhibiting this protease, **protease inhibitors** cause the release of immature noninfectious viral particles, blocking further spread of the infection [50, 51].

Examples of **HIV protease inhibitors** include **Norvir (ritonavir)** [52], **Kaletra (lopinavir)** [53] **(used together with ritonavir)**, **Reyataz (atazanavir)** [54], and **Crixivan (indinavir)**. (The last one, Crixivan, is no longer actively used.) These inhibitors mimic the natural peptide substrate of the HIV protease and bind tightly to its active site. **Atazanavir**, for example, binds to the active site of the HIV protease and prevents the cleavage of the protein substrate by occupying the substrate binding pocket. Like other aspartyl protease inhibitors, it interacts with the two aspartate residues on the enzyme's active site, thereby blocking the nucleophilic attack required for peptide bond hydrolysis.

The structure of the active site of the HIV protease and the positions of the aspartic acid residues differ from those of cellular proteases. This is important because it allows the drug to inhibit the viral protease without damaging normal cellular proteases [55–57].

- Integrase Inhibitors

Integrase inhibitors target another HIV enzyme, **HIV integrase**. This enzyme mediates the insertion of the viral DNA into the host genome. It prepares the ends of the viral DNA by cleaving it at specific sites, and it inserts this DNA into the host cell's DNA. **Bictegravir** is one of the three active ingredients in the combination drug Biktarvy [58**],** which includes **bictegravir, emtricitabine, and tenofovir alafenamide**. It is an **integrase strand transfer inhibitor (INSTI)**. It binds in the integrase active site, chelates essential Mg^{2+} ions, and thereby blocks the strand-transfer step of integration. By preventing the viral DNA from being covalently linked to the host DNA, it prevents integration of viral DNA into the host cell genome [59].

The different types of HIV inhibitors are often used in combination (combination antiretroviral therapy), and are quite effective at blocking HIV at various stages of the virus's life cycle [60].

3.2.2 Paxlovid

Paxlovid is [61] a **SARS-CoV-2 protease inhibitor**. It was authorized for emergency use in 2021 and was then FDA-approved in 2023. It is an oral antiviral drug combination (**nirmatrelvir** and **ritonavir**) for the **treatment of SARS-CoV-2 (COVID-19) infection**. The SARS-CoV-2 virus is an RNA virus (but not a retrovirus). It also depends on protease enzymes to cleave and activate large polyproteins into active proteins. Protease inhibitors are therefore effective drug targets for the treatment of COVID-19 infection.

Nirmatrelvir is a **3C-like (3CL) protease inhibitor** also known as 3CLpro or Mpro. It targets **3CL**, the main **protease** the virus uses. When the virus enters the host cell, its RNA is translated into a single polyprotein. This polyprotein must be cleaved into functional units for the virus to replicate. **3CL** is the enzyme that cleaves this viral polyprotein, resulting in functional proteins that are necessary for the virus to reproduce. **Nirmatrelvir** inhibits the 3CL enzyme's activity by mimicking its protein substrate. Thus, it acts as a **competitive inhibitor** and prevents the virus from producing its necessary proteins.

Paxlovid is a combination of both **nirmatrelvir** and **ritonavir**. **Ritonavir** is also an enzyme inhibitor, but it inhibits a host enzyme, not a viral enzyme. It is a potent inhibitor of the enzyme **cytochrome P450 (CYP) 3A4 (CYP3A4)**. **CYP3A4** has a role in metabolizing drugs. By inhibiting **CYP3A4**, **ritonavir** boosts the response to **nirmatrelvir** because it slows down the metabolism of the drug. Paxlovid reduces hospitalization and mortality and helps prevent transmission of the virus [62–64].

3.2.3 Tamiflu

Tamiflu (oseltamivir) [65] is used to treat patients with **influenza virus** infection [26]. It does so by inhibiting the viral enzyme **neuraminidase**. This is an enzyme that cleaves **sialic acid**, a sugar found on glycoproteins and glycolipids on the extracellular sides of cell membranes.

When flu viral particles form in an infected cell, new viral particles bud from the plasma membrane and acquire an envelope derived from the host cell's membrane. As part of this budding process, hemagglutinin (HA) proteins on the viral surface can bind the sialic acid, located on glycoproteins on the host cell's membrane. This may trap the virus in the cell at the cell surface and prevent its release. To solve this, the virus produces the enzyme **neuraminidase (NA)**, which cleaves the sialic acid from the host cell's glycoproteins. This breaks the interaction between the viral particle and the host membrane, allowing the virus to infect neighboring cells. **Tamiflu** prevents this cleavage by inhibiting **NA** activity. It is a **competitive inhibitor** that mimics sialic acid and binds to the active site of the NA enzyme. Thus, while it does not prevent an initial influenza infection, it prevents the virus from spreading to other cells [26, 66].

3.3 Vaccines for Preventing Viral Infections

In addition to treating viral infections, preventing viral infections may be even more important. Many different vaccines have been developed to prevent viral infections. **The COVID-19 vaccine** was designed to reduce the risk of severe disease, further infection by the SARS-CoV-2 virus, and transmission of the virus. This vaccine was already discussed in Chap. 7. It uses a new technology (mRNA vaccine) that could revolutionize how vaccines are made. Prior to this, many vaccines were made using inactivated virus or recombinant viral proteins. These include vaccines that prevent polio, MMR (measles, mumps, rubella), hepatitis (such as HBV and HAC), and influenza (flu). The idea is that inactivated viruses or proteins are injected. The body recognizes them as foreign substances and it mounts an immune response, activating T and B cells. An important result is the accumulation of memory B cells, which are ready to respond quickly when an infection with the same virus occurs [67].

The **Gardasil vaccine** [68] is an important vaccine. It is the only vaccine that is specifically designed to prevent certain types of cancer (although some other vaccines, such as the hepatitis B vaccine, can also reduce the risk of cancer, especially liver cancer). This vaccine protects against **human papillomavirus (HPV)**. The vaccine does not use live or inactivated viruses but uses recombinant proteins that correspond to parts of the HPV. HPV infection is quite common, but some forms of the virus can cause cancer, most commonly cervical cancer. Preventing HPV infection is thus highly effective against developing cervical and other cancers. The HPV vaccine is effective only before HPV infection occurs. Since HPV is largely transmitted by sexual contact, the vaccine is recommended for children at a young age, usually between ages 11 and 12 [69, 70].

FDA-Approved Drugs Discussed in This Chapter

Commonly used brand name	Generic name	FDA approval year	Main conditions	Main targets
Amtagvi	*Lifileucel*	2024	Melanoma	Tumor-infiltrating lymphocytes
Biktarvy	*Bictegravir/emtricitabine/tenofovir alafenamide*	2018	HIV	HIV integrase
Emtriva	*Emtricitabine*	2003	HIV	Reverse transcriptase
Epivir	*Lamivudine (3TC)*	1995	HIV	Reverse transcriptase
Gardasil	*HPV vaccine (recombinant)*	2006	HPV prevention	HPV (viruslike particles)
Herceptin	*Trastuzumab*	1998	HER2+ breast cancer	HER2
Humira	*Adalimumab*	2002	Inflammatory diseases (such as arthritis)	TNFα
Keytruda	*Pembrolizumab*	2014	Various cancers (such as melanoma, NSCLC, Hodgkin lymphoma)	PD-1
Leqembi	*Lecanemab-irmb*	2023	Alzheimer's disease	Amyloid beta
Norvir	*Ritonavir*	1996	HIV	HIV protease/CYP3A4
Paxlovid	*Nirmatrelvir + ritonavir*	2023	COVID-19	SARS-CoV-2 protease/CYP3A4

(continued)

Commonly used brand name	Generic name	FDA approval year	Main conditions	Main targets
Retrovir	*Zidovudine (AZT)*	1987	HIV	Reverse transcriptase
Reyataz	*Atazanavir*	2003	HIV	HIV protease
Tamiflu	*Oseltamivir*	1999	Influenza	Neuraminidase
Vemlidy	*Tenofovir alafenamide*	2015	HIV, hepatitis B	Reverse transcriptase
	Lopinavir	2000	HIV	HIV protease

Definitions

Antibodies Y-shaped molecules produced by B cells that bind to a specific target antigen.

Antibody-dependent cellular cytotoxicity (ADCC) In ADCC, natural killer (NK) cells recognize the Fc region on the antibody bound to the infected or abnormal cells, and then release cytotoxic granules to induce apoptosis in the targeted cell.

Antigen A molecule, usually recognized as foreign, which triggers an immune response.

B-cell receptor (BCR) Antibody made by B cells, and that remains bound to the B cell.

C3 convertase An enzyme in the complement pathway.

Checkpoint inhibitors Drugs that block inhibitory effects on the immune system, used to encourage the immune system to attack cancer cells.

Complement system A cascade of proteins enhancing immune responses, including opsonization and lysis. It involves a series of proteins in the blood.

Cytotoxic T lymphocyte (CTL) A T cell that kills infected or abnormal cells.

Degranulation Release of granule contents from immune cells, often cytotoxic or inflammatory.

Dendritic cells Antigen-presenting cells that activate adaptive immune responses.

Episome An extrachromosomal DNA molecule outside the host genome, usually circular.

Fab region The variable part of an antibody that binds antigen.

FasL-Fas pathway A pathway leading to apoptosis via receptor-ligand interaction.

Fc region The constant region of an antibody. This is the part of the antibody that binds to immune cell receptors.

Granzyme Proteins that trigger apoptosis in target cells.

IgG, IgA, IgM, IgE, IgD Classes of antibodies with different roles.

Integrase HIV enzyme that mediates the insertion of the viral DNA into the host genome.

Lectins Proteins that bind specific sugar molecules that trigger the lectin pathway of complement activation.

Macrophages Phagocytic cells that also release cytokines and recruit immune cells.

Major histocompatibility complex (MHC) A set of proteins that display antigens on the cell surface of cells, recognized by T cells. MHC1 displays endogenous antigen and is found on most nucleated cells; MHCII displays exogenous antigen and is found on antigen-presenting cells.

Membrane attack complex (MAC) A complex that can be formed in response to the complement system. It attaches to the membrane of a pathogen and makes pores in it so that it will lyse.

Memory B cell B cells that produce a certain antibody and remain in the body to respond to a repeated infection.

Neuraminidase An enzyme made by the influenza virus that allows it to infect the cell. It is the target of the drug Tamiflu.

Opsonization Coating of pathogens with components of the immune system, such as complement system proteins or antibodies, which tags them for phagocytosis by immune cells.

Pathogen An agent, often a microorganism, which can cause disease.

Pathogen-associated molecular patterns (PAMPS) Conserved microbial structures recognized by the innate immune system.

Perforin Substances that form pores on target cells.

Phagocytosis The mechanism that some immune cells use to engulf debris or pathogens.

Programmed cell death protein 1 (PD-1) A receptor on the surface of activated T cells that acts as an immune checkpoint.

Programmed death ligand 1 (PD-L1) A ligand found on some cells that binds to PD-1 on T cells. This interaction suppresses T-cell activity.

Reverse transcriptase Enzyme used by retroviruses to synthesize DNA from RNA.

T cell receptor (TCR) A receptor on T cells that binds specific antigens.

Study Questions

Multiple-Choice Questions

1. What type of immune cells are primarily responsible for producing antibodies?
   A. T cells
   B. Dendritic cells
   C. B cells
   D. Macrophages
2. Which complement pathway is activated by antibody-antigen complexes?
   A. Lectin pathway
   B. Alternative pathway
   C. Classical pathway
   D. Adaptive pathway
3. Which of the following drugs is an example of a checkpoint inhibitor?
   A. Humira
   B. Keytruda
   C. Tamiflu
   D. Retrovir
   E. Both A and B
4. What is the role of perforin in immune response?
   A. It neutralizes antigens.
   B. It forms pores in target cell membranes.
   C. It inhibits cytokine release.
   D. It activates the complement system.
   E. It activates B cells.

Fill-in-the-Blank Questions

5. The two main branches of the immune system are the ________ and ________ immune systems.
6. Retroviruses are RNA viruses but their RNA can be made into DNA due to the viral enzyme ____________
7. ________ is the process where immune cells engulf and digest pathogens.

Short-Answer Question

8. What kind of molecule is Herceptin and how does it treat some kinds of breast cancer?

Matching Question

9. **Match the immune cell with its primary function:**
   Cell Type Function
   A. Dendritic cell
   B. NK cell
   C. Plasma cell
   Function
   1. Presents antigens to T cells
   2. Induces apoptosis of infected cells
   3. Secretes antibodies

Answer Key

Multiple-Choice Questions

1. **Answer**: C
2. **Answer**: C
3. **Answer**: B
4. **Answer**: B

Fill-in-the-Blank Questions

5. **Answer**: Innate and adaptive
6. **Answer**: Reverse transcriptase
7. **Answer**: Phagocytosis

Short-Answer Question

8. **Answer**: Herceptin is a monoclonal antibody. It binds Her2, which is overproduced on some breast cancer cells, and triggers immune responses.

Matching Question

9. **Match the immune cell with its primary function**:
   Answers: A:1, B:2, C:3

References

1. Murphy K, Weaver C, Berg LJ. Janeway's immunobiology. 10th ed. New York: W.W. Norton & Company; 2022.
2. Abbas AK, Lichtman AH, Pillai S. Cellular and molecular immunology. 10th ed. Philadelphia: Elsevier; 2022.
3. Abbas A, Lichtman AH, Pillai S. Basic immunology. 7th ed. Philadelphia: Elsevier; 2023.
4. Marshall JS, Warrington R, Watson W, Kim HL. An introduction to immunology and immunopathology. Allergy Asthma Clin Immunol. 2018;14(Suppl 2):49. https://doi.org/10.1186/s13223-018-0278-1.
5. Medicine NLo. In brief: the innate and adaptive immune systems. 2023. Available from: InformedHealth.org.
6. Banchereau J, Steinman RM. Dendritic cells and the control of immunity. Nature. 1998;392:245–52. https://doi.org/10.1038/32588.
7. Chaplin DD. Overview of the immune response. J Allergy Clin Immunol. 2010;125(2 Suppl 2):S3–23. https://doi.org/10.1016/j.jaci.2009.12.980.
8. Mogensen TH. Pathogen recognition and inflammatory signaling in innate immune defenses. Clin Microbiol Rev. 2009;22(2):240–73, Table of Contents. https://doi.org/10.1128/cmr.00046-08.
9. Freeley S, Kemper C, Le Friec G. The "ins and outs" of complement-driven immune responses. Immunol Rev. 2016;274(1):16–32. https://doi.org/10.1111/imr.12472.
10. Ricklin D, Hajishengallis G, Yang K, Lambris JD. Complement: a key system for immune surveillance and homeostasis. Nat Immunol. 2010;11(9):785–97. https://doi.org/10.1038/ni.1923.
11. Sharpe AH. Mechanisms of costimulation. Immunol Rev. 2009;229(1):5–11. https://doi.org/10.1111/j.1600-065X.2009.00784.x.
12. Zheng SG. Regulatory T cells vs Th17: differentiation of Th17 versus Treg, are the mutually exclusive? Am J Clin Exp Immunol. 2013;2(1):94–106.
13. Krensky AM. Granulysin: a novel antimicrobial peptide of cytolytic T lymphocytes and natural killer cells. Biochem Pharmacol. 2000;59(4):317–20. https://doi.org/10.1016/S0006-2952(99)00177-X.
14. Tripani JA, Smyth MJ. Functional significance of the perforin/granzyme cell death pathway. Nat Rev Immunol. 2002;2(10):735–47. https://doi.org/10.1038/nri911.
15. Chiu ML, Goulet DR, Teplyakov A, Gilliland GL. Antibody structure and function: the basis for engineering therapeutics. Antibodies (Basel). 2019;8(4). https://doi.org/10.3390/antib8040055.
16. Alberts B, Heald R, Johnson A, Morgan D, Raff M, Roberts K, Walter P, Wilson J, Hunt T. Molecular biology of the cell. New York: W.W. Norton and Company; 2022.
17. Wöhner M, Nimmerjahn F. Cytotoxic IgG: mechanisms, functions, and applications. Immunity. 2025;58(6):1378–95. https://doi.org/10.1016/j.immuni.2025.05.008.
18. Dunkelberger JR, Song W-C. Complement and its role in innate and adaptive immune responses. Cell Res. 2010;20(1):34–50. https://doi.org/10.1038/cr.2009.139.
19. Merle NS, Church SE, Fremeaux-Bacchi V, Roumenina LT. Complement system part I – molecular mechanisms of activation and regulation. Front Immunol. 2015;6:262. https://doi.org/10.3389/fimmu.2015.00262.
20. Flint J, Racaniello VR, Glenn FR, Hatzilioannou T, Skalka AM. Principles of virology. 5th ed. Washington, DC: ASM Press; 2020.
21. Strauss JH, Strauss EG. Chapter 1 – Overview of viruses and virus infection. In: Strauss JH, Strauss

EG, editors. Viruses and human disease. 2nd ed. London: Academic Press; 2008. p. 1–33.
22. Dixit V, Sing A. Review article on DNA and RNA virus. Int J Pharmacol Biol Sci. 2020;10(3):210–8.
23. Traylen CM, Patel HR, Fondaw W, Mahatme S, Williams JF, Walker LR, et al. Virus reactivation: a panoramic view in human infections. Future Virol. 2011;6(4):451–63. https://doi.org/10.2217/fvl.11.21.
24. Louten J. Chapter 4 – Virus replication. In: Louten J, editor. Essential human virology. Boston: Academic Press; 2016. p. 49–70.
25. Chameettachal A, Mustafa F, Rizvi TA. Understanding retroviral life cycle and its genomic RNA packaging. J Mol Biol. 2023;435(3):167924. https://doi.org/10.1016/j.jmb.2022.167924.
26. Whalen K, Lerchenfeldt S, Giordano C. Lippincott illustrated reviews. Pharmacology. Wolters Kluwer; 2021.
27. FDA. HERCEPTIN (trastuzumab) approval overview. 1998. Available from: https://www.accessdata.fda.gov/scripts/cder/daf/index.cfm?event=overview.process&ApplNo=103792.
28. Maadi H, Soheilifar MH, Choi WS, Moshtaghian A, Wang Z. Trastuzumab mechanism of action; 20 years of research to unravel a dilemma. Cancers (Basel). 2021;13(14). https://doi.org/10.3390/cancers13143540.
29. FDA. HUMIRA (adalimumab) approval overview 1001. Available from: https://www.accessdata.fda.gov/scripts/cder/daf/index.cfm?event=overview.process&ApplNo=125057.
30. Horiuchi T, Mitoma H, Harashima S, Tsukamoto H, Shimoda T. Transmembrane TNF-alpha: structure, function and interaction with anti-TNF agents. Rheumatology (Oxford). 2010;49(7):1215–28. https://doi.org/10.1093/rheumatology/keq031.
31. Eisai Inc. LEQEMBI. LEQEMBI (lecanemab-irmb): Mechanism of action. Available from: https://www.leqembihcp.com/about-leqembi/mechanism-of-action.
32. Kothari M, Wanjari A, Acharya S, Karwa V, Chavhan R, Kumar S, et al. A comprehensive review of monoclonal antibodies in modern medicine: tracing the evolution of a revolutionary therapeutic approach. Cureus. 2024;16(6):e61983. https://doi.org/10.7759/cureus.61983.
33. Cleveland Clinic. Monoclonal antibodies. 2021. Available from: https://my.clevelandclinic.org/health/treatments/22246-monoclonal-antibodies.
34. National Cancer Institute. Immune checkpoint inhibitors. 2022. Available from: https://www.cancer.gov/about-cancer/treatment/types/immunotherapy/checkpoint-inhibitors.
35. Momtaz P, Postow MA. Immunologic checkpoints in cancer therapy: focus on the programmed death-1 (PD-1) receptor pathway. Pharmgenomics Pers Med. 2014;7:357–65. https://doi.org/10.2147/pgpm.S53163.
36. FDA. KEYTRUDA (pembrolizumab) approval overview. 2014. Available from: https://www.accessdata.fda.gov/scripts/cder/daf/index.cfm?event=overview.process&ApplNo=125514.
37. FDA. Amtagvi (lifileucel) approval overview. 2024. Available from: https://www.fda.gov/vaccines-blood-biologics/approved-blood-products/amtagvi?.
38. Shaik R, Royyala SA, Inapanuri B, Durgam A, Khan H, Unnisa A. Tumor infiltration therapy: from FDA approval to next-generation approaches. Clin Exp Med. 2025;25(1):254. https://doi.org/10.1007/s10238-025-01574-6.
39. Phililips C, National Cancer Institute. First TIL therapy gets FDA approval for advanced melanoma. 2024. Available from: https://www.cancer.gov/news-events/cancer-currents-blog/2024/fda-amtagvi-til-therapy-melanoma#:~:text=Enlarge,of%20cancer%2Dfighting%20immune%20cells.
40. Amaria RN, Komanduri KV, Schoenfeld AJ, Ramsingh G, Burga RA, Jagasia MH. Entering a new era of tumor-infiltrating lymphocyte cell therapy innovation. Cytotherapy. 2025;27(7):864–73. https://doi.org/10.1016/j.jcyt.2024.12.010.
41. Chan TW, Hein T, Tertius T, Goh YT. Pegcetacoplan: the first and only C3-targeted therapy for the treatment of adults with paroxysmal nocturnal hemoglobinuria. Expert Rev Hematol. 2025;18(1):11–20. https://doi.org/10.1080/17474086.2024.2440101.
42. Crass RL, Smith B, Adriaens S, Chapel S, Langdon G. Population pharmacokinetic and pharmacokinetic/pharmacodynamic analyses of pegcetacoplan in patients with paroxysmal nocturnal hemoglobinuria. Drugs R D. 2024;24(4):563–73. https://doi.org/10.1007/s40268-024-00500-7.
43. Chakravarthy U, Schwartz R, Guymer RH, Holz FG, Rachitskaya AV, Vujosevic S, et al. Visual function benefit after treatment with pegcetacoplan: microperimetry analysis from the Phase 3 Oaks Trial. Am J Ophthalmol. 2025;273:119–29. https://doi.org/10.1016/j.ajo.2025.02.012.
44. Lv Z, Chu Y, Wang Y. HIV protease inhibitors: a review of molecular selectivity and toxicity. HIV AIDS (Auckl). 2015;7:95–104. https://doi.org/10.2147/hiv.S79956.
45. FDA. Retrovir/zidovudine (azidothymidine (AZT)) approval overview. 1987. Available from: https://www.accessdata.fda.gov/scripts/cder/daf/index.cfm?event=overview.process&ApplNo=020518.
46. FDA. Emtriva (emtricitabine) approval overview. 2003. Available from: https://www.accessdata.fda.gov/scripts/cder/daf/index.cfm?event=overview.process&ApplNo=021500.
47. FDA. Vemlidy (tenofovir alafenamide) approval overview. 2015. Available from: https://www.accessdata.fda.gov/scripts/cder/daf/index.cfm?event=overview.process&ApplNo=208464.
48. Li G, Wang Y, De Clercq E. Approved HIV reverse transcriptase inhibitors in the past decade. Acta Pharm Sin B. 2022;12(4):1567–90. https://doi.org/10.1016/j.apsb.2021.11.009.

49. Holec AD, Mandal S, Prathipati PK, Destache CJ. Nucleotide reverse transcriptase inhibitors: a thorough review, present status and future perspective as HIV therapeutics. Curr HIV Res. 2017;15(6):411–21. https://doi.org/10.2174/1570162x15666171120110145.
50. Kim JG, Shan L. Beyond inhibition: a novel strategy of targeting HIV-1 protease to eliminate viral reservoirs. Viruses. 2022;14(6). https://doi.org/10.3390/v14061179.
51. Weber IT, Wang YF, Harrison RW. HIV protease: historical perspective and current research. Viruses. 2021;13(5). https://doi.org/10.3390/v13050839.
52. FDA. Norvir (ritonavir) approval overview. 1996. Available from: https://www.accessdata.fda.gov/scripts/cder/daf/index.cfm?event=overview.process&ApplNo=022417.
53. FDA. Kaletra (lopinavir) approval overview. 2000. Available from: https://www.accessdata.fda.gov/scripts/cder/daf/index.cfm?event=overview.process&ApplNo=21226
54. FDA. Reyataz (atazanavir) Approval overview. 2003. Available from: https://www.accessdata.fda.gov/scripts/cder/daf/index.cfm?event=overview.process&ApplNo=021567.
55. Berg JM, Tymoczko JL, Gatto GJ Jr, Stryer L. Biochemistry. 9th ed. New York: MacMillan Learning; 2019.
56. Smith JJ, Simons C. Enzymes and their inhibitors: drug development. Boca Raton: Taylor and Francis; 2004.
57. Ritter JM, Flower RJ, Henderson G, Loke YK, MacEwan DJ, Rang HP. Rang and Dale's pharmacology. 10th ed. Amsterdam: Elsevier; 2023.
58. FDA. Biktarvy (bictegravir/emtricitabine/tenofovir): Drugs@FDA approval overview. 2018. Available from: https://www.accessdata.fda.gov/scripts/cder/daf/index.cfm?event=overview.process&ApplNo=210251.
59. Engelman AN. Multifaceted HIV integrase functionalities and therapeutic strategies for their inhibition. J Biol Chem. 2019;294(41):15137–57. https://doi.org/10.1074/jbc.REV119.006901.
60. Ghosh AK. Four decades of continuing innovations in the development of antiretroviral therapy for HIV/AIDS: progress to date and future challenges. Glob Health Med. 2023;5(4):194–8. https://doi.org/10.35772/ghm.2023.01013.
61. FDA. Paxlovid (nirmatrelvir + ritonavir) approval overview. 2023. Available from: https://www.accessdata.fda.gov/scripts/cder/daf/index.cfm?event=overview.process&ApplNo=217188.
62. Marzi M, Vakil MK, Bahmanyar M, Zarenezhad E. Paxlovid: mechanism of action, synthesis, and in silico study. Biomed Res Int. 2022;2022:7341493. https://doi.org/10.1155/2022/7341493.
63. Hashemian SMR, Sheida A, Taghizadieh M, Memar MY, Hamblin MR, Bannazadeh Baghi H, et al. Paxlovid (nirmatrelvir/ritonavir): a new approach to Covid-19 therapy? Biomed Pharmacother. 2023;162:114367. https://doi.org/10.1016/j.biopha.2023.114367.
64. Zong K, Xu L, Luo C, Luo C, Liu B, Chen J, et al. Paxlovid reduces the 28-day mortality of patients with COVID-19: a retrospective cohort study. BMC Infect Dis. 2024;24(1):767. https://doi.org/10.1186/s12879-024-09482-y.
65. FDA. Tamiflu (oseltamivir) approval overview. 1999. Available from: https://www.accessdata.fda.gov/scripts/cder/daf/index.cfm?event=overview.process&ApplNo=021246.
66. Moscona A. Neuraminidase inhibitors of influenza. N Engl J Med. 2005;353(13):1363–73. https://doi.org/10.1056/NEJMra050740.
67. Hall HI, Wood K. Epidemiology of HIV in the United States. In: Smith J, editor. HIV Surveillance Report. Atlanta (GA): Centers for Disease Control and Prevention; 2021. p. 15–28. Available from: https://stacks.cdc.gov/view/cdc/252997
68. FDA. Gardasil (HPV vaccine (recombinant)) approval overview. 2006. Available from: https://www.fda.gov/vaccines-blood-biologics/vaccines/gardasil?.
69. Pallecaros A, Vonau B. Human papilloma virus vaccine—more than a vaccine. Curr Opin Obstet Gynecol. 2007;19(6):541–6. https://doi.org/10.1097/GCO.0b013e3282f1a91b.
70. Mayo Clinic Staff. HPV vaccine: who needs it and how it works. 2023. Available from: https://www.mayoclinic.org/diseases-conditions/hpv-infection/in-depth/hpv-vaccine/art-20047292#:~:text=Who%20is%20the%20HPV%20vaccine,should%20get%20the%20HPV%20vaccine.

Part VII

Advanced Technologies

These last few chapters in this part discuss some advanced techniques that scientists use to study human health and disease and to develop pharmacological therapies. The following topics will be introduced. In the first chapter in this part, Chap. 25, the gene editing technology and gene therapy methods are discussed, with a focus on *CRISPR-Cas9*. The newest CRISPR-based drug treatments are discussed, as well as potential future uses for the technology. The topic of *cellular therapy* is then addressed in Chap. 26, with a focus on stem cells. This topic centers on the idea of using whole cells to treat disease or injuries. The vast field of "omics" is also summarized in Chap. 27. This includes the technologies that scientists use to study whole genomes, transcriptomes, proteomes, and others, rather than focusing on individual genes and cells. Finally, a brief overview of *drug delivery mechanisms*, including the use of nanotechnology, will be discussed in Chap. 28.

CRISPR and Gene Therapy in Modern Medicine

25

Abstract

Pharmaceuticals, surgery, or dietary changes have long been used to treat disease. Another approach, however, is to modify human genes. This includes gene therapy, where endogenous genes are replaced, corrected, or modified. Scientists have developed several different methods for introducing gene changes into living cells and organisms. One of the most recent and prominent mechanisms for modifying genes is the CRISPR-Cas9 system. CRISPR is a powerful gene editing technology that is the focus of this chapter. CRISPR technology is still relatively new. One FDA-approved CRISPR-based treatment is currently available, which will be discussed in this chapter. More recently, CRISPR has been used to edit genes in vivo in an infant with a genetic disorder. Research is underway to develop more CRISPR-based treatments. CRISPR is also valuable for research, studying various aspects of human biology, developing disease models, and modifying other organisms, such as crops and farm animals. In addition to CRISPR, other mechanisms have been used for treating disease with gene therapy for some time. Several examples are discussed here.

Keywords

CRISPR · Casgevy · VERVE-101 · CPS I deficiency · Gene therapy

Abbreviations

AAV	Adeno-associated virus
ABE	Adenine base editing
ADA	Adenosine deaminase
Cas	CRISPR-associated protein
CCR5	C-C chemokine receptor 5
CPS1D	Carbamoyl phosphate synthetase I deficiency
CRISPR	Clustered regularly interspaced short palindromic repeats
crRNA	CRISPR RNA
DSB	Double-stranded break
GMO	Genetically modified organism
gRNA	Guide RNA
hATTR	Hereditary transthyretin-mediated amyloidosis
HbA	Hemoglobin A (adult hemoglobin)
HbF	Hemoglobin F (fetal hemoglobin)
HDR	Homology-directed repair
HSC	Hematopoietic stem cell
IV	Intravenous
IV	Intravenous
IVF	In vitro fertilization
LDL	Low-density lipoprotein
LDLR	LDL receptor
nCas9	Cas9 nickase
NHEJ	Nonhomologous end joining
PAM	Protospacer adjacent motif
PCSK9	Proprotein convertase subtilisin/kexin type 9
RBC	Red blood cell

A. Minden, *Biochemistry, Cell and Molecular Biology for Pharmacists and Health Professionals*,
https://doi.org/10.1007/978-3-032-20529-2_25

RNAi	RNA interference
RNP	Ribonucleoprotein
SCID	Severe combined immunodeficiency
SCID-ADA	Severe combined immunodeficiency due to adenosine deaminase deficiency
SDSA	Synthesis-dependent strand annealing
SMA	Spinal muscular atrophy
tracrRNA	Trans-activating CRISPR RNA

What You Will Learn

- The CRISPR Cas9 system for gene editing.
- The CRISPR locus including palindromic and spacer sequences.
- The biological role for the CRISPR locus as a bacterial adaptive immune system.
- The mechanism by which CRISPR can be modified to edit any gene.
- The first CRISPR disease treatment, CASGEVY, to treat hemoglobin disorders such as sickle cell disease.
- VERVE, another CRISPR treatment under investigation, to treat familial hypercholesterolemia.
- The first CRISPR personalized gene editing treatment to treat a baby with a urea cycle disorder.
- Other potential uses for CRISPR such as modifying farm animals or use in research.
- Gene drives to alter wild populations.
- Other mechanisms for altering genes that can be used for gene therapy such as the use of the adeno-associated viruses and several treatments that have already been developed using these methods.

Pharmaceuticals, surgery, or dietary changes have long been used to treat disease. Another approach, however, is to modify human genes. This includes gene therapy, where endogenous genes are replaced, corrected, or modified. Scientists have developed several different methods for introducing gene changes into living cells and organisms. Examples include viral vector-based gene replacement, antisense oligonucleotides (ASOs), and RNA interference (RNAi). New systems like zinc finger nucleases (ZFNs) and TALENs are also being developed for gene editing applications (though they will not be covered here). One of the most recent and prominent mechanisms for modifying genes is CRISPR. CRISPR is a powerful and relatively straightforward gene editing technology that is the focus of this chapter. CRISPR technology is still relatively new, and only one FDA-approved CRISPR-based treatment is currently available. However, research is underway to develop more CRISPR-based treatments. Beyond treating disease, CRISPR is also valuable for research, for studying various aspects of human biology, for developing models for disease, and for modifying other organisms such as crops and farm animals.

1 Introduction to CRISPR

The term CRISPR is based on a genetic locus present in some bacteria's genomes [1, 2]. CRISPR stands for clustered regularly interspaced short palindromic repeats, and it is the basis of a bacterial defense system. Scientists found that the CRISPR system can be modified and used to edit genes in almost any cell type. This is called the **RNA-programmable CRISPR-Cas9 system**, but it is often referred to as the **CRISPR-Cas9 system** or simply as **CRISPR**. This technology differs from previous gene editing methods because it is comparatively quick and straightforward. It allows scientists to edit genomes with more precision than previous methods. The CRISPR locus is described in the sections below. After describing the locus in bacteria, this chapter will cover its function in bacteria, followed by a discussion of how the system can be manipulated to edit any DNA. Although several different CRISPR-Cas systems exist in bacteria (as well as archaea), this chapter focuses one type of CRISPR system, called type II, which uses Cas9 and tracrRNA.

2 The CRISPR Locus

The CRISPR locus in bacteria consists of a leader sequence followed by the CRISPR array, which contains alternating short palindromic repeat sequences, and spacer sequences. Adjacent to the CRISPR array are the Cas genes. In type II systems, discussed here, the tracrRNA gene is also located nearby. Due to the presence of the short palindromic repeat sequences, this whole locus is called the clustered regularly interspaced short palindromic repeats (CRISPR) locus. These different components of the CRISPR locus are described [3–5] and discussed below.

2.1 The Palindromic Repeat Sequences

In the 1980s, scientists found that part of the bacterial genome contained palindromic DNA sequences of 20–49 nucleotides that were repeated multiple times. These sequences are called **palindromic repeat sequences**, or sometimes they are referred to as simply **repeat sequences** or **palindromic sequences**.

Palindromic sequences are sequences that contain palindromes. To clarify what palindromes are, they are sequences that read the same from left to right or from right to left. In written language, examples include "never odd or even" or "race car." For DNA, due to being double-stranded, a palindrome sequence is one where the 5′ to 3′ sequence on one strand is identical to the 5′ to 3′ sequence on the complementary strand. This means the sequence reads the same forward and backward—see the example below. Palindromes can impact DNA structure and allow for the formation of hairpin loops:

5'-GAATTC-3′
3'-CTTAAG-5′

2.2 Spacer Sequences

In addition to the palindromic repeat sequences, scientists found that there were sequences in between the palindromic repeats. These sequences were referred to as **spacer DNA** or **spacer sequences**. Unlike the palindromic repeat sequences, the spacer sequences differed from one another in sequence. Upon studying these spacer sequences, the scientists found that they matched sequences of foreign DNA, such as bacteriophage DNA, suggesting that they originated from viruses rather than bacteria. This was discovered in 2005 [6]. The spacer sequences and the palindromic repeat sequences form an array, where spacer sequences are located between the repeated palindromic sequences. The result is a **repeat-spacer-repeat** structure [7].

2.3 Upstream Regions: Leader Sequence, TracrRNA, and Cas Genes

The DNA just upstream of the array of palindrome and repeat sequences is called the **leader sequence**. Further upstream is a set of genes called the **Cas genes** (CRISPR-associated proteins), which are part of an **operon**. Even further upstream is a sequence that codes for the **tracrRNA** (trans-activating CRISPR RNA). The **Cas** genes can include genes encoding several different Cas proteins, such as Cas1, Cas2, Csn2, and Cas9. There may be differences in Cas proteins in various bacterial strains. This chapter will discuss primarily Cas9, but the other Cas proteins also have important functions. Cas9 has nuclease activity, and thus, it can cleave DNA. A diagram of the CRISPR locus is shown in Fig. 25.1a.

3 CRISPR as a Virus Defense Mechanism for Bacteria

3.1 Introduction to Viral Infection of Bacteria

Bacteriophages are viruses that infect bacteria (see Fig. 25.1b). These viruses bind to surface receptors on the bacteria and inject their genetic material (usually DNA) into the bacteria. Once this happens, the viral DNA uses the bacteria's machinery to replicate itself and make more virus.

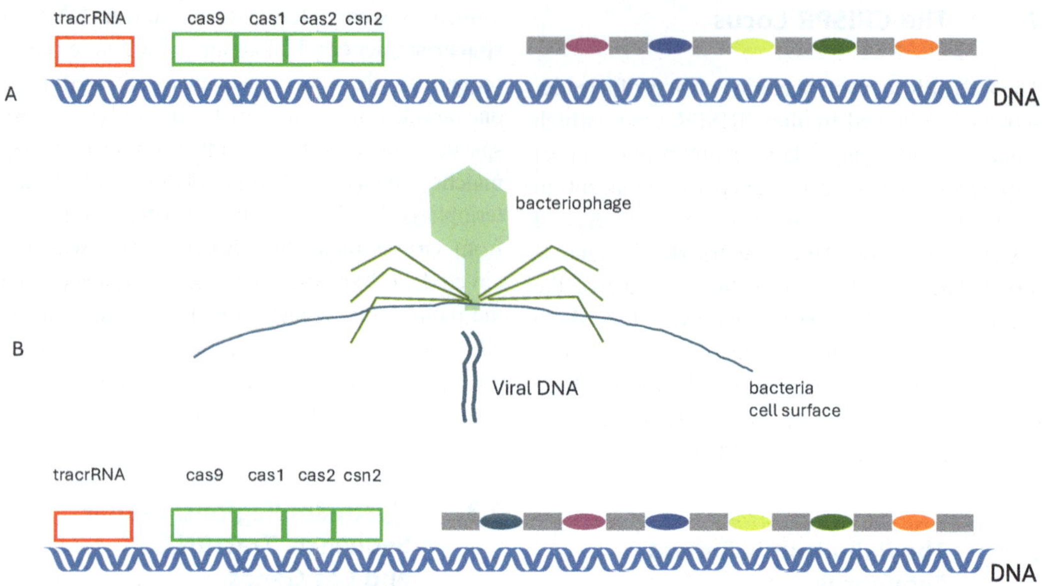

Fig. 25.1 The CRISPR locus, bacteriophage infection, and spacer acquisition. The CRISPR locus is shown in **A**. New spacer and repeat sequences are incorporated when a bacteriophage infection occurs, as seen in **B**. (Repeat sequences are in gray and spacer sequences are in different colors)

The new virus gets packaged, causes the cell to lyse, and can now infect more bacteria. This cycle of infection and lysis can cause serious harm to colonies of bacteria [8]. Bacteria have both non-specific and specific ways to prevent and combat some infections. An example of a nonspecific mechanism is modification of receptors on their surface, making it more difficult for the virus to bind to the cell surface and inject their genetic material [9]. We now know they also have more specific ways to combat infections from a particular phage. The CRISPR system is an important mechanism by which bacteria can "remember" an infection by a virus and prevent a second infection by that same virus [10].

3.2 How CRISPR Defends Bacteria from Repeat Infection by Viruses

The CRISPR locus helps provides bacteria with adaptive immunity against viruses. After the first infection, fragments of viral DNA are integrated into the CRISPR locus, in a process called spacer acquisition. During a second infection, these spacer sequences are transcribed and play a key role in destroying the invading virus. This system is discussed in 2, 4, 5, 11, and 12 and described below.

3.2.1 Spacer Acquisition

When the virus infects a bacterium, its DNA is inserted into the bacterium. If a bacterium gets infected with a new virus (one it has not been exposed to before), some of the viral DNA gets cleaved into pieces inside the cell. A short segment of DNA from the virus is integrated into the leader sequence of the CRISPR locus, the sequence just upstream of the repeat-spacer-repeat array. This new viral sequence now becomes a **spacer** sequence and integrates into the CRISPR locus. **Repeat** sequences are duplicated for each new spacer sequence added, thus maintaining the **repeat-spacer-repeat** structure (Fig. 25.1).

3.2.2 Transcription of the CRISPR Locus

The CRISPR locus (the repeat-spacer array) gets transcribed into an RNA sequence. This transcription can occur continuously, or it can be induced under certain conditions, such as during bacteriophage infection, depending on the bacterial strain and CRISPR locus [12, 13]. The resulting RNA sequence is called precursor crRNA (**pre-crRNA)** until it gets processed and cut into smaller fragments (Fig. 25.2).

3.2.3 Processing of the Pre-crRNA

After transcription (Fig. 25.2), the pre-crRNA gets processed. Several things happen during processing. The pre-crRNA is not translated. Instead, it is cleaved into individual crRNAs. The tracrRNA is also transcribed, and the Cas genes are transcribed to produce Cas proteins, including Cas9. Individual spacer/repeat pairs (in their RNA form) are separated from each other. Cas9, tracrRNA, and **RNase III** (an enzyme that can cleave RNA) all have roles in this process.

3.2.4 TracrRNA, Cas9 Protein, and the Ribonucleoprotein (RNP) Complex

The tracrRNA has sequences that are complementary to the palindromic **repeat** sequences, and it can thus anneal to them. The tracrRNA can also guide the Cas9 protein into position, mediated by RNA-binding domains within Cas9 [14]. Together, these can form a **ribonucleoprotein (RNP) complex**, which contains **Cas9**, one **palindromic repeat sequence, one spacer sequence**, and the **tracrRNA**. In this **complex**, the **palindromic repeat sequence + spacer sequence** can also be referred to as the **guide RNA**. This entire ribonucleoprotein complex will have a key role in cleaving foreign DNA, as discussed next.

3.2.5 After Reinfection by a Virus

The following occurs if the bacterium is infected by a virus that it has encountered before. When this virus inserts its DNA into the bacteria, the spacer RNA sequence in one of the **ribonucleoprotein complexes** will "recognize" it due to having a complementary sequence. This spacer RNA sequence will anneal to its complementary strand on the double-stranded viral sequence by complementary base pairing. This complementary sequence on the virus is now called the **protospacer**. Cas9, which is associated with the spacer sequence as part of the ribonucleoprotein complex, can cleave the viral DNA (Fig. 25.3). Cleavage involves creating a break on both strands of the DNA of the bacteriophage (a double-stranded break), thereby destroying this foreign viral DNA. While we focus on Cas9, other Cas proteins also have important functions, such as roles in selecting and acquiring the spacer sequences for the CRISPR locus [4].

3.2.6 The PAM Sequence

One more factor involved in the above sequence of events is the **PAM sequence** (see Figs. 25.2 and 25.3). The PAM sequence is the **p**rotospacer **a**djacent **m**otif. When the CRISPR system "chooses" a fragment of viral DNA to insert into the CRISPR locus, it will only select a sequence that is adjacent to a **PAM sequence**. The sequence adjacent to a PAM sequence is inserted into the CRISPR locus, but the **PAM sequence** itself is not inserted. Likewise, when a reinfection occurs, only a sequence adjacent to a PAM sequence will be recognized, incorporated into the ribonucleoprotein complex, and cleaved by Cas9. The PAM is a short sequence, two to six nucleotides long. The PAM probably serves a protective function. Since the PAM is not inserted into the CRISPR locus, the CRISPR locus is protected from being a target of its own CRISPR system because Cas9 will not cleave a nucleotide sequence that does not have a PAM [11].

3.2.7 Summary of CRISPR and Viral Infection

The CRISPR array contains palindromic repeats and spacer sequences. The spacer sequences consist of DNA from previous viral infections. The CRISPR array is transcribed into pre-crRNA. The pre-crRNA transcript is cleaved and processed into single spacer-repeat units. If there is a repeat viral infection, part of the DNA sequence of the virus will be complementary to one of the spacer sequences in the bacteria. In this case, as long as a PAM sequence is present, the complementary spacer sequence (which is in a complex with Cas9 and the tracrRNA in the RNP complex) anneals to the viral DNA. Cas9 then uses nuclease activity to create a double-stranded break in the

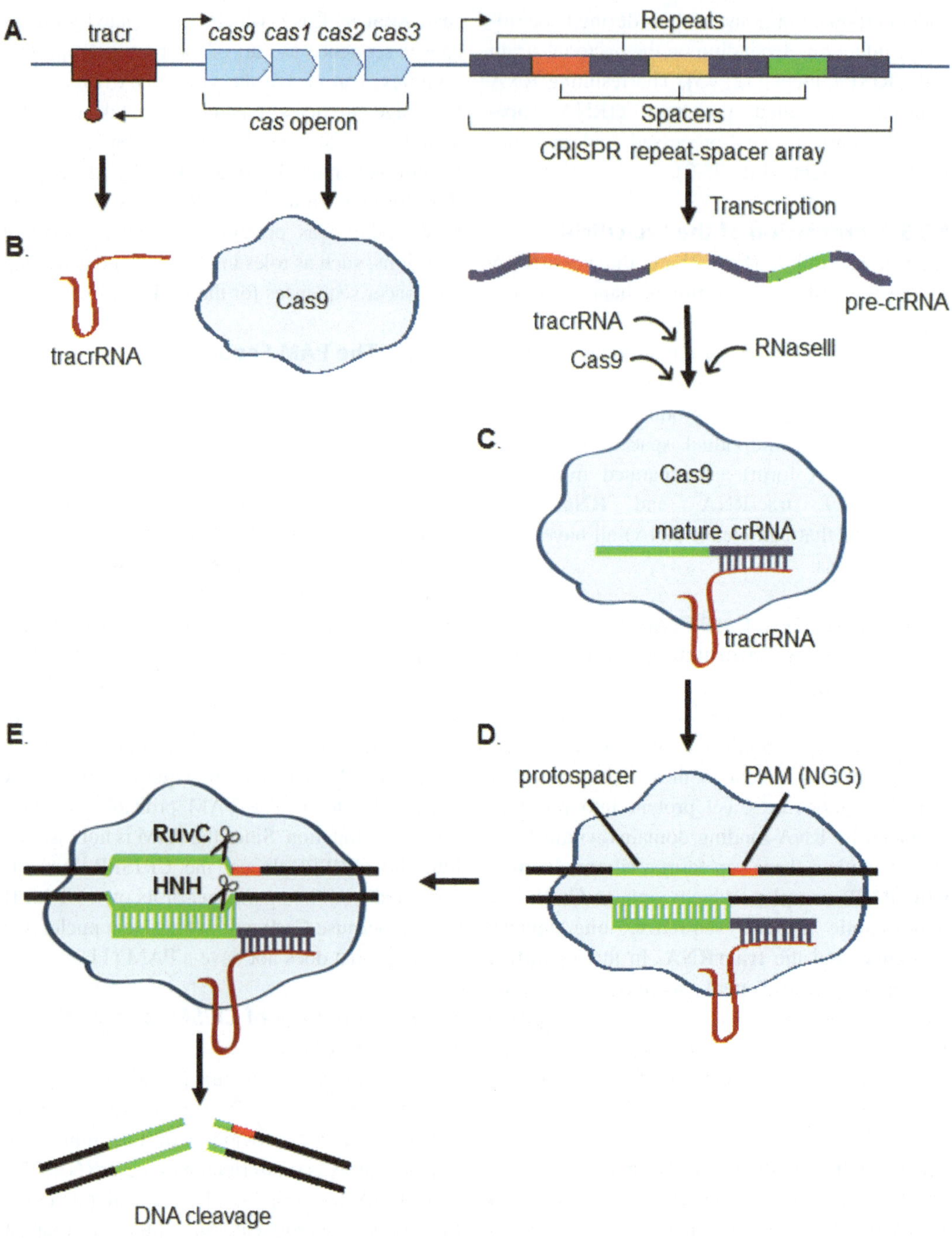

Fig. 25.2 Processing of the CRISPR locus. The CRISPR locus gets transcribed into a long precursor RNA, which is then processed into individual crRNAs. Cas proteins (encoded by the Cas genes) are translated separately. Cas, tracr, and the spacer-repeat sequence function together to cleave DNA. *Source:* Reproduced from: Gebre et al. (2018). Viruses. 2018 10(2):55. DOI: https://doi.org/10.3390/v10020055.Licensed under the Creative Commons Attribution 4.0 (CC BY 4.0) license

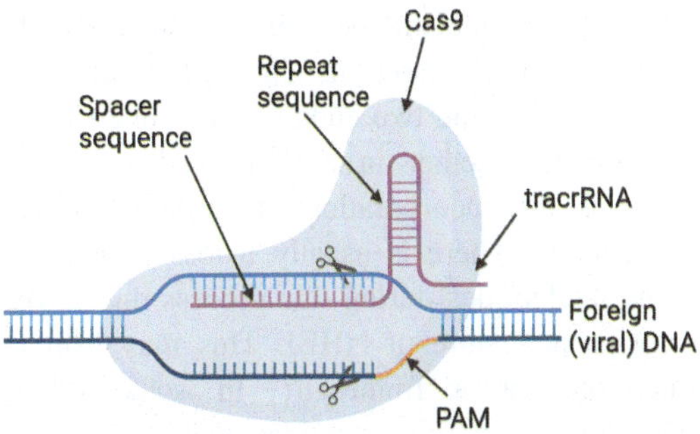

Fig. 25.3 CRISPR-Cas9. The structure of the ribonucleoprotein complex is shown here, along with the foreign DNA to be cleaved. See text for details. *Source:* Created with BioRender.com

viral DNA sequence. The cleavage of the viral DNA by the Cas9 nuclease activity deactivates and destroys the viral DNA. Thus, the bacterium is protected from being harmed by the virus due to the CRISPR locus [5].

4 CRISPR Can Be Manipulated for Use in Editing the DNA of Various Organisms

4.1 Introduction to the RNA-Programmable CRISPR-Cas9 System for Gene Editing

CRISPR is highly effective at cleaving and destroying bacteriophage DNA. Scientists reasoned that since the CRISPR system is based on recognizing specific sequences of viral DNA, it may be possible to manipulate the system so that it could modify or edit any DNA. Scientists began studying ways to modify the CRISPR system for this purpose, and they developed a method for using the system to modify DNA outside of the bacterial system. This method is called the RNA-programmable CRISPR-Cas9 system. Many scientists were involved in developing this breakthrough technology, and Jennifer Doudna (UC Berkeley) and Emmanuelle Charpentier (Max Planck Institute, Berlin) won the Nobel Prize in 2020 for discovering how the CRISPR-Cas9 system could be used as a gene editing tool [15].

4.2 The Guide RNA

The CRISPR system was modified so that it could be used to cleave DNA outside of bacteria. This could be done in vitro or in eukaryotic or other cells. This system, the **RNA-programmable CRISPR-Cas9 system,** is sometimes referred to as the **CRISPR-Cas9** system or as **CRISPR.** This system involves the generation of a synthetic **guide RNA** [16]. This synthetic guide RNA contains the targeting sequence (which is equivalent to the spacer sequence) and a scaffold (which is equivalent to tracrRNA + part of the palindromic **repeat** sequence). The targeting sequence in this guide RNA is complementary to a specific DNA sequence in the target sequence that the scientist wants to cleave or edit. (The synthetic **guide RNA** is sometimes referred to as **gRNA** or **sgRNA** [2].)

4.3 The Guide RNA Must Contain a PAM Sequence

The guide RNA must contain a PAM sequence because the PAM sequence is required for the CRISPR-Cas9 system to work and to cleave the DNA. What happens if scientists want to cleave a sequence that is not near a PAM? PAM sequences are small, and there are different PAM sequences that can be used. The most common PAM sequence is 5'-NGG-', where N can be any nucleotide. Thus, it is likely that the investigator can find a sequence adjacent to a PAM. If not, however, another solution is to use different Cas proteins from different bacterial strains, as some Cas proteins recognize different PAM sequences [5, 11].

4.4 Setting Up a CRISPR Study in the Lab

The CRISPR-Cas9 system can be used to modify gene sequences under many conditions. DNA modification can be done in vitro in a cell-free system, in cells, or even in embryos [5]. Here is an example of how the CRISPR-Cas9 system may be used in cultured animal cells to target a specific gene within the cell genomic DNA. A guide RNA (gRNA) is synthesized to include a single-stranded RNA sequence complementary to a specific region of the gene to be modified, fused to a tracrRNA sequence. This guide RNA is introduced into cells via transfection. Along with the guide RNA, cells are also transfected with a plasmid encoding the Cas9 protein. Alternatively, cells may be transfected with Cas9 mRNA or Cas9 protein. The cells are then cultured for several days. Within the cells, Cas9 associates with the guide RNA to form a ribonucleoprotein complex. This complex binds to the target DNA sequence via base pairing between the guide RNA and the genomic DNA. Cas9 then introduces a double-stranded break at the target site. Genomic DNA is then isolated from the cells and sequenced to verify that the intended alterations occurred. The types of alterations that may be introduced are discussed more in the next sections.

4.5 The Types of Changes Made in the DNA After Modification by the CRISPR-Cas9 System

4.5.1 Introducing a Double-Stranded Break (DSB)

The classic function of the CRISPR-Cas9 system is to introduce a double-stranded break (DSB) in DNA. If a cell-free system is being used (i.e., the CRISPR machinery and target DNA are mixed together in a tube), then the double-stranded breaks in the target gene will remain. However, within a cell, as described above, the double-stranded break will nearly always be repaired. The repair mechanism that usually occurs is nonhomologous end joining (NHEJ). This mechanism is error-prone, and thus the gene is likely to be disabled even though the repair mechanism takes place [11, 17]. NHEJ was covered in Chap. 23 and also briefly described below.

4.5.2 Nonhomologous End Joining (NHEJ)

Nonhomologous end joining (NHEJ) is the default mechanism that is usually used for repairing double-stranded breaks in cells. In NHEJ (also discussed in Chap. 23), after the DSB is made, the two ends of the broken DNA are joined together in a process that usually involves the incorporation of some additional nucleotides. There is usually a loss or gain of some nucleotides during this process due to the error-prone nature of NHEJ. This may lead to mutations or a frameshift. In some cases, incorporating new nucleotides may lead to an in-frame stop codon. This leads to a truncated and possibly ineffective protein product [11].

4.5.3 Homology-Directed Repair (HDR)

NHEJ [18] is the default cellular response for repair of DSBs such as those made by the CRISPR system, but it usually inserts random errors in the DS break site [19] (see Chap. 23). In many cases, that is the desired effect, because CRISPR is often used to disable a gene. However, sometimes, a more specific gene change is desired. In this case, the HDR method can be more useful. In contrast, the HDR method can be used for precise insertion of specific changes in the DNA. This method is challenging because NHEJ is the more dominant pathway, and it is necessary to stop the NHEJ method from occurring in order for the HDR method to proceed. To improve HDR efficiency, researchers may inhibit components in the NHEJ pathway, such as by inhibiting NHEJ-specific repair proteins, or they may synchronize cells so that cells are in S or G2 phase when HDR is more likely to occur [20].

For HDR, the cleaved DNA is mixed with a homologous DNA template. The template is homologous to the DNA at both sides of the break, but at the region where the break is, it contains a specific sequence of nucleotides to be added to the DNA. These nucleotides (or nucleotide) are to be added precisely to the break site. This sequence may be designed to contain a specific mutation that the investigator is trying to add to the DNA or a corrected sequence to repair a mutated DNA sequence [21].

HDR works on the principle of **homologous recombination (HR)** [22]. In CRISPR, the HDR method that is usually used is the SDSA method. HDR and the SDSA method were covered in more details in Chap. 23. In this HDR method,

CRISPR is used to generate a DSB. In this case, however, the reaction includes a homologous template. The template usually contains small changes to be incorporated into the break region. For the DNA with the DSB, the 5′ ends of the breaks are resected by nucleases to create single-stranded 3′ overhangs. The 3′ ends now invade the homologous template, which partly unwinds and denatures. These 3′ ends anneal to their complementary sequences in the homologous template. The strands of this homologous template will serve as a primer for DNA synthesis of each strand. DNA synthesis then takes place, and when it is complete, the newly synthesized strand dissociates from the template. The newly repaired DNA is now made. In the process, the original DNA with the DS break will incorporate specific changes that were present in the homologous template [2, 11, 21].

There are many reasons that researchers may want to use the HDR method to make specific targeted changes in a gene [20]. For example, a cancer cell line may have a mutant Ras gene such as RasV12, which contains a single nucleotide mutation. One could potentially use HDR to change this mutant Ras to wild-type Ras to determine what effect this has on cell growth. Alternatively, one might grow cells with normal Ras and want to use CRISPR to change wild-type Ras to a mutant Ras to study its effects in a certain cell type.

5 CRISPR-Cas9-Based Therapies

The CRISPR-Cas9 system is remarkable for being a relatively straightforward way to genetically engineer DNA in potentially any organism. It can be carried out more quickly than previous methods that have been used for gene modification. In fact, kits are available that have everything a lab needs to delete or mutate their gene of interest in a relatively short time [23]. Since its discovery as a gene modification tool, many studies have already used the CRISPR-Cas9 system to modify genes in animals, humans, and plants. A promising avenue is research into the possible use of the CRISPR system to treat human disease. Drugs made using the CRISPR system will almost always be considered biologics because they will involve treating patients with products of genes made from living cells. So far, one FDA-approved drug has been made using the CRISPR system. Below, we discuss this newest CRISPR-Cas9-based drug and other treatments in the pipeline or the research phase [2, 24].

5.1 CASGEVY: The First Approved CRISPR-Based Treatment

CASGEVY was approved in 2023 to treat sickle cell disease [25, 26]. Some background information about sickle cell disease is helpful before discussing how CASGEVY works.

5.1.1 Sickle Cell Diseases and Thalassemia

Sickle cell disease (such as **sickle cell anemia**) and **thalassemia** are inherited disorders caused by abnormalities in the gene that codes for the beta subunit of **hemoglobin**, the oxygen-carrying protein in red blood cells (RBCs). Hemoglobin transports oxygen in the RBCs from the lungs to the rest of the body. Hemoglobin is actually a protein complex that is made up of four subunits. Specifically, it is made up of four globin chains (see Chap. 4). There are two major categories of hemoglobin [27] (see Fig. 25.4a). **Fetal hemoglobin (HbF)** is present in the developing fetus. After birth, HbF is gradually replaced by **adult hemoglobin (HbA)** within the first year of life. Fetal and adult hemoglobin function similarly but have some important differences in their structures. **Fetal hemoglobin** is made of **two α-globin** chains and **two γ-globin** chains. In contrast, **adult hemoglobin** is made of **two α-globin** chains and **two β-globin** chains [28]. In both cases, the structure also contains an iron molecule, which is necessary for binding oxygen. RBCs no longer have a nucleus, but the globin genes are transcribed and translated in the erythroblasts, which are derived from hematopoietic stem cells and which mature into RBCs [29–31].

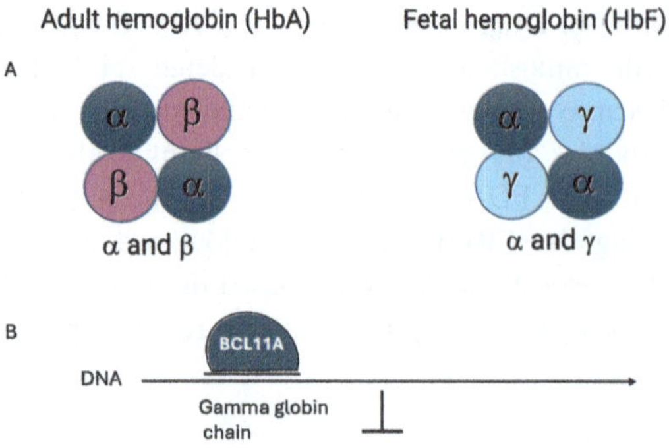

Fig. 25.4 A. Subunits of adult and fetal hemoglobin. Adult hemoglobin is composed of α-and β-subunits, while fetal hemoglobin is composed of α- and γ-subunits. B. The gene encoding γ-globin. The gene encoding γ-globin is negatively regulated. A major mechanism by which it is negatively regulated is by the transcription factor BCL11A

5.1.2 Sickle Cell Disease

Sickle cell disease is usually caused by **mutations** in the **beta-globin gene**, resulting in a mutated protein that is structurally abnormal. **Thalassemia** is also caused by mutations, but in this case, the result is absent or greatly reduced globin protein levels, and mutations may be in the alpha or beta subunits of hemoglobin, depending on the type of thalassemia [29]. **Sickle cell anemia** is the most common **sickle cell disease**. It is an inherited autosomal recessive disease. In this disease, there is a mutation in beta-globin (the beta-chain of hemoglobin) at a specific site (amino acid 6, glu → val). As a result, defective hemoglobin causes the red blood cells to form an abnormal shape (a sickle shape), which can cause the cells to get stuck in small vessels and reduce blood flow. This reduced blood flow can cause severe pain, interfere with the transport of oxygen, and lead to damage to the heart and other organs [32, 33].

The mutation in sickle cell anemia is found only in **adult hemoglobin**. Fetal hemoglobin does not contain a beta-chain at all. Instead, it has two gamma chains, and therefore, it completely lacks the beta-chain mutation. Fetal hemoglobin stops being produced after birth, but scientists reasoned that if they could cause its production to occur again after birth it may be able to compensate for the defective adult hemoglobin in patients with sickle cell anemia and other sickle cell diseases [32].

5.1.3 CRISPR and Treatment of Sickle Cell Disease

Casgevy (exagamglogene autotemcel) is made by CRISPR Therapeutics together with Vertex Pharmaceuticals. It was initially called CTXOO1. Casgevy is a new CRISPR-Cas9-based therapy designed to cause the production of fetal hemoglobin to occur even in adults. To understand how it works, it is helpful to understand the mechanism by which fetal hemoglobin expression is regulated.

- The mechanism by which fetal hemoglobin is regulated

Fetal hemoglobin is made up of α-and γ-globin units, while adult hemoglobin is made up of alpha and beta subunits. Fetal hemoglobin is gradually replaced after birth because the γ-globin gene becomes transcriptionally silenced and the β-globin genes become activated. This switch from γ-globin to β-globin is known as hemoglobin switching. The γ-globin genes are silenced in adults largely due to the **transcription regulator BCL11A**. After birth, BCL11A (along with several cofactors) binds to regulatory elements of the γ-globin gene and suppresses its expression. Thus, it **represses** the production of **fetal hemoglobin** after birth [32] (see Fig. 25.4b). (Although BCL11A is the key regulator of γ-globin gene repression, this switching process also involves other proteins such as other transcription factors and chromatin remodeling complexes [34]).

- The use of CRISPR to modulate gamma-globin expression

A CRISPR-Cas9 system was developed that specifically targets the regulatory element on the gene encoding BCL11A in red blood cell (RBC) precursors. The goal was to inactivate the BCL11A gene so that the γ-globin gene could continue to be expressed even in adults. Bone marrow-derived hematopoietic stem cells (HSPCs) were harvested from patients with sickle cell disease. The CRISPR-Cas9 system was used to disrupt the expression of the gene for BCL11A in these cells, ex vivo. This was accomplished by introducing a full ribonucleoprotein (RNP) complex, which was composed of Cas9 protein and a guide RNA that is complementary to a part of the BCL11A gene. Specifically, rather than targeting the coding sequence of BCL11A, the scientists targeted a regulatory enhancer within one of the introns of the BCL11A gene. This enhancer is called an erythroid-specific enhancer; it is located in exon 2 and positively regulates BCL11A gene expression specifically in erythroid cells (RBC precursors).

By inactivating this enhancer, they predicted that BCL11A would no longer be efficiently expressed in RBCs. The CRISPR system they used relied on the introduction of double-stranded breaks followed by NHEJ to introduce mutations that would disrupt this regulatory enhancer [35].

After CRISPR was used to edit hematopoietic stem and progenitor cells (HSPCs), some of the cells were analyzed to confirm the gene editing. The researchers optimized the conditions so that the BCL11A enhancer was disrupted in a high percentage of edited cells. These modified HSPCs cells were reinfused into patients following myeloablative conditioning. This is a procedure that eliminates (ablates) the patient's existing bone marrow cells before introducing the new and modified cells. This procedure comes with risks and involves the use of a chemotherapy drug.

As predicted and hoped for, many of the stem cells in these patients now produced fetal globin. Even better, they found that this new fetal hemoglobin could compensate for the abnormal adult hemoglobin in patients, and many sickle cell disease symptoms were alleviated [35].

- FDA approval of CASGEVY

The work described above became the **first approved CRISPR-Cas9-based treatment** for a **human disease**. The therapy, CASGEVY (exagamglogene autotemcel), was granted conditional marketing authorization in the UK in 2023 [36], and it was approved by the FDA in the USA a few weeks later for patients with sickle cell disease and later for patients with transfusion-dependent beta-thalassemia (a severe form of the disease where patients depend on transfusions) [37, 38]. It works by harvesting the patient's CD34+ hematopoietic stem and progenitor cells (bone marrow cells that can self-renew and give rise to all blood cell types including the RBCs) and editing them using CRISPR/Cas9 to disrupt the erythroid-specific enhancer of BCL11A, thereby reactivating production of fetal hemoglobin (HbF) in the erythroid lineage, resulting in adult red blood cells that have HbF. The success of this study illustrates the promise of using CRISPR-Cas9 technology. Challenges remain; however, because although the treatment is effective, at the current time, the treatment is limited by its high expense and requirement for intensive procedures. This is due to the challenge of obtaining patient HSCs, myeloablative conditioning, in which the patient receives chemotherapy before the stem cell transplant, editing the DNA, and reinjecting the blood cells [39–42].

5.2 VERVE-101: A Clinical Trial for a CRISPR-Cas9-Based Treatment to Treat Familial Hypercholesterolemia

5.2.1 PCSK9 and LDL

Currently, Casgevy is the only FDA-approved CRISPR-based treatment, but other promising treatments are under investigation. VERVE is a CRISPR-Cas9-based treatment still in clinical trials (as of 2024), and it is designed to treat familial hypercholesterolemia. It is based on the function of the protein PCSK9 (proprotein convertase subtilisin/kexin type 9). PCSK is found primarily in the liver and binds to the LDL receptor (Fig. 25.5). It was described in some detail in Chap. 9. Briefly, the LDL receptors (LDLRs) on the surface of liver cells remove LDL (low-density lipoprotein, the major carrier of cholesterol) from the bloodstream. A balance of the right amount of LDLRs on the liver cells is important for maintaining a low level of LDL in the bloodstream. PCSK9 is a secreted protein that is also produced in liver cells. PCSK9 binds to LDLRs on the surface of liver cells and leads to internalization of PCSK9-LDLR complexes. This leads to destruction of the LDLRs and decreases the number of LDLRs available to remove LDL from the bloodstream [43].

The amount of PCSK9 produced, therefore, has an impact on how cholesterol is processed. When the **levels of PCSK9 are high**, a large amount of LDL receptors gets internalized in the cells in the liver, often without even binding LDL. After this, the LDL receptors are destroyed in the lysosome. Consequently, this reduces the amount of LDL that can be taken up in the liver and increases the amount of LDL that remains in the bloodstream. This results in **high circulating LDL levels**. In contrast, when the levels of PCSK9 are low, most of the LDL receptor is not destroyed. Instead, more LDL receptors are present on the surface of liver cells, which bind and import LDL into the liver, where it is processed and removed from circulation. The LDL receptor then gets recycled back to the cell surface, where it can bind more LDL. This results in **low circulating LDL levels** [44].

5.2.2 VERVE-101: The CRISPR-Cas9 System that Inactivates the PCSK9 Gene

A CRISPR-Cas9 system called VERVE-101 is designed to disrupt the gene for PCSK9. It is being developed by Verve Therapeutics (Boston). PCSK9 is a regulator of the LDL receptor (LDLR) also discussed in Chap. 9. PCSK9 binds to LDLRs on the surface of liver cells, causing the receptor to internalize, and this lowers the amount of LDL receptors on the cell surface. As a consequence, high levels of PCSK9 can be associated with too much LDL, while lowering PCSK9 can be beneficial in lowering LDL levels (see Fig. 25.5).

The system for disrupting PCSK9 is a bit different from the system used for Casgevy, and it represents a significant advance in the CRISPR gene editing field. This system uses a modified Cas9 that was engineered in the lab, called a **base editor**. While Cas9 normally makes DS breaks, this modified Cas9 was engineered into a "nickase" that nicks only **one strand** of the DNA. This avoids the formation of DS breaks in the DNA and therefore avoids the triggering of the NHEJ process, which can lead to unwanted mutations. The first step in this modification of Cas9 involved mutating one of the two nuclease subunits of Cas9. This modified Cas9 was referred to as **Cas9 nickase (nCas9)**. In the second modification, this nCas9 was fused with a **deaminase enzyme**, which converts adenine (A) to inosine (I), which is interpreted as guanine (G) in single-stranded DNA. This is the final modified Cas9, and it is called a **base editor**.

When this base editor finds its target gene, it forms an R loop, which is a structure in which the target DNA binds the guide RNA, leaving the nontarget DNA (the complementary strand) single-stranded. Within this single-stranded

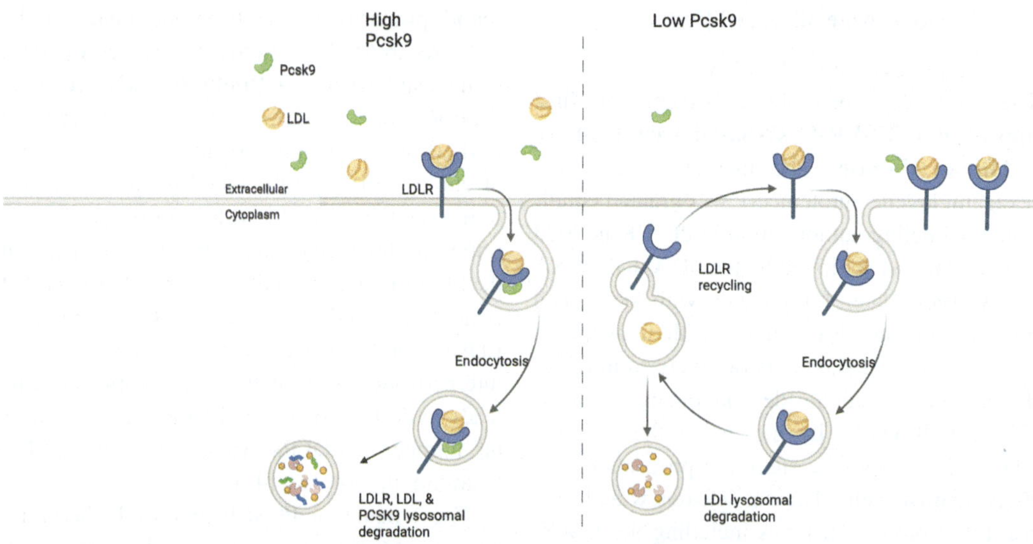

Fig. 25.5 PCSK9. PCSK9 binds to LDL receptor (LDLR) and targets it for lysosomal degradation, thereby lowering the number of LDL receptors available to clear LDL from the bloodstream. When PCSK9 levels are low, less LDLR is degraded. *Source:* Created with BioRender.com

exposed region, the A to I conversion takes place. This leads to a permanent change in the final gene. This is because during replication and repair of the modified DNA, the cell will read inosine as guanine (G) so that the base pair that was originally an AT base pair ultimately becomes a GC base pair.

Using **the base editor**, the researchers were able to add a precise mutation to the PCSK9 gene. Specifically, they designed the guide RNA so that a premature stop codon would be inserted into exon 1 of the PCSK9 gene, and as a result, no functional protein is made.

This modified CRISPR-Cas9 system was packaged in lipid-based nanoparticles. These nanoparticles contained the DNA sequence encoding the base editor and the guide RNA. It was delivered intravenously and taken up by the liver cells via LDL receptors.

In trials, a single dose of this treatment (delivery of the nanoparticles containing the CRISPR machinery) was given to patients enrolled in the studies. Studies were done in nonhuman primates and human patients with high LDL. Phase Ib trials have been initiated (2023). In addition to evaluating safety, patients have been evaluated for the reduction of PCSK9 protein and circulating LDL. LDL levels were reduced in up to 55% of participants who were given a high dose of treatment. While promising, some patients also have side effects, raising safety concerns. The next steps involve enrolling more patients, carrying out long-term follow-up studies, and studying the basis of any side effects [45–49].

5.3 The First Personalized Gene Editing Therapy Was Used to Treat a Urea Cycle Disorder

This year (2025) was a groundbreaking year for CRISPR technology because the **first personalized, in vivo CRISPR-Cas9 gene editing therapy** was administered to a patient. This treatment was used to treat a patient with CPS I deficiency (CPS1D). As discussed in Chap. 14, CPS1D, a mutation in the CPS I gene, disrupts the first step in the urea cycle. Since the gene mutation can be identified, this disease was considered to be a good candidate for CRISPR-Cas9 therapy. This therapy was done on an infant with CPS1D. The infant had two mutations in the CPS1D gene. The researchers focused on correcting one of the mutations using adenine base editing (ABE). This is a CRISPR technique that can convert an A-T base pair to a G-C base pair without actually cutting the DNA, thus reducing the risk of introducing more DNA damage. A modified Cas9 protein with a custom guide RNA was packaged in lipid nanoparticles. The particles were administered to the patient starting between 6 and 7 months of age, by IV infusion, and taken up by the patient's liver. After treatment, the targeted mutation was corrected in the liver. While this patient spent the first few months in the hospital under a serious restrictive diet, after the treatment, he was growing and continues to do well. This treatment is groundbreaking, as it is the first personalized gene editing treatment using CRISPR on a patient, and it is hopeful that this technique will be useful for other genetic disorders in the future.

The method used for this treatment was somewhat different from other CRISPR-Cas9 methodologies described in this chapter. Specifically, it used adenine base editing (ABE), which enables targeted base changes without generating double-stranded breaks in the DNA. Briefly, ABE can convert an **adenine-thymine (A-T) base pair** into a **guanine-cytosine (G-C) base pair**. This is done using a **modified Cas9 protein**, fused to a **deaminase enzyme** that chemically alters the adenine base. Because this approach avoids double-stranded DNA cleavage, it may reduce mutations that can occur in response to DNA break repair. It is also promising because it can lead to the insertion of precise genetic changes in the DNA [50, 51].

6 Other Potential Uses for CRISPR-Cas9 in Studying and Treating Human Disease

Some examples of other studies designed to assess the use of the CRISPR-Cas9 system for various purposes, including the treatment or prevention of other disorders, are discussed below.

6.1 CRISPR-Cas9 and HIV

HIV (human immunodeficiency virus) is the virus responsible for causing AIDS (acquired immunodeficiency syndrome). HIV enters cells by binding to CCR5 receptors, primarily on the cell surface of T cells (see Chap. 24, Fig. 24.4). HIV cannot enter cells that do not have functional CCR5 receptors. Scientists are therefore interested in using the CRISPR-Cas9 system to disrupt the CCR5 gene in human cells. One study to test this involved using the CRISPR-Cas9 system to induce an inactivating mutation in the CCR5 gene in human blastocyst stage embryos from IVF clinics. Disabling CCR5 was successful in a small percentage of the embryos. The purpose of this study was to determine whether CRISPR-Cas9 technology could be successful in embryos, with the long-term goal of developing strategies to make them HIV-resistant. The researchers for this study purposely worked with embryos that were not viable due to genetic abnormalities, but research on viable embryos is considered unethical [52, 53].

6.2 CRISPR and Modification of Human Embryos

Although it was considered a violation of scientific and ethical standards, in 2018 and 2019, three CRISPR-edited babies were made. These were made with the goal of developing babies that were resistant to HIV, using technology similar to that described above. The ability to genetically modify human embryos is remarkable, but the use of CRISPR to modify human babies has been outlawed. There is a worldwide moratorium on CRISPR editing on viable embryos while scientists address serious issues, including ethical issues, that this would entail [54].

6.3 Pigs as Organ Donors

The shortage of available human organs for transplants has led researchers to try transplanting animal organs into human patients. One of the best candidates is the pig due to various anatomical similarities between pig and human. So far, pig-to-human transplants have not lasted long, but scientists are studying ways to make pig organs more acceptable for human recipients and thus reduce the possibility of rejection. This requires genetic modification of pigs. While the genetic engineering of animals was previously relatively slow, using CRISPR makes it more efficient. Some researchers are studying ways to use the CRISPR-Cas9 system to modify the genes of pigs so that their organs will be less likely to be rejected by the human immune system [55].

6.4 Animal Models of Human Disease

CRISPR makes modifying genes of animals much more efficient than previous systems (which involved techniques such as homologous recombination in ES cells or microinjection techniques to make transgenic animals [56]). One interesting CRISPR study involves disrupting the SHANK3 gene in monkeys. This gene has been implicated in some forms of autism. While it is challenging to study autism in nonhumans, monkeys with the disrupted SHANK3 gene showed some signs of autism. Mutating individual portions of the gene may help scientists determine which parts of the gene and encoded protein may have the most important roles in autism [57].

6.5 CRISPR in Basic Research

One of the most important uses of the CRISPR-Cas9 system is in basic research. The CRISPR system allows scientists to turn certain genes off to study their functions in animal models or cultured cells. Scientists can use this method to study the functions of many different genes, such as those thought to be involved in cancer, cell cycle regulation, cell differentiation, regulation of transcription, development, metabolism, and others [11, 58].

7 CRISPR in Crops and Farm Animals

7.1 Gene Editing in Crops

CRISPR can be used for many purposes aside from developing drugs and studying disease, for example, it can be used to alter crops. Some bananas, for example, are susceptible to diseases caused by bacterial infections. This affects crop yield, economic loss, and, in some regions, problems related to food security. In some cases, CRISPR is being used to disrupt a gene that makes the bananas particularly sensitive to infection. Genetically edited bananas have enhanced resistance to infection. These types of studies are promising for increasing prop yield, although they are still in experimental stages [59].

7.2 CRISPR in Farm Animals

Pigs on farms sometimes succumb to infection by a porcine reproductive and respiratory syndrome virus (PRRSV). This is a serious problem, and pig deaths cost farmers millions of dollars each year. Vaccinations have failed to prevent the syndrome's spread. CD163 is a receptor for the virus, found on some pig immune cells, allowing the virus to enter cells. CRISPR was used to disrupt the CD163 gene in pigs. As a result, the pigs became resistant to exposure to the virus [60–62].

8 Gene Drives to Alter Wild Mosquito Populations

8.1 The Problem with Altering Genes in Wild Populations

An interesting potential use for CRISPR is in the creation of gene drives. Gene drives are a mechanism for altering a wild population, although they remain in experimental stages. While modulation of genes in individuals and some of their offspring could be carried out before the age of CRISPR editing, it would have been nearly impossible to modify wild populations. In traditional gene modification methods, only one allele of a gene is usually modified (a heterozygous mutation). Planned crossbreeding can generate homozygous deletions. However, if the heterozygous animal (or even a homozygous animal) mates with a normal animal in the wild, then the modified gene will be diluted with respect to the wild-type gene in the overall gene pool. This dilution will continue as the progeny mate and reproduce, and the mutation will eventually be present in only a tiny percentage of the animals or be selected against completely [63].

8.2 Gene Drive Mechanisms

CRISPR gives scientists a way around this by making gene drives (Fig. 25.6). By using gene drives, an animal that is heterozygous for a gene modification becomes homozygous. Not only that, but when they mate with a wild-type animal, the progeny becomes homozygous as well, and thus the modification can spread through the population. Gene drives work as follows and are also described in [64–66]:

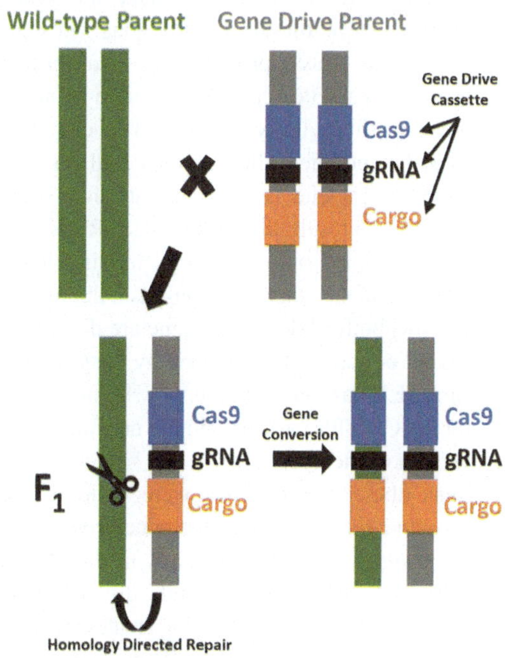

Fig. 25.6 Gene drives. Gene drives are a CRISPR-based mechanism that spreads a modified gene throughout wild populations. See text for details. *Source:* Croghan et al. (2023). Front. Agron. 5:1290781. doi:10.3389/fagro.2023.1290781. Image licensed under the Creative Commons Attribution 4.0 (CC BY 4.0) license

1. The scientists insert a "CRISPR cassette" into the germline cells of the animal (such as a mosquito). This CRISPR cassette is one long sequence that usually contains the Cas9 gene, the guide RNA (which includes a sequence complementary to a part of the gene being targeted), and flanking sequences that usually correspond to regions surrounding the gene to be targeted. This cassette is introduced along with the Cas9 protein or RNA and the guide RNA. This makes a double-stranded break in the target gene. As the cell tries to repair the cut, it uses HDR, using the cassette as the HDR template. Thus, the new allele now contains the entire CRISPR cassette, including Cas9 and guide RNA, disrupting or replacing the targeted gene. The animal is now heterozygous for the CRISPR cassette and the disabled gene. While this system is highly efficient, the efficiency of HDR can vary somewhat, according to species type and other factors.
2. At this point, the other chromosome still has the wild-type allele of the gene. In the germline cells, the guide RNA produced by the modified allele will recognize this wild-type sequence on the homologous chromosome. The Cas9 protein, also transcribed from the cassette, creates a double-stranded break in this wild-type allele. The cell then uses the modified allele, which still has the CRISPR cassette, as a template for HDR repair. As a result, the entire CRISPR cassette will be copied onto the corresponding wild-type site on the other allele, just as it did in step 1. Now, both alleles of the gene are disrupted, and all or part of it is replaced by the CRISPR cassette. Thus, the animal is now homozygous for the CRISPR cassette and the nonfunctional version of the gene.
3. Since these modifications were done in the germline, the offspring of the gene drive carrying parent will be homozygous in their germ cells for the mutation and the CRISPR cassette. When this offspring mates with a wild-type animal, its germ cells contribute one chromosome containing the CRISPR cassette and the disrupted gene. In the resulting offspring, the same sequence of events will occur in their germline cells, as described in step 2 above, and the other allele, the wild-type allele, will be replaced by the CRISPR cassette and the disabled gene. The system is designed so that this occurs primarily in the germline. In this way, the germ cells of the progeny animal become homozygous for the CRISPR cassette and the gene mutation, and the CRISPR cassette can be passed down to multiple generations.
4. This same process of cutting and copying will then occur as each new generation of animals mates. Thus, the gene drive will be inherited at a high rate. Over many generations, this allows the mutation to spread throughout a wild-type population.

8.3 Potential Uses for Gene Drives

One way gene drives may be beneficial is by inhibiting the spread of malaria by mosquitoes. Malaria is caused by microorganisms called plasmodia, which are parasites that are present in the saliva of some mosquitoes. Plasmodia can enter the body after a mosquito bite and cause malaria. The plasmodia are spread again to other individuals after another mosquito bite. This is a serious problem worldwide and is a major cause of death. Genetic modifications are being tested to make mosquitoes resistant to plasmodia. Once such a modification has been developed, gene drives could be used to spread the modification through the wild-type population of mosquitoes. The goal will be to replace most of the current population of mosquitoes with a modified strain that is resistant to spreading malaria [67]. Gene drives have not yet been used in the wild. This is due to many factors, such as ecological risks, ethical concerns, and some technical problems that are still being addressed [66].

9 Other Technologies Used for Gene Therapy

Although CRISPR is an up-and-coming technology for modifying human genes, various versions of gene therapy have been carried out for some time before CRISPR was developed. One important technology is viral vector-mediated gene replacement. This involves the use of viral vectors such as **adeno-associated viruses**

(AAV, a double-stranded DNA virus) [68, 69] or retroviruses, including lentivirus, which are RNA viruses that can integrate into the host cell genome [70]. **Lentiviruses** have advantages because they can also integrate into nondividing cells and lead to long-term gene expression. These types of delivery systems have been used in clinical trials and some approved therapies. Some types of gene therapy technologies are briefly discussed below:

Zolgensma (onasemnogene abeparvovec) [71] is an **AAV9-based therapy** for **spinal muscular atrophy (SMA)**. (AAV9 is a naturally occurring serotype of AAV) [72].) Zolgensma uses the AAV9 system to deliver a functional copy of the SMN1 gene to children with spinal muscular atrophy (SMA). It is delivered by intravenous infusion. It was FDA-approved in **2019**. AAV9 was used because it can cross the blood-brain barrier and its main target is motor neurons. In this treatment, the new gene remains in an episome, rather than integrating into the genome. Since neurons do not divide, this episomal DNA is relatively stable [69, 73, 74]. Zolgensma is FDA-approved; it does not cure the disease but has been shown to be successful in slowing down or stopping the disease's progression. Long-term data is still being collected, but results so far have been promising.

Strimvelis is a **gene therapy treatment** approved in Europe to treat severe combined immunodeficiency (**SCID**) that is caused by an inactivating mutant of the adenosine deaminase (ADA) gene. The treatment uses a type of retrovirus to deliver a functional ADA gene. In this treatment, the patient's bone marrow cells are harvested, and the cells are modified by infecting them with the retrovirus containing a functional ADA gene and are transfused back into patients [75]. This drug was approved in Europe but is not widely used because, although there has been some success, there have also been safety concerns. Other similar mechanisms are also being tested to improve results and safety for this type of treatment, such as using lentivirus instead of the gamma-retroviral vector [76–79].

Lyfgenia (lovotibeglogene autotemcel) [80] is a **cell-based gene therapy** that uses **lentivirus** for treating patients with **sickle cell disease**. This treatment was FDA-approved in **2023**. In sickle cell disease, patients have a **mutation in the beta-globin gene** as discussed above, and this mutation causes the RBCs to take on a sickle shape [81]. In this therapy, the patients' hematopoietic stem cells (HSCs) are removed and genetically modified to contain a modified copy of the **β-globin gene**. This beta-globin gene has a specific mutation (T87Q) that reduces hemoglobin polymerization and prevents sickling but otherwise functions normally. The gene is introduced using a **lentiviral** vector. The patient undergoes myeloablative chemotherapy to destroy most of their remaining hematopoietic stem cells, and the modified cells are infused back into the patient. The patient can now make beta-globin that functions normally and that can counteract sickling [82, 83].

Zynteglo (betibeglogene autotemcel) [84] is also a **cell-based gene therapy**, and it treats **transfusion-dependent thalassemia (beta-thalassemia major)**. It was FDA-approved in 2022. In this disease, **the β-globin gene (HBB) is mutated or missing**, resulting in a serious lack, or low level, of HBB. Patients cannot produce enough of the beta-chain of beta-globin. This can cause severe anemia and organ damage. Patients require frequent transfusions with healthy red blood cells. **Zynteglo** is a gene therapy treatment where the patients' hematopoietic stem cells (HSCs) are removed and modified by using a **lentivirus that contains a functional copy of the β-globin gene**. The patients' remaining stem cells are removed using myeloablative chemotherapy, and the new cells are infused back into the patient. The patient can now make healthy RBCs with functional hemoglobin [85, 86].

In addition to these technologies, other methods that have been discussed in earlier chapters are also forms of gene therapy. These include antisense oligonucleotides such as **Spinraza (nusinersen)** [87] for the treatment of **spinal muscular atrophy (SMA)** and RNAi technologies such as **Onpattro (patisiran)** [88] for treating **hereditary transthyretin-mediated (hATTR) amyloidosis** [89, 90], both of which were discussed in Chap. 7.

FDA-Approved Drugs Discussed in This Chapter

Commonly used brand name	Generic name	First FDA approval year	Main conditions	Drug type/mechanism
Casgevy	*Exagamglogene autotemcel*	2023	Sickle cell disease	CRISPR/Cas9, BCL11A gene
Lyfgenia	*Lovotibeglogene autotemcel*	2023	Sickle cell disease	Lentiviral gene therapy, HBB (beta-globin) with mutation, gene
Onpattro	*Patisiran*	2018	hATTR amyloidosis	RNA interference (RNAi), TTR gene
Spinraza	*Nusinersen*	2016	Spinal muscular atrophy (SMA)	Antisense oligonucleotide, SMN2 gene
Zolgensma	*Onasemnogene abeparvovec*	2019	Spinal muscular atrophy (SMA)	AAV9 gene therapy, SMN1 gene
Zynteglo	Betibeglogene autotemcel	2022	Beta-thalassemia major	Lentiviral gene therapy, HBB gene

Definitions

Adenine base editing (ABE) A CRISPR technique that can convert an A-T base pair to a G-C base pair without actually cutting the DNA, thus reducing the risk of introducing more DNA damage.

Base editor A modified Cas9 that alters bases without cutting DNA.

BCL11A A transcription factor that binds to the regulatory elements of the fetal gamma-globin gene and inhibits its expression after birth.

CRISPR-associated protein 9 (CAS9) A CRISPR-associated protein with nuclease activity (an endonuclease) that can cleave double-stranded DNA.

CAS9 nickase (nCAS9) A modified Cas9 that nicks only one strand instead of making a double-stranded break.

Casgevy The first FDA-approved CRISPR-based therapy, designed to treat sickle cell disease.

C-C chemokine receptor 5 (CCR5) A receptor on immune cells that serves as a co-receptor for HIV entry into cells.

CRISPR Clustered regularly interspaced short palindromic repeats, DNA sequences found in bacteria that are part of a defense mechanism against viruses.

Deaminase An enzyme that removes an amino group from a molecule; in base editing, adenine deaminase converts adenine (A) to inosine (I).

Double-stranded break (DSB) A break in both strands of the DNA double helix.

Erythroblast Cells that mature into red blood cells (RBCs).

Familial hypercholesterolemia A genetic disorder characterized by high LDL levels.

Guide RNA (gRNA) A synthetic RNA that contains the targeting sequence and a scaffold (which is equivalent to tracrRNA + part of a palindromic **repeat** sequence).

Hematopoietic stem cells (HSCS) Multipotent stem cells in the bone marrow that give rise to all blood cells.

Homology-directed repair (HDR) An accurate DNA repair mechanism that uses a homologous template to repair DSBs. Can be used with the CRISPR system for targeted genetic insertions.

Inosine (I) A nucleoside formed by deamination of adenosine, read as guanine (G) during replication.

Nonhomologous end joining (NHEJ) An error-prone DNA repair pathway that rejoins DSBs without a template.

Palindromic sequence DNA sequences that read the same forward and backward on complementary strands.

PAM sequence Protospacer adjacent motif, a short DNA sequence adjacent to the target sequence required for Cas9 to bind and cut the DNA. The requirement for the PAM prevents the CRISPR system from destroying the CRISPR locus itself.

Proprotein convertase subtilisin/kexin type 9 (PCSK9) A protein that binds LDL receptors, promoting their internalization and degradation.

Pre-crRNA The CRISPR locus RNA sequence before it gets cleaved into smaller pieces.

Protospacer Viral DNA sequence that is complementary to a spacer sequence in the CRISPR array.

Ribonucleoprotein complex (RNP) In the CRISPR system the RNP is a complex that contains Cas9 and the guide RNA (tracrRNA + crRNA).

Synthesis-dependent strand annealing (SDSA) A subtype of HDR sometimes used in CRISPR editing.

Sickle cell disease A genetic disorder caused by a point mutation in the β-globin gene, leading to abnormalities in hemoglobin, the oxygen-carrying protein in red blood cells.

Spacer sequence Segments of viral DNA inserted into the CRISPR locus, complementary to viral sequence.

Thalassemia A genetic disorder involving reduced or absent production of globin chains in hemoglobin.

TracrRNA Trans-activating CRISPR RNA. TracrRNA is involved in processing pre-crRNA and guiding Cas9.

Study Questions

Multiple-Choice Questions

1. What distinguishes fetal hemoglobin (HbF) from adult hemoglobin (HbA)?
   A. HbF has beta-globin chains instead of alpha.
   B. HbA has gamma-globin chains.
   C. HbF has gamma-globin chains instead of beta.
   D. HbA has only alpha chains.
2. Which CRISPR-based drug was the first to receive FDA approval?
   A. Spinraza
   B. Zynteglo
   C. CASGEVY
   D. VERVE-101
3. What is the role of the tracrRNA in the CRISPR-Cas9 system?
   A. It serves as a DNA template for editing DNA.
   B. It binds and stabilizes the spacer sequences during the viral infection.
   C. It guides Cas9 to the target DNA in the formation of the ribonucleoprotein complex.
   D. It degrades foreign DNA with its nuclease activity.
   E. In unwinds the foreign DNA.
4. What is a characteristic of the PAM sequence that prevents CRISPR from cleaving and inactivating its own CRISPR RNA?
   A. There is no PAM sequence in the CRISPR locus.
   B. The PAM sequence is only found in certain parts of the CRISPR locus.
   C. Only sequences without a PAM can be cleaved.
   D. The CRISPR array is modified by HDR to alter the sequence of the PAM.
   E. None of the above. The CRISPR array is cleaved each time the cell divides.
5. In the treatment of sickle cell disease, the drug Casgevy uses CRISPR to target:
   A. Cas9
   B. BCL11A
   C. CCR5
   D. PCSK9
   E. Beta-globin

True/False Question

6. Cas9 nickase generates double-stranded breaks.

Fill-in-the-Blank Questions

7. In CRISPR, a ____________ is a sequence of viral DNA incorporated into the bacterial genome for immune memory.
8. VERVE is an experimental CRISPR based treatment for the treatment of ____________.

Short-Answer Questions

9. Explain the main reason why base editors like those in VERVE-101, that involve a nickase, are considered safer than traditional CRISPR-Cas9 systems
10. Case Study
    A patient with familial hypercholesterolemia is enrolled in a clinical trial using CRISPR therapy. What is the likely target of the CRISPR-Cas9 system in this trial, and what is the intended genetic effect?

Answer Key

Multiple-Choice Questions

1. **Answer**: C
2. **Answer**: C
3. **Answer**: **C**
4. **Answer**: A
5. **Answer**: B

True/False Question

6. **Answer**: False

Fill-in-the-Blank Questions

7. Genome for immune memory.
   Answer: spacer
8. **Answer**: Familial hypercholesterolemia

Short-Answer Question

9. **Answer**: They avoid creating double-stranded DNA breaks. This reduces the risk of mutations due to mechanisms such as NHEJ.
10. **Case Study**
    Answer: The target is the PCSK9 gene. The idea is to target PCSK9, which is a secreted protein that binds to LDL receptors. By binding to LDL receptors, it reduces the uptake of LDL from the bloodstream. Targeting PCSK9 encourages increased uptake of LDL into the livre cells and lowers blood LDL.

References

1. Wang JY, Doudna JA. CRISPR technology: a decade of genome editing is only the beginning. Science. 2023;379(6629):eadd8643. https://doi.org/10.1126/science.add8643.
2. Doudna JA, Charpentier E. The new frontier of genome engineering with CRISPR-Cas9. Science. 2014;346(6213):1258096. https://doi.org/10.1126/science.1258096.
3. abm, Inc. CRISPR Cas9-Introduction. Available from: https://info.abmgood.com/crispr-cas9-introduction.
4. Karginov FV, Hannon GJ. The CRISPR system: small RNA-guided defense in bacteria and archaea. Mol Cell. 2010;37(1):7–19. https://doi.org/10.1016/j.molcel.2009.12.033.
5. Gebre M, Nomburg JL, Gewurz BE. CRISPR-Cas9 genetic analysis of virus-host interactions. Viruses. 2018;10(2):55. https://doi.org/10.3390/v10020055.
6. Mojica FJM, Díez-Villaseñor CS, García-Martínez J, Soria E. Intervening sequences of regularly spaced prokaryotic repeats derive from foreign genetic elements. J Mol Evol. 2005;60(2):174–82. https://doi.org/10.1007/s00239-004-0046-3.
7. Ishino Y, Krupovic M, Forterre P. History of CRISPR-Cas from encounter with a mysterious repeated sequence to Genome editing technology. J Bacteriol. 2018;200(7):e00580-17. https://doi.org/10.1128/jb.00580-17.
8. Nature Education. Bacteriophage (phage) definition. Scitable. Available from: https://www.nature.com/scitable/definition/bacteriophage-phage-293/
9. Oechslin F. Resistance development to bacteriophages occurring during bacteriophage therapy. Viruses. 2018;10(7):351. https://doi.org/10.3390/v10070351.
10. Zhang X, Zhang C, Liang C, Li B, Meng F, Ai Y. CRISPR-Cas9 based bacteriophage genome editing. Microbiol Spectrum. 2022;10(4):e0082022. https://doi.org/10.1128/spectrum.00820-22.
11. Jiang F, Doudna JA. CRISPR-Cas9 structures and mechanisms. Annu Rev Biophys. 2017;46:505–29. https://doi.org/10.1146/annurev-biophys-062215-010822.
12. Deltcheva E, Chylinski K, Sharma CM, Gonzales K, Chao Y, Pirzada ZA, et al. CRISPR RNA maturation by trans-encoded small RNA and host factor RNase III. Nature. 2011;471(7340):602–7. https://doi.org/10.1038/nature09886.

13. Sampson TR, Saroj SD, Llewellyn AC, Tzeng Y-L, Weiss DS. A CRISPR/Cas system mediates bacterial innate immune evasion and virulence. Nature. 2013;497(7448):254–7. https://doi.org/10.1038/nature12048.
14. Briner AE, Donohoue PD, Gomaa AA, Selle K, Slorach EM, Nye CH, et al. Guide RNA functional modules direct Cas9 activity and orthogonality. Mol Cell. 2014;56(2):333–9. https://doi.org/10.1016/j.molcel.2014.09.019.
15. The Nobel Prize. The Nobel Prize in chemistry 2020. 2025. Available from: https://www.nobelprize.org/prizes/chemistry/2020/popular-information/#:~:text=Emmanuelle%20Charpentier%20and%20Jennifer%20Doudna%20are%20awarded,plants%20and%20microorganisms%20with%20extremely%20high%20precision.
16. Bi Y, Sun L, Gao D, Ding C, Li Z, Li Y, et al. High-efficiency targeted editing of large viral genomes by RNA-guided nucleases. PLoS Pathog. 2014;10(5):e1004090. https://doi.org/10.1371/journal.ppat.1004090.
17. Synthego. What is CRISPR & why is it important. 2024. Available from: https://www.synthego.com/learn/crispr.
18. Stinson BM, Loparo JJ. Repair of DNA double-strand breaks by the nonhomologous end joining pathway. Annu Rev Biochem. 2021;90:137–64. https://doi.org/10.1146/annurev-biochem-080320-110356.
19. Asmamaw M, Zawdie B. Mechanism and applications of CRISPR/Cas-9-mediated genome editing. Biologics. 2021;15:353–61. https://doi.org/10.2147/btt.S326422.
20. Liao H, Wu J, VanDusen NJ, Li Y, Zheng Y. CRISPR-Cas9-mediated homology-directed repair for precise gene editing. Mol Ther Nucleic Acids. 2024;35(4):102344. https://doi.org/10.1016/j.omtn.2024.102344.
21. Addgene. CRISPR 101: homology-directed repair [Internet]. Watertown (MA): Addgene; [cited 2026 May 25]. Available from: https://blog.addgene.org/crispr-101-homology-directed-repair?utm
22. Elbakry A, Löbrich M. Homologous recombination subpathways: a tangle to resolve. Front Genet. 2021;12:2021. https://doi.org/10.3389/fgene.2021.723847.
23. Ledford H. Mail-order CRISPR kits allow absolutely anyone to hack DNA [Internet]. Scientific American. 2015 Jul 31 [cited 2026 May 25]. Available from: https://www.scientificamerican.com/article/mail-order-crispr-kits-allow-absolutely-anyone-to-hack-dna/
24. Laurent M, Geoffroy M, Pavani G, Guiraud S. CRISPR-based gene therapies: from preclinical to clinical treatments. Cells. 2024;13(10):800. https://doi.org/10.3390/cells13100800.
25. Singh A, Irfan H, Fatima E, Nazir Z, Verma A, Akilimali A. Revolutionary breakthrough: FDA approves CASGEVY, the first CRISPR/Cas9 gene therapy for sickle cell disease. Ann Med Surg (Lond). 2024;86(8):4555–9. https://doi.org/10.1097/ms9.0000000000002146.
26. FDA. Casgevy. 2024. Available from: https://www.fda.gov/vaccines-blood-biologics/casgevy.
27. Ahmed MH, Ghatge MS, Safo MK. Hemoglobin: structure, function and allostery. Subcell Biochem. 2020;94:345–82. https://doi.org/10.1007/978-3-030-41769-7_14.
28. Oneal PA, Gantt NM, Schwartz JD, Bhanu NV, Lee YT, Moroney JW, et al. Fetal hemoglobin silencing in humans. Blood. 2006;108(6):2081–6. https://doi.org/10.1182/blood-2006-04-015859.
29. American Society of Hematology. Sickle cell disease and thalassemia. 2008. Available from: https://www.hematology.org/about/history/50-years/sickle-cell-disease-thalassemia.
30. Alberts B, Heald R, Johnson A, Morgan D, Raff M, Roberts K, Walter P, Wilson J, Hunt T. Molecular biology of the cell. W.W. Norton and Company; 2022.
31. Cha HJ. Erythropoiesis: insights from a genomic perspective. Exp Mol Med. 2024;56(10):2099–104. https://doi.org/10.1038/s12276-024-01311-1.
32. Williams TN, Thein SL. Sickle cell anemia and its phenotypes. Annu Rev Genomics Hum Genet. 2018;19:113–47. https://doi.org/10.1146/annurev-genom-083117-021320.
33. Mayo Clinic. Sickle cell anemia. 2023. Available from: https://www.mayoclinic.org/diseases-conditions/sickle-cell-anemia/symptoms-causes/syc-20355876.
34. Sankaran VG, Xu J, Orkin SH. Advances in the understanding of haemoglobin switching. Br J Haematol. 2010;149(2):181–94. https://doi.org/10.1111/j.1365-2141.2010.08105.x.
35. Frangoul H, Altshuler D, Cappellini MD, Chen YS, Domm J, Eustace BK, et al. CRISPR-Cas9 gene editing for sickle cell disease and β-thalassemia. N Engl J Med. 2021;384(3):252–60. https://doi.org/10.1056/NEJMoa2031054.
36. Medicines and Healthcare Products Regulatory Agency (MHRA). MHRA authorises world-first gene therapy that aims to cure sickle-cell disease and transfusion-dependent β-thalassaemia. 2023. Available from: https://www.gov.uk/government/news/mhra-authorises-world-first-gene-therapy-that-aims-to-cure-sickle-cell-disease-and-transfusion-dependent-thalassemia?
37. FDA. FDA approves first gene therapies to treat patients with sickle cell disease. 2023. Available from: https://www.fda.gov/news-events/press-announcements/fda-approves-first-gene-therapies-treat-patients-sickle-cell-disease?
38. CRISPR Therapeutics. CRISPR Therapeutics announces U.S. Food and Drug Administration (FDA) approval of CASGEVY™ (exagamglogene autotemcel) for the treatment of transfusion-dependent beta thalassemia. Glove Newswire/Press Release; 2024. Found in: https://crisprtx.com/about-us/press-releases-and-presentations/crispr-therapeutics-

announces-u-s-food-and-drug-administration-fda-approval-of-casgevy-exagamglogene-autotemcel-for-the-treatment-of-transfusion-dependent-beta-thalassemia
39. Mohammadian Gol T, Zahedipour F, Trosien P, Ureña-Bailén G, Kim M, Antony JS, et al. Gene therapy in pediatrics—clinical studies and approved drugs (as of 2023). Life Sci. 2024;348:122685. https://doi.org/10.1016/j.lfs.2024.122685.
40. Vertex Pharmaceuticals. Casgevy. The power of precision. Available from: https://www.casgevyhcp.com/sickle-cell-disease.
41. Yen A, Zappala Z, Fine RS, Majarian TD, Sripakdeevong P, Altshuler D. Specificity of CRISPR-Cas9 editing in exagamglogene autotemcel. N Engl J Med. 2024;390(18):1723–5.
42. Hoy SM. Exagamglogene Autotemcel: first approval. Mol Diagn Ther. 2024;28(2):133–9. https://doi.org/10.1007/s40291-024-00696-z.
43. McKenney JM. Understanding PCSK9 and anti-PCSK9 therapies. J Clin Lipidol. 2015;9(2):170–86. https://doi.org/10.1016/j.jacl.2015.01.001.
44. Roth EM, Davidson MH. PCSK9 inhibitors: mechanism of action, efficacy, and safety. Rev Cardiovasc Med. 2018;19(S1):31–46. https://doi.org/10.3909/ricm19S1S0002.
45. Hooper AJ, Tang XL, Burnett JR. VERVE-101, a CRISPR base-editing therapy designed to permanently inactivate hepatic PCSK9 and reduce LDL-cholesterol. Expert Opin Investig Drugs. 2024;33(8):753–6. https://doi.org/10.1080/13543784.2024.2369747.
46. Horie T, Ono K. VERVE-101: a promising CRISPR-based gene editing therapy that reduces LDL-C and PCSK9 levels in HeFH patients. Eur Heart J Cardiovasc Pharmacother. 2023;10(2):89–90. https://doi.org/10.1093/ehjcvp/pvad103.
47. Verve Therapeutics. We are on a mission to protect the world from cardiovascular disease. 2025. Available from: https://www.vervetx.com/.
48. Gaudelli NM, Komor AC, Rees HA, Packer MS, Badran AH, Bryson DI, et al. Programmable base editing of a•T to G•C in genomic DNA without DNA cleavage. Nature. 2017;551(7681):464–71. https://doi.org/10.1038/nature24644.
49. Musunuru K, Chadwick AC, Mizoguchi T, Garcia SP, DeNizio JE, Reiss CW, et al. In vivo CRISPR base editing of PCSK9 durably lowers cholesterol in primates. Nature. 2021;593(7859):429–34. https://doi.org/10.1038/s41586-021-03534-y.
50. Musunuru K, Grandinette SA, Wang X, Hudson TR, Briseno K, Berry AM, et al. Patient-specific in vivo gene editing to treat a rare genetic disease. N Engl J Med. 2025;392(22):2235–43. https://doi.org/10.1056/NEJMoa2504747.
51. Children's Hospital of Pennsylvania (CHOP). World's first patient treated with personalized CRISPR gene editing therapy at children's hospital of Philadelphia. 2025. Available from: https://www.chop.edu/news/worlds-first-patient-treated-personalized-crispr-gene-editing-therapy-childrens-hospital#:~:text=CRISPR%20(clustered%20regularly%20interspaced%20short,many%20disease%2Dcausing%20variants%20exist.
52. Callaway E. Second Chinese team reports gene editing in human embryos. Nature. 2016; https://doi.org/10.1038/nature.2016.19718.
53. Kang X, He W, Huang Y, Yu Q, Chen Y, Gao X, et al. Introducing precise genetic modifications into human 3PN embryos by CRISPR/Cas-mediated genome editing. J Assist Reprod Genet. 2016;33(5):581–8. https://doi.org/10.1007/s10815-016-0710-8.
54. Normile D. Shock greets claim of CRISPR-edited babies. Science. 2018;362(6418):978–9. https://doi.org/10.1126/science.362.6418.978.
55. Sykes M, Sachs DH. Progress in xenotransplantation: overcoming immune barriers. Nat Rev Nephrol. 2022;18(12):745–61. https://doi.org/10.1038/s41581-022-00624-6.
56. Kalds P, Zhou S, Cai B, Liu J, Wang Y, Petersen B, et al. Sheep and goat Genome engineering: from random Transgenesis to the CRISPR era. Front Genet. 2019;10:750. https://doi.org/10.3389/fgene.2019.00750.
57. Tu Z, Zhao H, Li B, Yan S, Wang L, Tang Y, et al. CRISPR/Cas9-mediated disruption of SHANK3 in monkey leads to drug-treatable autism-like symptoms. Hum Mol Genet. 2019;28(4):561–71. https://doi.org/10.1093/hmg/ddy367.
58. Jacinto FV, Link W, Ferreira BI. CRISPR/Cas9-mediated genome editing: from basic research to translational medicine. J Cell Mol Med. 2020;24(7):3766–78. https://doi.org/10.1111/jcmm.14916.
59. Tripathi L, Ntui VO, Tripathi JN. Control of bacterial diseases of Banana using CRISPR/Cas-based gene editing. Int J Mol Sci. 2022;23(7):3619. https://doi.org/10.3390/ijms23073619.
60. Burkard C, Opriessnig T, Mileham AJ, Stadejek T, Ait-Ali T, Lillico SG, et al. Pigs lacking the scavenger receptor cysteine-rich domain 5 of CD163 are resistant to porcine reproductive and respiratory syndrome virus 1 infection. J Virol. 2018;92(16):e00415-18. https://doi.org/10.1128/jvi.00415-18.
61. National Hog Farmer. Cost of PRRS reaches $664 Million Annually. Available from: https://www.nationalhogfarmer.com/hog-health/cost-of-prrs-reaches-664-million-annually.
62. Augustyniak A, Pomorska-Mól M. Vaccination failures in pigs-the impact of chosen factors on the immunisation efficacy. Vaccines (Basel). 2023;11(2):230. https://doi.org/10.3390/vaccines11020230.
63. Oye KA, Esvelt K, Appleton E, Catteruccia F, Church G, Kuiken T, et al. Regulating gene drives. Science. 2014;345(6197):626–8. https://doi.org/10.1126/science.1254287.
64. Champer J, Buchman A, Akbari OS. Cheating evolution: engineering gene drives to manipulate the fate of wild populations. Nat Rev Genet. 2016;17(3):146–59. https://doi.org/10.1038/nrg.2015.34.

65. Gantz VM, Bier E. Genome editing. The mutagenic chain reaction: a method for converting heterozygous to homozygous mutations. Science. 2015;348(6233): 442–4. https://doi.org/10.1126/science.aaa5945.
66. Esvelt KM, Smidler AL, Catteruccia F, Church GM. Concerning RNA-guided gene drives for the alteration of wild populations. elife. 2014;3:e03401. https://doi.org/10.7554/eLife.03401.
67. Nolan T. Control of malaria-transmitting mosquitoes using gene drives. Philos Trans R Soc Lond Ser B Biol Sci. 2021;376(1818):20190803. https://doi.org/10.1098/rstb.2019.0803.
68. Zhao L, Yang Z, Zheng M, Shi L, Gu M, Liu G, et al. Recombinant adeno-associated virus 8 vector in gene therapy: opportunities and challenges. Genes Dis. 2024;11(1):283–93. https://doi.org/10.1016/j.gendis.2023.02.010.
69. Liu D, Li T, Liu L, Che X, Li X, Liu C, et al. Adeno-associated virus therapies: pioneering solutions for human genetic diseases. Cytokine Growth Factor Rev. 2024;80:109–20. https://doi.org/10.1016/j.cytogfr.2024.09.003.
70. Bulcha JT, Wang Y, Ma H, Tai PWL, Gao G. Viral vector platforms within the gene therapy landscape. Signal Transduct Target Ther. 2021;6(1):53. https://doi.org/10.1038/s41392-021-00487-6.
71. FDA. Zolgensma(onasemnogene abeparvovec) Approval overview. 2019. Available from: https://www.fda.gov/news-events/press-announcements/fda-approves-innovative-gene-therapy-treat-pediatric-patients-spinal-muscular-atrophy-rare-disease?
72. Lopez-Gordo E, Chamberlain K, Riyad JM, Kohlbrenner E, Weber T. Natural adeno-associated virus serotypes and engineered adeno-associated virus capsid variants: tropism differences and mechanistic insights. Viruses. 2024;16(3):442. https://doi.org/10.3390/v16030442.
73. Reilly A, Yaworski R, Beauvais A, Schneider BL, Kothary R. Long term peripheral AAV9-SMN gene therapy promotes survival in a mouse model of spinal muscular atrophy. Hum Mol Genet. 2024;33(6): 510–9. https://doi.org/10.1093/hmg/ddad202.
74. Zolgensma. The only gene therapy and only one-dose treatment for SMA. Available from: https://www.zolgensma.com/how-zolgensma-works.
75. Cicalese MP, Ferrua F, Castagnaro L, Rolfe K, De Boever E, Reinhardt RR, et al. Gene therapy for adenosine deaminase deficiency: a comprehensive evaluation of short- and medium-term safety. Mol Ther. 2018;26(3):917–31. https://doi.org/10.1016/j.ymthe.2017.12.022.
76. Ottaviano G, Qasim W. Current landscape of vector safety and genotoxicity after hematopoietic stem or immune cell gene therapy. Leukemia. 2025;39(6): 1325–33. https://doi.org/10.1038/s41375-025-02585-8.
77. Health EMASM. Strimvelis. Available from: https://www.ema.europa.eu/en/medicines/human/EPAR/strimvelis#:~:text=Strimvelis%20is%20a%20medicine%20used,delivering%20genes%20into%20the%20body.
78. Telethon F. Potential adverse reaction to gene therapy in a patient treated with Strimvelis for the treatment of ADA-SCID. 2020. Available from: https://www.fondazionetelethon.it/en/stories-and-news/news/from-telethon-foundation/potential-adverse-event-after-gene-therapy-for-ada-scid/.
79. Therapeutics O. Orchard therapeutics announces new england journal of medicine publication of HSC gene therapy data for ADA-SCID. 2025. Available from: https://ir.orchard-tx.com/news-releases/news-release-details/orchard-therapeutics-announces-new-england-journal-medicine.
80. FDA. Lyfgenia(lovotibeglogene autotemcel) Approval overview. 2023. Available from: https://www.drugs.com/history/lyfgenia.html?
81. Plus M. Sickle cell disease. 2024. Available from: https://medlineplus.gov/genetics/condition/sickle-cell-disease/#causes.
82. Lyfgenia. How does LYFGENIA work?. Available from: https://www.lyfgenia.com/what-is-lyfgenia/how-lyfgenia-works.
83. Administration USFaD. FDA approves first gene therapies to treat patients with sickle cell disease. 2023. Available from: https://www.fda.gov/news-events/press-announcements/fda-approves-first-gene-therapies-treat-patients-sickle-cell-disease.
84. Drugs.com. Zynteglo(betibeglogene autotemcel) Approval overview. 2022. Available from: https://www.drugs.com/history/zynteglo.html?
85. Zynteglo. What if you could let go of your transfusions?. Available from: https://www.zynteglo.com/.
86. Clinic C. Beta Thalassemia. 2022. Available from: https://my.clevelandclinic.org/health/diseases/23574-beta-thalassemia.
87. FDA. Spinraza (nusinersen) Approval Overview. 2016. Available from: https://www.accessdata.fda.gov/scripts/cder/daf/index.cfm?event=overview.process&ApplNo=209531.
88. FDA. ONPATTRO (patisiran) Approval Overview. 2018. Available from: https://www.accessdata.fda.gov/scripts/cder/daf/index.cfm?event=overview.process&ApplNo=210922.
89. Luigetti M, Romano A, Di Paolantonio A, Bisogni G, Sabatelli M. Diagnosis and treatment of hereditary transthyretin amyloidosis (hATTR) polyneuropathy: current perspectives on improving patient care. Ther Clin Risk Manag. 2020;16:109–23. https://doi.org/10.2147/tcrm.S219979.
90. Ottesen E. ISS-N1 makes the first FDA-approved drug for spinal muscular atrophy. Transl Neurosci. 2017;8: 1–6. https://doi.org/10.1515/tnsci-2017-0001.

26 Stem Cells and Cellular Therapies

Abstract

In addition to pharmaceuticals and surgery, cellular therapy represents another important approach for treating human disorders. Cellular therapy can take many forms, for example, it can involve replacing or supplementing abnormally functioning cells with healthy cells. An important type of cellular therapy is stem cell therapy. This chapter covers cellular therapy, with a focus on stem cells. It provides an overview of stem cells, including embryonic stem cells, and discusses where adult stem cells are located in the body. The major types of stem cells that are promising for disease treatment and regenerative therapy are then discussed, including embryonic stem cells, SCNT cells, iPS cells, and parthenogenetic stem cells. Examples of disorders for which stem cell therapy may hold promise are discussed, and the promises and challenges of stem cell use are addressed. In addition to stem cell therapy, other types of cellular therapy are discussed, such as CAR T-cell therapy and dendritic cell vaccines for the treatment of certain cancers.

Keywords

Embryonic stem cells · Adult stem cells · Somatic cell nuclear transfer (SCNT) technology · Induced pluripotent stem cells (iPS cells) · CAR T therapy

Abbreviations

ES cell	Embryonic stem cell
ESC	Embryonic stem cell
HSC	Hematopoietic stem cell
HSPC	HSCs + progenitor cells
iPSC	Induced pluripotent stem cell
MSC	Mesenchymal stromal cell, mesenchymal stem cell
Muse cell	Multi-lineage-differentiating stress-enduring cells
SCNT	Somatic cell nuclear transfer

What You Will Learn

- The basics of stem cells: what are stem cells and where are they found.
- The basics of embryogenesis. This is important because pluripotent stem cells come from an early embryo.
- The different potencies of stem cells: totipotent, pluripotent, multipotent, and unipotent.
- Different types of stem cells including embryonic stem cells, fetal stem cells, umbilical stem cells, and stem cells that are made in the lab such as SCNTs, iPS cells, and parthenogenetic stem cells.
- Potential uses for stem cells in medicine, for the treatment of conditions such as Parkinson's disease and MS or for regenerative medicine.
- Potential drawbacks and challenges to the use of stem cells.

A. Minden, *Biochemistry, Cell and Molecular Biology for Pharmacists and Health Professionals*,
https://doi.org/10.1007/978-3-032-20529-2_26

- Autologous stem cell therapy vs allogenic stem cell therapy.
- The pros and cons of using adult stem cells rather than embryonic stem cells for treating human conditions.
- Other types of cellular therapy besides stem cell therapy such as T-cell therapy, dendritic cell vaccines, and others.

This chapter covers cellular therapy, focusing on stem cells. This chapter is divided into four Sections: Section 1 provides an overview of key concepts that are important for understanding stem cells, including where stem cells are naturally found in the body. Section 2 discusses the major types of stem cells that are of value for research and therapy. Section 3 discusses the potential that these different types of stem cells have for treating disease, tissue regeneration, and biological research. Section 4 briefly discusses other types of cellular therapies that do not directly involve stem cells.

1 Introduction to Stem Cells

1.1 Introduction to Stem Cells in Medicine

1.1.1 Introduction to Cellular Therapy

In addition to pharmaceuticals and surgery, cellular therapy represents another important approach for treating human disorders. Cellular therapy can take many forms, for example, it can involve replacing or supplementing abnormally functioning cells with healthy cells. An important type of cellular therapy is stem cell therapy. All disorders are based at some level on problems with cells, but stem cell therapy may be more promising for some may disorders rather than others. For example, damage to certain types of cells can be permanent, limiting treatments options. An example is nerve damage, as most nerves do not have any capacity to self-renew. In such cases, stem cell therapy may be a promising mechanism for treating injury or damage to neurons. Similarly, heart attacks and strokes involve tissue damage that cannot necessarily be repaired or replaced and may be amenable to stem cell treatment. Another problem is that some diseases or injuries may be best treated by organ or tissue transplants, but this requires donors, who are in short supply. Even when donors are available, tissue rejection is a problem. In the future, stem cell treatments may be able to solve some of these types of problems. In addition to their therapeutic potential, stem cells are also valuable for basic research into understanding the basis of human disease and developing in vitro models [1–4].

1.1.2 What Are Stem Cells?

Stem cells are unique cells that **have the ability to self-renew** (to produce identical copies of themselves), as well as the ability to differentiate into other cell types. A common feature of many stem cells is asymmetric division, where one daughter cell remains a stem cell, replenishing the stem cell population, and the other can become a differentiated cell. Stem cells are present in early embryos as well as in adults. In the early embryo, stem cells contribute to all the cells in the body, and in adults, stem cells can serve as a reservoir for replacing terminally differentiated cells (Fig. 26.1) [5–7].

1.2 Cell Division and Cell Differentiation

A defining characteristic of stem cells is their ability to self-renew through cell division and their potential to differentiate into one or more specialized cell types. Many stem cells are capable of dividing asymmetrically, where one daughter cell can be undifferentiated and the other may become differentiated. A review of the concepts of cell division and cell differentiation, discussed below and in [8] and [9], is helpful for understanding how stem cells are unique.

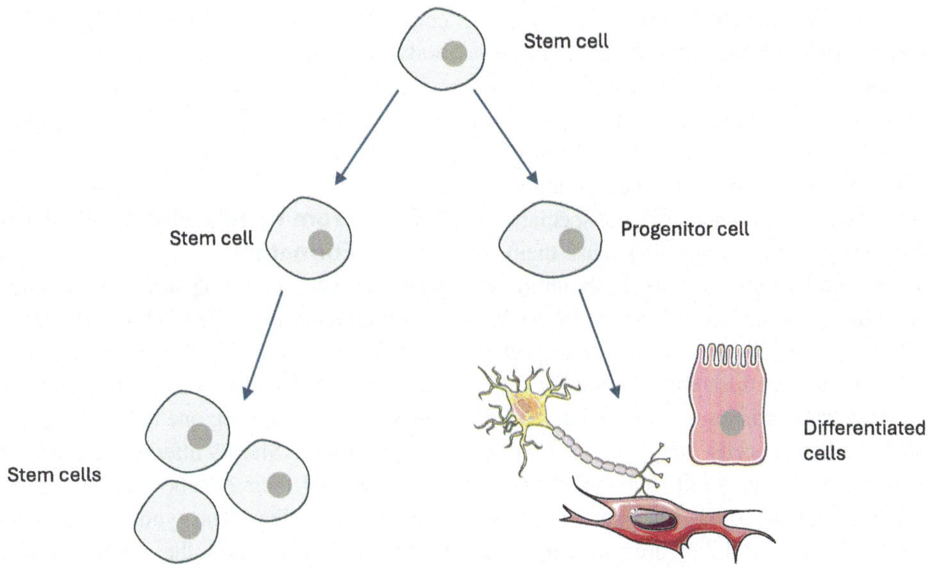

Fig. 26.1 Stem cells. Stem cells have the ability to divide asymmetrically, producing two daughter cells: one maintains the stem identity (stem cell renewal), while the other can become a progenitor cell, which can then further differentiate into a specialized cell type

1.2.1 Cell Division

Cell division is the process of one cell dividing into two daughter cells. (It can also be called cell proliferation or cell replication.) Cell division was discussed in depth in Chap. 20. During **cell division**, the cell passes the G1 checkpoint and enters S phase. In S phase, DNA is replicated. The cell then proceeds through G2 phase and into M phase. In M phase, the entire cell and all the chromosomes divide into two, and two daughter cells result.

1.2.2 Differentiated Cells

Cell **differentiation** is the process whereby a cell becomes a specialized cell. Our bodies have trillions of cells, but different cells have different specialized functions. Cells with specialized functions are called **differentiated cells**, for example, a liver cell is a differentiated cell that performs specialized functions such as regulating metabolism and detoxification. In contrast, neurons are cells that specialize in transmitting chemical and electrical signals to other nerves and to muscles. Neurons also have a unique shape that is completely different from liver cells or most other cells. These are only two examples of differentiated cells, but there are many others. Differentiated cells all express genes that enable them to carry out their specialized functions.

Some differentiated cells are **terminally differentiated**. These cells can no longer divide at all. Examples include most neurons and muscle cells. These cells are usually in the G_0 phase of the cell cycle. These cells are metabolically active and healthy, but they do not divide. Many of our nerve cells need to last our entire lives because they can never be replaced by cell division.

In contrast, some differentiated cells maintain the ability to divide. These cells are in G_0 phase when they are not dividing, but they can exit G_0 phase and re-enter the cell cycle upon appropriate signals. They may lose some of their differentiation characteristics when they divide but will regain them after cell division. Examples of differentiated cells that can also divide are hepatocytes (liver cells), endothelial cells, and fibroblasts.

1.2.3 Undifferentiated Cells

In contrast, **undifferentiated cells** do not carry out specialized functions. Stem cells are classic examples of undifferentiated cells. In the embryo, the cells in the inner cell mass of the blastocyst (discussed below) are undifferentiated stem cells. Undifferentiated cells do not carry out specialized functions. They express genes that cause them to maintain an undifferentiated state. Fully undifferentiated cells generally do not persist in adults. Some cells in the adult, such as adult stem cells, however, are partly undifferentiated, but they still show certain characteristics of differentiated cells.

Undifferentiated cells may be in various stages of the cell cycle. Adult stem cells that are not dividing may be quiescent, and they can remain in a G_0 state until they are called upon to divide. Embryonic stem cells usually grow rapidly; they are not usually in a G_0 state but progress rapidly through all phases of the cell cycle.

1.3 Early Embryogenesis and the Emergence of Stem Cells

Embryonic stem cells (ES cells) are a well-known type of stem cell that is present in the early embryo. ES cells have the classic characteristics of stem cells. They are undifferentiated, they have the ability to self-renew, and they have the ability to differentiate into specialized cell types. ES cells ultimately give rise to all the cells in the body. In order to understand where ES cells come from, it is helpful to first have a basic understanding of early embryonic development (Fig. 26.2a).

1.3.1 From Fertilization to Blastocyst Formation

To review early embryogenesis, we will start with the ovulated oocyte. Fertilization of this oocyte by sperm takes place in the fallopian tube, and it leads to the formation of a zygote, which initially contains two pronuclei, one from the egg and one from the sperm. Shortly after fertilization, the two nuclei and all their genetic material merge into one diploid nucleus. The zygote travels down the fallopian tubes toward the uterus. During this time, the zygote divides, and the cells then undergo a series of cleavages. This means that they are dividing but not increasing in size. Consequently, cells get smaller and smaller while they increase in number. This process results in the formation of a ball, or sphere, of cells. The stages after the zygote stage include the morula stage, which is a solid sphere of cells. This is followed, after more cell divisions, by the formation of the blastocyst. The blastocyst is different from the previous stages. It has a hollow, liquid-filled cavity called a blastocoel and an outer layer of cells that comprise the trophoblast. The trophoblast cells will eventually differentiate into the

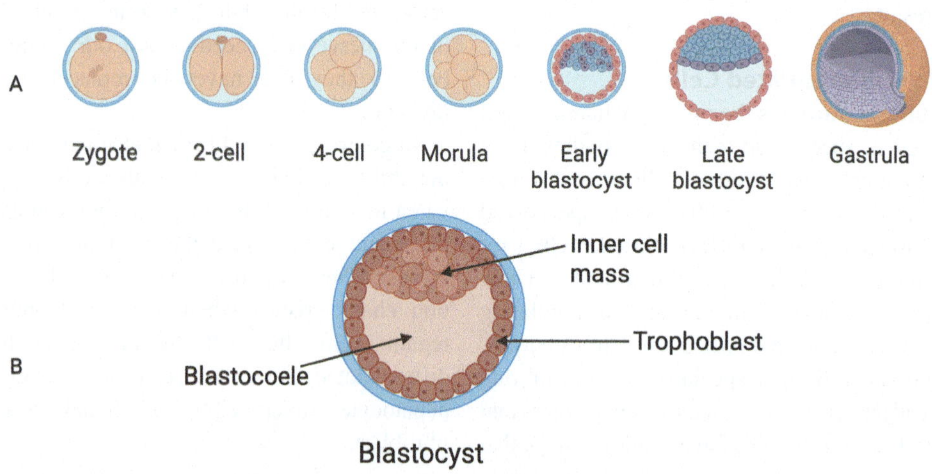

Fig. 26.2 Stages of early embryonic development. (**a**). Zygote to gastrula stagea. (**b**). Rgw vlastocyst. Pluripotent embryonic stem cells are found in the inner cell mass (ICM) of the blastocyst. **Source**: Created with BioRender.com

cells of the extraembryonic tissue, such as the placenta and yolk sac. Within the blastocoel (the cavity) are the inner cell mass (ICM) cells, which will eventually become the embryo [10].

The development of the blastocyst (Fig. 26.2b) marks an important stage in the embryo's development. Before this stage, all the cells are essentially in the same environment, having contact with each other and with the outside of the embryo. By the blastocyst stage, however, two groups of cells emerge to have two separate destinies. By this stage, the cells in the ICM, on the inside of the blastocyst, are in a different environment from the trophoblast cells on the outside of the structure. These **ICM** cells are the embryonic stem cells (ES cells), and they are **pluripotent**, which means they are undifferentiated but have the capacity to become any type of cell in the body (all three germ layers) [11].

1.3.2 From the Blastocyst Stage to the Gastrula Stage

The cells in the ICM are embryonic stem cells. They are defined as stem cells because they are undifferentiated, can self-renew, and have the capacity to differentiate into all cell types in the body. The timeframe during which ES cells are present is short. After the blastocyst is formed, it begins to invade the uterine epithelium, usually at around day 6–10 after fertilization. By day 16, the embryo has invaded the uterine epithelium and has developed into the **gastrula**. This is also a big step in terms of cell fate because the cells have now divided into cells with three different fates. Some cells will become endoderm, others mesoderm, and others ectoderm. The cells in the **endoderm** differentiate into soft tissues like the pancreas and liver. The **mesoderm** cells will become bone, blood, and muscle, including the heart. The **ectoderm** cells develop into skin and nerve cells. Thus, at this point, the cells in the embryo are no longer **pluripotent**. Instead, they are **multipotent** and have begun their commitment steps to becoming specific tissues. These multipotent cells do not have self-renewal properties like stem cells. Thus, at this point, the embryo has multipotent cells, but it no longer has embryonic stem cells. Thus, the period during which ES cells are present is relatively brief and is over by the gastrula stage [10, 12].

1.3.3 The Fetus

The next stage in which stem cells will appear is later in development, when organogenesis (the formation of the organs) is occurring. During this period, specialized tissues are starting to form. Many organs at this stage have their own populations of tissue-specific stem cells, which are **called fetal stem cells**. These cells are no longer pluripotent, but they maintain the capacity for self-renewal, and their daughter cells can take on the characteristics of specialized cell types within their lineage [13, 14].

2 Different Types of Stem Cells Used in Research and Disease Treatment

Embryonic stem cells have critical roles in the development of the embryo. Likewise, fetal and adult stem cells are important for the growing fetus and for maintaining terminally differentiated cells in the adult, respectively. When isolated, stem cells are of potential use for treating disease and for regenerative therapy. Different types of stem cells that can be of therapeutic use are discussed here. Some stem cells can be isolated directly from an embryo or from the body, such as from the bone marrow, but others are cells that are modified in the lab to take on stem cell characteristics. Important considerations for stem cells that may be used therapeutically include the source of the stem cells and their potency [1, 2]. Both of these are discussed below.

2.1 Potency of Stem Cells

An important characteristic that differentiates different types of stem cells from each other is potency. Stem cells are usually either pluripotent, multipotent, or unipotent, or more rarely, totipotent. **Pluripotent** stem cells have the capacity to differentiate into any cell in the body (all three germ layers), although they cannot differentiate into extraembryonic tissue such as the placenta or yolk sac. **Multipotent** cells, including multipotent stem cells, can differentiate into multiple types of cells in the body, but not into all cell types. Their potential is restricted to certain lineages. **Unipotent** stem cells can give rise to only one specific cell type but maintain the capacity for self-renewal. This type of cell is less common than multipotent stem cells. Examples include epidermal stem cells in the basal layer of the skin and limbal stem cells in the eye. These are different than progenitor cells, which are cells that are usually more committed to a specific cell type and have a more limited self-renewal capacity compared to stem cells. **Totipotent** stem cells can differentiate into any cell in the body, as well as the cells that contribute to the extraembryonic tissue such as the placenta and the yolk sac. This type of cell is found only in very early stages of embryonic development, such as the zygote and early cleavage stage. In animals, the quintessential example of a totipotent stem cell is a fertilized egg, which gives rise to all body cells and all the extraembryonic tissue that supports the fetus during its development [5].

2.2 Different Types of Stem Cells

Different types of stem cells are discussed below. Some of these are naturally occurring stem cells, and some of them are stem cells that are generated in the lab. Many of them are thought to have therapeutic potential.

2.2.1 Naturally Occurring Stem Cells

- Embryonic Stem (ES) Cells and Their Isolation

Embryonic stem (ES) cells are sometimes considered to be the most valuable type of stem cells for stem cell therapy because they are **pluripotent**. As described above, ES cells are in the ICM of the blastocyst [5, 15, 16]. ES cells can self-renew for a brief time in the embryo, and importantly, they can also self-renew when removed from the embryo and cultured in the lab.

ES cells can be isolated from human blastocysts and used in the lab or for potential treatments. Human blastocysts for the isolation of ES cells generally come from in vitro fertilization clinics and are stored in vitro in freezers (cryopreserved) until they are used. In animal studies, blastocysts can be isolated directly from the uterus of the animal. Once the blastocysts are obtained, researchers can isolate the ICM cells from the inside of the structure. One way to do this is with a fine needle under a microscope. There are other mechanisms as well, such as using various ways to disrupt the blastocyst or using microdissection methods [17, 18].

Once the ICM cells are isolated, they need to be cultured and maintained. Culturing these cells, which are the ES cells, is a delicate matter. The cells must be kept under conditions where they maintain their non-differentiated state in culture. If the right conditions are not used, the cells will start to differentiate prematurely, and they will no longer be pluripotent stem cells. One common way that scientists grow ES cells is on a layer of irradiated mouse fibroblast cells. These are called feeder cells. These cells secrete growth factors, which the ES cells need to maintain their undifferentiated state. Because the feeder cells are irradiated, they cannot grow and take over the ES cell culture. While feeder cells are a traditional method for growing ES cells, feeder-free methods are also being developed, which involve using defined media that can support the growth of ES cells [17, 18].

ES cells will generally grow in spherical clusters of cells when they are first cultured. These are called embryoid bodies (EBs). These embryoid bodies need to be dissociated and replated in the lab until a monolayer of cells forms. As the cells continue to grow, it is important to take care that they remain undifferentiated. After careful culturing under the right conditions, the ES cell line is firmly established in the lab. These ES cells are undifferentiated, but they have the ability to differentiate into specialized cell types. If conditions for maintaining pluripotency are not maintained, the cells can differentiate spontaneously, and various types of differentiated cells can sometimes be seen in the culture dish, for example, cardiomyocytes, which beat like heart cells, can be spotted readily under the microscope. However, the goal is to keep the cells in an undifferentiated pluripotent state until they are needed for experiments or for stem cell therapies [18].

- Fetal Stem Cells

Fetal tissue cells contain a mix of stem cells and progenitor cells. Fetal tissue cells can be isolated from fetuses that have already begun the development of organs (organogenesis). At this stage, the primitive organs have formed, but the embryo is still tiny and growing rapidly. This is an interesting time period because all organs contain rapidly dividing cells, even though in some cases the cells in their corresponding adult organs (such as the brain) cannot divide efficiently at all, and the organs may contain few or no stem cells. Since the fetus has already developed the basic structures of the organs, fetal tissue cells within these organs are more differentiated than embryonic stem cells. They are no longer pluripotent. Instead, these cells are often **multipotent**, but they can divide rapidly and efficiently. In the lab, fetal cells have been isolated from numerous tissues, either from humans or from animal studies. These tissues include the brain, the developing retina, the neural crest (a transient structure found in fetuses), bone marrow, the liver, the pancreas, and the gut. Fetal tissue cells used in the lab are isolated from aborted fetuses or animal studies [19, 20].

- Umbilical Cord Stem Cells

The umbilical cord is full of stem cells and progenitor cells. Stem cells are present in the umbilical cord blood and the umbilical cord tissue. The **umbilical cord blood** is a rich source of hematopoietic stem cells, which have the potential to differentiate into many types of cells. These include neuronal cells, oligodendrocytes, cardiomyocytes, endothelial cells, insulin-producing cells, and others. Stem cells from the **umbilical cord tissue** include cells from the **Wharton's jelly** (a gelatinous substance within the cord) and cells from the **umbilical vein**. The Wharton's jelly has large amounts of mesenchymal stromal cells, which are stem-like cells discussed later in this chapter. The umbilical vein is rich in endothelial cells and endothelial progenitor cells. Endothelial cells make up the lining of blood vessels. They are not stem cells but can maintain the endothelial cell population by replication and self-renewal [21, 22].

- Adult Stem Cells
  - Overview of adult stem cells

Even in adults, many tissues have a source of stem cells throughout life. Stem cells in the adult are not fully undifferentiated like ES cells, but they remain largely undifferentiated and typically express only a few lineage-specific differentiation markers. Adult stem cells often serve as a reservoir to replace cells that can no longer replace themselves by cell division. Some adult cells are terminally differentiated. For some terminally differentiated cells, stem cells are vital for their replacement when the cells die or become damaged. Unlike ES cells, adult stem cells are not pluripotent. Adult stem cells are often **multipotent**, although some are **unipotent**. A few examples of different types of adult stem cells [5, 7] are discussed below. Because adult stem cells are not pluripotent and partially differentiated, they may be less risky for use in treating patients. They carry a lower risk of forming teratomas or causing uncontrolled cell growth compared to pluripotent stem cells like ES cells or iPSCs. Challenges remain in finding the best ways to use adult stem cells for disease treatment.

- Hematopoietic stem cells (HSCs)

Hematopoietic stem cells (HSCs) are found in the bone marrow and are important examples of adult stem cells (Fig. 26.3). These stem cells give rise to cells in the blood and the immune system. HSCs are multipotent. They give rise to the lymphoid and myeloid progenitor cells. Lymphoid progenitor cells, in turn, give rise to T

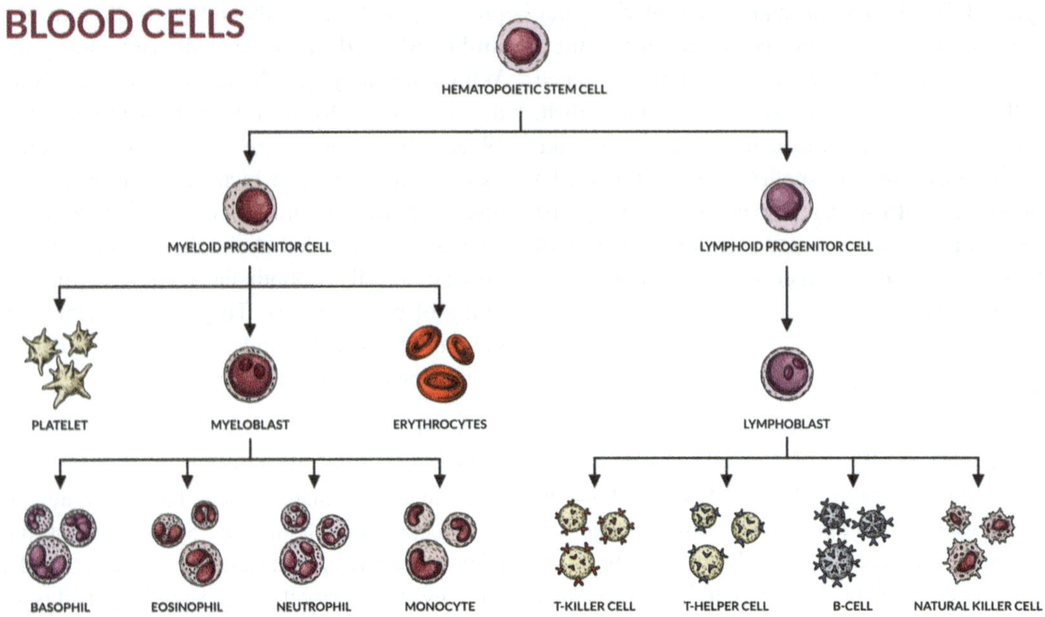

Fig. 26.3 Hematopoietic stem cells. Hematopoietic stem cells (HSCs) are an important example of adult stem cells, which give rise to the cells of the immune system

lymphocytes, natural killer cells, B lymphocytes, and dendritic cells. Myeloid progenitor cells give rise to many cells, such as macrophages, eosinophils, erythrocytes, mast cells, megakaryocytes, and platelets [5].

– Skin stem cells

Skin is another tissue that relies on adult stem cells (Fig. 26.4). The skin is a multilayered organ. The major layers are the **epidermis**, the dermis, and the hypodermis. The dermis and the hypodermis provide structural support and contain connective tissue. The epidermis, the layer that is described here, is the outer layer of the skin and has a protective function. The epidermis itself has several layers, also called strata. The top layer of the epidermis has cells that can no longer proliferate. These are the keratinized squames. These enucleated cells are no longer living cells but have vital protective functions. They cannot replicate and are constantly shed from the surface due to everyday wear and tear and injuries.

The basal layer of the epidermis is its deepest layer. It is the layer adjacent to the basal lamina, which is directly above the dermis. One population of adult stem cells is located in this basal layer. These cells can duplicate, and their daughter cells differentiate into mature skin cells that will replace the keratinized squames at the outermost layer when needed. They are involved in routine homeostasis (daily turnover). They constantly divide and replace keratinocytes that shed from the skin's surface. As the stem cells differentiate, they migrate through the different layers of the epidermis until they reach the outermost layer. At this point, the cells are terminally differentiated and become keratinized squames. These cells eventually flake off as well and will also need to be replaced by more stem cells [5, 8].

Another major skin stem cell population is at the hair follicle bulge region. The hair follicle stem cells (HFSCs) are normally quiescent but are activated by signals such as wounds, inflammation, or hair cycle signals. They can regenerate multiple skin cell lineages. They are strongly activated by wounds and injury and have important roles in repopulating the epidermis after an injury [28, 29].

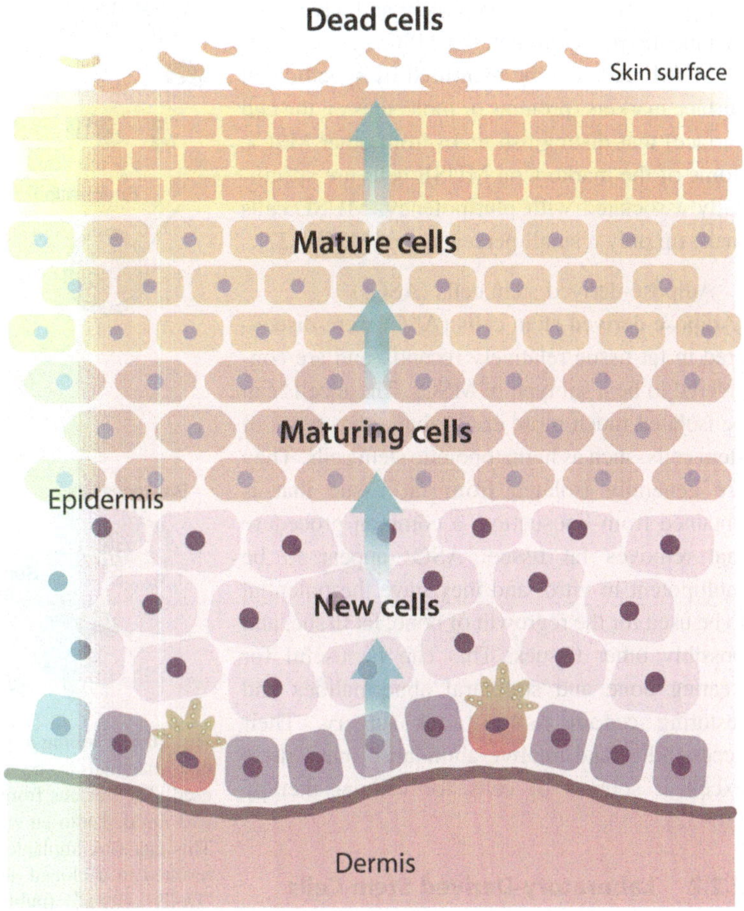

Fig. 26.4 Stem cells in the skin. The skin contains adult stem cells. One type of skin stem cell is located at the basal layer of the epidermis. These cells differentiate and migrate to outer layers as skin turnover occurs. **Source**: Illustration from Vecteezy.com (Pro license)

– Adult stem cells of the intestine

The colon's lining comprises large numbers of folds called crypts. At the upper parts of the crypts, sometimes called the neck or top, are highly differentiated cells, including enterocytes, goblet cells, and others. These cells have short life spans and are frequently shed from the tissue. They do not replace themselves by replicating. Instead, their replacement depends on stem cells. At the base of the crypts are stem cells, which are mostly undifferentiated but which have the ability to differentiate into the various differentiated cells of the colon. As they differentiate, they migrate upward to replace the cells at the upper regions of the crypts. A similar system operates in both the colon and the small intestine [5, 8].

– Mesenchymal stromal cells (MSCs)

Mesenchymal stromal cells (MSCs) are another type of cell with some stem-like properties. In fact, they are also sometimes called mesenchymal stem cells because of these properties. While MSCs can differentiate into multiple cell types in vitro, their stem cell capacity in vivo is still not certain. However, they have properties that make the valuable for disease treatment and regenerative medicine, largely because they secrete factors that promote tissue repair and because they have anti-inflammatory functions. They are present in bone marrow, and in other tissues such as adipose tissue and the umbilical cord [5, 23, 24, 25, 26].

– Muse cells

Multi-lineage-differentiating stress-enduring (**Muse**) cells were identified as a **subset of**

MSCs. They may be more potent and have more therapeutic potential than the MSCs.

MUSE cells can be identified by specific cell surface markers (proteins or molecules on the cell surface) that distinguish them from other MSCs. Some of the markers on MUSE cells are specifically associated with pluripotency. MUSE cells make up only a small percentage of MSCs [27].

– Adipose-derived stem cells (ASCs)

Adipose-derived stem cells (ASC) were discovered in fat tissue relatively recently and are considered to be a subtype of MSCs. These cells can be isolated much more easily than other types of stem cells, such as hematopoietic stem cells. They are generally isolated from fat tissue that is obtained from liposuction, a common procedure that removes fat tissue. ASCs appear to be multipotent in vitro, and they have the potential to be used for the regrowth of bone, fat tissue, and possibly other tissues. This can be useful for treating bone and structural abnormalities and restoring certain tissues after surgery. Their accessibility and relative abundance make them excellent sources of cells for cellular therapy [30–32].

2.2.2 Laboratory-Derived Stem Cells

Some stem cells do not occur naturally but have been developed or manipulated in the lab. These can be valuable for experimental research, and some of them have therapeutic potential. Major examples of these types of stem cells are discussed here.

- Somatic Cell Nuclear Transfer (SCNT) Stem Cells

– Overview of SCNT stem cells and methods

Somatic cell nuclear transfer (SCNT) stem cells are made in the lab. They are generated from an egg from one animal and a nucleus, including DNA, from a cell (usually a somatic cell) from another animal. These methods are the same methods used for cloning and were demonstrated for the first time in the cloning of Dolly the sheep in 1997 (Fig. 26.5) [5, 33, 34].

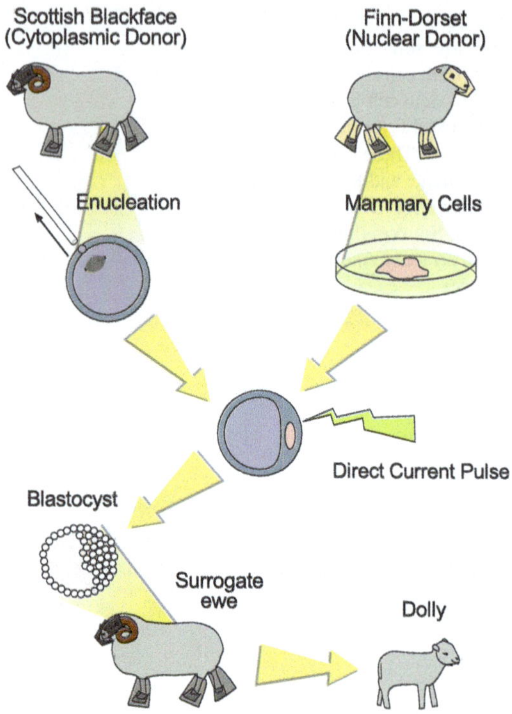

Fig. 26.5 Cloning of Dolly the sheep. The sheep was the first mammal to be cloned. The SCNT method was used. The nucleus from a somatic cell (a mammary cell) was inserted into an enucleated egg from another sheep. This egg was implanted into a host female ship, w hich resulted in a cloned sheep - Dolly the sheep. **Source**: "Dolly clone" (public domain image), Wikimedia Commons. Available at https://commons.wikimedia.org/wiki/File:Dolly_clone.svg

– Cloning of Dolly the sheep

In the cloning of Dolly the sheep, tissue was isolated from the donor sheep, sheep A, which was the sheep being cloned. A cell from this tissue was isolated, and the nucleus was removed and saved. At the same time, an egg cell was isolated from another sheep, sheep B. The nucleus of this egg was removed and discarded, and the nucleus from the somatic cell above, from sheep A, was inserted into the egg by injection with a fine needle. The egg, which now had the diploid DNA from sheep A, was activated artificially without sperm and was grown to the blastocyst stage in vitro. This blastocyst was implanted into a surrogate female sheep. This surrogate mother sheep gave birth to the resulting sheep, which was a replica, or clone, of the first sheep (sheep A). This sheep, named Dolly, was the first cloned mammal [35].

– Reproductive cloning vs therapeutic cloning
This SCNT technique is important for stem cell biology because the blastocyst obtained from the cloned cells is a good source of stem cells. ES cells can be isolated from the blastocyst and will be an exact genetic match of the original individual, in this case, the sheep, who donated the somatic cells. This brings us to the two terms, **reproductive cloning** and **therapeutic cloning**. Reproductive cloning is the technique described above for Dolly the sheep, where an entire animal is cloned. Therapeutic cloning instead involves the same method but stopping the study at the blastocyst stage, which is then used as a source of embryonic stem (ES) cells. When it comes to humans, reproductive cloning is banned worldwide, but therapeutic cloning might be a promising mechanism for isolating genetically matched stem cells from any individual [33].

– Generation of human SCNT stem cells
Since the cloning of Dolly the sheep, reproductive cloning has been done successfully in many different animals. These include cow (1998), mouse (1998), pig (2000), rabbit (2002), cat (2001), deer (2003), horse (2003), dog (Snuppy the hound) (2005), ferret (2006), camel (2009), and cynomolgus monkey (2018) [33]. For stem cell scientists, however, therapeutic cloning, not reproductive cloning, is the important goal. Generating human SCNT stem cells has been challenging. The first human SCNT stem cells were reported in 2004, but the work was found to be fraudulent [36]. One hurdle for generating human SCNT cells was that the human embryonic cells generated from this technique stopped dividing too early and did not reach the blastocyst stage. This had to do with premature exit from meiosis [37]. This problem existed in other primates as well. Conditions were optimized in primates such as macaques until primate SCNTs were made successfully [38]. In 2013, scientists made the first human SCNT stem cells. These stem cells were made where the donor nuclei were from skin cells of an infant. The scientists successfully made four pluripotent human stem cell lines from these nuclei. The infant donor had Leigh syndrome, a genetic neurologic disorder, and the goal was to develop a stem cell model that could be used to study this disorder [37]. While the first SCNT stem cells involved donor nuclei from infant cells, two subsequent studies successfully generated SCNT stem cells from adult donor cells, one from a 35-year-old and the other from a 75-year-old adult [33, 37, 39, 40].

– Telomere length and SCNT cells
Telomeres are found at the ends of chromosomes. They can be as long as 15,000 bases, and they prevent chromosomes from losing base pairs in the coding part of the genes when the DNA replicates. Telomeres shorten each time a cell replicates. Chromosomes from older individuals, therefore, have shorter telomeres than those from younger individuals. This raises the question of whether telomeres from SCNTs that were developed from older individuals will be shorter than telomeres from younger individuals. This could provide a disadvantage for the growth of the cells. Some animal studies have suggested that telomere length does not shorten during the SCNT process and may even modestly lengthen [41].

– Challenges associated with SCNT stem cells
The development and use of SCNTs are associated with ethical issues and legal restrictions in different parts of the world because they make use of the same techniques that are involved in cloning. Although reproductive cloning is outlawed throughout the world, therapeutic cloning laws vary from country to country. Furthermore, although SCNTs are a promising type of stem cell for future medical treatments, SCNTs may also have a downside. Evidence of this comes from studies of animals that have been reproduced by reproductive cloning. While cloned animals are genetically identical to the donor animal, in some cases, the health of the cloned animals is not perfect. There have been some cases where cloned mice were found to have damaged immune systems, increased cases of pneumonia and subsequent death, increased development of tumors, liver failure, fertility

problems, and shortened lifespans. While these problems may not apply to the SCNT stem cells themselves, it is important to take precautions and to study the cells carefully [42–44].

- Induced Pluripotent Stem Cells (iPS Cells)
  - Overview of iPS cells

Induced pluripotent stem cells (iPS cells; also referred to as iPSCs), are a type of stem cell made in the lab from somatic cells through a process called **cellular reprogramming**. Scientists reasoned that stem cells must express specific genes that cause them to be pluripotent and that if these genes can be added to somatic cells these cells could take on the characteristics of pluripotent stem cells. At the end of 2007, three groups, one in Japan and the other two in the USA, were successful in making this happen. They isolated human skin fibroblasts, which are somatic cells that are easy and unobtrusive to obtain. They experimented by testing different combinations of genes to determine which genes they could introduce into these that would induce pluripotency. After much work, they found that if they introduced four genes into these somatic cells, using the retroviral infection method, the cells were reprogrammed into pluripotent cells. These cells behaved like pluripotent stem cells in that they maintained an undifferentiated state but could also be induced to differentiate into any lineage. These cells are now called **induced pluripotent stem (iPS) cells** [45]. Shinya Yamanaka shared the Nobel Prize with Sir John Gurdon in 2012, for discoveries related to cellular reprogramming, which was critical for the development of iPS cells [46]. Gurdon's earlier work in 1962 had already challenged the prevailing idea of the time of that cellular differentiation was irreversible. He demonstrated in frogs that even mature adult cells can maintain their full developmental potential when their nuclei are transferred to enucleated eggs [47]. This idea of reversibility was key to the eventual generation of iPS cells.

iPS cells have many potential advantages. A significant advantage is that generating iPS cells completely abolishes the need to work with and destroy embryos. This solves many technical, legal, and ethical challenges. Furthermore, these cells can be made from any individual, resulting in a perfect genetic match with those who donated the fibroblasts to make the cells [5].

In addition, because iPS cells are **pluripotent**, they may be powerful tools for **regenerative medicine**, as they can be directed to become specific cell types needed to repair or replace damaged tissues. Importantly, since they can be potentially be derived from the patient receiving treatment, the risk of **immune rejection** is greatly reduced. This is an important example of autologous treatment, although currently this type of treatment is primarily in experimental stages.

  - The transgenes used to make the iPS cells

To generate these cells, the researchers studied stem cells and the genes they express to determine which, if any, of the genes could cause the somatic cells to become pluripotent. They tried many combinations before finding the right combination of four genes that resulted in the development of iPS cells. These four genes, called transgenes because they were introduced into cells, are Oct3/4, Sox2, c-Myc, and Klf4. Oct3/4 is a transcription factor that regulates many target genes implicated in the maintenance of pluripotency [48]. Sox2 is also a transcription factor, and it is necessary for embryonic development and for preventing ES cell differentiation [49]. c-Myc is a transcription factor with many cellular functions that are important for cell proliferation. Klf4 is a transcription factor whose overexpression inhibits the differentiation of ES cells. Later, a second study found that introducing two more genes was even more effective. These genes were Nanog, a transcription factor normally found in embryonic stem cells that has a role in promoting pluripotency, and Lin28, an mRNA-binding protein found in ES cells that has a role both in pluripotency and in differentiation [50, 51].

– iPS cell challenges

The generation of iPS cells was exciting and is an important milestone in stem cell research. Nevertheless, like all new technologies, the system still has challenges: First, the genes were originally introduced into the cells using a retroviral infection mechanism. Retroviruses are an effective way to introduce transgenes into cells but can also increase the risk of oncogenic transformation. Second, one of the genes is c-Myc, which is known to be oncogenic if it is overexpressed or activated. Thus, caution must be taken, and protocols must be optimized to ensure that iPS cells are safe for patients. Third, iPS cells take time to generate, and therefore, they are not necessarily a source of cells that can be available on short notice, as would be necessary for treating some conditions, such as a heart attack. There have also been some problems with growing some iPS cell lines. Some of the lines grow more slowly than they should or undergo cell senescence, and some of the cells develop mutations over time, which could increase the risk of cancer. The cells need to be used with caution, therefore, in clinical studies. While some clinical trials are being carried out using iPS cells, iPS cells are especially valuable as research tools, as they can be used to model organs and various disease cells in vitro [50, 52].

– iPS cells in research

iPS cells are quite valuable for developing in vitro models of diseases in the lab and for developing and studying organ models [7]. Somatic cells, such as fibroblasts, can be isolated from any patient with a genetic disease and developed into iPS cells. These cells can be differentiated into various cell types in the lab, resulting in an in vitro model for the disease, for example, iPS cells were used to study **liver disease**. In this study, somatic cells were isolated from patients with inherited liver disease. These cells were made into iPS cells and cultured. The scientists differentiated these iPS cells into hepatocytes, which are liver cells. Interestingly, the liver cells took on the same characteristics as the diseased livers of the patients. These cells could be used to study the disease further and test the effects of different drugs without needing to use biopsies to access the cells. This system can be used to study many different genetic diseases [53, 54].

• Parthenogenetic Stem Cells

Parthenogenetic stem cells are stem cells that come from blastocysts that developed from parthenogenetically activated eggs. Parthenogenetic egg activation involves the artificial activation of unfertilized eggs. This requires an understanding of the biological mechanisms behind egg activation. Researchers have found that several types of chemicals (such as ethanol, ionomycin, puromycin, and 6-DMAP (6-dimethylaminopurine)) can bring about activation of metaphase II arrested eggs in culture, even in the absence of sperm. Parthenogenetically activated eggs can form a blastocyst with diploid cells but cannot form an embryo. They are another potential source for therapeutic stem cells [55, 56].

2.2.3 Neural Stem Cells

Neural stem cells (NSCs) are given their own separate category here because they are a broad category of stem cells that can overlap with many of the types of stem cells described above. The term NSC can refer to different types of cells. Overall, NSCs are cells that can differentiate into neurons. They can come from various sources, all of which are often termed NSCs, for example, they can refer to iPS cells, ES cells, or MSCs that are differentiated into neuronal precursors in vitro. They can also refer to fetal stem cells isolated from the fetal nervous system. In some cases, they can, rarely, also refer to adult stem cells. While the nervous system is a part of the body generally recognized as having no or few adult stem cells, a small number of natural adult neural stem cells exist in some parts of the nervous system. Specifically, they are found in two regions of the brain: the subventricular zone (SVZ), where they give rise to olfactory neurons, and the subgranular zone of the dentate gyrus of the hippocampus (SGZ), where they give rise to some of the granule cells, which are a type of excitatory neuron [57–59].

3 Treatment of Diseases Using Cellular Treatment

3.1 Overview

A major goal for using stem cells in the treatment of human conditions is to encourage the stem cells to replace or repair damaged cells and tissues. The transplantation of hematopoietic stem cells (HSCs) has been used for years to treat blood disorders, including leukemia. In this type of therapy, the goal is to replace the patient's own cells with healthy cells from a donor. The possible uses of HSCs in stem cell therapy have expanded since their initial use, but the use of stem cells to treat other disorders is still in early stages. Some research and clinical studies in the stem cell field have been promising, and there is hope that stem cell therapy will increasingly become an effective way to treat certain human disorders. At present, most stem cell therapies are still in clinical trials. Most of these treatments involve the use of adult stem cells, most commonly hematopoietic stem cells. The only FDA-approved stem cell treatments at this time use adult hematopoietic stem cells. Other stem cells under investigation for treating human disease include the use of other types of stem cells, including ES cells, MSCs, adipose tissue stem cells, and NSCs [4, 60, 61].

3.2 The Promise and Challenges of Stem Cell Therapy

Stem cell therapy is still a relatively new, promising way to treat human disease and injuries or for regenerative therapy. For this reason, it has received a great deal of attention. Stem cell therapy also has risks, however. Before the promising benefits of using stem cells to treat human disorders are discussed, some of the challenges and problems associated with stem cell treatments are addressed below.

3.2.1 Cases of Insufficient Regulation

Historically, cells that come from a patient were not classified as drugs and were therefore not subjected to the regular FDA approval process. This allowed many clinics to offer unproven stem cell treatments without proper oversight. The law changed in 2017, but confusion over regulation persists, and some clinics still operate without proper oversight. Reports of patients experiencing no benefit from stem cell treatment or receiving serious side effects have raised concerns. The rise of "stem cell tourism," where patients travel across borders in search of treatment, has also increased the risk of encountering unregulated or unscientific treatments [5, 62–64].

3.2.2 Insufficient Education of Patients: Leading to Unrealistic Hope

Literature about various stem cell treatments for patients can be confusing and even misleading. Articles about potential stem cell therapies in the press do not always clarify which types of stem cells were used in the preclinical studies, what results can be realistically expected, and whether testing has gone further than the animal testing stage.

3.2.3 Technical Challenges with Using Stem Cells

While some stem cell therapies are already approved and in use, many technical challenges still remain to be addressed. Stem cells come from various sources, and determining which ones are the best for a particular purpose is one crucial challenge. Another challenge is stem cell differentiation. To use stem cells for disease treatment, it is desirable and sometimes necessary to start the differentiation process in vitro before introducing the cells into patients. This prevents cells from differentiating into the wrong lineage or staying undifferentiated, which can lead to risks of side effects or even cancer. Stem cell researchers, therefore, need to develop protocols for differentiating cells along different lineages.

Once the cells are differentiated, they generally need to be implanted into a host, and this can also be complex and involve multiple challenges [6].

Rejection by the immune system is another critical issue. Stem cells implanted into the patient are likely to be recognized as foreign and thus be rejected by the immune system. Immunosuppressants can reduce this problem, but they have side effects and may not completely eliminate the possibility of rejection. One solution may be to use **autologous** treatments, which means using stem cells that come directly from the patient and are, therefore, a genetic match. This is a promising idea but not always possible. The possibility of autologous treatments and the types of stem cells that might be used for such treatments are discussed at the end of this chapter [5].

One of the most serious risks associated with the use of pluripotent stem cells, such as embryonic stem cells (ESCs) and induced pluripotent stem cells (iPSCs), is the potential for uncontrolled growth, leading to tumor formation. If undifferentiated cells are transplanted into the body, they can form **teratomas**, which are benign but potentially harmful tumors composed of a variety of tissue types. This is particularly a risk if the stem cells are pluripotent. Ensuring that all transplanted cells are properly differentiated before implantation is critical to avoid this risk. This is one reason why it is usually desirable to initiate the differentiation process first before introducing cells into patients. Purification of the cells is also critical [65].

Interestingly, while teratoma formation is a safety concern, it also serves a purpose in stem cell research. When working with stem cells, it is often important to confirm that the stem cells are pluripotent. Scientists can use teratoma formation as a test to determine whether the stem cells they are working with are pluripotent. To test for pluripotency, scientists can inject stem cells into immunodeficient mice. Immunodeficient mice are used so that they do not have an immune system to reject the cells even if they are human cells. If the stem cells that are injected are indeed pluripotent, they may form teratomas that contain tissues from all three germ layers. For stem cells used for research purposes, this is a good way to test whether the cells are pluripotent [66].

Despite some of the challenges outlined above, stem cell therapy remains a promising area of research and clinical application. Some stem cell-based therapies are already FDA-approved, and others are under investigation or in preclinical stages.

3.3 Stem Cell Therapy within Stem Cell Medicine

The terms stem cell medicine and stem cell therapy are sometimes used interchangeably, but stem cell therapy is in fact, a subset of stem cell medicine. In this chapter, we will use the term **stem cell therapy** to refer to treating the patient with new stem cells to replace damaged or diseased cells or tissues. Another type of **stem cell medicine** is **regenerative medicine**. Regenerative medicine often involves pharmacologically activating the patient's own adult stem cells. An example of regenerative medicine is using a cytokine such as G-CSF (granulocyte colony-stimulating factor) to stimulate the patient's HSCs to produce more white blood cells, which can be done for many purposes, such as recovery of the immune system after chemotherapy [67]. The main focus of the chapter, however, will be **stem cell therapy** in which stem cells are used to replace damaged or diseased cells [68, 69].

3.4 Some Diseases and Conditions for Which Stem Cell Therapy Holds Promise

Stem cell therapies have been studied for the treatment of various conditions. Some of these studies are farther along than others. Below, we discuss some examples of various disorders for which scientists have studied the practicality and utility of stem cell therapy. This includes discussion of some approved treatments and some that are in the research or clinical trial stages.

3.4.1 Blood Disorders

Blood disorders include aplastic anemia, as well as cancers such as myeloma and leukemia. These diseases are often treated by hematopoietic stem cell transplant (HSCT). In these transplants, the patient's blood cells are destroyed using radiation or chemotherapy. The patients are then infused with HSCs from a donor. The donor is usually injected with cytokines or growth factors that encourage the proliferation and circulation of their HSCs. The donor's HSCs are then removed and administered to the patient. HSCT is the oldest known stem cell therapy. The procedure has risks because the patient's own blood cells are destroyed, and they may experience rejection of the donor cells or graft vs host responses. Nevertheless, HSCT is still used in some cases today, and it can lead to long-term remission from disease and has saved many lives [70, 71]. While HSCT is a well-established procedure for blood disorders, the reality of using stem cells for other disorders came later.

3.4.2 Parkinson's Disease (PD)

Parkinson's disease (PD) is one of the earliest non-hematological diseases for which stem cell therapy has been explored. It is thought to be a promising candidate for stem cell therapy because a single cell type, the dopamine neuron, is known to be disrupted in the disease [5].

- Pathophysiology of the Disease and Traditional Treatments

Parkinson's disease (PD) is caused specifically by the degeneration of **dopamine-producing nerve cells** in the substantia nigra in the brain, which is in the midbrain. Dopamine is a neurotransmitter that has important roles in controlling body movement. The result of the loss of dopamine is loss of muscle control, tremors, and stiffness. PD affects about 1% of people over age 65, and the problem is expected to grow. Traditional treatments for PD involve the use of the drug L-DOPA, which gets converted to dopamine. This can be effective for a while, but regulating the dose is complex, and the correct dose can change over time. Deep brain stimulation is a newer treatment involving transplanting electrodes into the brain. It is helpful for some patients but does not improve all symptoms, and results vary among different patients [72, 73].

- Stem Cell Therapy and Parkinson's

Stem cell therapy is another possible treatment option under investigation for Parkinson's disease. Some preclinical and clinical studies have been carried out. Different types of stem cells either have been tested or are under consideration for the treatment of PD, as described below.

– Fetal tissue cells

Some of the earliest stem cell studies for PD were done using fetal tissue cells, which contain a mixture of cells including stem and progenitor cells. In 1979, there were some successful studies in which fetal stem cells from rats were transplanted into rat models of PD. This led to clinical trials in human patients in the late 1990s, where human fetal cells were transplanted into the brains of PD patients. While some successful results were reported, other results were mixed, and ethical issues involved in using these cells limited the studies.

Using fetal cells in an attempt to treat PD has both advantages and disadvantages. The **advantages** are that fetal cells are a good source of dopamine neurons, yet, unlike adult cells, they still have a strong capacity for proliferation. Another advantage to using fetal stem cells is that they often do not induce an immune reaction.

Disadvantages of fetal cells include ethical concerns with using aborted fetuses, informed consent issues, and moratoriums that have been placed on their use at various times. Another disadvantage relates to the availability of fetal stem cells. The dopamine neuron stem cells needed for treating Parkinson's disease can be scarce, even within the fetus, and multiple fetuses may be necessary to obtain enough neurons for one patient. Even if one obtains enough of these cells, they can be contaminated with other neurons, including serotonin neurons.

In summary, while there have been some successes with fetal cells in clinical trials with

PD patients, results have been mixed. Not all studies have been successful, and rejection of the cells by the immune system has sometimes occurred [19].

– ES cells, MSCs, and MUSE cells

Mesenchymal stromal cells (MSCs) can differentiate into several cell types in vitro. Protocols have been developed to differentiate MSCs into dopamine neurons. These differentiated MSCs have been transplanted into **rat models** of PD, and the result was the relief of some of their symptoms [74, 75]. As discussed above, MUSE cells are a subset of MSCs. Some researchers are interested in using MUSE cells to treat PD due to some of the advantages these cells may have over MSCs. The development of dopamine-producing neurons from ES cells is also an area of investigation. The goal of all of these types of studies is to see whether transplantation of the stem cells into the brains of patients can slow the progression of the disease [76, 77].

– Induced pluripotent stem cells (iPS cells)

iPS cells have been tested in monkey models of PD in a study comparing autologous and allogeneic treatments. iPS cells were generated from fibroblasts from monkey models of PD. These iPS cells were differentiated into DA neurons. These neurons were transplanted back into the original monkeys (autologous) or into other monkeys (allogeneic), both of which were monkey models of PD. The treatment was more effective in the autologous recipients, who showed improved movement and decreased signs of depression. This study demonstrates the possible effectiveness of iPS cells in PD treatment and the advantages of using autologous therapies [78]. While there are no approved iPS stem cell therapies for PD yet, multiple clinical trials are in progress [79–81].

– Challenges remaining for using stem cell studies in Parkinson's disease

Using stem cells to treat PD appears promising, but many challenges remain. An important challenge is developing mechanisms for differentiating stem cells into functioning neurons. Other challenges involve developing the best methods for delivering the stem cells to the correct targets in the brain and learning how to get the stem cells to integrate into the brain correctly. Stem cells can potentially be tumorigenic, especially if they are not all differentiated. Developing strategies for minimizing this risk is critical. Other issues involve producing enough cells on a large scale and minimizing immunological responses [64, 82, 83].

3.4.3 Alzheimer's Disease (AD)

- Pathophysiology of Alzheimer's Disease

Alzheimer's disease (AD) is more complex and challenging to treat than PD because the disease affects multiple cell types in many parts of the brain. AD is the most common cause of dementia. The precise cause of the disease is still unknown. The brains of people with AD have an accumulation of the proteins amyloid-beta and tau. Amyloid-beta clumps together to form "plaques," and tau gets twisted into protein "tangles." One theory is that the tangles and plaques prevent nerve cells from communicating correctly and responding to nutrients. However, the roles for the plaques and tangles are complex, as recent studies have suggested that they might not always be the sole cause of the disease. Thus, AD remains poorly understood. It is a neurodegenerative disease, and over time, certain nerves die. There is currently no cure. Some drugs can temporarily help with some of the symptoms, but there are no drugs that prevent or delay the loss of neurons [84–86].

- Stem Cells and Alzheimer's Disease
  - Challenges for developing stem cell treatments for AD

Stem cell therapy for AD is an active area of research, but it is less advanced than the PD field, and much of the work on AD is being done in animal models. Using stem cells to treat AD has unique challenges. Unlike PD, many types of nerves are damaged, and multiple parts of the brain are affected in AD. Consequently, stem cells would have to differentiate into many types of neurons, travel to various areas of the brain, integrate into the correct places, and make the right connections. Even if this can be accomplished, the new neurons would need to function normally even once the disease has started, and it is unclear whether the new cells would be damaged by the high levels of tau and amyloid-beta protein already in the AD brain. Other challenges include the risk of rejection, tumor formation, and delivery challenges [87, 88].

– In vitro models of AD using iPS cells

Most cases of AD seem to be sporadic, but some cases may be familial and thus have a genetic component. Researchers have developed iPS cells from skin cells of patients with familial AD. These iPS cells were then differentiated into neurons and grown in culture. In some cases, these lab-grown neurons released beta-amyloid protein, similar to AD neurons. This can give scientists a tool for studying the disease in vitro and for testing drugs on cultured neurons [89].

– Animal models of AD using hematopoietic stem cells

Hematopoietic stem and progenitor cells (HPCS) are a subset of hematopoietic stem cells (HSCs), slightly more differentiated than HSCs. Both healthy HSCs and HSPCs were transplanted into the brains of an Alzheimer's (AD) mouse model. Some symptoms of AD were reversed in both studies, including memory and cognition problems. Neuroinflammation was reduced, and beta-amyloid accumulation was also reduced. The HSPCs migrated to the brain and replaced microglia, which are a cell type in the brain of myeloid lineage. Replacing these cells may have helped to lower neuroinflammation and clear β-amyloid plaques and toxins [90].

– Some recent clinical trials for AD

Clinical trials are being carried out to study the use of stem cells to reduce neuroinflammation and to repair neurons, for example, a study at **UTHealth Houston** is testing the use of **autologous adipose-derived mesenchymal stem cells (MSCs)**. This is a phase Ib/IIa clinical trial. In this study, patients are infused with their own MSCs. Scans are used to detect brain inflammation. This trial is being carried out in patients with early stages of the disease [91].

Another trial (by Regeneration Biomedical) uses adipose-derived stem cells that have high levels of Wnt pathway activity. The researchers specifically selected cells with high levels of Wnt pathway activity because this pathway is known to have a role in self-renewal and reduced differentiation. These cells are injected into the brains of patients. This is also a phase I trial. Researchers will assess the levels of markers such as tau, as well as cognitive functions [92].

Another trial, by **Longeveron**, a biotechnology company, tested a product called **Laromestrocel. This product** is derived from MSCs from a **donor**. This is a phase 2a trial, called CLEAR MIND. The cells are administered by IV infusions every month. Patients will be monitored and evaluated [93].

3.4.4 Multiple Sclerosis (MS)

• Pathophysiology of MS

Multiple sclerosis (MS) is a disease that affects the central nervous system (CNS). It is part of a family of diseases that involve the destruction of myelin in the myelin sheath. **Myelin sheaths** are made up of layers of myelin, which insulate the axons of nerves. In the CNS, myelin sheaths are formed by specialized neuroglial cells called oligodendrocytes (Fig. 26.6). MS is an autoimmune disease. In MS, the patient's immune system attacks the myelin sheaths surrounding nerve fibers. This leads to a breakdown in the transmission of nerve signals. Symptoms can be mild or severe, ranging from numbness to blindness and paralysis. Current treatments often involve immunosuppression. These treatments are effective but usually temporary [94, 95].

• Stem Cell Studies for MS

– Autologous hematopoietic stem cell transplant (aHSCT) clinical trials for MS

Autologous hematopoietic stem cell transplantation (aHSCT) is a promising therapy for MS, currently in clinical trials. This treatment involves the use of bone marrow-derived stem cells (HSCs) from the patient. These HSCs are harvested. The rest of the patients' immune cells are then intentionally destroyed using chemotherapy. Patients are then injected with their own

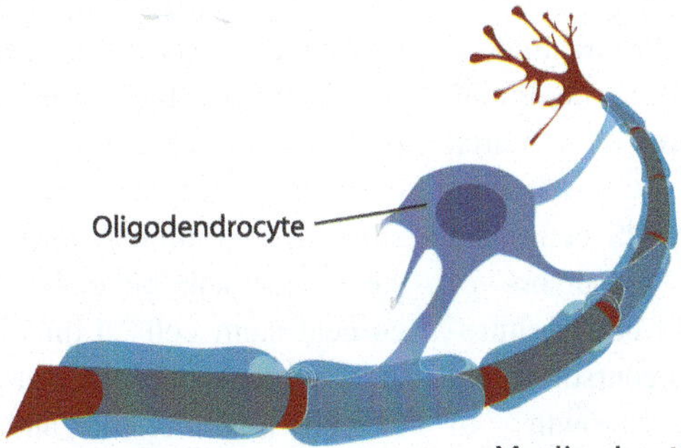

Fig. 26.6 Myelin sheaths. Myelin sheaths insulate the axons of nerves and allow rapid signal conduction. In the CNS, myosin is produced by specialized cells called oligodendrocytes. **Source**: Figure adapted from public domain image via Wikimedia Commons. https://commons.wikimedia.org/wiki/File:Neuron_with_oligodendrocyte_and_myelin_sheath.svg

HSCs, thus receiving an autologous hematopoietic stem cell transplant (aHSCT). This reinjection of HSCs apparently "resets" the body's immune system. It is interesting that autologous treatments are effective in MS patients, since the patients' own immune cells may already have the abnormality that is present in MS. Presumably, the treatment is effective because the harvested HSCs are at an earlier developmental stage, where they would not yet have developed abnormalities [96–98].

Two important clinical trials, **MIST** and **BEAT-MS,** have important pivotal roles in establishing the role of **autologous hematopoietic stem cell transplantation (aHSCT)** as a promising type of treatment for multiple sclerosis (MS). Both trials enrolled patients who did not respond well to other treatments. The **MIST trial** is an international phase III trial using aHSCT. Results so far indicate that the aHSCT was significantly more successful than standard treatment. It reduced disease progression and relapse. The MIST trial was completed in 2016, and results were positive, but due in part to the scale of the study and the risks involved, the treatment is still not FDA-approved. The **BEAT-MS trial** also uses aHSCT and compares it with the treatments that are effective in many patients. This study has also provided important evidence that aHSCT is a successful and promising treatment, but the study is still ongoing [97, 99]. aHSCT treatment for MS remains in clinical trials, but some patients are receiving this type of treatment through clinical trials or through centers offering it as an experimental therapy [100].

– MSCs

MSCs are also being tested in clinical trials for treating MS. MSCs are also often used in autologous treatments, although donor MSCs can also be used. In this type of treatment, MSCs are isolated from the patient or from a donor. They are isolated from bone marrow or blood, grown in culture, and multiplied. They are then reintroduced into patients in large amounts. MSCs are promising because they can suppress overactivity of the immune system, they release factors that support tissue repair, and some studies suggest they may differentiate into neuronal or glial cells, at least in vitro [101, 102]. Several studies using MSCs have shown that treatments are safe, and some studies have suggested improved neurological function. There are different mechanisms by which MSCs suppress the immune system. One important mechanism involves the secretion of immunosuppressive molecules which suppress T-cell proliferation [103–106].

– iPS cells

iPS cells have not yet been used in MS treatments, but research is being conducted to determine whether this is feasible. One area of research involves isolating somatic cells such as human skin cells, developing them into iPS cells, and then differentiating them into precursors of oligodendrocytes. The goal would be to reintroduce these oligodendrocyte precursors into patients to encourage the growth of oligodendrocytes, which in turn would make more myelin to counteract the destruction of myelin in MS [107].

3.4.5 Age-Related Macular Degeneration (AMD)

Age-related macular degeneration (AMD) is a vision disorder caused by the degeneration of the macula, a part of the retina (see Chap. 18). Retinal pigment epithelial (RPE) cells are primarily affected in this disorder. There are currently no approved stem cell treatments for AMD, but research is being carried out, and several clinical trials have been initiated. Replacing patients' **RPE cells** with healthy RPE cells is an active area of study and is considered to be a promising treatment avenue. Protocols for differentiating ES cells and iPS cells into RPE cells have been developed in several labs [5].

The National Eye Institute (NEI) launched a **clinical trial** for AMD, using **iPS cells** that are differentiated into **RPE cells**. The study was initiated in 2020, and patients will be followed for 15 years. Several other trials around the world have been launched. Clinical trials for AMD are in early stages, and they need to test for safety, tissue rejection, the development of tumors, as well as improved vision [108, 109].

3.4.6 Cornea Damage

Damage to the cornea is another eye problem that can abolish or reduce vision. It can be caused by foreign objects, chemical exposure, too much sunlight, or genetic abnormalities. Corneal transplants are one treatment option, but they are not always readily available. In addition, such transplants cannot be done at all if the limbus has been destroyed or is abnormal. The limbus is a zone of tissue that surrounds the iris and is located where the cornea (the transparent front part of the eye) meets the sclera (the white part of the eye). It contains stem cells that are responsible for continuously renewing the cornea. These stem cells can be called limbal stem cells or corneal stem cells. Limbal stem cell therapy is an important area of research, and there are several preclinical studies and clinical trials involving the use of limbal stem cells. These studies include isolating limbal stem cells from the good eye or from the eye of a donor and growing them in culture. These cells are then placed around the iris of the damaged eye. An important goal is to determine whether these transplanted stem cells can regenerate the different layers that make up the cornea [110].

3.4.7 Skin Damage and Burns

When there are severe burns to the skin, open wounds must be covered quickly and efficiently to prevent fluid loss and infection. Saline baths are one solution, but they have many drawbacks. Skin grafts are also a possibility, but the immune system often rejects them. Another possibility is creating artificial skin by tissue engineering. This is a slow process, and there is a need for faster treatment. A newer idea is to turn **iPS cells** into skin cells. These cells could be "banked" as a stock of skin cells ready for use if they are a close genetic match to the patient [111].

3.4.8 Organ Transplantation

Organ transplantations are sometimes the best option for treating certain disorders, but there is a shortage of organs available, and not all organs are replaceable. An alternative to whole organ transplants may be cell transplantation. In this proposed treatment, relevant cell types could be transplanted into a damaged organ to replace dysfunctional cells and subsequently regenerate the tissue. These cells could be delivered directly to the place in the body where they are required, such as the heart, muscle, pancreas, or other tissue. Alternatively, the cells may be administered intravenously through a blood vessel.

For such cell-based treatments to work, enough cells would need to be available to treat patients. Growing stem cells, amplifying them, and then differentiating them along the correct lineage could be a way to achieve this. Stem cells could also be modified with gene therapy to replace cells in an individual with a genetic abnormality.

Using stem cells instead of organ transplantation may be more promising for tissues that do not contain too many cell types that must operate together. Heart, for example, may be a promising candidate. In the heart, primarily one cell type, the heart muscle cell, dies following a myocardial infarction. New heart cells obtained by differentiation of pluripotent stem cells could be a promising way to replace the damaged cells. Other tissues, where multiple cell types are all damaged, may be more challenging to treat with stem cell therapy [60, 112, 113].

3.4.9 HIV

HIV (human immunodeficiency virus) is the virus that causes AIDS (acquired immunodeficiency syndrome). AIDs can be treated but not cured. The HIV virus infects immune cells, specifically T cells. It uses receptors to enter the cell, mainly CD4 and CCR5 (see Chap. 24). Once the virus infects the T cells, it can destroy the cells and, in turn, harm the immune system, which can eventually lead to death if infections are not controlled [114].

Patients who were HIV-positive and later developed blood cell cancers such as leukemia or lymphoma (independent of their HIV infection) received bone marrow transplants for their cancer treatment. A surprising finding was that a small percentage of these patients were also cured of HIV. This unexpected finding was traced back to mutations on both copies of the CCR5 gene in the donor stem cells. Thus, the patient's cells were replaced by the hematopoietic stem cells that did not have functional copies of the CCR5 gene. Consequently, the HIV could not infect the patient's new immune cells [115, 116].

More recently, similar results were obtained in an HIV-positive patient who had leukemia and was treated with umbilical cord stem cells rather than HSCs. The stem cells were from a donor who had banked their child's umbilical cord blood. These stem cells also had a deletion in the CCR5 gene. The patient was found to be HIV-negative after her treatment. She was followed for 5 years, during which time she remained HIV-negative and no longer needed to take antiviral drugs [117].

The studies described above indicate that stem cell therapy is a potential mechanism for curing HIV AIDS. While this is a promising result for the stem cell field overall, stem cell therapy is probably not appropriate for most patients with HIV because the procedure is invasive and risky, relative to today's antiviral drugs. However, it may have important implications for future treatments of other viral infections in cases where antiviral drugs are unavailable or ineffective [118].

3.4.10 Heart Failure

Heart failure is a condition where the heart does not pump as well as it should. As a result, the body does not get enough oxygen, and this can lead to organ damage and eventual death. There is no cure, though a heart transplant is sometimes possible in the most severe cases. Inflammation is a key feature of heart failure. Long-lasting inflammation can lead to blood vessel plaques, blood clots, heart attacks, and strokes [112].

To determine whether stem cell therapy may be effective for treating heart failure, a clinical trial was done where MSCs from healthy donors were injected into the myocardium heart muscle of patients with chronic heart failure. MSCs were used because they secrete anti-inflammatory factors, and an important goal was to reduce inflammation. Patients in this study showed a reduction in heart attack or stroke after 30 months, but the results were imperfect, and the rates of hospitalization were not sufficiently decreased. The results were also complicated because patients continued using other medications, making it difficult to interpret the results. Further testing is still considered to be warranted due to some initial success [119].

The **CATO trial** is currently in phase II, and it uses **umbilical cord-derived MSCs (UC-MSCs)**. Patients enrolled in the study have ischemic heart failure. The donor stem cells are introduced using intravenous (IV) infusion. Patients will be evaluated for improved cardiac function and reduction of inflammation and fibrosis [120].

3.4.11 Osteoarthritis (OA)

Osteoarthritis (OA) is a chronic joint disease. While some treatments for OA, such as exercise and certain medications, are available and these treatments help with managing symptoms, none of the current therapies regenerate degenerated tissues. Treatment with MSCs is considered a promising idea because these stem cells can differentiate into chondrocytes (specialized cells in cartilage tissue) and because of their anti-inflammatory properties.

Several different clinical trials that involve the use of MSCs have been initiated or are ongoing. MSCs in these studies are derived from different sources, for example, they can be derived from iPS cells that are differentiated into MSCs. They can also be derived from the umbilical cord, bone marrow, or adipose tissue. These MSCs are grown in vitro, followed by intra-articular injection (injection into the joint). The MSCs can then differentiate in vivo into the various tissues needed in the joint, such as chondrocytes (cartilage) and osteocytes (bone). MSCs can also secrete ECM, growth factors, and cytokines, which help regulate inflammation. An alternative approach to injecting undifferentiated MSCs is to pre-differentiate them in vitro into cells such as chondrocyte-like cells. This type of treatment is less common. Results from a number of these types of trials have been positive, with patients showing relief from certain symptoms [121–124].

3.4.12 COVID-19 and Acute Respiratory Distress Syndrome (ARDS)

COVID-19 infection can occasionally lead to severe inflammation, including what is known as a cytokine storm, an excessive immune response marked by the release of pro-inflammatory cytokines. This can result in acute respiratory distress syndrome (ARDS). This involves the activation of immune cells, the release of pro-inflammatory cytokines, and damage to the lungs. Severe ARDS has a high mortality rate despite improved treatments. In response to this, several clinical trials, such as those initiated by the School of Medicine at Mount Sinai, NY, initiated in 2020 [125], have been carried out to assess the effectiveness of MSCs in reducing inflammation and improving survival in patients with ARD. The rationale for these studies is that MSCs may have the potential to migrate to the lungs and release anti-inflammatory factors and immunomodulatory factors that can reduce cytokine storms and promote tissue repair. Preliminary results from some trials have shown that MSC therapy is safe and may be associated with improved oxygenation and reduced inflammatory markers. Current evidence as to the effectiveness of these methods is still inconclusive, however, and large-scale clinical trials are still needed [126–128].

3.5 Autologous Stem Cell Therapy vs Allogeneic Stem Cell Therapy

An important issue with any type of stem cell therapy is the possibility of rejection of the new cells by the patient's immune system. Using **autologous stem cell** therapy, a key approach in **personalized medicine** is one way to circumvent this problem. In autologous therapy, the patient's own stem cells are isolated, modified if necessary, and transplanted back into the same patient. The stem cells will thus be a perfect genetic match, and the chance for rejection by the immune system is minimal. This contrasts with **allogeneic** therapy, where the stem cells come from a donor. In allogenic therapy, there is a larger chance of rejection [129–131].

Autologous treatment has many advantages, but it is not always feasible, for example, it may not be feasible if the cells contain cancer cells. Likewise, if the patient has a genetic abnormality, using a donor may be preferable. Autologous treatment is almost impossible when using fetal stem cells, but it may be possible when using adult stem cells in certain cases. aHSCT for the treatment of MS is a classic example of autologous therapy. In this treatment, described above, HSCs from the patient are removed and readministered to the same patient. The resulting "reset" immune cells alleviated the autoimmune response in some MS patients.

Other adult stem cells may be useful for autologous therapies in the future. These may include umbilical cord stem cells that were saved from birth, or they may include adipose stem cells, which are easy to isolate and will be a perfect genetic match. SCNT cells could theoretically be used for autologous therapy. By making SCNT stem cells, the scientists would essentially clone the individual's own cells and develop them into stem cells. In reality, however, generating SCNTs is a long process using controversial methods and is not likely to be feasible in the near future. iPS cells may be a better option. They are made using the patient's own somatic cells that are made into pluripotent stem cells, and they do not rely on eggs or embryos. iPS cells are already being evaluated in some clinical trials. Parthenogenetic stem cells are also a possibility for autologous therapies, but they require invasive egg retrieval [5, 55].

Autologous therapy may not always be practical or timely in every situation. A more practical idea may be to collect banks of stem cells, both embryonic and adult, and record their genotypes. In this case, patients who need stem cell therapy may be able to access the best genetic match possible from these stores of cells, even though they will be unlikely to be a perfect match. This may also help solve the problem of speed, as treatment is often most successful if it is done quickly [31].

4 Other Types of Cellular Therapy

While this chapter has focused on stem cell therapy, other types of cellular therapies are also used to treat human disorders. Some examples are discussed below.

4.1 Chimeric Antigen Receptor T-Cell Therapy (T-Cell Therapy)

CAR T therapy involves isolating T cells from a patient who has cancer and altering them so that they produce a receptor that targets the cancer cells [132]. In this therapy, once they are isolated, the T cells are then modified to express a gene, engineered in the lab, containing a **chimeric antigen receptor (CAR)**. This cell surface receptor is designed to recognize specific proteins on cancer cells. The modified T cells are then amplified and transfused back into the patient. The goal is to develop T cells that attack specific types of cancer cells and destroy them. The CAR can be customized, but a common CAR is the CD19, which will be described below. CAR T therapy has been FDA-approved for some leukemias and lymphomas since 2017, and there has been some success in treating these types of cancers. While it is most often considered a cancer treatment, this type of therapy may also have potential use for treating other conditions, such as autoimmune diseases and viral infections [133].

Kymriah (tisagenlecleucel) [134] is **the first CAR T cell to be approved** (FDA-approved in **2017**), and it is still widely used. It is used to treat **pediatric ALL and adult DLBCL (diffuse large B-cell lymphoma)**. **Yescarta (axicabtagene ciloleucel)** [135] is also used to treat **DLBCL** as well as other **B-cell lymphomas**. For these two drugs, the T cells are engineered to express the antigen **CD19 (cluster of differentiation 19)**, which is found on early-stage and mature B cells but not on plasma cells, which are the final differentiated state of B cells. It is used as the target because it is expressed on malignant B cells in B-cell cancers. The gene for CD19 is delivered using a lentiviral vector (for Kymriah) or a retroviral vector (for Yescarta).

Carvykti (ciltacabtagene autoleucel) (FDA-approved in 2022) [136] is used to treat **multiple myeloma**. For this treatment, the antigen is BCMA (B-cell maturation antigen), which is found on plasma cells. BCMA is important in multiple myeloma. BCMA expression is high in multiple myeloma cells. Patients who receive CAR T therapy are often given immunoglobulin replacement therapy after the treatment to help protect their immune systems. This becomes necessary because the B cells, which produce antibodies, are destroyed due to the CAR T therapy. Overall, CAR T is an important new technology, and several other treatments have been approved using this approach [137, 138].

4.2 Dendritic Cell Vaccines for Cancer

Dendritic cells are immune cells that present antigens to T cells and activate a response. In dendritic cell vaccines, dendritic cells are isolated from the patient. These isolated dendritic cells are exposed to the patient's tumor antigen to activate them further. These activated dendritic cells are then reintroduced into the patient to enhance the immune response. The first dendritic cell treatment was FDA-approved in 2010 for the treatment of some types of prostate cancer [139].

4.3 Tumor-Infiltrating Lymphocyte (TIL) Therapy

Tumor-infiltrating lymphocytes (TILs) are lymphocytes that have already infiltrated a tumor and can recognize targets on the tumor. In TIL therapy, TILs are extracted from a patient's tumor, expanded in culture, and administered to the patient again by infusion. The purpose is to promote a stronger response. Sometimes, TILs can also be genetically modified in the lab in an effort to make them function even more effectively. Until recently, this type of therapy was mostly available in clinical trials, but one TIL therapy was approved by the FDA in 2024. This TIL therapy is for the treatment of melanoma [140–142].

4.4 Platelet-Rich Plasma (PRP) Therapy

This type of therapy is used for tissue regeneration and wound healing. In this procedure, the patient's blood is collected, and the plasma that is rich in platelets is isolated. Platelets are harvested and concentrated. Platelets are fragments of cells called megakaryocytes [143]. They secrete various molecules, such as cytokines, that can play a role in tissue repair. These concentrated platelets are transfused back into the patients at a location that has an injury or a location requiring tissue regeneration. The goal is to repair the damage or regenerate tissue. This type of treatment is sometimes used in orthopedics and for treating sports injuries [144].

4.5 Tissue Engineering and Skin Substitutes

In these procedures, cells are used to engineer artificial tissues or organs. An example is skin substitutes for the treatment of wounds or burns. **Apligraf** (FDA-approved since 1998) is a **skin substitute**. This skin substitute was constructed from a combination of living cells (including both stem cells and non-stem cells) and extracellular matrix. It is used in wound healing [145, 146].

FDA-Approved Drugs Discussed in This Chapter

Commonly used brand name	Generic name	First FDA approval year	Main conditions	Main target/ mechanism
Apligraf	*Bioengineered skin substitute*	1998	Chronic wounds and burns	Wound tissue
Carvykti	*Ciltacabtagene autoleucel*	2022	Multiple myeloma	BCMA antigen
Kymriah	*Tisagenlecleucel*	2017	Pediatric ALL, adult DLBCL	CD19 antigen
Yescarta	*Axicabtagene ciloleucel*	2017	DLBCL and other B-cell lymphomas	CD19 antigen

Definitions

Adult stem cell (Somatic stem cell) Stem cells that are present in adult tissues like bone marrow. They have roles in maintaining tissues composed of terminally differentiated cells and in repairing damaged cells and tissues.

Blastocyst An early-stage embryo. It has a hollow, liquid-filled cavity called a blastocoel and an outer layer of cells that comprise the trophoblast.

Bone marrow transplant Transplant of healthy hematopoietic stem cells from a donor into a patient with a disorder such as cancer or another disorder of their hematopoietic cells.

Differentiation The process of a cell becoming a differentiated cell, a cell with specialized functions.

Embryonic stem cell (ESC) Pluripotent stem cells that come from the inner cell mass (ICM) of the blastocyst.

Germ layer One of the primary tissue layers (ectoderm, mesoderm, endoderm) formed during embryonic development.

Hematopoietic stem cell (HSC) Multipotent stem cells that give rise to all blood cells, found in bone marrow.

Hematopoietic and progenitor cells (HSPCS) A broader term for HSCs that also includes more committed progenitor cells.

Induced pluripotent stem cell (iPSC) Stem cells that can be made in the lab from somatic cells that are reprogrammed to behave like stem cells.

Informed consent The process of ensuring that donors of tissues or cells understand the nature and implications of their donation.

Lineage The developmental path a stem cell follows to become a specific cell type.

Mesenchymal stromal cell (MSC) Mesenchymal stromal cells (MSCs) are also sometimes called mesenchymal stem cells. They are capable of differentiation into multiple cell types in vitro. They are promising for regenerative medicine largely because of their ability to secrete growth factors that promote cell survival, growth and tissue repair, and due to antiinflammatory properties.

Multipotent stem cell Multipotent stem cells have the ability to differentiate into multiple types of cells.

Pluripotent stem cells Pluripotent stem cells have the ability to differentiate into all cell types with the exception of extraembryonic tissue.

Regenerative medicine Medical treatments for repairing damaged tissues or organs.

Reprogramming In stem cell science, this can refer to introducing genes into somatic cells so that they function as undifferentiated stem cells.

Self-renewal The ability of a stem cell to divide and produce identical copies of itself over time.

Somatic cell Any of the nonreproductive cells in the body.

Teratoma A type of tumor containing a variety of tissue types.

Totipotent cell A totipotent stem cell can give rise to all cell types in the body including extraembryonic tissues. A fertilized egg is an example.

Unipotent stem cell A unipotent stem cell can differentiate into only one type of cell but still has self-renewal capacity.

Study Questions

Multiple-Choice Questions

1. Which type of cell has the ability to differentiate into all cell types of the body, including placental tissue?
   A. Multipotent stem cell
   B. Pluripotent stem cell
   C. Totipotent stem cell
   D. Unipotent stem cell
2. Which of the following stem cell therapies has received FDA approval for a specific disease treatment?
   A. ESC-derived neurons for Parkinson's
   B. Adult HSCs for leukemia
   C. iPSCs for diabetes
   D. MSCs for arthritis

3. Induced pluripotent stem cells (iPSCs) are reprogrammed from which type of cells?
   A. Embryonic cells
   B. Somatic (adult) cells
   C. Cancer cells
   D. Germline cells
4. The type of stem cells that are derived from the developing brain of a fetus are called:
   A. Embryonic stem cells
   B. Fetal stem cells
   C. Hematopoietic stem cells
   D. Parthenogenetic stem cells
   E. B or C
5. Which of the following best describes the function of stem cells in adult tissues?
   A. Repair DNA damage
   B. Produce hormones
   C. Maintain and repair the tissue
   D. Generate antibodies
6. Which feature is common to all types of stem cells?
   A. Ability to form placental tissue
   B. Ability to differentiate into all body tissues
   C. Ability to self-renew
   D. Ability to form tumors
7. An example of a totipotent cell is:
   A. A fertilized egg
   B. A blastocyst
   C. An adult stem cell
   D. A and B
8. Which of the following best describes **CAR T-cell therapy** in cancer treatment?
   A. It uses antibodies to block tumor growth signals.
   B. It involves reprogramming a patient's T cells to express a receptor that targets specific cancer cells.
   C. It delivers chemotherapy drugs directly into tumor cells via nanoparticles.
   D. It boosts natural killer (NK) cell activity using cytokines.

True/False Question

9. Isolation of some types of adult stem cells require isolation of cells from a blastocyst.

Fill in the Blank Questions

10. __________ stem cells are capable of giving rise to only one cell type but still have the ability to self-renew.
11. Stem cells that can give rise to all cells of the body but not placenta are called __________.
12. The inner cell mass of a blastocyst is the source of __________ stem cells.

Short Answer Question

13. A leukemia patient receives a bone marrow transplant. What type of stem cell is most relevant to this treatment, and what is its classification?

Matching Question

14. Match the stem cell type to its characteristic

Stem cell type
   A. Totipotent
   B. Pluripotent
   C. Multipotent
   D. Unipotent

Characteristic
   1. Can form only one type of cell
   2. Can form all cell types of the body but not placenta
   3. Can form all cell types, including placenta
   4. Can form several related cell types

Answer Key

Multiple-Choice Questions

1. **Answer**: C
2. **Answer**: B
3. **Answer**: B
4. **Answer**: B
5. **Answer**: C
6. **Answer**: C

7. **Answer**: A
8. **Answer**: B

True/False Question

9. **Answer**: False

Fill in the Blank Questions

10. **Answer**: Unipotent
11. **Answer**: Pluripotent
12. **Answer**: Embryonic

Short Answer Question

13. **Answer**: Hematopoietic stem cells; multipotent stem cells.

Matching Question

14. **Answer**: A-3, B-2, C-4, D-1

References

1. Hoang DM, Pham PT, Bach TQ, Ngo ATL, Nguyen QT, Phan TTK, et al. Stem cell-based therapy for human diseases. Signal Transduct Target Ther. 2022;7(1):272. https://doi.org/10.1038/s41392-022-01134-4.
2. Mousaei Ghasroldasht M, Seok J, Park H-S, Liakath Ali FB, Al-Hendy A. Stem cell therapy: from idea to clinical practice. Int J Mol Sci. 2022;23(5):2850.
3. Brianna, Ling APK, Wong YP. Applying stem cell therapy in intractable diseases: a narrative review of decades of progress and challenges. Stem Cell Investig. 2022;9(September 2022):4.
4. Dittmar T, Zanker KS, editors. Stem cell biology in health and disease. Springer; 2014.
5. Clarke M, Frampton J. Stem cells: biology and application. 1st ed. New York: Garland Science; 2020.
6. Zakrzewski W, Dobrzyński M, Szymonowicz M, Rybak Z. Stem cells: past, present, and future. Stem Cell Res Ther. 2019;10(1):68. https://doi.org/10.1186/s13287-019-1165-5.
7. National Institute of Health. Stem cell basics. Available from: https://stemcells.nih.gov/info/basics/stc-basics.
8. Alberts B, Heald R, Johnson A, Morgan D, Raff M, Roberts K, Walter P, Wilson J, Hunt T. Molecular biology of the cell. W.W. Norton and Company; 2022.
9. Cooper G, Adams K. The cell: a molecular approach. Sinauer Associates; 2022.
10. Rossant J, Tam PPL. Early human embryonic development: blastocyst formation to gastrulation. Dev Cell. 2022;57(2):152–65. https://doi.org/10.1016/j.devcel.2021.12.022.
11. Romito A, Cobellis G. Pluripotent stem cells: current understanding and future directions. Stem Cells Int. 2016;2016:9451492. https://doi.org/10.1155/2016/9451492.
12. Cleveland Clinic. Fetal development. 2024. Available from: https://my.clevelandclinic.org/health/articles/7247-fetal-development-stages-of-growth.
13. Abdulrazzak H, Moschidou D, Jones G, Guillot PV. Biological characteristics of stem cells from foetal, cord blood and extraembryonic tissues. J R Soc Interface. 2010;7(Suppl 6):S689–706. https://doi.org/10.1098/rsif.2010.0347.focus.
14. Mazo IB, Massberg S, von Andrian UH. Hematopoietic stem and progenitor cell trafficking. Trends Immunol. 2011;32(10):493–503. https://doi.org/10.1016/j.it.2011.06.011.
15. Mayo Clinic. Stem cells: what they are and what they do. 2024. Available from: https://www.mayoclinic.org/tests-procedures/bone-marrow-transplant/in-depth/stem-cells/art-20048117#:~:text=Embryonic%20stem%20cells.&text=These%20are%20pluripotent%20(ploo%2DRIP,repair%20diseased%20tissue%20and%20organs.
16. Burgess R. Stem cells: a short course. 1st ed. Hoboken: Wiley; 2016.
17. Danaher Life Sciences. Techniques in stem cell isolation. Available from: https://lifesciences.danaher.com/us/en/library/stem-cell-isolation-techniques.html#:~:text=Microdissection%20involves%20the%20use%20of,the%20embryo%20under%20a%20microscope.
18. Khan FA, Almohazey D, Alomari M, Almofty SA. Isolation, culture, and functional characterization of human embryonic stem cells: current trends and challenges. Stem Cells Int. 2018;2018:1429351. https://doi.org/10.1155/2018/1429351.
19. Ishii T, Eto K. Fetal stem cell transplantation: past, present, and future. World J Stem Cells. 2014;6(4):404–20. https://doi.org/10.4252/wjsc.v6.i4.404.
20. O'Donoghue K, Fisk NM. Fetal stem cells. Best Pract Res Clin Obstet Gynaecol. 2004;18(6):853–75. https://doi.org/10.1016/j.bpobgyn.2004.06.010.
21. Alatyyat SM, Alasmari HM, Aleid OA, Abdelmaksoud MS, Elsherbiny N. Umbilical cord stem cells: background, processing and applications. Tissue Cell. 2020;65:101351. https://doi.org/10.1016/j.tice.2020.101351.

22. Cells4Life. Umbilical cord tissue stem cells – more valuable than you think. 2016. Available from: https://cells4life.com/2016/08/umbilical-cord-tissue-stem-cells-more-valuable-than-you-think/.
23. Medical News Today. All you need to know about bone marrow. 2021. Available from: https://www.medicalnewstoday.com/articles/285666#transplants.
24. Li J, Wu Z, Zhao L, Liu Y, Su Y, Gong X, et al. The heterogeneity of mesenchymal stem cells: an important issue to be addressed in cell therapy. Stem Cell Res Ther. 2023;14(1):381. https://doi.org/10.1186/s13287-023-03587-y.
25. Murphy MB, Moncivais K, Caplan AI. Mesenchymal stem cells: environmentally responsive therapeutics for regenerative medicine. Exp Mol Med. 2013;45(11):e54. https://doi.org/10.1038/emm.2013.94
26. Caplan AI. Mesenchymal stem cells: time to change the name!. Stem Cells Transl Med. 2017;6(6):1445–1451. https://doi.org/10.1002/sctm.17-0051
27. Wakao S, Kushida Y, Dezawa M. Basic characteristics of muse cells. In: Dezawa M, editor. Muse cells: endogenous reparative pluripotent stem cells. Tokyo: Springer; 2018. p. 13–41.
28. Tan Y, Chang H. Hair follicle stem cells: the signaling hub of the skin. J Cell Immunol. 2024;6(1):7014.
29. Chen C-L, Huang W-Y, Wang EHC, Tai K-Y, Lin S-J. Functional complexity of hair follicle stem cell niche and therapeutic targeting of niche dysfunction for hair regeneration. J Biomed Sci. 2020;27(1):43. https://doi.org/10.1186/s12929-020-0624-8.
30. Brown SA, Levi B, Lequex C, Wong VW, Mojallal A, Longaker MT. Basic science review on adipose tissue for clinicians. Plast Reconstr Surg. 2010;126(6):1936–46. https://doi.org/10.1097/PRS.0b013e3181f44790.
31. Daley GQ. The promise and perils of stem cell therapeutics. Cell Stem Cell. 2012;10(6):740–9. https://doi.org/10.1016/j.stem.2012.05.010.
32. Tobita M, Orbay H, Mizuno H. Adipose-derived stem cells: current findings and future perspectives. Discov Med. 2011;11(57):160–70.
33. Matoba S, Zhang Y. Somatic cell nuclear transfer reprogramming: mechanisms and applications. Cell Stem Cell. 2018;23(4):471–85. https://doi.org/10.1016/j.stem.2018.06.018.
34. Wilmut I, Schnieke AE, McWhir J, Kind AJ, Campbell KH. Viable offspring derived from fetal and adult mammalian cells. Nature. 1997;385(6619):810–3. https://doi.org/10.1038/385810a0.
35. Loi P, Czernik M, Zacchini F, Iuso D, Scapolo PA, Ptak G. Sheep: the first large animal model in nuclear transfer research. Cell Reprogram. 2013;15(5):367–73. https://doi.org/10.1089/cell.2013.0032.
36. Hong S. The Hwang scandal that "shook the world of science". East Asian Sci Technol Soc. 2008;2(1):1–7. https://doi.org/10.1215/s12280-008-9041-x.
37. Tachibana M, Amato P, Sparman M, Gutierrez NM, Tippner-Hedges R, Ma H, et al. Human embryonic stem cells derived by somatic cell nuclear transfer. Cell. 2013;153(6):1228–38. https://doi.org/10.1016/j.cell.2013.05.006.
38. Liu Z, Cai Y, Wang Y, Nie Y, Zhang C, Xu Y, et al. Cloning of macaque monkeys by somatic cell nuclear transfer. Cell. 2018;172(4):881–7.e7. https://doi.org/10.1016/j.cell.2018.01.020.
39. Chung YG, Eum JH, Lee JE, Shim SH, Sepilian V, Hong SW, et al. Human somatic cell nuclear transfer using adult cells. Cell Stem Cell. 2014;14(6):777–80. https://doi.org/10.1016/j.stem.2014.03.015.
40. Kalmbach K, Robinson LG Jr, Wang F, Liu L, Keefe D. Telomere length reprogramming in embryos and stem cells. Biomed Res Int. 2014;2014:925121. https://doi.org/10.1155/2014/925121.
41. Wakayama T, Shinkai Y, Tamashiro KLK, Niida H, Blanchard DC, Blanchard RJ, et al. Cloning of mice to six generations. Nature. 2000;407(6802):318–9. https://doi.org/10.1038/35030301.
42. Volarevic V, Markovic BS, Gazdic M, Volarevic A, Jovicic N, Arsenijevic N, et al. Ethical and safety issues of stem cell-based therapy. Int J Med Sci. 2018;15(1):36–45. https://doi.org/10.7150/ijms.21666.
43. National Human Genome Research Institute. Cloning fact sheet. 2020. Available from: https://www.genome.gov/about-genomics/fact-sheets/Cloning-Fact-Sheet#:~:text=Researchers%20have%20observed%20some%20adverse%20health%20effects,aging%20and%20problems%20with%20the%20immune%20system.
44. Gouveia C, Huyser C, Egli D, Pepper MS. Lessons learned from somatic cell nuclear transfer. Int J Mol Sci. 2020;21(7):2314. https://doi.org/10.3390/ijms21072314.
45. The Nobel Prize. Press release: the Nobel Prize in physiology or medicine: 2012. 2012. Available from: https://www.nobelprize.org/prizes/medicine/2012/press-release/.
46. Lensch MW, Mummery CL. From stealing fire to cellular reprogramming: a scientific history leading to the 2012 Nobel prize. Stem Cell Rep. 2013;1(1):5–17. https://doi.org/10.1016/j.stemcr.2013.05.001.
47. Gurdon JB. The developmental capacity of nuclei taken from intestinal epithelium cells of feeding tadpoles. J Embryol Exp Morphol. 1962;10:622–40.
48. Geoghegan E, Byrnes L. Mouse induced pluripotent stem cells. Int J Dev Biol. 2008;52(8):1015–22. https://doi.org/10.1387/ijdb.082640eg.
49. Lewitzky M, Yamanaka S. Reprogramming somatic cells towards pluripotency by defined factors. Curr Opin Biotechnol. 2007;18(5):467–73. https://doi.org/10.1016/j.copbio.2007.09.007.
50. Han JW, Yoon YS. Induced pluripotent stem cells: emerging techniques for nuclear reprogramming. Antioxid Redox Signal. 2011;15(7):1799–820. https://doi.org/10.1089/ars.2010.3814.
51. Wang L, Su Y, Huang C, Yin Y, Chu A, Knupp A, et al. NANOG and LIN28 dramatically improve human cell reprogramming by modulating LIN41

and canonical WNT activities. Biol Open. 2019;8 (12):bio047225. https://doi.org/10.1242/bio.047225.
52. Kobold S, Bultjer N, Stacey G, Mueller SC, Kurtz A, Mah N. History and current status of clinical studies using human pluripotent stem cells. Stem Cell Rep. 2023;18(8):1592–8. https://doi.org/10.1016/j.stemcr.2023.03.005.
53. Rowe RG, Daley GQ. Induced pluripotent stem cells in disease modelling and drug discovery. Nat Rev Genet. 2019;20(7):377–88. https://doi.org/10.1038/s41576-019-0100-z.
54. Soto-Gutierrez A, Tafaleng E, Kelly V, Roy-Chowdhury J, Fox IJ. Modeling and therapy of human liver diseases using induced pluripotent stem cells: how far have we come? Hepatology. 2011;53 (2):708–11. https://doi.org/10.1002/hep.24143.
55. Brevini TA, Gandolfi F. Parthenotes as a source of embryonic stem cells. Cell Prolif. 2008;41(Suppl 1): 20–30. https://doi.org/10.1111/j.1365-2184.2008.00485.x.
56. International Stem Cell Corporation. Promising source for cell-based therapy. Parthenogenesis. Available from: https://www.internationalstemcell.com/science.
57. Shoemaker LD, Kornblum HI. Neural stem cells (NSCs) and proteomics. Mol Cell Proteomics. 2016;15(2):344–54. https://doi.org/10.1074/mcp.O115.052704.
58. Fu L, Zhu L, Huang Y, Lee TD, Forman SJ, Shih CC. Derivation of neural stem cells from mesenchymal stemcells: evidence for a bipotential stem cell population. Stem Cells Dev. 2008;17(6):1109–21. https://doi.org/10.1089/scd.2008.0068.
59. Maldonado-Soto AR, Oakley DH, Wichterle H, Stein J, Doetsch FK, Henderson CE. Stem cells in the nervous system. Am J Phys Med Rehabil. 2014;93(11 Suppl 3):S132–44. https://doi.org/10.1097/phm.0000000000000111.
60. Mummery CL, van de Stolpe A, Roelen B, Clevers H. Chapter 7 – Regenerative medicine: clinical applications of stem cells. In: Mummery CL, van de Stolpe A, Roelen B, Clevers H, editors. Stem Cells (Third Edition). Boston: Academic; 2021. p. 159–220.
61. Yu H, Commander CW, Stavas JM. Stem cell-based therapies: what interventional radiologists need to know. Semin Intervent Radiol. 2021;38(5):523–34. https://doi.org/10.1055/s-0041-1736657.
62. Schwartz ER. Ethics of modern stem cell research and therapy. Cham: Springer Nature; 2024.
63. U.S. Food and Drug Administration. Framework for the regulation of regenerative medicine products [Internet]. Silver Spring (MD): U.S. Food and Drug Administration; 2019 May 21 [cited 2026 May 25]. Available from: https://www.fda.gov/vaccines-blood-biologics/cellular-gene-therapy-products/framework-regulation-regenerative-medicine-products
64. Pereira LV. Stem cells: promise and reality. Singapore: World Scientific Publishing Co.; 2016.
65. Lezmi E, Benvenisty N. The tumorigenic potential of human pluripotent stem cells. Stem Cells Transl Med. 2022;11(8):791–6. https://doi.org/10.1093/stcltm/szac039.
66. Zhange WY, de Almeia PE, Wu JC. Teratoma formation: a tool for monitoring pluripotency in stem cell research. StemBook: Harvard Stem Cell Institute; 2012.
67. Link H. Current state and future opportunities in granulocyte colony-stimulating factor (G-CSF). Support Care Cancer. 2022;30(9):7067–77. https://doi.org/10.1007/s00520-022-07103-5.
68. McLennan A, Bates A, Elcock L, Turner P. BIOS instant notes in molecular biology. 4th ed. New York, NY: Garland Science; 2012.
69. Kumar A, Yun H, Funderburgh ML, Du Y. Regenerative therapy for the Cornea. Prog Retin Eye Res. 2022;87:101011. https://doi.org/10.1016/j.preteyeres.2021.101011.
70. Nervi B, Link DC, DiPersio JF. Cytokines and hematopoietic stem cell mobilization. J Cell Biochem. 2006;99(3):690–705. https://doi.org/10.1002/jcb.21043.
71. Juric MK, Ghimire S, Ogonek J, Weissinger EM, Holler E, van Rood JJ, et al. Milestones of hematopoietic stem cell transplantation – from first human studies to current developments. Front Immunol. 2016;7:470. https://doi.org/10.3389/fimmu.2016.00470.
72. Alexander GE. Biology of Parkinson's disease: pathogenesis and pathophysiology of a multisystem neurodegenerative disorder. Dialogues Clin Neurosci. 2004;6(3):259–80. https://doi.org/10.31887/DCNS.2004.6.3/galexander.
73. Poewe W, Seppi K, Tanner CM, Halliday GM, Brundin P, Volkmann J, et al. Parkinson disease. Nat Rev Dis Primers. 2017;3:17013. https://doi.org/10.1038/nrdp.2017.13.
74. Rodriguez-Pallares J, Garcia-Garrote M, Parga JA, Labandeira-Garcia JL. Dose-dependent effect of mesenchymal stromal cells co-grafted with dopaminergic neurons in a Parkinson's disease rat model. J Cell Mol Med. 2021;25(20):9884–9. https://doi.org/10.1111/jcmm.16900.
75. Hamedi H, Ghorbanian SH, Mirzaeian L, Abrari K, Mozdziak P, Ghorbanian MT. Intravenous transplantation of adipose-derived mesenchymal stem cells promoted the production of dopaminergic neurons and improved spatial memory in a rat model of Parkinson's disease. Cell J. 2023;25(5):317–26. https://doi.org/10.22074/cellj.2023.1972266.1161.
76. Parkinson's UK. Stem cell therapy for Parkinson's: Are we getting close? 2024. Available from: https://www.parkinsons.org.uk/research/blog/trials-treatments/stem-cell-therapy-parkinsons-are-we-getting-closer

77. Blue Rock Therapeutics. BlueRock Therapeutics phase I clinical trial for Parkinson's disease continues to show positive trends at 18 months 2024.
78. Tao Y, Vermilyea SC, Zammit M, Lu J, Olsen M, Metzger JM, et al. Autologous transplant therapy alleviates motor and depressive behaviors in parkinsonian monkeys. Nat Med. 2021;27(4):632–9. https://doi.org/10.1038/s41591-021-01257-1.
79. Healey N. Stem cell therapies advance in Parkinson's disease and beyond. Nat Med. 2025; https://doi.org/10.1038/d41591-025-00036-6.
80. Okano H. Clinical trials test the safety of stem-cell therapy for Parkinson's disease. Nature. 2025;641(8064):853–4. https://doi.org/10.1038/d41586-025-00688-x.
81. Brigham MG. Clinical trial launches for novel stem cell treatment for Parkinson's disease. 2024. Available from: https://www.massgeneralbrigham.org/en/about/newsroom/press-releases/clinical-trial-novel-stem-cell-treatment-for-parkinsons.
82. Stoker TB. Stem cell treatments for Parkinson's disease. In: Stoker TB, Greenland JC, editors. Parkinson's disease: pathogenesis and clinical aspects. Brisbane: Codon Publications; 2018.
83. Fan Y, Winanto NSY. Replacing what's lost: a new era of stem cell therapy for Parkinson's disease. Transl Neurodegener. 2020;9:2. https://doi.org/10.1186/s40035-019-0180-x.
84. National Institute of Aging. What happens to the brain in Alzheimer's disease. 2024. Available from: https://www.nia.nih.gov/health/alzheimers-causes-and-risk-factors/what-happens-brain-alzheimers-disease.
85. SciTechDaily. Amyloid plaques may not cause Alzheimer's: new research challenges decades-old theories. 2024. Available from: https://scitechdaily.com/amyloid-plaques-may-not-cause-alzheimers-new-research-challenges-decades-old-theories/.
86. Espay AJ, Herrup K, Kepp KP, Daly T. The proteinopenia hypothesis: loss of Aβ42 and the onset of Alzheimer's disease. Ageing Res Rev. 2023;92:102112. https://doi.org/10.1016/j.arr.2023.102112.
87. Duncan T, Valenzuela M. Alzheimer's disease, dementia, and stem cell therapy. Stem Cell Res Ther. 2017;8(1):111. https://doi.org/10.1186/s13287-017-0567-5.
88. Bhatti JS, Khullar N, Mishra J, Kaur S, Sehrawat A, Sharma E, et al. Stem cells in the treatment of Alzheimer's disease—promises and pitfalls. Biochim Biophys Acta (BBA) – Mol Basis Dis. 2023;1869(6):166712. https://doi.org/10.1016/j.bbadis.2023.166712.
89. Freude K, Pires C, Hyttel P, Hall VJ. Induced pluripotent stem cells derived from Alzheimer's disease patients: the promise, the Hope and the path ahead. J Clin Med. 2014;3(4):1402–36. https://doi.org/10.3390/jcm3041402.
90. Mishra P, Silva A, Sharma J, Nguyen J, Pizzo DP, Hinz D, et al. Rescue of Alzheimer's disease phenotype in a mouse model by transplantation of wild-type hematopoietic stem and progenitor cells. Cell Rep. 2023;42(8):112956. https://doi.org/10.1016/j.celrep.2023.112956.
91. UT Health Houston. Stem cell trial for early Alzheimer's disease begins at UTHealth Houston: The University of Texas Health Science Center at Houston (UTHealth Houston); 2025. Available from: https://www.uth.edu/news/story/stem-cell-trial-for-early-alzheimers-disease-begins-at-uthealth-houston#:~:text=A%20stem%20cell%20therapy%20trial,the%20brain%2C%20which%20cause%20dementia.
92. Stansfield N. Trial for regeneration biomedical's Alzheimer disease stem cell therapy doses first patient: CGTLive (Clinical & gene therapy live); 2024. Available from: https://www.cgtlive.com/view/trial-regeneration-biomedical-alzheimer-disease-stem-cell-therapy-doses-first-patient?utm_source.
93. Rash BG, Ramdas KN, Agafonova N, Naioti E, McClain-Moss L, Zainul Z, et al. Allogeneic mesenchymal stem cell therapy with laromestrocel in mild Alzheimer's disease: a randomized controlled phase 2a trial. Nat Med. 2025;31(4):1257–66. https://doi.org/10.1038/s41591-025-03559-0.
94. Genc B, Bozan HR, Genc S, Genc K, editors. Stem cell therapy for multiple sclerosis 2019; Cham: Springer.
95. National Institute of Neurological Disorders and Stroke. Multiple sclerosis 2024. Available from: https://www.ninds.nih.gov/health-information/disorders/multiple-sclerosis#:~:text=Multiple%20sclerosis%20(MS)%20is%20a,have%20a%20normal%20life%20expectancy.
96. HSCT Stops MS. HSCT for Multiple Sclerosis. Available from: https://www.hsctstopsms.com.
97. Burt RK, Balabanov R, Burman J, Sharrack B, Snowden JA, Oliveira MC, et al. Effect of Nonmyeloablative hematopoietic stem cell transplantation vs continued disease-modifying therapy on disease progression in patients with relapsing-remitting multiple sclerosis: a randomized clinical trial. JAMA. 2019;321(2):165–74. https://doi.org/10.1001/jama.2018.18743.
98. Northwestern Medicine. Stem cell transplants improve on current MS treatments 2019. Available from: https://news.feinberg.northwestern.edu/2019/01/22/stem-cell-transplants-improve-on-current-ms-treatments/.
99. Miller AE, Chitnis T, Cohen BA, Costello K, Sicotte NL, Stacom R. Autologous hematopoietic stem cell transplant in multiple sclerosis: recommendations of the National Multiple Sclerosis Society. JAMA Neurol. 2021;78(2):241–6. https://doi.org/10.1001/jamaneurol.2020.4025.

100. Muraro PA, Mariottini A, Greco R, Burman J, Iacobaeus E, Inglese M, et al. Autologous haematopoietic stem cell transplantation for treatment of multiple sclerosis and neuromyelitis optica spectrum disorder — recommendations from ECTRIMS and the EBMT. Nat Rev Neurol. 2025;21(3):140–58. https://doi.org/10.1038/s41582-024-01050-x.
101. Mareschi K, Novara M, Rustichelli D, Ferrero I, Guido D, Carbone E, et al. Neural differentiation of human mesenchymal stem cells: evidence for expression of neural markers and eag K+ channel types. Exp Hematol. 2006;34(11):1563–72. https://doi.org/10.1016/j.exphem.2006.06.020.
102. Jiang W, Xu J. Immune modulation by mesenchymal stem cells. Cell Prolif. 2020;53(1):e12712. https://doi.org/10.1111/cpr.12712.
103. Alanazi A, Alassiri M, Jawdat D, Almalik Y. Mesenchymal stem cell therapy: a review of clinical trials for multiple sclerosis. Regen Ther. 2022;21: 201–9. https://doi.org/10.1016/j.reth.2022.07.003.
104. Musiał-Wysocka A, Kot M, Majka M. The pros and cons of mesenchymal stem cell-based therapies. Cell Transplant. 2019;28(7):801–12. https://doi.org/10.1177/0963689719837897.
105. Bell E. Immunosuppression by mesenchymal stem cells. Nat Rev Immunol. 2008;8(3):165. https://doi.org/10.1038/nri2284.
106. Yagi H, Soto-Gutierrez A, Parekkadan B, Kitagawa Y, Tompkins RG, Kobayashi N, et al. Mesenchymal stem cells: mechanisms of immunomodulation and homing. Cell Transplant. 2010;19(6):667–79. https://doi.org/10.3727/096368910x508762.
107. Albert K, Niskanen J, Kälvälä S, Lehtonen Š. Utilising induced pluripotent stem cells in neurodegenerative disease research: focus on glia. Int J Mol Sci. 2021;22(9):4334. https://doi.org/10.3390/ijms22094334.
108. National Library of Medicine. ClinicalTrials.Gov. Autologous transplantation of induced pluripotent stem cell-derived retinal pigment epithelium for geographic atrophy associated with age-related macular degeneration. 2024. Available from: https://clinicaltrials.gov/study/NCT04339764.
109. Gupta S, Lytvynchuk L, Ardan T, Studenovska H, Sharma R, Faura G, et al. Progress in stem cells-based replacement therapy for retinal pigment epithelium: in vitro differentiation to in vivo delivery. Stem Cells Transl Med. 2023;12(8):536–52. https://doi.org/10.1093/stcltm/szad039.
110. Mass General Brigham. Mass Eye and Ear. Cell therapy that repairs cornea damage with patient's own stem cells achieves positive phase I trial results. 2023. Available from: https://masseyeandear.org/news/press-releases/2023/08/cell-therapy-that-repairs-cornea-damage-with-patients-own-stem-cells#:~:text=People%20who%20experience%20chemical%20burns,or%20showed%20any%20clinical%20benefit.
111. Zhou H, Wang L, Zhang C, Hu J, Chen J, Du W, et al. Feasibility of repairing full-thickness skin defects by iPSC-derived epithelial stem cells seeded on a human acellular amniotic membrane. Stem Cell Res Ther. 2019;10(1):155. https://doi.org/10.1186/s13287-019-1234-9.
112. Mummery A, Roelen B, Clevers H. Stem cells: scientific facts and fiction. 3rd ed. Academic; 2021.
113. Pawani H, Bhartiya D. Pluripotent stem cells for cardiac regeneration: overview of recent advances & emerging trends. Indian J Med Res. 2013;137(2): 270–82.
114. Woodham AW, Skeate JG, Sanna AM, Taylor JR, Da Silva DM, Cannon PM, et al. Human immunodeficiency virus immune cell receptors, coreceptors, and cofactors: implications for prevention and treatment. AIDS Patient Care STDs. 2016;30(7):291–306. https://doi.org/10.1089/apc.2016.0100.
115. Hütter G, Nowak D, Mossner M, Ganepola S, Müssig A, Allers K, et al. Long-term control of HIV by CCR5 Delta32/Delta32 stem-cell transplantation. N Engl J Med. 2009;360(7):692–8. https://doi.org/10.1056/NEJMoa0802905.
116. Gupta RK, Abdul-Jawad S, McCoy LE, Mok HP, Peppa D, Salgado M, et al. HIV-1 remission following CCR5Δ32/Δ32 haematopoietic stem-cell transplantation. Nature. 2019;568(7751):244–8. https://doi.org/10.1038/s41586-019-1027-4.
117. Hsu J, Van Besien K, Glesby MJ, Pahwa S, Coletti A, Warshaw MG, et al. HIV-1 remission and possible cure in a woman after haplo-cord blood transplant. Cell. 2023;186(6):1115–26.e8. https://doi.org/10.1016/j.cell.2023.02.030.
118. Kiem HP, Jerome KR, Deeks SG, McCune JM. Hematopoietic-stem-cell-based gene therapy for HIV disease. Cell Stem Cell. 2012;10(2):137–47. https://doi.org/10.1016/j.stem.2011.12.015.
119. Cardiology DDaI. DREAM-HF phase 3 trial results for Mesoblast cell therapy in heart failure published in J Am Coll Cardiol. 2023. Available from: https://www.dicardiology.com/content/dream-hf-phase-3-trial-results-mesoblast-cell-therapy-heart-failure-published-journal.
120. University of Louisville School of Medicine. UofL cardiologist leading clinical trial for high-potential new therapy for heart failure: University of Louisville School of Medicine; 2024. Available from: https://louisville.edu/medicine/news/uofl-cardiologist-leading-clinical-trial-for-high-potential-new-therapy-for-heart-failure.
121. Zhu C, Wu W, Qu X. Mesenchymal stem cells in osteoarthritis therapy: a review. Am J Transl Res. 2021;13(2):448–61.
122. Adelipour M, Lubman DM, Kim J. Potential applications of mesenchymal stem cells and their derived exosomes in regenerative medicine. Expert Opin Biol Ther. 2023;23(6):491–507. https://doi.org/10.1080/14712598.2023.2211203.
123. Carneiro DC, Araújo LT, Santos GC, Damasceno PKF, Vieira JL, Santos RRD, et al. Clinical trials with mesenchymal stem cell therapies for osteoarthritis: challenges in the regeneration of articular

cartilage. Int J Mol Sci. 2023;24(12):9939. https://doi.org/10.3390/ijms24129939.
124. Wu KC, Chang YH, Ding DC, Lin SZ. Mesenchymal stromal cells for aging cartilage regeneration: a review. Int J Mol Sci. 2024;25(23):12911. https://doi.org/10.3390/ijms252312911.
125. Mount Sinai. Mount Sinai Leading the Way in Innovative Stem Cell Therapy for COVID-19 Patients. 2020. Available from: https://www.mountsinai.org/about/newsroom/2020/mount-sinai-leading-the-way-in-innovative-stem-cell-therapy-for-covid19-patients-pr.
126. Durand N, Mallea J, Zubair AC. Insights into the use of mesenchymal stem cells in COVID-19 mediated acute respiratory failure. npj Regener Med. 2020;5(1):17. https://doi.org/10.1038/s41536-020-00105-z.
127. Whittaker Brown SA, Iancu-Rubin C, Aboelela A, Abrahams A, Burke E, Drummond T, et al. Mesenchymal stromal cell therapy for acute respiratory distress syndrome due to coronavirus disease 2019. Cytotherapy. 2022;24(8):835–40. https://doi.org/10.1016/j.jcyt.2022.03.006.
128. Zanirati G, Provenzi L, Libermann LL, Bizotto SC, Ghilardi IM, Marinowic DR, et al. Stem cell-based therapy for COVID-19 and ARDS: a systematic review. Npj Regen Med. 2021;6(1):73. https://doi.org/10.1038/s41536-021-00181-9.
129. Petrus-Reurer S, Romano M, Howlett S, Jones JL, Lombardi G, Saeb-Parsy K. Immunological considerations and challenges for regenerative cellular therapies. Commun Biol. 2021;4(1):798. https://doi.org/10.1038/s42003-021-02237-4.
130. Li C, Zhao H, Cheng L, Wang B. Allogeneic vs. autologous mesenchymal stem/stromal cells in their medication practice. Cell Biosci. 2021;11(1):187. https://doi.org/10.1186/s13578-021-00698-y.
131. Chang SH, Park CG. Comparing the benefits and drawbacks of stem cell therapy based on the cell origin or manipulation process: addressing immunogenicity. Immune Netw. 2023;23(6):e44. https://doi.org/10.4110/in.2023.23.e44.
132. Prazerekos P. Immunotherapy as a cancer treatment option in the future. Health Sci J. 2023;17(6):1035.
133. National Cancer Institute. FDA approval brings first gene therapy to fight aggressive lymphoma [Internet]. Bethesda (MD): National Cancer Institute; 2017 Oct 18 [cited 2026 May 25]. Available from: https://www.cancer.gov/news-events/cancer-currents-blog/2017/yescarta-fda-lymphoma?utm
134. FDA. Kymriah(Tisagenlecleucel). 2017. Available from: https://www.fda.gov/vaccines-blood-biologics/cellular-gene-therapy-products/kymriah?utm.
135. FDA. Yescarta(Axicabtagene ciloleucel). 2017. Available from: https://www.fda.gov/vaccines-blood-biologics/cellular-gene-therapy-products/yescarta?
136. FDA. Carvykti(Ciltacabtagene autoleucel) 2022. Available from: https://www.fda.gov/vaccines-blood-biologics/cellular-gene-therapy-products/carvykti?
137. Johnson and Johnson. CARVYKTI® is the First and Only BCMA-Targeted Treatment Approved by the U.S. FDA for Patients with Relapsed or Refractory Multiple Myeloma Who Have Received at Least One Prior Line of Therapy 2024. Available from: https://www.jnj.com/media-center/press-releases/carvykti-is-the-first-and-only-bcma-targeted-treatment-approved-by-the-u-s-fda-for-patients-with-relapsed-or-refractory-multiple-myeloma-who-have-received-at-least-one-prior-line-of-therapy.
138. Chekol Abebe E, Yibeltal Shiferaw M, Tadele Admasu F, Asmamaw Dejenie T. Ciltacabtagene autoleucel: the second anti-BCMA CAR T-cell therapeutic armamentarium of relapsed or refractory multiple myeloma. Front Immunol. 2022;13:991092. https://doi.org/10.3389/fimmu.2022.991092.
139. Nava S, Lisini D, Frigerio S, Bersano A. Dendritic cells and cancer immunotherapy: the adjuvant effect. Int J Mol Sci. 2021;22(22):12339. https://doi.org/10.3390/ijms222212339.
140. UT MD Anderson. TIL therapy: 6 things to know. 2021. Available from: https://www.mdanderson.org/cancerwise/what-is-tumor-infiltrating-lymphocyte-til-therapy%2D%2D6-things-to-know.h00-159460056.html#:~:text=Lymphocytes%2C%20made%20up%20of%20T,may%20recognize%20many%20unique%20targets.
141. National Cancer Institute. First cancer TIL therapy gets FDA approval for advanced melanoma. 2024. Available from: https://www.cancer.gov/news-events/cancer-currents-blog/2024/fda-amtagvi-til-therapy-melanoma#:~:text=Enlarge,of%20cancer%2Dfighting%20immune%20cells.
142. Kazemi MH, Sadri M, Najafi A, Rahimi A, Baghernejadan Z, Khorramdelazad H, et al. Tumor-infiltrating lymphocytes for treatment of solid tumors: it takes two to tango? Front Immunol. 2022;13:1018962. https://doi.org/10.3389/fimmu.2022.1018962.
143. Cleveland Clinic. Platelets. 2024. Available from: https://my.clevelandclinic.org/health/body/22879-platelets.
144. Everts P, Onishi K, Jayaram P, Lana JF, Mautner K. Platelet-rich plasma: new performance understandings and therapeutic considerations in 2020. Int J Mol Sci. 2020;21(20):7794. https://doi.org/10.3390/ijms21207794.
145. Eaglstein WH, Falanga V. Tissue engineering and the development of Apligraf, a human skin equivalent. Cutis. 1998;62(1 Suppl):1–8.
146. Organogeneseis. Unmatched. The gold standard proven to accelerate healing. 2024. Available from: https://www.apligraf.com/.

Introduction to Omics Technologies for Research, Diagnosis, and Drug Discovery

27

Abstract

The field of "omics" involves the study of different molecules, such as DNA, RNA, or proteins, that make up an individual, a tissue, or a cell type. Unlike some approaches that focus on individual or small groups of molecules, the field of omics analyzes the molecules on a large scale, for example, scientists may study all the genes expressed in a given cell type, rather than focusing on a few individual genes. Likewise, one might study all the proteins present in a cell, rather than focusing on studying a select few. The fields of omics encompass several subfields, and this chapter provides an overview of three fields: genomics, transcriptomics, and proteomics. Genomics is the study of the complete set of DNA (the genome) in the cell, tissue, or organism. It provides an overall picture of all the DNA in the genome, including mutations, deletions, and other structural variations. Transcriptomics is the study of the transcriptome, the full set of RNA transcripts in a given cell type or tissue. Studying transcriptomics can tell you which genes are expressed in a particular tissue or condition. Proteomics is the study of all the proteins in a cell or tissue. Different cells or tissues have distinct combinations of proteins, and proteomics can reveal how the set of proteins differs across situations. In addition to an overview of the basic mechanisms underlying these fields, this chapter discusses how the omics fields are contributing to modern medicine, including disease diagnosis and treatment.

Keywords

Genomics · Genetics · DNA sequencing · RNA sequencing · Proteomics

Abbreviations

cDNA	Complementary DNA
ENCODE	Encyclopedia of DNA Elements
ESI	Electrospray ionization
m/z	Mass-to-charge ratio
MS	Mass spectrometry
MSK-IMPACT	Memorial Sloan Kettering-Integrated Mutation Profiling of Actionable Cancer Targets
NGS	Next-generation sequencing
RNA-Seq	RNA sequencing
TCGA	The Cancer Genome Atlas

What You Will Learn

- An introduction to the basic concepts of omics studies, including genomics, transcriptomics, and proteomics
- Some history of genetics and genome science
- Some history of sequencing technologies, leading up to current technologies including next-generation sequencing technologies

A. Minden, *Biochemistry, Cell and Molecular Biology for Pharmacists and Health Professionals*,
https://doi.org/10.1007/978-3-032-20529-2_27

- The basic steps of Illumina-based next-generation sequencing
- RNA sequencing and transcriptomics to study differential expression
- Introduction to proteomics
- Use of omics techniques for personalized medical treatment, current treatments, present, and future

1 Introduction to Omics Studies

1.1 Overview of Omics Studies

The field of "omics" involves the study of different molecules, such as DNA, RNA, or proteins, that make up an individual, a tissue, or a cell type. Unlike some approaches that focus on individual or small groups of molecules, the field of omics analyzes the molecules on a large scale, for example, scientists may study all the genes that are expressed in a certain cell type, rather than focusing on a few individual genes. Likewise, one might study all the proteins present in a cell, rather than focusing on studying a select few. This type of study is considered **high-throughput** analysis. Omics studies involve isolating all molecules of interest (such as DNA, RNA, proteins, or other molecules) from cells or tissues, processing them, sequencing or analyzing them, and then using computational tools to sort, store, and analyze the data. This process helps to identify patterns and enables scientists to study how interactions between large groups of molecules may impact phenotype [1, 2].

Omics encompasses several subfields; each studies a particular set of biomolecules. Some major examples are as follows: **Genomics** is the study of the complete set of DNA (the genome) in the cell, tissue, or organism. It gives an overall picture of all the DNA in the genome, including any mutations, deletions, or other structural variations. **Transcriptomics** is the study of the transcriptome, which is the full set of RNA transcripts within a certain cell type or tissue. Studying transcriptomics can tell you which genes are expressed in a particular tissue or condition. **Proteomics** is the study of all the proteins in a cell or tissue. Different cells or tissues have a different combination of proteins, and proteomics can reveal how the set of proteins differs in different situations. **Metabolomics** investigates the metabolites present in a cell or organism under a specific condition. These include small molecules like sugars, amino acids, lipids, and nucleotides that result from cell metabolism. Metabolomics studies offer a real-time look into the cell's status and activity because it tells us what metabolic pathways are active and whether they appear to be functioning normally. **Epigenomics** studies the modifications on DNA and histones. **Lipidomics** detects the lipids in a system, and **glycomics** studies the sugars and carbohydrates in a system, especially those attached to proteins [3]. Integration of the different omics, such as combining transcriptomics, proteomics, and metabolomics, is also an emerging field to help scientists understand the large picture of what is taking place inside the cell [4]. Single-cell omics is also a tool that researchers increasingly use to study and analyze molecular data from single cells rather than the "bulk" omics approaches that average results across multiple cells in a tissue or culture [5].

1.2 Introduction to the Chapter

This chapter gives an overview of the first three foundational omics fields: genomics, transcriptomics, and proteomics. The chapters are broken down into six sections as follows: Section 1 provides a general introduction to omics studies and explores the connection between omics and the central dogma of molecular biology. Section 2 offers a brief history of genome science to help readers appreciate how rapidly this field has advanced. Section 3 provides an overview of the field of genomics and transcriptomics, Sect. 4 discusses transcriptomics, and Sect. 5 gives a brief overview of the techniques used in proteomics. Finally, Sect. 6 highlights some of the ways that the omics fields are contributing to modern medicine. Given the breadth of the omics field, this chapter is meant to provide an overview of the field to give learners interested in the topic a starting point for learning more.

1.3 Central Dogma of Molecular Biology

When studying "omics," it is helpful to review the **central dogma** of molecular biology, which describes the directional flow of genetic information (see Chap. 6). According to this dogma, DNA is the genetic material that codes for all the genes of the organism. DNA is transcribed into RNA, and RNA is translated into protein. Once proteins are made, they can be folded, modified, and combined with other proteins to carry out their functions [1, 6].

The complexity of the system increases at each step. The complete set of DNA in an individual is the **genome**, which is essentially identical in nearly every cell in the body, with the exception of some small changes in immune cells, where some genetic rearrangements occur. Studying the complete genome within a cell or organism is called **genomics**.

The transcription of DNA into RNA is called gene expression. Not all regions of the DNA are transcribed into RNA, because some of the DNA contains noncoding sequences. The DNA that is transcribed into RNA is the DNA that is "expressed." Here, things start to differ in different types of cells. Some cells express one set of RNA, and another cell type may express another set of RNA. Liver cells (hepatocytes), for example, may express a set of genes that code for proteins that are important for liver function, such as proteins involved in metabolism and detoxification. In contrast, hematopoietic stem cells may express genes that are important for giving rise to immune cells. Both cell types, however, may express some genes that are common to many cells, such as genes that encode proteins that are important for cell cycle progression. The complete set of RNA in a cell type or tissue is called a **transcriptome** [7]. The transcriptome not only varies from one cell type to another but also can vary at different times, such as at different developmental stages, during different phases of the cell cycle, or in response to different environmental conditions. Studying the transcriptome in a certain cell type, tissue, or condition is called **transcriptomics** [1].

RNA codes for protein. Proteins perform most of the functional tasks in the cell, through their roles as structural proteins, enzymes, signal proteins, transcription factors, and more. Certain RNA molecules also have important biological roles beyond their involvement in gene expression, but proteins are central to most biological processes. The complete set of proteins in a cell or tissue type at any given time is called the **proteome**, and studying the proteome is called **proteomics**. Studying proteomics is complex because not all RNA is translated into protein in every cell, and even within the same cell type, the proteome can show considerable variability. Some proteins are long-lasting in a cell, while others may be present only transiently. Likewise, not all the RNA is translated at the same time, so different proteins will be present at different times or under different conditions. Some proteins may be present when the cell is proliferating, for example, and others are present when the cell is quiescent. Furthermore, even if certain proteins are present in the cell, they are not necessarily active. The functions of some proteins may be regulated by posttranslational modifications, and some proteins may be targeted for destruction when they are no longer needed [1, 6, 8].

2 Some History of the Study of Genetics and Genomics

The fields of genetics and genomics are two related and connected fields, but the terms mean different things. **Genetics** is the study of individual genes and their roles in inheritance. **Genomics** is the study of how all genes in the genome operate collectively and how they interact with each other and the environment. The field of genomics emerged from studying the structure and function of DNA and how DNA encodes

RNA and protein, which are basic concepts in genetics. Before discussing modern-day genomics, this section reviews some history of genetics and early genomics research, which laid down the foundations for the field of modern genomics [9].

2.1 Chronological History of Genetics and Genome Science

The studies of genetics and genomics are still in their relative infancy. The structure of DNA was not even elucidated until 1953. Many of the major advances happened after that time, although significant studies about the mechanisms of inheritance had occurred before that. Here are some of the milestones in the study of genetics and genomics [10, 11]. They can be roughly divided into three overall eras.

The pre-genomic era: This era focused on genetics, heredity, the genetic material, and sequencing individual genes.

1865: (published in 1866) Gregor Mendel used pea plant studies to discover the basic mechanisms of inheritance. He is sometimes called "the father of modern genetics."

1869: Friedrich Miescher discovered "nuclein" in red blood cells, which turned out to contain nucleic acids.

1944: By studying bacteria, Oswald Avery, Colin MacLeod, and Maclyn McCarty found that DNA is the genetic material [12].

1951: Frederick Sanger sequenced the first protein, even before its relationship to DNA and the genetic code was known [13].

1952: Rosalind Franklin captured "photograph 51": an X-ray diffraction image that showed evidence of DNA's helical structure. This was key to discovering the structure of the DNA double helix.

1953: James Watson, Francis Crick, and Rosalind Franklin solved the structure of DNA.

1960s: The genetic code was elucidated. Multiple scientists were involved in this discovery. These include Marshall Nirenberg, Har Gobind Khorana, Severo Ochoa, and Heinrich Matthaei [14, 15].

1961. Francis Jacob and Jacques Monod proposed the operon model for bacteria, which lay the foundation for studying transcription regulation and the eventual study of transcriptomics [16].

1965: Robert Holley discovered the nucleotide sequence for the alanine tRNA (transfer RNA) [17, 18].

1972: Walter Fiers sequenced the gene for a protein in MS2, a bacteriophage that infects bacteria. It was the first sequence of a complete gene. This involved time-consuming techniques, much different from the techniques used later [11, 19].

1977: The Sanger sequencing method was developed by Frederick Sanger. It began to be used by many labs to sequence individual genes or portions of genes. Sanger sequencing became the major method used in labs for many years [20].

1977: The first complete genome was sequenced. This was the genome of the bacteriophage PhiX174, which infects bacteria. Frederick Sanger accomplished this using the Sanger method [20].

1980. Paul Berg, Walter Gilbert, and Frederick Sanger won the Nobel Prize for their work in DNA sequencing and recombinant DNA technology [21].

1982: The GenBank database was established, a public database of sequenced genes.

1983: The first human disease gene is mapped, the gene for Huntington's disease, by Nancy Wexler and colleagues [22].

1985: PCR (polymerase chain reaction) was invented. This allowed rapid DNA amplification, which facilitated analyzing and sequencing DNA.

1989: The cystic fibrosis disease gene was identified. This work was done by Francis Collins, Lap-Chee Tsui, and John Riordan [23, 24].

The genomic Era: The focus of this era was on sequencing whole genomes [25, 26].

1990: The start of the Human Genome Project (HGP). This was an international project funded largely by the US government and the Department of Energy and led by Francis Collins. The goal was to sequence the entire human genome. Around the same time, Craig Venter began a related project with private funding (Celera Genomics) [25–27].

1995: Craig Venter sequenced the first whole genome of a free-living organism, a bacterium called *Haemophilus influenzae*. This was done using Sanger sequencing and the shotgun method, where large pieces of DNA are broken up into smaller pieces and sequenced [28].

1999: The first human chromosome was sequenced (chromosome 22) as part of the HGP [29].

2000: A draft of the human genome was completed as part of the HGP, although more work continued after this. President Bill Clinton announced the completion of the draft sequence along with Francis Collins and Craig Venter [30, 31].

2001. The draft of the sequence of the human genome was published in two landmark papers, one by the Human Genome Project and the other by Celera Genomics [32, 33].

2003: Completion of the human genome sequence as part of the HGP [34].

2005: Elucidation of genes on the X chromosome, as part of the HGP [35].

Post-genomic era: This era is ongoing. It goes beyond sequencing to understand the functions and interactions between genes. Some key accomplishments are below:

2006: The Cancer Genome Atlas (TCGA) was started. The goal is to characterize the genomic, transcriptomic, and epigenetic changes in over 30 types of human cancers. TCGA is now one of the largest publicly available resources for cancer genomics. It has led to the discovery of many cancer-related genes and pathways [36].

2007: The ENCODE Project (Encyclopedia of DNA Elements) was started. The goal is to generate a resource of all functional elements in the human genome. Functional elements include many features such as transcription factor binding sites, promoters, enhancers, and others [37].

2007: The most widely used next-generation sequencing (NGS) technology was developed by Illumina [38].

2012: The CRISPR-Cas9 system was developed for gene editing [39].

2015: The "1000 Genomes Project" was completed. This involved sequencing the genomes of many people from all over the world. The main goal is to study genomic variations in different individuals and populations [40].

2015: The "Precision Medicine Initiative" was started. The goal is to study ways to develop personalized medical treatments based on individual genomes [41].

2016: The Human Cell Atlas was started. The goal is to study all human cell types. The project integrates genomic and transcriptomic and other data in diverse cell types [42].

2018: The "All of Us" Research Program was started. The goal is to collect genomic and other data from one million people in the USA to begin the creation of a health database [43].

2.2 History of Sequencing Technologies

The technologies for sequencing DNA have advanced significantly over the years. These technologies were instrumental in understanding genes and the genome. Below, some of the important sequencing methods are described, and some of these methods are discussed in more details in Sect. 3 of this chapter [11, 37, 44].

2.2.1 First-Generation Sequencing

1977: The Sanger sequencing method was developed by Frederick Sanger. This became the most commonly used sequencing method for many years. It is still sometimes used for sequencing individual genes or short DNA sequences.

1977: The Maxim and Gilbert sequencing method was developed but was never as commonly used as the Sanger method.

1986/87: Sanger DNA sequencing became automated and improved. Leroy Hood and Michael Hunkapiller were instrumental in this process [45].

2.2.2 Second-Generation Sequencing (Usually Called Next-Generation Sequencing (NGS))

This phase of sequencing technology included great advancements in sequencing and allowed scientists to sequence whole genomes much more quickly than previous methods.

1996: A method called **pyrosequencing** was developed and paved the way for the subsequent development of **NGS**. Pyrosequencing uses the **sequencing-by-synthesis method**, where one strand is used as a template, nucleotides are added one by one, and the complementary strand is synthesized. As nucleotides are incorporated, the released pyrophosphate is detected, ultimately leading to sequence information of the DNA strand.

2005: Pyrosequencing became more automated using a system called the 454 system (by Life Sciences). This is often considered to be the real beginning of **NGS**.

2007: Illumina, Inc. generated the Illumina system for NGS. This system became the most widely used NGS platform. Millions to billions of small DNA fragments are sequenced, and the results are compiled by bioinformaticians, which can result in sequencing information for entire genomes [46].

2.2.3 Third-Generation Sequencing

This generation of sequencing methods is still in progress. The goal is to develop long-read sequencing methods, which would improve on second-generation sequencing (which involves sequencing many small fragments of DNA and compiling the sequences together) [37, 46].

2.3 The Human Genome Project

2.3.1 The Overall Purpose of the Human Genome Project

The US Human Genome Project (HGP) [25, 26] was an international project funded mainly by the US Department of Energy and the National Institutes of Health. The project started in 1990. Celera Corporation, a private company founded by Craig Venter, also initiated a genome sequencing project at the same time. The timeframe for the HGP was meant to be 15 years, but it was actually finished in 13 years, in 2003. The main goals of the HGP were to sequence the entire genome, store the information in databases, identify all the human genes, and develop tools for analyzing the data.

2.3.2 Generation and Cloning of the DNA for the Human Genome Project

DNA for the Human Genome Project was collected from various individuals. Once collected, the DNA was cut into large pieces (about 150 kb long). These pieces were inserted into structures generated in the lab called bacterial artificial chromosomes, which were inserted into bacteria. In this way, the sequences were replicated in the bacteria when the bacteria replicated. This was done to amplify the DNA sequences. Once this was done, the DNA was cut into smaller pieces of about 1000 nucleotides in length. These fragments were sent to labs all over the world and sequenced. An important result of the project was the establishment of a "reference genome." This is a sort of blueprint or standard of the complete genome. Although DNA was collected from multiple individuals, one individual was the main source of the genetic information for this reference genome. Differences between individuals are small, and therefore, other genomes can be aligned and compared with this reference genome.

2.3.3 The Sanger Sequencing Method Was Used for the Human Genome Project

• Overview of the Sanger Sequencing Method

The Sanger sequencing method is the basis of the method used to sequence the genome in the HGP (Fig. 27.1). This method is also sometimes called the chain-termination method. The basics of the Sanger method are as follows:

1. Heat is used to denature double-stranded DNA to be sequenced into single strands.
2. This denatured DNA is divided equally into four tubes.
3. Each tube contains the DNA, a primer (a short nucleotide sequence complementary to a portion of the DNA), DNA polymerase, and all four deoxynucleotides (dNTPs): dTTP, dATP, dCTP, and dGTP.
4. To each tube individually is added a sample of a single dideoxynucleotide (ddNPT): either ddTTP, ddAPT, ddGTP, or ddCTP, each one fluorescently labeled.
5. DNA synthesis now takes place in each tube. DNA synthesis begins when the primer anneals to a specific sequence in the template DNA, and due to the presence of DNA polymerase and dNTPs, a new copy of DNA complementary to the template is synthesized. Occasionally, a ddNTP will be incorporated into the growing DNA chain. This is a key feature of Sanger sequencing. The ddNTPs have no OH group, so no further nucleotides can be added once a ddNTP is added to the chain. The ddNTPs are thus considered **termination nucleotides**.
6. In the original Sanger sequencing method, the DNA fragments from each reaction were separated by polyacrylamide gel electrophoresis, which separates all the newly synthesized DNA fragments by size. The type of gel used could separate sequences that differ by only one nucleotide in length. Each gel lane would have DNA from one of the four tubes. Bands could be visualized due to the fluorescently labeled ddNTPs. By reading through the bands on the gel, one could detect the

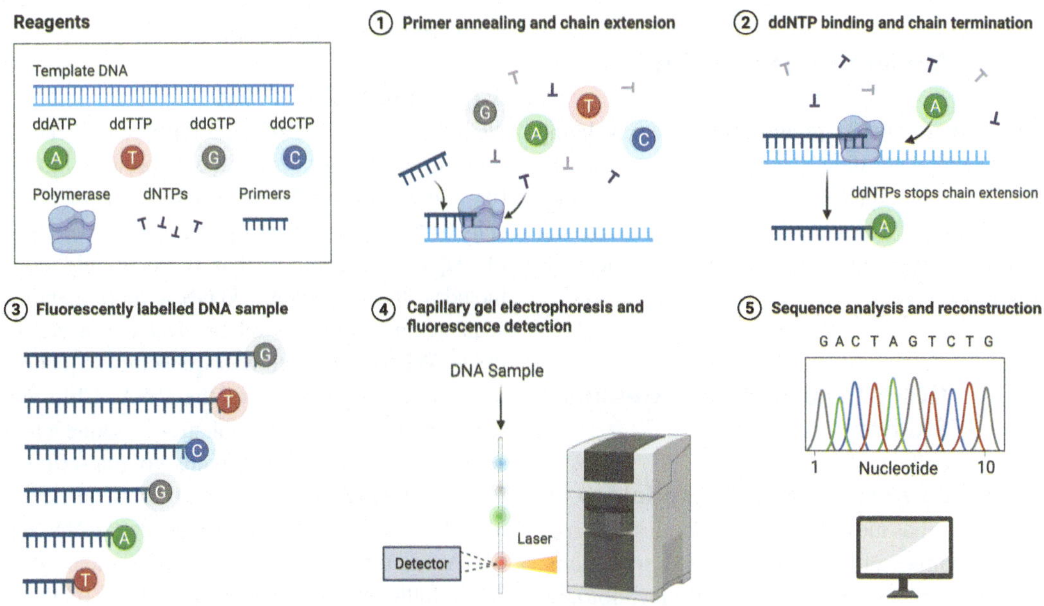

Fig. 27.1 Sanger sequencing. Sanger sequencing was one of the first sequencing methods developed. It is still sometimes used for smaller projects. It relies on the chain-termination method, described in the text. (*Source*: Created with BioRender.com)

sequence of the DNA. In more modern Sanger sequencing, this process is automated. The fragments are separated by capillary electrophoresis, where the fluorescently labeled ddNTPs are detected by a laser and analyzed by computer. The results are displayed as an electropherogram, with colored peaks corresponding to each nucleotide in the sequence [37, 47].

- How the Sanger Sequencing Method Was Used for the Human Genome Project

The above is the basic mechanism behind the Sanger sequencing method. However, it can be modified in various ways, including using radiolabeled ddNTPs instead of fluorescently labeled ddNTPs, and some parts of the process can be automated. The Human Genome Project was done using the Sanger sequencing method. For the Human Genome Project, however, the technique was modified and improved so that it did not rely on gel electrophoresis. Even with these improvements, the method was time-consuming and expensive. The genome project cost over 2 billion dollars [48].

3 Modern Genomics Studies

The study of modern **genomics** involves isolating **DNA** and determining the sequence of all the DNA within a cell or tissue. Genomics studies can involve various sequencing methods but often involve the use of **NGS**, which is discussed here.

3.1 NGS Is Used in Modern Genomics Studies

3.1.1 Overview of NGS Methodology

As science has moved into the genomics and even post-genomics eras, more advanced sequencing technologies are being used. New technologies have significantly reduced the time and expense of sequencing. While the Human Genome Project took 13 years and cost over 2 billion dollars, today, a whole genome can be sequenced within a few days for only a few thousand dollars [37, 44, 46, 49].

While Sanger sequencing is sometimes still used for small sequencing projects, large sequencing projects usually use high-throughput sequencing techniques, mainly **NGS** (see Fig. 27.2). NGS allows the sequencing of millions or more DNA molecules simultaneously. There are several platforms for NGS. These include Illumina MiSeq, Illumina HiSeq, 454 Sequencer, SOLiD system, and Ion Proton system. The term "platform" refers to the hardware, software, methodology, and biochemical reagents needed for sequencing. All these systems require a library, which is a pool of all DNA in the genome being sequenced, often isolated from cultured cells. The DNA in the library is usually ligated to adaptors on both ends.

The Illumina sequencing platform is one of the most commonly used platforms for sequencing DNA. In this system, the nucleic acid sequences are attached to a solid surface via adapters and amplified. The amplified sequences are then **sequenced** via a **sequence-by-synthesis method.** The raw data that is produced can then be analyzed by bioinformaticians [49, 50].

3.1.2 Subtypes of NGS

NGS can be used for different types of sequencing projects [46]. In all of these projects, DNA (or sometimes RNA) is usually isolated from cells or tissues, followed by sequencing. The type of sequencing analysis that is commonly done includes the following: **WGS** (whole genome sequencing) sequences all the DNA in a cell or tissue. **WES** (whole exome sequencing) sequences only the coding regions of the DNA. **Chip-Seq** (chromatin immunoprecipitation sequencing) sequences the DNA that binds to transcription factors. **Methyl-Seq** (methylation sequencing) identifies methylated DNA sequences. **RNA-Seq** (RNA sequencing) sequences RNA rather than DNA (although complementary DNA (cDNA), reverse transcribed from RNA, is usually sequenced in an RNA-seq project).

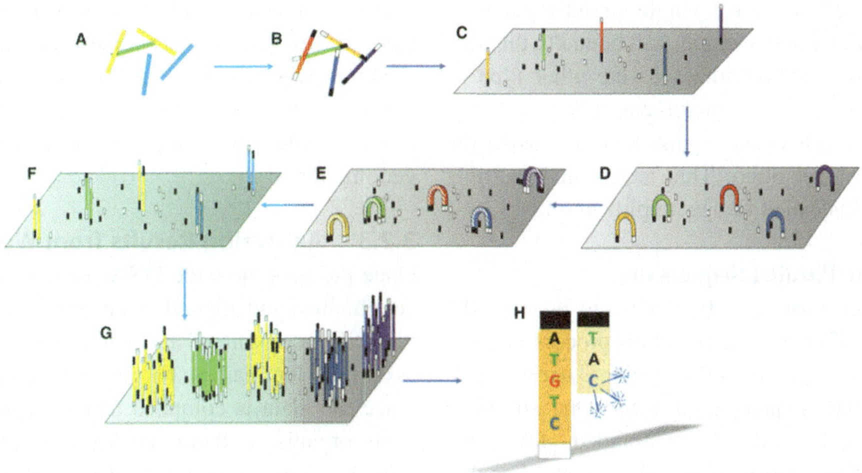

Fig. 27.2 Basic principles of NGS using the Illumina platform. In Illumina sequencing. DNA fragments are attached to a glass flow cell and amplified by bridge amplification to form clusters of identical sequences. The DNA in these clusters is sequenced by the sequence-by-synthesis method using fluorescein-labeled nucleotides, as described in the text. In this method, briefly, DNA is fragmented (**A**), and adapters are added to the fragments (**B**) Fragments are added to a glass flow cell that is coated with complementary oligonucleotides (**C**). DNA synthesis takes place, one strand is washed off, and the other end binds to its own complementary oligonucleotide on the flow cell, forming a U shape or bridge structure (**D**). Repeated cycles of bridge amplification lead to clusters of identical DNA fragments (**E**). The double-stranded DNA is denatured, leaving single-stranded DNA on the flow cell (**F**). Sequencing by synthesis, a form of massively parallel sequencing, is performed using fluorescently labeled nucleotides (**G**), and the sequence of each DNA fragment is determined by recording the fluorescent signal (**H**). (*Source*: Court [72]. Licensed under the Creative Commons Attribution 4.0 (CC BY 4.0) license)

WGS and RNA-seq methods are described below, though the basic concepts apply to the other sequencing methods [50–52].

3.2 Whole Genome Sequencing (WGS) and the Study of Genomics

3.2.1 Steps Involved in WGS (Whole Genome Sequencing)

WGS produces the complete DNA sequence of an organism's genome. Here, we discuss the basic steps involved in **WGS**, focusing on the steps that are specific to the Illumina system. While WGS is stressed here, this method is used for most forms of NGS [53].

Preparation of the DNA and the Beginning of DNA Replication

DNA is isolated from a cell or tissue and broken into small fragments, approximately 50–150 nucleotides long. This is the DNA library. Adapters (short DNA sequences) with known sequences are attached to the ends of the DNA fragments. Two different adapters are added: one type of adapter is added to the 5′ end and one to the 3′ end of the DNA. These double-stranded DNA fragments, containing their adapters, are denatured using heat.

Another set of oligonucleotides, the flow cell oligonucleotides (flow cell oligos), is now attached to a glass flow cell (or slide). These oligos are complementary to the two types of adapters used above. The DNA fragments are now also attached to the glass flow cell, where the adapters at their ends will hybridize to their matching (complementary) flow cell oligos. DNA polymerase and nucleotides are added, and DNA replication occurs on the glass flow cell. The hybridized flow cell oligos serve as the primers for DNA synthesis at this stage.

Bridge Amplification

After the DNA replicates, one strand of newly synthesized double-stranded DNA is washed off the flow cell, leaving a single strand remaining. The other end of the strand, still bound to the flow cell, bends and hybridizes to the other type of flow cell oligo via complementary base pairing. DNA synthesis occurs again, this time using the other flow cell oligo. This is repeated multiple times and is called bridge amplification.

Massively Parallel Sequencing

Many copies of each fragment are now on the flow cell after the amplification process is complete. This large amount of DNA is necessary to get accurate sequencing results. One strand is washed off, and single-stranded fragments remain. **Now, sequencing takes place**. For this step, a **sequencing primer** is annealed to each linear fragment. The primer is complementary to a region on one of the adapters that had been added to DNA fragments. DNA synthesis is initiated in the presence of fluorescently labeled nucleotides, which are incorporated as the DNA chain elongates. The computer records the incorporation of the fluorescently labeled nucleotides, generating sequence information for each strand.

3.2.2 Alignment to a Reference Genome

The process described above generates large amounts of data. The initial result will usually be millions of "reads," which consist of the nucleotide sequences of the short DNA fragments that were input into the system and processed. These reads are not arranged in any particular order, and the information still needs to be processed to be useful to the investigator. The sequences must be aligned to a reference genome to obtain valuable information about the sequences [53]. The reference genome is the complete sequence of the human genome, and this information is publicly available due to the Human Genome Project. Alternatively, if one is sequencing DNA from an organism other than humans, other reference genomes are also publicly available, such as that of the mouse and many different organisms. Using computers to align the short sequences to the reference genome is the first step in obtaining complete sequence information about the newly synthesized DNA [50]. (It is also possible to sequence a genome of an organism for which there is no reference genome, such as a rare organism whose genome has not been sequenced, but this involves de novo assembly, which is more complex than alignment with a reference genome [54]).

3.2.3 Analyzing Results from WGS

Once the genome-wide DNA sequencing results are obtained and aligned to the reference genome, the sequences can be annotated, which includes assigning the names to the sequenced genes. Once this data is compiled and the results have been organized, there are various things that scientists may look for in the sequences. This will depend on the purpose of the study, for example, in cancer research, sequencing data may be used to identify genetic abnormalities specific to a certain type of cancer cell. Researchers may look for certain genetic variations in the DNA of the disease cell. These may include variants such as SNVs (small nucleotide variants), indels (small insertions or deletions), SVs (structural variants: large insertions and deletions), or chromosomal rearrangements [55]. These studies might reveal, for example, that a group of genes associated with a certain signaling pathway or metabolic pathway carries activating mutations. The next steps may be to determine whether these mutations contribute to the symptoms and severity of the disease.

4 Transcriptomics

Transcriptomics involves studying expressed genes, those that are transcribed into RNA. This involves isolation, sequencing, quantifying, and studying all the RNA in a cell or tissue. Transcriptomics involves using similar techniques as above, but the RNA is sequenced instead of the DNA. RNA sequences are actually usually obtained indirectly in the form of cDNA. This is because DNA is more stable than RNA and thus less likely to become degraded during the sequencing process. In many cases, the RNA from two or more different cells or conditions is sequenced and compared with each other. This is called **differential expression (DE) analysis** [1, 56, 57].

4.1 The Basics of RNA Sequencing

The purpose of **RNA** sequencing (see Fig. 27.3) is to sequence the **transcriptome.** The transcriptome refers to all the transcripts in a cell. In other words, it is the complete set of expressed genes (transcribed into RNA). While the genome is essentially the same in all cells of an individual, the transcriptome will vary in different cells and under various conditions [1].

In an RNA sequencing study, the RNA is isolated from cells and then reverse-transcribed to cDNA using a **reverse transcriptase** enzyme. The resulting cDNA is fragmented, and it is sequenced using the basic protocol described above for DNA sequencing. The output contains the sequences that correspond to all the RNA in the cell, and the relative quantity of each sequence is indicated. The sequences are aligned to a reference genome for further analysis, as described for DNA sequencing [57].

One important difference between RNA sequencing and DNA sequencing is that in the latter sequencing the DNA from one type of cell, such as a blood cell or a skin cell, will provide information about the genome of the entire individual. Since different cells express (transcribe) different genes, however, every cell type in an individual will yield different results in an RNA sequencing study. Therefore, in RNA sequencing, the researcher will usually focus on studying a specific cell or tissue type for the analysis [57].

4.2 RNA Sequencing (RNA seq) and Differential Expression (DE) Analysis

RNA sequencing is often used to compare the transcriptomes of two or more cell or tissue types. This is called a differential expression (DE) study. After sequencing the RNA (actually

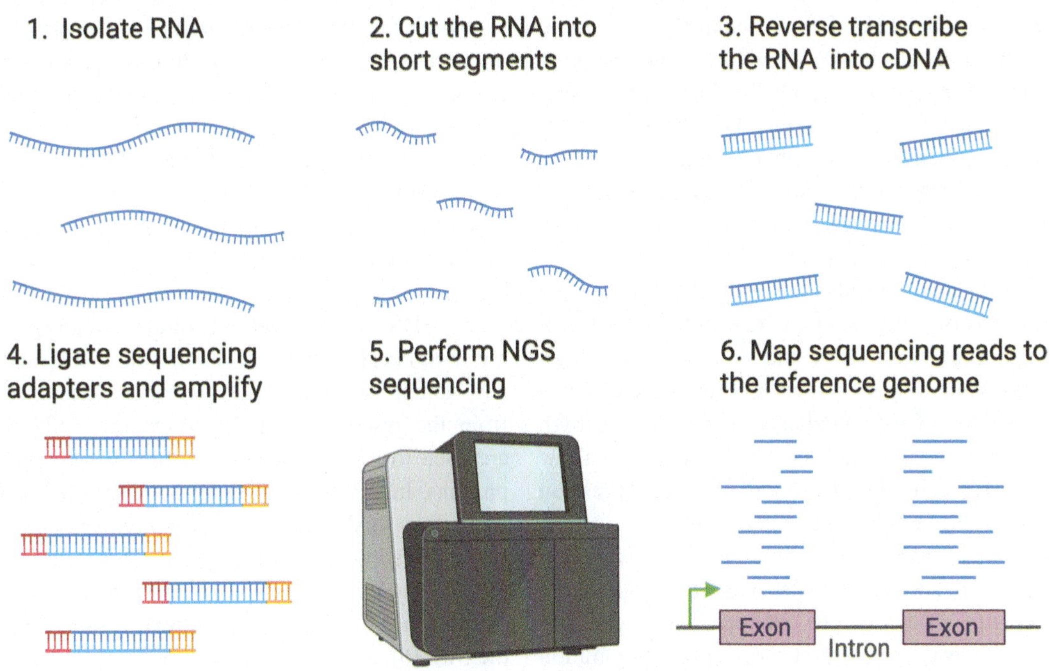

Fig. 27.3 Next Generation RNA sequencing (NGS) technology. RNA sequencing involves (1) the isolation of RNA, (2) cutting it into fragments (3) reverse transcription to cDNA (4), ligation to adapters, followed by (5) sequencing and (6) analyzing the results. More details are described in the text. (*Source*: Created with BioRender.com)

the reverse transcribed cDNA) from two or more cell types, the sequences are aligned to a reference genome and annotated. This reference genome can be a DNA genome. In fact, it can be the same reference genome used for genomics sequencing studies. The result after alignment and annotation is information about thousands of expressed genes. The genes that each sequence corresponds to are identified by name, and the relative quantity of cDNA corresponding to each gene is assessed. This quantity is often expressed as **FPKM** (fragments per kilobase of transcript per million mapped reads). This **relative quantity** is proportional to the number of copies of RNA present in the cell. Thus, RNA sequencing can be used not only to analyze sequencing information but also to assess the relative quantity of each expressed gene (i.e., the relative quantity of RNA produced). When comparing two or more cell types, researchers can determine whether there are expression-level differences in certain genes between the two cell types. This is called differential expression (DE). Many important results can result from these types of RNA sequencing DE studies, for example, one might find increased expression of growth-promoting genes in a type of cancer cell compared to normal cells. Alternatively, one might determine whether the treatment of cells with a certain drug leads to the regulated expression of certain genes or whether different gene expression patterns may be associated with disease severity [57].

One way to display the results of a differential expression study is with a **heat map,** A heat map is a chart with rows and columns. Each row represents a gene, and each column represents a condition. Color gradients (e.g., red for high expression and blue for low expression) allow researchers to detect patterns of gene expression under different conditions [58, 59].

After identifying differentially expressed genes, results can still be overwhelming. Large amounts of data can be obtained, and researchers need a way to interpret the results. A common way that researchers approach this is to perform **pathway analysis**. This is a computational method used to interpret the biological significance of the various gene expression changes that were identified in the sequencing study. This involves grouping the differentially expressed genes into known biological pathways, for example, researchers might find that most of the differentially expressed genes are involved in cell cycle regulation, immune response, apoptosis, or specific RTK signaling pathways. The main idea is to identify which cellular processes are most affected by the conditions that were tested. Pathway analysis can reveal patterns that aren't obvious just from looking at individual genes and can help researchers have a deeper understanding of the gene expression changes in different conditions [60]. Increasingly, pathway analysis and statistical modeling tools are also being used to identify patterns in large and complex datasets [61].

5 Proteomics

In proteomics, the identity and relative quantity of all the proteins in a cell, rather than the DNA or RNA, is assessed. The major method used in proteomics is **mass spectrometry** (often referred to as mass spec or MS). Proteomics experiments are often studies in which two or more conditions are compared. Reference proteomes are also available for comparison and alignment of proteomics studies [1, 62].

In a typical proteomics study, all the proteins in a cell or tissue are isolated. This starts with lysing the cells or tissues with detergents, commonly SDS (sodium dodecyl sulfate). Detergents are used because they can disrupt the membrane and solubilize proteins. Proteins are extracted from the lysed cells, and purification methods are used to remove other molecules besides the proteins. In most setups, proteins are then broken down into smaller peptides. This involves enzymatic digestion, often with trypsin. Chromatography techniques can be used to purify the peptides even further in order to get optimal results from the MS. [6]

Mass spectrometry is a technology that **separates ionized peptides based on their mass-to-charge ratio (m/z).** Peptides that were isolated during the proteomics workflow are ionized. In this ionizing process, the peptides will usually pick up one or more protons (H^+), causing them to obtain a positive charge. (This is typical in electrospray ionization, the method most commonly used in proteomics.) These ionized peptides are then introduced into a mass analyzer, where their motion is influenced by their m/z ratio. In earlier forms of mass spec, such as magnetic sector instruments, ion paths were bent using magnetic fields (still used for many types of studies), but more modern mass spectrometers mainly use electric fields and rely on analyzers such as quadrupoles, time of flight (ToF), Orbitraps, and ion traps. These analyzers allow faster, more sensitive, and higher-resolution measurements. For more information about these technologies, see [63].

In all mass spec systems, ions with a **lower m/z ratio** may travel faster or follow different paths than peptides with a higher m/z ratio, depending on the analyzer type, although some forms separate ions by velocity or oscillation frequencies. The instrument measures the m/z values of all the peptides in the sample. They can then be matched to peptide sequence databases to identify and quantify the proteins in the original sample [63–65].

Datasets from proteomics experiments are often compared against large datasets such as UniProt and PRIDE to identify and quantitate proteins [66]. Proteomics and MS data can be compared between different samples. The results can be used for many purposes, for example, one can detect whether there are differences in the levels of certain proteins under different conditions. Computational analysis and statistics are carried out to determine what changes are significant. Follow-up studies will involve determining the functions of proteins that are identified as significant in proteomics studies [67].

6 Omics Techniques in Personalized Medicine

An important goal of omics studies is to determine how each individual differs and what individuals share in common regarding their genetic and proteomics makeup. This information can be used to individualize drug treatment by predicting how different patients may respond to different types of treatments. This is potentially advantageous, rather than treating every patient with a certain disease in the same exact way. The methods scientists use to develop **personalized treatments** vary but can often involve the following.

6.1 Establishment of a Database

Genome sequencing techniques can improve personalized medicine. Some ongoing efforts, large-scale databases, are being created which store genetic information along with medical records and health information from many patients. By analyzing such databases, scientists and clinicians can identify patterns between specific genetic variants, disease risk, responses to specific treatments, and sensitivity to certain medications. This information can be used to improve patient care by enabling more accurate diagnoses and accurately predicting the right course of treatment, tailored to the patients' needs [55].

6.2 Personalized Cancer Medicine and the MSK-IMPACT Project

MSK-IMPACT (Integrated Mutation Profiling of Actionable Cancer Targets) is a precision oncology initiative developed by the Memorial Sloan Kettering Cancer Center (MSKCC). It involves sequencing DNA and RNA from tumors from **patients who have been diagnosed with different cancer types**. Normal tissue is also sometimes analyzed. This data uses a targeted gene panel to analyze hundreds of cancer-associated genes known to play roles in tumor development, tumor progression, or drug response. (A targeted gene panel analyzes selected types of genes rather than all genes, in this case, genes frequently known to be linked to cancer). This data is sometimes linked to clinical records, including treatments and outcomes. The goal of this database is to detect specific alterations, mutations, or expression variations that are associated with different tumor types as well as outcomes, in a population of individuals. This is called a **genomic profiling study**, and the information obtained from analyzing this data will help oncologists tailor treatment plans to the individual patient, for example, if a patient's tumor has a **biomarker** that is associated with response to a specific type of treatment, they may be directed toward that treatment. Alternatively, they may be directed toward appropriate clinical trials. Results from these studies are also used for research purposes. Information from the database can help researchers understand what types of genetic changes are typical in different types of cancer, with the goal of understanding the disease better and developing better treatments [68].

6.3 NGS and Personalized Medicine for COVID-19

One interesting way to use NGS information is to **predict how patients will respond to an infection, such as a viral infection**. The COVID-19 pandemic illustrated that a wide range of responses could occur in response to the same viral infection. While some COVID-19 patients had few or mild symptoms, others had severe, life-threatening symptoms. In the early days of the pandemic, it was challenging for physicians to know how to treat patients because it was almost impossible to predict how a patient would respond to the infection.

Genome sequencing might help alleviate this problem for COVID-19 and other viruses. Genome sequencing was performed on patients who had been infected with the COVID-19 virus and control patients who had not been infected. How severely patients reacted to the infection was recorded. Scientists analyzed the data to determine whether there were patterns of genomic differences between patients with severe responses to the virus and those with mild responses. They found that certain genetic variants seemed to be associated with severe reactions to the viral infection. Some of these variants are in immune system-related genes.

These types of studies can help physicians determine how to treat patients. Specifically, they can help determine which patients need urgent treatment and which are likely to recover on their own over time. They also allow scientists to better understand the biological mechanisms behind the different responses to infection [69].

6.4 Pros and Cons of Personalized Medicine

The fields of genomics and proteomics are quite promising for many reasons, including research into understanding the basis of human disease. As described in this section, omics studies are also groundbreaking tools for the advancement of personalized medicine. Personalized medicine has many potential advantages. An important advantage is the use of **targeted therapies** directed at the patient's precise genetic makeup. This targeted approach is likely to lead to **improved outcomes**. **Fewer side effects** are also expected compared to traditional broad-spectrum treatments. Understanding an individual's genetic makeup will also help prevent future disease by recommending lifestyle changes based on susceptibility to various disorders.

While there are many promising outcomes for personalized medicine, there may also be some drawbacks. One drawback is the potential **costs**, which may be high due to the technology involved. Although sequencing costs have decreased, large-scale studies are needed, as well as software, computers, and bioinformaticians dedicated to the work. Another potential issue is **privacy.** It will become important to learn how to deal with issues of confidentiality when large amounts of genetic data are obtained, stored, and shared. Along with privacy, there are ethical issues with regard to storing and sharing large amounts of genetic information, accessing expensive treatments that may be deemed necessary, and insurance issues. There is also the issue of overall **technical difficulty**. The technologies involved, including the complex technologies needed to analyze genomic data, may hinder the development of some personalized medicine strategies, although these techniques are likely to become increasingly streamlined and automated over time. Finally, there is the issue of the **efficacy of treatment**. The overall expectation is that personalized medicine will be more beneficial than more traditional methods of treating disease. In most cases, it is probably true. However, there may be situations where conventional methods are the most effective. In diseases like cancer, where cells and their genetic information can change rapidly, using a broad range of treatments may still have certain advantages [55, 70, 71]. Even as the technology for personalized medicine continues to advance, future disease treatments will likely consist of a balance between individualized therapies and more general approaches.

Definitions

Adapters Short synthetic DNA sequences that are ligated to the ends of the DNA fragments during library preparation. Enables attachment to the flow cell and amplification.

Biomarker A biological indicator of a condition, disease state, or response to treatment.

Bridge amplification A method used in Illumina sequencing to amplify DNA fragments on a flow cell.

Complementary DNA (cDNA) DNA synthesized from an RNA template by reverse transcription. It is used in RNA sequencing technology.

Differential expression A comparison of the genes that are expressed (transcribed into RNA) in one cell type or conditions, compared with another.

Electrospray ionization (ESI) A technique used in mass spectrometry to produce ions from molecules in liquid samples.

Encyclopedia of DNA Elements (ENCODE) A research project with the goal of generating a resource of all functional elements in the human genome. Functional elements include many features such as transcription factor binding sites, promoter, enhancers, and others.

Epigenomics The study of the epigenetic modifications in the DNA of the genome.

Flow cell A glass surface used in Illumina sequencing platforms to immobilize and amplify DNA fragments.

Gene annotation The process of identifying genes in DNA or RNA sequencing and assigning information such as their names.

Gene panel A group of selected genes and their sequences. Often, they are associated with specific functions, such as genes involved in cancer or a specific disease.

Genomics The study of the complete set of DNA (the genome) in the cell, tissue, or organism.

Heat map A data visualization method in transcriptomics or proteomics where color intensity represents expression levels.

Mass spectrometry (MS) A technology that separates ionized peptides based on their mass-to-charge ratio (m/z).

Metabolomics The study of the metabolites present in a cell under specific conditions.

MSK-IMPACT A precision oncology initiative developed by the Memorial Sloan Kettering Cancer Center (MSKCC). It involves sequencing DNA and RNA from tumors, from patients that have been diagnosed with different cancer types. Used to optimize treatments for patients.

Next-generation sequencing (NGS) A high-throughput method for sequencing large amounts of DNA or RNA, usually whole genomes or transcriptomes.

Pathway analysis A method to understand how differentially expressed genes or proteins are functionally related in biological pathways.

Proteomics The study of the identity and relative quantity of all the proteins in a cell or biological condition.

Reads (in sequencing) Short fragments of DNA or RNA that are sequenced.

RNA sequencing (RNA-seq) Sequencing of large amounts of RNA, usually from cells or tissues. The RNA is generally reverse transcribed into cDNA before sequencing.

Sanger sequencing A method for DNA sequencing before NGS was used. It is more time-consuming and not practical for large-scale sequencing projects, but it is still sometimes used for small projects. It involves the use of termination dideoxynucleotides.

The Cancer Genome Atlas (TCGA) A large-scale project to catalog genetic mutations responsible in cancer.

Transcriptomics The study of all RNA transcripts produced by a cell, tissue, or organism at a given time (the transcriptome).

Study Questions

Multiple-Choice Questions

1. What is the main focus of many transcriptomics studies?
   A. The complete DNA sequence of an organism
   B. The full set of RNA transcripts in a cell
   C. Protein expression patterns in disease
   D. Lipid metabolism analysis
2. Which of the following are parts of the Illumina NGS technology?
   A. Chain-termination method
   B. Sanger sequencing
   C. Bridge amplification
   D. Northern blotting
3. Which term describes the comparison of gene expression levels in two different conditions or cell types?
   A. Gene annotation
   B. Epigenomic mapping
   C. Differential expression
   D. RNA splicing
4. In the MSK-IMPACT project, sequencing is primarily used to:
   A. Discover new bacteria.
   B. Tailor cancer treatments to individual patients.
   C. Detect antibiotic resistance.
   D. Identify metabolic disorders.
5. What is the function of adapters in Illumina sequencing?
   A. To synthesize cDNA from RNA
   B. To identify which fragments should be sequenced
   C. To enable DNA fragments to attach to the flow cell
   D. To tag proteins for degradation
6. What is the main difference between the study of genomics and the study of genetics?
   A. Genomics focuses on the study of individual genes, while genetics examines all the genes in the cell.
   B. Genomics is the study of inherited traits, while genetics is concerned with the structure and function of DNA.
   C. Genomics analyzes the interactions of genes within an organism, whereas genetics looks at the interactions between groups of genes within an individual cell.
   D. Genomics involves the analysis of an organism's entire DNA, while genetics deals with the study of individual genes.
   E. Genetics studies only human genes, while genomics encompasses the study of genes in all living organisms.
7. What type of sequencing would you carry out if you wanted to find out what genes are expressed in a given cell type?
   A. Whole genome sequencing
   B. Whole exome sequencing
   C. Sanger sequencing
   D. Methyl sequencing
   E. RNA sequencing
8. In NGS, what does the reference genome refer to?
   A. A sequence that is already known and can be used for alignment
   B. A new alignment sequence that has been identified using the bridge amplification method
   C. An artificially derived sequence that allows for alignment
   D. A sequence from a virally infected cell

Fill in the Blank Questions

9. The __________ refers to all the RNA molecules transcribed in a particular cell or tissue.
10. __________ sequencing was used to complete the Human Genome Project.

Short Answer Question

11. What are two major benefits of personalized medicine using omics data?

Matching Question

12. Match each omics field to the biomolecule it studies:
    Term
    A. Genomics
    B. Transcriptomics
    C. Proteomics
    Biomolecule
    1. Proteins
    2. RNA
    3. DNA

Answer Key

Multiple-Choice Questions

1. **Answer:** B
2. **Answer:** C
3. **Answer:** C
4. **Answer:** B
5. **Answer:** C
6. **Answer:** D
7. **Answer:** E
8. **Answer:** A

Fill in the Blank Questions

9. **Answer:** Transcriptome
10. **Answer:** Sanger

Short Answer Question

11. **Answer**: (1) Targeted therapy tailored to genetic profile, (2) reduced side effects and improved treatment outcomes.

Matching Question

12. **Answers:**
    A → 3, B → 2, C → 1

References

1. Dai X, Shen L. Advances and trends in omics technology development. Front Med (Lausanne). 2022;9:911861. https://doi.org/10.3389/fmed.2022.911861.
2. National Human Genome Research Institute. About Genomics 2024. Available from: https://www.genome.gov/about-genomics.
3. Gutierrez Reyes CD, Alejo-Jacuinde G, Perez Sanchez B, Chavez Reyes J, Onigbinde S, Mogut D, et al. Multi omics applications in biological systems. Curr Issues Mol Biol. 2024;46(6):5777–93. https://doi.org/10.3390/cimb46060345.
4. Hasin Y, Seldin M, Lusis A. Multi-omics approaches to disease. Genome Biol. 2017;18(1):83. https://doi.org/10.1186/s13059-017-1215-1.
5. Tanay A, Regev A. Scaling single-cell genomics from phenomenology to mechanism. Nature. 2017;541(7637):331–8. https://doi.org/10.1038/nature21350.
6. Alberts B, Heald R, Johnson A, Morgan D, Raff M, Roberts K, Walter P, Wilson J, Hunt T. Molecular biology of the cell. New York: W.W. Norton and Company; 2022.
7. Nature Education. Transcriptome definition. Available from: https://www.nature.com/scitable/definition/transcriptome-296/#:~:text=A%20transcriptome%20is%20the%20full,Cell%20Origins%20and%20Metabolism.
8. Cui M, Cheng C, Zhang L. High-throughput proteomics: a methodological mini-review. Lab Investig. 2022;102(11):1170–81. https://doi.org/10.1038/s41374-022-00830-7.
9. National Human Genome Research Institute. Genetics vs. Genomics Fact Sheet 2018. Available from: https://www.genome.gov/about-genomics/fact-sheets/Genetics-vs-Genomics.
10. Unlocking Life's Code. A brief history: from Mendel to the Human Genome Project. Available from: https://www.unlockinglifescode.org/timeline.
11. Blog ETDU. A journey through the history of DNA sequencing. Available from: https://the-dna-universe.com/2020/11/02/a-journey-through-the-history-of-dna-sequencing/#:~:text=In%201972%2C%20Walter

%20Fiers%20was%20first%20to,al.%202019;%20Min%20Jou%20et%20al.%2C%201972.
12. National Human Genome Research Institute. 1944: DNA is /"Transforming Principle\". 2013. Available from: https://www.genome.gov/25520250/online-education-kit-1944-dna-is-transforming-principle.
13. HistoryofInformation.com. Frederick Sanger sequences the amino acids of insulin, the first of any protein. 2013. Available from: https://historyofinformation.com/detail.php?id=729#:~:text=Frederick%20Sanger%20Sequences%20the%20Amino%20Acids%20of,with%20no%20general%20rule%20for%20its%20assembly.
14. ASBMBToday. A mad race to the finish 2012. Available from: https://www.asbmb.org/asbmb-today/science/020112/a-mad-race-to-the-finish#:~:text=A%20conversation%20with%20Philip%20Leder,corresponded%20to%20which%20amino%20acid.
15. National Human Genome Research Institute. 1966: genetic code cracked. 2013. Available from: https://www.genome.gov/25520300/online-education-kit-1966-genetic-code-cracked#:~:text=Over%20the%20course%20of%20several,was%20the%20first%20codon%20deciphered.
16. Jacob F, Monod J. Genetic regulatory mechanisms in the synthesis of proteins. J Mol Biol. 1961;3(3):318–56. https://doi.org/10.1016/S0022-2836(61)80072-7.
17. Barciszewska MZ, Perrigue PM, Barciszewski J. tRNA—the golden standard in molecular biology. Mol Biosyst. 2016;12(1):12–7. https://doi.org/10.1039/c5mb00557d.
18. Holley RW, Apgar J, Everett GA, Madison JT, Marquisee M, Merrill SH, et al. Structure of a ribonucleic acid. Science. 1965;147(3664):1462–5. https://doi.org/10.1126/science.147.3664.1462.
19. Min Jou W, Haegeman G, Ysebaert M, Fiers W. Nucleotide sequence of the gene coding for the bacteriophage MS2 coat protein. Nature. 1972;237(5350):82–8. https://doi.org/10.1038/237082a0.
20. Sanger F, Nicklen S, Coulson AR. DNA sequencing with chain-terminating inhibitors. Proc Natl Acad Sci USA. 1977;74(12):5463–7. https://doi.org/10.1073/pnas.74.12.5463.
21. The Nobel Prize. The Nobel Prize in chemistry, 1980. 1980. Available from: https://www.nobelprize.org/prizes/chemistry/1980/summary/.
22. Chial H. Huntington's disease: the discovery of the Huntingtin gene. Nat Educ. 2008;1(1):71.
23. Michigan Medicine. A gene discovery that changed cystic fibrosis care, and genetic research, forever [Internet]. Ann Arbor (MI): Michigan Medicine; 2026 Apr 17 [cited 2026 May 25]. Available from: https://www.michiganmedicine.org/news-release/gene-discovery-changed-cystic-fibrosis-care-and-genetic-research-forever
24. Kerem B, Rommens JM, Buchanan JA, Markiewicz D, Cox TK, Chakravarti A, et al. Identification of the cystic fibrosis gene: genetic analysis. Science. 1989;245(4922):1073–80. https://doi.org/10.1126/science.2570460.
25. National Human Genome Research Institute. The Human Genome Project. 2024. Available from: https://www.genome.gov/human-genome-project.
26. Britannica. Human Genome Project. Available from: https://www.britannica.com/event/Human-Genome-Project#ref272803.
27. Whitelaw E. The race to unravel the human genome. EMBO Rep. 2002;3(6):515.
28. Fleischmann RD, Adams MD, White O, Clayton RA, Kirkness EF, Kerlavage AR, et al. Whole-genome random sequencing and assembly of Haemophilus influenzae Rd. Science. 1995;269(5223):496–512. https://doi.org/10.1126/science.7542800.
29. Dunham I, Shimizu N, Roe BA, Chissoe S, Hunt AR, Collins JE, et al. The DNA sequence of human chromosome 22. Nature. 1999;402(6761):489–95. https://doi.org/10.1038/990031.
30. National Human Genome Research Institute. International human genome sequencing consortium announces "working draft" of human genome. Bethesda: National Human Genome Research Institute (NHGRI); 2000. Available from: https://www.genome.gov/10001457/2000-release-working-draft-of-human-genome-sequence.
31. Clinton White House. President Clinton announces the completion of the first survey of the entire human genome. Washington, DC: The White House; 2000. Available from: https://clintonwhitehouse5.archives.gov/WH/EOP/OSTP/html/00628_4.html.
32. Lander ES, Linton LM, Birren B, Nusbaum C, Zody MC, Baldwin J, et al. Initial sequencing and analysis of the human genome. Nature. 2001;409(6822):860–921. https://doi.org/10.1038/35057062.
33. Venter JC, Adams MD, Myers EW, Li PW, Mural RJ, Sutton GG, et al. The sequence of the human genome. Science. 2001;291(5507):1304–51. https://doi.org/10.1126/science.1058040.
34. Jhala US, Canettieri G, Screaton RA, Kulkarni RN, Krajewski S, Reed J, et al. cAMP promotes pancreatic beta-cell survival via CREB-mediated induction of IRS2. Genes Dev. 2003;17(13):1575–80. https://doi.org/10.1101/gad.1097103.
35. National Human Genome Research Institute. Studies expand understanding of X chromosome. 2012. Available from: https://www.genome.gov/13514331/2005-release-studies-expand-understanding-of-x-chromosome#:~:text=BETHESDA%2C%20Md.%2C%20Wed.%2C%20March%2016%2C%202005%20%2D,the%20biological%20differences%20between%20males%20and%20females.&text=The%20new%20analysis%20confirmed%20the%20existence%20of%201%2C098%20protein%2Dcoding%20genes%20on%20the%20X%20chromosome.

36. National Cancer Institue. The Cancer Genome Project (TCGA). Available from: https://www.cancer.gov/ccg/research/genome-sequencing/tcga#:~:text=This%20joint%20effort%20between%20NCI%20and%20the,of%20genomic%2C%20epigenomic%2C%20transcriptomic%2C%20and%20proteomic%20data.
37. Lesk AM. Introduction to genomics. 3rd ed. Oxford: Oxford University Press; 2017.
38. Akintunde O, Tucker T, Carabetta VJ. The evolution of next-generation sequencing technologies. ArXiv. 2023.
39. Gostimskaya I. CRISPR-Cas9: a history of its discovery and ethical considerations of its use in genome editing. Biochemistry (Mosc). 2022;87(8):777–88. https://doi.org/10.1134/s0006297922080090.
40. Devuyst O. The 1000 Genomes Project: welcome to a new world. Perit Dial Int. 2015;35(7):676–7. https://doi.org/10.3747/pdi.2015.00261.
41. The White House of Barack Obama. The precision medicine initiative. 2016. Available from: https://obamawhitehouse.archives.gov/precision-medicine#:~:text=In%20his%202015%20State%20of%20the%20Union,how%20we%20improve%20health%20and%20treat%20disease.&text=The%20future%20of%20precision%20medicine%20will%20enable,microbiome%20composition%2C%20health%20history%2C%20lifestyle%2C%20and%20diet.
42. Human Cell Atlas. Human cell atlas. Available from: https://www.humancellatlas.org/.
43. Denny JC, Rutter JL, Goldstein DB, Philippakis A, Smoller JW, Jenkins G, et al. The "all of us" research program. N Engl J Med. 2019;381(7):668–76. https://doi.org/10.1056/NEJMsr1809937.
44. Pareek CS, Smoczynski R, Tretyn A. Sequencing technologies and genome sequencing. J Appl Genet. 2011;52(4):413–35. https://doi.org/10.1007/s13353-011-0057-x.
45. Adams J. DNA sequencing technologies. Nat Educ. 2008;1(1):193.
46. Satam H, Joshi K, Mangrolia U, Waghoo S, Zaidi G, Rawool S, et al. Next-generation sequencing technology: current trends and advancements. Biology (Basel). 2023;12(7). https://doi.org/10.3390/biology12070997.
47. Rodriguez R, Krishnan Y. The chemistry of next-generation sequencing. Nat Biotechnol. 2023;41(12):1709–15. https://doi.org/10.1038/s41587-023-01986-3.
48. MIT Biology. The Human Genome Project turns 20: here's how it altered the world. 2023. Available from: https://biology.mit.edu/the-human-genome-project-turns-20-heres-how-it-altered-the-world/.
49. Hu T, Chitnis N, Monos D, Dinh A. Next-generation sequencing technologies: an overview. Hum Immunol. 2021;82(11):801–11. https://doi.org/10.1016/j.humimm.2021.02.012.
50. Illumina. Introduction to NGS: learn how the technology works and what it can do for you. 2024. Available from: https://www.illumina.com/science/technology/next-generation-sequencing.html#:~:text=Applications%20of%20NGS,Identify%20novel%20pathogens.
51. NHS England. Whole exome sequencing. 2022. Available from: https://www.genomicseducation.hee.nhs.uk/genotes/knowledge-hub/whole-exome-sequencing/#:~:text=Whole%20exome%20sequencing%20(WES)%20is,whole%20genome%20sequencing%20(WGS).
52. Illumina. Illumina sequencing methods. 2024. Available from: https://www.illumina.com/techniques/sequencing.html.
53. Goodwin S, McPherson JD, McCombie WR. Coming of age: ten years of next-generation sequencing technologies. Nat Rev Genet. 2016;17(6):333–51. https://doi.org/10.1038/nrg.2016.49.
54. Xiao W, Wu L, Yavas G, Simonyan V, Ning B, Hong H. Challenges, solutions, and quality metrics of personal genome assembly in advancing precision medicine. Pharmaceutics. 2016;8(2). https://doi.org/10.3390/pharmaceutics8020015.
55. Tafazoli A, Guchelaar HJ, Miltyk W, Kretowski AJ, Swen JJ. Applying next-generation sequencing platforms for pharmacogenomic testing in clinical practice. Front Pharmacol. 2021;12:693453. https://doi.org/10.3389/fphar.2021.693453.
56. CD Genomics. What is differential gene expression analysis? Available from: https://www.cd-genomics.com/resource-differential-gene-expression-analysis.html#:~:text=limma%2Dvoom%20is%20a%20differential,microarray%20and%20RNA%2DSeq%20analyses.
57. Koch CM, Chiu SF, Akbarpour M, Bharat A, Ridge KM, Bartom ET, et al. A beginner's guide to analysis of RNA sequencing data. Am J Respir Cell Mol Biol. 2018;59(2):145–57. https://doi.org/10.1165/rcmb.2017-0430TR.
58. Rosati D, Palmieri M, Brunelli G, Morrione A, Iannelli F, Frullanti E, et al. Differential gene expression analysis pipelines and bioinformatic tools for the identification of specific biomarkers: a review. Comput Struct Biotechnol J. 2024;23:1154–68. https://doi.org/10.1016/j.csbj.2024.02.018.
59. Manzoor F, Tsurgeon CA, Gupta V. Exploring RNA-Seq data analysis through visualization techniques and tools: a systematic review of opportunities and limitations for clinical applications. Bioengineering. 2025;12(1):56. https://doi.org/10.3390/bioengineering12010056.
60. Reimand J, Isserlin R, Voisin V, Kucera M, Tannus-Lopes C, Rostamianfar A, et al. Pathway enrichment analysis and visualization of omics data using g:Profiler, GSEA, Cytoscape and EnrichmentMap. Nat Protoc. 2019;14(2):482–517. https://doi.org/10.1038/s41596-018-0103-9.
61. Libbrecht MW, Noble WS. Machine learning applications in genetics and genomics. Nat Rev Genet. 2015;16(6):321–32. https://doi.org/10.1038/nrg3920.

62. Al-Amrani S, Al-Jabri Z, Al-Zaabi A, Alshekaili J, Al-Khabori M. Proteomics: concepts and applications in human medicine. World J Biol Chem. 2021;12(5): 57–69. https://doi.org/10.4331/wjbc.v12.i5.57.
63. Zhang Y, Fonslow BR, Shan B, Baek M-C, Yates JR III. Protein analysis by shotgun/bottom-up proteomics. Chem Rev. 2013;113(4):2343–94. https://doi.org/10.1021/cr3003533.
64. Klein JB, M AK. Protein mass spectrometry made simple. J Am Soc Nephrol. 2018;29(6):1585–7. https://doi.org/10.1681/asn.2018030244.
65. Aebersold R, Mann M. Mass spectrometry-based proteomics. Nature. 2003;422(6928):198–207. https://doi.org/10.1038/nature01511.
66. Vizcaíno JA, Csordas A, del-Toro N, Dianes JA, Griss J, Lavidas I, et al. 2016 update of the PRIDE database and its related tools. Nucleic Acids Res. 2015;44(D1):D447–D56. https://doi.org/10.1093/nar/gkv1145.
67. ThermoFisher. Scientific Overview of mass spectrometry for protein analysis. Available from: https://www.thermofisher.com/us/en/home/life-science/protein-biology/protein-biology-learning-center/protein-biology-resource-library/pierce-protein-methods/overview-mass-spectrometry.html#:~:text=Mass%20spectrometry%20(MS)%20analysis%20of,modification%20and%20global%20protein%20dynamics.
68. Memorial Sloan Kettering Cancer Center. MSK-IMPACT: A comprehensive tumor sequencing test to detect targetable DNA mutations. Available from: https://www.mskcc.org/msk-impact.
69. Aschenbrenner AC, Mouktaroudi M, Krämer B, Oestreich M, Antonakos N, Nuesch-Germano M, et al. Disease severity-specific neutrophil signatures in blood transcriptomes stratify COVID-19 patients. Genome Med. 2021;13(1):7. https://doi.org/10.1186/s13073-020-00823-5.
70. Goetz LH, Schork NJ. Personalized medicine: motivation, challenges, and progress. Fertil Steril. 2018;109(6):952–63. https://doi.org/10.1016/j.fertnstert.2018.05.006.
71. Refractor. Personalized medicine: the pros, cons and concerns. Available from: https://newatlas.com/medical/personalized-medicine-benefits-concerns/.
72. Court D. Mitochondrial DNA in forensic use. Emerg Top Life Sci. 2021;5(3):415–26. https://doi.org/10.1042/ETLS20210204.

Drug Delivery and Nanotechnology 28

Abstract

Effective therapy depends not only on drug development and identifying the correct drug for a patient but also on selecting the right method for delivering drugs to the patient. This chapter will provide a brief overview of commonly used drug delivery mechanisms and introduce nanotechnology, a promising and rapidly evolving technology for drug delivery. The chapter discusses the major routes of delivery, including oral, injection, topical, and inhalation. It discusses some benefits of extended-release mechanisms. It then introduces the topic of nanotechnology-based delivery systems, including liposomes, lipid nanoparticles, and exosomes. It discusses how nanotechnology-based formulations can be combined with targeting mechanisms to deliver drugs to specific tissues or organs and provides examples of drugs delivered using nanotechnology systems. The chapter also discusses current challenges and limitations in the use of nanotechnology.

Keywords

Oral delivery · Topical delivery · Injections · Extended release · Nanoparticle delivery

Abbreviations

DOX	Doxorubicin
EPR	Enhanced permeability and retention
GI	Gastrointestinal
IV	Intravenous
LNP	Lipid nanoparticle
NP	Nanoparticle
PEG	Polyethylene glycol

What You Will Learn

- The major drug delivery routes, such as oral delivery, injection, topical delivery, inhalation, and extended release
- The principles of nanotechnology-based delivery systems, including the use of liposomes, lipid nanoparticles, and exosomes
- Mechanisms of drug targeting in nanotechnology-based formulations
- The COVID-19 vaccine as a key example of the use of lipid nanoparticle-mediated drug delivery
- Current challenges and limitations of nanotechnology

This book has explored a wide range of drug targets, therapeutic agents, and approaches to

A. Minden, *Biochemistry, Cell and Molecular Biology for Pharmacists and Health Professionals*,
https://doi.org/10.1007/978-3-032-20529-2_28

treating disease. In addition to the development of therapeutics, selecting the right method for delivering them is also essential. This chapter will give a brief overview of commonly used drug delivery mechanisms and introduce nanotechnology, which is a promising and relatively new drug delivery technology.

1 Traditional Methods of Drug Delivery

1.1 Orally Delivered Drugs and First-Pass Metabolism

Traditionally, many drugs have been designed to be delivered via **oral delivery**. Such drugs may be in a pill or liquid form, taken by mouth. These types of drugs are absorbed in the GI tract and must survive some harsh conditions, such as high acidity. Some drugs are delivered quite successfully in this way, although others have poor bioavailability and need to be delivered by other mechanisms [1–4]. Orally administered drugs are absorbed through the GI tract and then enter the **hepatic portal circulation**, which transports them directly to the **liver** before they enter **systemic circulation** [5]. This exposes them to **first-pass metabolism**, a process where compounds are significantly metabolized in the liver before reaching systemic circulation. This is the standard process for any orally delivered compounds, whether they are drugs or nutrients. Many drugs are delivered successfully this way, but oral delivery and first-pass metabolism can also affect drug efficacy.

All orally delivered drugs undergo some amount of metabolism in the liver, by liver enzymes, but some drugs are broken down more extensively than others. In some cases, metabolism of drugs in the liver is extensive and reduces their bioavailability and therapeutic efficacy too much. Such drugs may need to be delivered by other mechanisms, such as the mechanisms that are discussed below. On the other hand, some drugs are designed as prodrugs [6], which means they are inactive in their original forms until they are metabolized, often by liver enzymes. This may involve hydrolysis (which is usually associated with cleavage) or, in some cases, oxidation. These drugs are generally well-suited for oral delivery.

1.2 Injection

Some drugs are delivered via injection. This can be an **intravenous (IV) injection**, which is good for drugs that do not survive the digestive system, and it can lead to fast action. Other options are **intramuscular (IM)** or **subcutaneous (SC) injection**. Drugs delivered in this way are released more slowly than those delivered by IV. Drugs given by injection or intradermally (injection into the dermis, just under the outer skin layer) are often referred to as **parenteral drugs** [2].

1.3 Topical Delivery

Some drugs are delivered topically. They may be delivered, for example, via creams or patches (transdermal patches). This type of delivery is limited by the skin barrier but is effective for some treatments. Some topically delivered drugs are designed to treat skin conditions [7]. Others pass through the skin, enter blood vessels, and circulate throughout the body. This second type of topical delivery is referred to as **transdermal** drug delivery [8].

1.4 Inhalation Route

Another drug delivery mechanism is inhalation. This is good for delivering drugs to the lungs, such as those used to treat breathing problems like asthma [9].

1.5 Extended and Controlled Release

Some drugs use an extended-release formulation, and these can usually be delivered by one of the above mechanisms. **Extended release** is a subtype of **controlled release**, and it involves a design that allows the drug to be released gradually over time rather than all at once. This helps maintain a consistent level of the drug within the body [10]. How this works can depend on the route of administration. For oral delivery, capsules can be made in a way that they dissolve slowly in the GI tract [11]. For injectables, some drugs may be injected as parts of long-acting suspensions, such as in microspheres [12]. Microspheres are spherical particles often made of synthetic polymers that are larger than nanoparticles (described below) and which can encapsulate certain drugs.

2 Nanotechnology

Nanotechnology is an approach in drug delivery that only started being used clinically in the 1990s and has expanded and improved in recent years [13]. Nanotechnology has many advantages, and it can also be used with traditional methods of drug delivery. This technology is based on the encapsulation of drugs in nanoparticles (NPs) that are generated in the lab. Nanoparticles are small (submicron-sized) carriers. They generally range between 1 and 100 nm in size, although particles up to 1000 nm in size can also be considered nanoparticles [14, 15]. These particles can be used to encapsulate therapeutic agents. The encapsulation of drugs in nanoparticles provides a protective environment. This can allow the drug to have a longer half-life and to survive longer in various environments without degradation. Since drugs can act longer, the use of nanoparticles can sometimes minimize the dose or the need for repeat retreatments [16]. Nanotechnology has been most extensively developed for the treatment of cancer [17], although it has other uses as well, such as the delivery of mRNA vaccines [18].

Nanoparticles eventually release the encapsulated drug (sometimes referred to as the payload) in the body. In some cases, nanoparticles are designed in such a way that they facilitate the delivery of a drug to specific target tissues. Some nanoparticles can also release their encapsulated drug in a controlled manner. Nanoparticles can cross biological barriers, such as the blood-brain barrier (BBB). Overall, this technology can help with issues related to drug solubility, drug degradation, and nonspecific distribution within the body [4, 19–21].

Different systems exist for the construction of nanoparticles. **Liposomes, lipid nanoparticles, exosomes**, and **synthetic polymer-based carriers** are among the most commonly used nanoparticle systems, and some of these are discussed below.

2.1 Liposomes

Liposomes are spherical vesicles that have one or more phospholipid bilayers surrounding an aqueous core. In this way, they have biological characteristics similar to a simple cell, and some of the phospholipids are the same phospholipids found in cells. Liposomes can encapsulate both hydrophobic and hydrophilic drugs. Hydrophilic drugs can be enclosed within the aqueous interior of the sphere, and hydrophobic drugs can integrate into the lipid bilayer. Liposomes offer many advantages, including protecting their enclosed drugs from the environment. Furthermore, liposomes can be modified with targeting molecules on their surface, which can target them to specific destinations [4, 22] (see Fig. 28.1a, b).

A well-known example of a drug made using liposome technology is **Doxil**® (doxorubicin hydrochloride liposome injection) [23, 24]. Doxil consists of **doxorubicin** encapsulated in a pegylated liposome.

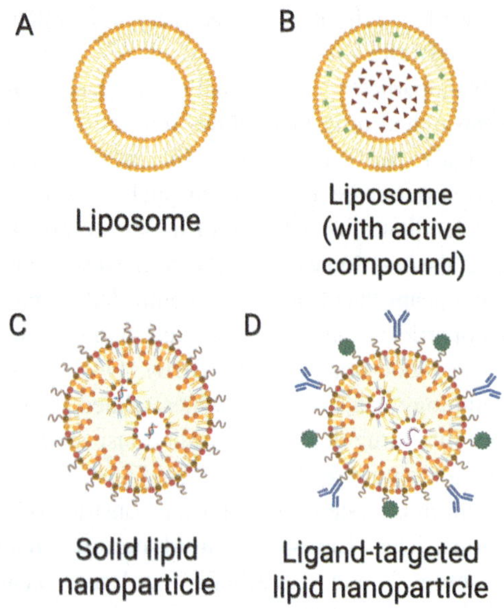

Fig. 28.1 Liposomes and lipid nanoparticles. Liposomes and lipid nanoparticles can be modified to carry and deliver drugs. They can also contain targeting molecules, such as antibodies, to target them to specific tissues. **A**. Liposome, **B**. Liposome with active compound, **C**. Solid lipid nanoparticle, **D**. Ligand targeted lipid nanoparticle. (*Source*: Created with BioRender.com)

PEGylation means that the PEG is covalently attached to lipid chains on the lipid surface, which creates a stabilized nanoparticle. (PEG is polyethylene glycol, a synthetic polymer.) PEGylation has several advantages, for example, it helps improve the stability of the molecule so that it lasts longer, and it helps reduce immunogenicity in some patients. Doxorubicin is a drug used for **cancer treatment** and discussed previously in this book. Encasing doxorubicin in a nanoparticle reduces the toxicity commonly seen with free doxorubicin because the drug tends to accumulate in tumors rather than healthy tissue. This is due to enhanced permeability and retention (EPR) effects, described more below. A disadvantage of the drug, however, is that it may accumulate in the skin, leading to some side effects and skin toxicity [25].

Another example of a liposome, but one that has been modified in a significant way, is a thermo-sensitive liposome called **ThermoDox®** (doxorubicin hydrochloride liposome injection, heat-activated). This is also a doxorubicin delivery system. It is still an experimental product, and its success has so far been mixed in clinical studies. ThermoDox is based on a **thermosensitive liposome**. This is a subtype of liposome that was developed to release its cargo in response to mild hyperthermia. Thus, it is temperature-sensitive delivery system. The idea is to heat tumor sites mildly, using hyperthermia-inducing techniques. This combination of elevating the heat at the site of the tumor and delivering a temperature-sensitive drug like ThermoDox is intended to encourage localized drug release at tumor sites [26].

2.2 Lipid Nanoparticles (LNPs)

Another class of lipid-based nanoparticles is the lipid nanoparticle (LNP). The LNP is similar to the liposome but also has some key differences. LNPs are usually made of ionizable lipids, cholesterol, and other lipids. Ionizable lipids are lipids that change their charge depending on the pH of their environment, and they are usually positively charged or neutral [27, 28]. LNPs often form a more condensed, less aqueous interior than liposomes. LNPs are especially useful for encapsulating nucleic acids because of the positive charges on the ionizable lipids. They are therefore good for gene-based therapies [29, 30] (see Fig. 28.1c).

2.3 Exosomes

Exosomes have some characteristics that differentiate them from liposomes and LNPs. Rather than being made entirely from scratch in the lab, exosomes are small extracellular vesicles naturally secreted by cells. They are about 30–150 nm in diameter. They contain proteins, lipids, and nucleic acids derived from the cells they came from. Since they derive from cells, they can be especially low in immunogenicity, and they are biocompatible in many situations. Exosomes are usually isolated from cultured cells in the lab, purified, and loaded with drugs or nucleic acids. They are especially effective at delivering nucleic acids such as DNA and RNA, making them promising for therapies such as siRNA-based therapies [31, 32].

2.4 Delivery and Targeting of Nanoparticles

Nanoparticles can be delivered using various delivery mechanisms. Most nanoparticles are delivered by injection, usually IV injection. This allows them to circulate and reach their targets efficiently. Researchers are also studying other ways to deliver nanoparticles, however, such as oral administration or inhalation. Once delivered, there are various mechanisms that are sometimes used to ensure that the particles are targeted to the correct destinations (see Fig. 28.1d). Some examples are as follows.

2.4.1 Ligand Receptor Interactions

Nanoparticles can be designed to release their contents at a specific target. This usually involves including certain molecules on the surface of the liposome, such as peptides or antibodies that recognize and bind to specific target receptors. This is called **active targeting**. Nanoparticles containing cancer therapies, for example, can be designed to target receptors that are overexpressed on cancer cells. This can be referred to as ligand-receptor interaction, and it can lead to receptor-mediated endocytosis and release of the drug into the target cell. For example, HER2-targeted liposomes are being investigated for delivering drugs to HER2-positive breast tumors. These nanoparticles consist of liposomes that have antibodies on their surface that bind to the HER2 receptor, and they contain drugs that kill rapidly dividing cells, such as doxorubicin. Because they target HER2, they deliver the drug specifically to HER2-positive cancer cells. This concentrates the drug at the appropriate place and minimizes damage to healthy tissue [33, 34].

2.4.2 Enhanced Permeability and Retention (EPR) Effect

Another way that nanoparticles can be targeted to cancer cells is the enhanced permeability and retention (EPR) effect. This has to do with tumor vasculature. Vessels that provide nutrients to tumors are often poorly organized and leaky. This is related to the rapid and disordered process of angiogenesis triggered by cancer cells. Nanoparticles, due to their small sizes, can usually readily pass through these vessels and accumulate within tumor tissues. The result is a high concentration of the drug in the tumor microenvironment compared to surrounding healthy tissues. This EPR effect is variable, however, among different tumor types and in individual patients, and it can also have some side effects [35]. Unlike active targeting, this type of targeting is called **passive targeting**, as it does not involve recognition of specific cellular markers. Instead, it is nonspecific and relies on general physiological features of the target, such as vascular features of the tumor, to promote nanoparticle accumulation [36].

2.5 The COVID-19 Vaccines

A well-known nanoparticle delivery system is the **COVID-19 vaccine**. In fact, the COVID-19 vaccines made by **Pfizer (COMIRNATY)** [37] and **Moderna (Spikevax)** [38] are currently the most widely used nanoparticle-based delivery systems worldwide. These two COVID vaccines use lipid nanoparticles (LNPs) to encapsulate mRNA. This is important because mRNA is not very stable, and it is protected within the nanoparticle. This nanoparticle-encased mRNA encodes a part of the virus (the spike protein). It is delivered by injection, and once released from the nanoparticle, the mRNA is translated into the spike protein. The spike protein then triggers the immune system to make antibodies [39]. This method for delivering mRNA is especially promising because the COVID-19 pandemic sparked a renewed interest in developing mRNA vaccines, and due to the fragile nature of RNA, nanotechnology is an excellent delivery system for this type of vaccine.

2.6 Challenges with Nanotechnology

Nanotechnology-based drug delivery systems are promising, but challenges also remain [40]. Challenges involve developing nanoparticles that are consistent in size and shape, that remain stable, and that do not have toxicity. Some nanoparticles are designed as **stealth nanoparticles**, which are modified to avoid detection by the immune system as much as possible and therefore to stay in circulation longer. PEGylation is one method for developing such stealth nanoparticles. Targeting nanoparticles to the correct target is also a challenge, and scientists are working on new ways to target NPs to disease cells. The overall cost of their production is also an important consideration, although as the technology becomes more advanced and standardized this may decrease. Another important formulation parameter is **encapsulation efficiency**, which refers to the percentage of the drug that is successfully enclosed within the liposome or nanoparticle [41]. Achieving high encapsulation efficiency is an important goal for developing lipid-encapsulated drugs [42]. Despite some challenges, nanoparticle technology represents an important mechanism for the delivery of drugs by providing a stable environment for drugs and in many cases specific drug targeting [16].

Definitions

Active targeting A drug delivery strategy where nanoparticles are designed with ligands that bind to specific cell receptors to target them to a specific location in the body.

Bioavailability The proportion of a drug that reaches the systemic circulation in its active form.

Controlled release Drug design that allows the drug to be released gradually over time rather than all at once.

Encapsulation efficiency The percentage of drug successfully contained within a carrier system, like a liposome or nanoparticle.

Enhanced permeability and retention effect (EPR effect) The phenomenon where nanoparticles accumulate in tumor tissue due to leaky blood vessels and poor lymphatic drainage.

Exosome Small extracellular vesicles naturally secreted by cells. They can be loaded with drugs or nucleic acids in the lab.

First-pass metabolism The process by which a drug is metabolized in the liver before it reaches systemic circulation.

Hydrophilic Hydrophilic molecules are water-attracting molecules.

Hydrophobic Hydrophobic molecules are water-resistant.

Inhalation route Drug administration through the lungs.

Intravenous (IV) administration Direct injection of a substance into a vein.

Ionizable lipids Lipids that change their charge depending on the pH of their environment. They are usually neutral at physiological pH

FDA-Approved Drugs Covered in This Chapter

Commonly used brand name	Generic name	First FDA approval date	Common conditions	Main targets
COMIRNATY	*BNT162b2 mRNA COVID-19 vaccine*	2020	COVID-19 prevention	SARS-CoV-2 spike
Doxil	*Pegylated liposomal doxorubicin*	1995	Cancer (e.g., ovarian, multiple myeloma)	Tumor tissue
Spikevax	*mRNA-1273 COVID-19 vaccine*	2020	COVID-19 prevention	SARS-CoV-2 spike

and become positively charged in acidic conditions.

Lipid nanoparticle (LNP) A type of nanoparticle made from ionizable lipids. It is often used to encapsulate and deliver nucleic acids such as mRNA and is central to the development of mRNA vaccines.

Liposome Spherical vesicles that have one or more phospholipid bilayers surrounding an aqueous core. Used to deliver drugs.

Nanoparticle Small, microscopic particles used to encapsulate, carry, and deliver drugs.

Parenteral administration Drug delivery that bypasses the gastrointestinal tract, administered by methods such as injection.

Passive targeting Passive targeting relies on natural physiological features (like the EPR effect) to direct nanoparticles to certain tissues.

PEGylation The modification of a nanoparticle with polyethylene glycol (PEG), which has advantages such as increasing the stability and reducing the immunogenicity of the particle.

Phospholipid A molecule with a hydrophilic head and hydrophobic tails that is the basis of membranes and also forms the basis of liposome bilayers.

Polyethylene glycol (PEG) A synthetic polymer used to coat drug carriers.

Portal circulatory system A system where blood from the intestine and other digestive organs flows to the liver via the portal vein, where it is processed before it returns to the heart and enters the systemic circulatory system.

Stealth nanoparticles Nanoparticles modified to evade immune detection, typically by PEGylation.

Systemic circulation The part of the circulation system that carries oxygenated blood from the heart to the rest of the body (except the lungs) and returns deoxygenated blood to the heart (heart → body → heart). (Other circulation systems are pulmonary circulation, coronary circulation, and portal circulation.)

Targeted delivery Delivery of a drug directly to a specific target tissue or cell type, minimizing side effects.

ThermoDox® An experimental drug based on encapsulation of doxorubicin in a thermosensitive liposome that releases its contents in response to heat.

Thermosensitive liposome Liposomes designed to release drugs at elevated temperatures, usually in response to localized heating (see ThermoDox).

Transdermal delivery A method where drugs are delivered through the skin, such as by using patches.

Study Questions

Multiple Choice Questions

1. Which of the following drug delivery methods is most likely to undergo first-pass metabolism?
   A. Intravenous injection
   B. Inhalation
   C. Oral delivery
   D. Transdermal delivery
2. What is an important function of PEGylation in drugs such as Doxil®?
   A. Avoid first-pass metabolism.
   B. Provide heat sensitivity.
   C. Reduce immunogenicity and increase stability.
   D. Promote protein degradation.
3. Which of the following best describes the enhanced permeability and retention (EPR) effect?
   A. Enhanced drug breakdown in the liver
   B. Active targeting of drugs of immune cells
   C. Nanoparticles accumulating in tumors due to leaky vasculature
   D. Controlled drug release via increased temperature
4. Why are lipid nanoparticles (LNPs) especially useful for delivering mRNA?
   A. They are composed of DNA.
   B. They are the perfect size for containing RNA.
   C. RNA is especially unstable and nanoparticles protect it.
   D. They are responsive to heat, which is especially important for RNA.

True/False Questions

5. Liposomes can only carry hydrophilic drugs.

Fill in the Blank Questions

6. Drugs delivered by __________ administration are absorbed in the GI tract and enter hepatic portal circulation before systemic circulation.
7. For orally delivered drugs, __________ metabolism occurs in the liver before a drug reaches systemic circulation.
8. __________ are small extracellular vesicles naturally secreted by cells and used for drug delivery.
9. The nanoparticle-based drug __________ is made from doxorubicin encapsulated in a PEGylated liposome, used in cancer treatment.

Short Answer Questions

10. Describe how ThermoDox® is designed to function.

Answer Key

Multiple-Choice Questions

1. **Answer**: C
2. **Answer**: C
3. **Answer**: C
4. **Answer**: C

True/False Questions

5. **Answer**: False. Liposomes can carry both hydrophilic and hydrophobic drugs.

Fill in the Blank Questions

6. **Answer**: Oral
7. **Answer**: First pass
8. **Answer**: Exosomes
9. **Answer**: Doxil®

Short Answer Questions

10. **Answer**: ThermoDox is designed to respond to heat. It is a thermosensitive delivery system. It releases its cargo in response to mild hyperthermia, and therefore, its release can be precisely controlled at the site of the tumor.

References

1. Alqahtani MS, Kazi M, Alsenaidy MA, Ahmad MZ. Advances in oral drug delivery. Front Pharmacol. 2021;12:618411. https://doi.org/10.3389/fphar.2021.618411.
2. Merck Manual Consumer Version. Drug administration. 2024. Available from: https://www.merckmanuals.com/home/drugs/administration-and-kinetics-of-drugs/drug-administration.
3. Whalen K, Lerchenfeldt S, Giordano C. Lippincott illustrated reviews. Pharmacology. Wolters Kluwer; 2021.
4. Ritter JM, Flower RJ, Henderson G, Loke YK, MacEwan DJ, Rang HP. Rang and Dale's pharmacology. 10th ed. Amsterdam: Elsevier; 2023.
5. Hall JE, Hall ME. Guyton and Hall textbook of medical physiology. 14th ed. Philadelphia: Elsevier; 2020.
6. Creative Proteomics. Prodrugs and their active metabolites. 2010. Available from: https://www.creative-proteomics.com/resource/prodrugs-and-their-active-metabolites.htm#:~:text=Enzymes%20responsible%20for%20converting%20prodrugs,other%20functional%20groups%20for%20activation.
7. Roberts MS, Cheruvu HS, Mangion SE, Alinaghi A, Benson HAE, Mohammed Y, et al. Topical drug delivery: history, percutaneous absorption, and product development. Adv Drug Deliv Rev. 2021;177:113929. https://doi.org/10.1016/j.addr.2021.113929.
8. Lambert PH, Laurent PE. Intradermal vaccine delivery: will new delivery systems transform vaccine administration? Vaccine. 2008;26(26):3197–208. https://doi.org/10.1016/j.vaccine.2008.03.095.

9. Mayo Clinic. Asthma medications: know your options. 2024. Available from: https://www.mayoclinic.org/diseases-conditions/asthma/in-depth/asthma-medications/art-20045557#:~:text=including%20the%20following.-,Inhaled%20corticosteroids,Ciclesonide%20(Alvesco).
10. UMP Pharmaceuticals. Drug delivery dynamics: a closer look at extended release vs. sustained release tablets. 2025. Available from: https://www.upm-inc.com/extended-release-vs-sustained-release.
11. DiFranco N. The guide to formulating oral extended-release tablets. Lubrizol Health; 2023. Available from: https://www.lubrizol.com/Health/Blog/2023/08/The-Guide-to-Oral-Extended-Release-Tablets.
12. News Medical Life Science. Sustained drug delivery: advancing long-acting injectables (LAIs) with biodegradable microspheres. 2024. Available from: https://www.news-medical.net/whitepaper/20240503/Sustained-drug-delivery-Advancing-long-acting-injectables-(LAIs)-with-biodegradable-microspheres.aspx
13. Afzal O, Altamimi ASA, Nadeem MS, Alzarea SI, Almalki WH, Tariq A, et al. Nanoparticles in drug delivery: from history to therapeutic applications. Nanomaterials (Basel). 2022;12(24). https://doi.org/10.3390/nano12244494.
14. Joseph TM, Kar Mahapatra D, Esmaeili A, Piszczyk Ł, Hasanin MS, Kattali M, et al. Nanoparticles: taking a unique position in medicine. Nanomaterials (Basel). 2023;13(3). https://doi.org/10.3390/nano13030574.
15. Khan I, Saeed K, Khan I. Nanoparticles: properties, applications and toxicities. Arab J Chem. 2019;12(7):908–31. https://doi.org/10.1016/j.arabjc.2017.05.011.
16. Gelperina S, Kisich K, Iseman MD, Heifets L. The potential advantages of nanoparticle drug delivery systems in chemotherapy of tuberculosis. Am J Respir Crit Care Med. 2005;172(12):1487–90. https://doi.org/10.1164/rccm.200504-613PP.
17. Zhu J, Lee H, Huang R, Zhou J, Zhang J, Yang X, et al. Harnessing nanotechnology for cancer treatment. Front Bioeng Biotechnol. 2025;12. https://doi.org/10.3389/fbioe.2024.1514890.
18. Khurana A, Allawadhi P, Khurana I, Allwadhi S, Weiskirchen R, Banothu AK, et al. Role of nanotechnology behind the success of mRNA vaccines for COVID-19. Nano Today. 2021;38:101142. https://doi.org/10.1016/j.nantod.2021.101142.
19. Wang Y, Minden A. The use of nanomedicine to target signaling by the PAK kinases for disease treatment. Cells. 2021;10(12). https://doi.org/10.3390/cells10123565.
20. Naqvi S, Panghal A, Flora SJS. Nanotechnology: a promising approach for delivery of neuroprotective drugs. Front Neurosci. 2020;14:494. https://doi.org/10.3389/fnins.2020.00494.
21. Wilczewska AZ, Niemirowicz K, Markiewicz KH, Car H. Nanoparticles as drug delivery systems. Pharmacol Rep. 2012;64(5):1020–37. https://doi.org/10.1016/S1734-1140(12)70901-5.
22. News Medical Life Science. C, B. What is a liposome? 2019. Available from: https://www.news-medical.net/life-sciences/What-is-a-Liposome.aspx#:~:text=A%20liposome%20is%20a%20spherical,become%20useful%20drug%20delivery%20systems.
23. FDA. Doxil (Pegylated liposomal doxorubicin) approval overview. 1995. Available from: https://www.accessdata.fda.gov/scripts/cder/daf/index.cfm?event=overview.process&ApplNo=050718.
24. Barenholz Y. Doxil®—the first FDA-approved nano-drug: lessons learned. J Control Release. 2012;160(2):117–34. https://doi.org/10.1016/j.jconrel.2012.03.020.
25. Lotem M, Hubert A, Lyass O, Goldenhersh MA, Ingber A, Peretz T, et al. Skin toxic effects of polyethylene glycol–coated liposomal doxorubicin. Arch Dermatol. 2000;136(12):1475–80. https://doi.org/10.1001/archderm.136.12.1475.
26. Regenold M, Bannigan P, Evans JC, Waspe A, Temple MJ, Allen C. Turning down the heat: the case for mild hyperthermia and thermosensitive liposomes. Nanomedicine. 2022;40:102484. https://doi.org/10.1016/j.nano.2021.102484.
27. Schlich M, Palomba R, Costabile G, Mizrahy S, Pannuzzo M, Peer D, et al. Cytosolic delivery of nucleic acids: the case of ionizable lipid nanoparticles. Bioeng Transl Med. 2021;6(2):e10213. https://doi.org/10.1002/btm2.10213.
28. AxisPharm. What are ionizable lipids. 2024. Available from: https://axispharm.com/what-are-ionizable-lipids/#:~:text=Ionizable%20Lipids%20Overview,nanoparticles'%20escape%20from%20the%20endosomes.
29. Mehta M, Bui TA, Yang X, Aksoy Y, Goldys EM, Deng W. Lipid-based nanoparticles for drug/gene delivery: an overview of the production techniques and difficulties encountered in their industrial development. ACS Mater Au. 2023;3(6):600–19. https://doi.org/10.1021/acsmaterialsau.3c00032.
30. BOSC Sciences, Best of Chemicals. Overview of lipid nanoparticle. 2022. Available from: https://www.bocsci.com/blog/overview-of-lipid-nanoparticle/.
31. Rajput A, Varshney A, Bajaj R, Pokharkar V. Exosomes as new generation vehicles for drug delivery: biomedical applications and future perspectives. Molecules. 2022;27(21). https://doi.org/10.3390/molecules27217289.
32. Li J, Wang J, Chen Z. Emerging role of exosomes in cancer therapy: progress and challenges. Mol Cancer. 2025;24(1):13. https://doi.org/10.1186/s12943-024-02215-4.

33. Tian H, Zhang T, Qin S, Huang Z, Zhou L, Shi J, et al. Enhancing the therapeutic efficacy of nanoparticles for cancer treatment using versatile targeted strategies. J Hematol Oncol. 2022;15(1):132. https://doi.org/10.1186/s13045-022-01320-5.
34. Nel J, Elkhoury K, Velot É, Bianchi A, Acherar S, Francius G, et al. Functionalized liposomes for targeted breast cancer drug delivery. Bioact Mater. 2023;24:401–37. https://doi.org/10.1016/j.bioactmat.2022.12.027.
35. Wu J. The enhanced permeability and retention (EPR) effect: the significance of the concept and methods to enhance its application. J Pers Med. 2021;11(8). https://doi.org/10.3390/jpm11080771.
36. Vagena I-A, Malapani C, Gatou M-A, Lagopati N, Pavlatou EA. Enhancement of EPR effect for passive tumor targeting: current status and future perspectives. Appl Sci. 2025;15(6):3189. https://doi.org/10.3390/app15063189.
37. FDA. Comirnaty [BNT162b2 (Pfizer-BioNTech)] approval overview. 2021. https://www.fda.gov/news-events/press-announcements/fda-approves-first-covid-19-vaccine
38. FDA. Spikevax [mRNA-1273 (Moderna)] approval overview. 2022. Available from: https://www.fda.gov/vaccines-blood-biologics/spikevax.
39. Schoenmaker L, Witzigmann D, Kulkarni JA, Verbeke R, Kersten G, Jiskoot W, et al. mRNA-lipid nanoparticle COVID-19 vaccines: structure and stability. Int J Pharm. 2021;601:120586. https://doi.org/10.1016/j.ijpharm.2021.120586.
40. Desai N. Challenges in development of nanoparticle-based therapeutics. AAPS J. 2012;14(2):282–95. https://doi.org/10.1208/s12248-012-9339-4.
41. Wallace SJ, Li J, Nation RL, Boyd BJ. Drug release from nanomedicines: selection of appropriate encapsulation and release methodology. Drug Deliv Transl Res. 2012;2(4):284–92. https://doi.org/10.1007/s13346-012-0064-4.
42. Wang JP, Huang ZR, Zhang C, Ni YR, Li BT, Wang Y, et al. Methodological advances in liposomal encapsulation efficiency determination: systematic review and analysis. J Drug Target. 2025:1–10. https://doi.org/10.1080/1061186x.2025.2484773.

Appendix A: Review of Some Key Biochemistry Concepts

This appendix reviews some basic concepts that are important in biochemistry.

Some Types of Chemical Bonds

Different types of bonds hold atoms together within molecules [1, 2]. Some important examples are reviewed here. Many bonds involve **electrostatic interactions**, meaning that they involve interactions between opposite charges. **Ionic bonds** are a classic example of bonds that involve electrostatic interactions. Ionic bonds form between oppositely charged ions, such as a negatively charged chloride ion and a positively charged sodium ion.

Hydrogen bonds are also an example of electrostatic interactions. These bonds form when a hydrogen atom covalently bonded to an electronegative atom (like oxygen or nitrogen) is attracted to another electronegative atom. These bonds are weaker than covalent bonds, but when molecules are held together by multiple hydrogen bonds (such as the double strands of DNA), this can strengthen the interaction.

Covalent bonds differ from ionic bonds because electrons are shared between atoms rather than transferred. These bonds are generally quite stable.

van der Waals interactions are weak, transient forces between molecules that are close to each other. They arise from temporary fluctuations in electron distribution that create instantaneous dipoles. (Dipoles are also discussed below.)

Polarity

A **polar molecule** is one in which one end of the molecule has a slightly positive charge and the other has a slightly negative charge. Polar molecules are usually **hydrophilic**, which means they are water-attracting.

Nonpolar molecules are generally uncharged and lack significant partial charges. They do not have significant charge differences across the molecule. These molecules tend to be **hydrophobic** (water-repelling).

A **dipole** is created when electrons are unevenly shared between atoms in a bond, resulting in partial positive and negative charges on opposite ends of the molecule or bond. Dipoles are the basis of polarity and help explain how molecules like water interact with each other and with ions or charged regions of other molecules.

Amphipathic molecules are molecules that have both a nonpolar region and a polar region. A classic example is a membrane phospholipid [1–3].

Electronegativity

Electronegativity is the tendency of an atom to attract shared electrons in a chemical bond. When two atoms in a bond have different electronegativities, the more electronegative atom pulls the electrons closer. This atom will thus have a **negative charge (δ^-)**. The other atom, which is less electronegative, will have a

A. Minden, *Biochemistry, Cell and Molecular Biology for Pharmacists and Health Professionals*, https://doi.org/10.1007/978-3-032-20529-2

partial positive charge (δ^+). This unequal sharing of electrons results in a **polar covalent bond**.

Examples of electronegative atoms are fluorine, oxygen, and nitrogen. Electronegativity is measured on a scale, which is called the Pauling scale, with fluorine being the most electronegative atom with a value of 3.98 [4].

The concept of electronegativity can be illustrated in a water molecule (H2O). Oxygen is more electronegative than hydrogen, meaning it will attract the shared electrons more strongly and thus will have a partial negative charge (δ^-), while the hydrogens will have partial positive charges (δ^+). This unequal distribution of charge creates a dipole, a molecule where one end has a partial positive charge and one end has a partial negative charge, and this helps water molecules interact with other molecules [2].

Electrophiles and Nucleophiles

Electrophiles are species that are electron-deficient. Thus, they attract electrons from electron-rich species, called **nucleophiles**. Electrophiles usually have a positive charge or a partial positive charge. As a result, they can accept an electron pair. Common electrophiles include cations like H^+ and molecules containing carbonyl groups, where the carbon atom is an electrophile [2].

Acids and Bases and Buffers

According to the Bronsted-Lowry definition, **acids** are substances that donate protons (H^+), while **bases** are substances that accept protons. This is the most commonly used definition in biological sciences, and it mainly involves reactions in aqueous-based solutions.

How acidic or basic a solution can be expressed as pH? The **pH scale** reflects the concentration of hydrogen ions in a solution. A solution with a lower pH is more acidic, and it has a higher concentration of hydrogen ions. A solution with a higher pH is more basic and has fewer hydrogen ions. **Buffers** are solutions that minimize pH changes by neutralizing added acids or bases. They may do this, for example, by consisting of a weak acid and its conjugate base so that they can donate or absorb protons.

Another definition of acids and bases is the Lewis theory. According to this model, the acid (called a Lewis acid) is defined as a substance that accepts an electron pair, and a Lewis base donates an electron pair. The Lewis theory is more general, and it can apply to reactions that are not necessarily aqueous [5, 6].

Some Common Functional Groups and Structures

Throughout the text, various functional groups are mentioned. Functional groups are groups that are parts of a larger molecule, which affect its functions. Students are expected to be familiar with most of these terms, but some functional groups and molecules that are frequently referred to in the textbook, as well as relevant bonds, are briefly described below. They are also described in more details in other biochemistry and organic chemistry textbooks, including 1, 2, and 5. Some functional groups are also shown in Fig. A.1 of Appendix A (Fig. A.1).

Basic Hydrocarbon Structures and Groups

An **alkane** is a hydrocarbon containing only single bonds (saturated). Thus, they contain only single bonds between carbons atoms and between carbon atoms and and hydrogen atoms. Examples are methane (CH_4), ethane (C_2H_6), and propane (C_3H_8).

An **alkyl group** is a **functional group** that consists of a simple hydrocarbon chain (or branched chain) made up of carbon and hydrogen. It is derived from an alkane, with one hydrogen atom removed, leaving one free bond.

An **aryl group** is a **functional group** derived from an aromatic ring system, such as a benzene ring, where one hydrogen is removed so that it

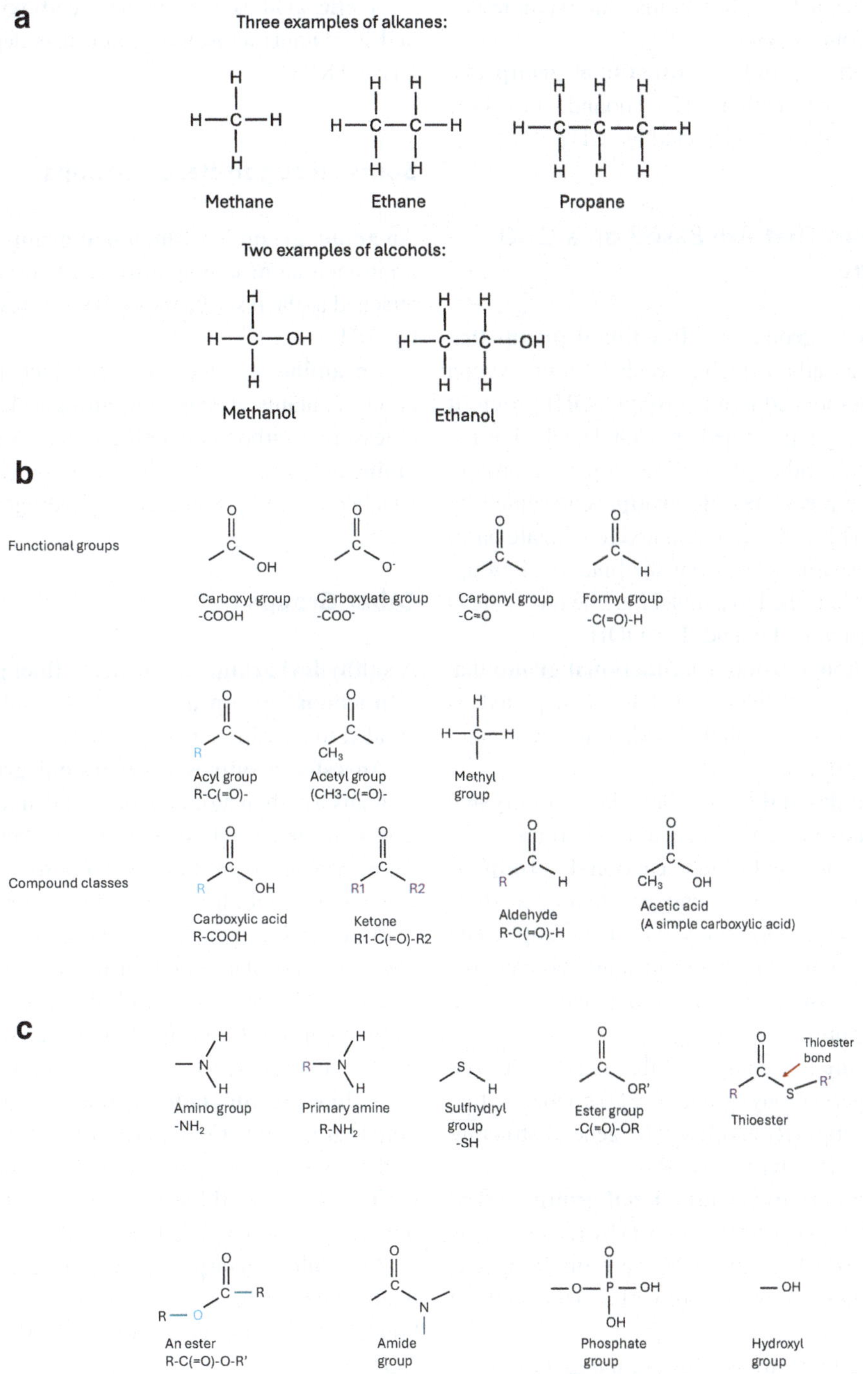

Fig. A.1 Some functional groups and important compound classes. (Source: Made by Polly Minden with permission)

can be attached to other atoms. An example is a phenyl group (C_6H_5-).

A **methyl group** is a **functional group** that contains one carbon (C) bound to three hydrogens. It can be depicted as -CH_3.

Molecules That Are Based on a C=O Structure

A **carboxyl group** is a **functional group** that contains a carbon double-bonded to an oxygen and single-bonded to a hydroxyl (-OH) group. It is commonly represented by -COOH, which is the protonated (acidic) form. When it loses a proton, it is called a **carboxylate group**, represented by -COO^-. This is the deprotonated (conjugate base) form. The carboxyl group is the functional group. When it is attached to a molecule, the molecule is called a **carboxylic acid**: R-COOH.

A **carbonyl group** is a **functional group** that contains a C double-bonded to O, depicted as C=O. C can be single-bonded to two other atoms or groups.

Aldehydes and ketones are classes of organic **compounds** that contain a carbonyl group.

In an **aldehyde**, the **carbonyl group** is attached to one carbon and one hydrogen (R-C(=O)-H). Within the aldehyde is the group -CHO (C double-bonded to an O and single-bonded to a H). This is also a functional group and is called a **formyl group**.

A **ketone** is a compound that is characterized by having a carbonyl group (C=O) bonded to two carbon groups (R_1 and R_2). The general structure of the ketone is R_1-C(=O)-R_2.

An **acyl group** is a **functional group** derived from carboxylic acid (R-COOH) by removing the hydroxyl (-OH) group, with the formula of R-C(=O). Thus, it is a carboxylic acid without the OH.

An **acetyl group** is a **functional group** that is a type of acyl group where R is a methyl group (CH_3) and C can bind to one more group. Essentially, it is a carbonyl group attached to a methyl group. It is depicted as CH_3-C(=O)-. It is a two-carbon unit.

Acetic acid is the simplest carboxylic acid, and it contains an acetyl group. It is depicted as CH_3-COOH.

Some Nitrogen-Based Groups

An **amino group** is a **functional group** in which a nitrogen atom is bound to two hydrogens and attached to the rest of a molecule. It is represented by -NH_2.

An **amine** is a molecule in which the compound contains at least one nitrogen bonded to at least one carbon containing group. A **primary amine** is a type of amine in which the nitrogen is attached to one carbon and two hydrogens.

Other Groups

A **sulfhydryl group**, also called a **thiol group**, is a **functional group** that contains a sulfur atom bonded to a hydrogen atom (–SH).

An **ester group** is a **functional group** that contains a carbon atom double-bonded to an oxygen atom and single-bonded to another oxygen atom, which is in turn bonded to a carbon-containing group. It is depicted as –COOR.

An **ester bond** is also known as an ester linkage. It is a covalent bond formed between a carboxylic acid and an alcohol. It connects a carbonyl group (C=O) to an oxygen atom, creating the structure R-C(=O)-O-R', an ester.

A **thioester functional group** contains a carbonyl group (C=O) bonded to a sulfur atom (rather than an oxygen atom), depicted as -C(=O)-S-R'. The **thioester bond** is the bond between the carbonyl carbon and the sulfur atom.

An **amide group** contains a carbon atom double-bonded to an oxygen atom and single-bonded to a nitrogen atom (–$CONH_2$ or –CONH–).

A **Schiff linkage** is also a term that is used in biochemistry. It is also called a **Schiff base**. It is a type of covalent bond that forms between a primary amine and a carbonyl group. This leads to a C=N double bond, the key component of the Schiff linkage, called an **imine**.

A **phosphate group** is a **functional group** consisting of a phosphorus atom bonded to four oxygen atoms, typically represented as $-PO_4^{2-}$ in its fully deprotonated form.

A **hydroxyl group** is a **functional group** that contains a hydrogen atom covalently linked to an oxygen. It is depicted as -OH.

An alcohol is a **molecule** that contains at least one hydroxyl (-OH) group attached to a carbon group. Examples of alcohols include methanol (CH_3OH) and ethanol (CH_3CH_2OH).

Oxidation and Reduction

Oxidation and reduction usually involve the transfer of electrons. Reactions involving the transfer of electrons are called redox reactions (short for reduction-oxidation reactions). In these reactions, hydrogen is usually the carrier of the electrons. Oxidation is the loss of electrons. This usually involves the loss of hydrogen atoms (which contain an electron) from the molecule. Reduction is the gain of electrons. This usually involves the gain of hydrogen atoms. Redox reactions are important during metabolism. During metabolism, for example, pyruvate can be changed to lactate in this reaction: Pyruvate + NADH + $H^+ \rightleftharpoons$ lactate + NAD^+. In this reaction, NADH is oxidized; it loses a hydride ion (H^-, a hydrogen atom with two electrons, one from hydrogen and one extra electron), and pyruvate is reduced as it gains electrons. Thus, in this reaction and in many metabolic reactions, NADH is the reducing agent, and NADH/NAD^+ is a redox pair. If the reverse reaction were to take place, NAD^+ would be considered the oxidizing agent, as it accepts electrons. Other common redox pairs in metabolic pathways are $FADH_2$/FAD and NADPH/$NADP^+$ [1].

Appendix B: Summary of Shuttles and Cycles Commonly Referred to in Metabolism

This short section reviews the shuttles and cycles frequently mentioned throughout the metabolism chapters. Please refer back to this appendix any time you need a refresher on these shuttles and cycles. Refer to the carbohydrate and lipid metabolism chapters in Part IV for more information and for figures representing these shuttles and cycles.

Definition of Shuttles and Cycles in Metabolism

Shuttles in this textbook refer to pathways whose primary function is to transport molecules or reduce equivalents between cellular compartments, usually within a single cell, for example, between the mitochondria and the cytoplasm. Their purpose is to transport, even though the pathway may often be cyclic. **Cycles**, in contrast, refer to metabolic pathways in which intermediates are chemically transformed and then regenerated. These pathways may occur within one cell (such as the Krebs cycle) or between different tissues (such as muscle and the liver in the Cori cycle). Their purpose is metabolic conversion.

Shuttles

Glycerol-3-Phosphate Shuttle

Function: Transfers electrons from cytosolic NADH to mitochondrial FAD.

Metabolites: Glycerol-3-phosphate ⇌ DHAP.

Tissues: The skeletal muscle and brain.

It is primarily discussed in Chap. 10 (see Fig. 10.13), glucose metabolism.

In the **glycerol-3-phosphate shuttle**:

1. In the cytosol, cytosolic **DHAP** is reduced to **glycerol-3-phosphate (G3P)** (not to be confused with glyceraldehyde-3-phosphate) by **cytosolic glycerol-3-phosphate dehydrogenase (GPD1)**. In the process, NADH is oxidized to NAD^+.
2. **Glycerol-3-phosphate** passes into the intermembrane space.
3. In the mitochondrial intermembrane space, **G3P is oxidized back to DHAP**. This is catalyzed by **mitochondrial glycerol-3-phosphate dehydrogenase**, which is on the outer surface of the inner mitochondrial membrane. In the process, FAD is reduced to $FADH_2$, and the electrons are transferred directly to CoQ in the ETC.
4. DHAP returns to the cytosol, completing the cycle [7].

The chemical reactions in this cycle are as follows:

1. **DHAP** in the **cytosol** is reduced to G3P:

   $$\mathrm{DHAP} + \mathrm{NADH} + \mathrm{H}^+ \rightarrow \mathrm{G3P} + \mathrm{NAD}^+$$

   Enzyme: **cytoplasmic glycerol-3-phosphate dehydrogenase (GDP1)**
2. G3P moves to the intermediate space (no transporter is needed; G3P can enter

A. Minden, *Biochemistry, Cell and Molecular Biology for Pharmacists and Health Professionals*,
https://doi.org/10.1007/978-3-032-20529-2

through the porins on the outer mitochondrial membrane).

3. On the mitochondrial inner membrane facing the intermembrane space, G3P is oxidized back to DHAP:

$$G3P + FAD \rightarrow FADH_2 + DHAP$$

   Enzyme: **mitochondrial glycerol-3-phosphate dehydrogenase (GPD2)**
4. **DHAP** returns to the **cytosol** (again, no transporter is needed).

Malate Aspartate Shuttle

Function: It transfers reducing equivalents of cytosolic NADH into the mitochondria. It relies cytosolic OAA which is reduced to malate to transfer electrons into the mitochondria. The shuttle gets its name because the transfer involves the interconversion of **malate** and **aspartate** across the mitochondrial membrane. OAA is reduced to malate, which crosses into the mitochondria, where it is reoxidized to OAA, producing mitochondrial NADH. Aspartate helps regenerate OAA in the cytosol.

The function of this shuttle is similar to the function of the glycerol-3-phosphate shuttle, but it uses different enzymes and a different part of the mitochondria (in this case, the mitochondrial matrix). Different tissues use different shuttles due to the availability of enzymes. This shuttle is more efficient than the glycerol-3-phosphate shuttle.

Key metabolites: Malate, aspartate, αKG, glutamate, oxaloacetate.

Tissues: Highly active in the liver, heart, and kidneys.

Importance: It preserves NADH for oxidative phosphorylation.

Other functions: Parts of this shuttle can also be used for other reactions, for example, it can be used for OAA movement across the membrane and for gluconeogenesis.

It is primarily discussed in Chap. 10 (see Fig. 10.13), glucose metabolism.

The **malate-aspartate shuttle** relies on a pool of cytosolic OAA to transfer electrons from cytosolic NADH to the mitochondria. In this shuttle system, the following occurs:

1. Electrons are transferred from cytosolic NADH to reduce cytoplasmic **oxaloacetate** to **malate** in a reaction catalyzed by **cytosolic malate dehydrogenase**. In this process, **NADH → NAD⁺**.
2. Malate then crosses into the mitochondria via a specific **malate transporter**, which is a **malate-α-ketoglutarate antiporter**. This is an antiporter; malate enters the mitochondria, while αKG is transported out of the mitochondria.
3. In the mitochondria, **malate** is reoxidized to **oxaloacetate** in a reaction catalyzed by **mitochondrial malate dehydrogenase**, and in the process, new **NADH** is generated in the mitochondria, which can enter the ETC.
4. The oxaloacetate cannot cross the mitochondrial membrane to return to the cytosol. It is therefore converted to **aspartate** via a transamination reaction (discussed in Chap. 14).
5. Aspartate can exit the mitochondria via a **glutamate-aspartate antiporter**. (Glutamate will enter the mitochondria as the aspartate exits.)
6. In the cytosol, **aspartate** can be transaminated back to **OAA**, completing the cycle [8]. Here is an overview of the chemical reactions that take place during this shuttle system:
   1. Reduction of **OAA** to malate in the **cytosol**:

$$OAA + NADH + H^+ \rightarrow NAD^+$$

      Enzyme: **cytosolic malate dehydrogenase**
   2. Transport of malate into the mitochondria This uses a transporter, which is a malate-α-ketoglutarate antiporter.
   3. Oxidation of malate to OAA in the mitochondria:

$$Malate + NAD^+ \rightarrow OAA + NADH + H^+$$

      Enzyme: **mitochondrial malate dehydrogenase**

4. Transamination of OAA to aspartate in the mitochondria:

$$\mathrm{OAA} + \mathrm{GLU} \rightarrow \mathrm{aspartate} + \alpha\mathrm{KG}$$

Enzyme: **mitochondrial aspartate aminotransferase**

5. Aspartate is shuttled out of the mitochondria.
This uses a glutamate-aspartate antiporter.
6. Transamination of aspartate back to **OAA** in the **cytoplasm**:

$$\mathrm{Aspartate} + \alpha\mathrm{KG} \rightarrow \mathbf{OAA} + \mathbf{GLU}$$

Enzyme: **cytosolic aspartate aminotransferase**

The Citrate Shuttle (Also Called Acetyl CoA Shuttle)

Function: Transports acetyl-CoA from the mitochondria to the cytosol for lipid synthesis, specifically fatty acid synthesis and cholesterol synthesis.
Key metabolites: Citrate, acetyl-CoA, OAA, malate, pyruvate, NADPH.
Tissues: Most active in the liver, adipose tissue, and lactating mammary glands.
Other names: Sometimes called the citrate pyruvate shuttle.
It is primarily discussed in Chap. 11 (see Fig. 11.7), fatty acid metabolism.

Fatty acids are made from **acetyl-CoA** in the cytosol. Much of the cell's **acetyl-CoA** is in the mitochondria, however, and acetyl-CoA cannot cross the mitochondrial membrane to enter the cytoplasm. To solve this, a shuttle system called the **citrate shuttle** (also sometimes called the **citrate-pyruvate cycle**) is used.

The citrate shuttle works as follows:

1. In the **mitochondria**, **oxaloacetate** (**OAA**; 4C) combines with **acetyl-CoA** (2C) to make citrate (with the release of free coenzyme A). This is also a normal step in the Krebs cycle, and it is catalyzed by **citrate synthase**.
2. **Citrate** is an intermediate in the **Krebs cycle**, but it can also cross the mitochondrial membrane and enter the cytoplasm. It crosses via a transporter called the **tricarboxylate transporter**. This is a **citrate malate antiporter**. It exchanges citrate for malate.
3. Once in the cytoplasm, **citrate (C6)** becomes cleaved in a reaction catalyzed by **ATP-citrate lyase (ACLY).** This cleavage leads to the formation of two molecules: (i) cytoplasmic **OAA**; 4C and (ii) **acetyl-CoA;** 2C. Acetyl-CoA can be used to synthesize FFAs, but cytoplasmic **OAA** can go on to complete the cycle, leading back to **OAA** formation, as follows:
4. This **cytoplasmic OAA** is first reduced to **malate** in a reaction catalyzed by **cytosolic malate dehydrogenase**. The resulting **malate** has two possible fates:
   (4a-1) Malate can enter the **mitochondria** through a transporter: the malate-a-KG antiporter.
   (4a-2) Once in the mitochondria, malate can be oxidized back to **OAA**, closing the cycle.
   Or
   (4b-1) Malate can instead form pyruvate in the cytoplasm in a reaction catalyzed by **malic enzyme (ME)**, which leads to the generation of NADPH.
   (4b-2) This **pyruvate** can enter the mitochondria via the **mitochondrial pyruvate transporter (MPC)**.
   (4b-3) The pyruvate can be carboxylated to form **OAA** by pyruvate carboxylase [9–11].

Reactions

1. In the **mitochondria**,

$$\mathbf{OAA} + \mathrm{acetyl} - \mathrm{CoA} + \mathrm{H2O} \rightarrow \mathrm{Citrate} + \mathrm{CoA} - \mathrm{SH}.$$

Enzyme: Citrate synthase

2. Citrate is transported out of the mitochondria into the cytosol via the **tricarboxylate transporter** (an antiporter that exchanges citrate for malate).

3. Cleavage of citrate (C6) to **OAA** (C4) and acetyl-CoA (C2):

$$\text{Citrate} + \text{CoA} + \text{ATP} \rightarrow \text{acetyl-CoA} + \text{OAA} + \text{ADP} + \text{Pi}$$

Enzyme: ATP-citrate lyase (ACLY)

4. Cytoplasmic OAA (C4) is reduced to malate:

$$\text{OAA} + \text{NADH} + \text{H}^+ \rightarrow \text{malate} + \text{NAD}^+$$

Enzyme:
cytosolic malate dehydrogenase

Two choices for malate:

(4a-1) Malate can enter the **mitochondria** through the malate-αKG-antiporter

(4a-2) Once in the mitochondria, malate can be converted back to mitochondrial OAA:

$$\text{Malate} + \text{NAD}^+ \rightarrow \textbf{OAA} + \text{NADH} + \text{H}^+$$

Enzyme: **malate dehydrogenase**

Or

(4b-1) Malate can form pyruvate in the cytoplasm:

$$\text{Malate} + \text{NADP}^+ \rightarrow \text{pyruvate} + \text{CO}_2 + \text{NADPH}$$

Enzyme: **malic enzyme (ME)**

(4b-2) Pyruvate can enter the **mitochondria** via the mitochondrial pyruvate transporter (MPC).

(4b-3) Pyruvate can become OAA:

$$\text{Pyruvate} + \text{CO}_2 + \text{ATP} \rightarrow \textbf{OAA} + \text{ADP} + \text{Pi}$$

Enzyme: **pyruvate carboxylase**

Cycles

The main cycles described here are the **glucose-alanine cycle** and the **Cori cycle.** We also discuss ketone body transport because, although it is not technically a cycle, it is a good example of how metabolites can be transported to different tissues where they have different functions.

The Glucose-Alanine Cycle

Function: Amino groups are transported from muscle to the liver (via alanine), facilitating their **elimination. Meanwhile**, glucose is regenerated via gluconeogenesis, providing energy for the muscle. **The glucose-alanine cycle** is an important link between protein metabolism and glucose metabolism and between muscle and the liver [12].

Main tissues involved: Muscle and the liver.

Other names: The Cahill cycle.

The main chapter is discussed in Chap. 14 (see Fig. 14.16), amino acid metabolism.

Here is how the cycle works:

Muscle

In **muscle**, proteins are broken down under certain conditions, such as fasting or heavy exercise. When proteins are broken down, amino acids lose their amino groups through **transamination** reactions catalyzed by **aminotransferases**. For many amino acids, the amino group is transferred to **αKG**, and this leads to the formation of **glutamate**. **Glutamate** then becomes the **nitrogen carrier.** This nitrogen (now on glutamate) needs to be excreted from the body via the **urea cycle**, which takes place in the liver. The nitrogen group, therefore, needs to travel from the muscle to the liver. In order to be transported, this nitrogen needs to be transferred to **alanine** (instead of glutamate), which is then transported to the liver. The transfer reaction is catalyzed by **alanine aminotransferase** (**ALT**). In this reaction, the amino group is transferred from **glutamate** to **pyruvate**, to form **alanine,** as in the reaction below:

$$\textbf{Glutamate} + \textbf{pyruvate} \rightleftharpoons \alpha\text{KG} + \textbf{alanine}$$

Enzyme: **alanine transaminase (ALT)**

Alanine is now the **nitrogen carrier**, and it will carry nitrogen to the **liver** where it can efficiently cross the membrane to be used for nitrogen disposal.

Liver
Once it reaches the liver, the **nitrogen** from alanine is then transferred to aKG to form **glutamate**, which is now the new nitrogen carrier. Meanwhile, **alanine** now becomes **pyruvate** in the following reaction.

Alanine + AKG ⇌ Pyruvate + glutamate

This reaction regenerates **pyruvate** and **glutamate**.

The role of the **glutamate** produced in this reaction is important. This glutamate will become deaminated, and the nitrogen group will enter the **urea cycle**, to be excreted in the urine.

The other product, **pyruvate**, is also an important part of this cycle. This pyruvate can lead to gluconeogenesis in the liver and thereby lead to more glucose production. This new **glucose** can now be released from the **liver** into the bloodstream and transported back to **muscle**.

Muscle
Back in the muscle, the **glucose** can enter **glycolysis**. As a result, **ATP** is generated, and more **pyruvate** is generated. **ATP** provides energy that the muscle needs, and **pyruvate** can reenter the **transamination** reaction (glutamine + pyruvate ⇌ αKG + alanine), thereby producing more **alanine** and keeping the cycle going.

Summary
In summary, the **glucose-alanine cycle** is important for nitrogen disposal, glucose homeostasis, and energy conservation in muscle, especially during fasting or intense muscle activity. It helps muscle safely remove nitrogen from the muscle by transporting it, via alanine, to the liver, where it enters the urea cycle for excretion, and this prevents toxic ammonia buildup in muscle. The cycle also helps maintain blood glucose levels even during fasting or intense exercise. It does so by providing the liver with pyruvate that can be used for gluconeogenesis, which produces glucose that can then be released into the bloodstream. This is also important during fasting of intense activity. The regenerated glucose returns to the muscle, where it can be used to generate ATP for muscle to continue supporting its activity. It also regenerates pyruvate so that the cycle can continue. Thus, the cycle both prevents nitrogen toxicity and supports glucose supply and ATP production.

The Cori Cycle (Lactate Cycle)

Function: Transport of lactate, produced by anaerobic glycolysis in muscle, to the liver, to produce glucose (gluconeogenesis). It is especially important when muscles are operating under **low oxygen** conditions, such as during intense exercise.
Main tissues involved: Muscle and the liver.
Other names: Lactate-glucose cycle.
It is primarily discussed in Chap. 10 (see Fig. 10.15), glucose metabolism.

Muscle
Muscle produces **lactate** from pyruvate during low oxygen conditions, catalyzed by **lactate dehydrogenase (LDH)**.

Liver
The lactate is then transported to the **liver** via the bloodstream. In the liver, **lactate** is converted back to **pyruvate** in a reaction that is also catalyzed by **LDH** in the following reaction: lactate + NAD^+ → pyruvate + NADH + H^+. This pyruvate then enters the **gluconeogenesis** pathway, and **glucose** is made. Gluconeogenesis could not happen in muscle because muscle lacks the necessary enzymes for gluconeogenesis. Instead, gluconeogenesis happens mostly in the liver.

Muscle
The glucose that is produced in the liver is sent back to the muscle via the bloodstream to replenish its supply, which is important because muscle

cannot carry out gluconeogenesis. In the muscle, this glucose undergoes glycolysis, where it makes a small amount of ATP (two ATPs per glucose molecule), which is especially valuable under low oxygen conditions.

In this way, the liver is helping the muscle to make more ATP since it is in a shortage due to its low oxygen. Glycolysis will also produce pyruvate, which is converted to more lactate, and this lactate can go back to the liver.

Summary
This makes a cycle, where muscle produces lactate, which is sent to the liver. The liver uses the lactate from the muscle to make more glucose, and this glucose is sent back to the muscle, where it can make more ATPs and more lactate. This cycle can continue temporarily while the muscle is in low oxygen conditions [10, 11, 13]. This cycle helps to provide a short-term energy supply to muscle under low oxygen conditions.

Ketone Body Metabolism

Notes: Ketone metabolism is not a true metabolic cycle and does not have a specific name. It is included here, though, because it is also a good example of how metabolites can be transported to different tissues where they have different functions.

Function: To provide energy to the brain (and some other tissues) when nutrients are low during prolonged **fasting or starvation** conditions, when glucose is in short supply, such as in a strict **no-carbohydrate diet**.

Main tissues involved: The liver and brain.

It is primarily discussed in Chap. 12, homeostasis.

When glucose levels become low during fasting or a strict low-carbohydrate diet, this can potentially be detrimental for the brain because the brain primarily relies on glucose as an energy source, and fatty acids do not readily cross the blood brain barrier. However, the brain can also switch to ketone bodies for energy during starvation as an alternative energy fuel. Therefore, transport of ketone bodies from the liver to the brain can supply the brain with a needed energy source during starvation.

Liver
If a fast or low-carbohydrate diet goes on for a long time, OAA in the liver is diverted towards gluconeogenesis. As a result, acetyl-CoA cannot efficiently enter the Krebs cycle and accumulates, leading to **ketone body** formation (**ketogenesis**). These ketone bodies get transported to the bloodstream, from which they can reach the brain because they can pass the blood-brain barrier. They also reach other tissues.

Brain
The brain lacks the enzyme for making ketone bodies, so during starvation, it relies on ketone bodies it obtains from the liver. In the brain, ketones can diffuse into the mitochondria and break down into **acetyl-CoA**. The brain, unlike the liver, still has plenty of mitochondrial OAA because it does not carry out gluconeogenesis and therefore does not use up its OAA. Thus, this **acetyl-CoA** can condense with **OAA** to form citrate, thus enter the Krebs cycle, and generate ATP for the brain. Consequently, the brain can generate ATP even in low-glucose and starvation conditions.

Appendix C: Exceptions to the Common Transamination and Oxidative Deamination Reactions

As described in Chap. 14, some amino acids do not undergo the same classic transamination reactions [14–16]. A few examples of these reactions are described briefly below.

Lysine

Lysine does not typically begin with the standard transamination to Glu. Instead, **lysine (Lys)** condenses with **αKG** and is reduced by **lysine α-ketoglutarate reductase (LKR)** to form an intermediate called **saccharopine** in a reaction catalyzed by **saccharopine dehydrogenase (SDS)** that then yields **α-aminoadipate semialdehyde (AASA) + Glu**. The resulting **AASA** is then further metabolized into Krebs cycle intermediates like acetyl-CoA or used in other metabolic processes.

Serine

Instead of going through the typical transamination pathway, serine is usually directly deaminated by the enzyme **serine dehydratase** to produce pyruvate and NH_3 (serine → pyruvate + NH_3). The pyruvate can then be converted to acetyl-CoA and enter the Krebs cycle, or it can enter glucose pathways [14]. The NH_3 can enter the urea cycle via CPSI in the liver, or it can be converted to Gln via **GS**.

Threonine

Threonine can be deaminated in different ways:

- *The threonine dehydratase pathway*
  Threonine in humans primarily undergoes non-oxidative deamination in a reaction catalyzed by **threonine dehydratase** (also called **serine/threonine dehydratase**) to form **α-ketobutyrate** and NH_3. **α-ketobutyrate** is then converted to **propionyl-CoA**. Propionyl-CoA undergoes multiple reactions to become **succinyl-CoA**, a Krebs cycle intermediate. The NH3 can follow the same path as the NH3 released from serine, described above.
- *Threonine Dehydrogenase Pathway*
  Threonine can, in some situations, also be oxidized to form **2-amino-3-ketobutyrate** in a reaction catalyzed by **threonine dehydrogenase (TDH)**. **2-amino-3-ketobutyrate** is then cleaved to form glycine and acetyl-CoA.

Glutamine, Proline, Arginine, Histidine, and Leucine, Isoleucine, and Valine

Here are a few other amino acids that undergo deamination in different ways [14, 17–19]. In each case, once glutamate is formed, it is deaminated as described above.

A. Minden, *Biochemistry, Cell and Molecular Biology for Pharmacists and Health Professionals*, https://doi.org/10.1007/978-3-032-20529-2

Glutamine (Gln) has two nitrogen groups. It is first deaminated by **glutaminase**. After its **first** nitrogen is removed, it is now **Glu**, which gets deaminated (Gln → Glu + NH_3).

Proline is oxidized to **Glu** in a two-step process, which involves opening of its ring structure and its conversion to Glu.

Arginine is hydrolyzed by **arginase,** resulting in **ornithine** and **urea**. Ornithine is then converted to **Glu** via a multistep reaction.

Histidine is converted to **urocanate** and then to **formiminoglutamate (FIGLU)**. FIGLU donates a one-carbon group to **tetrahydrofolate (THF)**. This leaves **Glu** remaining.

Leucine, isoleucine, and valine are branched-chain amino acids (BCAAs). These amino acids do undergo an essentially standard transamination reaction, but they are slightly different than those described in Chap. 14 and are therefore also included in this list.

The following reactions occur:

$$\text{BCAA} + \alpha\text{KG} \rightleftharpoons \text{BCKA} + \text{Glu}$$

The enzyme that catalyzes this reaction is **branched-chain aminotransferase (BCAT).** This is the reaction where the amino group is transferred to αKG, and the BCAA then becomes a branched-chain α-keto acid (BCKA).

In the second step, the BCKA undergoes oxidative decarboxylation and, after a series of reactions, results in **succinyl-CoA**. This is catalyzed by **branched-chain α-ketoacid dehydrogenase (BCKDH)** and several other enzymes [19].

References

Appendix A: Review of Some Key Biochemistry Concepts

1. Berg JM, Tymoczko JL, Gatto GJ Jr, Stryer L. Biochemistry. 9th ed. New York: MacMillan Learning; 2019.
2. Campbell MK, Farrell SO. Biochemistry. Pacific Grove: Brooks/Cole; 2007.
3. Alberts B, Heald R, Johnson A, Morgan D, Raff M, Roberts K, Walter P, Wilson J, Hunt T. Molecular biology of the cell. New York: W.W. Norton and Company; 2022.
4. Petrucci R, Herring G, Madura J, Bissonnette C. Petrucci's general chemistry: principles and modern applications. 12th ed. Harlow: Pearson Education; 2023.
5. Clayton J, Greeves N, Warren A. Organic chemistry. 2nd ed. Oxford/New York: OUP Oxford; 2012.
6. Pauling L. General chemistry. 3rd ed. New York: Dover Publications; 2014.

Appendix B: Summary of Shuttles and Cycles Commonly Referred to in Metabolism

7. Information NCfB. PubChem pathway summary for pathway PWY-6118, glycerol-3-phosphate shuttle. Source: BioCyc 2025. Available from: https://pubchem.ncbi.nlm.nih.gov/pathway/BioCyc:META_PWY-6118.
8. Koch J, Broeks MH, Gautschi M, Jans J, Laemmle A. Inborn errors of the malate aspartate shuttle – update on patients and cellular models. Mol Genet Metab. 2024;142(4):108520. https://doi.org/10.1016/j.ymgme.2024.108520.
9. Chandel NS. Navigating metabolism. 2nd ed. Cold Spring Harbor: Cold Spring Harbor Laboratory Press; 2020.
10. Berg JM, Tymoczko JL, Gatto GJ Jr, Stryer L. Biochemistry. 9th ed. New York: MacMillan Learning; 2019.
11. Salway JG. Metabolism at a glance. Hoboken: Wiley-Blackwell; 2017.
12. ScienceFacts.net. Glucose-alanine cycle. Available from: https://www.sciencefacts.net/glucose-alanine-cycle.html.
13. Hames BD, Hooper NM. BIOS instant notes in biochemistry. Taylor and Francis; 2011.

Appendix C: Exceptions to the Common Transamination and Oxidative Deamination Reactions

14. Berg JM, Tymoczko JL, Gatto GJ Jr, Stryer L. Biochemistry. 9th ed. New York: MacMillan Learning; 2019.
15. House JD, Hall BN, Brosnan JT. Threonine metabolism in isolated rat hepatocytes. Am J Physiol Endocrinol Metab. 2001;281(6):E1300–7. https://doi.org/10.1152/ajpendo.2001.281.6.E1300.
16. Darling PB, Grunow J, Rafii M, Brookes S, Ball RO, Pencharz PB. Threonine dehydrogenase is a minor degradative pathway of threonine catabolism in adult humans. Am J Physiol Endocrinol Metab. 2000;278

(5):E877–84. https://doi.org/10.1152/ajpendo.2000.278.5.E877.

17. Bender D. Amino acid metabolism. 3rd ed. Chichester: Wiley; 2012.
18. Chandel NS. Navigating metabolism. 2nd ed. Cold Spring Harbor: Cold Spring Harbor Laboratory Press; 2020.
19. Dimou A, Tsimihodimos V, Bairaktari E. The critical role of the branched chain amino acids (BCAAs) catabolism-regulating enzymes, branched-chain aminotransferase (BCAT) and branched-chain α-keto acid dehydrogenase (BCKD), in human pathophysiology. Int J Mol Sci. 2022;23(7). https://doi.org/10.3390/ijms23074022.

Index

A. Minden, *Biochemistry, Cell and Molecular Biology for Pharmacists and Health Professionals*, https://doi.org/10.1007/978-3-032-20529-2

N

S

The manufacturer's authorised representative in the EU is Springer Nature Customer Service Centre GmbH, Europaplatz 3, 69115 Heidelberg, Germany. If you have any concerns regarding our products, please contact ProductSafety@springernature.com

Printed and bound by CPI Group (UK) Ltd, Croydon, CR0 4YY
16/09/2026
02217815-0020